WASTEWATER TREATMENT FUNDAMENTALS I

LIQUID TREATMENT

2018

Water Environment Federation
601Wythe Street
Alexandria, VA 22314–1994 USA
http://www.wef.org

Association of Boards of Certification
2805 SW Snyder Blvd., Suite 535
Ankeny, IA 50023
http://www.abccert.org

ISBN: 978-1-57278-350-8

About WEF

The Water Environment Federation (WEF) is a not-for-profit technical and educational organization of 33,000 individual members and 75 affiliated Member Associations representing water quality professionals around the world. Since 1928, WEF and its members have protected public health and the environment. As a global water sector leader, our mission is to connect water professionals; enrich the expertise of water professionals; increase the awareness of the impact and value of water; and provide a platform for water sector innovation. To learn more, visit www.wef.org.

About ABC

The Association of Boards of Certification (ABC) was founded with the mission to advance the quality and integrity of environmental certification programs throughout the world. This charge has held strong through more than 40 years of providing knowledge and resources to nearly 100 certifying authorities representing more than 40 states, 10 Canadian provinces and territories, and several international and tribal programs. ABC believes in certification as a means of promoting public health and the environment while striving to give our members the necessary tools to ensure the knowledge and skills of their operators.

Prepared by

Sidney Innerebner, Ph.D., P.E., CWP, PO, A Industrial Wastewater and A Municipal Wastewater Operator in Colorado (Certified Water Professional), Grade IV Professional Wastewater Operator, Indigo Water Group, *Contractor*

With assistance and review provided by

Mike Bell, Supervisor, City of Hamilton, Hamilton, Ontario, Canada (Retired) (ABC)

Mark Cherniak

Nathan Coey, Pataskala Ohio Utility Director

Frank J. DeOrio, PO – Sr. Technical Director, O'Brien & Gere, Syracuse, New York (ABC)

Justin Elkins, City of Louisville, Colorado

John J. Fortin, P.E., Hazen and Sawyer, New York, New York

Sarah Galst

Mike Gosselin, PO (ABC)

Paul Krauth, Statepoint Engineering

Shaun Livermore, PO – Operations Manager, Poarch Band of Creek Indians Utility Authority, Atmore, Alabama (ABC)

Jorj Long

Michael P. Lutz, P.E., Principal Engineer, Dewberry Engineers, Inc., Denver, Colorado

Timothy Meloveck, PO, CWP – Utilities Supervisor, Town of Carbondale, Carbondale, Colorado (ABC)

Andy O'Neill, PO – Environment Specialist, Washington State Department of Ecology, Spokane, Washington (ABC)

Stacy J. Passaro, P.E., Passaro Engineering, LLC, Mount Airy, Maryland

Paul Pitt

Mark Poling

John R. Reynolds, PO – Contract Operator and Instructor, Sooke, British Columbia, Canada (ABC)

Kim Riddell, Alloway, Lima, Ohio

Joel C. Rife, P.E.

LeAnna Risso, PO – Process Control Manager, Clark County Water Reclamation District (ABC)

Kenneth Schnaars, P.E., Brown and Caldwell

Greg Seaman, President, O&M Solutions, LLC, Howell, New Jersey (ABC)

Ron Trygar

Jack Vanderland, Director of Wastewater Operator Training and Assistance Program, Virginia Department of Environmental Quality, Richmond, Virginia (Retired) (ABC)

Rich Weigand

Chris White, P.E., Hazen and Sawyer

Brian Woods, Clifton Sanitation District

Under the Direction of the Wastewater Treatment Fundamentals Steering Group of the **Technical Practice Committee**

Rudy Fernandez
John Hart
Cordell Samuels

Contents

Chapter 4 Primary Treatment of Wastewater 105

Chapter 5 Fundamentals of Biological Treatment. . . .175

Chapter 6 Wastewater Treatment Ponds 207

Association of Boards of Certification Formulas

Wastewater Treatment, Collection, Industrial Waste, & Wastewater Laboratory Exams

*Pie wheel format for this equation is shown at the end of the formulas.

$$\text{Alkalinity, mg/L as CaCO}_3 = \frac{(\text{Titrant Volume, mL})(\text{Acid Normality})(50\ 000)}{\text{Sample Volume, mL}}$$

$$\text{Amps} = \frac{\text{Volts}}{\text{Ohms}}$$

$$\text{Area of Circle*} = (0.785)(\text{Diameter}^2)$$

$$\text{Area of Circle} = (3.14)(\text{Radius}^2)$$

$$\text{Area of Cone (lateral area)} = (3.14)(\text{Radius})\sqrt{\text{Radius}^2 + \text{Height}^2}$$

$$\text{Area of Cone (total surface area)} = (3.14)(\text{Radius})(\text{Radius} + \sqrt{\text{Radius}^2 + \text{Height}^2})$$

$$\text{Area of Cylinder (total exterior surface area)} = [\text{End \#1 SA}] + [\text{End \#2 SA}] + [(3.14)(\text{Diameter})(\text{Height or Depth})]$$

Where SA = surface area

$$\text{Area of Rectangle*} = (\text{Length})(\text{Width})$$

$$\text{Area of Right Triangle*} = \frac{(\text{Base})(\text{Height})}{2}$$

$$\text{Average (arithmetic mean)} = \frac{\text{Sum of All Terms}}{\text{Number of Terms}}$$

$$\text{Average (geometric mean)} = [(X_1)(X_2)(X_3)(X_4)(X_n)]^{1/n}$$ The *nth* root of the product of *n* numbers

$$\text{Biochemical Oxygen Demand (seeded), mg/L} = \frac{[(\text{Initial DO, mg/L}) - (\text{Final DO, mg/L}) - (\text{Seed Correction, mg/L})][300\text{ mL}]}{\text{Sample Volume, mL}}$$

$$\text{Biochemical Oxygen Demand (unseeded), mg/L} = \frac{[(\text{Initial DO, mg/L}) - (\text{Final DO, mg/L})][300\text{ mL}]}{\text{Sample Volume, mL}}$$

$$\text{Blending or Three Normal Equation} = (C_1 \times V_1) + (C_2 \times V_2) = (C_3 \times V_3)$$

Where $V_1 + V_2 = V_3$; C = concentration, V = volume or flow; Concentration units must match; Volume units must match

$$\text{\# CFU/100 mL} = \frac{[(\text{\# of Colonies on Plate})(100)]}{\text{Sample Volume, mL}}$$

$$\text{Chemical Feed Pump Setting, \% Stroke} = \frac{\text{Desired Flow}}{\text{Maximum Flow}} \times 100\%$$

$$\text{Chemical Feed Pump Setting, mL/min} = \frac{(\text{Flow, mgd})(\text{Dose, mg/L})(3.785\text{ L/gal})(1\ 000\ 000\text{ gal/mil. gal})}{(\text{Feed Chemical Density, mg/mL})(\text{Active Chemical, \% express as a decimal})(1440\text{ min/d})}$$

$$\text{Chemical Feed Pump Setting, mL/min} = \frac{(\text{Flow, m}^3\text{/d})(\text{Dose, mg/L})}{(\text{Feed Chemical Density, g/cm}^3)(\text{Active Chemical, \% express as a decimal})(1440\text{ min/d})}$$

Circumference of Circle = (3.14)(Diameter)

$$\text{Composite Sample Single Portion} = \frac{\text{(Instantaneous Flow)(Total Sample Volume)}}{\text{(Number of Portions)(Average Flow)}}$$

$$\text{Cycle Time, min} = \frac{\text{Storage Volume, gal}}{\text{(Pump Capacity, gpm)} - \text{(Wet Well Inflow, gpm)}}$$

$$\text{Cycle Time, min} = \frac{\text{Storage Volume, m}^3}{\text{(Pump Capacity, m}^3\text{/min)} - \text{(Wet Well Inflow, m}^3\text{/min)}}$$

$$\text{Degrees Celsius} = \frac{(^\circ\text{F} - 32)}{1.8}$$

Degrees Fahrenheit = (°C)(1.8) + 32

$$\text{Detention Time} = \frac{\text{Volume}}{\text{Flow}}$$ *Units must be compatible*

Dilution or Two Normal Equation = $(C_1 \times V_1) = (C_2 \times V_2)$ *Where* C = Concentration, V = volume or flow; Concentration units must match; Volume units must match

Electromotive Force, V* = (Current, A)(Resistance, ohm - Ω)

$$\text{Feed Rate, lb/d*} = \frac{\text{(Dosage, mg/L)(Flow, mgd)(8.34 lb/gal)}}{\text{Purity, \% expressed as a decimal}}$$

$$\text{Feed Rate, kg/d*} = \frac{\text{(Dosage, mg/L)(Flow rate, m}^3\text{/d)}}{\text{(Purity, \% expressed as a decimal)(1000)}}$$

$$\text{Filter Backwash Rate, gpm/sq ft} = \frac{\text{Flow, gpm}}{\text{Filter Area, sq ft}}$$

$$\text{Filter Backwash Rate, L/(m}^2\cdot\text{s)} = \frac{\text{Flow, L/s}}{\text{Filter Area, m}^2}$$

$$\text{Filter Backwash Rise Rate, in./min} = \frac{\text{(Backwash Rate, gpm/sq ft)(12 in./ft)}}{\text{7.48 gal/cu ft}}$$

$$\text{Filter Backwash Rise Rate, cm/min} = \frac{\text{Water Rise, cm}}{\text{Time, min}}$$

$$\text{Filter Yield, lb/sq ft/hr} = \frac{\text{(Solids Loading, lb/d)(Recovery, \% expressed as a decimal)}}{\text{(Filter Operation, hr/d)(Area, sq ft)}}$$

$$\text{Filter Yield, kg/m}^2\cdot\text{h} = \frac{\text{(Solids Concentration, \% expressed as a decimal)(Sludge Feed Rate, L/h)(10)}}{\text{(Surface Area of Filter, m}^2\text{)}}$$

Flowrate, cu ft/sec* = (Area, sq ft)(Velocity, ft/sec)

Flowrate, m³/sec* = (Area, m²)(Velocity, m/s)

$$\text{Food-to-Microorganism Ratio} = \frac{\text{BOD}_5\text{, lb/d}}{\text{MLVSS, lb}}$$

$$\text{Food-to-Microorganism Ratio} = \frac{\text{BOD}_5\text{, kg/d}}{\text{MLVSS, kg}}$$

Force, lb* = (Pressure, psi)(Area, sq in.)

Force, newtons* = (Pressure, Pa)(Area, m^2)

$$\text{Hardness, as mg CaCO}_3\text{/L} = \frac{(\text{Titrant Volume, mL})(1000)}{\text{Sample Volume, mL}}$$

Only when the titration factor is 1.00 of ethylenediaminetetraacetic acid (EDTA)

$$\text{Horsepower, Brake, hp} = \frac{(\text{Flow, gpm})(\text{Head, ft})}{(3960)(\text{Pump Efficiency, \% expressed as a decimal})}$$

$$\text{Horsepower, Brake, kW} = \frac{(9.8)(\text{Flow, m}^3\text{/s})(\text{Head, m})}{(\text{Pump Efficiency, \% expressed as a decimal})}$$

$$\text{Horsepower, Motor, hp} = \frac{(\text{Flow, gpm})(\text{Head, ft})}{(3960)(\text{Pump Efficiency, \% expressed as a decimal})(\text{Motor Efficiency, \% expressed as a decimal})}$$

$$\text{Horsepower, Motor, kW} = \frac{(9.8)(\text{Flow, m}^3\text{/s})(\text{Head, m})}{(\text{Pump Efficiency, \% expressed as a decimal})(\text{Motor Efficiency, \% expressed as a decimal})}$$

$$\text{Horsepower, Water, hp} = \frac{(\text{Flow, gpm})(\text{Head, ft})}{3960}$$

Horsepower, Water, kW = (9.8)(Flow, m^3/s)(Head, m)

$$\text{Hydraulic Loading Rate, gpd/sq ft} = \frac{\text{Total Flow Applied, gpd}}{\text{Area, sq ft}}$$

$$\text{Hydraulic Loading Rate, m}^3\text{/(m}^2\cdot\text{d)} = \frac{\text{Total Flow Applied, m}^3\text{/d}}{\text{Area, m}^2}$$

Loading Rate, lb/d* = (Flow, mgd)(Concentration, mg/L)(8.34 lb/gal)

$$\text{Loading Rate, kg/d}^* = \frac{(\text{Flow, m}^3\text{/d})(\text{Concentration, mg/L})}{1000}$$

Mass, lb* = (Volume, mil. gal)(Concentration, mg/L)(8.34 lb/gal)

$$\text{Mass, kg}^* = \frac{(\text{Volume, m}^3)(\text{Concentration, mg/L})}{1000}$$

$$\text{Mean Cell Residence Time or Solids Retention Time, days} = \frac{(\text{Aeration Tank TSS, lb}) + (\text{Clarifier TSS, lb})}{(\text{TSS Wasted, lb/d}) + (\text{Effluent TSS, lb/d})}$$

Milliequivalent = (mL)(Normality)

$$\text{Molarity} = \frac{\text{Moles of Solute}}{\text{Liters of Solution}}$$

$$\text{Motor Efficiency, \%} = \frac{\text{Brake hp}}{\text{Motor hp}} \times 100\%$$

$$\text{Normality} = \frac{\text{Number of Equivalent Weights of Solute}}{\text{Liters of Solution}}$$

$$\text{Number of Equivalent Weights} = \frac{\text{Total Weight}}{\text{Equivalent Weight}}$$

$$\text{Number of Moles} = \frac{\text{Total Weight}}{\text{Molecular Weight}}$$

$$\text{Organic Loading Rate-RBC, lb SBOD}_5\text{/1000 sq ft/d} = \frac{\text{Organic Load, lb SBOD}_5\text{/d}}{\text{Surface Area of Media, 1000 sq ft}}$$

$$\text{Organic Loading Rate-RBC, kg SBOD}_5\text{/m}^2\cdot\text{d} = \frac{\text{Organic Load, kg SBOD}_5\text{/d}}{\text{Surface Area of Media, m}^2}$$

$$\text{Organic Loading Rate-Trickling Filter, lb BOD}_5\text{/1000 cu ft/d} = \frac{\text{Organic Load, lb BOD}_5\text{/d}}{\text{Volume, 1000 cu ft}}$$

$$\text{Organic Loading Rate-Trickling Filter, kg/m}^3\cdot\text{d} = \frac{\text{Organic Load, kg BOD}_5\text{/d}}{\text{Volume, m}^3}$$

$$\text{Oxygen Uptake Rate or Oxygen Consumption Rate, mg/L}\cdot\text{min} = \frac{\text{Oxygen Usage, mg/L}}{\text{Time, min}}$$

$$\text{Population Equivalent, Organic} = \frac{\text{(Flow, mgd)(BOD, mg/L)(8.34 lb/gal)}}{\text{0.17 lb BOD/d/person}}$$

$$\text{Population Equivalent, Organic} = \frac{\text{(Flow, m}^3\text{/d)(BOD, mg/L)}}{\text{(1000)(0.077 kg BOD/d}\cdot\text{person)}}$$

$$\text{Power, kW} = \frac{\text{(Flow, L/s)(Head, m)(9.8)}}{1000}$$

$$\text{Recirculation Ratio-Trickling Filter} = \frac{\text{Recirculated Flow}}{\text{Primary Effluent Flow}}$$

$$\text{Reduction of Volatile Solids, \%} = \left(\frac{\text{VS in} - \text{VS out}}{\text{VS in} - (\text{VS in} \times \text{VS out})}\right) \times 100\%$$

All information (In and Out) must be in decimal form

$$\text{Removal, \%} = \left(\frac{\text{In} - \text{Out}}{\text{In}}\right) \times 100\%$$

$$\text{Return Rate, \%} = \frac{\text{Return Flowrate}}{\text{Influent Flowrate}} \times 100\%$$

$$\text{Return Sludge Rate-Solids Balance, mgd} = \frac{\text{(MLSS, mg/L)(Flowrate, mgd)}}{\text{(RAS Suspended Solids, mg/L)} - \text{(MLSS, mg/L)}}$$

$$\text{Slope, \%} = \frac{\text{Drop or Rise}}{\text{Distance}} \times 100\%$$

$$\text{Sludge Density Index} = \frac{100}{\text{SVI}}$$

$$\text{Sludge Volume Index, mL/g} = \frac{(\text{SSV}_{30}\text{, mL/L})\text{(1000 mg/g)}}{\text{MLSS, mg/L}}$$

$$\text{Solids, mg/L} = \frac{\text{(Dry Solids, g)(1 000 000)}}{\text{Sample Volume, mL}}$$

$$\text{Solids Capture, \% (Centrifuges)} = \left[\frac{\text{Cake TS, \%}}{\text{Feed Sludge TS, \%}}\right] \times \left[\frac{\text{(Feed Sludge TS, \%)} - \text{(Centrate TSS, \%)}}{\text{(Cake TS, \%)} - \text{(Centrate TSS, \%)}}\right] \times 100\%$$

$$\text{Solids Concentration, mg/L} = \frac{\text{Weight, mg}}{\text{Volume, L}}$$

$$\text{Solids Loading Rate, lb/sq ft/d} = \frac{\text{Solids Applied, lb/d}}{\text{Surface Area, sq ft}}$$

$$\text{Solids Loading Rate, kg/m}^2\cdot\text{d} = \frac{\text{Solids Applied, kg/d}}{\text{Surface Area, m}^2}$$

Solids Retention Time: *see Mean Cell Residence Time*

$$\text{Specific Gravity} = \frac{\text{Specific Weight of Substance, lb/gal}}{\text{8.34 lb/gal}}$$

$$\text{Specific Gravity} = \frac{\text{Specific Weight of Substance, kg/L}}{\text{1.0 kg/L}}$$

$$\text{Specific Oxygen Uptake Rate or Respiration Rate, (mg/g)/h} = \frac{(\text{OUR, mg/L}\cdot\text{min})(\text{60 min})}{(\text{MLVSS, g/L})(\text{1 h})}$$

$$\text{Surface Loading Rate or Surface Overflow Rate, gpd/sq ft} = \frac{\text{Flow, gpd}}{\text{Area, sq ft}}$$

$$\text{Surface Loading Rate or Surface Overflow Rate, L/m}^2\cdot\text{d} = \frac{\text{Flow, L/d}}{\text{Area, m}^2}$$

$$\text{Total Solids, \%} = \frac{(\text{Dried Weight, g}) - (\text{Tare Weight, g})}{(\text{Wet Weight, g}) - (\text{Tare Weight, g})} \times 100\%$$

$$\text{Velocity, ft/sec} = \frac{\text{Flowrate, cu ft/s}}{\text{Area, sq ft}}$$

$$\text{Velocity, ft/sec} = \frac{\text{Distance, ft}}{\text{Time, sec}}$$

$$\text{Velocity, m/s} = \frac{\text{Flowrate, m}^3\text{/s}}{\text{Area, m}^2}$$

$$\text{Velocity, m/s} = \frac{\text{Distance, m}}{\text{Time, s}}$$

$$\text{Volatile Solids, \%} = \left[\frac{(\text{Dry Solids, g}) - (\text{Fixed Solids, g})}{(\text{Dry Solids, g})}\right] \times 100\%$$

$$\text{Volume of Cone}^* = (1/3)(0.785)(\text{Diameter}^2)(\text{Height})$$

$$\text{Volume of Cylinder}^* = (0.785)(\text{Diameter}^2)(\text{Height})$$

$$\text{Volume of Rectangular Tank}^* = (\text{Length})(\text{Width})(\text{Height})$$

$$\text{Water Use, gpcd} = \frac{\text{Volume of Water Produced, gpd}}{\text{Population}}$$

$$\text{Water Use, L/cap}\cdot\text{d} = \frac{\text{Volume of Water Produced, L/d}}{\text{Population}}$$

$$\text{Watts (AC circuit)} = (\text{Volts})(\text{Amps})(\text{Power Factor})$$

$$\text{Watts (DC circuit)} = (\text{Volts})(\text{Amps})$$

$$\text{Weir Overflow Rate, gpd/ft} = \frac{\text{Flow, gpd}}{\text{Weir Length, ft}}$$

$$\text{Weir Overflow Rate, L/m·d} = \frac{\text{Flow, L/d}}{\text{Weir Length, m}}$$

$$\text{Wire-to-Water Efficiency, \%} = \frac{\text{Water hp}}{\text{Motor hp}} \times 100\%$$

$$\text{Wire-to-Water Efficiency, \%} = \frac{(\text{Flow, gpm})(\text{Total Dynamic Head, ft})(0.746\ \text{kW/hp})(100\%)}{(3960)(\text{Electrical Demand, kW})}$$

Pie Wheels

- To find the quantity above the horizontal line: multiply the pie wedges below the line together.
- To solve for one of the pie wedges below the horizontal line: cover that pie wedge, then divide the remaining pie wedge(s) into the quantity above the horizontal line.
- Given units must match the units shown in the pie wheel.
- When US and metric units or values differ, the metric is shown in parentheses, e.g. (m^2).

Area of Circle

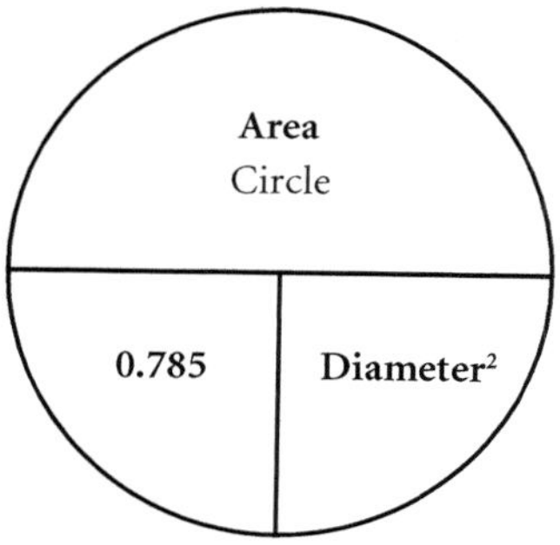

Area of Rectangle

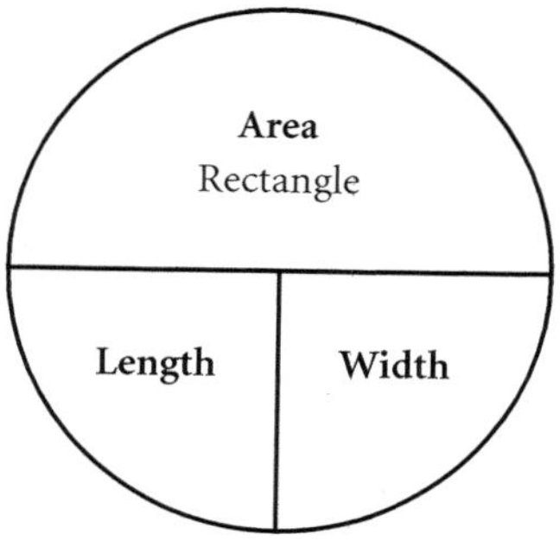

Area of Right Triangle

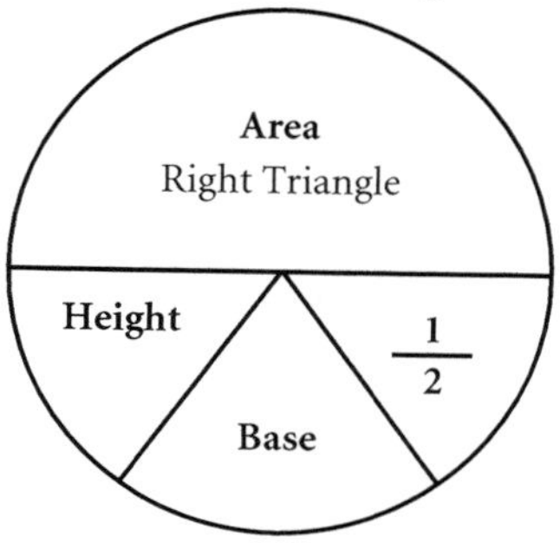

Electromotive Force (EMF), Volts

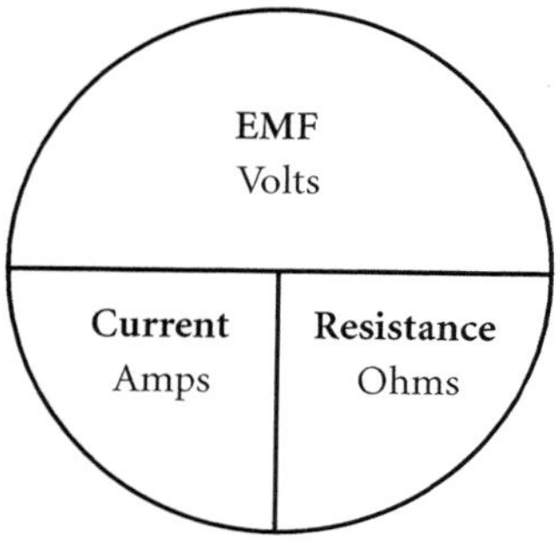

Feed Rate, lbs/day (kg/day)

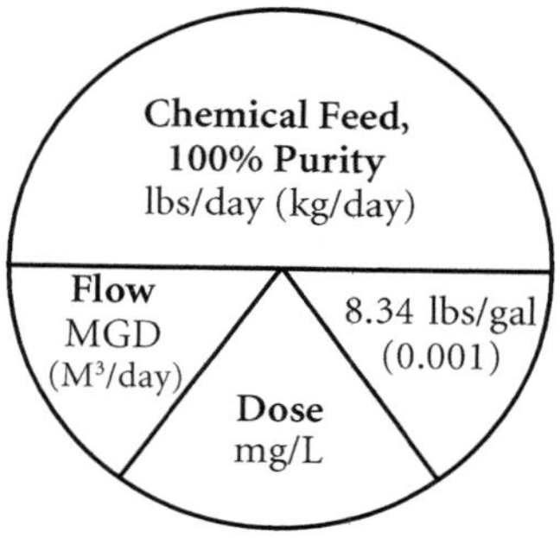

Flow Rate, ft^3/sec (m^3/sec)

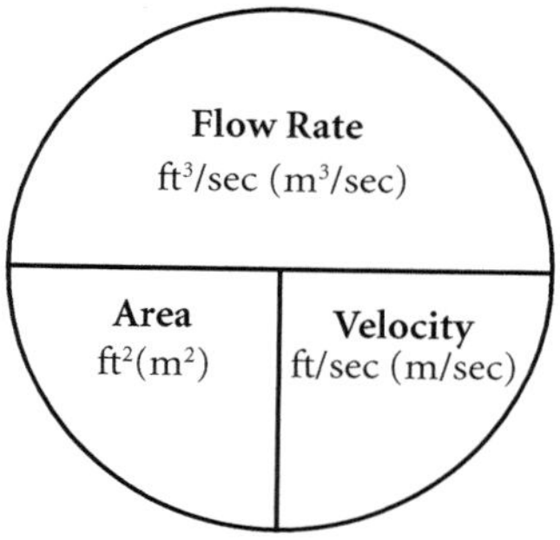

Force, lbs (Newtons)

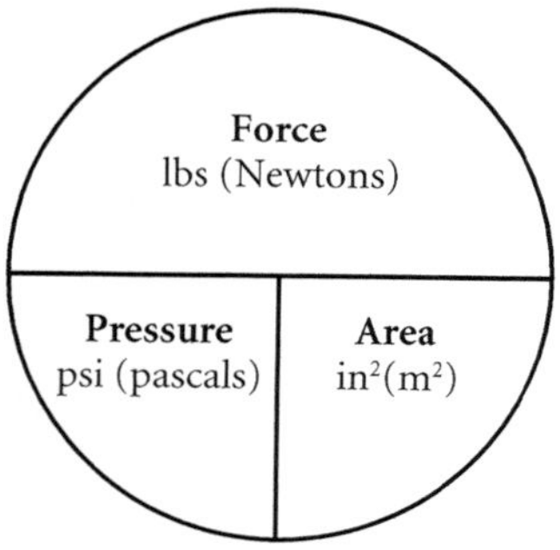

Loading Rate, lbs/day (kg/day)

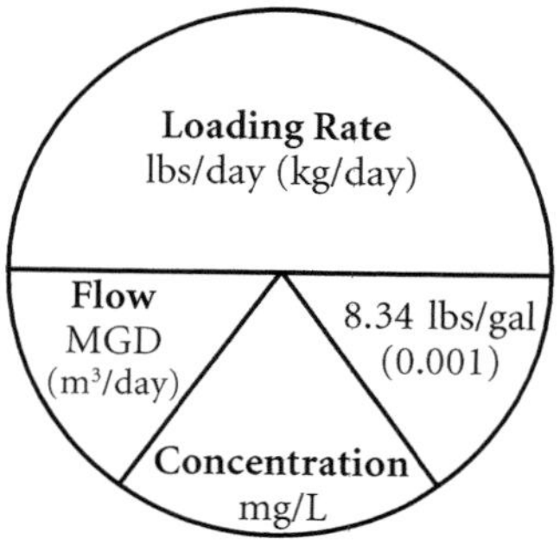

Mass, lbs (kg)

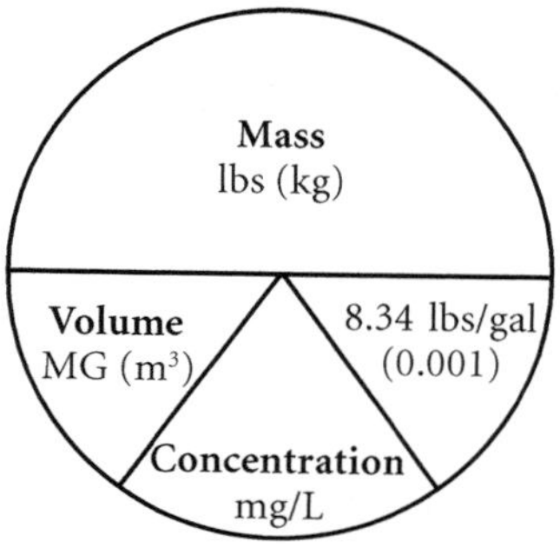

Volume of Cone

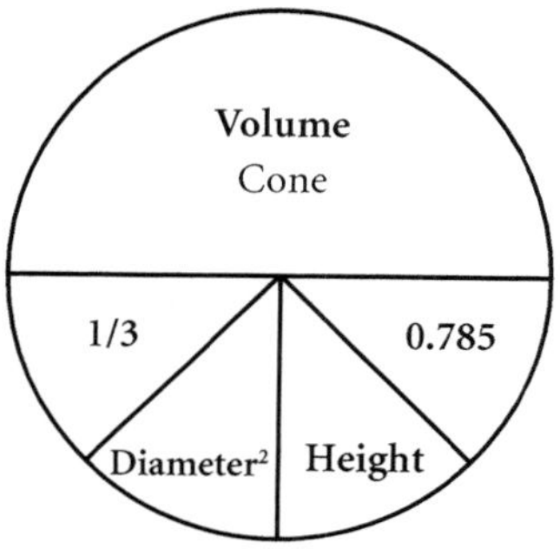

Volume of Cylinder

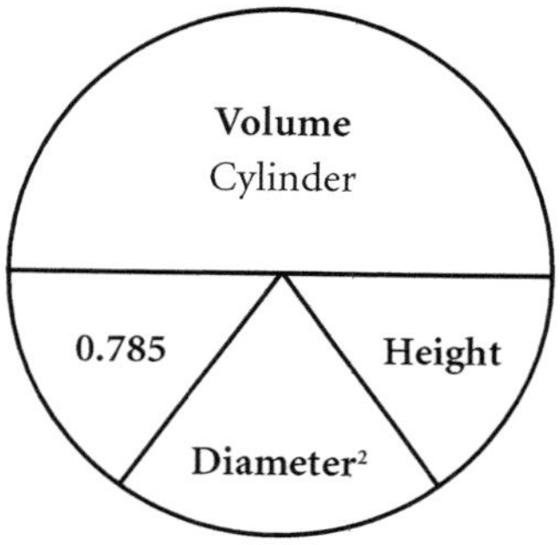

Volume of Rectangular Tank

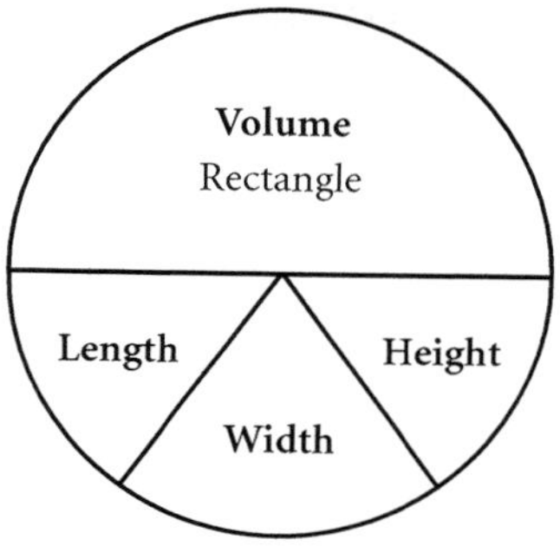

Conversion Factors

1 ac = 4046.9 m^2 or 43 560 sq ft

1 ac ft of water = 326 000 gal

1 atm = 33.9 ft of water
= 10.3 m of water
= 14.7 psi
= 101.3 kPa

1 cfs = 0.646 mgd
= 448.8 gpm

1 cu ft of water = 7.48 gal
= 62.4 lb

1 ft = 0.305 m

1 ft H_2O = 0.433 psi

1 gal (US) = 3.79 L
= 8.34 lb of water

1 gr/gal (US) = 17.1 mg/L

1 ha = 10 000 m^2

1 hp = 0.746 kW
= 746 W
= 33 000 ft lb/min

1 in. = 25.4 mm or 2.54 cm

1 L/s = 0.0864 ML/d

1 lb = 0.454 kg

1 m of water = 9.8 kPa

1 m^2 = 1.19 sq yd

1 m^3 = 1000 kg
= 1000 L
= 264 gal

1 metric ton = 2205 lb

1 mile = 5280 ft

1 mgd = 694 gpm
= 1.55 cfs
= 3.785 ML/d

Population equivalent (PE), hydraulic = 378.5 L/cap·d
= 100 gpd/cap

PE, organic = 0.077 kg BOD/cap·d
= 0.17 lb BOD/cap/d

1 psi = 2.31 ft of water
= 6.89 kPa

1 ton = 2000 lb

1% = 10 000 mg/L

π or pi = 3.14

CHAPTER 1

Introduction to Wastewater Treatment

Introduction

This chapter discusses the need for wastewater treatment, presents a brief overview of some of the pollutants in wastewater, reviews the various laws and regulations applicable to the permitting of water resource recovery facilities (WRRFs), describes the typical *unit processes* used to provide safe and acceptable treatment of wastewater, and discusses treatment and disposal alternatives for the solids removed from wastewater. To be consistent with the evolution of federal and state terminology, this manual uses the term *sludge* to refer to solids separated from wastewater during treatment and the term *biosolids* to refer to sludge after processing criteria have been achieved (i.e., at the outlet of the stabilization process).

Unit processes are a way of talking about distinct treatment steps within a treatment facility. An example of a unit process is a grit basin. A grit basin is a tank where the velocity of incoming wastewater is slowed down to approximately 0.3 m/s (1 ft/sec) to allow large particles of sand and gravel to settle by gravity.

LEARNING OBJECTIVES

Upon completing this chapter, you will be able to

- Understand the need for wastewater treatment.
- List some of the components of domestic wastewater.
- Identify major unit processes of domestic WRRFs.
- Understand the linkages between the liquid stream and solids handling sides of a WRRF. Label appropriately.
- Draw an example WRRF, clearly label the main unit processes, and give the function of each.
- Draw a typical natural treatment system and a typical mechanical treatment system.
- Describe the permitting requirements of the Clean Water Act (CWA) and the biosolids 503 regulations.

What's in a Name?

When the CWA was passed in 1972, it referred to publicly owned treatment works (POTWs). The term *POTW* didn't just include the treatment facility, but also all of the upstream infrastructure necessary to convey wastewater to the facility, including collection system pipes and lift stations. The CWA and legislation in many states still use the term *POTW*. Over the years, as our industry has worked to convey to the public the valuable services we provide, the name of a treatment facility has evolved from POTW to wastewater treatment plant or wastewater treatment facility to water reclamation facility and, finally, to water resource recovery facility (WRRF). The term *WRRF* was officially adopted by the Water Environment Federation in 2014 because it better reflects our goals as a profession. We no longer simply treat water to remove pollutants. A well-designed and operated facility may also reclaim effluent for irrigation and other uses; land-apply biosolids for beneficial reuse; produce methane gas, electricity, and heat; and even recover nutrients. Protecting public health and the environment and recovering valuable resources are the goals of a modern WRRF.

Why Treat Wastewater?

Rivers, streams, lakes, and the ocean all have some *assimilative capacity*; however, when groups of people or animals live close together, the amount of wastewater entering a river or lake in one place can easily overwhelm natural treatment processes. Overloading a waterbody with wastewater can cause

Assimilative capacity is the amount of pollutant that can be absorbed or treated by the environment without damaging the environment or having other negative consequences.

1. Low dissolved oxygen (DO) in the water,
2. Fish kills,

3. Algae blooms,
4. Spread of waterborne diseases like cholera, and
5. Violations of safe drinking water standards.

The objectives of treating wastewater are to clean the wastewater sufficiently to protect both public health and the environment. Water resource recovery facility personnel do this by reducing the concentrations of solids, organic matter, nutrients, *pathogens*, and other pollutants in wastewater.

A *pathogen* is an organism capable of causing disease. Pathogens include parasites, bacteria, and viruses.

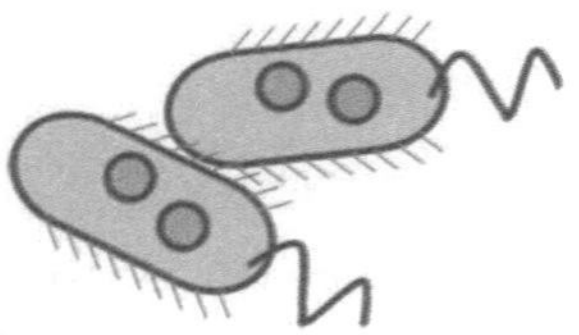

Components of Wastewater

Before going further, some terms that are used to describe the components of the influent wastewater should be defined. For this initial discussion, the definitions of these terms have been kept as simple as possible. Chapter 2 will discuss these terms in greater detail, including typical concentration ranges for municipal wastewater and how different constituents are related to one another.

WATER

Municipal wastewater contains various wastes; it typically consists of approximately 99.94% liquid and 0.06% solids. A typical U.S. city, including its private dwellings, commercial establishments, and industrial contributors, produces between 379 and 455 L/d·person (100 and 120 gpd/person) of wastewater—not including the water that infiltrates and exfiltrates the collection system. *Infiltration* refers to groundwater that enters the sewer through cracks in pipes and manholes. *Inflow* refers to groundwater, rain, and surface water that enter the sewer through direct openings such as open drain cleanouts. Exfiltration can occur when the groundwater level falls below the sewer. In this case, wastewater can seep out of pipes and into the ground. Per person water usage has been decreasing, particularly in the western United States, because of a combination of conservation, installation of water-saving fixtures, and modernization of sewer systems with plastic pipe. A survey of 12 western cities with 1188 homes measured median indoor water use between 204 and 241 L/d·person (54.0 and 63.8 gpd/person); however, this does not include contributions from commercial or industrial users (U.S. EPA, 2005).

SOLIDS

Raw wastewater contains a wide variety of solid material. Large debris such as rags, wipes, dentures, mop heads, and almost anything else you can imagine that might end up being flushed down a toilet or tossed into a floor drain—including the occasional diamond ring—shows up at the front end of a WRRF. These solids can damage equipment at the WRRF and plug pipes. Raw wastewater also contains smaller solids like sand and grit, dissolved salts, and solid organic material from toilets, washing machines, garbage disposals, and other sources. Solids are classified according to their size, whether they can pass through a 1.2-*μm* filter, and whether they are organic or inorganic. Organic solids can be burned or volatilized in a furnace at 550 °C. Inorganic solids or ash is what remains after a sample has been volatilized. Commonly used terms when discussing solids in wastewater are listed in Table 1.1.

The symbol *μm* stands for micron. There are 1000 μm in one millimeter. A human hair is 40 to 50 μm across, and bacteria may be only 0.5-μm wide and 1-μm long.

BIOCHEMICAL OXYGEN DEMAND

The *biochemical oxygen demand (BOD) test* was originally performed in England as a way to estimate the organic strength of wastewater. There are so many different organic compounds in wastewater that

The *BOD test* is conducted for 5 days at 20 °C because of historic convention. We think of BOD as a measure of organic material, but the test actually measures how much oxygen is needed to treat the wastewater.

Table 1.1 Types of Solids in Wastewater

Acronym	Term	Definition
TSS	Total Suspended Solids	Solids that cannot pass through a 1.2-μm filter
TVSS	Total Volatile Suspended Solids	Solids that cannot pass through a 1.2-μm filter AND are burned away when placed in a furnace at 550 °C
TDS	Total Dissolved Solids	Solids that are small enough to pass through a 1.2-μm filter. The sample must be dried completely before the dissolved solids can be seen with the naked eye.
TS	Total Solids	All of the solid material in a sample. This includes both organic and inorganic solids. TS = TSS + TDS
TVS	Total Volatile Solids	All of the solids in a sample that are burned away when placed in a furnace at 550 °C.

it would be very expensive and nearly impossible to measure them each individually. Instead, the BOD test measures how much oxygen is needed by the bacteria to stabilize the biodegradable organic material present in the wastewater. Biochemical oxygen demand allows regulators to estimate the effect the effluent from a WRRF might have on a river or lake. When excessive amounts of BOD are discharged to a river or lake, it can overwhelm the natural treatment capacity, consume most or all of the available oxygen, and result in fish kills and other environmental degradation. The treatment capacity of a WRRF is expressed in terms of how much flow it can treat in cubic meters per day (million gallons per day) and how much organic material it can process in kilograms per day (pounds per day).

The BOD test is conducted by taking a sample of wastewater, diluting it if necessary to bring it within the measureable range of the test, measuring the starting DO concentration, incubating the sample at 20 °C (68 °F) for a fixed period of time in the dark, and then measuring the ending DO concentration (Figure 1.1). The BOD concentration in milligrams per liter is calculated from the amount of oxygen consumed over the test period divided by the sample size used. The reason the test is conducted over a 5-day period is because the longest river in England is the Thames River. It takes approximately 5 days for water to flow from London to the ocean. Once the wastewater reached the ocean, it was thought that any remaining BOD did not matter. Similarly, the test is conducted at 20 °C because the warmest the Thames River got in the summer was 18.6 °C (65.5 °F). The test was standardized by rounding up to 20 °C.

The more organic material that the wastewater contains, the more oxygen the bacteria will use to consume and stabilize it. By definition, 1 kg of BOD will consume 1 kg of oxygen (1 lb of BOD consumes 1 lb of oxygen). The BOD is thought of in terms of organic strength, but the test really measures how much oxygen is needed to treat or stabilize the wastewater. The BOD can be either solid organic material, like bits of food from a garbage disposal, or dissolved organic material, like sugar in tea.

The BOD can be further divided into carbonaceous biochemical oxygen demand (CBOD) and the nitrogenous oxygen demand (NOD), as shown below. The amount of oxygen used in the BOD test that goes toward breakdown of organic material is the CBOD. Additional oxygen may be used by a special group of bacteria, the nitrifying bacteria, to convert ammonia to nitrite and nitrate. This reaction consumes 4.5 kg of oxygen for every 1 kg of ammonia converted to nitrate. When this reaction takes place in the BOD test, the additional oxygen demand is the NOD. If an analyst wants to measure only the CBOD, a substance that is toxic to the nitrifying bacteria will be added to the BOD test. For raw domestic wastewater (influent), the CBOD is typically approximately 85% of the total BOD (WEF, 2012).

$$BOD_{total} = CBOD + NOD$$

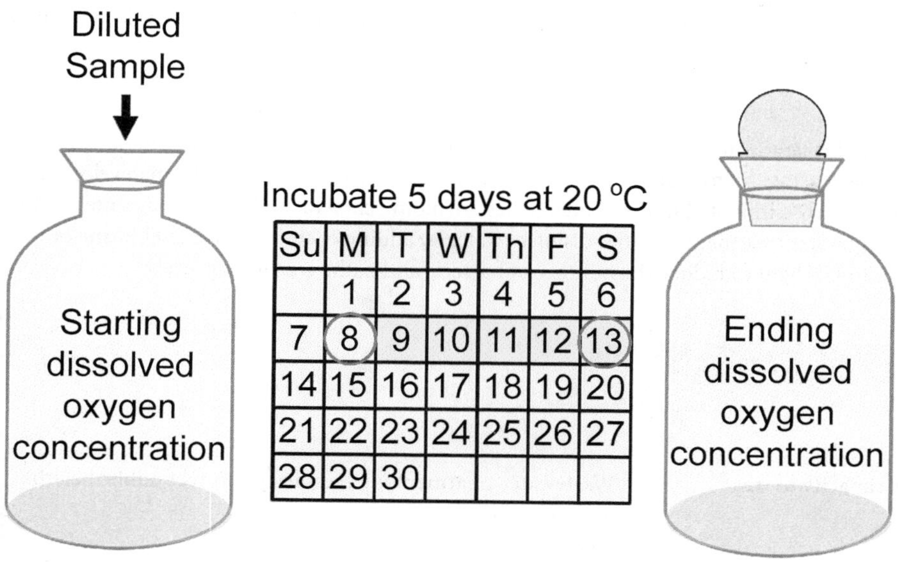

Figure 1.1 Biochemical Oxygen Demand (Courtesy of Indigo Water Group)

The BOD consists of both soluble (dissolved) and particulate fractions. If a wastewater sample is filtered through a piece of filter paper, some of the organic material will pass through the filter and some won't be able to pass through. The organic material that goes through the filter is the dissolved or soluble BOD. The organic material that remains on the filter paper is the particulate BOD. A can of soda contains a lot of sugar that is both organic and able to pass through a piece of filter paper because it is dissolved. The percentage of BOD that is soluble depends on the size of the collection system and the types of customers served. For domestic wastewater, including light commercial such as stores and schools, soluble BOD is typically between 20 and 40% of the total BOD (WEF, 2012).

NUTRIENTS

Influent wastewater also contains the macronutrients nitrogen and phosphorus as well as trace nutrients like iron and manganese. Nitrogen is present in many compounds in influent wastewater including urine, organically bound nitrogen (proteins and other compounds), and ammonia. Organically bound nitrogen can be soluble or particulate, whereas ammonia is only present as soluble. Phosphorus can also be particulate or dissolved. Phosphorus is present in proteins, urine, and detergents.

Ammonia can be toxic to fish and other aquatic life. It can be converted to nitrate during the treatment process, which mitigates the toxicity issue for aquatic life but does not reduce the amount of nitrogen going to the receiving stream or lake. Nitrate concentrations of 10 mg/L as nitrogen in drinking water supplies can cause blue baby syndrome or Methemoglobinemia in babies, the elderly, and persons with blood disorders. To better protect streams, lakes, and downstream water users, total nitrogen removal may be necessary. For total nitrogen removal, nitrate is converted to nitrogen gas biologically and then released to the atmosphere, which is approximately 78% nitrogen.

An *algae bloom* is a sudden overgrowth of algae.

Turbidity measures the cloudiness of water. A light is shined through the water and a detector, placed at a 90-deg angle to the light source, measures the amount of light scattered by suspended particles in the water.

When nutrients are discharged to rivers and lakes, they act as fertilizer and can cause *algae blooms*. Excessive amounts of algae can cause large fluctuations in pH and DO concentrations over a 24-hour period that can be harmful to fish and other aquatic life. Algae can also increase the *turbidity* of the water, making it more difficult for fish to find suitable places to lay their eggs. Although smaller increases in the amount of nitrogen and phosphorus discharged to a river or lake may not result in an algae bloom, they can still cause changes to the environment by altering the numbers and types of plant and animal species that are able to thrive under higher nutrient conditions.

FATS, OILS, AND GREASE

Fats, oils, and grease (FOG) in facility effluent can result in floating material in the receiving water. The FOG can enter the facility as discrete floatable particles, as emulsified material, or as a solution. Most of the FOG entering a domestic WRRF originates from plant and animal sources as food-based FOG. Think of cooking with vegetable oil or frying bacon. The FOG can also be from petroleum-based sources like industrial lubricants. The FOG is problematic in the collection system and in WRRFs because it can adhere to surfaces and create blockages.

BACTERIA AND PATHOGENS

Raw wastewater contains large numbers of bacteria, viruses, and protozoans. Some of these organisms are pathogenic or disease causing. Many waterborne illnesses that continue to be problematic in developing countries have been nearly eliminated in the United States, including cholera, typhoid, and dysentery. Most WRRFs and drinking water treatment facilities disinfect water by adding chemicals such as chlorine or exposing the water to UV light to reduce the number of bacteria and pathogens to safe levels.

TEST YOUR KNOWLEDGE

1. **Natural systems have enough assimilative capacity to treat wastewater from urban areas.**
   - ☐ True
   - ☑ False

2. **Wastewater treatment is necessary to protect public health and the environment.**
   - ☑ True
   - ☐ False

3. The term *water resource recovery facility* (WRRF) was recently adopted by Water Environment Federation because it better reflects our goals as a profession.
   - ☑ True
   - ☐ False

4. Most of the fats, oils, and grease present in influent wastewater come from petroleum products.
   - ☐ True
   - ☑ False

5. A pathogen is
   - a. A bacteria or virus found in wastewater
   - b. Any organism capable of causing disease
   - c. Unable to survive for long periods outside wastewater
   - d. Dependent on TSS to reproduce

6. Solids that are retained by a 1.2-μm filter paper and are burned away in a 550 °C furnace are:
   - a. Total dissolved solids (TDS)
   - b. Total volatile solids (TVS)
   - c. Total volatile suspended solids (TVSS)
   - d. Total non-volatile dissolved solids (TVDS)

7. The biochemical oxygen demand (BOD) test measures this:
   - a. Biodegradable organic material
   - b. Percentage of organic suspended solids
   - c. Quantity of live bacteria
   - d. Amount of oxygen needed to stabilize wastewater

8. Solids that are able to pass through a 1.2-μm filter paper and remain unchanged after spending time in a furnace at 550 °C may be described as
   - a. Dissolved and inorganic
   - b. Suspended and inorganic
   - c. Dissolved and organic
   - d. Suspended and organic

9. Some treatment facilities are required to remove ammonia as part of their discharge permits for this reason.
   - a. There is a safe drinking water limit for ammonia.
   - b. Ammonia reacts with organic matter in natural systems to form mustard gas.
   - c. To protect downstream agricultural users from over fertilizing crops.
   - d. Ammonia is toxic to fish and aquatic life.

10. The biochemical oxygen demand (BOD) test is typically performed as a 5-day test for this reason.
    - a. It only takes the Thames River 5 days to reach the ocean.
    - b. The bottles only hold enough dissolved oxygen for a 5-day test.
    - c. The bacteria only live for 5 days.
    - d. All of the organic material is consumed within 5 days.

11. By definition, how much oxygen is required to stabilize or treat 1 kg (lb) of BOD?
    - a. 1 kg (lb)
    - b. 2 kg (lb)
    - c. 3 kg (lb)
    - d. 4 kg (lb)

12. Which of the following pollutants is most likely to cause an algae bloom in a lake or river?
    - a. Total suspended solids (TSS)
    - b. Biochemical oxygen demand (BOD)
    - c. Phosphorus
    - d. Turbidity

The Clean Water Act

Water resource recovery facilities protect the natural environment from the wastewater generated in urban areas. Because uncontrolled releases of wastewater would degrade the water, land, and air on which life depends, the government has developed a comprehensive set of laws and regulations on safely treating and disposing of wastewater and sludge, which are summarized here. Anyone discharging wastewater to the environment must follow those laws and have an active discharge permit. The permitting process is explained in more detail within this section. Government agencies continually modify laws and regulations, so water and wastewater treatment professionals should review the current discharge rules when preparing and submitting the permit application for a specific facility. Operators should be familiar with the requirements of the discharge permit for their facility. It is a good idea to read through the entire discharge permit periodically to ensure that all requirements are being met. Some requirements for monitoring and reporting may be on a quarterly or annual basis and can be easily missed or forgotten.

Congress enacted the Water Pollution Control Act of 1948 in response to growing public concern about water pollution. When the Act was amended in 1972, this law became known as CWA. It established the basic structure for regulating discharges of pollutants to U.S. waters. The Act's four guiding principles are:

1. No one has the right to pollute U.S. waters, so a permit is required to discharge any pollutant;
2. Permits limit the types and concentrations of pollutants allowed to be discharged, and permit violations can be punished by fines and imprisonment;

3. Some industrial permits require companies to use the best treatment technology available regardless of the receiving water's assimilative capacity; and
4. Pollutant limits involving more treatment than technology-based levels, secondary treatment (for municipalities), or best practicable technology (for industries) are based on waterbody-specific water quality standards. Water quality based standards are set to protect aquatic life, downstream water supplies, and other uses such as agriculture and recreation.

The CWA's primary objective is to restore and maintain the chemical, physical, and biological integrity of U.S. waters. The CWA set two national goals: achieve fishable and swimmable waters (wherever possible) by 1983 and eliminate all pollutant discharges to navigable waters by 1985. The Act's 1981 and 1987 amendments reaffirm the importance of achieving these goals and give states the primary responsibility of implementing CWA through the permitting system established in Section 402.

Anyone discharging pollutants to U.S. waters must have a National Pollutant Discharge Elimination System (NPDES) permit. In some states, discharge permits are issued by the U.S. Environmental Protection Agency (U.S. EPA). Some states have primacy, which means that U.S. EPA has delegated their authority to issue discharge permits to the states. These permits are State Pollution Discharge Elimination System (SPDES) permits. Dischargers without permits or those who exceed their permit limits are violating the law and are subject to civil, administrative, or criminal penalties. Maximum federal criminal penalties are shown in Table 1.2 (Title 18, United States Code, Section 1001). The federal penalties shown in Table 1.2 apply when a person knowingly and willfully falsifies, conceals, or covers up a permit violation or misrepresents a permit violation with fraudulent statements. The permitting system depends on accurate self-monitoring and reporting. In addition to federal penalties, individual states may levy their own civil, administrative, or criminal penalties. It is worth noting that federal and state agencies understand that WRRFs will have occasional process upsets and discharge permit violations. A discharge permit violation typically does not result in a fine or penalty unless the discharger is a repeat offender or willful violator. The important thing for an operator is to fully disclose any violation, why it occurred, and how it will be prevented from occurring again in the future.

Concentrations of pollutants are typically expressed in µg/L (parts per billion) or mg/L (parts per million). Mass loading is the total amount of pollutant in kilograms or pounds.

The NPDES or SPDES permits typically specify the discharge location, the allowable discharge flows, the allowable *concentrations (mass loads) of pollutants* in the discharge, the limits of the mixing zone (if any), and facility-specific sampling, monitoring, and reporting requirements. The permit will also list the treatment capacity of the WRRF in terms of flow and organic load. Most discharge permits will require a WRRF to be in the planning stages for the next expansion when either the monthly flow or load reaches 80% of the permitted capacity and to be under construction when either the monthly flow or load reaches

Table 1.2 Maximum Federal Penalties for Discharge Permit Violations

Violation Type		Maximum Federal Fine ($/day per violation)	Imprisonment (Years)
Negligent*	Violation occurred because of an action (or inaction) taken by the operator that was not consistent with industry standards. In other words, the violation was not intentional, but resulted from poor operational practice.	$25,000	1
Knowing*	Violation was the direct result of an action (or inaction) taken by the operator even though the operator knew it would result in a permit violation.	$50,000	3
Knowing Endangerment*	Violation was the direct result of action (or inaction) taken by the operator even though the operator knew it would 1) result in a permit violation and 2) would have a negative environmental effect.	$250,000	15

*The terms *negligent violation, knowing violation,* and *knowing endangerment* are not explicitly defined in CWA.

95% of the permitted capacity. Before it can be discharged to U.S. waters, municipal wastewater must have received secondary treatment—or more stringent treatment if necessary to meet water quality standards. The secondary treatment standards are shown in Table 1.3 (40 CFR § 133.102). The effluent *pH* must be maintained between 6.0 and 9.0 standard units (SU). When establishing permit requirements for a WRRF, regulators may consider the following issues in addition to the minimum regulatory requirements:

pH is expressed in standard units (SU) and is the negative log of the hydrogen ion concentration. pH is discussed in more detail in Chapter 2.

1. Preventing disease,
2. Preventing nuisances,
3. Protecting drinking water supplies,
4. Conserving water,
5. Maintaining navigable waters,
6. Protecting waters for swimming and recreational use,
7. Maintaining healthy habitats for fish and other aquatic life, and
8. Preserving pristine waters to protect ecosystems.

Permit applications and required reports must be signed by an authority state-licensed individual of responsible charge, a principal executive officer, or ranking elected official. According to 40 CFR 122.22, all permits and reports submitted by a corporation must be signed by a responsible corporate officer; those submitted by a partnership or sole proprietorship must be signed by one of the general partners; and those submitted by a municipality, state, federal, or other public agency must be signed by a principal executive officer or ranking elected official.

U.S. EPA defines *interference* as a discharge that inhibits or disrupts the WRRF, its treatment processes or operations, or its sludge processes, use, or disposal and, therefore, causes a violation of any requirement of the WRRFs discharge permit.

Water resource recovery facilities designed to handle 19 000 m³/d (5 mgd) or more (and smaller ones with *interference* and pass-through problems) must establish pretreatment programs to regulate industrial and other nondomestic wastes discharged into sewers. A pretreatment program allows a WRRF to issue discharge permits to commercial and industrial users in their service area. An industrial user that discharges wastewater to the sewer is called an *indirect discharger*. A WRRF or industrial user that discharges treated wastewater directly to the receiving stream is called a *direct discharger*. Direct dischargers are permitted through U.S. EPA or a state agency (refer to Figure 1.2). Indirect dischargers are permitted by the WRRF that receives their wastewater.

If a WRRF is not designed to remove a particular pollutant from an *indirect discharger* or if a large enough quantity of a particular pollutant enters the WRRF, it may "pass through" the treatment process partially or completely untreated. The pretreatment program helps prevent these pollutants from entering the WRRF.

The CWA authorizes U.S. EPA to regulate the discharge of toxic chemicals to the environment. It also declares that U.S. policy is to prohibit toxic amounts of these pollutants from being discharged. Soluble toxics can threaten human health if they are in water used for drinking or swimming. Insoluble toxics can adsorb to sediment, be consumed by aquatic life, and thus enter the human food chain (via *bioaccumulation*).

The Act currently lists 126 priority pollutants (40 CFR 423, Appendix A) that must be controlled. These pollutants are divided into four classes: heavy metals and cyanide, volatile organic compounds, semivolatile organic compounds, and pesticides and poly-chlorinated biphenyls (PCBs). Also, certain nontoxic organic compounds in wastewater can become toxic chlorinated organic compounds, such as trihalomethanes, when the water is disinfected via chlorine.

Bioaccumulation is a gradual increase of a pollutant within an organism. The pollutant is taken up faster than it can be processed and eliminated. Bioaugmentation can concentrate pollutants further through the food chain. Algae might bioaccumulate selenium. Small fish eat the algae and the selenium is transferred to the fish; however, because the fish lives longer and eats a lot of algae, the selenium concentration in the fish will be higher than it was in the algae. A bigger fish might eat the smaller fish, increasing the selenium concentration further. Finally, a bear or a human might eat the fish. The selenium concentrates as it moves up the food chain.

Analysts use whole effluent toxicity testing to determine whether WRRF effluent is toxic to humans or the aquatic environment. The test is a bioassay in which sensitive aquatic organisms are put in a container of effluent and monitored for several days to see how well they survive. If the effluent is toxic, regulators require that WRRF staff systematically reduce the toxic components via a toxics reduction evaluation and appropriate WRRF or pretreatment-program modifications.

Table 1.3 Secondary Treatment Standards

	30-Day Average, mg/L	7-Day Average, mg/L	Minimum Percent Removal
BOD_5	30	45	85%
$CBOD_5$	25	40	85%
TSS	30	45	85%

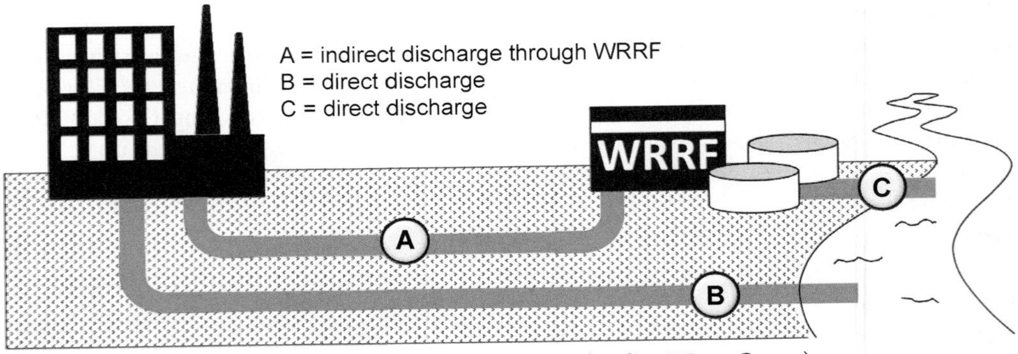

Figure 1.2 Direct and Indirect Dischargers (Courtesy of Indigo Water Group)

The CWA has helped improve water quality tremendously. Many lakes and rivers that were grossly polluted in the 1970s, when the CWA was enacted, now support aquatic life. In the 1960s, worst-case DO levels ranged from 1 to 4 mg/L in heavily polluted waterways; testing in 1985 and 1986 showed that DO levels had risen to between 5 and 8 mg/L.

TEST YOUR KNOWLEDGE

1. The Clean Water Act was promulgated in 1972.
   - ☑ True
   - ☐ False
2. All direct dischargers are required to have a discharge permit.
   - ☑ True
   - ☐ False
3. The discharge permit system depends on accurate self-reporting of effluent quality.
   - ☑ True
   - ☐ False
4. An operator will be fined and serve time in prison if the average effluent suspended solids concentration exceeds the limit given in the discharge permit.
   - ☐ True
   - ☑ False
5. The monthly discharge monitoring report must be signed by the town attorney.
   - ☐ True
   - ☑ False
6. U.S. EPA can set limits for heavy metals, cyanide, volatile organic compounds, and pesticides in direct discharge permits.
   - ☑ True
   - ☐ False
7. A WRRF has a 30-day monthly average BOD_5 limit in their permit of 30 mg/L. Two samples were collected in May with results of 28 and 36 mg/L. The operator should:
   a. Report only the first result to remain below the permit limit.
   b. Average the results together and report a permit violation.
   c. Alter the second result to read 26 mg/L and then average the results together.
   d. Go back to his office and work on his resume.
8. The secondary treatment standards set effluent limits for these parameters:
   a. BOD_5, $CBOD_5$, TSS, and pH
   b. FOG, BOD_5, and TSS
   c. Nitrogen and phosphorus
   d. Pathogenic organisms
9. ABC Corporation manufactures tires in Metro City. All of the process water they generate is discharged to the sewer and is conveyed to the city's water resource recovery facility (WRRF). What type of discharger is ABC Corporation and who issues their discharge permit?
   a. Direct, U.S. EPA
   b. Indirect, state
   c. Indirect, U.S. EPA
   d. Indirect, city WRRF

Treatment Goals

Water resource recovery facilities may be designed and operated to remove BOD and TSS, to convert ammonia to nitrate or nitrate to nitrogen gas, and/or to remove phosphorus. Specialized groups of bacteria are needed to remove ammonia, nitrate, and phosphorus. The details of the different microorganisms and

Table 1.4 Basic Wastewater Treatment Processes

Physical	Biological	Chemical
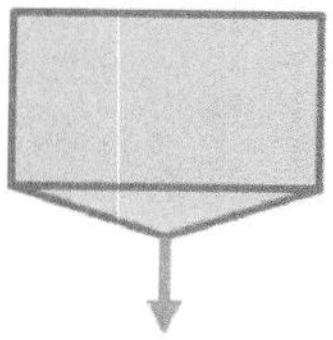	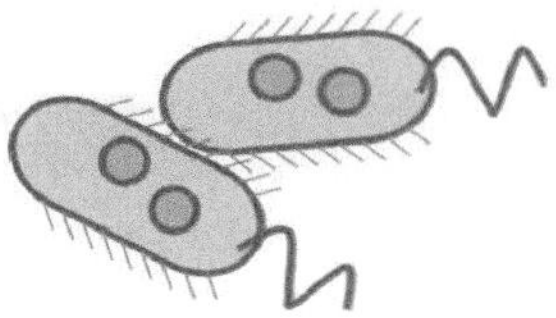	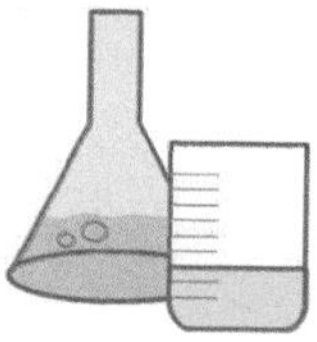
Physical treatment processes mechanically separate large solids from the wastewater. Screens are often the first treatment step in a WRRF. There are many different types of screens, with openings as small as a few millimeters or as large as 5 cm (2 in.). Screens remove rags, sticks, and other large debris. Particles that are too small to be screened, but are heavier than water, can be settled out using gravity. Gravity separation takes place in large, quiescent tanks including grit basins, primary clarifiers, and secondary clarifiers. The opposite of gravity separation is flotation. Some types of solids can be removed from wastewater by creating a bubble curtain of compressed air at the bottom of a tank. The bubbles carry the floatable material to the top of the tank where it is skimmed off. Filtration is another type of physical treatment process.	When wastewater is discharged to natural systems like rivers, lakes, and wetlands, bacteria and other organisms use the organic matter and nutrients in the wastewater as their food. These pollutants are turned into more bacteria, carbon dioxide, water, and other byproducts. As long as the assimilative capacity of the natural system isn't exceeded, these natural systems are capable of treating our wastes. Most natural treatment systems don't have large enough bacterial populations or enough oxygen to treat large amounts of waste quickly. Water resource recovery facilities use the same bacteria and organisms to break down wastes. The process is sped up by concentrating the number of bacteria available to do the work, by adding more oxygen than would be available in a natural system, and by keeping treatment temperatures as warm as possible.	Chemicals are used for a variety of purposes in wastewater treatment including: 1. Sodium hydroxide (caustic soda) and/or sodium carbonate/bicarbonate compounds (soda ash) to adjust the pH or alkalinity of the wastewater. 2. Ferric chloride for odor control; 3. Alum or aluminum compounds to coagulate solids and make particles stick together for easier removal; 4. Ferric chloride, alum, or other iron or aluminum salts to remove phosphorus; 5. Chlorine gas or bleach to disinfect treated wastewater; 6. Sodium bisulfite, sulfur dioxide, and other chemicals to remove chlorine before returning treated wastewater to the environment; and 7. Polymer for flocculation and to thicken and dewater biosolids.

biological reactions required to achieve each of these goals are discussed in detail in other chapters. Water resource recovery facilities may also be designed and operated to recover valuable resources, including recycled water, biosolids, biogas, heat, electricity, nitrogen, and phosphorus. The following sections will address some of the process goals for each of the design objectives previously cited.

Basic Wastewater Treatment Processes

Wastewater treatment processes can be grouped into three broad categories: physical treatment, biological treatment, and chemical treatment (refer to Table 1.4). Most WRRFs will use a combination of all three.

Liquid Treatment Processes

Wastewater treatment is a multistage process designed to clean water and protect natural waterbodies. Water resource recovery facilities may contain preliminary treatment, primary treatment, secondary treatment, tertiary treatment, and/or disinfection, as shown in the example activated sludge facility portrayed in Figure 1.3. Most WRRFs also have solids processing facilities that further stabilize the materials that are removed from the incoming wastewater. In this example, an anaerobic digester is shown; however, smaller facilities typically use aerobic digesters. The liquid and solids handling sides of a WRRF interact as solids and liquids are recycled between them. A particular WRRF may have one or more of these stages of treatment. There is tremendous variety in the processes used from one WRRF to another, partly because they have been constructed over more than 100 years, but also because treatment requirements can change depending on where the facility is located. A WRRF that discharges a large volume of treated wastewater

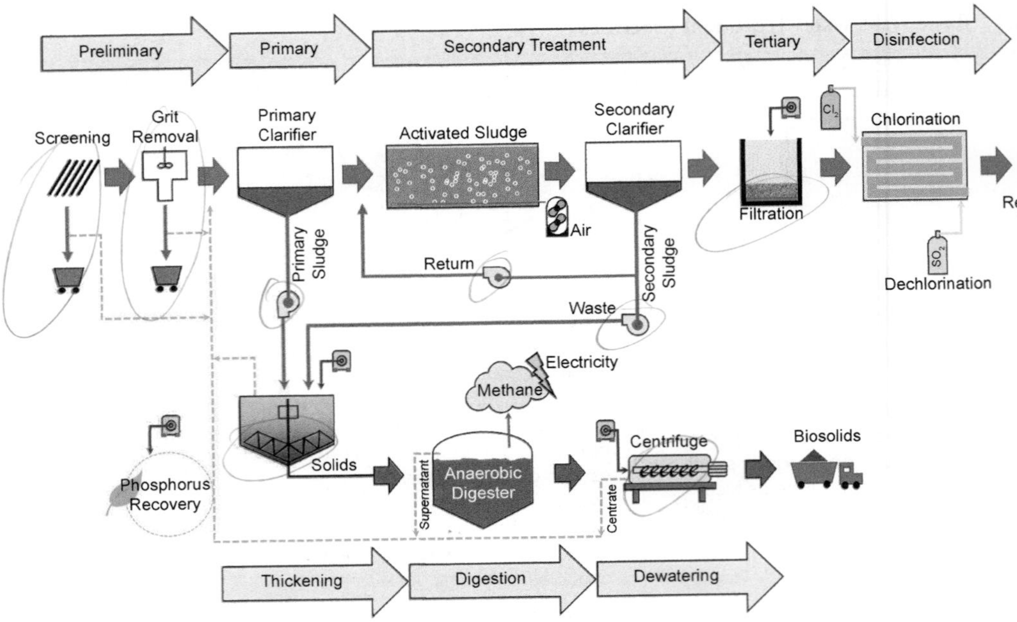

Figure 1.3 Treatment Process Overview (Courtesy of Indigo Water Group)

into a small, pristine trout stream will have more stringent ammonia limits in their discharge permit than a similarly sized facility that discharges through an ocean outfall or large river.

COLLECTION SYSTEM

A collection system is a network of pipes, conduits, tunnels, equipment, and appurtenances used to collect, transport, and pump wastewater. Typically, the wastewater flows through the network via conventional gravity sewers, which are designed so that the size and slope of each pipe will maintain flow toward the discharge point without pumping. Lift stations and pumping stations are used to move wastewater when gravity flow is not possible. Such lines under pumping pressure are referred to as *force mains*.

There are three principal types of municipal wastewater conveyance systems: sanitary sewers, storm sewers, and combined sewers. Sanitary sewers convey wastewater from residential, commercial, institutional, or industrial sources, as well as small amounts of groundwater infiltration and stormwater inflow. Storm sewers convey stormwater runoff and other drainage. Combined sewers convey both sanitary wastes and stormwater.

The type of collection system can profoundly affect WRRF operations. For example, the high flows and heavy sediment loads discharged by a combined system during and after a storm make effective treatment more challenging. Excessive infiltration and inflow in a poorly maintained sanitary system can have similar results. In warm climate conditions, long, flat, or extensive collection systems may discharge foul-smelling, septic wastewater to the facility unless the odors are controlled en route and will also have elevated levels of soluble organic constituents.

Today, municipalities rarely construct combined sewers, and most have made efforts to separate stormwater from sanitary wastewater. Their efforts may be inexpensive, such as requiring homeowners to disconnect roof drains from municipal sewer systems, or major construction projects, such as rerouting stormwater from sanitary sewers to nearby creeks and rivers.

PRELIMINARY TREATMENT

Preliminary treatment typically begins with removing larger materials such as wood, cardboard, rags, plastic, grit, grease, and scum that might damage downstream equipment or impair downstream operations. These materials may be removed with chemical addition, bar racks, screens, comminutors, and/or grit chambers. Preliminary treatment takes place at the WRRF headworks. Headworks may also include flow measurement, flow equalization, pumping, and/or odor control.

An example of a WRRF headworks is shown in Figure 1.4. Any given WRRF headworks may have all, some, or none of the unit processes shown in Figure 1.4. The individual unit processes may also be arranged in a different order, with flow measurement coming before screening and grit removal, for example. In the example headworks shown in Figure 1.4, the first unit process is a screen. Screens remove larger debris and may have openings as small as a few millimeters or as wide apart as 5 cm (2-in.). Figure 1.5 shows a manual bar screen with openings that are 2.5-cm (1-in.) apart. Screens must be cleaned or raked either automatically or manually by an operator or maintenance person several times during the day. This will stop large debris from entering the WRRF, but rags and stringy material may still pass. A manual screen with large openings is sometimes called a *trash rack*. Many different types of screens are used in wastewater treatment and will be discussed further in Chapter 3. The second unit process is a comminutor or grinder. The comminutor reduces the size of debris that was able to pass through the screen. The goal is to reduce the chance of damage to downstream equipment or processes and clogging of pipes.

Screens are typically followed by grit basins, which are channels or small tanks where the wastewater velocity or speed is decreased to approximately 0.3 m/s (1 ft/sec). Here, heavier particles like sand, grit, eggshells, and heavier organic particles settle to the bottom of the tank where they can then be removed, while lighter organic particles remain suspended in the wastewater and pass on to the next process. Figure 1.6 shows two parallel grit basins at the headworks of a WRRF that treats up to 6057 m^3/d (1.6 mgd) of flow. A slide gate is directing flow to one grit basin. Accumulated grit can be seen in the bottom of the dry grit basin. Vortex-type grit basins work by increasing the velocity through a circular basin to separate heavier grit particles from lighter, organic particles using centrifugal force. The volume of screenings and grit collected is highly dependent on the community served and the type of screening equipment used. Screenings and grit are typically sent to a landfill for disposal.

PRIMARY TREATMENT

Primary treatment follows preliminary treatment. Some facilities use a primary treatment process, such as a primary clarifier, to slow the water further and remove heavier organic material and other particles by gravity. Grease and floatable material collects on the water surface and is skimmed off. Wastewater remains in a primary clarifier for approximately 2 hours (WEF, 2017). Primary clarifiers may be rectangular, like the one shown in Figure 1.7, or circular, like the one shown in Figure 1.8. It is important to note that a primary clarifier can only remove settleable and floatable material, which means removal rates are limited by the characteristics of the incoming wastewater—that is, the percentages of particulate and soluble material. For domestic wastewater that is coming primarily from homes and light commercial activities like schools and stores, a primary clarifier can be expected to remove as much as 60 to 75% of suspended solids and between 20 and 35% of total BOD_5. It does not, however, remove *colloidal solids*, dissolved solids, soluble BOD_5, soluble phosphorus, or ammonia. Chemically enhanced primary treatment is a method to increase the removal efficiencies of suspended solids and BOD_5.

Colloidal solids are particles that are between 1 and 1000 μm in diameter. They are too small to settle to the bottom of the clarifier in the time allowed. Colloidal particles are measured in the total dissolved solids test, even though they are not truly dissolved.

Another type of primary treatment process is a dissolved air flotation thickener (DAFT). Dissolved air flotation thickeners work by adding compressed air to the bottom of a tank and using the resulting bubble curtain to lift and float solids. They are capable of achieving similar percentage removals for BOD_5 and TSS as primary clarifiers. Dissolved air flotation thickeners are more common in industrial treatment facilities than WRRFs.

The purpose of the primary clarifier or DAFT is to reduce the amount of BOD_5 and TSS going to the secondary treatment process. Reducing the load to the secondary treatment process will reduce its overall size and the amount of energy required to provide oxygen and operate downstream treatment process equipment. Generally, only larger facilities—greater than 1900-m^3/d (5-mgd) capacity—will have primary clarifiers or DAFTs because these processes are typically paired with anaerobic digestion for sludge stabilization.

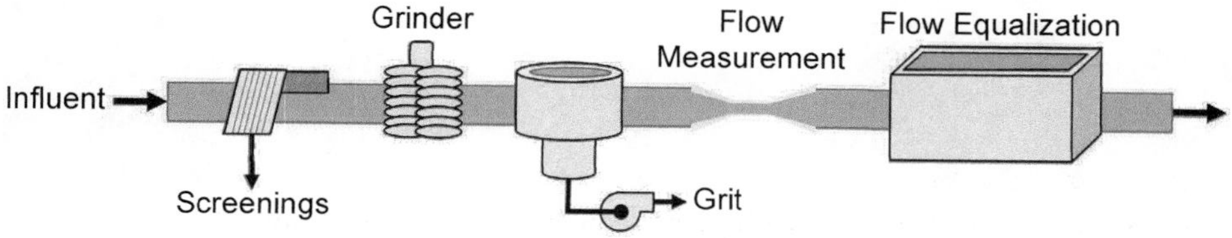

Figure 1.4 Typical WRRF Headworks

Figure 1.5 Manual Bar Screen (Reprinted with permission by Keith Radick)

Figure 1.6 Parallel Grit Basins (Reprinted with permission by Indigo Water Group)

Figure 1.7 Rectangular Primary Clarifier (Reprinted with permission by Indigo Water Group)

Figure 1.8 Circular Primary Clarifier (Reprinted with permission by Indigo Water Group)

TEST YOUR KNOWLEDGE

1. Another name for a grinder is a comminutor.
   - ☑ True
   - ☐ False

2. A trash rack is a manual bar screen with openings between the bars smaller than 5 mm (0.25 in.).
   - ☐ True
   - ☑ False

3. Which of the following treatment processes would be considered a physical treatment process?
   a. Grit basin
   b. Trickling filter
   c. Chlorine disinfection
   d. Anaerobic digestion

4. Sanitary sewers receive this type of flow.
   a. Stormwater
   b. Municipal wastewater
   c. Both stormwater and municipal wastewater
   d. Both municipal wastewater and industrial wastewater

5. Most municipalities have stopped constructing combined sewers and are removing existing combined sewers for this reason.
   a. Combined sewers deposit raw wastewater in rivers and lakes.
   b. Combined sewers affect WRRF operation during and after storm events.
   c. Combined sewers are difficult to keep clean and can generate odors.
   d. Combined sewers are expensive to construct due to larger pipe diameters.

6. Which technology are you likely to find in a WRRF headworks?
   a. Flow measurement
   b. Primary clarifier
   c. Ultraviolet disinfection
   d. Pond

7. The velocity of wastewater through a rectangular grit basin should be approximately ___________ to allow grit to settle while keeping lighter particles in suspension.
   a. 0.15 m/s (0.5 ft/sec)
   b. 0.3 m/s (1.0 ft/sec)
   c. 0.6 m/s (2.0 ft/sec)
   d. 1.5 m/s (5.0 ft/sec)

8. A primary clarifier is capable of removing:
   a. Soluble BOD_5
   b. Ammonia
   c. Total suspended solids
   d. Colloidal solids

9. A WRRF using ponds for secondary treatment is equipped with a manual bar screen that has openings 5-cm (2-in.) apart. What type of debris is most likely to be captured by this screen?
   a. Branches
   b. Rags
   c. Small rocks
   d. Paper

SECONDARY TREATMENT

By the time the wastewater reaches the secondary treatment process, nearly all of the larger particles that are capable of settling on their own have already been removed from the wastewater by screening, grit removal, and/or primary clarification. The particles that remain are very light and will not settle quickly on their own. Colloidal materials smaller than 1000 μm can take 2 years or longer to settle. For treatment to continue, the size of the remaining particles must be increased so that they can be efficiently removed. This can be done chemically or biologically. Water resource recovery facilities with *secondary treatment* remove at least 85% of influent TSS and BOD_5, resulting in effluent concentrations between 10 and 30 mg/L. Secondary treatment processes may also remove ammonia, nitrate, and phosphorus even though, technically, any treatment that goes beyond the secondary treatment standards is considered tertiary treatment. Secondary treatment is the heart of the liquid stream side of a WRRF and is discussed in detail in Chapters 5 through 9. The following sections give brief descriptions of some types of secondary treatment.

The *secondary treatment* standards (Table 1.3) require minimum percentage removals for BOD_5 and TSS. Occasionally, a WRRF might have difficulty meeting 85% removal because their influent BOD_5 and TSS concentrations have been diluted by inflow and infiltration.

Most secondary treatment processes involve biological treatment—typically, attached or suspended growth systems. Examples of secondary treatment types are shown in Figure 1.9; however, there are other types of secondary treatment not depicted. Biological treatment processes excel at converting smaller particles and soluble, biodegradable organic material in the raw wastewater into larger, heavier particles that can then be separated from the treated water by gravity. Biological treatment processes rely on a mixed population of microorganisms, oxygen, and trace amounts of nutrients to treat wastewater. The microorganisms

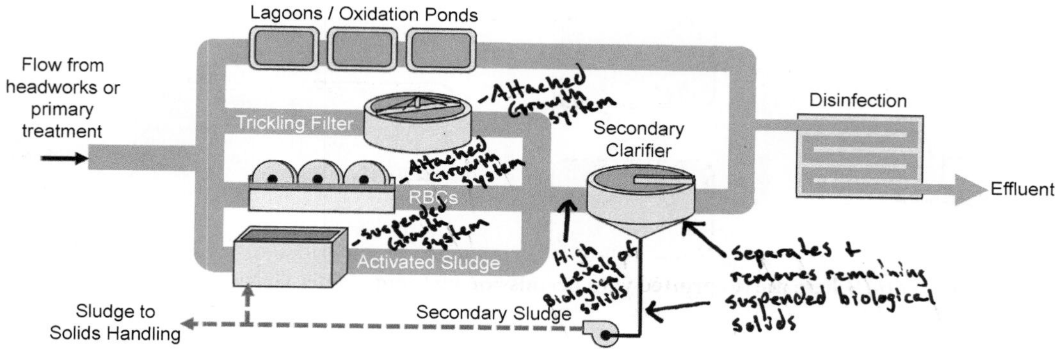

Figure 1.9 Secondary Treatment Alternatives (Reprinted with permission by Indigo Water Group)

consume organic material in the wastewater to sustain themselves and reproduce. They are naturally present in the influent and do not need to be added to the treatment process. Rapid treatment depends on having greater numbers of microorganisms in the treatment process than are typically present in natural systems.

In attached growth systems, such as trickling filters, packed towers, and rotating biological contactors, the microorganisms form a *biofilm* that is attached to the *supporting media*. In suspended growth systems, such as ponds and activated sludge processes, the microorganisms are drifting throughout the wastewater. Hybrid treatment processes combine fixed film and conventional suspended growth biological treatment processes. Hybrid processes generally consist of a fixed media for the microorganisms to grow on within an aerated biological treatment tank. One advantage of hybrid systems is that they allow for greater treatment within a given sized tank or the same level of treatment in a smaller tank. The important thing to remember is that all biological treatment processes use the same microorganisms and rely on the same underlying principles: Convert small particles into larger particles so they can be easily separated from the treated wastewater.

Biofilm is the slime layer that develops on almost any continuously wetted surface when food and nutrients are also available.

Supporting media provide a surface for microorganisms to colonize, much like the slime layer that grows in a toilet or shower that is not cleaned frequently. Media can be rock, concrete, or plastic.

Secondary treatment process effluent contains high levels of suspended biological solids that must be removed before the effluent is further treated or discharged to a receiving waterbody. Most WRRFs use secondary clarifiers to separate the solids from the liquid, as shown in Figure 1.9, although flotation, membranes, and other methods may be used.

PONDS

Ponds are one of the simplest forms of wastewater treatment. They have been in use for more than 3000 years, and currently make up approximately 50% of all WRRFs in the United States (U.S. EPA, 2011). However, they require a large facility footprint and thus are most commonly found in the Midwest or western parts of the country, where larger tracts of inexpensive land are more available. Ponds are also more common in rural communities than urban communities. Ponds are essentially natural treatment systems that have been re-created to isolate wastewater treatment from rivers and lakes.

Modern pond treatment systems typically consist of three or more separate ponds or cells that are interconnected with piping and valves so that any one pond can be taken out of service for cleaning and maintenance while the other two remain in service. During normal operating conditions, all three ponds will be in service. Raw wastewater may or may not be screened and degritted before being directed into the treatment process. The first two ponds in a pond system may be operated in *series* or in *parallel*. Microorganisms in the first two ponds treat the incoming wastewater. The third pond is the settling or polishing pond. Its purpose is to provide a quiet zone where the biological solids generated in the first two ponds can settle. Notice in Figure 1.10 that the pond system does not go to a secondary clarifier. The settling pond or polishing pond fulfills this purpose. Ponds may be lined with a synthetic liner, like the one shown in Figure 1.11, or simply have compacted clay bottoms and sides. This particular pond is equipped with baffle curtains to prevent short-circuiting and floating aerators to add oxygen. Many pond systems rely on wind and algae to supply oxygen instead of using mechanical aeration. Ponds are discussed in more detail in Chapter 6, along with other natural treatment systems.

Series operation means one after the other. *Parallel* operation means side by side.

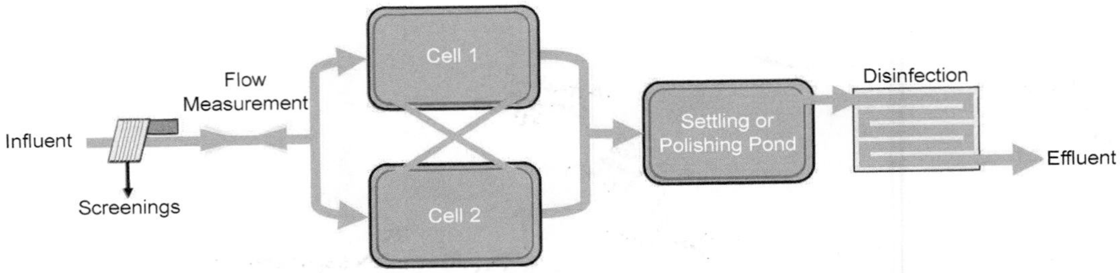

Figure 1.10 Pond Schematic (Reprinted with permission by Indigo Water Group)

ACTIVATED SLUDGE

The activated sludge process can be thought of as a modification to a pond treatment system. Pond systems are one-pass systems. The water enters at the front end of the process and passes through to the end of the process. Any biological solids that are generated in the pond either settle to the bottom of the polishing pond or pass into the final effluent. With a pond, the number of microorganisms available for treatment is limited because the solids spend the same amount of time in the system as the wastewater. The microorganisms don't have a lot of time to reproduce and increase their numbers. Consequently, ponds tend to be large compared to the amount of flow they are treating. The activated sludge process adds a recycle line that returns the biological solids that settle out in the clarifier back to the front end of the process. An example activated sludge process is shown in Figure 1.3. This relatively simple modification makes it possible to keep the biological solids in the system longer than the wastewater. Consequently, the number of microorganisms in the system increases, which means more treatment can be done in a smaller space. With a pond treatment system, the wastewater may remain in the ponds for as little as 12 days or longer than 180 days. With activated sludge, the wastewater remains in the system for 6 to 24 hours.

- In the activated sludge basin, incoming wastewater is fed to a complex mixture of bacteria and other microorganisms that form a community referred to as *biomass* or *mixed liquor suspended solids* (MLSS). Operators often refer to the MLSS as simply the "bugs", which is a shorthand reference to the bacteria in

Figure 1.11 Pond Treatment System with Synthetic Liner (Reprinted with permission by Indigo Water Group)

the process. The organic material in the influent wastewater becomes their food. As the bacteria consume the available food, they form large, floating colonies called *flocs* that will, under the right conditions, grow large enough (*flocculation*) and heavy enough that they can be separated from the treated wastewater by gravity. Non-biodegradable solids in the raw wastewater also become part of the floc particles through *bioflocculation* or are "swept out" in the effluent when the floc particles settle. Figure 1.12 shows an image of MLSS taken with a microscope at 200× magnification. The floc contains fibers, bacteria, inert material, and more complex microorganisms like the *Vaginicola* shown here. Individual floc particles have a life cycle of initial formation and growth. As the floc particles age, they accumulate dead bacteria and other inert material and increase in size. Collisions between smaller flocs yield larger flocs. The larger the floc particle gets, the more difficult it becomes for the bacteria at the center to rid themselves of waste products and gain access to nutrients and oxygen. Eventually, the floc will break into smaller flocs and the cycle will begin anew. The concentration of MLSS floc in the activated sludge basin is typically between 1200 and 3500 mg/L. The entire basin will be tan to dark brown with a small amount of white to light tan foam on the surface, as shown in Figure 1.13.

Flocculation means growing larger particles through collisions that help smaller particles stick together.

Bioflocculation uses a combination of flocculation and biological conversion and growth to agglomerate smaller particles into larger ones.

For most activated sludge processes, the MLSS will be conveyed to a secondary clarifier and allowed to settle, as shown in Figure 1.9. For the separation step to be efficient, the floc particles grown in the aeration basin must be large and dense. Process control is all about producing an MLSS that flocculates, settles, compacts, and ensures that the final effluent meets the requirements of the discharge permit (Wahlberg, 2016). Treated, clarified wastewater flows out through the top of the clarifier, whereas the MLSS settles to the bottom to form a sludge layer called the *blanket*. Operators use the settleometer test to predict how well the MLSS will settle and compact in the clarifier. A 2-L sample of MLSS is gently stirred in a wide-mouthed Mallory settleometer and then allowed to settle. The finished settleometer test shown in Figure 1.14 demonstrates what we hope to achieve in the secondary clarifier: a compact, well-defined sludge blanket with crystal clear water on top. Most of the settled MLSS in the sludge blanket is returned to the activated sludge process where it will be used to treat more influent wastewater. This is the return activated sludge. Excess MLSS is removed from the process entirely as waste activated sludge. In a sequencing batch reactor type activated sludge process, treatment and clarification take place in the same basin. Activated sludge is discussed in more detail in Chapter 8.

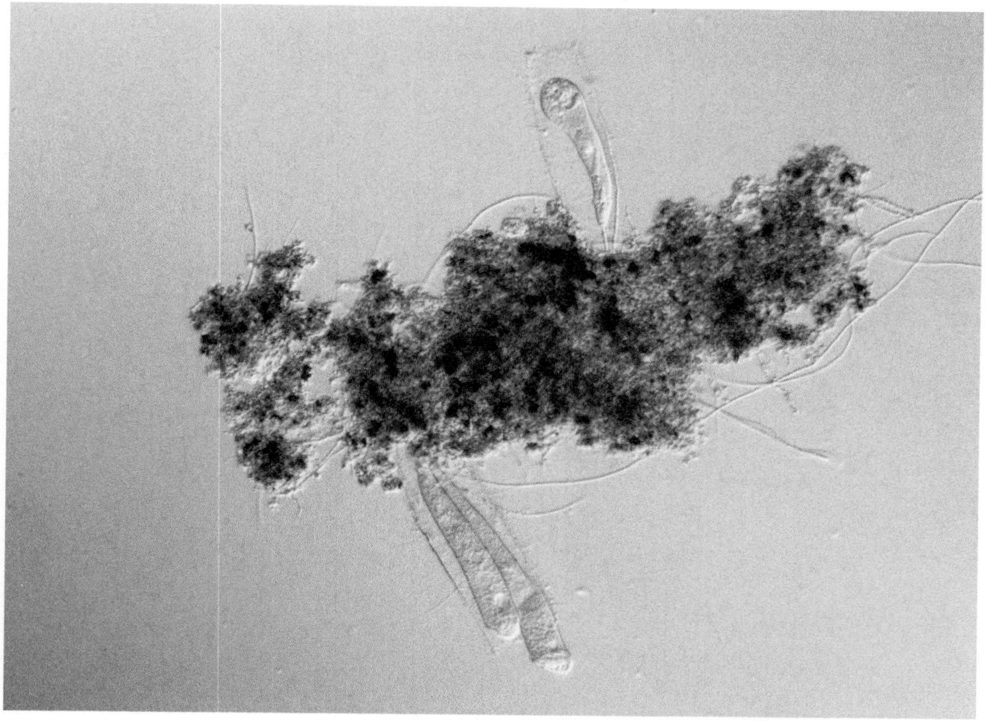

Figure 1.12 Activated Sludge at 200× Magnification (**Reprinted with permission by Richard Weigand, CET**)

Figure 1.13 Activated Sludge Basin (Reprinted with permission by Indigo Water Group)

TRICKLING FILTERS AND ROTATING BIOLOGICAL CONTACTORS

Trickling filters and rotating biological contactors (RBCs) are fixed-film processes. They use the same microorganisms to treat wastewater as ponds and activated sludge, but the microorganisms grow attached to a surface. A trickling filter consists of a tank filled with some kind of media—rock, plastic, wood, and so on—for the bacteria to grow upon. The media are supported by an underdrain system, which keeps the media inside the tank while allowing water to pass through. The underdrain system is also used to ventilate the trickling filter media. Wastewater is pumped to the top of the trickling filter where it is distributed by

Figure 1.14 Settleometer Test (Reprinted with permission by Indigo Water Group)

spray nozzles or a rotating arm, as shown in Figure 1.15. The water trickles down over the media and into the underdrain. Trickling filters are not submerged. The water trickling over the media takes up very little of the total space inside the filter. The trickling filter shown in Figure 1.15 is filled with natural rock media and is approximately 1.5-m (5-ft) deep.

Rotating biological contactors have been in full-scale use since the mid 1970s. They consist of multiple stacked plates or wheels of media that are submerged between 30 to 70% in wastewater (Figure 1.16). The wheels slowly rotate in and out of the wastewater. Rotating biological contactors are fixed-film processes, with the microorganisms forming a biofilm on the media. Unlike trickling filters, RBCs don't require pumping and don't require mechanical ventilation. Both trickling filters and RBCs produce excess biofilm that periodically sloughs off the media. Sloughed biofilm settles to the bottom of the secondary clarifier. In some facilities, trickling filters and RBCs have small hoppers to collect sloughed biofilm instead of true clarifiers. Unlike activated sludge, the sloughed biofilm is not returned to the trickling filters or RBCs. Rather, it is sent to the solids handling side of the WRRF for further processing. Fixed-film processes are discussed in more detail in Chapter 7.

PHYSICAL–CHEMICAL TREATMENT

Physical and chemical treatment processes are used to remove oil, grease, heavy metals, solids, and nutrients from wastewater. For example, screening, sedimentation, and filtration are used to physically separate solids from wastewater. Chemical coagulation and precipitation are used to promote sedimentation. Coagulation is also used to improve capture and removal of colloidal solids. Activated carbon adsorption is used to remove organic pollutants. Breakpoint chlorination and lime addition are used to reduce nitrogen and phosphorus concentrations, respectively.

There are many, many different types of secondary treatment. Ponds, activated sludge, trickling filters, and RBCs are some of the most prevalent.

ADVANCED WASTEWATER TREATMENT

Advanced wastewater treatment (AWT) processes are typically used to further reduce the concentrations of nutrients (nitrogen or phosphorus) and soluble organic chemicals in secondary treatment effluent. These processes may be physical, chemical, biological, or a combination. For example, membrane filtration—microfiltration, ultrafiltration, nanofiltration, and reverse osmosis—is used to remove organics, nutrients, and pathogens from wastewater. It traditionally was used for industrial wastewater treatment, but has been gaining popularity at municipal treatment facilities.

Permit requirements typically determine which, if any, AWT processes are used. Separate phosphorus reduction technologies are becoming more common because of more stringent discharge permit limitations. There are increasing requirements to both monitor and potentially treat microconstituents

Figure 1.15 Trickling Filter (Reprinted with permission by Indigo Water Group)

Figure 1.16 Rotating Biological Contactor (Reprinted with permission by Indigo Water Group via Upper Blue Sanitation District)

(sometimes called *compounds of emerging concern*). Microconstituents may include pharmaceuticals and personal care products that may also be referred to as *endocrine disruptor compounds*. This is of elevated interest in areas such as where WRRF effluent is being considered for potable water applications.

DISINFECTION

Disinfection inactivates or destroys pathogenic bacteria, viruses, and protozoan cysts typically found in wastewater. These pathogens cause waterborne diseases such as bacillary dysentery, cholera, infectious hepatitis, paratyphoid, poliomyelitis, and typhoid. Disinfection is not sterilization. It would be very expensive to inactivate or destroy every microorganism present in treated effluent. Instead, disinfection reduces the number of bacteria and pathogens to safe levels to prevent the spread of waterborne illnesses and to protect the environment. Most WRRFs use chlorine gas, sodium hypochlorite (bleach), or UV radiation to disinfect their treated effluent; however, these are not the only disinfection alternatives available.

The increased demand and interest in reuse of treated wastewater for irrigation and other purposes, and changes in requirements for toxics, including chlorine, to protect aquatic life, are altering disinfection policies and, subsequently, disinfection practices. Chemical disinfection processes involving chlorine historically dominated the wastewater treatment industry. However, concerns about chlorine safety and the requirement to dechlorinate some discharges have made other disinfection methods, such as ozonation and UV irradiation, more popular. Ozonation involves using ozone radicals to destroy pathogen cell walls. Ultraviolet irradiation involves using light energy to destroy an organism's genetic material. Both are more costly than chlorine disinfection, but are effective treatment alternatives. For UV irradiation to be effective, the light must be able to pass easily through the wastewater, which limits its application to systems with low turbidity and low total suspended solids effluent.

EFFLUENT DISCHARGE

The quality of effluent matters because where it will be discharged or how it will be reused influences the permit requirements set and the treatment processes needed to meet those requirements. Water resource recovery facility effluent can be discharged to a surface waterbody or wetlands, used to recharge groundwater aquifers via percolation through the ground or deep-well injection, or land applied. It may be considered an alternative water source (because of its nitrogen and phosphorus content) and used to irrigate golf courses, parks, plant nurseries, and farms. It also could be the source water for a constructed wetland, adding nutrients that support the aquatic environment, thereby enhancing wildlife habitat and public recreation. In addition, effluent can be used by industries as cooling or makeup water for certain chemical processes.

TEST YOUR KNOWLEDGE

1. Secondary treatment uses chemicals or bacteria to increase the size of particles in wastewater.
   - ☑ True
   - ☐ False

2. Water resource recovery facilities with secondary treatment typically remove more than 85% of influent BOD_5 and TSS.
   - ☑ True
   - ☐ False

3. Microorganisms in wastewater consume organic material in the wastewater to sustain themselves and reproduce.
   - ☑ True
   - ☐ False

4. Secondary treatment systems do not typically include a clarifier or other solids separation processes after biological treatment.
   - ☐ True
   - ☑ False

5. Ponds are typically unlined.
   - ☐ True
   - ☑ False

6. A well-functioning activated sludge process will be light tan to dark brown and have a small amount of white to tan foam on the surface.
   - ☑ True
   - ☐ False

7. A rock media trickling filter should be completely submerged with no free space between the rocks.
   - ☐ True
   - ☑ False

8. Biological treatment systems use the same microorganisms as natural systems, but are engineered to decrease treatment time by increasing the numbers of microorganisms in the treatment process.
   - ☑ True
   - ☐ False

9. Disinfection uses chemicals or UV light to sterilize treated wastewater. *Reduce to safe quantities
   - ☐ True
   - ☑ False

10. Which of the following treatment processes would be considered biological treatment?
    - a. Alum addition for phosphorus removal
    - b. Activated sludge
    - c. Belt filter press
    - d. Ultraviolet disinfection

11. A pond system is categorized as this type of treatment:
    - a. Primary
    - b. Suspended growth - cuz they floatin' around
    - c. Fixed growth
    - d. Physical

12. Where do the microorganisms in ponds, trickling filters, and activated sludge systems come from?
    - a. They are added by the operator.
    - b. They are naturally present in the influent wastewater.
    - c. They spontaneously generate from suspended solids.

13. This term is used to describe a collection of microorganisms growing on and attached to a media surface such as a rock.
    - a. Floc
    - b. Slime
    - c. Biofilm
    - d. Algae

14. In a pond treatment system, what is the purpose of the last pond in the series?
    - a. Increases the risk of short-circuiting
    - b. Removes the biological solids produced in the first two ponds
    - c. Warms the wastewater before discharge
    - d. Acts as a primary clarifier or grit basin - "polishing" pond

15. What is the primary difference between a pond treatment system and an activated sludge system?
    - a. Activated sludge recycles settled solids to the beginning of the process.
    - b. Pond treatment systems use specialized, cold-tolerant bacteria.
    - c. Activated sludge systems use algae for treatment.
    - d. Pond treatment systems perform better at higher elevation.

16. For an activated sludge system, which of the following statements is FALSE?
    - a. Activated sludge requires less time to treat wastewater than ponds.
    - b. Activated sludge is a suspended growth biological process.
    - c. Activated sludge uses fungus to treat wastewater.
    - d. Activated sludge holds the biological solids longer than the wastewater.

17. An example of a fixed-film treatment process is
    - a. Activated sludge
    - b. Pond
    - c. Rotating biological contactor
    - d. Clarifier

18. Which two methods of disinfection are most commonly used in domestic WRRFs?

a. Ozone and chlorine
b. Chlorine and UV light
c. Bleach and ozone
d. Ultraviolet light and boiling

19. Match the unit process to its place in the liquid treatment side.

A 1.	Collection system	a.	Pretreatment
D 2.	Grit basin	b.	Primary treatment
B 3.	Primary clarifier	c.	Disinfection
E 4.	Activated sludge	d.	Preliminary treatment
C 5.	Chlorine addition	e.	Secondary treatment

20. Draw a line from the liquid treatment type to its treatment goal.

1.	Preliminary treatment	a.	Reduce number of bacteria and pathogens
2.	Primary treatment	b.	Increase particle size for separation step
3.	Secondary treatment	c.	Protect downstream equipment
4.	Tertiary treatment	d.	Remove nitrogen and phosphorus
5.	Disinfection	e.	Decrease size and cost of secondary treatment

Solids Treatment Processes

The solid material that is removed in the liquid stream processes gets transferred to the solids treatment side of the WRRF for additional processing. Sludge stabilization processes transform *sludge* into *biosolids* by reducing the percentage of organic material, odors, pathogens, and biodegradable toxins, as well as binding heavy metals to inert solids so they will not leach into the groundwater. The resulting biosolids can be then be reused beneficially or disposed of safely. Typical solids treatment processes include thickening, stabilization, digestion, chemical stabilization, composting, dewatering, heat drying, and incineration. These processes are briefly described below. An example solids treatment process is shown in Figure 1.17. A particular WRRF may have some, all, or none of these processes in place. Similar to the liquid treatment side of the WRRF, each one of the solids treatment processes can be accomplished with a variety of different technologies.

Sludge encompasses any and all solid material—scum, settled solids, and floatables—that is removed from wastewater during treatment.

Biosolids refers to stabilized solids that have been thoroughly treated and meet regulatory criteria. Biosolids may be beneficially reused as a soil amendment and fertilizer or can be sent to a landfill.

Residuals refers to the solid material left over after liquid treatment is completed.

TYPES OF RESIDUALS

Wastewater *residuals* include primary sludge, secondary sludge, and chemical sludge, as well as screenings, grit, scum, and ash. Typically, screenings and grit are disposed of in landfills. Primary sludge consists of unprocessed (raw) organic and inorganic solids. This is material that was separated from the raw wastewater, typically by gravity, but has not received any additional treatment. Primary sludge typically has a

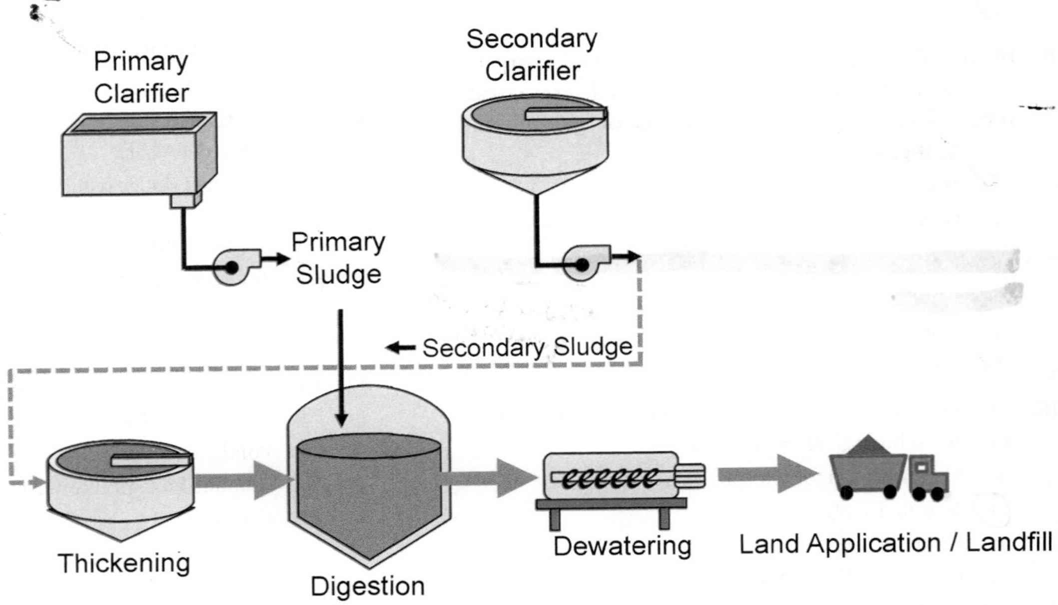

Figure 1.17 Solids Stream Process Overview (Reprinted with permission by Indigo Water Group)

concentration of 2 to 6% solids when removed from the primary clarifiers. Secondary (biological) sludge is composed largely of microorganisms grown during secondary treatment. Secondary sludge typically has a concentration of less than 1% solids and often goes through a thickening process before it is treated further. In some activated sludge processes, the secondary sludge is recycled to the primary clarifier where it is settled and co-thickened with the primary sludge. The concentration and characteristics of chemical sludge depend on the treatment chemicals (alum, ferric salts, or lime) used. Chemical sludge is generated at WRRFs that have tertiary treatment, such as phosphorus removal.

REGULATORY REQUIREMENTS FOR BIOSOLIDS

The use or disposal method for residuals depends on how much treatment they have received. Combustible residuals, such as screenings, may be incinerated or landfilled. Noncombustible residuals, such as grit, are typically landfilled. Primary and secondary sludge can be treated and broken down further using a variety of different treatment technologies to produce biosolids. Biosolids may be used as a soil amendment (land application), placed on a surface disposal site, incinerated, or sent to a landfill. Approximately 50% of all biosolids produced in the United States are beneficially reused through land application (U.S. EPA, 2016). Land application involves spreading biosolids on the soil surface or injecting it into soil. The material, which adds organic matter and nutrients to soil, must meet Class A or Class B standards to be land applied.

The U.S. EPA regulates all biosolids that are land applied, incinerated, or surface-disposed under Title 40 of the Code of Federal Regulations (CFR), Part 503 (58 FR 9248-9415). This set of rules is referred to as the *503 Regulations*. Sludge that is landfilled with other waste is regulated under 40 CFR 258. Individual states often have their own regulations governing the use and disposal of biosolids; however, the state regulations can never be more lenient than the federal regulations. Local regulations can be more stringent than federal requirements. Biosolids permits are issued to generators and to land appliers. Water resource recovery facility staff should assess local land availability, public acceptance, and transportation constraints before selecting a use or disposal alternative.

The 503 regulations set requirements for pathogen reduction; define two major classifications of biosolids, Class A and Class B; set limits for metals and organic compounds in biosolids; and set requirements for vector attraction reduction. Some of these requirements are listed in Table 1.5. Class A biosolids require more stabilization to reduce pathogens below detectable limits, but can be commercially marketed and made available for public takeaway or land applied without any pathogen-related restrictions. Class B biosolids can be land applied to private land without public access. Both Class A and Class B biosolids can be used as a soil amendment and fertilizer for growing crops. Recycling biosolids in this way is called *beneficial reuse*. The 503 regulations also contain rules and guidelines for the amount of biosolids that can be applied to land and where they can be applied. For example, biosolids can't be applied to land that is close to rivers and streams because of the potential for biosolids to be washed into the water when it rains or snows.

A *vector* transfers pathogens from one location to another. Examples of vectors are rats, birds, and other animals that may roll or dig in biosolids and then carry biosolids to other locations in their fur or feathers. Much of solids stabilization involves reducing the organic content of biosolids to make it unattractive as a food source for wildlife.

THICKENING

Thickening processes remove water to reduce the volume of liquid sludge, but the material retains the characteristics of a liquid (e.g., it flows). A thickened sludge typically contains between 1.5 and 8% solids. Thickening is intended to reduce the volume of sludge so sludge treatment, storage, and hauling processes, equipment, and costs can also be reduced.

Table 1.5 Regulatory Definitions for Class A and Class B Biosolids

	Class A	Class B
Fecal coliform bacteria, #/gram of total solids	1000	2 000 000
Salmonella, #/gram of total solids	4	
Treatment technique	Process to further reduce pathogens (PFRP)	Process to significantly reduce pathogens (PSRP)
Vector attraction reduction requirement	Must be met	Must be met

Thickening takes place before stabilization, whereas dewatering takes place after stabilization. Thickening and dewatering use many of the same technologies. Being specific with language helps wastewater professionals know which part of the process they are discussing.

There are three types of *thickening*:

1. Pre-thickening (thickening before stabilization and dewatering),
2. Post-thickening (thickening after stabilization, but before beneficial use), and
3. Recuperative thickening (thickening biosolids and returning them to the stabilization process).

Prethickening processes include gravity thickeners, DAFTs, centrifuges, gravity belt thickeners, and rotary drum thickeners. Gravity thickeners work best on primary and chemical sludges; they do not work well with combined sludges. Dissolved air flotation thickeners are typically used to thicken secondary sludge and can also be used to co-thicken mixed primary/secondary sludge. Mechanical thickeners, such as centrifuges, gravity belt thickeners, and rotary drum thickeners, are used on all types of sludges. They remove more water from sludges than a gravity thickener or DAFT does. The water removed during thickening is sent back to the liquid treatment side of the WRRF to a point before biological treatment, but after influent flow monitoring.

Outside and inside views of a rotary drum thickener are shown in Figure 1.18. This is one of many technologies that may be used for thickening and dewatering. A rotary drum thickener consists of a fine screen or trommel lying on its side. Sludge is mixed with a small amount of polymer to help the particles in the sludge stick together and form larger particles. The sludge and polymer mixture is then fed into one end of the trommel. The trommel screen rotates slowly at approximately 5 to 20 rpm to gently turn the sludge over inside, which allows free water to drain out of the sludge. An internal screw or other device along the drum length conveys sludge to the discharge chute. Wash water is applied both inside and outside of the drum periodically to clean solids from the screen openings. Depending on the type of sludge fed to the drum thickener—primary or secondary—the thickened sludge will be between 4 and 9% solids (WEF, 2017).

STABILIZATION

Stabilization via digestion or chemical stabilization reduces the sludge's pathogen content, transforming the sludge into biosolids suitable for beneficial use.

DIGESTION

Sludge digestion may be done aerobically (i.e., in the presence of oxygen) or anaerobically (i.e., in the absence of oxygen). In general, primary sludge is digested anaerobically, whereas secondary sludge may be digested either aerobically or anaerobically. Aerobic and anaerobic digestion both reduce the *volatile solids* and pathogen content of sludge, thereby reducing odors and producing an environmentally acceptable soil amendment. Digestion also makes the finished biosolids unattractive to rats and insects (vectors) as a potential food source by reducing the percentage of organic material remaining. In this way, digestion meets the vector attraction reduction requirement in the 503 regulations.

Volatile solids is a method of estimating the amount of organic material remaining in the sludge or biosolids. Higher volatile solids indicate larger amounts of organic material and are one indicator that the sludge-to-biosolids conversion is not complete.

Digesters are fed primary sludge, secondary sludge, or a mixture of the two. In the secondary treatment process, a mass of microorganisms is grown by feeding them the organic material (BOD_5) in the influent wastewater. Excess microorganisms are removed from the process and sent to thickening or directly to a digester, as shown in Figures 1.3 and 1.17. Once inside the digester, the microorganisms no longer have access to a food source. Think about what would happen if you suddenly had no access to food. You would gradually lose weight and eventually die. The same thing happens to the microorganisms in the digester through a process called *endogenous respiration*. The microorganisms use up their internally stored food supplies and, if available, consume the organic material in the primary sludge as well. The amount of food available per microorganism is very low so, over time, there is a net loss of live microorganisms and the percentage of volatile solids decreases.

Primary clarifiers are typically paired with anaerobic digesters because they don't require the addition of oxygen to stabilize the sludge. If a primary clarifier sent sludge to an aerobic digester, there isn't a net benefit to the WRRF. Remember that the purpose of primary treatment is to reduce the size and operating cost of secondary treatment. The biggest cost in both secondary treatment and aerobic digestion is the cost of supplying oxygen.

Aerobic digestion uses microbes in open or closed tanks or ponds to break down the sludge's organic matter into carbon dioxide, water, and ammonia. Ammonia is a component of proteins and is released when proteins break down. Solids remain in an aerobic digester for 40 days at 20 °C (68 °F) or longer when water temperatures are cooler (WEF, 2017). A well-operated aerobic digester can reduce more than 40% of volatile solids.

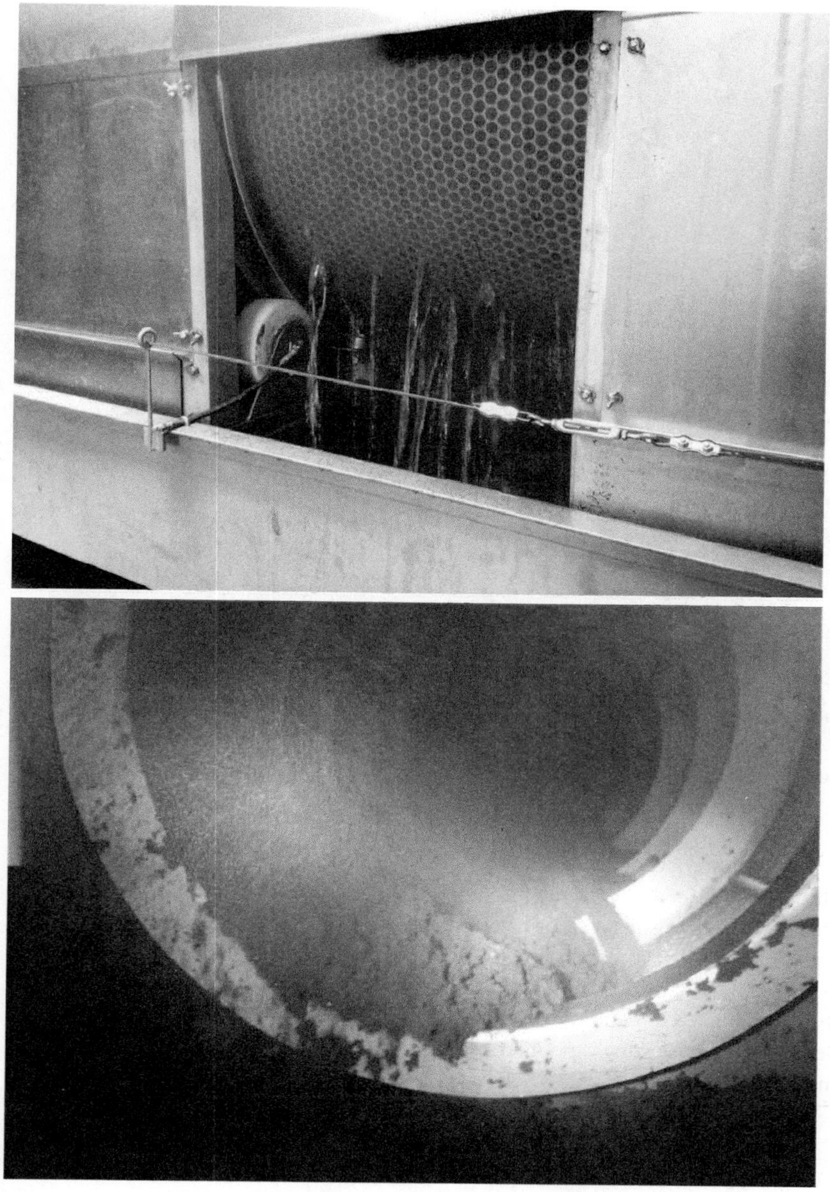

Figure 1.18 Outside and Inside Views of a Rotary Drum Thickener (**Reprinted with permission by Indigo Water Group**)

Anaerobic digestion uses microbes in a closed tank containing little or no oxygen (Figure 1.19). The tank is typically operated at temperatures between 35 and 38 °C (95 and 100 °F), but may be operated at more than 55 °C (131 °F) to further reduce pathogens and volatile solids. Solids remain in an anaerobic digester for 15 to 20 days or longer (WEF, 2017). Anaerobic digestion destroys more than 40% of volatile solids and produces an off-gas containing approximately 65% methane and 35% carbon dioxide, which may be collected and used as a fuel. Anaerobic digesters also produce water and ammonia. Both types of digestion can produce a Class A or Class B biosolids suitable for beneficial use, depending on the time, temperature, and flow configurations involved (40 CFR 503).

Digestion wants to reduce pathogens + volatile solids

very cool.

Aerobic Digester Retention Time
↳ 40 days @ 68°F (20°C)
*Longer if cooler temp

Anaerobic Digester Retention Time:
↳ 15-20 days @ 95°-100°F

CHEMICAL STABILIZATION

Chemical stabilization typically involves using lime to raise the sludge's pH to 12.0 for 2 hours and maintain it at pH 11.5 or greater for an additional 22 hours. This reduces pathogens and odors. Rules regarding chemical stabilization can be found in 40 CFR Part 503. This stabilization method can also produce a Class A or Class B biosolids suitable for beneficial use (40 CFR 503).

↳ Increases Alkalinity (Basic)

Figure 1.19 Anaerobic Digester (Reprinted with permission by Jorj Long)

COMPOSTING

Composting—specifically, windrow, aerated windrow, static pile, aerated static pile, and in-vessel—involves using microorganisms, a bulking agent (such as wood chips, leaves, or sawdust), and a controlled environment (typically 55 to 60 °C [131 to 140 °F]) to decompose organic matter in sludge, as well as to reduce its volume and odors. The moisture and oxygen levels are also controlled to minimize odors during the process. Biosolids are typically used in windrow systems, whereas sludge can be composted in aerated static piles or an in-vessel system. Composting is popular because the finished material is an excellent soil conditioner and often meets Class A biosolids standards.

DEWATERING

Dewatering takes place after sludge stabilization is complete. Dewatering further reduces the volume and weight of biosolids via air drying (on sand); vacuum-assisted drying beds; or mechanical dewatering equipment such as belt filter presses, centrifuges, plate-and-frame filter presses, and/or various types of horizontal or inclined screw presses. When used with polymers, belt filter presses and centrifuges can produce a biosolids "cake" that contains 15 to 25% solids. Sand-drying beds typically produce a cake containing between 10 and 50% solids. Vacuum-assisted beds produce a cake containing 10 to 15% solids (when polymers are used). Plate-and-frame presses can produce a cake containing 30 to 60% solids (when lime, ferric chloride, or fly ash is used). Vacuum filters can produce a cake containing 12 to 30% solids (when the biosolids have first been conditioned with lime, ferric chloride, or polymers). A few of these technologies are summarized in the following paragraphs.

BELT FILTER PRESS

Belt filter presses (BFPs) act to dewater solids using a combination of gravity and the process of squeezing water out of the biosolids. After water drains through the single gravity zone belt, biosolids are sandwiched between two porous belts and are subjected to gradually increasing amounts of pressure to force water out from between particles. There are five basic stages in the BFP process: chemical conditioning of feed sludge to release bound water, gravity drainage in the gravity zone, a low-pressure wedge zone to consolidate the solids, a high-pressure zone to squeeze water out from between sludge particles, and, finally, release of the solids from the press. Each of these zones is visible in Figure 1.20. Belt filter presses operate continuously.

Figure 1.20 Belt Filter Press (Reprinted with permission by Richard Weigand, CET)

CENTRIFUGE

Centrifuges use centrifugal force (typically 500 to 3000 times the force of gravity) to accelerate solids/liquid separation much in the same way that the spinning drum of a washing machine removes water from clothes (Figure 1.21). Centrifugal force throws particles to the inside shell of a spinning drum. Lighter liquid remains toward the center of the drum. Modern centrifuges use a scroll and bowl design. The bowl is the outer cylinder and the scroll rotates inside the cylinder to push solids out of the centrifuge. The bowl

Figure 1.21 Centrifuge (Reprinted with permission by the Sanitation Districts of Los Angeles County)

and scroll spin independently and each has their own drive motor. An important operating parameter with centrifuges is the differential speed or the difference between the scroll speed and the bowl speed. The bigger the difference in speeds, the faster solids will be removed from the centrifuge. Imagine using a drill to push a screw into a wall. If the wall is rotating at the same speed as the drill, the screw will not move into the wall. If there is a difference in speeds, the drill will be able to do its job.

HEAT DRYING

Heat drying processes, such as low-temperature heat drying, flash drying, rotary kiln drying, indirect drying, vertical indirect drying, direct–indirect drying, and infrared drying, reduce the volume of secondary sludge and destroy pathogens. They typically produce a commercially marketable biosolids. Digestion is typically not a prerequisite.

INCINERATION

Incineration typically involves firing biosolids at high temperatures in a multiple-hearth or fluidized-bed combustor, turning them into ash, and destroying volatile solids and pathogens in the process. Incineration is not widely practiced in the United States because of the availability of inexpensive land application sites and landfill space. Incineration degrades many organic chemicals, but can form others (e.g., dioxins) and, therefore, products of incomplete combustion must be controlled. Air emissions also must be controlled. Metals are not degraded; they concentrate in the ash. Most incinerated municipal sludge will produce nonhazardous ash, which can be landfilled (40 CFR 258). It also can be used as an aggregate in concrete or as a fluxing agent in ore processing. If inorganic or organic constituents exceed Appendix II to Part 258, disposal in a hazardous waste landfill will be required.

TEST YOUR KNOWLEDGE

1. Sludge stabilization processes like digesters transform sludge into biosolids.
   - ☑ True
   - ☐ False
2. Primary sludge typically contains between 2 and 6% total solids.
   - ☑ True
   - ☐ False
3. The beneficial use of biosolids is regulated under 40 CFR part 258.
   - ☐ True
   - ☑ False
4. The time required for aerobic digestion depends on the temperature of the sludge.
   - ☑ True
   - ☐ False
5. Put the following solids treatment processes into the correct order.
   a. Dewatering
   b. Secondary clarifier
   c. Thickening
   d. Digestion
6. Screenings and grit are typically:
   a. Sent to a landfill
   b. Dewatered and land applied
   c. Used for road base
   d. Digested anaerobically
7. Primary sludge consists of
   a. Microorganisms grown during treatment
   b. Rags, plastic, and other heavy materials
   c. Unprocessed, settleable organic and inorganic solids
   d. Grit and screenings
8. Secondary sludge consists of
   a. Microorganisms grown during treatment
   b. Rags, plastic, and other heavy materials
   c. Unprocessed, settleable organic and inorganic solids
   d. Grit and screenings
9. This type of biosolids may be made available for public takeaway.
   a. Class A
   b. Class B
   c. Class C
   d. Class D
10. The vector attraction reduction requirement in the biosolids 503 regulations
    a. Limits concentrations of heavy metals in biosolids
    b. Allows screenings and grit to be comingled with digested sludge
    c. Reduces the likelihood that rats and insects will be attracted to finished biosolids
    d. Prevents application of biosolids near streams and lakes

11. **Sludge thickening and biosolids dewatering are performed for this reason:**
    a. Reduces the total volume of sludge or biosolids
    b. Required by the discharge permit
    c. Reduces the total mass of sludge or biosolids
    d. Required by the 503 regulations

12. **All of the following statements about anaerobic digestion are true EXCEPT:**
    a. Reduces the amount of biosolids
    b. Meets vector attraction reduction requirements
    c. Typically paired with primary clarifiers
    d. Break down solids in the presence of oxygen

13. **Anaerobic digester gas contains approximately**
    a. 70% nitrogen and 22% oxygen
    b. 65% methane and 35% carbon dioxide
    c. 70% nitrogen and 30% carbon dioxide
    d. 65% methane and 22% oxygen

14. **Match the process to its defining characteristic.**

1. Aerobic digestion D	a. Uses lime to increase sludge pH
2. Anaerobic digestion C	b. Takes place prior to digestion
3. Thickening B	c. Breaks down sludge with oxygen
4. Dewatering E	d. Produces methane gas
5. Chemical stabilization A	e. Produces "cake" of up to 50% solids

- Got over excited. Answered 1+2 incorrectly without reading carefully whether it was aerobic or An-aerobic.

15. **This piece of equipment rapidly spins biosolids to remove water.**
    a. Belt filter press
    b. Rotary drum thickener
    c. Centrifuge
    d. Heat dryer

Example Water Resource Recovery Facilities

Figures 1.22, 1.23, and 1.24 present three examples of complete treatment processes. Figure 1.22 shows a natural treatment system—a three-cell pond system operated in series followed by a natural wetland. Natural treatment systems are the simplest and have the least amount of equipment associated with them. In this example, the pond is aerated by wind and wave action as well as oxygen produced from algae. There aren't any blowers or surface aerators to supplement the naturally occurring oxygen. Notice that a pond treatment facility does not have solids treatment processes. This is because the biological solids produced in the first two cells settle out in the settling or polishing pond (cell 3). These solids will be removed every few years for beneficial reuse or landfilling. This pond system is also lacking grit removal. Grit removal is less important for a pond than a mechanical treatment facility because there isn't a lot of downstream equipment like pumps that might be damaged or worn by grit. The grit is allowed to settle out in the first two cells.

Figure 1.23 shows a two-stage trickling filter treatment system. The level of complexity has increased substantially compared to a pond system. This WRRF has a complete headworks with both screening and grit removal. Two pumping stations and multiple recycle lines allow the operator flexibility in routing raw wastewater and recycle flows over both trickling filters. Waste solids are processed in an aerobic digester without pre-thickening. Notice that the liquid treatment side of this WRRF lacks a primary clarifier. The

By the end of this chapter, you should be able to draw an example WRRF, clearly label the significant unit processes, and give the function of each.

- Least amt. of equipment
- Grit can settle in 1st two ponds

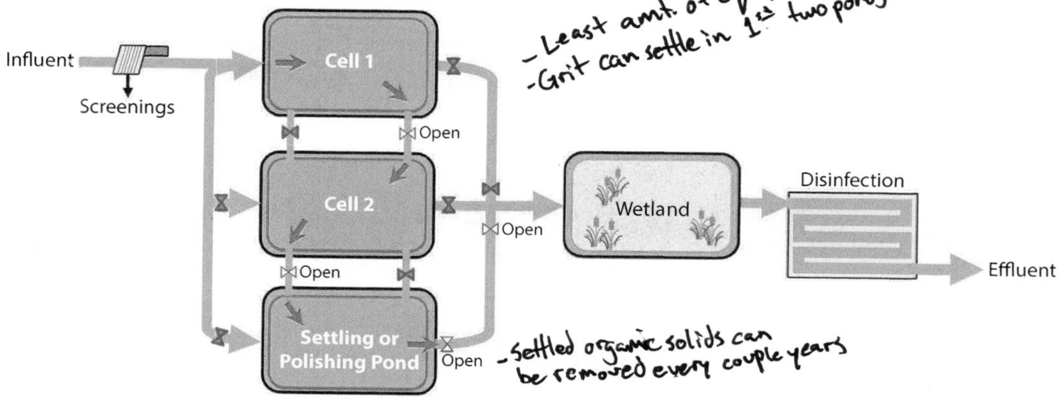

- Settled organic solids can be removed every couple years

Figure 1.22 Example of a Pond and Wetland Treatment Process (Reprinted with permission by Indigo Water Group)

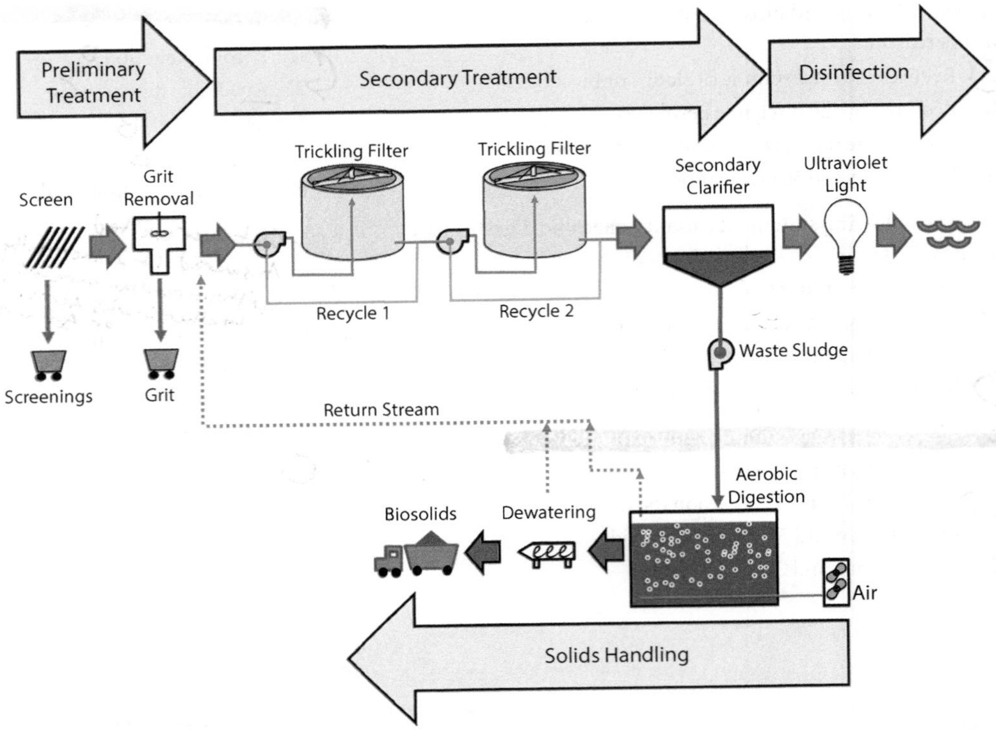

Figure 1.23 **Example of a Two-Stage Trickling Filter Treatment Process (Reprinted with permission by Indigo Water Group)**

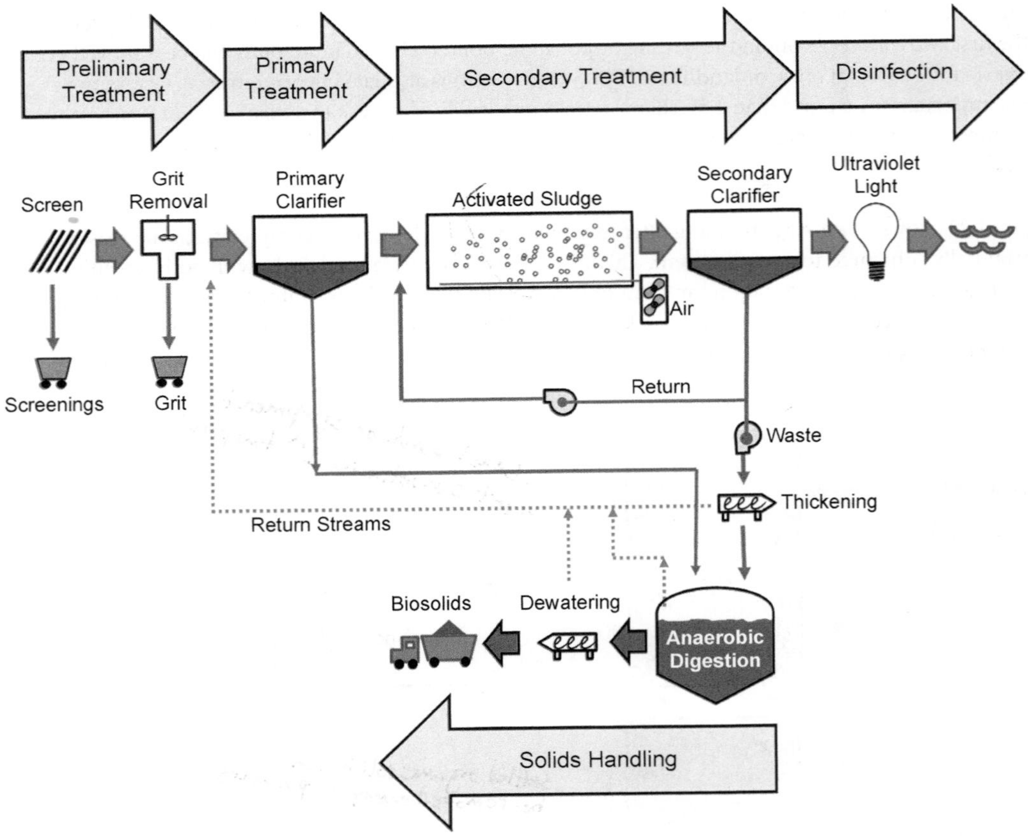

Figure 1.24 **Example of an Activated Sludge Treatment Process (Reprinted with permission by Indigo Water Group)**

trickling filters are naturally ventilated, which makes them very inexpensive to operate. Trickling filters are almost always preceded by primary clarifiers, even when the primary sludge will be sent to an aerobic digester. This is because trickling filters and rotating biological contactors can only remove soluble BOD. On the solids handling side of the WRRF, notice the blower providing oxygen to the digester and the recycle streams going back to the liquid side. The liquid and solids treatment processes of WRRFs affect each other, which sometimes requires careful control of both solids and recycle streams.

Figure 1.24 shows an activated sludge treatment process complete with primary clarification and anaerobic digestion. Note the differences between this WRRF and the two-stage trickling filter WRRF. Blowers are in place for the activated sludge process, which drives up operating costs. To minimize operating costs and reduce the size of the secondary treatment process, a primary clarifier has been included. There are three recycle streams connecting the solids side of the WRRF back to the liquid side. These three example facilities represent a few of the thousands of different combinations of unit processes that a wastewater operator might encounter in the field.

CHAPTER SUMMARY

	Liquid Treatment Processes	Solids Treatment Processes	
REGULATIONS	1. Clean Water Act amended 1972. 2. All dischargers must have a permit. 3. Sets secondary treatment standards of 30 mg/L each for BOD_5 and TSS and 25 mg/L for CBOD. 4. Direct dischargers are permitted by U.S. EPA or their state agency. 5. Indirect dischargers are permitted by the WRRF. 6. Depends on self-monitoring with monthly reports.	1. The Standards for the Use or Disposal of Sewage Sludge, also known as the 503 regulations, were promulgated in 1993. 2. Applies to the use or disposal of biosolids when they are land applied. 3. Sets standards for Class A and Class B biosolids. 4. Limits concentrations of heavy metals and other chemical compounds in biosolids. 5. Requires pathogen reduction and vector attraction reduction before land application.	REGULATIONS
PRELIMINARY	1. Physical treatment process. 2. Takes place at WRRF headworks. 3. Removes large debris to protect downstream processes and equipment. 4. May include screening, grit removal, flow measurement, and/or flow equalization. 5. Volume of screenings and grit produced is service-area dependent.	1. Physical treatment process. 2. Reduces the total volume of sludge by removing water. Polymer addition may be necessary. 3. Thickened sludges are typically between 2 and 8% total solids. 4. Takes place before sludge stabilization. 5. Many different technologies are available, including DAFTs, rotary drum thickeners, gravity belt thickeners, and others.	THICKENING
PRIMARY	1. Physical treatment process. 2. Decreases velocity of wastewater to remove settleable and floatable material. 3. May use gravity or flotation. 4. Reduces the size and cost of operation of secondary treatment processes. 5. Cannot remove colloidal material, soluble BOD, soluble phosphorus, or ammonia. 6. Removal rates depend on influent characteristics. For domestic wastewater, typically removes 20 to 35% of influent BOD_5 and 60 to 75% of influent TSS. 7. Paired with anaerobic digestion and other forms of sludge stabilization that do not require oxygen.	1. Biological or chemical treatment process. 2. Converts sludge to biosolids. 3. Reduces pathogens. 4. Biological treatment consists of aerobic or anaerobic digestion. Either method can reduce the total mass of biosolids by 40%. 5. Biological treatment can reduce the volatile solids content by as much as 50%. 6. Biological treatment converts primary and secondary sludge to carbon dioxide and water. In anaerobic digestion, biogas is also produced that is approximately 65% methane and 35% carbon dioxide. 7. Produces Class A or B biosolids.	STABILIZATION
SECONDARY	1. Biological treatment process. 2. Microorganisms consume organic material in wastewater and convert it into more microorganisms, carbon dioxide, and water. 3. Increases average particle size so biological solids can be separated from the treated wastewater by gravity. 4. Many variations in secondary treatment including fixed film processes (trickling filters and RBCs) and suspended growth processes (ponds and activated sludge). All processes depend on the same types of microorganisms. 5. Secondary effluent meets the secondary treatment standards of 30 mg/L of BOD_5 and TSS and 25 mg/L for CBOD.	1. Physical treatment process. 2. Reduces the total volume of biosolids for disposal. 3. Takes place after sludge stabilization. 4. Uses many of the same technologies as sludge thickening, but may also use belt filter presses, plate and frame presses, and centrifuges. Polymer addition is required. 5. Dewatered solids are referred to as "cake" and may contain as much as 50% total solids depending on the technology used.	DEWATERING

	Liquid Treatment Processes	Solids Treatment Processes	
TERTIARY	1. May be physical, biological, or chemical treatment. 2. Includes all treatment beyond the secondary treatment standards including removal of ammonia, nitrate, phosphorus, metals, and other constituents. 3. Includes technologies such as filtration and chemical precipitation.		
DISINFECTION	1. Reduces the numbers of bacteria and pathogens in the final effluent. 2. Disinfection is not sterilization. 3. Most WRRFs utilize chlorine gas, sodium hypochlorite (bleach), or UV light to disinfect; however, other technologies are available. 4. Protects public health and the environment.	*Does not sterilize or kill all bacteria	

References

Criteria for Municipal Solid Waste Landfills (1996) Code of Federal Regulations, Part 258, Title 40.

National Research Council (2002) *Biosolids Applied to Land: Advancing Standards and Practices;* National Academies Press: Washington, D.C.

Standards for the Use or Disposal of Sludge; Agency Response to the National Research Council Report on Biosolids Applied to Land and the Results of EPA's Review of Existing Sewage Sludge Regulations (2003) *Fed. Regist.*, **68** (68), 17379–17395.

U.S. Environmental Protection Agency (1999) *Biosolids Generation, Use, and Disposal in the United States;* EPA-530-R/99-009; U.S. Environmental Protection Agency: Washington, D.C.

U.S. Environmental Protection Agency (2003) Watershed Rule; 40 CFR 122, 124, and 130; U.S. Environmental Protection Agency: Washington, D.C.

U.S. Environmental Protection Agency (2005) *Onsite Wastewater Treatment Systems Manual;* EPA-625-R/00-008; U.S. Environmental Protection Agency, Office of Water, Office of Research and Development.

U.S. Environmental Protection Agency, Law & Regulations, Clean Water Act. http://www.epa.gov/r5water/cwa.htm (accessed April 2006).

U.S. Environmental Protection Agency (2000) Progress in Water Quality: An Evaluation of the National Investment in Municipal Wastewater Treatment; EPA/832-R/00/008; June.

U.S. Environmental Protection Agency (2011) *Principles of Design and Operations of Wastewater Treatment Pond Systems for Plant Operators, Engineers, and Managers;* EPA/600/R-11/088; U.S. Environmental Protection Agency, Office of Research and Development, National Risk Management Research Laboratory - Land Remediation and Pollution Control Division: Cincinnati, Ohio.

U.S. Environmental Protection Agency Frequently Asked Questions about Biosolids (2016); https://www.epa.gov/biosolids/frequent-questions-about-biosolids (accessed Oct 2016).

Water Environment Federation (2011) *Prevention and Control of Sewer System Overflows,* 3rd ed.; Manual of Practice No. FD-17; Water Environment Federation: Alexandria, Virginia.

Water Environment Federation (2012) *Basic Laboratory Procedures for the Operator-Analyst*, 5th ed.; Special Publication; Water Environment Federation: Alexandria, Virginia.

Water Environment Federation (2017) *Operation of Water Resource Recovery Facilities*, 7th ed.; Manual of Practice No. 11; Water Environment Federation: Alexandria, Virginia.

Wahlberg, E. (2016) Personal communication.

CHAPTER 2

Wastewater Characteristics

Introduction

Effective operation and control of a water resource recovery facility (WRRF) requires that the operator possess thorough knowledge of the composition of the influent, effluent, and internal process streams. To obtain that knowledge, the operator determines the characteristics of the raw wastewater and streams by collecting and analyzing representative samples throughout the facility. This chapter provides operators with information on wastewater characterization necessary to operate the facility effectively. Chapter 1 introduced two of the components that make up domestic wastewater: five-day biochemical oxygen demand (BOD_5) and various solids. Chapter 2 discusses additional wastewater components.

LEARNING OBJECTIVES

Upon completing this chapter, you will be able to

- List the significant sources of wastewater for a domestic WRRF.
- Evaluate the effect of activities in the service area on flow patterns at the WRRF.
- Explain how the size of the collection system and population served affect diurnal flow patterns.
- List and define additional components of domestic wastewater.
- Estimate flows and loads to a WRRF from population data. Calculate per capita generation rates.
- Convert nitrogen and phosphorus compounds to expressions as N and P, respectively.

Characterization of Wastewater

Characterization of facility influent, effluent, and internal process streams provides facility operators the information they need to properly control treatment processes. Depending on the size of the water resource recovery facility (WRRF) and the composition of the influent, wastewater characterization may require a few simple tests or several more complex tests in a well-equipped laboratory. As an alternative to maintaining a facility laboratory, an outside agency or private laboratory approved by the regulatory agency may provide the necessary analytical services. In either case, the operator must determine the specific information needed for each sample site. The following sections describe the various data that might be needed for process control and influent and effluent monitoring. The facility's National Pollutant Discharge Elimination System (NPDES) or State Pollutant Discharge Elimination System (SPDES) issued permit lists many of the pollutants to be tested.

SOURCES OF WASTEWATER

Wastewater can typically be categorized as originating from domestic, commercial, or industrial sources. In most treatment districts, sanitary wastewater is conveyed to treatment facilities in isolated piping networks or collection systems. This is the sanitary sewer system. Ground surface runoff induced by rain is stormwater. Potentially high volumes of stormwater typically collected in large catch basin drains are conveyed to specific locations throughout the treatment district. These largely untreated wastewaters may be piped directly to creeks, rivers, ponds, bays, or specially designed natural percolation areas. These intermittent, high flows are managed in their own piping system. This is the stormwater collection system. There are also combined sewer systems that carry both stormwater and wastewater. The combination of wastewater and stormwater can alter wastewater characteristics considerably. As a result, facility treatment processes are affected.

DOMESTIC

Domestic wastewater comes primarily from residential, commercial (nonindustrial), and institutional sources. Except for small WRRFs serving residential communities, most facilities treat some commercial and industrial wastewater. Wastewater with a predominantly domestic origin tends to be fairly uniform in composition. Composition varies somewhat among communities because of differing social, economic, geographic, and climatic conditions. Composition and quantities of domestic wastewater in some systems vary seasonally because of contributions from large institutions, such as colleges, or from resort areas where the population fluctuates widely.

• composition of the domestic waste water flowing to many WRRFs likely changed, due to the COVID outbreak + the quarantine.

INDUSTRIAL AND COMMERCIAL

Most municipal sewers convey wastewater from industrial and commercial sources and from domestic sanitary sources. *Industrial* and *commercial* wastewaters are typically monitored by the pretreatment group to verify that hazardous wastes are not discharged to the collection systems that could affect WRRF treatment. Industrial wastewater may contain substances derived from raw materials, intermediate products, byproducts, and end products of the industry manufacturing or production processes. Industrial wastewater changes with changing production mixes and schedules. This wastewater is more variable than domestic wastewater. Food-processing wastes, which typically contain high concentrations of soluble organic constituents, often cause extreme variations in facility loading because of the seasonal production associated with crop harvests. Food-processing wastes may also be nutrient-deficient (i.e., low in nitrogen or phosphorus), which can have an adverse effect on the biological treatment processes at the WRRF. Airports discharging significant amounts of glycol during the winter months may also affect facility secondary processes. Glycol is high in biochemical oxygen demand (BOD) and contributes to phosphorus nutrient deficiency. Commercial sources, such as retail businesses, primarily contribute domestic wastewater; other commercial sources, such as warehouses and distribution centers, may contribute variable wastewater from washing and other operations.

Industrial and *commercial* wastewaters are defined by the Federal Pretreatment Act as any non-domestic source of wastewater. The Act does not distinguish between industrial and commercial sources; however, most people think of heavy industry as being associated with processing of raw materials or manufacturing and light commercial as being associated with office buildings and warehouses. Examples of some heavy industries specifically regulated by the Act are sugar production, production of iron from ore, petroleum refining, and textile mills.

Industrial more variable, because of changes in operations

INFILTRATION AND INFLOW

Infiltration and inflow can affect hydraulic loadings. Infiltration enters sewers through leaky joints, cracks or holes in pipes, broken lateral lines from homes, and broken or missing clean-out caps. Groundwater is a common source of infiltration, particularly when the ground is saturated. Inflow results when storm or runoff water enters the system through direct openings such as leaky manhole covers or cracked casings. Illegal roof gutter and sump pump connections are also a source of inflow.

INFLUENT FLOW VARIATIONS

The facility operator needs to know whether the wastewater influent comes from separate or combined collection systems because stormwater flows from combined systems can adversely affect facility hydraulics. Even facilities served by separate sanitary sewer systems receive some extraneous water from inflow of surface runoff and infiltration of groundwater. Infiltration and inflow cause seasonal flow variations. These effects are influenced by the age, condition, and type of collection system. Combined systems cause significant changes in wastewater flow from runoff of stormwater or snowmelt. In addition to the hydraulic effects on the facility, changes may result from the organic matter and dissolved contaminants contained in runoff flushed from streets and other surfaces. The effects of hydraulic and pollutant loadings from combined sewers are especially pronounced during the first hours of a storm following a dry period.

Chapter 1 gave a broad overview of per capita generation rates for municipal wastewater. On an average day, each person in the *service area* will contribute between 204 and 455 liters per capita per day (L/cap·d) (54 and 120 gallons per capita per day [gpd/cap]). Wastewater is not generated at a constant rate all day long. More wastewater will flow through the collection system and into the WRRF at certain times of the day than others. Imagine a small community with fewer than 1000 residents. Most of the residents in this community will wake, shower, flush toilets, and cook breakfast within an hour or two of one another. They will go to work at about the same time and come home at about the same time to prepare dinner, run the dishwasher, and toss in a load or two of laundry. They will settle down for an evening of family time or television or playing cards, and then most of them will sleep for 7 or 8 hours before starting the whole cycle over again. Some members of the community will work a second shift or otherwise keep different hours. The patterns of daily living are reflected in how wastewater is generated and conveyed to the WRRF. At the WRRF, the influent flows will typically be the highest in the middle of the day and the lowest at night. This repeating pattern of changing flows over a 24-hour period is called a *diurnal flow pattern*.

The *service area* is the land area served by the collection system and WRRF. It could be as small as a single building or as large as a metropolitan area.

Diurnal means something is related to each day or that it takes place daily. The influent flow pattern is diurnal because the pattern will repeat every 24 hours.

Figure 2.1 shows an influent flow pattern for a community of approximately 20,000 residents. Notice that there are two distinct high points on the graph, the first appearing at about 10:00 A.M. and the second appearing at about 9:00 P.M. The first high point is larger. This is the peak hour flow or the highest hourly flow that occurs over a 24-hour period. Even though most of the residents in this community began going about their day at about the same time, the wastewater they generated did not make an instant appearance in the influent to the WRRF. It takes time for the flow to travel through the collection system. Engineers that design collection systems target a flow velocity between 0.61 and 1.22 m/s (2 and 4 ft/sec). This is equal to between 2.2 and 4.4 km/h (1.36 and 2.72 mph). This velocity range keeps solids from settling out in the collection system. The longer the collection system, the more time it will take for the wastewater to reach the WRRF. For very small communities with short collection systems, the peak hour flow will typically arrive before 9:00 A.M. For larger communities with longer collection systems, the peak hour flow may not occur until late morning or early afternoon. A second peak typically occurs later into the evening as residents return home to prepare their evening meals, wash dishes, and perhaps do laundry. Note the deep valley at either end of the diurnal graph where very little flow is coming into the WRRF. For a small community, influent flows can go nearly to zero between midnight and 7:00 A.M. This repeating pattern of peak hour flows and low flows is easier to see in Figure 2.2. Figure 2.2 shows a chart recorder with just over 3 days of influent flow recorded on it. Note the pronounced double peak pattern. The community this chart recorder was taken from serves about 500 residents.

Equipment maintenance is sometimes scheduled late at night when influent flows are lowest. Low flows can allow operators to take equipment out of service more easily without affecting WRRF operation.

As the service area gets larger and serves more residents and businesses, the shape of the diurnal flow pattern begins to change. The peaks become less pronounced and the diurnal pattern becomes flatter. Two things occur as the service area gets larger. First, the collection system gets larger and longer. Flow is input at many points along long pipes, an action that helps to spread out or *attenuate* the flow. Think about dumping a bucket of water on the top of a slide. The water will not arrive at the bottom of the slide in one big slug. It spreads out as it travels. The longer the slide, the greater the spread. The same type of thing happens in the collection system. Secondly, as the service area gets larger, the population within it tends to become more diverse. Bigger communities are more likely to have stores and other businesses that are open late into the night or even 24 hours per day. Residents may be working day shift, swing shift, or graveyard

Attenuate means to reduce the intensity of something.

↳ Regulate the flow

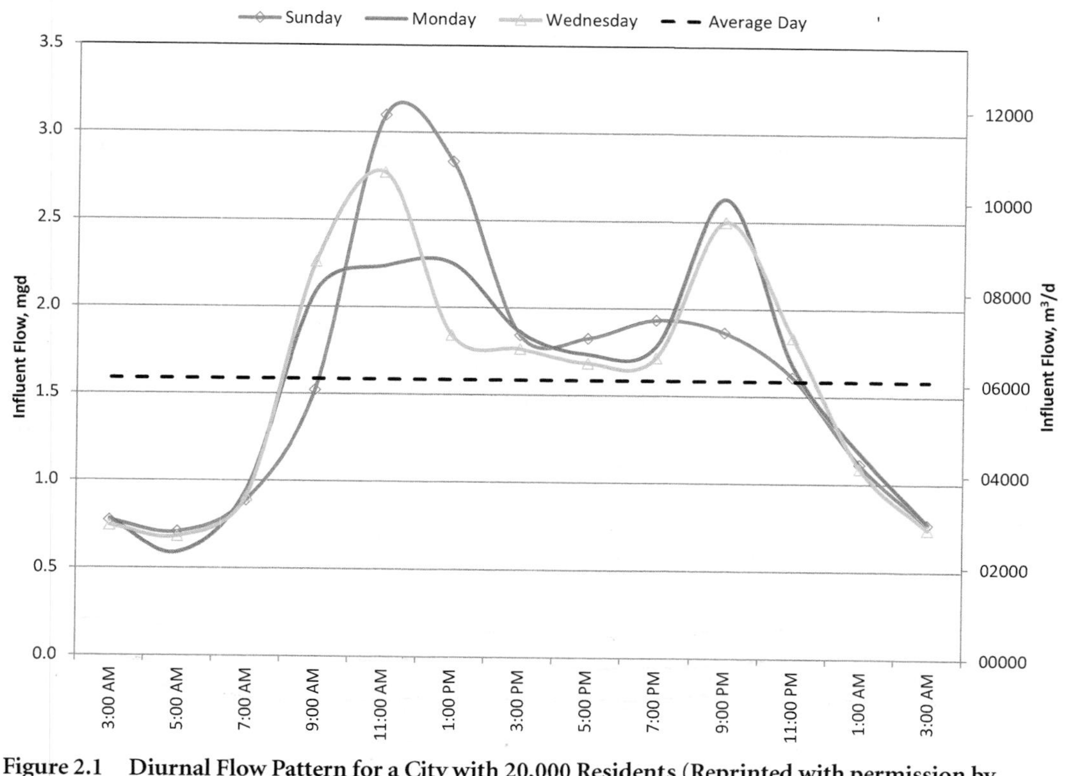

Figure 2.1 Diurnal Flow Pattern for a City with 20,000 Residents (Reprinted with permission by Indigo Water Group)

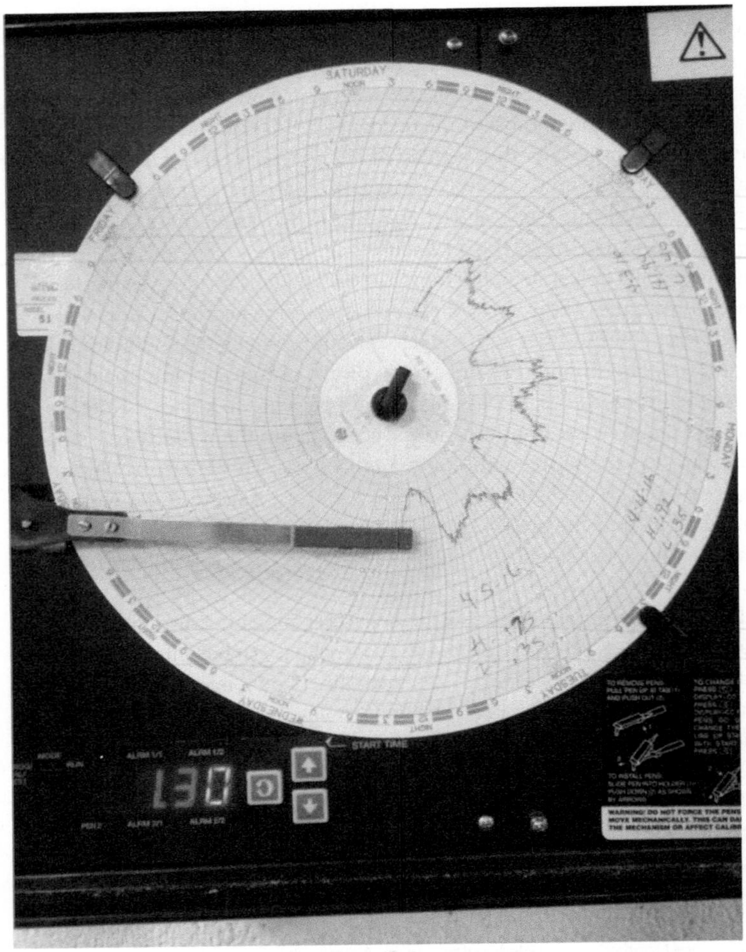

Figure 2.2 Influent Flow Measurements on a Chart Recorder (Reprinted with permission by Indigo Water Group)

shift. As a result, the peak hour peaks get smaller and the deep valley that occurred with smaller communities becomes less pronounced. Figure 2.3 shows some diurnal flow data for a large WRRF that serves about 250,000 residents. The diurnal pattern is still there, but it is not as pronounced. The peak hour flow is only about 1.4 times the average daily flow. In the previous example (Figure 2.1), the peak hour flow was about 2 times the average daily flow. Other influences on flow variation include the number and type of pumping stations, types of industries served, and population characteristics.

Look carefully at Figure 2.3. For this community, the peak hour flow arrives later in the day on Saturday and Sunday, and the total amount of flow has increased slightly. These changes are tied to residents waking later on the weekend and taking care of household chores that generate wastewater, including laundry and housekeeping. The evening peak— dinner time—is not shifted. For suburbs, flows may increase on the weekend simply because more residents are at home rather than commuting to workplaces outside of the service area. Conversely, flows may be higher at the city center during the week because of the presence of workers. The presence of large industries and storm events can further change flow patterns.

It is important for operators to be familiar with the flow patterns for their facilities. Operators should record the average daily influent flow and peak hour flow each day. For most WRRFs, the monthly average flow and maximum day flow must be reported on the monthly discharge monitoring report. Keeping daily records will help to alert the operator of any changes. Changes in influent flow may require an operator to respond by adjusting pumping cycles or chemical feed rates. A sudden, unexpected increase in flow could signal a new source of inflow and infiltration or it could indicate that a new subdivision or industrial user has become active. Figure 2.4 shows influent flow from a WRRF operating in an area experiencing

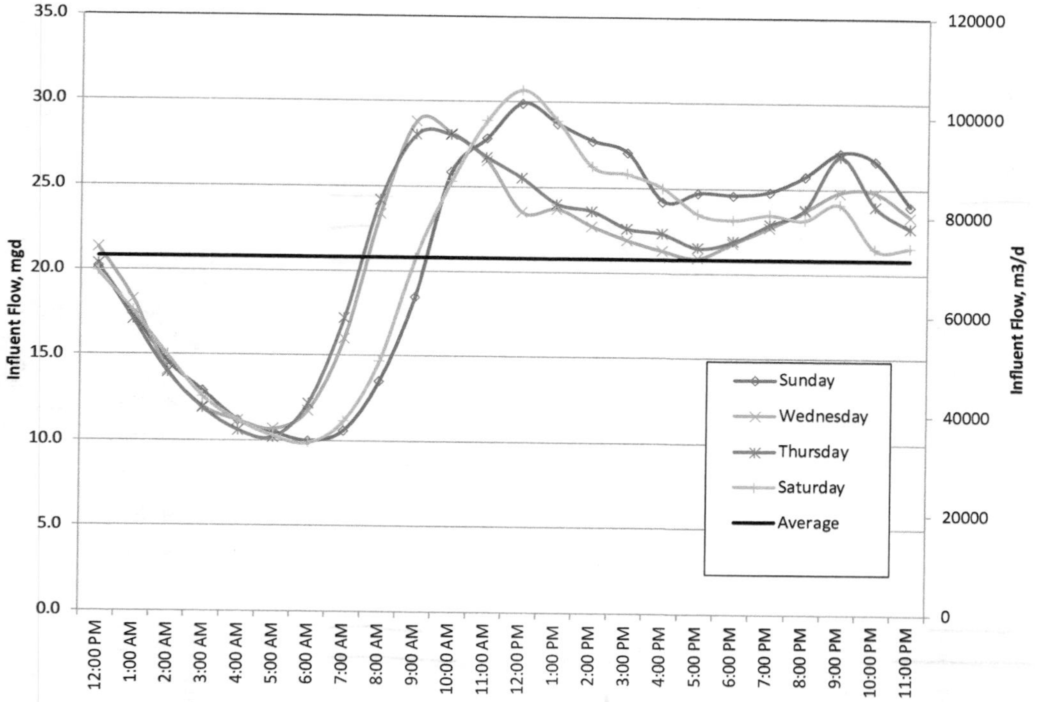

Figure 2.3 Diurnal Flow Pattern for a City with 250,000 Residents (Reprinted with permission by Indigo Water Group)

a drought. Between January and December, the WRRF influent flow steadily decreased from an average of about 71.9 ML/d (19 mgd) to an average of about 49.2 ML/d (13 mgd). Influent BOD concentrations increased as flow decreased, which indicates a loss of flow, but not a loss of load. A storm event in March 2004 caused the influent flow to increase suddenly and BOD concentrations to decrease temporarily. Many WRRFs have seasonal industries like tourism or crop processors in their service areas. Seasonal users can

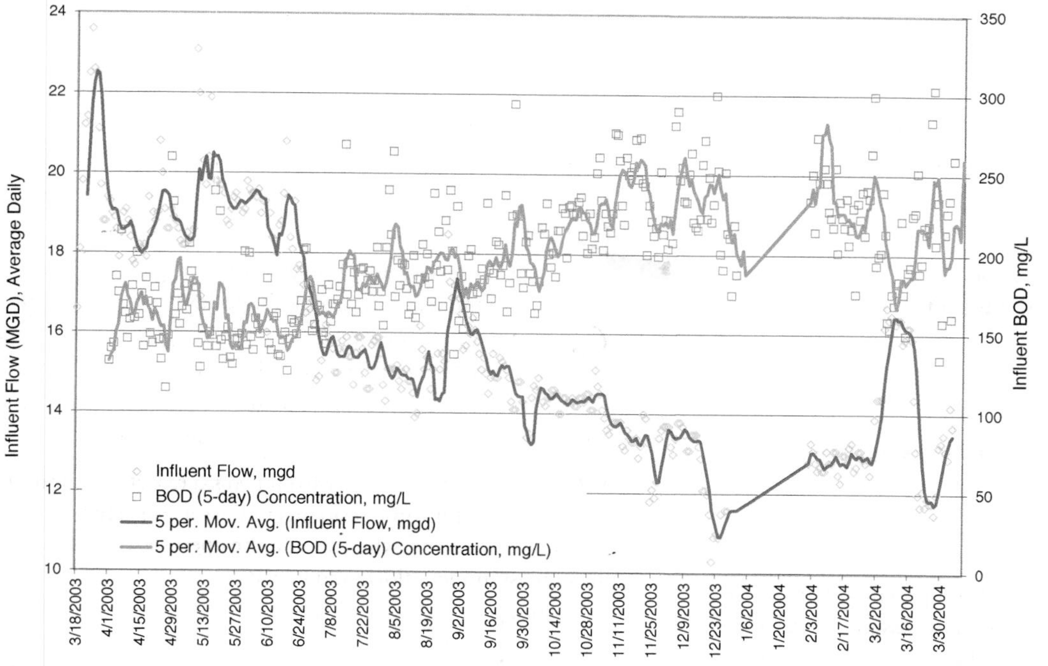

Figure 2.4 Effect of Inflow and Infiltration on Influent Concentrations (Reprinted with permission by Indigo Water Group)

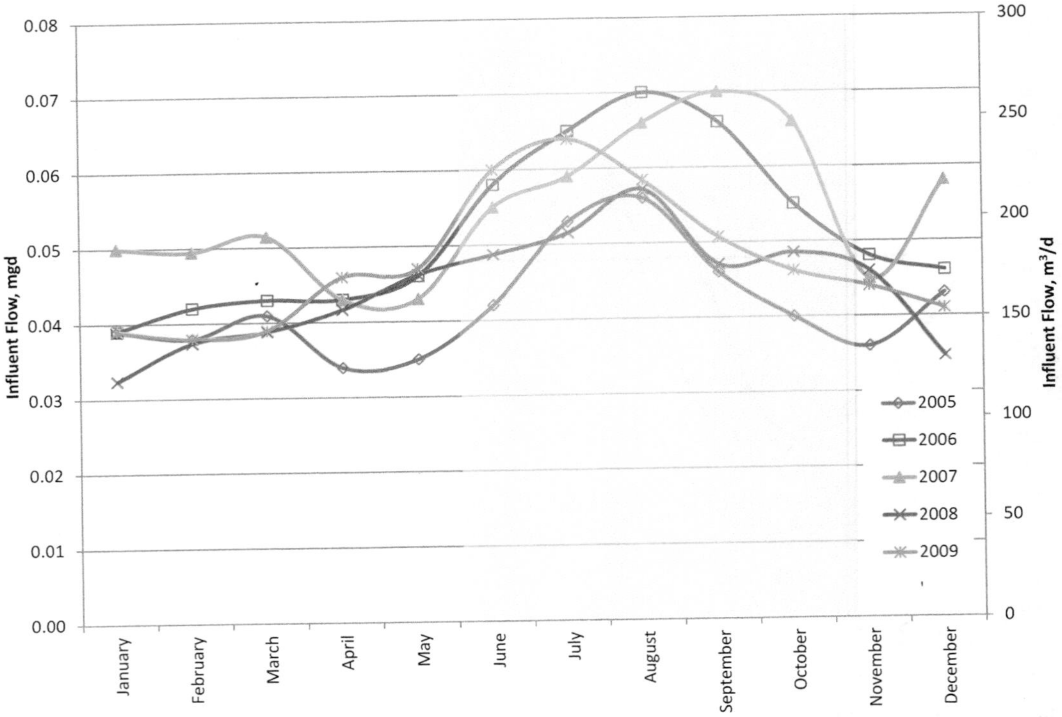

Figure 2.5 **Seasonal Flow Variations (Reprinted with permission by Indigo Water Group)**

cause sudden increases in the amount of flow and organic load to the WRRF. Schools and universities can also affect flow patterns and loading when in session. Both may affect operations. Figure 2.5 shows seasonal flow patterns for a small community that depends on summer tourism. Note that flows are substantially higher from May through November than they are during the rest of the year.

TEST YOUR KNOWLEDGE

1. The Federal Pretreatment Act defines industrial and commercial users differently from one another.
   - ☐ True
   - ☑ False

2. Influent wastewater consists of domestic, industrial, and commercial wastewater as well as contributions from inflow and infiltration.
   - ☑ True
   - ☐ False

3. Influent flow patterns reflect activities taking place in the service area.
   - ☑ True
   - ☐ False

4. An operator must take a piece of equipment out of service for maintenance. They will need to bypass pump wastewater around this piece of equipment during repairs. Assuming the WRRF has a typical diurnal flow pattern for domestic wastewater, when should maintenance be scheduled to minimize the amount of bypass pumping?

   "Normal 24 cycle" in other words

   a. Midmorning
   b. After lunch
   c. Afternoon
   d. Late evening

5. Which type of service area is likely to see the greatest variations in influent flow over a single day?
   a. Town with 500 residents
   b. City with separate domestic and storm sewers
   c. City with more than 50,000 residents
   d. Town without large commercial or industrial users

6. A WRRF receives wastewater from two ski resorts. Assuming the number of full-time residents in the service area is constant year-round, when should the operator expect to see the highest influent flows and loads?
   a. Spring
   b. Summer
   c. Fall
   d. Winter

INFLUENT CHARACTERISTICS

Raw wastewater is comprised of human wastes, ground-up vegetable matter from garbage disposals, trash, rags, grit, and other materials. Table 2.1 lists some of the many components of wastewater and their typical concentrations in untreated domestic wastewater. Solids and BOD were discussed in Chapter 1. Additional wastewater components are defined here. Domestic wastewater includes contributions from schools, stores, and other light commercial operations, but not flows from *significant or categorical* industrial users. Concentrations are described as low, medium, or high strength. Whether a particular wastewater is low or high strength reflects the makeup of the service area and how much water is being used by the population contained within that service area.

A *significant* industrial user is one that contributes 94 635 L/d (25 000 gpd) OR more than 5% of a WRRF's hydraulic or organic permitted capacity.

In the United States, the amount of water typically used can vary between 204 and 455 L/cap·d (54 and 120 gpd/cap). Flows higher than 455 L/cap·d (120 gpd/cap) should be investigated for contributions from inflow and infiltration. For communities that use less water, concentrations of organic matter and solids will be higher than for communities that use more water, assuming both communities receive the same number of kilograms (pounds) of organic matter and solids. Higher water use dilutes the strength. Commercial facilities such as schools, businesses, restaurants, and shopping centers may also increase the strength of the wastewater by discharging more kilograms (pounds) of organics and solids. Wastewater strength may also be affected by stormwater, whether it is coming from a combined system or as the result of significant inflow and infiltration. In this situation, wastewater strength can shift significantly between wet and dry weather.

A *categorical* industrial user is any industry listed in Title 40 of the *Code of Federal Regulations* (40 CFR) Parts 400–424 and 425–699. These industries have a high likelihood of affecting the way a WRRF operates and its ability to meet its discharge permit limits.

Industrial wastewater tends to be much higher in strength than domestic wastewater for certain components, particularly fats, oils, greases, heavy metals, and solvents. Table 2.2 gives concentration ranges for conventional pollutants (BOD, total suspended solids [TSS], oil and grease, total Kjeldahl nitrogen [TKN], and phosphorus [P]) for a wide variety of industrial wastewaters. If any of these industries discharge to the sewer system in significant amounts and without pretreatment, they can affect the overall makeup of the wastewater that arrives at the WRRF.

Table 2.1 Typical Concentrations for Selected Parameters in Raw Domestic Wastewater (Metcalf and Eddy/AECOM, 2014. Copyright © 2014 by McGraw-Hill Education, reprinted with permission.)

Parameter	Low Strength	Medium Strength	High Strength
Total solids, mg/L	537	806	1612
Total dissolved solids, mg/L	374	560	1121
Total suspended solids, mg/L	130	195	389
Fixed residue, mg/L	29	43	86
Volatile suspended solids, mg/L	101	152	304
Settleable solids, mg/L	8	12	23
Biochemical oxygen demand, 5-day, 20 °C, mg/L	133	200	400
Total organic carbon, mg/L	109	164	328
Chemical oxygen demand, mg/L	339	508	1016
Total nitrogen, mg/L as nitrogen	23	35	69
Total Kjeldahl nitrogen, mg/L as nitrogen	24	34	70
Organic nitrogen, mg/L as nitrogen	10	14	29
Free ammonia as nitrogen, mg/L	14	20	41
Nitrites as nitrogen, mg/L	0	0	0
Nitrates as nitrogen, mg/L	0	0	0
Total phosphorus, mg/L	3.7	5.6	11.0
Oil and grease, mg/L	51	76	153

Note: mg/L = g/m^3

Table 2.2 Typical Concentrations of Selected Parameters for some Industrial Wastewaters (WEF, 2008)

Industry Type	BOD_5 mg/L	TSS mg/L	Oil and Grease mg/L	COD mg/L	TKN mg/L as N	P mg/L as P
Dairies	1000–2500	1000–2000	300–1000		50–100	
Fruit and vegetable processing	300–1000	200–800				
Leather tanning and finishing	400–5900	710–8600	86–1600	1800–13 600	46–890	
Meat processing	1500–7200	360–3300	150–1800		24–310	35–82
Metal products and machinery	2000	1000	2300	11 300	600	170
Paint formulating	280–65 500	180–148 000	42–3400			
Synthetic rubber	9–420	15–770	1–200	50–2800		
Timber products	56–4000	400–1100	300	2600–19 300	0.17–4	0.3–3

TEMPERATURE

Wastewater temperatures are measured in degrees Celsius or degrees Fahrenheit. The two scales are related by the following equations.

Convert degrees Celsius to degrees Fahrenheit:

$$\text{Degrees Celsius} = (\text{Degrees Fahrenheit} - 32)\left(\frac{5}{9}\right)$$

Or

$$\text{Degrees Celsius} = \frac{(\text{Degrees Fahrenheit} - 32)}{1.8}$$

Convert degrees Fahrenheit to degrees Celsius:

$$\text{Degrees Fahrenheit} = (\text{Degrees Celsius})\left(\frac{9}{5}\right) + 32$$

Or

$$\text{Degrees Fahrenheit} = (\text{Degrees Celsius})(1.8) + 32$$

The degrees Celsius and degrees Fahrenheit scales cross at −40 degrees. Table 2.3 lists the temperature in both degrees Celsius and degrees Fahrenheit for a few reference points. It can be helpful to remember some of these as an easy way to compare temperatures.

A *temperate climate* has distinct seasonal changes in temperature: spring, summer, fall, and winter.

Wastewater is typically somewhat warmer than unheated tap water because wastewater contains heated water from dwellings and other sources. Because buried pipes convey wastewater long distances to the facility, the influent temperature typically approaches the temperature of the ground. Accordingly, summer wastewater temperatures exceed winter temperatures in *temperate climates*. Wastewater temperatures typically range between 10 and 20 °C (50 and 68 °F); however, temperatures may be higher or lower depending on where the WRRF is located. The temperature of the wastewater affects the operation of the WRRF.

Density is the amount of mass per unit volume. Water has a density of 998.2 kg/m³ (62.300 lb/cu ft) at a temperature of 20 °C (68 °F). As water cools, it becomes more dense. Water has a density of 1000 kg/m³ (62.424 lb/cu ft) at 4 °C (39 °F).

Cold water is *denser* than warm water. How fast particles settle in clarifiers is related to the difference in density between the particles and the water. As the water becomes denser with lower temperatures, the

Table 2.3 Reference Points in Degrees Celsius and Degrees Fahrenheit

Reference Point	Degrees Celsius	Degrees Fahrenheit
Freezing point of water at sea level	0	32
Typical winter wastewater temperature	10	50
Room temperature	20	68
Body temperature (humans)	37	98.6
Boiling point of water at sea level	100	212

density difference gets smaller and settling rates can decrease. For heavy, dense particles like sand and grit, this effect is so small that it is not even noticeable. For less dense organic material and the biological flocs produced in secondary treatment processes, changes in water temperature can have a big effect on settling rates.

In general, the rate of biological activity and chemical reactions depend on temperature. As temperature increases, microorganisms become more active and accelerate consumption of energy sources (food) and oxygen. Reaction rates approximately double with every 10 °C (18 °F) increase in temperature, until higher temperatures begin to *inhibit* biological activity. A biological treatment process operating at 10 °C may require twice as much time to do the same amount of treatment as the same process operating at 20 °C. Chemical reactions are similarly affected by temperature and will proceed faster at warmer temperatures.

Inhibit means to slow down or impede. When biological processes are inhibited, they are no longer functioning at their maximum rates.

A significant increase in influent temperature for a short period of time typically indicates the presence of an industrial discharge. A significant drop in temperature often indicates intrusion of stormwater or snowmelt.

DISSOLVED OXYGEN

Dissolved oxygen (DO) is simply the molecular oxygen present in water or wastewater. It is expressed as milligrams per liter. The maximum amount of DO that can be dissolved in water is the saturation concentration. The saturation concentration depends on water temperature, salinity, and atmospheric pressure. Colder water is capable of containing more DO than warmer water at the same salinity and pressure. Salinity is the amount of salt dissolved in the water. As salinity increases, the amount of DO the water can hold also decreases.

Atmospheric pressure is the weight of the air above a certain point. At sea level, the atmospheric pressure is equal to one atmosphere, which is the same as 101 kPa (14.7 psi). As one goes up in elevation on a plane or an elevator or in the mountains, there is less air above and less atmospheric pressure. Think about sitting on the bottom of a swimming pool. The weight of the water above you is similar to the weight of the atmosphere. There is more pressure at the bottom of the pool than at the top of the pool. At an elevation of 1609 m (1 mile), there is about 17% less atmosphere than at sea level. There is less pressure available to help hold dissolved gases in solution. Water exposed to higher atmospheric pressure is capable of containing more DO than the same water at lower pressure.

- Atmospheric pressure helps contain dissolved gases in a solution.

A container of pure water (zero salinity) at sea level can hold up to 11.29 mg/L of DO at 10 °C (50 °F), but only 9.09 mg/L at 20 °C (68 °F). If the salinity of the water (still at sea level) is increased to 45.2 parts per trillion (ppt), then the maximum amount of DO the water can contain will decrease to 8.45 mg/L at 10 °C and 6.96 mg/L at 20 °C. Keep in mind that just because water is capable of containing a certain concentration of DO does not mean that it will contain that concentration. A container of water at sea level with zero salinity and a temperature of 20 °C is capable of containing 9.09 mg/L, but it could contain any concentration up to that concentration or none at all.

When water freezes, oxygen and other gases dissolved in the water make the ice cloudy. Hot water will make clearer ice cubes because hot water contains less dissolved gases/oxygen than cold water.

Dissolved oxygen is critical for some types of biological treatment. Microorganisms in the treatment process breathe oxygen and exhale carbon dioxide. Some bacteria can use nitrate in place of DO. Oxygen is needed by bacteria to convert ammonia to nitrite and nitrate. Without oxygen, this reaction cannot take place. When oxygen is present, the water is aerobic. Another word for aerobic is *oxic*.

Wastewater becomes *septic* when all of the oxygen and nitrate has been used up by microorganisms. A group of bacteria called the *facultative anaerobes* can convert sulfate to hydrogen sulfide under septic conditions. Hydrogen sulfide has a strong, rotten egg smell. Some other sulfur compounds that can be produced under septic conditions include dimethyl disulfide (garlic) and methyl mercaptan (rotten cabbage).

COLOR

The color of wastewater depends on the amounts and types of dissolved, suspended, and colloidal matter present. Fresh wastewater is typically light brown to grayish in color. Wastewater becoming *septic* will be darker gray or even black. Other colors typically indicate the presence of industrial discharges. For example, green, blue, or orange discharges may emanate from plating operations; red, blue, or yellow discharges are often dyes; and white, opaque discharges often come from dairy wastes or latex paint. Knowledge of the types of industries contributing to the collection system and the colors of their discharges is of great value to the operator. Any time an unusual color or odor is observed in the WRRF influent, samples should be collected for analysis.

ODOR

Odor, a highly subjective parameter, can offer valuable information. The human nose, a sensitive odor-detecting system, can often smell wastewater constituents at low concentrations. Fresh wastewater typically produces a musty odor. Other odors, such as petroleum, solvents, or other abnormal scents, may indicate an industrial spill.

Because some of the compounds present in wastewater may be toxic, caution must be used when smelling wastewater, especially when smelling bottled samples. Detection of unusual odors in a facility, particularly in confined areas, requires exercising caution and strict adherence to safety procedures.

Decomposition of wastewater in the absence of oxygen can produce hydrogen sulfide, which has a distinctive rotten egg odor. When hydrogen sulfide is present, measures to increase the oxygen content of the liquid stream must be taken. In addition to indicating process problems, the presence of hydrogen sulfide raises concerns for other reasons. It is poisonous at relatively low levels, corrosive to concrete and metal, and potentially explosive. It is also important to note that hydrogen sulfide is particularly dangerous because it essentially paralyzes one's sense of smell when present at high concentrations. In this situation, one would initially note the rotten egg odor, but it would seem to quickly go away. If one interprets this lack of a rotten egg odor as no hydrogen sulfide present, it could quickly lead to one or more fatalities. Methane gas, which is even more explosive, may accompany hydrogen sulfide. Therefore, the presence of hydrogen sulfide and/or methane demand extreme caution and rigorous application of established safety procedures. This includes using appropriate gas detection equipment before entering a confined space and turning on ventilation equipment, if available.

TURBIDITY

When most people think of the word *turbidity*, they picture milky or cloudy water that contains a high concentration of suspended solids. A more correct definition is the tendency of a substance to scatter light. When light is scattered, it is bouncing off particles suspended in the substance and moving in all different directions. This is different from absorption, where light cannot fully penetrate a liquid, and transmittance, where some or all of the light passes through the substance. Turbidity in water is caused by suspended and colloidal material such as clay, silt, finely divided organic and inorganic matter, plankton, algae, and other microorganisms. Turbidity testing provides an indirect measurement of how much suspended and colloidal material there is in the flow stream, particularly at low solids concentrations.

Turbidity is measured with a special device called a *turbidimeter*. A rough representation of how a turbidimeter functions is shown in Figure 2.6. Notice that the light source is at a 90-deg angle relative to the detector. True turbidity is always measured at a 90-deg angle so the detector only sees light that has bounced and scattered. If the detector were placed opposite the light source on the other side of the sample, it would measure transmitted light instead. Turbidity is measured in nephelometric turbidity units (NTUs).

Turbidity does not directly correlate with suspended solids concentrations because (1) the size and quantity of particles can vary and (2) color can interfere with the turbidity measurement. It is possible for water to have low TSS and high turbidity. In this instance, the particles would be small and fine without a lot of total mass. Together, they cause turbidity, but would not result in a high TSS measurement. It is also possible for water to have high TSS and low turbidity. In this instance, the solids particles would be large, but there would be a lot of clear water between particles. The solids would have a lot of mass, but most of the light would be transmitted rather than scattered. The final effluent from a well operating, secondary treatment process may have a turbidity between 0.5 and 3 NTUs.

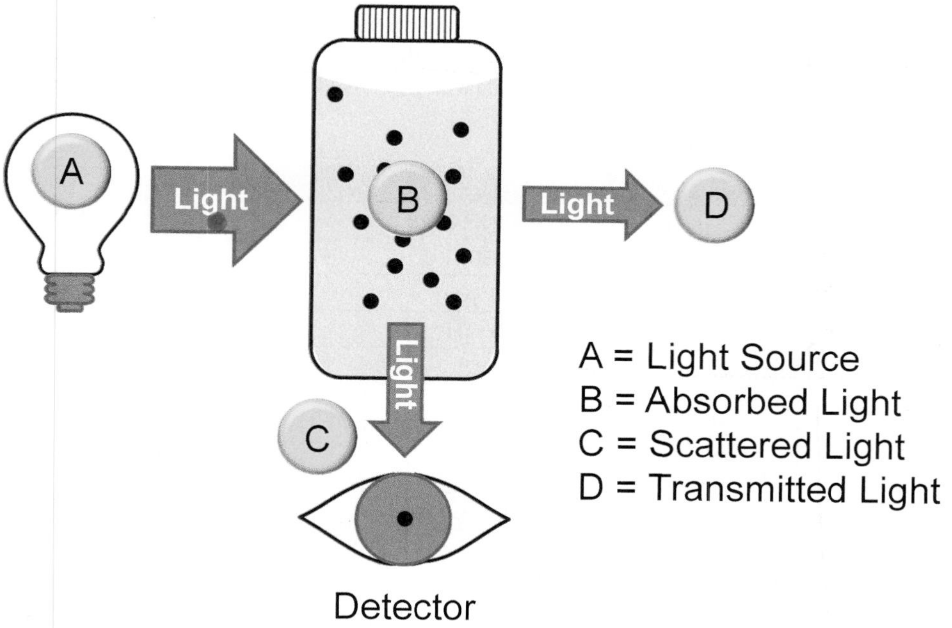

Figure 2.6 **Light Scatter with Turbidity (Reprinted with permission by Indigo Water Group)**

A relationship between TSS and turbidity can be established for a given system under a particular set of operating conditions. To do this, a series of samples would be analyzed for both TSS and turbidity. If conditions change significantly, the correlation must be reestablished.

CONDUCTIVITY

Conductivity measures the ability of an aqueous solution to carry an electrical current. The conductivity of domestic wastewater generally ranges from 50 to 1500 µS/cm, although some industrial wastewaters have conductivities higher than 10 000 µS/cm. The units *µS/cm* are microsiemens per centimeter. The conductivity of wastewater is related to the amount of total dissolved solids (TDS)—primarily salts—present in the water; however, conductivity and TDS are not the same thing. These are two different measurements for two distinctly different things: electrical conductivity and dissolved solids.

There are many commercially available meters that read both conductivity and TDS; however, the only true way of measuring TDS is by filtering the sample, evaporating the water, and physically weighing the dried solids that remain behind. Total dissolved solids meters actually measure conductivity and then use a conversion factor to estimate TDS. The conversion factor ranges between 0.47 and 0.85, depending on the manufacturer (HM Digital, 2008). Obtaining reliable data with conductivity monitoring equipment requires care of electrodes to prevent fouling and maintenance of adequate sample circulation. Conductivities greater than 10 000 to 50 000 µS/cm or less than 10 µS/cm can be difficult to measure.

pH

The term *pH* is traditionally used as a convenient representation of the concentration of hydrogen ions in a solution. The pH scale ranges from 0 to 14, with a neutral reading of 7. Readings below 7 are acidic and readings above 7 are basic. A reading of exactly 7 is defined as neutral. Raw wastewater typically has a pH near 7, although industrial and other nondomestic discharges can cause the pH to be much higher or lower. There are other conditions that can cause pH to deviate from the norm. Anaerobic conditions lower the pH of a wastewater. Low pH values, coupled with other observations, such as rotten egg odors (hydrogen sulfide) and black color, provide evidence of septic conditions in the collection system or within the treatment process.

The term *pH* is a mathematical abbreviation for the inverse log of the hydrogen ion concentration:

$$\text{pH} = -\log[\text{H}^+]$$

A *mole* is a unit of measurement used with atoms and molecules. Shoes come in a pair, paper comes in a ream, pencils come in a gross, and eggs come in a dozen. Atoms and molecules come in moles. One mole contains 6.022×10^{23} atoms or molecules. One millimole is one thousandth of a mole or 6.022×10^{20} atoms or molecules. Why such a huge number? Because atoms and molecules are so small that we need a large quantity of them in one place to measure them accurately. One mole of hydrogen atoms weighs 1.007 94 g.

Where $[H^+]$ is the actual hydrogen ion concentration in *millimoles* per liter (mmol/L) in a solution. The −log piece of the equation means that pH is expressed on a logarithmic or log scale. With a log scale, each segment of the scale increases by a factor of 10, so a sample with a pH of 4 has 10 times more hydrogen atoms than a sample with a pH of 5. For example:

pH	Hydrogen ion concentration, mmol/L
14	0.00000000000001
13	0.0000000000001
12	0.000000000001
11	0.00000000001
10	0.0000000001
9	0.000000001
8	0.00000001
7	0.0000001
6	0.000001
5	0.00001
4	0.0001
3	0.001
2	0.01
1	0.1

Because pH is expressed on a log scale, it takes a big change in hydrogen ion concentration to make a small change in the readout on the pH meter. Operators should be aware that pH measurements alone are rarely adequate for process control and that they should be paired with measurements of alkalinity. Often, by the time a pH change is noticeable, the process has degraded substantially and will be more difficult to bring back within normal operating range.

pH is important in biological and chemical treatment processes. Most microorganisms remain sufficiently active only with a narrow pH range, generally between 6.5 and 8.0. Outside of this range, pH can inhibit or completely stop biological activity. Chemical precipitation processes work best within narrow pH ranges that differ depending on what type of chemical is being used.

ALKALINITY

The alkalinity of water is its acid-neutralizing capability. In other words, how much acid water can absorb before there is a substantial change in pH. Alkalinity and pH are related, but do not measure the same thing. Alkalinity is buffering capacity. pH is the hydrogen ion concentration. Alkalinity is primarily composed of the following ions: carbonate (CO_3^-), bicarbonate (HCO_3^-), hydroxides (OH^-), and phosphates (PO_4^{-3}). The characteristics of the raw water supply influence alkalinity, which can be high in areas having hard water (typically associated with groundwater sources) or extremely low in areas having soft water.

In the laboratory, a known concentration of acid is gradually added to a sample while mixing continuously (Figure 2.7). The pH is monitored either directly with a pH meter or indirectly by adding special indicator solutions. Indicators change color at a specific pH. The first indicator used is phenolphthalein, which is pink above pH 8.3 and clear below. The second indicator used is bromocresol green methyl red, which is bright blue above pH 4.55 and red below. By definition, when the pH has reached 4.55, all of the alkalinity has been consumed and the result is reported as total alkalinity. Alkalinity is reported as milligrams per liter of calcium carbonate (mg/L $CaCO_3$).

NITROGEN

The nutrients nitrogen and phosphorus were introduced in Chapter 1. Both are essential for life. In wastewater, nitrogen occurs in four basic forms: organic nitrogen, ammonia, nitrite, and nitrate (Figure 2.8). The forms of nitrogen present in a wastewater indicate the level of treatment. Fresh wastewater typically has higher concentrations of organic nitrogen and ammonia than of nitrite and nitrate. Organic nitrogen

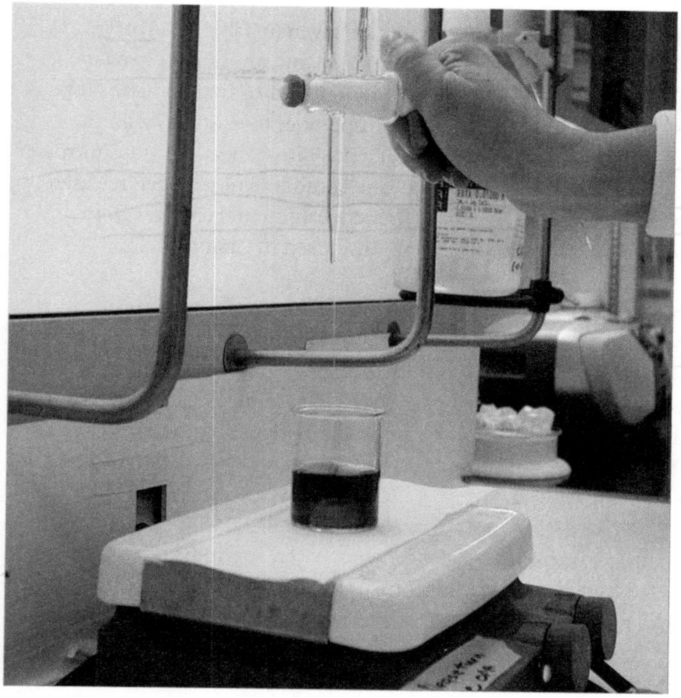

Figure 2.7 Titration for Total Alkalinity (Reprinted with permission by Indigo Water Group)

is found in proteins and other organic molecules. When organic material breaks down, ammonia is released. If conditions are suitable, different groups of bacteria in the treatment process gradually transform organic nitrogen into ammonia and then to nitrite, nitrate, and, finally, nitrogen gas. Nitrogen gas is not very soluble in water. Once produced in the process, it passes out of the wastewater and into the atmosphere. In addition to converting nitrogen from one form to another, some of the nitrogen will be incorporated to new bacteria. This is called *assimilative uptake*.

↳ when Nitrogen is taken in by bacteria

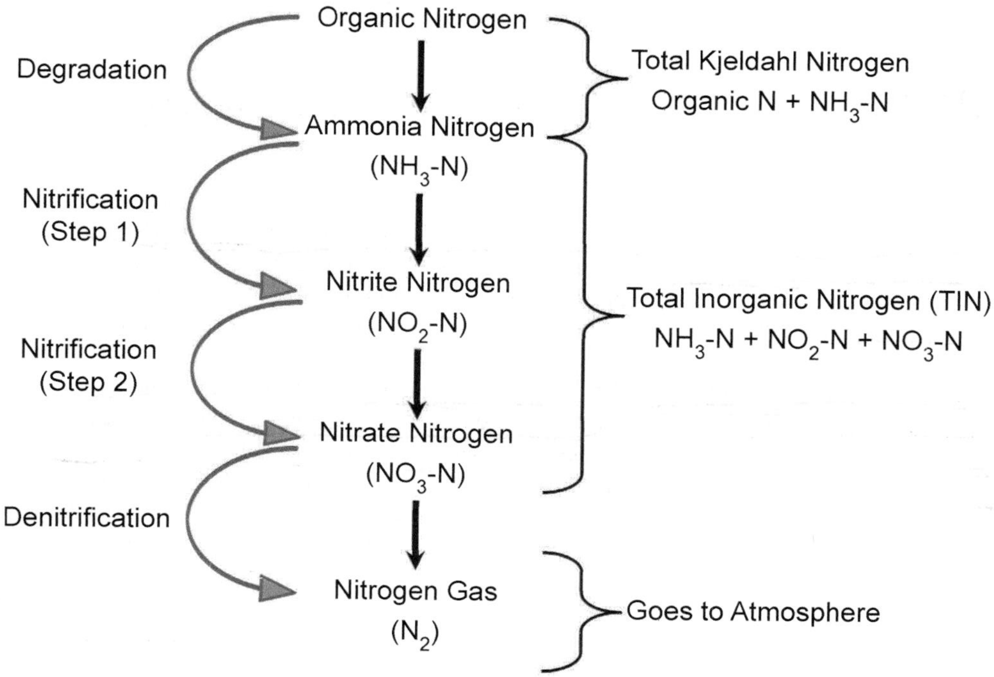

Figure 2.8 Nitrogen Species in Wastewater (Reprinted with permission by Indigo Water Group)

Typical ranges of nitrogen concentrations in raw domestic wastewater are given in Table 2.1. Total Kjeldahl nitrogen, pronounced "kell-doll", is the sum of the organic nitrogen and ammonia nitrogen (Figure 2.8). Some WRRFs are required to monitor for TKN in their influent and/or final effluent. Total Kjeldahl nitrogen is used by engineers when designing new facilities or upgrades to existing facilities. Total inorganic nitrogen (TIN) is the sum of the ammonia-nitrogen, nitrite-nitrogen, and nitrate-nitrogen (Figure 2.8). Many WRRFs have permit limits for TIN because even small amounts of nitrogen released to the environment can encourage the growth of algae and cause other environmental changes from *eutrophication*. The effects of nitrogen release on the environment are discussed in Chapter 1.

Eutrophication is an enrichment of nutrients in a natural system such as a lake or river. The increased nutrients can promote algae growth and change which species of plants and animals are able to thrive.

Nitrogen compounds are expressed as milligrams per liter as N (e.g., 1 mg/L of ammonia-nitrogen [NH_3-N]). This method of expression converts all of the different nitrogen compounds into the same currency—nitrogen. Organic nitrogen, ammonia, nitrite, and nitrate all contain other elements in addition to nitrogen, including hydrogen, oxygen, and carbon. Nitrogen is the element of concern from an environmental perspective. Expressing all of the nitrogen compounds as N focuses the attention on the primary pollutant. It also allows the different nitrogen compounds to be added and subtracted to find total nitrogen, TKN, or TIN.

Chapter 9 includes a review of basic *chemistry*.

Makes a common denominator: N, so we can add + subtract Nitrogen-containing compounds, to find key data such as TKN or TIN.

Expressing nitrogen compounds as N requires some knowledge of *chemistry*. The periodic table contains the atomic weights of different elements. Adding atomic weights for a particular compound like nitrate together gives the formula weight. An example follows showing how to convert 20 mg/L of nitrate into milligrams per liter of nitrate-nitrogen.

Step 1—Look up the atomic weights for nitrogen and oxygen. Nitrogen weighs 15 g/mol. Oxygen weighs 16 g/mol.

Step 2—Add the atomic weights together to find the formula weight.

1 Nitrogen = (1)(14) = 14 g/mol

3 Oxygen = (3)(16) = 48 g/mol

Total formula weight = 62 g/mol

Step 3—Use dimensional analysis to convert from mg/L nitrate to mg/L nitrate-nitrogen.

$$\frac{20\ \text{mg NO}_3}{\text{L}}\left|\frac{1\ \text{g}}{1000\ \text{mg}}\right|\left|\frac{1\ \text{mol NO}_3}{62\ \text{g NO}_3}\right|\left|\frac{1\ \text{mol N}}{1\ \text{mol NO}_3}\right|\left|\frac{14\ \text{g N}}{1\ \text{mol N}}\right| 1000\ \text{mg}/1\ \text{g} = 4.5\ \text{mg/L NO}_3 - \text{N}$$

PHOSPHORUS

Phosphorus, like nitrogen, assumes different forms in wastewater and serves as an essential element for biological growth and reproduction. Typical ranges of phosphorus concentrations in raw domestic wastewater are given in Table 2.1. An oversupply of nitrogen or phosphorus leads to excessive algae blooms and eutrophication. As a result of these effects, many WRRFs have effluent limits for phosphorus.

Phosphorus may be either soluble or particulate. The most common soluble form of phosphorus is orthophosphate. The most common forms of particulate phosphorus are polyphosphate and organic phosphate. Phosphorus may be expressed as either phosphate (PO_4^{-3}) or as phosphate as phosphorus (PO_4-P). Testing laboratories tend to consistently report all nitrogen compounds as N, but are not as consistent when reporting phosphorus compounds. It is crucial that operators be able to convert between milligrams per liter of PO_4^{-3} and milligrams per liter of PO_4-P. An example follows showing how to convert 3 mg/L of PO_4^{-3} to mg/L of PO_4-P.

Step 1—Look up the atomic weights for phosphorus and oxygen. Phosphorus weighs 31 g/mol. Oxygen weighs 16 g/mol.

Step 2—Add the atomic weights together to find the formula weight.

1 Phosphorus = (1)(31) = 31 g/mol

4 Oxygen = (4)(16) = 64 g/mol

Total formula weight = 95 g/mol

Step 3—Use dimensional analysis to convert from PO_4^{-3} to PO_4-P.

$$\frac{3\text{ mg }PO_4^{-3}}{\text{L}}\left|\frac{1\text{ g}}{1000\text{ mg}}\right|\left|\frac{1\text{ mol }PO_4^{-3}}{95\text{ g }PO_4^{-3}}\right|\left|\frac{1\text{ mol P}}{1\text{ mol }PO_4^{-3}}\right|\left|\frac{31\text{ g P}}{1\text{ mol P}}\right| 1000\text{ mg/1 g} = 0.98\text{ mg/L }PO_4\text{-P}$$

CHEMICAL OXYGEN DEMAND

Biochemical oxygen demand was discussed in Chapter 1. Biochemical oxygen demand is a measurement of the amount of oxygen consumed by microorganisms as they consume and stabilize organic compounds and ammonia-nitrogen in the wastewater. Wastewater contains additional organic compounds that cannot be consumed by the microorganisms within the 5-day BOD test or within most wastewater treatment processes. These organic compounds are non-biodegradable, but they can be broken down chemically. All of the organic compounds in a sample that can either be consumed by microorganisms or burned away chemically make up the chemical oxygen demand (COD). The COD test does not include oxygen demand from ammonia-nitrogen. Chemists define *COD* as the amount of a specified chemical that reacts with the sample under controlled conditions.

The COD test is conducted by combining the sample with potassium dichromate ($K_2Cr_2O_7$) and concentrated sulfuric acid. The samples are then heated to 150 °C for 2 hours. As the potassium dichromate reacts with organic compounds in the sample, it is converted to chromium ion (Cr^{+3}), which can then be measured. The concentration of chromium ion produced during the test is *directly proportional* to the amount of oxygen needed to break down the organic compounds present at the start of the test. Results are expressed in terms of oxygen equivalence. In other words, the COD test provides a measure of how much oxygen a sample will consume (oxygen demand), and it does so in only a few hours.

Directly proportional means that two variables increase or decrease together. *Inversely proportional* means that when one variable increases, the other decreases.

The COD test can be used to quickly estimate the BOD_5 of a sample. The correlation between BOD and COD varies from facility to facility. However, COD results will be equal to or higher than BOD results because all the organics are oxidized. For domestic wastewater without significant industrial contributions, the COD-to-carbonaceous biochemical oxygen demand (CBOD) ratio is typically between 1.9 and 2.2. Use of the COD test for process control requires that the BOD test must first be run in parallel with the COD test to determine this correlation. This allows BOD-to-COD ratios to be developed for each facility. These ratios vary across the facility from influent to effluent. The ratio changes as the wastewater moves through the treatment process and nearly all of the CBOD is consumed.

PRIORITY POLLUTANTS

Priority pollutants is a general term originally applied to a list of chemical compounds identified by U.S. Environmental Protection Agency (U.S. EPA) as being of significant concern because of their wide use and toxicity. This list has since been expanded and modified to include a listing of toxic pollutants and hazardous substances. U.S. EPA frequently revises this list based on its expanding knowledge of toxic compounds. The specific listing of these compounds is contained in the *Code of Federal Regulations*, Title 40 Part 122, Appendix D, "Tables of Toxic Pollutant and Hazardous Substances". Copies of the document can be obtained from U.S. EPA and should be retained in each facility's reference library. Upon request, U.S. EPA typically will include a facility on its mailing list to receive any updated listings.

The compounds on the toxic pollutant list typically come from industrial sources. The toxic compounds can be divided into two categories: toxic organic compounds (including organic solvents and many pesticides) and other toxic compounds (including heavy metals, cyanide and phenols). Analyses of most of the listed compounds require sophisticated instrumentation. Further, their concentrations in domestic wastewater are typically low (measured in parts per billion). Special sampling equipment may be required to prevent sample contamination.

Many of the listed toxic compounds, if they reach sufficient concentrations, inhibit biological activity. Industrial discharges of heavy metals, such as chromium, can result in process upsets or failure and can adversely affect the biosolids quality and process performance. The discharge of toxic compounds can impair receiving water environments because of their immediate toxicity and because many of the compounds, such as mercury and polychlorinated biphenyls, are bioaccumulative. This means that biological organisms remove the toxic substances from the water and concentrate them within their bodies. The concentration effect typically progresses up the food chain until high and possibly toxic levels are reached at the end of the chain in more complex organisms, including humans.

TEST YOUR KNOWLEDGE

1. When water use increases in a community, the influent wastewater concentrations decrease assuming that the population served has not changed.
   - ☑ True
   - ☐ False

2. An influx of stormwater from a combined sewer can increase average influent wastewater temperatures.
   - ☐ True
   - ☑ False

3. Septic conditions in the collection system upstream of the influent flow to a WRRF can cause the influent to be darker than usual and smell like rotten eggs.
   - ☑ True
   - ☐ False

4. Conductivity meters can also be used to accurately measure total dissolved solids.
   - ☐ True
   - ☑ False

5. Domestic wastewater includes contributions from
   a. Significant industrial users
   b. Categorical industrial users
   c. Homes and businesses
   d. Heavy industrial users

6. Fresh domestic wastewater will typically be
   a. White to pale yellow
   b. Light brown to grey
   c. Gray to black
   d. Orange, yellow, or blue

7. Hydrogen sulfide is a concern for all of these reasons EXCEPT
   a. Poisonous at low concentrations
   b. Corrodes concrete and metal
   c. Potentially explosive
   d. Smells strongly of garlic

8. Turbidity is a measurement of:
   a. Light scatter
   b. Cloudiness
   c. Solids concentration
   d. Organic matter

9. Alkalinity is a measurement of
   a. pH
   b. Buffering capacity
   c. Calcium carbonate concentration
   d. Hydroxide content

10. Which of the following pH values would be considered acidic?
    a. 4.6
    b. 7.1
    c. 8.3
    d. 9.4

11. If all of the alkalinity is consumed, what will the pH be?
    a. 1.2
    b. 4.5
    c. 7.0
    d. 8.3

12. An influent sample is analyzed for both COD and BOD. Which of the following statements must be true?
    a. BOD is equal to or greater than COD
    b. The BOD test was completed before the COD test
    c. COD is equal to or greater than BOD
    d. The COD test was performed at 20 °C

13. The laboratory reported a phosphorus concentration in the final effluent of 2.5 mg/L as PO_4^{-3}. What is this in milligrams per liter of PO_4-P?
    a. 0.25 mg/L PO_4-P
    b. 0.81 mg/L PO_4-P
    c. 2.5 mg/L PO_4-P
    d. 7.7 mg/L PO_4-P

P. 49
Go Back over; Don't understand.

PER CAPITA GENERATION RATES

Domestic wastewater has a similar chemical composition from one municipality to another. Large industrial users within a service area can skew the chemical composition; however, even the effects of significant industrial users are often averaged out over a large service area because their contributions are small relative to the total flow and load received by the WRRF. For smaller service areas, the effects of a single industrial or commercial discharger can be significant. Wastewater modeling and facility design both depend on being able to make reasonable assumptions about the composition of the influent wastewater when actual data are not available. The same assumptions can be invaluable to operators when determining if the flow and load measured at the facility influent are reasonable for the service area. Typical per capita (per person) generation rates in grams per person per day (pounds per person per day) are presented in Table 2.4. These generation rates are for typical municipal wastewater and should not be applied to industrial wastewaters or to municipal wastewaters that contain large percentages of industrial wastewater.

As an example, assume that a small wastewater facility serves a town with 5000 residents. The average daily flow for the hypothetical facility is 1.61 ML/d (0.425 mgd). In July, the influent composite sample results were 450 mg/L for BOD_5, 320 mg/L for TSS, and 45 mg/L for TKN. To find the per capita generation rate, divide the influent flow or mass by the number of persons in the service area. Refer to Table 2.4 to verify that the data provided are reasonable given what we know about the service area.

Step 1—Calculate the mass in kilograms or pounds

International System of Units for calculating mass:

$$\text{kg} = \frac{\left[\left(\text{Concentration, } \frac{\text{mg}}{\text{L}}\right)(\text{Volume, m}^3)\right]}{1000}$$

Alternatively, use dimensional analysis:

$$\frac{X\ \text{mg}}{\text{L}}\left[\frac{1\ \text{g}}{1000\ \text{mg}}\right]\left[\frac{1\ \text{kg}}{1000\ \text{g}}\right]\left[\frac{1000\ \text{L}}{1\ \text{m}^3}\right]\left[\frac{X\ \text{m}^3}{}\right] = \text{kg}$$

U.S. customary units for calculating mass:

$$\text{Pounds} = \left(\text{Concentration, } \frac{\text{mg}}{\text{L}}\right)(\text{Volume, mil. gal})\left(8.34\ \frac{\text{lb}}{\text{mil. gal}}\right)$$

Alternatively, use dimensional analysis:

$$\frac{X\ \text{mg}}{\text{L}}\left[\frac{1\ \text{g}}{1000\ \text{mg}}\right]\left[\frac{1\ \text{kg}}{1000\ \text{g}}\right]\left[\frac{2.204\ \text{lb}}{1\ \text{kg}}\right]\left[\frac{3.785\ \text{L}}{1\ \text{gal}}\right]\left[\frac{1\,000\,000\ \text{gal}}{1\ \text{mil. gal}}\right]\left[\frac{X\ \text{mil. gal}}{}\right] = \text{lb}$$

Step 2—Divide the flow or mass by population served to find the per capita generation rate.

Table 2.4 Per Capita Flows and Loads (Metcalf and Eddy/AECOM, 2014. Copyright © 2014 by McGraw-Hill Education, reprinted with permission.)

Parameter	Per Capita Generation Rate
Average daily flow	204 and 455 L/d (54 and 120 gpd/cap)
Chemical oxygen demand (COD)	110 – 295 g/d (0.30 – 0.65 lb/d)
Biochemical oxygen demand (BOD)	50 and 120 g/d (0.11 and 0.26 lb/d)
Total suspended solids (TSS)	60 and 150 g/d (0.13 and 0.33 lb/d)
Total Kjeldahl nitrogen as N (TKN)	9 and 18 g/d (0.020 and 0.040 lb/d)
Total phosphorus as P (TP)	1.5 and 4.5 g/d (0.003 and 0.010 lb/d)

~~Mass (lbs)~~ Loading Rate (lbs/day) = (Flow, MGD)(concent.)(8.34)

BOD = (.425 MGD)(450 mg/L)(8.34 lbs/gal) = 1,595 lbs/day

1595 lbs/day / 5000 cap = .319 lbs/day/cap (BOD)

TSS = (.425 MGD)(320 mg/L)(8.34 lbs) = 1,134.24 lbs/day

1,134.24 lbs/day / 5,000 cap = .23 lbs/day/cap

TKN = (.425 MGD)(45 mg/L)(8.34 lbs) = 159.5 lbs/day

159.5 lbs/day / 5000 cap = .032 lbs/day/cap

Overall:

- BOD: BAD - 122%
- TSS: GOOD - 50%
- TKN: GOOD - 60%

Table 2.5 Per Capita Flows and Loads Example

Parameter	Influent Total	Per Capita Day Rate	Expected Range
Flow	1.61 ML/d (0.425 mgd)	321.7 L/cap·d (85 gpd/cap)	204 and 455 L/d (54 and 120 gpd/cap)
BOD	450 mg/L or 723 kg/d (1595 lb/d)	0.15 kg/cap·d (0.32 lb/cap·d)	0.05 and 0.12 kg/d (0.11 and 0.26 lb/d)
TSS	320 mg/L or 514 kg/d (1134 lb/d)	0.10 kg/cap·d (0.23 lb/cap·d)	0.06 and 0.15 kg/d (0.13 and 0.33 lb/d)
TKN	45 mg/L or 72.6 kg/d (160 lb/d)	0.014 kg/cap·d (0.031 lb/cap·d)	0.09 and 0.022 kg/d (0.020 and 0.048 lb/d)

Note: To convert kilograms to grams, ~~divide~~ by a factor of 1000.

In the instance outlined in Table 2.5, all of the per person generation rates are within expected ranges except for BOD. The operator may want to investigate why the BOD load is higher than expected for domestic wastewater. It may be that the BOD sample was not taken correctly and is not representative or it may be that the BOD analysis was not performed correctly and the result was incorrectly reported as higher. Another possibility is that a discharger in the service area is contributing an excess of soluble BOD. Recall from Chapter 1 that BOD can be particulate or soluble. Because the TSS is within the expected range, but the BOD concentration is higher than expected, the additional BOD must be soluble BOD.

Per capita generation rates can be useful when estimating the effect of a new subdivision. Assume that the hypothetical town in the previous example is planning two new developments. The first subdivision will have 50 new homes and 130 residents. The second subdivision will have 520 new homes and 1250 residents. What is the expected increase in flow and load at the WRRF? If it is assumed that each new resident will generate 321.7 L (85 gal) of flow and 0.09 kg (0.20 lb) of BOD, then the new subdivisions will add 444 028 L (117 300 gal) of influent flow and 125 kg (276 lb) of BOD. Although these are rough estimates, they will help the operator know what to expect in the WRRF influent in the future.

Consider one more example. An operator at a particular facility is used to seeing influent BOD and TSS concentrations in the 280 to 320 mg/L range. This month, the influent concentrations are much lower, specifically, 178 mg/L for BOD and 213 mg/L for TSS. Is the laboratory analysis skewed or did something else happen to affect the BOD and TSS concentrations? This hypothetical town has 8300 residents. Flows are typically close to 2.95 ML/d (0.78 mgd), but this month the average daily flow for the day that the influent samples were collected was 4.23 ML/d (1.12 mgd). Refer to Table 2.4 to verify that the data provided are reasonable given what we know about the service area.

In this example, the per capita generation rates for BOD and TSS are within the expected range, but the per capita generation rate for flow is higher than expected (Table 2.6). The BOD and TSS results are reasonable. Higher influent flows have diluted the BOD and TSS load and have resulted in lower measured concentrations. The operator may want to review rainfall records to see if there was a significant amount of rain on the day or days before the samples were collected. A sudden increase in influent flows may indicate an inflow and infiltration problem in the collection system.

Table 2.6 Infiltration and Inflow Example

Parameter	Influent Total	Per Capita Day Rate	Expected Range
Flow	4.23 ML/d (1.12 mgd)	511 L/cap·d 135 gpd/cap	284 and 492 L/cap·d (70–130 gpd/cap)
BOD	178 mg/L or 753 kg/d (1660 lb/d)	0.09 kg/cap·d (0.20 lb/cap·d)	0.08 and 0.10 kg/cap·d (0.18–0.22 lb/cap·d)
TSS	213 mg/L or 904 kg/d (1992 lb/d)	0.11 kg/cap·d (0.24 lb/cap·d)	0.09–0.11 kg/cap·d (0.20–0.25 lb/cap·d)

pounds = (mg/L)(Flow, mgd)(8.34 lb/mil. gal).
kg/cap·d = kilograms per capita per day; L/cap·d = liters per capita per day.
lb/cap·d = pounds per capita per day; gpd/cap = gallons per capita per day.

CHAPTER SUMMARY

WASTEWATER SOURCES	1. Domestic is from residential, light commercial (nonindustrial), and institutional sources. It does not include grease trap waste or manufacturing wastes. 2. Industrial and commercial includes grease trap waste, manufacturing wastes, and other wastewater from industrial activity. 3. Industrial and commercial wastewater is more variable than domestic wastewater in composition and flow. 4. Infiltration and inflow is stormwater and groundwater that enters the collection system through cracks and other openings.
FLOW VARIATIONS	1. Domestic wastewater has a diurnal flow pattern. 2. The flow pattern reflects activities in the service area. 3. Influent flow is typically higher during the day than at night. 4. Smaller service areas generally experience greater variation in influent flows over a 24-hour period than larger service areas. 5. Concentrations of BOD, TSS, TKN, and other parameters are influenced by the quantity of water used in the service area. More water use dilutes the wastewater and, therefore, concentrations decrease.
INFLUENT CHARACTERISTICS	1. Biochemical oxygen demand, CBOD, COD, and TOC indirectly measure the amount of organic material in the wastewater by measuring the amount of oxygen needed to break them down. 2. Solids are classified by their size—soluble or particulate—and by whether they are organic (volatile) or inorganic (nonvolatile). 3. Temperature influences settleability, chemical reaction rates, and microbial activity. Treatment is faster at warmer temperatures. 4. Dissolved oxygen concentrations are influenced by water temperature, salinity, and atmospheric pressure. Dissolved oxygen is critical for some types of biological treatment. 5. Color is one indicator of wastewater freshness. Unusual colors may indicate industrial discharges. 6. In the absence of oxygen, some bacteria will convert sulfate to hydrogen sulfide and other sulfur compounds. Hydrogen sulfide smells like rotten eggs. 7. Turbidity is a measurement of light scatter by particles in the water. Although a relationship can be built between TSS and turbidity for a particular wastewater, TSS and turbidity cannot be used to predict one another without first building a correlation. 8. Conductivity measures the ability of a water-based solution to carry an electrical current. It is indirectly related to TDS. 9. pH measures the hydrogen ion concentration. The pH scale runs from 0 to 14, with a pH of 7 defined as neutral. pH less than 7 is acidic and pH greater than 7 is basic. pH is expressed on a log scale so pH 4 water has 10 times as many hydrogen ions than pH 5 water. 10. Alkalinity measures buffering capacity or how much acid a water can absorb without a change in pH. Alkalinity and pH are related, but one cannot predict the other. 11. Nitrogen is present in wastewater in four forms: organic, ammonia, nitrite, and nitrate. Microorganisms in the process convert nitrogen from one form to another. Nitrogen compounds are expressed a N. 12. Phosphorus and nitrogen are vital nutrients for supporting life. When excess quantities of either or both are present in natural systems, algae blooms and other environmental changes may result. Many WRRFs have final effluent limits for nitrogen and phosphorus.
PER CAPITA GENERATION RATES	1. Individual residents tend to produce about the same amount of flow, BOD, CBOD, TSS, TKN, and other constituents each day. 2. When the influent flow, concentration, and population are known, the per capita generation rates may be calculated in L/cap·d (gpcd) or kg/cap·d (lb/cap·d). 3. Per capita generation rates are useful for determining the presence of industrial discharges and the effect of future population growth on the WRRF influent.

References

HM Digital (2008) *TDS Meters, Conductivity, and Conversion Factors;* http://www.tdsmeter.com/what-is?id=0019 (accessed April 2017).

Metcalf and Eddy/AECOM (2014) *Wastewater Engineering Treatment and Resource Recovery*, 5th ed.; Tchobanoglous, G., Burton, F., Stensel, D.H., Abu-Orf, M., Bowden, G., Pfrang, W., Eds.; McGraw-Hill: New York.

Water Environment Federation (2008) *Industrial Wastewater Management, Treatment, and Disposal,* 3rd ed.; McGraw-Hill: New York; p 135.

Suggested Readings

Code of Federal Regulations, Title 40, Parts 136 and 503. Most Current Requirements for Sampling Containers, Holding Times, and Preservation Methods.

Water Environment Federation (2012) *Basic Laboratory Procedures for the Operator-Analyst*, 5th ed.; Water Environment Federation: Alexandria, Virginia.

CHAPTER 3

Preliminary Treatment of Wastewater

The purpose of preliminary treatment is to remove, reduce, or modify wastewater constituents in the raw influent that can cause operational problems with downstream processes or increase maintenance of downstream equipment. These constituents primarily consist of large solids and rags (screenings); abrasive inert material (grit); floating debris; and fats, oils, and grease (FOG). This chapter presents descriptions of, and operational considerations for, preliminary treatment processes. This chapter includes a separate section addressing the handling of hauled-in septic tank waste (septage).

LEARNING OBJECTIVES

Upon completing this chapter, you will be able to

- Describe the purpose of trash racks, bar screens, comminutors and grinders, and grit basins;
- Compare and contrast different screen types;
- Select a screen category given downstream equipment requirements;
- Place a screen or grit basin into service or take one out of service;
- Compare and contrast different types of grit basins;
- Calculate flow velocity for an open channel or full pipe;
- Determine the optimum number of grit basins to place into service to maintain a desired flow velocity;
- Troubleshoot common screening and grit removal process control and mechanical problems; and
- Describe potential effects of septage receiving on facility operations.

Introduction

Preliminary treatment typically uses bar racks, screens, comminutors, and/or grit chambers. Preliminary treatment takes place at the water resource recovery facility (WRRF) headworks. Headworks may also include flow measurement, flow equalization, pumping, and/or odor control and *septage* receiving. Flow measurement was discussed in Chapter 2.

Septage is domestic wastewater from cesspools, individual home septic systems, pit toilets, portable toilets, and campgrounds that is delivered to the WRRF by truck. Septage does not include grease trap waste or any other type of commercial or industrial waste.

An example WRRF headworks is shown in Figure 3.1. A particular WRRF headworks may have all, some, or none of the unit processes shown in Figure 3.1. The individual unit processes may be arranged in a different order, with, for example, flow measurement coming before screening and grit removal. In the example headworks shown in Figure 3.1, the first unit process is a screen. Screens remove larger debris and may have openings as small as a few millimeters or as wide apart as 5 cm (2 in.). Figure 3.2 shows a manual

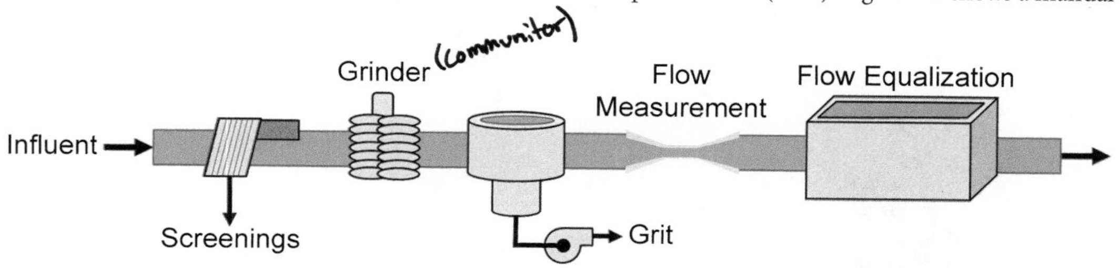

Figure 3.1 Typical WRRF Headworks (Reprinted with permission by Indigo Water Group)

Figure 3.2 Manual Bar Screen (Reprinted with permission by Keith Radick)

bar screen with openings that are 2.5-cm (1-in.) apart. Screens must be cleaned or raked either automatically or by an operator or maintenance person several times during the day. A bar screen with wide openings will stop large debris from entering the WRRF, but rags and stringy material may still pass through. A manual screen with large openings is sometimes called a *trash rack*. Many different types of screens are used in wastewater treatment and will be discussed in this chapter. The headworks may also include a comminutor or grinder. A comminutor, or grinder, reduces the size of any debris that passed through the screen. Comminutors and grinders are being phased out of many facilities in favor of screening. The goal is to reduce the chance of damage to downstream equipment or processes and clogging of pipes.

Grinders and comminutors reduce the size of debris entering the treatment facility, but they do not remove it from the wastewater. Shredded debris can still damage downstream equipment, clog pipes and pumps, and take up valuable treatment space in tanks.

Screens are typically followed by grit basins, which are channels or small tanks where the wastewater velocity, or speed, is decreased to about 0.3 m/sec (1 ft/sec). Here, heavier particles such as sand, grit, eggshells, and heavier organic particles settle to the bottom of the tank where they can then be removed, while lighter organic particles remain suspended in the wastewater and pass on to the next process. Figure 3.3 shows two parallel grit basins at the headworks of a WRRF that treats up to 6057 m^3/d (1.6 mgd) of flow. A *slide gate* is directing flow to one grit basin. Accumulated grit can be seen in the bottom of the dry grit basin. Several different types of grit basins are used in wastewater treatment and will be discussed in this chapter. Screenings and grit are typically sent to a landfill for disposal.

A *slide gate* is placed within a channel to direct flow. At smaller WRRFs, slide gates may be as simple as metal or fiberglass plates that are slid manually in and out of grooves mounted on the channel wall. At larger WRRFs, slide gates may be raised and lowered using hand wheels to move the gate up and down a threaded shaft. Proportional gates incorporate a weir that allows the operator to control the velocity of the water moving through the upstream process.

Figure 3.3 Parallel Grit Basins (Reprinted with permission by Indigo Water Group)

TEST YOUR KNOWLEDGE

1. All WRRF headworks include screening, grit removal, and flow measurement.
   - ☐ True
   - ☑ False

2. Screening and grit removal must be completed before flow measurement.
   - ☐ True
   - ☑ False

3. This device may be used to direct flow to or around a screen or grit basin.
   - a. Cogwheel
   - b. Slide gate
   - c. Weir block
   - d. Fraser valve

4. A bar screen may not prevent this type of material from entering the WRRF.
   - a. Sticks
   - b. Large rocks
   - c. Rags
   - d. Pallets

5. Grit basins typically remove sand, gravel, eggshells, and coffee grounds by
   - a. Placing wire mesh in the flow path as a strainer
   - b. Scooping the surface of the water
   - c. Introducing microorganisms to consume them
   - d. Decreasing water velocity and allowing them to settle

Screening

PURPOSE AND FUNCTION

As a WRRF's first treatment unit, screens protect facility equipment against damage such as clogging of pipes and pumps. The debris captured on the screen depends on the opening spacing and the size, configuration, and amount of debris. The type of screen selected will depend on which treatment processes are downstream and how much protection they require. Membranes, which are one type of downstream process, are easily damaged by debris and, therefore, require screens with very small openings upstream. Ponds, on the other hand, are much more tolerant of debris and grit and tend to have screens with much larger openings at their headworks. If the screen openings are too large, some debris will not be captured and may cause downstream difficulties. The passage of rags and debris into downstream processes is one of the biggest causes of equipment maintenance and failure because of jammed pump impellers, clogged sludge and scum pipelines, and imbalanced operation of rotating equipment. Floating material in downstream processes is unsightly and can be a safety hazard to operators attempting removal. As wastewater processes continue to advance, preventing damage from screeneable material becomes increasingly important. For these reasons, there is a trend toward installing screens with smaller openings. As screen openings become smaller, greater amounts of organic material are removed and it becomes important to wash and compact the screened material and return organic material to the wastewater flow. Screened material that is not washed sufficiently can putrefy and generate odors and hazardous gases. Removal of too much organic material with the screenings can also impede the ability of some biological treatment systems to remove nitrate.

Some bacteria in the treatment process are capable of converting nitrate, a pollutant, into nitrogen gas under certain conditions. This is called *denitrification*. The nitrogen gas is returned to the atmosphere, which is approximately 78% nitrogen. These bacteria require an organic food source to make the conversion. When too much organic material is removed with screenings and grit, denitrification can be impeded. Biological treatment fundamentals are discussed in Chapter 5.

Debris accumulated on the screens is removed at appropriate intervals by the use of either manual or automatic control methods. The cleaning rate depends on many variables including the type of collection system (separate or combined), demographics, diurnal flow patterns, condition of the collection system, and seasonal factors such as leaves during the fall. Debris removed from screens with wide openings typically consists of wood, tree limbs, rocks, and other large items. Debris removed from screens with smaller openings typically includes rags, food particles, bones, bottle caps, small plastic objects, leaves, paper, plastics, personal wipes, personal hygiene products, and other small objects. Screenings may be deposited directly into containers, washed and compressed before placement in containers, or loaded onto conveyors for transport elsewhere in the facility. Ultimately, the screenings are transported to a landfill or incinerator for final disposal. Increasingly, screening wash presses and vacuum containers are being used at WRRFs to remove the liquid from the screenings before transport, which reduces the volume and weight of the screenings.

-Too small screen openings can have an impedence on denitrification, because the organic material Nitrifying bacteria require, to turn Nitrate into Nitrogen gas, has been screened out.

DESIGN PARAMETERS AND EXPECTED PERFORMANCE

The quantity of screenings removed can vary significantly depending on the screen opening size and the activities in the service area. In 2008, Water Environment Federation (WEF) surveyed 328 WRRFs across the United States. Extreme variations in screening quantities were reported from less than 0.75 to 150 L/1000 m^3 (0.1 to 20 cu ft/mil. gal) (WEF et al., 2018). For screen openings between 25 and 50 mm (1 and 2 in.), for each 13-mm (0.5-in.) reduction of clear opening size, the volume of screenings will approximately double. In other words, a screen with 38-mm (1.5-in.) openings will remove about twice the amount of material as a screen with 50-mm (2-in.) openings. For screen openings smaller than 25 mm (1 in.), the volume of screenings removed increases rapidly as the opening size decreases (WEF et al., 2018).

There are four types of screening media that are typically used: bars, wedge wire, perforated plate, and mesh. Historically, bars have been the most commonly used because they are preferred for coarse screens and trash racks (Figure 3.2). Long, vertical or horizontal gaps between bars can allow the passage of long, thin objects like rags, sticks, and hair. Screens with large openings may pass larger objects, including construction debris and wood beams. If material is not removed from the screen regularly, the buildup of material and water behind the screen can force larger pieces of debris between the openings. Wedge wire is similar to bars, but is thinner and has smaller openings. The individual wires have a trapezoidal shape, with the smaller openings facing into flow, as shown in Figure 3.4. The trapezoidal shape helps prevent clogging because any object that is able to pass through the smaller initial opening should be able to move easily through the wider opening behind. Wedge wire allows the passage of long, thin material such as hair. Perforated plate media have round or oblong holes as small as 1 mm in diameter (Figure 3.5). Perforated plate improves capture of hair and other fine material that can pass through bar screens and wedge wire. Perforated plate is more likely to become clogged or "*blind off*", and will require more frequent cleaning. The last type of media is mesh. Mesh media are similar in appearance to a window screen, but are made from thicker, tougher wire.

"*Blinding off*" means that an excess of material has accumulated on the screen and that water is no longer able to move easily through the openings.

Screens are grouped into several categories based on the size of their openings. Coarse screens or trash racks have the largest openings, followed by bar screens, fine screens, and microscreens. Different publications use slightly different ranges for opening sizes when classifying screens.

The typical overall widths for a screen channel vary from 0.6 to 4.2 m (2 to 14 ft). The straining efficiency (screenings removal) of a screen and the hydraulic flow through the screen depend on the clear spacing between the bars. The smaller the spacing, the more debris is captured, with greater head loss developing across the screen. Head loss is the difference in water level between the upstream side of the screen and the downstream side of the screen, also known as *differential* (Figure 3.6). As material accumulates on the screen, it becomes more difficult for water to pass through the screen. As a result, water builds up behind the screen and can overflow the channel if the screen is not cleaned often enough. Screens with smaller openings require wider channels for a given flow to enable flow to pass through the screens. Screens are available in different widths, but each type of screen is limited by its mechanical components and has a maximum size available. To accommodate higher flowrates, most WRRFs will have multiple screens.

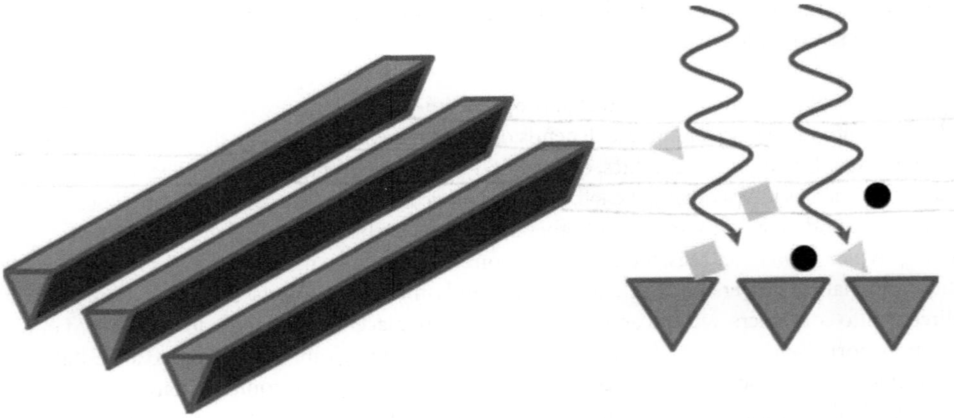

Figure 3.4 Wedge Wire Screen Material (Reprinted with permission by Indigo Water Group)

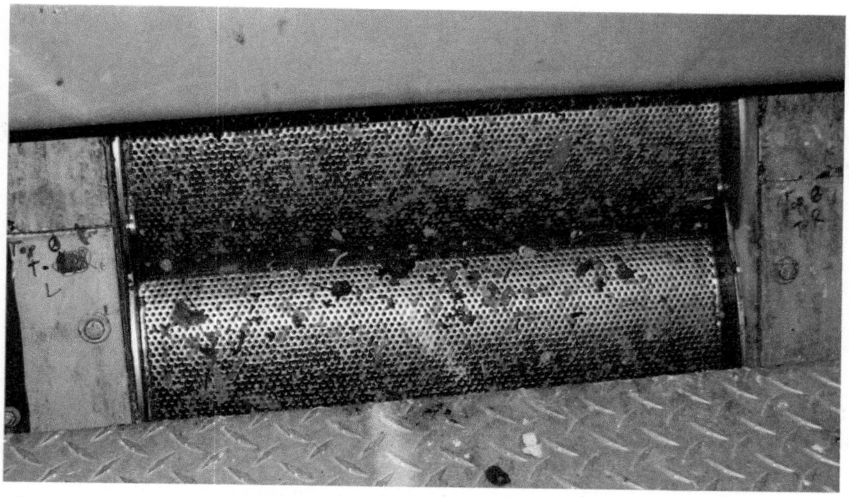

Figure 3.5 Perforated Plate Media (Reprinted with permission by Indigo Water Group)

Engineers typically assume screens will be approximately 50% blinded at any given time. This helps ensure the WRRF will have plenty of screening capacity under different operating conditions.

A WRRF headworks may have one or more screening channels and grit basins depending on the size and complexity of the facility. Thus, if one unit malfunctions, the other unit and channel may remain in service during repairs. A typical layout for a smaller WRRF is shown in Figure 3.7. The path of the influent during normal operation is shown in darker blue. The bypass channel is shown in lighter blue. Note that there is a bypass channel for both the screen and the grit basin. Slide gates are located upstream and downstream of each piece of equipment. The gates may be used to divert flow and allow a particular piece of equipment to be taken out of service for inspection, cleaning, and maintenance while leaving other pieces of equipment in service. In the example shown in Figure 3.7, it is possible to take either the mechanical screen or the grit basin out of service while leaving the other piece of equipment in service.

In this example, the flow path used for normal operation contains a mechanically cleaned bar screen followed by a vortex grit basin. If the mechanical bar screen must be taken offline for any reason (e.g., to remove stuck debris), the slide gates on either side of the mechanical bar screen can be closed and the slide gates on either side of the manual bar screen can be opened. Opening and closing gates will direct flow to the bypass channel. Slide gates fit into sleeves that are embedded in the channel. The top of the gate generally extends above the highest expected water level. If the gates are dropped to block off the bypass channel and the flow backs up in the channel because of a clogged mechanical bar screen, or because the operator is unavailable, the flow can exit over the top of the gate into the bypass channel. Larger facilities

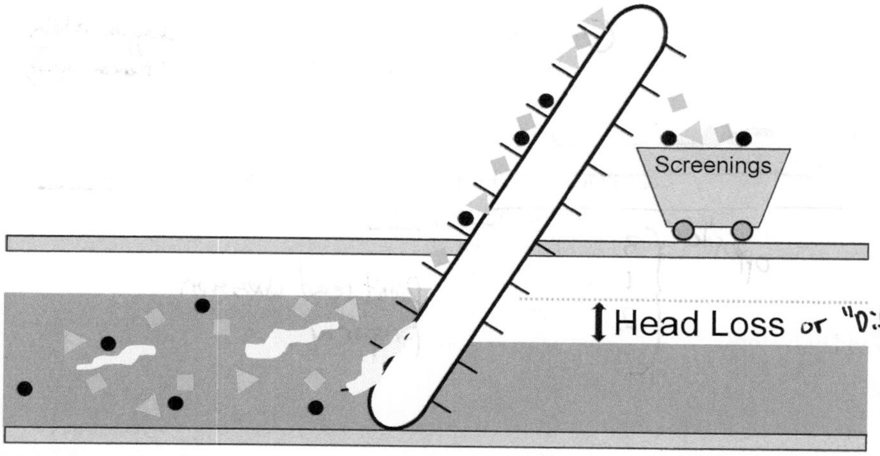

Figure 3.6 Head Loss across a Screen (Reprinted with permission by Indigo Water Group)

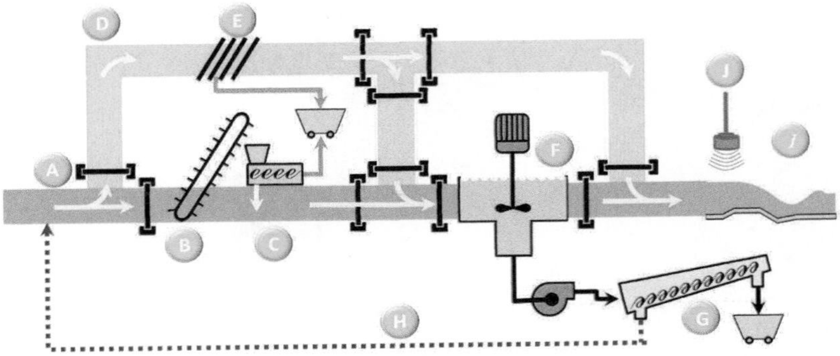

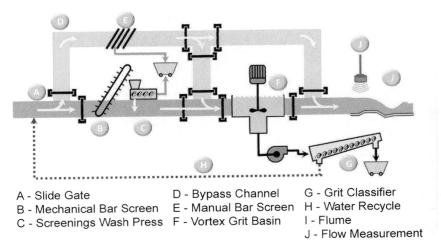

A - Slide Gate
B - Mechanical Bar Screen
C - Screenings Wash Press
D - Bypass Channel
E - Manual Bar Screen
F - Vortex Grit Basin
G - Grit Classifier
H - Water Recycle
I - Flume
J - Flow Measurement

Figure 3.7 Headworks Layout (Reprinted with permission by Indigo Water Group)

with multiple mechanical bar screens and grit basins may not have bypass channels because redundant equipment is available.

When multiple screens are available, influent flows should be equally distributed among all of the channels used for normal operation to ensure uniform loading among the screens. Uneven loading will require more frequent cleaning of some screens than others.

TEST YOUR KNOWLEDGE

1. Removal of grit, rags, and other debris is often necessary to protect downstream equipment from harm.
   - ☑ True
   - ☐ False

2. If too much organic material is removed at the WRRF headworks, the downstream biological process can be affected.
   - ☑ True
   - ☐ False

   De-Nitrifying bacteria ($NO_3 \rightarrow N_2$)

3. Head loss is the difference in water level between the upstream and downstream sides of a screen.
   - ☑ True
   - ☐ False

4. One consequence of allowing excess organic material to be removed along with rags and other inert debris is
   - a. Oxygen generation
   - b. Reduced capture of rags
   - c. Generation of odors
   - d. Reduced disposal costs

5. This type of screen has the widest openings and only stops the largest debris.
   - a. Trash rack
   - b. Manual bar screen
   - c. Mechanical bar screen
   - d. Step screen

6. A WRRF currently has a bar screen with 50-mm (2-in.) openings. Operators of the WRRF are considering replacing the screen with one that has 25-mm (1 in.) openings. How much should they expect the volume of screenings removed to change?
   - a. Volume will remain about the same
   - b. Screening volume will double
   - c. Screening volume will increase by a factor of 4
   - d. Screening volume will decrease by 50%

   Between 1"-2" every -.5" = 2x Volume

7. What is the most likely effect of receiving large quantities of grease on a perforated plate-type screen?
   - a. Blinding of the screen
   - b. Decreased cleaning frequency
   - c. Passage of hair and rags into the WRRF
   - d. Improved cog lubrication

   Clogging, reducing screening capacity

8. Place the following screen types in order from the smallest opening size to the largest.
   - 3 a. Fine screen
   - 1 b. Trash rack
   - 2 c. Bar screen
   - 4 d. Micro screen

   opposite — Didn't read Question properly

EQUIPMENT

COARSE SCREENS

Trash/Bar Racks

Coarse screens are classified as either trash racks or bar screens depending on the spacing between the bars. Trash racks, also called *bar racks*, are constructed of heavy, parallel rectangular or round steel bars with wide spacing set in a channel. They are intended to remove only the largest debris. Bars are sloped at an angle ranging from 30 to 45 deg from the vertical (Figure 3.2 and Figure 3.8). Bar racks may be cleaned manually or mechanically with robust steel rakes. The rake teeth fit between the bars and scrape accumulated debris to the top of the screen. The platform at the top of the screen in Figure 3.8 is perforated to allow water to drain from the screened material. Trash racks are mainly used in WRRFs that see high amounts of large debris, but may also be installed in a bypass channel (Figure 3.7). Bypass screens are used for emergency screening events such as when a mechanically cleaned screen must be taken out of service.

Bar Screens

Bar screens resemble bar racks except that they have smaller clear spacing between the bars (Figure 3.9). Bar racks and screens are designed as stationary structures built into concrete channels. Some mechanical units are designed with chains or cables and cogwheels attached to the submerged end of the screen and a pivot arrangement at the upper end to permit lifting of the submerged end for easy inspection and maintenance. Mechanical bar screens are typically positioned at a 65- to 70-deg angle for ease of operation of the raking mechanism (Figure 3.10). Mechanical cleaning reduces labor cost, improves flow conditions and screening capture, and reduces nuisances. For WRRFs with combined sewers, mechanically cleaned screens are better for handling large quantities of stormwater debris than manually cleaned screens. Many mechanically cleaned bar screen designs are currently used in WRRFs. These designs include chain/cable-driven, single rake, multiple rake, catenary, and continuous self-cleaning.

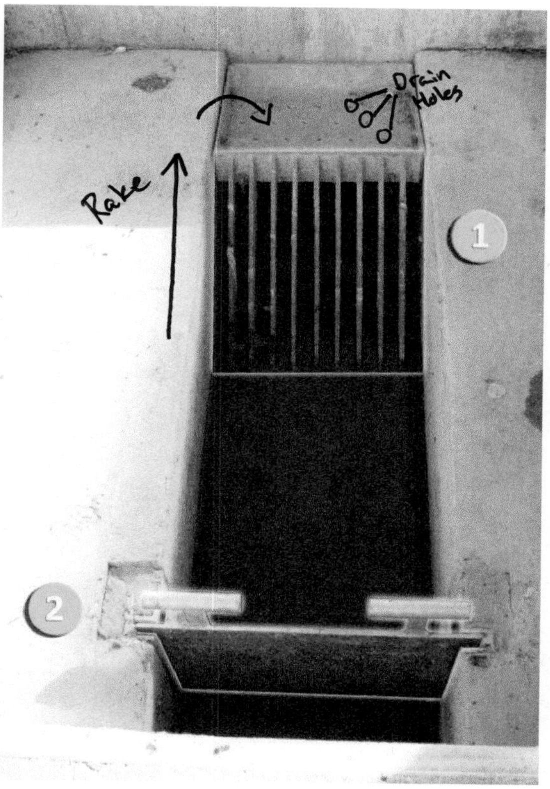

Figure 3.8 Manual Bar Screen with Slide Gate: (1) Manual Bar Screen and (2) Slide Gate (Reprinted with permission by Indigo Water Group)

Figure 3.9 Mechanical Bar Screen Openings (Reprinted with permission by Indigo Water Group)

Chain-Driven Screens

A *sprocket* is a wheel with teeth that is used to drive a chain or belt.

A common type of mechanically cleaned bar screen is the front-cleaning, chain-driven unit with the rake mechanism on the upstream side of the bar screen (Figure 3.11). These screens have two points of tension at the upper and lower *sprockets* that keep the chain taut. The typical front-cleaning unit includes a toothed hopper (rake) that swings out from the bar screen as it travels down to the bottom of the channel. After reaching the bottom of the channel, cables, chains, or cogwheels draw the toothed hopper toward the bar screen. The teeth of the hopper (rake) then slide between the bars and, during the upward movement, drag the screenings out of the wastewater. Think of dragging a fork across a barbecue grill. The tines of the fork

Figure 3.10 Three Mechanical Bar Screens Installed in Parallel (Reprinted with permission by Indigo Water Group)

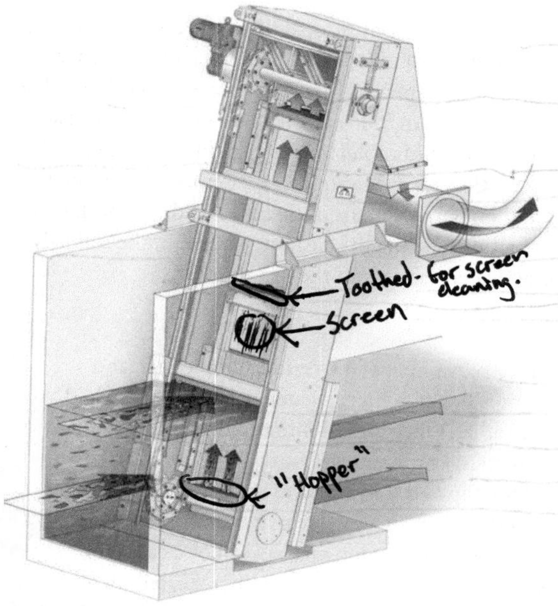

Figure 3.11 Front-Cleaning Screen (Courtesy of Huber Technology, Inc.)

slide easily into the openings between the grill bars and will remove captured debris from the surface of the bars. A variation of this type of screen is the rear-cleaning unit with the rake mechanism on the downstream side of the bar screen, the rake teeth of which protrude through the bars to the upstream side.

Single Rake Screens

A single rake or reciprocating rake bar screen simulates the movement of a person raking the screen (InfoBarScreens, 2009). A *reciprocating* rake bar screen consists of stationary components (frame, bar rack, dead plate, and pin rack) along which the rake assembly travels (Figure 3.12). The reciprocating rake bar

Reciprocating refers to something that moves backward and forward in a straight line.

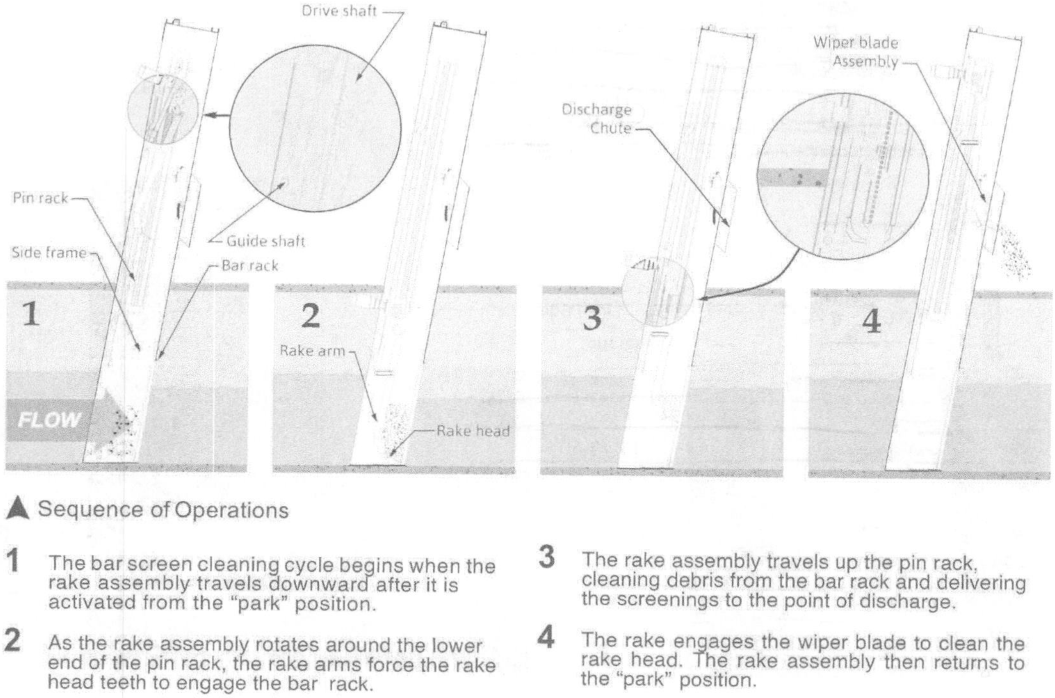

Drive shaft
Pin rack
Side frame
Guide shaft
Bar rack
1
FLOW
2
Rake arm
Rake head
Discharge Chute
3
Wiper blade Assembly
4
Sequence of Operations
1 The bar screen cleaning cycle begins when the rake assembly travels downward after it is activated from the "park" position.
2 As the rake assembly rotates around the lower end of the pin rack, the rake arms force the rake head teeth to engage the bar rack.
3 The rake assembly travels up the pin rack, cleaning debris from the bar rack and delivering the screenings to the point of discharge.
4 The rake engages the wiper blade to clean the rake head. The rake assembly then returns to the "park" position.

Figure 3.12 Reciprocating Rake Screen (Courtesy of Vulcan Industries, Inc.)

The *drive mechanism* is the pieces of the screen that physically move the rake across the screen.

A *cogwheel* is a wheel with teeth along its edge that are designed to interlock with another toothed mechanism. It is essentially a large gear.

screen typically goes through a single operation cycle (from top rest position, to downward stroke with rake in open position, to upward cleaning stroke with rake engaged in bar screen, to screenings discharge operation, and returning to top rest position) when initiated by timer or differential water-level signal. The rake assembly consists of a drive assembly, rake arms, rake, and cam followers, the latter controlling rake position. Although many *drive mechanisms* are available (chain and cable, hydraulic, and screw operated), the most popular design is the *cogwheel*. For these designs, the entire cleaning rake assembly, including the gear motor, is carriage-mounted on cogwheels that travel on a fixed pin or gear rack. The drive mechanism is typically designed to allow the rake to ride over obstructions during the cleaning stroke. In the unlikely event that the rake becomes jammed, a limit switch is activated to turn off the drive motor.

Catenary Screens

Catenary bar screens are similar to chain-driven screens. An example of a hybrid multiple rake and catenary screen is shown in Figure 3.13. Instead of having a flexible chain, a catenary screen uses rigid metal links or bars with joints that interlock with one another. Once locked into place, the links behave as a single, unbending unit along the full length of the screen. At the top and bottom of the screen, the links pivot at the joints, allowing them to bend around the top and bottom of the screen, as shown in Figure 3.13. One difference between a catenary screen and a chain-driven screen is that the catenary screen only has one point of tension: the sprocket at the top. The cleaning mechanism of a catenary screen consists of heavy tooth rakes held against the screen by the weight of its chain, rake, and rake counterweights. The term, *catenary*, comes from the catenary loop formed by the operating chain ahead of the screen. A curved transition piece at the base of the screen allows for efficient removal of solids captured at the bottom. Like reciprocating rake screens, all sprockets, shafts, and bearings are located out of the flow stream to reduce wear and corrosion and ease required maintenance. The submerged articulating chain joint surfaces, however, are subject to abrasive wear and fatigue failure. Because the cleaning rake is held against the bars primarily by just the weight of the chains, the rake can be pulled over large rags or solids preventing removal.

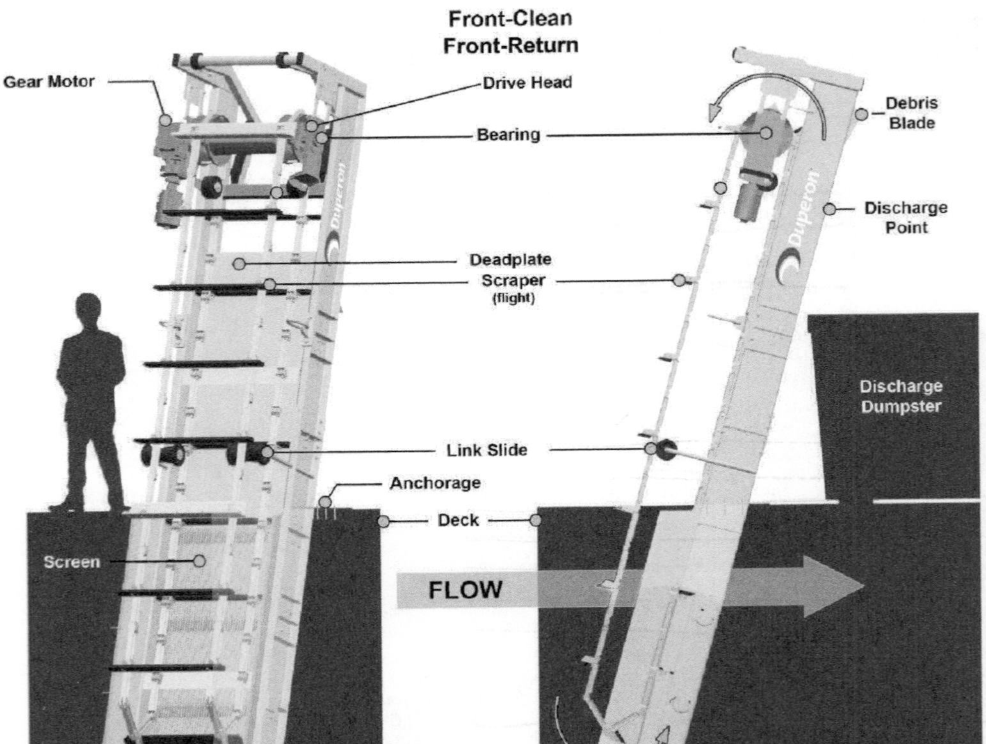

Figure 3.13 **Multiple-Rake Catenary Bar Screen** (Courtesy of Duperon Corporation)

Continuous Self-Cleaning Screens

Another type of mechanically cleaned bar screen is the continuous self-cleaning screen (Figure 3.14). This screen consists of a continuous belt of plastic or stainless steel elements that are pulled through the wastewater to provide screening along the entire submerged length of the screen. The screen openings are limited in both length and width to prevent the passage of long, thin materials. Continuous screens may have openings as small as 1 mm (0.039 in.) or greater than 72 mm (3 in.) (WEF et al., 2018). Solids are captured on the face of the screen and, as the belt rotates, teeth protrude on the upward movement to collect the screenings. Some of these screens include spray bars or brushes to improve the cleaning of the screening elements. Continuous screens have either a lower sprocket or a guide rail at the channel bottom to support the parts of the screen that are submerged.

COMMINUTORS, MACERATORS, AND OTHER GRINDING DEVICES

Comminutors and macerators grind or chop solids that could interfere with downstream wastewater treatment processes. A macerator is a slow-speed grinder. Note that comminutors increase the amount of inert material that will accumulate in the aeration basins and digesters, which effectively decreases treatment capacity. Other types of grinding devices are being used in pipes and channels that use twin shafts with cutting teeth (Figure 3.15). The cutting teeth are stacked onto rotating shafts that rotate toward each other. The rotation, along with the teeth, help pull debris between the shafts where it is cut into smaller pieces. These devices, typically preceded by a bar screen, do not remove the chopped solids from the wastewater flow and instead rely on subsequent treatment units for their removal. At some small facilities, grinding devices are used in place of bar screens. However, shredded rags can adversely affect the operation and maintenance of primary clarifiers, thickening units, and dewatering units. Because of the potential adverse downstream effects and the many other difficulties associated with the operation and maintenance of comminutors and other grinding devices, their use is limited.

Comminutors, macerators, and other grinding devices are often used in pumping stations, upstream of solids handling pumps, and in place of screens in some smaller WRRFs. This equipment can help to prevent blockages in pipes, pumps, and channels that could result in an overflow of raw wastewater. They are

Figure 3.14 Continuous Self-Cleaning Screen (Reprinted with permission by Indigo Water Group)

Figure 3.15 In-Channel Grinder (Muffin Monster) (Courtesy of JWC Environmental)

well suited to smaller facilities with limited staffing. Examples of places and associated conditions that may require a comminutor or macerator in the influent channel include the following:

- Pumping station in the collection system that receives shop rags and other debris that routinely clogs pumps and pipes. Pumping stations are not routinely staffed;
- Smaller WRRFs with narrow influent channels;
- Pond facility with excess solids storage capacity;
- A WRRF downstream of a prison complex or similar industry. Laundry, bed sheets, uniforms, and a host of other larger materials are often flushed by prisoners. This material can be difficult to remove, especially at smaller facilities. Blankets and uniforms can become stuck upstream of the screen and completely block the influent channel. Grinders can help to alleviate this issue; and
- Upstream of solids handling pumps and processes. Debris that enters the WRRF eventually ends up in the screenings dumpster or in the solids handling process. Once in a solids handling process, rags, plastic bags, and other debris can get tangled together, forming larger masses. Passing sludge through an in-line grinder can reduce the potential for damage to downstream equipment.

FINE SCREENS

Fine screens typically consist of wedge-wire, perforated plate, or closely spaced bars with openings of 0.5 to 6 mm (0.02 to 0.25 in.). These screens are freestanding units typically installed downstream from coarse screens. Mechanical cleaning of these screens is essential. Water sprays or brushes are typically used for cleaning these screens. Hot water provides better results than cold water for cleaning fine screens because it helps to remove grease that has adhered to the surface.

Screenings from fine screens typically contain higher concentrations of organic matter. Washer/compactors, either integral to the screen or standalone, must be used with fine screens. Some screening designs bag these screenings, while other fine-screening devices discharge directly to storage containers. Because of the higher organic content of these screenings, operators should transfer them more frequently to final disposal, before excessive odors are produced.

Continuous-Element Screens

Continuous-element screens consist of screens with an endless cleaning grid that is attached to a main drive via different configurations. Screenings are collected and conveyed to the top part of the screen and then discharged. Continuous-element screens come in a variety of forms; however, the most popular are the perforated-plate and belt-type technologies. Perforated-plate screens (Figure 3.5 and Figure 3.16) are constructed of plastic or stainless steel panels with orifices throughout the panel, and are provided in a stacked-panel configuration with rakes at regular intervals to prevent screenings *rollback*. Orifices are typically round for ease of construction. A proper cleaning mechanism (water spray and/or brushes) is required to remove buildup of organic material.

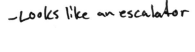

"*Rollback*" is when screenings fall or roll off the front of the screen instead of being carried up and out of the flow path.

Perforated-plate screens in a flow-through configuration typically require an inclination angle between 60 to 75 deg and have the greatest potential for screenings carryover. Although they have considerable head loss, continuous-element screens have one of the best screenings retentions in the industry. They are often used in instances where minimizing the screenings for downstream processes is important.

Stair Screens

Also referred to as *Step Screens®* (Figure 3.17), stair screens consist of long, thin parallel plates called *lamellas* made from stainless steel, with 3- to 6-mm (0.12- to 0.24-in.) openings between plates. The plates have a stair-stepped edge, as shown in Figure 3.18 and Figure 3.19. There are two sets of lamellas in a stair screen; most designs have one fixed and one moving set that rotates in and out of the screen to provide a step motion pattern that lifts the collected screenings upward until they are discharged on top of the screen (Figure 3.19). With each cycle, every other lamella moves while the others remain in place. The moving lamellas typically are connected by either chain drives or levers.

The thin lamellas are vulnerable to damage by large objects, rocks, broken glass, and grit. Sometimes, wedging of larger objects has been a problem with this design. Flexible lamellas at the bottom are used to prevent blockage or damage by large objects. Water flushing connections prevent accumulation of grit under the lamellas. Stair screens do provide a higher open area compared to other fine screens because they have open slots running from the bottom up to the top of the screen grid. This slot configuration allows passage of stringy solids, which can be minimized through "matted operation" in which the screen is operated with a mat of accumulated screenings created by using differential head control.

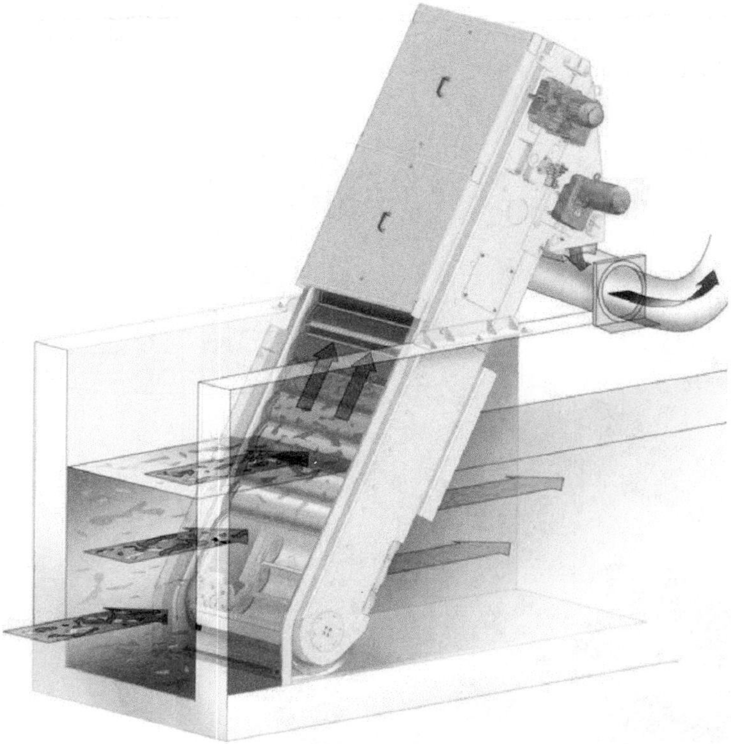

Figure 3.16 Perforated-Plate Screen (Courtesy of Huber Technology, Inc.)

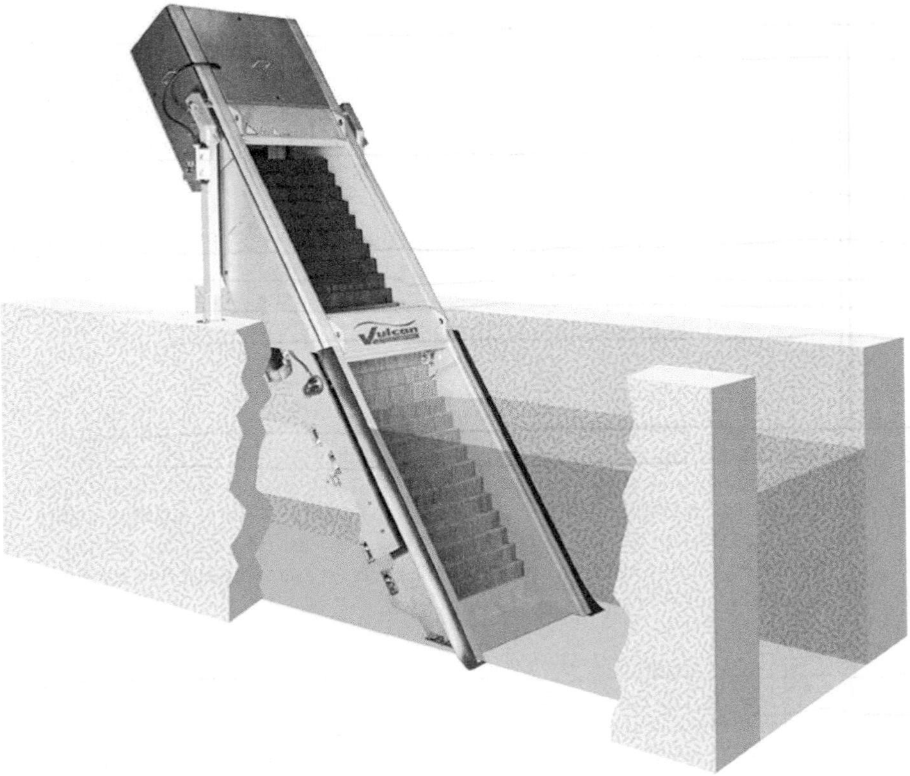

Figure 3.17 Stair Screen in a Channel (Courtesy of Vulcan Industries, Inc.)

Inclined Cylindrical Screens

Inclined cylindrical screens (Figure 3.20) consist of a cylindrical screening basket with an internal screenings removal mechanism that is typically a helical screw. The screen is typically installed at a 30- to 45-deg incline. These types of screens can either be "fixed" with screenings removed by the helical screw or "rotating" with screenings removed by a brush and spray bar. The screenings drop into a hopper feeding a

Figure 3.18 Stair Screen Debris (Courtesy of Huber Technology, Inc.)

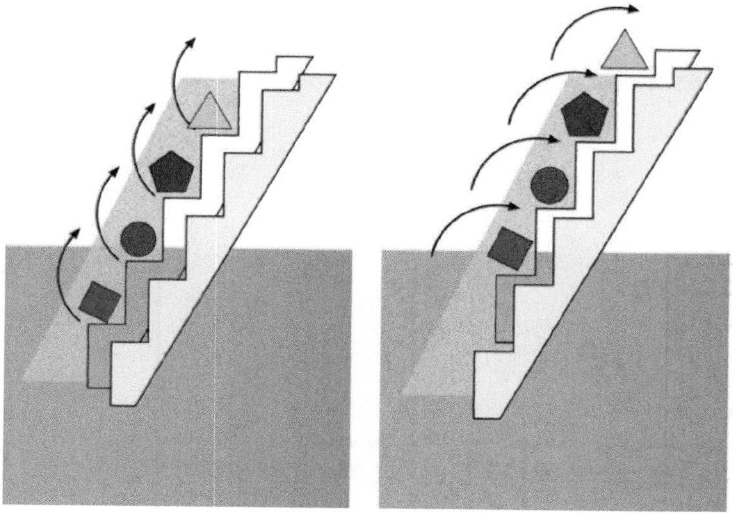

Figure 3.19 Stair Screen Operation (Courtesy of Huber Technology, Inc.)

screw and are conveyed through an inclined washer/compactor pipe. One of the biggest advantages of this screen is that it is supplied with an internal washer/compactor. However, it requires a larger footprint and shallower influent channel compared to other screens.

Table 3.1 summarizes the advantages and disadvantages of the screen types discussed in this chapter. There are many other screen types available for installation in WRRFs.

SCREENINGS PRESSES

Washing and compacting of removed screenings is a critical function of the screening process, particularly for fine screens. They are both recommended for coarse screens with openings less than 38 mm (1.5 in.) and for all fine screens and microscreens. Washer/compactors have two equally important functions.

Washers + Compactors Recommended for:

- Coarse Screens w/openings ≤ 1.5"
- All fine screens
- All micro screens

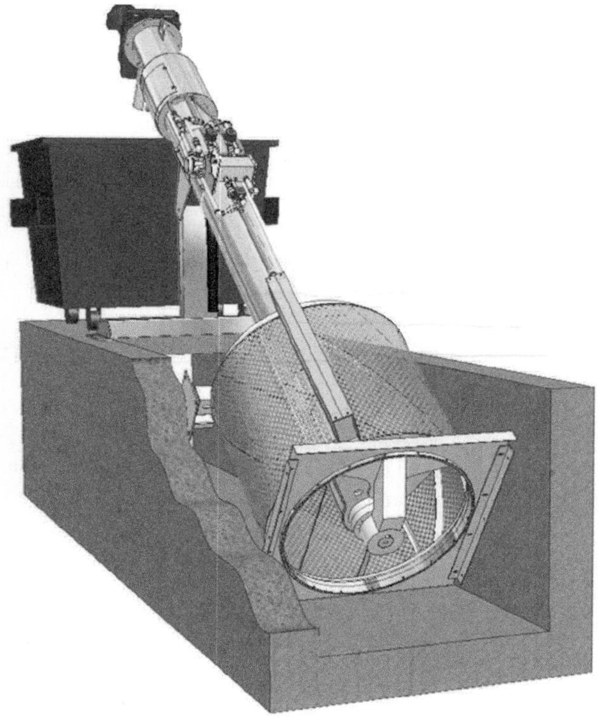

Figure 3.20 Inclined Cylindrical Screen (Courtesy of Lakeside Equipment Corporation)

Table 3.1 Advantages and Disadvantages of Various Screens (WEF et al., 2018)

Type of Screen	Advantages	Disadvantages
Coarse screens		
Chain or cable-driven screens	High screenings loading rate Insensitive to fats, oils, and grease	Submerged components
Reciprocating rake screens	No critical submerged components Allows a pivot design for servicing the unit above the channel	Low screening loading rate High overhead clearance, particularly for deep channels May be difficult to pivot screen out of the channel when connected directly to a washer compactor unit
Catenary screens	No critical submerged components	Large solids removal may be an issue
Continuous self-cleaning screen	Medium to low headroom required Allows a pivot design for servicing the unit above the channel	Several moving components May be difficult to pivot screen out of the channel when connected directly to a washer compactor unit
Fine screens		
Continuous element screens	Proven technology High screenings capture rate Allows a pivot design for servicing the unit above the channel	Numerous moving parts Some designs have components subject to wear and tear High head loss High potential for grease blinding Additional motor and/or spray bars for cleaning May be difficult to pivot screen out of the channel when connected directly to a washer compactor unit
Multiple rake screens	High screenings loading rate Low headroom requirements	Some designs have submerged components subject to wear and tear Medium head loss High potential for grease blinding
Stair screens	Low head loss Allows a pivot design for servicing the unit from above	Stringy solids can pass through the screen Not recommended when rocks or excessive grit loads are expected Requires considerable footprint for installation compared to others May be difficult to pivot screen out of the channel when connected directly to a washer compactor unit

Washing removes organic material and returns it to the wastewater flow. Compaction can reduce the water content of the screenings by up to 50% with a volume reduction of 60 to 85%, thereby reducing costs of storage and disposal. A typical performance specification for washer/compactors is to provide a minimum of 90% removal of suspended organics.

Screenings presses like the one shown in Figure 3.21 consist of a hopper section, an auger or ram, and a wedge section or compaction zone. Screened material is transferred from the screen into the hopper, where it then falls into the main housing of the wash press. An auger or ram then moves the screened material toward the wedge section or compaction zone. The auger is rotated by the drive unit. At the same time, wash water is sprayed over the screenings to help rinse away organic material. The auger stirs the screened material as it pushes it, making it easier for water to drain through the screenings and out the bottom of the wash press. The rinse water and organic material are returned to the main wastewater flow. In fine screening applications, separate wash tanks frequently are provided that mix and agitate the screenings in wash water. As the screenings are moved into the wedge or compaction zone, they are squeezed and compressed to remove water. Presses operate at a range of 4500 to 6000 kPa (1500 to 2000 psig) (WEF, 2016).

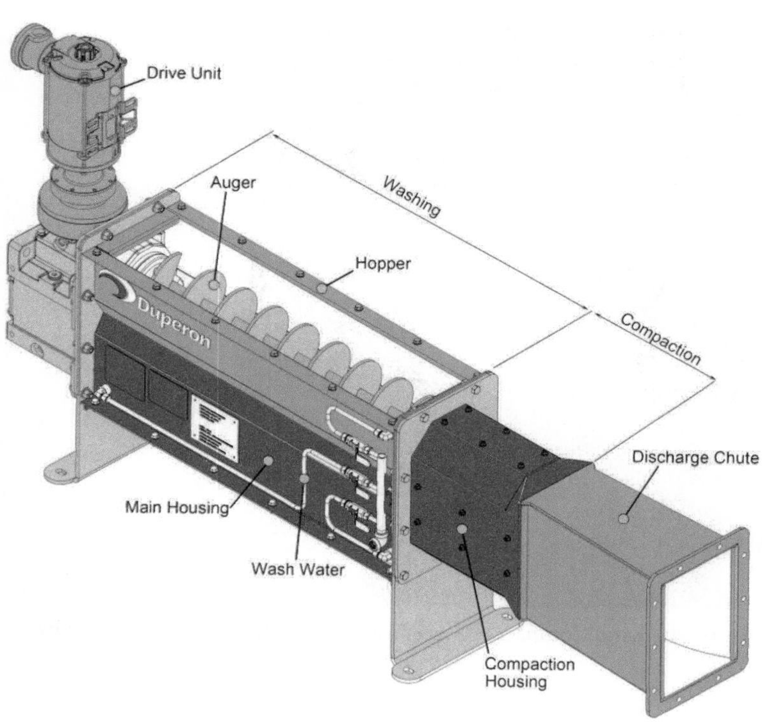

Figure 3.21 Screenings Wash Press (Courtesy of Duperon Corporation)

Washed and compressed screening material is shown in Figure 3.22 as it exits a wash press. Some systems deposit screened material into bags for odor control (Figure 3.23).

Sticks and other large objects in the screenings may cause mechanical breakdowns. Most wash presses have controls that sense jams, automatically reverse the mechanism, and actuate an alarm when a motor overloads. Washer/compactors can be equipped with grinders ahead of the auger to reduce significantly the quantity of screenings and provide maximum washing.

Figure 3.22 Washed and Compacted Screenings (Reprinted with permission by Indigo Water Group)

Figure 3.23 Bagging System for Washed and Compacted Screenings (Reprinted with permission by Indigo Water Group)

A *slurry* is a liquid that contains high concentrations of undissolved solids.

CONVEYING SYSTEMS

Belt conveyors are one method used in the screenings removal processes to transport debris from the point of removal to the point of collection and, eventually, disposal. Belt conveyor systems may be operated continuously or intermittently depending on loading conditions. An alternative to a conveyer, especially where screens are located below ground, is to flush the screenings into a grinder pump, which pumps the *slurry* up to ground level dewatering/screening units.

Screw augers are also used for screenings transport, particularly when an enclosed system is required for odor control. Both shafted and shaftless augers are used, but because of the abrasive nature of screenings, shafted augers may be preferred to avoid frequent replacement of the wear bars or liner. The biggest benefit of augers is the ability to function as the first part of a washer/compactor system before discharge to a compaction tube. Augers can have multiple introduction points. The auger should be sloped to a drain. A well-designed spray system is required along the auger and, particularly, at the drain point. The drain point must be accessible in the event of blockage.

TEST YOUR KNOWLEDGE

1. Trash and bar racks are installed at a 30- to 45-deg angle from the vertical.
   - ☐ True
   - ☐ False
2. Mechanical cleaning of bar screens increases labor costs because of increased maintenance.
   - ☐ True
   - ☐ False
3. Fine screens capture fewer organic solids than coarse screens.
   - ☐ True
   - ☐ False
4. Rollback occurs when screened material falls back down the face of the screen, possibly being returned to the wastewater flow.
   - ☐ True
   - ☐ False
5. A pond WRRF is most likely to have this type of screen in its headworks.
   - a. Manual bar screen
   - b. Perforated plate screen
   - c. Stair screen
   - d. Micro screen

6. Match the screen type to its identifying characteristic.

1. Trash rack
2. Manual bar screen
3. Single rake screen
4. Catenary screen
5. Continuous self-cleaning screen
6. Perforated plate screen
7. Stair screen
8. Inclined cylindrical screen

a. Single operating cycle
b. Interlocking links, no lower sprocket.
c. Efficiently removes hair and stringy material
d. Integrate wash press
e. Remove only largest debris
f. Rectangular openings limited in height and width
g. May be located in a bypass channel
h. Two sets of lamella lift solids up stairs

7. On a single rake screen, a limit switch

a. Sets the upper and lower limits of travel for the rake
b. Adjusts input line voltage
c. Modifies the minimum particle size captured
d. Turns off the drive motor if the rake is jammed

8. One disadvantage of using comminutors and grinders is

a. Reduced potential for clogged pipes and damaged equipment
b. Increased screenings disposal costs
c. Shredded material reduces treatment capacity downstream
d. More frequent overflows of the influent channel

PROCESS CONTROL

Some facilities have screens designed for manual cleaning. In this case, debris is removed with a hand rake, typically on a daily basis or several times each day. Because the amount of screenings will be different from WRRF to WRRF, operators will need to know what is normal for their facility and adjust the cleaning frequency as needed. If the screens are not cleaned often enough, excessive amounts of material will accumulate, which can cause water to build up in the channel. When the screenings are finally removed, a surge of wastewater can enter the WRRF all at once, which may impact the effectiveness of downstream equipment. Screenings should be removed from the face of the screen as often as necessary to ensure a reasonably free flow of wastewater. Unless otherwise specified by the manufacturer or the WRRF's operations and maintenance (O&M) manual, at a minimum, screens should be cleaned when the head loss across the screen reaches 7.6 cm (3 in.) (Bristow, 2008). During rainstorms, the cleaning frequency may have to be increased. In installations with multiple channels, additional screens may be placed into service during a storm event.

In general, bar rack installations use automated motor-operated rakes for cleaning because screens are typically unattended for long periods of time. Compared to manual cleaning, the use of a mechanical cleaning device tends to reduce labor costs, provide better flow conditions, and produce less nuisance conditions. Such devices are almost always used in facilities larger than 3.79 ML/d (1 mgd) and occasionally used in smaller facilities.

A *PLC* is a mini computer that controls a piece of equipment.

Screens may be controlled using timed cycles, head loss as measured by a level sensor immediately upstream of the screen, or by differential head loss. Most systems use a combination of these control methods. In these cases, the screen may normally operate based on head loss (either upstream or differential), but revert to a timed cycle if the amount of time between cycles is too long. This prevents debris that has accumulated on the screen from gradually working its way through the screen and into the WRRF. Clock-operated timing switches, level-sensing devices, or *programmable logic controllers* (PLCs) may control the motors automatically. In addition to automatic control, screens may be manually cycled by the operator using an on-off switch to activate the drive motor. Manual cycling may be needed to check proper operation during facility inspections.

Most of the equipment in a WRRF will have a mode selector switch that defines the equipment control mode. When the switch is in the "hand" position, the piece of equipment can be operated using field-mounted switches that move the rake up or down or to cycle the unit. When it is switched to "automatic", a timer, level switch, or PLC will control the equipment. When in automatic mode, equipment often may also be controlled remotely by the WRRF computer system.

With a clock-operated timing switch, the operator selects a fixed time interval for the screen to operate. For example, an operator may select a 10-minute interval on the timer, which would cause the screen to run through its operational sequence and clear accumulated debris once every 10 minutes.

With head loss control, level-sensing devices measure the water height in the channel immediately upstream of a screen. If the water level is rising, it means material is accumulating on the screen and head loss is increasing. The screen will cycle and clear debris based on an operator *setpoint*. For example, the operator may determine that an acceptable water level in the influent channel is 1 m (3.3 ft). Each time the water level reaches the setpoint of 1 m (3.3 ft), the screen will cycle to remove debris.

A *setpoint* is a limit determined by either the operator or the manufacturer. The setpoint determines how often a piece of equipment will operate.

With differential head loss control, level-sensing devices measure the water height both before and after the screen to determine the head loss. The operator selects the upstream water level or the amount of head loss that will trigger an operational sequence. For example, if the level indicator is set for 1 m (3.3 ft), then the screen will run each time the water level reaches 1 m (3.3 ft). Once the debris is cleared from the screen, the water level will drop. Finally, a PLC may be used to control the screen with a combination of timed cycles and level indicators or with some other piece of information.

Of the four control methods, the clock-operated timing switches are the most frequently used; however, more facilities are installing PLCs. Once experience determines the number of raking operations needed during the average day at a particular installation, the clock-operated timers are set to operate the raking mechanism at the required number of times each day. Timers work well, but cannot adjust to accommodate sudden large accumulations of debris that can occur during high flow periods. If a large amount of debris enters the WRRF at one time, as may occur during a storm event, a timed cycle could allow the debris to accumulate on the screen too long, which may result in a complete blockage and overflow. Therefore, level-sensing devices, such as *ultrasonic level systems*, are typically installed in parallel with the timers and are designed to override the timer control. Other types of level sensors include floats and bubbler tubes (WEF, 2016). The level-sensing system detects the increased head loss through the bar screen when debris clogs the bars, inducing an electrical switch to activate the raking mechanism. The circuit wiring gives the level sensor system precedence over the timer, so that the level sensor will activate the raking mechanism whenever an unusually large amount of debris collects on the screen. Controls are generally configured so that if a cleaning cycle is triggered by the level sensor, the clock timer cycle resets. At the next preset time interval after the level indicator activates the raking mechanism, the timer-operated switch will start the raking unless a large debris accumulation recurs. Controls may be configured with level-induced actuation as primary control with timer-induced override capability, or vice versa, with little difference in the system's function. In either case, timer actuation protects against debris accumulating for long periods resulting in breakthrough and adhesion challenges, and level actuation addresses sudden large debris accumulation on the bars.

An *ultrasonic level system* bounces a sound wave off the surface of the water. By measuring how long it takes for the sound to echo back, it is possible to calculate the height of the water. Ultrasonic level indicators are discussed in Chapter 2.

Some screening facilities have backup water level monitoring on the inlet side of the screen. The upstream-level alarm can be an ultrasonic sensor, bubbler system, submerged transducer, or a float switch. This device will alarm and activate a cleaning cycle if a high liquid level is sensed on the upstream side of the screen. Screening devices include a limit switch to ensure that the rake mechanism stops or parks outside the water after a cleaning cycle. Limit switches are provided to shut down the motor if excessive force is detected because of a rake jam or failure.

Shear pins are used in many different types of equipment to prevent an excessive buildup of force or pressure. The shear pin is designed to break before equipment is damaged.

Older screens may be equipped with a shear pin. *Shear pins* are used in many types of equipment to protect against an overtorque condition. *Torque* is the amount of force applied to make something rotate. If a piece of debris becomes jammed in the screen and the screen is unable to move, the motor will continue to operate and push the screen. Excess torque can build up, potentially damaging the screen. A shear pin connects the motor and gears to the screen. The pin is designed to break when a certain level of torque or pressure has been reached. When the shear pin breaks, the screen will no longer be connected to the motor. The motor will rotate freely, but the screen will no longer be forced against the obstruction. The motor will rotate at least one revolution where it will trigger a limit switch or other safety device that will shut down the motor and set off an alarm. The shear pin breaks so the screen does not.

Torque is the amount of force applied to make something rotate.

Variable-frequency drives are adjustable speed drives that are used to control motor speed and torque in motors that use AC power.

Most modern screens do not rely on shear pins. Instead, a power monitor controller, current monitor relay, or no-motion sensor is used. Power monitor controllers and current monitors are used in conjunction with *variable-frequency drives* (VFDs). Power monitor controllers monitor power use and current monitor relays monitor the current draw. Both power use and current draw are indirect measurements of the amount of torque being generated by the motor and gears. No-motion sensors, also called *zero speed switches*, monitor the starting and stopping points on the drive shaft. The drive shaft connects the motor and gears to the screen. A blockage or other stress on the screen should cause the drive shaft rotation to slow down and take longer to move between the starting and stopping points. Shaft rotation speed is another indirect measurement of the amount of torque.

Electrical current is measured in amps. Electrical power is measured in watts. [Power = (Volts) (Amps).] Electrical equipment that is doing more work requires more power. Power use and amp draw will both increase if the screen becomes stuck.

If the amount of torque exceeds the high torque limit set by the manufacturer (or the operator), the power monitor controller, current monitor relay, or no-motion sensor should stop the screen. The screen may then reverse direction for a few minutes to dislodge trapped debris. The screen may cycle through forward and reverse operation several times. If the blockage is cleared, the screen will return to normal operation. If the blockage cannot be cleared, the screen will stop completely and sound an alarm to alert the operator.

OPERATION

Under normal operation of a mechanically cleaned bar screen, the operator should visually check the equipment several times during a shift to ensure that it functions properly. Use of local manual controls allows immediate observation of the operation of the mechanically cleaned bar screen or other equipment, even though it may be inactive when inspection begins. Screens should be operated with manual switching until the accumulated solids are cleared. While the equipment is running, the operator needs to check for unusual noises, scraping of the screen, jerking motion of the drive mechanisms, overaccumulated screenings, and lubrication of the chain or drive mechanism.

The operator should avoid any overflows in the screening storage areas. Organic materials that are not transported off-site decay and result in offensive odors. Generation of hydrogen sulfide, a toxic gas, can create a hazardous atmosphere if the space is not adequately ventilated. Screenings are a significant source of odors. Check screenings containers daily and empty them as needed. Most importantly, workers must follow good hygiene practices because wastewater-soaked debris harbors pathogenic organisms. As such, hose down equipment weekly or more often to remove accumulated material and prevent odors. Remove grit, rocks, and other materials that may have settled in front of the screen at least once a week. Debris may be removed using a long-handled rake, shovel, or scoop only after the screen has been powered down and removed from service. Draining the channel first will make it easier to see and remove debris.

To place a screen into service,

- Start mechanically cleaned screens according to the manufacturer's recommendations,
- Open the downstream channel gate, and
- Open the channel inlet gate.

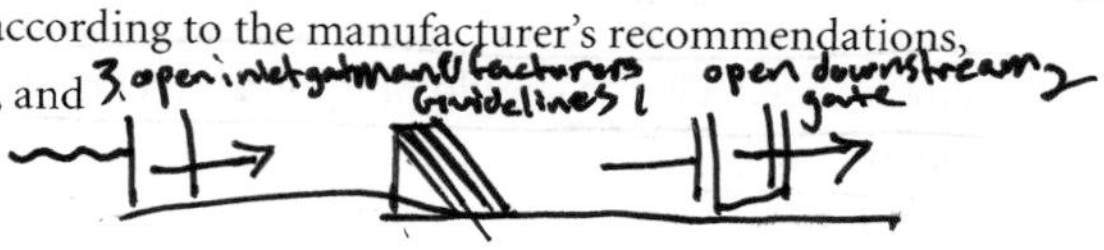

To remove a screen from service,

- Close the channel inlet gate;
- Close the downstream gate;
- Allow the mechanical bar screen to run and remove collected material;
- Shut down mechanically cleaned screens according to the manufacturer's recommendations;
- *Lockout and tagout* (LOTO) the screen and associated equipment;
- Drain the channel and remove debris. A portable sump pump is often used in smaller facilities. Larger facilities may be equipped with sumps and sump pumps. Water should be transferred to a point upstream of a second screen or to the screening bypass channel. Debris may be removed using a long-handled rake, shovel, or scoop only after the screen has been powered down and removed from service;
- Wash the screen to remove debris. Use hot water if available to help dissolve grease;
- Wash down the sides of the channel; and
- Drain the channel.

Lockout and tagout is a safety protocol used to isolate equipment and prevent its accidental operation. The LOTO procedures may include placing a physical lock on the power switch or circuit breaker for a piece of equipment. This ensures that no one can accidentally power up the piece of equipment while it is being worked on.

Verify proper operation of the screenings wash press each day by inspecting the screenings to see that they are compacted and dry. At the same time, verify that the wash water feed rate meets the manufacturer's specifications and check for leaks. To minimize buildup of grease and other materials, wash down the collection trough regularly with hot water, if available. At least once a week, thoroughly clean the screenings wash press, including the hopper and/or chutes.

MAINTENANCE

Equipment and systems described in this section benefit from planned maintenance procedures. The utility should include the headworks in its program of routing inspection, lubrication, and repair, guided by recommendations of the equipment suppliers. Keep in mind that screen channels, grit basins, and the headworks building can all be confined spaces. Be sure to follow confined-space safety protocols when conducting maintenance activities in confined spaces. Be sure to LOTO equipment before performing maintenance.

Trash/bar racks require daily inspection for visible and audible indications of possible malfunctions, as recommended by the preventive maintenance procedures in the O&M manuals. As a routine practice, the operator needs to observe all moving mechanisms to determine if the components are free of obstructions, properly aligned, moving at constant speeds, and producing no unusual vibrations. The operator should periodically verify that all control lamps are functional and that there are no burned-out light bulbs. The operator must listen to all moving mechanisms and recognize their normal operational sounds. For example, abnormal screeches could indicate lack of lubrication; thumps could indicate broken or loose components.

Preventive maintenance requires that worn parts and components be replaced before failure and that the proper lubrication of all moving parts be applied at the recommended intervals. The operator should always follow the manufacturer's recommendations on the type of lubricant to use because the lubricant's qualities are selected for the anticipated operating conditions and for compatibility with the materials used in the equipment.

Chains used in chain-driven bar screens tend to wear over time, which can cause the chain to appear stretched. The chain will gradually become looser over time. Figure 3.24 shows a single chain link and its associated barrel (on the left). When the two pieces rub together, it wears down both surfaces. Periodically, the operator may need to adjust the tensioners in accordance with the manufacturer's recommendations to compensate. Removal of a link may be necessary to ensure that the chain rides smoothly on the sprockets. To minimize equipment breakdown and maintain operational efficiency, badly worn parts should be replaced. Chain drives require frequent replacement of chains, sprockets, and other parts. Chain life also depends on how the unit is operated. For example, if the unit is operated continuously, then the chain life could be dramatically reduced. Grease should be applied weekly (or as required) for rake guides to ensure smooth and quiet operation.

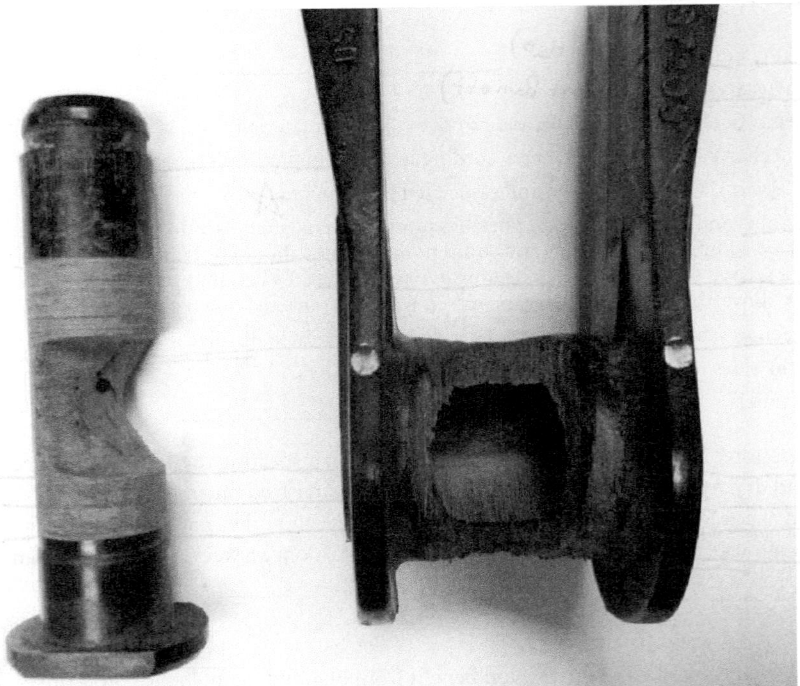

Figure 3.24 Worn Chain Link and Barrel (Printed with permission from Kenneth Schnaars)

Once a month, the entire screening area, screens, wash press, and conveyors should be thoroughly cleaned, including the sides of channels and the insides of covers. Inspect drainage holes on the screenings wash press for fouling and clean them with high-pressure water, as needed. Hot water should be used, if available, to remove grease. Monthly cleaning helps prevent the adherence of grease. Annually, or as recommended by the manufacturer, screening equipment should be thoroughly cleaned and inspected. Some items to look for include alignment of the rake mechanism, chain and belt tension, chain condition, loose screws and bolts, chipping paint, rust, loose or bent covers, worn and missing sprocket teeth, and any other annual inspection items recommended by the manufacturer. The screening channel should be dewatered for the annual inspection so all submerged components can be viewed. Check all nuts and bolts at this time and retighten them as necessary. Inspect and lubricate bearings annually or as recommended by the equipment manufacturer.

Once a month, seals on the main bearing of the screenings wash press should be inspected for leakage and wear. Inspect hydraulic units including all hoses and connections for wear and leaks. Inspect the screw conveyor guide bars for wear. Screenings and grit are extremely abrasive and will wear down both the guide bars and the edges of the screw conveyor. When worn, screenings will rotate in place rather than being pushed through and compacted. It will eventually become necessary to replace the guide bars and/or the compacting screw.

Table 3.2 Screening Equipment Troubleshooting Chart (U.S. EPA, 1978)

Indicator/Observations	Probable Cause	Check or Monitor	Solutions
1. Obnoxious odors, flies, and other insects	1. Accumulation of rags and debris	1. Method and frequency of debris removal	1. Increase frequency of removal and disposal to an approved facility.
2. Excessive grit in bar screen chamber	2. Flow velocity too low	2. Depth of grit in chamber, irregular chamber bottom, flow velocity	2. Remove bottom irregularity or reslope the bottom. Increase flow velocity in a chamber. Flush regularly with a hose.
3. Excessive screen clogging	3a. Unusual amount of debris in wastewater. Check industrial wastes. 3b. Inadequate cleaning frequency.	3a. Upstream conditions 3b. Cleaning frequency	3a. Identify source of waste causing the problem so discharge can be stopped. 3b. Increase cleaning frequency.
4. Mechanical rake inoperable, circuit breaker will not reset	4. Jammed mechanism	4. Screen channel	4. Remove obstruction. Adjust spring tension if appropriate.
5. Rake inoperative, but motor runs	5a. Broken shear pin 5b. Broken chain or cable 5c. Broken limit switch	5a. Inspect shear pin. 5b. Inspect chain. 5c. Inspect switch.	5a. Identify cause of break and replace shear pin. 5b. Replace chain or cable. 5c. Replace limit switch.
6. Rake inoperative, no visible problem	6. Defective remote control circuit	6. Check switching circuits.	6. Verify correct operation of the PLC. Replace circuit. Then replace circuit and motor.
7. Marks or metal against metal on screen binding	7. Screen needs adjustment	7. Operate screen through one cycle and listen for or observe metal-to-metal contact.	7. Manufacturer's adjustments recommended in equipment O&M manual.
8. Slow operation of grinders, screenings press, or belt conveyors. Screeching sounds during operation	8. Worn components	8. Operate through one cycle to identify components. Check maintenance logs.	8. Lubricate machine as indicated. Replace worn components.
9. Thumping sound during operation of screens, grinders, screenings press, or conveyors	9. Broken or loose components	9. Complete inspection of all components, nuts, and bolts.	9. Replace or tighten component.
10. Belt not tracking properly	10. Belt stretched; self-aligning idlers seized; tail pulley misaligned	10. Inspect belts, idlers, and tail pulley.	10. Adjust take-up; clean and lubricate self-aligning idlers; adjust tail pulley alignment.

TROUBLESHOOTING

The most common problems in the screening process are related to the following three categories:

1. Unusual operational conditions, such as sudden loads of debris that clog or jam the screening equipment; - Circumstancial
2. Equipment breakdown and component failure; and - Mechanical
3. Control failure. - Electrical

To understand the screening facility, the equipment used, and the limitations of the equipment, the operator should refer to the O&M manuals for the specific equipment used at the facility. After these manuals have been read and understood, efficient resolution of problems can become routine. Refer to Table 3.2 on the previous page for a troubleshooting chart related to screening equipment.

TEST YOUR KNOWLEDGE

1. **A WRRF with a combined sewer system is expecting a large storm. The manual bar screens are normally cleaned every hour. The operator should plan to decrease the cleaning frequency during and immediately after the storm.**
   - ☐ True
   - ☑ False

2. **A shear pin is designed to break under a certain amount of force or pressure to prevent additional damage to equipment.**
   - ☑ True
   - ☐ False

3. **Screens should be cleaned before the head loss across the screen reaches _______ or according to the screen manufacturer's recommendations.**
   - a. 2.5 cm (1 in.)
   - b. 7.6 cm (3 in.)
   - c. 12.7 cm (5 in.)
   - d. 17.8 cm (7 in.)

4. **This type of control mechanism is best suited for storm events or other conditions that produce highly variable amounts of screenable material.**
   - a. Clock timer - Doesn't compensate for debris surges
   - b. Level sensor
   - c. Particle counter
   - d. HOA switch

   Don't Make Sense

5. **A mechanical bar screen must be taken out of service for maintenance. Put the following steps in the correct order.**
   - 3 a. Shut down the screen according to the manufacturer's directions.
   - 1 b. Close the upstream (inlet) gate.
   - 2 c. Close the downstream (outlet) gate.
   - 6 d. Wash the screen to remove debris.
   - 5 e. Drain the channel and remove debris.
   - 4 f. Lockout and tagout equipment.
   - 7 g. Wash sides of channel and perform final draining of channel.

6. **At a minimum, how often should screens be inspected for visible and audible indications of possible malfunctions?**
   - a. Daily
   - b. Weekly
   - c. Monthly
   - d. Quarterly

Grit Removal

PURPOSE AND FUNCTION

Wastewater grit generally consists of fine, discrete, non-biodegradable particles that have a *settling velocity* greater than that of organic solids. Grit is generally defined as ranging in size from 0.050 to 1.0 mm, which is the same as 50 to 1000 µm (Metcalf and Eddy, Inc./AECOM, 2013). Such materials include sand, cinders, rocks, coffee grounds, seeds, fruit rinds, and other relatively *nonputrescible* organic and inorganic substances. Grit acts like liquid sandpaper that wears down and erodes equipment and pipes as it passes through and over them. Grit removal, an essential element of preliminary treatment, protects equipment from abrasion and accompanying wear. Grit removal prevents accumulation of these materials in downstream processes such as process tanks and digesters, where grit accumulation can result in loss of usable

Settling velocity is how fast a particle falls or settles through the water.

A *micron* is one-millionth of an inch. The average human hair is 70 to 100 µm in diameter and a grain of salt is about 70-µm across.

volume and treatment capacity. Settling of grit in primary clarifiers (Chapter 4) can cause the primary sludge concentration to increase so much that it becomes difficult to pump the sludge.

Putrescible substances are rapidly broken down or decayed by microorganisms.

Grit chambers are used to settle grit from the wastewater. The settled grit is then consolidated and separated from the wastewater using screw conveyors, chain and flight systems, bucket elevators, pumps, and cyclones. Finally, grit is rinsed by classifiers to remove unwanted organics.

DESIGN PARAMETERS AND EXPECTED PERFORMANCE

Variables influencing the quantity of grit include the type of collection system (combined or separate), presence of household garbage disposals, condition of the collection system, presence and types of industrial waste, and efficiency of grit removal. Often, grit accumulates in the collection system and is flushed into the WRRF in larger quantities during high flow (WEF et al., 2018). In the 2008 WEF member survey, grit quantities reported by WRRFs ranged from 3.7 to 148 L/1000 m^3 (0.5 to 20 cu ft/mil. gal) and averaged 37 L/1000 m^3 (5.0 cu ft/mil. gal). Because the amount of grit removed is different from one WRRF to another, operators will need to know what is normal for their system. Grit removal systems rely on differences in settling velocities to remove grit while allowing organic solids to pass on to the next treatment process. Settling velocity (i.e., how fast a particle sinks) depends on the density of the particle as well as its shape and size. Density is the weight per unit volume. For example, water has a density of 999 kg/m^3 (62.4 lb/cu ft) at 15.6 °C (60 °F). Sand has a density of about 1280 kg/m^3 (80 lb/cu ft) and whole kernel corn has a density of about 719 kg/m^3 (44.8 lb/cu ft). If the density of a particle is greater than the density of water, it will sink. If the density of a particle is less than the density of water, it will float. How fast a particle sinks or floats (settling velocity) is related to how different the density of the particle is from the density of water. A particle that is only slightly more dense than water will sink slowly, whereas particles that have high densities will sink rapidly. Round, compact solids sink faster than flat, feathery particles. Larger particles sink faster than smaller particles made of the same material.

To see settling velocity in action, place several centimeters (inches) of a mixture of fine sand and gravel into a 2-L bottle. Fill the rest of the bottle with water leaving at least 2.5 cm (1 in.) of airspace at the top. Shake the bottle vigorously. When the bottle is set on a counter and allowed to rest, the gravel will reach the bottom of the container, first forming a layer followed by the slow-settling sand.

When a particle enters the grit basin, gravity pulls it down in a straight line (Figure 3.25). The more dense and compact the particle is, the faster it will be pulled toward the bottom of the basin. At the same time, the water moving through the grit basin is pushing the particle forward (Figure 3.25). How fast the particles move forward is related to the *velocity* of the water. The result is that most of the particles entering the grit basin will have a diagonal path from the front of the basin toward the end of the basin (Figure 3.25). Particles that settle slowly or that have densities lower than the density of water will be carried though the basin before they have time to settle. Generally speaking, square and rectangular grit basins target a horizontal water velocity through the basin of about 0.3 m/s (1 ft/sec) to retain grit and pass on organics. Vortex grit basins target a water velocity of 0.45 to 1.1 m/s (1.5 to 3.5 ft/sec), depending on the manufacturer. If the flow velocity through the grit basin is too high, even very dense particles will be pushed through to the next unit process. Operators can make adjustments to the flow velocity by changing the cross-sectional area of a grit basin by raising or lowering the water level, taking basins in and out of service, and through the use of specialized weirs. Aerated grit basins use diffused air to maintain the desired velocities, whereas vortex-type grit basins use centrifugal force instead of gravity.

Velocity is speed. Velocity can be calculated in two different ways. Velocity is equal to the amount of flow divided by the cross-sectional area of the channel, pipe, or basin (eq 3.1). Velocity is also equal to distance traveled per unit of time. An example of velocity is a car driving along at 120 km/hr (75 mph).

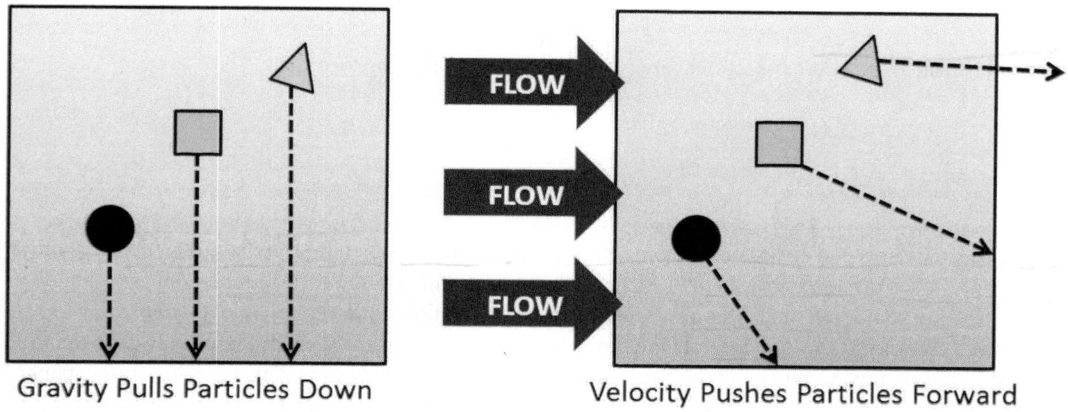

Figure 3.25 Settling Velocity (Reprinted with permission by Indigo Water Group)

TEST YOUR KNOWLEDGE

1. **The volume of grit removed for every m³ (mgd) of flow received is very consistent from one facility to the next.**
   - ☐ True
   - ☑ False

2. **As flow velocity through the basin increases, larger, denser particles will pass through to the next process.**
   - ☑ True
   - ☐ False

3. **All of the following characteristics will influence the settling velocity of a particle EXCEPT**
   a. Density
   b. Size
   c. Color
   d. Shape

4. **Gravity pulls grit particles straight down, but this pushes them forward toward the outlet of the grit basin:**
   a. Velocity
   b. Density
   c. Viscosity
   d. Conductivity

5. **Which of the following particles should settle the fastest through a column of water?**
   a. Corn kernel
   b. Coffee grounds
   c. Sand
   d. Gravel

6. **A grit basin has a flow velocity of 0.3 m/s (1 ft/sec) at 9 A.M. As flows increase throughout the day, grit basin performance will be affected in this way.**
   a. The percentage of organic material captured will increase.
   b. Grit removal efficiency will decrease.
   c. Velocity through the basin will decrease.
   d. Air consumption will increase.

EQUIPMENT

GRIT CHAMBERS

Detritus Tanks

–widely varying efficiency in grit/organic removal; dependent on influent flow

The *HRT* is the time required to fill or empty a container. If the container already contains water and water is both entering *and* leaving, then the HRT measures the amount of time an average bit of water remains in the container. HRT and detention time mean the same thing. HRT is more commonly used by engineers, though operators tend to use detention time. The ABC certification exam uses detention time.

Detritus is loose material or debris.

One of the earliest grit chambers was a constant-level, short *hydraulic retention time* (HRT) settling tank called a *detritus tank* (square tank degritter). Flow is distributed across the tank inlet to minimize velocity across the tank. There may be baffles at the tank inlet to minimize turbulence. Even flow distribution is maintained by proper adjustment of the deflector baffles. Because these tanks settle dense organics and grit, they require a grit-washing step to remove organic material and return it to the wastewater flow. Some designs incorporate a grit auger and a rake that remove grit from the tank. Detritus tanks are not recommended for WRRFs with widely varying flows. The velocity of the water through a detritus tank will vary with the amount of flow entering the facility. When flows increase in the middle of the day, some grit may pass through the basin. When flows decrease at night, some additional organic material will be captured with the grit.

Velocity Control Tanks

Keep Velocity of flow constant by varying the Area

Velocity is related to the cross-sectional area according to the following formula:

$$\text{Velocity} = \frac{\text{Flow}}{\text{Area}} \tag{3.1}$$

Additional information on weirs and flumes can be found in Chapter 2.

When flow increases and area stays the same, velocity will increase. When flow is changing, then area must also change to keep velocity constant. Horizontal flow grit chambers use proportional weirs or rectangular control sections (such as Parshall Flumes) to vary the depth of flow and keep the velocity of the flow stream at a constant 0.3 m/s (1 ft/sec). One type of proportional weir is shown in Figure 3.26. The weir is shaped so the area the water is moving through remains proportional to flow. This type of weir is capable of maintaining a nearly constant flow velocity through a channel. For the weir to operate correctly, debris must not be allowed to accumulate across the opening. Constant velocity can also be maintained by giving

Parshall Flume is what Solvay has ← measuring point

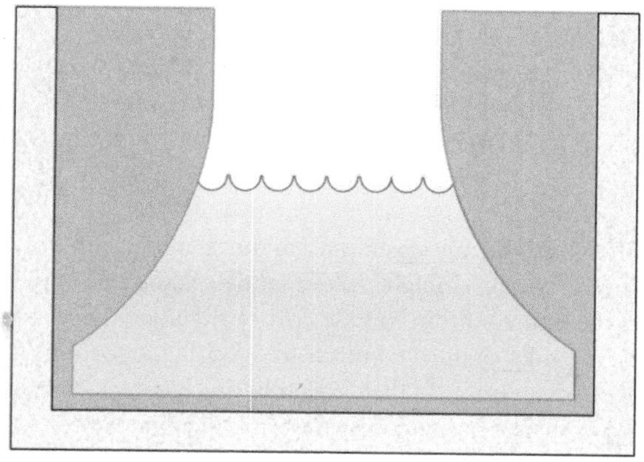

Figure 3.26 Flow-Proportional Weir

the grit channel a parabolic (u-shaped) or trapezoidal cross-sectional area. This channel shape functions similarly to a flow-proportional weir by maintaining a constant flow-to-area ratio.

Hydraulic retention times for horizontal flow grit chambers are typically between 45 and 90 seconds (U.S. EPA, 2003). Basin length is determined by the size and density of the average particle that must be removed. Slower settling particles require longer basins than faster settling particles (Figure 3.25).

Aerated Grit Basin

In aerated grit chamber systems, *diffused air* is introduced along one side of the basin near the bottom. The diffusers are attached to *headers* that are typically located at about 70% of the total depth of the basin to prevent them from becoming buried in grit (WEF et al., 2018). Airflow is generally between 4.7 and 12.4 L/(m·s) (3 and 8 cu ft/ft) and may be adjusted by the operator (GLUMRB, 2004). The air causes a spiral roll pattern perpendicular to the flow through the tank (Figure 3.27). The velocity (speed) of the roll pattern influences which particles will drop to the bottom and the particles that remain will be carried into the tank effluent. The velocity of the roll is governed by the shape of the grit basin and the amount of air introduced. Roll velocity is independent of flow through the tank, which allows this type of grit basin to operate effectively over a wide range of flows. Vortex-type grit basins are often used at facilities with highly variable influent flows. Smaller service areas tend to have higher peak hour to average day flows and can benefit from vortex grit basins. Heavier particles that settle are moved by the spiral flow across the tank bottom and then into a grit trough or hopper.

Aerated grit basins tend to be long and narrow with a length-to-width ratio between 3 : 1 and 8 : 1 (WEF et al., 2018). Longer basins tend to provide better grit removal. The width-to-depth ratio is nearly square (0.8 to 0.9) to ensure the roll pattern fills the basin (Figure 3.27). Aerated grit basins are typically designed

A good example of *diffused air* is an air stone in an aquarium creating fine air bubbles. Compressed air is fed through tubing from a compressor to the air stone. Diffused air in aerated grit basins is similar, but on a larger scale.

An air *header* is the piece of pipe that runs from the air source (i.e., compressor or blower) into the grit basin and along the bottom. Diffusers are typically mounted directly to the headers.

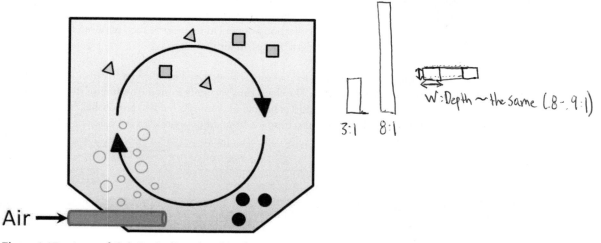

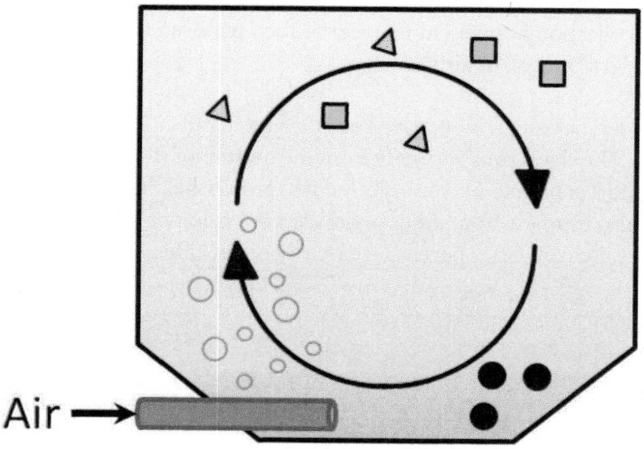

Figure 3.27 Aerated Grit Basin (Reprinted with permission by Indigo Water Group)

to remove 0.21-mm (65-mesh) or larger particles with an HRT of 2 to 5 minutes (Metcalf and Eddy/AECOM, 2013). Mesh measures the number of holes or openings in a 2.54 cm by 2.54 cm (1 in. by 1 in.) piece of screen. It is often used to describe the particle size that can be captured by different treatment technologies. Table 3.3 compares mesh, micron, inches, and millimeters.

Vortex Grit Basin

Vortex grit chambers are gravity-type grit chambers that swirl the raw wastewater in the chamber (Figure 3.28). Wastewater enters a vortex grit basin through a long, straight, inlet channel. At the end of the inlet flume, a ramp causes grit that may already be on the flume bottom to slide downward along the ramp until reaching the chamber floor, where it is captured. The inlet channel connects to the circular portion of the basin along one side, which creates a whirlpool or vortex effect in the circular basin (Figure 3.29). The velocity of the vertical roll pattern (whirlpool) allows denser particles to settle while keeping lighter organic material suspended. Vortex grit basins target a water velocity of 0.45 to 1.1 m/s (1.5 to 3.5 ft/sec), depending on the manufacturer.

Some vortex tank designs rely on natural hydraulics to achieve the proper rotational rate, whereas other designs use a slow, rotating-paddle-type mixer to enhance the natural hydraulics and achieve proper separation. The grit that settles in these tanks can be removed by an airlift pump or a non-clogging, recessed-impeller-type grit pump.

Multi-tray vortex systems are gravity-type chambers that use stacked trays to provide increased surface area to allow coarse and fine grit particle removal (Figure 3.30). Influent is distributed to each tray, and grit collects at the tank bottom. Grit slurry is pumped to a grit classification and dewatering system before disposal. Degritted effluent exits via a wall-mounted weir.

Hand-Cleaned Vs Mechanically Cleaned

Hand-cleaned grit chambers are used in the smallest facilities with flows generally less than 3.8 ML/d (1 mgd). Some have hopper-shaped bottoms for grit storage. Floor drains are provided to empty the tanks for manual grit removal by shoveling. An example of a manually cleaned grit chamber is shown in Figure 3.3.

Mechanically cleaned grit chambers are gravity-type units that are rectangular, circular, or square tanks. Mechanically cleaned grit chambers use a variety of different methods to push grit into a central hopper and remove it from the basin. Screw augers are the most commonly used method of moving grit along the bottom of the tank to the grit sump. Screw-type collectors or augers rotate to push the grit forward (Figure 3.31). Other methods include chain-and-flight collectors, chain-and-bucket collectors, clamshell buckets (Figure 3.32), and pumps.

In larger facilities with wider channels, chain-and-flight or screw-type collectors move the grit to a central sump for removal. A chain-and-flight system uses a series of boards, called *flights*, that span the width of the basin to push grit into a sump or hopper (Figure 3.33). Circular tanks include a circular collector mechanism that scrapes the grit to a central collection hopper (sump). Grit is then removed from the hopper by bucket elevators, screw conveyors, grit pumps, or airlift pumps.

Bucket elevators are simply buckets attached to chains that scoop grit from the grit chamber sump and remove it from the wastewater (Figure 3.34). The chain runs vertically from the bottom of the grit sump to the top of the basin, where grit is dumped into a container or is transferred to a grit washer and classifier. Buckets are often made from steel, but may also be made from other materials. Cast nylon or other strong, lightweight buckets reduce wear on the elevator chain. Grit buckets have wear shoes at either end; these

Table 3.3 Particle Sizes

Mesh	Microns	Inches	Millimeters
50	297	0.0117	0.297
70	210	0.0083	0.210

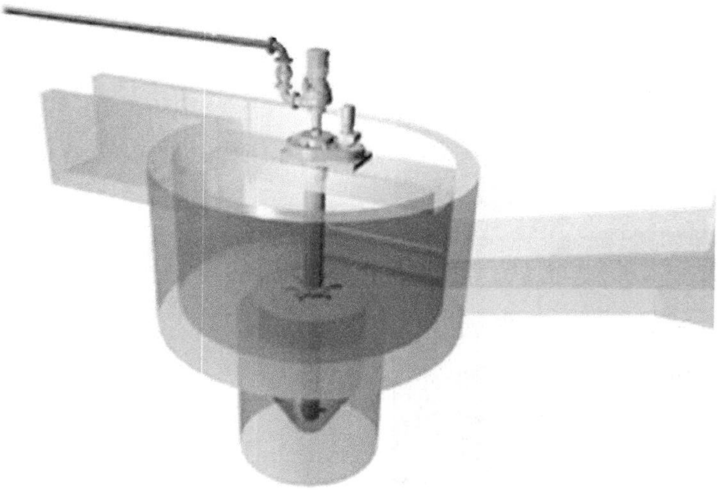

Figure 3.28 Hydraulic Forced Vortex Grit Basin (Courtesy of Smith & Loveless Inc.)

Figure 3.29 Vortex Grit Basin Installation (Reprinted with permission by Indigo Water Group)

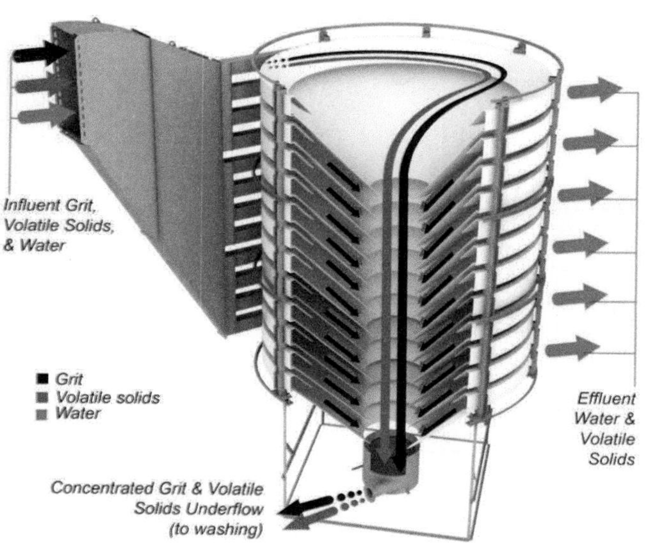

Figure 3.30 Multiple-Tray Grit Separator (Courtesy of Hydro International)

Figure 3.31 Grit Auger (Reprinted with permission by Indigo Water Group)

Figure 3.32 Clamshell Bucket (Getty Images)

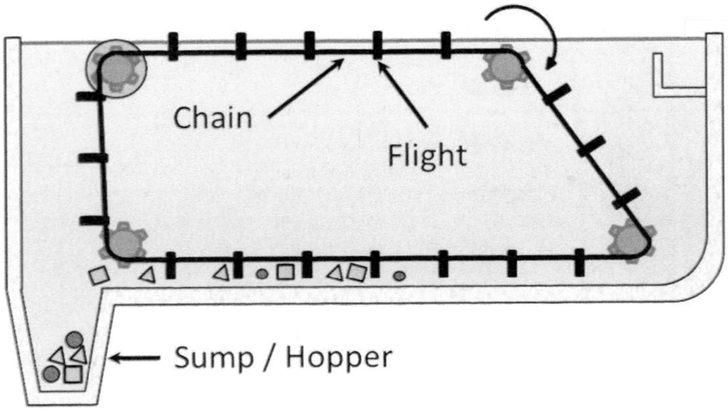

Figure 3.33 Chain-and-Flight Collection Mechanism (Reprinted with permission by Indigo Water Group)

Figure 3.34 Bucket Elevator (Reprinted with permission by Kim Riddell)

are pieces of metal that help the buckets move across the floor of the grit basin and reduce the amount of wear on the bucket itself. As the shoes wear down, they are replaced for less cost than replacing the entire bucket. A spray of water on the chain as it emerges from the wastewater helps prevent grit from becoming lodged in the chain joints. Polyurethane floats attached to the flights will reduce their effective weight, thus reducing the wear of the shoes and rails. As a precaution, however, grit must be kept below the height of the floats. Drain holes on the bottom of the buckets should be cleared occasionally so water can drain freely.

Preventative Maintenance

Scrapers or screw conveyers should be operated at low speed to push grit into the grit hopper while the bucket elevator should function at a rate fast enough to remove the collected grit. If scrapers or screw conveyors push grit into the hopper faster than the grit elevator can remove it, this may result in damage because of grit packing in the basin. Although scrapers, screw conveyors, and grit buckets are all separate pieces of equipment, there are some commercially available units that include both the scraper and grit elevator in one unit.

This avoids overloading the bucket

A grit pump is basically a centrifugal pump with a recessed impeller. Grit pumping can be more difficult than other solids pumping because the grit's sheer weight can clog pumps and its abrasive nature quickly deteriorates piping. To minimize these problems, the pumps are made of abrasion-resistant alloys. They are designed to be operated more frequently so large quantities of grit cannot accumulate. For more information on pumps, refer to Chapter 8 of *Operation of Water Resource Recovery Facilities* (MOP 11; WEF, 2016) and Chapter 6 of *Design of Water Resource Recovery Facilities* (MOP 8; WEF et al., 2018).

A grit pump is basically a centrifugal pump with a recessed impeller.

The removal of grit by pumps offers the advantage of eliminating the need for a bucket elevator and its submersed sprockets. However, blockages might occur in the pump suction line, in the intake area of the sump, or in the discharge piping. If this occurs, removal of the blockage may require reversing the pumping direction or blowing air into the sump. Some facilities contain water and/or air flushing connections on the suction and discharge side of the pumps to remove blockages.

- Grit pumping eliminates the need for bucket elevators.
- However blockages may occur throughout system.
- To remedy, the pump may be ran in reverse or air may be blown into the sump

A *tangent* is a line that touches a curve or circle in one place. Tangential feed means the feed is entering the cyclone along the edge or perimeter of the outer edge.

Orifice is another word for opening.

CYCLONES

Cyclones (Figure 3.35) use centrifugal force in a cone-shaped unit to separate grit and organics from the wastewater. A pump discharges a slurry of grit and organics into the cyclone at a controlled rate. The slurry enters the cyclone *tangentially* near its upper perimeter (A) and flows through an opening into the feed chamber. The feed chamber is the uppermost portion of the cyclone. The opening between the feed pipe and feed chamber is called the *feed orifice*. It is typically rectangular with the long dimension in line with the body of the cyclone. The feed velocity creates a vortex (whirlpool) in the classification zone (B). Larger, denser particles move downward producing a grit slurry at the lower, narrower opening (C), while a larger volume of slurry containing mostly volatile material is discharged through the upper port (D). The bottom of the cone section is the apex orifice (C). Cyclones capture approximately 95% of grit particles when functioning well. The grit stream falls into a grit washer, and the degritted flow leaves the cyclone through the opening near the top of the unit (D) and is returned to the treatment process. In some systems, a mechanical mixer is used to create the centrifugal effect.

The cyclone does not have any moving parts. The pump that moves grit slurry from the grit tank generates the initial velocity, which provides energy at the mouth of the cyclone. The fluid velocity starts the degritting process. The volume of pumped slurry and the resultant pressure at the cyclone are critical requirements specified by the cyclone manufacturer. Temperature, solids concentration, and other characteristics of the slurry may require changes in the sizes of the upper and lower orifices after installation and some initial operating experience. In some designs, these orifices are manually adjustable.

CLASSIFIERS

Grit classifiers effectively remove organics from the grit. Screw- and rake-type classifiers (Figure 3.36) can effectively remove organics from some grit slurries. Multiple papers have shown that screw classifiers alone do not effectively wash grit solids (Boldt, 2005; Kern and Pisano, 1998). To ensure a low organic content, however, ample dilution water is required to rinse the grit. Pumps typically provide sufficient dilution water, but bucket elevators may not, especially during periods of peak grit capture. Consequently, they may require supplementary liquid.

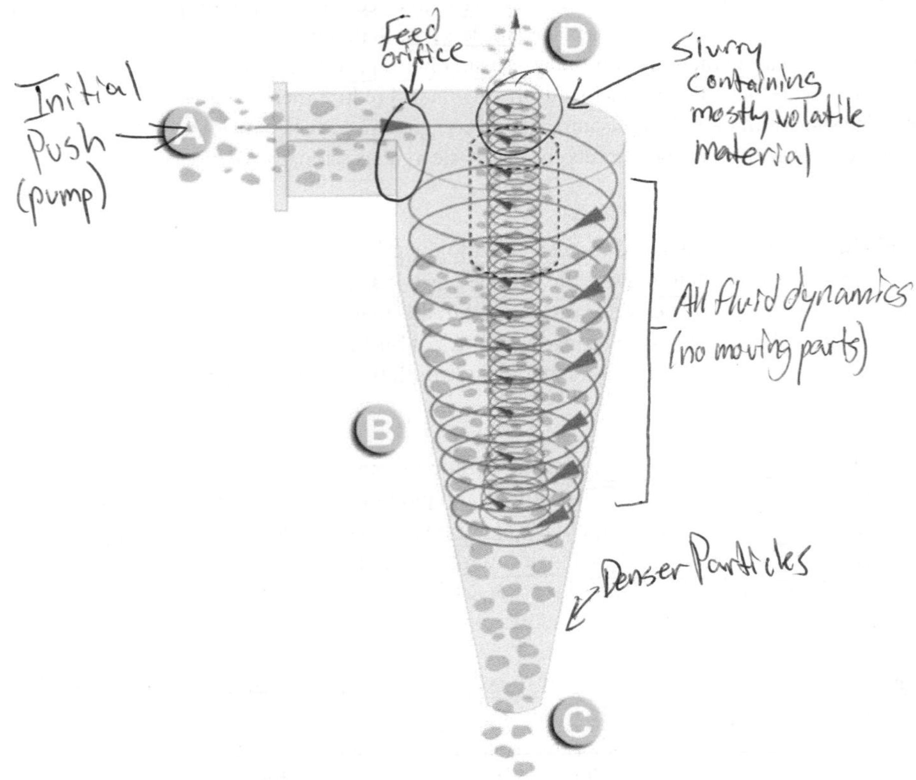

Figure 3.35 Hydrocyclone (Courtesy of Weir Minerals)

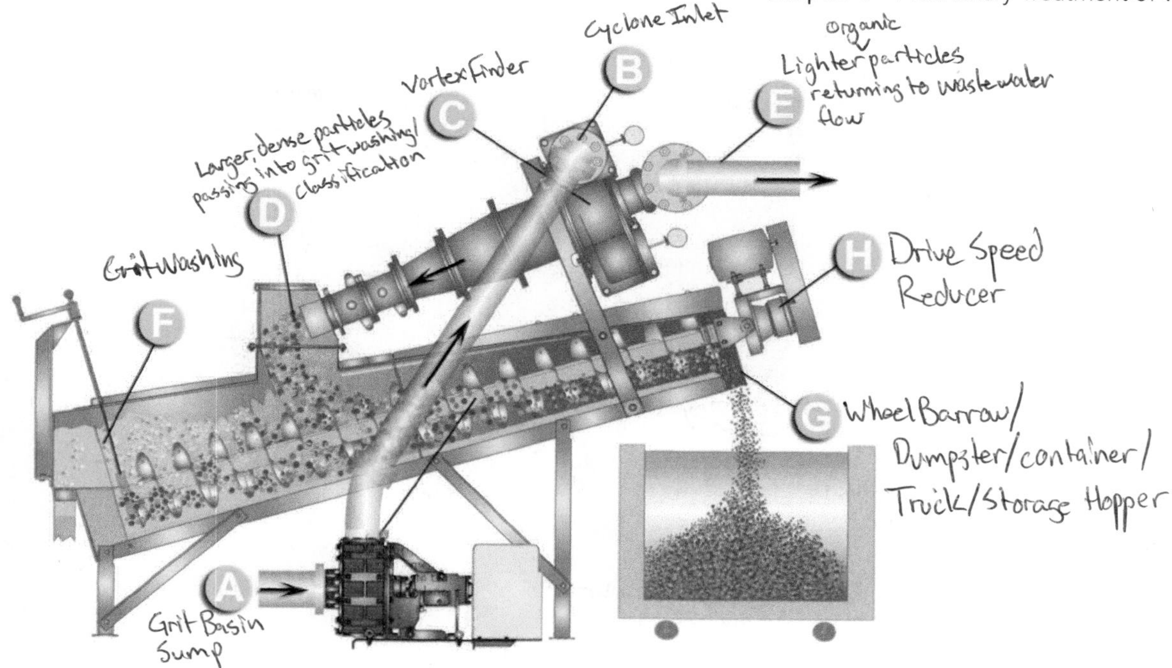

Figure 3.36 Grit Classifier (Courtesy of Weir Specialty Pumps)

The grit classifier in Figure 3.36 is paired with a grit cyclone. Grit is pumped from the grit basin sump (A) to the cyclone inlet (B). The vortex finder (C) is a cylinder within the cyclone that extends into the barrel and induces the spiral flow pattern through the cyclone. Larger, denser particles exit through the bottom of the cyclone where they pass into the grit washer classifier (D). Lighter, less dense organic particles exit the cyclone at the top and are returned to the wastewater flow (E). Grit washing occurs at the low end of the classifier in a well that has an adjustable weir to govern the depth of liquid above the settled grit (F). Most washers are equipped with spray nozzles along the conveyor to remove stray putrescibles. For optimum washing, some cyclones use an adjustable lower orifice to change the volume of liquid discharged with the grit. The classifier may be either a reciprocating rake device or a screw collector. Either type may include water sprays for grit washing. Both the reciprocating rake and the screw collector are inclined upward at an optimum angle for draining water and organic matter. The grit then falls into a wheelbarrow, dumpster, container, truck or storage hopper (G). The drive speed reducer is shown as (H) in the figure. A belt conveyor system may be used for transport of the washed and dewatered grit. Washers need regular lubrication and must be checked for wear. The screw conveyor will wear down over time from constant contact with grit and can develop chips and holes. The inside of the classifier will also wear. Eventually, a gap will develop between the conveyor and the wall of the classifier and grit will build up internally.

Free vortex-based grit classifier systems (Figure 3.37) use centrifugal force to wash organics from grit. Grit slurry is pumped to the cylindrical classifier, creating high centrifugal forces that concentrate grit particles at the outside wall (A). Grit particles then settle by gravity to the tank bottom (B). Lighter, less dense organic particles are swept up and out through the center of the vortex (C) as the vortex sweeps the grit particles to the bottom outlet, while rinse water is injected to provide a final rinse (D). Washed grit is discharged through a flow controller to a dewatering escalator unit for final disposal (E). Degritted effluent exits the top of the classifier carrying organics back to the treatment process (F). Free vortex classifiers of this type can produce output grit with less than 15% *volatile solids* while capturing 95% of grit particles as fine as 75 µm. Systems of this type can be effective for utilities with wide-ranging flow that see high organic content during low flows.

Volatile solids are organic solids.

Grit dewatering escalators (Figure 3.38) are typically used in conjunction with free vortex grit classifiers. Contrary to screw- and rake-type classifiers, dewatering escalator units are designed to capture all of the solids entering the clarifier. No washing is expected of this type of unit. Grit from the grit separator is discharged into a large hopper full of water that serves as a clarifier. The clarifier is essentially an unstirred tank where solids can settle to the bottom while lighter organics flow out with the degritted water. Grit

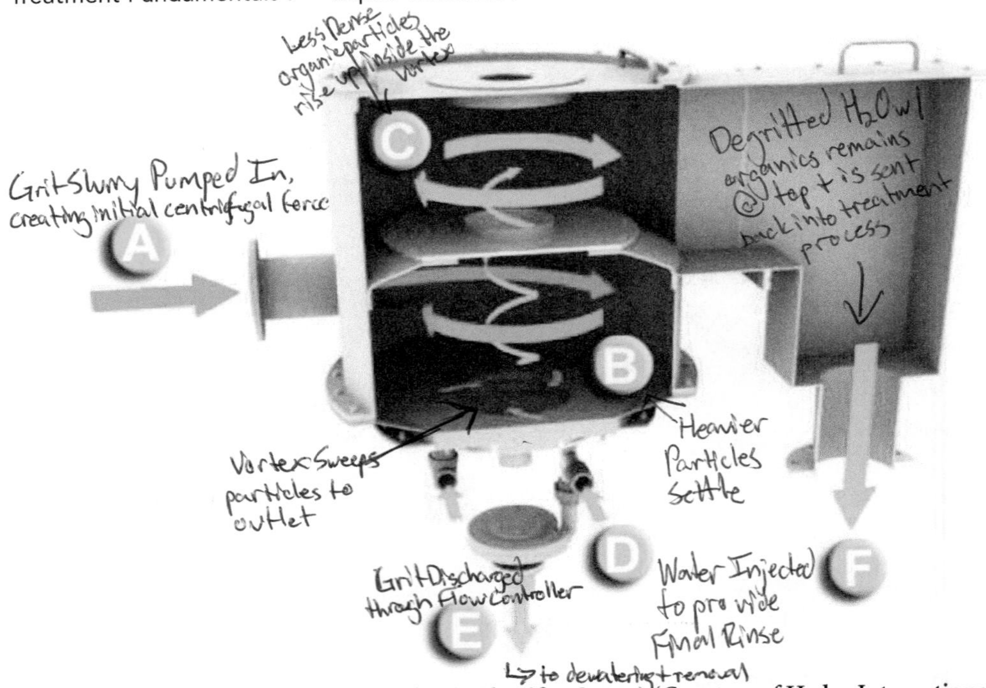

Figure 3.37 Free Vortex-Based Grit Classifier System (Courtesy of Hydro International)

dewatering escalators use a large clarifier surface area to achieve a low surface loading rate, which results in particles 75 μm and larger being retained. An inclined rubber cleated conveyor belt is used to slowly convey grit solids to the discharge point, while a wiper blade cleans each cleat as it passes. A chain-and-sprocket drive system moves the belt at 0.005 to 0.01 m/s (1 to 2 ft/min).

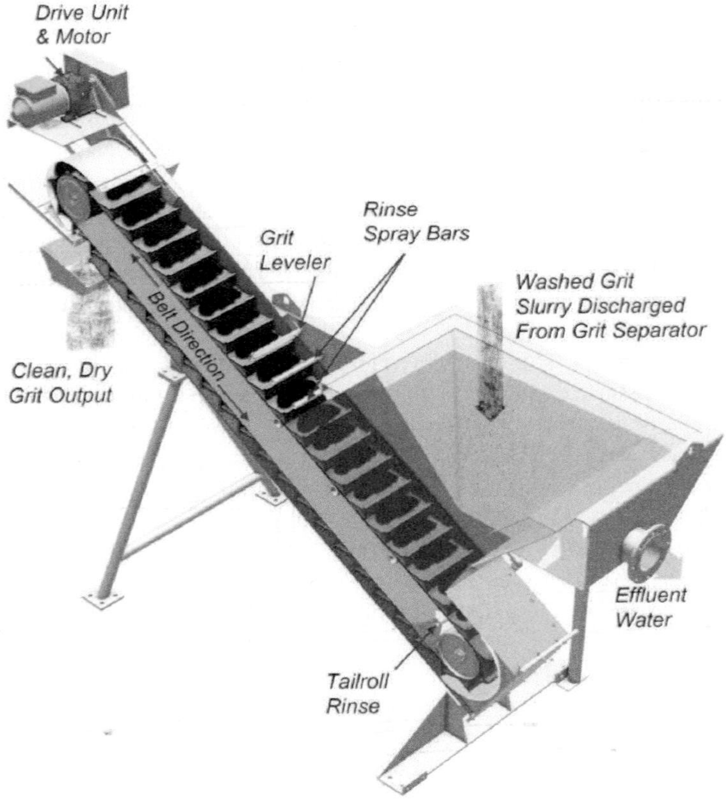

Figure 3.38 Grit Dewatering Escalator (Courtesy of Hydro International)

Conical grit washer technology uses a cone-shaped tank to capture grit. Rotating arms within the tank slowly mix the settled grit and a washing jet at the bottom of the unit, activated by a solenoid valve on a timer, vigorously washes it. Lighter material continuously overflows the unit and heavier organic material is removed at regular intervals from a midlevel overflow.

Washed grit settles to the bottom where a screw conveys and dewaters the output grit. Systems of this type are designed to capture 95% of grit particles 212 µm and larger while producing output grit with 5% volatile solids. The surface flow velocity, including the wash water, should be less than 25 m/h (0.02 ft/sec) and the weir overflow rate should be less than 15 m^2/h (160 sq ft/sec).

GRIT TRANSPORTATION WITHIN THE FACILITY

Grit removal equipment is typically located adjacent to the storage hoppers or containers. If this equipment is located elsewhere, conveyors may be used to transfer the grit to the loading area. Overhead clamshell bucket, chain-and-flight collectors, and augers can be used to remove grit from grit basins, but most installations use mechanical pumping. This is because grit slurry can be pumped directly to washing and dewatering equipment.

Grit can be conveyed directly to trucks, dumpsters, or storage hoppers. Containers should be covered to prevent odors during storage and hauling. Conveyors frequently are used for transporting grit from handling facilities to containers. Overhead storage hoppers that discharge to truck containers avoid the need to keep a truck at the facility.

TEST YOUR KNOWLEDGE

1. **Velocity is calculated by multiplying the flowrate by the cross-sectional area.** DIVIDING
   - ☑ True
   - ☐ False

2. **Grit classifiers return organic material to the wastewater flow and remove excess water from the grit.**
   - ☑ True
   - ☐ False

3. **A WRRF currently has one grit basin in service. The flow velocity through the basin is 0.8 m/s (2.6 ft/sec). How many grit basins should be in service to maintain a flow velocity between 0.24 and 0.3 m/s (0.8 and 1.0 ft/sec)?**
   a. 1
   b. 2
   c. 3
   d. 4

   Divide Current Velocity by Desired velocity max desired = 1 ft/sec 2.6/1.0 ; min desired = .8 ft/sec 2.6/.8

4. **Flow-proportional weirs are used in velocity control horizontal flow grit basins for this purpose:**
   a. Produce a constant discharge rate
   b. Reduce buildup of grit behind the weir
   c. Measure flowrate
   d. Maintain a constant velocity

5. **In an aerated grit basin, this process control parameter influences which particles will sink and which will continue on to the next process.**
   a. Velocity of the roll pattern
   b. Depth of the basin
   c. Vortex dimensions
   d. Flow-proportional weir setting

6. **This type of grit basin may use paddles to maintain the flow circulation pattern.**
   a. Detritus tank
   b. Aerated grit basin
   c. Vortex grit basin
   d. Velocity control

7. **An operator must bypass their aerated grit basin for several days while annual maintenance is performed on the diffusers. Which of the following statements is true?**
   a. Airflow should be set to 50% of maximum after basin dewatering.
   b. Grit may pass through and cause damage to downstream equipment.
   c. The organic loading to downstream processes will increase.
   d. Screenings will accumulate in the grit channel.

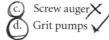

8. **Grit is typically removed from a vortex-type grit basin in this way.**
   a. Manual removal with shovels
   b. Chain-and-flight system
   c. Screw auger
   d. Grit pumps

PROCESS CONTROL

Grit removal efficiency is difficult to measure. Poor performance over an extended period of time will result in increased wear of downstream equipment, increased maintenance costs, and buildup of grit deposits in pipes, channels, and treatment processes, resulting in a loss of capacity. Therefore, a high level of early grit removal is a sound objective. However, the operator must consider grit-removal equipment wear vs increased removal efficiency and resultant downstream equipment longevity. Increased equipment run time will increase O&M costs.

Grit removal systems start at preset times and operate for preset times. For example, an operator may select setpoints for a grit pump to operate for 8 minutes once every 20 minutes. When selecting setpoints, keep in mind that the pump should cycle frequently enough to prevent the grit from compacting in the hopper. The cycle length, or total minutes of pump time, should be coordinated with the hopper size. For example, if the hopper holds 2 m^3 (528 gal) of grit and the pump has a capacity of 0.2 m^3/min (53 gpm), then the pump run time per cycle should be approximately 10 minutes. The time between pumping cycles is determined by how long it takes for the hopper to fill with grit. Hopper fill time depends on the influent flowrate and the amount of grit in the influent wastewater. Operators will need to observe conditions at their facilities to determine the optimum pump cycle and run time.

Grit should be removed from the grit collection area on a daily basis. Operators should visually inspect grit samples for low, high, and average flow conditions. Grit that is grey, greasy, or has a foul, rotten egg odor indicates the presence of excessive amounts of organic material. This indicates that flow velocities are too low or that some other adjustment needs to be made to either the grit removal or grit washing process. An accumulation of heavy, inorganic grit particles downstream of the grit basin indicates that flow velocities through the grit basin are too high at least some of the time. Operators should expect to see grit quality vary somewhat depending on the time of day that samples are collected.

Organic material in the grit can cause odor issues at the facility headworks. Operators should optimize grit basin operation and grit washing systems to minimize odors. Additionally, high organics/odors can lead to grit transportation and disposal issues. Grit, just like screenings, must pass the paint filter test to be acceptable for landfill disposal.

DETRITUS AND VELOCITY CONTROL TANKS

Detritus and velocity control tanks cannot be adjusted by the operator beyond placing more tanks into service or removing tanks from service. The goal of process control is to maintain a velocity of 0.3 m/s (1 ft/sec). Generally, detritus tanks and velocity control grit basins are intended to maintain an HRT of approximately 1 minute (WEF et al., 2018). Example velocity and HRT calculations are included in the calculations section of this chapter.

AERATED GRIT BASIN

Aerated grit chambers provide an HRT of 3 to 10 minutes (WEF et al., 2018). Compressed air, typically from the facility's process air supply, enters the grit chamber through diffusers at a controlled rate to induce the proper velocity of the spiral roll for the settling of grit. Air rates typically range from 5 to 12 L/s·m (3 to 8 cfm/ft) of tank length. Rates as low as 2 L/s·m (1 cfm/ft) have been used for shallow, narrow tanks. Rates higher than 8 L/s·m (5 cfm/ft) often are used for deep, wide tanks. The goal is to achieve air-induced rotation of the wastewater at approximately 0.6 m/s (2 ft/sec) near the tank entrance and 0.5 m/s (1.5 ft/sec) at the tank exit (WEF, 2007). The airflow rate should be adjusted to maximize removal of inorganic grit while minimizing entrapment of organic particles. Inconsistent air supply can result in erratic roll patterns, high percentages of organic matter in the removed grit, and/or low removal efficiencies.

When an aerated grit chamber is initially filled or after the tank is being returned to service, the operator should turn on the air when the wastewater reaches the diffusers. This will help to avoid fouling the diffusers. Grit should not be allowed to accumulate above the diffusers. Large grit accumulations complicate diffuser inspection, result in diffuser damage, and reduce the efficiency of the grit collection.

VORTEX GRIT BASIN

Operators should ensure that the tank is maintaining the proper swirl velocity. Swirl velocity can be adjusted by limiting the number of basins in service or by adjusting the paddle speed, if available. If excessive amounts of organic material are being captured with the grit, paddle speed should be increased. If excessive amounts of grit are found in downstream processes, it may be necessary to decrease paddle speed. Operators should inspect the pumping system during each shift to ensure that the equipment is not plugged. Cyclones can return grit, previously captured by the grit basin, to the influent channel. If excess liquid and uncaptured grit from these processes is returned to the influent stream downstream of the grit basin, it can accumulate in downstream channels and basins. Be sure to check all potential sources of grit at your facility before making a process control change.

GRIT COLLECTION MECHANISMS

Grit collection mechanisms can be damaged if excessive amounts of grit are allowed to accumulate or by rapidly flowing water. When initially filling grit tanks equipped with collection mechanisms, the influent gate should be opened slowly. Once the collection mechanism is submerged, the operator may then turn on the longitudinal collectors (flights), the screw collector (if any), bucket elevator or grit pump, and, if present, the grit washer and classifier. This approach avoids startup under load conditions after the grit builds up. Many grit collection mechanisms have a fixed operating speed and cannot be adjusted. Others are equipped with VFDs, which allow the drive and mechanism speed to be adjusted by the operator. Variable-frequency drives can decrease electrical use and operating costs. If possible, adjust the speed of the collection mechanism to maintain the grit level below the tops of the screw auger or flights. All components of the system are often interconnected electrically so that all will stop and an alarm will sound when any component fails to function. This feature helps prevent equipment damage and loss of the tank in use.

Buckets

During operation, buckets should be inspected frequently to check the volume of grit collected. If the buckets are filled, the elevator speed should be increased or an additional grit channel should be placed into service. If the grit volume is decreasing, the opposite actions may be appropriate. If it is not possible to adjust the operating speed, operators may run the buckets more frequently or increase total run time. For example, a grit bucket mechanism may be switched from operating once every 15 minutes to once every 10 minutes. The number of minutes of operation per operating cycle can also be adjusted.

Cyclones

If a cyclone is being started for the first time, has been out of service for more than a few weeks, or if vibration was the cause of shutdown, both orifices of the cyclone should be inspected before startup to ensure that they are clear. To view both orifices, lift up the apex end of the cyclone and look up the cyclone into the inlet area. The rectangular inlet orifice should be clearly visible. To start the system, the operator opens the inlet valve and starts the grit collector, pump, and washer. Cyclones are sized for a specific flowrate, pressure, and apex (outlet) size and will not function properly if one or more of these variables is out of range. Be sure to check the manufacturer's O&M manual and specifications for the cyclone at each facility. Pressures at the cyclone and the pump inlet need to be checked regularly. Operation should be routine, provided the proper volume of slurry is delivered to the cyclone at the proper pressure. Thus, the equipment needs regular observation. Excessive vibration of the cyclone may be caused by an obstruction in either the upper or lower port, and excessive flow at the lower end may result from a damaged liner or an obstruction of the upper port. Items needing regular inspection include bearings, pumps, valves, pipe work, cyclone lining (wear), and other items recommended by the manufacturer.

CALCULATIONS

Process control of grit removal systems may require an operator to calculate HRT or velocity. The formula for HRT, also referred to as *detention time*, is

$$\text{Detention Time} = \frac{\text{Volume}}{\text{Flow}} \tag{3.2}$$

Typical HRTs for different types of grit basins are as follows (Metcalf and Eddy/AECOM, 2013):

- Horizontal flow grit basin = 45 to 90 sec,
- Aerated grit basins = 2 to 5 min, and
- Vortex grit basins = 20 to 30 sec.

Velocity is related to the cross-sectional area the water is moving through according to the following formula:

$$\text{Velocity} = \frac{\text{Flow}}{\text{Area}} \tag{3.3}$$

Velocity is typically expressed as meters per second (m/s) or feet per second (ft/sec) for wastewater flowing through a collection system, pipe, conduit, or grit basin. For primary and secondary clarifiers, discussed in later chapters, velocity is expressed as $m^3/m^2 \cdot d$ or gpd/sq ft.

MAINTENANCE

Because grit removal equipment has many moving parts, many possible types of malfunctions and breakdowns can adversely affect the operation and efficiency of the device. The operator needs to learn the equipment and how to assess problems that will occur. The operator should always follow the manufacturer's O&M manuals for its equipment.

Additionally, the following items need to be checked regularly:

- Gates;
- Bolts on flights and elevator buckets, chains, and sprockets;
- Flight shoes and rails;
- Collector screws; and
- Shear pins.

The underwater equipment and chain idlers must be lubricated with the manufacturer's lubricant and at the recommended lubrication frequency. Gear drive units should be checked regularly to prevent contamination of lubricants by excessive water accumulation.

AERATED GRIT BASIN

Adequate ventilation is essential if the basin is enclosed; otherwise, the corrosive atmosphere will inevitably affect exposed electrical wiring and controls and concrete, as well as the risers and headers. Risers and headers should be inspected at least annually for corrosion or rupture. Some systems are equipped with removable headers and diffusers that can be lifted out of the basin as a single unit. Most systems, especially newer systems, have headers and diffusers that are fixed to the bottom of the tank. If the headers cannot be lifted, the basin must be drained for inspection. Bent or damaged diffusers and headers should be straightened or replaced as required. Diffuser (the device releasing the air into the wastewater) openings should be cleared regularly. If surface turbulence diminishes, the diffusers may need cleaning to remove rags or grit. Excessive turbulence in one area could indicate a missing or damaged diffuser or header. Rags should be removed whenever the headers are exposed.

If diffusers must be lifted to allow grit removal with a crane-operated bucket, the operator should maintain a reduced flow of air through the raised diffusers to prevent clogging. Crane-operated buckets remove grit only in large facilities (typically in facilities larger than 400 ML/d [100 mgd]). If the diffusers do not have to be raised during grit removal, an air volume reduction will usually be necessary to allow control of the bucket while it descends into the basin hopper. The wastewater flow can be maintained without restriction during the bucket removal of grit.

VORTEX GRIT BASIN

Operators should dewater and inspect the unit after 6 months of operation and then on an annual basis thereafter. Inspection requires the basin to be dewatered via the grit underflow pump. Operators should inspect for solids accumulation on the trays, support structure, and chamber floor. Accumulated solids can be hosed to the underflow collector for removal by the grit underflow pump. Larger solids should be removed manually.

CYCLONE DEGRITTERS AND GRIT WASHERS

Items requiring regular inspection include bearings, pumps, valves, pipework, the lining (wear), and the lining attachment to the casing. Washers need regular lubrication and must be checked for wear.

GRIT DEWATERING ESCALATOR

Operators should check the drive, belt, and wiper mechanism on a weekly basis.

TROUBLESHOOTING

To understand grit removal, the equipment used, and the limitations of the equipment, the operator should refer to the O&M manuals for the specific equipment used at the facility. After these manuals have been read and understood, efficient resolution of problems can become routine. Table 3.4 lists common grit equipment problems, their causes, and their remediation.

TEST YOUR KNOWLEDGE

1. **Increased equipment run time will increase O&M costs.**
   - ☐ True
   - ☐ False
2. **When grit pumps are used to remove grit from a hopper, a timed cycle of 10 minutes of operation once every 30 minutes should always be used.**
   - ☐ True
   - ☐ False
3. **When placing a rectangular grit basin with a chain-and-flight system into service, the basin should be filled completely before starting the mechanism.**
   - ☐ True
   - ☐ False
4. **A sample of grit is grayish in color and smells like rotten eggs. The operator should consider**
   a. Slowing down the vortex paddle
   b. Increasing flow velocity through the grit basin
   c. Decreasing the rinse water to the grit classifier
   d. Placing another grit basin into service
5. **The lower port of a grit cyclone is partially blocked by a piece of trash. The blockage will likely cause**
   a. Excessive vibration of the cyclone
   b. Increasing flow velocity through the grit basin
   c. Decreasing the rinse water to the grit classifier
   d. Liner delamination
6. **The headworks building at a new WRRF shows evidence of metal and concrete corrosion. The problem seems to be even worse in covered channels. The most likely cause is**
   a. Industrial discharges
   b. Grease buildup
   c. Inadequate ventilation
   d. Over-aeration
7. **An operator is used to seeing an even mixing pattern on the surface of the aerated grit basin. Today, the surface is fairly smooth and does not show signs of good mixing. The operator should**
   a. Increase the airflow rate to the basin
   b. Check the diffusers for rags
   c. Replace broken headers and diffusers
   d. Adjust the outlet weir

Additional Pretreatment Considerations

FLOW EQUALIZATION

Reducing flow fluctuations (diurnal and storm) in the collection system benefits the operation and performance of the WRRF. Some WRRFs are equipped with flow equalization basins as part of preliminary treatment. Flow equalization basins temporarily store influent wastewater during peak hour flow and storm events. The water level in the tank changes over the course of a day, filling up during peak hour flows and emptying again during periods of low flow. Ideally, the flowrate out of a flow equalization basin and into the WRRF will be at a constant rate. Facility flows through parallel units, such as screens and grit chambers, must balance to aid optimum performance of each unit.

Table 3.4 Grit Removal Equipment Troubleshooting Chart (U.S. EPA, 1978)

Indicators/Observations	Probable Cause	Check or Monitor	Solutions
1. Grit packed on collectors	1a. Collector operating at excessive speeds 1b. Bucket elevator or removal equipment operating at slow speeds	1a. Collector speed 1b. Removal system speed	1a. Reduce collector speed. 1b. Increase speed of grit removal from collector.
2. Vibration of cyclone degritter	2a. Obstruction in the lower port 2b. Obstruction in the upper port	2a. Flow from lower port 2b. Too much flow in lower end	2a. Remove obstruction. 2b. Reduce inlet flow.
3. Rotten egg odor in grit chamber	3. Hydrogen sulfide formation	3. Sample for sludge deposits, total and dissolved sulfides	3. Wash chamber and dose with hypochlorite (bleach).
4. Accumulated grit in chamber	4a. Submerged debris 4b. Flow velocity too low or broken chain or flight	4a. Inspect chamber for debris. 4b. Check equipment.	4a. Wash chamber daily. Remove debris. 4b. Repair equipment
5. Corrosion of metal and concrete	5. Inadequate ventilation	5. Ventilation and sample for sludge deposits, total and dissolved sulfides	5. Increase ventilation and perform annual repair and repainting.
6. Removed grit is gray in color, smells, and feels greasy	6a. Improper pressure on cyclone degritter 6b. Inadequate airflow rate 6c. Grit removal system velocity too low	6a. Discharge pressure on cyclone degritter. 6b. Check airflow rate. 6c. Use dye or floating objects to check velocity.	6a. Keep pressure at cyclone between 28 and 41 kPa (4 and 6 psi) by governing pump speed. Oil as O&M manual specifies. 6b. Increase airflow rate. 6c. Increase velocity in grit chamber.
7. Surface turbulence in aerated grit chamber is reduced	7. Diffusers covered by rags or grit	7. Diffusers	7. Clean diffusers and correct screens or other pretreatment steps to prevent recurrence.
8. Surface turbulence in aerated grit chamber is increased in one or more locations, but reduced in others	8. Broken diffusers or headers	8. Diffusers and headers	8. Repair or replace as needed.
9. Low recovery rate of grit	9a. Bottom scour at excessive speeds 9b. Too much aeration	9a. Velocity 9b. Aeration	9a. Maintain velocity near 0.3 m/s (1 ft/sec). 9b. Reduce aeration. Increase retention time by using more units or reducing flow to unit.
10. Overflowing grit chamber	10a. Pump surge problem 10b. Excessive influent flow	10a. Pumps 10b. Influent flowrate	10a. Adjust pump controls. 10b. Control inflow and infiltration.
11. Septic waste with grease and gas bubbles rising in basin	11. Sludge on bottom of basin	11. Grit basin bottom	11. Wash basin daily.
12. Grit pump efficiency decreased	12. Worn seals, shaft components, or impeller	12. Pump components	12. Replace seals; lubricate pump as specified; inspect and replace impeller as needed.
13. Decreased grit removal efficiency through cyclone	13. Inlet pressure is too low	13. Inlet pressure	13. Increase inlet pressure by adjusting the grit pump discharge.
14. Loud operation of cyclone	14. Torn or collapsed liner	14. Liner	14. Replace liner.

ODORS AND ODOR CONTROL

Odors often emanate from the headworks of a facility because of the long duration of wastewater transit through the collection system during hot weather. This may produce hydrogen sulfide and other anaerobic byproducts of decomposition that cause dangerous confined-space situations and complaints. Whenever possible, septic influent flows should be prevented in the collection system using appropriate techniques.

Otherwise, odor control with prechlorination, iron salts, potassium permanganate, hydrogen peroxide, nitrates, or preaeration may be necessary.

Within the WRRF, special odor control units use sodium hydroxide, sodium hypochlorite, activated carbon, or potassium permanganate to control releases of odor. These units typically require housing the influent structures, collecting the gases, and then treating them.

SEPTAGE MANAGEMENT

Septic waste is domestic wastewater collected from septic tanks, cesspools, portable toilets, pit toilets, and recreational vehicles. It is typically hauled to WRRFs in tanker trucks. Water resource recovery facilities can accept septage from domestic sources, but should not take grease trap waste or industrial septage because this is regulated differently than domestic wastewater. Domestic septage is regulated under the *Code of Federal Regulations* at 40 CFR Part 503. Domestic septage includes the liquid or solid material removed from a septic tank, cesspool, portable toilet, or similar system that only receives domestic septage. It does not include grease trap waste, shop pit wastes, car wash pit wastes, or dry cleaning waste residues. If these wastes are mixed with domestic septage, then the entire batch of septage becomes nondomestic septage and is no longer covered under the 503 regulations. Nondomestic septage, such as grease trap waste, is regulated under 40 CFR Part 257.

Septage tends to be much higher strength than typical domestic wastewater and may contain higher concentrations of FOG, heavy metals, toxic organics, trash, rocks, and debris. Biochemical oxygen demand (BOD) is typically 26 times more concentrated compared to typical raw domestic wastewater and may be as much as 300 times more concentrated (WEF, 1997). Total suspended solids (TSS) are typically 51 times more concentrated and can be as much as 374 times more concentrated (WEF, 1997). The increased load to the WRRF can be substantial, especially for smaller facilities. When accepting septage, operators should monitor and track incoming loads and calculate the total mass of BOD, TSS, and ammonia-nitrogen received (kilograms [pounds]). Tracking the total mass received will help to ensure that the amount of septage received does not exceed the WRRF's rated treatment capacity.

Many municipalities have enacted a waste hauler's ordinance that licenses and controls access for septage dischargers including sampling of waste loads. Each load received should have a tracking manifest that includes the following sections:

- Wastewater characterization section (e.g., BOD, TSS, FOG concentrations)
  - Amount of septage
  - Type of waste (municipal, commercial, industrial)
  - Location pumped from
- Generator's section
  - Generator's name and contact information
  - Certification statement (type, source, and volume)
- Hauler's section
  - Hauler's name and contact information
  - Hauler's permit number and vehicle license number
  - Pump-out date
  - Signed certification by hauler
  - Disposer/receiver's section
  - Septage receipt date
  - Sample identification number (if applicable)
  - Signature

Septage may be received at an upstream manhole, the facility headworks, or the solids handling process. Accepting septage at an upstream manhole makes for a simple and economical receiving station. Other advantages include allowing the septage to be diluted before reaching the WRRF. Disadvantages include increased odor potential near the manhole, increased opportunities for sewer blockages, increased line-cleaning needs downstream of the addition point, potential for hydrogen sulfide corrosion of the collection system at the addition point and immediately downstream, difficulty regulating and controlling access, and effects on traffic patterns and businesses. Accepting septage at the WRRF headworks gives the facility control of septage discharge into treatment processes as well as the ability to segregate the waste and hold it apart from the rest of the influent. This allows the operator an opportunity for flow equalization and sampling before addition. Disadvantages to this approach include increased odor potential at the facility (but greater potential for odor control) and additional staff time to manage facility access. Accepting septage into the solids handling process can be expensive because of receiving station requirements, but it reduces loadings on liquid processes and lessens the chance of a biological process upset. However, septage may affect the dewatering of processed solids. Adequate pretreatment using aeration, grit removal, and screening should precede treatment of septage wastes at the WRRF.

Some WRRFs are equipped with specially designed septage-receiving stations as part of the WRRF headworks or, sometimes, immediately upstream of the facility. Septage-receiving stations are equipped with screens to remove debris, one or more holding tanks, and septage pumps. The pumps allow operators to slowly introduce the septage to the influent, which decreases the effect of high BOD, TSS, and ammonia on downstream processes. Receiving stations are often equipped with card reader systems that allow authorized septic haulers to access the system and make deliveries unassisted. A septage-receiving station can dramatically reduce the amount of time required by WRRF staff and provide an easy paper trail for billing purposes.

Table 3.5 summarizes potential operational concerns that should be addressed before accepting septage.

Table 3.5 Operational Concerns Associated with Septage Receiving

Concern	Description
Hydraulic surge at smaller facilities	A single septic truck may hold up to 5000 gal of septage. Trucks can discharge their contents in under 30 min. If septage is not flow equalized, the instantaneous flowrate may be as high as 167 gal/min (0.24 mgd). At a smaller facility, this hydraulic surge can disrupt biological treatment and push solids from the secondary clarifier into the final effluent.
Increased organic and solids loadings	Sudden increases in organic and nutrient loads have the potential to pass through the WRRF partially treated or untreated. Excessive organic and nitrogen loading also can depress the dissolved oxygen concentration in the aeration basin if blower capacity is inadequate for the load.
Damage to downstream equipment	Septage often contains large debris such as rocks, rags, bits of metal, and other items that can damage downstream equipment. Prescreening septage before introducing it to the WRRF is recommended.
Potential foaming and/or toxicity in aeration basins	Septage contains volatile fatty acids and hydrogen sulfide and may contain heavy metals and organics that may harm biological treatment processes and/or affect biosolids disposal options. Volatile fatty acids and hydrogen sulfide can encourage the growth of certain types of filamentous bacteria in the activated sludge process. Concentrations of dissolved hydrogen sulfide as low as 1 mg/L have been shown to inhibit nitrification.
Additional operation and maintenance (O&M)	Accepting septage means increases in several areas of O&M such as ■ Staff time to manage receiving site access, sample collection, manifesting, recordkeeping, and disposal of grit, screenings, and biosolids; ■ Screenings and grit disposal; ■ Odors in headworks; ■ Scum in clarifiers; and ■ Solids handling and disposal.

Regulatory Considerations

To complete the handling of screenings and grit, the material must be transported for disposal in accordance with local, state, and federal regulations. Landfilling is by far the most common method of disposal; however, increasingly strict European regulations for landfilling are designed to require incineration and other methods of disposal. Screenings and grit must, at a minimum, pass the paint-filter liquids test before they can be accepted for disposal at a landfill. This test measures the amount of free water that leaches from the screenings by measuring the quantity of liquid from a representative screening sample that passes through a filter of Mesh No. 60. After the 5-minute test period, if any liquid passes through the filter, the material is deemed to contain free liquid. Some states may have additional testing and performance requirements. Operators should check with the landfill that will be receiving the screenings and grit to be sure all requirements are met.

Safety Considerations

It is important that O&M personnel follow safety guidelines specified by headworks equipment manufacturers and commonly practiced by the utility. This includes, but is not limited to, procedures describing appropriate equipment lockout, confined space entry, localized power interruption, and standby rescue personnel during maintenance activities. Operations personnel must familiarize themselves with and respect safe operations guidelines recommended by the equipment manufacturers and maintain a clean, well-ventilated, and well-lit environment to prevent accidents.

There is a tendency to view screens and rakes as nonhazardous because the equipment moves slowly. Although slow, the equipment is powerful. Never reach into a screen or attempt to dislodge equipment by hand. If debris becomes lodged in a screen, the screen should be taken out of service and powered off before any attempts are made to correct the situation. Use a rake or other long-handled tool to search for the item blocking the screen. Similarly, screened material should not be handled without thick, puncture-resistant gloves because syringes and other sharp objects may be present.

Because screening and grit equipment is located at the facility influent, the accumulation of potentially explosive or toxic gases is always a concern. The National Fire Protection Association 820 Standard, *Fire Protection in Wastewater Treatment and Collection Facilities* (2016), sets standards for fire protection in WRRFs and classifies spaces based on their risk of fire or explosion. Most headworks buildings will be classified as Class 1, Division 1. These are locations where ignitable concentrations of flammable gases might exist under normal operating conditions or when maintenance is performed. In Class 1, Division 1 areas, failure of electrical equipment could potentially be an ignition source. For this reason, any equipment installed in a WRRF headworks should be rated as explosion proof. Operating personnel must exercise extreme caution in these confined areas and explicitly follow all safety procedures for confined-space entry. Good safety practice demands continuous monitoring of hydrogen sulfide and methane levels in these confined areas and smoking must be prohibited.

Operators must follow good housekeeping procedures in the preliminary treatment area. Spills of screenings, grit, grease, and other such items should be regularly cleaned up to prevent slip and fall injuries. The operator should ensure that all safety guards are in place and that all safety devices are in good working order. The rotating equipment used in preliminary treatment is not commonly supplied with safety devices such as light curtains or power interruption trip cords (except some belt conveyors), so personnel must be trained to shut off and lockout equipment before attempting to free troublesome debris.

Proper barriers must be maintained around all equipment and tanks. For example, aerated grit tanks are turbulent and less buoyant than non-aerated water, which makes it more difficult for a worker that has fallen into the tank to extricate themselves. Screenings and grit contain pathogenic organisms and sharp objects that can result in illness or injuries. Facility personnel should wear gloves when working near these materials. Practice good personal hygiene by washing hands often and avoiding touching your face, eyes, and hair.

TEST YOUR KNOWLEDGE

1. The water level in flow equalization basins should fluctuate over a 24-hour period.
   - ☐ True
   - ☐ False

2. Because septic waste is already partially treated, it tends to be lower strength than typical domestic wastewater.
   - ☐ True
   - ☐ False

3. Septage can contain rocks and other debris, which can damage downstream equipment if it is not screened out.
   - ☐ True
   - ☐ False

4. Grease trap waste can be accepted and treated by most WRRFs under the 503 Regulations as long as it comes from restaurants and not categorical industrial users.
   - ☐ True
   - ☐ False

5. A septage hauling and receipt manifest should include all of the following information EXCEPT
   a. Source, type, and amount of septage
   b. Signed certification by the generator
   c. Hauler's commercial driver's license number
   d. Septage receipt date

6. Septage wastes received by WRRFs should be monitored and tracked to
   a. Proportion cost of service to customers
   b. Ensure treatment capacity is not exceeded
   c. Calculate percentage of industrial septage received
   d. Meet paint filter test requirements

7. This test measures the amount of free water in screenings and grit:
   a. Total suspended solids
   b. Mesh size
   c. Biochemical oxygen demand
   d. Paint filter

CHAPTER SUMMARY

	Screening Processes	Grit Removal Processes	
PURPOSE	■ Physically blocks entry of large debris (rags, sticks, bricks, wood, plastics) into the WRRF ■ Protects downstream equipment ■ Prevents blockages	■ Decreases velocity to about 0.3 m/s (1 ft/sec) ■ Larger, denser particles settle while lighter, less dense organic particles pass on to the next treatment process ■ Protects downstream equipment ■ Prevents blockages	PURPOSE
QUANTITY	■ Dependent on service area, type of collection system, and type of screen used ■ Varies between less than 0.75 to 150 L/1 000 m^3 (0.1 to 20 cu ft/mil. gal)	■ Dependent on service area and type of collection system ■ Varies between 3.7 to 148 L/1000 m^3 (0.5 to 20 cu ft/mil. gal) and averaged 37 L/1000 m^3 (5.0 cu ft/mil. gal)	QUANTITY
GENERAL	Three main types of screens: ■ Trash racks ■ Coarse screens ■ Fine screens	■ Three main grit removal systems: velocity-controlled grit systems, aerated grit systems, and vortex grit systems ■ All rely on differences in settling velocity to separate grit, organic material, and water ■ Settling velocity depends on the particle density, size, and shape	GENERAL
TRASH RACKS	■ Openings between 38 and 152 mm (1.5 and 6 in.) ■ Remove largest debris ■ May be used in bypass channels for short-term use	■ Earliest method of grit removal ■ Settle both grit and organics ■ Grit washing required ■ Square tanks decrease flow velocity ■ Grit removed manually or by screw augers, chain-and-flight systems, grit elevators, clamshell buckets, or pumps	DETRITUS TANKS
COARSE SCREENS	■ Bar screens ■ Manually or mechanically cleaned ■ Clear openings of 6 to 51 mm (0.25 to 2 in.) ■ Remove smaller material to further protect pumps and equipment ■ Stringy material can still pass through ■ Many different types: front clean screens, chain-driven screens, reciprocating screens, catenary screens, continuous self-cleaning, and other types	■ Velocity is calculated by dividing flow by cross-sectional area ■ Channels have parabolic or trapezoidal shaped cross-sectional areas to keep the flow to area ratio constant when flow is changing ■ Flow-proportional weirs may also be used to match cross-sectional area to flowrate ■ Grit removed manually or with screw conveyors, chain-and-flight collectors, and/or grit elevators	VELOCITY CONTROL TANKS
FINE SCREENS	■ Designed to remove smaller solids than coarse screens ■ Clear openings of 3 to 6 mm (0.12 to 0.24 in.) ■ Smaller openings capture more material ■ Screen face consists of wedge wire, perforated plate, or mesh ■ Stringy material captured by perforated plate and mesh ■ Many different types including continuous element screens, stair screens, inclined cylindrical screens, and other types	■ Diffused air introduced on one side of basin to create a spiral roll pattern ■ Velocity of roll pattern influences which solids settle ■ Velocity of roll pattern is independent of influent flow ■ Settled grit pushed into hopper by spiral flow pattern ■ Designed to achieve process goals at detention times of 4 to 5 minutes at maximum flowrate	AERATED GRIT BASIN
ADDITIONAL SCREENS	■ Wide variety of different screens available ■ All operate on the same basic principle—separation by size ■ See Chapter 11 of MOP 8, *Design of Municipal Wastewater Treatment Facilities* (WEF et al., 2018) for information on other types of screens	■ Circular basins ■ Flow introduced tangentially to create whirlpool effect ■ Velocities range from 0.3 to 1.1 m/s (1 to 3.5 ft/sec) ■ Denser particles settle to the bottom of the basin ■ Lighter, less dense particles overflow to the next process ■ Grit pumps pull grit from bottom hopper	VORTEX GRIT BASIN

<table>
<tr><td>PROCESS CONTROL</td><td>■ Cleaning cycles set as timed cycles or
■ Triggered by a difference in water level from one side of the screen to the other (head loss) or
■ Combination of both methods.
■ Minimum cleaning frequency is when head loss reaches 7.6 cm (3 in.)</td><td>Adjust process to maintain target velocity:
■ Velocity-controlled; adjust cross-sectional area
■ Aerated; adjust airflow
■ Vortex; adjust paddle speed
■ All types; limit the number of units in active service
Inspect grit:
■ Presence of organic material indicates velocities are too low.
■ Buildup of grit in downstream processes indicates velocities are too high.</td><td>PROCESS CONTROL</td></tr>
<tr><td>PROCESSING</td><td>■ Screenings deposited directly into dumpsters or
■ Washed and pressed to remove up to 90% of organic material and up to 50% of water content
■ Compaction at 4500 to 6000 kPa (1500 to 2000 psig) reduces volume by 60 to 85%</td><td>■ Grit deposited directly into dumpsters or trucks or
■ Concentrated with cyclones or
■ Washed and concentrated with grit classifiers;
■ Several different types of classifiers including paired cyclone with grit classifier, free vortex-based grit classifier, grit dewatering escalators, and conical grit washers</td><td>PROCESSING</td></tr>
<tr><td>MAINTENANCE</td><td colspan="2">■ Daily observation for visible and audible indications of malfunction
■ Preventive maintenance and planned replacement of worn parts
■ Regular cleaning and washdown to remove adhered material
■ Lubricate according to manufacturer's recommendations</td><td>MAINTENANCE</td></tr>
<tr><td>REGULATORY</td><td colspan="2">■ Grit and screenings must pass the paint filter test before being sent to landfill.</td><td>REGULATORY</td></tr>
</table>

CHAPTER EXERCISE

This exercise will draw on some of the knowledge acquired in this chapter. In this section, an operator must complete the following tasks:

- Determine how often a screenings dumpster must be emptied,
- Select a screen type given information about the downstream equipment, and
- Determine how many grit basins should be in service to maintain a target velocity.

WRRF Information:

Current Screen Type – 2.0-in. bar screen.

New downstream process does not tolerate stringy material such as rags and hair. The new screen should effectively remove these materials.

There are four, trapezoidal, velocity control grit channels placed to operate in parallel. Place one or more of the grit basins into service to maintain a target velocity as close as possible to 0.3 m/s (1.0 ft/sec).

Parameter	SI Units	English Units
Influent Flow	10.598 ML/d	2.8 mgd
Screenings Produced	0.075 m^3/ML	10 cu ft/mil. gal
Screenings Container Volume	1.133 m^3	40 cu ft

Grit Basin Length	4.57 m	15 ft
Grit Basin Width	0.457 m	1.5 ft
Grit Basin Height / Depth	0.457 m	1.5 ft

CHAPTER EXERCISE SOLUTIONS

1. Determine how often the screenings container must be emptied.

Step 1—Find the total volume of screenings produced each day.

SI

$$\frac{10\ 598\ \text{m}^3\ \text{influent flow}}{\text{d}}\left[\frac{1\ \text{ML}}{1000\ \text{m}^3}\right]\left[\frac{0.0748\ \text{m}^3\ \text{screenings produced}}{1\ \text{ML influent flow}}\right]=0.79\ \text{m}^3/\text{d}$$

English

$$\frac{2.8\ \text{mil. gal influent flow}}{\text{d}}\left[\frac{10\ \text{cu ft screenings produced}}{1\ \text{mil. gal influent flow}}\right]=28\ \text{cu ft}$$

Step 2—Determine how long it will take to fill the screenings container.

SI

$$1.13\ \text{m}^3\ \text{available space}\left[\frac{1\ \text{day}}{0.79\ \text{m}^3\ \text{screenings produced}}\right]=1.43\ \text{days}$$

English

$$40\ \text{cu ft available space}\left[\frac{1\ \text{day}}{28\ \text{cu ft screenings produced}}\right]=1.43\ \text{days}$$

Step 3—Convert time in days to hours.

$$1.43\ \text{days}\left[\frac{24\ \text{hours}}{1\ \text{day}}\right]=34.3\ \text{hours}$$

2. Select a new screen that will meet the needs of the downstream process. The downstream process is sensitive to rags and stringy material. Look back through the screen descriptions. One of them is better at removing stringy material than the others. Select the continuous element, perforated plate screen for best performance.

3. Determine how many grit basins should be in service to maintain a flow velocity as close as possible to 0.3 m/s (1 ft/sec). If the velocity is too low, excessive amounts of organic material will be removed with the grit. If the velocity is too high, downstream processes may not be protected.

Step 1—Find the formula for calculating velocity.

$$\text{Velocity}=\frac{\text{Flow}}{\text{Area}}$$

Step 2—Determine which units velocity, flow, and area must be in to solve the equation. The target velocity is given in meters per second or feet per second. The units selected for flow and area must work together to produce the desired units for velocity. Look at the equations in both SI and English units below. See how the units on the right side of the equation cancel out to leave units that match the left side of the equation.

SI

$$\text{Velocity, }\frac{\text{m}}{\text{s}}=\frac{\text{Flow, m}^3/\text{s}}{\text{m}^2}$$

English

$$\text{Velocity, }\frac{\text{ft}}{\text{s}}=\frac{\text{Flow, cfs}}{\text{Area, sq ft}}$$

Step 3—Assume all of the flow is going through a single grit basin. Find the area of one grit basin. The formula for the area of a rectangle is length multiplied by width. This is deceptive because the problem gives dimensions for length, width, and height or depth of channel. The area of importance for velocity through a grit basin is the cross-sectional area. The cross-sectional area, shown in green on the figure below, is the area that the water is flowing through or past. The cross-sectional area will be calculated using the dimensions given for width and height. These are the length and width for this rectangle. Note that the area used for velocity is not the entire cross-sectional area of the channel, but only the area that the water is flowing through. If the question had given both a channel depth and a water depth, then the water depth would be used to find the cross-sectional area of the water.

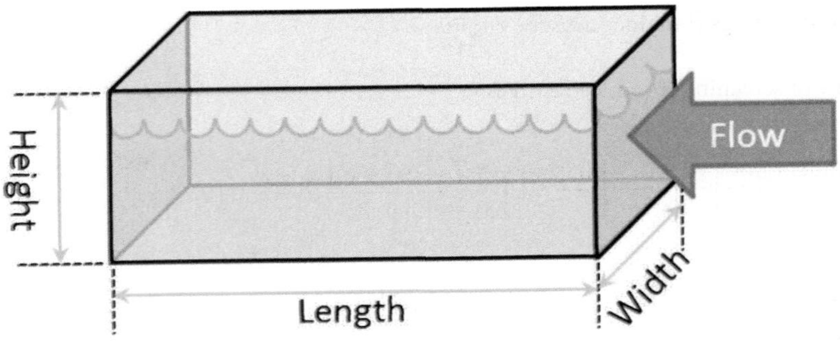

SI

$$\text{Area} = (\text{Length})\ (\text{Width})$$

$$\text{Area} = (0.457\ \text{m})\ (0.457\ \text{m})$$

$$\text{Area} = 0.209\ \text{m}^2$$

English

$$\text{Area} = (\text{Length})\ (\text{Width})$$

$$\text{Area} = (1.5\ \text{ft})\ (1.5\ \text{ft})$$

$$\text{Area} = 2.25\ \text{sq ft}$$

Step 4—Convert the influent flowrate to compatible units.

SI

$$\frac{10.598\ \text{ML}}{\text{day}}\left[\frac{1\ \text{day}}{24\ \text{hours}}\right]\left[\frac{1\ \text{hour}}{60\ \text{minutes}}\right]\left[\frac{1\ \text{minute}}{60\ \text{seconds}}\right]\left[\frac{1\,000\,000\ \text{L}}{1\ \text{ML}}\right]\left[\frac{1\ \text{m}^3}{1000\ \text{L}}\right] = 0.123\ \text{m}^3/\text{s}$$

English

$$\frac{2.8\ \text{mil. gal}}{\text{d}}\left[\frac{1\ \text{day}}{24\ \text{hours}}\right]\left[\frac{1\ \text{hour}}{60\ \text{minutes}}\right]\left[\frac{1\ \text{minute}}{60\ \text{seconds}}\right]\left[\frac{1{,}000{,}000\ \text{gal}}{1\ \text{mil. gal}}\right]\left[\frac{1\ \text{cu ft}}{7.48\ \text{gal}}\right] = 4.33\ \text{cfs}$$

Step 5—Calculate the velocity with a single grit basin in service.

SI

$$\text{Velocity}, \frac{\text{m}}{\text{s}} = \frac{\text{Flow, m}^3/\text{s}}{\text{Area, m}^2}$$

$$\text{Velocity}, \frac{\text{m}}{\text{s}} = \frac{0.123\frac{\text{m}^3}{\text{s}}}{0.209\ \text{m}^2}$$

$$\text{Velocity}, \frac{\text{m}}{\text{s}} = 0.59$$

English

$$\text{Velocity}, \frac{\text{ft}}{\text{s}} = \frac{\text{Flow, cfs}}{\text{Area, sq ft}}$$

$$\text{Velocity}, \frac{\text{ft}}{\text{s}} = \frac{4.33 \text{ cfs}}{2.25 \text{ sq ft}}$$

$$\text{Velocity}, \frac{\text{ft}}{\text{s}} = 1.93$$

Step 6—Compare the velocity with one basin in service to the desired velocity. The desired velocity is 0.3 m/s (1 ft/sec) and the velocity with one basin in service would be 0.59 m/s (1.93 cfs). This is approximately double the desired flowrate. The operator should place two grit basins in service.

References

Boldt, J. (2005) Eliminating Grit Deposition Problems Through Objective Grit System Design: A Case Study at the Fox Lake NRWRP, Illinois. Paper presented at the Illinois Water Environment Association Conference, March 9–13.

Bristow, L. (2008) Racks, Screens, Comminutors, and Grit Removal. In *C. S. University, Operation of Wastewater Treatment Plants: A Field Study Program*, Vol 1, 7th ed.; California State University at Sacramento Office of Water Programs: Sacramento, California; pp 57–100.

Great Lakes-Upper Mississippi River Board (2004) *Recommended Standards for Wastewater Facilities;* Health Research Inc., Health Education Services Division: Albany, New York.

InfoBarScreens (2009) About Reciprocating Rake Bar Screens. www.infobarscreens.com (accessed Dec 2016).

Kern, J. W.; Pisano, W. (1998) Vortex Grit Chambers: Process, Performance, and Modifications. *Proceedings of the 71st Annual Water Environment Federation Technical Exposition and Conference*; Orlando, Florida, Oct 3–7; Water Environment Federation: Alexandria, Virginia.

Metcalf and Eddy, Inc./AECOM (2013) *Wastewater Engineering: Treatment and Resource Recovery,* 5th ed. McGraw Hill: New York.

National Fire Protection Association (2016) *Fire Protection in Wastewater Treatment and Collection Facilities;* NFPA 820; National Fire Protection Association: Quincy, Massachusetts.

U.S. Environmental Protection Agency (1978) *Field Manual for Performance Evaluation and Troubleshooting at Municipal Wastewater Treatment Facilities;* U.S. Environmental Protection Agency: Washington, D.C.

U.S. Environmental Protection Agency (2003) Wastewater Technology Fact Sheet: Screening and Grit Removal; U.S. Environmental Protection Agency Office of Water, Municipal Technology Branch: Washington, D.C.

Water Environment Federation (1997) *Septage Handling*; Manual of Practice No. 24; Water Environment Federation: Alexandria, Virginia.

Water Environment Federation (2016) *Operation of Water Resource Recovery Facilities*, 7th ed.; WEF Manual of Practice No. 11; Water Environment Federation: Alexandria, Virginia.

Water Environment Federation; American Society of Civil Engineers; Environmental and Water Resources Institute (2018) *Design of Water Resource Recovery Facilities,* 6th ed.; WEF Manual of Practice No. 8/ASCE Manuals and Reports on Engineering Practice No. 76; Water Environment Federation: Alexandria, Virginia.

Suggested Readings

Arasmith, S. (2011) Pumps and Pumping www.acrp.com (accessed Aug 2017).

Gravette, B.; Strehler, A.; Finger, D.; Palepu, S. (2000) Troubleshooting a Grit Removal System. *Proceedings of the 73rd Annual Water Environment Federation Technical Exposition and Conference* [CD-ROM]; Anaheim, California, Oct 14–18; Water Environment Federation: Alexandria, Virginia.

U.S. Environmental Protection Agency (1987) *Preliminary Treatment Facilities, Design and Operational Considerations;* EPA 430/9-87-007; U.S. Environmental Protection Agency: Washington, D.C.

U.S. Environmental Protection Agency (1994) *Guide to Septage Treatment and Disposal*; EPA 625/R-9-402; U.S. Environmental Protection Agency: Washington, D.C.

Water Environment Federation; American Society of Civil Engineers; Environmental and Water Resources Institute (2018) Chapter 11—Preliminary Treatment. In *Design of Water Resource Recovery Facilities*, 6th ed.; WEF Manual of Practice No. 8/ASCE Manuals and Reports on Engineering Practice No. 76; Water Environment Federation: Alexandria, Virginia.

CHAPTER 4

Primary Treatment of Wastewater

Introduction

Primary treatment involves the separation and removal of readily settleable suspended solids (sludge) and readily floatable material (scum) by providing a *quiescent* tank where the velocity of the wastewater is reduced to a fraction of a meter per second (foot per second). Settleable solids sink to the bottom of the tank, where they are collected and removed. Settleable solids consist of dense, inorganic particles, like sand and grit, as well as dense organic solids. Removing grit and other abrasive material helps protect downstream equipment from wear and damage. Floatable materials rise to the top of the tank where they are skimmed off and removed. Floatable material is generally organic and includes grease, oils, plastics, and soaps. Removal of floatables from wastewater minimizes operational problems from the buildup of scum and improves the overall appearance and smell of the wastewater. Removing both settleable and floatable materials reduces the mass of total suspended solids (TSS), chemical oxygen demand (COD), and biochemical oxygen demand (BOD) that passes on to secondary treatment. Having fewer solids and organics pass through to the secondary treatment processes decreases the overall size and amount of energy needed in that process.

Quiescent means calm or undisturbed. Gravity separation processes work most efficiently when currents are minimized.

Often, the terms *clarification* and *thickening*, or *sedimentation*, are used to describe gravity separation processes, depending on whether the treatment focus is on the clarified liquid (effluent) or the thickened solids (sludge). Although the primary purpose of a primary clarifier is to remove settleable and floatable solids, many primary clarifiers are deliberately operated to produce thickened sludge. Older publications sometimes refer to primary clarifiers as *sedimentation basins*.

LEARNING OBJECTIVES

Upon completing this chapter, you will be able to

- Describe the purpose of primary clarification;
- Categorize influent parameters according to whether they can be removed by primary clarification;
- Label all components of circular and rectangular primary clarifiers and describe the function of each;
- Inspect and maintain clarifier equipment;
- Place a primary clarifier into service or remove one from service;
- Calculate hydraulic detention time and surface overflow rate;
- Determine the optimum number of clarifiers to place into service to maintain a desired surface overflow rate and detention time;
- Estimate the quantity of sludge produced given primary influent and effluent TSS concentrations;
- Calculate sludge pumping time required for a given sludge thickness and volume;
- Collect process control samples, conduct settleable solids analysis, and evaluate results;
- Anticipate seasonal changes and make appropriate process control changes; and
- Troubleshoot common mechanical and process control problems.

Purpose and Function

Primary treatment follows preliminary treatment. Primary treatment is designed to remove materials heavier than water by gravity and those lighter than water by flotation. Some facilities use a primary treatment process such as a primary clarifier to slow the water further and remove dense organic material and other particles by gravity. Grease and floatable material collect on the water surface and are skimmed off. A simple model of a primary clarifier can be constructed using some soil, vegetable oil, and a large jar filled with water. When the soil and vegetable oil are added to the jar full of water, most of the soil will sink to the bottom and most of the oil will collect on the surface. Some particles will be too light to sink and will remain hanging in suspension. If the model clarifier is stirred, the settleable solids will rapidly return to the bottom and the oil will return to the top. Primary clarifiers may be rectangular like the one shown in Figure 4.1 or circular like the one shown in Figure 4.2. Operation of rectangular and circular clarifiers with identical parameters in Winnipeg, Canada, resulted in no significant differences in performance (Ross and Crawford, 1985).

The purpose of the primary clarifier is to reduce the amount of COD or biochemical oxygen demand (5-day test) (BOD_5) and TSS going to the secondary treatment process. Reducing the load to the secondary treatment process reduces its overall size and the amount of energy required to provide oxygen and to operate downstream equipment. One kilogram (pound) of BOD_5 entering an *aerobic* secondary treatment process consumes one kilogram (pound) of oxygen. Generally, only larger facilities (i.e., greater than 1900-m^3/d [5-mgd] capacity) will have primary clarifiers because these processes are typically paired with *anaerobic* digestion for sludge stabilization. An anaerobic digester breaks down organic material in the absence of oxygen. An aerobic digester requires oxygen to break down organic material. If primary sludge is transferred to an aerobic digester, energy use increases rather than decreases because the sludge is still being broken down (digested) with oxygen and additional energy will be spent on clarifier operation and pumping.

Aerobic means molecular oxygen is present. Nitrate and nitrite may also be present.

Anaerobic means neither molecular oxygen nor nitrite or nitrate are present.

Figure 4.1 Rectangular Primary Clarifier (Reprinted with permission by Indigo Water Group)

Figure 4.2 Circular Primary Clarifier (Reprinted with permission by Indigo Water Group)

TEST YOUR KNOWLEDGE

1. **Primary clarifiers are capable of removing 100% of the influent TSS and BOD_5.**
   - ☐ True
   - ☐ False

2. **Which of the following components is most likely to be in the sludge at the bottom of a primary clarifier?**
   a. Ammonia
   b. Soluble BOD
   c. Grease
   d. Settleable solids

3. **The main purpose of primary treatment is to**
   a. Remove grit and larger debris to protect downstream equipment.
   b. Reduce the loading to the secondary treatment process.
   c. Convert particulate BOD into soluble BOD for easier treatment.
   d. Generate electricity by reducing the organic load to the digester

Theory of Operation

Clarifiers rely on differences in density to separate particles from the surrounding wastewater. Particles that are more *dense* than water are pulled by gravity to the bottom of the clarifier. Particles that are less dense than water float to the top of the clarifier. Gravity is not the only force influencing the movement of particles. Drag is created by the particles as they move through the water. They rub against the water, which creates friction between the water molecules and the particles. Drag slows the particles down. Think of skydivers jumping out of an airplane (Figure 4.3). Skydivers are much more dense than the surrounding air, but how fast they fall depends on how much of their body contacts the air (surface area). A skydiver that spreads himself or herself out flat will fall slower than a skydiver that pulls his or her body into a tight ball. More surface area means more friction and more drag. This is why parachutes are so effective—they have lots of surface area. Mass and density also influence how fast particles will settle. Mass is how much something weighs. Density is how much mass there is per volume. In the skydiver example, the mass of the skydiver does not change, whether they are curled into a ball or are spread out. By spreading themselves out, skydivers spread their mass over a larger area, decreasing their density and increasing the amount of drag. Now, imagine two skydivers with the same density (mass per volume), but one of the skydivers weighs 80 kg (176 lb) and the other weighs 120 kg (264 lb). Which one will fall faster? The heavier skydiver will fall faster. How fast particles settle in a primary clarifier depends on their shape (surface area and drag), density, and total mass.

Density is the amount of mass per volume. Water has a density of 998.2 kg/m^3 (8.34 lb/gal or 62.3 lb/cu ft) at 20 °C (68 °F). As water cools, it becomes more dense. Water has a maximum density of 1000 kg/m^3 (62.4 lb/cu ft) at 4 °C.

Specific gravity is the ratio of the density of any substance divided by the density of water. Water has a specific gravity of 1. If the specific gravity is greater than 1, then the substance is denser than water.

Flocculate means to stick together. Particles that flocculate will tend to bump into one another and stick together, which forms larger particles. Flocculation can occur naturally or chemicals can be added to the process to help particles flocculate.

Suspended particles may be classified as granular or flocculent. Granular particles (sand and silt) settle at a constant velocity, with no change in size, shape, or weight. Ideally, most granular particles will be removed upstream in the grit chambers. Flocculent particles (organic matter and flocs formed by coagulants or biological growths) tend to *flocculate* during settling, with changes in size, shape, and relative density. The clusters formed by flocculation ordinarily settle more rapidly than individual particles. Settleable solids, including portions of the granular and flocculent material, settle under quiescent conditions within a reasonable time. Finely divided nonsettleable solids and colloidal materials are too fine to settle within the time the water remains in the clarifier.

Currents in the clarifier push particles in different directions. The most significant current is the velocity of the water as it moves through the tank and over the effluent weirs. This type of velocity has a special name: surface overflow rate (SOR). The SOR pushes particles up and out of the clarifier along with the wastewater. For the clarifier to effectively settle solids, the SOR (upward velocity) must be smaller than the settling velocity, which is influenced by the downward force caused by gravity. For particles that are less dense than water, the SOR pushes them to the surface at an accelerated rate.

As addressed in Chapter 3, grit particles settle to the bottom of the grit basin when the velocity of the wastewater is decreased to about 0.3 m/s (1 ft/s). In the primary clarifier, the velocity is slowed down much more to allow additional particles to settle. The water is moving so slowly that is it no longer useful to express velocity in meters per second (feet per second). Instead, the SOR, which can be thought of as the flow velocity through the tank, is expressed as $m^3/m^2 \cdot d$ (gpd/sq ft). Currents can also be created by the movement of the sludge collection mechanism or wind on the clarifier surface. Temperature and density differences in clarifiers can also cause currents to form. These currents, along with SOR, can prevent particles from settling to the bottom before the water moves out of the tank.

Calculations for SOR and detention time are given in the process control section of this chapter.

Primary clarifier performance depends, in part, on tank surface area and tank volume. For any influent flow, tank surface area controls the SOR and tank volume controls the hydraulic retention time (HRT). Hydraulic retention time is also referred to as *hydraulic detention time* (HDT) or, simply, detention time. The HDT can be (1) the time required to fill a vessel or (2) the time required to empty a vessel, or (3) the average amount of time that water remains in a vessel. Although the rate of removal of granular particles settling at uniform velocities depends almost entirely on the tank surface area, the rate of removal of flocculent particles settling at variable velocities depends on the detention time. More detention time gives flocculant particles more time and opportunities to collide with one another and form larger particles.

Figure 4.3 Skydivers Create Drag to Slow Their Fall (Getty Images)

Sludge that settles to the bottom of the clarifier builds up in a layer called a *blanket*. The concentration of total solids in the clarifier blanket depends, in part, on how fast the sludge is removed from the clarifier. When solids are removed quickly, they do not have an opportunity to compact, so the solids concentration in the sludge will be lower. All sludges have a maximum concentration that can be achieved and longer detention times will not increase the solids concentration beyond this maximum. If a sample of wastewater is allowed to sit undisturbed, a layer of settled solids will form on the bottom of the container. When solids are allowed to remain in the clarifier for longer periods of time, the sludge blanket will have time to compress and water will be squeezed out of the spaces between particles. The result is a thicker sludge with a higher total solids concentration. Settling and compaction are essentially complete within about 2 hours. If solids are left in the clarifier too long, however, microorganisms can begin to break them down and may cause the sludge to become septic. *Septic* sludge generates *hydrogen sulfide* and is generally darker brown to black in color than nonseptic sludge. Very thick sludges can be difficult or impossible to pump depending on the type of sludge pump used.

Wastewater becomes *septic* when all of the oxygen and nitrate have been used up by microorganisms. A group of bacteria called the *facultative anaerobes* can convert sulfate to hydrogen sulfide under septic conditions. *Hydrogen sulfide* has a strong, rotten egg smell.

TEST YOUR KNOWLEDGE

1. **Clarifiers rely on density differences to separate particles from wastewater.**
   - ☐ True
   - ☐ False
2. **Two particles have the same density. The particle with more surface area will settle faster than the particle with less surface area.**
   - ☐ True
   - ☐ False
3. **Primary sludge will continue to compact and achieve higher total solids concentrations for as long as it remains in the clarifier.**
   - ☐ True
   - ☐ False
4. **Which of the following particles should reach the bottom of the clarifier first?**
   a. Grease
   b. Feather
   c. Sand
   d. Lettuce
5. **What phenomena in a clarifier push particles up and out of the clarifier?**
   a. Drag
   b. Surface tension
   c. Gravity
   d. Surface overflow rate
6. **Surface overflow rate is comparable to**
   a. Detention time
   b. Velocity
   c. Weir loading rate
   d. Density
7. **The sludge layer that forms at the bottom of the clarifier is called the**
   a. Schmutzdecke
   b. Float
   c. Blanket
   d. Emulsion
8. **A potential negative consequence of leaving primary sludge in the clarifier for an extended period of time might be**
   a. Hydrogen sulfide gas
   b. Higher solids concentrations
   c. Lower solids concentrations
   d. Breakdown of grit particles

Design Parameters

Primary clarifiers are generally designed to meet the following criteria (WEF, 2005):

- Hydraulic detention time of 1.5 to 2.5 hours,
- Average SORs of 24.4 to 48.9 $m^3/m^2 \cdot d$ (600 to 1200 gpd/sq ft),
- Peak hour SORs of 102 to 122 $m^3/m^2 \cdot d$ (2500 to 3000 gpd/sq ft),
- Sludge blanket depth of 0.6 to 0.9 m (2 to 3 ft), and
- Sludge concentration of 3 to 6% total solids.

Expected Performance

Primary clarifier performance is limited by SOR, HDT, and the characteristics of the incoming wastewater (i.e., the percentages of particulate and soluble material). For domestic wastewater that is coming primarily from homes and light commercial activities like schools and stores, a primary clarifier can be expected to remove 50 to 70% of TSS, 25 to 40% of total BOD_5, and 5 to 10% of total phosphorus (WEF et al., 2018). Particulate nitrogen will also be removed to some extent. Primary clarifiers do not, however, remove *colloidal solids*, dissolved solids, soluble BOD_5, soluble phosphorus, or ammonia. Chemically enhanced primary treatment is a method to increase the removal efficiencies of suspended solids and BOD_5.

Colloidal solids are particles that are between 1 and 1000 µm in diameter. They are too small and light to settle to the bottom of the clarifier in the time allowed. Colloidal particles are measured in the total dissolved solids test even though they are not truly dissolved.

Understanding primary clarifier performance requires an understanding of the underlying relationships between different influent wastewater constituents. Domestic wastewater is wastewater that comes primarily from homes, schools, offices, stores, and other nonmanufacturing businesses. Figure 4.4 shows the graphical relationship between BOD_5 and TSS. For domestic wastewater, 20 to 40% of the influent BOD_5 is soluble. The primary clarifier cannot remove it. Soluble BOD will pass on to secondary treatment. The remaining BOD_5 is particulate, but only some of those particles will be large enough and dense enough to settle in the clarifier. Consider a typical domestic wastewater that contains 200 mg/L of BOD_5 that is 30% soluble. This means 60 mg/L is soluble and 140 mg/L of the BOD_5 is particulate. If half of the particulate BOD_5 is settleable (i.e., large enough and dense enough), then the primary clarifier will be able to remove a maximum of 70 mg/L of BOD_5. How does this look as percent removal?

$$\text{Percent Removal} = \left(\frac{\text{In} - \text{Out}}{\text{In}}\right) \times 100 \qquad (4.1)$$

$$\text{Percent Removal} = \left(\frac{200\text{ mg/L} - \left(60\frac{\text{mg}}{\text{L}}\text{ soluble} + 70\frac{\text{mg}}{\text{L}}\text{ particulate}\right)}{200\text{ mg/L}}\right) \times 100$$

$$\text{Percent Removal} = 35\%$$

The influent TSS can either be inorganic (sand and silt) or organic. Looking at Figure 4.4, one can see the relationship between volatile suspended solids (VSS) (organic) and particulate BOD_5. For domestic wastewater, these components are often nearly equal to one another. Solid material that is organic can be

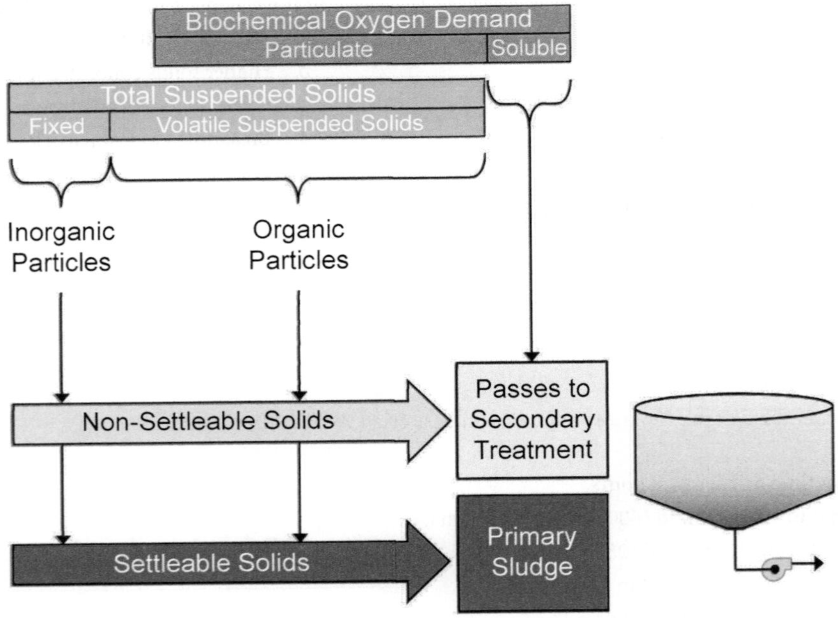

Figure 4.4 Primary Clarifier Removal Efficiency (Reprinted with permission by Indigo Water Group)

measured both as VSS and as BOD_5. In other words, organic solids contribute to BOD_5. When organic solids end up in the sludge, BOD is also removed from the process. Not all of the VSS will be large enough and dense enough to settle in the clarifier. Some VSS will pass on to secondary treatment.

Many water resource recovery facilities (WRRFs) use primary clarifiers for thickening primary sludge as well as for separating solids from wastewater. In these cases, the primary clarifiers may be larger and/or deeper to accommodate sedimentation and thickening. Some facilities use their primary clarifiers as sludge fermenters to break down complex organic material into simpler organic compounds called *volatile fatty acids* (VFAs). Volatile fatty acids are necessary for biological phosphorus removal.

TEST YOUR KNOWLEDGE

1. **A well-operated primary clarifier will always remove 40% of total BOD_5.**
   - ☐ True
   - ☐ False
2. **Primary clarifier performance is limited by SOR, HDT, and the characteristics of the incoming wastewater.**
   - ☐ True
   - ☐ False
3. **Biodegradable organic solids are measured by both the BOD and VSS laboratory tests.**
   - ☐ True
   - ☐ False
4. **The influent to a primary clarifier contains 190 mg/L of TSS. The effluent contains 125 mg/L of TSS. Find the percent removal.**
   a. 26.1%
   b. 34.2%
   c. 65.7%
   d. 87.4%
5. **The influent to a primary clarifier contains 250 mg/L of BOD. Fifty milligrams per liter is soluble and 200 mg/L is particulate. If the clarifier removes 50% of the particulate BOD, what will the total BOD concentration be in the primary effluent?**
   a. 50 mg/L
   b. 100 mg/L
   c. 125 mg/L
   d. 150 mg/L
6. **A typical HDT for a primary clarifier would be**
   a. 30 minutes
   b. 1.5 hours
   c. 3 hours
   d. 6 hours

Equipment

Primary clarifiers can be rectangular, circular, or square. In the circular and square types, the wastewater typically enters at the center and flows toward the outside edge, with the settled solids scraped or otherwise transported to the center. In some circular tank designs (called *peripheral feed clarifiers*), wastewater enters at the outer edge and flows inward. In the rectangular type, wastewater flows from one end of the tank to the other, and scrapers move the settled sludge to the inlet end. Imhoff tanks perform the dual function of settling and anaerobic digestion; however, these tanks represent old technology and are no longer allowed in most states.

The sludge collection process entails moving settled solids to a point in the clarifier where it is drawn off. Primary clarifiers, except for those in a few small facilities relying on manual removal, use mechanical collection. One of the following types of equipment may be used:

- Rotating scrapers (circular and square basins),
- Flight and chain (rectangular basins), or
- Traveling bridge with screens (rectangular basins).

Typically, pumping, but sometimes gravity flow, removes sludge from primary clarifiers. Positive-displacement-type pumps such as piston, progressive cavity, or diaphragm pumps are typically used because of the high solids concentrations. However, some facilities also use torque-flow, centrifugal-type pumps, or hose (peristaltic) pumps. For additional information on pumps, see Chapter 8, "Pumping of Wastewater and Sludge", in Water Environment Federation's Manual of Practice No. 11, *Operation of Water Resource Recovery Facilities*. For gravity removal, sludge flows from the collection point to a separate vault, with a lower liquid surface level than the primary tank.

IMHOFF TANKS

Imhoff tanks, which are seldom used, contain two compartments where settleable solids fall through a slot in the bottom of the upper chamber and travel to the lower compartment. The solids are digested in the lower compartment of the same unit. As stated previously, some state regulatory agencies do not allow the design of Imhoff tanks for new or expanded facilities.

CIRCULAR CLARIFIER COMPONENTS

Circular clarifiers are used extensively in primary, secondary, and tertiary treatment. Circular clarifiers have earned a reputation for being the most trouble-free with respect to the sludge-collection mechanisms. Square, hexagonal, and octagonal tanks are somewhat similar to circular tanks in how water moves through them; however, settled sludge can build up in the corners and degrade performance. If *fillets* are present in the corners and simple collection mechanisms are used, these alternate shapes are nearly equal to circular tanks. For purposes of this chapter, tanks of these shapes are essentially considered equivalent to circular tanks.

Fillets are extra fill material used to reduce sharp corners. Filling the space with grout or cement smooths over the corner and prevents solids from collecting in the space.

A typical primary clarifier is shown in Figure 4.5, with all significant components labeled. Flow enters the clarifier through the bottom, coming up through the center pipe. An energy-dissipating inlet and feedwell slow the influent current and direct flow toward the bottom of the clarifier, as shown in Figure 4.6. These barriers prevent the influent wastewater from traveling straight across the clarifier surface to the outlet. Short-circuiting occurs when flow passes rapidly through the clarifier by bypassing a majority of the tank volume. The feedwell prevents short-circuiting by forcing the influent current down. As the water moves through the clarifier, settleable solids fall to the bottom where the rake mechanism pushes the sludge toward the sludge discharge pit, also called the *sludge hopper*. The rake mechanism and skimmer arm are rotated slowly through the clarifier by a drive unit located on the center platform. Treated water flows down through the feedwell, across the tank, under the scum baffle, and over the weirs into the effluent launder. The scum baffle prevents floating material from passing over the weir and into the effluent launder. A skimmer arm rotates with the rake mechanism and skims floatable material from the surface and pushes it into the scum trough.

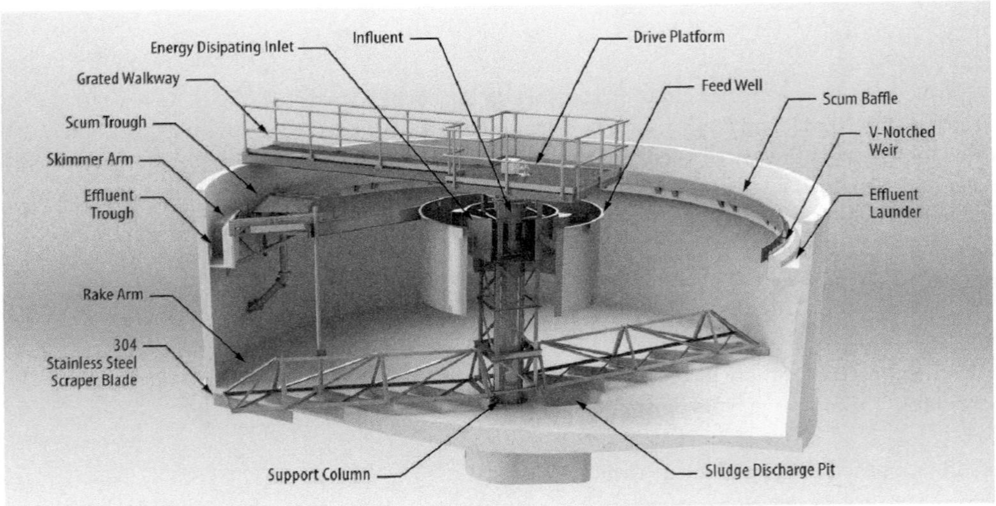

Figure 4.5 Circular Primary Clarifier Schematic (Reprinted with permission by Monroe Environmental [Monroe, MI] www.mon-env.com)

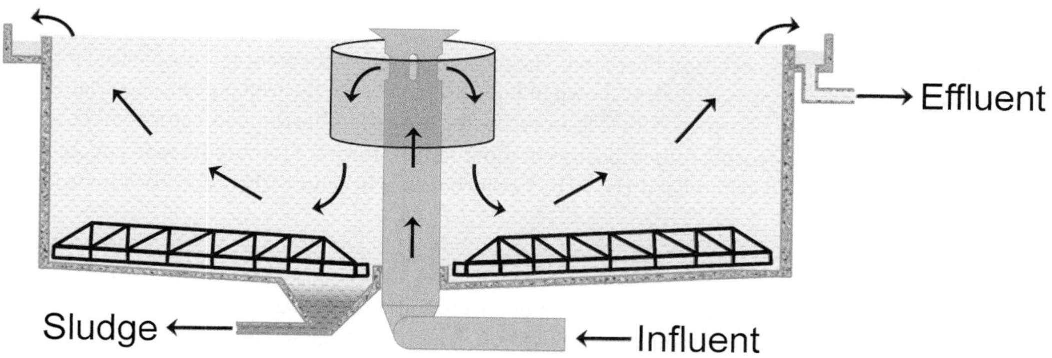

Figure 4.6 Flow Pattern through a Circular, Primary Clarifier (U.S. EPA, 1975)

DIAMETER AND DEPTH

Circular clarifiers used in primary and secondary treatment have tank diameters that range from 3 m (10 ft) to greater than 100 m (300 ft). However, for most facilities, diameters are kept to less than 50 m (150 ft) to avoid the adverse effects of wind on the surface. Depths of approximately 3 to 4 m (10 to 12 ft) are most common for primary treatment. Most primary clarifiers have gradually sloped bottoms so the clarifier is deeper in the center than it is at the outer wall. The steepness of the slope is not standardized for primary clarifiers and can vary considerably from one facility to another. Tank depth is measured at the outer wall of the tank. This is called the *sidewater depth.*

FREEBOARD

It has been common practice for U.S. engineers to use 0.5 to 0.7 m (1.5 to 2.0 ft) of freeboard for most clarifier designs, including rectangular clarifiers. Freeboard is the distance between the top of the clarifier wall and the water surface. The origin of this range is uncertain, but has been common practice for many years (WEF, 2005). This much freeboard allows for downstream hydraulic problems (such as a pump failure or partial pipe inlet blockage) that may cause flow to back up into the clarifier.

TEST YOUR KNOWLEDGE

1. **Fillets are used to fill in the corners of square and octagonal clarifiers.**
   - ☐ True
   - ☐ False

2. **Imhoff tanks settle primary sludge in the upper compartment and digest it in the lower compartment.**
   - ☐ True
   - ☐ False

3. **Generally, 1.5 m (5 ft) of freeboard is provided. This allows excess water to be stored when there is a blockage or downstream equipment malfunction.**
   - ☐ True
   - ☐ False

4. **This device helps prevent floatable material from escaping over the weir.**
   a. Energy dissipater
   b. Rake
   c. Scum baffle
   d. Wear strip

5. **In a circular primary clarifier, the sludge hopper is typically located**
   a. Along the outer edge of the clarifier
   b. Near the center of the clarifier
   c. Adjacent to the scum baffle
   d. Below the scum box

6. **Most primary clarifiers are smaller than 100 m (300 ft) in diameter for this reason:**
   a. Wind can create currents in larger clarifiers
   b. Larger sludge collection mechanisms are not available
   c. Difficulty in reaching equipment for cleaning
   d. Skimmer arm weight cannot be supported

Torque is force generated by twisting. The drive in the center of the clarifier generates torque, which then rotates the entire clarifier mechanism.

INLET PIPE AND PORTS

Circular primary clarifiers are fed from the center. The most common configuration is for the feed pipe to enter through the bottom of the clarifier, as shown in Figure 4.6; however, the feed pipe in some clarifiers enters through the side of the clarifier. With side entry, the feed pipe runs horizontally across the clarifier tank to the feedwell. Center feeding causes the flow to move radially outward toward the weir, and, in many tanks, there is a doughnut-shaped roll pattern formed, which results in some surface flow back toward the center.

PIPE SIZE AND VELOCITIES

Most circular clarifiers in the United States are equipped with the drive mechanism located on the top of the center column. The center-feed pipe must then serve a dual role of bringing influent to the tank and transmitting rotational *torque* from the drive into the bottom foundation. Generally, the influent velocity at peak-hour flow should not exceed 1.4 m/s (4.6 ft/sec). More clarifiers may be placed into service to keep the influent velocity within this range.

Most center feed columns have four rectangular opening ports. They are often submerged, although some clarifiers may show the top several centimeters of the ports exposed. In many center feed inlet designs, the inlet ports discharge freely into the inlet feedwell. In some, however, deflectors are constructed just outside of each port to break up the jetting velocities to the inlet baffled area. As shown in Figure 4.7, center feed tanks can be fed with horizontal pipes or vertical pipes that discharge freely at their end. Sometimes, these pipes are equipped with a bell-mouth outlet that reduces the release velocity into the tank center.

INLET AND FEEDWELL

A standard center feed inlet configuration for a circular tank is shown in Figure 4.8. It remains one of the most common ways of feeding center-feed primary clarifiers. The diameter of the feedwell is typically equal to 15 to 25% of the tank diameter; however, feedwell diameters do not typically exceed 10 to 15 m (32 to 50 ft), regardless of the tank size. Center wells may be 1- to 2.5-m (3- to 8-ft) deep (Metcalf and Eddy, Inc./AECOM, 2014). Likewise, downward flow velocities leaving the feedwell are often limited to 0.7 m/min (2.3 ft/s). Energy dissipaters reduce velocity through ports to 0.75 m/s (maximum flow) and 0.30 to 0.45 m/s (average flow) (Metcalf and Eddy, Inc./AECOM, 2014).

The top elevation of the feedwell generally extends above the water surface at peak hour flow. A few ports are typically cut into the top portion of the baffle to allow scum and other floatable material to move from the feedwell into the tank proper. There are typically four of these openings spaced equally around the baffle. Floatable material may also pass over the top of the baffle during peak hour flow.

A typical feedwell extends downward from as little as 30% to as much as 75% of the tank depth. It is also common that the center feedwell bottom edge be located approximately 0.3 m (1 ft) below the bottom of the center-feed pipe ports. It must be low enough so that the flow jetting out of the ports does not get below the baffle and short-circuit across the top of the clarifier. In conventional tanks, the feedwell rotates with the sludge scraper mechanism. In other tanks, it remains stationary. The feedwell can be supported from the bridge or from the sludge collection mechanism.

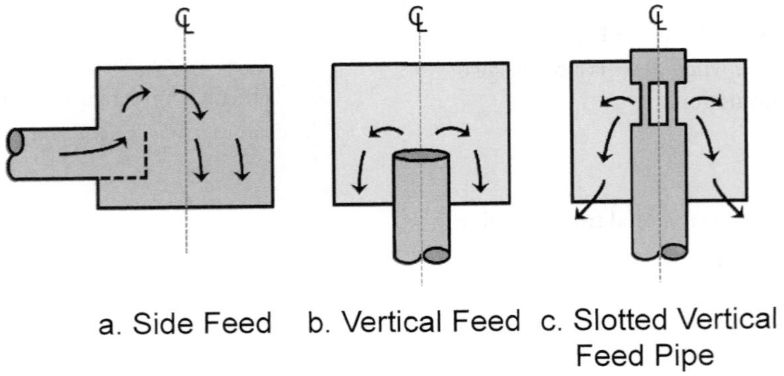

Figure 4.7 Various Center Feed Inlet Designs

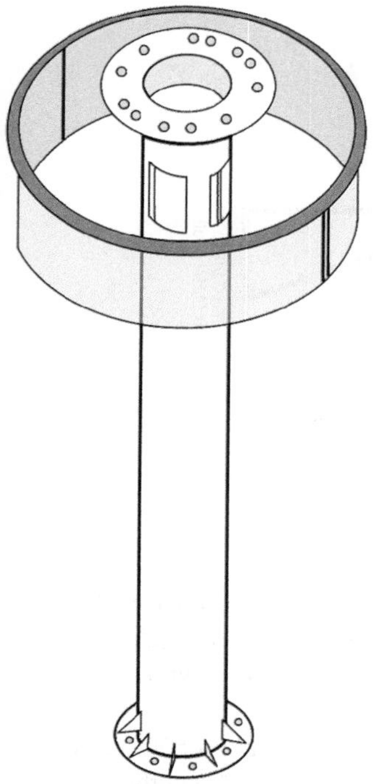

Figure 4.8 Center-Feed Configuration

TEST YOUR KNOWLEDGE

1. **Most circular, primary clarifiers use a center-feed, peripheral-rim, take-off flow pattern.**
   - ☐ True
   - ☐ False
2. **The feedwell directs flow down toward the bottom of the clarifier while increasing flow velocity.**
   - ☐ True
   - ☐ False
3. **A WRRF has three primary clarifiers, but only two of them are currently in service. The influent velocity is currently 2 m/s (6.6 ft/s). The operator should**
   a. Record the influent velocity on the log sheet.
   b. Place the third clarifier into service.
   c. Adjust the feedwell baffles.
   d. Take a clarifier out of service.
4. **The purpose of the feed well in a primary clarifier is to**
   a. Prevent short-circuiting
   b. Channel the flow to the effluent weir
   c. Aerate and mix the influent wastewater
   d. Collect scum and sludge

OUTLETS

Outlets for most circular feed clarifiers consist of a single perimeter v-notch weir that overflows into an effluent trough. Alternatives to this include cantilevered or suspended double weir troughs and submerged-orifice collector tubes. A relatively small number of circular tanks have been constructed with submerged orifices and are not discussed further in this text.

Peripheral Weir

There are two common designs for peripheral weir outlets for circular clarifiers. In the first, a concrete trough is constructed on the inside of the tank wall. The weir plate is then bolted at the top of the inward face of the trough wall. This is referred to as an *inboard launder* (Figure 4.9 [a]). Floatable material can rise up underneath an inboard launder and become stuck. Weirs and launders require cleaning at regular

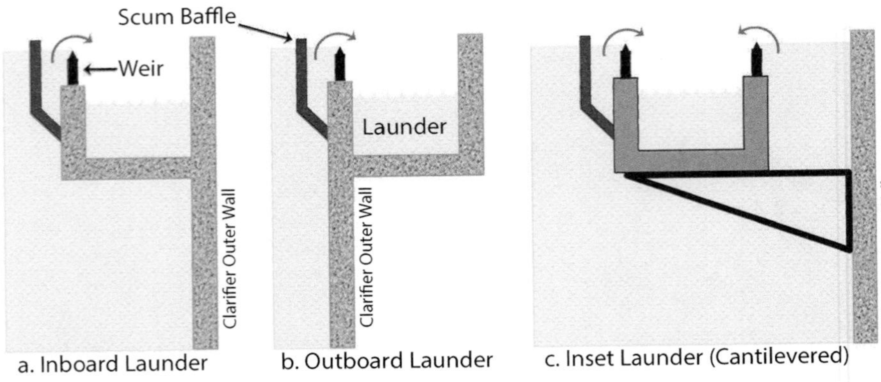

Figure 4.9 Inboard, Outboard, and Inset Launders (**Reprinted with permission by Indigo Water Group**)

intervals. The other common arrangement for a perimeter weir is to have the weir plate bolted to the inside of the tank wall. A concrete effluent trough is then constructed outside the tank wall. This is referred to as an *outboard launder* (Figure 4.9 [b]). Inset launders (Figure 4.9 [c]) are discussed in the next section.

The most common type of weir plate involves the placement of 90-deg v-notches at 152- or 304-mm (6- or 12-in.) intervals. A typical weir plate is shown in Figure 4.10, with the weir plate (B) in the foreground and the scum baffle (A) shown behind. The weir shown in Figure 4.10 is on a secondary clarifier; however, the appearance in primary clarifiers is identical. This weir placement allows a balance of relatively low increases in water elevation when flows increase as a result of diurnal and other changes. V-notch-type weirs are also forgiving of imperfect leveling of the weir plate. Some clarifiers have square notches, which are more prone to partial notch blockage by leaves, algae strings, and other surface debris. A rectangular weir plate is shown in Figure 4.11, with the weir plate (B) in the foreground and the scum baffle (A) behind.

Cantilevered Double or Multiple Launders

A typical cantilevered double launder is shown in Figure 4.12. This type of launder increases the total weir length for a given diameter of tank. This reduces the weir loading rate. The weir loading rate is the amount of flow passing through the clarifier per length of weir. Inset launders can reduce the effects of some types

Figure 4.10 V-Notch Weir Plate with Scum Baffle on a Secondary Clarifier: (A) Scum baffle and (B) Weir (Reprinted with permission by South Platte Water Renewal Partners)

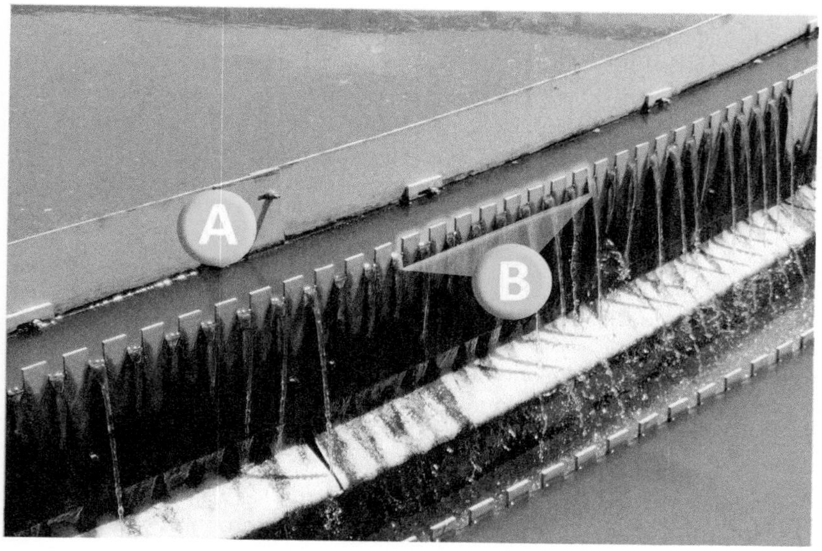

Figure 4.11 Rectangular Weir on a Primary Clarifier: (A) Scum baffle and (B) Weir plate (Reprinted with permission by South Platte Water Renewal Partners)

of short-circuiting in clarifiers; however, floatable material can become trapped between the inset launder and the wall of the clarifier. If the trapped material is not removed, it can putrify and generate odors.

The effluent troughs (launders) should be kept clean, neat, and tidy. For primary clarifiers, stringy material may be caught in the weirs. To clean the weirs and launders, operators often lean over the tank wall and hose these areas clean. Some clarifiers are equipped with brushes that are attached to the end of the skimmer arm. As the arm rotates, the brushes help to remove debris from the weirs. They are not a substitute for regular maintenance.

SLUDGE REMOVAL SYSTEMS

Effective removal of sludge from circular tanks is vital to process performance. There are two basic types of sludge removal systems, namely plows and hydraulic suction. Plows are used for all types of sludge encountered in wastewater treatment, whereas hydraulic suction is primarily limited to activated sludge secondary clarifiers.

Figure 4.12 Inset Launder with Dual Scum Baffles (Reprinted with permission by South Platte Water Renewal Partners)

Scrapers

There are several basic scraper designs that are used. Figure 4.13 shows four different types of scrapers. The multiblade plows, using straight scraper blades, have been used most extensively in the United States. The designs using curved blades are commonly referred to as *spiral plows* and have been used for decades in Europe. The number of spiral blade clarifiers in the United States is increasing. The straight, multiblade design (C) remains the most widely used mechanism for primary clarifiers. The scrapers plow furrows of primary sludge progressively toward the centrally located sludge hopper.

The conventional plow type (Figure 4.14) consists of scrapers that drag the tank floor at a tip speed of approximately 1.8 to 3.7 m/min (6 to 12 ft/min). Plows are located at an angle to force primary sludge toward the hopper, typically at the center of the tank, as the device rotates. The rotating element of the device can be driven from either the center platform or the outside tank wall. Torque must be sufficient to move the densest primary sludge expected. Several revolutions are required to move the solids from the tank perimeter to the center hopper.

More recently, deeper, tapered, segmented, spiral-type sludge scrapers (Figure 4.13 [B]) have been favored over conventional plow-type scrapers. The continuously tapered spiral-shaped scraper blades allow slightly faster operating tip speeds of 2.4 to 4.3 m/min (8 to 14 ft/min) and move solids to the center hopper in as little as one revolution. This enables the facility operator to increase sludge transport capacity and improve concentrations. Spiral-type scrapers have been successfully used in clarifiers up to 68 m (225 ft).

Hoppers

Settled sludge is typically scraped into a hopper where it is removed by gravity or pumping. For circular tanks, the hopper is typically located in the center of the tank. The hopper, up to 3-m (10-ft) deep, typically has steep sides. Until recent years, most U.S. clarifiers using scraper mechanisms were equipped with trapezoidal hoppers. Depending on the tank size, these hoppers are typically several meters deep and have walls with slopes of at least 60 deg above horizontal. This is still the most common type of hopper for primary clarifiers.

Drive Location

Most clarifiers in the United States have the drive mechanism located in the center of the clarifier at the top of the center column. For smaller tanks with side-feed pipes, the drive mechanism may be located on or under a fixed bridge that spans the full width of the tank. In Europe, it is common to have a drive located at the tank wall. Locating the drive at the tank wall is less common in the United States. This powers rubber-tired wheels that ride up on top of the tank wall and rotate the bridge that spans the tank diameter and is pivoted in the middle.

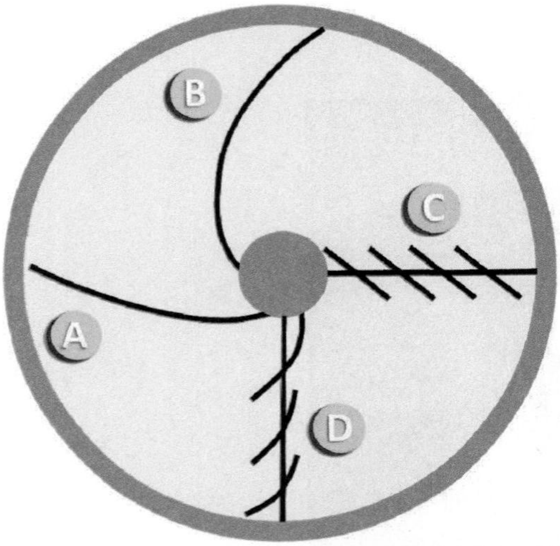

Figure 4.13 Scraper Configurations: (A) Nierskratzer spiral plow, (B) Logarithmic spiral plow, (C) Straight window shade plow, and (D) Curved window shade plow

Figure 4.14 Conventional Straight Plows (Reprinted with permission by Keith Radick)

SKIMMING SYSTEMS

The presence of scum and floatable material on the surface of clarifiers is a common problem in most municipal WRRFs. In primary clarifiers, the main contributors consist of grease and oils, plastics, leaves, rags, hair, and other materials. It has become common practice to remove floating materials from the surfaces of primary and secondary clarifiers. For circular tanks, a variety of skimming mechanisms have been designed and operated with varying degrees of capacity and success. The most common system used for center-feed tanks, a revolving skimmer, is shown in Figure 4.15 (highlighted in green). This design has been used for many years and is considered by many to be the standard, especially in primary clarifiers. It features a rotating skimmer arm and wiper that travels along the outer edge of the tank next to a scum baffle. It moves the floatables onto a beach or egress ramp connected to a scum removal box (Figure 4.16). During each rotation, floating solids are pushed toward the egress ramp (1), where they leave the water surface, go up the ramp, and drop into the scum box (2). The skimmer blade then passes over the scum box and dips back into the water to repeat its rotation. The scum baffle (3) prevents scum from being pushed toward the effluent weir (4). An exposed scum beach is shown in Figure 4.17. Most primary clarifiers have one rotating blade per clarifier, whereas secondary clarifiers may have two or even four such rotating blades. Some scum boxes are also equipped with an automatic flushing valve located on the centermost end of the box (Figure 4.18). The valve is mechanically actuated with each pass of the skimmer, which results in a water flush of the solids into the box hopper bottom and discharge pipe. The flush volume and duration are typically adjustable by the operator. Figure 4.19 shows the flush valve in both the open and closed positions. When the valve is open, water enters the valve (A) and flows into the scum removal box (B). Once the skimmer has passed by the scum removal box, the valve automatically closes, stopping the flow of flush water.

The scum trough often extends several meters (feet) from the scum baffle toward the center of the tank. Some designs extend this to the center feed well, thereby obtaining full radius skimming. For shorter scum troughs, some system is generally provided to move the floatables toward the outer scum baffle. Another method of moving floatables out toward the scum baffle is the use of water surface sprays.

Figure 4.15 Rotating Skimmer Arm (Reprinted with permission by Indigo Water Group)

Figure 4.16 Scum Beach, Scum Baffle, and Weir: (1) Scum beach, (2) Drop to scum box, (3) Scum baffle, and (4) Weir (Reprinted with permission by South Platte Water Renewal Partners)

Figure 4.17 Exposed Scum Beach: (1) Scum beach, (2) Drop to scum box, (3) Scum baffle, and (4) Weir (Reprinted with permission by Indigo Water Group)

Figure 4.18 Automatic Flushing Device for Scum Removal Box (Reprinted courtesy of City of Columbia, Metro WWTP. Photo by Adrian Martin.)

Figure 4.19 **Operation of Scum Box Flushing Mechanism: (A) Valve and (B) Scum removal box (Reprinted courtesy of City of Columbia, Metro WWTP. Photo by Adrian Martin.)**

TEST YOUR KNOWLEDGE

Label the components of this circular primary clarifier. Check your answers against Figure 4.5.

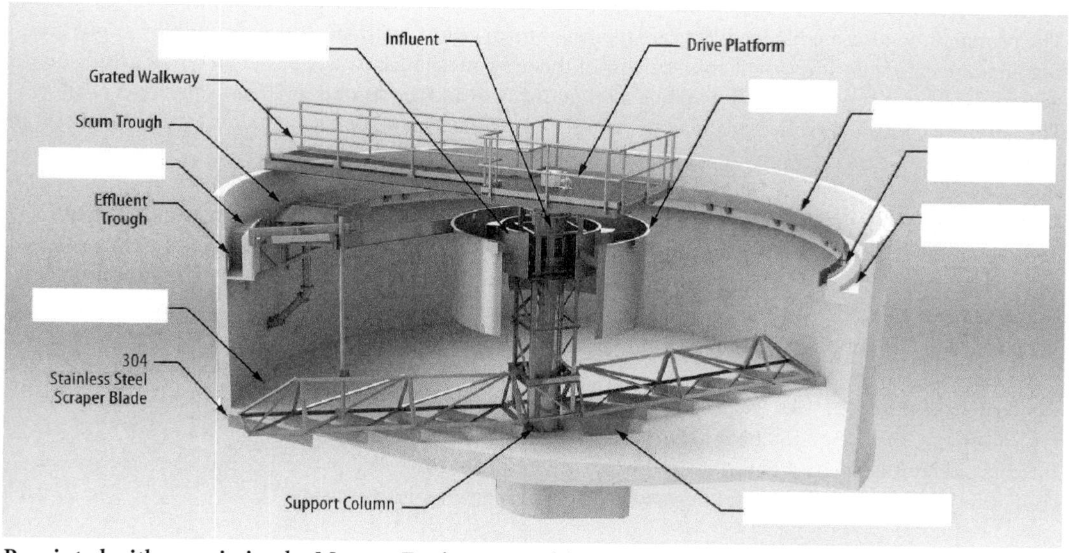

Reprinted with permission by Monroe Environmental (Monroe, MI) www.mon-env.com

1. Weirs for circular clarifiers generally consist of rectangular openings spaced at 0.3-m (1-ft) intervals.
   - ☐ True
   - ☐ False
2. The plows on a circular clarifier scraper mechanism are angled to direct sludge to the center hopper.
   - ☐ True
   - ☐ False
3. The drive for the sludge collection mechanism is typically mounted on the wall of a circular clarifier.
   - ☐ True
   - ☐ False
4. This type of launder projects into the clarifier center and is mounted on the outer wall.
   a. Inset launder
   b. Outboard launder
   c. Cantilevered launder
   d. Inboard launder
5. One potential disadvantage of an inboard launder is
   a. Weir length is increased compared to an inset launder
   b. Lack of a mounting location for the weir plate
   c. Floatable material can become trapped under the launder
   d. Difficulty in reaching both weir plates for cleaning

6. One advantage of using an inset or cantilevered launder is
   a. Increases total weir length
   b. Difficulty reaching weir plates for cleaning
   c. Floatable material can become trapped
   d. Increases weir loading rate

7. Scum and other floatable material
   a. Passes under the scum baffle and into the effluent.
   b. Is pushed by the skimmer arm onto the egress ramp.
   c. Consists primarily of grease and sand.
   d. Should be removed upstream in preliminary treatment.

WALKWAYS AND PLATFORMS

Most circular clarifiers are equipped with a single walkway that extends from the perimeter of the tank to the center area. Some clarifiers have walkways that extend the full diameter. Walkways are typically a minimum of 1-m (3-ft) wide. On circular clarifiers, the center platforms for clarifiers with center drives are generally large enough to give operators a minimum of 1.4 m (4 ft) of clearance around the mechanism for the operator to work. Platforms that are 2.4 m by 2.4 m (8 × 8 ft) or larger are common. Access holes and trap doors may be present at convenient locations to facilitate maintenance, operation, sampling, and observation activities.

DRIVES

The clarifier drive unit rotates the sludge collection mechanism and skimmer arm. Figure 4.20 shows examples of both shaft drive and cage drive units. A motor located at the top of each drive is coupled to the gear reducers below. The drive units for circular tanks typically consist of three sets of reducers that transition speeds from the motor to the rotating mechanism. These are commonly referred to as *primary, intermediate,* and *final reducers.* Worm gears, cycloidal speed reducers, and cogged gears have been used by different manufacturers. More details on gears can be found in the information box later in this section. The primary and intermediate reducers are typically worm gears. The final reducer is typically a spur gear. Bearings are extremely important components of the drive mechanisms. The principal types include one where steel balls run on hardened strip liners set in cast iron and the second involves forged steel raceways. The latter are commonly called *precision drives.* Precision bearings can be seen in both drive mechanisms in Figure 4.20.

In the United States, there are two common forms of clarifier drive mechanisms: bridge-supported styles and center pier (column) supported styles. Bridge-supported drives are used in full-span bridge clarifiers, typically less than 15 m (50 ft) in diameter. The access bridge supports the center drive. The output flange of the drive attaches to the rotating torque tube (drive shaft), which rotates the collector mechanism. Typical drive configurations include a primary and final gear reduction unit.

Center-pier-supported styles are used on half-span-bridge, center column support clarifiers that are typically larger than 15 m (50 ft) in diameter. The center column supports the center drive, and the rotating spur gear attaches to the drive cage, which rotates the collector mechanism. Typical drive configurations include a primary, intermediate, and final gear reduction unit.

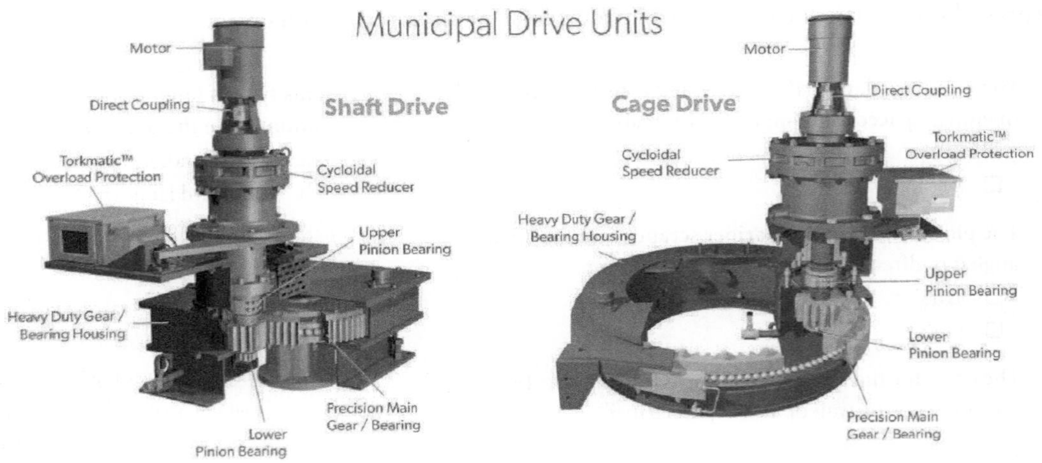

Figure 4.20 Clarifier Drive Units (Courtesy of WesTech Engineering, Inc.)

Another type of drive used more commonly in Europe is the rim-drive mechanism. It features a motor, gearbox, and drive wheel that runs on top of a circular tank wall. There are a few units operating in the United States; however, because of their small number, they are not discussed further in this text.

The drive and the clarifier mechanism for a circular clarifier must be protected from overload conditions. If the mechanism becomes restricted on a piece of debris, hung up on the floor, or cannot move the sludge because the solids concentration is too high or for any other reason, the clarifier drive must be protected by setting off a high-torque alarm or by shutting down the collector mechanism to prevent damage to the drive and/or the mechanism. Although the mechanism moves slowly, there is a high amount of torque in the system, which is capable of twisting or otherwise damaging the entire clarifier mechanism. Most suppliers of clarifier drive packages provide drive torque monitoring devices as part of the drive mechanism. Torque is sensed on the scraper arms and is transmitted to the drive unit. High-torque and high-high torque warning, alarm, and shutoff switches are typically installed at each clarifier drive mechanism. Torque indication can typically be read from a scale, which is expressed as a percentage of the maximum torque load. The overload system can operate using either lateral displacement or free rotation (Weidler, 2017).

The lateral displacement method may be used with drive mechanisms that include a primary reducer, chain-and-sprocket set, intermediate reducer, and final spur gear set (Weidler, 2017). The intermediate reducer, typically a worm gear and gear combination, is designed to slip or shift when too much torque is applied. When the worm gear shifts, it displaces or puts tension on a spring mechanism, which then triggers one or more limit switches inside the gearbox. The shift is small, typically +/− 9.5 mm (3/8 in.) (Weidler, 2017). A limit switch is a switch that is activated or tripped by the movement of another component. The limit switches in Figure 4.21 are inside the overload protection box. Most drives will have two limit switches. If the high-torque warning switch senses a high-torque condition (typically when the torque load reaches 40 to 50% of the maximum design drive torque), it generates a high-torque alarm. This alarm is typically displayed on the facility operator's main control panel. If the high-high torque switch senses that the torque load has reached 80 to 85% of the maximum design drive torque, it generates a high-high torque alarm. If the drive is not turned off after the actual torque reaches the high-high level, the high-high torque switch will turn off the drive motor to protect it from overload and to protect the sludge collection mechanism from damage.

A third level of protection, a shear pin, is provided as part of many drive mechanisms. A shear pin is a metal rod that is designed to break under a certain amount of load or torque. In the event that the limit switches do not engage, the shear pin will break. Depending on the design, a broken shear pin may allow the drive motor to continue to operate, but stop the collector from turning or, if the shear pin device

Figure 4.21 Clarifier Drive Mechanism Showing Location of Shear Pin: (A) Torque sensor and (B) Torque switch box (Reprinted with permission by Kenneth Schnaars)

Figure 4.22 Chain and Sprocket Assembly with Shear Pin (Reprinted with permission by Kenneth Schnaars)

operates in conjunction with a limit switch with a trip, may shut down both the drive and the collection mechanism. A chain-and-sprocket assembly-type clarifier is shown in Figure 4.21. The cover has been removed from the chain-and-sprocket assembly in Figure 4.22 to show the location of the shear pin assembly (highlighted in green).

The second method is the rotational reducer-type system (Weidler, 2017). With this system, the primary and intermediate reducers are allowed to rotate with the mechanism. As they rotate, a plunger or rod is depressed. The rod interacts with the limit switches. As torque increases and the clarifier mechanism slows down, the rotation of the primary and intermediate reducers also slows, which triggers the high-torque and high-high torque alarms.

Rectangular clarifiers use a different shutdown device. They also use both shear pins and limit switches. The type of limit switch used and whether the shear pin or limit switch has primary control depends on both the age of the equipment and the manufacturer. An example of clarifier drive limit switches is shown in the upper right corner of Figure 4.23 adjacent to the load indicator.

In the event that the clarifier mechanism shuts down or the shear pin breaks, the operator should remove the clarifier from service and determine the cause of the problem. Possible causes for an overload condition

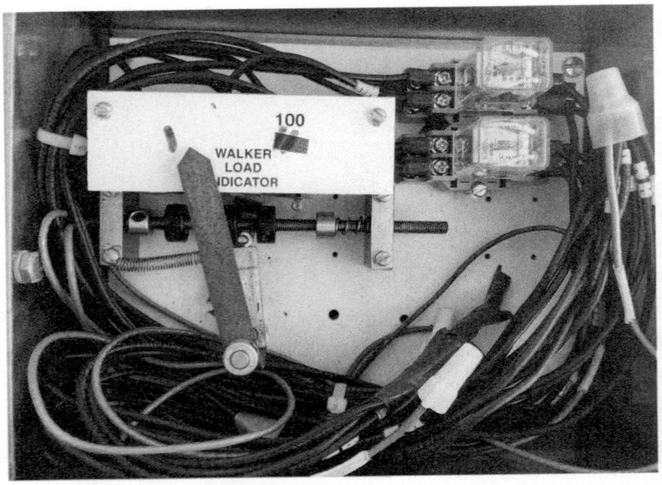

Figure 4.23 Clarifier Drive Limit Switches (Reprinted with permission by Manuel Freyre, City of Northglenn, Colorado)

include debris stuck in the sludge collection mechanism, excessively high total solids concentrations in the sludge blanket, a failure of the sludge removal system (pumps), or the sludge collection mechanism being caught on the clarifier floor. It is extremely important that the cause of the overload be determined and corrected. Under no circumstance should an operator install a larger, more robust shear pin. The shear pin breakage is a warning that there is something wrong with the collection mechanism. Forcing continued operation is likely to result in equipment damage.

Gears, Gears, Gears!

Much of the equipment used in wastewater treatment relies on motors and gears to function. Gears transmit power from one machine to another. If two or more gears are meshed together so that their teeth align, turning one gear will turn the other gear as well. A gearbox is a housing that holds one or more gears within it. Figure 4.24 illustrates several different types of gears. A gear with teeth that stick straight out from the edge of the gear is called a *spur gear*. Several of the gears shown in Figure 4.24 are spur gears. If two gears are placed next to one another, the larger gear is simply called the *gear* while the smaller of the two is called the *pinion gear*. An intermediate gear that transfers power from one gear to another gear is called an *idler gear*. Idler gears are used when the pinion gear needs to rotate in the same direction as the gear. A worm gear looks similar to a screw and is typically one piece of a larger shaft. Worm gears are used to transfer power from a horizontal shaft to another gear or piece of equipment that is set perpendicular to the worm gear. A sprocket looks similar to a gear, but is not a true gear. A sprocket is a toothed wheel that is used in combination with a belt or a chain. The teeth of a sprocket fit through openings in the chain or belt to pull it along.

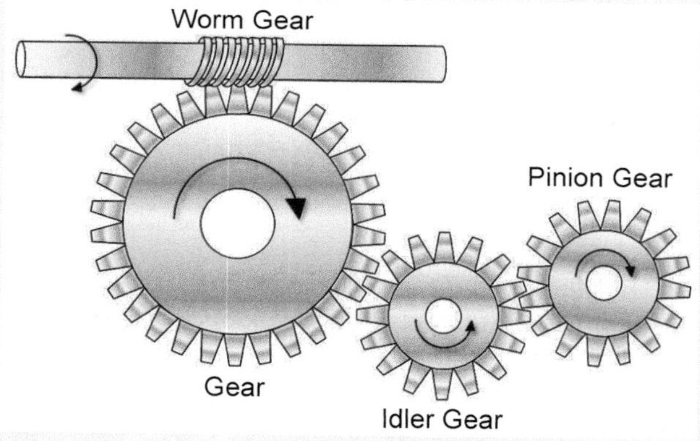

Figure 4.24 Types of Gears (Reprinted with permission by Indigo Water Group)

Torque is a force that tends to produce rotation. Imagine having a ruler that is nailed down in the center. If you push on the ruler at its pivot point, nothing happens. If you push on the ruler some distance from the center, the ruler will rotate around the pivot point. The force or power that was applied to make the ruler rotate is called *torque*. When power is applied to a gear, the resulting torque can be used to move other gears or pieces of equipment. Power, torque, and speed are related to one another by the following equation:

$$\text{Power} = \text{Torque} \times \text{Speed}$$

Imagine a larger gear connected to a smaller gear (pinion). If power is applied to the larger gear, it will rotate at a certain speed and cause the pinion gear to rotate. The smaller pinion gear will rotate faster than the larger gear because it has fewer teeth along its perimeter. Because the amount of power remains about the same between the two gears and the smaller pinion gear is rotating faster, the smaller pinion gear will have less torque than the larger gear.

Worm gears and cycloidal reducers are two types of reducers. Gear reducers work by connecting larger gears to smaller gears. Gear reducers create a decrease in speed and an increase in torque. In a clarifier drive unit, the motor may rotate at 1500 to 1800 rpm. The motor is connected to the sludge collection mechanism through a series of reducers. This allows the mechanism to turn very slowly, about 0.04 rpm, while generating enough torque to push the sludge around the bottom of the clarifier and into the hopper. The load applied to the rotating rake arm is the continuous operating torque or running torque.

TEST YOUR KNOWLEDGE

1. **Clarifier drive units typically consist of a motor and three sets of reducers.**
   - ☐ True
   - ☐ False
2. **A sprocket is a toothed wheel that is used in combination with a chain or belt.**
   - ☐ True
   - ☐ False
3. **Gears transfer power from one machine to another.**
   - ☐ True
   - ☐ False
4. **A clarifier drive mechanism has a broken shear pin. The operator should**
   a. Replace the shear pin and restart the drive.
   b. Determine and correct the source of the overload condition.
   c. Decrease the drive motor speed.
   d. Adjust the gear-to-pinion teeth ratio
5. **The clarifier mechanism is hung up on the clarifier floor. The resulting torque load reaches 45% of the maximum design drive torque. As a result, the worm gear shifts inside the gearbox causing**
   a. Destruction of the spur gear
   b. Drive motor to shut down
   c. Rake speed to decrease
   d. Overtorque alarm to trigger
6. **When two gears are placed next to one another, the smaller gear is called the**
   a. Gear
   b. Idler
   c. Pinion
   d. Worm
7. **This type of gear may be used when it is necessary to connect a horizontal shaft to a perpendicular gear.**
   a. Sprocket
   b. Worm
   c. Idler
   d. Pinion
8. **Two gears are operating together. The smaller gear is spinning at 2000 rpm. The larger gear spins at 500 rpm. How much more torque does the larger gear have?**
   a. Both gears have the same torque
   b. Twice as much
   c. Four times less
   d. Eight times as much
9. **When a reducer is placed between a motor and a piece of equipment**
   a. Speed is reduced and torque is increased.
   b. Speed is increased and torque is reduced.
   c. Speed is reduced and torque is reduced.
   d. Speed is increased and torque is increased.

RECTANGULAR TANKS

A typical rectangular primary clarifier is shown in Figure 4.25, with all significant components labeled. Flow enters the clarifier at one end. An inlet channel with multiple openings is typically used to distribute flow across the width of the tank. Baffles break up the influent current, further distribute flow across the tank width, and direct flow toward the bottom of the clarifier. The baffles help prevent the influent wastewater from traveling straight across the clarifier surface to the outlet and short-circuiting. As the water moves through the clarifier, settleable solids fall to the bottom where the sludge collection mechanism pushes the sludge toward the sludge hopper. With a chain-and-flight type sludge collection system (Figure 4.25), the flights are moved slowly through the clarifier by a drive unit located on the clarifier wall. These drive units are similar to those found in circular clarifiers. Treated water flows around the inlet baffle, across the tank, under the scum baffle, and over the weirs into the effluent channel. Scum and floatable materials will rise to the clarifier surface. Baffles and skimming devices prevent this undesirable material from passing over the weir and into the effluent channel. Primary clarifiers typically use rotating skimmers to remove floatable material from the surface.

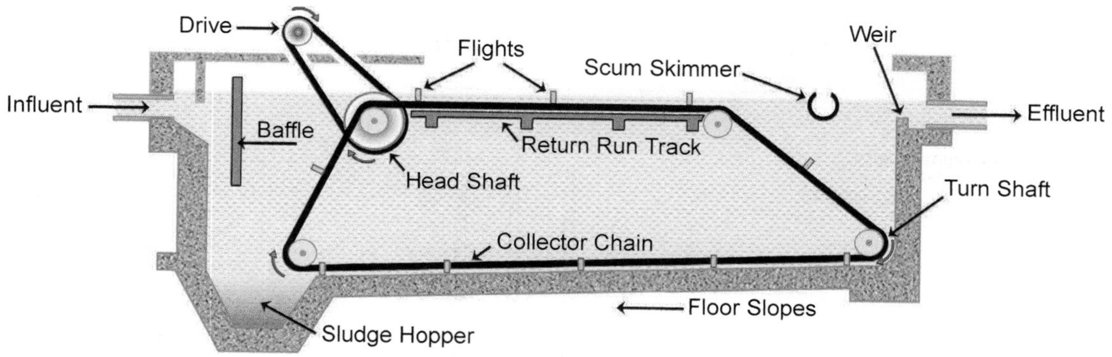

Figure 4.25 Rectangular Primary Clarifier Schematic

Length, Width, and Depth

A rectangular tank, either a single unit or one of several adjacent units, typically has a length several times its width. Typically, the length-to-width ratio is greater than 5 : 1 (WEF, 2005). With a long, narrow tank, the water tends to move forward through the tank as if it were in a very large pipe with few opportunities for short-circuiting. This type of flow pattern is called *plug flow*. Rectangular clarifiers range from 15 to 90 m (50 to 300 ft) in length and 3 to 24 m (10 to 80 ft) in width (WEF et al., 2018). The width of the tank is limited by the available sizes of sludge removal equipment. Most primary tanks at mid- to large-sized facilities are 12.5-m (41-ft) wide to accommodate the width of two, parallel, chain-and-flight mechanisms at 6 m (20 ft) each, with a 0.3-m (1-ft) gap between them. Depths typically range from 3 to 4.9 m (10 to 16 ft). Most tanks are between 3- and 3.7-m (10- and 12-ft) deep. Recall that the tank surface area, or the length times the width, is important because it determines the SOR. The SOR is the velocity of the water upward through the tank.

Freeboard

Freeboard in rectangular tanks is similar to circular tanks, with 0.5 to 0.7 m (1.5 to 2.0 ft) of freeboard for most clarifier designs.

Inlet Conditions

Flow is introduced to rectangular clarifiers by spanning the width of the clarifier with a short open channel or by a *manifold* piping system. In both of these options, the flow is directed to multiple inlet openings in the tank. The multiple inlet ports are situated and sized to uniformly distribute flow over the width of the clarifier. For example, in a 6-m- (20-ft-) wide tank, there are typically three to four inlet ports. Inlet ports are typically spaced at 2- to 3-m (6.5- to 10-ft) intervals. The size of the ports maintains an inlet velocity of 3 to 9 m/min (10 to 30 ft/min) to distribute flows across the rectangular tank (Metcalf and Eddy, Inc./AECOM, 2014). Sometimes, an inlet baffle is placed in the flow path of the inlet stream. It may be a solid target baffle to deflect the flow or a perforated baffle to break up any jetting action and disperse the flow.

A *manifold* system consists of multiple pipes all connected to a central pipe similar to the way that tines on a fork are connected to the base of the fork.

Outlets (Weirs)

The two types of effluent collectors that are commonly used are surface launders (overflow weirs) and submerged launders (outlet tubes), which consist of pipes with openings. Surface launders are similar in appearance to the v-notch weir and launder arrangement used for circular clarifiers. With rectangular tanks, surface launders can be placed down the length of the tank (longitudinal), across the tank (transverse), or in a grid arrangement (Figure 4.26). Figure 4.27 shows transverse weirs installed in a primary clarifier. Uniform withdrawal along the length of the weir is important to prevent localized regions of high velocity. Weirs must be kept clean and level to distribute flow evenly. Some clarifiers have submerged launders or outlet tubes in place of surface overflow weirs (Figure 4.28). Submerged launders consist of outlet tubes with evenly spaced openings along their length that discharge to a common manifold. Submerged launders can make it easier to remove scum from the surface and to avoid passing floatable material on to the secondary treatment process. Algae growth is less of a problem with submerged launders than surface overflow weirs. Submerged launders can, however, collect rags and debris that are not easily removed without dewatering the tank.

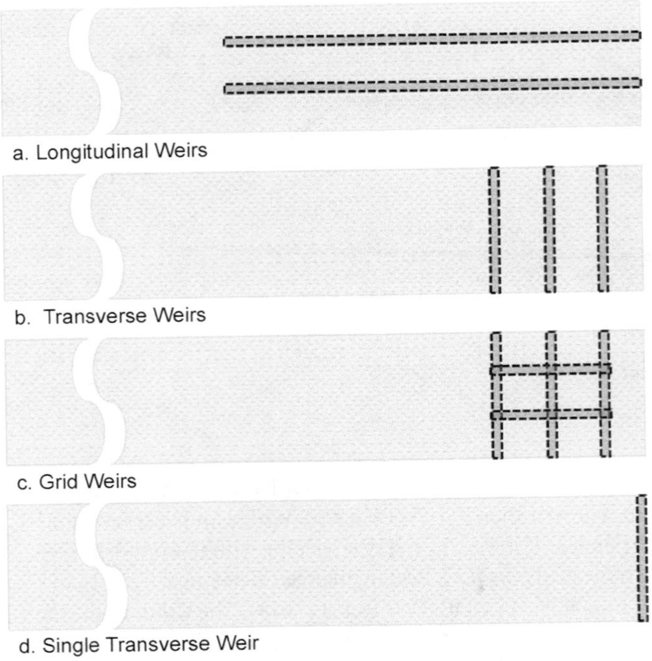

Figure 4.26 Plan Views of Typical Surface Weir Configurations

Figure 4.27 Transverse Weirs on a Primary Clarifier (Reprinted with permission by Kenneth Schnaars)

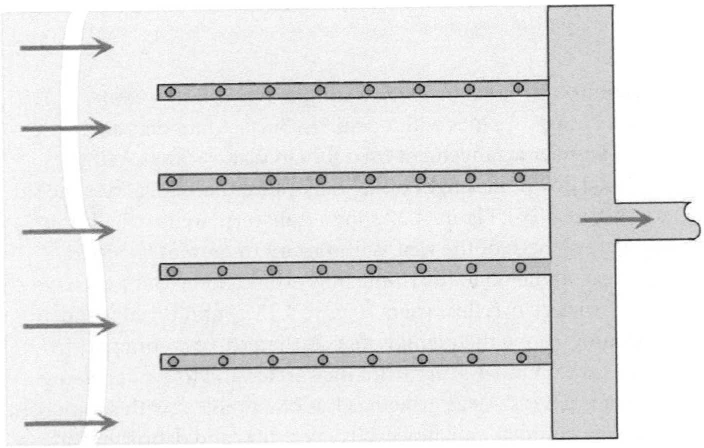

Figure 4.28 Submerged Launders (Adapted from Larsen, 1977)

Sludge Removal Systems

Chain-and-Flight

Primary sludge collection equipment for rectangular tanks typically includes either a chain-and-flight system or a traveling bridge. A chain-and-flight system (Figure 4.25) consists of two, parallel, endless loops of chains with cross scrapers (flights) attached at approximately 3-m (10-ft) intervals. The flights are long, narrow boards that span the width of the tank, as shown in Figure 4.29. Typical flights measure 5 to 6 m (16 to 20 ft), depending on the width of the clarifier. The sprocket wheels are mounted on rotating shafts. The flights move slowly along the clarifier floor, scraping the settled sludge into the sludge hopper. At the same time, on their return path near the surface, the partially submerged flights serve as skimming devices to push any floating solids or foam to a skimmer pan or trough. This requires the use of four rotation points or sprockets (Figure 4.25). If the flights are not used to move the skimmings on the surface, only three rotation points are required and the chain forms a long triangle shape. Sometimes, a five-sprocket system is used with the extra sprocket helping to guide and hold down the flights in the bottom, midtank area of long tanks. Typically, chain-and-flight collectors move approximately 5 to 15 mm/s (1 to 3 ft/min) and are driven by a single drive unit located on the wall. Higher operating speeds push sludge toward the hopper faster and help ensure the flights are not buried by the settling sludge. If the sludge blanket becomes too deep, the flights can become stuck. Single tanks can have either a single or double hopper. Multiple tanks or tanks with more than one chain-and-flight assembly often use a cross collector in a transverse trough.

Some primary clarifiers are so large that it becomes necessary to place two or more chain-and-flight systems side by side in the same tank. Each chain-and-flight system may have its own sludge hopper. Alternatively, sludge may be collected in a single, larger trench called a *cross-collector* that runs the full width of the basin. Sludge that is pushed into the cross-collector trench may be removed with an auger/screw conveyor or a three-sprocket chain-and-flight system like the one shown in Figure 4.30.

The sludge removal equipment in rectangular clarifiers has greater maintenance requirements than similarly sized circular tanks because drive bearings in rectangular tanks are submerged. Historically, cast iron chains and cedar wood flights were used. Today, stainless steel or nonmetallic (plastic) chains and fiberglass flights are used almost exclusively. Nonmetallic chain, flights, and sprockets are currently used more often

Figure 4.29 Chain-and-Flight Sludge Collection Mechanism: (A) Submerged flight pushing sludge, (B) Skimming flight, (C) Chain, (D) Wear strips, (E) Wear shoes, and (F) Rotating trough scum skimmer (Reprinted with permission by Kenneth Schnaars)

Figure 4.30 Cross-Collector (Reprinted with permission by Kenneth Schnaars)

in new installations and in retrofits of existing tanks. The nonmetallic equipment is lighter, making it easier to maintain, and requires less power to operate. Also, the nonmetallic equipment is less susceptible to corrosion. Other facilities prefer metallic-type chains for their collector mechanisms because they are less likely to decay in sunlight and are less affected by temperature or stretching. In some tanks, flat water sprays angled toward the liquid surface move the scum and grease to the skimming device. As plastic chain wears and stretches over time, it is normal maintenance to remove a link approximately every year on long clarifiers.

Plastic or metal wear shoes (Figure 4.29) fixed to the bottoms of the flights allow them to slide on rails near the surface and wear strips on the bottom of the clarifier so that the chain does not bear the full weight of the flights. In this way, less power is used to move the flights. Wear shoes and wear strips must be replaced periodically. An adjustable rubber scraper is typically attached to the bottom edge and side of at least some of the flights to provide complete sludge scraping and prevent unwanted stationary sludge deposits. A floor slope of 1% is common to help the sludge move toward the hopper.

Traveling Bridge

Dewater means to empty a tank by removing water and sludge.

Traveling bridge systems were developed to solve the problem of having to *dewater* the chain-and-flight system for routine maintenance. While these systems are easier to maintain, they generally allow the settled sludge to accumulate to a greater extent before moving the sludge to the sludge hopper for withdrawal.

A traveling bridge (Figure 4.31) consists of a scraper-blade mechanism that can be raised or lowered. The scraper-blade mechanism is mounted on a bridge or carriage that travels approximately 1.8 m/min (6 ft/min) back and forth on tracks or rails mounted on top of the tank. The weight of the scraper blade is often supported by wheels to minimize the frictional resistance. As it travels away from the hopper at approximately 3.7 m/min (12 ft/min), the mechanism, largely out of the water, acts as a skimmer, pushing floating material toward the scum removal mechanism. As it reaches the end of the tank, the mechanism drops to the floor of the tank, reverses direction, and travels toward the hopper end of the tank, pushing settled primary sludge to a hopper. Traveling-bridge collectors cannot be used with covered tanks.

Hoppers

Transverse means to extend across something. In this case, the transverse trench is extending across multiple primary clarifiers.

The settled solids may collect in a single hopper, multiple hoppers, or a *transverse* trench with a hopper at one end. The typical hopper for rectangular clarifiers is an inverted pyramid with a rectangular opening on top. The sides are steep, with slopes of 52 deg to prevent solids from accumulating on the upper walls (WEF, 2005). A single rectangular tank may have two or more withdrawal hoppers, each equipped with a

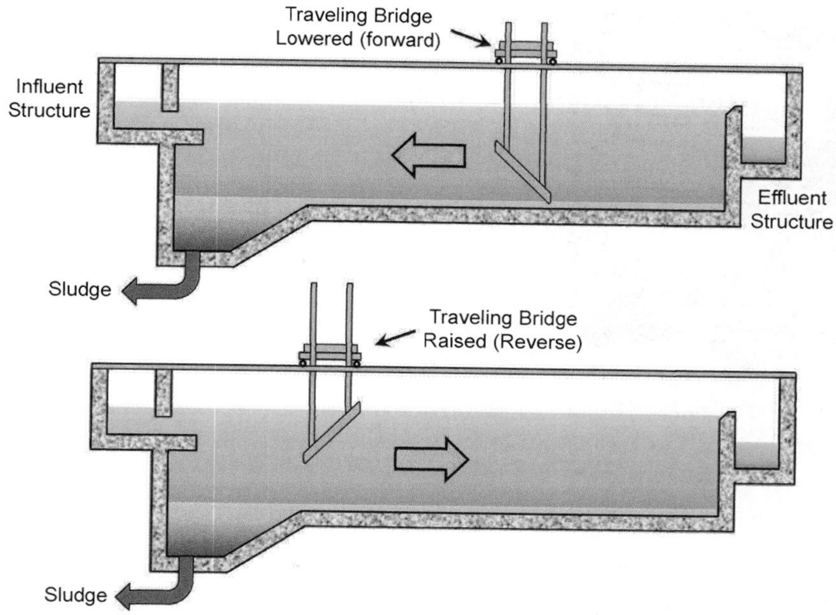

Figure 4.31 Traveling Bridge

withdrawal pipe. The hopper for rectangular tanks is typically located at the inlet end of the tank to minimize travel time of particles to the hopper. The transverse trench may be equipped with a flight-and-chain collector, referred to as a *cross collector* or a *screw conveyor*, that moves the sludge to the hopper.

Skimming System

Scum collection typically has been located on the effluent end of the rectangular primary clarifier tank. Some clarifiers, however, have located scum collection on the influent end of the clarifier tank to decrease the travel distance to the collection point and ensure rapid removal of all floatables. A slotted pipe or rotating helical wiper skimmer is used in rectangular primary tanks and is typically operated by hand or on a timer. The material is then pumped or conveyed to another process, such as a digester, concentrator, or holding basin. Methods used to convey scum, grease, and other floating material include pneumatic ejectors, positive-displacement pumps, or recessed impeller centrifugal pumps.

A rotating trough skimmer is shown in Figure 4.32. A skimmer board is connected to the sludge removal mechanism through a hinged, counter-weighted assembly. It pushes floatable material toward a fixed, rotating trough that turns into position as the skimmer board approaches and trips a trigger switch. When

Figure 4.32 Rotating Trough Scum Skimmer (Reprinted with permission by Jim Myers & Sons, Inc, (JMS))

Figure 4.33 Rotating Trough Scum Skimmer Drive: (1) Motor, (2) Sprocket, (3) Chain, and (4) Skimmer sprocket (Reprinted with permission by Jim Myers & Sons, Inc, (JMS))

the board reaches the rotating trough, it ducks under the trough, and its counterweights return it to the surface to continue rotation around the tank. This device has an advantage of offering full-radius scum removal. Separate flushing is generally not required, although some clarifiers have a deeper cut opening at the inner end of the rotating trough to take on more water, which moves the floatables into the collector box at the end of the trough. The trough rotation can be programmed to trip every time the skimmer arm approaches or can skip some of the cycles to concentrate the amount of floatable material removed. The trough movement is controlled with a small motor and gear system. Figure 4.33 shows an example rotating trough skimmer drive with a small motor at the top (1) rotating a sprocket (2) that is connected to the skimmer (4) with a chain (3).

TEST YOUR KNOWLEDGE

Label the following components of a rectangular clarifier. Check your answers against Figure 4.25.

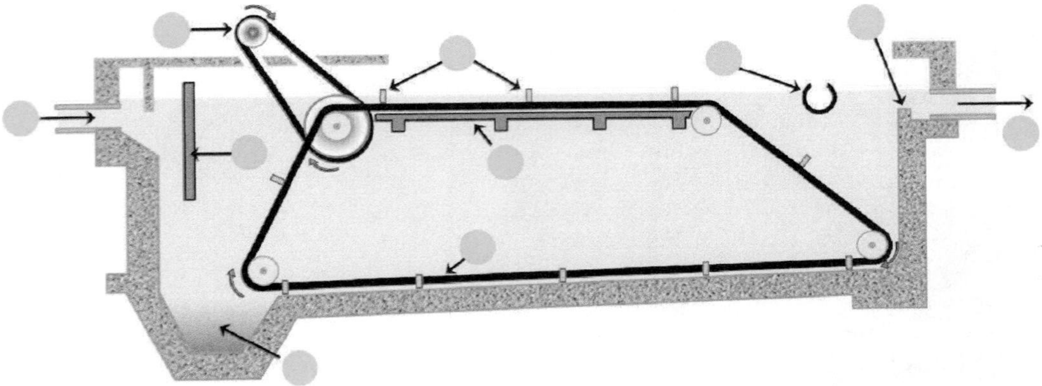

1. The flow in a rectangular clarifier is radial.
   - ☐ True
   - ☐ False

2. A chain-and-fight system consists of two, continuous, parallel loops of chains.
   - ☐ True
   - ☐ False

3. **A perforated inlet baffle may be used to break up jetting action and disperse flow in a rectangular clarifier.**
   - ☐ True
   - ☐ False

4. **This type of sludge collection mechanism is commonly used with rectangular clarifiers.**
   a. Niersfratzer spiral
   b. Chain and flight
   c. Window shade plow
   d. Logarithmic spiral plow

5. **When a high-high torque alarm is triggered, this may also occur.**
   a. Tooth loss on the pinion gear
   b. Collector mechanism may speed up
   c. Load indicator shows zero
   d. Shear pin breaks

6. **A minimum of ________ sprockets is needed to support a chain-and-flight system that removes both settleable and floatable material.**
   a. 1
   b. 2
   c. 3
   d. 4

7. **The drive unit for a rectangular clarifier is located**
   a. On the clarifier wall
   b. Underneath the inlet baffle
   c. Parallel to the scum skimmer
   d. In the influent manifold

8. **The inlet manifold**
   a. Is typically 25% of the tank width
   b. Distributes flow evenly across the tank
   c. Requires a minimum of 10 openings
   d. Incorporates a waterfall cascade

9. **One disadvantage of submerged launders over transverse weirs is**
   a. Diminished algae growth
   b. Easier to remove scum from the clarifier surface
   c. Trapped rags and debris can be difficult to remove
   d. Higher flow velocities

10. **Plastic or metal wear shoes at either end of a clarifier flight**
   a. Support the weight of the flights
   b. Do not require replacement
   c. Compensate for stretching of the chain
   d. Assist in cleaning the scum dipper

11. **Historically, cast iron chains and wooden flights were used. Today, most systems use stainless steel or plastic with fiberglass flights for this reason.**
   a. Plastic chains are less likely to decay in sunlight.
   b. Cast iron is more expensive.
   c. U.S. Environmental Protection Agency requires the use of food-grade materials.
   d. These materials are lighter and easier to maintain.

12. **A transverse collection trench may be used in this situation.**
   a. Longitudinal trench will not fit the space
   b. Length-to-width ratio is less than 3:1
   c. Multiple chain-and flight systems used in parallel
   d. Expected sludge concentrations exceed 10%

13. **The trough rotation on a rotating skimmer**
   a. Typically operates on a timed cycle
   b. Turns in the direction of flow towards the weir
   c. May be triggered by an approaching skimmer arm
   d. Cannot be used with submerged launders

GATES AND VALVES

Water resource recovery facilities use slide gates, canal gates, and valves to direct flow. These may also be used to isolate a treatment unit and remove it from service for cleaning and maintenance. If there are multiple treatment units, such as primary clarifiers, they are often manifolded together like the primary clarifiers shown in Figure 4.34. Here, influent flow is directed to a splitter box with connections to each clarifier. A gate or valve is used to either allow flow into a particular clarifier or to prevent it from reaching that clarifier. Primary effluent from all of the clarifiers is brought together in a joiner box. The joiner box is also equipped with multiple gates or valves that can be opened to allow flow into the joiner box or closed to prevent flow from backing up into an offline clarifier.

Gates are used in open channels and splitter/joiner boxes. They are used to cover the entrances to channels and pipes. At smaller WRRFs, slide gates may be as simple as metal or fiberglass plates that are slid in and out of grooves mounted on the channel wall. Slide gates can be difficult to maneuver in larger channels. The force of the water moving through the channel puts pressure on the gate and pushes it against the channel grooves. Larger WRRFs typically use canal gates (Figure 4.35). A canal gate is similar to a slide gate, but incorporates a threaded stem and handwheel. The handwheel, which looks like a car steering wheel, is rotated to raise or lower the gate. The handwheel acts like a lever and increases the amount of force that an operator can exert, making it easier to raise and lower the gate. Canal gates typically have rising stems. This means that as the gate opens below, the stem rises up through the handwheel. Rising stem gates and valves make it easy for an operator to see from a distance which gates are open and closed. Many canal gates are

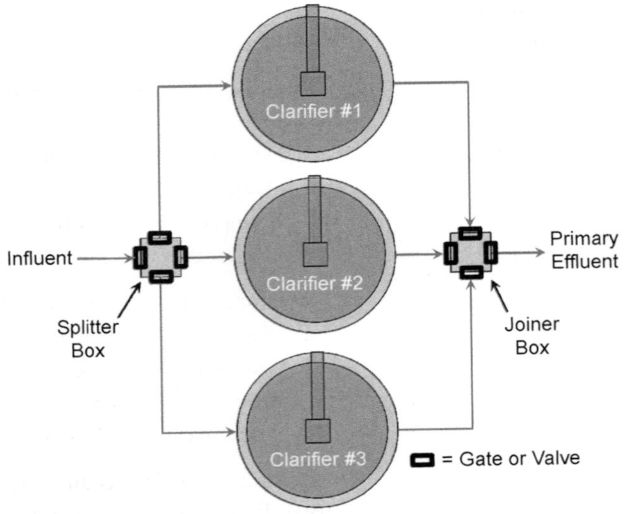

Figure 4.34 Manifolding of Multiple Clarifiers (Reprinted with permission by Indigo Water Group)

equipped with transparent covers for the rising stems that help to keep out dirt and moisture and help prevent rust.

A third option for taking equipment and processes in and out of service is isolation valves. Isolation valves are located within pipelines and are used in both smaller and larger WRRFs. One type of valve commonly used for this purpose is a gate valve (Figure 4.36 and Figure 4.37). Ball valves may also be used as isolation valves. Gate valves have a gate that is completely removed from the flow path when open. This is a big advantage in wastewater treatment where rags and other solids could become stuck inside the valve. The gate is stored inside the valve bonnet when the valve is open. Gate valves are available in both rising stem and nonrising-stem configurations. With a nonrising stem, the gate moves up the threads inside the bonnet and there is no indication from the outside whether the gate is open or closed. Gate valves are intended to operate in either the full-open or full-closed position. They should not be operated partially opened because this puts stress on the gate.

Canal gates and valves require regular maintenance to prevent them from rusting and/or freezing in place. At a minimum, canal gates and valves should be exercised once a year. To exercise a valve or gate, completely open and close it several times. The stem should be lubricated according to the manufacturer's

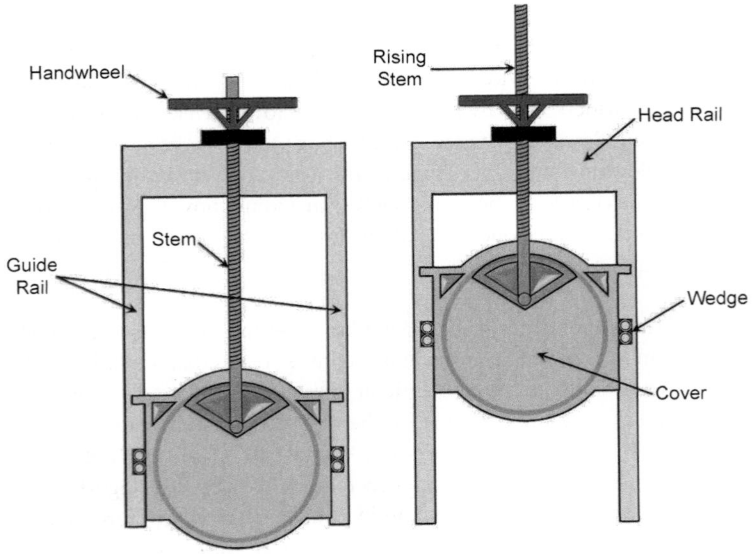

Figure 4.35 Canal Gate (Reprinted with permission by Indigo Water Group)

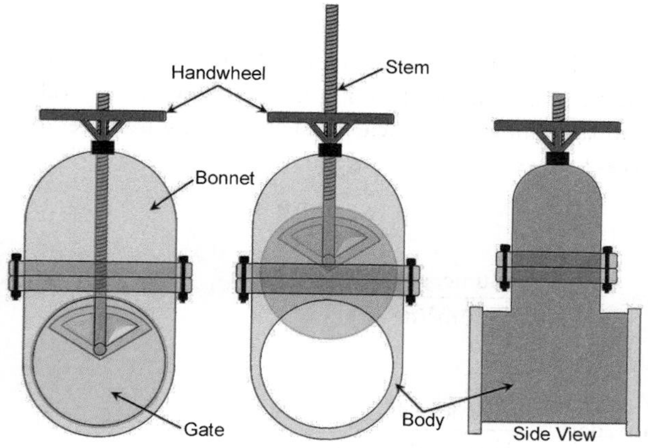

Figure 4.36 Gate Valve Diagram (Reprinted with permission by Indigo Water Group)

directions at least annually, although many manufacturers recommend more frequent lubrication. For more information on different types of valves, refer to *The Valve Primer* (1997).

Process Variables

VARIABLES AFFECTING SETTLEABLE SOLIDS REMOVAL

Primary clarifiers rely on differences in density to separate particles from the surrounding wastewater.

SURFACE OVERFLOW RATE

Surface overflow rate and other currents in the clarifier push particles in other directions, including up and out of the clarifier. For particles to settle, the downward force from the pull of gravity and the resultant settling velocity must be greater than the upward force applied as a result of SOR (upward velocity) and other currents. Surface overflow rate is expressed as follows:

In International Standard units:

$$\text{Surface Overflow Rate, } \frac{m^3}{m^2 \cdot d} = \frac{\text{Flow, } m^3/d}{\text{Surface area, } m^2} \qquad (4.2)$$

In U.S. customary units:

$$\text{Surface Overflow Rate, } \frac{\text{gpd}}{\text{sq ft}} = \frac{\text{Flow, gpd}}{\text{Surface area, sq ft}} \qquad (4.3)$$

Figure 4.37 Gate Valve Exterior (Reprinted with permission by Indigo Water Group)

HYDRAULIC DETENTION TIME

Hydraulic detention time is the amount of time that the water remains within the clarifier. The detention time should be sufficient to allow nearly complete removal of settleable solids. A longer hydraulic period would not improve removal and might actually impair removal efficiencies by allowing the wastewater to become septic. The formula for detention time is as follows:

In International Standard units:

$$\text{Detention Time, days} = \frac{\text{Volume, m}^3}{\text{Flow, m}^3\text{/d}} \tag{4.4}$$

In U.S. customary units:

$$\text{Detention Time, days} = \frac{\text{Volume, gal}}{\text{Flow, gpd}} \tag{4.5}$$

To find the surface area of a circular clarifier, use the formula of a circle.

$$\text{Area of a Circular Clarifier} = 0.785 \times \text{Diameter} \times \text{Diameter}$$

or

$$\text{Area of a Circular Clarifier} = \pi \times (\text{Radius}^2) \tag{4.6}$$

Why are there two different formulas for finding the area of a circle or the volume of a cylinder? When the diameter of a circle is doubled, the area increased by a factor of 4. A 30-cm (12-in.) pipe carries 4 times as much flow as a 150-cm (6-in.) pipe when both pipes are running full and at the same velocity. Because the radius is half the diameter, the formula must be adjusted to compensate. The 0.785 factor is simply π or 3.14 divided by 4. Both formulas give the same answer, so use the one you are most comfortable with.

To find the surface area of a rectangular or square clarifier, simply multiply the length of the clarifier by its width.

To find the volume of a clarifier, multiply the surface area by the depth.

$$\text{Volume of a Circular Clarifier} = 0.785 \times \text{Diameter} \times \text{Diameter} \times \text{Height}$$

or

$$\text{Volume} = 0.785 \times (\text{Diameter})^2 \times \text{Height} \tag{4.7}$$

$$\text{Volume of a Rectangular or Square Clarifier} = \text{Length} \times \text{Width} \times \text{Height} \tag{4.8}$$

Because the dimensions of primary clarifiers are fixed, the SOR and detention time will vary with flow, resulting in variable removal efficiencies. Consequently, the number of units required for clarification may vary if the flow changes significantly.

Example Calculations for Surface Overflow Rate and Detention Time

A primary clarifier is 22.9 m (75 ft) in diameter and has a sidewater depth of 3.66 m (12 ft). The average daily influent flow to the facility it 18.9 ML/d (5 mgd). Calculate the SOR and the HDT. Compare your answers to accepted design criteria.

In International Standard units:

Step 1—Find the surface area of the clarifier using eq 4.6.

$$\text{Area} = 0.785 \times (\text{Diameter})^2$$

$$\text{Area} = 0.785 \times (22.9\text{ m})^2$$

$$\text{Area} = 411.66\text{ m}^2$$

The more traditional formula for area may also be used.

$$\text{Area} = \pi \times (\text{radius})^2$$

$$\text{Area} = (3.14) \times (11.45\ \text{m})^2$$

$$\text{Area} = 411.66\ \text{m}^2$$

Step 2—Convert the influent flow from ML/d to m^3/d.

$$\frac{18.9\ \text{ML}}{\text{d}} \left| \frac{1\,000\,000\ \text{L}}{1\ \text{ML}} \right| \left| \frac{1\ \text{m}^3}{1000\ \text{L}} \right| = 18\,900\ \text{m}^3/\text{d}$$

Step 3—Find the SOR using eq 4.2.

$$\text{Surface Overflow Rate}, \frac{\text{m}^3}{\text{m}^2\cdot\text{d}} = \frac{\text{Flow, m}^3/\text{d}}{\text{Surface area, m}^2}$$

$$\text{Surface Overflow Rate}, \frac{\text{m}^3}{\text{m}^2\cdot\text{d}} = \frac{18\,900\ \text{m}^3/\text{d}}{411.66\ \text{m}^2}$$

$$\text{Surface Overflow Rate}, \frac{\text{m}^3}{\text{m}^2\cdot\text{d}} = 45.9$$

The design criteria for primary clarifiers have an SOR between 24.4 and 48.9 $m^3/m^2 \cdot d$. The calculated value for this clarifier is within the accepted range.

Step 4—Find the volume of the clarifier using eq 4.7.

$$\text{Volume} = 0.785 \times (\text{Diameter})^2 \times \text{Height}$$

$$\text{Volume} = 0.785 \times (22.9\ \text{m})^2 \times 3.66\ \text{m}$$

$$\text{Volume} = 1506.7\ \text{m}^3$$

Step 5—Find the detention time using eq 4.4.

$$\text{Detention Time, days} = \frac{\text{Volume, m}^3}{\text{Flow, m}^3/\text{d}}$$

$$\text{Detention Time, days} = \frac{1506.7\ \text{m}^3}{18\,900\ \text{m}^3/\text{d}}$$

$$\text{Detention Time, days} = 0.08$$

Step 6—Convert days to hours.

$$\frac{0.08\ \text{days}}{} \left| \frac{24\ \text{hours}}{1\ \text{day}} \right| = 1.9\ \text{hours}$$

The design criteria for primary clarifiers have detention times between 1.5 and 2.5 hours. The calculated value for this clarifier is within the accepted range.

In U.S. customary units:

Step 1—Find the surface area of the clarifier using eq 4.6.

$$\text{Area} = 0.785 \times (\text{Diameter})^2$$

$$\text{Area} = 0.785 \times (75\text{ ft})^2$$

$$\text{Area} = 4415.6\text{ sq ft}$$

The more traditional formula for area may also be used.

$$\text{Area} = \pi \times (\text{radius})^2$$

$$\text{Area} = (3.14) \times (37.5\text{ ft})^2$$

$$\text{Area} = 4415.6\text{ sq ft}$$

Step 2—Convert the influent flow from million gallons to gallons per day.

$$\frac{5\text{ mil. gal}}{\text{d}} \left| \frac{1\,000\,000\text{ gal}}{1\text{ mil. gal}} \right| = 5\,000\,000\text{ gal/d}$$

Step 3—Find the SOR using eq 4.3.

$$\text{Surface Overflow Rate, } \frac{\text{gpd}}{\text{sq ft}} = \frac{\text{Flow, gpd}}{\text{Surface area, sq ft}}$$

$$\text{Surface Overflow Rate, } \frac{\text{gpd}}{\text{sq ft}} = \frac{5\,000\,000\text{ gpd}}{4415.6\text{ sq ft}}$$

$$\text{Surface Overflow Rate, } \frac{\text{gpd}}{\text{sq ft}} = 1132.3$$

The design criteria for primary clarifiers have an SOR rate between 600 and 1200 gpd/sf at average daily flow. The calculated value for this clarifier is within the accepted range.

Step 4—Find the volume of the clarifier using eq 4.8.

$$\text{Volume} = 0.785 \times (\text{Diameter})^2 \times \text{Height}$$

$$\text{Volume} = 0.785 \times (75\text{ ft})^2 \times 12\text{ ft}$$

$$\text{Volume} = 52\,987.2\text{ cu ft}$$

Step 5—Convert volume in cu ft to mil. gal.

$$52\,987.2\text{ cu ft} \left| \frac{7.48\text{ gal}}{1\text{ cu ft}} \right| \left| \frac{1\text{ mil. gal}}{1\,000\,000\text{ gal}} \right| = 0.4\text{ mil. gal}$$

Step 6—Find the detention time using eq 4.5.

$$\text{Detention Time, days} = \frac{\text{Volume, mil. gal}}{\text{Flow, mgd}}$$

$$\text{Detention Time, days} = \frac{0.4\ \text{mil. gal}}{5\ \text{mgd}}$$

$$\text{Detention Time, days} = 0.08$$

Step 7—Convert days to hours.

$$\frac{0.08\ \text{days}}{} \left| \frac{24\ \text{hours}}{1\ \text{day}} \right| = 1.9\ \text{hours}$$

The design criteria for primary clarifiers have detention times between 1.5 and 2.5 hours. The calculated value for this clarifier is within the accepted range.

INFLUENT CHARACTERISTICS

As discussed earlier in this chapter in the section entitled "Theory of Operation", the characteristics of the influent wastewater will determine what percentage of the influent TSS, COD, and BOD is potentially removable by the clarifier. The clarifier can only remove readily settleable and floatable material. If an industrial user discharges a large quantity of soluble BOD, it will pass through the clarifier unchanged. This will cause the percent removal of BOD through the clarifier to decrease even when the clarifier performance has been optimized.

SETTLING VELOCITY

Particle Characteristics

Gravity pulls particles down at a certain rate. This is the settling velocity. Settling velocity is affected by particle size and shape and the difference in density between the particle and the surrounding wastewater. A dense particle settles more quickly than a light one, a particle with a large surface-area-to-weight ratio settles slowly, and a particle with an irregular shape settles slowly because of its high frictional drag. The size, shape, and density of particles entering a primary clarifier depend on whether they are inorganic or organic and on both the sewer network and the types of upstream unit processes. Inorganic particles tend to be dense and compact. They settle quickly. Organic particles tend to be less dense and may have open, irregular shapes. Organic particles may be partially broken down in the collection system, which reduces their size. Longer or warmer collection systems will encourage more breakdown of organic material than shorter or colder collection systems. Some preliminary treatment devices, such as comminuting and pumping, fragment the particles, reducing their ability to settle.

Freshness of the Wastewater

Solids in stale (septic) wastewater settle less readily than those in fresh wastewater because biological degradation of the stale wastewater reduces the particle sizes. The collection system affects the freshness of the wastewater. The time elapsed while wastewater flows through the sewer system depends on the physical characteristics of the system, such as sewer slope and length, number of pumping stations, and the method of operating the collection system. Regarding the latter, many raw wastewater pumping stations store flows in the sewer, thus extending flow time through the entire sewer system. This extended storage time diminishes the freshness of the wastewater. Large contributions of industrial wastewater may also affect the freshness of wastewater or the percentage of soluble vs particulate BOD. Soluble BOD cannot be removed by the primary clarifiers. An increase in the amount of soluble BOD may make it appear as though removal efficiency is decreasing. Another factor that affects freshness of wastewater is temperature. Typically, the warmer the wastewater temperature, the more likely it is to become septic.

As discussed in the section entitled "Primary Sludge Fermentation", soluble BOD and VFAs are key ingredients for biological phosphorus removal. Soluble BOD and VFAs are produced in the collection system, especially when the wastewater is older and warmer. Stale or septic wastewater may be beneficial for phosphorus removal, but wastewater that is too old can cause settleability problems.

Temperature

As wastewater cools, its density increases. Water achieves its maximum density at 4 °C (39.2 °F). Density does not change drastically when water is cooled from 20 °C (68 °F) to 10 °C (50 °F), a typical summer to winter water temperature change for many WRRFs; however, the change is big enough to affect how fast particles settle in the primary clarifier. Most of the organic particles in wastewater have densities that are fairly close to the density of water, so even small changes in water density can have big effects on clarifier performance. Colder water also becomes more *viscous*, which increases drag. Increased drag slows the settling velocity even more. Water resource recovery facilities that experience changes in influent wastewater temperatures from summer to winter should expect clarifier performance to be better in the summer than in the winter. As wastewater warms, it becomes less dense and viscosity decreases. Particles settle faster. At a water temperature of 27 °C (80 °F), the settling rate exceeds that at 10 °C (50 °F) by nearly 50% (WEF et al., 2018).

If something is *viscous*, it is thick, sticky, and gooey because of internal friction. Water becomes more viscous at colder temperatures because the water molecules are rubbing up against one another more often than they do at warmer temperatures when the water is less dense.

Warm weather increases the rate of biological activity, thus diminishing the freshness of the wastewater in the collection system and promoting gasification in the settling basins and slow settling. On the other hand, the lower viscosity of warm water, compared to that of cold water, allows particles to settle faster.

Temperature differences, and the associated density differences between the tank contents and the incoming flow, may create density currents within the basin. These currents can interfere with effective particle settling characteristics, thereby impairing settling performance.

ENHANCING SETTLING VELOCITY

Primary clarifier performance can be improved by encouraging more flocculation to take place between particles. Influent wastewater is sometimes gently mixed and aerated immediately upstream of the primary clarifier to encourage flocculation. Chemicals, including alum and ferric, may be used to remove more finely divided suspended and colloidal material than is possible with gravity settling alone. These chemicals react with constituents in the wastewater and with the water, or in combination with other added chemicals to form a heavy, flocculent precipitate. The settling precipitate traps the suspended and colloidal particles and *adsorbs* them on the floc surface. Some treatment facilities install chemical-feed equipment strictly for the addition of chemicals to the wastewater influent flow to improve settling. Other WRRFs use chemicals for odor control or phosphorus removal and, in the process, receive the additional benefit of improved solids capture in the primary clarifiers. However, most facilities with primary clarifiers do not use any chemicals to enhance settling.

When something is *absorbed*, it is taken up inside. Sponges absorb water. When something is *adsorbed*, it adheres to the outer surface. Road dust adsorbs onto a vehicle.

VARIABLES AFFECTING REMOVAL OF FLOATABLES

Efficiency of grease and scum removal depends on tank configuration, wastewater characteristics, and the type and condition of removal mechanism. Each of these are described below.

TANK CONFIGURATION

Primary tank baffling at the outlet must be continuous and deep enough to prevent grease and scum from traveling underneath the baffles. Grease removal efficiencies vary slightly depending on the equipment and the system used.

WASTEWATER CHARACTERISTICS

Two wastewater characteristics that influence grease and scum removal efficiency are temperature and pH. At summer wastewater temperatures or at a pH less than 7 (acidic), grease and scum may tend to stay in suspension, and some may enter the settled sludge instead of floating to the surface.

TEST YOUR KNOWLEDGE

1. **Hydraulic detention time can be defined as the amount of time, on average, water remains in a vessel.**
   - ☐ True
   - ☐ False

2. **As SOR increases, the velocity of water through the clarifier decreases.**
   - ☐ True
   - ☐ False

3. **A primary clarifier has a diameter of 36.6 m (120 ft) and a depth of 4.57 m (15 ft). Find the volume of the clarifier.**

____

4. **The influent flow to a primary clarifier is 30.3 ML/d (8 mgd). If the clarifier holds 2.27 ML (0.6 mil. gal), what is the HDT in hours?** _______________

5. **Find the SOR for a primary clarifier with a diameter of 36.6 m (120 ft) and an influent flow of 30.3 ML/d (8 mgd).**

____

6. **All of the following variables can affect the settling velocity of a particle EXCEPT**
   a. Size
   b. Shape
   c. Density
   d. Color

7. **A long, warm collection system may have this effect on the influent wastewater:**
   a. Decreases particle size
   b. Increases ratio of inorganic to organic particles
   c. Decreases ammonia concentration
   d. Increases opportunities for flocculation

8. **As water temperatures decrease in a primary clarifier,**
   a. Particles will settle slower
   b. Odor potential increases
   c. Particles will settle faster
   d. Water becomes less viscous

9. **A small WRRF decides to accept hauled septage waste from a nearby campground. On days when septage is received**
   a. Percent removal of TSS increases
   b. Percent removal of soluble BOD increases
   c. Percent removal of both TSS and BOD decreases
   d. Percent removal is unaffected

10. **Cold water affects settling velocity because**
   a. Cold water contains more dissolved gases
   b. There is less drag on particles
   c. Microbial activity increases as temperatures drop
   d. Water is denser at colder temperatures

11. **These conditions tend to entrain grease and prevent it from being skimmed off in the primary clarifier.**
   a. Presence of ducks
   b. Low pH, high temperatures
   c. Lack of turbulence in the sewer
   d. High pH, low temperatures

Process Control

The factors affecting primary clarifier performance are summarized in Table 4.1. Not all of these factors are under the control of the operator.

Good process control involves seven operational aspects: hydraulic considerations, solids handling, skimming of floatables, recycle flow management, odor control, housekeeping, and chemical addition (if applicable).

HYDRAULIC CONSIDERATIONS

The two hydraulic control variables for the primary influent are the flowrate and the number of tanks in service. Both will affect the SOR and DT. Because most facilities must accept flow as it enters the facility, flow control is limited. Some facilities have *equalization basins* to smooth diurnal flow variations. If equalization basins are available, influent flow can be diverted during periods of high flow, such as a storm event. Later, the stored wastewater may be pumped back to the primary clarifiers at a steady rate. Facility recycle flows should be timed to smooth rather than aggravate variable hydraulic loading of the primary treatment units.

Both the SOR and detention time should be calculated for average daily flow and peak hour flow for process control. Verify that the SOR and detention time are within the accepted ranges given under the section on primary clarifier design criteria. If the SOR is too high or the detention time is too low, additional tanks should be placed into service. If the SOR is too low or the detention time is too high, tanks should be removed from service. The following operational considerations apply to removing primary tanks from service:

- Remove a tank from service less frequently than monthly,
- Rotate the removal of tanks from service,
- Keep idle tanks filled with facility effluent water, and
- Exercise mechanisms of idle tanks daily.

Partially buried clarifiers and other large tanks can be pushed out of the ground by groundwater. While there is not a lot of upward pressure, tanks have a lot of surface area. When you combine a small amount of pressure with a lot of area, you get force. If the force is more than the weight of the empty tank, it can "float" upward. While a tank full of final effluent may grow algae that will have to be removed before putting the clarifier back into service, it is preferable to accidentally float a tank. Keeping the clarifier mechanism submerged also helps protect it from UV radiation, which can damage squeegees and other nonmetal components.

Table 4.1 Process Control Variables for Primary Clarifiers

Parameter	Monitor	Adjustment
Influent wastewater characteristics	Percentage of settleable material vs soluble	No adjustment; Monitor only
Surface overflow rate	Upward velocity	Number of basins in service
Hydraulic retention time	Settling time Flocculation of smaller particles into larger particles	Number of basins in service
Solids loading rate	Total mass of solids entering the clarifier	Number of basins in service
Chemical addition	Flocculation Odor control Phosphorus precipitation	Chemical dosage to maximize flocculation and minimize chemical usage
Preliminary treatment	Carryover of grit and screening material	Upstream equipment issues such as pipe and equipment plugging
Sludge collection mechanism speed if connected to a variable frequency drive	Solids buildup in one or more areas Clarifier blanket depth	Collection mechanism Note that most collection mechanisms operate at constant speed and cannot be adjusted by the operator.
Sludge withdrawal rate	Sludge concentration Clarifier blanket depth; septicity and odor	Pump cycle time and duration
Nature and amount of internal recycle streams	Total mass of solids entering the clarifier Fine particles	Timing and duration of recycle streams and/or adjustments to the primary sludge withdrawal rate

Equalization means to even out the load, typically over a 24-hour period. Some facilities are equipped with flow equalization tanks. Flow, in this case supernatant, flows into the equalization tank as it is produced or received. Flow is taken out of the tank at a constant rate. The water level in the tank will rise and fall because sometimes flow goes in faster than it is removed and sometimes flow is removed faster than it is replaced.

If the primary tank detention times are short because of a shortage of tanks, then the operator's flexibility for tank maintenance is limited because the tank downtime must be kept to a minimum. Some procedural options include the following:

- Use all available tanks for primary influent settling so that one primary tank at a time can be temporarily removed from service.
- Reduce downtime of tanks by a comprehensive preventive maintenance program.
- Minimize the effect of recycle flows by altering their timing or time of day, redirecting them to another portion of the facility and eliminating nonessential flows.
- Add chemicals to increase removal of settleable solids, while one tank is removed from service.
- Use flow equalization basins, if available, for temporary storage of influent flow.

Ensuring equal flows among multiple tanks and uniform flow distribution across the width of each rectangular tank (or around the perimeter of circular tanks) are important hydraulic objectives. Good flow distribution will minimize variations in sludge blankets and aid overall performance. Clean, level weirs and properly adjusted influent gates help equalize flows to each tank. Because flows vary throughout the day, the gates may require several adjustments during the day to maintain good flow distribution. Influent flow distribution boxes, with an upflow pipe and influent weirs for each tank, have been installed in New York City and Nassau County, New York. The use of influent weirs allows the gate to be left wide open and accounts for flow variations throughout the day.

SOLIDS HANDLING

Two variables, sludge thickness (or solids concentration) and quantity, affect handling operations. The quantity depends on thickness and the suspended solids loading and removal efficiency of the primary tanks. These two variables and their interaction with the two sludge handling activities—collection and pumping—are discussed below.

SLUDGE THICKNESS (SOLIDS CONCENTRATION)

The desired thickness of the sludge removed from primary clarifiers depends on the type of solids-handling facilities downstream. If the sludge is pumped to thickeners, a lower sludge concentration (0.5 to 1.0%) may be permissible. If the sludge is pumped to digesters or a dewatering process, a thicker concentration is desirable.

Most primary sedimentation tanks can be operated to consistently produce a thickened solids concentration of 3 to 6% solids by allowing a 0.6- to 0.9-m (2- to 3-ft) blanket of solids to build up and compact the primary sludge (WEF, 2005). Higher concentrations can be achieved by some facilities, but this often causes problems with the sludge collection and/or pumping systems. The solids concentration achievable ultimately depends on the characteristics of the wastewater entering the primary clarifier. Wastewater that contains more grit and dense solids will yield a higher solids concentration. Wastewater with very little grit and fewer dense solids may produce sludge that is only 1 to 2% total solids. The process control goal is to find a blanket depth or level that provides as thick a sludge as can be achieved for that wastewater without adversely affecting tank removal efficiency, overloading collector equipment, or allowing decomposition in the bottom of the clarifier.

SLUDGE QUANTITY

The operator may estimate the volume of sludge to be removed from the primary tank by measuring the primary influent and effluent suspended solids concentrations. Solids that do not pass through to the final effluent must become part of the sludge blanket. The mass of sludge in the clarifier blanket can be calculated using the following equations:

In International Standard units:

$$\text{Dry Solids Removed, } \frac{\text{kg}}{\text{d}} = \frac{\left(\text{Flow, } \frac{\text{m}^3}{\text{d}}\right)\left(\text{Concentration, } \frac{\text{mg}}{\text{L}}\right)}{1000} \tag{4.9}$$

In U.S. customary units:

$$\text{Dry Solids Removed, } \frac{\text{lb}}{\text{d}} = (\text{Flow, mgd})\left(\text{Concentration, } \frac{\text{mg}}{\text{L}}\right)\left(8.34\frac{\text{lb}}{\text{mil. gal}}\right) \tag{4.10}$$

If the total solids concentration is less than 7% and the specific gravity of the sludge is unknown, use a value of 1.0 for the specific gravity. Raw primary sludge typically has a specific gravity of 1.0 plus 0.010 for every 1% of total solids (U.S. EPA, 1987).

If the concentration of the settled sludge is known, then the volume to be removed from the bottom of the clarifier can be calculated using the following equations:

In International Standard units:

$$\text{Wet Sludge Removed, } \frac{\text{kg}}{\text{d}} = \text{Dry Solids Removed, kg/d} \times \left(\frac{100}{\text{Dry solids in sludge, \%}}\right) \tag{4.11}$$

For liquids with a specific gravity of 1.0 (equal to water), 10 000 mg/L is equal to 1%. Consider that 1 mg/L is the same as 1 ppm. If you take 1 000 000 and multiply by 1% (0.01), the result is 10 000.

In U.S. customary units:

$$\text{Wet Sludge Removed, } \frac{\text{lb}}{\text{d}} = \left(\frac{\text{Dry solids removed, lb/d} \times 100}{\text{Dry solids in sludge, \%}}\right) \tag{4.12}$$

In International Standard units:

$$\text{Wet Sludge Removed, } \frac{\text{m}^3}{\text{d}} = \frac{\text{Wet sludge removed, kg/d}}{998.2\frac{\text{kg}}{\text{m}^3} \times \text{Specific gravity of sludge}} \tag{4.13}$$

In U.S. customary units:

$$\text{Wet Sludge Removed, gpd} = \frac{\text{Wet sludge removed, lb/d}}{8.34\frac{\text{lb}}{\text{gal}} \times \text{Specific gravity of sludge}} \tag{4.14}$$

From the above equations, the operator may derive the approximate duration of pumping for a known pump capacity. Lack of daily analytical information may prevent daily use of the above equations; nonetheless, use of estimated values will provide an appraisal of wet sludge volumes valuable to the operator as a general guideline. The operator may then adjust the guideline based on daily observations of the sludge blanket levels and the guideline number determined weekly using eqs 4.9 through 4.14. Of course, chemical addition will generate more material for removal than that indicated by these equations.

Operators should calculate the amount of solids into and out of the primary tank to ensure that the system is operating efficiently. Calculating the balances in and out of the tank provides a useful tool for operators to monitor the system to ensure that it is within operating parameters.

Example Calculation for Estimating Sludge Pumping Time

A primary clarifier receives an influent flow of 18.9 ML/d (5 mgd). The primary influent TSS concentration is 250 mg/L and the primary effluent concentration is 125 mg/L. The settled sludge contains 5% total solids. The primary sludge pump has a capacity of 378.5 L/min (100 gpm). How many minutes will the sludge pump need to run over a 24-hour period?

In International Standard units:

Step 1—Convert the influent flowrate in ML/d to m³/d.

$$\frac{18.9\text{ ML}}{\text{d}}\left|\frac{1\,000\,000\text{ L}}{1\text{ ML}}\right|\left|\frac{1\text{ m}^3}{1000\text{ L}}\right| = 18\,900\text{ m}^3/\text{d}$$

Step 2—Find the dry solids removed using eq 4.9.

$$\text{Dry Solids Removed, }\frac{\text{kg}}{\text{d}} = \frac{\left(\text{Flow, }\frac{\text{m}^3}{\text{d}}\right)\left(\text{Concentration, }\frac{\text{mg}}{\text{L}}\right)}{1000}$$

$$\text{Dry Solids Removed, }\frac{\text{kg}}{\text{d}} = \frac{(18\,900\text{ m}^3/\text{d})\left(250\frac{\text{mg}}{\text{L}} - 125\text{ mg/L}\right)}{1000}$$

$$\text{Dry Solids Removed, }\frac{\text{kg}}{\text{d}} = 2363$$

Step 3—Find the wet solids removed in kg/d using eq 4.11.

$$\text{Wet Sludge Removed, }\frac{\text{kg}}{\text{d}} = \text{Dry solids removed, kg/d} \times \left(\frac{100}{\text{Dry solids in sludge, \%}}\right)$$

$$\text{Wet Sludge Removed, }\frac{\text{kg}}{\text{d}} = 2363\text{ kg/d} \times \left(\frac{100}{5\%}\right)$$

$$\text{Wet Sludge Removed, }\frac{\text{kg}}{\text{d}} = 47\,260$$

Step 4—Convert wet solids in kilograms per day to volume in m^3/d using eq 4.13.

$$\text{Wet Sludge Removed, }\frac{m^3}{d}=\frac{\text{Wet sludge removed, kg/d}}{998.2\dfrac{kg}{m^3}\times\text{Specific gravity of sludge}}$$

$$\text{Wet Sludge Removed, }\frac{m^3}{d}=\frac{47\ 260\text{ kg/d}}{998.2\dfrac{kg}{m^3}\times 1.0}$$

$$\text{Wet Sludge Removed, }\frac{m^3}{d}=47.34$$

Step 5—Convert wet solids in m^3/d to liters per day.

$$\frac{47.34\ m^3}{d}\left|\frac{1000\ L}{m^3}\right|=47\ 340\text{ L/d}$$

Step 6—Use the detention time formula (4.4) to find the total pump time. Units may be different than the ones used in formula (4.4) as long as the units on the top of the equation agree with the units on the bottom of the equation.

$$\text{Detention Time}=\frac{\text{Volume}}{\text{Flow}}$$

$$\text{Detention Time}=\frac{47\ 340\text{ L sludge}}{378.5\text{ L/min pump flow}}$$

$$\text{Detention Time}=125\text{ minutes}$$

In U.S. customary units:

Step 1—Find the dry solids removed using eq 4.10.

$$\text{Dry Solids Removed, }\frac{lb}{d}=(\text{Flow, mgd})\left(\text{Concentration, }\frac{mg}{L}\right)\left(8.34\frac{lb}{MG}\right)$$

$$\text{Dry Solids Removed, }\frac{lb}{d}=(5\text{ mgd})\left(250\frac{mg}{L}-125\text{ mg/L}\right)\left(8.34\frac{lb}{MG}\right)$$

$$\text{Dry Solids Removed, }\frac{lb}{d}=5212.5$$

Step 2—Find the wet solids removed in pounds per day using eq 4.12.

$$\text{Wet Sludge Removed, }\frac{lb}{d}=\frac{\text{Dry solids removed, }\dfrac{lb}{d}\times 100}{\text{Dry solids in sludge, \%}}$$

$$\text{Wet Sludge Removed, }\frac{lb}{d}=\frac{5212.5\times 100}{5\ \%}$$

$$\text{Wet Sludge Removed, }\frac{lb}{d}=104\ 250$$

Step 3—Convert wet solids in pounds per day to volume in gallons per day using eq 4.14.

$$\text{Wet Sludge Removed, gpd} = \frac{\text{Wet sludge removed, lb/d}}{8.34\frac{\text{lb}}{\text{gal}} \times \text{Specific gravity of sludge}}$$

$$\text{Wet Sludge Removed, gpd} = \frac{104\ 250\ \text{lb/d}}{8.34\frac{\text{lb}}{\text{gal}} \times \text{Specific gravity of sludge}}$$

$$\text{Wet Sludge Removed, gal} = 12\ 500$$

Step 4—Use the detention time formula (eq 4.5) to find the total pump time. Units may be different than the ones used in formula (4.5) as long as the units on the top of the equation agree with the units on the bottom of the equation.

$$\text{Detention Time} = \frac{\text{Volume}}{\text{Flow}}$$

$$\text{Detention Time} = \frac{12\ 500\ \text{gal of sludge}}{100\ \text{gpm pump flow}}$$

$$\text{Detention Time} = 125\ \text{minutes}$$

COLLECTION

The collectors may run continuously or intermittently, depending on the type of facility, equipment, and characteristics of the wastewater. If the facility is designed to pump a thin sludge to thickeners or to a degritting unit, the collectors must run continuously. If the primary tanks are square or circular, the sludge collection mechanisms are typically run continuously because sludge movement to the hoppers requires more time than that for rectangular tanks. Continuous operation avoids the possibility of overloading the mechanism with accumulated solids. Regardless of tank type or facility design, continuous collector operation allows easy operation of an automatic withdrawal system. Because of its many advantages, continuous sludge collector operation is preferred for most facilities.

Intermittent operation of the collectors may be necessary if the primary tanks are used to thicken the waste sludge from the secondary process or if primary sludge is pumped directly to digestion or dewatering units. As a good practice for intermittent operation of rectangular tanks, the operator may start the collector mechanisms approximately 1 hour before sludge is drawn and then turn off the collector when pumping stops. This procedure helps avoid overloading the collector mechanisms or packing the sump.

SLUDGE PUMPING

Positive-displacement pumps such as piston, progressive cavity, or diaphragm pumps are typically used to remove sludge from the primary clarifiers. Piston pumps are rarely installed in new facilities, but many remain in service in older facilities because of their reliability. Many facilities use centrifugal torque flow pumps or hose pumps for primary sludge. Others use centrifugal pumps. The type of pump selected depends, in part, on the expected sludge solids concentration. Centrifugal pumps work well with sludge that is less than 5% total solids, while positive displacement pumps are better suited to higher total solids concentrations. Pump capacity should be coordinated with sludge hopper size, whether the clarifier is circular or rectangular. The pump should evacuate one hopper volume each pumping cycle. The control sequence may use automatic timers and the timing cycle should be adjustable so that the pump is not started until the hopper has had a chance to refill with sludge. The operator will need to periodically monitor the sludge blanket in the clarifier and primary sludge concentration. If the sludge pump discharge line is equipped with a sampling spigot, it may be used to collect a sludge sample in a bucket or other container for inspection. Be sure to let the sludge pumps run long enough that the sludge coming out of the sample spigot is representative of current operating conditions. The pumping time can be increased in response to a buildup of the sludge blanket or decreased in response to a decrease in the sludge solids concentration. If facility influent is prone to septicity or other site-specific conditions that prevent sludge from settling well, the primary clarifiers may be operated with continuous withdrawal of dilute primary sludge to minimize thickening, maximize removal, and prevent anaerobic decomposition of settled solids. Anaerobic or septic conditions will result in resolubilization of BOD and can cause rising sludge blankets and poor removal of solids.

For more information on different types of pumps, refer to Chapter 8 of MOP 11. Another excellent reference for centrifugal pumps is *Pumps & Pumping* by ACR Press.

With several pumps, each pump may withdraw sludge from a single hopper simultaneously. A single pump should withdraw from only one hopper at a time to prevent unequal sludge withdrawals from the individual tanks. Pumping for short durations at frequent intervals is a good practice, and pumping for long periods at infrequent intervals should be avoided because excessive sludge to an anaerobic digester can upset the process. Ideally, pumping approaches continuous operation to evenly feed downstream processes. Some facilities vary the amount of sludge pumped with diurnal flow or historical solids loading patterns.

The proper solids concentration of sludge pumped (see Chapter 8) from primary clarifiers varies from facility to facility. The operator monitors solids concentrations from sample ports in the discharge lines and measures sludge levels before and after pumping to ensure that the proper pumping frequency and duration have been selected. Some techniques used to measure sludge blankets include using a pole to "feel" the height of the blanket, sample aspiration at various depths, core sampling, and electronic light-penetration measurement. Routine monitoring of the torque indicator, when available, will give the operator an indication of the sludge blanket depth and density. By observing and recording daily torque readings, the operator can determine whether the rate of primary sludge pumping needs to be adjusted. Using this observation in conjunction with other operational parameters such as sludge blanket depth, volume of sludge pumped, and visual/olfactory observations (black or septic color, increased odor, and rising/floating/clumping sludge) should prompt the operator to adjust the sludge pumping rate.

Many facilities use automatic sludge pumping controlled with either a timer, a programmable logic controller (PLC), or a timer with a density meter. With the first and second options, preset times operate the pumps for each tank at the selected frequency and duration. With the third option, the timer turns the pump on and the density meter shuts the pump off. In this option, sludge thickness governs the pumping time. With any type of automatic system, routine level measurements are necessary to ensure the operator that sludge buildups are not occurring. Typical WRRFs designed today do not include density meters because of their expense and maintenance needs. In addition, WRRFs designed today use centralized computer systems to control sludge pumping and/or grease removal.

PUMPING APPURTENANCES

Appurtenances include cleanouts in grease lines, sludge lines, pumps, flushing systems, and line-cleaning devices. Flushing of lines, pumps, and sumps uses high-pressure water or air. Sludge lines may also be cleaned with steam or hot water. Line cleaning equipment includes hydraulically propelled tools (pigs) of metal or plastic and special fittings to insert and catch the cleaning device.

Safety Tip! Biological activity in sludge continues even when the sludge is within pumps and pipelines. This activity can generate hydrogen sulfide and methane gas. Sludge should never be allowed to sit undisturbed in pipelines and pumps. It is extremely important to be sure to flush all lines and pumps to be sure they are clean before taking them out of service. This includes pump rotation where a pump may only be out of service for several days. Failure to properly clean and store equipment can result in excessive gas buildup, which may explode.

Cleanouts, typically provided in sludge lines and pumps, need to be checked if pumping trouble occurs. All lines should be equipped for cleaning with hydraulically propelled tools or balls because regular sludge line cleaning reduces jamming at valves or fittings. The frequency of line cleaning depends on velocities in the pipe, the grease content of the sludge, grit in the pipe, chemical constituents in the line, and temperature. As the weather becomes colder, grease coatings become thicker and tougher, which means that line cleaning may be necessary as often as twice monthly.

Flushing systems are used to clean sludge pumps, suction lines, and discharge piping. If a flushing system is provided, a few minutes of weekly flushing helps to keep lines clean. Whenever sludge becomes difficult to pump, the system should be flushed. Facility water connections are typically provided at the end of the suction line to clear the sludge pump. At some facilities, compressed air is also used for cleaning a packed sump or suction line.

TEST YOUR KNOWLEDGE

1. Sludge blanket depths should generally be kept below 0.9 m (3 ft).
   - ☐ True
   - ☐ False

2. Managing the collection system can help reduce odors at the WRRF.
   - ☐ True
   - ☐ False

3. Given the following information, calculate the mass of dry solids that will end up in the clarifier blanket over a 24-hour period. Primary influent TSS is 280 mg/L and primary effluent TSS is 168 mg/L. The average daily flow through the WRRF is 12.5 ML/d (3.3 mgd).

   ____

4. A primary clarifier captures 2671 kg (5885 lb) of solids. If the total solids concentration in the sludge is 4.2%, how much volume will need to be pumped? Express your answer as m^3 (gal).

   ____

5. A primary clarifier produces 20.2 m^3 (5333 gal) of primary sludge at a concentration of 4.2% total solids. If the sludge pump has a capacity of 435 L/min (115 gpm), how many minutes will the sludge pump need to operate each day?

   ____

6. Thickened sludge concentrations greater than 6% total solids
   a. Are easily achieved in primary clarifiers.
   b. Can cause problems with sludge collection and pumping.
   c. Deliver excess water to anaerobic digesters.
   d. Would be considered "thin" sludge.

7. The primary sludge concentration is approaching 8% total solids. A high-torque alarm has sounded for the drive mechanism. The operator should
   a. Replace the shear pin with a stronger model.
   b. Decrease the clarifier mechanism speed.
   c. Increase the sludge pumping rate to decrease the total solids concentration.
   d. Remove a clarifier from service to decrease detention time.

8. A WRRF begins adding ferric chloride to its influent channel for odor control. Primary sludge production
   a. Increases
   b. Decreases
   c. Is unaffected
   d. Doubles

9. When sludge pumping is intermittent
   a. The total solids concentration in the sludge tends to be lower
   b. Sludge lines should be flushed with water
   c. The risk of septicity and hydrogen sulfide generation is lower
   d. Pump cycle times should be coordinated with sludge hopper size

10. This week, the percentage of BOD removal over the clarifier has dropped from 40 to 25%; however, the percent removal of TSS is unchanged. What is the most likely cause?
   a. Recycling of solids and BOD from the anaerobic digester
   b. Industrial discharge of soluble BOD
   c. Temperature changes affecting settling velocities
   d. Collection mechanism is broken

FLOATABLE SOLIDS SKIMMING

Grease, fats, oils, plastics, and other floatables must be consistently removed. Skimmings and floatables removed from primary tanks are pumped to various disposal facilities in WRRFs. Some facilities transfer their skimmings to concentrators or dissolved air flotation thickeners. The concentrated skimmings are then transferred to an incinerator or trucked off-site for disposal. Other facilities transfer skimmings to dewatering facilities such as belt presses, whereas others pump their skimmings to anaerobic digesters; this practice of pumping is gradually being phased out because excessive grease in digesters causes scum buildup and foaming. The removal frequency depends on the amount of floatables in the incoming wastewater, flow variations throughout the day, and wastewater temperature. Because of their design, circular tanks are skimmed continuously; most rectangular tanks remove skimmings periodically. Rectangular tanks need skimming at least once a day and more frequently whenever grease or floatables appear in the primary effluent, secondary influent channels, or the final clarifiers. Rectangular tanks typically use scum troughs that are manually operated, or they can be automatically operated by a timer or PLC. Control of odors may also require more frequent skimming. The skimming process must capture enough wastewater to allow scum conveyance through pumps, screw conveyors, and pipelines.

RECYCLE FLOW MANAGEMENT

Recycle flow management can improve primary clarifier performance by minimizing hydraulic and solids loadings while reducing the potential for septicity. Good recycle flow management includes the following actions:

- Identify each recycle flow and its duration and quality;
- Assess the effect of each recycle on gravity settling performance; and
- Eliminate, reduce, redirect, or smooth the recycle flows as indicated by the evaluation.

INTERACTION WITH OTHER UNIT OPERATIONS

Process treatment objectives require all process units to be operated in an interrelated manner, and primary clarification units are no exception. A properly operated primary clarifier helps with the operation of downstream units. Conversely, proper operation of the primary clarifier depends, to a large degree, on the sound operation of several unit operations, both upstream and downstream. Entry-level operators may wish to return to this section once they are more familiar with the different types of treatment processes that can be present elsewhere in the WRRF.

UPSTREAM UNITS

The upstream units affecting primary clarification are described below, including comminuting (grinding), bar screens, grit removal, and influent pumping.

Comminuting and grinding operations reduce floating and entrained solids concentrations, changing their settling properties. Improper comminution leads to the carryover of rags and debris that clog and tangle the sludge collecting and pumping mechanisms of primary tanks.

Bar screens remove debris larger than the screen openings. This debris would otherwise interfere with other facility processes. Inefficient removal of such debris results in carryover of rags and debris that will clog sludge pumps and tangle in primary tank collectors. Fine bar screens can fragment organic solids, altering their settling properties.

The grit chamber removes discrete inorganic particles. If grit enters the primary tanks, pumping and other problems will ensue. The abrasive grit causes excessive wear of collector mechanisms, pump impellers, and sludge lines. In addition, a buildup of grit in the primary tank will reduce its effective volume for sludge storage and reduce the tank's detention time.

At influent pumping stations, safe operation depends on careful operation of the primary tanks. When primary tanks are drained, large quantities of septic sludge may enter the wet well of the pumping station. This situation could result in dangerous accumulations of hydrogen sulfide or combustible gases.

DOWNSTREAM UNITS

The downstream units interacting with the primary clarifiers include secondary treatment processes, solids handling systems, and recycle flows from liquid and solids processes. These units are described below.

SECONDARY PROCESSES

Improper primary tank operation may overload the secondary process with solids and BOD. Primary settling removes settleable solids and suspended solids and their associated BOD efficiently—if the process is properly designed and operated. If settleable solids are allowed to pass through to the secondary system, the amount of organic material passing through will also increase. Remember, the solids entering the WRRF are mostly organic. More organics in the secondary process increases energy usage because the blowers must provide more oxygen to treat them. More solids passing to the secondary process will increase the amount of secondary sludge produced, which may adversely affect the operation of the secondary clarifiers. Grease carryover from primary tanks may interfere with the activated sludge or fixed-film (trickling filter or rotating biological contactor) biological system operations, possibly impairing final effluent quality. If a primary tank is being operated as a *fermenter* or there is a separate thickener/fermenter, care must be taken to ensure proper recycle rates and solids retention times. If these are not properly controlled, it could hinder biological phosphorus removal.

Fermenters break down organic material in the absence of oxygen and nitrate. Beer and wine are both made by fermentation. When a primary clarifier is used as a fermenter, the sludge is allowed to remain in the clarifier blanket long enough for the bacteria in the sludge to start breaking it down. Because there is no oxygen or nitrate in the sludge blanket, the breakdown is incomplete. A group of smaller, simpler organic compounds called *VFAs* are produced. Volatile fatty acids are important in biological phosphorus removal and can encourage the growth of some nuisance organisms in the activated sludge process.

SOLIDS HANDLING SYSTEMS

Operation of primary tanks may affect solids-handling systems in several ways. For WRRFs without prethickening of the removed primary sludge, pumping of thin sludge may reduce facility digester capacity (either aerobic or anaerobic), cause hydraulic overloading of concentration tanks and thickening processes, and increase sludge heating costs. The continuous pumping of dilute sludge could result in the failure of the digestion process. If sludge becomes septic, downstream solids-handling processes, including thickening and dewatering, may be adversely affected with resultant odor problems or poor operating efficiencies.

Pumping of toxic primary sludges to aerobic or anaerobic digesters may result in digester failure. The operator typically can prevent process failure caused by toxicity only by the following:

- Watching for obvious physical changes in the wastewater such as color, odor, or consistency;
- Altering unit process operations as appropriate, if changes are detected;
- Notifying their supervisor or industrial pretreatment personnel if industrial contributors are suspected; and
- Watching for abrupt changes in the chemistry of the anaerobic digestion process, such as pH, alkalinity, gas production, methane content, and volatile acid level.

RECYCLE FLOWS

Recycle flow sources include thickening, conditioning, dewatering, digestion, secondary and tertiary processes, and cooling waters. These are described below.

Thickeners or dewatering units, when properly operated, capture and concentrate the solids. After thickening, the sludge concentration is higher. The water removed from the sludge should be low in BOD and suspended solids and should not impair primary clarification. The extracted water goes by different names depending on what type of thickening or dewatering process was used. Belt filter presses produce filtrate. Centrifuges produce centrate. Gravity thickeners and digesters produce supernatant. Polymers are typically added to dewatering processes to improve solids capture. Use of polymers improves return-flow quality and generally will not affect primary tank performance.

Excess water must be removed from aerobic and anaerobic digesters to make room for more solids. Operators decant the digester by allowing the solids in the digester to settle. The water can then be removed from the digester. This decanted water is the supernatant. Supernatant from digesters may contain high concentrations of TSS, especially when the digesting sludge settles poorly. If too many solids are sent to primary settling basins, they can overwhelm the sludge pumps and the blanket depth will increase. This could cause septicity, odors, and solids carryover. If the addition of high solids supernatant liquor to primary tanks cannot be avoided, the supernatant loading should be *equalized* to reduce its effects. Waste activated sludge from secondary processes is sometimes pumped to the primary clarifiers for thickening with the primary sludge. Note that co-settling waste activated sludge in primary clarifiers reduces the allowable SOR. Caution must be used when waste activated sludge is co-settled with primary sludge. If biological phosphorus removal is used, high concentrations of phosphorus will be released and recycled to the secondary process, which may result in failure of the biological phosphorus removal process. This practice can cause pumping problems, septicity, and solids carryover if the extra solids are not accounted for during removal from the primary tank. If waste sludge must return to the primary tank, an even pumping rate, preferably limited to low- or average-flow periods, is necessary. Also, sending waste activated sludge to the primary clarifiers will reduce the solids concentration of the primary sludge, resulting in longer pumping times to the solids-handling facilities. It is important that operations personnel take into account the quantity of waste activated sludge and primary sludge that accumulate in a primary clarifier because this combined sludge must be removed from the tank. If the total co-settled quantity of sludge is not taken into account in the sludge withdrawal process, the primary tanks will become overloaded with sludge and sludge may be carried over to the secondary treatment process, thus hindering that process. Some facilities have reported advantages to the co-settling of waste activated and primary sludge. These advantages have included improved suspended solids removal, reduced odors, and improved anaerobic digester operation.

Tertiary processes, such as sand filters, return backwash flows to the primary tanks, which increases the hydraulic loading on the tank and could reduce the settling efficiencies. Return of cooling waters primarily affects hydraulic loading.

TEST YOUR KNOWLEDGE

1. **Chemical addition for odor control and phosphorus removal upstream of the primary clarifiers can increase the percentage of BOD and TSS removed by the primary clarifier.**
   - ☐ True
   - ☐ False

2. **Recycle flows from other WRRF processes are best returned to the primary clarifiers during peak hour influent flow.**
   - ☐ True
   - ☐ False

3. **The surface of a primary clarifier has a buildup of scum and foam and smells strongly of rotten eggs. The operator should**
   a. Adjust the sludge pumping frequency
   b. Remove the scum baffle
   c. Increase the skimming frequency
   d. Verify proper operation of the sludge pump

4. **A poorly functioning grit basin or bar screen**
   a. Retains excessive quantities of grit
   b. Reduces odors upstream of the primary clarifier
   c. Generates soluble BOD
   d. May pass debris on to the primary clarifier

5. **A WRRF has two primary clarifiers. One must be taken out of service for cleaning and maintenance. While this is being done, the SOR in the remaining clarifier will be excessively high. What is the likely effect to overall facility operations?**
   a. Longer HRTs through the remaining clarifier
   b. Increased BOD and TSS loading to the secondary treatment process
   c. Reduced efficiency of the grit basin and screening equipment
   d. Increased BOD and TSS loading to the solids handling process

Operation of Primary Clarifiers

Under normal operating conditions for a primary clarifier, the operator should visually check the equipment at least once per shift to ensure that it is functioning properly. Verify that the sludge collection mechanism is moving smoothly. There should not be any unusual noises, scraping sounds, or jerking motion of the sludge collector or skimmer. Verify proper operation of the primary sludge pumps (if present). Use of local controls allows immediate observation of the operation of the pumps, even though they may be inactive when inspection begins.

PLACING A PRIMARY CLARIFIER INTO SERVICE

Step 1—Remove any debris from the tank floor and sludge hopper.

Be sure to follow all required safety procedures when filling, draining, or cleaning a tank.

Step 2—Inspect all equipment for wear and proper operation. Some items to check include the following:

With a chain-and-flight system, verify that the wear shoes, wear strip, and rails are in good condition. Replace worn parts as needed. Verify that the flights are level and parallel to the floor and that squeegees (if present) are in good condition. Adjust as necessary. Inspect the chains. Verify that both chains have the same number of links. Adjust chain tension according to the manufacturer's directions. This may require removal of a link if the chain has stretched. Check all bolts and tighten as needed.

With a circular clarifier mechanism, verify that the plows are level and parallel to the floor and that squeegees (if present) are in good condition. Adjust as necessary. The plows should ride easily over the basin floor. Check all bolts and tighten as needed.

Step 3—Be sure the sludge collection drive mechanism is properly lubricated. For chain-and-flight mechanisms, each flight should be manually lifted to ensure it has not adhered to the bottom of the tank. Flights with metal wear shoes can rust into place.

Step 4—Drain condensate from the clarifier drive. Verify proper lubrication levels. Check drive alignment.

Step 5—Start the sludge collection mechanism in a dry tank. With a circular mechanism, observe several full rotations. Check for scraping, binding, or jerking. With rectangular tanks, it may be necessary to add enough water to cover the flights and lower the sprockets before starting the mechanism. Some rectangular tanks use water to lubricate the lower shaft bearings rather than oil.

Step 6—Adjust plows or flights if necessary.

Step 7—Be sure ancillary equipment is in good operating condition.

Step 8—Open slide gates or valves at the clarifier outlet followed by the clarifier inlet. Flow should be able to move easily into and out of the basin. Continue operating the clarifier mechanism.

Step 9—When the sludge hopper is full, start the primary sludge pumps and allow them to operate for one full cycle. Verify that the pumps are functioning properly.

Step 10—Continue filling the tank and place all equipment into automatic operation.

REMOVING A PRIMARY CLARIFIER FROM SERVICE

Tanks should be removed from service annually for inspection of submerged equipment and the tank itself. At this time, operators should also check for cracks in the concrete, spalling, and evidence of corrosion.

Step 1—Shut off the influent flow to the clarifier. Divert flow to other clarifiers if possible.

Step 2—Drain the clarifier tank using the sludge withdrawal system. A sump pump may be used to remove remaining liquid from the sump. When draining a tank, immediately flush it completely. If sludge does not drain quickly, spray lime, calcium hypochlorite, or potassium permanganate on the sludge surface to reduce odors. Because even a clean tank can produce odors, flushing the tank with a chlorine solution or keeping the tank floor covered with a low concentration of chlorine solution will reduce odors.

Step 3—If the tank is not completely empty, locate and mark submerged tripping hazards such as the sludge pit or other drains.

Step 4—Wash down the interior of the clarifier using a high-pressure hose and squeegee to push solids into the sludge pit. Be sure to wash down the mechanism.

Step 5—Flush dedicated sludge lines and pumps with water. Sludge should not be allowed to sit for long periods of time in sludge lines or pumps. Gas produced from the anaerobic breakdown of primary sludge can build pressure and damage equipment.

GENERAL HOUSEKEEPING AND ODOR CONTROL

Adoption of the following regular practices will increase removal efficiencies, reduce odors, and provide better working conditions for the operator:

- Establish a housekeeping schedule for the primary treatment area, including galleries, stairwells, control rooms, and related buildings, and assign responsibility for each item to a specific employee;
- Remove scum routinely and with increased frequency during warm weather;
- Clean scum removal equipment regularly; otherwise, obnoxious odors and an unsightly appearance will result;
- Remove sludge before it can bubble or float;
- Wash weirs and other points where floatables and slime collect regularly;
- Wash down all spills and grease coatings;
- If the wastewater is septic, add chemicals in the collection system or at the facility, as appropriate, to reduce sulfides;
- If tanks are covered for odor control, keep plates and access hatches in place, except when operations or maintenance require their removal;
- The splashing of primary effluent into weir troughs and effluent channels can result in the release of hydrogen sulfide. If possible, try to minimize the splashing of primary effluent into the channel or weirs;
- Regularly remove accumulations from the inlet baffles and effluent weirs with a hose or a broom with stiff bristles. Only experience can determine the necessary frequency;
- Avoid hosing down motors and enclosed control devices; and
- Repaint surfaces, as necessary, for surface protection and appearance.

Data Collection, Sampling, and Analysis

PROCESS CONTROL DATA

The testing conducted for primary treatment influent, effluent, and sludge varies between facilities. In deciding which tests are essential, the operator must consider permit requirements, downstream treatment units, and anticipated industrial wastes. Typical tests for evaluating primary clarifiers include settleable solids, suspended solids, total solids, volatile solids, COD or BOD_5, pH, and oil and grease concentrations. Each of these laboratory tests is described in detail in the 5th edition of *Basic Laboratory Procedures for the Operator-Analyst* (WEF, 2012). It is highly recommended that operators obtain this reference for both day-to-day operations and as an additional reference when studying for certification exams.

Keep in mind that primary clarifiers are not typically found in facilities smaller than 19 ML/d (5 mgd) because they tend to be paired with anaerobic digesters. Larger facilities are more complex and have more resources to devote to process monitoring.

VISUAL INSPECTION

The surface of a primary clarifier should appear grey to yellowish brown. Figure 4.38 shows the surface colors of four different, well-functioning primary clarifiers. The wastewater in the clarifier should have an earthy, slightly sour smell, but should not smell like rotten eggs. Hydrogen sulfide, which smells like rotten eggs, is indicative of septic conditions. Septicity may result when the sludge is left in the clarifier too long. Check for bubbles and floating sludge on the clarifier surface. When sludge turns septic, microorganisms in the sludge blanket produce hydrogen sulfide and methane gas. These gases form bubbles in the sludge blanket that can rise to the clarifier surface, sometimes carrying pieces of sludge with them. As the sludge collection mechanism moves through the sludge blanket, it can releases accumulated gases. The easiest place to see whether gas bubbles are rising to the surface will be 0.3 to 0.6 m (1 to 2 ft) behind the skimmer arm. To alleviate septicity, increase the sludge pumping frequency.

SETTLEABLE SOLIDS TEST

The settleable solids test may be the best indicator of primary clarifier performance and is preferred over simple percent removal calculations for TSS, COD, and BOD. The settleable solids test estimates the removal efficiency of primary and secondary clarifiers. The test should be conducted at both average flow and peak hour flow. The test is conducted using an Imhoff cone by following these steps (Figure 4.39):

- Step 1—Fill the Imhoff cone to the 1-L mark with a well-mixed sample. Occasionally, settleable material in a given wastewater sample may exceed the 40-mL/L graduation on the side of the cone. In these instances, a 500-mL sample volume should be used.
- Step 2—Settle for 45 minutes.
- Step 3—Gently stir the upper portion of the sample with a glass rod to dislodge any material clinging to the tapered sides of the cone.
- Step 4—Settle for an additional 15 minutes.
- Step 5—Read the volume of settleable matter in milliliters per liter. If floatable material is present on the top of the water at the end of the test, it should not be included as part of the settleable material measurement.

The settleable solids test should be conducted using both primary clarifier influent and primary clarifier effluent. Running the test on the primary influent reveals how much settleable material is present in the influent. Running the test on the primary effluent reveals how efficiently the clarifier removed those settleable solids. Ideally, the primary effluent sample will not show any settleable material in the base of the cone at the end of the test. If settleable material is passing through the primary clarifier, then treatment process efficiency could be improved.

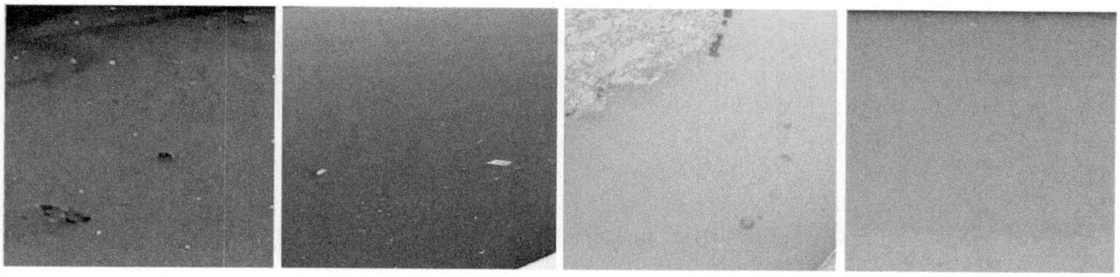

Figure 4.38 Primary Clarifier Surface Color

Figure 4.39 Settleable Solids Test Using an Imhoff Cone (Reprinted with permission by Indigo Water Group)

Recall that if the soluble BOD concentration in the influent increases, it will pass through to the primary effluent. An operator who is only monitoring primary influent and effluent BOD concentrations would see a drop in percent removal across the clarifier and wrongly conclude that something was wrong with the clarifier and that process control adjustments needed to be made. The settleable solids test, on the other hand, would show that the clarifier was continuing to perform to the best of its ability given the characteristics of the influent wastewater.

TOTAL SUSPENDED SOLIDS

Suspended solids are typically analyzed on composited samples, whereas settleable solids can be determined from composite or grab samples. These tests are typically performed daily at large facilities (flows higher than 19 ML/d [5 mgd]) and at least twice weekly at smaller facilities. Samples of the influent and effluent of the tank are taken at points where the wastewater is well mixed. Test results may be used to calculate the percent removal over the clarifier, the mass of solids loading entering downstream treatment units, and the quantity of sludge to be pumped.

TOTAL AND VOLATILE SOLIDS IN SLUDGE

These tests typically are performed daily on manually composited samples removed from the tanks. The results provide information on the solids loading on digesters and dewatering units. Exact knowledge of the percentage of moisture in the sludge enables the operator to check visual observations on the degree of compaction, the need for changing sludge removal rates and quantities, and the amount of excess water entering the digestion and dewatering units. A sudden increase in the amount of volatile solids could indicate an industrial discharge. A sudden increase in the percentage of inorganic (nonvolatile) solids could indicate a failure of upstream grit removal equipment or a problem in the collection system that is allowing more grit and silt to enter the facility. Primary sludge from domestic wastewater should have a volatile solids concentration between 75 and 80% (U.S. EPA, 1987).

CHEMICAL OXYGEN DEMAND AND BIOCHEMICAL OXYGEN DEMAND

Tests for COD or BOD_5 on composite samples of influent and effluent are typically performed at least once a week. Tests on the tank effluent provide an excellent measure of the organic loading on downstream biological processes. Soluble COD or BOD tests on the influent and effluent can serve as indicators of septic conditions in the tank. If the soluble COD or BOD increases across the clarifier, then it is likely that sludge is remaining in the clarifier long enough for bacteria to break down the organic material.

TEST YOUR KNOWLEDGE

1. **All rectangular clarifier mechanisms may be safely operated in a completely dry tank.**
   - ☐ True
   - ☐ False

2. **When placing a circular clarifier into service, the mechanism should be operated for several revolutions in a dry tank to verify correct operation.**
   - ☐ True
   - ☐ False

3. **When a primary clarifier is taken out of service, it is acceptable to leave sludge in pipelines and pumps.**
   - ☐ True
   - ☐ False

4. **A rectangular clarifier with a chain-and-flight sludge collection mechanism will be placed into service. Because the plastic chain is several years old, the operator may need to**
   a. Lubricate the chain with a lightweight machine oil
   b. Ensure the chain has not rusted to the sprockets
   c. Remove a link from each chain to compensate for wear
   d. Replace the wear shoes on each flight

5. **When removing a tank from service, it should be immediately flushed with water for this reason.**
   a. Prevent generation of odors
   b. Dried sludge is harder to remove
   c. Continued operation of water lubricated bearings
   d. Tank may be returned to service faster

6. **A sample is collected from the primary clarifier effluent for the settleable solids test. After settling for an hour, the tip of the cone is empty. The operator concludes**
   a. Excessive grit is entering the clarifier
   b. The sludge collector speed should be increased
   c. Settleable solids are passing on to the secondary process
   d. Process efficiency cannot be improved

7. **Routine laboratory testing shows that the percentage of volatile solids in the primary sludge has decreased over the last week from 80% down to 70%. What is the most likely cause?**
   a. Increased BOD loading to the WRRF
   b. Failure of upstream grit removal equipment
   c. Sludge blanket is deeper than 1.2 m (4 ft)
   d. Hydraulic retention time should be decreased

Maintenance

The elements of a maintenance program include preventive maintenance, established written procedures for emergency maintenance, and a system to ensure an adequate supply of essential spare parts. Primary clarifiers represent a tremendous investment in machinery, including equipment that typically operates underwater. Protection of this investment requires at least an annual inspection of the tanks and appurtenant equipment. A thorough inspection includes the following procedures:

- Completely dewater the basin once each year to allow for a complete inspection of all equipment;
- Inspect all mechanical equipment for wear and corrosion. Replace as necessary and apply a protective coating to chains, sprockets, guide rails, shaft bearings, bearing brackets, and other equipment parts;
- Lubricate all metal parts;
- Check flights for chipping or breaking, missing or loose wear shoes, or missing bolts. Replace these items as necessary;
- In rectangular clarifiers, check baffle boards for warping or breakage and replace them as required. Tighten all bolts;
- In rectangular clarifiers, inspect wear strips. Repair any locations where the strip is separating from the basin floor;
- In circular clarifiers, inspect plows for bent or damaged pieces and replace them as required. Tighten all bolts;
- Adjust the tension of chains, both longitudinal and crosscollector. Note that a nonmetallic chain has tension and slack requirements different from those of a steel chain, and nonmetallic chain stretches much more during its initial break-in period, thus requiring frequent checks;
- Check all suction lines and sumps for debris collection and clogging. Clean all lines before filling tanks;
- Check all mechanical parts of the skimming unit for wear or corrosion. Counterbalance the chain for buckets and trolley or carriage (adjustment may be required); and
- Inspect all concrete above and below the waterline, particularly blind areas such as sumps and anchorage areas for mechanical equipment. Patch concrete and caulk and seal joints as necessary.

The most likely circular primary clarifier equipment problems include the following:

- Failure of the overload protection, resulting in equipment damage;
- Oil seal leak or failure; and
- Failure of the needle bearings (roller bearing) or worm and pinion gear, possibly resulting from excessive wear or lack of lubrication.

The equipment of circular tanks requires periodic lubrication (turntable, worm drive, and lower gear drive), oil changes, and inspection (gear wear and gears and bearings for alignment). Necessary maintenance activities include brushing and cleaning gear housings, removing and cleaning oil pump strainers, and replacing skimmer blades as necessary. A sample drive maintenance log (courtesy of WesTech) is presented in Table 4.2. All utilities should develop their own maintenance logs for routine inspection, lubrication, and repair, guided by recommendations of their equipment suppliers.

One of the most important elements in a preventive maintenance program is a schedule for protective coating and painting. Ferrous metal above the waterline requires regular repainting every 3 years; ferrous metal below the waterline requires repainting every 5 to 10 years. Paint manufacturer's recommendations will be helpful in establishing a painting program.

Table 4.2 Sample Drive Maintenance Log

	Interval	Initials	Date
Break-In Maintenance Requirements			
Drain and Fill Oil Cavity/Cavities (Before Operating)	0 hours		
Drain and Replace Oil	500 hours		
Preventive Maintenance Requirements			
Grease Cyclo Reducer	M		
Grease Upper Bearing	M		
Grease Main Bearing	W		
Oil Main Gear/Lower Bearing	W		
Oil Torque Box Plunger	W		
Inspection Requirements			
Inspect Fasteners for Tightness	M/A		
Visually Inspect Drive Mech. for Wear	W/A		
Test Torque Box Limit Switches	W		
Check/Drain Condensate	W		
Check/Drain Particulates	S		
Inspect/Repair Drive Unit Paint	A		
Inspect Torque Control Device	A		

A - Annually, S - Semiannually, M - Monthly, W - Weekly, D - Daily

Note: This maintenance information is provided by WesTech Engineering Inc. as an example. It is not intended to be utilized as-is by any plant. Each plant should develop its own schedule and log, specific to the equipment at the facility.

Suggested maintenance items beyond the annual inspection are listed below; however, this list is not comprehensive and different manufacturers may recommend different frequencies for particular maintenance items.

Daily

- Conduct facility walkthrough.
- Verify proper operation of all equipment.
- Remove debris from channels and surface of clarifier.
- Follow housekeeping procedures described in the section entitled "Operation of Primary Clarifiers" to prevent odors and keep equipment running smoothly.

Weekly

- Drain condensation from drive unit. Oil and water do not mix. Condensation will find its way into the oil reservoir for the drive unit. Most units have a valve on the bottom of the reservoir. Water, being denser than oil, will collect on the bottom near the valve. Open the valve and allow it to drain until oil begins to come out. Close the valve. Top off the oil as needed.
- Grease bearings and reducers.
- Check drive oil and change if necessary.
- Lubricate torque plunger according to the manufacturer's instructions.
- Wash down clarifier weirs and launders to remove algae. A stiff bristled brush may be required. Some facilities paint their weirs with an epoxy to make algae removal easier.
- Inspect the weir to ensure that it is level and flow is distributed evenly across the weir length. Adjust as needed. If the weir is not level, the amount of water passing over some parts of the weir will be higher than others. This can cause clarifier performance to deteriorate as velocities increase through some sections of weir.
- Wash down influent and effluent channels.

Monthly

- Verify the skimmer arm is level and at the correct height. The skimmer arm should project above the height of the water at peak hour flows while remaining partially submerged.
- Examine wipers and squeegees for tears and flexibility. Replace as needed.
- On circular clarifiers with flush valves, verify operation of the flush valve.
- Grease bearings on the drive.

Semiannually

- Drain and refill the main gear oil.

EMERGENCY MAINTENANCE

Some emergency maintenance considerations are presented below. When a problem occurs, stop the flow and drain the tank. Notify all concerned staff of the tank draining schedule and of the possible presence of dangerous gases in both the primary clarifier and in any areas receiving primary sludge. While draining, continue to operate the sludge collection and removal equipment for as long as possible. The clarifier mechanism may be safely operated in a dry tank as long as there are not any obstructions and the mechanism is moving smoothly. The primary sludge pumps must be turned off once the sludge hopper can no longer be kept full. Continuing to operate the primary sludge pumps after this point may draw air into the pumps, which may damage the pumps. A portable sump pump placed in the sludge collection trough may be used to continue removing sludge and water.

To clear remaining sludge from a rectangular tank, stop the chain-and-flight mechanism and support the flights above the floor on blocks. This will allow the operator to wash sludge toward the sump with a hose and squeegee. Flushing valves in the effluent walls will expedite cleaning by allowing primary effluent to flush the inoperable bays. After work is completed, check the floor and sump areas for debris such as broken flights, tools, and broken pins. Check for proper operation of all mechanical equipment before returning the tank to service. Inspect flights to be sure they are adjusted parallel to the tank floor and that the wear shoes are in good condition.

To avoid disastrous results, never use a shear pin larger than the correct size.

SPARE PARTS

The spare parts inventory must include a stock of the parts necessary to maintain primary tank operation. The stock of significant spare parts typically includes flights and drive chains for rectangular tanks and wiper blades, skimmer assembly, turntable gears, and a motor for circular tanks. Rectangular tanks also use wearing shoes on the collector flights. It is recommended that a stock of wearing shoes be kept for spare parts. There should be enough wearing shoes to equip one tank and have an excess of 10% for miscellaneous replacements. The following list is a suggested stock of spare parts:

- Motors;
- Gear reducers;
- Sprockets, chains, and flights (enough to equip one tank);
- Wall brackets (enough to equip one tank);
- Chain pins, six per strand of chain; and
- Shear pins, 10 per tank.

When determining the stocking of spare parts, operators must also take into consideration the fail occurrence of a piece of equipment and the delivery time of a piece of equipment. For example, if a piece of equipment fails frequently, then additional spare parts for that equipment must be in stock at the treatment facility. In addition, if there is a long delivery lead time for a piece of equipment, that should also be taken into consideration when ordering and stocking spare parts. Equipment suppliers should be consulted for complete spare parts recommendations.

TEST YOUR KNOWLEDGE

1. **Clarifiers should be drained and all equipment and concrete thoroughly inspected annually.**
   - ☐ True
   - ☐ False

2. **Condensate must be drained from clarifier drive units.**
   a. Daily
   b. Weekly
   c. Monthly
   d. Annually

3. **Weirs should be kept clear of debris, level, and clean to prevent**
   a. Uneven flow distribution
   b. Retention of fine particulates
   c. Unsightly growth of algae
   d. Clogging of the flush valves

4. **Each facility should create its own equipment maintenance logs. What is the best source of information for determining the type and frequency of necessary maintenance?**
   a. Books like this one that provide general information
   b. Online training classes that provide general information
   c. Manufacturer's operations and maintenance manuals
   d. Chilton's guide

Troubleshooting

To remedy difficulties experienced with the primary treatment units, the operator must know the characteristics of the collected primary sludge, design shortcomings of the primary tanks and equipment, and indicators of operational problems. Brief descriptions and tabulations are given below concerning each of these three aspects of troubleshooting.

The nature of an operational problem and its severity are frequently temperature dependent. In colder weather, sludge will be more difficult to pump, sludge lines will collect grease more quickly, and skimming quantities will increase; however, septicity and odor will likely diminish.

Table 4.3 presents a troubleshooting guide that will enable the operator to identify problems with primary clarification and determine their solutions.

Table 4.3 Troubleshooting Guide for Primary Clarifiers (U.S. EPA, 1978)

Indicators/Observations	Probable Cause	Check or Monitor	Solutions
Floating sludge	Excessive sludge accumulating in the tank		Remove sludge more frequently or at a higher rate.
	Scrapers worn or damaged	Inspect scrapers.	Repair or replace as necessary.
	Return of well-nitrified waste activated sludge	Effluent nitrates	Reduce age of returned sludge or move point of waste sludge recycle.
	Sludge withdrawal line plugged	Sludge pump output	Flush or clean line.
	Damaged or missing inlet baffles	Damaged baffles	Repair or replace baffles.
Black and odorous septic wastewater or sludge	Sludge collectors worn or damaged	Inspect sludge collectors	Repair or replace as necessary.
	Improper sludge removal pumping cycles	Sludge blanket	Increase frequency and duration of pumping cycles until sludge blanket decreases.
	Inadequate pretreatment of organic industrial wastes	Pretreatment practices	Pre-aerate waste. Locate source of industrial discharge. Require pretreatment of industrial wastes.
	Wastewater decomposing in collection system	Retention time and velocity in collection lines	Inspect collection system for septic conditions. Increase pump station pumping frequency. Control septic conditions in collection system by adding chemicals or providing aeration and mixing.
	Recycle of excessively strong digester supernatant	Digester supernatant quality and quantity	Improve sludge digestion to obtain better quality supernatant. Reduce or delay withdrawal until quality improves. Select better quality supernatant from another digester zone. Discharge supernatant to ponds, aeration tank, or sludge drying bed.
	Sludge withdrawal line plugged	Sludge pump output	Clean line.
	Delivery of septage wastes (portable toilets)	Collect samples from septage haulers. If it is not possible to sample all trucks, random sampling may be used.	Regulate or curtail acceptance of hauled septage.
	Insufficient run time for sludge collectors	Review operation logs	Increase run time or run continuously.
Scum overflow	Frequency of removal inadequate	Scum removal rate	Remove scum more frequently.
	Heavy industrial waste contributions	Influent waste	Limit industrial waste contributions.
	Worn or damaged scum wiper blades	Wiper blades	Clean or replace wiper blades.
	Improper alignment of skimmer	Alignment	Adjust alignment.
	Inadequate depth of scum baffle	Scum bypassing baffle	Increase baffle depth.
Sludge difficult to remove from hopper	Excessive grit, clay, and other easily compacted material	Operation of grit removal system	Improve operation of grit removal unit.
	Low velocity in withdrawal lines	Sludge removal velocity	Increase sludge pumping rate. Verify pump capacity.
	Pipe or pump clogged	Pump output. Pump discharge pressure.	Backflush clogged pipe lines and pump sludge more frequently.
	Excessive solids loading	Mass of solids (kb or lb) entering the clarifier	Provide more even flow distribution in all tanks, if multiple tanks. If only one tank is available, increase sludge pumping frequency.

(*continues*)

Table 4.3 Troubleshooting Guide for Primary Clarifiers (U.S. EPA, 1978) (*Continued*)

Indicators/Observations	Probable Cause	Check or Monitor	Solutions
Undesirably low solids content in sludge	Hydraulic overload	Influent flowrate	Provide more even flow distribution in all tanks, if multiple tanks.
	Short-circuiting of flow through tanks	Dye or other flow tracers	(See "Short-circuiting of flow through tanks".)
	Overpumping of sludge	Frequency and duration of sludge pumping; suspended solids concentration	Reduce frequency and duration of pumping cycles.
Short-circuiting of flow through tanks	Uneven weir settings	Weir settings	Change weir settings.
	Damaged or missing inlet line baffles	Damaged baffles	Repair or replace baffles.
Surging flow	Poor influent pump programming	Pump cycling	Modify pumping cycle.
Excessive accumulation of solids in inlet channel	Velocity too low	Velocity	Increase velocity or agitate with air or water to prevent decomposition.
Settleable solids present in primary effluent. Test for settleable solids using an Imhoff cone.	Hydraulic overloading	Flow	Use available tankage, shave peak flow, chemical addition to flocculate solids.
	Short-circuiting		(See "Short-circuiting of flow through tanks".)
	Poor sludge removal practices	Monitor pumping duration and sludge levels.	Frequent and consistent pumping
	Recycle flows	Inventory flows—quality and quantity	(See "Rectangular Tanks" section in this chapter.)
	Density currents wind or temperature related	Monitor wastewater temperature and wind.	Eliminate storm flows from sewer system. Install wind barrier.
Excessive growth on surfaces and weirs	Accumulations of wastewater solids and resultant growth.	Inspect surfaces.	Frequent and thorough cleaning of surfaces.
Erratic operation of sludge collection mechanism	Broken shear pins, damaged collector	Shear pins and sludge collector	Repair or replace damaged parts.
	Rags and debris entangled around collector mechanism	Sludge collector	Remove debris. Improve influent screening.
	Excessive sludge accumulation	Sludge blanket depth	Increase frequency of pumping sludge from tank.
	Inadequate preventive maintenance program		Annual draining and inspection of tank
Broken scraper chains and frequent shear pin failure	Improper flight alignment	Flight alignment	Realign flights.
	Ice formation on walls and surfaces	Inspect walls and surfaces.	Remove or break up ice formation.
	Excessive loading on mechanical sludge scraper	Sludge loading	Operate collector for longer period and/or remove sludge more often.
	Inadequate preventive maintenance program		Annual draining and inspection of tank
Excessive corrosion on unit	Septic wastewater	Color and odor of wastewater	Paint surfaces with corrosion-resistant paint or coating.
Noisy chain drive	Moving parts rub stationary parts	Alignment	Tighten and align casing and chain. Remove dirt or other interfering matter.
	Chain does not fit sprockets		Replace with correct parts.
	Loose chain		Adjust chain to manufacturer's recommended tension.
	Faulty lubrication	Lubrication	Lubricate properly.
	Misalignment or improper assembly	Alignment and assembly	Correct alignment and assembly of drive.
	Worn parts		Replace worn chain or bearings. Reverse worn sprockets before replacing.

(*continues*)

Table 4.3 Troubleshooting Guide for Primary Clarifiers (U.S. EPA, 1978) (*Continued*)

Indicators/Observations	Probable Cause	Check or Monitor	Solutions
Rapid wear of chain drive	Faulty lubrication	Lubrication	Lubricate properly.
	Loose or misaligned parts	Alignment	Align and tighten entire drive.
Chain climbs sprockets	Chain does not fit sprockets		Replace chain or sprockets.
	Worn-out chain or worn sprockets		Replace chain. Reverse or replace sprockets.
	Loose chain		Tighten. Remove links as necessary.
	Alignment		Align sprocket and chain.
Stiff chain	Faulty lubrication	Lubrication	Lubricate properly.
	Rust or corrosion		Clean and lubricate.
	Misalignment or improper assembly	Alignment and assembly	Correct alignment and assembly of drive.
	Worn-out chain or worn sprockets		Replace chain. Reverse or replace sprockets.
Broken chain or sprockets in chain-drive system	Shock or overload	Solids loading rate	Avoid shock and overload or isolate through couplings.
	Wrong size chain or chain that does not fit sprockets		Replace chain. Reverse or replace sprockets.
	Rust or corrosion		Replace chain. Correct corrosive conditions.
	Misalignment	Alignment	Correct alignment.
	Interferences		Make sure no solids interfere between chain and sprocket teeth. Loosen chain if necessary for proper clearance over sprocket teeth.
Oil seal leak	Oil seal failure	Oil seal	Replace seal.
Bearing or universal joint failure	Excessive wear		Replace joint or bearing.
	Lack of lubrication	Lubrication	Lubricate joint and/or bearing.
Binding of sludge pump shaft	Improper adjustment of packing		Adjust packing.

DESIGN SHORTCOMINGS

The operator needs to recognize design shortcomings that affect settling performance and possible remedies. Some of the remedies may be carried out by facility personnel at a low cost.

Table 4.4 presents a list of design shortcomings observed for primary clarifiers and associated equipment and possible remedies for those shortcomings.

TEST YOUR KNOWLEDGE

1. **Primary sludge concentrations are between 1.5 and 3%. What action can the operator take to increase the sludge concentration?**
   a. Decrease pumping rate
   b. Increase collector speed
   c. Remove a clarifier from service
   d. Clean the sludge lines

2. **The settleable solids test shows a significant amount of settlable material in the primary effluent. What action can the operator take to reduce effluent settleable solids?**
   a. Adjust recycle flows to coincide with peak hour flow
   b. Decrease collector speed
   c. Place additional clarifiers into service
   d. Adjust sludge withdrawal rate

3. The shear pin on a chain-and-flight system has broken three times over the previous week. The collector is operated intermittently. To prevent additional breakages, the operator could
   a. Install a stronger shear pin
   b. Increase the collector operating time
   c. Decrease sludge withdrawal rate
   d. Remove a clarifier from service

4. Over the past month, it has become progressively more difficult to remove sludge from the hopper. Loading to the WRRF has not changed. The upstream headworks equipment is functioning well. What is the most likely cause?
   a. Decreased grit content in the sludge
   b. Velocity in sludge lines is too high
   c. Squeegees on flights need to be replaced
   d. Buildup of grease or other material in the sludge lines

Table 4.4 Common Design Shortcomings and Methods for Compensating (U.S. EPA, 1978)

Shortcoming	Solution
Poor flotation of grease	Pre-aeration of wastewater to increase grease buoyancy
Scum overflow	Move scum collection system away from outlet weir.
Sludge hard to remove from hopper because of excessive grit	Install grit chamber or eliminate sources of grit entering the system.
Short-circuiting of flow through tank causing poor solids removal	Modify hydraulic design and install appropriate baffles to disperse flow and reduce inlet velocities.
Heavy wear and frequent breakage of scrapers and shear pins because of grit	Install grit chamber.
Grease particles adhered to flights	Lower return flights to below the water surface or install water sprays to remove grease into scum troughs.
Inadequate removal of heavy grease loading	Install flotation or evacuator equipment.
Septic conditions resulting from overloading	Divert or provide alternate disposal for other facility process wastes (i.e., centrates and supernates) that are typically recirculated to the primary clarifier.
Excessive corrosion because of septic wastewater	Coat all surfaces with proper paint or other coating and/or install cathodic protection.
Consistent problems with thermal currents in the clarifier	Install flow equalization and mixing basin ahead of the clarifier.
Poor scum removal because of wind	Install a wind barrier to protect the tank from wind effects.
Septic wastewater	Improve hydraulics of the collector system to reduce accumulation of solids.

Records and Reports

Maintaining good performance of the primary units requires records of operation, sampling and analysis, and maintenance. Operating records should include the following items:

- Number of tanks in operation;
- Equipment run times;
- Sludge blanket levels;
- Pumping time;
- Operator and shift;
- Housekeeping checklist;
- Chemicals used (if applicable);
- Suspended solids concentrations of the primary influent and effluent (and waste activated sludge, if co-settling is used at the facility);
- The COD or BOD_5 concentrations of the primary influent and effluent;
- Total solids and total volatile solids (TVS) concentrations in the sludge;
- Flowrates of the influent, effluent, and sludge;
- pH; and
- Temperature.

Maintenance records typically include summaries of preventive, predictive, and corrective maintenance performed. In addition, these records must give the time, date, equipment, and employee for each task.

Safety Considerations

It is important that operations and maintenance personnel follow safety guidelines specified by primary equipment manufacturers and commonly practiced by the utility. This includes, but is not limited to, procedures describing appropriate equipment lockout, confined space entry, localized power interruption, and standby rescue personnel during maintenance activities. Operations personnel must familiarize themselves with and respect safe operations guidelines recommended by the equipment manufacturers and maintain a clean, well-ventilated, and well-lit environment to prevent accidents. Some safety considerations, especially for primary clarifiers, are as follows:

- Apply confined space entry procedures relating to pump pits, ejector pits, and enclosed or partially enclosed primary tanks;
- Use extreme caution to avoid falls during entry to and exit from drained tanks;
- Require hard hats and provide good footing to avoid slips and falls while working on tank floors;
- Provide constant ventilation of covered tanks and prohibit smoking in their vicinity;
- Practice meticulous personal hygiene when work requires contact with sludges, wastewater, and scum;
- Exercise caution to prevent burns or eye injuries from chemical feed pumps and tanks;
- Keep access plates and hatches in place;
- Keep tank areas well lit;
- Lockout/tagout inlet gates or valves before entering an empty tank;
- Lockout/tagout electrical equipment before any maintenance inspection of an empty tank; and
- Relocate light fixtures, switches, valves, and so on, that are difficult to access easily or safely.

CHAPTER SUMMARY

<table>
<tr><td>PURPOSE</td><td colspan="2">■ Removes readily settleable material by gravity
■ Removes floatable material
■ Reduces COD, BOD_5, and TSS load to the secondary treatment process
■ Reduces size and operating cost of secondary treatment process</td></tr>
<tr><td>THEORY</td><td colspan="2">■ Differences in density used to separate particles from water
■ Gravity pulls particles down
■ Surface overflow rate pushes particles up and out
■ Particle size and shape affects settleability
■ Particles are granular or flocculant. Flocculant particles can increase in size through the process.
■ Detention time, in part, determines the amount of flocculation that can take place
■ Chemicals may be added to enhance performance
■ Changes in flowrate will change SOR and DT</td></tr>
<tr><td>DESIGN PARAMETERS</td><td colspan="2">■ Detention time of 1.5 to 2.5 hours
■ Average SOR of 24.4 to 48.9 $m^3/m^2{\cdot}d$ (600 to 1200 gpd/sq ft)
■ Peak hour SOR of 102 to 122 $m^3/m^2{\cdot}d$ (2500 to 3000 gpd/sq ft)
■ Sludge blanket depth of 0.6 to 0.9 m (2 to 3 ft)
■ Sludge concentration of 3 to 6% total solids</td></tr>
<tr><td>EXPECTED PERFORMANCE</td><td colspan="2">■ Highly dependent on influent characteristics
■ Typical removal rates for domestic wastewater are 50 to 70% TSS, 25 to 40% BOD, and 5 to 10% total phosphorus.
■ Cannot remove soluble or colloidal material</td></tr>
<tr><td>EQUIPMENT TANK</td><td>Circular Clarifiers
■ Circular tank up to 100 m (300 ft) in diameter
■ Depths of 3 to 4 m (10 to 12 ft)
■ Floor slope of 1 on 12
■ 0.5 to 0.7 m (1.5 to 2 ft) freeboard typical</td><td>Rectangular Clarifiers
■ Rectangular tank up to 90 m (300 ft) in length and 24-m (80-ft) wide
■ Length-to-width ratio typically greater than 5:1
■ Depths up to 4.9 m (16 ft), but depths of 3 to 3.7 m (10 to 12 ft) more common
■ Floor slope of 1%
■ 0.5 to 0.7 m (1.5 to 2 ft) freeboard typical</td></tr>
<tr><td>FLOW PATTERN</td><td>■ Three patterns: (1) center feed, peripheral take-off, (2) peripheral feed, center take-off, and (3) peripheral feed, peripheral take-off
■ Feedwell dissipates influent flow and directs downward to prevent short-circuiting</td><td>■ Flow enters at one end and travels lengthwise across to the effluent launder
■ An influent channel or manifold may be used to distribute flow across the width of the tank
■ Inlet baffles dissipate influent flow and direct flow downward to prevent short-circuiting</td></tr>
<tr><td>SLUDGE REMOVAL</td><td>■ Rotating sludge scraper with four different blade arrangements
■ Straight multiblade window shade plow most common
■ Sludge hopper typically located near center of tank</td><td>■ Two options: chain-and-flight system or traveling bridge
■ Chain-and-flight system consists of a series of boards (flights) pulled by chains across the floor of the clarifier. The same flights may also be used for scum skimming.
■ Chain-and-flights come in 3-, 4-, or 5-sprocket versions.</td></tr>
</table>

SLUDGE REMOVAL	■ Most common sludge hopper shape is trapezoidal with steep sides ■ Sludge is typically pumped out of the hopper and sent to the solids handling side of the facility for additional processing.	■ Large tanks may have more than one chain-and-flight system placed side-by-side. Each system may have a separate sludge hopper or they may discharge into a larger cross-collector. ■ Cross-collector troughs use screw conveyors or chain-and-flight systems to move sludge to a central hopper. ■ A traveling bridge has a scraper mechanism that can be raised or lowered. ■ Pumps transfer sludge to solids handling.
CLARIFIER DRIVE	■ Typically located at clarifier center, but occasionally on outer wall ■ Motor is coupled to the clarifier mechanism through a series of reducers ■ Reducers, like worm gears, increase torque and decrease speed. ■ Torque monitoring needed to prevent equipment damage if the mechanism becomes stuck ■ Limit switches may be used to sound an alarm and/or shut down the drive to prevent equipment damage. ■ Shear pin may also be present	■ Located on outer wall ■ Similar to drives used with circular clarifiers ■ Shear pin and limit switches used to prevent equipment damage in an overload situation
WEIRS	■ Weir plate typically has 90-deg v-notch weirs spaced at 152- or 304-mm (6- or 12-in.) intervals. ■ Inboard launder projects into the clarifier from the outer wall. ■ Outboard launder projects outward from the outer wall. ■ Inset launder is located several feet within the clarifier tank. ■ All three launder and weir arrangements follow the circumference or perimeter of the circular tank. ■ Scum baffle set inside the weir prevents floatable material from passing into the final effluent.	■ May utilize surface overflow weirs similar to those in circular tanks or submerged launders (outlet tubes) ■ Overflow weirs may be longitudinal, transverse, or grid. ■ Submerged launders have larger openings at regularly spaced intervals.
SKIMMING SYSTEM	■ Revolving skimmer with scum beach and drop to scum box ■ Revolving skimmer is typically joined to the rotating sludge scraper.	■ Ducking skimmers are most common with rectangular tanks. ■ Rotating skimmers use a rotating trough to periodically remove scum from the clarifier surface. ■ Flights push floatable material toward the ducking skimmer. ■ Approach of the flight triggers an operating cycle where the skimmer rotates to expose its openings to the oncoming scum.
PROCESS VARIABLES	■ Gravity pulls particles down while SOR pushes particles up and out. ■ Hydraulic detention time aids in particle retention. ■ Settling velocity is determined by particle size, shape, and density. ■ Flocculation can increase the size of organic particles and help to remove some colloidal material. ■ Particle size tends to be larger in fresher, nonseptic wastewater. ■ Particles settle faster in warmer, less dense water. ■ Chemicals can be added to enhance flocculation and improve percent removal. ■ Low pH and high temperatures increase grease solubility, which decreases removal of floatable material.	

PROCESS CONTROL	■ Variables within the control of the operator include SOR, DT, sludge withdrawal rate, and the timing of recycle streams. ■ Tanks should be placed into or be taken out of service to maintain the SOR and DT within accepted limits. ■ Solids thickness of 3 to 6% total solids can be achieved with a blanket depth of 0.6 to 0.9 m (2 to 3 ft). ■ Sludge quantity and required pumping times can be estimated from process control data. ■ Sludge collectors are generally operated continuously. ■ Sludge pumping cycles should be coordinated with hopper volume to maximize sludge concentrations. ■ Upstream unit processes can affect primary clarifier operation when they fail to capture grit and screenings. ■ Primary clarifiers can affect operation of the secondary processes when they fail to remove settleable solids. An increased load to the secondary process increases operating costs. ■ Primary clarifiers can affect operation of anaerobic digesters when the organic concentration (VSS concentration) increases or when sludge is not transferred consistently. Either may result in an upset condition.
PROCESS CONTROL DATA	■ Use the settleable solids test to determine process performance. ■ Monitor influent and effluent TSS and COD or BOD to calculate percent removal and loading to the secondary treatment process. ■ Monitor influent soluble COD or BOD as an indicator of industrial discharges. ■ Monitor total solids and TVS in the primary sludge to determine percent volatile solids. Low volatile solids could indicate a failure of upstream headworks equipment. High volatile solids could indicate an industrial discharge. ■ Monitor total solids and TVS to determine solids and organic loading rates to solids handling processes like digesters.
MAINTENANCE	■ Daily observation for visual and audible indications of malfunction ■ Regular maintenance should be performed in accordance with the manufacturer's recommendations. ■ Dewater the clarifier and perform a complete inspection of all equipment annually. ■ Perform other maintenance tasks daily, weekly, monthly, or as required by the manufacturer. ■ Maintain an inventory of spare parts.

CHAPTER EXERCISE

This exercise will draw upon some of the knowledge acquired in this chapter. In this section, an operator must complete the following tasks:

- Estimate primary effluent BOD, TSS, and total phosphorus concentrations from laboratory data.
- Calculate percent removal for BOD.
- Determine how many primary clarifiers should be in service.
- Estimate total sludge pumping time and set a pumping schedule.

Water Resource Recovery Facility Information:

The influent flow to the WRRF is 28.4 ML/d (7.5 mgd). The influent contains 315 mg/L of TSS, of which 50% is settleable; 280 mg/L of BOD, of which 84 mg/L is soluble; and 7 mg/L of total phosphorus, of which 50% is soluble. Assume that 50% of the particulate BOD and 50% of the particulate phosphorus are settleable. There are three primary clarifiers. Each clarifier is 24.4 m (80 ft) in diameter and 3.66-m (12-ft) deep. Each clarifier has a dedicated sludge pump with a capacity of 265 L/min (70 gpm). The settled sludge has an average total solids concentration of 3.5%.

CHAPTER EXERCISE SOLUTIONS

Estimate Primary Effluent Concentrations:

When faced with a lot of complex information, it can be helpful to start by drawing a picture like the one below. The soluble BOD and soluble phosphorus pass through the clarifier unchanged.

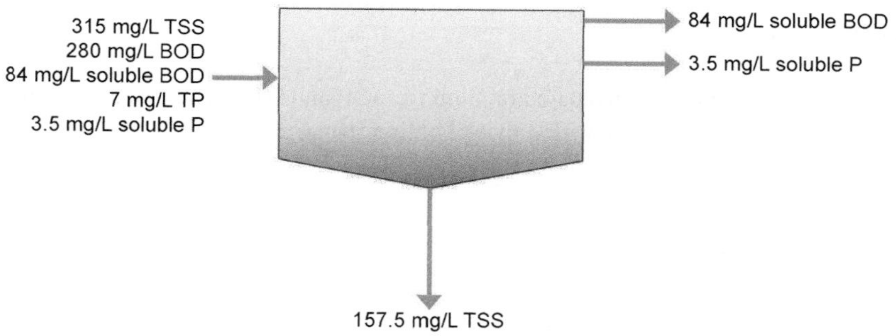

Fifty percent, or half, or the TSS are settleable.

$$\left(315\frac{\text{mg}}{\text{L}}\text{TSS}\right)(50\%) = 157.5 \text{ mg/L}$$

The same is true for the particulate BOD. Because the total BOD includes the soluble BOD, the soluble BOD must be subtracted from the total before the percent removal can be applied.

$$\left(280\frac{\text{mg}}{\text{L}} - 84\frac{\text{mg}}{\text{L}}\right)(50\%) = 98\frac{\text{mg}}{\text{L}} \text{ of Particulate BOD}$$

$$\text{Total BOD} = \text{Particulate} + \text{Soluble}$$

$$\text{Total BOD} = 98\frac{\text{mg}}{\text{L}} + 84 \text{ mg/L}$$

$$\text{Total BOD} = 182 \text{ mg/L}$$

Find the effluent phosphorus using the same method.

$$\left(7\frac{\text{mg}}{\text{L}} - 3.5\frac{\text{mg}}{\text{L}}\right)(50\%) = 1.75\frac{\text{mg}}{\text{L}} \text{ of Particulate P}$$

$$\text{Total P} = \text{Particulate} + \text{Soluble}$$

$$\text{Total P} = 3.5\frac{\text{mg}}{\text{L}} + 1.75 \text{ mg/L}$$

$$\text{Total P} = 5.25 \text{ mg/L}$$

Calculate Percent Removal for BOD:

$$\text{Percent Removal} = \left(\frac{\text{In} - \text{Out}}{\text{In}}\right) \times 100$$

$$\text{Percent Removal, BOD} = \left(\frac{280\frac{\text{mg}}{\text{L}} - 182\frac{\text{mg}}{\text{L}}}{280 \text{ mg/L}}\right) \times 100$$

$$\text{Percent Removal, BOD} = 35$$

Determine the Number of Clarifiers:

The number of clarifiers in service should be selected based on both the SOR and HDT. The goal is to keep both within acceptable ranges. Start by finding the SOR and detention time with only one basin in service.

In International Standard units:

Step 1—Find the surface area of the clarifier using eq 4.6.

$$\text{Area} = 0.785 \times (\text{Diameter})^2$$

$$\text{Area} = 0.785 \times (24.4\ \text{m})^2$$

$$\text{Area} = 467.36\ \text{m}^2$$

Step 2—Convert the influent flow from ML/d to m^3/d.

$$\frac{28.4\ \text{ML}}{\text{d}} \left| \frac{1\,000\,000\ \text{L}}{1\ \text{ML}} \right| \left| \frac{1\ \text{m}^3}{1000\ \text{L}} \right| = 28\,400\ \text{m}^3/\text{d}$$

Step 3—Find the SOR using eq 4.2.

$$\text{Surface Overflow Rate, } \frac{\text{m}^3}{\text{m}^2\cdot\text{d}} = \frac{\text{Flow, m}^3/\text{d}}{\text{Surface area, m}^2}$$

$$\text{Surface Overflow Rate, } \frac{\text{m}^3}{\text{m}^2\cdot\text{d}} = \frac{28\,400\ \text{m}^3/\text{d}}{467.36\ \text{m}^2}$$

$$\text{Surface Overflow Rate, } \frac{\text{m}^3}{\text{m}^2\cdot\text{d}} = 60.8$$

The design criteria for primary clarifiers have the SOR between 24.4 and 48.9 $m^3/m^2 \cdot d$. The calculated value for this clarifier is too high. If two clarifiers are in service, then the SOR will be cut in half to 30.4 $m^3/m^2 \cdot d$, which is well within the desired range. Based on SOR, the operator should place two clarifiers into service.

Step 4—Find the volume of the clarifier using eq 4.7.

$$\text{Volume} = 0.785 \times (\text{Diameter})^2 \times \text{Height}$$

$$\text{Volume} = 0.785 \times (24.4\ \text{m})^2 \times 3.66\ \text{m}$$

$$\text{Volume} = 1710.5\ \text{m}^3$$

Step 5—Find the detention time using eq 4.4.

$$\text{Detention Time, days} = \frac{\text{Volume, m}^3}{\text{Flow, m}^3/\text{d}}$$

$$\text{Detention Time, days} = \frac{1710.5\ \text{m}^3}{28\,400\ \text{m}^3/\text{d}}$$

$$\text{Detention Time, days} = 0.06$$

Step 6—Convert days to hours.

$$\frac{0.06 \text{ days}}{} \left| \frac{24 \text{ hours}}{1 \text{ day}} \right| = 1.4 \text{ hours}$$

The design criteria for primary clarifiers have DTs between 1.5 and 2.5 hours. The calculated value for this clarifier is within the accepted range; however, the SOR will require two clarifiers to be in service.

In U.S. customary units:

Step 1—Find the surface area of the clarifier using eq 4.6.

$$\text{Area} = 0.785 \times (\text{Diameter})^2$$

$$\text{Area} = 0.785 \times (80 \text{ ft})^2$$

$$\text{Area} = 5024 \text{ sq ft}$$

Step 2—Convert the influent flow from million gallons to gallons per day.

$$\frac{7.5 \text{ mil. gal}}{\text{d}} \left| \frac{1\,000\,000 \text{ gal}}{1 \text{ mil. gal}} \right| = 7\,500\,000 \text{ gpd}$$

Step 3—Find the SOR using eq 4.3.

$$\text{Surface Overflow Rate, } \frac{\text{gpd}}{\text{sq ft}} = \frac{\text{Flow, gpd}}{\text{Surface area, sq ft}}$$

$$\text{Surface Overflow Rate, } \frac{\text{gpd}}{\text{sq ft}} = \frac{7\,500\,000 \text{ gpd}}{5024 \text{ sq ft}}$$

$$\text{Surface Overflow Rate, } \frac{\text{gpd}}{\text{sq ft}} = 1492.8$$

The design criteria for primary clarifiers have SORs between 600 and 1200 gpd/sf. The calculated value for this clarifier is too high. If two clarifiers are in service, then the SOR will be cut in half to 746 gpd/sq ft, which is well within the desired range. Based on SOR, the operator should place two clarifiers into service.

Step 4—Find the volume of the clarifier using eq 4.8.

$$\text{Volume} = 0.785 \times (\text{Diameter})^2 \times \text{Height}$$

$$\text{Volume} = 0.785 \times (80 \text{ ft})^2 \times 12 \text{ ft}$$

$$\text{Volume} = 60\,288 \text{ cu ft}$$

Step 5—Convert volume in cubic feet to million gallons.

$$\frac{60\,288 \text{ cu ft}}{} \left| \frac{7.48 \text{ gal}}{1 \text{ cu ft}} \right| \left| \frac{1 \text{ mil. gal}}{1\,000\,000 \text{ gal}} \right| = 0.45 \text{ mil. gal}$$

Step 6—Find the detention time using eq 4.5.

$$\text{Detention Time, days} = \frac{\text{Volume, mil. gal}}{\text{Flow, mgd}}$$

$$\text{Detention Time, days} = \frac{0.45\ \text{mil. gal}}{7.5\ \text{mgd}}$$

$$\text{Detention Time, days} = 0.06$$

Step 7—Convert days to hours.

$$\frac{0.06\ \text{days}}{} \left| \frac{24\ \text{hours}}{1\ \text{day}} \right| = 1.4\ \text{hours}$$

The design criteria for primary clarifiers have detention times between 1.5 and 2.5 hours. The calculated value for this clarifier is within the accepted range; however, the SOR will require two clarifiers to be in service.

Estimate Total Sludge Pumping Time and Set a Pumping Schedule:

In International Standard units:

Step 1—Convert the influent flowrate in ML/d to m^3/d.

$$\frac{28.4\ \text{ML}}{\text{d}} \left| \frac{1\,000\,000\ \text{L}}{1\ \text{ML}} \right| \left| \frac{1\ \text{m}^3}{1000\ \text{L}} \right| = 28\ 400\ \text{m}^3$$

Step 2—Find the dry solids removed using eq 4.9.

$$\text{Dry Solids Removed, } \frac{\text{kg}}{\text{d}} = \frac{\left(\text{Flow, } \frac{\text{m}^3}{\text{d}}\right)\left(\text{Concentration, } \frac{\text{mg}}{\text{L}}\right)}{1000}$$

$$\text{Dry Solids Removed, } \frac{\text{kg}}{\text{d}} = \frac{\left(28\ 400\ \frac{\text{m}^3}{\text{d}}\right)\left(157.5\frac{\text{mg}}{\text{L}}\right)}{1000}$$

$$\text{Dry Solids Removed, } \frac{\text{kg}}{\text{d}} = 4473$$

Step 3—Find the wet solids removed in kilograms per day using eq 4.11.

$$\text{Wet Sludge Removed, } \frac{\text{kg}}{\text{d}} = \text{Dry solids removed, kg/d} \times \left(\frac{100}{\text{Dry solids in sludge, \%}}\right)$$

$$\text{Wet Sludge Removed, } \frac{\text{kg}}{\text{d}} = 4473\ \text{kg/d} \times \left(\frac{100}{3.5\%}\right)$$

$$\text{Wet Sludge Removed, } \frac{\text{kg}}{\text{d}} = 127\ 800$$

Step 4—Convert wet solids in kg/d to volume in m^3/d using eq 4.13.

$$\text{Wet Sludge Removed, } \frac{m^3}{d} = \frac{\text{Wet sludge removed, kg/d}}{998.2\frac{kg}{m^3} \times \text{Specific gravity of sludge}}$$

$$\text{Wet Sludge Removed, } \frac{m^3}{d} = \frac{127\ 800\ \text{kg/d}}{998.2\frac{kg}{m^3} \times 1.0}$$

$$\text{Wet Sludge Removed, } \frac{m^3}{d} = 128$$

Step 5—Convert wet solids in cubic meters per day to liters per day.

$$\frac{128\ m^3}{d}\left|\frac{1000\ L}{m^3}\right| = 128\ 000\ \text{L/d}$$

Step 6—Use the detention time formula (eq 4.4) to find the total pump time. Units may be different than the ones used in formula (eq 4.4) as long as the units on the top of the equation agree with the units on the bottom of the equation.

$$\text{Detention Time} = \frac{\text{Volume}}{\text{Flow}}$$

$$\text{Detention Time} = \frac{128\ 000\ \text{L of sludge}}{265\ \text{L/min pump flow}}$$

$$\text{Detention Time} = 483\ \text{minutes}$$

The primary sludge pump only needs to operate for 483 minutes out of each day. If the pumping time is distributed over a 24-hour period, the total number of minutes per hour can be calculated.

$$\frac{483\ \text{minutes}}{}\left|\frac{1\ \text{day}}{24\ \text{hours}}\right| = 20\ \text{minutes per hour}$$

Because there will be two clarifiers in service and each clarifier has its own sludge withdrawal pump, the operator should set each pump to operate for 10 minutes out of each hour.

In U.S. customary units:

Step 1—Find the dry solids removed using eq 4.10.

$$\text{Dry Solids Removed, } \frac{lb}{d} = (\text{Flow, mgd})\left(\text{Concentration, } \frac{mg}{L}\right)\left(8.34\frac{lb}{MG}\right)$$

$$\text{Dry Solids Removed, } \frac{lb}{d} = (7.5\ \text{mgd})(157.5\ \text{mg/L})\left(8.34\frac{lb}{MG}\right)$$

$$\text{Dry Solids Removed, } \frac{lb}{d} = 9851.6$$

Step 2—Find the wet solids removed in pounds per day using eq 4.12.

$$\text{Wet Sludge Removed, } \frac{\text{lb}}{\text{d}} = \frac{\text{Dry solids removed, } \frac{\text{lb}}{\text{d}} \times 100}{\text{Dry solids in sludge, \%}}$$

$$\text{Wet Sludge Removed, } \frac{\text{lb}}{\text{d}} = \frac{9851.6 \times 100}{3.5\%}$$

$$\text{Wet Sludge Removed, } \frac{\text{lb}}{\text{d}} = 281\ 474$$

Step 3—Convert wet solids in pounds per day to volume in gallons per day using eq 4.14.

$$\text{Wet Sludge Removed, gpd} = \frac{\text{Wet sludge removed, lb/d}}{8.34\frac{\text{lb}}{\text{gal}} \times \text{Specific gravity of sludge}}$$

$$\text{Wet Sludge Removed, gpd} = \frac{281\ 474\ \text{lb/d}}{8.34\frac{\text{lb}}{\text{gal}} \times \text{Specific gravity of sludge}}$$

$$\text{Wet Sludge Removed, gal} = 33\ 750$$

Step 4—Use the detention time formula (eq 4.5) to find the total pump time. Units may be different than the ones used in formula (eq 4.5) as long as the units on the top of the equation agree with the units on the bottom of the equation.

$$\text{Detention Time} = \frac{\text{Volume}}{\text{Flow}}$$

$$\text{Detention Time} = \frac{33\ 750\ \text{gal of sludge}}{70\ \text{gpm pump flow}}$$

$$\text{Detention Time} = 482\ \text{minutes}$$

The primary sludge pump only needs to operate for 482 minutes out of each day. If the pumping time is distributed over a 24-hour period, the total number of minutes per hour can be calculated.

$$\frac{482\ \text{minutes}}{} \left| \frac{1\ \text{day}}{24\ \text{hours}} \right| = 20\ \text{minutes per hour}$$

Because there will be two clarifiers in service and each clarifier has its own sludge withdrawal pump, the operator should set each pump to operate for 10 minutes out of each hour.

References

Larsen, P. (1977) On the Hydraulics of Rectangular Settling Basins. *Report 1001*, Department of Water Resources Engineering, University of Lund: Sweden.

Metcalf and Eddy, Inc./AECOM (2014) *Wastewater Engineering: Treatment and Reuse*, 4th ed.; McGraw Hill: New York.

Ross, R. D.; Crawford, G. V. (1985) The Influent of Waste Activated Sludge on Primary Clarifier Operation. *J. Water Pollut. Control Fed.*, **57**, 1022–1026.

Stojkov, B. T. (1997) *The Valve Primer;* Industrial Press, Inc.: New York.

U.S. Environmental Protection Agency (1978) *Field Manual for Performance Evaluation and Troubleshooting at Municipal Wastewater Treatment Facilities;* U.S. Environmental Protection Agency: Washington, D.C.

U.S. Environmental Protection Agency (1987) *Design Manual: Dewatering Municipal Wastewater Sludges;* EPA-625-1/87-014; Office of Research and Development, Center for Environmental Research Information: Cincinnati, Ohio.

Water Environment Federation (2005) *Clarifier Design,* 2nd ed.; Manual of Practice No. FD-8; Water Environment Federation: Alexandria, Virginia.

Water Environment Federation (2012) *Basic Laboratory Procedures for the Operator-Analyst*, 5th ed.; Water Environment Federation: Alexandria, Virginia.

Water Environment Federation (2016) *Operation of Water Resource Recovery Facilities,* 7th ed.; WEF Manual of Practice No. 11; Water Environment Federation: Alexandria, Virginia.

Water Environment Federation; American Society of Civil Engineers; Environmental and Water Resources Institute (2018) *Design of Water Resource Recovery Facilities,* 6th ed.; WEF Manual of Practice No. 8/ASCE Manuals and Reports on Engineering Practice No. 76; Water Environment Federation: Alexandria, Virginia.

Weidler, J. (2017) Hydraulic Overload Systems Provide Increased Protection for Clarifier Drives. In *WaterWorld.* http://www.waterworld.com/articles/print/volume-20/issue-2/editorial-focus/hydraulic-overload-systems-provide-increased-protection-for-clarifier-drives.html (accessed Aug 2017).

Suggested Readings

Heinke, G. W.; Tay, A. J.-H.; Qazi, M. A. (1980) Effects of Chemical Addition on the Performance of Settling Tanks. *J. Water Pollut. Control Fed.*, **52**, 2946.

Stojkov, B. T. (1997) *The Valve Primer;* Industrial Press, Inc.: New York.

Water Environment Federation (2012) *Basic Laboratory Procedures for the Operator-Analyst*, 5th ed.; Water Environment Federation: Alexandria, Virginia.

Water Environment Federation (2016) Chapter 8–Pumping of Wastewater and Sludge and Chapter 12–Maintenance. In *Operation of Water Resource Recovery Facilities,* 7th ed.; WEF Manual of Practice No. 11; Water Environment Federation: Alexandria, Virginia.

CHAPTER 5

Fundamentals of Biological Treatment

Introduction

Biological treatment depends on a healthy community of bacteria and other microorganisms to produce a final, treated effluent that is clean and clear with very little biochemical oxygen demand (BOD) and total suspended solids (TSS) remaining. Some biological treatment processes also remove nitrogen and phosphorus by encouraging the growth of specialized bacteria. Several different types of biological secondary treatment processes will be described in the upcoming chapters, including lagoons, wetlands, trickling filters, rotating biological contactors, and activated sludge. These processes were described briefly in Chapter 1. This chapter describes some of the different types of microorganisms and environmental conditions present in those processes. Anaerobic bacteria—bacteria that thrive in the absence of oxygen, nitrite, and nitrate—will not be discussed in detail in this chapter because, although important in anaerobic digestion, they do not play a significant role in most secondary treatment systems.

The presence of microorganisms within biological treatment processes cannot be observed with the naked eye. Operators can only observe some of the physical effects of microbial activity such as color change, odor, bubbling of gases, floc size, sludge settling characteristics, and foaming. These direct observations of the treatment process are supplemented by measurements of physical and chemical characteristics, including BOD, TSS, pH, alkalinity, dissolved oxygen (DO) concentration, ammonia, nitrite, nitrate, phosphate, oxidation–reduction potential, and other parameters. An operator can run a water resource recovery facility (WRRF) effectively based on these observations and physical and chemical measurements alone. However, understanding why changes occur or do not occur in a biological treatment process requires understanding what different groups of microorganisms need to live and grow as well as the conditions that can threaten their survival or decrease their activity. Understanding the microbiology behind the process makes it easier to optimize and troubleshoot secondary treatment processes.

LEARNING OBJECTIVES

Upon completing this chapter, you will be able to

- Describe how dissolved and particulate organic matter are removed by biological treatment processes;
- Predict the fate of dissolved organic material based on its characteristics;
- Categorize the main groups of microorganisms and their functions in secondary treatment processes;
- Compare and contrast carbon and energy sources for heterotrophic and autotrophic bacteria;
- Compare growth rates of heterotrophic and autotrophic bacteria;
- Define and explain the purpose of aerobic, anoxic, and anaerobic conditions in biological treatment processes;
- Predict the behavior of heterotrophic bacteria, nitrifying autotrophic bacteria, and phosphate accumulating bacteria under aerobic, anoxic, and anaerobic conditions;
- Understand the terms *yield, decay, maximum growth rate, half-saturation coefficient,* and *substrate*;
- Compare bacterial growth rates at different wastewater temperatures and explain how growth rates affect treatment process capacity; and
- Explain the effect of substrate concentrations on bacterial growth rates.

Biological Treatment

Biological treatment was developed in the early part of the 20th century. Biological treatment relies on naturally occurring microorganisms to break down organic matter contained in wastewater into simple substances, mainly carbon dioxide and water. Some microorganisms use the organic matter present in wastewater as food to grow and reproduce. Others rely on ammonia or other materials to grow and reproduce. These microorganisms are retained at much higher concentrations in biological treatment processes than would occur in soil or natural bodies of water. The high concentrations of microorganisms maintained in treatment processes allow the waste to be broken down much faster than would otherwise occur in natural environments. Once treatment is complete, the *biomass* is separated from the treated wastewater, which can then be safely discharged as effluent to a stream, river, lake, wetland, or sea.

Biomass refers to the mass of microorganisms in biological treatment processes.

Physical and Chemical Requirements for Biological Treatment

Understanding the relationships between different wastewater components and whether or not they are biodegradable can help predict how wastewater characteristics will change as the wastewater moves through the treatment process. Chapters 3 and 4 discussed some of the mechanical processes (screens, grit basins, and primary clarifiers) that are used to physically remove larger debris and settleable solids. The particles that remain in the wastewater will not settle on their own within a reasonable amount of time. During secondary treatment, microorganisms consume biodegradable organic matter and nutrients to grow and reproduce, cleaning the wastewater in the process. Specialized bacteria convert ammonia to nitrite and nitrate or uptake phosphorus. Still other bacteria convert nitrate to nitrogen gas. The microorganisms form *biofilms* or *flocs*, depending on the type of secondary treatment process used. Biofilms and flocs are communities of microorganisms. Biofilms grow attached to surfaces while floc particles grow suspended in the surrounding wastewater. These communities of microorganisms incorporate biodegradable matter and convert it into biomass. Some of the microorganisms in biofilms and flocs produce exopolymer, a sticky substance that helps the bacteria stick together. Biofilms and flocs remove non-biodegradable particles from wastewater when they become trapped in the exopolymer around the microorganisms. Biofilms and flocs are denser than water and can be separated from the treated wastewater by gravity.

Biofilm formation is discussed in Chapter 7, "Fixed Film Treatment". Floc formation is discussed in Chapter 8, "Activated Sludge".

Figure 5.1 depicts some of the different wastewater components that will be discussed in the next several sections. The bars in the graph are sized proportionally to one another to show typical percentages of various wastewater constituents found in the influent of WRRFs receiving domestic wastewater without significant industrial contributions. The example shown in Figure 5.1 was created from average data collected from many different facilities (Pasztor et al., 2009). Because of differences between service areas

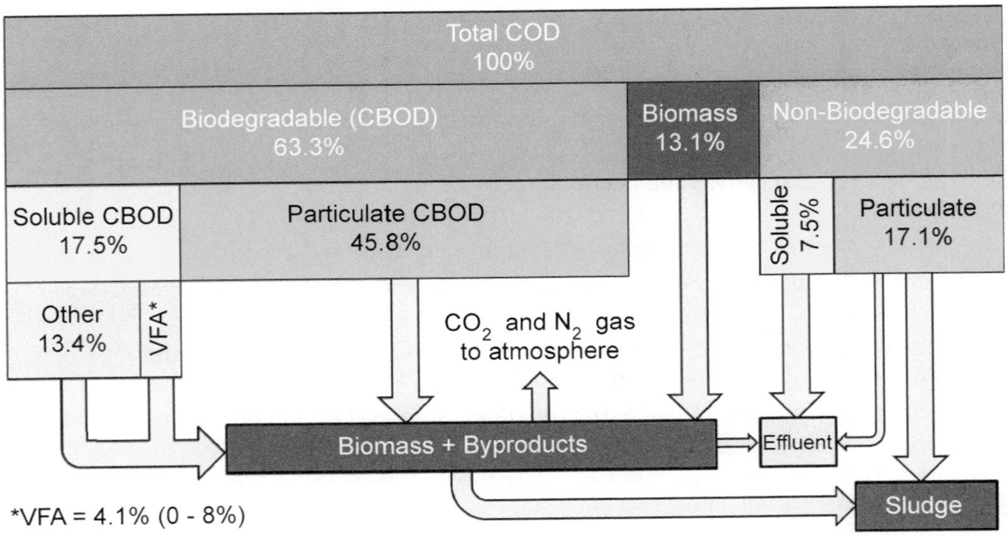

Figure 5.1 Organic Wastewater Fractions (COD = chemical oxygen demand, CBOD = carbonaceous biochemical oxygen demand, and VFA = volatile fatty acids) (adapted from Pasztor et al., 2009)

and small errors in laboratory tests, sampling, and subsampling, influent from a particular facility will not match these average percentages exactly.

BIODEGRADABLE WASTE

Bacteria are responsible for most of the biodegradation that takes place in WRRFs. Domestic wastewater contains organic matter consisting of carbohydrates, proteins, and fats that are biodegradable. Most of the organic matter in wastewater is biodegradable, but some is not. If something is biodegradable, it can be broken down into simpler components through the action of living organisms.

NON-BIODEGRADABLE AND INERT WASTE

In addition to biodegradable organic matter, wastewater contains some organic matter and chemicals that degrade very slowly or are not biodegradable at all. Non-degradable organic chemicals are referred to as *recalcitrant.* Non-degradable inorganic matter, such as salt, silt, sand, and grit, are referred to as *inert.* Slowly biodegradable waste may not be completely treated by the microorganisms during normal detention times within biological treatment processes and may not be measured in the standard, 5-day biochemical oxygen demand (BOD_5) test. The slowly biodegradable waste and inert waste that is not physically removed in the treatment process is discharged with the treated wastewater effluent. Physical removal methods include settling and filtration.

If a waste is *recalcitrant*, it is not responsive to treatment.

An *inert* material is chemically non-reactive.

Secondary treatment processes remove biodegradable material by turning it into biomass. Inorganic particles are removed when they attach to and become part of biofilms and flocs.

BIOCHEMICAL AND CHEMICAL OXYGEN DEMAND

Biochemical oxygen demand measures how much oxygen is needed by the microorganisms in the process to break down or stabilize the waste materials present in the wastewater. In the United States, the BOD test is commonly conducted over 5 days and is referred to as the *BOD_5 test.* Other countries sometimes use a 7-day test, called *BOD_7*, or a much longer test conducted over 20 or 30 days known as the *$BOD_{ultimate}$*. In Chapter 1, we learned that BOD can be further divided into carbonaceous biochemical oxygen demand (CBOD) and nitrogenous oxygen demand (NOD) as shown in eq 5.1. Carbonaceous biochemical oxygen demand is the amount of oxygen needed to stabilize organic compounds. Nitrogenous oxygen demand is the amount of oxygen needed to convert ammonia to nitrite and nitrate. Nitrogenous oxygen demand is not shown in Figure 5.1. A specialized group of bacteria, the nitrifying bacteria, make this conversion so it is part of the biochemical oxygen demand. If an analyst wants to measure only CBOD, a substance that is toxic to the nitrifying bacteria is added to the sample in the BOD test. For raw domestic wastewater (influent), the CBOD is typically about 85% of the total BOD (WEF, 2012).

$$BOD_{Total} = CBOD + NOD \tag{5.1}$$

Chapter 1 gives a more extensive description of BOD and how the test is conducted.

Carbonaceous biochemical oxygen demand consists of both soluble and particulate fractions. Soluble CBOD includes any CBOD that can pass through a piece of filter paper. In practice, some very small particulates will also pass through the filter paper and be measured as soluble. An example of soluble CBOD is sugar dissolved in tea. Soluble BOD includes both soluble CBOD and NOD. Particulate CBOD remains on the filter paper after filtering. For domestic wastewater, including light commercial dischargers such as stores and schools, soluble CBOD is typically between 20 and 40% of the total CBOD (WEF, 2012).

Enzymes are molecules produced by cells that help them carry out specific biochemical functions. Amylase is an enzyme found in saliva that converts starch into sugar. Enzymes are catalysts.

Not all of the organic matter entering the treatment process is readily biodegradable. Some of the organic material is actually composed of live bacteria and other microorganisms (biomass). Moreover, some of the organic material cannot be broken down because the bacteria lack the necessary *enzymes.* Some of the organic material can be broken down; however, because it is so complex, the bacteria need more than the 5 or 7 days given in the BOD test to process the organic material. All of these organic components can be chemically *oxidized.* They may, given enough time, consume oxygen if they are discharged to the environment. Other substances in wastewater that can consume oxygen include ferrous iron and sulfides. While these organic and inorganic substances are not measured in the BOD_5 or BOD_7 tests, they can be measured in the chemical oxygen demand (COD) test.

In chemistry, *oxidation* refers to adding oxygen to a compound. Iron metal combines with oxygen to form iron oxide or rust. The iron is oxidized. A substance can also be oxidized if it loses hydrogen or loses electrons.

The COD test uses sulfuric acid, potassium dichromate, and high temperatures to break down organic matter. It does not matter if the organic matter is biodegradable or not. Chemical oxygen demand includes CBOD, non-biodegradable organic material, and biomass (eq 5.2). The chemicals in the COD test do not react with ammonia and will not convert it to nitrite or nitrate, so the NOD is not included in the COD test result. There are a few types of complex organic compounds that cannot be broken down even in the COD test.

$$\text{COD} = \text{CBOD} + \text{Biomass} + \text{Non-biodegradable organics} + \text{Inorganic demand} \quad (5.2)$$

The COD concentration will always be greater than or equal to the CBOD concentration because of non-biodegradable organic and inorganic matter that can be chemically *oxidized*. For domestic wastewater without a large industrial component, the ratio of COD to $CBOD_5$ is typically within the range of 1.8 to 2.2, but can vary from one WRRF to another. In other words, the influent COD is typically about twice the influent $CBOD_5$. The COD-to-CBOD ratio typically remains fairly consistent over time for a WRRF, which allows either parameter to be estimated based on measurements of the other. A facility may collect COD and $CBOD_5$ data for its influent and effluent over time to better refine the calculation for its wastewater. If a new industry begins discharging, the ratio of COD to $CBOD_5$ could change; therefore, it is a good idea to periodically run both COD and $CBOD_5$.

The average percentages of biodegradable and non-biodegradable COD in wastewater are shown in Figure 5.1. Wastewater that has a high biodegradable content will have a relatively low ratio of COD to CBOD. A high ratio of COD to CBOD indicates that the wastewater contains relatively high amounts of non-biodegradable and slowly biodegradable wastes.

A nanometer (nm) is one billionth of a meter. A micrometer or micron (µm) is one millionth of a meter. A single human hair may be between 17- and 181-µm wide.

Colloidal means a mixture of fine particles that cannot be separated by gravity settling or ordinary filtering.

When something is *adsorbed*, it adheres to the surface. When something is absorbed, it is incorporated beneath the surface. Dirt adsorbs to the outside of a car, but a sponge absorbs water.

PARTICULATE ORGANIC MATTER

About 60% of the organic matter in wastewater is in particulate form (Figure 5.1). Slightly less than half of the particulate organic matter is large enough to settle out of suspension. If the WRRF has an efficiently functioning primary clarifier, the settleable solids and settleable organics will be removed upstream of the secondary treatment process. Particles between 1 and 100 µm remain in *colloidal* suspension. During treatment, colloidal particles become *adsorbed* onto biomass. Microorganisms can break down adsorbed biodegradable particles into soluble organic matter, which can then be used as a food source. Non-biodegradable particles, which are adsorbed onto biomass, are removed when the biomass is wasted or discharged from the system. In Figure 5.1, both the soluble and particulate biodegradable organic matter is turned into additional biomass. A small amount of biomass will escape into the final effluent; however, most of the biomass and adsorbed, non-biodegradable particulate matter will eventually end up in the sludge.

Chemical oxygen demand, CBOD, BOD, TSS, and total volatile solids (TVS) are linked. Organic solids (TVS) are also measured as COD and CBOD. Chemical oxygen demand results can be used to estimate CBOD results.

SOLUBLE ORGANIC MATTER

Some of the soluble organic matter in the influent wastewater is readily biodegradable and will be rapidly consumed by microorganisms. Soluble biodegradable organic matter can be measured in two ways: as soluble CBOD or as readily biodegradable COD (rbCOD). To determine rbCOD, the influent and effluent soluble COD are both measured. If soluble COD is present in the final effluent, it is assumed to be non-biodegradable (Figure 5.1). The microorganisms were unable to process it, which is why it ended up in the final effluent. The difference between the influent soluble COD and effluent soluble COD is the soluble COD that the bacteria were able to consume (eq 5.3), as follows:

$$\text{rbCOD} = \text{Soluble COD}_{\text{influent}} - \text{Soluble COD}_{\text{effluent}} \quad (5.3)$$

The influent rbCOD can vary significantly in temperate climates, with ranges from 25 to 125 mg/L. Microorganisms absorb and oxidize the influent rbCOD very rapidly. Readily biodegradable COD includes volatile fatty acids (VFAs), which are necessary for biological phosphorus removal. Volatile fatty acids are short chain organic acids such as *acetic acid*, propionic acid, butyric acid, isobutyric acid, valeric acid, and isovaleric acid. From a chemistry perspective, VFAs are organic acids that contain five or fewer carbon atoms. Municipal wastewater influent typically contains 30 to 70 mg/L of VFAs because of *anaerobic* biodegradation of organic matter in the wastewater collection system pipelines. Long, warm, flat collection systems will generate more VFAs than shorter, colder collection systems with steeper slopes. Essentially, the more time the wastewater spends in the collection system, the more VFAs will be generated. Some operators add sodium or calcium nitrate to their collection systems for odor control, which can reduce the concentration of VFAs and soluble CBOD in the WRRF influent.

Acetic acid goes by a more common name: *vinegar*. It has a pungent, sour odor. Propionic, valeric, and isovaleric acids contribute to body odor and give sweat its sour smell. Butyric acid smells like rancid butter or vomit. Volatile fatty acids are needed at low concentrations for biological phosphorus removal and are used in denitrification. In high concentrations, they can be a source of odors.

SOLIDS

Solid material in the influent can be classified by size and by whether it is organic or inorganic. Figure 5.2 shows the different types of solids and their relative proportions in the influent of WRRFs receiving domestic wastewater without significant industrial contributions. The different compounds that make up the volatile suspended solids (VSS) and volatile dissolved solids can also be measured as either COD or CBOD. Most of the dissolved, fixed solids are salts that cannot be removed by biological treatment processes. They pass through to the final effluent. Particulate and soluble volatile solids that are biodegradable will become biomass and end up in the sludge. Soluble, volatile solids that are non-biodegradable will also pass through to the final effluent. Non-volatile suspended solids will become part of the sludge.

Biodegradable organic matter, particulate matter, and nutrients can be removed by biological treatment. Soluble, non-biodegradable matter cannot be removed and passes through to the final effluent.

TYPICAL RATIOS FOR DOMESTIC INFLUENT WASTEWATER

Typically, raw domestic wastewater ratios fall within the following ranges (EnviroSim Associates, 2006):

BOD : TSS = 0.82 to 1.43
COD : CBOD = 1.80 to 2.20
Soluble CBOD : CBOD = 0.20 to 0.40
CBOD : total Kjeldahl nitrogen = 4.2 to 7.1
CBOD : total phosphorus = 20 to 50

The ratio of COD to CBOD will change as the wastewater moves through the process. This is because nearly all of the CBOD will be removed during treatment as well as the majority of particulate COD. However, soluble, non-biodegradable COD will pass through the process unchanged, ending up in the final effluent. As an example, imagine the influent to a WRRF has a COD of 500 mg/L and a CBOD of 250 mg/L. This is an influent COD-to-CBOD ratio of 2 to 1. If 5% of the influent COD is both soluble and non-biodegradable, then the final effluent will contain 25 mg/L of soluble COD. The effluent will also contain some particulate COD. The effluent CBOD, on the other hand, is typically much lower. If the effluent CBOD is 10 mg/L and the total COD in the effluent is 35 mg/L, then the ratio of COD to CBOD will be 3.5 to 1. The influent COD-to-CBOD ratios are fairly consistent from one WRRF to another, but the final effluent ratios can be quite different. It all depends on the amount of soluble, non-biodegradable COD in the influent.

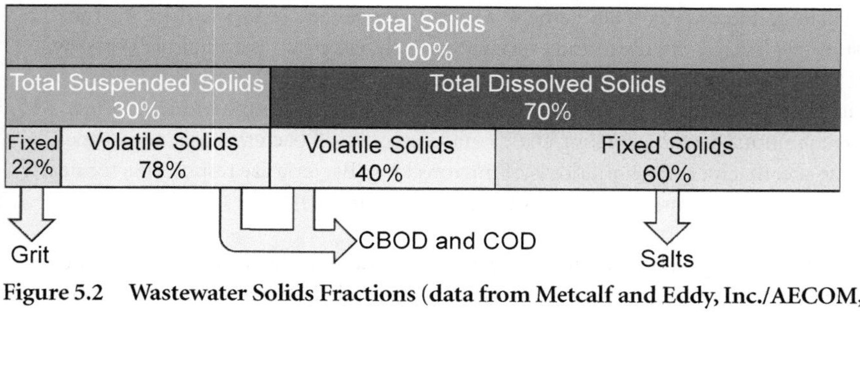

Figure 5.2 Wastewater Solids Fractions (data from Metcalf and Eddy, Inc./AECOM, 2013)

TEST YOUR KNOWLEDGE

1. Biological secondary treatment processes contain bacteria in much higher concentrations than natural systems.
   - ☐ True
   - ☐ False
2. Individual bacteria are easily observed with the naked eye.
   - ☐ True
   - ☐ False
3. The 5-day CBOD test measures all of the biodegradable material present in the wastewater.
   - ☐ True
   - ☐ False
4. Liquid may be adsorbed into a sponge.
   - ☐ True
   - ☐ False
5. One goal of biological treatment systems is to
   a. Increase the numbers of pathogenic organisms
   b. Release ammonia and phosphorus
   c. Convert BOD into biomass
   d. Decrease overall particle size
6. Starch can be broken down by bacteria into simple sugars. The starch could be described as
   a. Hydrolytic
   b. Biodegradable
   c. Inert
   d. Inorganic
7. Which of the following substances can be measured with the COD test, but not the BOD test?
   a. Biodegradable organics
   b. Ammonia
   c. Non-biodegradable organics
   d. Sulfates
8. Substances that are both soluble and non-biodegradable will be
   a. Incorporated to biomass
   b. Adsorbed onto sludge particles
   c. Outgas as carbon dioxide
   d. Discharged into the final effluent
9. A sand particle that was not dense enough to be captured by the primary clarifier passes into the biological secondary treatment process. Where is the sand particle most likely to end up once treatment is complete?
   a. Biomass
   b. Sludge
   c. Effluent
   d. Atmosphere
10. The difference between the soluble COD in the WRRF influent and effluent is
   a. Readily biodegradable COD
   b. Ammonia
   c. Nitrogenous oxygen demand
   d. Isobutyeric acid

Microbiology

The biomass in biological treatment processes may contain as many as 300 different types of microorganisms. Microorganisms of greatest significance to wastewater professionals can be classified into four groups: bacteria, protozoa, metazoa, and viruses. Each of these groups plays a key role in the complex world of wastewater biology. Individual microorganisms that are important in particular secondary treatment processes will be discussed in the chapters covering those processes. Readers that are interested in a more in-depth discussion of microbiology may refer to *Wastewater Biology: The Microlife* (WEF, 2017) and *Wastewater Biology: The Life Processes* (WEF, 1994).

All cells are either prokaryotes or eukaryotes. Eukaryotes keep their genetic material, DNA, separated from the rest of the cell with a membrane. Think of the membrane as a flexible bag. Prokaryotes do not have a membrane separating their DNA from the rest of the cell. Fungi, protozoa, metazoa, plants, and animals are all eukaryotes.

BACTERIA

Bacteria are *prokaryotes*, a diverse group of microorganisms. These small, single-celled organisms are found in tremendous numbers in polluted water. They come in a variety of sizes and shapes and are the most prolific microorganisms on earth, essentially found everywhere. Each bacterium is a single cell varying in size from about 0.2 to 2 µm, which is too small to be seen with the naked eye. Powerful microscopes are needed to view bacteria at 400 to 1000 times magnification. Most bacteria are spherical, but some are rod or spiral shaped. Filamentous bacteria are long chains made up of small bacterial cells, sometimes surrounded by a tubular sheath, and can be hundreds of microns long. Bacteria are responsible for nearly all of the biological treatment that occurs in secondary treatment systems.

Describing the cell will provide the groundwork for understanding how and why it works the way it does. Figure 5.3 shows a typical bacterial cell with significant components labeled. Bacteria have a rigid outer

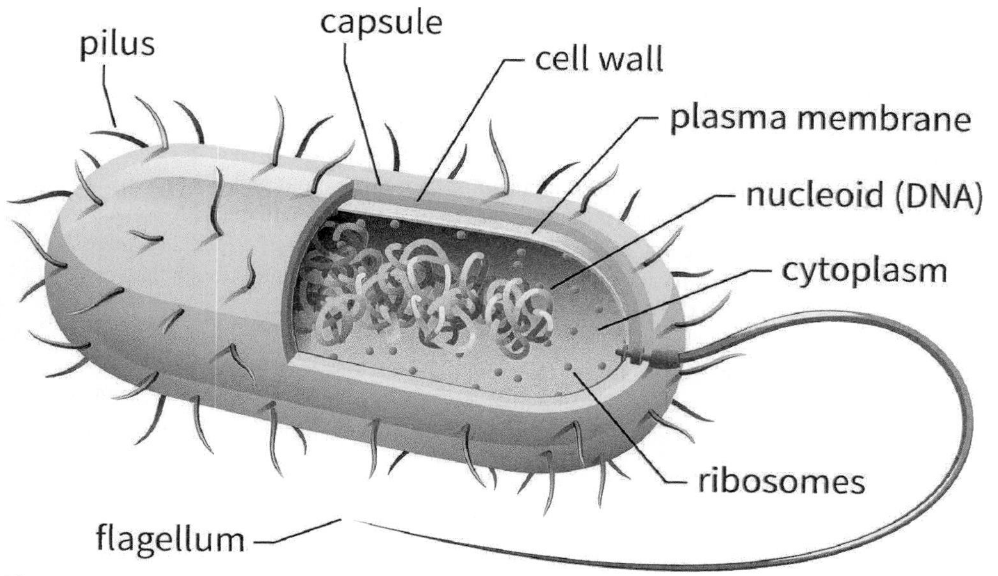

Figure 5.3 Bacteria Cell Structure (Getty Images)

coating that gives them structure and maintains their shape. This is the cell wall. Bacteria also have an inner, flexible membrane called the *periplasmic membrane* or *cell membrane.* This dual-layered covering has been compared to a balloon inside a box. The cell membrane is very important because it controls the intake of food and other nutrients and the discharge of waste products. It keeps "in" what should be inside (e.g., *enzymes*, nutrients, and food) and keeps "out" what should be outside (e.g., excess water). The box is the cell wall. The cell wall provides the structural support and maintains the shape of the cell. Much of the cellular contents are large protein molecules, known as enzymes, that are manufactured by the cell. Other cellular contents may include granules of polyphosphate, sulfur, or stored organic material.

Periplasmic translates to "around the plasm". Cytoplasm is the mixture of water, proteins, salts, and other compounds that make up the liquid filling of a cell.

Related to *enzyme*—a catalyst is a substance that participates in a chemical reaction, but is not used up by that reaction. A molecule of amylase, an enzyme found in saliva, attaches to starch molecules and breaks them into sugars. The amylase is not used up or changed by the reaction and can immediately attach to a new starch molecule to repeat the process. Enzymes often speed up chemical reactions. They can also cause chemical reactions that cannot take place on their own.

BACTERIA CELL ELEMENTAL COMPOSITION

The organic fraction of chemical elements in microbial cells can be represented as $C_5H_7O_2N$. A more complete chemical formula for the composition of a bacterial cell is $C_{60}H_{87}O_{23}N_{12}P$. As shown in Table 5.1, carbon, hydrogen, oxygen, nitrogen, phosphorus, and sulfur account for 96% of the dry mass of bacteria. Other elements account for 4%. The microbial cell average biomass contains 50% carbon, 12% nitrogen, and 2% phosphorus, by weight. If one or more of the components listed in Table 5.1 is not present in the influent wastewater in high enough concentrations, the bacteria will not be able to grow and reproduce at their maximum rate. Early researchers reported that a ratio of organic carbon (BOD_5) to nitrogen (N) to phosphorus (P) of 100 : 5 : 1 was needed for efficient treatment of wastewater. Later studies showed that these minimum ratios depended on the microorganism, process, type of wastewater being treated, and other environmental factors (WEF, 1994).

Table 5.1 Elemental Composition of Bacterial Cells (Gerardi, 2006; WEF, 1994)

Elements	Percent	Elements	Percent
Carbon	45–55	Potassium	0.8–1.5
Oxygen	16–22	Sodium	0.5–2.0
Nitrogen	12–16	Calcium	0.4–0.7
Hydrogen	7–10	Chlorine	0.4–0.7
Phosphorus	2–5	Magnesium	0.4–0.7
Sulfur	0.8–1.5	Iron	0.1–0.4
		Trace elements	0.2–0.5
Total	96		4.0

CHEMICAL TRANSPORT ACROSS THE CELL MEMBRANE

Bacteria absorb chemicals through their cell membrane. Waste products are eliminated from the cell by movement through the cell membrane to the external environment. Many chemicals cross the cell membrane by diffusion, moving from regions of high concentration outside the cell to regions of low concentration inside the cell. Diffusion is driven by the difference in concentration and does not require an expenditure of energy. Pores, or openings, in the cell membrane are approximately 0.008 mm in size. The pores in the cell membrane are so small that most molecules cannot pass through the membrane unless they are actively pumped across by the cell. Only water molecules; ions such as chloride (Cl^-), potassium (K^+), and sodium (Na^+); alcohols; and some small fat-soluble substances can pass through the pores passively (Brock and Madigan, 1991; Gerardi, 2006).

The cell membrane also contains carrier molecules that help transport substances across the cell membrane. Carrier molecules are proteins that can bind to specific substances and assist their diffusion through the cell membrane without an expenditure of cellular energy. Chemical transport by carrier molecules is known as *facilitated diffusion*.

Hydrophilic means water loving. Hydrophilic compounds are easily dissolved in water. *Hydrophobic* means water fearing. Oil is hydrophobic. When mixed together, oil and water separate into two distinct layers.

Some *hydrophilic* chemicals are actively transported through the cell membrane. Bacteria have an inner membrane made from lipids. Lipids are fats and oils. Most of the CBOD that is soluble in water is not very soluble in fat. Think about salad dressing. The oil separates and floats on top of the water. If you add spices, they end up in the water. Vigorous shaking will mix the oil and water and distribute the spices, but, if the salad dressing is allowed to sit undisturbed, the spices will go into the water and the oil and water will separate into two distinct layers again. Another simple kitchen experiment to try (or think about) is to make a circle on the counter with lard or vegetable shortening. Fill the circle with a small amount of water. Then, add a few drops of food coloring. After a few minutes, the food coloring will disperse and the water will be uniformly colored, but the lard or vegetable shortening will not change color. Why not? The food coloring is water soluble, but not fat soluble. If we want the food coloring to move to the other side of the barrier that we made with lard or shortening, we have to cut a channel through the barrier. The lipid membrane is a barrier that CBOD cannot move through very easily. To get past the lipid membrane, the CBOD must be pumped across. Pumping is active transport. Active transport requires a specific carrier molecule for each type of chemical and an expenditure of cellular energy to convey chemicals through the cell membrane.

Soluble CBOD in the wastewater is rapidly transported across the cell membrane by the bacteria using active transport. To feed on particulate CBOD, the bacteria must first adsorb the particulate CBOD and then excrete enzymes. The enzymes convert particulate CBOD to soluble CBOD outside the cell. Only then can the organic molecules be transported across the cell membrane. The breakdown step followed by active transport is comparable to having to cut up your steak into smaller pieces before you can eat it. In this case, the knife and fork are doing the work of enzymes. Bacterial cells can absorb and metabolize soluble CBOD quickly, and they can adsorb particulate CBOD quickly. However, it takes time to convert particulate CBOD to soluble CBOD outside the cell. If the steak came to the table already cut into bite-sized pieces, you would be able to eat it quickly and with very little effort. When it comes to the table as a large steak, it takes time to cut it up before you can eat it.

In biology, *substrate* can have several different meanings. It can be the underlying surface that an organism grows on, like plastic or wood. It can be the material an organism uses to grow, like ammonia, oxygen, or sugar. Think of substrates in terms of the resources a microorganism needs to survive, grow, and reproduce.

The degree of *ionization* of *substrates*, nutrients, and toxic wastes strongly affects chemical transport into bacterial cells. Ionization occurs when a molecule or atom gains or loses electrons. For example, dissolved ammonia has the chemical formula NH_4^+, with the positive charge denoting that the molecule has one more proton than electrons. This is the ammonium ion. Ammonia may also be present as NH_3 with no charge. At low pH, NH_4^+ dominates and, at high pH, NH_3 dominates. Therefore, pH is a critical factor for microorganisms. Wastes produced by bacteria may significantly change the pH within a biological treatment process. A change in pH may result in different bacteria becoming dominant within a biological treatment process, which may affect treatment efficiency.

Polysaccharides are molecules made of many (poly) sugars (saccharides) that have been chemically bonded together. Starch, cellulose, and glycogen are three types of polysaccharides.

EXTERNAL SLIME LAYER

Many bacterial cells produce extracellular polymeric substances (EPS), which are secreted through the cell wall onto the exterior surface of the cell. The EPS consist mainly of *polysaccharides*. Extracellular polymeric substances develop a protective coating on the outside of the cell wall (called a *glycocalyx*). The protective coating may be either a distinct gelatinous capsule or a diffuse and irregular slime or gel layer. The external slime layer or gelatinous capsule minimizes the loss of water and nutrients from the cell. It can also protect the cell from toxins.

The slime or gel layer allows the bacteria to adhere to solid surfaces and to form colonies, *biofilms*, and *floc particles*. This ability to form larger particles, which can settle out of the water by gravity, is essential to many wastewater treatment processes, especially the activated sludge process. All bacteria produce multiple adhesive chemicals, which allow them to switch from attached growth to suspended growth forms under different environmental influences.

Biofilm is created when bacteria adhere to a surface forming a slime layer.

Floc is a form of bacterial growth found in activated sludge systems. The bacteria form colonies several hundred microns (μm) in diameter that are suspended and mixed in the wastewater.

Large or suspended organic particles cannot pass through the cell wall and membrane. However, bacteria can adsorb large organic particles onto the slime or gel on the outside of the cell wall. Enzymes embedded within the slime or gel break down large, organic particles into smaller chemicals that can be ingested and used as food. Different kinds of bacteria produce and secrete different enzymes that determine which chemicals they can adsorb and ingest as a food source.

CELLULAR STORAGE PRODUCTS

Bacterial cells store excess food and energy internally in the form of organic and inorganic molecules. The primary organic storage molecules include glycogen, starch, and poly-β-hydroxybutyrate (PHB). Inorganic storage molecules include long chains of polyphosphate, known as *volutin granules*, and insoluble elemental sulfur that is stored as sulfur granules. Bacteria use these storage molecules to provide energy when external sources of food are limited or absent.

Only soluble materials can be absorbed by bacteria. Solids must be broken down by enzymes first. Bacteria need energy to absorb most materials. The cell membrane restricts what can enter and leave the cell.

MICROBIAL GROWTH AND REPRODUCTION

When a bacterial cell grows to full size, it reproduces by dividing into two daughter cells through *binary fission* (Figure 5.4). Rapid growth and cell division of the daughter cells result in exponential growth in the number of bacteria. Two daughter cells divide into four, four cells divide into eight, eight cells divide into 16, and so on. Under ideal conditions, some bacteria can double their numbers in as little as 20 minutes.

Microorganisms generally exhibit a pattern of sequential growth phases, which have different growth rates (Figure 5.5). When a new food source becomes available, microorganisms often have a lag phase, during which no growth is observed initially. During the lag phase, microorganisms begin producing specific

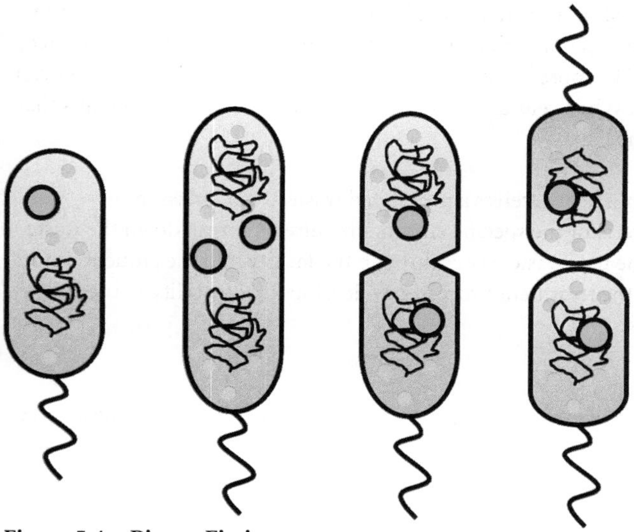

Figure 5.4 Binary Fission

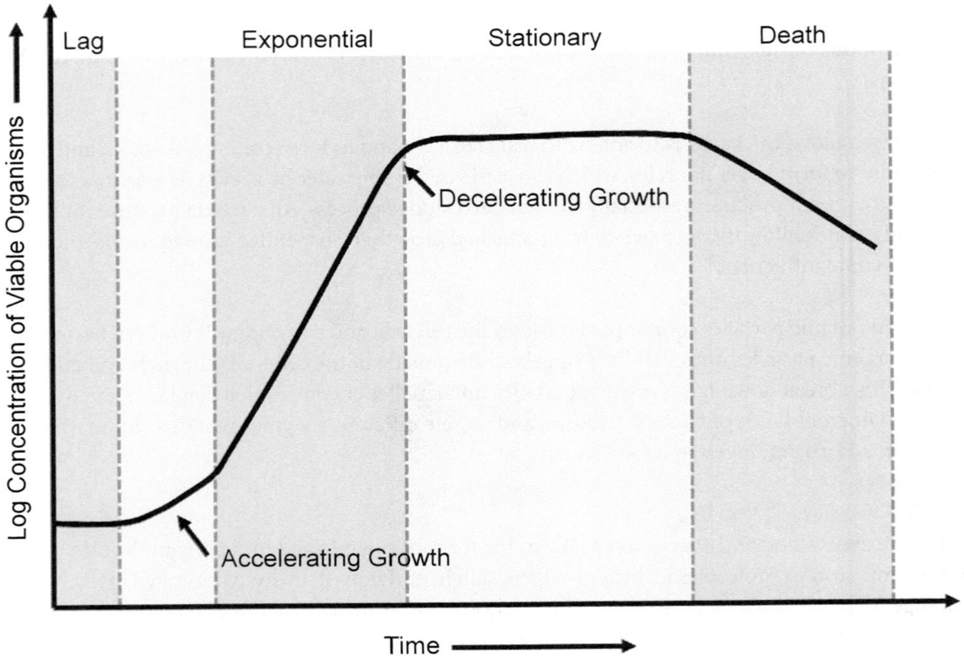

Figure 5.5 Bacterial Growth Curve

enzymes required to break down and oxidize the new food source. The lag phase is followed by a period of rapidly increasing growth called the *acceleration phase*. The initial lag and acceleration phases reflect the time required for the microorganisms to produce or activate the specific enzymes required to break down the new food source. Formation of a specific enzyme under the influence of its *substrate* is known as *enzymatic adaptation*. When the enzyme system for the specific food source is fully activated, microbial growth reaches the maximum growth rate, which remains constant until the food concentration decreases.

The acceleration phase is followed by an exponential growth phase, during which the microorganisms reproduce at their maximum rate. In this stage of development, resources are plentiful. As the number of bacteria increases, the amount of resources available to each bacterium gradually decreases and fewer bacteria will have enough resources to reproduce. Resources and energy are first used to maintain existing cell mass. Bacteria can only reproduce if excess resources and energy are available. This period of gradually declining growth rate is called the *deceleration phase*. Eventually, the number of bacteria in the system will level out to match the available food and energy coming into the system. This is the stationary phase. The stationary phase is shown as a flat line in Figure 5.5. In the stationary phase, the number of new bacteria formed is roughly equivalent to the number of bacteria that die. The population stabilizes. The bacteria in secondary treatment processes will be in the stationary phase most of the time. If the amount of resources coming into the process decreases, there will be more live microorganisms than the available resources can support and some of them will begin to die. When cells die, they lyse (break apart), releasing contents that are then available as food for other organisms.

When a WRRF receives waste from a new source or receives a prohibited waste, the biomass in the treatment process may lack the type of bacteria or the specific enzymes required to break down the waste. Discharge permit violations may result as the new waste passes through the facility into the effluent. If the new waste is biodegradable, the biomass in the treatment process may develop the capability to break it down over a period of several days or weeks.

Bacteria in secondary treatment processes are typically in the stationary phase because food and other resources are limited.

PROTOZOA

An operator can observe protozoa and metazoa under the microscope. These *indicator organisms* can also give insight to the health of the treatment system. Protozoa are unicellular eukaryotes that include amoebae, flagellates, and ciliates (Figure 5.6). The name *protozoa* derives from the Greek *protos*, "first", + *zoia*, "animals". Many microbiologists prefer the name protista to the historical name *protozoa*, but either name can be found in many microbiology books or manuals. In wastewater treatment facilities, protozoa account for over 90% of the total nonbacterial biomass. Protozoa, like bacteria, are typically found in large numbers in polluted water and wastewater as well as natural systems like streams and lakes. Many of them are quite mobile, swimming rapidly through their microscopic world. Others attach themselves to tiny particles and extend into the water around them. Protozoans common in secondary treatment processes include flagellates, amoebae, and ciliates. Ciliates can be crawling, free-swimming, or stalked protozoa. They are typically the dominant protozoa in wastewater systems in terms of number of species. Ciliates account for 10% of the biomass in suspended growth systems (10^4 ciliates per milliliter) and 20% of the total biomass in biofilms (Pauli et al., 2001).

An *indicator organism* "indicates" or tells something about the process by its presence or absence. For example, a large number of flagellates in an activated sludge system is not normal except during system startup, when organic loading is very high, or if a toxic substance entered the treatment process. The presence of a lot of flagellates "indicates" an abnormal condition.

Protozoa are predators to bacteria and algae, which are their food sources. Protozoa are generally larger than bacteria, and each protozoan can consume hundreds of bacteria per hour. Some protozoa, such as amoebae, can engulf particles of food and whole bacteria. Some protozoa, such as paramecium, have a mouth-like opening to ingest food. All protozoa digest food in specialized compartments called *vacuoles*. The large numbers of bacteria consumed by protozoa as food reduces the volume of waste biomass generated in WRRFs by approximately 20%.

In suspended growth systems, the bacterial colonies are suspended in the wastewater. They float and sink freely. An example of suspended growth is activated sludge. In fixed film systems, the bacteria grow on a surface and do not move with the water. As described previously, the slime layer formed is called a biofilm. Trickling filters are an example of a fixed film process.

Absorption of colloidal particles by protozoa improves the settleability of suspended particles. In both activated sludge basins and clarifiers, the ciliate population has sufficient time to filter the entire wastewater volume several times, which helps clarify the water by removing bacteria and colloidal particles. Experiments have demonstrated that when protozoa are absent from the treatment process, high numbers of suspended bacteria remain in the effluent. After reintroduction of protozoa to the treatment process, the suspended bacteria population decreases by 70% (Pauli et al., 2001). Filter-feeding ciliates also substantially reduce the numbers of pathogenic bacteria in wastewater treatment facilities.

Some protozoans are associated with waterborne disease. *Giardia* and *Cryptosporidium* are two protozoans responsible for significant outbreaks of illness in the United States.

METAZOA

Metazoa is a term that technically applies to all animals composed of more than one cell. From a practical standpoint, it refers to larger, more complex animals than those considered protozoa. Included in the metazoan group are organisms known as *rotifers*, so named because of their rotating crowns of hairlike cilia. Rotifers are fairly common in some wastewater treatment processes and play an active role in the breakdown of organic wastes. Although typically somewhat larger than bacteria and protozoa, they are still microscopic in size.

Many other metazoa must be considered in the discussion of wastewater microbiology. Nematodes, water bears, bristleworms, water fleas, and seed shrimp are all metazoa that may take up residence in a wastewater treatment process.

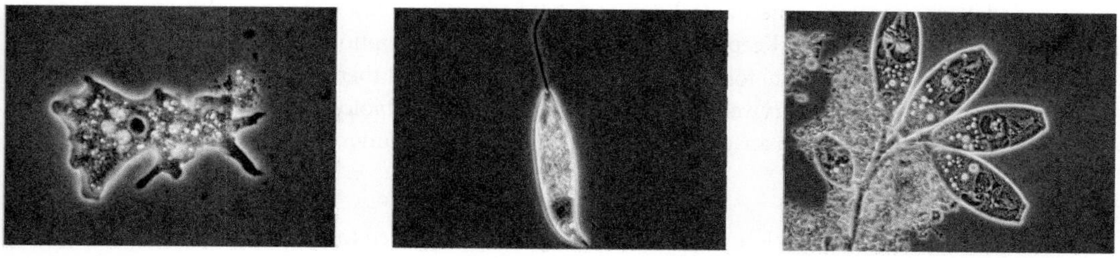

a) Amoeba b) Flagellate c) Ciliate

Figure 5.6 Protozoans in Wastewater (Reprinted with permission by Steven C. Leach, Novozymes)

Protozoa and metazoa are indicator organisms. Observing them under the microscope can give operators information about the biological process.

TEST YOUR KNOWLEDGE

1. The cell membrane serves as a selective barrier that only allows some substances to pass through.
   - ☐ True
   - ☐ False
2. Enzymes are catalysts.
   - ☐ True
   - ☐ False
3. Protozoans are predators that can consume large quantities of free-swimming bacteria.
   - ☐ True
   - ☐ False
4. Metazoa are multicellular animals and are typically smaller than both protozoans and bacteria.
   - ☐ True
   - ☐ False
5. A bacterium can move ________ across its cell membrane without expending energy.
   a. Soluble CBOD
   b. Water
   c. Particulate COD
   d. Phosphate ions
6. Extracellular polymeric substances:
   a. Help the cell adhere to surfaces and other bacteria
   b. Form a protective coating around the cell
   c. Consist of sugars chemically joined together
   d. All of the above.
7. When a bacterium reproduces, how many daughter cells are produced from each parent cell?
   a. 1
   b. 2
   c. 3
   d. 4
8. A single bacterium enters the treatment process and has access to unlimited amounts of food, energy, oxygen, and the other substances it needs to grow and reproduce. If the bacterium can reproduce every 20 minutes, how many bacteria will there be after 1 hour?
   a. 2
   b. 4
   c. 8
   d. 16
9. During the lag phase of bacterial growth:
   a. No growth is observed
   b. Growth rate rapidly increases
   c. Growth rate and death rate are about equal
   d. Death rate exceeds growth rate

Bacteria in Secondary Treatment Processes

The bacteria in secondary treatment processes can be broken into a few significant groups depending on their energy source, carbon source, oxygen requirements, and their preferred growth patterns. Over the next several sections, the terms *heterotroph, autotroph, aerobic, facultative, anaerobic*, and others will be defined. Many wastewater professionals initially enter the profession as mechanics or have a mechanical background. For this reason, a mechanical approach has been used to describe the needs and functions of the different groups of bacteria. Keep in mind that the following descriptions comparing microorganisms to engines and factories are useful for learning basic concepts, but that there are many factors that contribute to the growth and survival of different microorganisms in biological treatment processes. The anaerobic bacteria, important in anaerobic digestion, have their own unique physiology and have not been included in this discussion.

Bacteria are somewhat predictable and, at a basic level, can be compared to miniature combustion engines. For an engine to function, it requires both a fuel source and an oxygen source (Figure 5.7). The oxygen source is used to chemically burn fuel to release energy. The technically correct term for this process is *oxidation*. The byproducts of combustion when burning organic fuel with oxygen are carbon dioxide (CO_2)

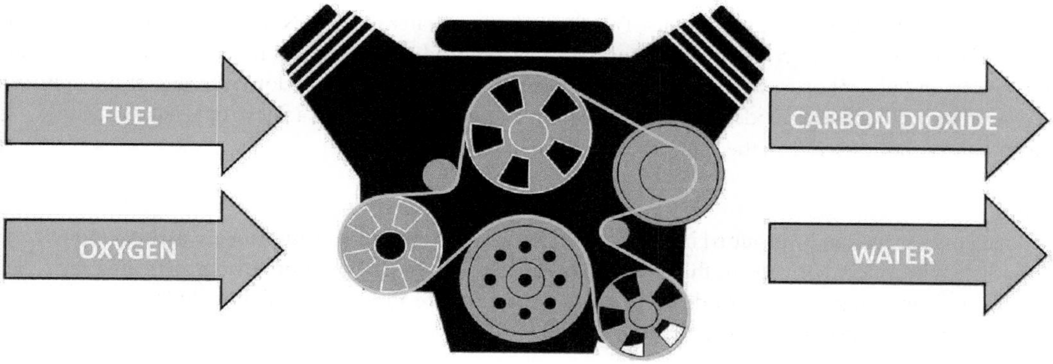

Figure 5.7 Combustion Engine (Reprinted with permission by Indigo Water Group)

and water (H_2O). In the example in eq 5.4, the organic fuel is ordinary table sugar ($C_6H_{12}O_6$). Burning organic compounds releases stored chemical energy, much in the same way that burning gasoline provides energy to operate a car.

$$\text{Sugar} + \text{Oxygen} \rightarrow \text{Carbon dioxide} + \text{Water}$$
$$C_6H_{12}O_6 + O_2 \rightarrow CO_2 + H_2O \tag{5.4}$$

Bacteria are a little more complicated than simple combustion engines because the bacteria have another goal: to produce more of themselves. It is more useful to think of bacteria as factories with an engine at their heart (Figure 5.8). The factory needs fuel and an oxygen source to run its engine to produce energy. To build more bacteria, the factory also needs a ready supply of spare parts. The bacteria will require large quantities of carbon, oxygen, nitrogen, and phosphorus as well as a host of other materials like iron, manganese, and potassium in smaller quantities. Carbon, nitrogen, and phosphorus are called *macronutrients* because they are needed in large quantities. Iron, manganese, potassium, and other vital elements are called *micronutrients*. Recall from Table 5.1 that bacterial cells contain about 50% carbon, 14% nitrogen, and 3% phosphorus by dry weight. The overall biochemical reaction that occurs for all of the bacteria in biological treatment processes may be expressed as follows:

$$\text{Fuel} + \text{Nutrients} + \text{Bacteria} + \text{Oxygen Source} \rightarrow \text{New Bacteria} + \text{Products} \tag{5.5}$$

Domestic wastewater typically has all of the ingredients needed to sustain life. Unless a WRRF receives a significant amount of industrial wastewater, there should be plenty of nitrogen and phosphorus available. Typical medium-strength domestic wastewater contains about 200 mg/L of $CBOD_5$, 40 mg/L of total

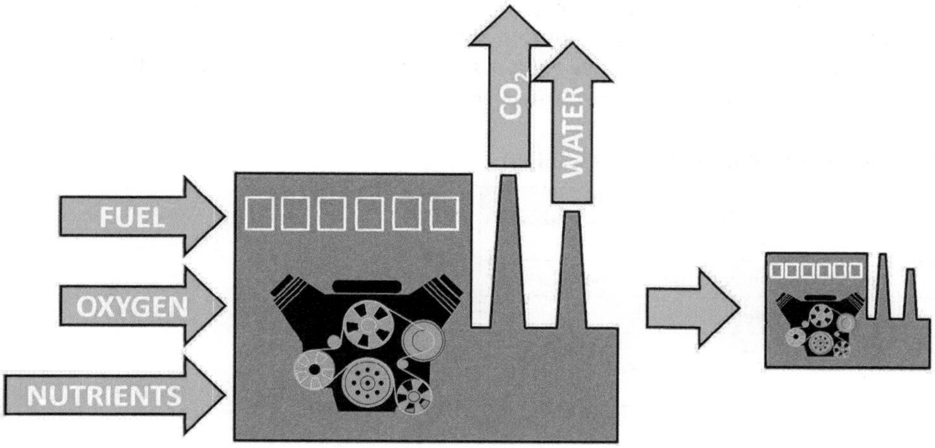

Figure 5.8 Bacterial Factory (Reprinted with permission by Indigo Water Group)

nitrogen, and 7 mg/L of total phosphorus. This example wastewater has a C:N:P ratio of 100:20:3.5—an excess of nutrients (nitrogen and phosphorus). Wastewater produced by certain industries may be deficient in nitrogen and phosphorus. When the concentration of these nutrients is too low, nitrogen or phosphorus may need to be added to provide the correct ratio for microbial growth and achieve optimal biological treatment. Without nutrients, the bacteria are unable to reproduce.

ENERGY AND CARBON SOURCES

All living organisms can be grouped into broad categories according to how they obtain carbon and energy from the environment. Organisms that use carbon dioxide as their carbon source for growth are called *autotrophs*. Plants, algae, and many different kinds of bacteria are autotrophs. All other living organisms are heterotrophs, which derive the carbon they need for growth from organic compounds.

Living organisms can further be divided into categories based on the way they obtain energy. All plants and algae and a few bacterial species get their energy from sunlight through photosynthesis and are called *phototrophs*. All other organisms get their energy from the oxidation of chemical substances and are called chemotrophs. If those chemical substances are organic, the organisms are *chemoorganotrophs*. If those chemicals are inorganic, the organisms are called *chemolithotrophs*.

These long, scientific words can be a little intimidating, but, if they are broken down into their base components, it is a little easier to keep them straight. Many scientific words are Greek or Latin. From Table 5.2, a lithoautotroph is a rock-eating self-nourisher and a photoheterotroph is a light-eating other-nourisher.

HETEROTROPHIC BACTERIA

The bacteria in a WRRF, which feed mainly on organic (carbonaceous) matter, are called *heterotrophs*. The word *heterotroph* means other-nourishing and derives from the Greek words *hetero* (other) and *trophe* (from trephine, meaning to nourish). *Heterotroph* literally translates to an organism that gets its nourishment from others. The bacteria do not literally consume one another whole, but they do get their energy and carbon from organic compounds that were produced by other microorganisms and not themselves. All animals and most bacteria are heterotrophs. This large group of different types of bacteria obtains its carbon and its energy from organic molecules like carbohydrates, proteins, and fats in the wastewater. These compounds are measured collectively as CBOD. Heterotrophs require DO or an oxidized inorganic chemical such as nitrate (NO_3^-), nitrite (NO_2^-), or sulfate (SO_4^{-2}) to process CBOD. Heterotrophic bacteria are typically the dominant microorganisms in most environments, including in the biomass in wastewater treatment processes.

AUTOTROPHIC BACTERIA

Autotrophs are microorganisms that obtain their carbon from inorganic chemicals such as carbon dioxide (CO_2), carbonate (CO_3^{-2}), and bicarbonate (HCO_3^-). Most plants, algae, and many different kinds of bacteria are autotrophs. *Auto* means self and troph means nourishing. An autotroph is literally self-nourishing. Autotrophs cannot use organic compounds (CBOD) as an energy or carbon source. Instead, they obtain their energy from inorganic sources. Autotrophs build or synthesize all of their own organic molecules. Although there are many different types of autotrophic bacteria, some of the most important ones in wastewater treatment are the nitrifying bacteria and the sulfate-reducing bacteria.

Table 5.2 Organism Descriptors and Their Meanings

Description	Meaning
Auto	Self (inorganic carbon)
Hetero	Other (organic carbon)
Troph	Feed or nourish
Photo	Light
Chemo	Chemical
Organo	Organic
Litho	Rock

OXYGEN REQUIREMENTS

There are three different environments that can be present in wastewater treatment processes: aerobic (oxic), anoxic, and anaerobic. Aerobic environments contain DO and may or may not contain nitrite or nitrate. Anoxic environments contain nitrite and/or nitrate, but do not contain DO. Anaerobic environments do not contain DO, nitrite, or nitrate. The presence or absence of oxygen, nitrite, and nitrate influences which bacteria are active and how they process their fuel.

OBLIGATE AEROBES

Microorganisms that grow only in the presence of oxygen are called *obligate*, or *strict, aerobes*. Obligate aerobes cannot survive without oxygen. Humans are obligate aerobes. Humans must have oxygen to breathe and do not survive very long without it. Some heterotrophs require an oxygen source for *respiration* and growth. For heterotrophs using DO, the overall biochemical reaction that occurs may be expressed as follows:

$$\text{Organic matter} + \text{Nutrients} + \text{Bacteria} + O_2 \rightarrow \text{New bacteria} + CO_2 + H_2O \quad (5.6)$$

Respiration is the extraction of energy from organic or inorganic compounds. In aerobic processes, oxygen is combined with other molecules to extract energy. In anoxic and anaerobic processes, nitrate or sulfate takes the place of oxygen.

Most of the nitrifying bacteria are obligate aerobes and must have DO. They function well in the aerobic or oxic environments, but cannot function in either anoxic or anaerobic environments. Most of the nitrifying autotrophs will become dormant for the time they are in either anoxic or anaerobic conditions.

OBLIGATE ANAEROBES

Microorganisms that grow only in the absence of oxygen are *obligate*, or *strict, anaerobes*. Examples of obligate anaerobes include *Actinomyces, Bacteroides*, and *Clostridium*. These microorganisms cannot use oxygen or nitrate for respiration. Instead, they partially break down organic matter to get some energy and then use the breakdown products for respiration. This type of anaerobic respiration is called *fermentation*. The byproducts of fermentation are organic acids and alcohols. In secondary treatment systems, fermentation is important for generating the VFAs that are needed for biological phosphorus removal.

In anaerobic digestion, two groups of anaerobic bacteria break down organic matter in primary and secondary sludge to form methane and carbon dioxide. The first group of anaerobes is the acid formers, also known as the *saprophytic microorganisms*. The acid formers form VFAs through fermentation. The second group of anaerobes is the methanogens. The methanogens convert VFAs into methane. The methanogens are extremely slow growing and are very sensitive to changes in pH and temperature. The methanogens do not play a significant role in most secondary treatment processes.

FACULTATIVE ANAEROBES

Some heterotrophic bacteria are capable of surviving under different environmental conditions because they are adaptable. Facultative heterotrophic bacteria prefer to use DO but can switch to a *chemically bound* form of oxygen like nitrite (NO_2^-) or nitrate (NO_3^-) when oxygen is not available. When facultative heterotrophic bacteria use nitrite or nitrate as an oxygen source instead of DO, the byproducts are nitrogen gas (N_2), carbon dioxide (CO_2), and water (eq 5.7). This process is called *denitrification*. Most species of denitrifying bacteria belong to three genera including *Alcaligenes, Bacillus*, and *Pseudomonas*. Using nitrate instead of oxygen is not as efficient for the bacteria. For facultative heterotrophs using nitrate as an oxygen source, the overall biochemical reaction that occurs may be expressed as follows:

Chemically bound means the oxygen is combined with other elements like nitrogen, sulfur, or carbon.

$$\text{Organic matter} + \text{Nutrients} + \text{Bacteria} + NO_X \rightarrow \text{New organisms} + N_2 + CO_2 + H_2O \quad (5.7)$$

If facultative, heterotrophic bacteria are placed into an aerobic secondary treatment process where they will have access to a fuel source (CBOD), an oxygen source (DO), and the spare parts (nitrogen, phosphorus, and other micronutrients) they require to build more of themselves, the byproducts will be carbon dioxide and water. When the same facultative, heterotrophic bacteria are subjected to anoxic conditions in a secondary treatment process, they no longer have access to DO. They will have access to nitrate. The CBOD is also present as are nutrients. In anoxic secondary processes, the facultative, heterotrophic bacteria will continue to grow and reproduce, but the byproducts will now be nitrogen gas, carbon dioxide, and water. If this same group of organisms is placed into anaerobic conditions, they will not be able to function. There is no nitrate or oxygen for them to use to burn fuel. The heterotrophs will be dormant for the time they are in the anaerobic basin.

Living organisms are classified by the type of carbon they use (organic or inorganic); where they get their energy (sunlight, organic compounds, or inorganic compounds); and their oxygen requirements (aerobic, facultative, or anaerobic).

SPECIALIZED GROUPS OF BACTERIA

NITRIFYING BACTERIA

Nitrifying bacteria use ammonia (NH_3) or nitrite (NO_2^-) as their fuel and DO as their oxygen source. The inputs for autotrophic, nitrifying bacteria will be either ammonia or nitrite as fuel plus DO and spare parts (Figure 5.9). The outputs from nitrifying autotrophic bacteria will be nitrite (NO_2^-), nitrate (NO_3^-), carbon dioxide (CO_2), and water (H_2O).

The spare parts list is a little bit different for the autotrophic bacteria than it is for the heterotrophic bacteria because their fuel source, which is ammonia or nitrite, does not contain any carbon. Carbon is a necessary building block of all life on earth. The autotrophs use the inorganic carbon compounds, carbonate (CO_3^{-2}) and bicarbonate (HCO_3^-), as their carbon source instead of CBOD. Carbonate and bicarbonate are two components of alkalinity. They are formed when carbon dioxide gas is dissolved in water. For every 1 mg/L of ammonia that is converted to nitrate, 7.14 mg/L of alkalinity will be consumed. For the nitrifying bacteria, the overall biochemical reaction that occurs may be expressed as follows:

$$NH_3^+ + O_2 + HCO_3^- + \text{Nitrifying bacteria} \rightarrow \text{New bacteria} + NO_3^- + CO_2 + H_2O \quad (5.8)$$

Chapter 2 includes an in-depth discussion of alkalinity.

In the previous paragraph, the conversion of ammonia to nitrate was shown as a single step. There are actually two different groups of nitrifying bacteria: the ammonia-oxidizing bacteria (AOB) and the nitrite-oxidizing bacteria (NOB). The ammonia-oxidizers convert ammonia (NH_3) to nitrite (NO_2^-). The nitrite-oxidizers then convert the nitrite to nitrate (NO_3^-), as shown in Figure 5.10. Both use inorganic carbonate and bicarbonate as a source of carbon to build the organic molecules needed for growth. Ammonia oxidizing bacteria include the following: *Nitrosomonas, Nitrosococcus, Nitrosospira, Nitrosovibrio,* and *Nitrosolobus* (Koops and Pommerening-Röser, 2001). *Nitrosomonas* appears to be the most important genus in activated sludge and trickling filter systems receiving relatively high specific loadings of ammonia-N (≥20 mg/L). *Nitrosospira* is the most important genus in biofilm systems receiving very low ammonia-N (≤1 mg/L) concentrations (Grady and Filipe, 2000). Nitrite-oxidizing bacteria include the following: *Nitrobacter, Nitrospira, Nitrococcus,* and *Nitrospina*. Members of the genus *Nitrospira* appear

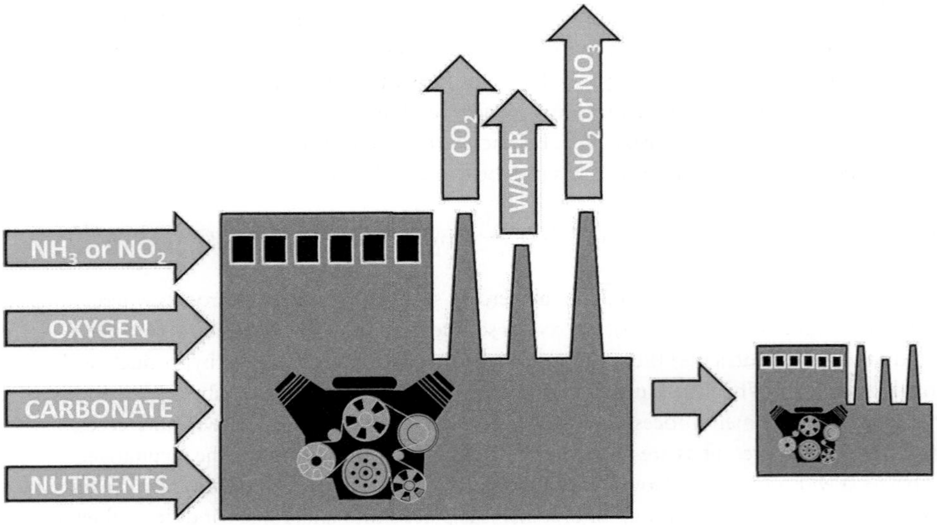

Figure 5.9 Nitrifying Autotrophic Bacterial Factory (Reprinted with permission by Indigo Water Group)

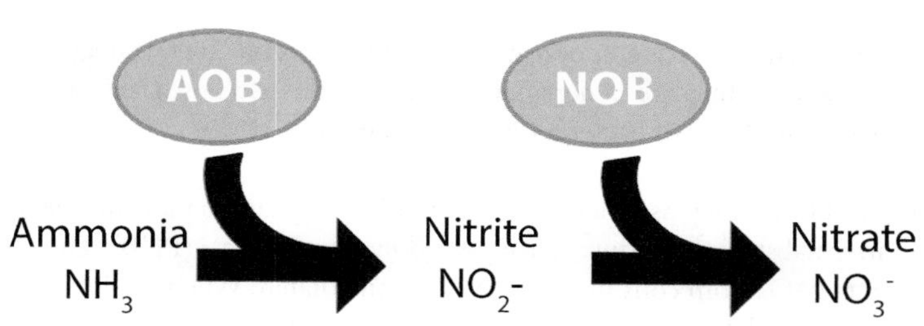

Figure 5.10 Nitrification as a Two-Step Process

to be the most important NOB in wastewater treatment systems, although *Nitrobacter* are also present. *Nitrococcus* and *Nitrospina* are marine species that do not live in wastewater treatment systems. Nitrifying bacteria are discussed in more detail in the chapter on biological nutrient removal.

Many different species of bacteria can convert ammonia to nitrite (AOB) and nitrite to nitrate (NOB). They are all autotrophs, or, more specifically, chemolithotrophs.

COMPARING HETEROTROPHS AND AUTOTROPHS

Which group of bacteria grows faster, the heterotrophs or the nitrifying autotrophs? The heterotrophs use energy-rich organic compounds as their fuel source, whereas the nitrifying autotrophs use relatively energy poor ammonia as their fuel source. Higher-energy-content fuel means the heterotrophs can grow much faster than the autotrophs. Under ideal conditions, a heterotrophic bacterium may only need 20 minutes to reproduce, whereas a nitrifying bacterium may need between 7 and 13 hours. For this reason, nitrifying bacteria will be present in the biomass in significant numbers only when the biomass is retained in a WRRF for several days. Fast-growing heterotrophs tend to outcompete autotrophs for nutrients and, therefore, are the dominant microorganisms in most environments.

PHOSPHATE ACCUMULATING ORGANISMS

Phosphate accumulating organisms (PAOs) are a specialized subgroup of facultative heterotrophic bacteria. When these bacteria are cycled between anaerobic and either anoxic or aerated conditions, they will pick up and store excess phosphorus. This process is called *luxury uptake* because the bacteria are incorporating more phosphorus than they need for growth. The PAOs are able to store up to 40% of their dry weight as phosphorus. Phosphorus is removed from the wastewater when excess biomass is removed from the system.

Understanding why luxury uptake occurs begins with understanding what happens to CBOD under anaerobic conditions. In an anaerobic secondary treatment process, some of the CBOD is broken down through fermentation by anaerobic bacteria into soluble CBOD and simpler organic molecules called *VFAs*. Volatile fatty acids are chemical compounds that contain 5 carbons or less such as acetic acid (2 carbons), propionic acid (3 carbons), and butyric acid (4 carbons). Acetic acid is more commonly known as *vinegar*. Volatile fatty acids are a preferred source of carbon and energy by heterotrophic bacteria, including the PAOs, because these compounds are easily absorbed into the bacteria. The PAOs have a logistical problem: when they are under anaerobic conditions, they are exposed to VFAs, but without oxygen, nitrite, or nitrate present, they cannot access them. Picking up "fuel" takes energy, which is unavailable to most facultative heterotrophs in the absence of oxygen, nitrite, and nitrate.

The PAOs have a creative solution to this problem. They build a chemical battery from phosphorus. Most cells, including human cells, use a compound called *adenosine triphosphate* (ATP) to store energy. Adenosine triphosphate consists of three phosphate molecules chemically bound together. The chemical bond between phosphorus atoms stores energy until the cell needs it. The ATP is continuously formed and consumed within cells as energy is stored and recovered. The PAOs take ATP to the next level and form an energy-rich compound called *polyphosphate*, which strings together large numbers of phosphate molecules. Polyphosphate is a chemical battery. When the PAO needs energy, it breaks the chemical bonds between phosphorus molecules to recover the stored energy and temporarily releases phosphorus. This strategy allows the PAOs to pick up and store VFAs when they are under anaerobic conditions. Because they have a battery pack, they are able to outcompete other types of bacteria for the available VFAs.

When the PAOs are under anaerobic conditions, they use energy stored in polyphosphate to pick up VFAs. The VFAs are then combined into an internal storage product called *poly-β-hydroxybutyrate* (PHB). The lack of oxygen, nitrite, or nitrate prevents PAOs from consuming the VFAs immediately. When the PAOs move into either anoxic or aerobic conditions, they immediately begin to consume the PHB and rebuild their polyphosphate battery. Cycling the PAOs between anaerobic and anoxic or aerated conditions is required for luxury uptake to occur. Without exposure to an anaerobic zone, there is no incentive for the PAOs to store phosphorus. Phosphate accumulating organisms are discussed in more detail in the chapter on biological nutrient removal.

Chemical reduction is the opposite of oxidation. If a compound is reduced, it loses oxygen, gains hydrogen, or gains electrons and becomes more negatively charged.

SULFATE-REDUCING BACTERIA

Sulfate-*reducing* bacteria (SRB) are a large group of microorganisms that live in many environments. There are 220 known species within 60 different genera. They can tolerate extreme pH values ranging from as low as 2 to as high as 10. Sulfate-reducing bacteria obtain energy by oxidizing organic compounds under anaerobic conditions using sulfate (SO_4^{-2}) as an oxygen source. Sulfate-reducing bacteria produce carbon dioxide (CO_2) and hydrogen sulfide (H_2S) as waste products. Hydrogen sulfide smells like rotten eggs and is extremely toxic. In addition to sulfate, SRBs can reduce other sulfur compounds including thiosulfate ($S_2O_3^{-2}$), sulfite (SO_2^{-3}), and elemental sulfur (S^o) to sulfide (S^{-2}). Most SRB are anaerobes, but some can switch to aerobic respiration when oxygen is available. *Desulfovibrio desulficans* is one SRB found in WRRFs that produces hydrogen sulfide.

SULFUR-OXIDIZING BACTERIA

The sulfur-oxidizing bacteria are a diverse group of microorganisms that are capable of oxidizing sulfur as an energy source. Reduced sulfur compounds that are used by sulfur-oxidizing bacteria include hydrogen sulfide (H_2S), elemental sulfur (S^o), and thiosulfate ($S_2O_3^{-2}$). The final product of sulfur oxidation is sulfate (SO_4^{-2}) or sulfuric acid (H_2SO_4), which can corrode concrete and steel. Sulfur-oxidizing bacteria include *Thiobacillus* and *Beggiatoa*. Some sulfur-oxidizing bacteria can also oxidize ferrous iron (e.g., *Thiobacillus ferrooxidans and Sulfolobus* species). *Thiothrix*, common in wastewater treatment facilities, and *Sulfolobus*, confined to sulfur-rich hot springs, transform hydrogen sulfide to elemental sulfur.

Sulfur-oxidizing and sulfate-reducing bacteria both live in wastewater collection systems and WRRFs. Both H_2S and H_2SO_4 can be formed within pipelines and treatment processes under alternating aerobic and anaerobic conditions.

Some of the specialized groups of bacteria in wastewater include the following:

- Nitrifying bacteria, which convert ammonia to nitrite and nitrate;
- Phosphate accumulating bacteria, which uptake phosphorus;
- Anaerobic bacteria, which break down CBOD to VFAs;
- Sulfur-reducing bacteria, which convert sulfate to hydrogen sulfide; and
- Sulfur-oxidizing bacteria, which convert hydrogen sulfide and sulfur to sulfuric acid.

SUMMARY

Table 5.3 summarizes the characteristics of anaerobic, anoxic, and aerobic conditions and lists some of the biological reactions that can take place under each set of conditions. Within limits, the behavior of different groups of bacteria can be predicted from their environment. Facultative heterotrophs will use oxygen when it is available and nitrate when it is not. Nitrifying autotrophs will convert ammonia to nitrite and nitrate when oxygen is available. Figure 5.11 illustrates some of the interrelationships between heterotrophic and nitrifying autotrophic bacteria under aerobic and anoxic conditions.

Table 5.3 Anaerobic, Anoxic, and Aerobic Conditions

Environmental Condition	Defining Characteristics	Possible Biological Reactions
Anaerobic	No oxygen present No nitrite or nitrate present Sulfate may be present	CBOD broken down into soluble CBOD and VFAs. Heterotrophs are dormant. Nitrifiers are dormant. PAOs release phosphorus. PAOs uptake VFAs. Sulfate reduced to hydrogen sulfide.
Anoxic	No oxygen present Nitrite and/or nitrate present Sulfate may be present	Heterotrophs convert CBOD to biomass. Heterotrophs convert NO_2^- and NO_3^- into nitrogen gas. Nitrifiers are dormant. PAOs consume stored VFAs. PAOs uptake phosphorus.
Aerobic/Oxic	Oxygen present Nitrite and nitrate may also be present Sulfate may be present	Heterotrophs convert CBOD to biomass. Nitrifiers convert NH_3 to NO_2^- and NO_3^-. PAOs consume stored VFAs. PAOs uptake phosphorus. Hydrogen sulfide oxidized to sulfate.

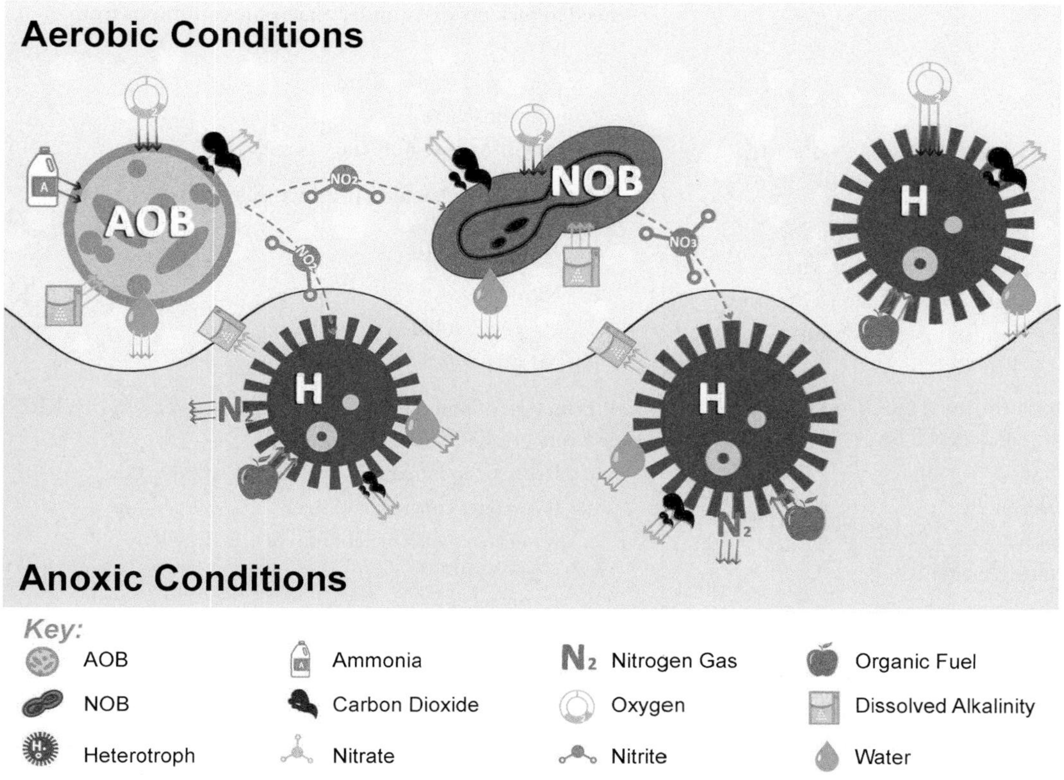

Figure 5.11 Summary of Requirements for Autotrophic and Heterotrophic Bacteria

TEST YOUR KNOWLEDGE

1. **Sugar is combined with oxygen to produce carbon dioxide and water. In this reaction, the sugar is being oxidized.**
   - ☐ True
   - ☐ False

2. **Autotrophic bacteria consume CBOD and produce new bacteria, carbon dioxide, and water.**
   - ☐ True
   - ☐ False

3. **Autotrophic bacteria grow faster than heterotrophic bacteria because their fuel source is energy rich.**
   - ☐ True
   - ☐ False

4. **The phosphate accumulating organisms must be cycled between anaerobic and anoxic or aerobic conditions to perform luxury uptake of phosphorus.**
   - ☐ True
   - ☐ False

5. **Heterotrophic bacteria obtain both their energy and their carbon from**
   a. Carbonate and bicarbonate ion
   b. DO
   c. Biodegradable organic material
   d. Sulfate

6. **Algae are classified as**
   a. Autotrophs
   b. Phototrophs
   c. Heterotrophs
   d. Taxitrophs

7. **Match the environmental condition to its description.**

1. Anaerobic	a. Dissolved oxygen present
2. Anoxic	b. Nitrate present, no DO present
3. Aerobic	c. Neither oxygen nor nitrate present

8. **When heterotrophic bacteria consume CBOD in the presence of oxygen, the byproducts will be**
   a. Nitrate, nitrite, and water
   b. Nitrate and carbon dioxide
   c. Carbon dioxide and water
   d. Carbon dioxide and nitrogen gas

9. **Obligate aerobes require _______ for respiration.**
   a. Nitrate
   b. Oxygen
   c. Ammonia
   d. CBOD

10. **A heterotrophic bacterium is placed under anoxic conditions. This bacterium will**
    a. Convert CBOD to CO_2 and H_2O using oxygen
    b. Convert NH_3^- to NO_2^- and NO_3 using oxygen
    c. Convert CBOD to N_2, CO_2, and H_2O using nitrate
    d. Convert NH_3 to N_2, CO_2, and H_2O using nitrate

11. **The process of consuming CBOD with nitrate is called**
    a. Denitrification
    b. Nitrification
    c. Fermentation
    d. Hydrolysis

12. **The NOB**
    a. Consume only soluble CBOD
    b. Require an inorganic carbon source
    c. Prefer VFAs
    d. Convert ammonia to nitrite

13. **Phosphate accumulating bacteria obtain the energy they need to pick up VFAs under anaerobic conditions from**
    a. VFAs
    b. Adenosine triphosphate
    c. Polyphosphate
    d. Poly-β-hydroxybutyrate

14. **Sulfate-reducing bacteria produce this toxic gas under anaerobic conditions.**
    a. Carbon dioxide
    b. Nitrogen gas
    c. Hydrogen sulfide
    d. Nitrous oxide

15. **Production of sulfuric acid in collection systems and WRRFs is a concern because**
    a. It is toxic to humans at low concentrations in air.
    b. It corrodes concrete and steel.
    c. It interferes with metabolism of CBOD at low concentrations.
    d. It consumes DO.

Microbial Growth Rates

Biological secondary treatment process design and operation are based on the growth kinetics of microorganisms. Kinetics is how fast or slow chemical and biological reactions take place. Kinetics determine how much time is required to treat wastewater. The rest of this chapter defines several key terms and concepts that will be used in later chapters: maximum specific growth rate (μ_{max}), saturation coefficient, half-saturation coefficient (K_s), yield, and decay. The following are key concepts to understand by the end of this chapter:

- Each microorganism in the process has a maximum growth rate it can achieve;
- Lack of carbon, oxygen, or nutrients can limit growth;
- Growth rate increases with resource availability until the maximum rate is reached;
- If more than one resource is limiting growth, the effects add up;
- If excess resources are available, growth is limited by the maximum growth rate;
- Environmental conditions (pH, temperature, toxicity) can also limit growth; and
- Yield is fixed. Bacteria can only grow and reproduce when food is available.

Do not worry about the math in this section. Understanding the concepts is more important.

MONOD GROWTH CURVE

Each kind of bacteria will grow at its maximum rate only when the water surrounding the bacteria has an optimum pH and contains sufficient amounts of food, oxygen source, nutrients, and trace elements. Microbial growth rates are often lower than the maximum growth rates because one or more of the substances they need to grow and reproduce is not available in sufficient quantities. When a substance is not present in sufficient quantities to support growth, it is said to be limiting. The most common limiting factors include (a) exhaustion of nutrients, (b) accumulation of toxic byproducts, and (c) changes in ion concentrations, especially pH (Monod, 1949).

Monod kinetics relates microbial growth rates to the concentration of a limiting substrate in an aqueous environment. Microbial growth rates are represented by an equation proposed by Jacques Monod (1942, 1949). The Monod equation states that when substrate concentrations are low, the growth rate of the microorganisms will be proportional to the concentration of the limited substrate. In other words, the growth and reproduction microorganism will be restricted by the availability of resources. The Monod equation also states that when resources are plentiful, the growth rate of the microorganisms will be maximized. However, the microorganism cannot grow faster than its maximum rate when resources are available in excess.

Think back to the factory analogy used earlier in this chapter. Imagine that an assembly line within that factory is producing automobiles. The assembly line needs to have all of the parts available to keep producing cars. What would happen to the speed of the assembly line if a key component was limited? If the automobile factory only has access to 60 radiators on any given day, then the maximum number of automobiles that can be produced is 60. If the automobile factory had access to as many parts as it needed, then the number of automobiles it can produce is limited only by the speed of the assembly line. The speed of the assembly line is the maximum growth rate. It represents the speed at which a particular microorganism is capable of running its internal cellular assembly lines. The maximum substrate utilization rate is related to the maximum growth rate. When the microorganism is growing at its maximum rate, it will use up resources—specifically, CBOD, ammonia, oxygen, nitrate, and nutrients—at the maximum rate. If the assembly line is only capable of producing 100 new automobiles each day, then the maximum number of radiators it can use each day is also 100.

The Monod equation generates a characteristic growth curve, shown on the graph in Figure 5.12. The microbial growth rate, shown as μ, is determined by the concentration of the limiting substrate until the maximum growth rate (μ_{max}) is reached. The growth rate has units of grams of VSS produced per gram of VSS added to the process per day (g VSS/g VSS·d). It is typically shown in mathematical equations simply as "per day" (or d^{-1}).

The Monod equation is written as follows:

$$\mu = \mu_{max} \times \left[\frac{S}{K_s + S} \right] \tag{5.9}$$

Where

μ = specific growth rate of the microorganisms (pronounced mew), per day;
μ_{max} = maximum possible growth rate of the microorganisms, per day;
S = concentration of substrate, mg/L; and
K_s = half-saturation coefficient, mg/L.

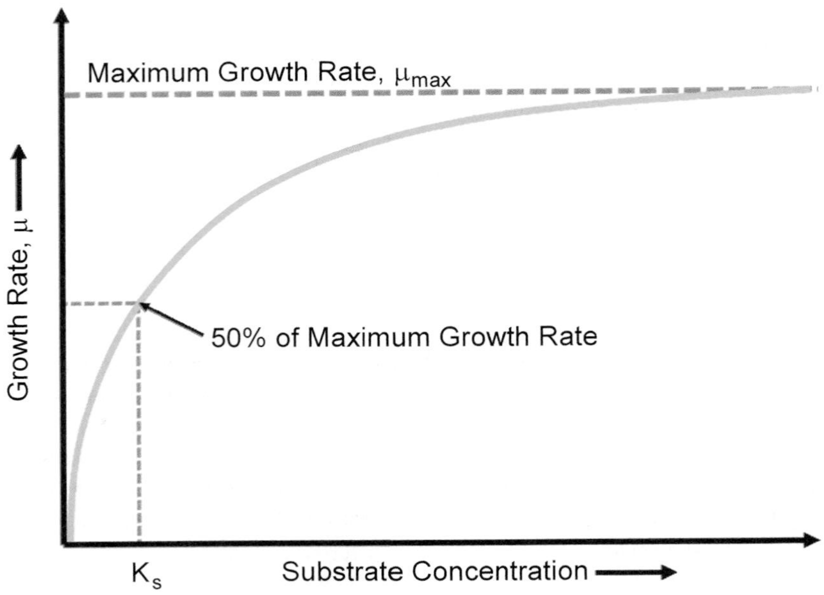

Figure 5.12 Graph of the Monod Equation

MAXIMUM SPECIFIC GROWTH RATE

The maximum specific growth rate is an important factor that, in part, determines biological treatment capacity. It is the fastest that a particular microorganism can grow and reproduce. Microorganisms respond to higher concentrations of substrate (food) by increasing their growth rate up to the maximum growth rate (μ_{max}). When the substrate is scarce (low food concentration), microorganisms cannot grow as fast as they could if more food were available. In contrast, an abundance of substrate allows each microorganism to consume more food and increase their growth rate. Increasing the amount of substrate reduces competition for food and allows the microorganisms to increase their growth rate proportional to the available food source.

Saturated means that there is more substrate available than can be immediately used.

In Figure 5.12, the growth rate curve increases rapidly from the origin in response to the increasing substrate concentration. It gradually flattens out and approaches the horizontal line, which indicates the maximum growth rate (μ_{max}). When the microorganisms are *saturated* by substrate and attain the maximum growth rate, the growth rate remains constant as substrate concentrations increase further. When surplus substrate is available, growth becomes limited by the number of microorganisms in the system instead of by the amount of substrate available.

The maximum growth rate (μ_{max}) and the half-saturation coefficient (K_s) used in the Monod equation are empirical coefficients. *Empirical* means they must be measured by experimentation. Growth rates are measured at several different concentrations of the limiting substrate and the results are plotted on a graph. The maximum growth rate (μ_{max}) and half-saturation coefficient (K_s) are then estimated by fitting a curved line to the graph.

Each species of bacteria has a different maximum growth rate (μ_{max}) and half-saturation coefficient (K_s) for every limiting nutrient and for the surrounding environmental conditions. The maximum growth rate for the biomass in a secondary treatment process is simply the average of the maximum growth rates for each of the different types of bacteria present in the biomass. The average maximum growth rate in a secondary treatment process can be determined for the biomass in the process using the same methods as those for a single type of bacteria and a single limiting substrate. When engineers design secondary treatment processes, they use a μ_{max} between 3 and 13.2 g VSS/g VSS·d for heterotrophic bacteria and a μ_{max} between 0.20 and 0.90 g VSS/g VSS·d for the nitrifying autotrophic bacteria (Metcalf and Eddy, Inc./AECOM, 2013). The big differences in maximum growth rate make it clear that the heterotrophic bacteria can reproduce much more rapidly than the nitrifying autotrophic bacteria.

The Monod equation represents a stable microbial culture. Changing environmental conditions or changing wastewater characteristics may cause different microorganisms to displace previously dominant species in the treatment process. For example, an influent wastewater with high BOD and low ammonia will favor the heterotrophs and increase their numbers rapidly. An influent that is low in BOD and high in ammonia will favor the nitrifying autotrophs instead. The changing makeup of the population of microorganisms in the process can change the μ_{max} and K_s values, thereby affecting process performance.

Every microorganism has a maximum growth rate. They cannot grow faster than this even when excess resources are available.

SATURATION COEFFICIENT

The saturation coefficient is the concentration of a resource that a microorganism needs to grow at its maximum rate. Higher concentrations of the resource will not continue to increase growth rates. If an assembly line can produce a maximum of 75 cars a day, then having access to 75 radiators is saturation. The assembly line cannot use more than 75 radiators. If more radiators are available, the assembly line will not produce more cars. This is the saturation coefficient. When resources are available at the saturation coefficient or higher, the microorganism will grow at its maximum rate. Below the saturation coefficient concentration, the assembly line will be limited and production of new microorganisms will slow down. The saturation coefficient is typically expressed as milligrams per liter (mg/L).

HALF-SATURATION COEFFICIENT

The half-saturation coefficient, or half-velocity coefficient, shown as K_s, is the concentration of a resource needed to achieve exactly half of the maximum growth rate. The half-saturation coefficient is typically calculated from measurements of the maximum growth rate and is rarely measured directly. It is typically expressed in milligrams per liter (mg/L).

The values of the maximum growth rate (μ_{max}) and half-saturation constant (K_s) are highly dependent on the microorganisms and substrates in the biological system. Readily biodegradable substrates have high μ_{max} values and low K_s values. In other words, soluble BOD, which is easy for the bacteria to take up, will help the bacteria to grow very quickly. They do not need to waste time and energy breaking down particulate BOD before they can bring it through the cell membrane. In contrast, slowly biodegradable substrates have low μ_{max} values and high K_s values. Therefore, the percentages of readily biodegradable and slowly biodegradable substrates in the wastewater affect the μ_{max} and K_s values for each WRRF.

Many wastewater treatment systems are required to meet stringent discharge limits, which are close to or less than the half-saturation coefficient concentrations (Shaw, 2015). For this reason, wastewater treatment systems with low effluent limits operate at low substrate concentrations and at growth rates that are much lower than the maximum growth rate (μ_{max}). Looking at Figure 5.12, these facilities are operating toward the left side of the characteristic Monod growth curve. In other words, the bacteria are often growing much slower than they would under ideal conditions.

For facultative heterotrophic bacteria, the K_s ranges between 25 and 100 mg/L of BOD, with a typical value of 60 mg/L of BOD (Metcalf and Eddy/AECOM, 2003). This means that when the BOD concentration is in this range, the growth rate of the microorganisms is about half of maximum. Compare this to an effluent BOD limit of 30 mg/L for most mechanical treatment facilities. For the nitrifying bacteria, the K_s for ammonia-nitrogen is 0.50 mg/L (U.S. EPA, 2010). When the ammonia-nitrogen concentration is higher than 1 mg/L in the activated sludge process, it becomes possible for the AOB to grow at their maximum rate. As the concentration of ammonia-nitrogen gets lower and lower, its the AOB growth rate slows down. There is a similar effect with DO concentrations. The K_s for oxygen for the nitrifying bacteria ranges from 0.13 to 0.47 mg/L (Manser et al., 2005). Ultimately, the maximum growth rate of the microorganisms will be limited by the availability of any or all of the resources they need as well as water temperature.

Although operators will likely never need to calculate growth rates, an example calculation is included here to demonstrate the effects of substrate concentration on growth rate. In this example, the bacteria reproduce at a maximum rate of 3 times per day and the half-saturation coefficient (K_s) is 0.7 mg/L. The substrate concentration will be varied, from 0.5 mg/L up to 12.5 mg/L (Table 5.4).

$$\mu = \mu_{max} \times \left[\frac{S}{K_s + S}\right]$$

$$\mu = 3\ \text{d}^{-1} \times \left[\frac{0.5\ \text{mg/L}}{0.7\ \text{mg/L} + 0.5\ \text{mg/L}}\right]$$

$$\mu = 3\ \text{d}^{-1} \times \left[\frac{0.5\ \text{mg/L}}{1.2\ \text{mg/L}}\right]$$

$$\mu = 1.25\ \text{d}^{-1}$$

The amount of substrate available is low, so the growth rate is also low. If the same calculation is performed with progressively increasing substrate concentrations, the growth rate achieved begins to get very close to the maximum growth rate. In Table 5.4, one can see that when the substrate concentration is equal to the half-saturation coefficient (K_s) of 0.7 mg/L, the resulting growth rate is equal to 1.5 per day, which is exactly half of the maximum growth rate. As the substrate continues to increase, the actual growth rate gets closer and closer to the maximum growth rate. However, even when the substrate concentration is higher than twice the half-saturation coefficient (K_s), the growth rate can never be higher than the maximum growth rate. Look what happens when the same calculation is performed using a concentration of 12.5 mg/L and when K_s is 0.7 mg/L. No matter how high the substrate concentration goes, the microorganism cannot grow faster than its μ_{max}.

$$\mu = \mu_{max} \times \left[\frac{S}{K_s + S}\right]$$

Table 5.4 Effect of Substrate Concentration on Growth Rate

S, mg/L	μ, g VSS/g VSS·d
0.7	1.50
1.0	1.76
1.5	2.05
2.0	2.22
2.5	2.34
3.0	2.43
3.5	2.50
4.0	2.55
4.5	2.60
5.0	2.63
5.5	2.66
6.0	2.69
6.5	2.71
7.0	2.73
7.5	2.74
12.5	2.84

$$\mu = 3\ \text{d}^{-1} \times \left[\frac{12.5\ \text{mg/L}}{0.7\ \text{mg/L} + 12.5\ \text{mg/L}}\right]$$

$$\mu = 3\ \text{d}^{-1} \times \left[\frac{12.5\ \text{mg/L}}{13.2\ \text{mg/L}}\right]$$

$$\mu = 2.84\ \text{d}^{-1}$$

When resources are limited—like they are in most secondary treatment processes—the growth rate will be proportional to the concentration of the resource. When resources are not limited, the growth rate will be very close to the maximum growth rate, but never higher.

EFFECT OF MULTIPLE LIMITING SUBSTRATES

Growth rates in wastewater treatment processes are often limited by two or more substrates at the same time. When growth rates are limited by more than one substrate or growth factor (e.g., organic matter and oxygen are both required by heterotrophic bacteria), multiple Monod terms can be multiplied together to estimate the combined effect of two or more substrates, as shown by the following equation:

$$\mu = \mu_{max} \times \left[\frac{S_1}{K_{s1} + S_1}\right] \times \left[\frac{S_2}{K_{s2} + S_2}\right] \tag{5.10}$$

Think back to the factory analogy used earlier in this chapter. The assembly line for autotrophic nitrifying bacteria requires DO and ammonia. In this example, the K_{s1} represents the half-saturation coefficient for DO and S_1 is the DO concentration in the process. The K_{s2} term represents the half-saturation coefficient for ammonia and S_2 is the ammonia concentration in the process.

The nitrifying bacteria need ammonia and DO to grow. The K_s for ammonia is 0.5 mg/L and the K_s for DO is 0.47 mg/L. What happens when one or both are limited? For the nitrifying bacteria, we will use a μ_{max} of 1 per day (U.S. EPA, 2010). In the first example, ammonia will be plentiful at 5 mg/L as N, but DO will be in short supply at 0.3 mg/L. The subscripts have been changed to make it easier to see which pieces of the equation are associated with ammonia and which are associated with DO.

$$\mu = \mu_{max} \times \left[\frac{S_{NH_3-N}}{K_{sNH_3-N} + S_{NH_3-N}}\right] \times \left[\frac{S_{DO}}{K_{sDO} + S_{DO}}\right]$$

$$\mu = 1\ \text{d}^{-1} \times \left[\frac{5\ \text{mg/L}}{0.5\ \text{mg/L} + 5\ \text{mg/L}}\right] \times \left[\frac{0.3\ \text{mg/L}}{0.47\ \text{mg/L} + 0.3\ \text{mg/L}}\right]$$

$$\mu = 1\ \text{d}^{-1} \times \left[\frac{5\ \text{mg/L}}{5.5\ \text{mg/L}}\right] \times \left[\frac{0.3\ \text{mg/L}}{0.77\ \text{mg/L}}\right]$$

$$\mu = 1\ \text{d}^{-1} \times [0.91] \times [0.39]$$

$$\mu = 0.35\ \text{d}^{-1}$$

Because there was plenty of ammonia, the second term was calculated as 0.91 and is very close to 100% of the maximum growth rate. There was not enough DO, so the third term calculated out to be 0.39, which is pretty low. The overall growth rate is reduced from 1 per day to 0.35 per day. What if both ammonia and DO had been limiting? Imagine if the ammonia term had also calculated out as 0.39. Then, the overall growth rate would be reduced even further.

$$\mu = 1\ \text{d}^{-1} \times [0.39] \times [0.39]$$

$$\mu = 0.15\ \text{d}^{-1}$$

Remember, when working on a complicated algebra problem, follow the order of operations: (1) parentheses and other grouping symbols, (2) exponents, (3) multiplication and division from left to right, and (4) addition and subtraction from left to right.

It seems odd that the growth rate would continue to drop, but it makes sense if you think about it in terms of the population of nitrifying bacteria instead of individuals. The further resources are reduced, the more likely it becomes that an individual bacterium will be lacking something essential. The percentage of bacteria in the whole population that have enough resources to reproduce goes down and so does the overall growth rate. If a single bacterium is missing two essential resources, then the resource that is scarcest will determine the maximum growth rate. Think back to our factory producing cars. If the factory has 10 production lines and there is a limited supply of both radiators and wheels, some of those lines will get enough radiators and wheels to produce some cars. Other lines will get a lot of wheels and only a few radiators. Overall, the output of the entire factory decreases.

BIOMASS YIELD

Yield is the amount of new microorganisms produced per amount of substrate added to the system (eq 5.11). Yield is how many cars the factory can produce. Growth rate is how fast those cars are produced. Substrates include organic compounds, ammonia, VFAs, oxygen, and nitrate. The availability of each resource affects how efficiently the microorganisms can grow and reproduce. Heterotrophic bacteria produce an average of 0.40 g of VSS, a surrogate measurement for new bacteria, for every 1 g of COD added to the process when oxygen is available (Metcalf and Eddy and AECOM, 2003). For secondary treatment processes receiving domestic wastewater, this is equivalent to about 0.6 g VSS/g BOD. The COD and BOD that do not become new biomass become carbon dioxide and water. Yields are very consistent for WRRFs accepting domestic wastewater. It is not possible to produce more new bacteria than the amount of food in the influent will support. Put another way, if a person consumes a large number of donuts, he or she can gain weight. The same person cannot gain weight without eating the donuts. Without food, there cannot be an increase in biomass.

Yields are somewhat lower when oxygen is not available and the facultative heterotrophs are forced to substitute nitrate or nitrite for oxygen (denitrification), requiring the bacteria to use more energy. Using nitrate or nitrite is less efficient for the bacteria than oxygen.

The AOB and NOB produce an average of only 0.15 and 0.05 g, respectively, of VSS for every 1 g of ammonia converted to nitrate (U.S. EPA, 2010). The yield is much lower for nitrifiers than it is for the heterotrophs, which means their growth is slower. The amount of new cells produced is essentially constant when the energy and oxygen sources are known. Yield is unaffected by changes in water temperature between 4 °C and 20 °C, but rates change substantially (Sayigh and Malina, 1978). As temperatures increase, the bacteria will grow faster, but it still is not possible to grow more biomass without resources. Yield can be calculated as follows:

$$\text{Yield} = \frac{\text{Mass of new cells formed}}{\text{Mass of substrate removed}} \tag{5.11}$$

The yield is fixed. You cannot grow microorganisms without food, nutrients, and other resources.

BIOMASS DECAY RATE

Decay accounts for losses from the system caused by microorganisms consuming internal storage products, dying, and being eaten by other microorganisms. The amount of solids lost each day is a fraction of the total solids in the system (i.e., typically 0.10 g VSS lost per gram of VSS in the system per day for the heterotrophs and 0.17 g VSS lost per gram of VSS in the system per day for the nitrifying autotrophs) (Metcalf and Eddy/AECOM, 2003; U.S. EPA, 2010). In other words, about 10% of the heterotrophs in the process will die or be reduced in size each day. New microorganisms are produced each day (yield) and some get smaller and/or die each day (decay). The difference between the yield and the decay is the net yield or observed yield. This is an important concept for secondary treatment process control. The amount of COD or BOD entering the process each day dictates the amount of new cells that can potentially be produced. When the amount of COD or BOD decreases, fewer new microorganisms are produced. It is not possible to build up a population of live microorganisms without COD or BOD, or, in the case of the nitrifying bacteria, ammonia. Decay can be calculated as follows:

$$\text{Decay} = \frac{\text{Mass of cells lost per day}}{\text{Mass of cells in system}} \tag{5.12}$$

Both the maximum specific growth rate and the amount of decay depend on temperature. When the treatment process is warmer, the bacteria grow faster than when it is colder (Metcalf and Eddy, Inc./ AECOM, 2003). From a practical standpoint, it means that treatment takes less time when the water is warm. Decay also depends on temperature. The amount of bacteria that die or lose mass because they are consuming internal storage products slows down at colder temperatures.

MONOD KINETICS

The yield, decay, maximum specific growth rate, maximum substrate utilization rate, and saturation coefficient are related to one another through the following equations (Metcalf and Eddy, Inc./ AECOM, 2003):

$$\mu = Y \times \frac{kS}{K_s + S} - k_d \tag{5.13}$$

Alternatively

$$Y = \left[\frac{K_s + S}{kS}\right][\mu + k_d] \tag{5.14}$$

Where

μ = specific growth rate, g VSS/g VSS·d;
Y = Yield, g VSS/g BOD or COD;
k = maximum utilization rate, g COD/g VSS·d;
S = substrate concentration, mg/L BOD or COD;
K_s = substrate concentration at one-half the maximum specific substrate utilization rate, mg/L; and
K_d = decay, g VSS/g VSS·d.

This particular example relates the growth rate of heterotrophic bacteria to the availability of BOD or COD. Similar equations exist for the nitrifying autotrophic bacteria and are presented in the chapter on biological nutrient removal. Operators will not need to calculate growth rates for microorganisms, but it is useful to show the mathematical relationship so we can understand what happens to the growth rate as the available resources diminish. Equation 5.14 can be rewritten as follows:

$$\text{Growth} = (\text{Potential growth})(\text{Resource availability}) - \text{Decay} \tag{5.15}$$

If the amount of resources available to the microorganism is less than 100% of what is needed, then the potential growth will be reduced. It is possible for oxygen sources to be limited as well or for there to be a limitation on the quantity and types of spare parts (carbon and nutrients). If the amount of resources available to each bacteria begins to decrease, they will not be able to grow or reproduce as quickly. If available resources continue to decrease, the bacteria will begin to use up stored resources. This is called *endogenous respiration* and is similar to what happens when you go on a diet. Eventually, the bacteria run out of stored resources and will lyse or break open. Their remains then become food for other bacteria in the process.

The next four chapters, Wastewater Treatment Ponds, Fixed-Film Treatment, Activated Sludge, and Nutrient Removal, will build on the concepts developed in this chapter. Understanding how microorganisms feed, grow, and multiply is critical to understanding biological treatment processes.

TEST YOUR KNOWLEDGE

1. Every type of bacteria has a maximum growth rate that can be achieved when resources are present in excess and environmental conditions are favorable.
   - ☐ True
   - ☐ False

2. When resources are limited, the growth rate of the bacteria will be proportional to the concentration of the most limited substrate.
   - ☐ True
   - ☐ False

3. The saturation coefficient for a particular heterotrophic bacteria is 20 mg/L of BOD. The growth rate will increase rapidly at BOD concentrations above 20 mg/L.
   - ☐ True
   - ☐ False

4. There can be only one limiting substrate at any time.
   - ☐ True
   - ☐ False

5. The amount of new microorganisms produced each day is dependent on the amount of food entering the treatment process.
   - ☐ True
   - ☐ False

6. Assuming that a bacterium has all the nutrients and environmental conditions it needs to grow and reproduce, when will growth rates be fastest?
   a. When the water temperature reaches 10 °C
   b. After all of the BOD has been consumed
   c. When the water temperature is 22 °C
   d. Before ammonia concentrations reach 15 mg/L

7. This term is used to describe how fast or slow a chemical or biological reaction takes place.
   a. Monod
   b. Kinetics
   c. Saturation coefficient
   d. Growth

8. A common limiting factor for bacterial growth in secondary treatment processes is
   a. Excess BOD
   b. pH near neutral
   c. Abundant nutrients
   d. Accumulation of toxic byproducts

9. This equation predicts the growth rate of bacteria based on the availability of resources.
   a. Yield
   b. Monod
   c. Decay curve
   d. μ_{max} parametric

10. The influent BOD to a WRRF suddenly decreases by 30%. The number of new bacteria produced in the secondary treatment process
    a. Is dependent on the influent flowrate
    b. Will increase according to the BOD-to-ammonia ratio
    c. Is unaffected by the change
    d. Will decrease to match the influent BOD supply

11. The half-saturation coefficent of BOD for a particular bacteria is 15 mg/L and the maximum growth rate is 12 g VSS/g VSS·d. If the concentration of BOD in the process is 15 mg/L, what is the growth rate of the bacteria?
    a. 6 d^{-1}
    b. 12 d^{-1}
    c. 15 d^{-1}
    d. 24 d^{-1}

CHAPTER SUMMARY

	Microbiology Fundamentals
PHYSICAL AND CHEMICAL REQUIREMENTS FOR BIOLOGICAL TREATMENT	■ Relies on naturally occurring microorganisms ■ Secondary treatment processes have higher concentrations of microorganisms than natural systems. ■ Some components of wastewater are biodegradable. ■ Organic matter can be particulate or soluble. ■ Biodegradable wastes are transformed into new microorganisms. They are removed from the treatment process as sludge. ■ Particulate, non-biodegradable wastes will settle or adhere to microorganisms in the treatment process. In both cases, these wastes are eventually removed from the treatment process as sludge. ■ Small amounts of microorganisms and particulate, non-biodegradable wastes pass through the WRRF and into the final effluent. ■ Soluble, non-biodegradable wastes pass through to the final effluent. ■ The ratio of COD to CBOD changes through the WRRF.
MICROBIOLOGY	■ Bacteria, protozoans, metazoans, and viruses are all present in secondary treatment processes. ■ Bacteria have a cell membrane that limits what can pass into and out of the bacteria. ■ Only soluble matter can pass through the cell membrane. ■ Particulate BOD is converted to soluble BOD by enzymes secreted by the bacteria. ■ Bacteria secrete EPS that form a protective coating and help groups of bacteria stick together and adhere to surfaces. ■ Bacteria reproduce by binary fission. ■ Most secondary treatment processes operate in the stationary phase of bacterial growth.
	Bacteria
ENERGY	■ Heterotrophs obtain their energy from organic compounds. ■ Autotrophs obtain their energy from inorganic compounds including ammonia, nitrite, and sulfate.
CARBON	■ Heterotrophs obtain their carbon from organic compounds. ■ Autotrophs obtain their carbon from dissolved carbonate and bicarbonate. This is alkalinity.
OXYGEN	■ Three different environments potentially exist in all WRRFs: anaerobic, anoxic, and aerobic (oxic). ■ *Anaerobic* means the absence of DO, nitrate, and nitrite. Sulfate may be present. ■ *Anoxic* means the absence of DO. Nitrate and/or nitrite are present. Sulfate may also be present. ■ *Aerobic*, or *oxic*, means that DO is present. Nitrate, nitrite, and sulfate may also be present. ■ Obligate aerobes must have oxygen. ■ Many bacteria are facultative and can use oxygen, nitrate, nitrite, or sulfate.
NITRIFYING BACTERIA	■ Obtain their carbon from alkalinity ■ Obtain their energy from ammonia or nitrite ■ Ammonia oxidizing bacteria (AOB) convert ammonia to nitrite. ■ Nitrite oxidizing bacteria (NOB) convert nitrite to nitrate. ■ Can be paired with facultative heterotrophs to convert ammonia to nitrogen gas for complete nitrogen removal

PHOSPHATE ACCUMULATING ORGANISMS	■ Specialized facultative, heterotrophic bacteria ■ Store excess phosphorus to build a chemical energy source ■ Preferred carbon source is VFAs ■ Volatile fatty acids are produced by other bacteria through fermentation under anaerobic conditions.
SULFUR BACTERIA	■ Sulfate-reducing bacteria convert sulfate to hydrogen sulfide under anaerobic conditions. ■ Sulfur-oxidizing bacteria convert hydrogen sulfide to sulfuric acid under aerobic conditions. ■ Hydrogen sulfide is a toxic gas that smells like rotten eggs. It is a safety concern in collection systems and WRRFs. ■ Sulfuric acid can corrode concrete and metal.
	Microbial Growth Rates
YIELD	■ The amount of new organisms produced per amount of substrate added to the system. ■ Heterotrophic bacteria have a yield of 0.40 g VSS for 1 g COD consumed. ■ Nitrifying autotrophic bacteria have a yield of 0.10 g VSS for 1 g NH_3-N converted to NO_3-N.
DECAY	■ Microorganisms lost to death and predation ■ Decay is a fraction of the total biomass in the system. ■ Decay is temperature dependent.
MAXIMUM SPECIFIC GROWTH RATE	■ Fastest possible rate of reproduction ■ Each species of microorganism has its own maximum specific growth rate. ■ Nitrifying autotrophs grow much slower than heterotrophs. ■ Temperature dependent
SATURATION COEFFICIENT	■ Concentration of a resource that a microorganism needs to grow at its maximum rate ■ Higher concentrations will not increase the growth rate. ■ The half-saturation coefficient is the concentration of a resource that produces exactly half the maximum growth rate.
MONOD KINETICS	■ The yield, decay, maximum specific growth rate, and half-saturation coefficient are related to one another by the Monod equation. ■ The Monod equation can be used to predict growth rates of microorganisms if environmental conditions are known, including temperature and availability of resources.

References

Brock, T. D.; Madigan, M. T. (1991) *Biology of Microorganisms*, 6th ed.; Prentice Hall: New Jersey.

EnviroSim Associates, Ltd. (2006) Influent Specifier (Raw) 2_2.xls. Spreadsheet for fractionating domestic wastewater into components required for Biowin modeling.

Gerardi, M. H. (2006) *Wastewater Bacteria*; Wiley & Sons: Hoboken, New Jersey.

Grady Jr., C. P. L.; Filipe, C. D. M. (2000) Ecological Engineering of Bioreactors for Wastewater Treatment, In *Environmental Challenges*, Belkin, S., Ed.; Springer: Dordrecht, Netherlands.

Koops, H.-P.; Pommerening-Röser, A. (2001) Distribution and Ecophysiology of the Nitrifying Bacteria Emphasizing Cultured Species. *FEMS Microbiol. Ecol.*, **37** (1), 1–9.

Manser, R.; Gujer, W.; Siegrist, H. (2005) Consequences of Mass Transfer Effects on the Kinetics of Nitrifiers. *Water Res.*, **39** (19), 4633–4642.

Metcalf & Eddy, Inc./AECOM (2003) *Wastewater Engineering: Treatment and Resource Recovery*, 4th ed.; McGraw-Hill: New York.

Metcalf & Eddy, Inc./AECOM (2013) *Wastewater Engineering: Treatment and Resource Recovery*, 5th ed.; McGraw-Hill: New York.

Monod, J. (1942) *Recherches Sur la Croissance Des Cultures Bacteriennes*; Librairie Scientifique: Paris, France (in French).

Monod, J. (1949) The Growth of Bacterial Cultures. *Annu. Rev. Microbiol.*, **3**, 371.

Pasztor, I.; Thury, P.; Pulai, J. (2009) Chemical Oxygen Demand Fractions of Municipal Wastewater for Modeling of Wastewater Treatment. *Int. J. Environ. Sci. Technol.*, **6** (1), 51–56.

Pauli, W.; Jax, K.; Berger, S. (2001) *Protozoa in Wastewater Treatment: Function and Importance, The Handbook of Environmental Chemistry Vol. 2 Part K Biodegradation and Persistence, Chapter 3*; Springer-Verlag: Heidelberg, Germany.

Sayigh, B. A.; Malina, J. F. (1978) Temperature Effects on the Activated Sludge Process. *J. Water Pollut. Control Fed.*, **50** (4), 678–687.

Shaw, A. R. (2015) Investigating the Significance of Half-Saturation Coefficients on Wastewater Treatment Processes. Doctor of Philosophy thesis, Illinois Institute of Technology: Chicago, Illinois.

U.S. Environmental Protection Agency (2010) *Nutrient Control Design Manual;* EPA-600/R-10-100; U.S. Environmental Protection Agency, Office of Research and Development: Cincinnati, Ohio.

Water Environment Federation (1994) *Wastewater Biology: The Life Process;* Water Environment Federation: Alexandria, Virginia.

Water Environment Federation (2012) *Basic Laboratory Procedures for the Operator-Analyst*, 5th ed.; Water Environment Federation: Alexandria, Virginia.

Water Environment Federation (2017) *Wastewater Biology: The Microlife*, 3rd ed.; Water Environment Federation: Alexandria, Virginia.

Suggested Reading

Gerardi, M. H.; Zimmerman, M. C. (2005) *Wastewater Pathogens*; Wiley & Sons: Hoboken, New Jersey.

CHAPTER 6

Wastewater Treatment Ponds

Introduction

Wastewater treatment ponds are commonly used to treat municipal wastewater from small communities with populations of fewer than 20,000 residents; however, they are capable of treating a variety of wastewaters from domestic wastewater to complex industrial wastes. Treatment ponds typically provide secondary treatment and are capable of achieving effluent 5-day biochemical oxygen demand (BOD_5) and total suspended solids (TSS) concentrations below 30 mg/L. They can operate in various climatic conditions, from the tropics to the arctic. Ponds may be combined with other treatment processes, or operate as standalone systems. More than 50% of the water resource recovery facilities (WRRFs) currently operating in the United States are pond systems (U.S. EPA, 2011).

Advantages of pond systems include that they are

- Relatively inexpensive to operate,
- Simple to operate compared to other treatment processes,
- Better able to accommodate shock loads (because of their size),
- Effective at removing BOD, and
- Effective at removing pathogens.

Disadvantages of pond systems include that they

- Require more land than other types of treatment processes,
- Are less efficient in cold climates,
- May generate odors during certain times of year or if operated improperly,
- Inconsistently remove nutrients (nitrogen and phosphorus), and
- Inconsistently remove TSS.

Readers unfamiliar with wastewater microbiology should review Chapter 5 before continuing with ponds.

LEARNING OBJECTIVES

Upon completing this chapter, you will be able to

- List and describe the three types of wastewater ponds;
- Describe the interrelationships between bacteria, algae, and predator organisms;
- Explain the three mechanisms for nitrogen removal in ponds;
- Evaluate the effects of water temperature, sunlight, nutrient availability, and other operational parameters on population dynamics;
- Discuss the causes of fall and spring turnover in facultative ponds;
- Inspect and maintain pond components;
- Determine when to use series versus parallel operation;
- Calculate hydraulic detention time and organic loading rate;
- Place a pond into service or remove a pond from service;
- Collect process control samples and evaluate results; and
- Troubleshoot common process control problems.

Purpose and Function

Ponds, also called lagoons, are one of the simplest forms of wastewater treatment and have been in use for thousands of years. Ponds require more land than other types of secondary treatment and are more common in rural than urban communities. Like all biological secondary treatment processes, ponds rely on bacteria and other microorganisms to treat wastewater by converting biochemical oxygen demand (BOD) in the influent into biomass, carbon dioxide, water, and other byproducts. Gravity settling separates solids and biomass from the treated wastewater. Wastewater ponds systems may or may not have a headworks to remove screenable materials and/or grit before the wastewater enters the ponds.

Wastewater treatment ponds are generally classified according to the environmental conditions and biological processes that occur within them. Using this classification system, the main types of ponds are aerobic, anaerobic, and facultative. Aerobic ponds have dissolved oxygen (DO) distributed throughout their depth. Anaerobic ponds do not contain DO and facultative ponds have both aerobic and anaerobic layers. Facultative ponds may also have an anoxic layer. Aerobic and facultative ponds are further classified according to how oxygen is added to them. Aerated ponds use mechanical equipment to provide oxygen. Unaerated ponds rely on algae and wind and wave action to provide oxygen. Ponds may also be classified according to how often they discharge: continuous, intermittent, or zero discharge. Zero discharge ponds are also called total containment ponds or evaporation ponds.

Plants, including algae, use energy from sunlight to convert carbon dioxide and water into organic molecules like sugars. This process is called *photosynthesis*.

Don't worry about memorizing depth and detention time ranges for different types of ponds. The ranges for each vary from one design manual to the next and each state regulatory agency has its own requirements. Remember that aerobic ponds are shallow, anaerobic ponds are fairly deep, and that facultative ponds are in between.

AEROBIC PONDS

Aerobic ponds, also called oxidation ponds, contain DO throughout their entire depth. They are typically 0.3- to 0.9-m (1- to 3-ft) deep, which allows sunlight to penetrate to the bottom of the pond (WEF, 2016). An example of a mixed aerobic pond is shown in Figure 6.1. These shallow ponds are often mixed to expose all of the algae in the system to sunlight and prevent solids from settling. Most of the DO in the pond is produced by algae during *photosynthesis*. Dissolved oxygen may also be added to the system from wind, wave action, or mechanical aeration devices. In this oxygen-rich environment, aerobic bacteria dominate and are the main consumers of BOD_5. These systems often have short hydraulic detention times, with 4 to 6 days being typical for high-rate systems and 10 to 40 days for low-rate systems (U.S. EPA, 1975). Aerobic ponds are typically limited to warm, sunny climates and are most commonly used in the southern United States and similar climates. Aerobic ponds are used infrequently as a standalone system, but are now coming into favor as a nutrient polishing step following other mechanical processes.

ANAEROBIC PONDS

Anaerobic ponds are the opposite of aerobic ponds. They receive such heavy organic loading that DO, nitrite, and nitrate are not available. In this environment, anaerobic bacteria dominate and are the main stabilizers of the waste. The byproducts of anaerobic treatment are organic acids, methane gas, and carbon dioxide. Anaerobic treatment also produces hydrogen sulfide and other odorous compounds. Aerobic

Figure 6.1 Aerobic Pond Racetrack Model, Utah State, Logan Utah (Reprinted with permission by Paul Krauth P.E.)

water may be recycled from a downstream aerobic process to the influent end of an anaerobic pond. The recycled water creates a thin aerobic layer at the top of the pond, which helps control odors.

Anaerobic ponds are typically 2.5- to 4.5-m (8- to 16-ft) deep and have detention times of 20 to 50 days (U.S. EPA, 2011). Long detention times are needed because the bacteria responsible for anaerobic breakdown of BOD_5 grow very slowly. Anaerobic ponds often develop a crust of floating scum over their surface when left undisturbed. The crust prevents air from entering the wastewater, keeps the anaerobic pond anaerobic, and helps contain odors. Anaerobic ponds may also be covered with a flexible membrane to capture methane gas, which may be burned for heating or to generate electricity (Figure 6.2).

Anaerobic ponds are often used for treatment of strong industrial and agricultural wastes or as a pretreatment process. Pretreatment is treatment that occurs upstream of the main treatment process. It does not provide complete treatment, but does reduce wastewater strength. The wastewater BOD_5 concentration from an industrial or agricultural process may be as a high as 3000 or 5000 mg/L. Pretreatment with an anaerobic pond can reduce BOD_5 and TSS down to 100 to 300 mg/L, similar to domestic wastewater. Pretreated wastewater is then sent to a secondary treatment process such as another pond, trickling filter, rotating biological contact, or activated sludge process, or may be discharged to the sanitary sewer. Anaerobic ponds are not widely used for secondary treatment of municipal wastewater in the United States and will not be discussed further in this chapter.

FACULTATIVE PONDS

The most common type of pond is the facultative pond. These ponds have an upper aerobic layer overlying a deeper anaerobic layer. The anaerobic layer contains settled sludge. Unaerated facultative ponds are typically 1.1- to 2.5-m (3.5- to 7-ft) deep with detention times ranging anywhere from 25 to more than 180 days. Like aerobic ponds, DO is supplied by algae (photosynthesis) and wind and wave action. Some facultative ponds also use mechanical aeration devices in addition to algae and wind to supply DO and may be either partially or completely mixed (Figure 6.3). They are still considered to be facultative because DO doesn't penetrate the full depth. Mechanically aerated ponds are typically deeper than unaerated ponds. They are 1.8- to 6-m (6- to 20-ft) deep, with detention times ranging from 5 to 30 days. Aerated ponds are capable of treating higher BOD loads than unaerated ponds, are less susceptible to odors, and generally require less land because they are both deeper and smaller. Aerated pond systems typically include a final, unaerated facultative pond that serves as a clarifier to reduce TSS concentrations prior to discharge. The remainder of this chapter is devoted to facultative pond systems.

Figure 6.2 Covered Anaerobic Pond (Reprinted with permission by Indigo Water Group)

Figure 6.3 Facultative Pond with Subsurface Aerators (Reprinted with permission by Paul Krauth P.E.)

DISCHARGE FREQUENCY

Pond systems are also classified according to the duration and frequency of their effluent discharges. Discharge patterns include

- Total containment ponds (zero discharge),
- Controlled discharge ponds,
- *Hydrograph* controlled release ponds, and
- Continuous discharge ponds.

A *hydrograph* is a graph showing the amount of water flowing through a stream or river over a period of time.

Total containment ponds, also called evaporation ponds, can only be used in climates in which the evaporation rate exceeds the precipitation rate on an annual basis. In other words, the pond must lose more water from evaporation and other losses than will enter the pond as wastewater, rain, or snow. Theseponds are designed with large surface areas to encourage evaporation. Total containment ponds are sometimes used in place of more complex treatment when it would be too difficult or expensive to meet discharge permit limits for nutrients, metals, or other parameters. A variation of the total containment pond is one designed for seasonal storage of reclaimed wastewater. Reclaimed wastewater is wastewater that has been treated to high enough standards that it may be used for a variety of purposes such as irrigation and fire suppression. Reclaimed water may be used instead of *potable* water for some purposes. Reclaimed water is also called recycled water or reuse water. These ponds tend to have very long hydraulic detention times and store water for months at a time. Reuse ponds accumulate water during times when it can't be used for irrigation. When the water can be used, it is pumped directly from the pond. It does not get discharged to a natural lake, stream, or river nor is it retreated before use.

Potable water is water that is safe for humans to drink. Potable water meets treatment and quality standards are described in the Safe Drinking Water Act.

Controlled discharge ponds discharge only when stream conditions are satisfactory. For example, a pond system may have low effluent ammonia limits in its discharge permit during some months each year, but not others. Pond discharges are timed to occur during the months without low ammonia limits or when stream flows are high. Hydrograph-controlled release ponds, a variation of the controlled discharge concept, limit the amount of effluent discharged based on the amount of flow in the receiving stream. More effluent is discharged when receiving stream flows are higher. Operators of hydrograph release ponds often set their effluent discharge rates equal to a percentage of receiving stream flow and stop discharging completely when the stream flow falls below a critical level. With both controlled discharge and hydrograph discharge ponds, the operator must keep careful watch on pond water levels and release water before the ponds overflow. With continuous discharge ponds, the effluent is discharged at the same rate (less evaporation and *seepage*) as the influent wastewater flow.

Seepage is water that passes through the bottom of the pond and into the ground.

TEST YOUR KNOWLEDGE

1. All wastewater pond systems have a headworks to remove grit and rags before the wastewater enters the pond.
   - ☐ True
   - ☐ False

2. Membrane covers on anaerobic ponds capture methane gas for reuse.
   - ☐ True
   - ☐ False

3. Wastewater pond systems can reliably remove nutrients from their effluent.
   - ☐ True
   - ☐ False

4. This type of pond treatment system is best suited to high-strength industrial waste.
   a. Aerated
   b. Aerobic
   c. Anaerobic
   d. Facultative

5. The most commonly used type of wastewater pond system is
   a. Aerated
   b. Aerobic
   c. Anaerobic
   d. Facultative

6. This type of wastewater pond is shallow enough for sunlight to penetrate the entire pond.
   a. Aerobic
   b. Anoxic
   c. Facultative
   d. Anaerobic

7. How often does an evaporation pond typically discharge effluent to a receiving stream, river, or lake?
   a. Continuously
   b. Weekly
   c. Semiannually
   d. Never

Theory of Operation—Facultative Ponds

Modern wastewater pond systems typically consist of three or more separate ponds that are interconnected with piping and valves so that any one pond may be taken out of service for cleaning and maintenance while the other two remain in service. There are many pond systems in the United States that consist of only one or two ponds; however, most regulatory agencies now require a minimum of three ponds. Multiple ponds, also called cells, give operators more operational flexibility and decrease the likelihood of *short-circuiting*. Figure 6.4 presents typical flow schematics for a three-pond system. Normally, the three ponds are operated in series with raw wastewater entering the first pond, flowing through to the second pond, followed by the third pond. Ponds may also be operated in parallel. In parallel operation, the influent wastewater is divided among multiple ponds as shown in Figure 6.4. When the ponds are operated in series, the first pond in the series receives raw, influent wastewater and is referred to as the primary pond. When ponds are operated in parallel, two or more ponds receive untreated influent wastewater. In this case, all of the ponds receiving untreated wastewater are considered to be primary ponds. In Figure 6.4, the ponds labeled as (1) are primary ponds.

Short-circuiting occurs when wastewater moves quickly from the inlet to the outlet of a treatment process. Short-circuiting can result in the discharge of untreated or partially treated wastewater. Depending on the inlet and outlet structure locations, short-circuiting can still occur even in multiple pond systems. In general, the more ponds in series, the better the treatment efficiency becomes.

The first pond or ponds operate much like primary clarifiers. Here, heavier, denser materials settle out and accumulate on the pond bottom. In aerobic and facultative ponds, microorganisms in the upper portions of the first and second ponds use oxygen to convert BOD into new biomass, carbon dioxide, and water. By the time the wastewater reaches the last pond in the series, very little BOD remains. The last pond in the treatment process is sometimes called a tertiary pond, settling pond, or polishing pond. Here, the biomass produced upstream and dead algae settle to the bottom of the pond. Live algae are buoyant and can remain near the surface of the pond. Live algae contribute to the amount of TSS in the final effluent. Treated wastewater from the polishing pond is often *disinfected* before being discharged to the environment. Disinfection is not always needed to meet discharge permit limits for *fecal coliform* and *E. coli* bacteria. This is especially true for ponds with long detention times.

Disinfection uses chlorine, ultraviolet light, or other methods to reduce the number of bacteria and other microorganisms in the effluent. Disinfection is covered in Chapter 10—Disinfection.

Most discharge permits set limits on the number of *fecal coliform* and *E. coli* bacteria that can be discharged to the environment. If these bacteria are present, it is likely that disease causing microorganisms are also present.

AEROBIC LAYER

Facultative ponds rely on a variety of different bacteria and algae to treat wastewater. Take a moment to study Figure 6.5 carefully. Examine the connections among the different groups of microorganisms. They each rely upon the others to obtain the materials they need to grow and reproduce. Maintaining balance among these different groups is the key to successful pond operation. In the upper portion of the pond,

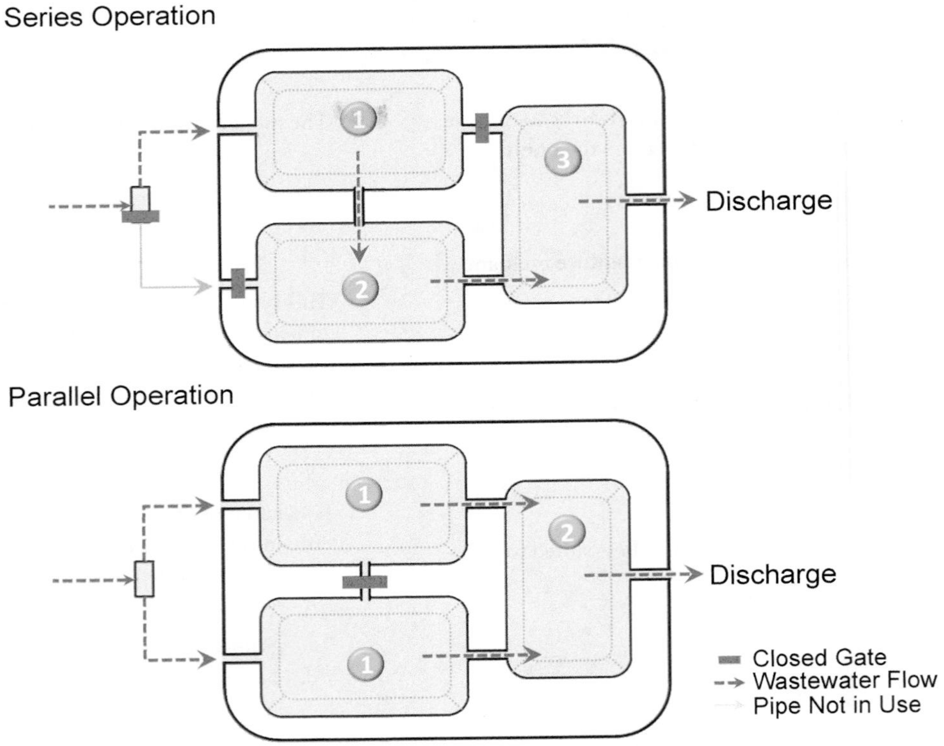

Figure 6.4 Three-Pond Treatment System. For Series Operation: (1) Primary Pond; (2) Secondary Pond; and (3) Tertiary, Settling, or Polishing Pond. For Parallel Operation: (1) Primary Pond and (2) Tertiary, Settling, or Polishing Pond.

The AOB and NOB are collectively called the nitrifying bacteria. These specialized bacteria are introduced in Chapter 5—Fundamentals of Biological Treatment and are discussed in depth in Chapter 9—Nutrient Removal.

algae use energy from sunlight to convert carbon dioxide (CO_2) and water (H_2O) into organic molecules like sugar ($C_6H_{12}O_6$). This process is called photosynthesis. Oxygen is produced as a byproduct. The oxygen is then used by heterotrophic bacteria to consume BOD_5 in the influent wastewater and convert it into new bacteria, carbon dioxide, and water. As the BOD_5 is broken down, ammonia (NH_3) and phosphate (PO_4^{-3}) are released. Some of the released nitrogen and phosphate, along with nitrogen and phosphate in the influent wastewater, is used as fertilizer by the algae. Some is taken up by bacteria to grow and build new bacteria. When conditions are right, ammonia oxidizing bacteria (AOB) and nitrite oxidizing bacteria (NOB) will use oxygen to convert ammonia to nitrite (NO_2^-) and nitrate (NO_3^-). In the process, the AOB and NOB produce acid and use up alkalinity. For the AOB and NOB to convert ammonia, warmer water temperatures and longer hydraulic detention times are needed. Facultative ponds often remove ammonia through nitrification in the summer, but not in the winter.

ANOXIC LAYER

Some heterotrophic bacteria can use nitrite and nitrate in place of oxygen to consume BOD_5. They only do this when oxygen is unavailable. Algae and wind keep oxygen concentrations high near the top of the pond during the day, but concentrations decrease from the top to the bottom of the pond. Dissolved oxygen concentrations will also decrease on cloudy days and at night when algae don't have access to sunlight. Deep in the pond where sunlight can't reach, oxygen may not be present at all. When the AOB and NOB in the aerobic layer are active, nitrite and nitrate are produced and part or all of the pond may become anoxic. Nitrite and nitrate will be converted to nitrogen gas when BOD is available. The nitrogen gas produced forms small bubbles that travel up through the water to the atmosphere. This is one way that ponds can achieve total nitrogen removal.

ANAEROBIC LAYER

Settleable solids from the influent along with dead algae and bacteria from the treatment process settle to the pond bottom to form a sludge blanket. Oxygen, nitrite, and nitrate are not available in the settled sludge. It is anaerobic. Different groups of bacteria work together to break down the organic solids in the

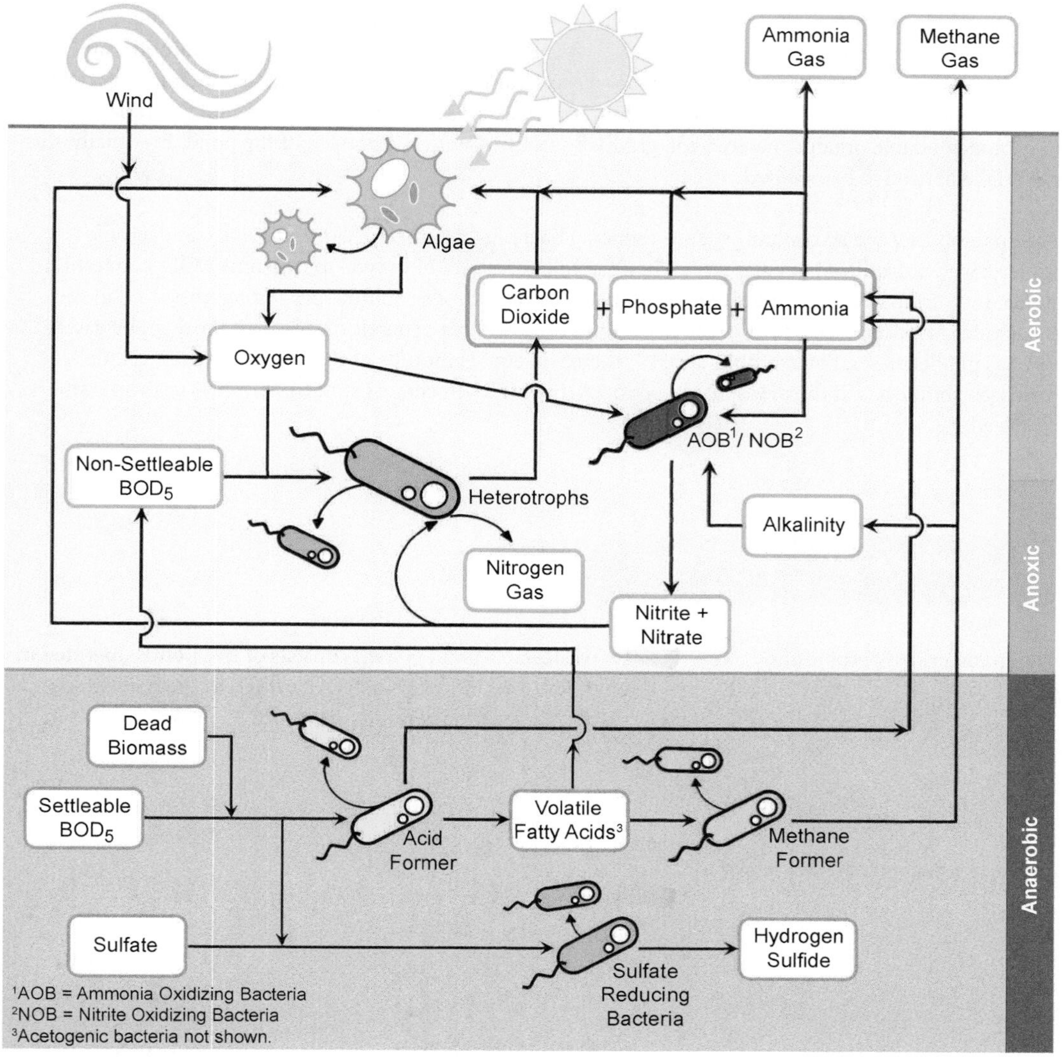

Figure 6.5 Biological Reactions within Facultative Ponds (Reprinted with permission by Indigo Water Group)

sludge anaerobically: the acid formers (*acidogenic bacteria*), the acetic acid formers (*acetogenic bacteria*), and the methane formers (*methanogens*). The acid formers are sometimes referred to as the *saprophytic* bacteria. Anaerobic breakdown is typically shown as a two-step process converting BOD_5 into methane and carbon dioxide (Figure 6.5). In reality, anaerobic breakdown is more complex and involves at least three different groups of bacteria and four distinct steps (Bajpai, 2017). Each group produces byproducts that are used by the next group.

In the first step, extracellular *enzymes* (enzymes operating outside the cells) break down solid, complex organic compounds (carbohydrates, fats, and proteins) into simpler, soluble organic compounds (fatty acids, alcohols, carbon dioxide, and ammonia). The technical term for this step is hydrolysis. In the second step, acidogenesis, the acid formers absorb these soluble organic compounds and break them down further to extract energy. The end products are *volatile fatty acids* (VFAs) and carbon dioxide. Volatile fatty acids are a group of soluble organic acids that contain five or fewer carbon atoms. The VFAs produced in the sludge blanket include acetic, proprionic, and butyric acid. Acetic acid is more commonly known as vinegar. Both acetic and proprionic acid smell like vinegar, whereas butyric acid smells like rancid butter orvomit. Some of the soluble organics and VFAs produced in the first two steps are released from the sludge blanket and travel up into the overlying water. Next, another group of bacteria, the acetogenic bacteria, convert some of the soluble organics and VFAs from step two into acetate, hydrogen gas, and

Saprophytic organisms obtain carbon and energy from dead organisms or decaying organic matter.

Enzymes can be thought of as chemical scissors that cut larger molecules into smaller molecules. Enzymes can only make cuts into particular molecules. For example, the enzyme amylase can break down starch, but it cannot break down fat.

Volatile fatty acids are one component of soluble BOD.

When solids must be *removed* from a pond, be sure to contact the state or local regulatory agency, as they may require permits for the disposal of pond solids.

Early study materials from U.S. EPA show anaerobic breakdown of organic material with only two steps and two groups of bacteria: the acid formers and the methane formers. Other manuals leave out the first step (hydrolysis), during which organics are broken down outside of the cell. They show the breakdown as three steps: acidogenesis, acetogenesis, and methanogenesis. Showing the process as four distinct steps is the most correct.

carbon dioxide. This is acetogenesis. In the fourth and last step, the methane-forming bacteria (*methanogens*) convert these byproducts into methane, carbon dioxide, and alkalinity. Ammonia and phosphate are released during all four steps. Over time, approximately 60% of all the organic material that ends up in the sludge blanket will be turned into methane and carbon dioxide. Inorganic material and non-biodegradable organic material will gradually accumulate at the bottom of the pond. Eventually, this material will need to be *removed*.

The anaerobic layer also contains sulfate-reducing bacteria (Figure 6.5). These facultative anaerobic bacteria were introduced in Chapter 5. They use sulfate instead of oxygen to consume BOD; converting sulfate (SO_4^{-2}) to hydrogen sulfide (H_2S) in the process. Hydrogen sulfide may remain dissolved in the wastewater or be released as a gas. The percentage of each form depends on pH with more in the gas form as pH decreases. For example, at pH 7, there will equal amounts of dissolved hydrogen sulfide and hydrogen sulfide gas. If the pH is increased to 7.4, then the percentages shift to only 26% gas and 74% dissolved.

TEST YOUR KNOWLEDGE

1. **Most of the dissolved oxygen in unaerated aerobic and facultative ponds comes from**
   a. Surface aerators
   b. Wind
   c. Photosynthesis
   d. Respiration

2. **These organisms are primarily responsible for BOD removal in wastewater ponds:**
   a. Algae
   b. Bacteria
   c. Protozoans
   d. Ducks

3. **The first pond in a wastewater pond system is similar to what unit process in a mechanical treatment facility?**
   a. Headworks
   b. Primary clarification
   c. Trickling filter
   d. Disinfection

4. **The minimum number of ponds required by most regulatory agencies to prevent short-circuiting is**
   a. One
   b. Two
   c. Three
   d. Four

5. **The main gas produced in any type of wastewater pond system by bacteria is**
   a. Carbon dioxide
   b. Methane
   c. Nitrogen
   d. Oxygen

6. **A wastewater pond system consists of five ponds operated in parallel. Raw influent wastewater is fed to the first two ponds. How many primary ponds are there?**
   a. One
   b. Two
   c. Three
   d. Four

7. **Match the organism to its function.**

1. Algae	a. Convert ammonia to nitrite and nitrate
2. Bacteria	b. Consume BOD and release hydrogen sulfide
3. AOB and NOB	c. Produce oxygen
4. Acid formers	d. Convert VFAs into methane
5. Sulfate-reducing bacteria	e. Convert settleable BOD and biomass into VFAs
6. Methanogens	f. Consume BOD and release nutrients

8. **About half of the organic material that settles to the bottom of the pond will ultimately be converted into**
   a. Algae and bacterial biomass
   b. Methane and carbon dioxide
   c. Carbon dioxide and water
   d. Volatile fatty acids and oxygen

POND BIOLOGY

Wastewater ponds contain many different types of bacteria, algae, and other organisms. Not all of them are shown in Figure 6.5. Different types of bacteria are discussed in Chapter 5.

ALGAE

Algae is a term used to describe a large, diverse group of photosynthetic organisms. Some algae exist as individual cells while others grow in long chains called filaments. Some algae can propel themselves through the water, while others grow attached to surfaces or simply drift with the current. All algae are phototrophs. They obtain their energy from sunlight and carbon from carbon dioxide. They cannot use BOD as either an energy or carbon source. Like bacteria and other microorganisms, algae need nitrogen, phosphorus, and other nutrients to grow and reproduce. Algae use phosphate, ammonia, nitrite, and nitrate. These nutrients are available in the influent wastewater and are released into the pond during treatment as BOD is consumed by heterotrophic bacteria (Figure 6.5).

Algae have two operational modes: photosynthesis and respiration (Figure 6.6). They switch from one mode to the other depending on whether or not sunlight is available. Photosynthesis occurs when sunlight is available. Algae take in carbon dioxide and water to produce organic compounds like sugars. Oxygen is a byproduct of this reaction. The organic compounds are either stored or used to generate more algae (reproduction). When sunlight isn't available, algae operate their cellular machinery in reverse. They take in oxygen and use it to burn the organic compounds formed when sunlight was available. The byproducts of this reaction are carbon dioxide and water. This process is called respiration. In essence, algae store excess energy from sunlight within the chemical bonds of organic compounds. Later, when they need the energy, the stored organic fuel is burned to release it. Over a 24-hour period, the concentrations of dissolved carbon dioxide and oxygen can change dramatically within a pond. These changes affect pond pH and alkalinity concentrations. Dissolved oxygen concentrations and pH will be the highest in late afternoon and the lowest right before the sun comes up in the morning. Pond pH and alkalinity are discussed in more detail in the section on process variables within this chapter.

Algae are grouped into three major categories based on the type or types of chlorophyll and other *pigments* they contain: green, brown, or red. Green and brown algae are common to wastewater ponds. Red algae occur infrequently. Chlorophyll and other photosynthetic pigments are large organic molecules that react

Pigments are compounds that are colored.

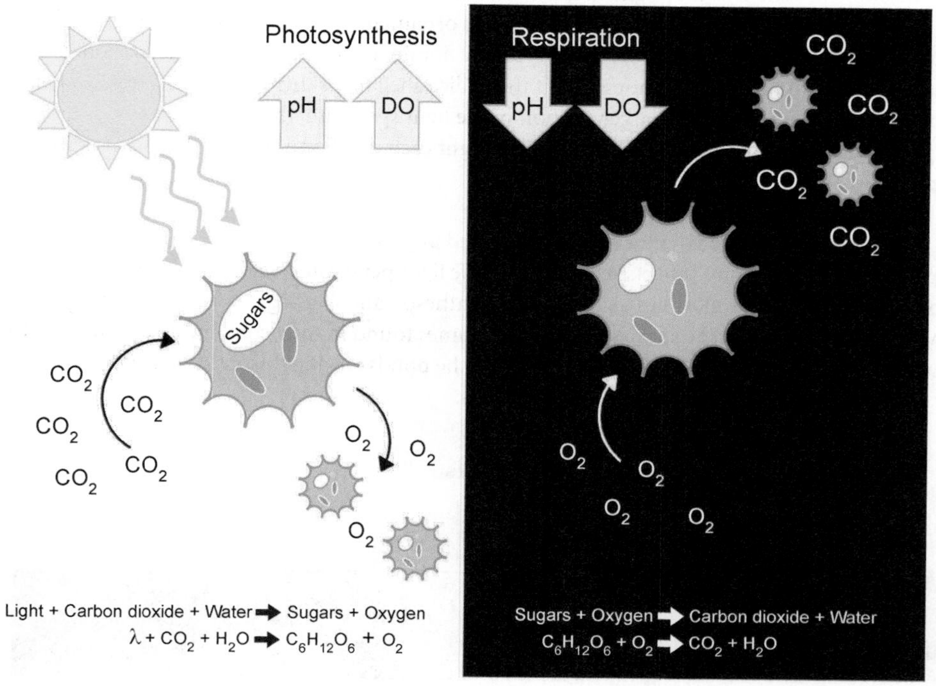

Figure 6.6 Photosynthesis and Respiration (Reprinted with permission by Indigo Water Group)

with sunlight. Think of these compounds as chemical antennae or solar panels that the algae use to absorb light energy. Chlorophyll and other pigments absorb light within narrow ranges. The colors of light that are not absorbed reflect back. The reflected light is what we see. Figure 6.7 shows the light spectrum. The numeric labels along the top of the graph are the wavelengths in *nanometers* (nm) associated with each color. Think of it as a radio dial that the algae can tune into. Green algae appear green because they contain chlorophyll types a and b. These types of chlorophyll absorb red and blue light. If you cover up the red and blue ends of the light spectrum with your fingers, only the green and yellow light remain. This is what makes green algae appear green. Two examples of the many types of green algae found in facultative ponds are *Scenedesmus* (Figure 6.8) and *Volvox* (Figure 6.9). Pond systems with an abundance of green algae will be bright green and sparkling (Figure 6.10).

One *nanometer* is equal to one billionth of a meter. A sheet of paper is about 100 000 nm thick.

Growth rates of green algae are affected by temperature and the availability of light, inorganic carbon, and nutrients. Think back to the factory analogy in Chapter 5. Green algae can only achieve their maximum growth rates when all of the inputs to the factory are present in excess, including sunlight. In the northern hemisphere, wastewater ponds typically experience *blooms* of green algae in the spring and summer as water temperatures increase and the days get longer. Ponds often become turbid resulting from the large number of green algae present. The algae absorb the incoming sunlight and block it from reaching further down into the pond. As a result, rapidly growing green algae often crowd out slower growing types of algae and bacteria. Green algae are a food source for other microorganisms, snails, fish, and crustaceans. When the amount of algae lost to predation is greater than the amount of algae grown, the green algae population in the pond decreases.

The sudden appearance of a large number of algae or cyanobacteria is called a *bloom*.

Brown and red algae contain other pigments in addition to chlorophyll. These pigments assist with photosynthesis by capturing light energy at other wavelengths, which they then pass on to chlorophyll. Brown algae contain chlorophyll a, chlorophyll c, and carotenoids (red and yellow pigments) to capture light energy for photosynthesis. This allows brown algae to collect light over a broader range than green algae. Use your fingers to cover up the light spectrum in Figure 6.7 at the wavelengths used by brown algae: chlorophyll a (430 and 662 nm), chlorophyll c (450 nm), and carotenoids (400 to 500 nm). Only orange and green light remain. Blended together, the reflected light appears brown.

Brown algae include many different species of marine algae and diatoms. These single-celled algae come in a wide variety of shapes and sizes and may be *flagellated*. Diatoms have an outer shell made from silica. Their patterned silica shells scatter light causing diatoms to appear very bright and transparent when viewed through a microscope, like tiny glass Christmas ornaments. A diatom commonly found in facultative ponds is *Navicula* (Figure 6.11). Diatoms typically dominate pond systems when sunlight is less plentiful and the water is cooler. It is thought that their silica shell helps them survive in cooler water. Because diatoms can absorb light from a larger portion of the light spectrum, they are better equipped to survive than green algae when there are fewer hours of sunlight each day. Pond systems that are dominated by brown algae and diatoms will appear brownish.

A *flagellum* is a tail or whip that some microorganisms use to swim.

Red algae get their color from a photosynthetic pigment called phycoerythrin. This pigment absorbs blue and green light (470 to 550 nm) and reflects red. Because blue light penetrates water to a greater depth than other colors, these pigments allow red algae to photosynthesize and live at greater depths than most other algae. Most red algae are marine species, but are sometimes found in total containment ponds due to their high salt content. Water continuously evaporates from the ponds leaving dissolved salts behind. Over time, the salt concentration in an evaporative pond increases.

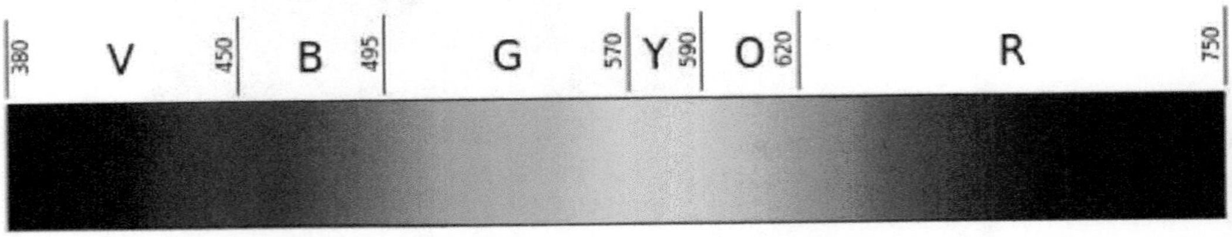

Figure 6.7 Light Spectrum (Reprinted with permission by Gringer)

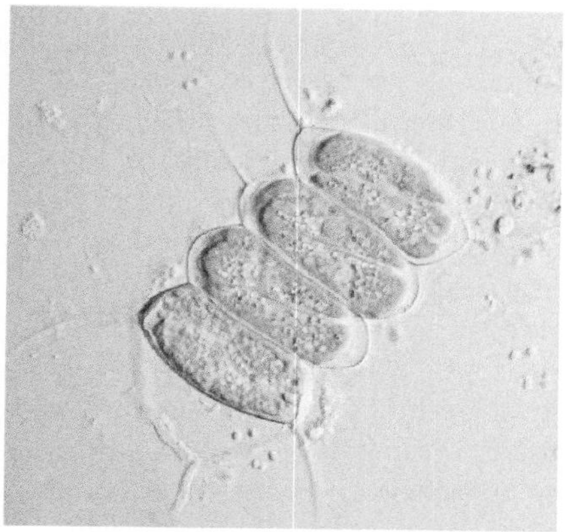

Figure 6.8 Green Algae *Scenedesmus* (Reprinted with permission by Frank Fox, www.mikro-foto.de)

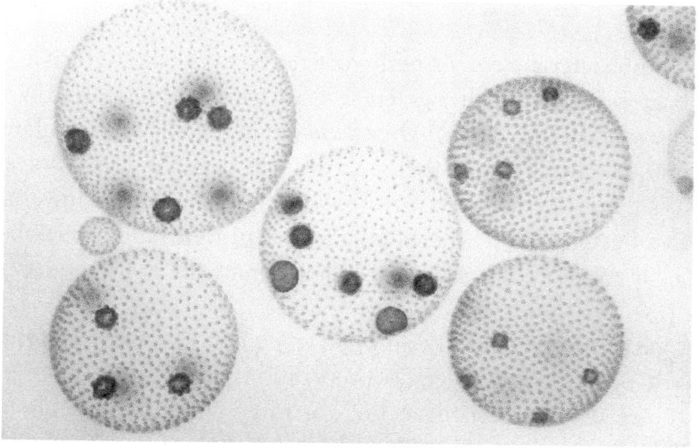

Figure 6.9 Green Algae *Volvox* (Reprinted with permission by Frank Fox, www.mikro-foto.de)

Figure 6.10 Healthy Pond with an Abundance of Green Algae (Reprinted with permission by Paul Krauth P.E.)

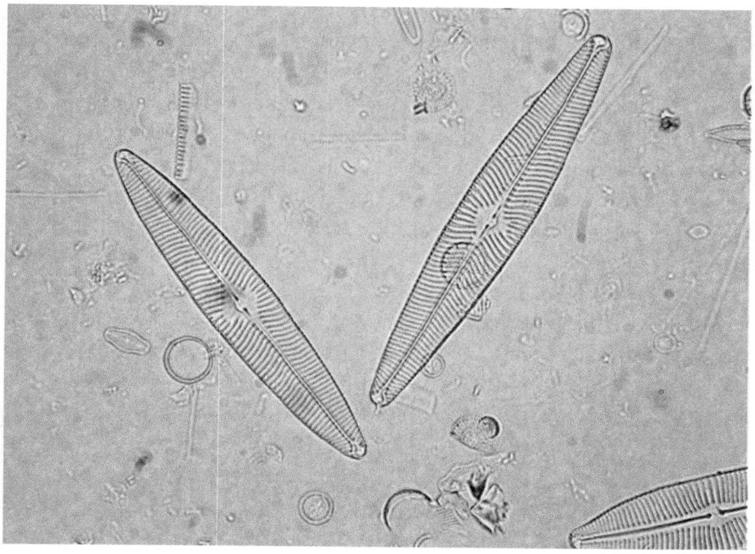

Figure 6.11 Brown Algae, the Diatom *Navicula radiosa* (Reprinted with permission by Kristian Peters)

CYANOBACTERIA (BLUE-GREEN ALGAE)

Cyanobacteria were formerly referred to as the blue-green algae, but they are not true algae. These bacteria are often blue-green, as their name implies, but can also be blue, green, reddish purple, or brown (WDNR, 2006). They grow as single cells or long filaments. Cyanobacteria are phototrophs and, like algae, obtain their energy from sunlight, use dissolved carbon dioxide, and produce oxygen as a byproduct of photosynthesis. Cyanobacteria have lower maximum growths rate than green algae (Mur et al., 1999). When sunlight and nutrients are plentiful, green algae grow faster than both brown algae and cyanobacteria and will dominate the pond biology (Table 6.1).

Low light availability could be the result of turbid water, overcast days, or fewer hours of sunlight each day.

Cyanobacteria tend to dominate when conditions are unfavorable for green algae including: *low light availability*, low nitrogen concentrations, and high predation pressure. The first two conditions slow the growth of green algae, but don't have the same effect on the cyanobacteria. Like brown algae, cyanobacteria contain chlorophyll a, carotenoids, and other pigments that allow them to harvest light energy all across the light spectrum (Mur et al., 1999). They are efficient photosynthesizers in low light conditions. Most species of cyanobacteria use inorganic nitrogen and phosphorus present in the wastewater. However, some species are able to use nitrogen gas directly from the atmosphere (Williams and Burris, 1952). They convert nitrogen gas directly into ammonia. This ability enables them to grow faster than green algae when wastewater nitrogen concentrations are limiting, but phosphorus is not. Predators consume green algae and reduce their numbers, but typically don't consume cyanobacteria (WDNR, 2006). For facultative ponds in the northern hemisphere, low light, low nitrogen, and high predation tend to occur together in the autumn and early winter months. The net effect is to bring the growth rates of cyanobacteria and algae closer together, or in some cases, to reduce the growth of algae below the growth rate of cyanobacteria. Blooms of cyanobacteria are most likely to occur in late autumn while the water is still fairly warm, but can occur at any time. Low nitrogen concentrations in ponds can cause blooms to occur in early and mid-summer. Cyanobacteria blooms have even been observed under the ice in frozen ponds (WDNR, 2006).

Cyanobacteria are problematic in both naturally occurring ponds and wastewater pond systems for two reasons. Many species of cyanobacteria produce unpleasant, odorous, or toxic compounds. Exposure to some of these compounds can cause stomach cramps, diarrhea, vomiting, and even seizures (WDNR, 2006). Cyanobacteria are responsible for many of the taste and odor problems in drinking water reservoirs. Most species of cyanobacteria are buoyant and can form large, floating mats like the ones shown in Figures 6.12 and 6.13. These mats prevent sunlight from reaching deeper into the pond and prevent air from reaching the pond surface. Without access to sunlight, algae and diatoms are unable to produce oxygen. Biochemical oxygen demand removal may be reduced if not enough oxygen is available.

PURPLE SULFUR BACTERIA

Elemental sulfur is pure sulfur without any other elements like oxygen attached.

The purple sulfur bacteria are another group of photosynthetic bacteria. They require both sunlight and anaerobic conditions to survive. During photosynthesis, these bacteria convert hydrogen sulfide into either *elemental sulfur* (S) or sulfate (SO_4^{-2}). Hydrogen sulfide is produced in the anaerobic layer by the sulfate-reducing bacteria (Figure 6.5). Normally, purple sulfur bacteria are found in a thin layer between the sludge blanket and aerobic layer; deep enough to be fully anaerobic, but shallow enough for blue light to reach them. Wastewater ponds can become entirely anaerobic either because they are receiving too much BOD or because not enough oxygen is being added to keep up with the organic load. A bloom of purple sulfur bacteria may occur in an anaerobic pond and turn the entire pond dark pink to purple (Figure 6.14).

Table 6.1 Growth Rates for Cyanobacteria and Algae at 20 °C (68 °F) (created with information from Carr and Whitton [Eds.])*

Type	Time to Double Population, hours
Cyanobacteria	17–80
Diatoms	12.5–30
Green algae (single cell)	10.5–18.5

*Light must be available in excess.

Figure 6.12 Cyanobacteria Bloom (Reprinted with permission by Paul Krauth P.E.)

Figure 6.13 Closeup View of Floating Cyanobacteria Mat (Reprinted with permission by Paul Krauth P.E.)

Figure 6.14 Purple Sulfur Bacteria Bloom (Reprinted with permission by Paul Krauth P.E.)

PROTOZOANS AND INVERTEBRATES

Although bacteria and algae are primarily responsible for the wastewater treatment that occurs in ponds, protozoans and invertebrates are an important part of pond biology. Rotifers, daphnia, midge, and mosquito larvae are relatively slow growing and typically only occur in pond systems with hydraulic detention times longer than 10 days. These organisms feed on bacteria and algae. They are too small to be seen with the naked eye and must be viewed through a *microscope*. Rotifers feed on free-swimming bacteria and algae. Predation by rotifers encourages bacteria and other particles to *flocculate* into larger, denser masses because individual bacteria are easier prey than larger groups. There is safety in numbers. Larger, denser particles settle to the bottom of the pond and become part of the sludge blanket. Rotifers are associated with lower turbidity and lower TSS in the final effluent.

Flocculation means growing larger particles through collisions that help smaller particles stick together.

Daphnia, also called water fleas, are 0.2 to 5 mm (0.01 to 0.2 in.) in length (Figure 6.15). Daphnia get their nickname from their resemblance to the insect by the same name (fleas) and by the way they move through the water in short jerky bursts. Daphnia are not insects, but are tiny crustaceans that belong to the same family as crayfish, krill, and crabs. They are easily seen with the naked eye and can often be observed swimming close to the surface of a pond. They grow well at water temperatures above 10 °C (50 °F). Daphnia are tolerant of poor water quality and low DO, but are extremely sensitive to heavy metals, like copper and zinc, pesticides, detergents, chlorine, and other dissolved toxins (Clare, 2002).

Daphnia feed primarily on single-celled green algae and are, in turn, consumed by fish and salamanders. Predation by daphnia also encourages flocculation and settling of particles. This results in clearer water, better light penetration, increased algae growth, and more DO production. The number of daphnia will generally be proportional to the amount of green algae as the two populations balance one another. A rotifer or daphnia bloom often follows immediately after an algae bloom. The increased food supply supports a larger number of predators. The rotifers and daphnia consume the algae, which decreases their food supply and can result in die-off. Without predators, the algae population may bloom again setting up a cycle of boom-and-bust conditions for the two populations. If too much algae is lost from predation, low DO conditions may result.

Daphnia use *hemoglobin* to pull DO out of the water and into their bodies. Hemoglobin turns bright red in the presence of oxygen. Under low DO conditions, daphnia produce more hemoglobin than normal to help maximize capture of the DO that is available. The extra hemoglobin turns the daphnia bright red. Thus, daphnia can act as a biological DO probe in a pond system. Red clouds of daphnia are often seen in pond systems in the early morning when DO concentrations are at their lowest (Figure 6.16). Red daphnia clouds should not be seen late in the afternoon when algae should be producing the most oxygen.

Hemoglobin is a protein that contains iron. It is the same protein used by human red blood cells to carry oxygen. When oxygen is present, hemoglobin appears red. In the absence of oxygen, hemoglobin appears blue.

Mosquitos can be problematic for some ponds. Mosquitos lay floating rafts of eggs in stagnant water, often at the edge of the water or attached to plant stems and leaves. The eggs hatch into larvae (worms) that are submerged in the water, but attached to the water surface. Mosquito larvae breathe air. Mosquito larvae proliferate in ponds where shoreline vegetation is not removed. Vegetation protects eggs and larvae from waves and helps hide them from predators. Mosquitos spread many diseases including encephalitis, malaria, and yellow fever (U.S. EPA, 2011). Many regulatory agencies require removal of shoreline vegetation to reduce mosquito populations.

A Winogradsky column is a miniature pond system. Build one using a graduated cylinder, pond sediment, some eggshells, a bit of raw egg, and pond water. Place the finished column in a sunny location. Over time, the column will develop colored layers as the green and brown algae, sulfate reducing bacteria, purple sulfur bacteria, and other groups of microorganisms bloom and recede. Directions can be found on many different websites.

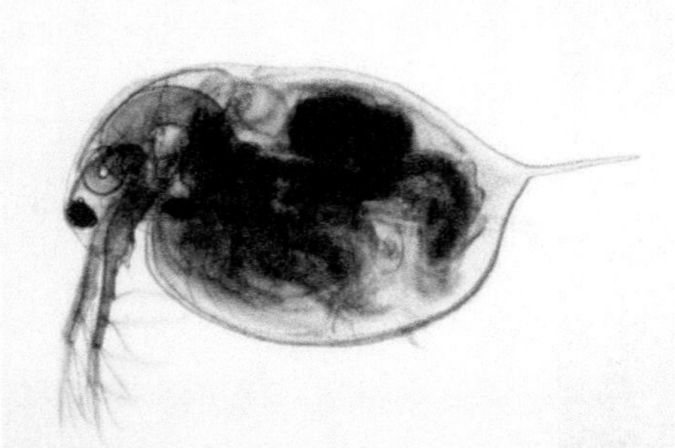

Figure 6.15 Daphnia (Water Flea) (Source: *Are We Underestimating Species Extinction Risk?* PLoS Biology Vol. 3/7/2005, e253 doi:10.1371/journal.pbio.0030253)

Figure 6.16 Daphnia Bloom (Reprinted courtesy of Paul Krauth P.E.)

TEST YOUR KNOWLEDGE

1. **The pH in an unaerated, facultative pond will be lowest**
   a. Early morning
   b. Before lunch
   c. Late afternoon
   d. At night

2. **Algae obtain carbon for growth and reproduction from**
   a. BOD
   b. Volatile fatty acids
   c. Dissolved carbon dioxide
   d. Sugar

3. **During photosynthesis, algae produce**
   a. Nitrogen gas
   b. Oxygen
   c. Carbon dioxide
   d. Methane

4. **Dissolved oxygen concentrations will be highest in an unaerated, facultative pond**
   a. On a bright, sunny day
   b. After a daphnia bloom
   c. After 3 days of rain
   d. In the middle of the night

5. **This type of algae is rarely seen in wastewater treatment ponds:**
   a. Brown
   b. Green
   c. Blue-green
   d. Red

6. **A healthy pond with high dissolved oxygen should appear**
   a. Brownish
   b. Pink to dark purple
   c. Grey to light green
   d. Green and sparkling

7. **Ponds tend to turn from green to brown as the days get shorter and less sunlight is available because**
   a. Lack of dissolved oxygen prevents algae growth
   b. Diatoms can capture light over more of the light spectrum
   c. Presence of large numbers of daphnia and rotifers
   d. Colder water temperatures

8. **A sudden increase in the amount of algae in a pond is called a**
   a. Burst
   b. Flowering
   c. Bloom
   d. Sparkle

9. **This unique ability of cyanobacteria, also called the blue-green algae, allows them to prosper when green algae cannot:**
   a. Use nitrogen gas from the atmosphere
   b. Recycle phosphorus internally
   c. Reproduce faster under ideal conditions
   d. Obtain carbon from BOD

10. Cyanobacteria blooms are most likely to occur in the

a. Spring
b. Summer
c. Autumn
d. Winter

11. Cyanobacteria are problematic in natural and wastewater pond systems because

a. They increase phosphorus concentrations
b. Some species produce toxins
c. They support daphnia blooms
d. Ponds may become overoxygenated

12. A small community hosts an annual festival that doubles or triples the town's population. During the festival, the surface of the wastewater pond turns from sparkling green to dark pink. What is the most likely cause?

a. Increased BOD loading
b. Growth of cyanobacteria
c. Flowering of green algae
d. Portable toilet chemicals

13. Red clouds of tiny, darting organisms are observed at the edge of a pond in the early morning. Which of the following must be true?

a. This pond recently experienced a bloom of blue-green algae.
b. Dissolved oxygen concentrations are low.
c. Shoreline vegetation should be trimmed to remove breeding habitat.
d. Hydrogen sulfide generation has increased in the sludge blanket.

14. A bloom of daphnia or rotifiers is most likely to occur

a. When dissolved oxygen concentrations are lowest
b. After a storm event or several overcast days
c. Immediately after a bloom of green algae
d. When toxic compounds are present in the influent

15. A facultative pond turns dark purple. Which of the following statements must be true?

a. Daphnia populations have bloomed in response to a bloom of green algae.
b. Sludge blanket nitrite and nitrate concentrations have dropped below a critical level.
c. Anaerobic conditions have caused a bloom of purple sulfur bacteria.
d. Pond pH and DO are both increasing.

Design Parameters

Design of aerobic and facultative wastewater treatment ponds is based on the minimum expected water temperature, organic loading rate, availability of DO, and the need to store both treated water and settled sludge. These variables determine the total pond volume, minimum hydraulic detention time, and amount of surface area required. Surface area is related to the need for DO. Pond depth is determined partly by the need to store settled solids as they break down, but is limited by the availability of DO deeper in the pond. Storage requirements often determine the total volume needed for intermittent discharge ponds.

Wastewater ponds, like all biological treatment processes, are sensitive to changes in water temperature. For every 10 °C decrease in water temperature (18 °F decrease), the growth rate of the bacteria decreases by approximately 50%. Slower growth rates mean that the bacteria need more time to consume the same amount of BOD. Algae also grow slower at colder temperatures so less oxygen is produced. As a result, ponds that operate at colder temperatures must be larger than those operated at warmer temperatures. Larger ponds have more surface area, which cools the water faster. This means that ponds in cold climates tend to be much larger than ponds receiving similar organic loads in warm climates.

The amount of BOD_5 that an aerobic or facultative pond can process is dependent on the amount of oxygen available. Every one kilogram (one pound) of BOD_5 that enters the pond will require a minimum of one kilogram (one pound) of oxygen. In practice, the oxygen needed will be greater than this because the new bacteria produced will eventually break down in the treatment process and become BOD_5 themselves. Dead algae also contribute to the BOD_5 load. Think about a food chain where the influent BOD_5 is passed along from one bacterium to another bacterium to protozoans and other microorganisms. Each of these steps requires some additional oxygen. The BOD_5 test is only 5 days long, but the wastewater will remain in most facultative pond treatment systems for much longer than 5 days.

Dissolved oxygen may be added to aerobic and facultative ponds by three different mechanisms. First, oxygen is produced by algae during photosynthesis. Algae are present in the upper layers of pond systems, but only to depths where sunlight fully penetrates. For most ponds, this will limit algae growth to the upper 0.6- to 1.5-m (2- to 5-ft) of water depth. Second, oxygen dissolves into the water from the air above the pond (oxygen transfer). Wind creates waves and turbulence, which increases oxygen transfer. Third, mechanical devices can be used to inject air directly into the pond or create turbulence or sprays to increase oxygen transfer from the air. A diffused air system is shown in Figure 6.17. This system uses a blower to inject air into the water through *diffusers*. One type of floating aerator is shown in Figure 6.18. The type pictured is an aspirating aerator. It creates a current by injecting a mixture of air and water into the pond. It is similar to blowing bubbles into a glass of liquid with a straw.

Diffusers contain numerous tiny holes that help turn a steady stream of air into many, tiny bubbles. An example of a ceramic air diffuser is the air stone used in a home aquarium. Diffusers may be porous, like an air stone, or be perforated. Diffusers are categorized as coarse bubble (typical bubble size > 5 mm [0.2 in.]) or fine bubble (bubble size < 5 mm [0.2 in.]).

Facultative ponds that rely on algae and oxygen transfer from the atmosphere will have less oxygen available than mechanically aerated ponds. As a result, the amount of organic load they can process is lower. Many unaerated facultative ponds have been upgraded to increase their treatment capacity by adding mechanical aeration. These ponds are shallow with large surface areas to maximize the amount of oxygen transferred from the atmosphere and the percentage of water exposed to sunlight for algae growth. Naturally aerated facultative ponds are usually only 1.1- to 2.1-m (3.5- to 7-ft) deep (U.S. EPA, 1975). Increasing the pond depth does not result in a lot of additional treatment because DO concentrations decrease with depth. Mechanically aerated ponds are deeper; typically between 1.8- and 6-m (6- and 20-ft) deep. Mechanical aeration devices add oxygen and provide mixing. Mixing often prevents algae from growing in the aerated cells in large quantities as they are unable to remain in the uppermost layer of the pond. Mechanical aeration may be used 24 hours a day, which maintains DO concentrations at night when concentrations decrease in naturally aerated ponds. The additional oxygen reduces the amount of treatment time required to treat the same amount of organic load.

Additional information on aeration systems may be found Chapter 8 of *Design of Water Resource Recovery Facilities* (WEF et al., 2018) and in Chapter 8 of *Energy Conservation in Water and Wastewater Facilities* (WEF, 2009a).

There are at least five different methods used by engineers to design wastewater ponds and determine the volume and surface area needed. The simplest and most conservative method is based on the amount of BOD_5 applied per surface area of pond. Loading rates for different types of ponds are given in Table 6.2. Organic loading rates to ponds are expressed in kilograms of BOD_5 per hectare per day (kg BOD_5/ha·d) or pounds of BOD_5 per acre per day (lb BOD_5/ac/d). Take a moment to examine Table 6.2. Notice that acceptable BOD_5 loading rates for naturally aerated facultative ponds decrease with temperature. With mechanical aeration added, hydraulic detention times decrease and loading rates increase.

The load to a wastewater pond system should be distributed between ponds to ensure that the primary pond(s) do not become anaerobic. Loading rates to the primary cell(s) of naturally aerated ponds are usually limited to 40 kg/ha·d (35 lb/ac/d) or less where the average air temperature is below 0 °C (32 °F). In mild climates where the air temperature is greater than 15 °C (59 °F), the BOD_5 loading rate to the primary

Figure 6.17 Diffused Air Aeration (Reprinted with permission by Indigo Water Group)

Figure 6.18 Surface Aerator/Aspirator (Reprinted with permission by Indigo Water Group)

cell(s) may be as high as 100 kg/ha·d (89 lb/ac/d). For mechanically aerated ponds, operators usually have some flexibility to increase or decrease the amount of aeration provided. Still, heavy influent BOD loads can cause the DO concentrations to fall to unacceptably low levels even in mechanically aerated ponds. Operators can reduce the loading to the primary cells and increase DO concentrations by switching from series to parallel operation. Series operation is preferred to minimize the possibility of short-circuiting, but switching to parallel operation during periods of heavy organic loading can prevent ponds from becoming anaerobic.

Table 6.2 Naturally Aerated Facultative Pond BOD_5 Loading Rates (U.S. EPA, 1975; WEF, 2009b; WEF, 2016)

Pond Type	Hydraulic Detention Time, days	Depth, m (ft)	BOD Loading Rate, kg/ha·d (lb/ac/d)
Aerobic (oxidation)			
Low rate	10–40	0.5–0.9 (1.5–3)	67–135 (60–120)
High rate	4–6	0.3–0.5 (1–1.5)	90–179 (80–160)
Naturally aerated facultative ponds*			
<0 °C (<32 °F)	80–180	1.5–2.1 (5–7)	11–22 (10 to 20)
0–15 °C (32–59 °F)	40–60	1.2–1.8 (4–6)	22–45 (20 to 40)
>15 °C (>59 °F)	25–40	1.1 (3.5)	45–90 (40 to 80)
Partial-mix aerated facultative ponds	5–30	1.8–6 (6–20)	34–112 (30–100)
Anaerobic ponds	20–50	2.4–4.9 (8–16)	224–560 (200–500)

*Facultative pond depth is based on average winter air temperature.

TEST YOUR KNOWLEDGE

1. **BOD_5 loading to facultative ponds is limited by**
   a. Surface area
   b. Available oxygen
   c. Sludge storage volume
   d. Hydraulic detention time

2. **A mechanically aerated facultative pond system has three cells operated in series. Dissolved oxygen concentrations in the primary pond have been dropping steadily and are now below 1 mg/L even in the late afternoon. The operator should**
   a. Bypass flow to the polishing pond
   b. Decrease aerator run time
   c. Consider switching to parallel operation
   d. Add phosphate to grow green algae

3. **Facultative ponds are capable of processing more BOD load when**
   a. Wastewater is colder
   b. Less oxygen is provided
   c. Wastewater is warmer
   d. Cyanobacteria are present

Expected Performance

Facultative ponds are capable of consistently meeting secondary treatment standards for BOD_5 of less than 30 mg/L (30-day average) and 45 mg/L (7-day average). They do not consistently meet secondary treatment standards for TSS. The settling pond at the end of the treatment process removes bacteria, dead algae, and other particles by gravity settling, but cannot remove live algae. Algae must remain in the upper portion of the pond to have access to sunlight. Different species of algae have evolved different methods of staying near the surface. As a result, they end up being discharged with the treated wastewater. Effluent TSS concentrations of 40 to 100 mg/L are common during periods of maximum algae growth. Pond systems sometimes have multiple outlet pipes that allow operators to discharge water from different depths; however, it can be difficult to avoid both algae and stirring up settled solids in ponds that are less than 2.1 m (7 ft) deep. Controlled discharge and hydrograph controlled release pond operators often schedule pond discharges around periods of maximum algae production.

The U.S. Environmental Protection Agency (U.S. EPA) recognizes that high effluent TSS from a pond system is frequently a result of algae growth and not because of incomplete treatment. U.S. EPA adopted a regulation in 1977 that allows states to adjust the maximum allowable TSS concentrations for facultative ponds. The rule was updated in 1984. Many states have since adopted alternative TSS limits for ponds (Table 6.3). These alternative discharge limits only apply to facilities that use ponds for secondary treatment.

Total suspended solids may also be lost during spring and fall turnover. Turnover occurs in the spring when warmer water temperatures increase the activity of anaerobic bacteria in the sludge blanket. Because the anaerobes are nearly inactive below 10 °C (59 °F), organic material tends to accumulate in the sludge blanket during the winter. Increased gas production in the warming sludge blanket causes bubbles of carbon dioxide and methane to form. As they leave the sludge blanket, they can carry settled sludge back to the top of the pond. Fall turnover occurs because of temperature differences between the top and bottom of the pond. As air temperatures decrease, the top of the pond begins to cool. The bottom of the pond is insulated by the surrounding earth and does not cool as fast as the top of the pond. The pond becomes stratified. Cold water is denser than warm water. At some point, the colder, heavier water will sink to the bottom of the pond and the warmer, less dense water will rise. The two layers swap positions. The swap can occur suddenly, stirring up the settled sludge and bringing it to the surface. Noxious odors and ammonia will be released with the rising sludge in both the spring and fall. It can take several days or weeks for the sludge to resettle. Turnover is beyond the control of the operator in unaerated, unmixed ponds. If possible, avoid discharging treated wastewater until the effluent quality returns to normal.

Nitrogen removal ranges from 40 to 95% (Crites et al., 2006). Pond systems remove ammonia in three ways: assimilative uptake, volatilization, and biological nitrification. Assimilative uptake is the ammonia taken up by algae and bacteria as part of growth and reproduction. The ammonia is incorporated into

Table 6.3 State-Specific Adjusted TSS Requirements (49 FR 37005, September 20, 1984 [Table 5-4 in U.S. EPA, 2010])

Location	Alternate TSS limitation (30-day average) (mg/L)
Alabama	90
Alaska	70
Arizona	90
Arkansas	90
California	95
Colorado	
■ Aerated ponds	75
■ All others	105
Connecticut	None
Delaware	None
District of Columbia	None
Florida	None
Georgia	90
Guam	None
Hawaii	None
Idaho	None
Illinois	37
Indiana	70
Iowa	
■ Controlled discharge, 3 cell	Case-by-case but not greater than 80
■ All others	80
Kansas	80
Kentucky	None
Louisiana	90
Maine	45
Maryland	90
Massachusetts	None
Michigan: Controlled seasonal discharge	
■ Summer	70
■ Winter	40
Minnesota	40
Mississippi	None
Missouri	80
Montana	100
Nebraska	80
North Carolina	90
North Dakota	
■ North and east of Missouri R.	60
■ South and west of Missouri R.	100
Nevada	90
New Hampshire	45
New Jersey	None
New Mexico	90
New York	70
Ohio	65
Oklahoma	90
Oregon	
■ East of Cascade Mountains	85
■ West of Cascade Mountains	50
Pennsylvania	None
Puerto Rico	None
Rhode Island	45
South Carolina	90
South Dakota	120
Tennessee	100
Texas	90
Utah	None
Vermont	55
Virginia	
■ East of Blue Ridge Mountains	60
■ West of Blue Ridge Mountains	78
■ East slope counties	Case-by-case application of 60/78 limits
Virgin Islands	None
Washington	75
West Virginia	80
Wisconsin	80
Wyoming	100
Trust Territories and N. Marianas	None

proteins and other compounds and becomes part of the new biomass. Ultimately, nitrogen removed by assimilative uptake ends up in the settled sludge at the bottom of the pond. As the sludge breaks down anaerobically, some of this ammonia will be re-released into the water above. Assimilative uptake accounts for 5 to 15% of the total ammonia removed.

When something is *volatilized*, it is converted into a gas.

Ammonia is lost by *volatilization*. Ammonia exists in two forms in wastewater: ammonium ion (NH_4^+) and free ammonia (NH_3). Free ammonia is a dissolved gas that can be transferred from the pond to the atmosphere. Like nitrogen gas, free ammonia is not very soluble in water. The percentage of NH_4^+ and NH_3 changes depending on pH. At pH 9.25, there will be equal amounts of each form. As pH decreases, the percentage of NH_4^+ increases. Only free ammonia can be volatilized, so more ammonia will be lost to the atmosphere when pH is higher than when pH is lower. Volatilization from ponds is a slow process and may only account for 3 to 5% of the total amount of ammonia removed when the pH is below 8 (Gross et al., 1998), but can be up to 80% of the total ammonia removed when the pH is above 8 (Rockne and Brezonik, 2006). Ammonia losses from volatilization can trigger cyanobacteria blooms, especially in pond systems with intermittent mechanical aeration. When the aerators are on, the water is agitated, which increases volatilization. Then, when the aerators are turned off, the pond surface is not mixed and algae growth can occur. Cycling between the two conditions produces the low nitrogen, no competition environment that is ideal for cyanobacteria growth.

WINTER OPERATING CONDITIONS

In cold climates, ponds may freeze over in the winter. Photosynthetic activity can continue beneath the ice and may continue to provide some dissolved oxygen. Snow cover on the ice, however, will limit algal activity and anaerobic conditions may develop. With prolonged cold weather, biodegradation of organic matter will slow. Settleable organic matter will accumulate on the bottom of the pond(s). Spring turnover will reintroduce the stored organic matter throughout the entire pond and cause a high bacterial oxygen demand. Anaerobic conditions and odor problems often result. If sufficient hydraulic detention time and dissolved oxygen are not available, effluent quality will be degraded temporarily.

NUTRIENT REMOVAL

In Chapter 5, the nitrifying bacteria were introduced. These bacteria convert ammonia to nitrite and nitrate, but grow slowly. Generally, ponds will only remove ammonia through biological nitrification when hydraulic detention times are long and the water is warm. Requirements for nitrifier growth are discussed in detail in Chapter 9—Nutrient Removal. In most of the United States, ponds will remove ammonia and nitrate through biological nitrification during the summer, but not the rest of the year. Pond systems in the southernmost portions of the United States and Hawaii can remove ammonia year-round.

Phosphorus removal is low, typically less than 40%. Phosphorus removal mechanisms include assimilative uptake followed by settling.

TEST YOUR KNOWLEDGE

1. **Given the following information, what is the most likely cause of high TSS in the final effluent? Pond is located in Minnesota. It is early spring. Air temperatures have increased from 5 °C (41 °F) to 15 °C (59 °F) over the last 5 weeks. The pond is free of ice, but has not yet achieved its summer green sparkle. Effluent BOD_5 is less than 10 mg/L. Effluent TSS is 120 mg/L.**
   a. Increased algae production
   b. Short-circuiting
   c. Incomplete treatment
   d. Pond turnover

2. **When is green algae growth most likely to contribute to high effluent TSS?**
   a. Early spring
   b. Summer
   c. Fall
   d. Winter

3. **A facultative pond has a pH of 8.5 and an abundance of green algae. The hydraulic detention time is less than 10 days. The influent ammonia concentration is 25 mg/L as N and the final effluent concentration is 8 mg/L as N. What is the primary method of ammonia removal under these operating conditions?**
   a. Assimilative uptake
   b. Conversion to nitrite and nitrate
   c. Volatilization to atmosphere
   d. Particles settling into sludge

4. **For biological nitrification to take place in a wastewater pond, which of the following conditions must be met?**
   a. Long HRT and warm temperatures
   b. Short HRT and warm temperatures
   c. Short HRT and cold temperatures
   d. Long HRT and cold temperatures

Equipment

PHYSICAL CONFIGURATION OF TREATMENT PONDS

Ponds are usually constructed entirely of compacted earth. By using cut and fill to remove the original topsoil, ponds are built by enclosing an area with earthen dikes (Figure 6.19). The finished pond is typically partly below and partly above the surrounding ground level. Length-to-width ratios for complete-mix ponds are near 1 : 1, with the width of the pond being nearly equal to the length. Plug flow ponds have length to width ratios of 3 : 1 or greater. In plug flow, the water moves across the pond from one end to the other in the same way that water moves through a pipe. With plug flow, the water at the outlet end of the pond should not mix very much with the water at the inlet end of the pond. The goal is to maximize the hydraulic detention time and reduce the likelihood of short-circuiting. In practice, the actual hydraulic detention times are shorter than the calculated hydraulic detention times for both complete-mix and plug-flow ponds.

DIKE CONSTRUCTION

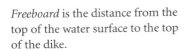

Freeboard is the distance from the top of the water surface to the top of the dike.

The outside slopes of the dikes are either 3 to 1 (horizontal to vertical) or flatter to permit grass mowing. Interior slopes are typically steeper, ranging from 2 to 1 to 3 to 1. However, ponds with surface aerators may have interior slopes of 4 to 1 to minimize erosion. Steep slopes on the interior of the pond minimize the amount of shallow water where cattails and other aquatic plants can grow. Larger ponds include a minimum of 0.6 to 0.9 m (2 to 3 ft) of *freeboard* that provides stability to the dike and room for ice accumulation in colder climates (Townshend and Knoll, 1987). Dikes must be protected to prevent damage from heavy rain or snow, waves, vehicle traffic, trees and shrubs, and burrowing animals. Each of these can compromise the strength and integrity of a dike; potentially causing portions of the dike to collapse. Collapse of a dike or a portion of a dike can release the contents of the pond.

WEATHER AND WAVE PROTECTION

Dikes are also called berms.

Earthen *dikes* that are not protected will erode over time. Damage from heavy rain or snow often begins as barely noticeable channels funneling stormwater into the pond (Figure 6.20). Without intervention, the channels become wider and deeper with each storm event until they destabilize the dike. The most common method of weather erosion protection for large dikes is grass. The roots of the grass help stabilize the dike by holding the soil in place with interlocking roots. Grass is effective for protecting the tops and outer slopes of dikes. For grass covers to be effective, they must cover the dike without large bald areas. Operators should mow and reseed as necessary to maintain healthy cover. A thin layer of gravel may be used instead of grass on the outer slopes and tops of dikes.

The *fetch* is the distance that the wind blows over the water. Bigger ponds have a longer fetch and bigger waves.

Ponds that are greater than 4 ha (10 ac) or are located in windy areas typically have *riprap* along the shoreline to protect the dikes from wave erosion. Riprap is loose stone that is placed over the surface. Examples of riprap along the shoreline of a pond and around an inlet pipe are shown in Figure 6.21. Riprap may be held in place with chicken wire. Riprap should extend from 0.3 m (1 ft) below the minimum water surface to at least 0.3 m (1 ft) above the maximum water surface (Kays, 1986; Zickefoose and Hayes, 1977). Wave height in a pond depends on wind speed and *fetch* length. The longer the fetch, the larger the waves and the more erosion they can cause. The size of the individual stones used in riprap also depends on the fetch length (Kays, 1986). Generally, larger waves require larger stones that are more difficult to dislodge. Riprap varies from river run rocks that are 15 to 20 cm (6 to 8 in.) across to quarry boulders that weigh 7 to

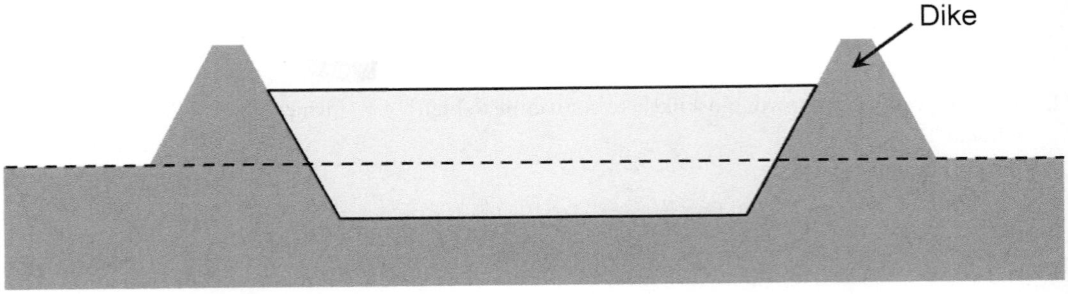

Figure 6.19 Cut and Fill Pond Construction

Figure 6.20 Dike Erosion at the Edge of a Pond (Reprinted with permission by Indigo Water Group)

Figure 6.21 Riprap: (a) Along the Edge of a Pond and (b) Around an Inlet Pipe (Reprinted with permission by Indigo Water Group)

14 kg (15 to 30 lb) apiece. River rock that is all one size can be unstable and should be mixed with smaller material and be carefully placed and pressed into the dirt. Because river rocks are smooth, wave action can loosen them and cause them to slip down steeper-sloped dikes more easily than rougher rock. Broken concrete pavement can also be used for riprap. Operators should inspect the shoreline daily and replace riprap as needed when bare areas develop.

Riprap can make it difficult for operators to remove weeds and control rodent populations. Weeds can be seen among the riprap in Figure 6.21(a). Asphalt, concrete, geotextile fabric, and low grasses can also be used to provide protection from wave action. Low grasses only grow to a certain height and do not require mowing.

VEHICLE PROTECTION

Operators are often required to drive along the tops of dikes as part of conducting daily rounds and inspections. This is particularly true for larger ponds. Well-designed dikes are wide enough at the top to allow vehicle wheels to pass over every part of the surface. Narrow dikes force vehicles to follow the same path each time, which can wear bald areas in grass and gravel and cause ruts to develop (Figure 6.22). Ruts often create runoff erosion problems in areas of high rain intensity. Although operators can't change the width of a dike after it is constructed, they can take actions to minimize damage. With wide dikes, operators should vary driving patterns as much as possible to prevent ruts and bald areas from developing. With narrow dikes, a smaller vehicle such as a golf cart may be used for daily inspections. Operators should avoid driving on dikes after storm events when the ground may be soft and vary the route followed each day to give grass a chance to recover. Gravel may be added to the tops of dikes can help minimize the impact of vehicles.

TREES AND SHRUBS

Trees and shrubs should never be allowed to grow within or through a dike or within a pond. Grass roots help stabilize the soil surface and prevent erosion. Larger roots from trees and shrubs can penetrate the entire width of a dike and provide a path for water to move through the dike. Larger roots can push the dike apart from the inside and damage pond liners; potentially allowing partially treated wastewater to contaminate the groundwater under the pond. Trees and shrubs also provide habitat for rodents and other animals that can dig holes in the dikes.

RODENT PROTECTION

Wastewater ponds attract a variety of wildlife including muskrats, nutria, rabbits, gophers, minks, and badgers. All of these animals live in underground dens. Given an opportunity, they will dig deeply into dikes to make their homes. Dikes should be inspected daily for evidence of burrowing animals.

Figure 6.22 Grassy Dike Showing Vehicle Wear (Reprinted with permission by Paul Krauth P.E.)

Preventing damage from animals begins with making the pond as unfriendly as possible. Ground cover other than short grass should be eliminated, cattails and other plants removed, and riprap be kept in good repair. Animals need food to eat, places to hide from predators, and access to the soft dirt underneath the riprap. If these things are unavailable, animals are less likely to attempt to make their homes in the dike. Wastewater pond operators sometimes joke that the key to operating a pond is to mow the grass, mow the grass, mow the grass, and, when you are done with that, mow the grass. A flat, green lawn with nowhere to hide doesn't make a good home. The pond shown in Figure 6.23 has an abundance of cattails, reeds, and rushes around the outer edge that provide ideal habitat for mosquitos, crawfish, and rodents. This situation should never be allowed to occur.

If burrows are found, filling in the burrow with dirt may be sufficient to deter the animal and send it looking for a new home. Burrows that reappear may require trapping or other means to remove and relocate or exterminate the responsible animal. Contact the local animal control agency for assistance if needed. Muskrats prefer a partially submerged tunnel. Varying pond water depth can discourage muskrat infestations (Crites et al., 2006; Zickefoose and Hayes, 1977). Raising and lowering the water level over a several-week period will discourage them from burrowing into the dike.

LINERS

The primary reason for sealing ponds is to prevent *seepage*. Seepage affects treatment by causing fluctuations in the water depth and can cause pollution of groundwater. Changes in water depth also cause changes in the hydraulic detention time and pond surface area. Operators should be familiar with liner construction because they are often required to make patches and other repairs.

Seepage is water that leaves the pond by passing through the pond liner and dikes.

Traditionally, ponds have been lined with compacted, natural materials such as bentonite and clay, but may also be lined with asphalt, soil cement, or synthetic, waterproof materials. Bentonite is a naturally occurring clay that expands when wet, absorbing several times its dry weight in water. There are several different methods for creating bentonite liners (U.S. EPA, 2011). One method uses a mixture of water and bentonite spread over the pond bottom. The bentonite settles to form a thin liner. Another technique is to apply bentonite over a gravel bed. The bentonite settles into the gravel layer and seals the spaces between pieces of gravel. Another option is to spread the bentonite in a 2.5- to 5-cm (1- to 2-in.) layer and then cover it with a 20- to 30-cm (7- to 12-in.) layer of soil and gravel. Bentonite may also be mixed with sand or some of the existing (native) soil. The bentonite soil mixture is then layered over the bottom and interior dike walls followed by compaction. With all of these methods, the bentonite expands when the pond is filled with water to create a nearly water-tight seal. Bentonite liners can develop permanent cracks if allowed to dry out. Seepage losses through buried bentonite liners are approximately 0.2 to 0.25 $m^3/m^2 \cdot d$ (5 to 6 gpd/sq ft).

Figure 6.23 Cattails and Other Weeds around a Small Pond (Reprinted with permission by Indigo Water Group)

This figure is for thin blankets and represents approximately a 60% improvement over ponds with no lining. Poor quality bentonite deteriorates rapidly in the presence of hard water, and it also tends to erode in the presence of currents or waves. Buried bentonite liners last between 8 and 15 years (U.S. EPA, 2011).

Asphalt, known as pitch in its natural form, was used to waterproof boats long before being used in road surfaces and pond liners. Seepage losses through hot mix asphalt (HMA) liners are typically less than 0.2 $m^3/m^2 \cdot d$ (0.5 gpd/sq ft), making them one of the most watertight liners available. A well-maintained HMA liner should last for at least 15 years. Hot mix asphalt is made from a mixture of approximately 6.5% asphalt cement and 93.5% stone (Schlect, 1990). Pond liners are typically constructed by layering 5 to 7.5 cm (2 to 3 in.) of asphalt over 20 cm (8 in.) of crushed stone. Hot mixed asphalt liners are more flexible than concrete. In an empty pond, rising groundwater can push the liner upward, sometimes causing it to balloon and rupture. *Pressure relief valves* (PRVs) are often included to protect the liner from groundwater damage. When the pond is full, water pressure from above keeps these valves closed. When the pond is empty, pressure from groundwater forces the valves open to allow groundwater into the pond. The amount of water in the pond will equalize so the pressure above and below the valve is the same. Pressure relief valves can be a source of frustration for operators when trying to drain a pond completely. To empty the pond, the pumping rate must be greater than the groundwater inflow rate.

Emergent plants grow in water, but pierce the surface. Examples of emergent plants include cattails, bulrushes, water lilies, and water willow.

Bentonite and asphalt linings are easily damaged by plant roots, crawfish, and rodents. The roots of *emergent* plants can penetrate the pond lining and allow wastewater to seep into groundwater. Operating water depths should be kept at 1.5 m (5 ft) or more to avoid the growth of emergent plants. The slope of the dike going into the pond means there will always be an area around the edge of the pond where the water is less than 1.5-m (5-ft) deep. Operators must remove these plants before they can become established. The pond shown in Figure 6.24 has multiple issues including an abundance of vegetation growing around the edge and a cyanobacteria bloom. Plants growing in the middle of the pond suggest that sludge has been allowed to accumulate and the water depth is now less than 1.5 m (5 ft). An abundance of plant roots has likely compromised the liner.

Impermeable means that water cannot pass through.

Synthetic, plastic liners give the highest level of protection against seepage. They are essentially *impermeable*. These liners are most often used when the underlying native soil can't be improved enough to reduce seepage to acceptable levels. Because of their high cost, they are most often used with smaller ponds. Figure 6.25 shows a three-cell facultative pond with a synthetic liner. Plastic lining material is sold in large sheets that can be combined during installation to cover larger surfaces. The ground surface under the liner must be free of stones and debris that might damage the liner. Like asphalt liners, plastic liners in an empty pond can be pushed up and out of place by rising groundwater. Plastic liners typically have

Figure 6.24 Vegetation and Cyanobacteria Bloom in a Small Pond (Reprinted with permission by Indigo Water Group)

Figure 6.25 Pond System with Synthetic Liner (Reprinted with permission by Indigo Water Group)

vents, openings in the liner, that are covered with additional liner material to create a flap or tent over the opening. When the pond is full of water, the weight of the water holds the flap closed so wastewater cannot enter the ground. When the pond is empty, groundwater can flow through the opening in the liner instead of becoming trapped between the liner and the bottom of the pond. Plastic liners may be covered with a layer of soil or gravel to help protect them or may be left exposed (Figure 6.26). Exposed plastic liners can be damaged by sunlight, animals, equipment, excessive foot traffic, and vandalism. Some liner materials on the market today, such as chlorosulfonated polyethylene, are resistant to ultraviolet light damage. U.S. EPA recommends fencing ponds to keep animals and vandals out (U.S. EPA, 2011). Operators should inspect synthetic liners for cracks and breaks and make any necessary repairs before damage becomes extensive.

INLET, OUTLET, AND TRANSFER CONFIGURATIONS

Historically, influent wastewater entered ponds through a single pipe (Figure 6.26) that was typically located near the center of the pond. Several studies have since shown this is not the best location and that

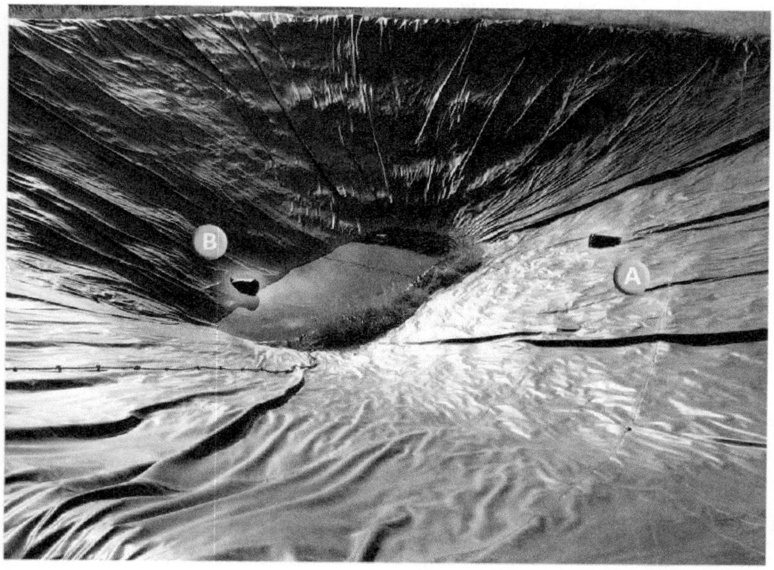

Figure 6.26 Exposed Synthetic Pond Liner with (A) Inlet Pipe and (B) Outlet Pipe (Reprinted with permission by Indigo Water Group)

it can contribute to short-circuiting. In modern pond designs, the inlets and outlets are located as far from one another as possible. Inlets for small ponds may consist of a single pipe, whereas larger ponds are more likely to have inlet diffusers. A single pipe inlet or outlet often consists of two pipes that intersect to form a T-shape (Figure 6.27). Flow may also go the opposite direction; in though the top. This arrangement minimizes disturbance to the sludge blanket and helps prevent sludge from entering the pipe. A diffuser consists of a length of pipe that may be perforated along its length or have multiple outlet ports to help distribute the influent wastewater over a larger area. Distributing the flow helps minimize the chance of short-circuiting and ensures the entire pond volume is being used for treatment. Inlet and outlet openings are typically located some distance above the expected sludge blanket depth, but below the pond surface. Inlets and outlets are generally staggered to further minimize opportunities for short-circuiting. In other words, the inlet to the first pond may be closer to the bottom, but the connecting pipe to the next pond is closer to the top of the pond. This forces the water to move through a larger volume, taking a serpentine path, rather than moving straight across from inlet to outlet. Many ponds, especially larger ponds, are equipped with multiple pipes at different depths to give operators flexibility at both the inlet and outlet of each pond. Operators should inspect inlet and outlet structures daily to check for blockages and debris.

Stratification occurs when layers of cooler and warmer water form in the pond. This is a consequence of the water near the top of the pond gaining or losing heat in response to air temperature and the tendency of the ground and dikes to act as insulation for the water near the bottom of the pond. A pond located near Denver, Colorado, may reach 22 °C (71.6 °F) toward the end of summer, but drop down as low as 0.5 °C (32.9 °F) in winter. The influent wastewater for most treatment systems, on the other hand, is more consistent and generally stays between 10 °C (50 °F) and 20 °C (68 °F) even in very cold climates. The influent wastewater gains heat from showers, dishwashers, and other uses. If the pond is cold and the influent wastewater is warm by comparison, then the influent wastewater will have a tendency to stay near the surface of the pond. If both the inlet and outlet pipes are near the surface, the warm wastewater won't mix very much with the cold pond water. Instead, the warm water will form a current from inlet to outlet and short-circuiting will occur. If the inlet pipe is near the bottom and the outlet pipe is near the top, the water is forced to mix on its way through the pond. In the summer, the incoming wastewater may be cooler than the pond water and tends to sink to the bottom.

Transfer and outlet structures between ponds and at the final effluent generally allow the water level to be lowered at a rate of less than 300 mm/wk (1 ft/wk) when the system is receiving normal flow. Pond depth (Middlebrooks et al., 1982) is typically controlled with manhole-type structures containing either valved piping, adjustable stop log-type overflow gates, or weir gates. *Gate valves* are often used in pipes between ponds. These valves are designed to be operated in the full open or full closed position. They are often used to isolate equipment and processes. Gate valves may be buried deep within dikes inside of small enclosures called valve boxes. Valve boxes are usually just larger than the valve housed and are connected to the ground surface by a long, narrow pipe. They may be made from concrete, fiberglass, plastic, or metal. The boxes are topped with a small, usually metal, lid. A metal lid enables an operator to locate the valve box with a metal detector if it becomes covered with dirt or vegetation and can't be easily seen. To open or close the valve, an operator lowers a valve key into the box and connects it to the *operating nut* on the top of the valve. Valve keys are similar in appearance to a tire iron, but much longer and thicker.

Gate valves have a sliding plate or gate that can be lowered to fill the entire opening or be raised completely out of the pipe. These valves are less likely to clog than other types of valves.

An *operating nut* is the piece on the top of the valve that is connected to the mechanism inside the valve. Look at a water spigot on an exterior faucet. The round piece that one grabs onto and turns to open and close the valve is called the hand wheel. The metal rod that comes up through the valve into the center of the hand wheel is the valve stem and the top of the valve stem is the operating nut.

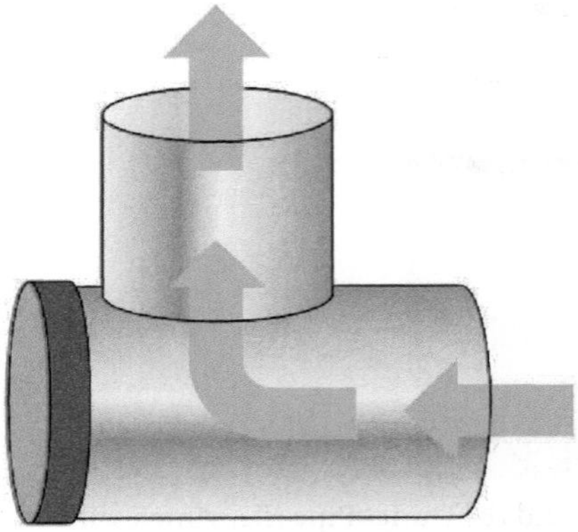

Figure 6.27 Inlet, Outlet, or Transfer Pipe Arrangement (Reprinted courtesy of Indigo Water Group)

Stop logs, also called flashboard risers, allow operators to increase or decrease the water level in the pond by adjusting the height of the effluent weir. The effluent weir height, in turn, controls the water level in the pond. Stop logs are typically located in a manhole or outlet structure immediately upstream of the pond outlet pipe as shown in Figure 6.28. Stop logs are boards that are cut to fit together by overlapping the two edges of the boards. The fit is sometimes called a tongue and groove. The overlap helps to form a nearly watertight seal. Some amount of seepage between boards is acceptable since the seeped water joins the water passing over the top of the stop log assembly. Stop logs are now available in wood, fiberglass, resin composites, and stainless steel. The stop logs fit inside channels set within the outlet structure. Water level in the pond is increased or decreased by adding or removing individual stop logs, respectively, from the outlet structure. Stop logs give operators coarse control over water level since each stop log has a set thickness. For example, if an operator wants to lower the pond level by 18 cm (7 in.) he or she will need to remove some combination of stop logs whose thickness adds up to 18 cm (7 in.). The desired combination of stop log widths may not always be available.

Weir gates have replaced stop logs in many facilities because they offer finer control and are less likely to become stuck. Weir gates, like the one shown in Figure 6.29, have a rectangular gate (A) that moves up and down within a channel (B). The top of the gate is flat and level. Water flows over the top of the gate. The height of the gate is adjusted using a hand wheel (C). Notice that a single hand wheel operates both sides of the gate. This ensures that the weir remains level and that the gate does not become stuck in its channel. The sight glasses (D) show the height of the gate stems and enable an operator to see from a distance where the gate opening is set. Unlike stop logs, an operator can choose any water level desired that is within the range of the gate. Weir gates may also be used for flow measurement when combined with flow measurement devices such as ultrasonic level indicators.

Pipes and other structures that penetrate through dikes are typically installed in a trench with a layer of crushed rock or gravel at the bottom called bedding. The gravel helps ensure that the pipe is level. The gravel creates a path for water to move easily through the dike, potentially causing erosion within the dike. To prevent this type of seepage, seepage collars are placed around each dike penetration. A seepage collar consists of a flat plate that extends outward in all directions from the pipe or structure. After installation, they are similar in appearance to a piece of paper (seepage collar) with a pencil (pipe) poked through the center. Seepage collars typically extend at least 0.7 m (2 ft) from the pipe or other structure and reach well below any gravel bedding. They may be constructed from rubber, concrete, corrugated metal, foam, or other materials.

BAFFLING

Individual ponds or cells are often subdivided into smaller compartments by adding baffles. Baffles help reduce short-circuiting by forcing the water to follow a path through the pond. Baffles may be constructed

Figure 6.28 Stop Log or Flashboard Riser Assembly (**https://webpages.uidaho.edu/larc380/new380/pages/qualityWetlands3.html**)

Figure 6.29 Weir Gate (Courtesy of RW Gate Company)

A *bollard* is a short, thick post. Bollards are often made from a partially buried piece of pipe filled with concrete.

from earth, stone, concrete, or synthetic materials. Earth and stone baffles are similar in appearance to the outer dikes of the pond. Baffles made from synthetic materials are more like thin curtains that hang down into the pond. The upper edge of the curtain contains a series of sealed compartments, each containing a floatation device. Picture a bunch of pillow cases sewn together, end to end, with a Styrofoam pillow or pool noodle inside of each one. Each end of the top edge of the curtain is tethered to a *bollard* or other structure to hold it in place. The upper portion of a baffle curtain can be seen in Figure 6.30 highlighted in green. The bottom of the curtain is weighted down with chains or some other type of ballast to keep the curtain stretched between the top and bottom of the pond. Synthetic baffle curtains may extend over the entire width or length of the pond. Openings in the curtain, called windows, allow water to flow from one side of the curtain to the other.

Baffle curtains can be damaged by high winds and should be kept well away from any mechanical aeration devices. Operators must ensure that baffle curtains and aeration devices are securely moored to prevent damage to both. Daily checks are recommended.

POND RECIRCULATION AND CONFIGURATION

Oxygen demand in a facultative pond will be highest in the primary cell. Pond systems are often equipped with recirculation pumps that allow operators to transfer water from the outlet of one pond back to the inlet of another pond (Figure 6.31). Ponds that follow the primary pond tend to have less BOD_5, lower TSS, greater light penetration, more algae, and more DO. Recirculation transfers DO back to the primary pond or ponds where it is needed most. Recirculation can also help to increase the flow rate through the ponds, which aids in mixing. Note that recirculation only increases the speed of the water as it moves through the process, but does not affect the overall treatment time. In other words, increasing recirculation will not cause water to leave the treatment system faster.

MECHANICAL AERATION DEVICES

A wide variety of mechanical aeration devices are used in facultative ponds, including several different types of diffused aeration and floating aeration devices. Some floating aerators commonly used with pond systems include high-speed aerators and aspirating aerators.

Figure 6.30 Top of Baffle Curtain (Reprinted with permission by Indigo Water Group)

HIGH-SPEED AERATORS

High-speed aerators are typically mounted on floats held in place by *guy wires* anchored at the edge of the pond. Power is brought to the aerator by a cable that can be submerged in the wastewater. The motor is connected directly to the mixer *impeller*, both on a *vertical axis*. The impeller splashes the wastewater into the air, which causes oxygen to move from the air into the water droplets. Aeration is controlled by an operator-adjusted timer depending on daily or seasonal flow and load variation. The float rises and falls with the liquid depth in the pond. The submergence of the impeller is relatively constant. Operation of these units must be carefully monitored during cold weather, as ice can result in equipment damage.

A *guy wire* is a cable that has been pulled tight to help support a piece of equipment.

The *impeller* is the rotating end of the mixer.

Equipment components that are aligned vertically are stacked one above the other. Horizontally aligned pieces of equipment are side-by-side.

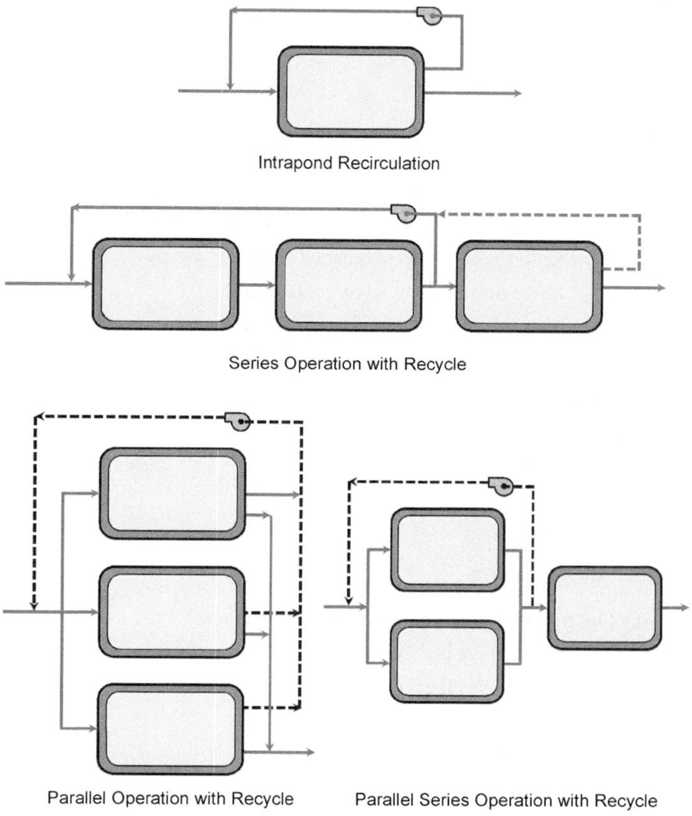

Figure 6.31 Intrapond Recirculation Patterns (U.S. EPA, 1983)

ASPIRATING AERATORS

Aspirating aerators are available in two different configurations. An example of the first configuration is shown in Figure 6.18. It consists of a shaft mounted inside a long tube. The top part of the shaft is connected to a motor above the surface of the pond. The bottom part of the shaft is connected to a submerged impeller. At the top of the tube, close to the motor, there are openings called air intake ports. When the impeller rotates, water is pulled down through the tube. This creates a low-pressure zone inside the tube that draws air in through the air intake ports. The air becomes entrained, or held within, the pumped water. Alternatively, the device may use a submersible pump attached to a vertical air tube that is open to the atmosphere.

STATIC TUBE AERATORS

Static tube aerators use a blower to generate pressurized air. The air moves from the blower to the bottom of the pond through a long pipe called an aeration header. The aeration header has openings (orifices) spaced at regular intervals along its length. Above each orifice is a tube, about 0.30 m (12 in.) in diameter and 0.75 m (30 in.) in height. Inside each tube are static mixers (plates or tines) that break the air into smaller bubbles as it rises. These are typically coarse-bubble systems.

TEST YOUR KNOWLEDGE

1. **Calculated detention times and actual detention times are nearly identical for pond systems.**
   - ☐ True
   - ☐ False

2. **Dike slopes are steeper on the outside of the pond than on the inside of the pond.**
   - ☐ True
   - ☐ False

3. **In general, larger ponds require larger rocks for shore protection.**
   - ☐ True
   - ☐ False

4. **An operator notices several large bald patches on the top of a dike where the grass no longer grows. It is almost winter and heavy snows are likely. The operator should**
   a. Mow the surrounding grass and reseed.
   b. Place a layer of gravel over the bald areas.
   c. Wait until spring to reseed the grass.
   d. Postpone further dike inspections until spring.

5. **Vegetation should be removed from the shoreline and grass kept neatly trimmed to**
   a. Maintain the appearance of the facility.
   b. Protect the pond liner and prevent leaks.
   c. Make the pond inhospitable to burrowing animals.
   d. Increase mosquito and midge larvae to feed fish.

6. **Match the term to its definition.**

1. Freeboard	a. Plant that grows in shallow water
2. Riprap	b. Clay mineral used for lining ponds
3. Fetch	c. Distance from top of water to top of dike
4. Emergent plant	d. Water cannot pass through
5. Bentonite	e. Distance wind travels over the water
6. PRV	f. Loose stone placed to prevent erosion
7. Impermeable	g. Prevents ballooning of pond liner

7. **Riprap should extend at least _______ m (ft) above and below the minimum and maximum water levels.**
   a. 0.3 m (1 ft)
   b. 0.7 m (2 ft)
   c. 1.0 m (3.3 ft)
   d. 1.3 m (4.3 ft)

8. **When driving vehicles on the top of pond dikes, operators should ________________ to minimize erosion.**
   a. Keep tires within established ruts
   b. Drive on wet dikes to recompact soil
   c. Minimize use of gravel on dike sides
   d. Vary driving patterns

9. **Dikes should be inspected at least ________________ for evidence of burrowing animals.**
   a. Daily
   b. Weekly
   c. Monthly
   d. Quarterly

10. **The primary reason for lining a pond is to**
    a. Discourage burrowing
    b. Reduce weed growth
    c. Prevent seepage
    d. Eliminate mosquitos

11. An operator needs to empty a pond for repairs. They calculate the volume of the pond and determine that their dewatering pump should be able to completely empty the pond in 9 days. The water level dropped rapidly in the beginning, but after two weeks, it seems the pump isn't moving any water at all. The water level just won't go below 0.7 m (2 ft). What is the most likely cause?
    a. Water is passing through the dike from the adjacent pond.
    b. Groundwater is entering the pond through an open PRV.
    c. Area of pond increases with depth, so it appears the pumping rate slowed.
    d. Inlet valve is in the full open position.

12. A facultative pond has cattails growing up in the middle of the pond. What must be true?
    a. Mosquito breeding ground has been eliminated.
    b. There are tears in the plastic liner.
    c. The bentonite clay liner has been compromised.
    d. Water depth at that location is less than 1.5 m (5 ft).

13. This type of liner provides the highest level of protection for groundwater.
    a. Bentonite
    b. Natural soil
    c. Hot mix asphalt
    d. Synthetic, plastic

14. An inlet diffuser on a pond
    a. Is typically only perforated at one end
    b. Distributes the influent flow across the pond
    c. Requires a riser pipe or T-structure
    d. Would be located at the center of a newly constructed pond

15. A facultative pond has several inlet pipes at different locations and depths. It is winter and the water temperature within the pond is 5 °C (41 °F) and the influent wastewater temperature is 15 °C (59 °F). To minimize the possibility of short-circuiting, which inlet pipe should be used?
    a. Inlet near the center of the pond across from the outlet
    b. Inlet nearest the top of the water surface and furthest from the outlet
    c. Inlet closest to the pond bottom and furthest from the outlet
    d. Inlet at the bottom of the pond and near the center

16. Pond outlet structures often have two pipes put together to form a T-shape to
    a. Prevent settled sludge from being drawn into the pipe
    b. Eliminate thermal stratification of the pond
    c. Reduce short-circuiting by decreasing mixing
    d. Provide operational flexibility to the operator

17. This device is placed around a pipe penetrating a dike to prevent water from traveling from one side of the dike to the other.
    a. Bollard
    b. Bedding wall
    c. PRV
    d. Seepage collar

Process Variables

Process variables important in wastewater pond systems include: organic loading rate, dissolved oxygen, hydraulic detention time, pH, alkalinity, temperature, and percent removal.

ORGANIC LOADING RATE

The organic loading rate is the mass of BOD_5 (kilograms or pounds) applied to the first pond in the series divided by the surface area (hectares or acres) of the first pond. One hectare is equivalent to 10 000 m². One acre is equivalent to 43 560 sq ft. Acceptable organic loading rates for different types of pond systems are given in Table 6.2. The organic loading rate may be calculated using formulas 6.1 through 6.5. Use the surface area of all the ponds in the system to calculate the overall system loading rate. To find the loading rate to the primary pond(s), only use the surface area for ponds receiving raw influent wastewater.

$$\text{Organic Loading Rate} = \frac{\text{Mass of BOD}_5\text{ applied (kg or lb)}}{\text{Surface Area of Pond(s) (ha or ac)}} \qquad (6.1)$$

The organic loading rate may also be calculated using carbonaceous BOD_5 or chemical oxygen demand.

Mass may be calculated as either kilograms or pounds.

In International Standard units for calculating mass:

$$\text{kg/d} = \frac{\left[\left(\text{Concentration, }\frac{\text{mg}}{\text{L}}\right)\left(\text{Flow, }\frac{\text{m}^3}{\text{d}}\right)\right]}{1000} \qquad (6.2)$$

Alternatively, use dimensional analysis:

$$\frac{X\text{ mg}}{\text{L}}\left|\frac{1\text{ g}}{1000\text{ mg}}\right|\left|\frac{1\text{ kg}}{1000\text{ g}}\right|\left|\frac{1000\text{ L}}{1\text{ m}^3}\right|\left|\frac{X\text{ m}^3}{\text{d}}\right| = \text{kg/d} \tag{6.3}$$

In U.S. customary units for calculating mass:

$$\frac{\text{lb}}{\text{d}} = (\text{Concentration, mg/L})(\text{Flow, mgd})\left(8.34\frac{\text{lb}}{\text{mil. gal}}\right) \tag{6.4}$$

Alternatively, use dimensional analysis:

The "X"s shown in the dimensional analysis equations indicate unknown numbers. These numbers come from the problem being solved. Formulas for area are given in the front of the book in the ABC formula sheet.

$$\frac{X\text{ mg}}{\text{L}}\left|\frac{1\text{ g}}{1000\text{ mg}}\right|\left|\frac{1\text{ kg}}{1000\text{ g}}\right|\left|\frac{2.204\text{ lb}}{1\text{ kg}}\right|\left|\frac{3.785\text{ L}}{1\text{ gal}}\right|\left|\frac{1\,000\,000\text{ gal}}{1\text{ mil. gal}}\right|\left|\frac{X\text{ mil. gal}}{\text{d}}\right| = \text{lb/d} \tag{6.5}$$

Example Calculations for Organic Loading Rate

A naturally aerated facultative pond system contains three ponds. The ponds are currently being operated in series. All three ponds are the same size: 250-m (820-ft) wide × 750-m (2461-ft) long × 2.5-m (8-ft) deep. The influent flow averages 4.5 ML/d (4500 m³/d) (1.19 mgd) and has a BOD_5 concentration of 200 mg/L. Calculate the organic loading rate to the primary pond. Compare your answer to the design criteria in Table 6.2 and in the design criteria section. Should the pond be switched to parallel operation if the average air temperature drops below 0 °C (32 °F)?

In International Standard units:

Step 1—Find the surface area of the pond using the equation for area given in the formula sheet in the front of the book.

$$\text{Area} = \text{Length} \times \text{Width}$$
$$\text{Area} = 250\text{ m} \times 750\text{ m}$$
$$\text{Area} = 187\,500\text{ m}^2$$

Step 2—Organic loading rates for ponds are expressed as kg/ha·d. Currently, the area is in square meters. Convert square meters to hectares.

$$\frac{187\,500\text{ m}^2}{}\left|\frac{1\text{ ha}}{10\,000\text{ m}^2}\right| = 18.75\text{ ha}$$

Step 3—Find the mass of BOD_5 entering the primary pond using formula 6.2 or 6.3.

$$\text{kg/d} = \frac{\left[\left(\text{Concentration, }\frac{\text{mg}}{\text{L}}\right)\left(\text{Flow, }\frac{\text{m}^3}{\text{d}}\right)\right]}{1000}$$

$$\text{kg/d} = \frac{\left[\left(200\frac{\text{mg}}{\text{L}}\right)\left(4500\frac{\text{m}^3}{\text{d}}\right)\right]}{1000}$$

$$\text{kg/d} = 900$$

Step 4—Find the organic loading rate using formula 6.1.

$$\text{Organic Loading Rate} = \frac{\text{Mass of BOD}_5\text{ applied (kg or lb)}}{\text{Surface Area of First Pond(s) (hectares or acres)}}$$

$$\text{Organic Loading Rate} = \frac{900\text{ kg/d BOD}_5}{18.75\text{ ha}}$$

$$\text{Organic Loading Rate} = 48\text{ kg/ha}\cdot\text{d}$$

The recommended loading rate for primary ponds operating with air temperatures below 0 °C (32 °F) is 40 kg/ha·d. The calculated loading rate is higher than the recommended loading rate. The operator should switch from series to parallel operation. The switch will double the surface area (two primary ponds) and reduce the organic loading rate to 24 kg/ha·d.

In U.S. customary units:

Step 1—Find the surface area of the pond using the equation for area given in the formula sheet in the front of the book.

$$\text{Area} = \text{Length} \times \text{Width}$$

$$\text{Area} = 2461\text{ ft} \times 820\text{ ft}$$

$$\text{Area} = 2\,018\,020\text{ sq ft}$$

Step 2—Organic loading rates for ponds are expressed as lb/ac/d. Currently, the area is in square feet. Convert square feet to acres.

$$\frac{2\,018\,020\text{ sq ft}}{}\left|\frac{1\text{ ac}}{43\,560\text{ sq ft}}\right| = 46.3\text{ ac}$$

Step 3—Find the mass of BOD_5 entering the primary pond using formula 6.4 or 6.5.

$$\frac{\text{lb}}{\text{d}} = (\text{Concentration, mg/L})(\text{Flow, mgd})\left(8.34\frac{\text{lb}}{\text{mil. gal}}\right)$$

$$\frac{\text{lb}}{\text{d}} = (200\text{ mg/L})(1.2\text{ mgd})\left(8.34\frac{\text{lb}}{\text{mil. gal}}\right)$$

$$\frac{\text{lb}}{\text{d}} = 2001.6$$

Step 4—Find the organic loading rate using formula 6.1.

$$\text{Organic Loading Rate} = \frac{\text{Mass of BOD}_5\text{ applied (kg or lb)}}{\text{Surface Area of First Pond(s) (ha or ac)}}$$

$$\text{Organic Loading Rate} = \frac{2001.6\text{ lb/d BOD}_5}{46.3\text{ ac}}$$

$$\text{Organic Loading Rate} = 43.2\text{ lb/ac/d}$$

The recommended loading rate for primary ponds operating with air temperatures below 0 °C (32 °F) is 35 lb/ac/d. The calculated loading rate is higher than the recommended loading rate. The operator should switch from series to parallel operation. The switch will double the surface area (two primary ponds) and reduce the organic loading rate to 21.6 lb/ac/d.

OXYGEN

Bacteria use oxygen to convert BOD_5 into new bacteria, carbon dioxide, and water. In naturally aerated facultative ponds, most of the dissolved oxygen comes from algae during photosynthesis. Mechanically aerated ponds obtain most of their oxygen from the mechanical aerators. How much oxygen can be added to a pond depends on three factors: temperature, barometric pressure, and salinity. As water temperature increases, the amount of dissolved gas the water can hold decreases. Think about how a can of soda behaves when it is cold versus when it is hot. No one minds opening a cold can of soda directly from the refrigerator. Some carbon dioxide gas is released, but not too much. If the same can of soda is opened after it has been sitting in a warm car for a few hours, the rush of escaping gas turns the soda into a fountain. Table 6.4 gives a few examples of the saturation concentration for oxygen in clean water at sea level. As barometric pressure decreases, the amount of dissolved gas the water can hold also decreases. Barometric pressure is the weight of the atmosphere over a particular location. Small changes in barometric pressure occur when the weather changes. When the weatherman talks about high and low pressure zones on the map, they are talking about barometric pressure. Barometric pressure also changes with elevation. The higher you go, the less atmosphere there is above you. Barometric pressure is lower in Denver than it is in San Francisco. Less pressure on the liquid also means there is less pressure holding dissolved gases in. Table 6.5 gives a few examples of how barometric pressure affects oxygen solubility. Salinity also has an effect. As salinity increases, the amount of dissolved gas the water can hold decreases.

Before going further, two terms must be defined: saturated and supersaturated. A water sample that is saturated with DO contains the highest concentration the water can hold for that temperature, barometric pressure, and salinity. A water sample that is supersaturated contains more DO than it should be able to hold. If the supersaturated water is stirred or shaken, the excess oxygen will be forced out of the water and DO concentration will decrease. Supersaturated conditions can occur in naturally aerated facultative ponds, especially in the late afternoon, when algae produce an abundance of oxygen.

Dissolved oxygen concentration can reach 200 to 300% of saturation (15 to 25 mg/L) from algae photosynthesis by mid-afternoon due to algal photosynthesis. Mechanical aeration devices both mix and aerate the wastewater, which releases excess oxygen and other dissolved gases to the atmosphere. Operating mechanical aeration devices when the pond is supersaturated in oxygen due to algae may strip oxygen to the atmosphere during the daytime. A quick calculation can determine whether a pond is supersaturated with oxygen. Three pieces of information will be needed: dissolved oxygen concentration in the pond, system elevation, and water temperature.

Example Calculation for Oxygen Saturation

A treatment pond is located in the foothills outside of Salt Lake City at an elevation of 1200 m (3937 ft). By late afternoon, the water temperature in the pond has reach 20 °C (68 °F) and the oxygen concentration is 8.2 mg/L. Calculate the saturation concentration of DO under these conditions. Compare the result to the in-pond oxygen measurement to determine whether oxygen will be removed by turning on the mechanical aerators.

Step 1—Look up the saturation concentration at sea level in Table 6.4. For this facility, the saturation concentration at 20 °C (68 °F) is 9.09 mg/L.

Table 6.4 Dissolved Oxygen Solubility Changes with Temperature at Sea Level

Temperature, °C (°F)	Oxygen Solubility, mg/L
0 (32)	14.62
5 (41)	12.77
10 (50)	11.29
15 (59)	10.08
20 (68)	9.09
25 (77)	8.26

Table 6.5 Effect of Barometric Pressure (Altitude) on Oxygen Solubility

Elevation, m (ft)	Saturation at Sea Level, %
0 (0)	100
300 (984)	96.4
600 (1969)	93.1
900 (2953)	89.8
1200 (3937)	86.6
1500 (4921)	83.5
1800 (5906)	80.4
2100 (6890)	77.5
2400 (7874)	74.6
2700 (8858)	71.9
3100 (10 171)	68.3

Step 2—Find the percent saturation adjustment for the facility elevation in Table 6.5. For an elevation of 1200 m (3937 ft), the percent saturation is 86.6% of sea level.

Step 3—Multiply the saturation concentration at sea level by the percent saturation for barometric pressure.

$$\text{Saturation Concentration, } \frac{\text{mg}}{\text{L}} = (\text{Saturation at Sea Level, mg/L})(\text{Saturation, \%})$$

$$\text{Saturation Concentration, } \frac{\text{mg}}{\text{L}} = (9.09\text{ mg/L})(0.866) \tag{6.6}$$

$$\text{Saturation Concentration, } \frac{\text{mg}}{\text{L}} = 7.87$$

The saturation concentration for the operating temperature and elevation is lower than the DO concentration in the pond. The pond is supersaturated for oxygen. Turning on mechanical aerators at this time will cause the DO concentration to decrease.

HYDRAULIC DETENTION TIME

Hydraulic detention time is the amount of time that the water remains in the treatment process. The required system hydraulic detention time is dictated by the disinfection method used (if any). When natural disinfection is used (no additional disinfection methods used), most regulatory agencies require a minimum of 180 days. When additional disinfection methods are used, then the total system detention time can be shorter and based on the minimum pond temperature expected during the year. Required system detention time ranges from 5 to 180 days, depending on temperature. Although the calculation is straightforward, the actual detention time is often 50 to 60% of the calculated detention time resulting from incomplete mixing. An example of the hydraulic detention time calculation may be found in Chapter 4—Primary Treatment.

Hydraulic detention time (HDT) is the same as hydraulic retention time (HRT). Engineers tend to use HRT, whereas the ABC certification exams and operators tend to use HDT.

In International Standard units:

$$\text{Detention Time, days} = \frac{\text{Volume, m}^3}{\text{Flow, m}^3\text{/d}} \tag{6.7}$$

In U.S. customary units:

$$\text{Detention Time, days} = \frac{\text{Volume, gal}}{\text{Flow, mgd}} \tag{6.8}$$

More detailed explanations of alkalinity and pH are given in Chapter 2.

ALKALINITY AND pH

Photosynthesis and respiration by algae can cause large swings in pH and alkalinity. pH measures the concentration of hydrogen ions. Alkalinity measures buffering capacity. Alkalinity in wastewater pond systems is composed of the following ions: carbonate (CO_3^{-2}), bicarbonate (HCO_3^-), hydroxides (OH^-), and phosphates. Carbon dioxide (CO_2), which is taken up by algae during photosynthesis and released during respiration, is an acid. During the day, algae remove carbon dioxide from the water, lower the amount of acid, and increase alkalinity. At night, algae release carbon dioxide, increase the amount of acid, and decrease alkalinity. These relationships were shown earlier in the chapter in Figure 6.6.

Figure 6.32 shows a typical relationship between pH, CO_2, HCO_3^-, CO_3^{-2}, and OH^- concentrations (Sawyer et al., 1994). For this system, the total alkalinity is 100 mg/L as $CaCO_3$. When CO_2 concentrations increase, pH falls. When CO_2 concentrations decrease, pH increases.

The pH increases during the day and decreases at night. As long as the amount of carbon dioxide produced and consumed remains about the same, the pH will remain within a fairly narrow range as shown in Figure 6.33. In the northern hemisphere, the days get longer during the summer. Longer days promote more photosynthesis and less respiration. Over time, more and more carbon dioxide is pulled out of the water, alkalinity increases, and pH rises from day to day. It isn't unusual for facultative ponds to see pH values over 9.5 on a summer afternoon. Operators should keep close watch on pH to ensure that it does not exceed discharge permit limits. Many pond systems feed dilute sulfuric acid to reduce pond pH immediately prior to discharge.

PERCENT REMOVAL

Percent removal is useful for assessing pond performance and for discharge permit compliance. Composite samples give information about average performance. Grab samples provide a quick snapshot of how the pond is performing at the moment the sample is collected. Percent removal is calculated according to the following formula:

$$\text{Removal, \%} = \left(\frac{\text{In} - \text{Out}}{\text{In}}\right) \times 100 \tag{6.9}$$

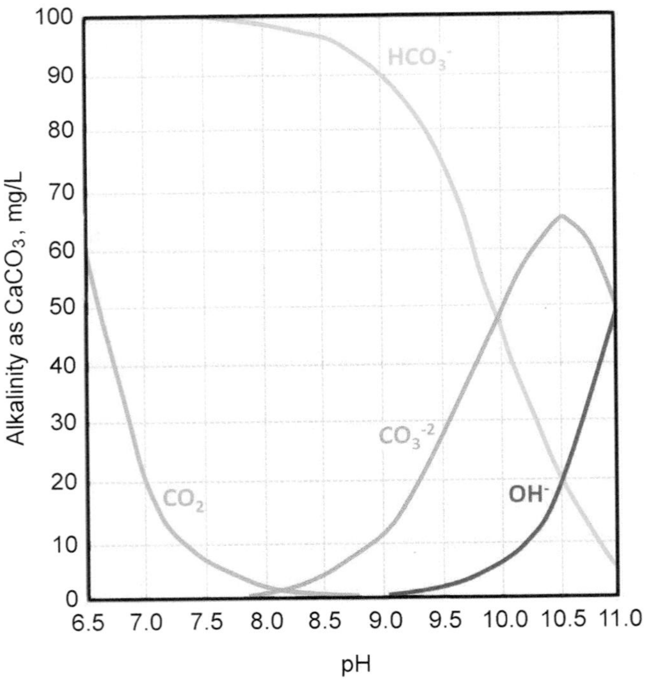

Figure 6.32 Relationship between Alkalinity and Carbon Dioxide Concentration (Sawyer et al., 1994. Copyright © 1994 by McGraw-Hill Education, reprinted with permission.)

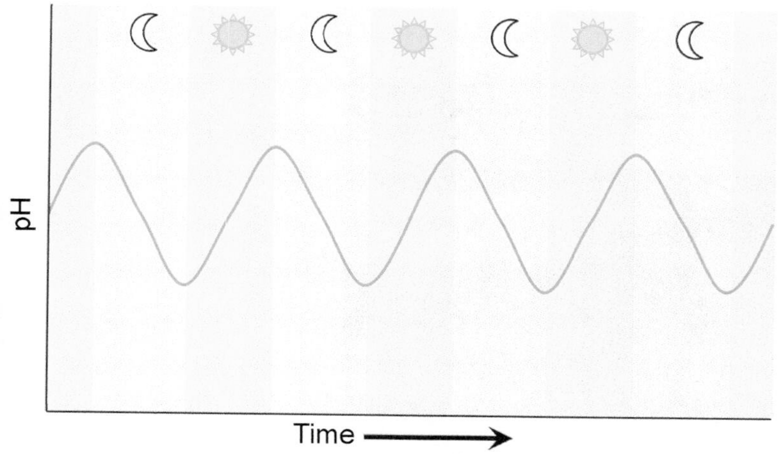

Figure 6.33 **pH Cycle when Light and Dark are Nearly Equal (Reprinted with permission by Indigo Water Group)**

Example Calculation for Percent Removal

The influent wastewater contains 225 mg/L of BOD_5. A sample taken at the end of the primary pond contains 117 mg/L of BOD_5. Find the percent of BOD_5 removed in the primary pond.

$$\text{Removal, \%} = \left(\frac{\text{In} - \text{Out}}{\text{In}}\right) \times 100$$

$$\text{Removal, \%} = \left(\frac{225\frac{\text{mg}}{\text{L}} - 117\frac{\text{mg}}{\text{L}}}{225\frac{\text{mg}}{\text{L}}}\right) \times 100$$

$$\text{Removal, \%} = 48\%$$

SLUDGE ACCUMULATION

Sludge accumulates in the bottoms of ponds. Settled sludge includes grit, inert material, and settleable organics from the influent wastewater as well as biomass and algae grown in the treatment process. The largest amounts of sludge tend to accumulate at the inlet end of the primary pond because this is where most of the rapidly settling particles in the influent wastewater will settle out. All of the ponds in a treatment process, including the polishing pond, will accumulate sludge. Accumulated sludge takes up treatment space in the pond, can come to the surface during turnover events, and produces soluble BOD.

Approximately 60% of the organic material in the sludge blanket is eventually converted into carbon dioxide and methane gas by anaerobic bacteria. Some of the organic material isn't broken down easily and will remain. Inert materials can't be broken down at all. They will also remain. Eventually, the accumulated sludge must be removed. Figure 6.34 shows sludge being removed from a facultative pond that is being permanently taken out of service. Normally, sludge removal is accomplished while the pond is still full of water and may be done while the pond continues to operate. Hoses are lowered into the sludge blanket. Sludge pumps are then used to remove the sludge. Operators should check with their local regulatory agency before removing sludge and comply with all regulatory requirements for testing and disposal.

Sludge accumulation is difficult to estimate accurately. There are many factors that affect how fast sludge accumulates in a pond and how much space the settled sludge will occupy, including:

- BOD_5 loading rate to the pond. Ponds with higher loading rates accumulate sludge faster.
- Quantity of inert material in the influent. Ponds without screening and/or grit removal will accumulate sludge faster.

Figure 6.34 Sludge Removal during Pond Closure (Reprinted with permission by Industrial Service Group)

- Temperature in the sludge blanket. Warmer temperatures result in faster breakdown of organics and slower sludge accumulation.
- Seasonal temperature variations. During the winter, sludge breakdown slows. Ponds in colder climates will accumulate sludge during the winter months that may not be fully broken down when warmer weather returns. Ponds in warm climates break down sludge year-round and accumulation is slower.
- Size of the pond. Larger ponds that are lightly loaded will accumulate the same mass (kilograms or pounds) of sludge per person served as a smaller pond, but it will be spread over a larger area and won't be as deep.
- Concentration of solids in the sludge blanket. Sludge concentrations in ponds can vary considerably from very dilute to very concentrated. Primary clarifier sludge, which is similar in composition to primary pond sludge, typically contains between 1 and 6% total solids (Metcalf and Eddy, Inc. /AECOM, 2014). More concentrated sludge (higher percent solids) takes up less space than less concentrated sludge.

Data from multiple studies on sludge accumulation in facultative ponds show sludge accumulations ranging from 0.04 to 0.18 m^3 (1.4 to 6.4 cu ft) of settled sludge per person served per year (Harris, 2003). Most of these studies did not include the concentration of the settled sludge. Remember, volume isn't mass. How much room the sludge takes up will depend on its mass and concentration. For example, a sludge mass of 1 kg (2.204 lb) will take up twice as much space at 3% total solids than it would at 6% total solids. Depending on the size of the pond, sludge depth may increase between 0.5 and 11 cm (0.2 and 4.3 in.) each year (Harris, 2003). Naturally aerated facultative ponds are lightly loaded and cover large areas. Sludge accumulation in this type of pond may not even be noticeable for many years. These large ponds often accumulate sludge for 10 or 20 years before it needs to be removed. Completely mixed, aerated ponds are much smaller, so the same mass of sludge will form a thicker sludge blanket. In this type of pond, sludge may need to be removed from the settling pond every other year. Measuring the sludge blanket depth at least annually is the most accurate method for determining how much sludge has accumulated in the pond and how fast it is accumulating.

TEST YOUR KNOWLEDGE

1. In general, sludge accumulates faster in ponds located in Michigan than in ponds located in Alabama.
   - ☐ True
   - ☐ False

2. Algae growth causes pond pH to increase during the day.
   - ☐ True
   - ☐ False

3. **The oxygen saturation concentration at 20 °C (68 °F) is lower at sea level than it is in the mountains.**
   - ☐ True
   - ☐ False

4. **A pond system with three cells is currently operated in series with one primary pond. Over the last few weeks, temperatures have decreased and the pond has turned from green to brown. The operator notices red streaks at the surface of the primary pond that persist through late afternoon. What corrective action should the operator take to eliminate the red streaks?**
   a. Add diatoms to consume daphnia.
   b. Lower the pond level in the primary pond.
   c. Switch from series to parallel operation.
   d. Increase loading to the primary pond.

5. **The influent flow to a primary pond contains 250 mg/L of BOD_5. The influent flow rate is 5000 m^3/d (1.32 mgd). How many kilograms (pounds) per day of BOD_5 are entering the primary pond?**
   a. 149 kg/d (330 lb/d)
   b. 1120 kg/d (2468 lb/d)
   c. 1250 kg/d (2752 lb/d)
   d. 1650 kg/d (3637 lb/d)

6. **A wastewater pond system has four ponds that are all the same size. The design criteria for the facility state that the maximum organic loading rate to the primary pond should be kept below 40 kg/ha·d (36 lb/ac/d). If all of the influent BOD_5 load is sent to one pond, the loading rate will be 65 kg/ha·d (58.5 lb/ac/d). How many ponds should receive raw influent wastewater to keep the system within its design parameters?**
   a. 1
   b. 2
   c. 3
   d. 4

7. **The dissolved oxygen concentration will be the highest in a facultative pond when**
   a. The water is cold and sunlight is plentiful
   b. The water is warm and all mechanical aerators are operating
   c. The water is cold and all mechanical aerators are operating
   d. The water is warm and sunlight is plentiful

8. **A primary pond has a surface area of 18.75 ha (46.3 ac). Find the loading rate when the influent flow is 6000 m^3/d (1.6 mgd) and the BOD concentration is 180 mg/L.**
   a. 28.8 kg/ha·d (26.0 lb/ac/d)
   b. 57.6 kg/ha·d (51.9 lb/ac/d)
   c. 115.2 kg/ha·d (103.8 lb/ac/d)
   d. 172.8 kg/ha·d (155.7 lb/ac/d)

9. **This chemical may be used to decrease effluent pH prior to discharge.**
   a. Sodium hydroxide (NaOH)
   b. Calcium carbonate ($CaCO_3$)
   c. Sodium bicarbonate ($NaHCO_3$)
   d. Sulfuric acid (H_2SO_4)

10. **Sludge accumulation in a wastewater pond system will be greatest at the**
    a. Inlet to primary pond
    b. Outlet of primary pond
    c. Inlet to settling pond
    d. Outlet of settling pond

11. **Sludge blanket depth should be checked**
    a. Monthly
    b. Quarterly
    c. Annually
    d. Every 5 years

Process Control

Pond operators have limited control over facultative pond operation. Pond operation essentially consists of the following tasks:

1. **Daily observation of pond appearance.** Most of the following colors and conditions have been addressed throughout this chapter. During the warmer months, primary cells should be highly productive and have a green color indicating high pH and DO. Secondary or final ponds should have a high DO and produce an effluent that meets permit requirements. In the late fall and winter, biological activity will decrease and the color of the ponds will change to brown and then gray. Pond wastewater color indicates the condition of the pond:
   a. Dark sparkling green: good; high pH and DO (Figure 6.10).
   b. Dull green to yellow: pH and DO are decreasing; blue-green algae are becoming established (Figure 6.24).
   c. Gray to black: bad; anaerobic conditions.
   d. Tan to brown: more brown algae than green algae; may also be caused by silt or dike erosion.
   e. Red or pink: presence of purple sulfur bacteria (anaerobic conditions) (Figure 6.14) or red algae (aerobic conditions). Daphnia blooms can also cause a red color (Figure 6.16).

f. Milky appearance: indicates system is approaching or already is septic, typically from being overloaded.

2. **Balancing the influent organic load with the amount of oxygen available.** Maintaining balance often requires operators to either reduce the BOD_5 loading rate to the primary cell or to increase the amount of oxygen available. Primary cell loading rates may be reduced by feeding raw influent wastewater to multiple ponds in parallel operation. If mechanical aerators are available, increasing either the speed or total run time will increase the amount of oxygen unless the pond is already supersaturated with oxygen. Recycling treated wastewater from downstream ponds back to the primary pond inlet also increases oxygen in the primary pond.
3. **Monitoring the influent wastewater** and calculating loading rates for the primary pond and overall system. Switch to parallel operation when loading rates to the primary ponds are high enough to deplete the available oxygen or when loading rates to the primary pond exceed the system design. Series and parallel flow paths for two- and three-pond systems are shown in Figures 6.4 and 6.35, respectively. Symptoms of low dissolved oxygen conditions include the appearance of red daphnia and/or foul odors.
4. **Monitoring the final effluent** to ensure that it meets the limits in the discharge permit for BOD_5, TSS, pH, alkalinity, and other parameters. For controlled discharge and hydrograph release ponds, treated effluent should only be released after testing is complete and water quality is at its best. Controlling TSS concentrations may require operators to control the amount of algae in the pond. Algae control is discussed in the next section.
5. **Monitoring sludge buildup and removing it as needed.** Operators should check sludge depth at least annually. Sludge depth can vary across the pond and is often deeper near the inlet end. U.S. EPA recommends testing blanket depth in many different locations across the pond, working from side to side in a grid pattern.

Be aware that anaerobic ponds, ponds with deep sludge blankets, and facultative ponds that have become anaerobic can produce hazardous levels of hydrogen sulfide gas. If hydrogen sulfide is a concern, atmospheric monitoring should be conducted before and during sludge measurement and removal.

CONTROL OF ALGAE GROWTH

Facultative ponds rely on algae for oxygen production, but algae also contribute to effluent TSS. Excessive algae growth is responsible for most discharge permit violations of pond systems (Richard, 2003). For algae to grow, they must have access to sunlight and nutrients. For algae to accumulate in a pond, the growth rate must exceed the death rate. Control methods for algae either decrease access to sunlight or increase the death rate through predation or other means. If control methods are too successful, there may not be enough algae remaining in naturally aerated ponds to produce the oxygen needed by the bacteria. The pond may become anaerobic. Control methods are often used only in the settling or polishing pond after BOD removal is complete.

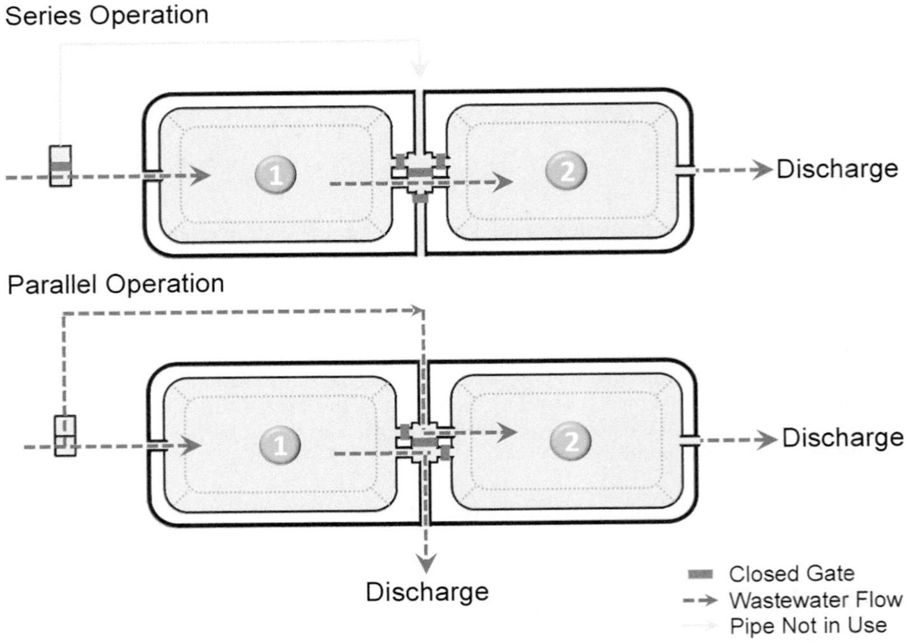

Figure 6.35 Series and Parallel Operation for a Two-Pond System: (1) Primary Pond and (2) Tertiary, Settling, or Polishing Pond

DYES

Dyes have been applied to small ponds to control algae growth; however, U.S. EPA has not approved dyes for use in municipal or industrial wastewater ponds. Aquashade, a mixture of blue and yellow dye, is marketed as a means of controlling algae in backyard garden pools, large business parks, and residential development ponds. It works by blocking those regions of the light spectrum that algae depend upon. The product is registered with U.S. EPA for these uses. However, future approval of the use of dyes in wastewater ponds is unknown.

Dyes are not approved for use in municipal or industrial wastewater ponds.

DUCKWEED COVER

Duckweed is a small aquatic plant with 1 to 3 leaves that are about 3-mm (0.125-in.) long. They grow in dense, floating colonies consisting of hundreds or thousands individual plants. Duckweed is sometimes used to control algae growth in ponds because its leaves block sunlight. These small plants are fast growers that can quickly cover the surface of a pond. Duckweed coverage is rarely complete, especially in large ponds, because the wind tends to push them to one side of the pond. If complete coverage does occur, the pond may become anaerobic.

FABRIC STRUCTURES AND FLOATING COVERS

Pond systems in Colorado and other locations have constructed trellis-like structures to suspend greenhouse fabrics and opaque materials to shade small wastewater ponds. Floating covers and plastic balls can also be used to reduce the amount of sunlight available to algae.

DAPHNIA

Daphnia predation is a form of biological control. Live daphnia can be purchased and added directly to the pond. Daphnia eat algae.

BARLEY STRAW

Barley straw has been used for algae control in ponds since the early 1900s. When placed into a well-aerated pond, the barley straw decays, releasing compounds that inhibit the growth of new algae. The same compounds won't kill existing algae, so it is important to place barley straw early in the spring before algae have had an opportunity to bloom. Barley straw is an algistat, not an algicide. Ideally, barley straw should be put into position as soon as the pond surface has thawed. When applied to water that is 10 °C (50 °F) or colder, it may take 6 to 8 weeks for the straw to decay enough to inhibit algae growth. For water above 20 °C (68 °F), barley straw may be effective in as little as one week (Newman, 2017). For best results, fresh barley straw should be added several weeks before the previous application is used up. A pond may have several applications of barley straw in various stages of decay. Continuous decay and dosing is needed to keep algae under control. Recommended dosage rates vary between 150 and 200 kg/ha (90 and 225 lb/ac) (Newman, 2017; Zhou et al., 2005).

Floating booms of barley straw are easily constructed using plastic mesh or snow fencing (Figure 6.36). The straw is loosely packed within the mesh so water can circulate freely. Zip ties or cord may be used to close the fencing around the straw. Empty plastic containers may be added to the finished boom to help keep the barley at the pond surface near the algae. Once boom construction is complete, a boat is used to drag it into place (Figure 6.37). Additional cord and stakes hold the finished boom in place (Figure 6.38).

ULTRASOUND

Ultrasound devices have been used to control algae in golf course ponds, large residential ponds, water treatment storage ponds, and wastewater ponds. Ultrasound systems work by sending high-frequency sound waves through the water. The sound waves cause irreversible damage to algae cells, chlorophyll loss, and cell death. Although initial results can be observed after 2 weeks of treatment, it may take 6 to 12 weeks to eliminate an algae bloom.

Operation of Wastewater Treatment Ponds

Daily operation of pond systems typically consists of checking pond DO and pH levels followed by dike, liner, and equipment inspections. For many pond systems, daily operation that does not involve compliance sampling or equipment, dike, or liner repairs can be completed in less than 2 hours.

Figure 6.36 Barley Straw Boom Construction (Reprinted with permission by Paul Krauth P.E.)

Figure 6.37 Boom Placement (Reprinted with permission by Paul Krauth P.E.)

Figure 6.38 Barley Straw Boom across a Facultative Pond (Reprinted with permission by Paul Krauth P.E.)

PLACING A POND INTO SERVICE

The optimal season for startup to avoid low temperatures and possible freezing is either spring or summer. If it is necessary to start the system in the late fall or winter, the water level should be brought to 0.75 to 1 m (2.5 to 3 ft) and no discharge allowed until late spring (Zickefoose and Hayes, 1977). The procedure outlined below is generally for facultative ponds and is intended to minimize weed growth and anaerobic conditions during system startup:

Step 1—Fill the primary cell to the 0.6-m (2-ft) level, using fresh water.

Step 2—Start the addition of wastewater to the primary cell.

Step 3—Maintain the pH above 7.5.

Step 4—Check DO daily.

Step 5—Begin introducing fresh water to successive cells when the water level in the primary cell reaches 1 m (3 ft).

Step 6—Add fresh water to a depth of 0.6 m (2 ft).

Step 7—Begin transferring water from the previous cell using only the top draw-off. To avoid transferring settled solids to the next cell, do not draw off water from the bottom 45 cm (18 in.).

Step 8—Do not allow the water level in the previous cell to fall to less than 1 m (3 ft).

Step 9—To equalize the water depths in all the cells, do not discharge until all the cells are filled. If available, use an effluent box with gates or valves to recycle the effluent to any cell in the system to raise the water level. Raise the water level in each cell in 15-cm (6-in.) increments until cells are at their operating depth.

Step 10—Implement the design operating strategy once a healthy growth of organisms has formed in the systems, the appropriate chemical and biological tests have been made, and state permission has been received.

REMOVING A POND FROM SERVICE

Ponds are rarely removed from service. Typically, facultative ponds remain in operation until they are permanently decommissioned. A pond may need to be removed from service to repair or replace a damaged liner. To begin, flow is diverted away from the pond and sent to other ponds by closing off the inlet

Be sure to follow all required safety procedures when filling, draining, or cleaning a tank.

structure. Next, the outlet structure should be opened as far as possible to allow as much water as possible to pass through. A portable pump can then be used to pump the remaining water to the inlet end of another pond. Some ponds have outlet pipes near the bottom of the pond that may be used to drain them.

TEST YOUR KNOWLEDGE

1. **Dyes that block sunlight are approved by U.S. EPA for use in wastewater ponds to control algae.**
   - ☐ True
   - ☐ False

2. **Sludge blanket depth should be monitored at multiple locations throughout all ponds.**
   - ☐ True
   - ☐ False

3. **Algae provide oxygen to aerobic and facultative ponds and**
   a. Consume organic matter
   b. Contribute to effluent TSS
   c. Can be killed with barley straw
   d. Grow symbiotically with duckweed

4. **A three-pond system is in parallel operation, but dissolved oxygen concentrations in the primary cells are still too low. If equipment is available, the oxygen concentration could be increased by**
   a. Reducing surface aerator speed
   b. Increasing recycle flow from the settling pond
   c. Adding barley straw booms
   d. Decreasing surface aerator run time

5. **Duckweed can outcompete algae for**
   a. Nitrogen
   b. Phosphorus
   c. Oxygen
   d. Sunlight

6. **Which of the following algae control methods is an algistat?**
   a. Barley straw
   b. Ultrasound
   c. Daphnia
   d. Rotifers

Data Collection, Sampling, and Analysis

Regular testing of several parameters is essential to properly evaluate the performance of a pond system. System performance data are also required to satisfy the requirements of the WRRF's National Pollutant Discharge Elimination System (NPDES) discharge permit. Table 6.6 presents significant parameters and suggested frequencies of their measurement. Arrangements for competent, well-managed laboratory facilities for regular monitoring of BOD_5, suspended solids, and coliform counts are particularly important for determining the performance of the pond system. Because each pond system will have slightly different requirements, a site-specific standard data sheet should be developed to enter the required data. The process control and regulatory compliance tests used with pond systems are common to other treatment processes and will not be discussed in this chapter.

Detailed descriptions and procedures for laboratory tests commonly used by wastewater reclamation facility operators may be found in *Basic Laboratory Procedures for the Operator-Analyst* (WEF, 2012).

Maintenance

The elements of a maintenance program include preventive maintenance, established written procedures for emergency maintenance, and a system to ensure an adequate supply of essential spare parts. By design, wastewater ponds require less maintenance than a mechanical WRRF, but they cannot be expected to provide treatment without any maintenance. If ponds are subjected to deferred maintenance, they will deteriorate such that costly repairs and possible reconstruction will be required.

There are two types of maintenance, corrective and preventive. Corrective maintenance, or emergency repairs, is reactive. Corrective maintenance deals with issues after they have occurred. Examples of corrective maintenance include: replacing a seized motor or pump bearings, replacing broken drive belts, repairing/replacing valves, and repairing pond dikes. Preventive maintenance, or planned maintenance, is proactive. Preventative maintenance keeps equipment in good repair to prevent problems before they occur. Examples of preventive maintenance include: routine lubrication of motors and pumps, replacing or rebuilding equipment on scheduled intervals, and routine visual inspections. Each system is unique, so specific maintenance schedules and spare parts inventories are dependent on the equipment at the system.

Table 6.6 Recommended Sampling and Analysis for Pond Systems

Parameters	Measurements Frequency (Minimum/Ideal)[a]			
	Unaerated Aerobic Ponds	Facultative Ponds	Anaerobic Ponds	Aerated Ponds
Flow				
Facility influent	(1/day)/continuous	(1/day)/continuous	(1/day)/continuous	(1/day)/continuous
Facility effluent				
Continuous discharge	1/week	1/week	Capability only	Capability only
Short-period discharge	—	1/day	—	—
Recirculation and cell to cell	Not required	Not required	Not required	Not required
pH (in each pond cell)	(1/week)/(1/day)	(1/week)/(1/day)	1/day	(3/week)/(1/day)
Dissolved oxygen (in each pond cell)				
Except during short-period discharge	(1/week)/(1/day)	(1/week)/(1/day)	—	1/day
During short-period discharge	3/day	3/day	—	—
Liquid temperature in each pond cell	(1/week)/(1/day)	(1/week)/(1/day)	1/day	1/day
Liquid level				
Except during short-period discharge	1/week	1/week	—	—
During short-period discharge	1/day	1/day	—	—
Biochemical oxygen demand				
Facility influent	(2/month)/(1/week)	(2/month)/(1/week)	1/week	1/week
Unit effluent	—	—	1/week	1/week
Facility effluent				
Except during short-period discharge	(2/month)/(1/week)	(2/month)/(1/week)	1/week	1/week (after settling tank, if any)
During short-period discharge	—	5/week (each a composite of 3/day)	—	—
Suspended solids				
Facility influent	(2/month)/(1/week)	(2/month)/(1/week)	1/week	1/week
Unit effluent	—	—	1/week	1/week (after settling tank, if any)
Facility effluent				
Except during short-period discharge	1/week	1/week	1/week	1/week (after settling tank, if any)
During short-period discharge	—	5/week (each a composite of 3/day)	—	—
Coliforms (total or fecal)				
Facility influent	1/month	1/month	2/month	2/month
Unit effluent	—	—	—	—
Facility effluent				
Except during short-period discharge	1/week	1/week	1/week	1/week
During short-period discharge	—	—	—	—
Dissolved solids (influent and effluent)	Optional/(1/week)	Optional/(1/week)	(2/month)/(1/week)	Optional/(1/week)
Chlorine residual in effluent (when chlorine is used as specified in effluent discharge permit)				
Automatic control	1/week	1/week	2/week	2/week
Manual control	1/day	1/day	Not recommended	1/day

(continues)

Table 6.6 Recommended Sampling and Analysis for Pond Systems (*Continued*)

Parameters	Measurements Frequency (Minimum/Ideal)[a]			
	Unaerated Aerobic Ponds	Facultative Ponds	Anaerobic Ponds	Aerated Ponds
Sludge level				
In pond	1/(3 months)	1/(3 months)	1 month	1/(3 months) if followed by a settling pond; 1/ month otherwise.
In settling pond following pond	—	—	—	1/month
Nitrogen and phosphorus species	Not required	Not required	Not required	Not required
Volatile acids and alkalinity	—	—	Not ordinarily required	—
Heavy metals and toxic organic compounds	Not ordinarily required	Not ordinarily required	Not ordinarily required	Not ordinarily required
Ice cover, % or area (seasonal requirements)				
Continuous discharge	2/week	2/week	—	2/week
Short-period discharge	1/day	1/day	—	—
Ice thickness (when 100% covered)	1/month, safety permitting	1/month, safety permitting	1/month, safety permitting	—
Weather				
Ambient air temperature	Daily	Daily	Daily	Daily
Cloud cover	—	—	—	—
Precipitation	—	—	—	—
Wind direction	Daily[b]	Daily[b]	Daily[b]	Daily[b]
Wind speed	—	—	—	—
Humidity	—	—	—	—
Descriptive summary	—	—	—	—
Power consumption (cumulative)				
Surface aerators or rotary blowers	—	—	—	Weekly
Pumps and miscellaneous equipment	Monthly	Monthly	Monthly	Monthly
Air flow (for diffused air systems)	—	—	—	Weekly
Air pressure (on discharge header of diffused air system)	—	—	—	Weekly

[a]Sampling frequencies listed are intended to apply to domestic wastes; units receiving substantial proportions of industrial wastewaters may require other parameters or frequencies.
[b]Records of nearby weather stations may be used.

Individual equipment operations and maintenance manuals should be the basis for a system's maintenance schedules. This schedule should be reviewed and revised annually or when new equipment is installed. Refer to Table 6.7 for an operation and maintenance checklist.

ROUTINE PHYSICAL MAINTENANCE

PERIMETER FENCING

Fences should be inspected daily and kept in good repair. Unattended systems should have all gates closed and locked. Warning signs stating the type of system should be legible and in good repair. As these signs are often targeted by vandals, it is recommended to have a large inventory on hand.

ACCESS AND DIKE ROADS

Both access and dike roads should be inspected monthly and regraveled or patched as needed to ensure year-round access. Access roads should be snow plowed on an as-needed basis. If ruts develop, regrading may be necessary.

Table 6.7 Example Operation and Maintenance Checklist (Zickefoose and Hayes, 1977)

Operation and Maintenance	Frequency					
	Daily	Weekly	Monthly	Quarterly	Yearly	As Needed
Facility Survey						
Drive around perimeters of ponds taking note of the following conditions:						
Any buildup of scum on pond surface and discharge outlet boxes		X				
Signs of burrowing animals	X					
Anaerobic conditions: noted by odor and black color, floating sludge, large number of gas bubbles	X					
Water-grown weeds	X					
Evidence of dike erosion		X				
Dike leakage		X				
Fence damage		X				
Ice buildup in winter					X	X
Evidence of short-circuiting	X					
Pretreatment						
Clean inlet and screens, and properly dispose of trash	X					X
Check inlet flow meter and float well	X					
If discharge is once or twice per year, the discharge permit may require observations of the following:						
Odor		X				
Aquatic plant coverage of pond		X				
Pond depth		X				
Dike condition		X				
Flow (influent)		X				
Rainfall (or snowfall)		X				
Note: Each state has requirements for data collected prior to and during discharge that are defined in the pond system discharge permit.						
If discharge is continuous, the discharge permit may require the following information:						
Weather	X					X
Flow	X					X
Condition of all cells	X					X
Depth of all cells	X					X
Pond effluent						
DO and pH grab sample	X					X
Cl^- residual	X					X
BOD_5 & TSS run on composited sampled						X
Microbial tests						X
Cl^- used and remaining, kg (lb)	X					X
Other tests and frequency information will be defined in the individual permit						
Mechanical Equipment						
Check mechanical equipment and perform scheduled preventive maintenance on						
Pump stations:						
Remove debris	X					
Check pump operation		X				
Run emergency generator			X			
Log running times	X					
Clean floats, bubblers, or other control devices		X				
Lubricate						X

(continues)

Table 6.7 Example Operation and Maintenance Checklist (Zickefoose and Hayes, 1977) (*Continued*)

Operation and Maintenance	Frequency					
	Daily	Weekly	Monthly	Quarterly	Yearly	As Needed
Flow measuring devices:						
Check and clean floats, etc.	X					
Verify accuracy			X			
Valves and gates:						
Check to see if set correctly		X				
Open and close to be sure they are operational			X			

To *exercise* a gate or valve, it should be fully opened and fully closed several times. This helps lubricate the mechanism and ensures that it won't rust shut.

TRANSFER STRUCTURES

Most pond systems have valves and or gates to transfer wastewater between ponds (cells). Most gates and valves can/will rust or deteriorate/freeze. Valves and gates should be *exercised* fully weekly. Lubrication should be done based upon manufacturer's recommendations. Leaking slide gates are a very common issue operators face with pond systems. Valves and gates should be locked or the handles removed to minimize any vandalism attempts.

DIKE MAINTENANCE

Dike maintenance was discussed earlier in this chapter and is reiterated here. To minimize erosion on pond embankments, a dense growth of grass should be established and maintained. The grass should be kept neatly mowed on embankments to ensure a dense growth for erosion control and to reduce cover and food for burrowing animals. Small ungulates (sheep and goats) have been used in place of mowing, but they should *not* be used in systems with exposed flexible membrane liners.

Any deep rooting plant or trees should be removed as soon as they are noticed during the monthly dike inspection, as their roots may compromise dike structural integrity. In larger ponds (cells), internal banks are often protected from erosion by stone riprap, concrete, or asphalt liners; these also should be checked monthly.

BURROWING ANIMALS

Catastrophic dike collapse can be caused by burrowing animals. Minks and muskrats usually construct a submerged entrance making identifying their burrows more difficult. Inspect dikes for evidence of burrowing daily. Removal of these animals should be done immediately after noticing them. Trapping or shooting are the most common methods of removal. Note that harvesting any of these animals may be subject to state or local game regulations. Additionally, muskrats use vegetation for food and nest construction. Keeping the inner pond banks clear of weeds and vegetation will discourage their nesting.

MECHANICAL AERATION SYSTEMS

Surface aerators (particularly those on timer or DO controllers) should be started on a weekly basis to ensure operation. The flow pattern should be observed to ensure floating debris has not clogged or partially clogged the aerator. Lubrication should be done in accordance with the manufacturer's recommendations. At that time, the aerator should be cleared of any attached debris. Record equipment run times, amperage, and other parameters as recommended by the manufacturer.

Diffused aerators should have their flow pattern observed on a weekly basis to ensure the aerator is not clogged or ruptured. The valves to the aeration lines (if so equipped) should be exercised through the complete range monthly. Lubrication of the blowers should be done in accordance with the manufacturer's recommendations.

TREATMENT PROCESS MAINTENANCE

VEGETATION CONTROL

Extensive growth of vegetation interferes with treatment by reducing water circulation, reducing/excluding sunlight, and adding to organic loading when the vegetation dies. Vegetation can cause stagnant water, providing ideal breeding areas for insects. These "dead zones" also trap grease and scum resulting in odors.

Aquatic plants are those that grow in water with their roots either under water or in the mud along the shore. Drowning, burning, chemical treatment, and physical removal are all effective aquatic plant controls. A 1.5-m (5-ft) pond depth should eliminate these plants from growing on the pond (cell) bottom. Often, pond water levels are drawn down in the fall, which dries out some of these aquatic plants. Dried plants may be burned away. Be sure to comply with local regulations and obtain necessary permits before conducting a controlled burn. Application of commercial *herbicides* can be effective in controlling aquatic plants. Table 6.8 lists some herbicides used to control aquatic plants. Effects on the pond microbiology and receiving stream should be the first concern when choosing any herbicide. Federal, state, and/or local regulations may apply to any herbicide application. Check with regulatory agencies before application even if the herbicide has been permitted in the past. Be aware that any chemicals added to the treatment process, including herbicides, are often added to the discharge permit. Usage of chemicals today may increase testing and compliance costs in the future. Physical removal is the most common method used. Pulling plants (manually or mechanically) either from the bank or a boat is very effective provided they are pulled out by the roots. If only the leaves are removed, the plants will grow back quickly. The most important factor in controlling vegetation is to keep ahead of the growth. When these plants are controlled at the shoreline, they will be prevented from spreading into the pond where they may have to be removed by boat.

Burning of vegetation, while legal in many states and counties, contributes to air pollution. Ash may also contribute to water pollution. Consider composting solids to produce a reusable and potentially saleable product instead.

Herbicides work because they are toxic to plants. When they are discharged to the environment through final effluent, they can have long-term effects even in extremely low concentrations. They should be used only as a last resort.

SCUM AND FLOATING MATERIALS

Large floating mats of sludge, scum, grease, and algae can become septic, creating odors and insect breeding areas. Mats can form when settled sludge is carried back to the surface by gas bubbles. Cyanobacteria also form floating mats (Figure 6.12). Operators should inspect the pond surface and outlet structures for scum buildup weekly. Agitation of these mats will allow trapped gases to escape and solids to settle to the bottom of the pond. A high-pressure hose from a portable pump from shore or a boat motor can provide the agitation.

Along with these mats, other floating material such as plastics, papers, and leaves, can block sunlight into the pond. Decreased access to sunlight will decrease algae DO production and may affect BOD_5 removal. Accumulations of floating trash need to be physically removed before causing operational problems. Screening at the headworks is the most effective method; however, not all pond systems have influent screens. Skimming from shore or a boat is another method that can be used to remove debris.

INSECTS

Elimination of desirable breeding habitats is the most effective control method for insects. It should include vegetation control around the ponds (cells), as well as removal or breaking up of floating mats or accumulations at the shoreline. Buildings, valve boxes, and other structures should be checked monthly for any insect nests.

Table 6.8 Herbicides Used Around Wastewater Treatment Ponds (Zickefoose and Hayes, 1977)*

Chemical	Marketed By	To Control
Dalapon	Dow	Cattails
Silvex	Dow	Willows and emergent weeds
Copper sulfate	—	Filamentous algae
Endo-thal	Ortho	Suspended weeds
Simazine (use qualified ticket sprayer)	—	Weeds

*These herbicides should only be used as permitted by the appropriate regulatory agencies.

TEST YOUR KNOWLEDGE

1. **An example of preventative maintenance is**
   a. Replacing a broken drive belt
   b. Repairing pond dikes
   c. Routine lubrication of motors
   d. Replacing a seized motor

2. **Valves and gates should be exercised**
   a. By completely opening and closing them several times
   b. During quarterly inspections for surface rust
   c. Any time leakage is observed around the stem
   d. Immediately before placing a pond into service

3. **Burrowing animals must be removed**
   a. As soon as they are noticed
   b. Prior to breeding season
   c. Only if dikes and liners are impacted
   d. In early spring before the pond thaws

4. **Before using any herbicide to control vegetation at a WRRF**
   a. Plants should be cut as short as possible to increase effectiveness.
   b. Burrowing animals should be relocated to other areas.
   c. A test application should be tried on one-quarter of the pond.
   d. Operators should contact their regulatory agency for guidance.

5. **Once a regulatory agency has approved an herbicide for use, it may be used again in the future.**
   ☐ True
   ☐ False

6. **An effective method for resettling floating sludge is to spray it with a high pressure hose.**
   ☐ True
   ☐ False

Troubleshooting

Poor effluent quality may be caused by (Barsom, 1973):

- Overloading;
- Low ambient temperatures;
- Ice formation;
- Toxic materials in influent;
- Nutrient deficiencies;
- Short-circuiting;
- Loss of liquid volume due to deposition, leakage, and evaporation;
- Aeration equipment malfunction;
- Mixing or agitation equipment malfunction;
- Operating levels too deep for light to penetrate the full depth;
- Interference of light penetration by high turbidity, algal mats, duckweed cover, or scum;
- Blockage of light by plant growth on dikes; and
- Groundwater contamination caused by leakage through the bottom or sides.

When not specifically designed for, anaerobic conditions in a pond can cause odor problems. Other possible nuisance problems are

- Foaming and spray problems in aerated ponds,
- Insect propagation, and
- Aesthetics.

In addition, erosion of the dikes, the presence of burrowing animals, and the growth of rooted woody plants can destroy the structural stability of the pond walls, and thereby create a hazardous condition. Table 6.9 presents a troubleshooting guide for ponds that will enable the operator to quickly identify problems and their possible solutions.

Table 6.9 Troubleshooting Guide for Wastewater Treatment Ponds (Zickefoose and Hayes, 1977)

Indicators/Observations	Probable Cause	Solutions
Weeds	Poor circulation/maintenance, insufficient water depth	Pull weeds by hand if new growth. Mow weeds with a sickle bar mower.
		Lower water level to expose weeds, then burn with gas burner.
		Allow the surface to freeze at a low water level, raise the water level and the floating ice will pull the weeds as it rises.
		Increase water depth to above tops of weeds.
		Use riprap; caution: if weeds get started in the riprap, they will be difficult to remove but can be sprayed with acceptable herbicides; get appropriate regulatory agencies' approval to use herbicides.
		To control duckweed, use rakes or push a board with a boat, then physically remove duckweed from pond.
Burrowing animals	Bank conditions that attract animals; high population in area adjacent to ponds	Remove food supply such as cattails and burr reed from ponds and adjacent areas.
		Muskrats prefer a partially submerged tunnel; if the water level is raised, the tunnel will extend upward and if lowered sufficiently, the muskrat may abandon the tunnel completely; they may be discouraged by raising and lowering the level 0.15 to 0.20 (6 to 8 in.) over several weeks; trapping or other methods of animal removal may be used if permitted by all appropriate regulatory agencies.
High weed growth, brush, trees, and other vegetation	Poor maintenance	Periodic mowing is the best method.
		Sow dikes with a mixture of fescue and blue grasses on the shore and short native grasses elsewhere; it is desirable to select a grass that will form a good sod, drive out tall weeds, bind the soil, and outcompete undesirable growth.
		Spray with approved weed control chemicals.
		Some small animals, such as sheep, have been used; may increase fecal coliform, especially to the discharge cell; some regulatory agencies advise against using animals for vegetation control due to concerns for liner and dike integrity.
		Practice rotation grazing to prevent destroying individual species of grasses; an example schedule for rotation grazing in a three-pond system would be to graze each pond area for 2 months over a 6-month grazing season.
Undesirable scum formations	Pond bottom is turning over with sludge floating to the surface; poor circulation and wind action; high amounts of grease and oil in influent will also cause scum	Use rakes, a portable pump to get a water jet, or motor boats to break up scum formations; broken scum usually sinks.
		Any remaining scum should be skimmed and disposed of by burial or hauled to landfill with approval of regulatory agency.

(*continues*)

Table 6.9 Troubleshooting Guide for Wastewater Treatment Ponds (Zickefoose and Hayes, 1977) (*Continued*)

Indicators/Observations	Probable Cause	Solutions
Odors that are a nuisance	The odors are generally the result of overloading, long periods of cloudy weather, poor pond circulation, industrial wastes, or ice melt; use parallel feeding to primary cells to reduce loading; also, weeds and sometimes riprap will serve as areas for organic material accumulation resulting in odors	Apply chemicals such as sodium nitrate, or 1,2-dibromo-2,2-dichloroethyl dimethylphosphate to introduce oxygen; application rate: 5 to 15% sodium nitrate/mil. gal [mil. gal × 3785 = m^3]; repeat at a reduced rate on succeeding days; use 112 kg/ha (100 lb sodium nitrate/ac) for the first day, then 56 kg/ha·d (50 lb/ac/d) thereafter if odors persist; apply in the wake of a motorboat.
		Install supplementary aeration such as floating aerators, caged aerators, or diffused aeration to provide mixing and oxygen; daily trips over the pond area in a motorboat also help; note: stirring the pond may cause odors to be worse for short periods but will reduce total length of odorous period.
		Recirculate pond effluent to the pond influent to provide additional oxygen and to distribute the solids concentration; recirculate on a 1:6 ratio.
		Eliminate septic or high-strength industrial wastes.
Cyanobacteria (blue-green algae)	Blue-green algae is an indication of incomplete treatment, overloading, or poor nutrient balance	Use three applications of a solution of copper sulfate. Note that the use of copper sulfate is no longer acceptable in many regions. Operators should check with their regulatory agencies to obtain approval before use. Most streams and lakes have maximum concentration limits on copper and other heavy metals. The use of copper sulfate can cause a discharge permit violation.
		If the total alkalinity is above 50 mg/L, apply 1200 kg/m^3 (10 lb/mil. gal) of copper sulfate; if alkalinity is less than 50 mg/L, reduce the amount of copper sulfate to 600 kg/m^3 (5 lb/mil. gal). Some states do not approve the use of copper sulfate because it is toxic to certain organisms and fish; also, extensive use of copper sulfate may cause a buildup of copper in the sludge, making sludge disposal a problem.
		Break up algal blooms by motorboat or a portable pump and hose; motorboat motors should be air cooled as algae may plug water-cooled motors.
		Provide shading, increase DO, add plankton and algae-eating fish.
Mosquitoes, midges	Poor circulation and maintenance	Keep pond clear of weeds and allow wave action on bank to prevent mosquitoes from hatching.
		Keep ponds free of scum.
		Stock pond with *Gambusia* (mosquito fish).
		Spray with larvacide as a last resort; check with state regulatory officials for approved chemicals.

(*continues*)

Table 6.9 Troubleshooting Guide for Wastewater Treatment Ponds (Zickefoose and Hayes, 1977) (*Continued*)

Indicators/Observations	Probable Cause	Solutions
High algal suspended solids in pond effluent	Weather or temperature conditions that favor particular population of algae	Draw off effluent from below the surface by use of a good baffling arrangement.
		Use multiple ponds in series.
		Intermittent sand filters and submerged rock filters may be used but will require modification and the services of a consulting engineer.
		Provide shading, increase DO, add plankton and algae-eating fish.
		In some cases, alum dosages of 20 mg/L have been used in final cells, used for intermittent discharge, to improve effluent quality; dosages at or less than this level are not toxic.
Lightly loaded ponds that may produce filamentous algae and penetration	Overdesign, low seasonal flow	Increase the loading by reducing the number of cells in use.
		Use series operation.
A low, continued downward trend in DO	Poor light penetration, low detention time, high BOD loading, or toxic industrial wastes (daytime DO should not drop below 3.0 mg/L during warm months)	Remove weeds, such as duckweed, if covering greater than 40% of the pond.
		Reduce organic loading to primary cells by going to parallel operation.
		Add supplemental aeration (surface aerators, diffusers, or daily operation of a motorboat).
		Add recirculation by using a portable pump to return final effluent to the head works.
		Apply sodium nitrate [see "Odors that are a nuisance" (above) for rate].
		Determine if overload is owing to an industrial source and eliminate it.
Overloading, which results in incomplete treatment of the wastewater; overloading problems can be detected by offensive odors or a yellow, dull green, or gray color; laboratory tests showing low pH, DO, and excessive BOD loading per unit area should be considered	Short-circuiting, industrial wastes, poor design, infiltration, new construction (service area expansion), inadequate treatment, and weather conditions	Bypass the cell and let it rest.
		Use parallel operation.
		Apply recirculation of pond effluent.
		Look at possible short-circuiting.
		Install supplementary aeration equipment.
Decreasing trend in pH; pH should be on the alkaline side, preferably about 8.0 to 8.4	A decreasing pH is followed by a drop in DO as the green algae die off; this is most often caused by overloading, long periods of adverse weather, or higher animals such as Daphnia or rotifers feeding on the algae	Bypass the cell and let it rest.
		Use parallel operation.
		Apply recirculation of pond effluent.
		Check for possible short-circuiting.
		Install supplementary aeration equipment if problem is persistent and results from overloading.
		Look for possible toxic or external causes of algae die off and correct at source.
Odor problems, low DO in parts of the pond, anaerobic conditions, and low pH found by checking values from various parts of the pond and noting them on a plan (drawing) of the pond; differences of 100% to 200% may indicate short-circuiting; after recording the readings for each location, the areas not receiving good circulation are characterized by a low DO and pH	Poor wind action because of trees or poor arrangement of inlet and outlet locations; may also be due to shape of pond, weed growth, or irregular bottom	Cut trees and growth at least 150 m (500 ft) away from pond if in direction of prevailing wind.
		Install baffling around inlet location to improve distribution.
		Add recirculation to improve mixing; provide new inlet-outlet locations including multiple inlets.
		Clean out weeds.
		Fill in irregular bottoms; investigate potential for sludge accumulation in various parts of the pond.

(*continues*)

Table 6.9 Troubleshooting Guide for Wastewater Treatment Ponds (Zickefoose and Hayes, 1977) (*Continued*)

Indicators/Observations	Probable Cause	Solutions
Facultative pond that turned anaerobic, resulting in high BOD, suspended solids, and scum in the effluent in continuous discharge ponds, unpleasant odors, the presence of filamentous bacteria, yellowish-green or gray color, and short-circuiting placid surface, indicating anaerobic conditions	Overloading, short-circuiting, poor operation, or toxic discharges	Change from series to parallel operation to divide load; helpful if conditions exist at a certain time each year and are not persistent.
		Add supplemental aeration if pond is continuously overloaded.
		Change inlets and outlets to eliminate short-circuiting.
		Add recirculation (temporary-use portable pumps) to provide oxygen and mixing.
		In some cases, temporary help can be obtained by adding sodium nitrate at rates suggested for odor control.
		Eliminate sources of toxic discharge.
High BOD concentrations that violate NPDES or other regulatory agency permit requirements; visible dead algae	Short detention times, poor inlet and outlet placement, high organic or hydraulic loads, and possible toxic compounds	Check for collection system infiltration and eliminate at source.
		Use portable pumps to recirculate the water.
		Add new inlet and outlet locations.
		Reduce loads from industrial sources if greater than design level.
		Prevent toxic discharges; convert to parallel operation.
Fluctuating DO, fine pin floc in final cell effluent, and frothing or foaming	Shock loading, overaeration, industrial wastes, and floating ice	Control aeration system by using timeclock to allow operation during high-load periods; monitor DO to set. up schedule for even operation, holding approximately 1 mg/L or more.
		Vary operation of aeration system to obtain solids that flocculate or clump together in the secondary cell but are not torn apart by excessive aeration.
		Locate industrial wastes that may cause foaming or frothing and eliminate or pretreat such wastes as slaughterhouse, milk, and some vegetable wastes.
Ice interfering with the operation of aerated ponds	Floating ice	Operate units continuously during cold weather to prevent freezing damage or remove completely if not a type that will prevent freeze-up.
Odors in anaerobic ponds, caused by hydrogen sulfide (rotten egg) odors or other disagreeable conditions due to sludge in a septic condition	Lack of cover over water surface and insufficient load to have complete activity that eventually forms scum blanket	Use straw cast over the surface or polystyrene planks as a temporary cover until a good surface sludge blanket has formed.
		Recirculate aerobic pond effluent to anaerobic pond.
		Distribute over anaerobic pond by spraying to establish thin layer of aerobic water.
Low pH in anaerobic ponds; a pH below 6.5 accompanied by odors, results from acid bacteria working in the anaerobic condition	Acid formers working faster than methane formers in an acid condition	The pH can be raised by adding a lime slurry of 100 lb of hydrated lime/50 gal (500 kg/200 L) of water at a dosage rate of 1 lb of lime/10 000 gal (120 g/10 000 L) in the pond; the slurry should be mixed while being added; the best place to put in the lime is at the entrance to the pond so that it is well mixed as it enters the pond.
Groundwater contamination	Leakage through bottom or sides	Potential major capital improvements.

Pond Safety

This section was lightly edited from Section 9.5 in U.S. EPA (2011).

Operators and others conducting activities around treatment ponds must proceed with caution and make safety and public health a priority. Treatment ponds must be used for their designed purpose only, not for public recreation. In some areas, treatment ponds are the only sizeable body of water and have attracted people for recreation. Incidents of boating, ice skating, waterfowl hunting, and even swimming in ponds have been reported. Recreational use should be discouraged and safety practices encouraged for several important reasons. The potential for contamination or infection from pathogenic organisms exists when a person comes into contact, especially full body contact, with wastewater in a treatment pond. People can drown in treatment ponds. Clay and synthetic liners become very sticky and slick when water is added. Should a person fall into, or jump into, a pond, the presence of liners can make it extremely difficult to get out. To discourage the use of ponds for recreation, the entire area should be fenced and warning signs displayed.

The same personal hygiene rules apply to ponds as other wastewater treatment processes. Don't eat, drink, or smoke out near the treatment processes. Save those activities for the break room or office. Be sure to thoroughly wash hands before eating. Consider washing your face as well. Wear waterproof gloves such as latex, nitrile, or other rubber when handling samples and performing laboratory tests. Wear safety glasses and steel toed shoes to protect eyes and toes.

CHAPTER SUMMARY

PURPOSE AND FUNCTION	■ Natural treatment system. ■ Three types classified by environmental conditions and the types of biological processes occurring within them: aerobic, anaerobic, and facultative. ■ Facultative ponds are the most common type. ■ Used for secondary treatment. ■ Naturally or mechanically aerated. ■ Mechanically aerated ponds treat a greater mass of BOD_5 in less space than naturally aerated ponds. ■ Continuous, intermittent, or zero discharge. ■ Anaerobic ponds primarily used by industry for pretreatment.
THEORY OF OPERATION—FACULTATIVE PONDS	■ Typically consist of three ponds operated in series or in parallel. ■ Single, two, and multiple pond systems also exist. ■ Treatment improves with more ponds in series. Reduces opportunities for short-circuiting. ■ Primary ponds receive raw influent. ■ Tertiary, settling, or polishing ponds act as clarifiers to settle solids prior to discharging treated wastewater. ■ Uppermost pond layer is aerobic. Bottom of pond contains sludge and is anaerobic. Anoxic conditions may exist in some ponds. ■ Algae produce oxygen during the day, which is used by bacteria to consume BOD_5. ■ Algae consume oxygen and produce CO_2 at night and in low light conditions. ■ Cycles of taking up and releasing CO_2 increase pH, DO, and alkalinity during the day. pH, DO, and alkalinity decrease at night. ■ Aerobic bacteria release nutrients as they break down BOD_5. ■ Anaerobic bacteria convert settleable BOD_5 and settled biomass into CO_2, methane, soluble BOD_5, and nutrients. ■ Some nutrients are incorporated into biomass. Nitrifying bacteria convert ammonia to nitrite and nitrate. Ammonia volatilizes to the atmosphere when pH is high. Most nutrients pass through to the final effluent. ■ Pond colors reflect the dominance of different microorganisms and other environmental conditions.
DESIGN PARAMETERS—FACULTATIVE PONDS	■ Pond size is determined by coldest expected water temperature, BOD_5 load, and availability of oxygen. ■ Ponds treating colder water must be larger than ponds treating warmer water. ■ In naturally aerated ponds, most of the dissolved oxygen is produced by algae. ■ Wind and wave action transfers oxygen from the atmosphere. ■ BOD_5 loading rates are limited for both the entire system and to the primary pond. ■ Switch from series to parallel pond operation when DO concentrations can't be maintained in primary ponds. ■ Parallel operation typically used in winter.
EXPECTED PERFORMANCE—FACULTATIVE PONDS	■ Facultative ponds meet secondary treatment standards for BOD_5. ■ Facultative ponds often discharge TSS in excess of 30 mg/L resulting from algae growth. ■ Many states have alternative discharge limits for TSS for ponds. ■ Spring and fall turnover can result in solids loss, increasing effluent TSS. ■ Nutrient removal is inconsistent. ■ Nitrogen removal is primarily through assimilative uptake and volatilization. ■ Phosphorus removal is primarily through assimilative uptake.

EQUIPMENT AND ROUTINE MAINTENANCE

- Constructed from earth. Cut and fill used to build dikes.
- Length to width ratios vary. 1 : 1 for complete mix. 3 : 1 or greater for plug flow ponds.
- Earthen dikes sloped for stability. Kept at 3 : 1 or flatter to permit mowing. Steeper inside the pond.
- Grass, riprap, and gravel used to prevent erosion.
- Tops of dikes are typically wide enough for a road. Reduce erosion by varying driving habits and by staying off wet, muddy dikes.
- Trees, shrubs, and burrowing animals can compromise dike stability. Remove as soon as they are noticed.
- Keep grass mowed and riprap in good repair to prevent erosion and deter animals.
- Liners prevent wastewater from contaminating groundwater.
- Liners may be natural earth, clay, asphalt, concrete, or synthetic waterproof materials.
- Inlet, outlet, and transfer structures allow water to flow into, out of, and between ponds.
- Inlet, outlet, and transfer structures may be a single pipe or a diffuser with multiple openings.
- Gate valves are used to close pipes to take ponds in and out of service.
- Stop logs, flashboard risers, or weir gates are used to control pond water level.
- Baffles (walls or curtains) are used to divide ponds into smaller sections and to direct the flow of water through the pond. Baffles reduce short-circuiting.
- Algae, recirculation of oxygen-laden water, and/or mechanical aeration devices provide oxygen for treatment.
- Briefly operate mechanical aeration systems weekly when not in use to keep parts functional and lubricated.

PROCESS VARIABLES

- Organic loading rate expressed as kg BOD_5/ha·d or lb BOD_5/ac/d.
- Oxygen must be matched to the organic load or the pond will become anaerobic. Oxygen concentrations increase during the day and decrease at night and on cloudy days.
- Supersaturation can occur when excess oxygen, produced by algae, increases the oxygen concentration beyond the saturation concentration for the current operating temperature and pressure. Turning on mechanical aeration devices when the pond is supersaturated will strip oxygen from the water. The oxygen concentration will decrease.
- Hydraulic detention time required depends on water temperature and availability of oxygen. Naturally aerated ponds may require 180 days or longer to meet secondary treatment standards.
- Alkalinity and pH change diurnally in response to carbon dioxide uptake and production by algae.
- Sludge accumulates in the bottoms of all ponds. It contains settleable solids from the influent wastewater and biomass. Eventually, approximately 60% of the organic matter will be converted to carbon dioxide and methane.
- Differences in water temperature between the influent and the pond can cause short-circuiting.

PROCESS CONTROL

- Daily observation of pond appearance. Color is a key indicator.
- Balance the influent organic load with the amount of oxygen available. Operators may:
  - Switch between series and parallel operation.
  - Recirculate oxygenated water from one pond to another.
  - Increase surface aerator run speed or time.
- Monitor influent and effluent parameters.
- Monitor sludge buildup and remove excess sludge as needed.
- Control algae growth, especially in polishing ponds, to keep effluent TSS concentrations below discharge permit limits. Methods of algae control include limiting growth by restricting access to light, introducing predators like daphnia, barley straw, and ultrasound.

SAMPLING AND ANALYSIS

- Minimum and ideal measurement frequencies given.
- Dissolved oxygen in each pond should be measured daily along with water temperature.
- Sampling frequencies for BOD_5, TSS, and fecal coliforms are set forth in the discharge permit. Minimum frequencies are recommended for process control.
- Sludge blanket depth should be measured at least annually.

References

Bajpai, P. (2017) Basics of Anaerobic Digestion Process. *Anaerobic Technology in Pulp and Paper Industry*; Springer Singapore; Singapore.

Barsom, G. (1973) *Lagoon Performance and the State of Lagoon Technology*; EPA-R2-73-144; U.S. Environmental Protection Agency, Office of Research and Monitoring: Washington, D.C.

Carr, N. G.; Whitton, B. A. (Eds.) (1982) The Biology of Cyanobacteria, Botanical Monographs, Volume 19. In *The Biology of Cyanobacteria*, J. H. Burnett, H. G. Baker, H. Beevers, F. R. Whatley, Eds.; University of California Press: Berkeley and Los Angeles, California.

Clare, J. (2002) *Daphnia: An Aquarist's Guide.* July. https://www.caudata.org/daphnia/#cult2 (accessed Aug 2017).

Crites, R. W.; Middlebrooks, E. J.; Reed, S. C. (2006) *Natural Wastewater Treatment Systems*; CRC Press: Boca Raton, Florida.

Gross, A., Boyd, C. E.; Wood, C. W. (1998) Ammonia Volatilization from Freshwater Fish Ponds. *J. Environ. Qual.*, **28** (3), 793–797.

Harris, S. M. (2003) *Wastewater Lagoon Troubleshooting*; H&S Environmental: Mesa, Arizona.

Hathaway, J. M.; Hunt, W. F. (2009) An Evaluation of the Dye Branch Wetlands; North Carolina State University, Biological and Agricultural Engineering.

Kays, W. B. (1986) *Construction of Linings for Reservoirs, Tanks, and Pollution Control Facilities*; Wiley & Sons: New York.

Metcalf and Eddy, Inc./AECOM (2014) *Wastewater Engineering: Treatment and Resource Recovery*, 5th ed.; McGraw Hill: New York.

Middlebrooks, E. J.; Middlebrooks, C. H.; Reynolds, J. H.; Watters, G. Z.; Reed, S. C.; George, D. B. (1982) *Wastewater Stabilization Lagoon Design, Performance and Upgrading.* Macmillan Publishing Co., Inc.: New York.

Mur, L. R.; Skulberg, O. M.; Utkilen, H. (1999) Cyanobacteria in the Environment. In *Toxic Cyanobacteria in Water: A Guide to Their Public Health Consequences, Monitoring, and Management*; Chorus, I., Bartram, J., Eds.; World Health Organization: Geneva, Switzerland.

Newman, J. (2017) *Water Quality: Barley Straw.* http://extension.psu.edu/natural-resources/water/ponds/barley-straw (accessed Aug 15, 2017).

Richard, M. (2003) *Microbiological and Chemical Testing for Troubleshooting Lagoons.* http://www.lagoons online.com/trouble-shooting-wastewater-lagoons.htm (accessed Aug 18, 2017).

Rockne, K. J.; Brezonik, P. L. (2006) Nutrient Removal in a Cold-Region Wastewater Stabilization Pond: Importance of Ammonia Volatilization. *J. Environ. Eng.*, **132** (4), 451–459.

Sawyer, C. N.; McCarty, P. L.; Parkin, G. F. (1994) *Chemistry for Environmental Engineering*; McGraw Hill: New York.

Schlect, E. D. (1990) Oregon and Washington Fish Hatcheries Lined with Asphalt. *Asphalt*, **4** (3).

Townshend, A. R.; Knoll, H., Eds. (1987) Cold Climate Sewage Lagoons:. *Proceedings of the 1985 Workshop, Winnipeg, Manitoba*; Rep. EPS 3/NR/1; Environment Canada: Ottawa, Ontario, Canada.

U.S. Environmental Protection Agency (1975) *Wastewater Treatment Lagoons*; EPA-430/9-74-001, MCD-14; U.S. Environmental Protection Agency: Washington, D.C.

U.S. Environmental Protection Agency (1983) *Municipal Wastewater Stabilization Ponds*; EPA 625/1-83-015; U.S. Environmental Protection Agency, Office of Research and Development Laboratory, Center for Environmental Research Information: Cincinnati, Ohio.

U.S. Environmental Protection Agency (2011) *Principles of Design and Operations of Wastewater Treatment Pond Systems for Plant Operators, Engineers, and Managers*; EPA/600/R-11/088; U.S. Environmental Protection Agency Office of Research and Development, National Risk Management Research Laboratory—Land Remediation and Pollution Control Division: Cincinnati, Ohio.

Water Environment Federation (2012) *Basic Laboratory Procedures for the Operator-Analyst*, 5th ed.; Water Environment Federation: Alexandria, Virginia.

Water Environment Federation (2009a) *Energy Conservation in Water and Wastewater Treatment Facilities*; Manual of Practice No. 32; Water Environment Federation: Alexandria, Virginia.

Water Environment Federation (2009b) *Natural Systems for Wastewater Treatment*, 3rd ed.; Manual of Practice No. FD-16; Water Environment Federation: Alexandria, Virginia.

Water Environment Federation (2016) *Operation of Water Resource Recovery Facilities*, 7th ed.; WEF Manual of Practice No. 11; Water Environment Federation: Alexandria, Virginia.

Water Environment Federation; American Society of Civil Engineers; Environmental and Water Resources Institute (2018) *Design of Water Resouce Recovery Facilities*, 6th ed.; WEF Manual of Practice No. 8; ASCE Manuals and Reports on Engineering Practice No. 76; Water Environment Federation: Alexandria, Virginia.

Williams, A. E.; Burris, R. H. (1952) Nitrogen Fixation by Blue-Green Algae and Their Nitrogenous Composition. *Am. J. Bot.*, **39** (5), 340–342.

Wisconsin Department of Natural Resources (2006) *Blue-Green Algae*. http://dnr.wi.gov/lakes/bluegreen algae/ (accessed July 25, 2017).

Zhou, J.; Corley; C. E.; Zhu, L.; Broeckling, L.; Renth, R.; Kluge, T. (2005) Barley Straw to Improve Performance of Small Wastewater Treatment Lagoon System. *Proceedings of the 78th Annual Water Environment Federation Technical Exhibition and Conference* [CD-ROM]; Washington, D.C., Oct 29–Nov 2; Water Environment Federation: Alexandria, Virginia; pp 3672–3684.

Zickefoose, C., Hayes, R. B. J. (1977) *Operations Manual: Stabilization Ponds*; U.S. Environmental Protection Agency, Municipal Operations Branch, Office of Water Programs Operations: Washington, D.C.

Suggested Readings

H&S Environmental (2003) *Wastewater Lagoon Troubleshooting: An Operator's Guide to Solving Problems and Optimizing Wastewater Lagoon Systems*; H&S Environmental: Mesa, Arizona.

U.S. Environmental Protection Agency (1983) *Design Manual: Municipal Wastewater Stabilization Ponds*; EPA-625/1-83-015; U.S. Environmental Protection Agency: Cincinnati, Ohio.

U.S. Environmental Protection Agency (2011) *Principles of Design and Operations of Wastewater Treatment Pond Systems for Plant Operators, Engineers, and Managers*, EPA/600/R-11/088; U.S. Environmental Protection Agency: Cincinnati, Ohio.

CHAPTER 7

Fixed-Film Treatment

Introduction

Trickling filters and rotating biological contactors (RBCs) are fixed-film processes, which are used at many water resource recovery facilities (WRRFs). Fixed-film biological treatment processes remove dissolved organics and finely divided organic solids from wastewater. Removal occurs primarily by converting soluble and colloidal material into a biological film. This biological film, or biofilm, is attached to solid surfaces of support media. Fixed-film processes may also remove ammonia, nitrite, and nitrate. They retain a large amount of biomass in the biofilm, which provides process stability and tolerance for temperature changes and toxic substances. Both types of fixed-film processes have low energy and low to moderate maintenance requirements, especially when compared to activated sludge processes.

LEARNING OBJECTIVES

Upon completing this chapter, you will be able to

- Understand biofilm growth and decay,
- Explain the principles underlying mass transport through a biofilm,
- Label all components of trickling filters and RBCs and describe the function of each,
- Determine the direction of airflow through a trickling filter,
- Place a trickling filter or RBC into service or remove one from service,
- Inspect and maintain trickling filter and RBC equipment,
- Calculate organic and hydraulic loading rates (HLRs) for trickling filters and RBCs,
- Make process adjustments to maintain a desired biofilm thickness or weight in trickling filters and RBCs,
- Determine the recycle ratio required to maintain a desired Spülkraft (SK) value for a trickling filter,
- Interpret influent biochemical oxygen demand (BOD) data and adjust the size of stage 1 treatment for an RBC in response,
- Collect process control samples and evaluate results,
- Implement corrective actions to minimize the effects of biofilm predators, and
- Troubleshoot common mechanical and process control problems.

Purpose and Function of Fixed-Film Processes

Trickling filters and RBCs are fixed-film processes. They use the same microorganisms as pond systems and activated sludge to treat wastewater, but the microorganisms grow attached to a surface. Fixed-film processes are also referred to as *attached growth processes. Support media* provides a surface for microorganisms to attach to and grow. They gradually form a slime layer called a *biofilm.* A thin layer of wastewater flowing over the biofilm either intermittently or continuously wets the surface of the biofilm and brings food and nutrients to the microorganisms and carries away waste products. Excess or dead biomass, which *sloughs* off of the media surface is carried away by the wastewater flowing over the biofilm and through the *void spaces* of the support media. Sloughed biomass is removed from the trickling filter or RBC effluent by gravity settling in a clarifier.

The microorganisms that grow in biofilms convert soluble organic matter into biomass solids. Both trickling filters and RBCs can remove carbonaceous matter measured as biochemical oxygen demand (BOD) as well as ammonia, nitrite, and nitrate. These biofilms typically do not remove phosphorus beyond *assimilative uptake.* Biological treatment fundamentals are discussed in Chapter 5.

Biofilm is the slime layer that develops on almost any continuously wetted surface when food and nutrients are also available.

Sloughing is the loss of biofilm from a fixed-film process. Small pieces of biofilm are continuously washed out of the filter. Occasionally, larger sloughing events may occur.

Support media, or simply media, is the underlying rock, wood, plastic or any other material that the biofilm grows upon.

Void is another word for empty space. Void spaces are the openings between pieces of media in the filter.

All microorganisms are made up of carbon, nitrogen, phosphorus, and other elements. When they grow and reproduce, these elements are removed from the wastewater and become part of their biomass. This is *assimilative uptake.*

Figure 7.1 Rock Media Trickling Filter (Reprinted with permission by Indigo Water Group)

A trickling filter consists of a tank filled with some kind of media—rock, plastic, wood, etc.—for the bacteria to grow upon. The media is supported by an underdrain system, which keeps the media inside the tank while allowing water to pass through. The underdrain system is also used to ventilate the trickling filter media. Wastewater is pumped to the top of the trickling filter where it is distributed by spray nozzles or a rotating arm as shown in Figure 7.1. The water trickles down over the media and into the underdrain. Trickling filters are not submerged. The water trickling over the media takes up very little of the total space inside the filter. The trickling filter shown in Figure 7.1 is filled with natural rock media and is approximately 1.5-m (5-ft) deep.

Rotating biological contactors have been in full-scale use since the mid-1970s. Rotating biological contactors consist of multiple stacked plates or wheels of media that are submerged between 30 to 70% in wastewater (Figure 7.2) (Metcalf and Eddy, Inc., 2003). Most are submerged approximately 40% (U.S. EPA, 1984). The wheels slowly rotate through the wastewater. Unlike trickling filters, RBCs don't require pumping and often don't require mechanical ventilation. Both trickling filters and RBCs produce excess biofilm that periodically sloughs off the media. Sloughed biofilm settles to the bottom of the secondary clarifier. In some facilities, trickling filters and RBCs have small hoppers to collect sloughed biofilm instead of true clarifiers. The sloughed biofilm is not returned to the trickling filters or RBCs. It is removed from the process for further processing.

Raw domestic and industrial wastewater typically contains settleable solids, floatable materials, and other debris. Failure to remove these solids before the wastewater enters fixed-film processes can interfere with their ability to absorb oxygen, plug the filter media, result in high solids yield, or create other problems. Therefore, fixed-film processes are typically preceded by screening and grit removal processes. Primary treatment processes should be used to reduce the fixed-film process load.

TEST YOUR KNOWLEDGE

1. In fixed-film processes, the microorganisms grow attached to some type of support media.
   - ☐ True
   - ☐ False

2. Some fixed-film processes convert ammonia to nitrite and nitrate.
   - ☐ True
   - ☐ False

Figure 7.2 Rotating Biological Contactor (Reprinted with permission by Indigo Water Group via Upper Blue Sanitation District)

3. **For a trickling filter to function properly, all of the media must be submerged.**
   - ☐ True
   - ☐ False

4. **Fixed-film processes don't typically remove _______ beyond assimilative uptake.**
   a. BOD
   b. TSS
   c. Nitrogen
   d. Phosphorus

5. **Excess biofilm leaves a fixed-film process by**
   a. Mechanical wasting
   b. Scraping
   c. Sloughing
   d. Power washing

6. **The primary difference between trickling filters and RBCs is**
   a. RBCs rotate through the wastewater.
   b. Only trickling filters can remove ammonia.
   c. Sloughing rates are lower in RBCs.
   d. Trickling filters do not have influent pumps.

Theory of Fixed-Film Operation

Biofilm formation begins as microorganisms in the wastewater attach to a surface. Over time, these microorganisms reproduce and excrete *exopolymer*, which forms a slime layer on the media surface (Figure 7.3). Eventually, the bacteria and microorganisms form colonies that protrude away from the underlying media. Biofilms are often shown in diagrams as thin and flat, but in reality, biofilms are full of protrusions, valleys, open spaces, and channels. As new microorganisms grow, the thickness of the biofilm increases. Excess biofilm eventually sloughs off the media surface. The sloughed biomass and suspended particles, which were not absorbed by the biofilm are separated by gravity from the liquid discharged from the process and are sent to waste sludge processing.

Exopolymer is a sticky substance made up primarily of sugar molecules that helps the microorganisms adhere to the media and to one another.

When fixed-film processes are used for BOD removal, the microbial population consists of various species of heterotrophic bacteria with smaller populations of protozoa and fungi. If these processes are used primarily for ammonia removal, autotrophic nitrifying bacteria predominate, with smaller numbers of heterotrophic bacteria.

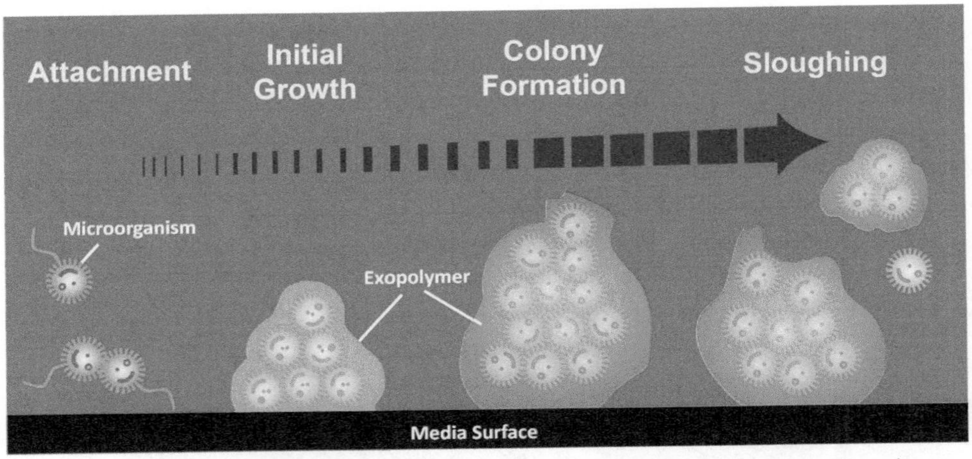

Figure 7.3 Biofilm Growth Cycles (Reprinted with permission by Indigo Water Group)

If something is *adsorbed*, it adheres to the surface. If something is absorbed, it is drawn up, through, or into something else. For example, dirt is adsorbed onto the surface of a car, but a sponge absorbs water.

An *enzyme* is a special type of protein that cells use to speed up chemical reactions. Amylase, an enzyme found in saliva, breaks down starch into simple sugars.

Diffusion describes the natural movement of substances from areas of high concentration to low concentration.

As wastewater passes over the biofilm, particles in the wastewater are *adsorbed* onto the biofilm surface. *Enzymes*, secreted by bacteria, break down particulate biodegradable organic material, gradually converting particulate organic matter into soluble organic matter. Soluble organic matter, dissolved oxygen (DO), ammonia, nitrite, nitrate, phosphorus, and other nutrients in the wastewater *diffuse* directly into the biofilm where they are used by the microorganisms (Figure 7.4). Waste products diffuse out of the biofilm into the passing wastewater. The concentrations of the different wastewater components change throughout the depth of the biofilm. For most wastewater components, a gradient exists where the concentration goes from higher to lower from one side of the biofilm to the other. Wastewater components move naturally by diffusion from areas of higher concentration to areas of lower concentration. The movement of the different components is called *mass transfer*. The mass of food, nutrients, oxygen, waste products, and other substances that can be transferred throughout the depth of the biofilm limits the numbers and types of microorganisms present as well as the thickness of the biofilm.

The outside layer of the biofilm has the greatest exposure to food, nutrients, and oxygen from the passing wastewater. As a result, the microorganisms in this portion of the biofilm will grow the fastest. Oxygen must pass through this outer layer to reach microorganisms deeper in the biofilm and is often consumed

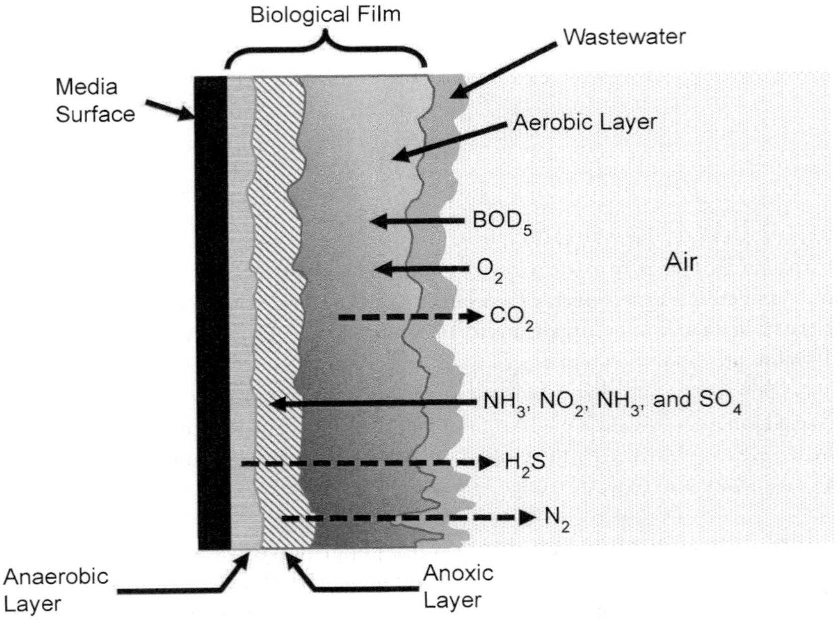

Figure 7.4 Biofilm Diagram (Reprinted with permission by Indigo Water Group)

before reaching the inner layers. Aerobic, anoxic, and anaerobic zones develop within biofilms (Figure 7.4). To review the definitions of *aerobic, anoxic,* and *anaerobic,* revisit Chapter 5. Microorganisms in biofilms can consume BOD, convert ammonia to nitrite, convert nitrite to nitrate, convert nitrite and nitrate to nitrogen gas, and uptake phosphorus. In deep biofilms, anaerobic zones form where sulfate may be converted to hydrogen sulfide. The types of microorganisms that develop within biofilms are controlled by the food source provided and the environmental conditions within the process.

Biofilms cannot perform enhanced biological phosphorus removal. Some phosphorus is taken up for biological growth (assimilative uptake).

Biofilms have limited capability to adsorb suspended solids. Only soluble chemicals can diffuse through the biofilm and be taken up by the microorganisms. Primary treatment to remove suspended solids is typically required prior to treatment in fixed-film processes. Biodegradable solids will be consumed by the microorganisms. Inert non-biodegradable matter and particles, which are not adsorbed from the wastewater flowing over the biofilm, pass through the fixed-film process. Settleable particles will be captured downstream in secondary or final clarifiers.

Nitrification converts ammonia to nitrite and nitrate. Denitrification converts nitrite and nitrate to nitrogen gas. For more details, review the chapters on Fundamentals of Biological Treatment and Nutrient Removal.

Under conditions of low DO, nutrient deficiencies, or low pH values, organisms that either remove BOD slowly or exhibit poor settling characteristics can dominate the fixed-film reactor. These organisms are predominately *filamentous bacteria* and sometimes fungi and are a nuisance. Eliminating filamentous bacteria typically involves identifying the source of the nuisance and eliminating the condition that allows them to dominate the system.

Filamentous bacteria are bacteria that grow together in long chains. A few filamentous bacteria that may be present in fixed-film systems include *Beggiatoa, Thiothrix,* and *Sphaerotilus natans*. These filaments thrive in low DO conditions.

BIOFILM STRUCTURE

Biofilms consist of a layered, porous matrix of living and dead cells plus debris that is held together and attached to a solid surface with exopolymer. Exopolymer is also referred to as exopolymeric substances (EPS). The EPS consists mainly of polysaccharides (sugars) with lower amounts of proteins and fats. The EPS helps the biofilm adhere to the media and to keep the microorganisms attached to the biofilm. The amount of EPS increases with the age of the biofilm and may account for 50 to 90% of the total organic carbon of mature biofilms. Near the outer surface, the biofilm will be dominated by living cells and EPS, while the inner layers nearer the surface of the support media will be dominated by inert solids and dead biomass.

Biofilms are composed of a large and diverse population of living organisms including bacteria, protozoa, algae, fungi, worms, and even insect larva. Cells within the biofilm can form single species *microcolonies* simply by remaining attached following cell division. Single-species microcolonies can rapidly protrude away from the support media surface to separate themselves from competing cells and gain better access to nutrients at the biofilm surface. Figure 7.5 shows green microcolonies of nitrifying bacteria (left) growing

Microcolonies are small groups of bacteria or other microorganisms growing together in a clump.

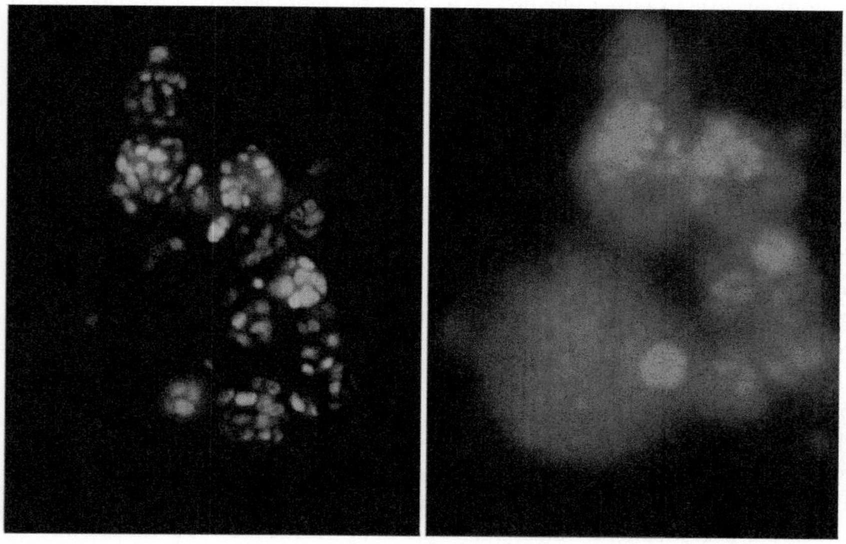

Figure 7.5 Nitrifying Bacteria Growing in Biofilm (Reprinted with permission by Indigo Water Group)

out of a red biofilm (right). The biofilm in Figure 7.5 has been treated with specialized fluorescent dyes. As the bacteria within the biofilm compete for the same resources (DO and nutrients), growing higher and wider above the surrounding surface gives the cells within the protruding microcolonies a competitive advantage over neighboring bacteria.

Microorganisms that are able to grow rapidly accumulate on and near the outer surface of the biofilm. Slower growing microorganisms accumulate closer to the media as faster growing microorganisms cover them. When the BOD concentration is high, heterotrophic bacteria will grow faster than autotrophic bacteria. The heterotrophs will grow over the nitrifying bacteria and deplete the DO within the deeper layers of the biofilm, resulting in decreased nitrification rates. When the BOD concentration is low, the growth rate of the heterotrophs slows down, oxygen penetrates deeper into the biofilm, and the number and activity of nitrifying bacteria increase. In general, the soluble BOD (sBOD) concentration must fall below about 20 mg/L before nitrifying bacteria will be the dominant bacteria in a fixed-film process.

BIOFILM THICKNESS

The biofilm thickness and density both increase over time. Water flowing over the biofilm creates shear forces and turbulence that cause pieces of the biofilm to break away. This gradual loss of small pieces of biofilm is called *abrasion* or *erosion*. This forces the biofilm to consolidate to a higher density over time. Think of pruning a bush or a tree. When branch length is limited by regular pruning, new growth fills in the empty spaces between existing branches, which makes the bush or tree more compact overall. More leaves and branches grow in a smaller volume. The biomass density in a biofilm is 5 to 10 times greater near the support media than it is near the outer surface of the biofilm (Zhang and Bishop, 1994).

The thickness of the biofilm (perpendicular to the surface of the support media) depends both on the shear created from wastewater passing over the biofilm and the availability of resources near the media surface. Shear is controlled in trickling filters by the amount of water sent to the filter and the speed of the distributor arms. It is controlled by the rotation speed of the media wheel in RBCs. When resources are no longer available to the microorganisms in the lowest layers of the biofilm, those microorganisms may die and detach from the support media. Detachment of biofilm from the support media allows living cells to be carried by flowing water from nutrient-depleted areas to new environments downstream and re-attach themselves at locations where the environmental conditions are favorable for growth. Most of the detached biofilm ends up leaving the process and going to the secondary clarifier, where it is removed by gravity.

BIOFILM DETACHMENT AND SLOUGHING

Biofilm formation is determined by the balance between the microbial attachment and growth rates versus the rate of biomass detachment. Growth rate is determined by the availability of resources (Chapter 5). Increasing the BOD or ammonia load to a fixed-film process will produce more heterotrophic bacteria and nitrifying bacteria, respectively. Biofilm detachment can occur continuously by erosion of small particles from the surface of the biofilm or by intermittent sloughing of relatively large pieces of biofilm. With sloughing, particle sizes often exceed the thickness of the biofilm (Figure 7.6). During sloughing, pieces of the biofilm can be removed down to the surface of the support media.

Biofilm growth and sloughing can be compared to your own head of hair. Each day, some hairs are lost when you shower or comb your hair. Because there are so many individual hairs on your head, the few that are lost each day aren't noticeable. New hairs grow to replace the ones that are lost. As long as the growth of new hair balances with the loss of hair, you'll maintain a full head of hair. A healthy biofilm process functions in the same way. Small losses each day are replaced by new growth and the amount of biomass that remains in the process is nearly constant from day to day.

A *substrate* is any food, nutrient, or other substance that the microorganism needs to survive and reproduce. Examples of substrates include BOD, oxygen, nitrogen, and phosphorus.

The type and rate of biofilm detachment depends on the HLR, DO concentration, soluble *substrate* concentration, and other factors. Detachment from biofilms exposed to constant hydraulic loading (shear stress) occurs primarily by erosion. A sudden increase in HLR can cause significant increases of both the total suspended solids (TSS) concentration and the average particle size in the trickling filter effluent (Choi and Morgenroth, 2003). Think about cleaning a driveway with a hose. If a small amount of water is allowed to flow gently down the driveway, smaller debris like grass clippings will be pushed along, but larger debris will remain on the driveway. If the amount of water is suddenly increased, there

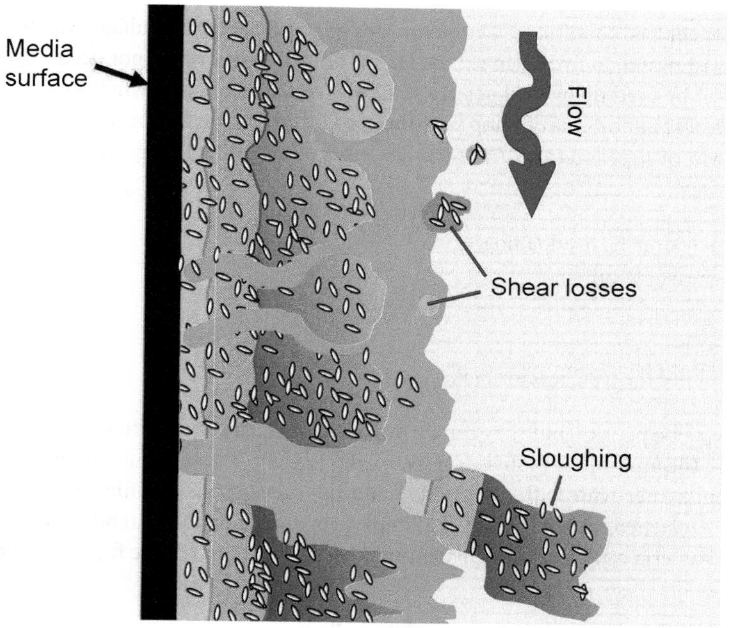

Figure 7.6 Shear Losses and Sloughing (Reprinted with permission by Indigo Water Group)

will be enough force to move more debris all at the same time. Higher hydraulic loading rates in fixed-film processes encourage continuous erosion and discrete sloughing events, which produce smoother, more compact, and thinner biofilms (Picioreanu et al., 1999).

As biofilms age, inert residues of decayed cells accumulate near the surface of the support media, which decreases the adhesive strength of the biofilm. When low DO concentrations (<0.5 mg/L) persist, hollow cavities form at the base of the biofilm, which also reduces the adhesive strength of the biofilm. The accumulation of inert residues of decayed cells and development of biofilm cavities at the surface of the support media contribute to biofilm sloughing (Laspidou and Rittmann, 2004). The biomass sloughed from trickling filters has a *specific gravity* of approximately 1.025, which results in good settling characteristics in the secondary clarifier. For comparison, the specific gravity of suspended growth (activated sludge) solids is approximately 1.005 (Metcalf and Eddy, Inc., 2003).

Specific gravity is the ratio of the density of a substance to the density of water. Water has a specific gravity of 1. Density is the amount of mass (weight) per volume. If a substance has a specific gravity greater than 1, it is more dense than water.

BIOFILM DENSITY AND MASS

Biofilm density over the surface of the support media is non-uniform and varies across the thickness of the biofilm. Biofilm porosity and composition affect how easily soluble compounds are able to diffuse through the biofilm depth. As colonies form, die, detach, and are replaced, biofilms develop pores and channels, which aid transport of DO, nutrients, and waste products through the biofilm.

Pores are openings in the biofilm that can penetrate partially or completely through the biofilm.

Near the outer surface, the biofilm is less dense, with pores approximately 2 to 3 μm in diameter. Near the surface of the support media, the biofilm density increases to 5 to 10 times higher than in the surface layer and pore size decreases to approximately 0.3 to 0.4 μm in diameter (Zhang and Bishop, 1994). As biofilms age, the void spaces between microbial clusters or colonies gradually fill with new cells and inert residues of decayed cells, making the biofilm more uniform with a smoother surface.

FILAMENTOUS BACTERIA

Filamentous bacteria are long chains of individual bacteria that grow together. *Beggiatoa* and *Thiothrix*, two types of filamentous bacteria, commonly grow in both trickling filters and RBCs. These filaments compete with heterotrophic organisms for oxygen and space on trickling filter and RBC media surfaces. *Beggiatoa* and *Thiothrix* utilize hydrogen sulfide and elemental sulfur as energy sources. Over time, they can form large mats that look like white or gray hair on the media surface. When the mats are peeled away from the media, the surface underneath is typically black and smells like rotten eggs. *Beggiatoa* and *Thiothrix* mats may also appear on clarifier weirs downstream from a fixed-film process.

Hydrogen sulfide is formed under anaerobic conditions. Certain groups of bacteria convert sulfate, which is always present in the incoming wastewater, to hydrogen sulfide (H_2S). This conversion will not occur if oxygen is available. The oxygen can be used up by the bacteria at the surface of the biofilm when too much BOD is available. Then, anaerobic conditions develop near the media surface. Hydrogen sulfide is generated, which supports the growth of *Beggiatoa* and *Thiothrix* as well as other sulfur-loving bacteria. Sulfides may result from the following:

- Low DO concentrations caused by organic overloading,
- Low DO concentrations during warm weather,
- Septic wastes,
- Industrial discharges, and
- Anaerobic deposits on the bottom of the RBC reactor tank.

Some filamentous bacteria will always be present, but they can be problematic in large quantities. Mats of filamentous bacteria can increase the amount of biomass on the media (weight) while simultaneously reducing BOD removal. The filaments intertwine with one another and can prevent thick biofilm from sloughing off of the media surface. Detached biofilm can be held in place by filaments that extend into the surrounding biofilm. Filamentous bacteria can cause sloughed biomass to resist settling in the final clarifier.

Macroinvertebrates include some of the larger, non-microscopic organisms found in trickling filters. Snails, filter flies, and worms are macroinvertebrates. Macroinvertebrates are more commonly referred to as *macrofauna*.

MACROFAUNA IN BIOFILMS

The biofilms in both trickling filters and RBCs are inhabited by a variety of *macrofauna* including filter flies, insect larvae, snails, and worms. These organisms feed on the biofilm and reduce the amount of excess biomass sloughed from the biofilm. Possible benefits associated with the presence of macrofauna include reduced sludge production, improved sludge settleability, and biofilm thickness control (WEF et al., 2010). Trickling filters with macrofauna achieve higher percentages of BOD removal and lower effluent BOD concentrations than trickling filters that do not have these invertebrates (Williams and Taylor, 1968). However, biofilm predation by rapidly growing macrofauna populations can degrade treatment performance significantly by removing large amounts of the biomass within a few days or weeks. Macrofauna may have the following detrimental effects:

- Grazing of nitrifying biofilm;
- Plugging of process piping;
- Damaging pumps;
- Damaging belts on gravity-belt thickening and belt-press dewatering equipment;
- Organic snail bodies remaining in the effluent stream, which may increase effluent fecal counts by shielding bacteria from disinfection processes;
- Shells remaining in the effluent stream, which may increase effluent fecal counts by shielding bacteria from disinfection processes;
- Exerting additional solids loading on secondary clarifiers; and
- Accumulating in downstream processes, thereby reducing treatment capacity.

FILTER FLIES

Trickling filters provide an ideal habitat for filter flies (Figure 7.7). The biofilm provides an abundant source of food for adult flies and larvae. Adult filter flies are 3 mm (1/8 in.) long. The emergence of large numbers of adult flies often becomes a serious nuisance during warm weather, particularly with *Psychoda* and *Anisopus*. The *Psychoda* species are found in many WRRFs.

Filter flies live for 8 to 24 days. They breed only once, often within hours of emerging from their pupal casings. Females lay 30 to 100 eggs on dry surfaces just above the water surface. The eggs hatch within 48 hours. The fly larval stage lasts 10 to 15 days. The pupal stage lasts for 20 to 40 hours. An adult fly will emerge from the pupal casing. The development time from egg to adult is 7 to 28 days, depending on temperature and food availability. The short life cycle and high reproduction rate can result in enormous numbers of filter flies within a few weeks.

Filter fly larvae are unable to absorb oxygen through water and must breathe through a small tube (spiracle). They must reach the water surface regularly to obtain oxygen. For this reason, alternating wet and dry

Figure 7.7 Filter Fly (*Psychoda*) (Creative Commons Wikimedia. Photograph by Sanjay Acharya.)

media surfaces inside trickling filters are ideal places for filter flies to deposit their eggs and for the larvae to grow. Typically, filter flies accumulate along the outer edge of the filter where it is damp, but they are not flushed away. During the pupal stage, the insect remains submerged near the water surface, breathing through a spiracle.

Filter fly larvae can consume large amounts of biomass in a few weeks. The effect of fly larvae on heterotrophic biofilm is not significant because of the high growth rates of *heterotrophic* bacteria. Filter fly larvae can cause significant reductions in treatment performance and poor effluent quality in nitrifying biofilms because of the slow growth rates of *autotrophic* bacteria.

Chapter 5 contains additional information on *heterotrophic* and *autotrophic* bacteria.

SNAILS

Snails belonging to genus *Physa*, commonly known as the pouch snail, frequently inhabit trickling filters and RBCs (Figure 7.8). The number of snails observed in trickling filters tends to be highest for rock media, less for plastic dump media, and least for plastic sheet media. Rock media and plastic dump media have lower HLRs than plastic sheet media and contain more dry media areas. Both factors may enable snails to reproduce more efficiently in rock and plastic dump media. Members of the genus *Physa* are among the

Figure 7.8 *P. gyrina* (Snail) (WEF, 2010)

most common and widespread of all freshwater snails. Snails belonging to genus *Physa* range in size from 3 mm to 15 mm and have a life span of 15 to 18 months. *P. gyrina*, common to trickling filters, is a lung breather that inhales oxygen and exhales carbon dioxide. Although these snails prefer a moist environment, they can't tolerate being submerged completely for extended periods of time.

Egg development occurs mainly in spring, but population spikes also occur during the winter. Each snail produces eight to 25 eggs embedded in a gelatinous mass, which adheres to the filter media. Eggs are laid below water. None of the eggs in the same cluster hatch at the same time. *Physa* can typically withstand less than 16 to 24 hours of anoxic conditions, dislike saline conditions, require a minimum of 2 mg/L of DO, and require calcium carbonate (alkalinity) for shell development (Boltz et al., 2008). Snails are most vulnerable directly after hatching.

Snails are biofilm predators. They can consume large amounts of biomass in a few days or weeks. The loss of biomass can cause significant reductions in treatment performance and poor effluent quality in nitrifying biofilms because of the slow growth rates of autotrophic bacteria. In essence, the snails consume the biofilm faster than it can regrow. Snail infestations have completely stripped trickling filter media of biofilm in some cases.

The quantities of snail shells produced in trickling filters can hinder performance of downstream aeration basins, secondary clarifiers, and sludge handling equipment. These process units may need to be taken out of service for cleaning and maintenance to remove piles of shells. Figure 7.9 shows accumulated snail shells and grit in the bottom of an activated sludge basin. Snail shells are highly abrasive and can damage pumps, piping, and other equipment over time. A small portion of the snails have air entrained inside their shells, which causes them to be disconnected from their shell. This means a fraction of the snails in trickling filter effluent may float and work their way into the WRRF final effluent. In the final effluent, snail bodies can increase BOD. High flushing intensity is ineffective to control snail populations in the trickling filters.

WORMS

Numerous worms may inhabit a trickling filter including nematodes, annelids, and other kinds. Beneficial conditions for aquatic worms include DO concentrations greater than 2 mg/L, pH between 5 and 9, low ammonia concentrations, and temperatures between 15 and 25 °C. These conditions prevail in trickling

Figure 7.9 Snails and Grit Accumulation Below Diffusers in an Activated Sludge Basin (diffuser width is 1.2 m [4 ft]; diffusers are the shorter pieces extending out from the longer air pipeline along the bottom of the tank) (Reprinted with permission by the City of Boulder, Colorado)

filters with low organic loading rates and in the lower portions of trickling filters with moderate organic loading rates. High rate and roughing filters don't have high enough DO concentrations to support their growth.

The family of *Tubificidae* worms includes both free-swimming and *sessile* organisms, which attach themselves to the media. Sessile *Tubificidae* are typically red in color, up to 20-cm (8-in.) long, and less than 2 mm (0.08 in.) in diameter. They burrow into the biofilm with their posterior end projecting out of the biofilm to uptake oxygen. Sessile *Tubificidae* populations can reach very high densities of 400 000 worms/m^2 (37 000 worms/sq ft) or more. *Tubificidae* tolerate anoxic conditions for up to 25 days by switching to anaerobic metabolism.

While some operators dislike finding *Tubificidae* in their treatment processes because they are unsightly, they can be beneficial. *Tubifex* worms, one type of *Tubificidae*, reduce chemical oxygen demand (COD) concentrations in trickling filter effluent and reduce trickling filter biomass yield (sludge production) substantially compared to trickling filters operating without worms (Rensink and Rulkens, 1997). Decreased sludge production tends to increase nitrite, nitrate, and phosphate concentrations in the wastewater. Worms do not affect nitrification. *Sessile Tubificidae* bioaccumulate zinc and cadmium, which increases the concentrations of these metals in waste sludge and decreases these metals in the effluent. Worms produce compact feces, which have a low *sludge volume index* (approximately 60 mL/g), which improves trickling filter sludge settleability (Elissen, 2007).

Tubificidae are true worms and, although they are often red, they aren't the same organism as the red worms often found in lagoons and activated sludge processes. Red worms, also called *blood worms*, are actually midge fly larvae, not worms. Midge flies look similar to mosquitoes, but don't bite.

Sessile means fixed in one place. Sessile worms live attached to a surface.

The *sludge volume index* is a measure of how well sludge settles and compacts in a standard laboratory test. The sludge volume index test is discussed in the activated sludge chapter.

ALGAE

Trickling filters and RBCs without covers will grow algae on the media surfaces that are exposed to sunlight. Thick mats of algae can develop on the upper surface of the exposed media and can cause media plugging and *ponding*. Some algae species such as *Cladophora glomerata* and *Vaucheria* develop into thick patches or mats, which can contribute to plugging and ponding. *C. glomerata* grows during the spring and summer and becomes dormant during cold weather. *Vaucheria* growth increases during cooler and wetter periods (Dölle and Peluso, 2015).

Ponding describes a situation where water cannot pass through the media and accumulates on the surface of the trickling filter. Ponding is discussed further in the troubleshooting section.

TEST YOUR KNOWLEDGE

1. **Only soluble wastewater components can diffuse into the biofilm and be utilized by the microorganisms.**
   - ☐ True
   - ☐ False
2. **In natural systems, substances move from areas of low concentration to areas of high concentration.**
   - ☐ True
   - ☐ False
3. **The amount of EPS present in biofilms decreases with biofilm age.**
   - ☐ True
   - ☐ False
4. **The first step in biofilm formation is**
   a. Attachment
   b. Initial Growth
   c. Colony Formation
   d. Sloughing
5. **Microorganisms in the biofilm excrete ________ to help convert particulate BOD into soluble BOD.**
   a. Exopolymeric substances
   b. Enzymes
   c. Poisons
   d. Digestive acids
6. **Which of the following substances can diffuse through a biofilm?**
   a. Total suspended solids
   b. Colloidal BOD
   c. Dissolved oxygen
   d. Food particles
7. **Microorganisms in this portion of the biofilm have the greatest access to resources and will grow the fastest.**
   a. Outermost layer
   b. Anoxic layer
   c. Anaerobic layer
   d. Media surface

8. This substance, secreted by bacteria, helps the biofilm attach to the supporting media.

a. Enzymes
b. Exopolymer
c. Carbon dioxide
d. Lipocolonic glucosamine

9. A fixed-film process receives a heavy organic load. Which of the following statements is true?

a. Oxygen will easily penetrate the full depth of the biofilm.
b. The biofilm will contain high numbers of nitrifying bacteria.
c. Growth will be limited due a lack of soluble BOD.
d. Heterotrophic bacteria will overgrow the nitrifying bacteria.

10. Sloughed biofilm from a trickling filter or RBC

a. Is recycled to the front of the treatment process
b. Is removed by gravity in a secondary clarifier or hopper
c. Is typically less dense than the surrounding wastewater
d. Consists primarily of live microorganisms

11. Filter flies require this condition for egg laying and reproduction.

a. Low organic loading rates
b. High flushing intensities
c. DO greater than 2 mg/L
d. Dry surfaces

12. Which of the following statements about snail infestations is NOT true?

a. Consume large quantities of biofilm
b. May contribute to effluent BOD
c. Increase TSS removal
d. Shells can damage equipment

13. These sessile organisms can be found in the biofilms of lightly loaded trickling filters.

a. Filter flies
b. Snails
c. Worms
d. Rotifers

Trickling Filters

Trickling filters are characterized by their HLR, organic loading rate (OLR), treatment objectives, and location in the treatment process. The HLR is the amount of water sent to the trickling filter and is expressed as cubic meters of flow per square meter per day ($m^3/m^2 \cdot d$) or gallons per minute per square foot (gpm/sq ft) or gallons per day per square foot (gpd/sq ft). Most trickling filters have the ability to recycle treated effluent back to the beginning of the process. This helps to maintain a constant HLR and control biofilm thickness. The organic loading rate is the mass of contaminant, either BOD or ammonia-nitrogen (NH_3-N), applied to the trickling filter. For BOD, it is expressed as kilograms per cubic meter per day ($kg/m^3 \cdot d$) or pounds per thousand cubic feet per day (lb/d/1000 cu ft). Loading rates for NH_3-N are expressed as kilograms per square meter of media surface area per day ($kg/m^2 \cdot d$) or pounds per thousand square feet of media surface area per day (lb/d/1000 sq ft). Trickling filters may be designed to remove only carbonaceous BOD (CBOD), only NH_3-N, or a combination of CBOD and NH_3-N. In general, the more lightly loaded a trickling filter is, the more likely that some or all of the NH_3-N will be removed.

DESIGN PARAMETERS FOR TRICKLING FILTERS

Historically, trickling filters have been classified as standard rate, intermediate rate, high rate, or roughing based on their OLR. In the first three categories—standard, intermediate, and high-rate filters—the filter removes all or essentially all of the BOD applied. In the fourth category (the roughing filter), a substantial amount of BOD is allowed to pass through the filter. Roughing filters are typically combined with another biological treatment step such as activated sludge, RBC, or another trickling filter where a substantial amount of BOD removal occurs. Notice that both hydraulic and organic loading rates are higher for plastic media than rock media (see Table 7.1). This is because plastic media have more surface area per volume and larger openings between and through the media. Operators do not need to memorize design or operating ranges for different types of filters. Ranges differ slightly from one manual to the next. Operators should be familiar with the design parameters for their own facilities and be able to calculate the different design and operating variables. The important things to remember are:

- Higher OLRs and HLRs may be used with plastic media than rock media.
- As organic loading rates increase, the ability to nitrify decreases.
- At very high organic loading rates, the amount of BOD and TSS in the trickling filter effluent will increase.

Table 7.1 summarizes generally accepted criteria defining each operational mode. Media types are described in the equipment section. Notice that both hydraulic and organic loading rates are higher for plastic media than for rock media.

Table 7.1 Historic Trickling Filter Categories

Design Characteristics	Low or Standard Rate	Intermediate Rate	High Rate	Super Rate	Roughing
Media	Rock	Rock	Rock	Plastic	Rock/Plastic
HLR, $m^3/m^2 \cdot d$ (gpd/sq ft) excluding recirculation	1.0 to 3.7 (25 to 90)	3.7 to 9.4 (90 to 230)	9.4 to 36.7 (230 to 900)	14.3 to 85.6 (350 to 2100)	57.0 to 171 (1400 to 4200)
Organic loading rate, kg $BOD_5/m^3 \cdot d$ (lb BOD_5/d/1000 cu ft)	0.07 to 0.22 (5 to 15)	0.24 to 0.48 (15 to 30)	0.48 to 2.4 (30 to 150)	0.6 to 3.2 (37.5 to 200)	>1.5 (>100)
Recirculation	Minimum	Typically	Always	Typically	Not normally required
Filter Flies	Many	Varies	Few	Few	Few
Sloughing	Intermittent	Intermittent	Continuous	Continuous	Continuous
Depth, m (ft)	1.8 to 2.4 (6 to 8)	1.8 to 2.4 (6 to 8)	0.9 to 2.4 (3 to 8)	≤12 (≤40)	0.9 to 6 (3 to 20)
BOD removal, % (after clarification)	80 to 90	50 to 70	40 to 80	65 to 85	40 to 85
Effluent quality	Well nitrified	Some nitrification	No nitrification	Limited nitrification	No nitrification

Low-rate filters, also known as *standard-rate filters*, are used for loadings of less than 0.22 kg five-day biochemical oxygen demand (BOD_5)/100 $m^3 \cdot d$ (15 ppd BOD_5/1000 cu ft). The low organic loading rates applied to standard-rate filters result in low biomass yields and effluent BOD_5 typically less than 25 mg/L. Low-rate filters often reduce effluent ammonia concentrations during warm weather. Low-rate rock media filters range in depth from 1.8 to 2.4 m (6 to 8 ft). Most low-rate filters are circular with rotary distributors. However, a number of rectangular rock filters remain in operation. Sloughed solids from a low-rate filter are typically well-*digested*, so these filters discharge fewer solids than higher rate filters. Solids *yields* of 0.5 kg TSS per kg of incoming BOD_5 are not uncommon (WEF, 2007).

When solids are well-*digested*, it means that little readily biodegradable material remains.

Chapter 5 includes an extensive discussion on yield.

Intermediate-rate filters are used for loadings between 0.24 and 0.48 kg/$m^3 \cdot d$ (15 to 30 lb BOD_5/d/1000 cu ft). Higher hydraulic loadings increase the velocity of wastewater flowing over the biofilm, resulting in more uniform sloughing of excess biomass and thinner biofilms than in low-rate filters. These filters recirculate trickling filter effluent to keep the media wet and scour excess biomass from the media. The biological solids that slough from an intermediate trickling filter are not as well-digested as those from a low-rate filter. Yields ranging from 0.6 to 0.8 kg TSS/kg BOD_5 are common, depending on the filter media type.

High-rate and super-high-rate filters receive total BOD_5 loadings ranging from 0.48 to 3.2 kg/$m^3 \cdot d$ (30 to 200 lb BOD_5/d/1000 cu ft). Achieving secondary effluent quality with high-rate filters reliably is less predictable than with low- or intermediate-rate filters. High-rate filters are often combined with second-stage processes to meet low secondary effluent water quality standards.

Roughing filters are somewhat unique in that they are not designed to provide complete removal of BOD. This allows roughing filters to be operated at extremely high loading rates. Roughing filters typically have a design load ranging from 1.6 to 4.8 kg $BOD_5/m^3 \cdot d$ (100 to 300 lb BOD_5/d/1000 cu ft). Because much of the BOD passes all the way through the tower, the microorganisms in the biofilm are exposed to an excess of BOD and nutrients throughout the entire depth of the tower. This means that all of the microorganisms can grow at their maximum rate provided that environmental conditions (pH and DO) are suitable. It also means that removal rates are maximized over the entire tower. In other types of trickling filters, the microorganisms in the lower portion of the tower are substrate limited and can't grow at their maximum rate. Only a portion of these trickling filters is operating at maximum efficiency.

Since the 1980s, most trickling filters have been designed with plastic media rather than rock media. Plastic media allow for higher organic loading rates and often achieve high percentages of ammonia removal. Modern trickling filters can be better classified as roughing, carbon oxidation, combined carbon oxidation and nitrification, and tertiary nitrification filters. Although the terminology has changed to better reflect the treatment goals of the different types of filters, comparing the ranges in Table 7.1 and Table 7.2 shows clear parallels. Carbon oxidation filters are comparable to high-rate filters and carbon oxidation and nitrification filters are comparable to intermediate-rate filters. Tertiary filters follow secondary treatment and clarification. Table 7.2 summarizes characteristics and average performance ranges of these trickling filter categories (WEF, 2010).

Trickling filters can be built with several different process configurations and flow patterns such as single-stage, two-stage, and series or parallel operation. As noted previously, trickling filters are almost always preceded by grit removal, screening, and primary clarification to minimize the amount of solid material entering the filter. Single-stage trickling filter systems have two or more filters that are operated in parallel (side-by-side) (Figure 7.10). The OLRs and HLRs applied to each trickling filter are the same. One filter can be removed from service while maintaining flow to the other filter(s). Removing a filter from service will increase the organic load to the remaining filters. Each filter may have its own recycle line or a single recycle line may pull water from the combined effluent and recirculate it to a point just downstream of the primary clarifier.

Two-stage trickling filter systems have two filters, which operate in series (Figure 7.11). Two-stage systems may have an intermediate clarifier installed between the two trickling filters. Recirculation is almost universally practiced at two-stage facilities. In the system shown in Figure 7.11, the operator is able to recycle treated wastewater from the effluent of either trickling filter or to send effluent from the second trickling filter to the inlet of the first trickling filter. The recirculation configuration can be varied to produce the best effluent for the particular conditions of wastewater strength and other characteristics. Series operation sometimes increases the overall BOD removal performance because of increased efficiency in the higher loaded first stage. In a two-stage system, the second trickling filter may also remove ammonia.

Table 7.2 Modern Trickling Filter Categories

Operating Characteristics	Trickling Filter Function			
	Roughing	Carbon Oxidation	Carbon Oxidation and Nitrification	Tertiary Nitrification
Media type[a]	VF	Rock, RA, XF, or VF	Rock, RA, XF	Rock, RA, XF
Wastewater source	Primary Effluent	Primary Effluent	Primary Effluent	Secondary Effluent
Hydraulic loading, $m^3/m^2 \cdot d$ (gpm/sq ft)	52.8–1278 (0.9–2.9)	14.7–88.0 (0.25[b]–1.5)	14.7–88.0 (0.25[b]–1.5)	35.2–88.0 (0.6[b]–1.5)
Hydraulic loading, gpd/sq ft	1300–4180	360–2160	360–2160	860–2160
BOD_5 loading, $kg/m^3 \cdot d$ (lb BOD_5/d/1000 cu ft)	1.6–3.52 (100–220)	0.32–0.96 (20–60)	0.08–0.24 (5–15)	N/A
NH_3-N, $kg/m^2 \cdot d$ (lb NH_3-N/d/1000 sq ft)	N/A	N/A	0.2–1.0 (0.04–0.2)	0.5–2.4 (0.1–0.5)
$CBOD_5$ removal, percent[a]	40–70	N/A	N/A	N/A
Effluent $CBOD_5$, mg/L[c]	Percentage	15–30	<10	<10
Effluent NH_3-N, mg/L[b]	No removal		<3	0.5–3.0
Media depth, m	0.90–6.10 (3–20)	<12.2 (<40)	<12.2 (<40)	<12.2 (<40)
Predation	Negligible	Beneficial	Detrimental (nitrifying biofilm)	Detrimental

[a]Media types include VF = vertical flow, RA = random pack, and XF = cross-flow; applicable to shallow trickling filters.
[b]Concentration remaining in the clarifier effluent stream.
[c]Conversions:

gpm/sq ft × 58.674 = $m^3/m^2 \cdot d$ (cubic meters of flow per day per square meter of TF plan area).
ppd BOD_5/d/1000 cu ft × 0.0160 = $kg/d \cdot m^3$ (kilograms per day per cubic meter of media).
ppd NH_3-N/d/1000 sq ft × 4.88 = $g/m^2 \cdot d$ (grams per day per square meter of media).

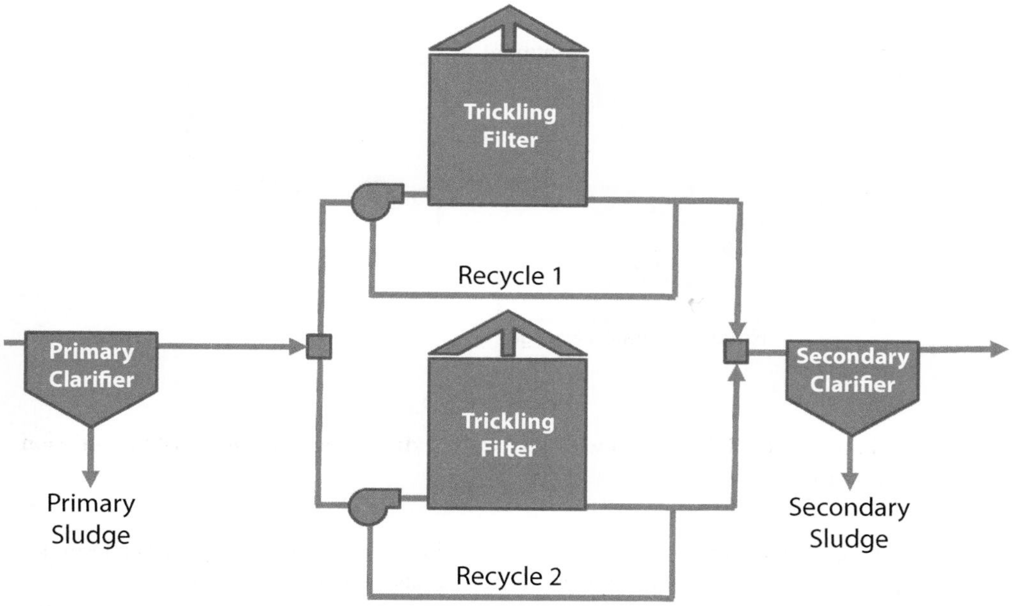

Figure 7.10 Single-Stage Trickling Filter in Parallel Operation (Reprinted with permission by Indigo Water Group)

Alternating double filtration consists of two trickling filters operating in series. The first-stage trickling filter receives a high organic load, which causes the biofilm to grow at a high rate. The second-stage trickling filter receives effluent from the first stage, which has a much lower organic load that results in a lower biological growth rate and less biofilm development. The order of the trickling filters is exchanged every 3 to 7 days by changing control valve positions. When the configuration is reversed, the first-stage trickling filter becomes the second stage and vice versa. The filter positions are rotated to encourage more uniform biofilm development over the full depth of the media in both trickling filters. The much lower organic load applied in the second-stage position causes excess biofilm to slough, which helps prevent media clogging. Nitrifying trickling filters (NTFs) operating in the alternative double filtration mode instead of a single-stage NTF achieved improved ammonia removal because of virtual elimination of dry spots on the media and better utilization of the media surface area (Aspegren, 1992).

Alternating double filtration can accommodate high organic loadings by using the first-stage filter as a roughing filter followed by a second-stage filter with much lower organic loadings. The cost of double pumping of the entire facility flow is a potential disadvantage. However, operating in series effectively doubles the HLRs, which may reduce or eliminate the need for recirculation pumping to control biofilm thickness and prevent media clogging. Operating trickling filters in parallel requires recirculation of trickling filter effluent at rates that may equal or exceed the cost of double pumping for series operation.

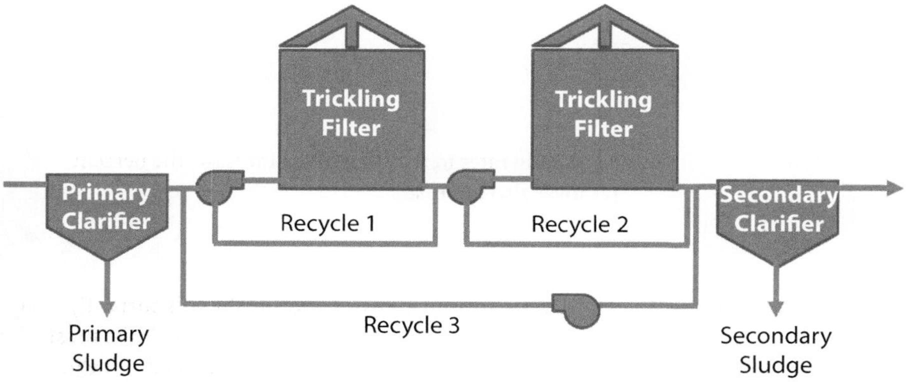

Figure 7.11 Two-Stage Trickling Filter (Reprinted with permission by Indigo Water Group)

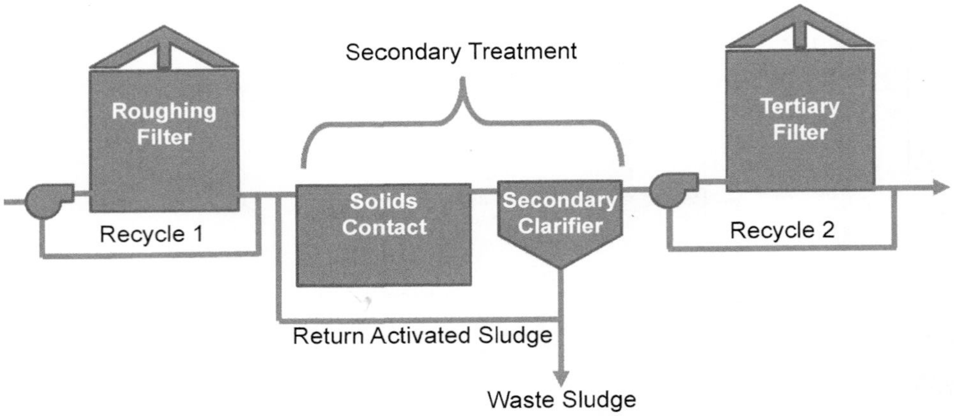

Figure 7.12 Trickling Filter Solids Contact with Tertiary Trickling Filter (**Reprinted with permission by Indigo Water Group**)

The main disadvantage of double filtration is the need for additional interconnecting piping and valves and increased operator attention to routinely alternate the first- and second-stage filters.

Trickling filters can be combined with other treatment processes. Figure 7.12 is a schematic of South Platte Water Renewal Partners in Englewood, Colorado. An upstream roughing filter removes approximately 70% of the incoming BOD load. The remaining BOD and sloughed biofilm pass into a small activated sludge process. It is so small that it has a special name: solids contact. In the solids contact process, the remaining BOD is removed and the sloughed solids plus new biomass produced in the solids contact basin are gently mixed to help the particles stick together and form larger particles. Coupling a roughing filter with a solids contact process decreases the amount of energy needed to provide oxygen to the microorganisms. After separating the solids from the treated wastewater in a secondary clarifier, the water passes to another trickling filter. This is a tertiary trickling filter. Because all of the BOD has already been removed, the nitrifying bacteria dominate in this type of filter. Ammonia is converted to nitrite and nitrate in the tertiary filter. Trickling filters can be used in combination with other processes as well, including RBCs.

EXPECTED PERFORMANCE FROM TRICKLING FILTERS

Trickling filter performance is limited by OLRs and HLRs and characteristics of the incoming wastewater, including the percentages of particulate and soluble material and temperature. For domestic wastewater that is coming primarily from homes and light commercial activities like schools and stores, a trickling filter can remove 85 to 95% of the BOD, TSS, and ammonia applied to the process. Trickling filters are capable of meeting *secondary treatment standards* and reducing ammonia concentrations below 2 mg/L as N. Settling to remove sloughed biofilm and other solids is needed before these effluent limits can be met. Expected performance for different loading conditions is shown in Table 7.2.

Secondary treatment standards require the TSS and BOD_5 in the final effluent to be less than 30 mg/L as a monthly average and below 45 mg/L as a 7-day average.

TEST YOUR KNOWLEDGE

1. Trickling filters are classified according to their organic loading rate.
   - ☐ True
   - ☐ False

2. An intermediate rate trickling filter may remove both BOD and ammonia when lightly loaded.
   - ☐ True
   - ☐ False

3. As loading rates to trickling filters increase, the percent removal also increases.
   - ☐ True
   - ☐ False

4. The organic loading rate to a trickling filter is currently near 1.5 kg/m^3·d (100 lb BOD_5/d/1000 cu ft). What is the most likely effect of reducing the loading rate to 0.07 kg/m^3·d (5 lb BOD_5/d/1000 cu ft)?

a. Effluent TSS will increase
b. Soluble BOD will pass through
c. Effluent ammonia will decrease
d. Biofilm growth will increase

5. **Trickling filters are capable of removing all of the following wastewater components EXCEPT:**
   a. CBOD
   b. TSS
   c. Fecal coliforms
   d. Ammonia

6. **Treated wastewater is often recycled back to the front of a trickling filter process to**
   a. Increase the organic loading rate
   b. Control biofilm thickness
   c. Decrease wetting rates
   d. Capture sloughed biofilm

7. **One advantage of plastic media over rock media is**
   a. Allows higher loading rates
   b. Locally available building material
   c. Shallower media beds
   d. Improved TSS removal efficiency

8. **Two-stage, alternating double filtration trickling filters processes regularly switch which filter receives primary effluent for this reason.**
   a. Ensure adequate removal of phosphorus
   b. Promote uniform biofilm growth in both filters
   c. Improve hydraulic wetting rates
   d. Decrease biofilm sloughing events

9. **A tertiary trickling filter, placed after the secondary clarifier, is intended to remove**
   a. BOD
   b. TSS
   c. Ammonia
   d. Nitrate

TRICKLING FILTER EQUIPMENT

Trickling filters are essentially tanks full of some type of media that provide some way to both distribute the wastewater over the media and collect it after treatment is complete. Media may consist of rocks, loose plastic shapes, or plastic sheeting. A diagram of a trickling filter is shown in Figure 7.13. Water is pumped through the inlet pipe to the top of the trickling filter where it is distributed over the media. Figure 7.13 shows a rotating distributor; however, fixed nozzle distributors are also available.

Wastewater is pumped into the trickling filter through the inlet pipe (A) and into the center well (B). It then flows through the arms of the rotary distributor. Rotary distributors may have two or four arms. Water rushing out of the outlet orifices (C) on the trailing edge (D) help to push the distributor so that it rotates, evenly distributing wastewater over the entire surface area of the trickling filter. Speed retarder orifices (E), also called *brake jets*, may be opened on the leading edge of the distributor arms. These orifices are set at a higher elevation. As flow to the trickling filter increases, the amount of water leaving through the speed retarder orifices increases. Wastewater flowing through these orifices pushes against the rotation of the distributor arm and may be used to control its speed. Some trickling filters are equipped with motors in the center to help rotate the distributor arm (not shown). The wastewater percolates down through the filter media (F). This particular trickling filter is filled with river rock, a common, locally available building material in many communities. Modern trickling filters are typically filled with plastic media. The media are supported and held in place by a support structure (G) and perimeter wall (H). Wastewater flows through the support structure and into the underdrain system (I). The underdrain system both collects treated wastewater and provides ventilation channels to distribute airflow evenly through the media. Most trickling filters filled with rock media have ceramic clay tile support systems, which have rectangular drain and ventilation channels, which are approximately 7.6-cm (3-in.) wide by 7.6- to 15-cm (3- to 6-in.) high. Trickling filters filled with plastic media typically have support systems consisting of short *stanchions* or piers and beams, which provide greater height between the bottom of the media and the trickling filter floor, which allows higher HLRs and improves ventilation.

Stanchions are adjustable columns used to support the underdrain system.

Note that the wastewater is dripping from the media and that neither the media nor the underdrain system are completely submerged. During normal operation, the trickling wastewater will form a thin layer over the media with most of the space between pieces of media being open space. An underdrain channel (J), outlet box (K), and outlet pipe (L) direct the treated wastewater out of the trickling filter and onto the next treatment process. The outlet gate (M) may be closed to take the filter out of service. There is a similar gate on the inlet pipe (A); however, it is not shown here. The floor of the underdrain system (N) is sloped toward the underdrain channel (J) to help treated wastewater and sloughed biomass exit the process.

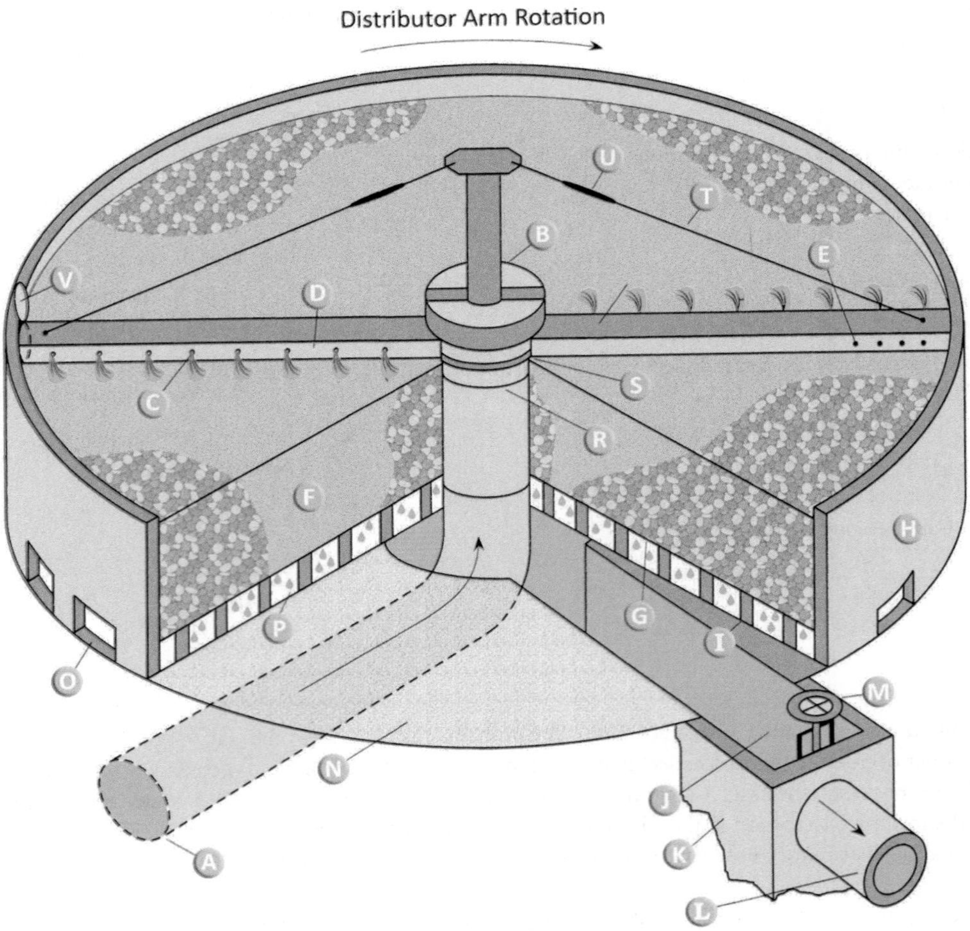

Figure 7.13 Trickling Filter Diagram (U.S. EPA, 1970)

Naturally ventilated trickling filters are equipped with ventilation ports (O) along the bottom perimeter. If present, the ventilation ports allow air to flow through the open spaces in the underdrain system between the media support system (G and I) and the floor of the trickling filter (N). This space is called the *plenum* (P) in trickling filters filled with plastic media. Trickling filters filled with rock media do not have a plenum, only the shallow drain channels through the ceramic clay tile blocks.

A *diurnal* cycle changes over a 24-hour period.

Depending on the difference in temperature between the outside air and wastewater, the airflow through naturally ventilated trickling filters may be down through the media and out through the ventilation ports or through the ventilation ports and up through the media. Airflow direction often changes *diurnally*. The direction of airflow is discussed further in the section on process variables. When the temperature of the outside air and wastewater is the same, the ventilation airflow through the media will stop or be very low and may cause the DO concentration in the wastewater to be depleted. Low or zero DO concentrations in the wastewater may cause odors during periods of low ventilation. Some trickling filters are ventilated with forced air from a fan or blower to maintain a constant ventilation airflow through the media. This is mechanical ventilation. Mechanically ventilated trickling filters do not have ventilation ports through the perimeter wall. Trickling filters are sometimes covered with domes to help retain heat and control odors. Covered trickling filters use mechanical ventilation.

The distributor consists of other components including the distributor base (R), distributor bearings (S), stay rod (T), stay rod turnbuckle (U), arm dump gate (V), and splash plates (not shown). The distributor base supports the weight of the distributor mechanism and provides a conduit for the wastewater to reach the distributor arms. The bearings support the weight of the distributor mechanism and the weight of the water flowing through it. The bearings allow the mechanism to rotate smoothly with a minimum amount of friction. The stay rods (T) help to support the weight of the distributor arms and keep them level and parallel to the media surface. Stay rods are also called *guy wires*. The stay rods will expand and contract with temperature and may require seasonal adjustment to keep the distributor arms level.

The stay rod turnbuckles (U) are used to adjust the length of the stay rods. A turnbuckle consists of a mechanical coupling with female threads inside each end. The threads on opposite ends of the turnbuckle are threaded in opposite directions (left-hand threads in one end and right-hand threads in the other). Each stay rod consists of two parts. One end of each rod is screwed into the end of the turnbuckle. Rotating the turnbuckle in one direction will unscrew both portions of the turn rod and increase the overall length. Rotating the turnbuckle in the other direction shortens the overall length. The adjustment of the turnbuckle is limited by the length of the turnbuckle and the number of threads on each piece of the stay rod.

The dump gate (V) is normally closed, but may be opened by the operator to help flush accumulated debris out of the arm. It may also be used to wash down the insides of the trickling filter walls. However, the dump gate must be normally closed to prevent wastewater short-circuiting down the inside face of the trickling filter perimeter walls instead of through the media. Dump gates are also called *flush gates*.

TEST YOUR KNOWLEDGE

1. **All trickling filters are equipped with motors, located at the top of the inlet column, to rotate the distributor arm.**
   - ☐ True
   - ☐ False
2. **In a naturally ventilated trickling filter, the direction of airflow may change during the day.**
   - ☐ True
   - ☐ False
3. **Orifices on the leading edge of a trickling filter arm**
   a. May be used to reduce arm speed
   b. Help push the distributor arm around
   c. Should all be in the full open position
   d. Rinse the outer wall of the trickling filter
4. **The underdrain system of a trickling filter**
   a. Helps distribute wastewater evenly over the media
   b. Must be cleaned daily to prevent plugging
   c. Collects treated wastewater and aids in ventilation
   d. Slopes away from the underdrain channel
5. **The open space below the underdrain system in a plastic media trickling filter is called the**
   a. Channel
   b. Ventilator
   c. Stanchion
   d. Plenum
6. **This device is used to level the distributor arms of a trickling filter.**
   a. Stanchion
   b. Turnbuckle
   c. Shims
   d. Dump gate

BIOFILM SUPPORT MEDIA

Biofilm support media may consist of rock, wood, plastic, textiles, or other materials. Support media should provide a large amount of surface area for biofilm growth in a relatively small volume. The support media must contain open spaces (voids) between media surfaces to allow both wastewater and air to flow freely through the entire volume of the trickling filter. Characteristics of different media types are listed in Table 7.3.

Rock Media

Crushed stone has historically been the most widely used media for trickling filters because it is inexpensive and readily available. Crushed slag, rounded gravel, and river rock have also been used. Rock media consist of large aggregates that are 6-cm (2.5-in.) minimum in diameter and a uniform size to provide uniform void spaces through the media. One issue with rock media is that it is nearly impossible to control

Table 7.3 Biofilm Support Media Characteristics*

				Plastic Random Media		Corrugated Plastic Sheet Media	
Parameter	River Rock	Slag	Redwood	Pall Ring	Bio-Pac	XF	VF
SSA, m^2/m^3	62	46	46	91	98	100, 138, and 223	102 and 131
(sq ft/cu ft)	(19)	(14)	(14)	(28)	(30)	(30, 42, and 68)	(31 and 40)
Void space, percent	50	60	94	95	92	95	95
Media size							
Diameter, cm (in.)	2.4 to 7.6	7.6 to 12.8	N/A	9	19	N/A	N/A
Height, cm (in.)	(0.08 to 0.25)	(0.25 to 0.42)	1	(3.5)	(7.5)	N/A	N/A
			(0.375)	9	5		
				(3.5)	(2)		
Module size	N/A	N/A	0.04 × 0.009 × 0.9	N/A	N/A	0.61 × 0.61 × 1.22	0.61 × 0.61 × 1.22
W, H, L, m (ft)			(0.125 × 0.03 × 3)			(2 × 2 × 4)	(2 × 2 × 4)
Material	Mineral	Mineral	RW	PP	PP		PVC
Dry weight, kg/m^3	1442	1600	160	44	27	24 to 45	24 to 45
(lb/cu ft)	(90)	(100)	(10)	(2.75)	(1.7)	(1.5 to 2.8)	(1.5 to 2.8)

*SSA = specific surface area; RW = redwood; VF = vertical-flow; XF = cross-flow; W, H, L = width × height × length. PVC = polyvinyl chloride plastic; PP = polypropylene plastic.

how the rocks settle into the trickling filter structure. Smaller rocks may fill some of the spaces between larger rocks and reduce the flow of wastewater and air. Narrow spaces between rocks can become overgrown with biofilm or filled with debris, which can blind off portions of the trickling filter. Rock media should be as nearly all the same size as possible, with no more than 10% by weight being elongated or flat or both. Smaller rocks should not exceed 10% by weight of the total.

Surface area is the amount of space available for biofilm growth. In general, greater surface area per volume equals more treatment capacity.

Void spaces are the open spaces between pieces of media.

Rock media have relatively little *surface area* per unit of trickling filter volume compared to plastic media, which limits the amount of biofilm that can be grown. Stone or rock media have a relatively low percentage of *void spaces* and are more prone to plugging than structured plastic media. This limits the amount of flow and organic load that a rock media filter can process. Round rock has a higher percentage of void space than crushed or angular rock. Rock media are extremely heavy and require more structural support than other types of media. Their weight is one of the factors limiting the maximum depth of rock media trickling filters. Most loose stone aggregates have a dry weight density of approximately 1282 kg/m^3 (80 lb/cu ft) compared to a density of 32 to 48 kg/m^3 (2 to 3 lb/cu ft) for plastic media (WEF, 2016).

Most rock trickling filters are 1.8- to 2.4-m (6- to 8-ft) deep. This depth range was determined to be optimal by experiments conducted in the early 1900s. Increasing the depth of the trickling filters did not improve performance. The 6- to 8-ft depth of nearly all rock media trickling filters is associated with limited ventilation rates produced by natural draft and an increased tendency for water to pond on the media surface (WEF, 2010). These performance limitations are consistent with early experimental results and with the design standards adopted by most states.

Redwood Media

Commercial wood media consist of rough sawn redwood slats stacked horizontally on wood spacers. Redwood media are no longer manufactured, but can still be found in many existing trickling filters (Figure 7.14). Redwood media are similar in appearance to pallets (Figure 7.15). The slats are 1-cm (3/8-in.) thick, by 3.3-cm (1.5-in.) wide and either 0.9- or 1.2-m (35.5- or 47.5-in.) long. The horizontal spacing between edges of the slats is 1.7 cm (11/16 in.) (Trickling Filter Floor Institute, 1970). Wood media were produced in modules (Del-Pak) to provide structure and make installation easier.

Figure 7.14 Trickling Filter with Redwood Slat Media (Reprinted with permission by Kenneth Schnaars)

The use of redwood as filter media declined because of low availability, low ratio of media surface area to volume, and higher costs than alternative materials. Additionally, the wood slats sag in the center of the modules because of the weight of the biofilm. Redwood media have been replaced with plastic media in some existing filters.

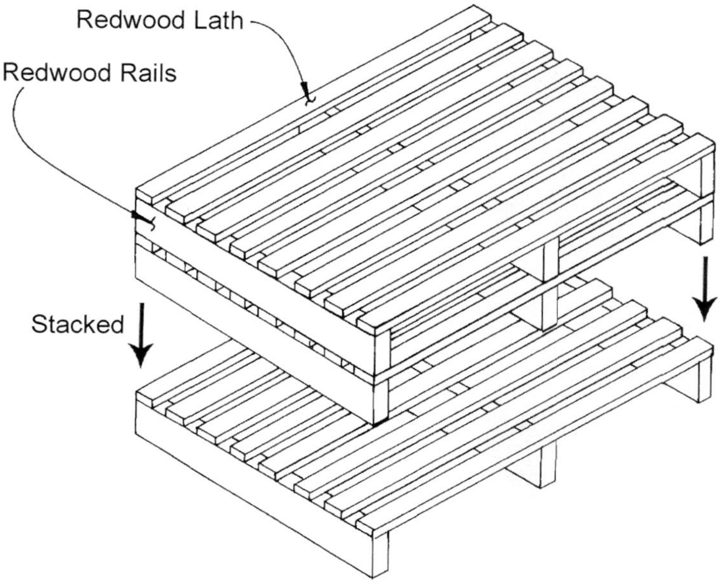

Figure 7.15 Redwood Media Module (U.S. EPA, 1977)

Plastic media trickling filters, which can have media depths up to (12 m) 40 ft, are sometimes called *biotowers* because of their depth and to distinguish them from shallower rock trickling filters.

The *footprint* is amount of surface area a process occupies. A tall narrow process has a smaller footprint than a short, wide process even though both may contain the same volume.

Plastic Media

In the 1950s, *plastic media* began to be used instead of rock media. Plastic has many advantages over rock, including having larger void spaces, providing more surface area for biofilm growth per unit of trickling filter volume, and being lighter. Large void spaces and more surface area per volume allow higher OLRs and HLRs without clogging. Larger void spaces also minimize many operational problems associated with rock media such as plugging, uncontrolled sloughing, odors, and filter flies. Increased surface area is advantageous because greater masses of biofilm can be grown in a relatively small volume. Large void spaces allow higher ventilation rates, which helps to keep the biofilm aerobic and reduces odor potential. Because plastic media are much lighter than stone or wood, it is possible to build trickling filters as tall as 12-m (40-ft) high. Tall trickling filters have a smaller *footprint* than shallow trickling filters.

Random Media

Plastic random media consist of molded shapes with internal bracing to help the media hold its shape. Three examples of random media, Bio-Pac (Raschig USA, Inc., Arlington, Texas), Bio-Rings (Raschig USA, Inc.), and Pall-Ring (The Pall Ring Company, U.K.), are shown in Figure 7.16, Figure 7.17, and Figure 7.18, respectively. Random media are installed by dumping them into the trickling filter vessel. For this reason, they are sometimes called *dump media*. Random media are light, have a high amount of surface area per volume, and are easy to install. The void spaces within random media are uniform, but the void spaces between pieces of media vary in size. Similar to rock media, small void spaces between some pieces of media tend to plug. Water and air distribution through random media may be irregular, leading to non-uniform media wetting and ventilation.

Vertical-Flow Media

Vertical-flow (VF) media are made with flat plastic sheets inserted between corrugated sheets to form larger blocks of media (Figure 7.19). Single media blocks are typically 0.6-m wide × 0.6-m tall × 1.2-m long (2 ft × 2 ft × 4 ft). The stacked sheets form channels, which are continuous from top to bottom of each module and do not intersect any other channels. Vertical-flow media provide a nearly equal distribution of wastewater while minimizing potential plugging at higher organic loading rates. This type of media is typically used in roughing filters and high-rate filters.

Cross-Flow Media

Cross-flow (XF) media consist of vacuum-formed plastic sheets with corrugations and shallow ridges or flutes that structurally stiffen the individual sheets. Individual sheets are similar in appearance to an egg flat with its peaks and valleys (Figure 7.20). Multiple sheets are assembled into modules, which measure 0.6 m × 0.6 m × 1.2 m (2 ft × 2 ft × 4 ft) (Figure 7.21). During assembly, alternating corrugated sheets are rotated 180 deg, which makes the angled corrugations run in opposite directions and creates the void

Figure 7.16 Bio-Pac Random Media (Reprinted with permission by Indigo Water Group)

Figure 7.17 Bio-Rings Random Media (Reprinted with permission by Indigo Water Group)

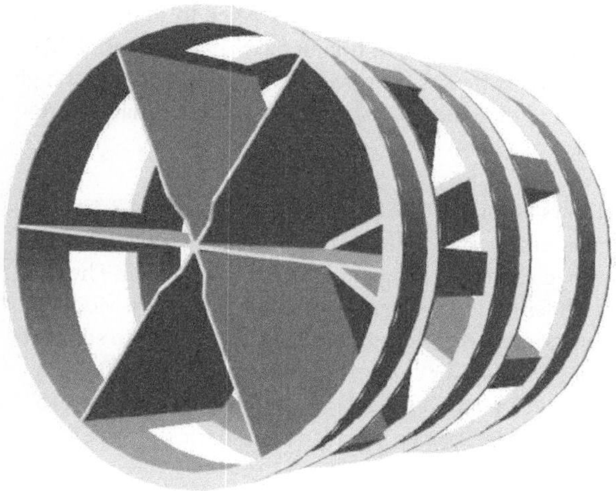

Figure 7.18 PallPAK100 Random Media (Reprinted with permission by Pall Ring Company)

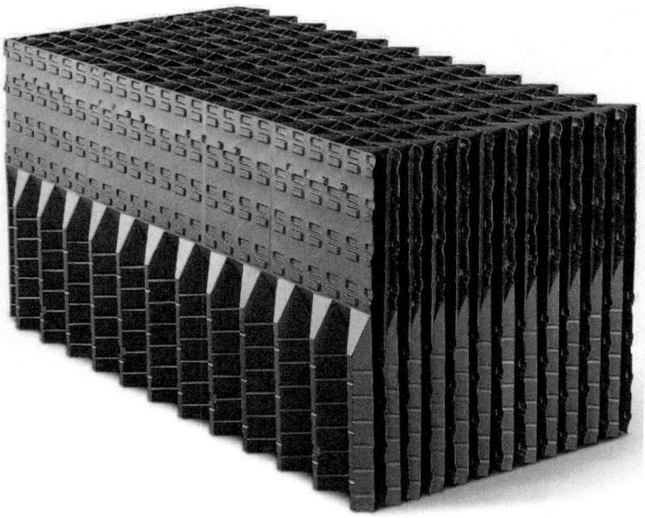

Figure 7.19 Vertical-Flow Media (Courtesy of Brentwood Industries, Inc.)

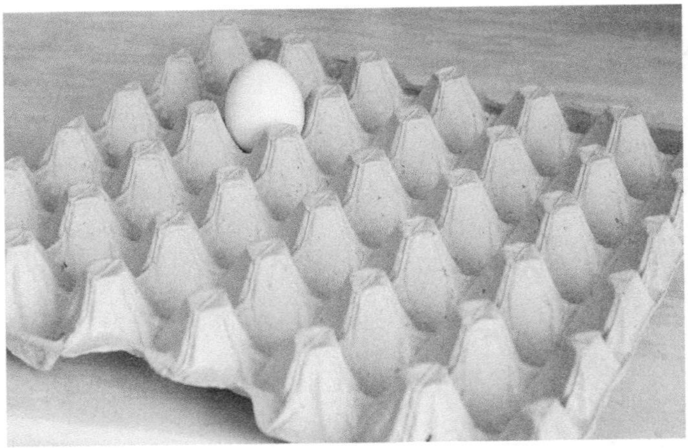

Figure 7.20 Empty Egg Flat Resembles Cross-Flow Media (Getty Images)

space within the modules. The intersections of the angled corrugations between each sheet are solvent- or heat-welded together. As wastewater trickles down over the media, each of these intersections becomes a place where wastewater splits into multiple streams and is joined by previously split streams. The overall effect both mixes the wastewater and creates areas of turbulence, which help pull oxygen from the air into the wastewater. The top and bottom of the modules have a hexagonal or honeycomb shape (Figure 7.22). The geometry of the intersecting angled corrugations and stiffening ribs give the lightweight modules high compressive strength, which makes the media strong and difficult to crush. Cross-flow media have more uniform (better) flow distribution than other media.

Cross-flow media modules are installed in parallel rows to form 0.6-m (2-ft) high layers. Figure 7.23 shows layers of plastic sheet media being installed. Several layers are stacked one upon another to fill the trickling filter vessel. The edges of the media blocks are trimmed so they fit snuggly against the outer wall. The thickness of the plastic sheets used in the media modules typically varies within different layers. Modules installed in the bottom layers use the thickest sheets and have the highest compressive strength to support the weight of upper layers. Modules with thinner sheets and lower compressive strength are installed in the upper layers where the overhead weight is lower. The top layer contains modules with thick sheets to support foot traffic and the impact of water falling onto the upper surface plane of the trickling filter.

Figure 7.21 Cross-Flow Media (Reprinted with permission by Brentwood Industries, Inc.)

Figure 7.22 Cross-Flow Media Openings (Reprinted with permission by Indigo Water Group)

Specific Surface Area

The ratio of media surface area to media volume is called the *specific surface area* (SSA). The SSA measures the amount of surface area available for biofilm growth per unit of media. For example, rock media have SSA between 46 and 60 m^2/m^3 (14 and 19 sq ft/cu ft). This means that for every 1 m^3 (1 cu ft) of media, there will be between 46 and 60 m^2 (14 and 19 sq ft) of area available for biofilm growth. Plastic packing has a SSA between 91 and 223 m^2/m^3 (28 and 68 sq ft/cu ft). Plastic media, because they are mostly empty space, can have 3.5 times or more surface area than the same volume of rock. More biofilm growth area means that more BOD or ammonia can be treated in a much smaller space. With all media types, effective upstream screening, grit removal, and primary treatment are essential to prevent clogging and to ensure efficient treatment.

Figure 7.23 Installation of Plastic Sheet Media (Reprinted with permission by South Platte Water Renewal Partners)

TEST YOUR KNOWLEDGE

1. **Match the media type to its defining characteristic.**

a. Rock	1. Also called dump media
b. Redwood	2. Used in roughing filters
c. Random	3. Low specific area to volume ratio
d. Vertical-Flow	4. Rotated sheets mix and aerate
e. Cross-Flow	5. Narrow slats like pallets

2. **Which type of media has the highest SSA?**
   a. River Rock
   b. Redwood
   c. Cross-Flow
   d. Random Pack

3. **An advantage of rock media is**
   a. Larger void spaces than random media
   b. Light weight facilitates taller towers
   c. Potential for blinding
   d. Inexpensive and locally available

4. **Voids in trickling filter media**
   a. Should be avoided in most circumstances
   b. Allow the flow of wastewater and air
   c. Are responsible for short-circuiting
   d. Consume less than 25% of the total volume

5. **As the SSA of trickling filter media increases**
   a. More biofilm can be grown in a smaller volume.
   b. The likelihood of blockages decreases.
   c. BOD and ammonia loading rates must decrease.
   d. TSS removal efficiency increases.

Poured-in-place means that the tank was built on site using forms and poured concrete. Precast concrete is manufactured elsewhere and is delivered to the site as a finished piece.

CONTAINMENT TANKS AND STRUCTURES

Trickling filters with rock media and random media typically have *poured-in-place* structural concrete tanks, which contain the media and the wastewater being treated. Steel tanks built with bolted steel panels are also used to contain rock and random media. Concrete and steel tanks support the weight of the rock or random media, biofilm, and wastewater within the tank. Some concrete and steel tanks are built as water retaining structures that can be completely filled with water.

Containment structures for trickling filters with self-supporting corrugated sheet media are often built with precast concrete, masonry, fiberglass, and other materials. These containment structures may have a waterproof membrane, lining, or coating on the interior walls to retain moisture and protect the structure from water damage. Most of these containment structures are not capable of being filled with water.

The wall of the containment structure may extend 1.2 to 1.5 m (4 to 5 ft) above the top of the filter media. The wall height above the media prevents spray from staining the exterior sides of the structure, reduces wind effects that may reduce wastewater temperatures or stall the distributors, and provides a structural base for domed covers. Domed covers help to retain heat, protect plastic media from UV light degradation, and keep out windblown debris. Covers can also be used to contain odors and direct air to air scrubbers or other odor control equipment.

Ventilation ports allow air to flow through the media, which provides oxygen to the biofilm. Many trickling filters have ventilation ports located at the base of the filter tower below the bottom of the media and around the full perimeter of the containment structure (Label #2, Figure 7.24). Trickling filter ventilation ports may have louvers or removable closure panels. These louvers or closure panels may be partially closed during cold weather to reduce the airflow rate and minimize cooling of the wastewater. Containment structures with ventilation ports at the base of the filter tower cannot be filled with water.

Trickling filters equipped with forced air ventilation systems typically do not have ventilation ports at the base of the structure. Low-pressure fans and air ducts supply air to the filter underdrain system (Label #1, Figure 7.24). Air flows upward through the media and out the top of the filter. Air always flows in the same direction in a mechanically ventilated trickling filter. Containment structures for trickling filters with forced air ventilation can be designed to be filled with water, which requires greater structural support to withstand the additional weight from the water.

Concrete and steel tanks, which are structurally strong enough to be completely filled with water, can be operated to completely submerge the media. Periodic immersion of the media controls nuisance

Figure 7.24 Ventilation Ports (Reprinted with permission by Kenneth Schnaars)

organisms such as filter flies, which require dry surfaces within the media to lay their eggs. Complete submergence of the media in wastewater also distributes organic matter uniformly to the entire media surface during the immersion period. The periodic increase in organic matter at lower levels of the media helps to grow biofilm near the bottom of the trickling filter where very little organic matter would be available when the media are not submerged.

Tanks that can be completely filled with water do not have ventilation ports located at the base of the filter tower. Instead, vertical vent pipes around the inner circumference of the walls at the tank perimeter supply air to the underdrain system.

UNDERDRAINS

Trickling filters have underdrain systems, which consist of drainage channels (rock and redwood media) or an open plenum space (plastic media) below the support media. An underdrain system for a cross-flow media trickling filter is shown in Figure 7.25. The underdrain system conveys the trickling filter effluent to a channel or pipeline that then carries it to a secondary clarifier or other treatment processes. Most trickling filters have an effluent channel in the center, which divides the filter floor into two halves that slope toward the effluent channel. In Figure 7.25, the underdrain channel (A) can be seen toward the bottom of the picture. Concrete piers (B) and media support beams may also be seen as well as ductwork for the ventilation system (C) being prepared for installation. The central effluent channel is typically offset to one side of the central support column for the rotary distributor. To facilitate cleaning and inspection, the center channel may extend through the filter wall at one or both ends and be of sufficient size to permit entrance by personnel. The underdrain system supports the media and allows air to flow through the trickling filter media.

Underdrain support systems for plastic media consist of short piers, columns, or field-adjustable stanchions, support beams, and fiber-glass-reinforced plastic (FRP) grating spanning between columns or support beams. These underdrain support systems create an open plenum approximately 0.3- to 0.6-m (1- to 2-ft) high under the media. Underdrain support systems may be constructed from concrete as shown in Figure 7.25 or FRP grating as shown in Figure 7.26. Historically, redwood or pressure-treated wood have also been used to support media, but are no longer used.

The large underdrain plenum space in plastic media trickling filter systems allows access for inspection of the underdrain support system. Inspection of the underdrain support structure should be conducted using

Figure 7.25 Underdrain Channel for a Cross-Flow Media Trickling Filter (Reprinted with permission by South Platte Water Renewal Partners)

a remote-controlled video camera so that operators do not need to enter the plenum space. If operators must enter the plenum space for inspection, confined space entry procedures must be used.

Vitrified clay underdrain blocks have been used in most rock trickling filters. The underdrain blocks are laid in staggered rows at right angles to the center drain channel and cover the entire floor area. The underdrain blocks were laid on a thin layer of dry mortar directly on the filter floor. The mortar would be wetted after the blocks were laid to bond them to the concrete floor and form a continuous underdrain system. Vitrified clay underdrain blocks are manufactured currently, and many remain in service in existing rock and plastic media trickling filters.

Figure 7.26 Plastic Media Support System with Adjustable Stanchions and FRP Grating (Courtesy of Brentwood Industries, Inc.)

A standard-size, clay underdrain block is illustrated in Figure 7.27. Clay underdrain blocks come in different sizes and configurations, but all have a few traits in common. The blocks are mostly hollow with two or three channels that run through the length of the block. Blocks either have slots cut across the width of the top of the block as shown in Figure 7.27 or have a series of shorter cuts across the top to make a grating pattern. Water passes through these openings and into the channels. Blocks are placed so the channels align between blocks to create a smooth path for the water to follow out of the filter. The bottom of the drainage channels in each filter block is curved to form a trough, which minimizes potential for solids buildup in the underdrain system.

Standard, clay underdrain blocks have hydraulic capacity up to 47 $m^3/m^2 \cdot d$ (1150 gpd/sq ft). Standard blocks have channel depths of 10 to 13 cm (4.25 to 5.125 in.). Drainage openings or slots no more than 3.8-cm (1.5-in.) wide penetrate at least 20% of the top surface of the filter blocks and form a continuous underdrain system. High-rate blocks provide the same hydraulic capacity but have deeper channel depths of 18 to 19 cm (7.25 to 7.5 in.) to increase air circulation into the filter. Vent blocks were also installed to connect pipe riser vents, which extend from the underdrain to above the media surface (Trickling Filter Floor Institute, 1970).

ROTARY DISTRIBUTOR

Most trickling filters use a rotary distributor mechanism to apply wastewater to the surface of the support media. A rotary distributor consists of a stationary central column, which supports two or more horizontal rotating arms positioned several inches above the filter media. The wastewater is fed through the center column into the horizontal rotating arms. The wastewater discharges through orifices located along one side of each of the rotating arms and is distributed over the media. Distributors with four arms can be configured so that two arms distribute low flows and all four arms distribute high flows, permitting an operating range with peak flows up to about five times the minimum.

Large rotary distributors have guy or stay rods, which support the rotating arms and allow adjustment of the distributor arms to keep them level and maintain an even distribution of wastewater over the media. The stay rods in Figure 7.29 can be seen running from the top of the center column out to the arms. Rotary distributors have quick-opening dump gates at the end of each arm to permit easy flushing of the mechanism (Figure 7.28). Dump gates are also called flush gates.

Hydraulically driven rotary distributors on a trickling filter use the force of the wastewater flowing out of the orifices along the back side of the distributor arm to rotate the mechanism, similar to a rotary lawn irrigation sprinkler. Rotation of the distributor is driven by the force of the water flowing through the orifices. On large distributors, approximately 1 revolution per minute (rpm) is a normal speed. The rotational speed of a hydraulically driven rotary distributor can usually be controlled by adjusting the amount of flow through orifices located on the front side of each rotating arm. As the water exits the orifices, it hits a splash plate, which forces the water up and out from the arm (Figure 7.30). This helps to distribute the wastewater evenly over the media surface and minimize erosion of the media. Splash plates may be made of aluminum, plastic, or other materials. A plastic splash plate is shown in Figure 7.31. Splash plates wear out and can become bent over time, but are easily replaced.

Hydraulic refers to fluid. *Hydraulically* driven means driven by the force of a fluid, in this case water.

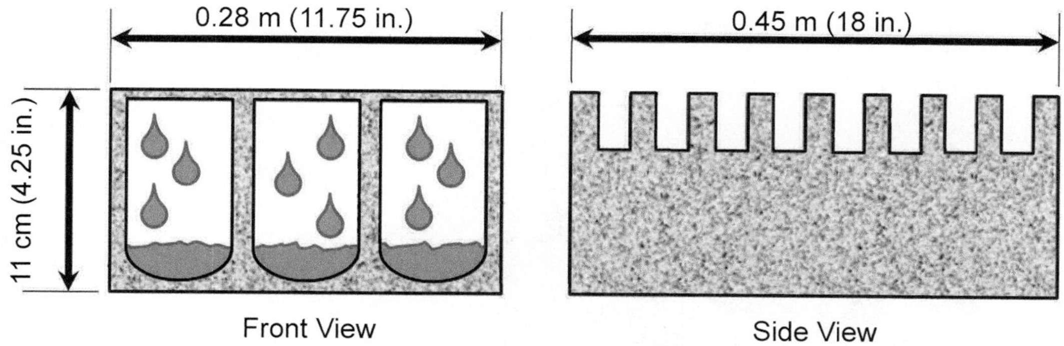

Figure 7.27 **Clay Tile Underdrain Block (Reprinted with permission by Indigo Water Group)**

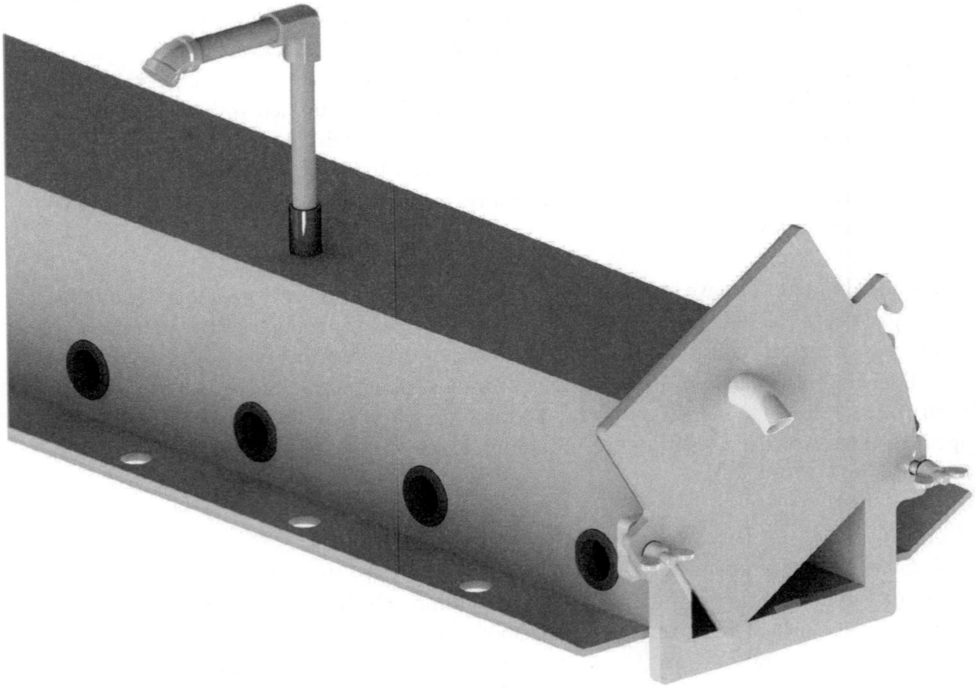

Figure 7.28 Dump Gate at End of Distributor Arm (Reprinted with permission by DBS Manufacturing)

One disadvantage of hydraulically driven rotary distributors is that high winds or partial blockage of some orifices by biofilm, debris, or ice during freezing weather may affect the rotational speed of the distributor. Mechanically driven rotary distributors use electric motors and gears to rotate the distributor mechanism. Motor-driven rotary distributors can maintain constant speed under all operating conditions.

Oscillations are side-to-side movements. In this case, oscillations refer to a rocking or wavelike movement of the entire distributor.

Rotary distributors (see Figure 7.32) can develop dynamic instability and *oscillations* of the mechanism. A small downward deflection of one of the arms increases the amount of water flowing into the arm. The additional weight of this water causes further deflection. When the water spills out rapidly, the empty arm

Figure 7.29 Electrically Driven Rotary Distributor (Courtesy of WesTech Engineering, Inc.)

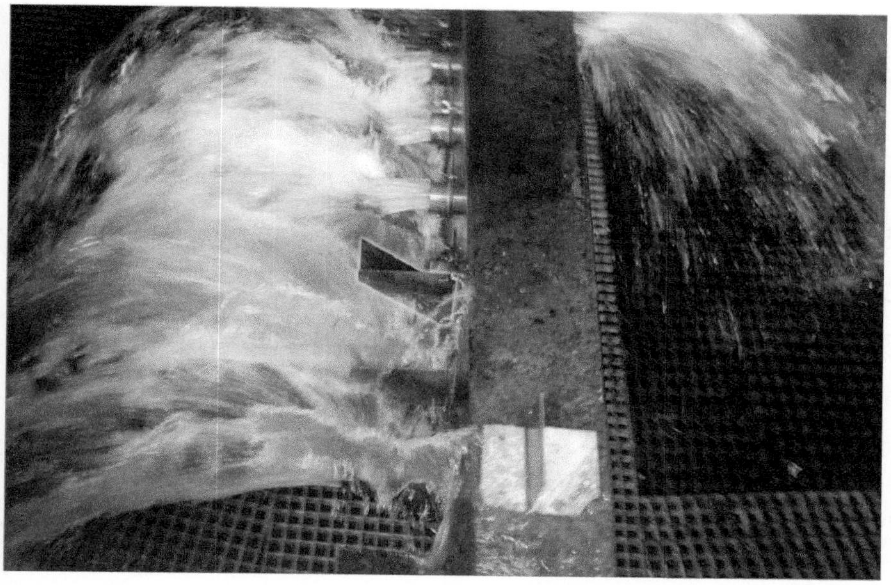

Figure 7.30 Rotary Distributor Arm with Splash Plates (Reprinted with permission by South Platte Water Renewal Partners)

will deflect upward and cause the opposing arm to deflect downward and repeat the cycle. The resulting oscillation creates structural stress on the members supporting the arms, which can cause structural damage or failure of the rotary distributor. Operators must adjust the turnbuckles on the guy- or stay rods to keep the arms level, maintain good balance, and prevent dynamic instability and oscillations of the mechanism.

Bearings

Rotary distributors are supported on spherical roller bearings, sealed tapered roller bearings, or non-sealed tapered roller bearings. Spherical roller bearings tolerate deflection and misalignment better than other types. The bearings ride on removable races (tracks) in an oil bath. The oil typically contains oxidation and corrosion inhibitors.

Figure 7.31 Plastic Splash Plate (Reprinted with permission by South Platte Water Renewal Partners)

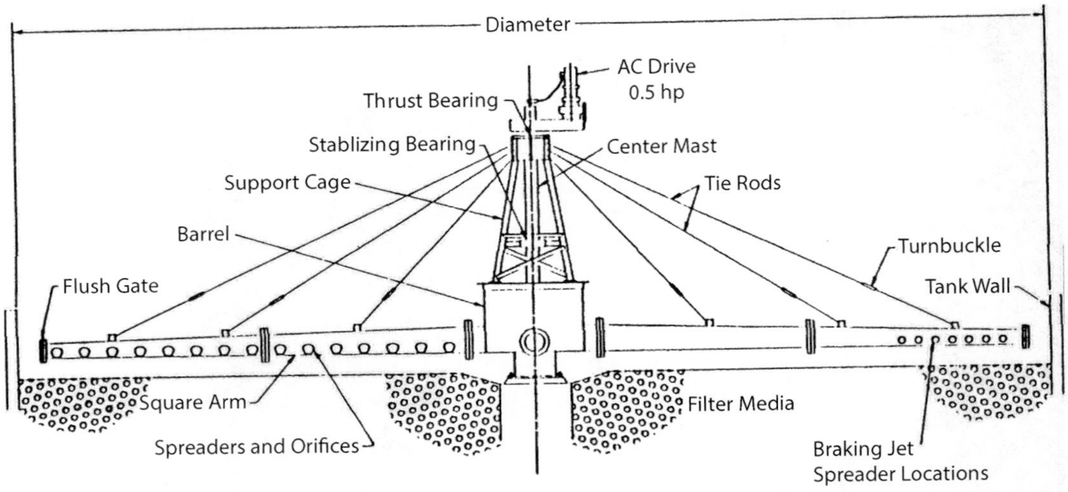

Figure 7.32 Rotary Distributor Diagram (Courtesy of WesTech Engineering, Inc.)

The bearings may be located in the base of the center column or at the top of the mast (Figure 7.33 and Figure 7.34). Bearings located at the top of the mast are better protected from corrosion or water intrusion than bearings located in the base of the center column. The bearing located closest to the motor in a mechanically driven distributor is the thrust bearing. Rotary distributors are equipped with mechanical seals at the center column to prevent leakage and protect the bearings.

Seals

Distributors require a seal to prevent water from leaking between the fixed inlet column and the rotary section. Older seal designs include water traps, mercury seals, or packed mechanical seals. Modern mechanical seals, with a double neoprene seal with a stainless steel seal ring, require no maintenance and need less head than the seal-less design. Older distributors can be upgraded with new mechanical seals.

FIXED NOZZLE DISTRIBUTOR

Fixed nozzle distributors consist of a pipe network or grid suspended above the trickling filter media. Nozzles or orifices located at a uniform spacing along the pipe network or grid distribute the wastewater over the media. The fixed nozzles are positioned to create an overlapping spray pattern, which distributes

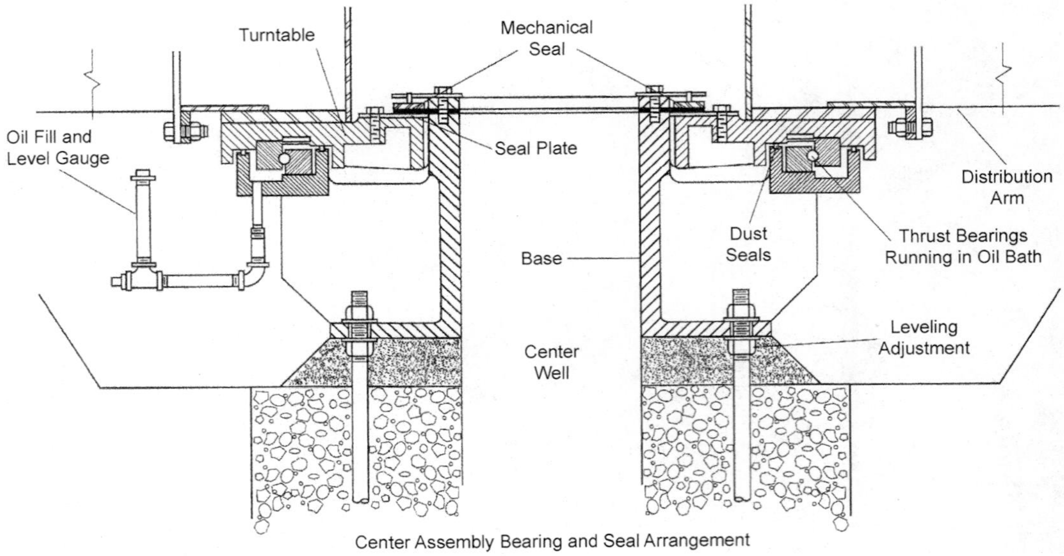

Figure 7.33 Center Assembly Bearing and Seal Arrangement (U.S. EPA, 1970)

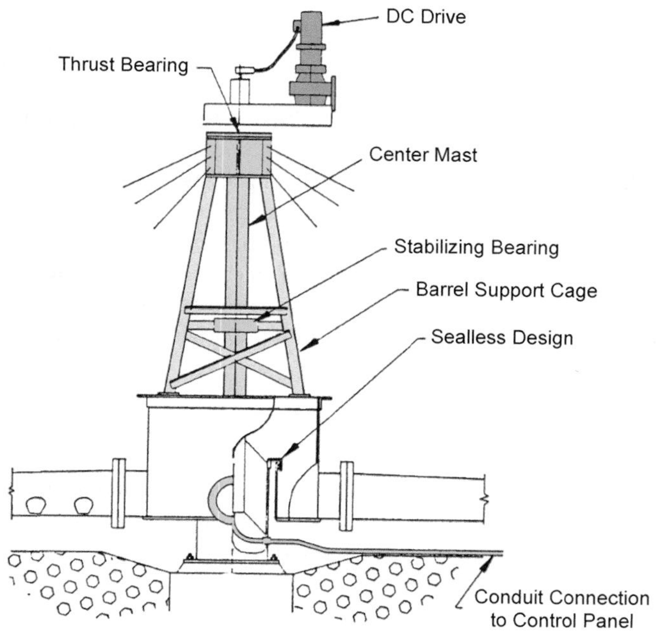

Figure 7.34 Electrically Driven Distributor Diagram (WEF et al., 2010)

the water over the whole media surface area. Trickling filters with fixed nozzle distributors are often square or rectangular. Two major disadvantages of fixed nozzle distributors are that they are difficult to access for cleaning and maintenance and that they require high flowrates. At lower flowrates, some of the nozzles in a grid system may not dispense water.

Fixed nozzle distributors are typically used only at smaller facilities. Fixed nozzle distributors typically operate intermittently. Dosing tanks are used at some facilities to distribute wastewater intermittently. Automatic siphons regulate the flowrate from dosing tanks, which eliminates the need for automatic control valves. The flow to the fixed nozzle distributors starts when the dosing tank is filled to the high water level and stops when the water surface in the dosing tank decreases to the minimum level. The dosing cycle varies in response to the facility influent flowrate.

MEDIA SURFACE PROTECTION

The biofilm growing on the support media creates a very slippery surface. To reduce the risk of slipping or falling and to protect from excessive wear, access areas should be covered with non-slip fiberglass grating. A walkway made of fiberglass grating from the entry point to the central column of the rotary distributor reduces the risk of slipping on the wet media when inspecting and maintaining bearings and seals as well as splash plates along the distributor arms.

Biofilm will grow on the grating and make it slippery also, but the grating provides a more stable surface and distributes the weight of personnel or other point loads over a larger area, reducing structural stress on the media.

Corrugated plastic sheet media are strong enough to carry the weight of wet biofilm, but the material can buckle or collapse when additional weight or force is suddenly applied to a small area. Sudden impacts on the media will damage PVC modules. Fiberglass grating over the media protects the plastic sheet media from impacts and damage by frequent foot traffic.

TRICKLING FILTER PUMPING STATION

The trickling filter pumping station lifts primary effluent and recirculated filter effluent, if any, to the distributor, which spreads the water across the top of the media. The pumps may be vertical column, submersible, or dry pit centrifugal pumps. Some WRRFs pump primary effluent to a siphon dosing tank instead of pumping directly to the distributor. A small number of WRRFs have sufficient slope across the facility site to use gravity flow to the trickling filter distributor, which eliminates the need for a trickling filter pumping station.

Trickling filters are typically elevated to allow the filter effluent to flow by gravity back to the trickling filter pumping station wet well and on the secondary clarifier or other downstream process. Most trickling filters have the ability to recycle some of their treated effluent back to the beginning of the process. The trickling filter effluent returns to the pumping station wet well where it mixes with primary effluent and is recirculated to the filter. The remainder of the trickling filter effluent (which is not recirculated) flows by gravity to the secondary clarifier or other downstream treatment units. The percentage of recirculated effluent may be controlled with a weir, which eliminates the need for a valve to control the amount of effluent returning to the pumps. The secondary clarifier effluent launder weirs typically control the water surface elevation in the trickling filter pumping station wet well.

SECONDARY CLARIFIER

Trickling filter effluent typically contains between 100 to 250 mg/L TSS. Separation of the TSS from the trickling filter effluent in the secondary clarifier is the key to good process performance. Effluent quality depends largely on being able to settle sloughed biofilm and other solids. Trickling filter performance is typically not limited by soluble BOD removal. Secondary clarifier operation is discussed in the activated sludge chapter.

TEST YOUR KNOWLEDGE

1. **Ventilation ports are typically located around the perimeter of the trickling filter and below the underdrain system.**
   - ☐ True
   - ☐ False

2. **Bearings located at the top of the mast are more likely to corrode than bearings located in the base of the center column.**
   - ☐ True
   - ☐ False

3. **The outer walls of a trickling filter often extend 1.2 to 1.5 m (4 to 5 ft) above the top of the media bed for this reason:**
   - a. Prevent maintenance workers from falling
   - b. Protects the distributor from interference by wind
   - c. Provides a staging area for new biofilm growth
   - d. Prevents wind from blowing media out of structure

4. **Plastic media trickling filters are often covered with domes to**
   - a. Maintain humidity in the plenum
   - b. Encourage the growth of algae
   - c. Reduce oxygen transfer
   - d. Protect media from UV light degradation.

5. **The louvers on trickling filter ports**
   - a. May be partially closed during cold weather to decrease airflow
   - b. Require weekly exercising to prevent them from rusting shut
   - c. Are only used with mechanically ventilated trickling filters
   - d. Must be closed prior to flooding the trickling filter

6. **Most hydraulically driven rotary distributors on trickling filters rotate at this rate:**
   - a. 1 rpm
   - b. 3 rpm
   - c. 5 rpm
   - d. 10 rpm

7. **Splash plates on a rotary distributor**
   - a. Must be replaced quarterly
   - b. Operate in the fully open or fully closed position
   - c. Prevent trash and debris from clogging the filter
   - d. Distribute wastewater over the media

8. **One disadvantage of hydraulically driven rotary distributors on trickling filters is**
   - a. Low energy usage and cost
   - b. Orifices are more prone to blockage
   - c. High winds may slow or stop the rotation
   - d. Slower rotation speeds impede biofilm growth

9. **Instability and oscillations of a rotary distributor may occur if**
   - a. Arms are not kept level and balanced.
   - b. Biofilm distribution on media is uneven.
   - c. Wastewater flows are split evenly between arms.
   - d. Roller bearing plate becomes worn.

PROCESS VARIABLES FOR TRICKLING FILTERS

DISSOLVED OXYGEN

Dissolved oxygen is needed to sustain the aerobic microorganisms in the fixed-film process. As water flows over the biofilm on the media, oxygen is transferred to both the water and biofilm. When water is recycled in a fixed-film process, the presence of a high concentration of DO in the fixed-film underflow or treated effluent does not necessarily mean that this same concentration is available in the trickling filter's interior. Without adequate oxygen, BOD and ammonia removal efficiency will decline and the biofilm may generate odors.

ORGANIC LOADING RATE

The *organic loading rate* is the mass of BOD_5 applied to the filter divided by the volume of filter media. It is expressed as kilograms of BOD_5 applied per cubic meter of filter media per day ($kg/m^3 \cdot d$) or pounds of BOD_5 applied per day per 1000 cubic foot of media per day (lb BOD_5/d/1000 cu ft). The organic loading may be calculated using either of the following two formulas:

The *organic loading rate* may also be calculated using carbonaceous BOD (CBOD) or chemical oxygen demand (COD). Some facilities prefer to use COD for process control.

In International Standard units:

$$\text{Organic Loading Rate, } \frac{\text{kg BOD}_5}{\text{m}^3 \cdot \text{d}} = \frac{\text{Organic Load, kg BOD}_5/\text{d}}{\text{Volume, m}^3} \tag{7.1}$$

In U.S. customary units:

$$\text{Organic Loading Rate, } \frac{\text{lb BOD}_5}{\text{d} \cdot 1000 \text{ cu ft}} = \frac{\text{Organic Load, lb BOD}_5/\text{d}}{\text{Volume, 1000 cu ft}} \tag{7.2}$$

Mass may be calculated as either kilograms or pounds.

In International Standard units for calculating mass:

$$\text{kg/d} = \frac{\left[\left(\text{Concentration, } \frac{\text{mg}}{\text{L}}\right)(\text{Flow, m}^3/\text{d})\right]}{1000} \tag{7.3}$$

To find kilograms, use the volume in cubic meters in place of flow in cubic meters per day.

Alternatively, use dimensional analysis:

$$\frac{X \text{ mg}}{\text{L}} \left| \frac{1 \text{ g}}{1000 \text{ mg}} \right| \frac{1 \text{ kg}}{1000 \text{ g}} \left| \frac{1000 \text{ L}}{1 \text{ m}^3} \right| \frac{1\,000\,000 \text{ L}}{1 \text{ ML}} \left| \frac{X \text{ ML}}{\text{d}} \right| = \text{kg/d} \tag{7.4}$$

An "*X*" shown in dimensional analysis indicates an unknown number. Think of this as your starting value. When converting 20 mg/L into kilograms per day or pounds per day, the number 20 replaces the first "*X*". Flowrate replaces the second "*X*". Formulas for filter area and volume are given in the front of the book on the ABC formula sheets.

In U.S. customary units for calculating mass:

$$\text{Pounds per day} = \left(\text{Concentration, } \frac{\text{mg}}{\text{L}}\right)(\text{Flow, mgd})\left(8.34 \frac{\text{lb}}{\text{mil. gal}}\right) \tag{7.5}$$

To find pounds, use the volume in million gallons in place of flow in million gallons per day.

Alternatively, use dimensional analysis:

$$\frac{X \text{ mg}}{\text{L}} \left| \frac{1 \text{ g}}{1000 \text{ mg}} \right| \frac{1 \text{ kg}}{1000 \text{ g}} \left| \frac{2.204 \text{ lb}}{1 \text{ kg}} \right| \frac{3.785 \text{ L}}{1 \text{ gal}} \left| \frac{1\,000\,000 \text{ gal}}{1 \text{ mil. gal}} \right| \frac{X \text{ mil. gal}}{\text{d}} \left| = \frac{\text{lb}}{\text{d}} \right. \tag{7.6}$$

Where recirculation is used, the organic matter remaining in the trickling filter effluent is added to the incoming BOD, which increases the organic loading to the filter. However, this added organic load from recycle is often omitted from calculations.

Example Calculations for Organic Loading Rate

A rock trickling filter has a diameter of 24.4 m (80 ft). The media depth is 1.8 m (6 ft) and the SSA is 49 m^2/m^3 (15 sq ft/cu ft). The average daily influent flow to the facility is 5.7 ML/d (1.5 mgd) and the recycle ratio is 1.5. The primary effluent BOD_5 concentration is 120 mg/L. The treated water leaving the filter has a BOD_5 less than 30 mg/L. Calculate the organic loading rate applied to the filter. Compare your answers to the accepted design criteria in Table 7.1 and Table 7.2.

In International Standard units:

Step 1—Find the volume of the trickling filter using the equation for volume given in the formula sheet in the front of the book.

$$\text{Volume} = (0.785)(\text{diameter})^2(\text{height})$$

$$\text{Volume} = (0.785)(24.4\ \text{m})^2(1.8\ \text{m})$$

$$\text{Volume} = 841.2\ \text{m}^3$$

Volume may also be found using a more traditional formula, also found in the formula sheet in the front of the book.

$$\text{Volume} = (\pi)(\text{radius})^2(\text{height})$$

$$\text{Volume} = (3.14)(12\ \text{m})^2(1.8\ \text{m})$$

$$\text{Volume} = 841.2\ \text{m}^3$$

Why are there two different formulas for finding the area of a circle or the volume of a cylinder? When the diameter of a circle is doubled, the area increases by a factor of 4. A 30-cm (12-in.) pipe carries 4 times as much flow as a 15-cm (6-in.) pipe when both pipes are running full and at the same velocity. Because the radius is half the diameter, the formula must be adjusted to compensate. The 0.785 factor is simply π (3.14) divided by 4. Both formulas give the same answer, so use the one you are most comfortable with.

Step 2—Convert the influent flow from ML/d into m^3/d.

$$\frac{5.7\ \text{ML}}{\text{d}}\left|\frac{1\,000\,000\ \text{L}}{1\ \text{ML}}\right|\left|\frac{1\ \text{m}^3}{1000\ \text{L}}\right| = 5700\ \text{m}^3/\text{d}$$

Step 3—Find the mass of BOD_5 applied to the filter.

$$\text{kg/d} = \frac{\left[\left(\text{Concentration, }\frac{\text{mg}}{\text{L}}\right)(\text{Flow, m}^3/\text{d})\right]}{1000}$$

$$\text{kg/d} = \frac{\left[\left(120\frac{\text{mg}}{\text{L}}\right)(5700\ \text{m}^3/\text{d})\right]}{1000}$$

$$\frac{\text{kg}}{\text{d}} = 684$$

Step 4—Find the organic loading rate.

$$\text{Organic Loading Rate, }\frac{\text{kg BOD}_5}{\text{m}^3\cdot\text{d}} = \frac{\text{Organic Load, kg BOD}_5/\text{d}}{\text{Volume, m}^3}$$

$$\text{Organic Loading Rate, }\frac{\text{kg BOD}_5}{\text{m}^3\cdot\text{d}} = \frac{684\ \text{kg BOD}_5/\text{d}}{841.3\ \text{m}^3}$$

$$\text{Organic Loading Rate, }\frac{\text{kg BOD}_5}{\text{m}^3\cdot\text{d}} = 0.81$$

The design criteria for trickling filters has the organic loading rate at 0.48 to 2.4 kg/m³·d for high-rate trickling filters (historic category) and between 0.32 and 0.96 kg/m³·d for carbon oxidation (modern category). This trickling filter should remove nearly all of the BOD_5 load, but will not convert ammonia to nitrite and nitrate.

In U.S. customary units:

Step 1—Find the volume of the trickling filter using the equation for volume given in the formula sheet in the front of the book.

$$\text{Volume} = (0.785)(\text{diameter})^2(\text{height})$$

$$\text{Volume} = (0.785)(80\text{ ft})^2(6\text{ ft})$$

$$\text{Volume} = 30\,144\text{ cu ft}$$

Step 2—Divide the volume by 1000 to get the volume in thousands of cubic feet.

$$\frac{30\,144\text{ cu ft}}{1000} = 30.144 \times 1000\text{ cu ft}$$

Step 3—Find the mass of BOD_5 applied to the filter.

$$\text{Pounds} = \left(\text{Concentration, }\frac{\text{mg}}{\text{L}}\right)(\text{Volume, mil. gal})\left(8.34\,\frac{\text{lb}}{\text{mil. gal}}\right)$$

$$\text{Pounds} = \left(120\frac{\text{mg}}{\text{L}}\right)(1.5\text{ mil. gal})\left(8.34\,\frac{\text{lb}}{\text{mil. gal}}\right)$$

$$\text{Pounds} = 1501.2$$

Step 4—Find the organic loading rate.

$$\text{Organic Loading Rate, }\frac{\text{lb BOD}_5}{\text{d}\cdot 1000\text{ cu ft}} = \frac{\text{Organic Load, lb BOD}_5\text{/d}}{\text{Volume, 1000 cu ft}}$$

$$\text{Organic Loading Rate, }\frac{\text{lb BOD}_5}{\text{d}\cdot 1000\text{ cu ft}} = \frac{1501.2\text{ lb BOD}_5\text{/d}}{30.144 \times 1000\text{ cu ft}}$$

$$\text{Organic Loading Rate, }\frac{\text{lb BOD}_5}{\text{d}\cdot 1000\text{ cu ft}} = 49.8$$

The design criteria for trickling filters has the organic loading rate at 30 to 150 lb BOD_5/d/1000 cu ft for high-rate trickling filters (historic category) and between 20 and 60 lb BOD_5/d/1000 cu ft for carbon oxidation (modern category). This trickling filter should remove nearly all of the BOD_5 load, but will not convert ammonia to nitrite and nitrate.

PERCENT REMOVAL

Percent removal is useful for assessing trickling filter performance under different operating conditions. Composite samples of the primary clarifier effluent and trickling filter effluent give information about average performance. Grab samples pulled from the same locations may be used to determine performance under peak load or other conditions. Samples collected to determine the concentration entering the trickling filter should not include recycle flow. Process performance may be evaluated by comparing removal of soluble or total BOD_5 or soluble or total $CBOD_5$. Removal percentage is calculated according to the following formula:

$$\text{Removal, \%} = \left(\frac{\text{In} - \text{Out}}{\text{In}}\right) \times 100 \tag{7.7}$$

Example Calculations for Percent Removal

The trickling filter in the above example had a feed concentration of 120 mg/L of BOD_5 from the primary effluent and discharged less than 30 mg/L of BOD_5.

$$\text{Removal, \%} = \left(\frac{\text{In} - \text{Out}}{\text{In}}\right) \times 100$$

$$\text{Removal, \%} = \left(\frac{120 - 30}{120}\right) \times 100$$

$$\text{Removal, \%} = 75\%$$

RECIRCULATION

A portion of the trickling filter effluent is often recirculated to the trickling filter inlet to maintain a minimum hydraulic loading on the biofilm. The trickling filter receives a blend of primary effluent and recirculated filter effluent. The recirculation ratio is the ratio of recycled filter effluent flow to the primary effluent flow.

$$\text{Recirculation Ratio} = \frac{\text{Recirculated Flow}}{\text{Primary Effluent Flow}} \tag{7.8}$$

The recirculation ratio is typically in the range 0.5 to 4.0, which is equivalent to 50% to 400% of the primary effluent flow. Recirculation may be accomplished by conveying a portion of the trickling filter effluent to a point upstream of the filters. The most common recirculation patterns for trickling filters are shown in Figure 7.35.

The facility details given at the beginning of the process variables section gave a primary effluent flow of 5.7 ML/d (1.5 mgd) and a recirculation ratio of 1.5. The recirculation flow is equal to 150% of the primary effluent flow. Equation 7.7 may also be used to find the recirculation flowrate if the recirculation ratio is known.

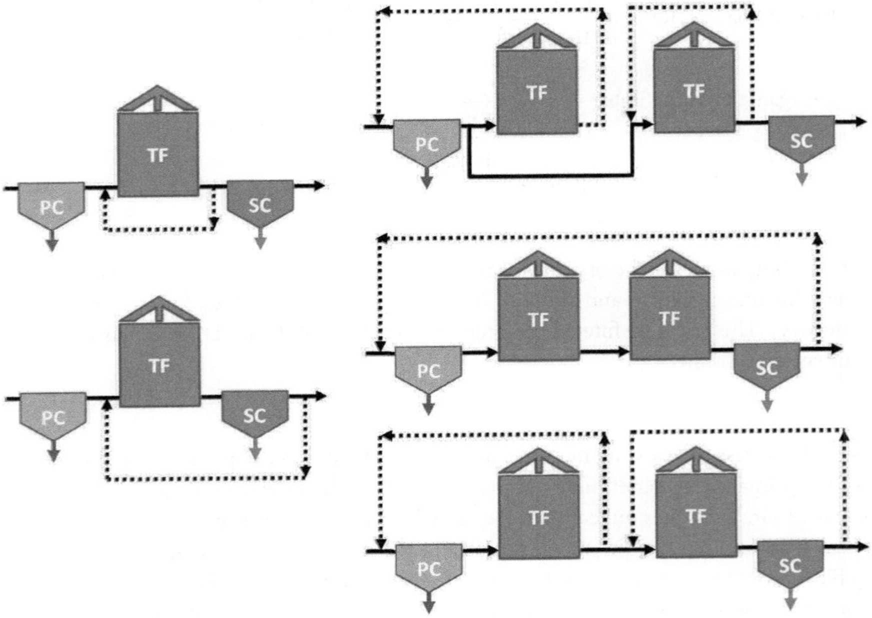

Figure 7.35 Recirculation Patterns for Single-Stage and Two-Stage Trickling Filters (WEF, 2007)

Example Calculations for Recirculation Ratio and Total Flow Applied

In International Standard units:

$$\text{Recirculation Ratio} = \frac{\text{Recirculated Flow}}{\text{Primary Effluent Flow}}$$

$$1.5 = \frac{\text{Recirculated Flow}}{5.7\ \text{ML/d}}$$

$$8.55\ \text{ML/d} = \text{Recirculated Flow}$$

In U.S. customary units:

$$\text{Recirculation Ratio} = \frac{\text{Recirculated Flow}}{\text{Primary Effluent Flow}}$$

$$1.5 = \frac{\text{Recirculated Flow}}{1.5\ \text{mgd}}$$

$$2.25\ \text{mgd} = \text{Recirculated Flow}$$

The total flow going to the filter is the sum of the incoming flow and the recirculated flow.

$$\text{Total Flow} = \text{Primary Effluent Flow} + \text{Recirculated Flow} \qquad (7.9)$$

In International Standard units:

$$\text{Total Flow} = \text{Primary Effluent Flow} + \text{Recirculated Flow}$$

$$\text{Total Flow} = 5.7\frac{\text{ML}}{\text{d}} + 8.55\ \frac{\text{ML}}{\text{d}}$$

$$\text{Total Flow} = 14.25\ \text{ML/d}$$

In U.S. customary units:

$$\text{Total Flow} = \text{Primary Effluent Flow} + \text{Recirculated Flow}$$

$$\text{Total Flow} = 1.5\ \text{mgd} + 2.25\ \text{mgd}$$

$$\text{Total Flow} = 3.75\ \text{mgd}$$

Recycling filter effluent increases the hydraulic loading on the biofilm, which has the following benefits:

- Reduced BOD_5 concentration of the wastewater being applied to the filter;
- Reduced opportunities for snail and filter fly breeding;
- Maintain distributor movement during low flows;
- Produce hydraulic shear to encourage solids sloughing and prevent ponding;
- Dilute toxic wastes, if present;
- Reseed the filter's microbial population;
- Provide uniform flow distribution;
- Prevent filters from drying out; and
- Provide additional oxygen to the upper portions of the filter.

Recirculation increases the hydraulic loading and dilutes the BOD concentration applied to the biofilm. Recirculation does not change the organic loading rate. Biofilm thickness is directly related to the hydraulic loading of the wastewater. Recirculation increases the hydraulic loading, which keeps the biofilm relatively thin.

Water can hold a maximum amount of DO. When the water contains the maximum amount, it is saturated. The *saturation concentration* depends on water temperature, atmospheric pressure, and salinity. Additional details are given in Chapter 2.

As wastewater percolates down through the trickling filter media, it absorbs oxygen from the air and transfers it to the microorganisms in the biofilm. Oxygen use decreases as the BOD is consumed so the upper portions of the filter consume more oxygen than the lower portions of the filter. Excess oxygen often remains in the treated wastewater as it exits the filter. It is not uncommon for the filter effluent to have the DO at or near the *saturation concentration*. Recycling this high oxygen water helps provide oxygen to the uppermost portions of the filter where the organic loading is the highest.

Recirculation may be constant or intermittent and at a steady or fluctuating rate. Recirculation during periods of low influent flow to the WRRF keeps rotary distributors in motion and prevents drying of the biofilm. Recirculation in proportion to the facility flow reduces the concentration of the wastewater applied to the filter. In this scenario, the operator would maintain a constant recirculation ratio and the total flow to the filter would vary during the day. The recirculation ratio may be varied to complement changing influent flows to maintain a constant HLR and wetting rate of the biofilm. In this scenario, the operator would adjust the amount of recycle flow so that the total flow to the filter remains relatively constant over the course of the day. Maintaining a constant HLR gives the greatest amount of control over biofilm thickness and the sloughing event. However, recirculation at a constant rate consumes more energy. Some facilities operate at high recirculation rates for 2 or 3 hours per week to promote sloughing of excess biomass on a regular basis rather than allowing the biofilm to build up and slough spontaneously under uncontrolled conditions.

HYDRAULIC LOADING RATE

The hydraulic loading applied to a filter is the total volume of water, including recirculation, over the surface area of the trickling filter. It is expressed as cubic meters per day per square meter ($m^3/m^2{\cdot}d$) or as gallons per day per square foot of filter surface area (gpd/sq ft). The HLR indicates whether the amount of water applied to the trickling filter is sufficient to completely wet the media surface. Biofilm requires a wet surface to develop and grow as well as a steady supply of wastewater to provide nutrients. Dry media surfaces do not contribute to treatment and provide locations where nuisance organisms and biofilm predators such as filter flies or snails can develop. The hydraulic loading may be calculated using either of the following two formulas:

In International Standard units:

$$\text{HLR}, \frac{\text{m}^3}{\text{m}^2 \cdot \text{d}} = \frac{\text{Total Flow Applied, m}^3/\text{d}}{\text{Area, m}^2} \quad (7.10)$$

In U.S. customary units:

$$\text{HLR}, \frac{\text{gpd}}{\text{sq ft}} = \frac{\text{Total Flow Applied, gpd}}{\text{Area, sq ft}} \quad (7.11)$$

Example Calculations for Hydraulic Loading Rate

Find the HLR for the trickling filter system described at the beginning of the process variables section. Then, compare your answers to the design criteria given in Table 7.1 and Table 7.2.

In International Standard units:

Step 1—Find the surface area of the trickling filter in square meters using the formula for area of a circle given on the formula sheet in the front of the book.

$$\text{Area} = (0.785)(\text{diameter})^2$$

$$\text{Area} = (0.785)(24.4\ \text{m})^2$$

$$\text{Area} = 467.4\ \text{m}^2$$

Step 2—Convert the total flow from megaliters to cubic meters per day.

$$\frac{14.2\text{ ML}}{\text{d}}\left|\frac{1\,000\,000\text{ L}}{1\text{ ML}}\right|\left|\frac{1\text{ m}^3}{1000\text{ L}}\right| = 14\,200\text{ m}^3/\text{d}$$

Step 3—Find the HLR.

$$\text{HLR, }\frac{\text{m}^3}{\text{m}^2\cdot\text{d}} = \frac{\text{Total Flow Applied, m}^3/\text{d}}{\text{Area, m}^2}$$

$$\text{HLR, }\frac{\text{m}^3}{\text{m}^2\cdot\text{d}} = \frac{14\,200\text{ m}^3/\text{d}}{467.4\text{ m}^2}$$

$$\text{HLR, }\frac{\text{m}^3}{\text{m}^2\cdot\text{d}} = 30.4$$

The design criteria for trickling filters has the HLR at 9.4 to 36.7 $m^3/m^2{\cdot}d$ for high-rate trickling filters (historic category) and between 14.7 and 88.0 $m^3/m^2{\cdot}d$ for carbon oxidation (modern category). The calculated answer is reasonable.

In U.S. customary units:

Step 1—Find the surface area of the filter in square feet using the formula for area of a circle given on the formula sheet in the front of the book.

$$\text{Area} = (0.785)(\text{diameter})^2$$

$$\text{Area} = (0.785)(80\text{ ft})^2$$

$$\text{Area} = 5024\text{ sq ft}$$

Step 2—Convert the total flow from million gallons per day to gallons per day.

$$\frac{3.75\text{ mil. gal}}{\text{d}}\left|\frac{1\,000\,000\text{ gal}}{1\text{ mil. gal}}\right| = 3\,750\,000\text{ gal}$$

Step 3—Find the HLR.

$$\text{HLR, }\frac{\text{gpd}}{\text{sq ft}} = \frac{\text{Total Flow Applied, gpd}}{\text{Area, sq ft}}$$

$$\text{HLR, }\frac{\text{gpd}}{\text{sq ft}} = \frac{3\,750\,000\text{ gpd}}{5024\text{ sq ft}}$$

$$\text{HLR, }\frac{\text{gpd}}{\text{sq ft}} = 746.4$$

The design criteria for trickling filters has the HLR at 230 to 900 gpd/sq ft for high-rate trickling filters (historic category) and between 360 and 2160 gpd/sq ft for carbon oxidation (modern category). The calculated answer is reasonable.

FLUSHING INTENSITY

The water entering the trickling filter is applied to the growing biofilm differently depending on the rotational speed of the distributor. Think about holding a bucket full of water. The bucket is equipped with a nozzle that limits the amount of water that is dispensed. If you spin around one time quickly while pouring the water out on a driveway, the amount of water reaching the driveway will be small. If you spin around very slowly while pouring the water, much more water will reach the driveway even though it is coming out of the bucket at the same rate as when you spun around quickly. If you keep spinning, eventually the bucket will empty whether you spin fast or slow. What if the driveway was covered with dirt? Which scenario—a single fast or a single slow spin—would do a better job removing the dirt? A single slow spin would work better. Biofilm responds in a similar way. When the water is sprinkled lightly over the trickling filter by a rapidly rotating distributor, there is very little shear to break off pieces of biofilm. The biofilm is able to grow thicker. When the distributor is slowed down, more water is dumped on each patch of media. There is more shear and pieces of the biofilm break off more easily. The biofilm will be thinner.

In trickling filters, the amount of water dispersed by the distributor arm with each rotation is referred to as the *flushing intensity* or SK. Spülkraft is expressed as millimeters of water per pass (mm/pass). Spülkraft depends on the total flow entering the filter, the number of distributor arms, and the rotational speed. It is calculated using either of the following formulas.

In International Standard units:

$$\text{SK, mm/pass} = \frac{\text{HLR}, \dfrac{\text{m}^3}{\text{m}^2 \cdot \text{h}} \times 1000 \dfrac{\text{mm}}{\text{m}}}{N_a \times \omega_d \times 60 \dfrac{\text{min}}{\text{h}}} \quad (7.12)$$

In U.S. customary units:

$$\text{SK, in/pass} = \frac{\text{HLR}, \dfrac{\text{cu ft}}{\text{sq ft} \cdot \text{hr}} \times 12 \dfrac{\text{in.}}{\text{ft}}}{N_a \times \omega_d \times 60 \dfrac{\text{min}}{\text{hr}}} \quad (7.13)$$

Where

SK = Spülkraft (mm/pass or in./pass),
N_a = number of arms on the distributor, and
ω_d = distributor rotational speed (rpm).

The typical hydraulically driven distributor in North America operates in the range of 51 to 254 mm/pass (2 to 10 in./pass). Table 7.4 lists recommended ranges for rotary distributors. Higher dosing rates are recommended for higher organic loading rates to enhance biofilm thickness control. Besides a normal operating dosing rate, it may be beneficial to periodically use a higher flushing dosing rate for 5 to 10% of the 24-hour operating period. To increase the flushing intensity, decrease the rotational speed of the distributor.

Table 7.4 Operating and Flushing Dose Rates for Distributors (WEF, 2010)*

Organic Loading Rate (OLR) kg/m³·d (lb BOD_5/d/1000 cu ft)	Operating Dosing Rate mm/pass (in./pass)	Flushing Dosing Rate mm/pass (in./pass)
<0.4 (<25)	25 to 75 (1 to 3)	100 (4)
0.8 (50)	50 to 150 (2 to 6)	150 (6)
1.2 (75)	75 to 225 (3 to 9)	225 (9)
1.6 (100)	100 to 300 (4 to 12)	300 (12)
2.4 (150)	150 to 450 (6 to 18)	450 (18)
3.2 (200)	200 to 600 (8 to 24)	600 (24)

*Actual values are site-specific and vary with media type.

Example Calculations for Flushing Intensity

Find the flushing intensity for the trickling filter system described at the beginning of the process variables section. Assume that the distributor has two arms and that it takes 3.5 minutes (0.3 revolutions per minute) to make one complete revolution. Check to see if the calculated results are reasonable for the OLR received.

In International Standard units:

Step 1—Convert the HLR from $m^3/m^2{\cdot}d$ to $m^3/m^2{\cdot}h$.

$$\frac{30.4\ \text{m}^3}{\text{m}^2\cdot\text{d}}\left|\frac{1\ \text{d}}{24\ \text{h}}\right| = 1.27\frac{\text{m}^3}{\text{m}^2\cdot\text{h}}$$

Step 2—Calculate the flushing intensity.

$$\text{SK, mm/pass} = \frac{\text{HLR},\dfrac{\text{m}^3}{\text{m}^2\cdot\text{h}}\times 1000\dfrac{\text{mm}}{\text{m}}}{N_a\times\omega_d\times 60\dfrac{\text{min}}{\text{h}}}$$

$$\text{SK, mm/pass} = \frac{1.27\dfrac{\text{m}^3}{\text{m}^2\cdot\text{h}}\times 1000\dfrac{\text{mm}}{\text{m}}}{2\times 0.3\times 60\dfrac{\text{min}}{\text{h}}}$$

$$\text{SK, mm/pass} = 35.3$$

In U.S. customary units:

Step 1—Convert the total flow from gpd/sq ft to cu ft/sq ft/hr.

$$\frac{746.4\ \text{gal}}{\text{sq ft}\cdot\text{d}}\left|\frac{1\ \text{cu ft}}{7.48\ \text{gal}}\right|\left|\frac{1\ \text{d}}{24\ \text{hr}}\right| = 4.16\frac{\text{cu ft}}{\text{sq ft}\cdot\text{hr}}$$

Step 2—Calculate the flushing intensity.

$$\text{SK, in./pass} = \frac{\text{HLR},\dfrac{\text{cu ft}}{\text{sq ft}\cdot\text{hr}}\times 12\dfrac{\text{in.}}{\text{ft}}}{N_a\times\omega_d\times 60\dfrac{\text{min}}{\text{hr}}}$$

$$\text{SK, in./pass} = \frac{4.16\dfrac{\text{cu ft}}{\text{sq ft}\cdot\text{hr}}\times 12\dfrac{\text{in.}}{\text{ft}}}{2\times 0.3\times 60\dfrac{\text{min}}{\text{hr}}}$$

$$\text{SK, in./pass} = 1.4$$

The OLR for this system is 0.81 $kg/m^2{\cdot}d$ (49.8 lb BOD_5/d/1000 cu ft). From Table 7.4, the recommended flushing intensity is 50 to 150 mm/pass (2 to 6 in./pass). The operator for this system should decrease the rotational speed or increase the recycle ratio to bring the flushing intensity within the recommended range.

TEST YOUR KNOWLEDGE

1. The organic loading rate to a trickling filter is the mass of BOD_5 per square meter (square foot) of surface area.
   - ☐ True
   - ☐ False

2. Trickling filters may be operated in parallel or in series.
   - ☐ True
   - ☐ False

3. When the diameter of a pipeline is doubled, the cross-sectional area
   a. Increases by 50%
   b. Increases by 100%
   c. Increases by 200%
   d. Increases by 400%

4. One benefit of increasing recycle flows during times of low primary effluent flow followed by decreasing recirculation flows when primary effluent flows increase is
   a. Maintains constant HLR
   b. Increases biofilm thickness
   c. Maintains constant organic loading rate
   d. Reduced pumping costs

5. When the rotational speed of the distributor is increased
   a. Biofilm thickness increases.
   b. The SK value increases.
   c. The HLR will match the OLR.
   d. Recycle flows must decrease.

6. All of the following methods may be used to increase the flushing intensity for a trickling filter EXCEPT:
   a. Close all of the orifices on two arms of a four-arm rotary distributor.
   b. Decrease the rotational speed by opening orifices on the leading edge of the arms.
   c. Increase the recirculation ratio to increase the total HLR.
   d. Place an additional trickling filter into service to increase surface area.

7. Find the surface area of a trickling filter that is 24.4 m (80 ft) in diameter and 1.8-m (6-ft) tall.
   a. 116.8 m^2 (1256 cu ft)
   b. 210.2 m^2 (7536 cu ft)
   c. 467.4 m^2 (5024 cu ft)
   d. 841.3 m^2 (30144 cu ft)

8. The primary effluent flow is 5.7 ML/d (1.5 mgd) and the BOD concentration is 120 mg/L. Find the total mass of BOD in kilograms (pounds) going to the trickling filter when a recycle ratio of 1.5 is used.
   a. 684 kg (1501.2 lb)
   b. 1020 kg (2251.8 lb)
   c. 1368 kg (3002.4 lb)
   d. 1704 kg (3753 lb)

9. The primary effluent flowrate is 30.3 ML/d (8 mgd). The operator would like to increase the total flow to the trickling filter to 60.6 ML/d (16 mgd). What recycle ratio is needed to reach the desired total flowrate?
   a. 0.5
   b. 1
   c. 1.5
   d. 2

10. A trickling filter that is 41.1 m (135 ft) in diameter and 7.3-m (24-ft) deep receives 22.7 ML/d (6 mgd) of primary effluent. The influent BOD concentration is 180 mg/L and the recycle ratio is 2. Find the organic loading rate to the trickling filter in kg/m^3·d (lb BOD/d/1000 cu ft).
    a. 0.11 kg/ m^3·d (6.6 lb BOD_5/d/1000 cu ft)
    b. 0.42 kg/ m^3·d (26.2 lb BOD_5/d/1000 cu ft)
    c. 0.84 kg/ m^3·d (52.5 lb BOD_5/d/1000 cu ft)
    d. 1.68 kg/ m^3·d (104.9 lb BOD_5/d/1000 cu ft)

11. Calculate the HLR to a trickling filter given the following information. Surface area is 1326 m^2 (14 306.625 sq ft). Influent flow is 30.3 ML/d (8 mgd) and the recycle ratio is 2.
    a. 22.9 m^3/m^2·d (559.2 gpd/sq ft)
    b. 34.2 m^3/m^2·d (838.8 gpd/sq ft)
    c. 68.5 m^3/m^2·d (1677.5 gpd/sq ft)
    d. 272.6 m^3/m^2·d (6710.2 gpd/sq ft)

12. Find the SK value for a 24.4-m (80-ft) diameter trickling filter receiving a total flow of 117 ML/d (31 mgd). The filter has a rotary distributor with 4 arms. The distributor completes .3 revolutions per minute.
    a. 36.19 mm/pass (1.4 in./pass)
    b. 144.9 mm/pass (5.7 in./pass)
    c. 289.52 mm/pass (11.4 in./pass)
    d. 579.04 mm/pass (22.8 in./pass)

VENTILATION

Natural Draft Ventilation

Most trickling filters rely on natural draft to circulate air through the media and underdrain system. Natural draft occurs because of differences in air density caused by differences in air temperature and humidity between the air inside and outside of the trickling filter.

When a difference in temperature exists between any two materials in contact with one another, heat flows from the warm material to the cool material. The *heat transfer rate* is directly related to the difference in temperature. As the temperature difference between the two materials increases, the heat transfer rate also increases. Anyone who has stood on ice knows that the feet get cold faster than when standing on pavement or grass.

The *heat transfer rate* is how quickly heat moves from one area to another area.

Inside a trickling filter, a difference in temperature between the wastewater and the air causes heat to transfer from one to the other. When the wastewater is warmer than the air, heat is transferred from the water to the air. This increases the air temperature inside the trickling filter. The warmer air inside the trickling filter expands and rises because of its lower density than the cooler outside air. The rising air inside the trickling filter flows out the top of the media and draws fresh outside air into the bottom of the trickling filter.

When the wastewater is warmer than the outside air, air movement will be upward though the tower.

When the outside air is warmer than the wastewater, heat is transferred from the air to the water inside the trickling filter. As a result, the air cools and sinks because of higher density than the warm outside air. The sinking, denser air inside the trickling filter flows out the bottom of the media and draws fresh outside air into the top of the trickling filter.

When the wastewater is colder than the outside air, air movement will be down though the tower.

During cold weather, natural draft causes air to flow upward through the trickling filter media. Oxygen transfer from the air into the water decreases the oxygen concentration in the air as it moves upward through the media. The lowest oxygen concentration in the air occurs at the top of the trickling filter. This is where the highest organic loading and maximum oxygen demand also occur. As a result, oxygen transfer from the air into the water is reduced during upward airflow (Schroeder and Tchobanoglous, 1976). The difference between the DO concentration in the water and the oxygen concentration in the air is smaller at the top of the tower than it is at the bottom of the tower. Just as heat moves from areas of high concentration to low concentration, DO moves from areas of high concentration to low concentration. The bigger the difference in concentration, the faster the transfer takes place. At the same time, cold water is capable of holding more DO than warm water. This is true for all gasses, which is why no one ever opens a can of soda that has been sitting in a hot car. The net effect is that upward airflow during cold weather has little or no effect on oxygen transfer into the water and biofilm.

Airflow in naturally ventilated trickling filters slows down and can even stagnate when the wastewater and air have nearly the same temperature. Natural draft ventilation may be inadequate to meet the oxygen demand in this situation. Such conditions often occur for short periods each day as the air temperature rises in the early morning and falls in the evening. Natural draft ventilation may also be inadequate during the spring and fall seasons when average daily air temperature is nearly the same as the wastewater temperature. Inadequate ventilation will cause oxygen depletion and development of anaerobic layers inside the biofilms (near the media surface) and poor trickling filter performance.

The amount of water vapor in the air (humidity) affects air density. Humid air is denser than dry air. Inside the trickling filter, water evaporates into the air and saturates the air with water vapor. The dense humid air inside the trickling filter tends to sink. During cold weather, the sinking effect of humid air is overcome by the upward flow of warming air from contact with the relatively warm water. During warm weather, the sinking effect of humid air increases the downward airflow rate.

Mechanical Ventilation

Many trickling filters use air supply and/or exhaust fans to mechanically induce airflow. Mechanical ventilation provides a constant controlled airflow rate, which is not affected by daily or seasonal temperature differences between the air and wastewater. The ventilation or air exchange rates required are based on the organic loading rate and *oxygen-transfer efficiency* from the air into the water, which ranges from 2 to 10%.

Oxygen-transfer efficiency is a measurement of how easily oxygen can be dissolved into water. The more efficient the transfer, the higher the DO concentrations will be in the water.

Mechanical ventilation systems may operate with upward or downward airflow direction through the media. Downward airflow provides fresh air with a high oxygen concentration at the top of the trickling filter where the maximum oxygen demand occurs. Downward airflow carries humid air down into media. This can help reduce winter ice formation above the media and outside the trickling filter where the saturated air would otherwise condense and freeze. Downward airflow in covered trickling filters reduces the amount of water vapor in the air and condensation of water inside the enclosure. In contrast, upward airflow increases the amount of water vapor in the air to the point that the interior air may resemble a steam room or sauna and visibility is greatly reduced. Covered trickling filters with mechanical ventilation create a controlled exhaust air stream, which can be directed to an odor control system.

Air pressure losses through synthetic trickling filter media are typically low, often less than 1 mm (0.01 in.) of water column per meter (feet) of media depth (Grady et al., 1999). Because there is very little back-pressure in a trickling filter, a small fan may be used instead of a blower. The fan-discharge pressure typically ranges from approximately 20 to 30 mm (0.5 to 0.75 in.) of water column. Trickling filters at a medium-size facility may only require fan motors of 3.7 to 4.6 kW (5 to 7.5 hp). Activated sludge processes, on the other hand, may have multiple blowers operating at 111 to 222 kW (150 to 300 hp) each.

To distribute the airflow uniformly through the trickling filter media, covered trickling filters with mechanical ventilation systems typically have air supply ducts inside the enclosure above the rotary distributor. Air ducts are not usually installed under the media in the underdrain plenum where access for maintenance would be very limited and biofilm growth on ductwork could interfere with air distribution.

Media, Underdrain, and Biofilm Effects on Ventilation

Airflow is impeded by narrow, twisting air passages through the trickling filter media. The orifices and drainage slots in vitrified clay underdrain blocks also restrict airflow. Biofilm growing on the media surface decreases the open area through void spaces in the media and may also partially block the orifices and drainage slots in the underdrain blocks. Excessively thick biofilm can completely block portions of the media and underdrain blocks. To maintain sufficient airflow, trickling filters must be operated to eliminate excessively thick biofilm.

Rock trickling filters, which rely on natural draft to circulate air through the media, typically have vertical vent stacks installed at the trickling filter perimeter. The vent stacks allow air to flow into the underdrain system when air flows upward through the media. The vent stacks exhaust air from the underdrain system when air flows downward through the media. Rock media trickling filters typically have a minimum vent cross-sectional area of 15% of the trickling filter cross-sectional area. The underdrain channels are sized for less than 50% submergence at maximum HLRs to provide adequate space for air movement.

TEMPERATURE

During cold weather, biological reaction rates decrease within the biofilm and the activity of macroinvertebrates also decreases. When low temperatures prevail, worms and fly larvae move from the surface layers into the interior of the media and fungal growths may increase. The reduction of biofilm *predation* by macroinvertebrates at low temperatures may result in a thicker biofilm, which may cause partial or complete clogging of the media. When wastewater temperatures increase, the accumulated excess biomass tends to slough off fairly rapidly. This can increase the concentration of suspended solids in the effluent. This spring season sloughing has been attributed to cell death within the biofilm and to increased activity of the macroinvertebrate populations in response to increasing temperatures.

Predation might seem like an odd word choice, but snails, filter flies, and worms all graze on the biofilm. These predators help keep the biofilm a healthy thickness.

Evaporation within the trickling filter tends to decrease the temperature of the wastewater as it passes through. In a well-ventilated trickling filter, the temperature of the wastewater may decrease by a few degrees Celsius across the filter during extreme cold weather. To achieve good treatment during extreme cold conditions, monitor influent and effluent temperatures. If the temperature decreases by 1 °C or more, 25 to 50% of the air inlets should be closed to reduce the ventilation rate and prevent excessive cooling.

In cold climates, trickling filters are often covered. During very cold weather, ice may accumulate on the media surface or walls of exposed trickling filters and BOD removal may be impaired. Wind will blow water droplets from the distributor spray and snow, if present, to the downwind side of the filter. Ice deposits may build up there. In very cold areas, covers can reduce or eliminate ice formation, reduce heat losses, and improve overall performance.

Attempting to conserve heat in the wastewater as it proceeds through treatment may sometimes reduce the adverse effects of cold temperatures. The following are some ideas on how this can be accomplished:

- Removing a unit from service,
- Covering the fixed-film process,
- Reducing recirculation,
- Using forced rather than natural ventilation,
- Reducing the settling tank's hydraulic retention time (HRT),
- Operating in parallel rather than in series,
- Adjusting orifice and splash plates to reduce spray,
- Constructing windbreaks to reduce wind effects,
- Intermittently dosing the filter,
- Opening dump gates or removing splash plates from distributor arms, and
- Covering open sumps and transfer structures.

pH AND ALKALINITY

Trickling filters are biological treatment processes. The pH entering and leaving the process should be near pH 7. If the pH drops below 6.5, consider adding a *base* to the influent flow for pH adjustment. For trickling filters that remove ammonia, the effluent alkalinity should be at least 100 mg/L as $CaCO_3$. When pH and alkalinity are too low, biological activity will be inhibited.

A *base* is a chemical that has a pH higher than 7. Commonly used bases include sodium hydroxide (caustic or caustic soda), sodium carbonate (soda ash), calcium oxide (lime), and calcium hydroxide (slaked lime).

MEDIA PLUGGING AND PONDING

Water accumulating on the upper surface of the media, called *ponding*, occurs when the flow of wastewater through the media slows down or stops completely. Ponding may be caused by excessively high organic loading rates, accumulation of debris such as trash and grit, mats of filamentous bacteria within the media, irregularly sized media, and algae growth on the filter surface.

Excessive organic loading rates can cause rapid growth of the biofilm followed by large sloughing events. Sloughed material can become trapped in the void spaces and cause water to pond above it. Thick biofilms can grow to completely fill the void spaces in rock and redwood media. Plastic media, with their larger void spaces, are less susceptible to plugging from excessive biofilm growth. Operating with a high flushing intensity or high recirculation rate can help prevent plugging by keeping the biofilm thin and can remove biomass trapped in the media voids. Biomass trapped in the media voids may also be removed by temporarily stopping the flow of primary effluent to the filter while continuing to recirculate filter effluent through the media. Isolating the trickling filter eliminates the food source. After several hours or days, sloughing will increase, reducing the excess biomass. An isolated trickling filter may also be treated with a high concentration of chlorine (sodium hypochlorite or bleach) or high pH to remove excess biomass. When chlorinating, add chlorine to the filter influent for several hours to maintain a chlorine residual of 1 to 2 mg/L in the filter effluent. However, loss of too much biomass may lead to poor performance until new growth restores the biofilm after chemical treatment.

Ponding may also be caused by an accumulation of trash, fibrous debris, or mats of filamentous bacteria in the media voids. Debris on the upper media surface should be removed daily before it is carried deeper into the media. Debris that becomes trapped inside the media may initiate plugging as sloughed biomass is trapped above the debris. Inorganic, non-degradable debris trapped in the media voids that cannot be removed by high flushing intensity may require excavation of the media to remove it.

CAUTION! Chlorine removes biomass because it is toxic. Extreme caution should be used when dosing with chlorine. If too much is used, all of the biomass can be destroyed. It can take weeks for a damaged trickling filter to fully recover.

Ponding can also be caused by media that are not sufficiently uniform in size. In non-uniform media, the large voids between large pieces of media can be filled by small pieces of media and create narrow channels where sloughing biomass or debris can be trapped. Rock media may break down over time, especially when subjected to temperature extremes. The smaller rocks and fines will then fill the pore spaces between larger rocks. If non-uniform media causes plugging, media replacement may be necessary. Grit and other particulate matter can also fill the voids if not captured upstream by grit removal and primary clarification.

PROCESS CONTROL FOR TRICKLING FILTERS

Trickling filters have very few operational parameters that are directly under the control of the operator. Items that can be controlled include: management of upstream processes to remove particulates, recirculation ratio, and distributor rotation speed. Changes to the recirculation ratio adjust the HLR and help to ensure adequate wetting of the entire media surface. The distributor rotation speed changes the SK value and influences biofilm thickness. Operators should adjust each of these parameters as needed to keep the trickling filter within accepted design and operating values (Table 7.1 and Table 7.2) and to control biofilm thickness and sloughing.

For trickling filters equipped with mechanical ventilation, operators may be able to adjust the amount of airflow. For trickling filters with natural ventilation, the louvres on the ventilation ports may be opened to varying degrees to increase or decrease airflow through the filter. In cold weather, it can be beneficial to minimize airflow to prevent heat loss. Operators cannot control the influent flow or organic load coming into the WRRF, but opportunities may exist to take units offline or operate in stages (parallel or series filters). Trickling filters are not taken in and out of service very often because it can take days to weeks to bring one back into service.

When adjusting the treatment process or attempting to correct a problem, make only one change at a time and wait approximately one week to allow the biological process to stabilize so the effectiveness of each change can be evaluated. Operators should keep detailed operational records and graph results from sampling and process control calculations. Trend charts will help operators see small changes before they can develop into big problems.

CONTROL OF MACROFAUNA

Several strategies have been applied to manage macrofauna accumulation and/or development in trickling filters including physical, chemical, or a combination of physical and chemical applications. The key is application of a condition that is either toxic to the animals or creates an environment not conducive to their accumulation. Trickling filter flooding is an effective method for controlling filter flies and snails.

Adult filter flies are difficult to drown and are not affected by contact with most toxic chemicals, including chlorine. Filter fly eggs are highly resistant to chemical agents and can withstand periods of dehydration. However, filter fly larvae can be drowned, which disrupts the fly life cycle. The most effective filter fly control method is to submerge the media for 3 to 6 hours. Flooding prevents completion of the filter fly life cycle, which provides sufficient control of filter flies. Some trickling filters are not designed to hold water and cannot be flooded. After flooding, drain the water slowly to the facility influent to prevent peak loading on the facility. Increasing the flushing intensity (SK) and/or increasing the recirculation rate will help wash filter fly larvae out of the filter media.

Submergence of trickling filter media in a basic solution (pH 9 to 10) for 4 hours controls snails and other macrofauna. Treatment with a basic solution removed 76% of filter fly larvae at pH 9 and 99% at pH 10 in nitrifying trickling filters at Englewood, Colorado (Parker et al., 1997). Subsequent research trials demonstrated that flooding and backwash for 4 hours at pH 9 reduced the number of snails by two-thirds and re-established high nitrification efficiency (Parker, 1998).

Exposure of biofilm to 150 mg/L ammonium chloride (NH_4Cl) solution at pH 9.2 results in 100% snail mortality. The Truckee Meadows Wastewater Treatment Plant in Reno Sparks, Nevada, uses this snail control method (Gray et al., 2000). Ammonia-rich anaerobic digester centrate is directed to a nitrifying trickling filter recirculation pumping station. Sodium hydroxide is added to the recirculation flow to raise the pH to 9 (range 9.0 to 9.5) (Lacan et al., 2000). This control method is typically applied once per month. During the treatment cycle, the solution is recirculated through the isolated trickling filter for approximately 2 hours to achieve 100% mortality of adult snails and their larvae. After chemical treatment is completed, the chemical solution is returned to the head of the WRRF and the nitrifying trickling filter is then flushed with secondary effluent in the "recirculation mode" for 10 hours prior to placing the unit in service.

Physical removal of snails can be achieved by (1) screening trickling filter effluent or underflow, (2) gravity separation in low-velocity channels and removal with a dedicated pumping circuit, and (3) accelerated gravity separation using grit removal equipment.

Snail shell removal from the secondary clarifier underflow by a free *vortex classifier* has been implemented at several WRRFs. Snail shells in secondary clarifier effluent have been removed by subsequent settling in a low-velocity sedimentation basin at Montgomery, Alabama. Snail shells deposited in the basin are pumped to a static screen where they fall by gravity into a collection bin. Snail shells captured by settling in aeration basins have been removed by grit pumps and separated by grit classifiers (Tekippe et al., 2006).

Vortex classifiers are discussed in the preliminary treatment chapter.

TEST YOUR KNOWLEDGE

1. **Water entering a trickling filter has a temperature of 20 °C (68 °F). Air will rise through the filter on a hot summer day when the air temperature is 29 °C (85 °F).**
   - ☐ True
   - ☐ False

2. **Mechanical ventilation uses blowers to force air through the trickling filter media.**
   - ☐ True
   - ☐ False

3. **If the biofilm is allowed to become too thick, it will impede airflow through the filter.**
   - ☐ True
   - ☐ False

4. **During extreme cold weather conditions, this should be done to reduce heat loss from a trickling filter.**
   a. Increase recycle flows.
   b. Close 25 to 50% of air inlets.
   c. Heat the primary effluent.
   d. Double airflow.

5. **If some of the voids in trickling filter media are blocked by debris or excessive biofilm growth, this condition may result in**
   a. Dosing
   b. Ponding
   c. Abrasion
   d. Flooding

6. **This type of trickling filter media is least susceptible to plugging.**
   a. Rock
   b. Redwood
   c. Random
   d. Vertical-flow

7. **This may be used to remove excess biofilm from trickling filter media.**
   a. Sodium pentothal
   b. Sodium hypochlorite
   c. Sodium chloride
   d. Hydrochloric acid

OPERATION OF TRICKLING FILTERS

DAILY OPERATION

After normal operation has been established, perform daily observation of the following items:

- Roughness or vibration of the distributor arms,
- Leakage past the distributor turntable seal,
- Any indication of ponding,
- Plugged orifices,
- Odors,
- Splash beyond the filter media,
- Filter flies or snails, and
- Visually check the equipment at least once per shift to ensure that it is functioning properly. Verify that the trickling filter pumps operate normally and that the distributor mechanism rotates smoothly.

Recirculation

Recirculate trickling filter effluent during low-flow periods of the day and night to maintain minimum biofilm wetting rates, minimize filter fly development, and scour excess biomass from the media. Recirculation of filter effluent, rather than *secondary clarifier* effluent, does not affect the hydraulic loading on the secondary clarifier. Reduce or stop recirculation during high-flow periods, if necessary, to avoid hydraulic overloading of secondary clarifiers. The recirculation rate should be adjusted to maintain a DO concentration from 3 to 6 mg/L in the effluent from rock media and from 4 to 8 mg/L from synthetic media.

Operation of *secondary clarifiers* is discussed in Chapter 8—Activated Sludge.

Because of the large volumes of water distributed over trickling filters, large plastic storage *tubs* work well for checking the hydraulic distribution. Be sure that the distributor arm can pass easily over the tubs.

Hydraulic Distribution

Periodically check the distribution of wastewater over the filter. Place *tubs* of the same size on the media surface at several points along the radius of a circular filter. The distributor arm should then be run long enough to almost fill the pans. Stop the distributor and measure the amount of water collected in each pan. The amount of water in each pan should not differ from the average by more than 5%. If some pans collect more water than others, the orifices and/or the stay rod turnbuckles should be adjusted.

Sloughing

Sloughing is a natural part of all biofilm processes. The goal from a process control standpoint is to encourage nearly constant sloughing of smaller particles from the biofilm surface to keep the biofilm healthy and to avoid sudden detachment of large portions of biofilm. These goals are met when biofilm thickness is managed by manipulating both the HLR and flushing intensity. Thinner biofilms are associated with sloughing of smaller particles, whereas thick biofilms can detach in large sheets. Uncontrolled sloughing of large quantities of biofilm from the trickling filter media can degrade effluent quality. Excessive solids in the effluent increase chlorine demands and reduce disinfection effectiveness. To control excessive sloughing, slow down the rotation of the distributor arms by opening more reverse orifices. Calculate the Spülkraft factor and verify that it is within the recommended range for the current OLR.

Upstream Processes

If mechanical screening equipment fails, the bypass bar rack must be cleaned frequently to remove debris and prevent it from entering the trickling filter and potentially plugging the media. Frequent skimming of the primary clarifiers also will help to reduce plugging of the orifices in the rotary distributor arm.

Grit must be removed upstream of the rock trickling filters. If grit reaches the rock media, it will be trapped in and block the small voids between the media. Accumulation of grit and other debris will plug the media, reduce ventilation, and cause ponding. When this occurs, the filter must be taken out of service for cleaning. Rock media may be washed off with high-pressure water. Grit must also be removed from the underdrain. Grit has less impact on plastic, corrugated sheet media because of the large void spaces, which allow most of the grit to pass through. Some grit particles will adhere to the biofilm.

PLACING A TRICKLING FILTER INTO SERVICE

If possible, plan to bring trickling filters online during warmer weather. Biofilm grows faster at higher temperatures. Waiting for warmer weather will reduce the amount of time required to develop a mature biofilm throughout the filter. Biofilm development won't be visibly observable for several days and it may take several weeks before a mature biofilm develops through most of the filter.

Figure 7.36 shows a trickling filter system with two trickling filters being operated in parallel. This is only one of many different ways that a trickling filter system may be configured; however, startup procedures for trickling filters in different arrangements will be similar. Adapt the following steps for your facility. In Figure 7.36, each filter has its own pumping station that pumps both primary effluent and recycle flows. Some facilities have a single, common pumping station that feeds multiple filters. Some facilities may have separate pumps for primary effluent and recycle flows. The operator of this system has the option of recycling trickling filter effluent directly from the discharge of each filter back to that filter's pumping station (Recycle 1 or 2) or recycling secondary clarifier effluent back to a junction box upstream of both filters (Recycle 3). The green circles on Figure 7.36 indicate locations of valves or slide gates. The *junction box* is shown between valve locations 1a and 1b on Figure 7.36.

A *junction box* may also be called a *splitter box* or a *joiner box*. These are typically small concrete structures where process piping can be joined together, different wastewater sources can be mixed, and combined flows can be divided among different process units.

For startup of a new filter or one that has been offline long enough for the biofilm to die, recycling directly from the filter effluent returns sloughed biofilm to the top of the tower. The returned biofilm contains bacteria and other microorganisms that may reattach to the filter media and helps speed up biofilm development throughout the filter. In this example, trickling filter #1 will be placed into service.

Step 1—Before placing a new trickling filter or a trickling filter that has been offline for an extended period of time into service, operators should inspect each of these areas and remove debris, if found: 1) surface of trickling filter media, 2) interiors of fixed-nozzle distributors and individual orifices on rotary distributor arms, 3) underdrain system, 4) influent and effluent channels, and 5) pumping station wet well. Debris in

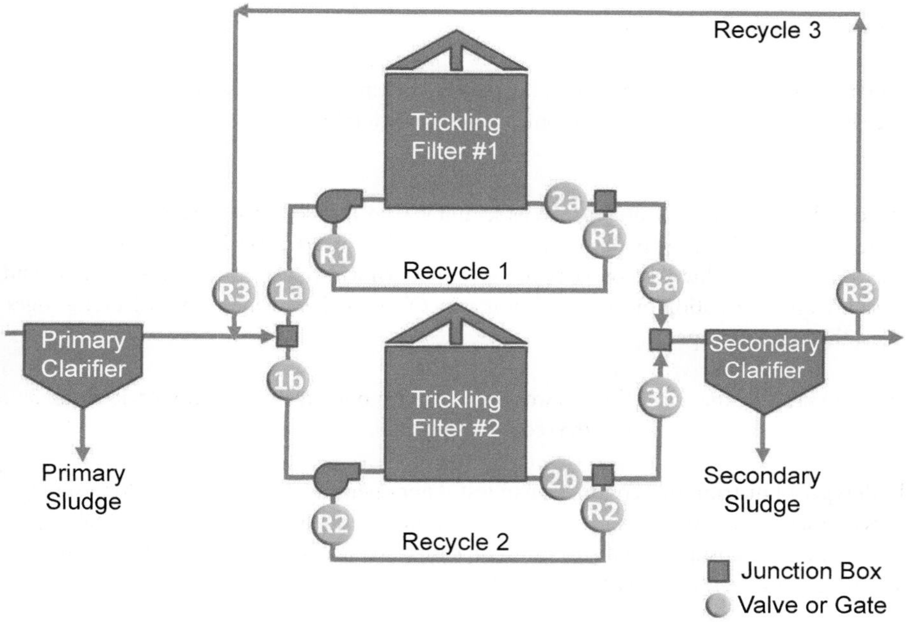

Figure 7.36 Example Parallel, Single-Stage Trickling Filter Process Diagram (Reprinted with permission by Indigo Water Group)

any of these locations could find its way into the filter media and create a blockage and/or ponding. Debris in the underdrain system and effluent channel may be recycled back to the beginning of the process.

Step 2—Exercise all valves and slide gates that will be in service. If a valve or gate is sticking or needs some other repair, that should be known before wastewater is flowing through the trickling filter. Start with valves and gates on the downstream side of the filter and work backward toward the front of the process.

Step 3—Ensure the arms on rotary distributors are level and even. Use the stay rods and turnbuckles to level the arms (if needed). Ensure that the dump gates on the ends of the arms are closed.

Step 4—Perform any manufacturer recommended maintenance, including checking the oil in the motor and distributor bearings. If water is present or the oil is very dark, replace it according to the manufacturer's recommendations prior to startup. It is extremely important that all equipment be fully serviced prior to startup as trickling filters tend to remain in service for years at a time.

Step 5—Check painted surfaces for peeling or rust. Repaint as needed.

Step 6—Inspect concrete structures for spalling, cracking, and other signs of decay. Repair as needed.

Step 7—Open the valve or gate on the trickling filter outlet pipe. In Figure 7.36, this is location 2a. All other valves and gates should be closed at this time.

Step 8—Open the valves or gates that will allow trickling filter effluent to be recycled directly to the front of the process. In Figure 7.36, these are both of the locations labeled as R1.

Step 9—Open the valve or gate at the first junction box. Primary effluent can now flow through the junction box to the pumping station for trickling filter #1.

Step 10—Start the pumps. It will take some time before water makes it all the way through the filter media and starts coming back through the recycle line. It may be necessary to pump at a lower rate initially to avoid emptying the pumping station wet well. Once steady flow has been established, the pump rate may be gradually increased.

Step 11—Verify proper operation of the distributors.

For fixed-nozzle distributors, check to see that the spray pattern is even and all media are being wetted evenly. Check to see if wastewater is being sprayed onto or past the containment wall. Adjust nozzles as needed.

For rotary distributors, listen for unusual noises such as scraping. Check to see if wastewater is flowing from all of the orifices and that the spray pattern coming off the splash plates is wetting the media evenly. Adjust splash plates as needed to achieve an overlapping spray pattern without spraying wastewater beyond the containment wall. Time the rotation of the distributor arm. Open or close orifices on the leading edges of the arms to adjust the rotational speed.

Never enter a trickling filter while it is operating or attempt to stop a rotating distributor with your body. If adjustments are needed, stop the pump and wait for the distributor to stop before making adjustments. Follow good safety practices including lockout/tagout, and confined space entry, as appropriate.

Step 12—When you are satisfied that everything is working correctly, open the discharge gate, location 3a in Figure 7.36, to allow filter effluent to flow to the secondary clarifier.

Step 13—With both types of distributors, conduct a pan test. After operating the distributors for a short period of time, check to see if the depth of water in each pan is about the same. If it isn't, additional adjustments should be made to ensure even distribution of wastewater across the media surface. If the HLR will vary, conduct several pan tests over the range of expected HLRs.

If the trickling filter will be operated at different flowrates, several pan tests should be conducted over the range of expected flows.

Step 14—If desired, switch the recycle to pull flow from the secondary clarifier effluent rather than directly from the trickling filter effluent. First, open the gates or valves on both ends of the secondary clarifier recycle line. They are shown in Figure 7.36 as locations R3. After flow has been established through the secondary clarifier recycle line, close the gates or valves on the direct recycle line. Close the downstream gate or valve first so water can drain out of the pipeline into the pumping station wet well.

Step 15—Calculate the organic loading rate, HLR, and Spülkraft value for the current and anticipated peak and average operating conditions. Compare them to the design criteria in your facility's operations and maintenance manual or, if one is not available, against the accepted design criteria in Table 7.1 and Table 7.2. If the operating parameters are out of range, adjust the recycle ratio and arm speed to bring the HLR and Spülkraft value within range. If the organic loading rate is too high, consider placing an additional trickling filter into service.

A trickling filter that has just been placed into service will not provide treatment. This takes several days to weeks as the biofilm needs time to develop. The same wastewater can be continuously recycled over the trickling filter until the effluent quality begins to improve. During this time, primary effluent will need to be diverted to another treatment process. As effluent quality begins to improve, gradually increase the amount of primary effluent sent to the trickling filter and open the discharge gate, location 3a in Figure 7.36, to allow filter effluent to flow to the secondary clarifier. If it isn't possible to bring the process online gradually without discharging partially treated effluent, heavy doses of chlorine may be used to chemically treat the remaining organic material. Be sure to check with your regulatory agency before using heavy doses of chlorine or discharging any effluent during startup that does not meet the limits in your discharge permit.

Steps for starting up pumps, clarifiers, and other equipment can be found in other chapters or in *Operation of Water Resource Recovery Facilities* (WEF, 2016).

REMOVING A TRICKLING FILTER FROM SERVICE

Step 1—If the trickling filter has its own pumping station or is served by dedicated pumps, gradually decrease the pumping rate. This will gradually shift the primary effluent flow to any trickling filters and clarifiers remaining in service.

Step 2—Stop the primary effluent and recycle pumps.

Step 3—After the distributor has stopped moving, open the dump gates on the end of each arm. Be sure to follow all safety procedures including those for lockout/tagout and confined space entry before performing any work. Lockout/tagout procedures ensure the equipment will not be restarted while maintenance is performed. If the distributor begins to move while dump gates are being opened, serious injury may occur.

Step 4—Restart the pumps for a few minutes. This will help push any debris and sloughed biofilm out of the distributor arms.

Step 5—Stop the primary effluent and recycle pumps.

Step 6—Close the pump discharge valves, gates and valves on the recycle pipeline, and gates and valves on the trickling filter influent and effluent channels. This will isolate the filter and prevent wastewater from other processes from backing up into the outlet channel. Be sure to follow procedures for confined space entry and lock-out/tag-out to prevent injuries.

Step 7—Clean the distributor and media surface and perform any necessary maintenance including replacing the oil in the motor and distributor bearings.

Step 8—Hose down the interior walls of the containment structure, the distributor arm, stay rods, and other equipment. Take care to avoid wetting the motor and its electrical components.

Step 9—Reopen the outlet channel gate or valve to allow rinse water to drain to the secondary clarifier. Reclose the outlet channel gate or valve. Use a portable sump pump to remove any remaining water from the outlet channel to prevent breeding of filter flies and other insects. Remove grit and any debris remaining in the channel.

All valves and gates should be exercised at least once each year. To exercise a valve or gate, it should be fully opened and fully closed several times. Lubricant may be applied to exposed surfaces at this time to prevent rusting and seizing.

TEST YOUR KNOWLEDGE

1. **One indication that the recirculation ratio is within the right range is an effluent DO concentration between 3 and 8 mg/L.**
   - ☐ True
   - ☐ False

2. **Taking grit basins and primary clarifiers out of service for several weeks will have no effect on trickling filter operation.**
   - ☐ True
   - ☐ False

3. **Biofilm generally grows faster at higher temperatures.**
   - ☐ True
   - ☐ False

4. **If the results from the pan test differ by more than 5%,**
   a. Orifices on the leading edge should be closed.
   b. The stay rods may need to be adjusted.
   c. Recycle ratio should be increased.
   d. Influent water temperature is too high.

5. **After making adjustments to the recycle ratio or distributor speed,**
   a. Calculate the HLR and SK value.
   b. Verify proper placement of the splash plates.
   c. Measure biofilm thickness with calipers.
   d. Decrease the effluent chlorine concentration.

6. **When placing a trickling filter into service, pumps should be started at a lower pump rate because**
   a. Media may be crushed by sudden increase in flow.
   b. Debris could be forced into the distributor nozzles.
   c. Avoids emptying the pumping station wet well.
   d. Prevents excessive stress on column bearings.

7. **How long does it take for biofilm to fully develop and provide treatment after a trickling filter is placed into service?**
   a. 24 hours
   b. 1 to 3 days
   c. Several weeks
   d. Months

8. **When removing a trickling filter from service,**
   a. Gradually decrease pumping rates before stopping the pumps.
   b. Use your hands to slow and stop the distributor.
   c. Notify your regulatory agency.
   d. Add chlorine to the distributor arms.

DATA COLLECTION, SAMPLING, AND ANALYSIS FOR TRICKLING FILTERS

The testing conducted for trickling filters varies between facilities. In deciding which tests are essential, the operator must consider treatment objectives, discharge permit requirements, downstream treatment units, and anticipated industrial wastes. Laboratory tests are described in detail in *Basic Laboratory Procedures for the Operator–Analyst* (WEF, 2012).

For trickling filters that remove BOD, trickling filter performance can be assessed by sampling the influent and effluent for BOD. Trickling filter influent samples should be collected from a location after the primary clarifier. Influent samples should not include recycle flows (if any). Influent samples should be analyzed for total BOD_5. Trickling filter effluent samples may be collected from either the trickling filter effluent or the secondary clarifier effluent. If samples are collected from the trickling filter effluent, they should be allowed to settle for at least 30 minutes prior to running the BOD test. Sloughed solids contain BOD. Remember, biological treatment processes convert soluble BOD into particulate BOD (biomass). Allowing the particulates to settle mimics the behavior of the secondary clarifier and will be a better predictor of final effluent quality. When sending samples to a contract laboratory, request a settled BOD test or a soluble BOD test for trickling filter effluent.

CBOD or COD may also be used for process control testing.

Trickling filter effluent should be monitored daily for dissolved oxygen. Record results in the facility log book. Dissolved oxygen concentrations should be greater than 2 mg/L and will often be much higher. Low DO concentrations are a concern as they could indicate excessive organic loading or inadequate ventilation. Either condition may result in anaerobic conditions, production of hydrogen sulfide, and the growth of certain filamentous bacteria.

Trickling filters that remove ammonia should be monitored for additional parameters including temperature, pH, alkalinity, ammonia, nitrite, and nitrate. Nitrification, the conversion of ammonia to nitrate, consumes 7.14 mg/L of alkalinity for every 1 mg/L of NH_3-N converted. If too much alkalinity is lost, the pH will decrease. Alkalinity will change faster than pH. If only one analysis can be completed, alkalinity is preferred over pH. Corrective action should be taken if the trickling filter effluent pH falls below 6.5 or if alkalinity falls below 100 mg/L as $CaCO_3$. Alkalinity and pH may be increased by adding sodium bicarbonate, lime, sodium hydroxide, or other *base* to the trickling filter influent.

Substances with a pH lower than 7 are acidic, higher than 7 are basic, and equal to 7 are neutral. Adding a chemical that is a *base* will increase pH. See Chapter 2 for more information on alkalinity and pH.

Monitoring the influent and effluent ammonia, nitrite, and nitrate concentrations will allow an operator to calculate percent removal. If the concentration of either parameter increases in the trickling filter effluent, the influent results can be used to determine if the increase is the result of a change in filter performance or a change in influent concentration.

Operating data that should be recorded daily or at least any time an operational change is made include: number of trickling filters in service, rotational speed of distributor, influent flow, recycle flow, and any maintenance activities. When changes are made to the recycle flow, both the HLR and Spülkraft value should be calculated and recorded. When changes are made to the distributor speed, the Spülkraft value should be calculated and recorded. The OLR should be calculated whenever BOD data are available. Verify that the calculated results are within the design parameters for the facility. Calculations should be done at least monthly even when no process changes have occurred.

An introduction to the nitrifying bacteria may be found in Chapter 5—Fundamentals of Biological Treatment. Ammonia removal is discussed in depth in Chapter 9—Nutrient Removal.

MAINTENANCE OF TRICKLING FILTERS

The elements of a maintenance program include preventive maintenance, established written procedures for emergency maintenance, and a system to ensure an adequate supply of essential spare parts. Planned maintenance will vary from facility to facility, depending on unique design features and equipment installed. Although this chapter cannot address all of these items, a summary of the most common and important maintenance tasks is listed in Table 7.5. The information provided in Table 7.5 is not equipment- or facility-specific. Consult and follow the manufacturer's literature and operating instructions.

Proper procedures need to be followed when shutting down a trickling filter for maintenance, especially for covered trickling filters in warm climates. The filter needs to be flushed for an extended period of time. Proper ventilation needs to be maintained throughout the filter to prevent heat buildup. The filter inside temperature needs to be monitored.

Table 7.5 Planned Maintenance for Trickling Filters (WEF, 2016)

Rotary Distributors
■ Observe the distributor daily. Make sure the rotation is smooth and that spray orifices are not plugged.
■ Lubricate the main support bearings and any guide or stabilizing bearings according to the manufacturer's instructions. Change lubricant periodically, typically twice a year. If the bearings are oil-lubricated, check the oil level, drain condensate weekly, and add oil as needed.
■ Time the rotational speed of the distributor at one or more flowrates. Record and file the results for future comparison. A change in speed at the same flowrate indicates bearing trouble.
■ Flush distributor arms monthly by opening end shear gates or blind flanges to remove debris. Drain the arms if idle during cold weather to prevent damage via freezing.
■ Clean orifices weekly with a high-pressure stream of water or with a hooked piece of wire.
■ Keep distributor arm vent pipes free of ice, grease, and solids. Clean in the same manner as the distributor arm orifices. Air pockets will form if the vents are plugged. Air pockets will cause uneven hydraulic loading in the filter, and non-uniform load and excessive wear of the distributor support bearing.
■ Make sure distributor arms are level. To maintain level, the vertical guy wire should be taken up during the summer and let out during the winter by adjusting the guy wire tie rods. Maintain arms in the correct horizontal orientation by adjusting horizontal tie rods.
■ Periodically check distributor seal and, if applicable, the influent pipe to distributor expansion joint for leaks. Replace as necessary. When replacing, check seal plates for wear and replace if wear is excessive. Some seals should be kept submerged even if the filter is idle or their life will be severely shortened. Verify proper seal maintenance by checking the manufacturer's recommendations.
■ Remove ice from distributor arms. Ice buildup causes non-uniform loads and reduces main bearing life.
■ Paint the distributor as needed to guard against corrosion. Cover bearings when sand- blasting to protect against contamination. Check oil by draining a little oil through a nylon stocking after sandblasting. Ground the distributor arms to protect bearings if welding on distributor and lock out the drive mechanism at the main electrical panel.
■ Adjust secondary arm overflow weirs and pan test wastewater distribution on filter as needed. The pan test is described in the section on Hydraulic Distribution under Trickling Filter Operation.
Fixed-Nozzle Distributors
■ Observe spray pattern daily. Unplug blocked nozzles manually or by increasing hydraulic loading. Flush headers and laterals monthly by opening end plates. Adjust nozzle spring tension as needed.
Filter Media
■ Observe condition of filter media surface daily. Remove leaves, large solids and plastics, grease balls, broken wood lath or plastic media, and other debris. If ponding is evident, find and eliminate the cause. Keep vent pipes open, and remove accumulated debris. Store extra plastic media out of sunlight to prevent damage from UV rays. Observe media for settling. After they are installed, media settle because of their own weight and the weight of the biofilm and water attached to their surface. Settling should be uniform and should stabilize after a few weeks. Total settling is typically less than 0.3 m (1 ft) for random plastic media, less for plastic sheet media, and nearly zero for rock. If settling is non-uniform or excessive, remove some of the media for inspection.
■ Observe media for hydraulic erosion, particularly in regions where reversing jets hit the media.
Underdrains
■ Flush out periodically if possible. Remove debris from the effluent channels.
Media Containment Structure
■ Maintain spray against inside wall of filter to prevent filter fly infestation and to prevent ice buildup in winter.
■ Practice good housekeeping. Keep fiberglass, concrete, or steel outside walls clean and painted, if applicable. Keep grass around structures cut, and remove weeds and tall shrubs to help prevent filter fly and other insect infestations. Remember, using insecticides around treatment units may have adverse effects on water quality or the biological treatment units.
Filter Pumps
■ Check packing or mechanical seals for leakage daily. Adjust or replace as needed. Lubricate pump and motor bearings as per manufacturer's instructions. Keep pump motor as clean and dry as possible. Periodically check shaft sleeves, wearing rings, and impellers for wear; repair or replace as needed. Perform speed reducer, coupling, and other appurtenant equipment maintenance according to manufacturer's instructions.
Appurtenant Equipment
■ Maintain piping, valves, forced draft blowers, and other appurtenant equipment according to the manufacturer's instructions.

DISTRIBUTOR BEARINGS AND SEALS

Distributor bearings typically ride on removable races (tracks) in a bath of oil. The oil, typically specified by the manufacturer, is selected to prevent oxidation and corrosion and to minimize friction. Because the oil level and condition are crucial to the life of the equipment, they need regular checking in accordance with the manufacturer's recommendations (typically weekly). A common procedure is to check the oil by draining approximately 0.6 L (1 pint) into a clean container. It is important to collect the sample from the lowest point as this is where the water, if any is present, will collect. If the oil is clean and free of water, it may be returned to the unit. If the oil is dirty, it should be drained and refilled with a mixture of approximately one part oil and three parts solvent. Kerosene is often used as the solvent; however, the manufacturer's recommendations for a solvent should be followed. This will help remove additional impurities and water. The distributor should be operated for a few minutes and then stopped. Drain the oil-solvent mixture and refill the distributor with clean oil.

If water is found in the oil, either the seal fluid is low or the gasket in the mechanical seals requires replacement.

Bearings often are delivered packed in a special grease to prevent damage during transportation. This packing grease must be removed and replaced with the proper grease before startup.

CAUTION! The underdrain of a trickling filter is a confined space. Never enter a confined space without following confined space entry procedures.

UNDERDRAINS

Trickling filter underdrain maintenance consists of flushing the lines or channels to remove sludge deposits or debris every 3 to 6 months. Some plastic media filters have underdrain channels that are large enough for an operator to enter and inspect the underdrain. A closed-circuit television camera may be used to inspect both rock and plastic media underdrain systems.

TEST YOUR KNOWLEDGE

1. **How often should the condensate be drained from oil-lubricated bearings in a rotary distributor?**
   a. Daily
   b. Weekly
   c. Monthly
   d. Quarterly

2. **Last week, the distributor made 1 complete revolution in 55 seconds. This week, the same distributor is making 1 complete revolution every 70 seconds. The HLR has not changed. What should the operator check?**
   a. Oil levels
   b. Guy wires
   c. Bearings
   d. Dump gates

3. **Distributor orifices should be cleaned weekly by**
   a. Removing each splash plate
   b. Spraying with high-pressure water
   c. Soaking in chlorine solution
   d. Flooding the distributor arms

4. **Ice buildup on the distributor arms should be removed as quickly as possible to avoid:**
   a. Uneven loads and bearing stress
   b. Condensate accumulation in the oil
   c. Water ponding up inside the arms
   d. Snapping the guy wires

5. **Debris should be removed from the top of the trickling filter daily to**
   a. Keep the operators busy
   b. Decrease solids loading to the filter
   c. Ensure it does not get stuck in nozzles
   d. Prevent blockages and ponding

6. **Maintaining a neat and clean landscape by weeding and keeping grass short**
   a. Increases the likelihood that windblown debris will enter the filter
   b. Enhances natural draft through the trickling filter
   c. Helps prevent filter flies and other insect infestations
   d. Extends the life of concrete and fiberglass

7. **Oil samples should be collected by draining from the bottom of the oil reservoir because**
   a. If water is present, it will be at the bottom.
   b. Pulling from the top can entrain air in the sample.
   c. Sample collected from the top may not be well-mixed.
   d. The oil reservoir will be mostly empty during normal operation.

TROUBLESHOOTING TRICKLING FILTERS

Even though the trickling filter process is considered one of the most trouble-free means of secondary treatment, the potential for operating problems exists. Potential operating problems and corrective actions are listed in Table 7.6. The source of mechanical problems is often obvious; however, less obvious causes of problems may stem from operations, design overload, influent characteristics, and other non-equipment-related items.

Good records and data associated with the trickling filter are essential in locating, identifying, and applying the proper corrective measure to solve problems. Tracking soluble BOD, suspended solids, pH, temperature, and other parameters is necessary for operators to observe operational trends. Mechanical parameters that should be recorded in the facility log book include distributor rotation speed and recycle ratios.

SAFETY CONSIDERATIONS FOR TRICKLING FILTERS

It is important that operations and maintenance personnel follow safety guidelines specified by primary equipment manufacturers and commonly practiced by the utility. This includes, but is not limited to, procedures describing appropriate equipment lockout, confined space entry, localized power interruption, and standby rescue personnel during maintenance activities. Operations personnel must familiarize themselves with and respect safe operations guidelines recommended by the equipment manufacturers and maintain a clean, well-ventilated, and well-lit environment to prevent accidents. Some safety considerations, especially for trickling filters, are listed below.

- Work on distributors may proceed only after the arms have been stopped and locked in place and the distributor pump or control valve electrical switch has been disengaged and locked out on the electrical panel;
- Never attempt to stop a moving distributor with your hands or other body parts;
- The filter medium should not be walked on because it will be slippery. Plastic grating is often placed as a permanent walking surface to provide safe access to the distributor;
- Covered trickling filters have special safety considerations because they are considered confined spaces. The possibility exists for the atmosphere under the dome to contain little oxygen or much hydrogen sulfide or ammonia. Maintenance in these areas must include proper confined-space entry procedures; and
- Keep access plates and hatches in place.

SECTION EXERCISE

This exercise will draw on some of the knowledge acquired about trickling filters. In this section, an operator must complete the following tasks:

- Label a diagram of a trickling filter;
- Match the component of a trickling filter to its function;
- Given information for a trickling filter process, determine the recycle ratio needed to maintain a target SK value; and
- Determine the most likely cause of a trickling filter malfunction.

1. Label the following items in the image below. Speed retarder orifice, dump gate, underdrain channel, ventilation ports, and distributor bearings.

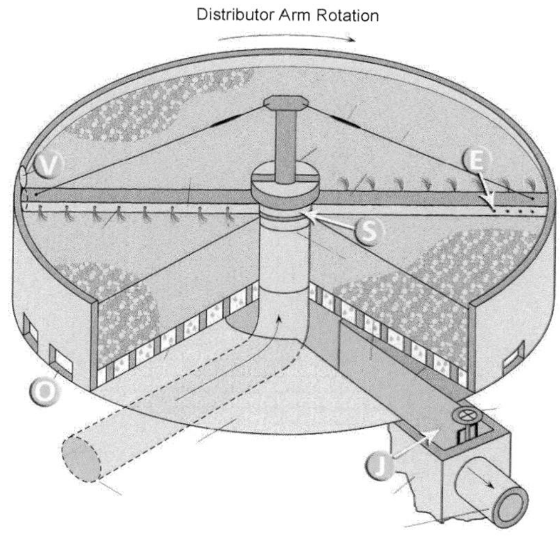

Table 7.6 Troubleshooting Trickling Filters (adapted from U.S. EPA, 1978)

Problem/Possible Cause	Corrective Action
Operations	
Increase in Secondary Clarifier Effluent Suspended Solids	
Clarifier hydraulically overloaded	Check clarifier surface overflow rate; if possible, reduce flow to clarifier to less than 35 $m^3/m^2 \cdot d$ (900 gal/d/sq ft) by reducing recirculation or putting another clarifier into service Expand facility
Denitrification in clarifier	Increase clarifier sludge withdrawal rate Increase loading on trickling filter to prevent nitrification; skim floating sludge from entire surface of clarifier or use water sprays to release nitrogen gas from sludge so sludge will resettle
Excessive sloughing from trickling filter because of changes in wastewater	Increase clarifier sludge withdrawal rate Check wastewater for toxic materials, changes in pH, temperature, BOD, or other constituents Identify and eliminate source of wastewater causing the upset Enforce sewer-use ordinance
Equipment malfunction in secondary clarifier	Check for broken sludge-collection equipment and repair or replace broken equipment
Short-circuiting of flow through secondary clarifier	Level effluent weirs Install clarifier center pier exit, baffles, effluent weir baffles, or other baffles to prevent short-circuiting
Increase in Secondary Clarifier Effluent BOD	
Increase in effluent suspended solids	See corrective actions for "Increase in secondary clarifier effluent suspended solids"
Excessive organic loads on filter	Calculate loading Reduce loading by putting more trickling filters in service Increase BOD removal in primary settling tanks by using all tanks available and minimizing storage of primary sludge in tanks Eliminate high-strength sidestreams in facility Expand facility
Undesirable biological growth on media	Perform microscopic examination of biological growth Chlorinate filter to kill off undesirable growth
Objectionable Odors from Filter	
Excessive organic load causing anaerobic decomposition in filter	Calculate loading Reduce loading by putting more trickling filters into service Increase BOD removal in primary settling tanks by using all tanks available and minimizing storage of primary sludge in tanks Encourage aerobic conditions in treatment units ahead of the trickling filter by adding chemical oxidants (e.g., chlorine, potassium permanganate, or hydrogen peroxide) or by preaerating, recycling facility effluent, or increasing air to aerated grit chambers Enforce industrial waste ordinance, if industry is source of excess load Scrub trickling filter offgases Replace rock media with plastic media Expand facility
Insufficient ventilation	Increase hydraulic loading to wash out excess biological growth; remove debris from filter effluent channels and underdrains; remove debris from top of filter media Unclog vent pipes Reduce hydraulic loading if underdrains are flooded Install fans to induce draft through filter Check for filter plugging caused by breakdown of media

(continues)

Table 7.6 Troubleshooting Trickling Filters (adapted from U.S. EPA, 1978) (*Continued*)

Problem/Possible Cause	Corrective Action
Operations	
Ponding on Filter Media	
Excessive biological growth	Reduce organic loading Slow down distributor to increase SK value Increase hydraulic loading to increase sloughing Flush filter surface with high-pressure stream of water Chlorinate filter influent for several hours; maintain 1 to 2 mg/L residual chlorine on the filter Flood filter for 24 hours Shut down filter until media dries out Enforce industrial waste ordinance if industry is source of excess load
Poor media	Replace media
Poor housekeeping	Remove debris from filter surface, vent pipes, underdrains, and effluent channels
Filter Flies (*Psychoda*)	
Insufficient wetting of filter media (a continually wet environment is not conducive to filter fly breeding and a high wetting rate will wash fly eggs from the filter)	Increase hydraulic loading Unplug spray orifices or nozzles Use orifice opening at end of rotating distributor arms to spray filter walls
Filter environment module conducive to filter fly breeding	Flood filter for several hours each week during fly season Chlorinate the filter for several hours each week during fly season; maintain a 1- to 2-mg/L chlorine residual on the filter
Poor housekeeping	Keep area surrounding filter mowed; remove weeds and shrubs
Icing	
Low wastewater temperature	Decrease recirculation Remove ice from orifices, nozzles, and distributor arms with a high-pressure stream of water Reduce number of filters in service, provided effluent limits can still be met Reduce retention time in pretreatment and primary treatment units Construct windbreak or covers
Maintenance	
Rotating Distributor Slows Down or Stops	
Insufficient flow to turn distributor	Increase hydraulic loading Close reversing jets
Clogged arms or orifices	Flush out arms by opening end plates; flush out orifices; check function of preliminary treatment processes, especially screening equipment
Clogged distributor arm vent pipe	Remove material from vent pipe by rodding or flushing Remove solids from influent wastewater
Bad main bearing	Replace bearing
Distributor arms not level	Adjust guy wires at tie rods
Distributor rods hitting media	Level media Remove some media
Dirt in Main Bearing Lube Oil	
Worn bearing dust seal	Replace seal
Worn turntable seal or seal plate	Replace seal; inspect seal plate and replace if worn
Condensate not drained regularly or oil level too low	Check oil level, drain condensate, and refill if needed

(*continues*)

Table 7.6 Troubleshooting Trickling Filters (adapted from U.S. EPA, 1978) (*Continued*)

Problem/Possible Cause	Corrective Action
	Operations
Water Leaking from Distributor Base	
Worn turntable seal	Replace seal
Leaking expansion joint between distributor and support column	Repair or replace expansion joint influent piping
Broken Top Media	
Foreign material	Flush out with a high-pressure stream of clean water Rod out with wire or hook Disassemble and clean
Secondary Clarifier Sludge Collector Stopped	
Torque overload setting exceeded	Reduce sludge blanket; withdraw excess sludge Check if skimmer portion of collector hung up on scum trough; free and repair or adjust skimmer Drain tank and remove foreign objects
Loss of power	Reset drive unit circuit breaker if tripped (after cause for trip is identified and corrected) Reset drive unit, motor control center, or facility main circuit breakers as necessary when power is restored to plant after interruption Check drive motor for excessive current draw; if current excessive, determine reason Check drive motor overload relays; replace if bad or undersized
Failure of drive unit	Check drive chains and shear pins; replace as necessary and use proper size shear pin, or damage will occur Check and replace worn gears, couplings, speed reducers, or bearings as needed; lubricate and provide preventive maintenance for units as per manufacturer's instruction
Recirculation Pumps Delivering Insufficient Flow	
Excessive head	Open closed or throttled valves Unplug distributor arms, headers, and laterals Unplug distributor nozzles and orifices Unplug distributor vent lines
Pump malfunction	Adjust or replace packing or mechanical seals Adjust impeller to casing clearance; replace wear rings if worn excessively Replace or resurface worn shaft sleeves Check impeller for wear and entangled solids; remove debris; replace impeller if necessary Check pump casing for air lock Release trapped air Lubricate bearings as per manufacturer's instructions Replace worn bearings
Pump drive motor failure	Lubricate bearings as per manufacturer's instructions Replace worn bearings Keep motor as clean and dry as possible Pump and motor misalignment; check vibration and alignment Redesign as needed Burned windings; rewind or replace motor Check drive motor for excessive current draw; if current draw is excessive, determine reason Check drive motor overload relays; replace if bad or undersized; reset drive motor, motor control centers, or facility main circuit breakers after cause for trip is identified and corrected, or when power is restored after interruption

2. Match the trickling filter component to its function:

1. Speed retarder orifices	A. Allow air movement into filter
2. Filter media	B. Distributes air across filter area
3. Ventilation ports	C. Used to level distributor arms
4. Stay rod	D. Slows speed of distributor arm
5. Dump gate	E. Isolates trickling filter
6. Plenum	F. Used to remove debris from distributor
7. Outlet gate	G. Provides surface for biofilm growth

3. WRRF Information: A trickling filter is 36.6 m (120 ft) in diameter and 7.3-m (24-ft) deep. The total media volume is 7676.7 m^3 (271 296 cu ft). The filter surface area is 1051.6 m^2 (11 304 sq ft). The filter receives an average daily flow from the primary clarifier of 14.4 ML/d (3.8 mgd). There are 4 distributor arms. The distributor speed is fixed at 1.5 rpm. How much recycle flow is needed to maintain a flushing intensity of 77.1 mm/pass (3 in./pass)?
4. A naturally ventilated trickling filter is performing well, but one area on the filter surface is covered with standing water. What is the most likely cause and what should the operator do to correct the problem?

SECTION EXERCISE SOLUTIONS

1. E = Speed retarder orifices, V = Dump gate, J = Underdrain channel, O = Ventilation ports, S = Distributor bearings
2. 1. D, 2. G, 3. A, 4. C, 5. F, 6. B, 7. E
3. To solve this problem, the operator needs to know the total HLR needed to obtain the desired flushing intensity. Once that is known, subtract the primary effluent flow from the total flow to get the amount of recycle flow required.

   In International Standard units:

   Step 1—Find the total HLR required.

$$\text{SK, mm/pass} = \frac{\text{HLR}, \dfrac{\text{m}^3}{\text{m}^2 \cdot \text{d}} \times 1000 \dfrac{\text{mm}}{\text{m}}}{N_a \times \omega_d \times 60 \dfrac{\text{min}}{\text{h}}}$$

$$77.1 \text{ mm/pass} = \frac{\text{HLR}, \dfrac{\text{m}^3}{\text{m}^2 \cdot \text{d}} \times 1000 \dfrac{\text{mm}}{\text{m}}}{4 \times 1.5 \times 60 \dfrac{\text{min}}{\text{h}}}$$

$$77.1 \text{ mm/pass} = \text{HLR}, \frac{\text{m}^3}{\text{m}^2 \cdot \text{d}} \times 2.78$$

$$27.7 = \text{HLR}, \frac{\text{m}^3}{\text{m}^2 \cdot \text{d}}$$

   Step 2—Use the HLR equation to find the total flow applied to the filter.

$$\text{HLR}, \frac{\text{m}^3}{\text{m}^2 \cdot \text{d}} = \frac{\text{Total Flow Applied, m}^3/\text{d}}{\text{Area, m}^2}$$

$$27.7 \frac{\text{m}^3}{\text{m}^2 \cdot \text{d}} = \frac{\text{Total Flow Applied, m}^3/\text{d}}{1051.6 \text{ m}^2}$$

$$29\ 129 \frac{\text{m}^3}{\text{d}}$$

Step 3—Convert the primary effluent flow to cubic meters per day.

$$\frac{14.4\ \text{ML}}{\text{d}}\left|\frac{1\ 000\ 000\ \text{L}}{1\ \text{ML}}\right|\left|\frac{1\ \text{m}^3}{1000\ \text{L}}\right| = 14\ 400\ \text{m}^3/\text{d}$$

Step 4—Find the amount of recycle flow required.

$$\text{Total Flow} - \text{Primary Effluent Flow} = \text{Recycle Flow}$$

$$29\ 129\ \frac{\text{m}^3}{\text{d}} - 14\ 400\ \frac{\text{m}^3}{\text{d}} = 14\ 729\ \frac{\text{m}^3}{\text{d}}$$

In U.S. customary units:

Step 1—Find the total HLR required.

$$\text{SK, in./pass} = \frac{\text{HLR,}\ \frac{\text{cu ft}}{\text{sq ft}\cdot\text{d}} \times 12\frac{\text{in.}}{\text{ft}}}{N_a \times \omega_d \times 60\frac{\text{min}}{\text{hr}}}$$

$$3\ \text{in./pass} = \frac{\text{HLR,}\ \frac{\text{cu ft}}{\text{sq ft}\cdot\text{d}} \times 12\frac{\text{in.}}{\text{ft}}}{4 \times 1.5 \times 60\frac{\text{min}}{\text{hr}}}$$

$$1080 = \text{HLR,}\ \frac{\text{cu ft}}{\text{sq ft}\cdot\text{d}} \times 12$$

$$90 = \text{HLR,}\ \frac{\text{cu ft}}{\text{sq ft}\cdot\text{d}}$$

Step 2—Convert the HLR from cu ft/sq ft·d to gpd/sq ft.

$$\frac{90\ \text{cu ft}}{\text{sq ft}\cdot\text{d}}\left|\frac{7.48\ \text{gal}}{1\ \text{cu ft}}\right| = 673.2\frac{\text{gpd}}{\text{sq ft}}$$

Step 3—Use the HLR equation to find the total flow applied to the filter.

$$\text{HLR,}\ \frac{\text{gpd}}{\text{sq ft}} = \frac{\text{Total Flow Applied, gpd}}{\text{Area, sq ft}}$$

$$673.2\frac{\text{gpd}}{\text{sq ft}} = \frac{\text{Total Flow Applied, gpd}}{11\ 304\ \text{sq ft}}$$

$$7\ 609\ 853 = \text{Total Flow Applied, gpd}$$

$$7\ 609\ 853\ \text{gpd} = 7.6\ \text{mgd}$$

Step 4—Find the amount of recycle flow required.

$$\text{Total Flow} - \text{Primary Effluent Flow} = \text{Recycle Flow}$$

$$7.6\ \text{mgd} - 3.8\ \text{mgd} = \text{Recycle Flow}$$

$$3.8\ \text{mgd} = \text{Recycle Flow}$$

4. Standing water indicates an area where water movement has slowed or stopped. The most likely cause is a blockage in the media below. The operator should make a note of the location. It may be necessary to remove media from the filter to find the blockage and remove it.

Rotating Biological Contactors

Rotating biological contactors consist of closely spaced parallel plastic discs mounted on a central horizontal shaft (Figure 7.37). The shaft is supported by bearings at each end, which are mounted on the walls of the basin. Biofilm grows on the surface of the parallel plastic discs (biofilm support media). Approximately 40% of the volume of the RBC support media is immersed in wastewater and the remaining media volume is exposed to the air. This level of submergence keeps the shaft and bearings above the water level. The media rotates continuously through the wastewater to keep the media surface completely wet, which allows the biofilm to grow over the entire media surface. Rotating the RBC media between wastewater and air exposes the microorganisms in the biofilm to food, nutrients, and oxygen. Continuous media rotation also creates shearing forces, which help to remove excess biomass from the media and keeps sloughed biomass solids in suspension until they are removed from the RBC tank. Faster rotation speeds result in thinner biofilms. The biological suspended solids produced in the RBC are typically separated from the treated RBC effluent by a secondary clarifier. Some smaller, WRRFs have a hopper rather than a true clarifier.

Rotating biological contactors have been extensively used at hundreds of locations in the United States to treat municipal and industrial wastewater. It is estimated that more than 600 RBC facilities are now used for industrial and municipal wastewater treatment. Most of the facilities are designed and used for BOD removal and a few for both BOD and nitrogen removal.

DESIGN PARAMETERS FOR ROTATING BIOLOGICAL CONTACTORS

When RBCs were initially introduced for wastewater treatment during the late 1970s and early 1980s, mechanical problems and organic overloading occurred frequently. By the mid-1980s, both equipment manufacturers and consulting engineers had developed standards that minimized most of the problems, but some systems are still mechanically unsound.

The flow pattern for RBCs resembles that for most other biological treatment processes (Figure 7.38). Preliminary treatment removes settleable solids, grease, and other floatable material upstream of the RBCs. A secondary clarifier follows RBCs to remove sloughed solids from the treated wastewater. Solids that settle in the secondary clarifier may be pumped directly to a solids handling process or be recycled to the primary clarifier for *cosettling* with the primary sludge. Primary sludge is pumped directly to a solids handling process.

Cosettling means both the primary and secondary sludge are being settled in the same clarifier. Secondary sludge is typically pumped back to the primary clarifier.

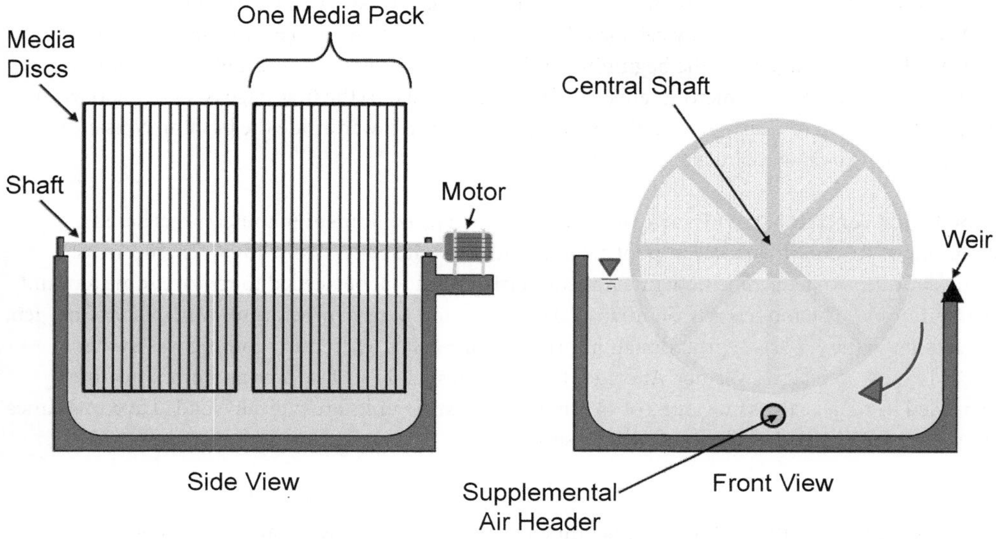

Figure 7.37 Rotating Biological Contactor (adapted from Wikimedia Commons, submitted by Milton Beychok)

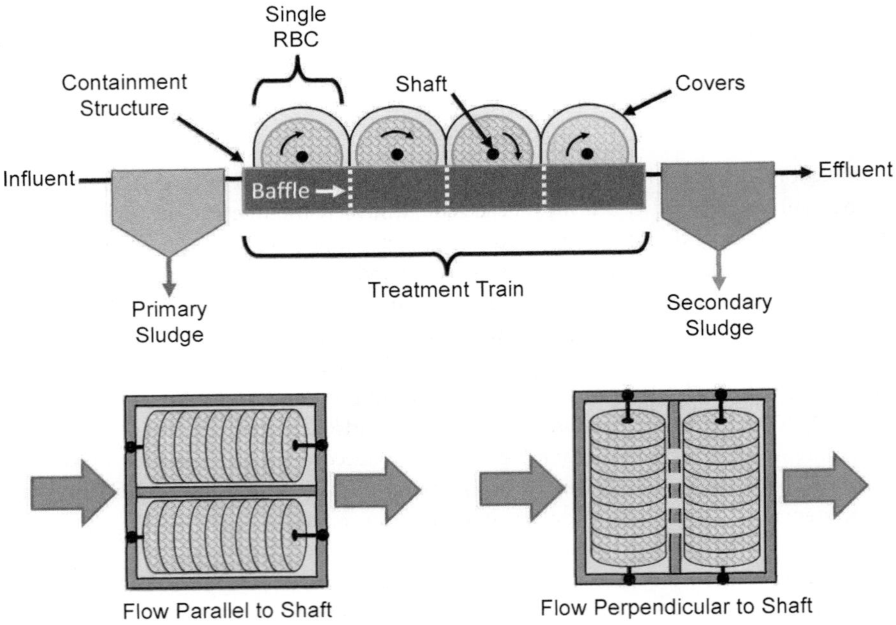

Figure 7.38 RBC Process Flow Schematics (Adapted from U.S. EPA, 1978, and WEF, 2016)

The term "shaft" typically is used to describe both the metal support and the filter media discs. Individual RBCs may be used alone or in combination with other units. It is typical to see multiple RBCs arranged in series as shown in Figure 7.38. A series of RBCs treating wastewater as a unit are called a *train*. Individual RBCs in a single train that are separated by internal baffles are called *stages*. The first stage receives the highest organic load. Each subsequent RBC receives a lower organic load than the RBC before it. Rotating biological contactors may be placed parallel to one another with wastewater entering the containment structure either parallel or perpendicular to the RBC shafts. Feeding of wastewater parallel to the shaft (end flow) is typically only done at very small facilities (bottom left of Figure 7.38). When flow is perpendicular to the RBC shafts, the RBC should rotate against the direction of flow. Baffles divide the containment vessel into compartments and help to prevent *short-circuiting* through the basin.

Short-circuiting occurs when the water in the basin is not well-mixed. This can result in currents or paths forming where water bypasses most of the tank volume and quickly reaches the effluent pipeline.

Loading rates for RBCs are typically based on the mass of BOD going to the RBC units divided by the media's surface area. With trickling filters, the organic loading rate was calculated as the mass of BOD_5 applied per volume of trickling filter media. The organic load may be based on either soluble BOD_5 or total BOD_5. Organic loading is typically calculated for all units online and also for the first stage. Because the organic loading is the highest at the beginning of the process, this is where the highest amount of oxygen will be consumed by the microorganisms. If the organic load to the first stage is too high, oxygen concentrations may drop to nearly zero in the wastewater and even lower inside the biofilm. Nuisance organisms like *Beggiatoa* may populate the biofilm as a result.

Rotating biological contactor systems were originally designed by equipment manufacturers based on observations of RBC performance in pilot systems and at municipal WRRFs. Today, engineers use mathematical models to determine how much media surface area will be needed to remove some amount of BOD or ammonia. The models rely on microbial growth rates, water temperature, availability of oxygen, and other factors. Table 7.7 lists typical design information for RBCs when the incoming wastewater temperature is 13 °C (55 °F) or warmer. At colder temperatures, more surface area and more treatment time are needed to achieve the same levels of treatment. First-stage units are typically loaded at 3 to 4 times the recommended BOD load of the total units in service.

EXPECTED PERFORMANCE FROM ROTATING BIOLOGICAL CONTACTORS

Similar to trickling filters, RBC performance is limited by organic and HLRs, characteristics of the incoming wastewater, and temperature. For domestic wastewater, an RBC process can be expected to remove between 75 and 90% of the BOD_5 applied. In a study of RBC performance, U.S. EPA found the

Table 7.7 Typical Design Information for Rotating Biological Contactors (Metcalf and Eddy, Inc., 2003. Copyright © 2003 by McGraw-Hill Education, reprinted with permission.)

Parameter	Units	BOD_5 Removal*	BOD_5 Removal and Nitrification*	Separate Nitrification*
HLR	m³/m²·d (gpd/sq ft)	0.08–0.16 (2–4)	0.03–0.08 (0.8–2)	0.04–0.10 (1–2.5)
Organic Loading Rate (soluble BOD_5)	g $sBOD_5$/m²·d (lb $sBOD_5$/d/1000 sq ft)	4–10 (0.8–2)	2.5–8 (0.5–1.6)	0.5–1.0 (0.1–0.2)
Organic Loading Rate (total BOD_5)	g BOD_5/m²·d (lb BOD_5/d/1000 sq ft)	8–20 (1.6–4.0)	5–16 (1–3.3)	1–2 (0.2–0.4)
Maximum 1st Stage Organic Loading (soluble BOD_5)	g $sBOD_5$/m²·d (lb $sBOD_5$/d/1000 sq ft)	12–15 (2.5–3)	12–15 (2.4–3)	N/A
Maximum 1st Stage Organic Loading (total BOD_5)	g BOD_5/m²·d (lb BOD_5/d/1000 sq ft)	24–30 (4.9–6.1)	24–30 (4.9–6.1)	N/A
Ammonia (NH_3-N) Loading	g N/m² (lb NH_3-N/d/1000 sq ft)	N/A	(0.14–0.31)	No typical value
HRT	hours	0.7–1.5	1.5–4	1.2–3
Effluent BOD_5	mg/L	15–30	7–15	7–15
Effluent NH_3-N	mg/L		<2	1–2

*Wastewater temperatures above 13 °C (55 °F).

average monthly effluent concentration for the RBCs studied was 18 mg/L (U.S. EPA, 1984). There is no RBC equivalent of a roughing trickling filter where a substantial amount of BOD is intentionally passed through.

Rotating biological contactor units can achieve organic carbon removal and *nitrification*. Nitrification may occur in the latter stages, which have lower organic loading. In general, the soluble BOD_5 concentration must be lower than 20 mg/L before nitrification can take place. Rotating biological contactor systems have the potential to achieve denitrification, but this mode of operation is not typical. For denitrification to occur, a low DO concentration must be maintained in the first RBC unit, which may cause adverse effects on the biofilm and reduce BOD removal. Expected performance for different loading conditions is shown in Table 7.7. As with trickling filters, the more lightly loaded the process is, the more likely that ammonia removal will be achieved in addition to removal of BOD.

Nitrification converts ammonia to nitrite and nitrate. Denitrification converts nitrate to nitrogen gas. For more details, review the chapters on Fundamentals of Biological Treatment and Biological Nutrient Removal.

TEST YOUR KNOWLEDGE

1. **What percentage of an RBC is typically submerged during normal operation?**
   a. 20%
   b. 40%
   c. 60%
   d. 80%

2. **For an RBC system that uses cosettling, the solids that settle in the secondary clarifier are**
   a. Sent directly to solids handling
   b. Typically greater than 10% total solids
   c. Sent to the primary clarifier
   d. Applied to the RBC in stage 1

3. **An RBC system has two treatment trains with four stages each. Each stage contains one RBC unit. How many units are there altogether?**
   a. 2
   b. 4
   c. 8
   d. 16

4. **Loading rates are based on the mass of BOD (kilograms or pounds) per**
   a. Square meters or square feet of media
   b. Number of RBCs per train
   c. Cubic meters of 1000 cubic feet of media
   d. Population equivalent

5. If the organic loading rate to the first stage of treatment in an RBC is too high,

a. Treatment efficiencies will increase.
b. Biofilm development will be inhibited.
c. Mass sloughing may occur.
d. Oxygen concentrations will decrease.

6. An operator notices large patches of white biofilm on an RBC in the first stage of a treatment train. This indicates

a. Snail predation
b. Lack of oxygen
c. Organic underloading
d. Nitrification

7. An RBC that is lightly loaded

a. May remove both BOD_5 and NH_3-N
b. Requires a longer HRT than a highly loaded RBC
c. Produces poor quality effluent high in BOD_5
d. Performs better at colder water temperatures

EQUIPMENT FOR ROTATING BIOLOGICAL CONTACTORS

Rotating biological contactor systems typically include the following equipment items and may also include instrumentation:

- Tankage (containment structure),
- Baffles,
- Filter media,
- Cover,
- Drive assemble, and
- Inlet and outlet piping.

Figure 7.39 illustrates some of the various equipment components that are typically used in the RBC process. The names for individual equipment components may differ slightly; depending on the manufacturer.

REACTOR BASIN

The reactor basin contains the wastewater and the portion of the media, which is submerged. Containment structures or tanks for RBC equipment may consist of metal tanks for small pilot facilities or single-shaft units. However, multi-shaft units almost always include tanks made of concrete basins. Figure 7.40 shows construction of concrete tanks for a series of three, parallel treatment trains. For this system, each train will ultimately contain five RBCs with four media packs each. The end of the reactor basin may be equipped with an adjustable overflow weir. During periods of low flow, the height of the overflow weir or the weirs of the downstream secondary clarifier will maintain the water level in the tank.

BAFFLES

Internal baffling or weir structures separate the stages of the RBC reactors. Baffling used to separate one shaft from another may be made of either concrete or wood. In Figure 7.40, the baffle walls separating each RBC unit are visible. Removable baffles are often used to allow process changes after the facility is constructed. Baffles help prevent short circuiting and may be used to restrict the flow of wastewater from one step of the process to the next. This can help maintain high organic loading rates at the first RBC unit (baffles in) or even the load out over more RBC units (some baffles out). Each baffled compartment is referred to as a stage. With all of the baffle walls in place, the system shown in Figure 7.40 will have 5 stages of treatment. If the baffles are removed from in between the first and second RBC unit, then the first stage of treatment will contain 2 RBC units. The wastewater within a single stage has the same concentration throughout.

COVERS

Rotating biological contactor units are almost always covered or located inside a building. Rotating biological contactor covers retain heat and prevent ice accumulation on the shaft and drive units during freezing weather, prevent deterioration of the plastic media resulting from ultraviolet (UV) radiation from exposure to sunlight, and prevent algae growth on the media. If left uncovered, algae growth on the surface of the media can clog the media and reduce air movement. Rotating biological contactor covers typically have an arched shape and are made of fiberglass for durability (Figure 7.41). Another approach involves housing a number of RBC units in a building (Figure 7.2).

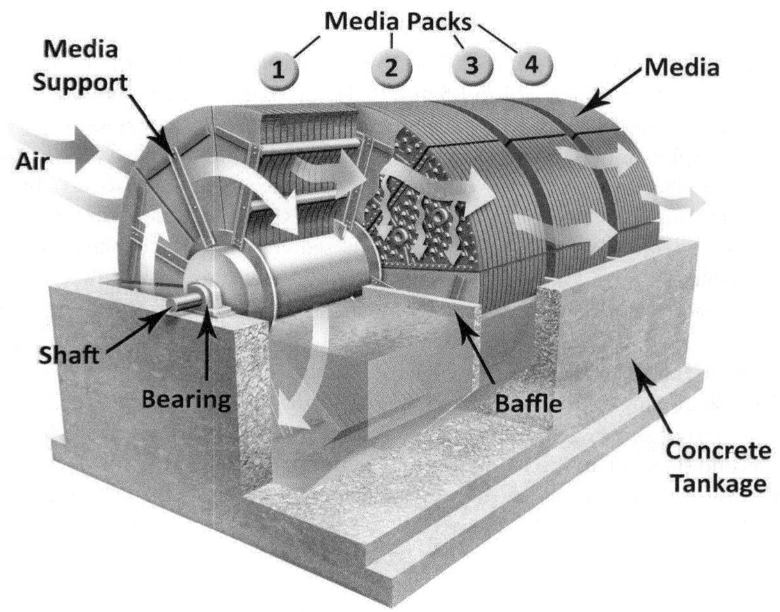

Figure 7.39 Rotating Biological Contactor Components (Courtesy of Walker Process Equipment Division of McNish Corporation)

Figure 7.40 Installation of RBC Equipment (Courtesy of Walker Process Equipment Division of McNish Corporation)

Figure 7.41 RBC Mounted on a Horizontal Shaft (left) and RBC Covers (right) (WEF et al., 2010)

BIOFILM SUPPORT MEDIA

Rotating biological contactors use plastic media made of high density polyethylene (HDPE) (or polyvinyl chloride or polystyrene). Single sheets with thickness of 1 to 1.5 mm (0.04 to 0.06 in.) can be as large as 3.6 m (12 ft) in diameter. The sheets, bonded and assembled onto the horizontal shafts are typically 7.6-m (25-ft) long. Corrugations, dimples, or honeycomb patterns molded into the plastic stiffen the thin sheets and serve as spacers to keep the sheets separated. Corrugations increase the available surface area by 15 to 20% and create a winding path through the media. As a result, water takes longer to drain from the media when it rotates. This increases the amount of time the water is exposed to the air, which increases the amount of oxygen transferred into the water (U.S. EPA, 1984). Some RBC media have radial and concentric passages in the media to facilitate movement of wastewater and air over the surface area.

The amount of media surface area and the SSA in each RBC unit varies depending on the treatment requirements for the system. Each shaft (single RBC) of standard density media typically provides approximately 9300 m^2 (100 000 sq ft) of surface area for biofilm attachment. Standard density media is normally used in the first and second stages of an RBC train for BOD removal. Higher density media are typically used for ammonia removal as the biofilms formed by the nitrifying bacteria tend to be thinner. Higher density media may have between 11 200 and 16 700 m^2 (120 000 and 180 000 sq ft) of SSA per shaft (U.S. EPA, 1984).

High-density media is used in the middle and final stages of the RBC train where the organic loading has been reduced and thinner biofilm develops. High density media have the space between the repeating plastic sheets reduced by 33%, which increases the available surface area by 50% compared to standard density media. The biofilm thickness on high-density media should be limited to approximately 1.3 mm (0.05 in.) to ensure that the weight of the biofilm doesn't overload the shaft and bearings.

For a tertiary nitrification process receiving secondary effluent, an RBC train could use high-density media in the first and second stages as well as in subsequent stages. Some manufacturers also produce media with intermediate and extra high densities. Intermediate density media is used in the middle stages of the RBC train. A standard RBC system with four units in series could use all four media densities in ascending order through the four units.

SHAFT

The media are supported on a single shaft with bearings at each end. The shaft may be cylindrical, hexagonal, or square, depending on the manufacturer. The standard shaft length is 8.2 m (27 ft) with a maximum of 7.6 m (25 ft) occupied by media. Shorter shafts are available from some manufacturers. The media are attached to the shaft by one of several methods specific to each manufacturer. Media can be supported by radial steel arms that are bolted onto the central shaft (Figure 7.42) or bolted onto clips welded to the shaft. The plastic media sections or wedges can be field assembled for easy replacement of media without removing the heavy 8.2-m (27-ft)-long shaft (Figure 7.43).

BEARINGS

Rotating biological contactor shafts are supported on spherical roller bearings, sealed tapered roller bearings, or non-sealed tapered roller bearings. For most RBCs, the bearings supporting the RBC shaft and the drive assembly are mounted on the reactor basin walls. A bearing can be seen in Figure 7.44 highlighted in green. Spherical roller bearings tolerate deflection and misalignment better than other types of bearings. Every RBC shaft has at least one bearing designed to accommodate *thermal expansion* as the shaft heats and cools. Most RBC shafts have one expansion bearing and one non-expansion bearing.

When metals heat up, they expand slightly. This *thermal expansion* must be accommodated. Otherwise, the small changes in shaft length caused by temperature changes could damage the bearings.

DRIVE MECHANISM

The RBC discs may be rotated by either air or mechanical drive units. Air-driven RBC units have a blower and coarse bubble air diffusers at the bottom of each RBC shaft. Air diffusers are similar in operation to the air stones used in home aquariums. Air is pushed through the diffusers with pressure generated by the blower. The diffuser is full of small openings that break the airflow up into a lot of smaller bubbles. Coarse bubble diffusers produce larger bubbles than fine air diffusers. Air cups pinned to the edge of the plastic disc trap air bubble released from the air header (Figure 7.2 and 7.45). As the bubbles rise and are caught in the cups, they cause the RBC shaft to rotate. The bubbles are released when the cups move above the

Figure 7.42 Shaft and Supports for RBC Media (Courtesy of Walker Process Equipment Division of McNish Corporation)

Figure 7.43 Partial Installation of RBC Media (Courtesy of Walker Process Equipment Division of McNish Corporation)

Figure 7.44 RBC Bearing and Shaft (Reprinted with permission by Indigo Water Group via Upper Blue Sanitation District)

Figure 7.45 Air Cups on Exterior of RBC Media (Reprinted with permission by Indigo Water Group via Upper Blue Sanitation District)

water surface. Approximately 4.2 to 11.3 m^3/min (150 to 400 standard cubic feet per minute [scfm]) are provided per standard shaft size. The amount of air needed depends on the size of the RBC, media density, revolutions per minute, wastewater temperature, biofilm thickness (weight), and other conditions.

Air-drive units typically operate between 1.0 and 1.4 rpm, but may vary from one RBC to another in the same treatment train (U.S. EPA, 1984). The first and second stages of air driven RBC units typically operate at rotational speeds of 1.4 rpm, or *tip speeds* of 0.27 m/s (53 ft/min). The latter stages of an RBC train, where oxygen demand is lower, often operate at rotational speeds of 1.0 rpm and tip speeds of 0.19 m/s (38 ft/min). The rotational speed of air driven RBCs can be varied by adjusting the airflow. Keep in mind that adjusting the air to one RBC often shifts excess air to the other RBCs in the system.

Air-driven RBCs are sometimes called aerated biological contactors (ABCs).

Tip speed is how fast a point on the outside edge of the RBC wheel is traveling.

Air-driven RBC units operate at lower rotational speeds than mechanically driven RBCs because the additional aeration provided by the diffused air system reduces the amount of oxygen, which must be transferred to the biofilm from the air. The diffused air bubbles rising through the media also increase the shear forces, which remove excess biomass from the media, reducing the need for hydraulic shearing. The shearing force of the bubbles rising across the surface of the media maintains a thinner, primarily aerobic biofilm than that in mechanical drive RBCs (McCann and Sullivan, 1980). The thinner biofilm decreases the weight carried by the shaft, which extends the useful life of the equipment. Air-drive systems eliminate the high shaft torque of mechanical drives by distributing the turning force along the entire length of the shaft instead of applying the force at one end of the shaft.

The advantages of air-driven units are that less torque is applied to the shaft, the biomass tends to be thinner (sheared by the air), and the wastewater may contain slightly more DO. In some cases, diffused air has been used in mechanically driven RBCs to reduce solids accumulation in the bottom of the tank.

Positive-displacement or centrifugal blowers may be used; however, centrifugal blowers are easier to control by adjusting the throttle on the blower suction valve. A major problem with air-driven units has been loping or unbalanced rotation. In extreme cases, rotation has stopped and mechanical rotation, cleaning, or the use of high air rates has been necessary to re-establish rotation. Other disadvantages of air-drive units are higher power use and the need to balance and adjust the airflow to the diffusers.

Mechanical-drive systems consist of an electric motor, speed reducer, and belt- or chain-drive components for each shaft. Multistage speed reducers are also used instead of belt or chain drives. The motors, typically rated at 3730 to 5590 W (5 to 7.5 hp) per shaft, may be equipped to allow changing shims or sprocket sizes and installation of a variable frequency drive to vary rotational speed. Mechanical-drive units typically operate at 1.2 to 1.6 rpm. Rotating a 25-foot-long shaft supporting a heavy biomass requires high torque. The high twisting stress in the shaft can cause metal fatigue of the shaft and connections to the shaft.

An explanation of gear reducers and torque may be found in Chapter 4—Primary Treatment of Wastewater.

Mechanically driven RBC units normally have constant speed drives to ensure consistent rotational speed to apply appropriate hydraulic shear forces to control biofilm thickness and maintain an aerobic environment for the biomass on the media. The rotational speed may be varied in each stage depending on the organic loading conditions and oxygen requirement at each stage. Constant speed drives can be converted to variable speed drives to increase operational flexibility.

A diffused aeration system can be installed in the reactor basin of mechanically driven RBCs to increase oxygen transfer. Air from a blower system is discharged into the wastewater. Diffusers break the air stream into fine bubbles. The smaller, or finer, the bubbles are, the greater the oxygen transfer.

Uneven biofilm growth can cause an unbalanced weight distribution in the RBC, resulting in uneven shaft rotation (loping). An unbalanced weight distribution increases torque loads on the shaft and mechanical drive. In air-driven RBC units, an unbalanced weight distribution causes the rotational speed to vary cyclically during each rotation. Rotational speed decreases when the heavier portion of the media is lifted above the water surface. When the heavier portion of the media rotates past the top center of the drum and begins to descend, the rotational speed accelerates. A loping condition often accelerates and, if not corrected, may lead to inadequate treatment and an inability to keep the RBC rotating. Loping and jerking of the media can also damage the gears. Figures 7.46 through 7.48 show broken teeth and a broken gear caused by uneven loading and loping.

Figure 7.46 Broken Teeth on RBC Gear (Reprinted with permission by Kenneth Schnaars)

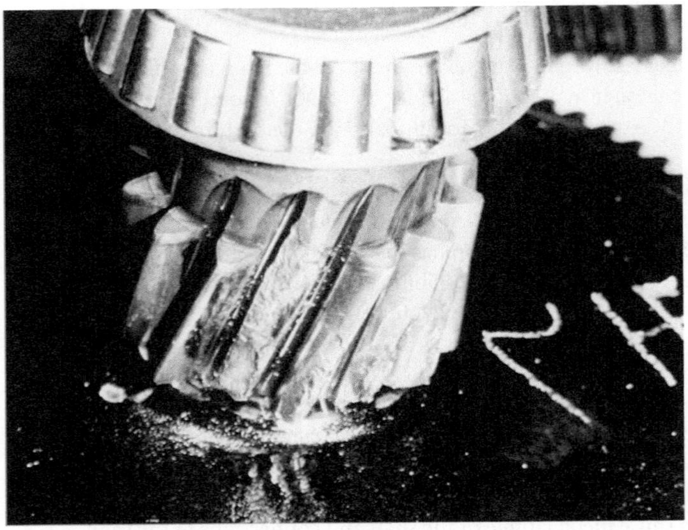

Figure 7.47 Broken Teeth (Reprinted with permission by Kenneth Schnaars)

Figure 7.48 Broken Gear (Reprinted with permission by Kenneth Schnaars)

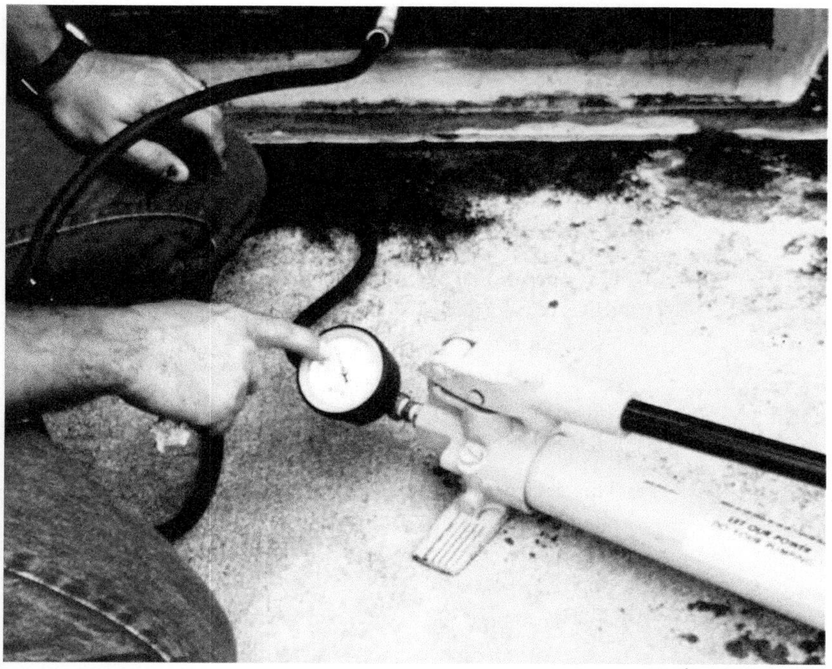

Figure 7.49 Hand-Operated Hydraulic Pump Connected to Load Cell to Measure Media Weight (Reprinted with permission by Kenneth Schnaars)

LOAD CELLS

Many RBCs are equipped with load cells, which measure the total weight on the shaft (see Figure 7.48). A load cell device is typically installed under the shaft support bearing on the idle end of the shaft, which is the opposite end from the motor and drive. A load cell has a hand operated hydraulic pump that may be used to lift the bearing from its base (see Figure 7.49). This generates hydraulic pressure, which can be measured and converted to shaft weight. The total weight can be used to estimate the total biomass and biofilm thickness on the media surface. Electronic strain gauge load cells provide continuous measurements of the shaft load without lifting the bearing off its base.

When field-measured power consumption exceeds the normal range, the operator should investigate whether the higher power consumption is being caused by equipment problems, heavier than normal biofilm growth, or both. Potential equipment problem areas include misalignment, inadequate lubrication, excessive rotational speed, excessive belt tension or belt slippage, and general wear and deterioration of the drive components.

TEST YOUR KNOWLEDGE

1. **An RBC treatment train may contain different media densities across the train.**
   - ☐ True
   - ☐ False
2. **Metals will contract slightly when heated and expand slightly when cooled. This change in length with temperature is why at least one RBC bearing must accommodate temperature changes.**
   - ☐ True
   - ☐ False
3. **Either a motor or air may be used to rotate RBCs.**
   - ☐ True
   - ☐ False
4. **Baffles between RBC units**
   - a. Reduce hydraulic detention time
   - b. Separate treatment stages
   - c. Cannot be moved
   - d. Are always made from concrete

5. **Removing the baffles between the first and second sequencing batch reactor (SBR) units**
   a. Discourages hydraulic short-circuiting
   b. Promotes full nitrification in stage 2
   c. Distributes the influent load
   d. Decreases oxygen availability

6. **RBCs are typically covered to**
   a. Prevent algae growth
   b. Reduce losses of volatile organics
   c. Keep out vandals
   d. Protect the distributor arm

7. **The amount of media surface area per volume is the**
   a. Weighted ratio allowance
   b. SSA
   c. Turndown ratio
   d. Media density factor

8. **The individual plastic sheets of media in an RBC are corrugated for this reason:**
   a. Increases flexibility of modules
   b. Reduces opportunities for biofilm bridging
   c. Increases available surface area
   d. Decreases oxygen transfer

9. **Air-driven RBCs**
   a. Rotate 4 to 6 times per minute
   b. Capture air bubbles in cups
   c. Rotate faster than motor-driven units
   d. Have thicker biofilm than motor-driven units

10. **One operational problem associated with air-driven RBCs is**
    a. Harmonics
    b. Shaft stress
    c. Loping
    d. Shedding

11. **This device is used to measure the total weight on the shaft.**
    a. Load cell
    b. Strain hoist
    c. Load bearing
    d. Tension cable

PROCESS VARIABLES FOR RBCS

ORGANIC LOADING RATE

The *organic loading rate* may also be calculated using soluble BOD, CBOD, or COD.

The *organic loading rate* is the mass of BOD_5 applied to the RBC divided by the surface area of media. It is expressed as grams of BOD_5 applied per square meter per day (g/m²·d) or pounds of BOD_5 applied per 1000 square feet of media per day (lb/d/1000 sq ft). The organic loading rate may be calculated using either of the following two formulas:

In International Standard units:

$$\text{Organic Loading Rate, } \frac{\text{g}}{\text{m}^2 \cdot \text{d}} = \frac{\text{Organic Load, g BOD}_5\text{/d}}{\text{Surface Area of Media, m}^2} \quad (7.14)$$

In U.S. customary units:

$$\text{Organic Loading Rate, } \frac{\text{lb BOD}_5}{\text{1000 sq ft} \cdot \text{d}} = \frac{\text{Organic Load, lb BOD}_5\text{/d}}{\text{Surface Area of Media, 1000 sq ft}} \quad (7.15)$$

Methods for calculating mass in kg or lb are given in equations 7.3 through 7.6.

The amount of organic load that can be effectively processed by an RBC is heavily influenced by the availability of DO. 2.2 kg (1 lb) of oxygen is needed to treat 2.2 kg (1 lb) of BOD_5. If there isn't enough oxygen available, biofilm growth and BOD removal will be limited. Keeping loading rates within the typical design ranges given in Table 7.7 will help to ensure that there is enough oxygen available.

Example Calculations for Organic Loading Rate

An RBC process has two treatment trains with four stages each. The trains are operated in parallel. The influent flow to the facility averages 13.6 ML/d (3.6 mgd). After primary clarification, the wastewater contains 80 mg/L of total BOD_5 and 50 mg/L of soluble BOD_5. All eight RBC units are made from high-density media with a surface area of 13 935 m² (150 000 sq ft) per wheel. The surface area information was provided by the manufacturer. Find the total BOD_5 load to the entire RBC process and find the soluble BOD_5 load to the first stage of the process. Compare your answers to the typical design information given in Table 7.7.

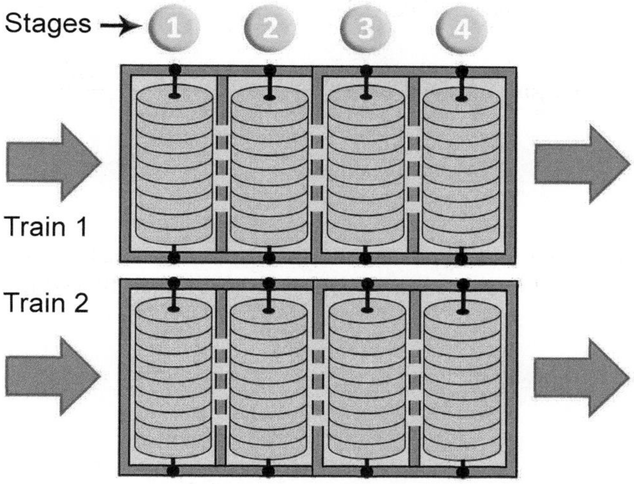

Reprinted with permission by Indigo Water Group

In International Standard units:

Step 1—Convert the influent flow from megaliters per day to cubic meters per day.

$$\frac{13.6\text{ ML}}{\text{day}}\left|\frac{1\,000\,000\text{ L}}{1\text{ ML}}\right|\left|\frac{1\text{ m}^3}{1000\text{ L}}\right| = 13\,600\text{ m}^3/\text{d}$$

Step 2—Find the mass of total BOD_5 going to the RBC process.

$$\text{kg total BOD}_5 = \frac{\left[\left(\text{Concentration, }\frac{\text{mg}}{\text{L}}\right)(\text{Flow, m}^3/\text{d})\right]}{1000}$$

$$\text{kg total BOD}_5 = \frac{\left[\left(80\frac{\text{mg}}{\text{L}}\right)(13\,600\text{ m}^3/\text{d})\right]}{1000}$$

$$\text{kg total BOD}_5 = 1088$$

Step 3—Convert kilograms of total BOD_5 to grams.

$$\frac{1088\text{ kg total BOD}_5}{}\left|\frac{1000\text{ g}}{1\text{ kg}}\right| = 1\,088\,000\text{ g}$$

Step 4—Find the total surface area available for biofilm growth in the entire process.

$$\frac{13\,935\text{ m}^2}{\text{RBC unit}}\left|\frac{4\text{ RBC units}}{\text{treatment train}}\right|\left|\frac{2\text{ treatment trains}}{\text{process}}\right| = 111\,480\text{ m}^2$$

Step 5—Find the loading rate of total BOD_5 over the entire process.

$$\text{Organic Loading Rate}, \frac{g}{m^2 \cdot d} = \frac{\text{Organic Load, g BOD}_5/d}{\text{Surface Area of Media, m}^2}$$

$$\text{Organic Loading Rate}, \frac{g}{m^2 \cdot d} = \frac{1\,088\,000 \text{ g BOD}_5/d}{111\,480 \text{ m}^2}$$

$$\text{Organic Loading Rate}, \frac{g}{m^2 \cdot d} = 9.8$$

From Table 7.7, a typical organic loading rate for total BOD_5 over the entire RBC process is between 8 and 16 g $BOD_5/m^2{\cdot}d$. The calculated result is reasonable. Now find the soluble BOD loading to the first stage of the process.

Step 1—Find the mass of soluble BOD_5 going to the RBC process.

$$\text{kg soluble BOD}_5 = \frac{\left[\left(\text{Concentration}, \frac{mg}{L}\right)(\text{Flow, m}^3/d)\right]}{1000}$$

$$\text{kg soluble BOD}_5 = \frac{\left[\left(50 \frac{mg}{L}\right)(13\,600 \text{ m}^3/d)\right]}{1000}$$

$$\text{kg soluble BOD}_5 = 680$$

Step 2—Convert kilograms of soluble BOD_5 to grams.

50 kg total BOD_5	1000 g	= 680 000 g
	1 kg	

Step 3—Find the total surface area available for biofilm growth in the first stage.

13935 m²	1 RBC unit	2 treatment trains	= 27 870 m²
RBC unit	treatment train	process	

Step 4—Find the soluble BOD_5 loading to the first stage of the process.

$$\text{Organic Loading Rate}, \frac{g}{m^2 \cdot d} = \frac{\text{Organic Load, g soluble BOD}_5/d}{\text{Surface Area of Media, m}^2}$$

$$\text{Organic Loading Rate}, \frac{g}{m^2 \cdot d} = \frac{680\,000 \text{ g soluble BOD}_5/d}{27\,870 \text{ m}^2}$$

$$\text{Organic Loading Rate}, \frac{g}{m^2 \cdot d} = 24.4$$

From Table 7.7, a typical organic loading rate for soluble BOD_5 to the first stage of an RBC process is between 12 and 15 g $BOD_5/m^2{\cdot}d$. The operator of this RBC system needs to take action to reduce the loading to the first stage. They could bypass some of the primary effluent flow to the second stage by removing some baffles or put another RBC train into service. Bypassing will help even out the load in the short term.

In U.S. customary units:

Step 1—Find the mass of total BOD_5 going to the RBC process.

$$\text{Total BOD}_5\text{, lb} = \left(\text{Concentration}, \frac{\text{mg}}{\text{L}}\right)(\text{Flow, mgd})\left(8.34\frac{\text{lb}}{\text{mil. gal}}\right)$$

$$\text{Total BOD}_5\text{, lb} = \left(80\frac{\text{mg}}{\text{L}}\right)(3.6\text{ mgd})\left(8.34\frac{\text{lb}}{\text{mil. gal}}\right)$$

$$\text{Total BOD}_5\text{, lb} = 2402$$

Step 2—Find the total surface area available for biofilm growth in the entire process.

$$\frac{150\,000\text{ sq ft}}{\text{RBC unit}}\left|\frac{4\text{ RBC unit}}{\text{treatment train}}\right|\left|\frac{2\text{ treatment trains}}{\text{process}}\right| = 1\,200\,000\text{ sq ft}$$

Step 3—Divide the total media area by 1000 to get 1200 × 1000 sq ft.

Step 4—Find the loading rate of total BOD_5 over the entire process.

$$\text{Organic Loading Rate}, \frac{\text{lb BOD}_5}{1000\text{ sq ft}\cdot\text{d}} = \frac{\text{Organic Load, lb BOD}_5/\text{d}}{\text{Surface Area of Media, 1000 sq ft}}$$

$$\text{Organic Loading Rate}, \frac{\text{lb BOD}_5}{1000\text{ sq ft}\cdot\text{d}} = \frac{2402\text{ lb BOD}_5/\text{d}}{1200\times1000\text{ sq ft}}$$

$$\text{Organic Loading Rate}, \frac{\text{lb BOD}_5}{1000\text{ sq ft}\cdot\text{d}} = 2$$

From Table 7.7, a typical organic loading rate for total BOD_5 over the entire RBC process is between 1 and 3.3 lb BOD_5/1000 sq ft/d. The calculated result is reasonable. Now find the soluble BOD loading to the first stage of the process.

Step 1—Find the mass of total BOD_5 going to the RBC process.

$$\text{Pounds soluble BOD}_5 = \left(\text{Concentration}, \frac{\text{mg}}{\text{L}}\right)(\text{Flow, mgd})\left(8.34\frac{\text{lb}}{\text{mil. gal}}\right)$$

$$\text{Soluble BOD}_5\text{, lb} = \left(50\frac{\text{mg}}{\text{L}}\right)(3.6\text{ mgd})\left(8.34\frac{\text{lb}}{\text{mil. gal}}\right)$$

$$\text{Soluble BOD}_5\text{, lb} = 1501$$

Step 2—Find the total surface area available for biofilm growth in the first stage.

$$\frac{150\,000\text{ sq ft}}{\text{RBC unit}}\left|\frac{1\text{ RBC unit}}{\text{treatment train}}\right|\left|\frac{2\text{ treatment trains}}{\text{process}}\right| = 300\,000\text{ sq ft}$$

Step 3—Divide the total media area by 1000 to get 300 × 1000 sq ft.

Step 4—Find the loading rate of soluble BOD_5 to the first stage of the process.

$$\text{Organic Loading Rate}, \frac{\text{lb BOD}_5}{1000\ \text{sq ft}\cdot\text{d}} = \frac{\text{Organic Load, lb soluble BOD}_5/\text{d}}{\text{Surface Area of Media, 1000 sq ft}}$$

$$\text{Organic Loading Rate}, \frac{\text{lb BOD}_5}{1000\ \text{sq ft}\cdot\text{d}} = \frac{1501\ \text{lb soluble BOD}_5/\text{d}}{300 \times 1000\ \text{sq ft}}$$

$$\text{Organic Loading Rate}, \frac{\text{lb BOD}_5}{1000\ \text{sq ft}\cdot\text{d}} = 5$$

From Table 7.7, a typical organic loading rate for soluble BOD_5 to the first stage of an RBC process is between 2.4 and 3 lb soluble BOD_5/1000 sq ft/d. The operator of this RBC system needs to take action to reduce the loading to the first stage. They could bypass some of the primary effluent flow to the second stage by removing some baffles or put another RBC train into service. Bypassing will help even out the load in the short term.

PERCENT REMOVAL

The treatment efficiency of an RBC can be determined by calculating the percent removal of soluble or total BOD_5 or soluble or total $CBOD_5$ across the process. An example calculation is shown in the section on Process Variables for Trickling Filters.

$$\text{Removal, \%} = \left(\frac{\text{In} - \text{Out}}{\text{In}}\right) \times 100 \qquad (7.16)$$

HYDRAULIC LOADING RATE

The HLR for an RBC is the influent wastewater flow per unit of media surface area per day. It is expressed as either cubic meters of flow per square meter of media per day ($m^3/m^2 \cdot d$) or gallons per day per square foot (gpd/sq ft). The HLR is much less important to RBC systems than the organic loading rate; however, if the flow through the RBC tank is too high, there may not be enough time to complete treatment. At excessively high HLRs, biofilm may be stripped from the media. The hydraulic loading rate may be calculated using either of the following two formulas:

In International Standard units:

$$\text{HLR}, \frac{\text{m}^3}{\text{m}^2\cdot\text{d}} = \frac{\text{Flow, m}^3/\text{d}}{\text{Suface Area of Media, m}^2} \qquad (7.17)$$

In U.S. customary units:

$$\text{HLR}, \frac{\text{gpd}}{\text{sq ft}} = \frac{\text{Flow, gpd}}{\text{Surface Area of Media, sq ft}} \qquad (7.18)$$

HYDRAULIC DETENTION TIME

Hydraulic retention time is also referred to as the *hydraulic detention time* (HDT) or simply *detention time.* You may also see this as *HRT*. The HDT can be 1) the time required to fill a vessel or 2) the time required to empty a vessel, or 3) the average amount of time that water remains in a vessel.

With RBCs, only a small portion of the wastewater in the RBC tank is exposed to the biofilm on the media at any given time. It takes many rotations of the wheel through the wastewater to eventually reduce BOD_5 and/or NH_3-N concentrations. This takes time. Depending on the type of treatment desired, the HDT may be between 45 minutes and 4 hours. More time is required for systems that remove ammonia and even more time is needed for systems that remove both BOD_5 and NH_3-N. The detention time is largely out of the control of the operator as tank sizes were determined by the engineer during design. However, operators may have the ability to place more RBCs into service or take RBCs out of service. Detention time should be calculated if there is sudden increase in flow, if effluent BOD_5 and/or ammonia concentrations are increasing, or to help determine how many treatment trains should be in service. The formula for detention time is:

In International Standard units:

$$\text{Detention Time, days} = \frac{\text{Volume, m}^3}{\text{Flow, m}^3\text{/d}} \tag{7.19}$$

In U.S. customary units:

$$\text{Detention Time, days} = \frac{\text{Volume, gal}}{\text{Flow, gpd}} \tag{7.20}$$

Example detention time calculations may be found in the Primary Treatment chapter.

Hydraulic detention time is the amount of time the water spends in the treatment process. Solids retention time (SRT) is the amount of time the solids spend in the treatment process. For trickling filters and RBCs, the SRT is much longer than the HRT.

BIOFILM THICKNESS AND ROTATIONAL SPEED

A study of biofilms in RBCs measured biofilm thicknesses of 0.04 cm (0.016 in.) after 3 days of growth and up to 0.32-cm (0.125-in.) thick after 9 days of growth (Zahid and Garnczarczyk, 1994). The biofilm on the first stages is typically 0.15- to 0.33-cm (0.06- to 0.13-in.) thick (WEF, 2010). The thickness of the biofilm is controlled by the availability of BOD, nutrients, and shear. Shear is created by the discs being rotated through the water. Higher rotational speeds increase shear forces, which scour excess biomass from the media surface. Scouring velocity varies across the entire media surface, with the highest shear forces occurring at the outer edges of the media and the lowest shear forces occurring at the central shaft. As a result, the biofilm thickness may be thin at the outer edges of the media and much thicker in the interior of the media. The interior of the media cannot be observed without disassembling some of the media. Rotational speed should be adjusted to keep biofilms from becoming too thick and heavy while staying within the operating parameters recommended by the manufacturer. Rotational speeds range from 1 to 6 rpm.

Although less energy is required to rotate RBCs at slower speeds, slower speeds will reduce the amount of oxygen that is pulled from the air into the wastewater. The movement of oxygen from air to water is called the *oxygen-transfer efficiency*.

OXYGEN-TRANSFER RATE

The oxygen-transfer rate determines the biofilm oxygen concentration. Oxygen from the air must diffuse into the water, then across into the liquid layer at the biofilm surface, and then into the biofilm. Oxygen is transferred into the water remaining on the biofilm surface when the media are rotated through the air. How much water remains on the biofilm depends on how fast the media are rotating and on the surface structure of the biofilm. Thicker, rougher biofilms will retain more water than thinner, smoother biofilms.

Higher RBC rotational speeds increase oxygen transfer into the biofilm by increasing turbulence in the reactor tank and by lifting more wastewater into the air. This increases the amount of wastewater draining across the biofilm surface back to the reactor tank. The cycle of lift, air exposure, and drainage transfers oxygen from the air into the water and biofilm.

SECONDARY EFFLUENT RECIRCULATION

Secondary clarifier effluent may be pumped back into the influent of the RBC reactor tanks to increase DO concentrations. However, recycling treated secondary clarifier effluent will also decrease the hydraulic detention time in the RBC reactor tanks, which could reduce treatment efficiency. Recycling effluent may allow a larger portion of the biofilm to nitrify by diluting the influent organic concentration.

TEMPERATURE

Rotating biological contactors will perform best when wastewater temperatures are between 13 and 29 °C (55 and 85 °F). Biological reaction rates decrease with colder water temperatures. As temperatures drop, a greater mass of microorganisms and/or more media surface area may be needed to achieve the same level of treatment. Operators may observe a seasonal shift in how the biofilm appears on the RBC media. In spring and summer when the water is warmer, the biofilm will appear to move toward the earlier stages.

The last stages of the media may not have a visible biofilm. In fall and winter when the water is cooler, biofilm development may be more evenly distributed across all stages.

During winter operation, keep RBC covers in place and close access hatches to minimize heat loss and ice accumulation. During summer, inspection hatches may be opened to enhance airflow through the media.

PH AND ALKALINITY

Rotating biological contactors are biological treatment processes. The pH entering and leaving the process should be near pH 7. If the pH drops below 6.5, consider adding a base to the influent flow for pH adjustment. For RBCs that remove ammonia, the effluent alkalinity should be at least 100 mg/L as $CaCO_3$. When pH and alkalinity are too low, biological activity will be inhibited.

TEST YOUR KNOWLEDGE

1. **The OLR to an RBC is the mass of BOD_5 divided by the volume of media.**
   - ☐ True
   - ☐ False

2. **Ammonia removal requires less HDT than BOD_5 removal.**
   - ☐ True
   - ☐ False

3. **The OLR that can be treated in an RBC is limited by**
   a. Influent flow
   b. Oxygen availability
   c. Rotation speed
   d. Beggiatoa

4. **Because RBCs don't effectively treat particulates, this parameter is often used to calculate percent removal.**
   a. CBOD
   b. BOD
   c. sBOD
   d. COD

5. **An RBC process contains two treatment trains with four RBC units each. Each unit contains 13 935 m² (150 000 sq ft) of media surface area. Find the organic loading rate if 680 kg (1501 lb) of soluble BOD_5 is applied to the entire process in a 24-hour period.**
   a. 5.2 g sBOD/m²·d (1.0 lb sBOD/d/1000 sq ft)
   b. 6.0 g sBOD/m²·d (1.3 lb sBOD/d/1000 sq ft)
   c. 19.5 g sBOD/m²·d (4.2 lb sBOD/d/1000 sq ft)
   d. 23.6 g sBOD/m²·d (4.8 lb sBOD/d/1000 sq ft)

6. **The influent total BOD_5 concentration to an RBC is 80 mg/L. The effluent soluble BOD_5 is 8 mg/L. What is the percent removal?**
   a. 10%
   b. 72%
   c. 90%
   d. 95%

7. **The rotational speed of an RBC is increased slightly. Which of the following statements is true?**
   a. Biofilm thickness will increase
   b. Oxygen concentration will increase
   c. Bearing temperature will decrease
   d. HLR will increase

PROCESS CONTROL FOR ROTATING BIOLOGICAL CONTACTORS

Rotating biological contactors have very few operational parameters that are directly under the control of the operator. Items that can be controlled include: management of upstream processes to remove particulates, rotational speed, placement of baffles, and airflow (if available). Some facilities may be able to alternate their RBCs for parallel or series operation and/or recycle secondary clarifier effluent back to the beginning of the process. Operators should adjust each of these parameters as needed to keep the RBC within accepted design and operating values (Table 7.7) and to control biofilm thickness and sloughing. Operators should keep detailed operational records and graph results from sampling and process control calculations. Trend charts will help operators see small changes before they can develop into big problems.

VISUAL INSPECTION OF THE BIOFILM

The most important element of process control is daily shaft inspection by a trained operator. Principal observations typically include the biomass condition in each stage and the DO concentrations exiting the individual stages.

Observations about the first-stage biomass are the most critical. A healthy biomass on the first stage tends to be light brown, while biofilm on later stages tends to have a gold or reddish sheen. Lightly loaded units may be nearly devoid of visible biomass. A white or gray biomass indicates domination by certain types of filamentous bacteria such as *Beggiatoa*, *Thiothrix*, or *Lepothrix*. These bacteria are an indicator of low DO conditions—an unhealthy sign. *Beggiatoa* are described in more detail earlier in this chapter. A heavy, shaggy biomass in the first stage indicates an organic overload, which can be caused by an insufficient number of RBCs, industrial waste, or the effect of sidestreams returning organic loads from other treatment processes.

Excessive biofilm growth can fill the gaps between the media completely. This condition is called *bridging*. Bridging can prevent water from draining from the media, reduce oxygen transfer, and increase the weight on the support media and shaft (Mba, 2007). If biomass bridging occurs, operators should remove the excess biofilm and then adjust operating conditions to maintain biofilm thickness within normal range. Adjustments may include removing baffles to distribute the organic load over more RBC units, placing another treatment train into service, or increasing the rotational speed.

SHAFT OPERATING WEIGHT

An operating weight exceeding the shaft capacity can cause an RBC shaft or bearing failure or damage the plastic media. Operators must check the load cells on the RBC shafts prior to startup to obtain a baseline weight for the equipment. Measurements should be taken daily thereafter to verify that the biomass weight does not exceed the shaft structural loading capacity. If the operating weight is high, operators must adjust operating conditions to increase the removal of excess biomass from the media surfaces to reduce the weight within normal range. Remove excess biomass with either a pressured hose or increase the shaft speed.

Standard-density RBC media have a media weight of approximately 4537 kg (10000 lb) (Table 7.8). A biofilm that is 3.5-mm (0.15-in.) thick and has a specific gravity of 1.0 has a biomass weight of 35 390 kg (78 000 lb).

ORGANIC LOADING RATES

The organic loading to the first stage of any RBC system is limited by the oxygen transfer capability of the system. Exceeding this oxygen-transfer capability will often result in the proliferation of sulfide-oxidizing organisms and reduced BOD removal. When the first-stage organic loading exceeds 20 to 30 g/m^2·d (6.4 ppd BOD_5/1000 sq ft), white or gray patches develop on the disc surfaces, indicating large numbers of sulfide-oxidizing *Beggiatoa* or *Thiothrix* spreading across the media surface. This loading corresponds approximately to a soluble BOD_5 loading in the range of 12 to 20 g/m^2·d (2.6 to 3.8 ppd/1000 sq ft).

The organic loading to the first stage can be reduced by step feeding to distribute the influent flow to subsequent RBC units. The organic loading to the first stage can be reduced by placing additional RBC trains in service (if available). Recirculation of RBC effluent can reduce the organic loading to the first stage and also increase the DO in the wastewater applied to the first stage.

Table 7.8 Rotating Biological Contactor Media Density (developed with data from U.S. EPA, 1984b)*

	RBC Media Density			
Parameter	Standard	Intermediate	High	Extra High
Surface area per RBC unit m^2	9290	11 148	13 935	16 722
(sq ft)	(100 000)	(120 000)	(150 000)	(180 000)
SSA, m^2/m^3	118	143	180	215
(sq ft/cu ft)	(36.4)	(43.6)	(54.5)	(65.5)
Approximate media weight, kg	4536	5443	6803	8164
(lb)	(10 000)	(12 000)	(15 000)	(18 000)

*RBC unit standard media volume = 2750 cu ft (25 feet length 12 feet diameter).

DISSOLVED OXYGEN CONCENTRATION

For CBOD removal, a minimum DO concentration of 0.5 to 1.0 mg/L is needed at the end of the first stage, and at least 2 to 3 mg/L is needed at the end of the last stage of the RBC system. When the DO concentration falls below the minimum level, operators may need to adjust one or more of the following process control variables:

- Increase number of RBC stages and trains in service,
- Increase supplemental aeration to provide more oxygen,
- Start step feeding to reduce the first-stage organic loading,
- Increase recirculation to reduce the first-stage organic loading, and/or
- Increase RBC rotational speed to provide more oxygen.

SOLIDS REMOVAL

Suspended solids may accumulate in the RBC reactor basin. Periodically stop rotation of the RBC and extract a sample from the bottom of the reactor basin. Use either a dipper or a sample pump. Resuspend settled solids by injecting compressed air or pressurized flushing water into the bottom of the basin and drain them from the reactor basin. Be aware that introduction of air on one side can cause rotation of the media and shaft. If the RBC is equipped with an aeration system, solids deposition may be reduced by increasing the airflow rate.

CONTROL OF MACROFAUNA

Rotating biological contactor biofilms are populated by many of the same nuisance organisms as trickling filter biofilms, including filter flies, insect larvae, snails, worms, and filamentous bacteria. Control strategies are similar to those for trickling filters. Snails in RBCs can be effectively controlled by recirculation of a basic solution at a pH of 10 for a 24-hour period. Freshwater snails in RBCs can be effectively controlled by recirculation of either a salt (sodium chloride) solution or an ammonium chloride solution. Sodium or calcium hypochlorite solution (bleach) may be metered into the RBC system at a concentration of 60 to 70 mg/L for a 2- to 3-day period to help minimize snail populations.

TEST YOUR KNOWLEDGE

1. **The biomass in the first stage of an RBC treatment process is thick and shaggy. This may indicate**
   a. High rotation speed
   b. Organic overloading
   c. Insufficient aeration
   d. Septic conditions

2. **Healthy RBC biomass in the first stage of a process removing BOD_5 should be**
   a. Gray to black
   b. Red to gold
   c. Light brown
   d. White

3. **Load cell results are 10% higher than the rated shaft capacity. The operator should**
   a. Reduce rotational speed
   b. Manually remove excess biofilm
   c. Increase organic loading
   d. Seed the system with snails

4. **All of the following methods may be used to reduce the organic loading rate in the first stage of an RBC process EXCEPT**
   a. Remove baffles between stages
   b. Step-feed primary effluent to later stages
   c. Recirculate RBC effluent to stage 1
   d. Add chlorine to the influent channel

5. **An RBC treatment train consists of four stages. The DO concentration leaving the last stage is below 1 mg/L. The operator should**
   a. Increase the number of RBC stages and trains in service
   b. Replace any missing baffles between stages
   c. Decrease RBC rotational speed
   d. Resuspend settled solids in the RBC tank

6. **Low DO conditions may encourage the growth of this nuisance organism.**
   a. Snails
   b. Worms
   c. *Beggiatoa*
   d. Rotifers

OPERATION OF ROTATING BIOLOGICAL CONTACTORS

DAILY OPERATION

After normal operation has been established, perform daily observation of the following items:

- Visually inspect the biofilm.
    - Thick, shaggy biofilm in the first stage may indicate organic overloading.
    - White biomass indicates the presence of *Beggiatoa.*
    - Excessive sloughing may indicate toxicity.
    - Determine if biofilm predators are present in excess.
- Monitor DO concentrations at the end of each stage of treatment. This should be done in the middle of the day when the highest organic load is entering the process. For some facilities, peak load may occur during a different time of day.
- If multiple trains are in use, verify flow is distributed evenly between trains.
- Visually check equipment at least once per shift to ensure that it is functioning properly. Listen for any unusual noises as they may signal abnormal conditions.
- Rotational speed should be smooth with no evidence of loping. Record time to complete one revolution and time to complete each quarter revolution. The time (seconds) required for each one-quarter turn must be the same for each quadrant. If the time required for each quarter revolution differs, the biofilm weight distribution within the media is becoming unbalanced. Biofilms can grow unevenly when some portions of the media have more exposure time to the wastewater than others.
- Monitor shaft operating weight with the load cell.
- Shaft bearings should be cool.
- Inspect air drive (if present).
    - Blower output pressure is normal.
    - Bubble pattern in tank beside media is even. Turbulent areas can indicate a broken air header or diffuser.
- Inspect mechanical drive (if present).
    - Motor temperature.
    - Smooth operation of chain drive.
    - Verify oil level in speed reducer and chain drive is at correct level.
    - Check power consumption.
- Inspect ventilation systems for proper function. Excessive humidity or condensation could indicate inadequate airflow.
- Check for buildup of solids in bottom of RBC tank.

PLACING A ROTATING BIOLOGICAL CONTACTOR INTO SERVICE

If possible, plan to bring RBCs online during warmer weather. Biofilm development won't be visibly observable for several days and it may take several weeks before a mature biofilm develops through most of the filter. Startup procedures specific to your equipment may be found in the manufacturer's operations and maintenance manual (the following example is from Walker Process Equipment).

Step 1—When bringing a new unit into service, operators should ensure that all equipment is functioning properly before sending any wastewater to the process. Be sure to complete manufacturer-recommended maintenance before starting the unit.

Step 2—Start the RBC and allow it to complete one revolution. Look for movement of the chain casing and listen for any unusual noises. If adjustments are needed, stop the RBC.

Step 3—Start the RBC and allow it to rotate in the dry tank for a short period of time. While it is rotating, listen for any unusual noises. Verify motor amperage is within the manufacturer's recommendations. Watch to see that the media rotates smoothly. Time and record the number of revolutions per minute. After 15 minutes, check the temperature of the shaft bearings. Confirm that the bearing temperature is below the maximum recommended by the manufacturer. Absent a manufacture's recommendation, the bearings should not be warmer than 93 °C (200 °F). Do not allow the RBC to rotate for more than 15 to 20 minutes in a dry tank as the full weight of the RBC on the bearing can increase wear. When the tank is full, much of the RBC weight will be supported by the water.

Step 4—After proper operation of all equipment has been confirmed, the tank may be filled with wastewater by opening the upstream slide gate or valve. Continue careful observation as the tank fills. If further adjustments need to be made, the RBC may be stopped while the tank fills.

An RBC that has just been placed into service will not provide treatment. It takes several days to weeks for the biofilm to develop. The same wastewater can be left in the tank until the effluent quality begins to improve. The RBC must be rotated continuously to prevent uneven development of biofilm. During this time, primary effluent will need to be diverted to another treatment process. As effluent quality begins to improve, gradually increase the amount of primary effluent. Open the downstream slide gate or valve to allow RBC effluent to flow to the secondary clarifier. If it isn't possible to bring the process online gradually without discharging partially treated effluent, heavy doses of chlorine may be used to chemically treat the remaining organic material. Be sure to check with your regulatory agency before using heavy doses of chlorine or discharging any effluent during startup that does not meet the limits in your discharge permit.

Steps for starting up pumps, clarifiers, and other equipment can be found in other chapters or in *Operation of Water Resource Recovery Facilities* (WEF, 2016).

REMOVING A ROTATING BIOLOGICAL CONTACTOR FROM SERVICE

Shutdown procedures specific to your equipment may be found in the manufacturer's operations and maintenance manual (items 1 and 2 in the following example come from Walker Process Equipment). When removing an RBC from service, be aware of the following:

If the rotation of the RBC is stopped, it will develop an unbalanced load of biofilm. An RBC assembly standing idle in wastewater promotes rapid biomass growth in the submerged areas and decreased biomass growth in the non-wetted areas. These varying rates of growth cause imbalance when restarted. It is imperative that the RBC unit not be left standing idle in wastewater for more than 1 or 2 hours. If an RBC unit must sit idle and the tank cannot be drained, the RBC shaft must be rotated a quarter of a turn every 4 to 6 hours. This will help keep the biomass wet and reduce the amount of imbalance when restarting. Failure to rotate the RBC will result in extremely high stresses during restart, with possible damage to the drive, bearings, and RBC shaft.

If the tank is drained, the buoyancy effect of the water will be lost. The resulting additional load on the shaft will reduce the lift of the shaft if it is kept rotating. If the tank will be drained, stop the rotation of the RBC as the tank is being drained. Lockout and tagout power to the RBC so it cannot be accidentally restarted.

Step 1—Close the upstream slide gate or valve to stop the flow of wastewater.

Step 2—Stop the rotation of the RBC.

Step 3—Lockout and tagout power to the unit.

Step 4—Drain the tank. A portable sump pump may be required.

Step 5—Hose down the RBC media to remove as much biofilm as possible. The biofilm may be sprayed with a dilute bleach solution and allowed to soak. This will kill some of the biofilm and make it easier to remove.

Step 6—Remove any remaining water and solids from the tank.

Step 7—Close the downstream slide gate or valve to prevent water from backing up into the tank from other processes.

DATA COLLECTION, SAMPLING, AND ANALYSIS FOR ROTATING BIOLOGICAL CONTACTORS

Data collection, sampling, and analysis for RBCs is nearly identical to that recommended for trickling filters. Rotating biological contactors that remove BOD should monitor influent and effluent $sBOD_5$, $CBOD_5$, or COD. The DO concentration should be measured and recorded daily for each stage of treatment. Rotating biological contactors that remove ammonia should also be monitored for temperature, pH, alkalinity, ammonia, nitrite, and nitrate. Operating data that should be recorded daily, or at least any time an operational change is made, include: number of RBCs in service, rotational speed, influent flow, load cell test results, and any maintenance activities. The HDT should be calculated at least monthly or more often if influent flows are inconsistent. The OLR should be calculated whenever BOD data are available, but at least monthly. Verify that the calculated results are within the design parameters for the facility.

MAINTENANCE FOR ROTATING BIOLOGICAL CONTACTORS

Like any treatment process, the RBC system demands routine attention or operations and maintenance problems will occur. Chain drives, belts, sprockets, rotating shafts, and other moving parts need inspection and maintenance according to the manufacturer's instructions or with guidance from the design engineer. Although these requirements will vary among facilities, the RBC maintenance guide given in Table 7.9 provides many of the typically required maintenance procedures for RBCs. Operators may wish to add an additional column for comments and/or maintenance frequencies recommended by the manufacturer that don't fit into one of the existing columns.

Table 7.9 Rotating Biological Contactor Maintenance Schedule

		Frequency				
Function	Task or Activity	Daily	Weekly	Monthly	Quarterly	Annually
Mechanical	Check for excessive shaft deflection	X				
	Check shaft load cells for excessive weight	X				
	Check shaft alignment			X		
	Check shaft bearings, listen for unusual noise			X		
	Check speed reducer for oil leaks	X				
	Check oil levels per manufacturer specifications		X			
	Check all drive belts and chains			X		
	Lubricate bearings per manufacturer specifications					X
	Lubricate equipment per manufacturer specifications					X
	Check carbon steel components for corrosion				X	
	Repair protective coatings as needed				X	
Electrical	Verify motors are not overheating	X				
	Check motor torque loading		X			
	Check motors for high amperage draw		X			
Operation	Check for uniform and continuous shaft rotation	X				
	Check for ice formation in winter	X				
	Check for excessive growth, clogged media	X				
	Check for damaged media sections	X				
	Check for ponding or channeling	X				
	Check for odors	X				
	Check for excessive solids deposits in reactor basin		X			
	Check submerged aeration system in reactor basin					X
Maintenance	Perform general and scheduled maintenance	X	X	X	X	X
	Perform cover maintenance					X
	Align and tighten chain and belt drives				X	
	Adjust shaft bearings if needed				X	
	Change oil or lubricants per specifications					X
	Grease electric motor bearings if appropriate					X

Routine maintenance should include the inspection of shafts and replacement of broken air cups or media that might otherwise jam or interfere with shaft rotation. Housekeeping should include the removal of grease balls via a net device. The manufacturer may provide advice on making field repairs to media that become separated. Unbonded surfaces may sometimes be repaired by melting the plastic with a heated metal rod or other manufacturer-recommended product.

MECHANICAL DRIVE SYSTEMS

Shaft bearings should be inaudible above the splashing. A screwdriver or metal rod can be used to transmit bearing noise to the operator's ear. Vibration meters can also sense noise (vibrations). Drive motors, which need daily inspection, should run cool enough to touch with a bare hand (less than 60 °C [140 °F]). If motor amperage readings are recorded, they should be taken and logged weekly. During daily observations of the belt drive, a squealing noise is the first indication that a problem has occurred. Because belts are often sold as a set, the whole set should be replaced with identical belts from the same manufacturer.

AIR-DRIVE SYSTEMS

Air-drive systems require more careful monitoring and attention than mechanical drive units because shaft speed and balance must be maintained via indirect air lift. Rotation speed should be checked daily and compared with manufacturer recommendations. Shaft balance also requires periodic checks to ensure that excess biomass has not built up on one side of the discs. For air-drive RBCs, each shaft must be timed for quarter revolutions once or twice each week. The time (seconds) required for each one-quarter turn must be the same for each quadrant. If the time required for each quarter revolution differs, the biomass weight distribution within the media is becoming unbalanced. When a shaft starts to go out of balance, the heavy side tends to stay on the bottom longer, which gives it more time in contact with the food supply. As a result, the biomass on the heavy side becomes even heavier and the weight imbalance increases.

On a monthly basis, or more frequently, the air volume supplied to each operating shaft should be increased to 150% of normal to remove excess biomass from the media surfaces. Stripping excess biomass keeps the biofilm thin and minimizes the potential for the shafts to become unbalanced. Daily monitoring of blower discharge pressure will help indicate the possible presence of clogged diffusers or other interferences.

If a shaft imbalance is detected early, excess biomass can be removed by air purging. If the shaft is badly out of balance, the unit must be shut down and the tank must be drained. The isolated unit can be dried out or be treated chemically to allow the biomass to die and slough from the media.

TEST YOUR KNOWLEDGE

1. **Sudden, excessive sloughing of biofilm from a trickling filter or RBC may indicate**
   a. BOD_5 increase
   b. Toxicity
   c. Snail infestation
   d. Excessive aeration

2. **Rotating biological contactor rotation should be monitored per quarter turn to determine if _______ is occurring.**
   a. Sloughing
   b. Chain slippage
   c. Loping
   d. Vibration

3. **If an RBC must be stopped for a period of time while remaining in a full tank,**
   a. Manually rotate the unit one-quarter turn every 4 to 6 hours.
   b. Use a power washer to hydraulically scrub all biofilm from the media.
   c. Immediately shut off the blower and diffused air headers to prevent clogging.
   d. Open the cover doors to increase airflow over the exposed media surface.

4. **Allowing an RBC to sit idle in a tank full of wastewater**
    a. Is acceptable if idle time is less than 8 hours
    b. Controls the growth of *Beggiatoa* and *Thiothrix*
    c. Is part of a feast and famine control strategy
    d. May result in uneven biofilm growth

5. **Dissolved oxygen concentrations should be measured at the end of each stage of treatment in an RBC treatment train**
    a. Daily
    b. Weekly
    c. Monthly
    d. Quarterly

6. **Excessive accumulation of biofilm on an RBC may cause**
    a. Media delamination
    b. Increased rotation speed
    c. Shaft deflection
    d. Lower load cell readings

7. **Air-drive systems should be operated with higher airflow rates some percentage of the time to**
    a. Remove settled solids from diffusers
    b. Scour excess biofilm from the media
    c. Increase media buoyancy
    d. Reduce blower strain from low airflow

TROUBLESHOOTING OF ROTATING BIOLOGICAL CONTACTORS

When properly designed and operated, RBCs can provide trouble-free secondary treatment. However, some RBC facilities constructed during the late 1970s and early 1980s required significant troubleshooting or facility modifications to achieve the desired treatment level without operating problems.

Troubleshooting operational problems begins with maintaining good operating records and collecting data associated with the RBC process. Tracking total and soluble BOD_5, suspended solids, organic nitrogen, ammonia-nitrogen, pH, alkalinity, DO, and other parameters is necessary to recognize trends that may have an adverse effect on the RBC system. The sampling frequency may have to be increased beyond what is required by the discharge permit to ensure representative data.

Equipment failures (e.g., broken shafts or failed filter media) were common occurrences in many of the early RBC installations. Most of these problems were resolved through equipment warranty or performance specifications. Many of the problems associated with the early designs have been mitigated by using more conservative design practices, improving equipment design and manufacturing practices, shaft-weighing devices, and using supplemental aeration to improve biomass uniformity. Table 7.10 is a troubleshooting guide for other problems associated with the design, operation, and maintenance of RBCs.

SAFETY CONSIDERATIONS FOR ROTATING BIOLOGICAL CONTACTORS

It is important that operations and maintenance personnel follow safety guidelines specified by primary equipment manufacturers and commonly practiced by the utility. This includes, but is not limited to, procedures describing appropriate equipment lockout, confined space entry, localized power interruption, and standby rescue personnel during maintenance activities. Operations personnel must familiarize themselves with and respect safe operations guidelines recommended by the equipment manufacturers and maintain a clean, well-ventilated, and well-lit environment to prevent accidents. Some safety considerations, especially for RBCs, are listed below:

- Work on RBCs may proceed only after media rotation has been stopped. The RBC must be locked in place and the electrical switch must be locked out and tagged out on the electrical panel.
- Never attempt to stop a moving RBC with your hands or other body parts.
- Never work underneath an RBC unless additional supports have been provided to secure the full weight of the RBC.
- Covered RBCs have special safety considerations, because they are considered confined spaces. The possibility exists for the atmosphere under the dome to contain little oxygen or much hydrogen sulfide or ammonia. Maintenance in these areas must include proper confined space entry procedures.

Table 7.10 Troubleshooting Guide for Rotating Biological Contactors (U.S. EPA, 1978)

Indicators/Observations	Probable Cause	Check or Monitor	Solutions
Decreased treatment efficiency	Organic overload	Check peak organic loads; if less than twice the daily average, should not be the cause	Improve pretreatment or expand facility
	Hydraulic equalization	Peak hydraulic loads less than twice the daily average, should not be the cause	Flow equalization; eliminate source of excessive flow; balance flows between reactors
	pH too high or too low	Desired range is 6.5 to 8.5 for secondary treatment, 8 to 8.5 for nitrification	Eliminate source of undesirable pH or add an acid or base to adjust pH; when nitrifying, maintain alkalinity at seven times the influent ammonia concentrations
	Low wastewater temperatures	Temperature less than 15 °C (59 °F) will reduce efficiency	If available, place more treatment units in service
	Snails stripping biofilm from media	Observe presence of snails on media or at reactor basin water surface	Remove unit from service and add caustic, salt, ammonium chloride, or hypochlorite to remove snails and eggs from media
Excessive sloughing of biomass from discs	Toxic materials in influent	Determine material and its source	Eliminate toxic material if possible; if not, use flow equalization to reduce variations in concentration so biomass can acclimate
	Excessive pH variations	pH below 5 or above 10 can cause sloughing	Eliminate source of pH variations or maintain control of influent pH
Development of white biomass over most of disc area	Septic influent or high hydrogen sulfide concentrations	Influent odor	Pre-aerate wastewater or add sodium nitrate, hydrogen peroxide, or ferrous sulfate; supplemental aeration may also help, especially in the first stages
	First stage is organically overloaded	Organic loading on first stage	Adjust baffles between first and second stages to increase fraction of total surface area in first stage
Solids accumulating in reactors	Inadequate pretreatment	Determine if solids are grit or organic	Remove solids from reactors and improve grit removal or primary settling
Shaft bearings run hot or fail	Inadequate maintenance	Maintenance schedules and practices	Lubricate bearings per manufacturer's instructions
Motors running hot	Inadequate maintenance	Oil level in speed reducer and chain drive	Lubricate bearings per manufacturer's instructions
	Chain drive alignment incorrect	Alignment	Adjust and correct alignment

SECTION EXERCISE

This exercise will draw on some of the knowledge acquired about RBCs. In this section, an operator must complete the following tasks:

- Label a diagram of an RBC.
- Match the components of an RBC to their function.
- Given information for an RBC process, determine the most likely reason for the described condition, and suggest a corrective action.

1. Label the following items in the image below. Media Support, Media, Tankage, Baffle, Shaft, and Bearing.

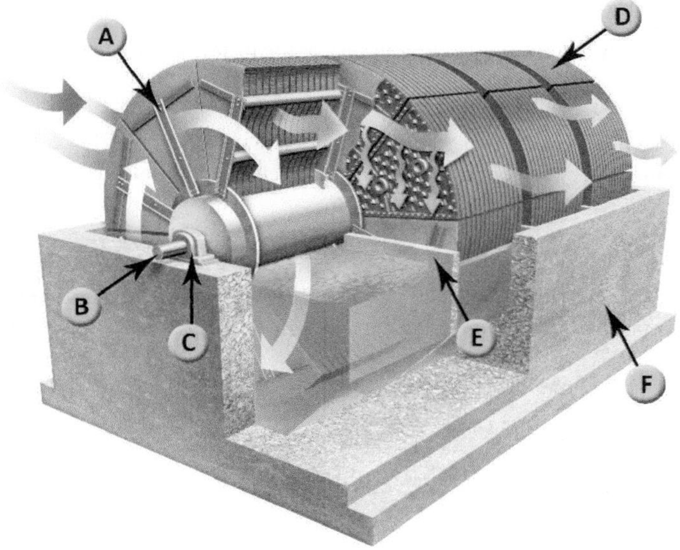

2. Match the RBC components to their function.

1. Media	a. media support
2. Central shaft	b. prevents algae growth
3. Mechanical drive	c. surface for biofilm growth
4. Baffles	d. support shaft weight
5. Covers	e. capture air bubbles for rotation
6. Bearings	f. constant speed media rotation
7. Expansion bearing	g. prevents shaft damage from temperature changes
8. Air cups	h. weighs shaft and biomass
9. Load cell	i. separates treatment stages

3. An RBC facility for a small town has been unstaffed for a 3-day weekend because the circus is in town. The system usually operates very well with very little variation in performance or influent characteristics. Upon returning to the facility on Tuesday morning, the operators find the following: 1) thick, shaggy biofilm on the SBRs in the first stage of treatment, 2) patches of white biofilm, and 3) the RBC in stage 2 appears to be stuck and the media on the upper portion of the RBC has dried. What caused each of these problems and what should the operators do to fix them?

SECTION EXERCISE SOLUTIONS

1. A = Media support, B = Shaft, C = Bearing, D = Media, E = Baffle, F = Tankage
2. 1 = c, 2 = a, 3 = f, 4 = i, 5 = b, 6 = d, 7 = g, 8 = e, 9 = h
3. The thick, shaggy biofilm in the first stage of treatment indicates organic overloading. It could be a result of the circus coming to town because it probably drew visitors from neighboring communities. Several different things could be done to decrease the OLR to the first stage of the system. The operators could 1) remove the baffles between RBC stages to increase the amount of biofilm exposed directly to the influent wastewater, 2) step-feed primary effluent to later stages in the process, 3) place additional treatment trains into service (if available), or 4) recycle treated wastewater to help dilute the influent wastewater. The operators should also take action to reduce the amount of biofilm on the RBCs in stage 1. If the RBCs are air-driven, increasing the airflow to 150% of normal will help remove excess biofilm. If the RBCs are mechanically driven, increasing the rotation speed will have the same effect.

Notes:

Loading may be up for other reasons, too. The operators should collect samples of the influent to see if the organic loading to the WRRF has returned to normal. They should also collect samples of the primary clarifier effluent just in case the loading to the RBCs went up because the clarifier is not working efficiently.

Patches of white biofilm indicate the presence of *Beggiatoa*. *Beggiatoa* is an indicator of low DO conditions and organic overloading. Increasing the rotation speed and adding supplemental air are two methods that could be used to increase DO concentrations.

The RBC in stage 2 is stuck in place. If the RBCs are air-driven, then uneven biofilm development and loping are the most likely cause. If the RBCs are mechanically driven, it is possible that either the motor failed or became disconnected from the gear reducer. Before the RBC can be restarted, excess biofilm must be removed. In the situation described here, it will be difficult to remove excess biofilm evenly. The operators should drain the tank and completely clean the RBC with high-pressure water before placing it back into service.

CHAPTER SUMMARY

<table>
<tr><th></th><th>Trickling Filters</th><th>Rotating Biological Contactors</th></tr>
<tr><td>THEORY</td><td colspan="2">■ Microorganisms convert non-settleable organic solids and soluble organic matter into new microorganism and waste products.
■ Process may remove BOD_5, NH_3-N, NO_2-N, and NO_3-N.
■ Microorganisms grow attached to media as a biofilm.
■ Mass transfer, the movement of substances through the biofilm, limits biofilm growth and thickness.
■ Biofilm loss (sloughing) is influenced by availability of resources and hydraulic shear.
■ Snails and filter flies consume biofilm. Snails have other negative effects on operation and downstream processes.
■ Filamentous bacteria can reduce treatment efficiency.</td></tr>
<tr><td>OVERVIEW</td><td>■ Containment structure filled with some type of material that provides a surface for biofilm growth.
■ Wastewater is pumped to the top of the media bed and allowed to trickle down through the media.
■ Treated wastewater is collected in an underdrain system and is sent to a secondary clarifier where the solids are removed from the treated water.</td><td>■ Wheels of plastic media provide a surface for biofilm growth.
■ Wheels are rotated through a shallow tank of wastewater to alternately expose the biofilm to wastewater (food) and air (oxygen).
■ Treated wastewater overflows at the end of the tank and is sent to a secondary clarifier where solids are removed from the treated wastewater.</td></tr>
<tr><td>DESIGN PARAMETERS</td><td>■ Based primarily on organic loading rate.
■ Historic classifications are roughing, super-rate, high-rate, intermediate-rate, and low- or standard-rate filters.
■ Modern classifications accommodate the higher hydraulic and organic loading rates possible with plastic media.
■ Modern classifications are roughing, carbon-oxidizing, combined BOD and nitrification, and nitrification only.
■ Roughing filters must be followed by another treatment process for complete removal of BOD_5.
■ Filters may be operated in parallel or in series. They may also be combined with other biological treatment processes like activated sludge.</td><td>■ Designed by engineers and manufacturers based on field experience and kinetic models.
■ Based primarily on organic loading rate. Loading calculations use total BOD_5 and soluble BOD_5.
■ Loads are calculated for the overall RBC process and for the first stage of treatment. Loading to the first stage is typically 4 times greater than overall loading.
■ Organic loading rate is limited by the amount of oxygen that can be transferred from the air to the wastewater by the rotating biofilm.
■ Rotating biological contactors receiving high organic loading remove BOD_5 only. As loading rates decrease, RBCs may also remove ammonia.
■ RBCs may be operated in parallel or in series.
■ Multiple RBCs operated in series are a treatment train.
■ Baffles may be used to separate individual RBCs in a treatment train. Each segmented portion of the treatment train is called a treatment stage. A stage may have multiple RBC units.</td></tr>
</table>

EXPECTED PERFORMANCE	■ Depends on the organic loading rate. Percent removal of BOD_5 decreases with increasing loading. ■ Possible to achieve < 10 mg/L of soluble BOD_5 and < 3 mg/L NH_3-N. ■ Conversion of NH_3-N to NO_2-N and NO_3-N when soluble BOD_5 is less than 20 mg/L.

EQUIPMENT	Trickling Filters	RBCs
MEDIA	■ Rock, redwood slats, or plastic. ■ Random media consist of loose pieces randomly packed together. ■ Structured media consists of rectangular blocks made from plastic sheets. Surface area per volume of media varies depending on the type and manufacturer. ■ Voids allow air and wastewater to move through the media. ■ Media must have a high surface area available for biofilm growth per volume of media. This is the SSA. ■ Media is not submerged. ■ Media is supported by an underdrain system. ■ Underdrain may be constructed from clay tile, redwood beams, concrete beams, or plastic beams.	■ Plastic media wheels that may be up to 3.6 m (12 ft) in diameter. ■ Standard shaft length is 8.2-m (27-ft) long with 7.6 m (25 ft) occupied with media. ■ Standard RBC media contain 9290 m^2 (100 000 sq ft) of media surface per 7.6-m (25-ft) long shaft. Higher density media with more surface area are available. ■ Shafts may be cylindrical, hexagonal, or square depending on the manufacturer. ■ Media are mounted to the shaft in sections. Damaged media may be replaced by removing individual sections. ■ Media are rotated through the wastewater with either a mechanical drive unit or may be air driven. Air-driven units rely on air bubbles rising from the bottom of the tank to push one side of the media wheel toward the surface.
CONTAINMENT	■ Square or rectangular tanks. ■ Concrete or steel. ■ May or may not be water tight. ■ May have ventilation ports along the bottom perimeter to allow for natural ventilation.	■ Rectangular tanks approximate the outer dimensions of the individual RBC units. ■ Two bearings, one on each end of the RBC, are mounted on the wall of the containment structure. The bearings help support the weight of the shaft. ■ The bearing opposite the motor will be equipped with a load cell. The load cell is used to measure the weight of the media and biofilm. Operators must ensure that the total weight doesn't exceed the rated capacity of the shaft. ■ Tanks may be divided by permanent or adjustable baffles. Removable baffles allow the operator to adjust the size of the first stage of treatment. ■ Hydraulic retention time in the RBC tank varies between 45 minutes and 4 hours depending on treatment objectives. More time is needed to remove ammonia and for combination systems.
VENTILATION	■ Natural or mechanical ■ Direction of airflow in naturally ventilated systems changes depending on water and air temperature differences.	■ Biofilm is exposed to air directly with each rotation of the media. ■ Supplemental air may be provided by injecting air into the bottom of the RBC with a blower and air diffusers.

EFFLUENT	■ Underdrain collects treated wastewater. Floor slopes to channel in center. ■ Space below media is called a *plenum* in plastic media filters. ■ Underdrain helps distribute air over filter cross-sectional area.	■ Treated wastewater flows out over the end of the RBC tank. ■ Baffles between stages help to prevent short-circuiting.
WASTEWATER DISTRIBUTION	■ Wastewater moved to the top of the media bed by a pumping station. ■ Rotary distributor or fixed nozzle. ■ Rotary distributors have splash plates to fan water out. ■ Biofilm thickness is influenced by distributor speed. Faster speeds equate to thicker biofilms. ■ Distributor speed controlled either hydraulically or with a motor and gear reducer. ■ Speed of hydraulically driven distributors controlled by number of open orifices on the leading and tailing edges of the distributor arm. ■ Debris is removed from rotary distributor arms through the dump gates at the end of the arms. ■ Treated effluent may be recycled to the beginning of the process to help maintain a constant HLR. ■ Verify even distribution of wastewater with a pan test.	■ Wastewater flows from one end of the RBC tank to the other. Baffles separate long treatment trains into treatment stages. ■ Wastewater is not typically recycled in RBC processes. ■ Recycle can help dilute high-strength wastewater. ■ If the first stage in an RBC treatment train is organically overloaded, baffles may be removed from between the first stage and second stage of treatment. Step-feeding may also be used to bypass primary effluent to other stages in the treatment train.
PROCESS VARIABLES	■ Dissolved oxygen concentration ■ Organic loading rate—mass of BOD_5 per volume of media; influences biofilm thickness. ■ Recirculation ratio—influences total HLR and flushing intensity; returns DO to the beginning of the process. ■ Hydraulic loading rate—includes both primary effluent flow and recycle flow; total flow over the cross-sectional area. ■ Flushing intensity or SK influences biofilm thickness. ■ Temperature ■ pH and Alkalinity ■ Ponding—water pooling on the surface of the trickling filter; occurs when there is a blockage below.	■ Organic loading rate—mass of BOD_5 per square meter or square foot of media surface area; influences thickness of biofilm. ■ Hydraulic loading rate—volume of wastewater treated per square meter or square foot of media surface area. ■ Hydraulic detention time—the amount of time the wastewater spends in the RBC tank. ■ Rotational speed—influences thickness of biofilm; faster wheel speeds decrease biofilm thickness; adds oxygen to the wastewater. ■ Temperature ■ pH and Alkalinity ■ Media bridging; caused by an overgrowth of biofilm.

<table>
<tr>
<td>PROCESS CONTROL</td>
<td><ul>
<li>Variables under the direct control of the operator include management of upstream processes to remove particulates, the recirculation ratio, and HLR.</li>
<li>Adjust the recycle ratio to influence the HLR. Goal is to maintain complete biofilm wetting at all times.</li>
<li>Adjust the distributor speed to influence the flushing intensity (SK value) and biofilm thickness. Goal is to encourage continuous sloughing of small quantities of biofilm. Adjustments to both the recycle ratio and arm speed may be needed to obtain the desired flushing intensity.</li>
<li>Control of macrofauna to prevent excessive losses of biofilm, especially in systems designed for ammonia removal.</li>
</ul></td>
<td><ul>
<li>Variables under the direct control of the operator include management of upstream processes to remove particulates and the rotational speed of the RBC.</li>
<li>Operators may also adjust the size of treatment stages by adding or removing baffles.</li>
<li>Some facilities are able to control airflow or can alternate between parallel and series operation.</li>
<li>Daily inspection of the biofilm is the most critical element of process control. Biofilm checks should include: appearance (color and texture), weight (load cell), and DO concentrations leaving each stage.</li>
<li>Adjustments should be made to keep RBC weight within accepted parameters and maintain DO concentrations leaving each stage of treatment.</li>
<li>Periodically check for presence of sloughed solids in the bottom of the RBC tank.</li>
<li>Control of macrofauna to prevent excessive losses of biofilm.</li>
</ul></td>
</tr>
<tr>
<td>PROCESS CONTROL DATA</td>
<td colspan="2"><ul>
<li>Measure total and soluble BOD_5 in trickling filter and RBC influent samples. Samples should not include recycle flow.</li>
<li>Measure settled or soluble BOD_5 in effluent samples.</li>
<li>Record effluent DO concentrations daily. For RBCs, this should be done for each stage of treatment.</li>
<li>For systems removing ammonia, also monitor effluent pH, alkalinity, ammonia, nitrite, and nitrate in both the influent and effluent. Influent samples should not include recycle flows.</li>
<li>Record operating data including number of units in service, distributor speed, RBC rotational speed, recycle flow, influent flow, and results of process control calculations.</li>
</ul></td>
</tr>
<tr>
<td>MAINTENANCE</td>
<td colspan="2"><ul>
<li>Daily observation for visual and audible indications of malfunction. Make sure rotation is smooth and that spray orifices are not plugged.</li>
<li>Maintenance should only be performed after components have come to a complete stop, been locked out and tagged out, and are locked into place.</li>
<li>NEVER try to stop a rotating distributor or RBC with your hands or any other body part.</li>
<li>Check for water in drive oil by collecting a sample from the bottom of the oil reservoir.</li>
<li>Regular maintenance should be performed in accordance with the manufacturer's recommendations.</li>
<li>Maintain an inventory of spare parts.</li>
</ul></td>
</tr>
</table>

References

Aspegren, H. (1992) *Nitrifying Trickling Filters, A Pilot Study of Malmö, Sweden*; Malmö Water and Sewage Works: Malmö, Sweden.

Boltz, J. P.; Goodwin, S. J.; Rippon, D.; Daigger, G. T. (2008) A Review of Operational Control Strategies for Snail and Other Macrofauna Infestations in Trickling Filters. *Water Pract.*, **2** (4) 1–16.

Choi, Y. C.; Morgenroth, E. (2003) Monitoring Biofilm Detachment Under Dynamic Changes in Shear Stress Using Laser-Based Particle Size Analysis and Mass Fractionation. *Water Sci. Technol.*, **47** (5), 69–76.

Dölle, K.; Peluso, C. (2015) Earthworm and Algae Species in a Trickling Filter. *J. Adv. Biol. Biotechnol.*, **3** (3), 132–138.

Elissen, H. J. H. (2007) *Sludge Reduction by Aquatic Worms in Wastewater Treatment with Emphasis on the Potential Application of Lumbriculus variegatus.* PhD thesis, Wageningen University, Wageningen, Netherlands. ISBN 978-90-8504-777-3.

Grady, L. E.; Daigger, G. T.; Lim, H. (1999) *Biological Wastewater Treatment,* 2nd ed.; Marcel Dekker: New York.

Gray, R.; Ritland, G.; Chan, R.; Jenkins, D. (2000) Escargot...Going...Gone, A Nevada Facility Controls Snails with Centrate to Meet Stringent Total Nitrogen Limits. *Water Environ. Technol.,* **12** (5), 80–83.

Lacan, I.; Gray, R.; Ritland, G.; Jenkins, D.; Resh, V.; Chan, R. (2000) The Use of Ammonia to Control Snails in Trickling Filters. *Proceedings of the 73rd Annual Water Environment Federation Technical Exposition and Conference,* Anaheim, California, Oct 14–18; Water Environment Federation: Alexandria, Virginia.

Laspidou, C. S.; Rittmann, B. E. (2004) Evaluating Trends in Biofilm Density Using the UMCCA Model. *Water Res.*, **38**, 3362–3372.

Mba, D. (2003) Mechanical Evolution of the Rotating Biological Contactor into the 21st Century. *Proc. Inst. Mech. Eng., Part E,* **217** (3), 189–219.

McCann, K. J.; Sullivan, R. A. (1980) Aerated RBC's—What Are the Benefits. *Proceedings of the First National Symposium/Workshop on Rotating Biological Contactor Technology*; Champion, Pennsylvania, Feb 4–6.

Metcalf and Eddy, Inc. (2003) *Wastewater Engineering: Treatment and Reuse,* 4th ed.; McGraw Hill: New York.

Parker, D. S. (1998) Establishing Biofilm System Evaluation Protocols. WERF Workshop: Formulating a Research Program for Debottlenecking, Optimizing, and Rerating Existing Wastewater Treatment Plants. *Proceedings of the 71st Annual Water Environment Federation Technical Exposition and Conference,* Orlando, Florida, Oct 3–7; Water Environment Federation: Alexandria, Virginia.

Parker, D. S.; Jacobs, T.; Bower, E.; Stowe, D. W.; Farmer, G. (1997) Maximizing Trickling Filter Nitrification Through Biofilm Control: Research Review and Full Scale Application. *Water Sci. Technol.,* **36,** 255–262.

Picioreanu, C.; van Loosdrecht, M. C. M.; Heijnen, J. J. (1999) Multidimensional Modeling of Biofilm Structure. *Proceedings of the 8th International Symposium on Microbial Ecology*; Atlantic Canada Society for Microbial Ecology: Halifax, Canada.

Rensink, J. H.; Rulkens W. H. (December 1997) Using Metazoa to Reduce Sludge Production. *Water Sci. Technol.*, **36**, (11), 171–179.

Schroeder, E. D.; Tchobanoglous, G. (1976) Mass Transfer Limitations on Trickling Filter Design. *J. Water Pollut. Control Fed.,* **48,** 771–775.

Tekippe, T. R.; Hoffman, R. J.; Matheson, R. J.; Pomeroy, B. (2006) A Simple Solution to Big Snail Problems—A Case Study at VSFCD's Ryder Street Wastewater Treatment Plant. *Proceedings of the 79th Annual Water Environment Federation Technical Exposition and Conference,* Dallas, Texas, Oct 21–25; Water Environment Federation: Alexandria, Virginia.

Trickling Filter Floor Institute (1970) *Handbook of Trickling Filter Design*; Public Works Journal Corporation: Ridgewood, New Jersey.

U.S. Environmental Protection Agency (1970) *Operation of Wastewater Treatment Plants: A Field Study Training Program*; U.S. Environmental Protection Agency, Office of Water Programs, Division of Manpower and Training: Cincinnati, Ohio.

U.S. Environmental Protection Agency (1978) *Field Manual for Performance Evaluation and Troubleshooting at Municipal Wastewater Treatment Facilities;* EPA Contract No. 68-01-4418; U.S. Environmental Protection Agency: Washington, D.C.

U.S. Environmental Protection Agency (1984a) *Rotating Biological Contactors (RBCs) Checklist for a Trouble-Free Facility*; 905R84120; U.S. Environmental Protection Agency: Washington, D.C.; May.

U.S. Environmental Protection Agency (1984b) *Design Information on Rotating Biological Contactors*; EPA-600/2-84-106; U.S. Environmental Protection Agency, Municipal Environmental Research Laboratory: Cincinnati, Ohio; June.

U.S. Environmental Protection Agency (1984c) *Review of Current RBC Performance and Design Procedures*; EPA-430/9-84-008; U.S. Environmental Protection Agency, Office of Water, Program Operations: Washington, D.C.; September.

Water Environment Federation (2007) *Operation of Municipal Wastewater Treatment Plants,* 6th ed.; Manual of Practice No. 11; Water Environment Federation: Alexandria, Virginia.

Water Environment Federation (2010) *Biofilm Reactors;* Manual of Practice No. 35; Water Environment Federation: Alexandria, Virginia.

Water Environment Federation (2012) *Basic Laboratory Procedures for the Operator–Analyst*, 5th ed.; Water Environment Federation: Alexandria, Virginia.

Water Environment Federation (2016) *Operation of Water Resource Recovery Facilities*, 7th ed.; Manual of Practice No. 11; Water Environment Federation: Alexandria, Virginia.

Water Environment Federation; American Society of Civil Engineers; Environmental and Water Resources Institute (2010) *Design of Municipal Wastewater Treatment Plants,* 6th ed.; WEF Manual of Practice No. 8/ASCE Manuals and Reports on Engineering Practice No. 76; Water Environment Federation: Alexandria, Virginia.

Williams, N.V.; Taylor, H. M. (1968) The Effects of Psychoda alternate (Say) (Diptera) and Lumbricillus rivalis (Levinson) (Enchytraeidae) on the Efficiency of Sewage Treatment in Percolating Filters. *Water Res.*, **2**, 139–150.

Zahid, W.; Ganczarczyk (1994) Structure of RBC Biofilms. *Water Environ. Res.,* **66**, 100–106.

Zhang, T. C.; Bishop, P. L. (1994) Density, Porosity, and Pore Structure of Biofilms. *Water Res.,* **28**, 2267–2277.

Suggested Readings

Arasmith, S. E. E.; Scheele, M.; Zentz, K. (2000) *Pumps and Pumping,* 8th ed.; ARC Publications, Inc.: Albany, Oregon.

U.S. Environmental Protection Agency (1973b) *Procedural Manual for Evaluating the Performance of Wastewater Treatment Plants;* U.S. Environmental Protection Agency, Office of Water Programs: Washington, D.C.

U.S. Environmental Protection Agency (1978) *Field Manual for Performance Evaluation and Troubleshooting at Municipal Wastewater Treatment Facilities*; U.S. Environmental Protection Agency: Washington, D.C.

U.S. Environmental Protection Agency (1992) *Rotating Biological Contactors*; EPA/540/S-92/007; U.S. Environmental Protection Agency: Washington, D.C.

USFilter (1998) *Applying the Rotating Biological Contactor Process*; Bulletin No. USF 315-13A6, 5M-10/98; Envirex Products: Waukesha, Wisconsin.

Water Environment Federation (2012) *Basic Laboratory Procedures for the Operator–Analyst*, 5th ed.; Water Environment Federation: Alexandria, Virginia.

CHAPTER 8

Activated Sludge

Introduction

The activated sludge process is the most widely used biological treatment process for reducing the concentration of organic pollutants in wastewater. Well-established design and operational standards based on empirical data and scientific basis have evolved over the years. As a result, our understanding of the process has advanced from a system originally designed simply for removal of solids and organic material to one that now removes nutrients such as nitrogen and phosphorus. New process configurations and technologies continue to evolve. Despite these advances, poor process performance can still present problems for many water resource recovery facilities (WRRFs).

LEARNING OBJECTIVES

Upon completing this chapter, you will be able to

- Describe the components of an activated sludge process and the functions of each.
- Describe the relationship between activated sludge microbiology and sludge settleability.
- Explain the importance of balancing the growth of floc formers and filament formers.
- Identify microorganisms common to activated sludge processes including filamentous bacteria, protozoa, metazoa.
- List at least three groups of microorganisms present in activated sludge and describe the conditions that promote their growth.
- Evaluate information on types of microorganisms present to determine underlying operating conditions.
- Compare and contrast complete-mix, plug-flow, and batch operation.
- Inspect and maintain equipment associated with the activated sludge process.
- Compare and contrast gould sludge age, mean cell residence time (MCRT), solids retention time (SRT), and solids retention time aerobic ($SRT_{aerobic}$).
- Calculate process control variables, including MCRT, SRT, $SRT_{aerobic}$, and food-to-microorganism ratio (F/M).
- Determine whether MCRT, SRT, or $SRT_{aerobic}$ is the most appropriate control variable given facility data.
- Select a target sludge age to meet treatment objectives at a particular water temperature.
- Calculate theoretical maximum return activated sludge/waste activated sludge (RAS/WAS) concentrations from settleometer test results.
- Calculate actual RAS/WAS concentrations using influent flow and RAS flow. Compare against theoretical maximum thickness to optimize RAS flowrate.
- Predict the effect of increasing or decreasing sludge age on other process control variables including mixed liquor suspended solids (MLSS) concentration, mixed liquor volatile suspended solids (MLVSS) concentration, F/M, and wasting rate.
- Select a target dissolved oxygen (DO) concentration to prevent filamentous bulking and/or maximize ammonia removal rates.
- Describe how hydraulic and solids loading parameters are calculated for secondary clarifiers and the relative importance of each.
- Explain how the maximum solids loading rate to a secondary clarifier depends on sludge settling characteristics.
- Collect process control samples, conduct testing, and evaluate results.
- Start up a new activated sludge process, place a basin into service, or take one out of service.

- Troubleshoot common activated sludge and secondary clarifier process control and mechanical problems.
- Discuss differences between different types of activated sludge processes (complete mix, step feed, oxidation ditch, pureox, etc.). Understand that they are all based on the same underlying biological principles.

Purpose and Function

By the time the wastewater reaches the secondary treatment process, many of the larger particles that are capable of settling on their own have already been removed from the wastewater by screening, grit removal, and/or primary clarification. Regardless of whether or not a facility has primary clarifiers, many of the particles that remain in the wastewater when it reaches the activated sludge process won't settle quickly on their own. For treatment to continue, the size and density of the remaining particles must be increased so they can be efficiently removed.

Flocculation means growing larger particles through collisions that help smaller particles stick together. *Bioflocculation* uses a combination of flocculation and biological conversion and growth to combine smaller particles into larger ones.

The activated sludge process and other biological treatment processes excel at converting smaller particles and soluble, biodegradable organic material into larger, heavier particles through *bioflocculation*. In the *activated sludge* basin, incoming wastewater is fed to a complex mixture of bacteria and other microorganisms known as the MLSS. Operators often refer to MLSS as simply the "bugs", which is a shorthand reference to the bacteria in the process. The organic material in the influent wastewater becomes their food. As the bacteria consume the available food, they form large colonies called *flocs* that will, under the right conditions, grow large enough and dense enough that they can be separated from the treated wastewater by gravity. The flocs are kept suspended in the activated sludge process with either mixers, aeration, or a combination of mechanical mixing and aeration. Non-biodegradable solids in the raw wastewater also become part of the floc particles through bioflocculation. Individual floc particles have a life cycle of initial formation and growth. As the floc particles age, they accumulate dead bacteria and inert material and they increase in size. The larger the floc particle gets, the more difficult it becomes for the bacteria at the center to rid themselves of waste products and gain access to nutrients and oxygen. Eventually, the floc will break into smaller flocs and the cycle will begin anew.

Activated sludge is a suspended growth treatment process because the microorganisms are not attached to a surface.

Return activated sludge refers to the settled solids from the bottom of the secondary clarifier that are collected and sent back to the beginning of the activated sludge process.

Waste activated sludge is the excess solids that are removed from the treatment process. Waste activated sludge is typically transferred to solids handling.

For most activated sludge processes, the MLSS will be conveyed to a separate secondary clarifier and allowed to settle. For the separation step to be efficient, the floc particles grown in the activated sludge basin must be large and dense. Process control is all about producing an MLSS that flocculates, settles, compacts, and produces a final effluent that meets discharge permit limits for organics, solids, ammonia, and other parameters (Wahlberg, 2016). Treated, clarified wastewater flows out through the top of the clarifier while the MLSS settles to the bottom to form a sludge layer called the *blanket*. Most of the settled MLSS is returned to the activated sludge process where it will be used to treat more influent wastewater. This is the *return activated sludge* (RAS). Excess MLSS is removed from the process entirely as *waste activated sludge* (WAS). In a sequencing batch reactor (SBR)-type activated sludge process, treatment and clarification take place in the same basin.

TEST YOUR KNOWLEDGE

1. **The floc particles that make up the MLSS include both live and dead microorganisms.**
   - ☐ True
   - ☐ False
2. **The goal of activated sludge process control is to produce an MLSS that flocculates, settles, compacts, and meets the requirements of the discharge permit.**
   - ☐ True
   - ☐ False
3. **Soluble biochemical oxygen demand (BOD) is removed by ____________________.**
   a. Bioflocculation
   b. Conversion into MLSS
   c. Settling in the clarifier
   d. Offgassing

4. **Which of the following components cannot be removed through bioflocculation?**
   a. Non-biodegradable organic solids
   b. Biochemical oxygen demand
   c. Toilet paper fibers
   d. Ammonia-nitrogen

5. **Waste activated sludge**
   a. Is the returned settled sludge to the activated sludge basin
   b. Allows floc particle size control by breaking up larger floc
   c. Is the excess sludge removed from the activated sludge process
   d. Converts large particles into food for the bacteria

6. **The purpose of the secondary clarifier is to**
   a. Separate the solids from the treated wastewater
   b. Promote denitrification
   c. Manipulate the MLSS concentration
   d. Reduce soluble BOD

Theory of Operation

In its simplest form, the activated sludge process consists of an activated sludge basin where wastewater is combined with oxygen and a mixed population of bacteria and other microorganisms to consume the organic material in the wastewater (Figure 8.1).

ACTIVATED SLUDGE BASIN

The activated sludge basins are the heart of the process. This is where treatment takes place. The microorganisms use organic material, *BOD*, to grow, reproduce, and form larger, heavier particles called flocs through bioflocculation. Historically, activated sludge basins were referred to as aeration basins or A-basins because air is added to the process; however, not all activated sludge basins are aerated. Air, pure oxygen, or mechanical mixers may be used to provide oxygen to the bacteria, mix the activated sludge basin, keep the flocs in suspension, and blend the microorganisms with the influent wastewater.

Biochemical oxygen demand is a way of measuring the organic strength of wastewater. Organics are not measured directly. Instead, the amount of oxygen the microorganisms need to consume the organic material is measured. Additional information on BOD can be found in Chapter 1.

The mixture of microorganisms and wastewater remains in the activated sludge basin for 4 to 36 hours depending on the level of treatment required. The tanks are sized to provide sufficient *hydraulic detention time* (HDT) for treatment of the carbonaceous BOD (CBOD) and ammonia (if nitrification is a requirement) and to ensure proper flocculation of the microorganisms. Depending on the requirements for effluent quality, the basin may be subdivided or compartmentalized to achieve specific *biochemical* reactions.

The *HDT* is the time required to fill or empty a container with water. If the container already contains water and water is both entering and leaving, then the HDT is the amount of time an average bit of water remains in the container. Hydraulic detention time is also called the hydraulic retention time (HRT).

The contents of the activated sludge basin consist of microorganisms and biodegradable and non-biodegradable suspended, colloidal, and soluble organic and inorganic matter. The solids are referred to as the *MLSS* and the organic fraction is called the *MLVSS*. If you looked at a sample of MLSS under the microscope, you would see a world of different organisms living together in compact clumps suspended in the wastewater. The activated sludge sample shown in Figure 8.2 has several well-developed floc particles (A) as well as filamentous bacteria (B), stalked ciliates (C), and a rotifer (D). In fact, MLSS borrows its name from the mining industry in which a liquor is any high suspended solids mixture or slurry. In activated sludge, it is called a *mixed liquor* because of the variety of different organisms living in it. Microorganisms consist primarily of organic matter (70 to 80%) and are often measured as MLVSS;

Biochemical refers to chemical reactions caused by bacteria and other microorganisms in the treatment process.

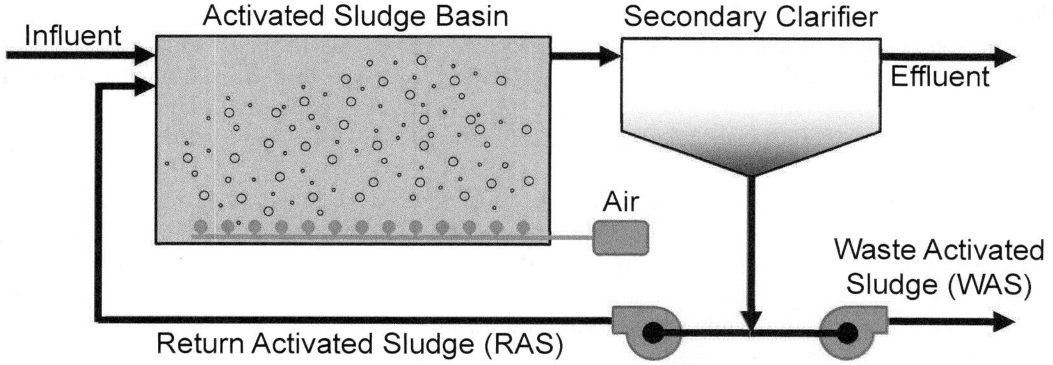

Figure 8.1 Activated Sludge Process (Reprinted with permission by Indigo Water Group)

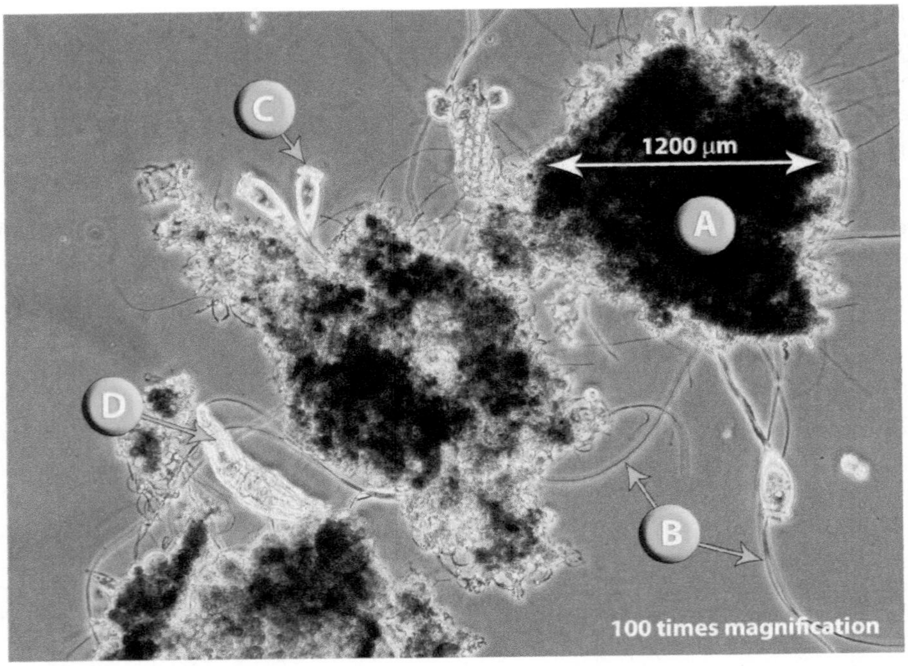

Figure 8.2 Activated Sludge Floc at 200× Magnification (Reprinted with permission by Toni Glymph-Martin)

however, it must be emphasized that a fraction of the MLVSS represents inert organic matter including organisms that are no longer viable (i.e., living and active).

SECONDARY CLARIFIER

Next, the mixture passes into the secondary clarifier. The secondary clarifier is not aerated or mixed. The velocity of water through the clarifiers is very low. The floc particles produced in the activated sludge basin are heavier than water and will gradually sink to the bottom of the clarifier to form a layer called the *sludge blanket*. The clarified, treated wastewater flows out over the top of the clarifier. This is the effluent. Secondary clarifiers are typically smaller than the activated sludge basins, with the water remaining in the clarifier for approximately 2 hours.

Clarifiers rely on differences in density to separate particles from the surrounding wastewater. Particles that are more dense than water are pulled by gravity to the bottom of the clarifier. Particles that are less dense than water float to the top of the clarifier. Gravity is not the only force influencing the movement of particles. Drag is created by the particles as they move through the water. They rub against the water, which creates friction between the water molecules and the particles. Drag slows the particles down. Currents in the clarifier push particles in different directions. The biggest current is the velocity of the water as it moves through the tank and over the effluent weirs. This type of velocity has a special name: surface overflow rate (SOR). Surface overflow rate pushes particles up and out of the clarifier along with the wastewater. For the clarifier to effectively settle solids, the SOR (upward velocity) must be smaller than the settling velocity, which is influenced by the downward force because of gravity. For particles that are less dense than water, SOR pushes them to the surface at an accelerated rate. Wind, temperature gradients, and the movement of the clarifier mechanism can also create currents in clarifiers that push particles in different directions.

Particles settle differently in primary clarifiers than they do in secondary clarifiers. The clarifiers are essentially the same, but the particles are different. Wastewater coming into primary clarifiers has a low solids concentration (120 to 400 mg/L) compared to the mixed liquor entering secondary clarifiers (1000 to 4000 mg/L). Fewer collisions occur between particles in primary clarifiers, so flocculation is limited and particles tend to settle independently at their own speeds (Chapter 4). Secondary clarifiers, on the other hand, are designed to encourage numerous collisions between particles. The MLSS particles interact with each other to flocculate, increase particle size, and capture smaller particles as they settle. These differences in solids concentration and flocculation cause something very interesting to happen in secondary clarifiers: the MLSS settles as if it were one large mass instead of a bunch of smaller, individual floc particles. This behavior can be observed by taking a sample of MLSS and placing it into a clear container like the ones shown in Figure 8.3. A clear interface (line) forms between the treated water on top and the settling particles. The interface forms quickly. By the time the MLSS has been allowed to settle for even 5 minutes, the division can be easily seen. Once the interface forms, all of the floc particles settle at the same rate, so it looks almost as if the MLSS were a solid plug being drawn down through the bottom of the container.

Sludge that settles to the bottom of the clarifier builds up in a layer called a *blanket*. Although you can't see it, a second interface (line) forms at the bottom of the container as soon as the first MLSS particles land upon it. More particles land on top of the first particles and more particles land on those, stacking solids on the bottom of the container. The interface at the top of the container and the interface at the bottom of the container move toward each other until settling and compaction are complete. How fast the interfaces move toward one another depends, in part, on the concentration of the MLSS. In the beginning, it takes the same amount of time for the upper interface to go from 1000 mL down to 950 mL as it does to go from 950 mL down to 900 mL. As the MLSS gets more concentrated, the particles interfere with one another more and the interface gradually falls slower and slower.

The concentration of solids in the clarifier blanket depends, in part, on how fast the sludge is removed from the clarifier. When solids are removed quickly, they don't have an opportunity to compact as much, so the solids concentration in the sludge will be lower. All sludges have a maximum concentration that can be achieved and longer detention times will not increase the solids concentration beyond this maximum. If a sample of wastewater is allowed to sit undisturbed, a layer of settled solids will form on the bottom of the container. When solids are allowed to remain in the clarifier for longer periods of time, the sludge blanket will have time to compress and water will be squeezed out of the spaces between particles. The result is a thicker sludge with a higher total solids concentration.

RETURN ACTIVATED SLUDGE

The activated sludge basin and secondary clarifier are connected by the RAS line. This pipeline typically runs from the bottom of the secondary clarifier to the front of the activated sludge basins, as shown in Figure 8.1. The purpose of the RAS line is to take the settled, active, biological solids (MLSS) from the bottom of the clarifier and return them to the treatment process. The RAS flow may be between 25 and 125% of the influent flow to the WRRF.

Figure 8.3 Side-by-Side Settleometers after 10 Minutes and 15 Minutes of Settling (Reprinted with permission by Indigo Water Group)

Many different names were proposed for the *activated sludge process*. The term "activated" was used because "active"—live—biomass is recycled in the process. Because the return activated sludge (RAS) line is used to return the biomass, it is what makes activated sludge "activated".

Wastewater treatment pond systems are discussed in Chapter 6—Wastewater Treatment Ponds.

The *SRT* is the amount of time the solids remain in the treatment process. For a lagoon, the HDT and SRT are the same. For an activated sludge process, the SRT is much longer than the HDT.

Loading rates to activated sludge processes are expressed as kilograms per cubic meters per day (kg/m³) or pounds per thousand cubic feet per day (lb/d/1000 cu ft).

Sludge age is a general term for the SRT in an activated sludge process.

Yield is discussed in Chapter 5—Fundamentals of Biological Treatment.

The RAS line is what differentiates an *activated sludge process* from a wastewater treatment pond system. A wastewater treatment pond system is a natural treatment system consisting of one or more ponds that may or may not be mechanically aerated. Wastewater enters a pond system bringing with it a population of microorganisms. As the microorganisms consume the organic material in the wastewater, they reproduce and form heavier particles, similar to an activated sludge process. Typically, the last pond in a pond treatment process serves as a settling pond. Settling ponds, also called *polishing ponds*, perform the same function in pond systems as secondary clarifiers do for activated sludge processes. They separate the treated wastewater from the solids. With a pond system, the settled solids remain in the bottom of the settling pond where they are broken down further by bacteria before eventually being removed by the operator. They are not recycled. Because the lagoon is a pass-through system, the wastewater and microorganisms spend the same amount of time in the process. The HDT and *SRT* are the same. This limits the total number of microorganisms available for treatment to the number of microorganisms entering with the raw wastewater plus the number added through reproduction. Reproduction is limited by the amount of CBOD and oxygen available, water temperature, and the amount of time the wastewater is allowed to spend within the lagoon. Other environmental factors can also affect reproduction. As a result, the concentration of microorganisms in a lagoon system is relatively low, which extends the time required for treatment, increases the size of the process, and limits the amount of BOD that can be processed.

Adding a RAS line allows the operator to keep the solids in the activated sludge process longer than the water. Wastewater still moves through the system in a single pass, but the solids are continuously recycled, which gives the microorganisms in the process more time to grow and reproduce. As a result, the concentration of microorganisms in an activated sludge process is relatively high, which reduces the time required for treatment and increases the amount of BOD that can be processed. A typical, aerated lagoon system producing a final effluent with less than 30 mg/L of BOD and total suspended solids (TSS) may need an HDT between 5 and 20 days to treat a 5-day biochemical oxygen demand (BOD_5) load of 1682 kg/km²·d (15 lb/d/ac) (WEF et al., 2010). An activated sludge process typically has an HDT between 4 and 36 hours—much shorter—and is able to treat BOD_5 loads up to 641 kg/m³·d (40 lb/d/1000 cu ft) (U.S. EPA, 1974, 1978b). Reducing the HDT reduces the size of the treatment process. The fundamental difference between the two treatment systems is the RAS line.

For an activated sludge process, the SRT has a special name: *sludge age*. Sludge age is measured in days and may be as short as 2 days or as long as 30 days. This is the amount of time that an average floc particle will spend cycling between the activated sludge basin and the clarifier before being removed from the process. Floc particles are removed by the operator in the WAS, but can also pass through the secondary clarifier to the final effluent. Typically, the amount of floc in the final effluent is very small—less than 30 mg/L. The amount of MLSS leaving the system each day determines sludge age. Sludge age is a critical concept for activated sludge processes and will be discussed in much more detail throughout this chapter.

WASTE ACTIVATED SLUDGE

Waste activated sludge is excess sludge that is intentionally removed from the process. For every 1 kg (2.2 lb) of BOD that is added to the activated sludge basin, the bacteria feeding on it will produce about 0.6 kg of new bacteria (Metcalf and Eddy, Inc./AECOM, 2014). This is equivalent to about 0.4 kg of new bacteria for every 1 kg of chemical oxygen demand (COD). The *yield* will vary somewhat depending on the type of organic material available and whether oxygen, nitrite, or nitrate is being used by the bacteria. In addition, inert material is being added to the activated sludge basins every day by the influent wastewater. If some of those solids aren't removed from the treatment process periodically, the MLSS concentration will gradually increase. Eventually, the amount of MLSS going to the secondary clarifiers will exceed the ability of the clarifiers to effectively settle it and MLSS will accumulate in the clarifier. If enough solids accumulate, they will be carried over the clarifier weirs, potentially causing a discharge permit violation.

To maintain a stable treatment process, MLSS must be removed on a regular schedule. The MLSS can be removed from the bottom of the clarifier or from the activated sludge basin. In Figure 8.1, the WAS line is shown pulling MLSS from the bottom of the secondary clarifier. The MLSS removed directly from the activated sludge basin is renamed as *WAS*. Some clarifiers have separate pipelines for RAS and WAS. In other cases, WAS is pumped out of the RAS pipeline. Regardless of whether it is removed from the activated sludge basin, from the bottom of the clarifier through a dedicated WAS line, or from the RAS line, the WAS is typically transferred to solids-handling processes. Operators determine how much WAS to remove each day by calculating the system sludge age and then adjusting the mass of WAS removed to maintain a target sludge age. Sludge age is discussed in detail later in this chapter. Sludge age is essentially how long the MLSS remains in the process before leaving in the WAS or final effluent. It is the principal mechanism used for process control.

SEQUENCING BATCH REACTORS

An SBR is a fill-and-draw activated sludge system in which both steps of treatment and clarification take place in the same basin. For conventional systems, there are five steps that are carried out: fill, react, settle, decant, and idle (Figure 8.4). Settling occurs when the air and mixers are turned off. Treated wastewater is discharged from the top of the SBR. This is called the *decant*. The decant is typically approximately 20% of the total basin volume. Because the MLSS remains in the basin during all cycles, SBRs do not have RAS lines or pumps. Waste activated sludge is removed from the bottom of the basin during the idle phase.

ANAEROBIC, ANOXIC, AND AEROBIC BASINS OR ZONES

The activated sludge process shown in Figure 8.1 is a conventional activated sludge system. It contains one or more aerated basins followed by one or more secondary clarifiers. A conventional activated sludge process is intended to remove CBOD and TSS, but may also remove ammonia by converting it to nitrite and nitrate. Many activated sludge systems contain other types of basins with different environmental conditions and may also contain additional recycle flows between the basins. Figure 8.5 shows a three-stage activated sludge system that contains an anaerobic zone (A), anoxic zone (B), and aerated zone (C). This activated sludge process is intended to remove BOD and TSS, convert ammonia to nitrite, convert nitrite to nitrate, convert nitrate to nitrogen gas, and remove phosphorus. The different environments in each basin encourage different groups of bacteria to behave in particular ways. Major groups of bacteria are discussed in Chapter 5.

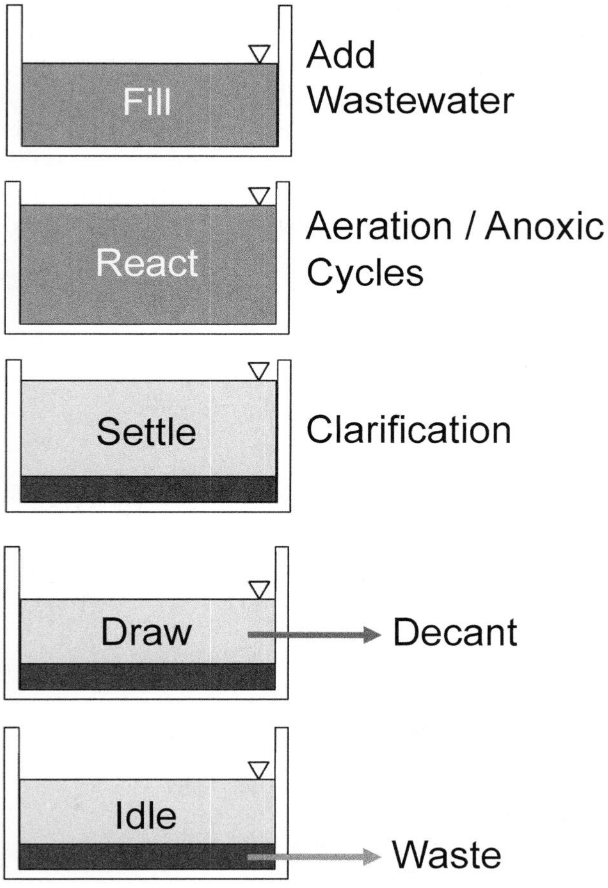

Figure 8.4 Sequencing Batch Reactor (Reprinted with permission by Indigo Water Group)

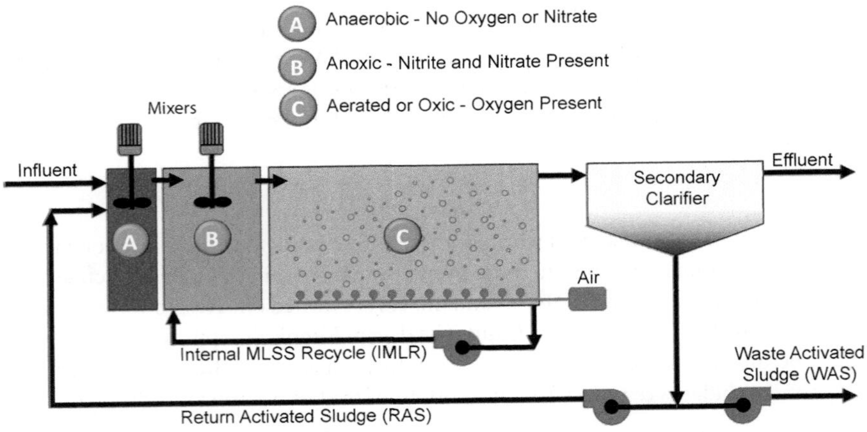

Figure 8.5 Three-Stage Biological Nutrient Removal Activated Sludge System (Reprinted with permission by Indigo Water Group)

The first basin in this example is the anaerobic basin. *Anaerobic* literally means without oxygen. In wastewater treatment, this term is used to describe zones or tanks or times during treatment where or when the wastewater does not contain DO, nitrite, or nitrate. The next basin is anoxic. *Anoxic* also means without oxygen, but specifically refers to zones or tanks or times in the treatment process where or when the wastewater does not contain DO, but does contain nitrite and/or nitrate. In wastewater treatment, it is important to distinguish between anaerobic and anoxic because different types of treatment are able to take place under each set of conditions. Both the anaerobic and anoxic zones contain mixers to keep the MLSS in suspension. The last and largest basin is the aerated basin. *Oxic* or *aerobic* means that the wastewater in the tank or zone contains DO and often indicates that oxygen is being added to the wastewater during treatment. An oxic or aerobic zone may also contain nitrite and nitrate. This three-stage activated sludge process contains an additional recycle line that runs from the end of the aerated basin back to the beginning of the anoxic basin. It is labeled as the *internal mixed liquor recycle* (IMLR) in Figure 8.5. Its purpose is to bring the nitrite and nitrate that are produced in the aerated zone back to the anoxic zone where they can be converted to nitrogen gas. There are many variations on the activated sludge process that contain more or fewer anaerobic, anoxic, and aerated basins switched around in different ways and with or without internal recycle lines connecting different types of basins. The important thing to remember is that the same fundamental operating principles apply to all of them.

The name *"internal mixed liquor recycle (IMLR)"* makes it clear that the MLSS is being moved around within the activated sludge basins. It does not go through the clarifiers and is completely separate from the RAS.

SUMMARY

In summary, the activated sludge process

- Combines influent wastewater with a mixture of microorganisms and oxygen;
- The microorganisms consume the organic material, reproduce, and form flocs;
- The mixture of microorganisms and inert material is called the MLSS;
- The fraction of microorganisms is estimated by measuring the volatile (organic) portion of the MLSS. This is the MLVSS;
- Once treatment is complete, the mixture of MLSS and treated water pass into the secondary clarifier. The clarifier is not mixed or aerated. Here, the solids settle to the bottom and treated water passes out over the top;
- A portion of the settled solids are returned to the activated sludge basins. This is the RAS;
- A portion of the settled solids are removed from the system. This is the WAS;
- Waste activated sludge may be removed directly from the activated sludge basin, the RAS line, or through a dedicated WAS line;
- The amount of MLSS removed each day determines the sludge age of the system. Sludge age is an important parameter for activated sludge process control; and
- Portions of activated sludge processes may be aerobic, anoxic, or anaerobic. The presence or absence of oxygen, nitrite, and nitrate help determine the types of microbial activity.

TEST YOUR KNOWLEDGE

1. **Floc particles may be kept in suspension in the activated sludge process with air or mixers.**
   - ☐ True
   - ☐ False

2. **As the MLSS settles, larger particles settle faster than smaller particles, reaching the bottom of the settleometer first.**
   - ☐ True
   - ☐ False

3. **The organic fraction of activated sludge floc is referred to as**
   a. MLSS
   b. Bioflocculated TSS
   c. MLVSS
   d. Bioslurry volatile fraction

4. **The fundamental difference between a wastewater treatment pond system and activated sludge is**
   a. Activated sludge systems discharge to polishing ponds.
   b. Pond systems contain larger populations of microorganisms.
   c. Only pond systems remove both BOD_5 and TSS.
   d. Activated sludge systems have a RAS line.

5. **Sludge age refers to**
   a. The length of time the solids remain in the activated sludge process
   b. How many years an activated sludge process has been in operation
   c. The ratio of the return to WAS pumping rates
   d. How often the return activated sludge pumping rate is adjusted each week

6. **One effect of returning sludge to the beginning of the activated sludge process is**
   a. Increases total time required to complete treatment
   b. Increases concentration of microorganisms in the process
   c. Links hydraulic and SRT to increase efficiency
   d. Controls the system sludge age by removing excess biomass

7. **Anaerobic and anoxic zones use ___________ to keep the MLSS in suspension.**
   a. Fine bubble diffusers
   b. Mixers
   c. Draft tubes
   d. Recycle pumps

8. **One reason air is added to some activated sludge basins is to**
   a. Prevent freezing
   b. Keep the MLSS in suspension
   c. Circulate waste products
   d. Maintain water temperature

9. **The settled solids in the bottom of the clarifier are called the**
   a. Blanket
   b. Septic layer
   c. Schmutzdecke
   d. Scum

10. **What is the most likely outcome if excess solids are not removed from the activated sludge basin on a regular basis?**
   a. Treatment efficiency will increase.
   b. Sludge settleability will improve.
   c. Solids may be lost to the final effluent.
   d. Oxygen use will decrease.

11. **The purpose of the RAS line is to**
   a. Transfer settled solids to solids handling processes
   b. Return settled solids to the activated sludge basin
   c. Manipulate MLSS concentrations
   d. Control sludge age

Chapter 5 discusses some of the fundamental principles that apply to different types of biological treatment processes.

ACTIVATED SLUDGE MICROBIOLOGY

Bacteria are grouped according to their carbon source (*autotroph* versus heterotroph) and by their oxygen usage (*aerobic, facultative*, or *anaerobic*). Bacteria may also be divided into groups depending on their preferred growth pattern. In the activated sludge process, bacteria may grow in clumps or colonies called "flocs" or they can grow in long chains called "filaments". Filaments may be long and straight, bent, curved, coiled, or branched like a tree. Flocs and two different kinds of filaments can be seen in Figure 8.2. Flocs are held together with exopolymer. Exopolymer is a sticky substance produced by some bacteria. It is made up primarily of sugar molecules and helps the microorganisms stick together. Flocs formed primarily from floc-forming bacteria and exopolymer settle relatively fast in the secondary clarifier because they are compact, round, and dense. A small number of filaments helps to hold the floc together and provides structure, which allows floc size to increase. Flocs with some filaments settle slower than flocs with fewer filaments. Too many filaments can produce feathery, low-density flocs, flocs with filaments that protrude from the edges, and/or flocs that are connected by bridges made of filaments. All three conditions can cause settling

Autotrophs obtain their carbon from inorganic carbon dioxide dissolved in the wastewater. Inorganic carbon is measured as alkalinity. Heterotrophs must have organic carbon. Organic carbon is measured in the BOD test.

Aerobic bacteria must have dissolved oxygen. *Facultative bacteria* may use dissolved oxygen, nitrite, nitrate, and, in some cases, sulfate. Oxygen is toxic to *anaerobic bacteria*.

Which came first, the chicken or the egg? The difficulty in learning about the activated sludge process partially stems from needing to understand its microbiology while simultaneously understanding how different process variables affect the microbiology. It's nearly impossible to talk about one without talking about the other. After finishing the section on process control, you may wish to review this section on microbiology.

problems in the secondary clarifier by making the floc less dense and creating more drag. Successful operation of an activated sludge treatment process depends on balancing the growth of floc-forming and filament-forming bacteria. Good floc will settle slowly enough to capture smaller floc particles as it settles, but not so slowly that it won't settle and compact well in the clarifier.

TYPES OF FILAMENTOUS BACTERIA

There are 20 to 25 different types of filamentous microorganisms found in activated sludge processes treating municipal wastewater (Jenkins et al., 2003). About 10 to 12 types are responsible for the majority of settling problems in domestic WRRFs (Jenkins et al., 2003). Table 8.1 lists many of the different filament types and the conditions that support their growth. Don't worry if some of the terms in the table don't make sense; as you work your way through this chapter, these terms will be explained in detail. Some of

Table 8.1 Conditions Associated with Filament Growth in Activated Sludge (Jenkins et al., 2004; Richard, 2003)

Filament Type	Causes Foaming	Causes Sludge Bulking	Low DO	High Sludge Age	Low F/M	High Organic Acids	Fat, Oil, and Grease (FOG)	Septicity (H2S)	Nutrient Deficiency		Low pH (less than 6.5)
									N	P	
Beggiatoa spp.		X						X			
Fungi		X									X
Haliscomenobacter hydrossis			X							X	
Microthrix parvicella	X	X		X			X				
Nostocoida limicola I						X		X			
Nostocoida limicola II						X		X			
Nostocoida limicola III						X		X		X	
Nocardioforms	X						X				
Sphaerotilus natans		X	X							X	
Thiothrix I		X				X		X	X		
Thiothrix II		X				X		X	X		
Type 0041		X		X	X						
Type 0092		X		X		X		X			
Type 021N		X				X		X		X	
Type 0411		X				X		X			
Type 0581		X		X		X		X			
Type 0675		X		X	X						
Type 0803		X		X	X						
Type 0914		X		X		X		X			
Type 0961		X				X		X			
Type 1701		X	X								
Type 1851		X		X	X						
Type 1863	X	X					X				
Actinomycetes		X									

the filamentous microorganisms in Table 8.1 are named with their proper scientific names for genus and species (e.g., *Nostocoida limicola I*). Others are designated with a type and number (e.g., Type 021N). As odd as it may seem, the reason some filaments have only a number to identify them is because we don't yet know what genus and species they belong to. Many of these organisms can't be grown in the laboratory, which makes them difficult to study.

Filaments, and other microorganisms, grow in response to specific conditions in the activated sludge process. If a particular microorganism is observed in abundance, the environmental conditions in the process favor its growth. Environmental conditions include pH, DO concentration, the presence or absence of vital nutrients, the availability of preferred food sources, and other factors. Activated sludge samples collected from 525 WRRFs in the 1970s and 1980s most commonly contained nocardioforms, Type 1701, Type 021N, Type 0041, *Thiothrix spp., Sphaerotilus natans,* and *Microthrix parvicella* (Jenkins et al., 2003). These filaments grow in response to high concentrations of fats, oil, and grease (FOG), low DO concentrations, septicity, and high sludge age.

Encouraging or preventing the growth of a particular microorganism depends on controlling the environmental conditions in the process. For example, *Haliscomenobacter hydrossis* (*H. hydrossis*) and *Sphaerotilus natans* (*S. natans*) are two filaments that grow well in low DO conditions. The growth of many floc-forming bacteria is limited when the DO concentration is too low. As a result, when DO in the activated sludge process is chronically low, these two filaments will perform better than the floc formers, eventually dominating the system. To help control the growth of these two filaments, an operator could increase the DO concentration in the basin. Increasing DO concentrations should boost the growth rate of the floc formers and limit the growth of the low DO filaments, assuming the growth of the floc formers is not being limited in some other way.

The names of filaments can be shortened by using the first letter of their genus followed by their species. For example, *Haliscomenobacter hydrossis* is shortened to *H. hydrossis*.

Bulking sludge settles slowly and doesn't compact well in the bottom of the clarifier. Bulking can be caused by many different things, including filaments.

Even when the right environmental conditions exist to support growth, a particular filament or microorganism still won't be able to accumulate in the process unless it is allowed to remain long enough to grow and reproduce. Bacteria, filaments, protozoa, metazoa, and other microorganisms are entering the WRRF with the influent wastewater all the time. If they leave the process through the WAS or effluent as fast as they enter, their numbers will not increase. Some microorganisms reproduce quickly, in a matter of hours, whereas others require days. If an organism requires 2 days to reproduce in the process, but the sludge age is only 1.8 days, then their numbers will not increase during treatment. An operator may see the occasional nematode (worm) under the microscope because it came in with the influent wastewater. An operator won't see a lot of worms unless they have time to reproduce and the conditions are right to support their growth. This is true for every organism that can live in activated sludge.

Old sludge filaments (filaments that only appear at longer sludge ages) may be able to compete better when five-day biochemical oxygen demand (BOD_5) concentrations are low. Chapter 5 discusses specific growth rate. Essentially, when the microorganisms in the process have access to all the BOD_5 they can use, they will grow at their maximum rate. If they don't, their growth rate will be related to the concentration of BOD_5 available. For heterotrophic bacteria, the maximum growth rate occurs when BOD_5 is greater than 50 to 200 mg/L, depending on the type of bacteria. Many filamentous bacteria are able to maintain their maximum growth rates at much lower BOD_5 concentrations. Filaments typically grow much slower than floc formers. But, when resources are limited, the growth rates of filaments and floc formers get closer and closer together. Eventually, if the BOD_5 gets low enough, the filaments will grow at a faster rate than the floc formers (Lou and de los Reyes III, 2008). Figure 8.6 illustrates the differences in growth rates between these two essential groups. To the left of the dashed line, old sludge filaments will dominate the activated sludge floc. To the right, floc formers will dominate. For reasons that are explained later in this chapter, BOD_5 concentrations in the activated sludge basin decrease with increasing sludge age. Lower BOD_5 concentrations, lower growth rates for floc formers, and time to reproduce all play a role in the appearance and abundant growth of old sludge filaments.

It is now known that some filamentous microorganisms can grow as individual cells or small groups within the floc, growing only as filaments when the environment favors growth in its filamentous form. This helps explain why blooms of filaments can occur quickly, often causing settling problems.

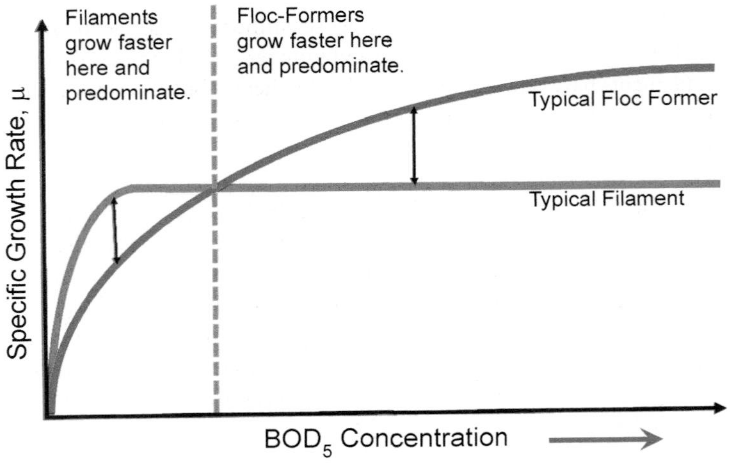

Figure 8.6 Growth Rates for Floc Formers and Filament Formers (Reprinted from *Water Research*, Vol. 7, Chudoba et al., "Control of Activated Sludge Filamentous Bulking—II. Selection of Microorganisms by Means of a Selector", pp 1389–1406, Copyright © 1973, with permission from Elsevier)

PROTOZOA AND METAZOA

As bacteria grow in the system, different types of protozoans and metazoans grow along with them, including flagellates, amoeba, free-swimming ciliates, attached (stalked) ciliates, crawler ciliates, suctoria, rotifers, worms, and tardigrades.

Operators will find two other books helpful in deepening their understanding of activated sludge microbiology and improving facility operations. The first is *Wastewater Biology: The Microlife* (WEF, 2017b). The second is the *Manual on the Causes and Control of Activated Sludge Bulking, Foaming, and Other Solids Separation Problems* (Jenkins et al., 2003).

SECTION EXERCISE

1. Label the diagram below. Check your answers against Figure 8.5.

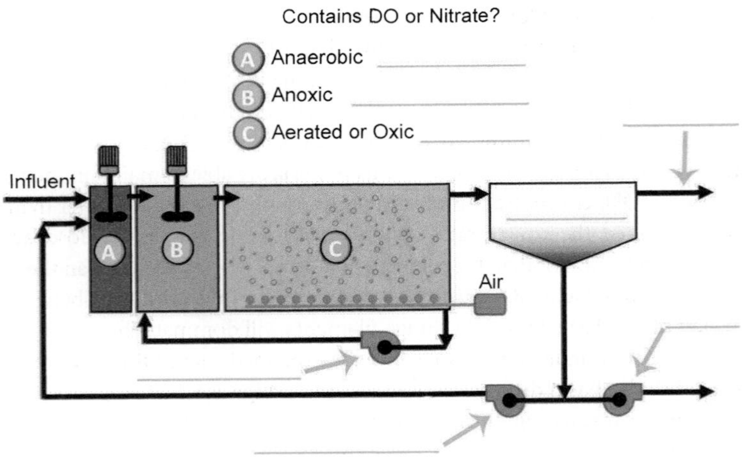

2. In your own words, describe the purposes of RAS and WAS. Check your answers by reviewing this section. ______________________
____
____
____
____

TEST YOUR KNOWLEDGE

1. **Ideally, filament-forming bacteria will outnumber floc-forming bacteria.**
   - ☐ True
   - ☐ False

2. **Match the following terms to their definitions.**

a.	Heterotroph	1.	Utilizes inorganic carbon
b.	Autotroph	2.	Uses DO, nitrite, or nitrate
c.	Aerobic	3.	Grows in clumps
d.	Facultative	4.	Utilizes organic carbon
e.	Anaerobic	5.	Grows in long chains
f.	Floc former	6.	Cannot use DO, nitrite, or nitrate
g.	Filament	7.	Utilizes DO

3. **While some filamentous bacteria provide structure and support to floc particles, too many filaments**
   a. Will increase floc particle settling velocity
   b. Require excessive quantities of nutrients
   c. Can cause settling problems in the clarifier
   d. Increase floc density and oxygen usage

4. **Which of the following filaments indicates a low DO condition?**
   a. *M. parvicella*
   b. Nocardioforms
   c. *N. limicola*
   d. *H. hydrossis*

5. **Match these commonly occurring filaments to their preferred environmental condition.**

a.	Nocardioforms	1.	Septicity
b.	*S. natans*	2.	High sludge age
c.	*Beggiatoa*	3.	FOG
d.	*M. parvicella*	4.	Nutrient deficiency
e.	Type 021N	5.	Low DO

6. **The growth of the filament *M. parvicella* has been linked to influent FOG. Even if FOG is present, this must also be true for *M. parvicella* to accumulate in the process.**
   a. Minimum time to reproduce has been met
   b. Dissolved oxygen concentration is low
   c. Hydrogen sulfide is present in excess
   d. pH is less than 6.5

Design Parameters

The activated sludge process may be designed and operated to remove CBOD, to convert ammonia to nitrite and nitrate, to remove nitrogen compounds, and/or to remove phosphorus. The different biological reactions are discussed in Chapter 5.

Activated sludge operating modes are characterized by the sludge age, volumetric loading rate, F/M, and flow pattern. The volumetric loading rate is the mass of BOD_5 added per volume of activated sludge basin. It is similar to the organic loading rate calculation for trickling filters. The F/M is the mass of BOD_5 added to the activated sludge basin each day divided by the mass of MLVSS in the basin. The flow pattern is how the water moves through the process. Table 8.2 shows typical ranges for these parameters in conventional, high-rate, and low-rate (often called *extended aeration*) systems. There are many different types of activated sludge systems in each of these broad categories. Flow patterns and different system types are discussed at the end of this chapter in the section on Process Variations.

Conventional systems provide BOD_5 removal efficiencies of 85 to 95% and typically carry MLSS concentrations varying from 1000 to 3000 mg/L. At conventional loading rates, some ammonia removal may occur, especially in warmer climates or when the system is operating well below its design capacity. Conventional systems are typically not intended to remove ammonia, nitrite, or nitrate.

Low-rate systems are typically used for low flows and are characterized by high oxygen requirements and low sludge production rates. These systems are considered to be more stable than conventional systems and thus require less operational attention. Typically, BOD removal efficiencies range from 75 to 95% and nitrification is complete. However, these systems may suffer from higher effluent suspended solids concentrations because of poor flocculation (pinpoint floc) and clarifier *denitrification*.

Denitrification is the conversion of nitrite and nitrate to nitrogen gas. Nitrogen gas bubbles can carry settled floc back to the surface of the clarifier.

High-rate systems are often used as pretreatment processes in staged biological treatment systems and are also used where only CBOD removal is required. They are characterized by low oxygen requirements and somewhat higher than normal sludge generation compared to conventional facilities. The process may

Table 8.2 Typical Process Loading Ranges for the Activated Sludge Process

Loading Range	Sludge Age, days	Volumetric loading, kg BOD_5/m^3 (lb BOD/1000 cu ft)	F/M, kg/kg·d (lb/d/lb)
High rate	1 to 3	1.60 to 16.0 (100 to 1000)	0.5 to 1.5
Conventional	5 to 15	0.32 to 0.64 (20 to 40)	0.2 to 0.5
Low Rate	20 to 30	0.16 to 0.40 (10 to 25)	0.05 to 0.15

produce BOD removal efficiencies over a wide range, from less than 50% to as high as 95% depending on loading rates and waste characteristics. High-rate activated sludge systems can be compared to roughing filters. Sludge settling can be a problem during instances of high loading when flocculation does not effectively occur.

Secondary clarifiers have hydraulic capacity and solids-handling capacity. The hydraulic capacity determines the amount of water that can pass through the clarifier while maintaining performance. There are two different variables for hydraulic capacity: SOR and weir overflow rate. The SOR is the amount of flow per clarifier surface area. It essentially measures the velocity of the water through the clarifier and includes only the influent flow. The RAS flow is not included in the SOR calculation. The weir overflow rate is the amount of flow per foot of weir. Many states limit the weir overflow rate differently for facilities smaller than 3.8 ML/d (1 mgd) than for larger facilities. The solids handling capacity, called the *solids loading rate* (SLR), is the mass of MLSS that can be effectively settled. When calculating the SLR, the influent flow and the RAS flow are combined to find the mass of MLSS going to the clarifier. The SLR is dependent on how well the MLSS settles and compacts. When the MLSS is settling poorly, the SLR must be decreased. If either the hydraulic or solids handling capacity of the clarifier is exceeded, the clarifier may not be able to effectively settle the incoming MLSS. Table 8.3 shows typical ranges for these parameters.

Expected Performance

With typical municipal wastewater, a well-designed and operated activated sludge system should achieve a CBOD effluent quality of 5 to 15 mg/L. Effluent suspended solids should also typically be less than 15 mg/L. The secondary treatment standards set forth by the U.S. Environmental Protection Agency (U.S. EPA) require all mechanical treatment facilities, like activated sludge, to meet effluent BOD and TSS limits of 30 mg/L or less as a 30-day average. Additionally, mechanical treatment facilities are required to remove a minimum of 85% of all influent BOD and TSS. To achieve consistent BOD and TSS concentrations less than 5 mg/L, some type of tertiary treatment may be required.

Table 8.3 Design Criteria for Secondary Clarifiers (GLUMRB, 2014; Metcalf and Eddy, Inc./AECOM, 2014; WEF, 1998)

Parameter	Average Daily Flow	Peak Hourly Flow
Solids loading rate, kg/m²·d (lb/d/sq ft)	100 to 150 (20 to 30)	240 (48)
Surface overflow rate, m³/m²·d (gpd/sq ft)	16.3 to 28.6 (400 to 700)	40.8 to 65.0 (1000 to 1600)
Weir overflow rate, m³/m·d (gpd/ft)	N/A	125 (10 000)[a] 375 (30 000)[b]

[a]For facilities smaller than 3.79 ML/d (1 mgd).
[b]For facilities larger than 3.79 ML/d (1 mgd).

Activated sludge processes may be designed to remove ammonia, nitrite, nitrate, and phosphorus to extremely low concentrations. Ammonia-nitrogen is often undetectable (below 1 mg/L) in the final effluent of some types of activated sludge processes. Nitrite-nitrogen concentrations are typically below 0.25 mg/L. Nitrate-nitrogen concentrations below 10 mg/L are easily achieved without the addition of chemicals. Biological phosphorus removal can reduce effluent phosphorus concentrations below 1 mg/L. Refer to the chapter on nutrient removal for additional information.

TEST YOUR KNOWLEDGE

1. **Compared to high-rate activated sludge processes, low-rate processes**
   a. Typically have higher organic loading rates
   b. Typically have longer sludge ages
   c. Remove less ammonia than high-rate systems
   d. Are often less stable than conventional systems
2. **Activated sludge processes are characterized by**
   a. Sludge age, loading rate, and F/M
   b. Presences of flagellates, ciliates, and rotifers
   c. Percentage removal of ammonia, nitrite, and nitrate
   d. Average MLVSS concentration
3. **Surface overflow rate in a secondary clarifier is an example of**
   a. Downward velocity
   b. Hydraulic capacity
   c. Loading capacity
   d. Sludge settleability
4. **If the solids loading capacity of a secondary clarifier is exceeded**
   a. The hydraulic loading rate may be increased to compensate.
   b. SOR must be kept below 100 kg/m^2·d (20 lb/d/sq ft)
   c. The clarifier may not be able to settle the incoming MLSS.
   d. The effluent suspended solids concentration should decrease.
5. **The BOD_5 concentration in the final effluent from an activated sludge process (after clarification) should be**
   a. Less than or equal to 30 mg/L
   b. About 50% of the influent BOD_5 concentration
   c. Directly proportional to the ammonia concentration
   d. Independent of operational mode

Equipment

Equipment associated with activated sludge processes includes: activated sludge basins, secondary clarifiers, aeration systems, mixers, and pumps. Secondary clarifiers are nearly identical to primary clarifiers, with some exceptions. Differences between primary and secondary clarifiers are discussed in this section. Aeration systems, mixers, and pumps are common to many treatment processes. A brief overview of each has been included here. Readers are encouraged to review the following references for more information on aeration systems and pumps:

- *Pumps and Pumping* (ACR Publications, 2011)
- Chapter 8, Pumping of Wastewater and Sludge in *Operation of Water Resource Recovery Facilities* (WEF, 2016)
- Chapter 5, Facility Hydraulics and Pumping in *Design of Water Resource Recovery Facilities* (WEF et al., 2018)
- Chapter 12, Suspended-Growth Treatment, Section 8, Oxygen-Transfer Systems in *Design of Water Resource Recovery Facilities* (WEF et al., 2018)

ACTIVATED SLUDGE BASINS OR BIOLOGICAL REACTORS

Activated sludge basins are often constructed of reinforced concrete, although some package facilities (e.g., extended air) may use steel. The basins are typically rectangular to accommodate common-wall construction for multiple basins, although some installations may have circular or oval tanks (the oxidation ditch is one example). A minimum of two basins is desirable, even for small facilities, to accommodate any shutdown for occasional maintenance. Each basin should be furnished with inlet and outlet gates or valves so they can be isolated and removed from service. Proper drains or sumps should be provided for rapid dewatering (approximately 8 to 20 hours). In addition to aeration equipment, basins are often equipped with a froth-control system to control foaming. Froth-control systems consist of water sprays that help break up foam and/or push it to a central area for removal or to the basin outlet. Inlet design should ensure that flow to the basin is as uniform as possible to avoid severe short-circuiting of influent and to provide for intermixing of RAS with influent wastewater. Where parallel multiple basins are used, a method for proper flow splitting should be provided to ensure equal flow to each basin. In step-feed systems, flow control is particularly important. In areas where high groundwater conditions can occur, deep tanks should be equipped with groundwater relief valves.

BASIN CONFIGURATION

Basin configuration deals primarily with the hydraulic characteristics of the process. Continuous-flow systems are often categorized as "ideal plug flow" or "ideal complete mixed" systems, although most operate in a less-than-perfect flow regime somewhere between the two.

Ideal Complete Mix

Ideal completely mixed flow implies that the composition of the mixed liquor is the same throughout the basin volume. The influent wastewater immediately and completely mixes with the basin contents so that the concentration of a given component is the same as the effluent concentration. As a result, the DO, soluble BOD, nitrogen species, pH, temperature, and other characteristics are identical throughout the basin. Completely mixed flow is difficult to achieve, although the use of square or round basins with intense mixing can get very close.

The advantages of the complete-mix process include the dilution of shock loads (e.g., high loads or toxic chemicals) and its relative simplicity compared to other configurations. Shock loads can lead to anaerobic conditions or biomass toxicity. There are several disadvantages to completely mixed basins. First, they are often plagued by filamentous bulking problems. This can be overcome, in part, by using *selectors*, which provide short-term conditioning of the RAS and influent wastewater ahead of the activated sludge basin. Theoretically, complete-mix processes also suffer because the removal efficiency of the wastewater constituents is lower than in plug flow systems, requiring longer basin detention times and/or the use of basins in series to achieve comparable effluent quality. These systems are often used for low-rate applications or for treating industrial or industrial–municipal wastewater where large variations in load are anticipated.

Selectors are small anaerobic, anoxic, or aerobic tanks placed at the beginning of an activated sludge process. Influent flow and RAS are directed into the selector. The goal is to dramatically increase the F/M for a short period of time before the water flows into the main part of the process. The growth of many different filaments is inhibited by selectors. These small basins are called *selectors* because they "select" for the desired biology.

Ideal Plug Flow

In an ideal plug flow system, water entering the basin will move evenly along the basin length with very little mixing (Figure 8.7). Plug flow is how water moves through a pipe. There is virtually no mixing between the water at the end of the pipe and the water in the middle of the pipe. A particular bit of water will remain in the basin for a time equal to the theoretical *detention time*. As a result, the concentration of BOD and the oxygen usage will decrease along the basin length. The MLSS and MLVSS will increase along the basin length as biomass is produced. This type of flow is approximated by long, narrow basins with length-to-width ratios greater than 10. That is, the basins are at least 10 times longer than they are wide. Plug flow may also be simulated by placing multiple basins in series or by folding rectangular basins. Like the complete-mix configuration, it is not practical to produce a true plug flow system.

Detention time calculations have been demonstrated in previous chapters. The theoretical detention time is equal to the volume of the basin divided by the flowrate. In practice, the actual detention time is always something less than the theoretical.

This configuration results in high organic load and oxygen usage at the inlet to the basin because the concentrated microorganisms in the RAS are blended with the influent wastewater in a small percentage of the total basin. With a complete-mix basin, the influent wastewater is evenly distributed throughout the basin, which dilutes the overall concentration of BOD. With plug flow, there is a concentration gradient—highest at the influent end and lowest at the discharge end. Theoretically, the plug flow basin delivers the highest removal rate per unit volume. It is also less susceptible to filamentous bulking provided that sufficient DO is present at the inlet.

BASINS-IN-SERIES

The *basins-in-series* configuration is described as two or more completely mixed basins operating in a series configuration. It can be approximated with baffles or by folding basins, as shown in a Figure 8.7. This configuration, with three or more basins-in-series, attempts to simulate a plug flow condition that has a kinetic advantage over mixed-flow systems, as mentioned previously. This configuration also seems to mitigate filamentous bulking.

Series operation places one treatment unit after the other. The second treatment unit receives the effluent from the first treatment unit. Parallel operation has the treatment units operating side by side with each receiving the same influent flow.

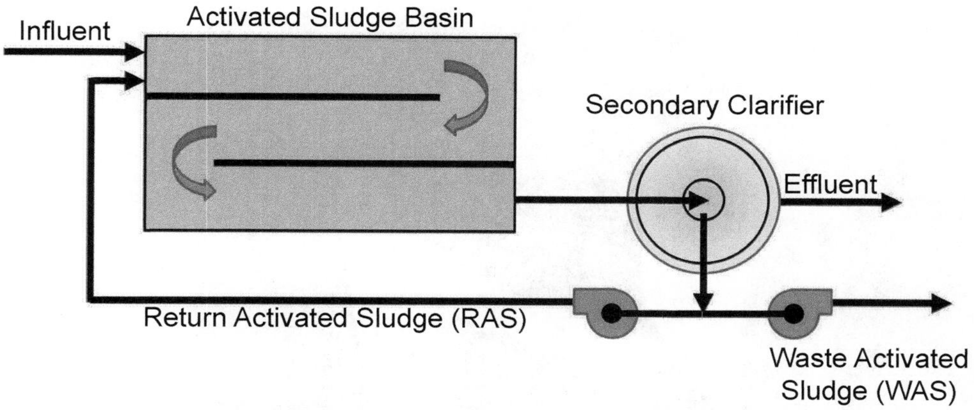

Figure 8.7 Plug Flow Activated Sludge (Reprinted with permission by Indigo Water Group)

BATCH OPERATION

Sequencing batch reactors are discussed at the beginning of this chapter. This type of activated sludge system uses a fill-and-draw system to treat wastewater in small batches. Both treatment and clarification take place in the same tank. At the end of a treatment cycle, the MLSS is allowed to settle and treated water is decanted off the top of the basin. Sludge is wasted from the settled sludge blanket. Batch systems behave like ideal plug flow systems with respect to biological process performance; the reaction time for batch systems is interchangeable with the space–time for the plug flow basin.

OXIDATION DITCH

In the typical oxidation ditch, mixed liquor is pumped around an oval or circular pathway by brushes, rotors, or other mechanical aeration devices as well as pumping equipment located at one or more points along the flow path (Figure 8.8). The mixed liquor is typically moved at a velocity of 0.24 to 0.37 m/s (0.8 to 1.2 ft/sec) in the channel. Most oxidation ditches are designed as low-rate processes with long detention times (approximately 24 hours). Oxidation ditches are considered to be completely mixed. Ditches may be designed with a single channel or multiple interconnected channels set inside each other (Figure 8.9). They are widely used for small- to medium-sized communities.

SECONDARY CLARIFIERS

Separation of MLSS from the liquid stream is vital to the operation and performance of activated sludge systems. This is typically achieved by gravity in secondary clarifiers although membranes can also be used. Secondary clarifiers, both circular and rectangular, are nearly identical to primary clarifiers (Chapter 5). There are a few variations in circular secondary clarifiers that are not used in primary clarifiers, including

- Peripheral feed, peripheral take-off flow path,
- Presence of a floc well outside of a smaller center well,

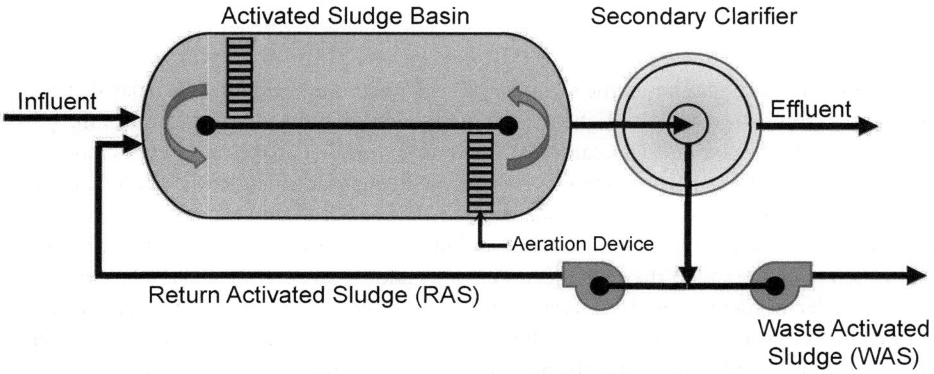

Figure 8.8 Oxidation Ditch with Surface Aerators (Reprinted with permission by Indigo Water Group)

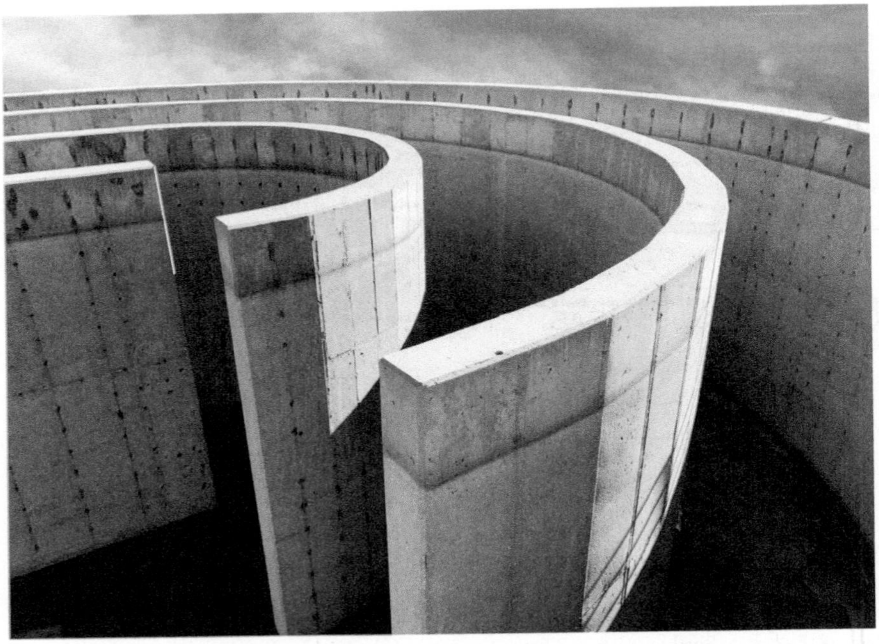

Figure 8.9 Oxidation Ditch with Multiple Channels (Reprinted with permission by Indigo Water Group)

- Presence of interior baffles,
- Equipment to control algae growth on weirs and launders,
- Hydraulic suction method of settled sludge removal,
- Riser pipe mechanisms for settled sludge removal, and
- Differences in depth and floor slope.

Peripheral refers to the outer edge of the clarifier.

PERIPHERAL FEED, PERIPHERAL TAKE-OFF FLOW PATH

Circular clarifiers can be fed by several different inlet configurations. Most clarifiers are fed from the center (Figure 8.10 [a]), however, there are two basic *peripheral* feed alternatives (Figure 8.10 [b, c]). In the first and second alternative, the influent enters the clarifier on one side and then passes into a wide channel that follows the outer wall of the clarifier (Figure 8.11). Holes in the bottom of the channel allow influent to flow into the clarifier as shown in Figure 8.10 (b) and (c). The channel helps distribute the influent flow over the clarifier surface. Peripheral feed clarifiers may have either center take-off as shown in Figure 8.10 (b) or peripheral take-off as shown in Figure 8.10 (c).

The location of the feed point determines how water will move through the clarifier. Center feeding causes the flow to move radially outward toward the weir, and, in many tanks, there is a doughnut-shaped roll pattern formed, which results in some surface flow back toward the center. The opposite pattern is formed by peripheral feed devices.

FLOCCULATING CENTER FEED WELL

In many facility designs, the MLSS arrives at the secondary clarifier without being fully flocculated. The floc can be broken up to some degree by the higher flow velocities found in the pipelines connecting the activated sludge basin to the clarifier. This can result in pieces of floc that are not large enough or dense enough to settle well in the clarifier and may increase the effluent suspended solids. To address this problem, many secondary clarifiers contain a flocculation zone. The goal is to provide an area of gentle mixing within the clarifier to help reflocculate the MLSS and improve settleability. The flocculation zone may be created by increasing the size of the center feed well or by adding an outer flocculator center well that surrounds the smaller, influent center well. Two flocculation center wells are shown in Figure 8.12. The center feed well helps to slow down the influent flow. The flocculation well keeps the MLSS together for reflocculation. Some flocculation wells, like the one shown at the bottom of Figure 8.12, include mechanical flocculators.

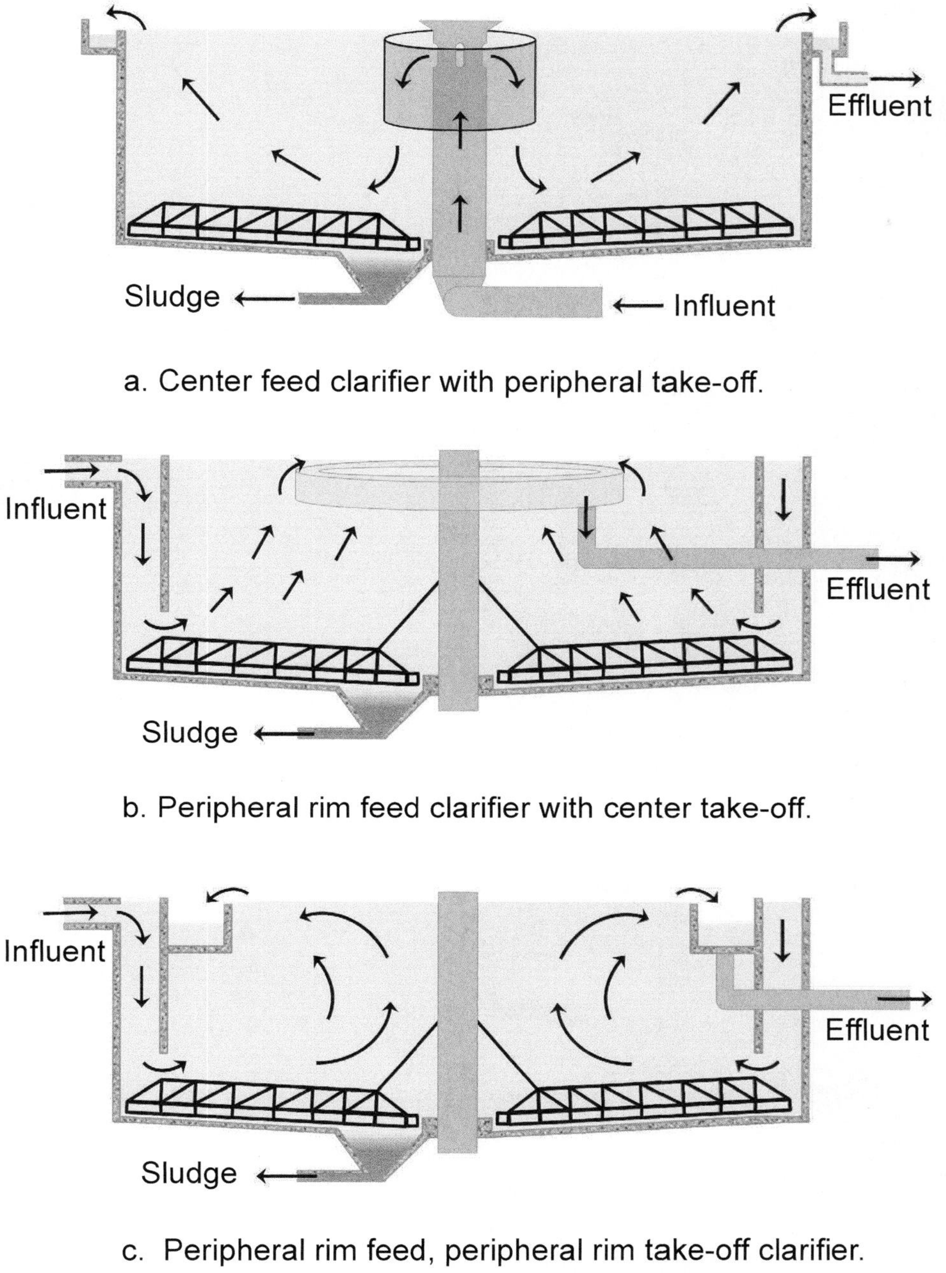

Figure 8.10 Flow Patterns through Circular Secondary Clarifiers by (U.S. EPA, 1978a)

The size of the flocculation center well has been inconsistent from one clarifier to another. Floc wells that provide 20 minutes of HDT appear to work best and can reflocculate over 90% of the floc particles entering the clarifier (Parker et al., 1971; Wahlberg et al., 1994). If the floc well is too big, the incoming floc particles will be too spread apart and reflocculation may not occur. Many clarifiers have floc wells that are 30 to 35% of the clarifier diameter regardless of the recommended HDT (WEF, 2005). The flocculating center well typically extends about halfway down into the depth of the clarifier, although shallower ones have been used.

PRESENCE OF INTERIOR BAFFLES

For many years, circular clarifiers were constructed without interior baffles, except for the inlet well. Center-feed activated sludge clarifiers often create an updraft of suspended solids along the outer wall. Under normal operating conditions, flow enters the clarifier, passes through the center well and flocculating

Figure 8.11 Peripheral Feed Clarifier Channel (Reprinted with permission from Indigo Water Group)

Figure 8.12 Secondary Clarifier Flocculator Center Wells (Reprinted with permission by Indigo Water Group)

well, and is then forced down toward the bottom of the clarifier. The current may continue across the bottom of the clarifier floor until it hits the outer wall. At this point, the only place for the current to go is upward. Most of the time, the current will dissipate before reaching the outer wall. During periods of high flow, the current can be strong enough to disturb the blanket and push solids up and out along the wall. The clarifier edges will show a rolling pattern of billowing sludge that is not seen toward the center of the clarifier. After hitting the clarifier wall, the current turns back toward the center of the clarifier. Solids now have an opportunity to resettle. Inset launders (discussed in Chapter 4) try to avoid capturing solids disturbed by currents by moving the weirs toward the center of the clarifier. The hope is that the current will return any disturbed solids back to the bottom of the clarifier before they reach the weirs. Another option is to install baffles along the outer wall of the clarifier as shown in Figure 8.13. The baffles help redirect the current back toward the bottom of the clarifier and prevent solids from reaching the weirs. The Stamford baffle is typically made of fiberglass and bolts directly to the clarifier wall. A traditional Stamford baffle projects out at a 45-deg angle; however, modified versions are now available that use a 30-deg angle. The McKinney baffle is typically an extension of an inboard launder and is made of concrete. These types of baffles have been shown to reduce effluent TSS by as much as 70% over clarifiers without baffles. Both types of baffles can trap air bubbles and floating sludge beneath them.

WEIR COVERS

Water leaving the secondary clarifier should have less than 30 mg/L of TSS. The clarity of the water allows sunlight to penetrate deeply into the clarifier. Nitrogen and phosphorus present in the wastewater help to support the growth of algae on almost all surfaces of secondary clarifiers. Algae accumulation on the effluent weirs and launders can be especially problematic (Figure 8.14). If not removed regularly, algae can blind off sections of the weir, resulting in unbalanced flows over the length of the weir. High flowrates in one area can pull settling solids up and out of the clarifier.

Most WRRFs remove algae manually by having operators scrub the weirs with brushes and high-pressure washers. When water temperatures are above about 15 °C (59 °F), algae removal may be needed weekly. Weir covers, like the ones shown in Figure 8.15, are sometimes installed to prevent sunlight from reaching the weirs. Each section of the weir cover opens to allow access for sampling, cleaning, and inspection of the clarifier. Another option is to install a brush system that cleans the weirs with each rotation of the clarifier mechanism (Figure 8.16). Preventing algae growth and continuous removal can help minimize the amount of algae that collects on downstream UV disinfection systems.

SLUDGE REMOVAL SYSTEMS

Effective removal of sludge is vital to process performance. There are two basic types of sludge removal systems, namely plows and hydraulic suction. Plows are used for all types of sludge encountered in wastewater treatment, whereas hydraulic suction is primarily limited to activated sludge secondary clarifiers.

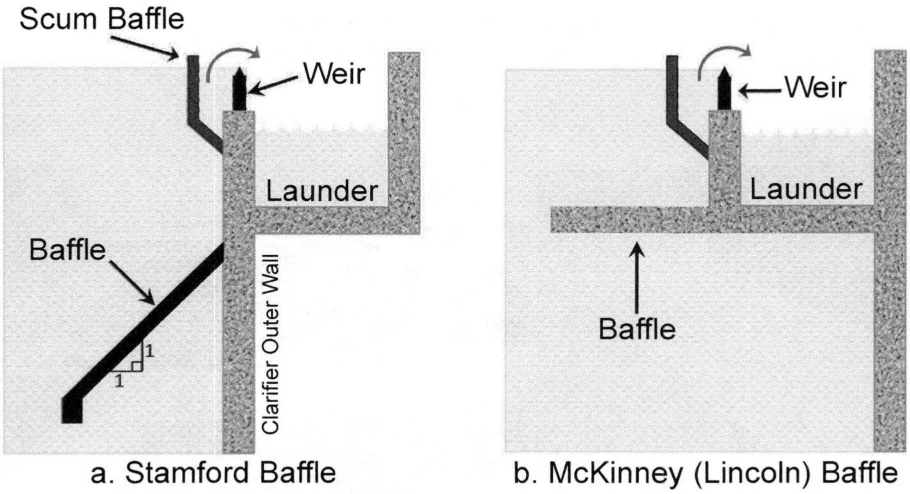

Figure 8.13 Stamford and McKinney Baffles (Reprinted with permission by Indigo Water Group)

Figure 8.14 Inset Weir with Algae Growth (Reprinted with permission by Indigo Water Group)

The multi-blade plows, using straight scraper blades, have been used most extensively in the United States (Figure 8.17). Designs using curved blades, commonly referred to as *spiral plows*, have been used for decades in Europe and are gaining popularity in the United States. For more information on plow-type sludge removal systems, refer to Chapter 4.

For activated sludge systems that remove ammonia, denitrification in the clarifier blanket can cause solids to float and effluent quality to degrade. Hydraulic suction sludge removal systems can remove sludge more rapidly than plow-type systems. Hydraulic suction mechanisms lift solids from across the entire tank radius. There are two fundamentally different types of hydraulic suction removal. The first, commonly called an *organ pipe* or *riser pipe*, has a separate collector pipe for each suction inlet opening (Figure 8.18). V-shaped plows direct the sludge to the multiple riser pipes. The other type has a single or double arm extending across the full radius of the tank. The arm is tubular and has a number of openings along its length (Figure 8.19). It is commonly referred to as a *manifold design*, but is also known as a "header", "tubular", or "Tow-Bro", in recognition of Townsend and Brower, who developed it. Both types of hydraulic suction removal work by using pumps or adjustable valves to create a difference in water level (hydraulic head) between the sludge blanket and the location where the sludge is discharged. Large hydraulic head differences result in faster sludge removal.

Figure 8.18 shows a typical riser pipe clarifier design. The horizontal runs of the riser pipes are stacked vertically, one on top of the other (A). The riser pipes may also be stacked side-by-side, horizontally, which

Figure 8.15 Weir Covers on a Secondary Clarifier (Reprinted with permission by Indigo Water Group)

Figure 8.16 Clarifier Weir Brush Cleaning System (Courtesy of Ford Hall Company, Inc.)

reduces tank stirring by the mechanism. Also visible in Figure 8.18 are a flocculating inlet well (B) and Stamford baffles (C). Each riser pipe is fitted with an adjustable, *telescoping weir*, movable sleeve, or ring arrangement that allows the operator to adjust the flow independently for each suction inlet. Sludge moves through the riser pipes to a sludge collection box (Figure 8.19). If the telescoping weirs in the sludge collection box are above the water level and can discharge freely, then the weir on each riser pipe must be adjusted separately in order to change the sludge withdrawal rate from the tank. If the telescoping weirs are submerged, then changing the level of sludge in the sludge box will change the flowrate through

A *telescoping weir* consists of a smaller tube fitted inside of a larger tube. The smaller tube can be moved in or out to adjust the total length of the pipe.

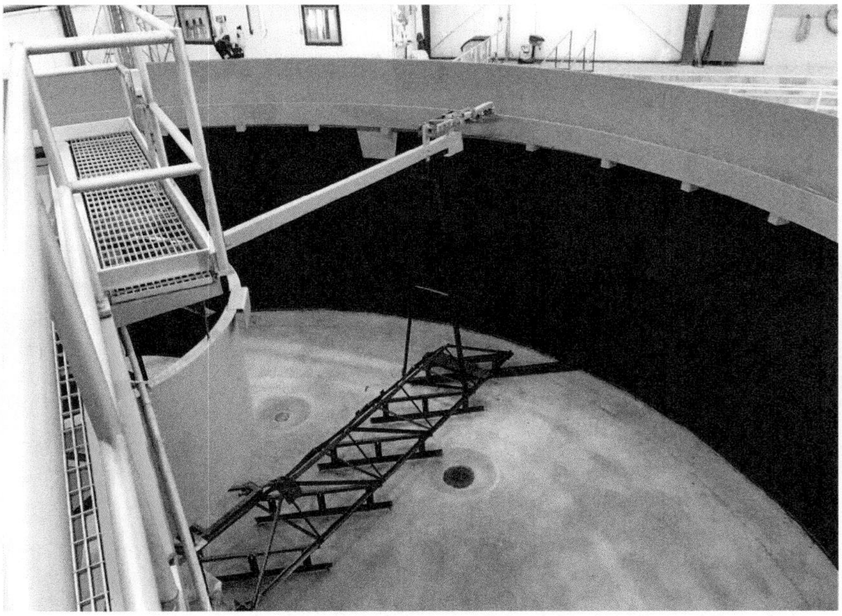

Figure 8.17 Scraper Type Sludge Collection Mechanism (Reprinted with permission by Indigo Water Group)

Figure 8.18 Organ Pipe Hydraulic Suction Sludge Collection Mechanism (Reprinted with permission by Kenneth Schnaars)

all of the riser pipes at the same time. The sludge level in the sludge collection box may be decreased by increasing the pumping rate out of the box.

The manifold-type hydraulic suction mechanism or Tow-Bro has multiple openings along its length (Figures 8.20 and 8.21). Some clarifiers have a single tube, whereas others have two tubes opposite each other so the mechanism crosses the entire tank diameter. The openings are sized and spaced during manufacturing to obtain a near-optimum pattern of collecting solids from the floor. The openings at the far end of the tube are the largest while the openings toward the center of the clarifier are the smallest. As the mechanism rotates, the flow through each opening is proportional to the percentage of the clarifier floor traveled by that section of the mechanism. As a result, solids are removed from the entire clarifier floor at the same rate. The main advantage of the manifold is that sludge may be pumped directly from the manifold or from a RAS wet well. This allows suction of relatively dense sludge.

Figure 8.19 Sludge Box with Telescoping Weirs (Reprinted with permission by Indigo Water Group)

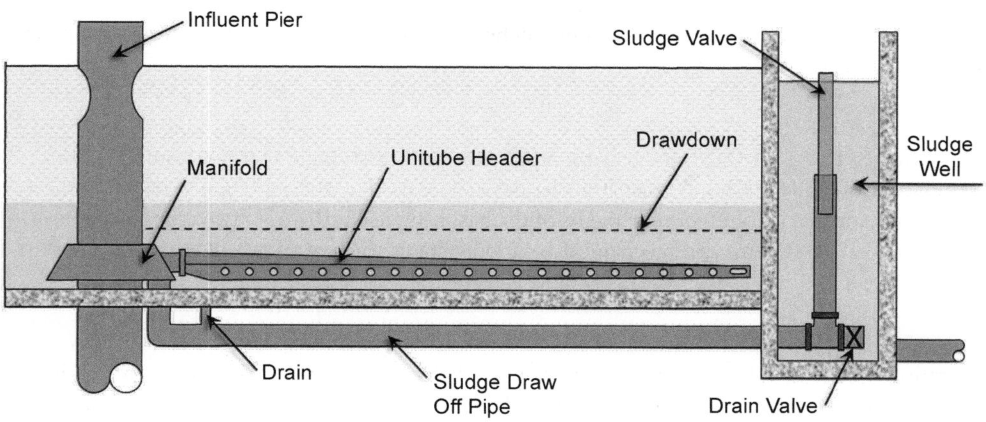

Figure 8.20 Manifold Hydraulic Suction Type Sludge Collector (Courtesy of Evoqua Water Technologies LLC)

DEPTH AND FLOOR SLOPE

Before the early 1980s, circular clarifiers used for primary and secondary clarification often had depths of 2.4 to 3 m (8 to 10 ft). Performance data showed that deeper secondary clarifiers worked better in activated sludge systems than shallower clarifiers. Deeper clarifiers had lower effluent suspended solids concentrations and were more resistant to upsets from peak hydraulic loads. In a 1984 survey (Tekippe, 1984), many of the largest engineering firms were designing and building secondary clarifiers with depths of 4 to 5 m (13 to 16 ft). Depths of up to 6 m (20 ft) have been used (Metcalf and Eddy, Inc./AECOM, 2014).

The floors of both primary and secondary clarifiers typically slope toward the sludge collection hopper. Unlike primary clarifiers where the floor slope can vary considerably from one facility to the next, most secondary clarifiers with plot or spiral sludge collection mechanisms have a constant floor slope of 1 on 12. In other words, a clarifier that is 12 m across from the outer wall of the clarifier to the center would have a drop of 1 m (foot) across that distance. The true origin of this particular slope is uncertain, but has received widespread use for decades.

For hydraulic suction clarifiers, a 1 on 12 slope may also be used. However, because it is not necessary to move the sludge across the floor, relatively flat floors are acceptable. Tow-bro clarifiers use a completely flat bottom. Often, a bottom slope of 1 or 2% is provided to make it easier to drain and clean the tank. Some

Figure 8.21 Field Installation of Tow-Bro Sludge Removal System (Courtesy of Ovivo USA)

design engineers actually prefer to reverse the slope (higher in the middle) and provide a gutter along the perimeter of the tank.

AERATION SYSTEMS

The supply of oxygen to the activated sludge basin represents the largest single energy consumer in an activated sludge facility (50 to 90%). Oxygen-transfer devices are used to not only supply oxygen to the process, but also to mix the aerobic compartments of the basin. Typically, there are two types of aeration devices: diffused-aeration systems and mechanical-aeration systems. Each of these will be discussed in the following subsections.

DIFFUSED AERATION

Diffused aeration is defined as the injection of air or oxygen below the water level. The air or oxygen is supplied by low-pressure blowers with pressures typically up to 210 kPa absolute (30 psia) or 105 kPa gauge (15 psig). Jet aerators, which combine gas injection with mechanical pumping or mixing, are also included in this category.

Diffusers include both porous and nonporous devices. They are similar to the air stones used in aquariums. Details about these devices can be found in *Design of Water Resource Recovery Facilities* (WEF et al., 2018). The porous diffusers, often referred to as *fine-pore* or *fine-bubble* diffusers, are highly efficient and are currently the most widely used of the diffused-air systems (Figure 8.22). They are typically produced from ceramic materials, porous plastics, or perforated membranes. Nonporous systems include an array of diffusers that have larger openings ranging from holes in pipes to specially designed valved openings. Nonporous systems are also referred to as *coarse-bubble diffusers* and are less efficient than fine-bubble. Diffusers are placed near the floor of the activated sludge basin and may be configured in a grid arrangement, along one or both longitudinal sides (Figure 8.23). Space is left below the diffusers to accommodate accumulation of grit and debris. Diffusers may be placed in uniform densities or arranged in *tapered configurations.* The diffuser type, pattern of diffuser layout, basin geometry, submergence, and airflow rate all influence *oxygen-transfer* performance. Maintenance of porous diffusers is greater than for the large-orifice diffusers. Clogging and fouling of the fine pores require occasional cleaning, which can be provided by automatic gas-cleaning systems in situ or by draining the basins and cleaning with high-pressure hosing or acid spritzing (WEF et al., 2010). Diffusers must be replaced every 3 to 5 years.

In *tapered configurations*, more diffusers are placed at the front of the activated sludge basin than at the discharge end. This ensures more oxygen is available where the BOD_5 concentration is highest and where the microorganisms will consume the most oxygen.

Oxygen transfer is the movement of oxygen from the air bubbles (gas) into the water. Air bubbles in water are not dissolved oxygen.

JET AERATION

Jet aeration systems combine aeration with mechanical pumping. These systems may be arranged with two headers running the full length of a basin, as shown in Figure 8.24, or have a central hub arrangement, as shown in Figure 8.25. A blower is typically used to provide pressurized air to the upper header while

Figure 8.22 Fine-Bubble, Flexible-Membrane Diffuser (Reprinted with permission by Indigo Water Group)

Figure 8.23 Fine-Bubble Diffusers at the Bottom of an Activated Sludge Basin (Reprinted with permission by Indigo Water Group)

MLSS is simultaneously pumped through the bottom header. Submersible pumps are typically used, but some installations use external pumps. In the case of a circular system, the air line enters at the top of the hub. Nozzle assemblies, also called *jets*, are spaced evenly along the liquid header, as shown in Figure 8.24 and Figure 8.25. Each jet assembly contains two nozzles nested one within the other, as shown in Figure 8.25. The air header is connected to the liquid header by air transfer ducts, as shown in Figure 8.26 and Figure 8.27. As MLSS is pumped through the header and exits through the inner nozzle, it mixes with the incoming air in the outer nozzle. The intense mixing and high degree of turbulence help to produce air bubbles that are similar in size to those produced by fine-bubble aeration systems.

Figure 8.24 Jet Aeration System (Reprinted with permission by Fluidyne Corp.)

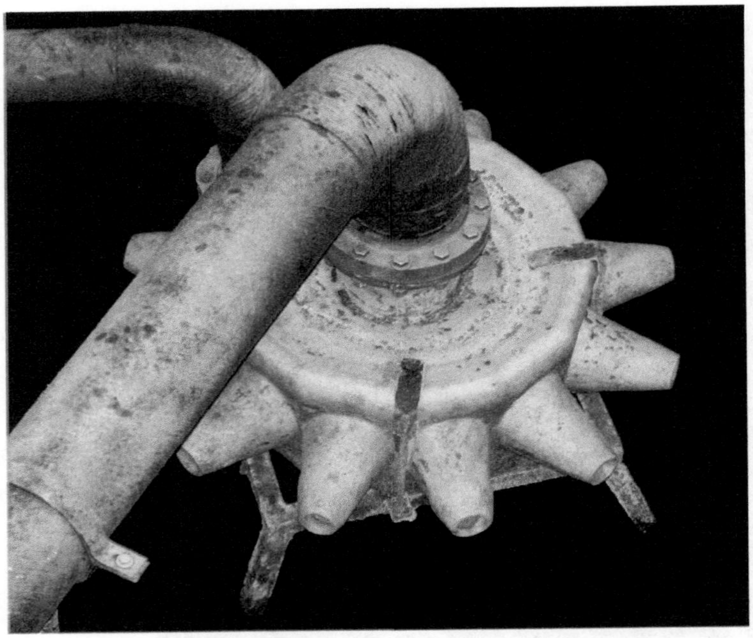

Figure 8.25 Circular Jet Aerator (Reprinted with permission by Indigo Water Group)

The recirculation pump is typically a constant speed pump and the aeration capacity is adjusted by modulating the airflow rate. Airflow is not required to operate the recirculation pump. Operators may temporarily turn off the blower while continuing to mix the MLSS with the reciculation pump to encourage denitrification.

AIR DELIVERY

The three basic components of the air-delivery system are air filters or conditioners, blowers, and piping. Air filters remove particulates such as dust from the inlet air to the blowers and protect both the blowers and diffusers from mechanical damage or clogging. The degree of air cleaning depends on the inlet air quality, the type of blower, and the diffusers.

Today, many types of dynamic and positive-displacement blowers are used. The dynamic, or centrifugal, blowers are considered constant-pressure machines and are the most efficient and easiest to operate at variable airflow rates, are quieter than positive-displacement blowers, and require less maintenance. Their

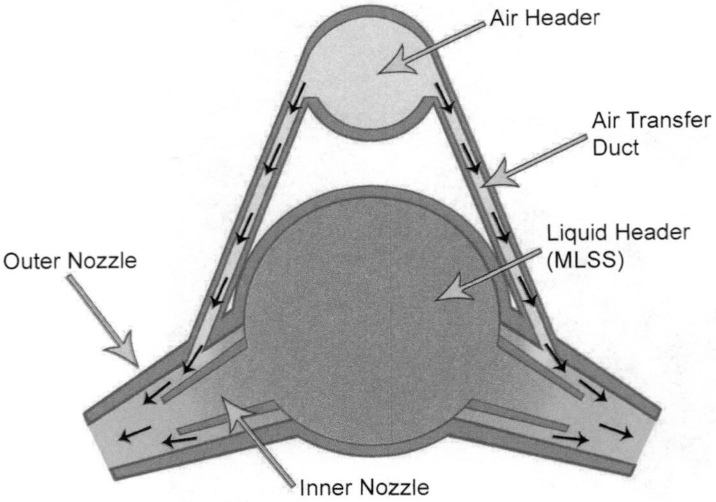

Figure 8.26 Jet Aeration Header Cross-Section (Reprinted with permission by Indigo Water Group)

Figure 8.27 Fiberglass Jet Aeration Header Showing Connections Between Air and Liquid Headers (Reprinted with permission by Mixing Systems, Inc. [www.mixing.com])

disadvantages include a limited operating pressure range and reduced delivered air volumes with any increase in backpressure caused by diffuser clogging. The positive-displacement blower is a constant-volume device capable of operating over a wide range of discharge pressures. It has a lower initial cost and requires relatively simple control procedures. Centrifugal blowers have higher efficiency and are favored at higher airflows. More details on blowers can be found in literature by WEF et al. (2018) and U.S. EPA (1989).

Piping connects the blowers to vertical drop pipes called *downcomers* or drop legs that are typically located along the longer sides of the basin. At the bottom of the basin, the downcomer pipes tee off into horizontal pipes or laterals that sit 0.5 to 1.5 m (1 to 3 ft) off the bottom of the tank. The pipes are connected to the diffusers. Valves in the air system are used to control airflow to individual basins or sectors of basins.

DISSOLVED OXYGEN CONTROL

Activated sludge basin DO may be controlled by adjusting individual airflow control valves located on the downcomers. In this approach, basin DO is measured at several locations and the corresponding air control valve is adjusted manually or automatically to maintain the desired DO concentration. Manual adjustment is simple and inexpensive, but can result in overaeration and underaeration. Automatic control modulates the airflow on a continuous basis to meet demand.

In conjunction with basin DO control, aeration blowers may be operated in several modes that may be classified as manual or automatic control. In manual control, the operator opens or closes the inlet valves on centrifugal blowers. Typically, this is done on an as-needed basis and results in periodic overaeration. Automatic control of blower capacity (blower speed or number of online blowers) may be achieved based on main air header pressure. In the most common automatic mode, air capacity is varied around an operator-set air header pressure. For example, increased air demand will result in reduced air header pressure, which will trigger an upward adjustment of the airflow to reestablish the preset header pressure.

MECHANICAL AERATION

Mechanical aeration systems used with activated sludge include surface splash-type aeration devices and submerged mixers with or without draft tubes. Draft tubes look like wide, short pipes. The draft tube surrounds the portion of the mixer that is below water level. Draft tubes help the water to pass through the aerator at a minimum speed (velocity) and reduce interference from waves created by other aeration devices in the same basin. Surface aerators commonly used with activated sludge can be grouped into the following general categories:

- Radial flow, low-speed aerators; and
- Horizontal rotors.

Each system is widely used and has distinct applications. Surface aerators are typically float-, bridge-, or platform-mounted and some may be equipped with submerged draft tubes for deep basin applications (Figure 8.28). The radial low-speed aerators have been increasingly used recently and are considered good mixing devices that are more efficient than high-speed aerators. Horizontal rotors, designed in several configurations, are used in oxidation ditches (Figure 8.29). The rotors agitate the liquid surface, lifting and mixing the MLSS to transfer oxygen from the air into the MLSS. Horizontal rotors also push the MLSS around the oxidation ditch and keep the MLSS in suspension. There are several different oxidation ditch designs that combine fine-bubble diffusers with a large mixer instead of using rotors.

MIXING

In the activated sludge process, mixing is important for maintaining the MLSS in suspension. In most applications, the aerators serve as both oxygen-transfer devices and mixers. With the exception of the effluent end of plug flow basins, oxygen requirements typically control aerator design and operation. In other words, more air is needed by the microorganisms than is needed to keep the MLSS in suspension. However, at the effluent end of plug flow basins and in some low-loaded, completely mixed systems, mixing may be the controlling factor, especially during low-flow conditions. In most cases, the aeration system is used for mixing in these situations; however, use of low-speed, submerged propeller, horizontal mixing devices (such as banana blade mixers) to move mixed liquor horizontally along the basin is another option.

Figure 8.28 Surface Aerator Mounted Under Platform in Activated Sludge Basin (Reprinted with permission by Indigo Water Group)

Figure 8.29 Horizontal Rotor in an Oxidation Ditch (Reprinted with permission by Envirodyne Systems Inc.)

Both submerged propeller or turbine mixers have been used for mixing anoxic and anaerobic compartments of activated sludge basins. These devices mix without breaking the water surface and are capable of maintaining biological solids in suspension at minimal energy inputs. The number and placement of these units is critical to effective suspension of MLSS.

RETURN AND WASTE ACTIVATED SLUDGE SYSTEMS

The RAS system pumps the settled sludge, thickened in the clarifier, from the clarifier back to the activated sludge basins. Most RAS pumping stations use centrifugal pumps for this application. For intermediate-sized facilities, the pumps are often connected directly to the sludge withdrawal pipes. In larger facilities, a wet well may be provided. In this case, sludge is transferred from the clarifier to the wet well (typically by gravity) and then RAS is pumped from the wet well back to the front of the activated sludge basins. Screw pumps are also used in many facilities in the United States and Europe. Return activated sludge pumps may be equipped with variable frequency drives to allow the operator to increase and decrease the RAS pumping rate over a wide range of flows, typically between 25 and 125% of the WRRF design flow. Many WRRFs flow pace their RAS pumps to maintain the RAS at constant percentage of the influent flow.

All activated sludge processes must have a WAS system to remove excess MLSS from the system. Waste sludge may be wasted from the clarifier under flow (clarifier blanket) or directly from the activated sludge basins. Most WAS pumping stations use centrifugal pumps, but diaphragm pumps, plunger pumps, and progressing cavity pumps are also used. The RAS and WAS pumping systems may include some type of flow measurement device such as a magnetic flow meter (Chapter 2).

RECIRCULATION PUMPING

Activated sludge systems that must remove ammonia, *nitrite*, and nitrate often contain internal mixed liquor recycle (IMLR) pumps. These pumps cycle nitrite and nitrate-rich MLSS from the end of aerated zones back to anoxic zones, as shown in the three-stage process in Figure 8.5. Mixed liquor recycling is typically accomplished by low-head, submersible nonclog pumps, propeller pumps, or nonclog vertical turbine pumps. Constant-speed pumps are typically used because it is not essential that recycle flows match or follow influent flows. Recycle pumps are typically large enough to pump between 50 and 400% of the WRRF's rated flow capacity.

Nitrite is an intermediate product and does not typically accumulate in the activated sludge process. Effluent nitrite concentrations are typically below 0.25 mg/L as N.

SECTION EXERCISE

1. The effluent from the clarifier shown in the picture below is high in TSS. The operator watches the clarifier carefully during high and low flow conditions. It does not appear to be short-circuiting, but there are a lot of tiny floc particles that don't seem to want to settle. What is missing from this clarifier that could, if added, help flocculate those tiny floc particles and improve effluent quality?

Reprinted with permission by Indigo Water Group

2. There are three organ-pipe-style clarifiers in operation. The flow distribution is even between the clarifiers so they are each receiving the same mass (kilograms or pounds) of MLSS. Two of the clarifiers have a very low blanket, but the third clarifier is full of settled sludge. Even though troubleshooting has not been discussed, what do you think could be wrong with the third clarifier based on what you know about how it works?

____

____

____

SECTION EXERCISE SOLUTIONS

1. Flocculating center well
2. Could be a blockage in one or more of the organ pipes preventing sludge withdrawal or the sludge pump connected to this clarifier stopped working.

TEST YOUR KNOWLEDGE

1. **Water resource recovery facilities typically have multiple activated sludge basins and clarifiers so they can be taken out of service for cleaning and maintenance.**
   - ☐ True
   - ☐ False

2. **Blowers used with activated sludge processes are low-pressure.**
   - ☐ True
   - ☐ False

3. **Having air bubbles in water is the same as having DO.**
   - ☐ True
   - ☐ False

4. **Water moves through this type of activated sludge basin in the same way that water moves through a pipe.**
   a. Complete mix
   b. Oxidation ditch
   c. Sequencing batch reactor
   d. Plug flow

5. **One advantage of complete-mix-type activated sludge systems is**
   a. More complicated operation
   b. Dilution of shock loads
   c. Prone to filamentous bulking
   d. Lower oxygen demand

6. **Selectors are useful for**
   a. Diverting flow between basins
   b. Thickening WAS
   c. Controlling the growth of some filaments
   d. Decreasing total treatment time

7. **The holes in the bottom of a peripheral feed clarifier channel**
   a. Prevent short-circuiting of MLSS to the weirs
   b. Must have their o-rings replaced weekly
   c. Are linked to floc-well performance
   d. Distribute the flow over the clarifier surface

8. **This modification to secondary clarifiers provides a zone of gentle mixing to improve sludge settling.**
   a. Perimeter channel
   b. Flocculating center feed well
   c. Inlet channel mixers
   d. Riser pipes

9. **A Stamford baffle**
   a. Is typically made of concrete
   b. Attaches to the scum baffle
   c. Redirects currents downward
   d. May increase effluent TSS

10. **Algae growth on secondary clarifier weirs should be controlled because**
    a. Excessive growth can result in unbalanced flows.
    b. The roots can eat into the metal v-notches of the weir plate.
    c. Algae absorb vital nutrients from the process.
    d. Its messy appearance offends visitors to the facility.

11. **An operator wants to increase the sludge withdrawal rate from an organ-pipe style clarifier with submerged telescoping valves. He or she should**
    a. Decrease the sludge pumping rate
    b. Lower the water level in the sludge collection box
    c. Manually adjust each riser pipe telescoping weir
    d. Shift influent flow to a different secondary clarifier

12. **With organ-pipe-style clarifiers, stacking the pipes vertically**
    a. Increases access to the tops of the riser pipes for sampling
    b. Interferes with placement of Stamford baffles
    c. Limits the adjustment of the telescoping weirs
    d. Increases clarifier mixing over horizontally stacked pipes

13. **Which of the following statements about Tow-Bro type sludge collection mechanisms is false?**
    a. Openings along the manifold are largest at the clarifier edge.
    b. May use either a sludge wet well or pump directly from clarifier.
    c. Removes sludge from the entire clarifier floor at the same rate.
    d. Requires routine adjustment of the telescoping weirs.

14. **In a plug flow activated sludge basin with tapered aeration, where would you expect to find the most diffusers?**
    a. At the influent end of the basin
    b. Near the middle of the basin
    c. At the effluent end of the basin
    d. Equally spaced along the basin length

15. **Nonporous aeration systems are**
    a. More efficient than fine-bubble
    b. Have larger openings than fine-bubble
    c. Most widely used type of diffused air
    d. Consist of a header with nozzles

16. **Jet aeration systems are classified as**
    a. Fine bubble
    b. Coarse bubble
    c. Surface aeration
    d. Draft tubes

17. **The nesting cones of a jet aeration header**
    a. Inject air through the inner cone
    b. Create turbulence for oxygen transfer
    c. Require both air and MLSS flow at all times
    d. Typically point up toward the basin surface

18. **The vertical drop pipes that connect the blowers to the diffusers at the bottom of the activated sludge basins are called**
    a. Laterals
    b. Diffusers
    c. Draft tubes
    d. Downcomers

19. **Surface aerators and rotors transfer oxygen into the MLSS by**
    a. Injecting air through nozzles
    b. Agitating the water surface
    c. Conveying air supplied by a blower
    d. Recirculating MLSS over diffusers

20. **Return activated sludge pumps are often equipped with ____________ to allow the operator to adjust the flowrate.**
    a. Upstream wet wells
    b. Magnetic flow meters
    c. Variable frequency drives
    d. Diaphragms

The F/M is the mass of BOD_5 added per mass of MLVSS in the activated sludge basin.

Process Variables for the Activated Sludge Basin

Activated sludge is extremely flexible and can be manipulated to meet a variety of process goals. Operators can influence MLSS microbiology, settleability, and final effluent quality by adjusting five process variables: the number of basins and clarifiers in service, the quantity of WAS removed from the system each day, the RAS, the availability of oxygen, and, in some systems, IMLR flows. These five process variables give operators an immense amount of control over sludge settleability, the level of treatment achieved, and final effluent quality. Other factors that can affect activated sludge performance include temperature, pH, alkalinity, and the characteristics of the influent wastewater.

HYDRAULIC DETENTION TIME

Hydraulic retention time is also referred to as the HDT or simply detention time (DT). You may also see this as *HRT*. The HDT can be 1) the time required to fill a vessel, 2) the time required to empty a vessel, or 3) the average amount of time that water remains in a vessel. Operators have limited control over the HDT, but opportunities exist for placing additional basins and clarifiers into service or removing them from service.

The formula for DT is

In International Standard units:

$$\text{DT, days} = \frac{\text{Volume, m}^3}{\text{Flow, m}^3\text{/d}} \tag{8.1}$$

In U.S. customary units:

$$\text{DT, days} = \frac{\text{Volume, gal}}{\text{Flow, gpd}} \tag{8.2}$$

Example DT calculations may be found in the Primary Treatment chapter. Detention times for activated sludge basins and clarifiers are in hours, not days. Remember to convert your answers from days to hours.

RETURN ACTIVATED SLUDGE

Return activated sludge is arguably a clarifier process control variable rather than an activated sludge basin control variable. It is included in this section to keep the descriptions of RAS and WAS together and, hopefully, minimize confusion surrounding the functions of each of these important process control variables. Return activated sludge, along with the influent flow, pushes MLSS from the activated sludge basin into the secondary clarifier. The mass of MLSS (kilograms or pounds) entering the secondary clarifier is the SLR. The SLR may be calculated according to the following formula:

In International Standard units:

$$\text{SLR, kg/m}^2\cdot\text{d} = \frac{(Q_{\text{inf}} + Q_{\text{RAS}}, \text{m}^3\text{/d})\left(\text{Concentration, } \frac{\text{mg}}{\text{L}}\right)}{(1000)(\text{Clarifier Surface Area, m}^2)} \tag{8.3}$$

In U.S. customary units:

$$\text{SLR, lb/d/sq ft} = \frac{(Q_{\text{inf}} + Q_{\text{RAS}}, \text{mgd})\left(\text{MLSS, } \frac{\text{mg}}{\text{L}}\right)\left(8.34 \frac{\text{lb}}{\text{mil.gal}}\right)}{\text{Clarifier Surface Area, sq ft}} \tag{8.4}$$

Where:

Q_{inf} = Influent flow;
Q_{RAS} = Return activated sludge flow; and

8.34 lb/mil. gal = Conversion factor. Water weighs 8.34 lb/gal; however, the 8.34 used in this equation is a conversion factor used to convert milligrams per liter and million gallons to pounds. The conversion of 1 mg/L into pounds per million gallons is shown below.

$$\frac{1\ \text{mg}}{\text{L}}\left|\frac{1\ \text{g}}{1000\ \text{mg}}\right|\left|\frac{1\ \text{kg}}{1000\ \text{g}}\right|\left|\frac{2.204\ \text{lb}}{1\ \text{kg}}\right|\left|\frac{3.785\ \text{L}}{1\ \text{gal}}\right|\left|\frac{1\,000\,000\ \text{gal}}{1\ \text{mil. gal}}\right| = 8.34\ \frac{\text{lb}}{\text{mil. gal}} \tag{8.5}$$

Assuming that the clarifier surface area (number of clarifiers in service) and MLSS concentration remain constant, increasing either the influent flow or the RAS flow will increase the SLR to the secondary clarifier. The SLR is a critical operating parameter for secondary clarifiers. If the SLR is too high, the clarifier won't be able to settle the solids and a blanket will build up in the clarifier, eventually washing out over the top of the weirs. Acceptable SLR for secondary clarifiers are between 122 to 195 kg/m^2·d (25 to 40 lb/hr/sq ft) (WEF, 2018).

The allowable SLR is affected by how well the sludge settles. When the MLSS settles poorly, the maximum allowable SLR to the clarifier will be lower than when the MLSS is settling well. This idea is discussed in more detail in the section on Secondary Clarifier Process Control.

Increasing the RAS flow pushes solids into the clarifier faster, but it also removes solids from the bottom of the clarifier faster and returns them to the activated sludge basin. Within limits, the RAS flow controls how long the MLSS remains in the clarifier and controls the blanket depth at the bottom of the clarifier. The RAS flow also controls the solids concentration in the sludge blanket. If the RAS flow is increased, the solids spend less time in the clarifier and have less time to settle and compact. The RAS concentration goes down. If the RAS flow is decreased, the solids remain in the clarifier longer and the solids have more time to settle and compact. The RAS concentration goes up. The effect of RAS on clarifier solids detention time, blanket depth, and RAS concentration is limited by how well the sludge flocculates, settles, and compacts. The clarifier cannot magically thicken the MLSS indefinitely. At some point, turning the RAS flow down more will not continue to increase the RAS concentration.

There is a persistent myth that increasing the RAS flowrate effectively decreases the amount of treatment time in the activated sludge basin. This is not the case. Turning up the RAS does move water between the clarifier and activated sludge basins and back again faster, but the water spends the same amount of time in the treatment process overall. The time per pass decreases, but total time does not. If treatment time really did decrease, then increasing the RAS flowrate would cause treated wastewater to flow out of the secondary clarifier and into the final effluent faster, which does not happen. Increasing or decreasing the RAS rate cannot change the water level in the activated sludge basins by more than a few millimeters. The water level in the activated sludge basins is typically controlled by the height of the secondary clarifier weirs.

The RAS does not typically affect the *MLSS concentration* in the activated sludge basin to an appreciable degree. The RAS simply moves solids between the clarifier and the activated sludge basins. Changing the RAS flow can only affect the MLSS concentration in the activated sludge basin if large quantities of solids are being stored in the clarifier blanket compared to the quantity of MLSS in the activated sludge basin. Blanket depths in secondary clarifiers should be kept below approximately 0.6 m (2 ft) whenever possible. Storing solids and maintaining deep blankets in the secondary clarifiers increases the likelihood that solids will be washed out of the clarifier and into the final effluent, especially during periods of high influent flow. Deep blankets should be avoided.

Sudden changes in the *MLSS concentration* from one day to the next are almost always a result of sampling and analysis errors. Although it is possible for the MLSS concentration to increase suddenly, it is unlikely. Remember that the yield is nearly constant. Samples with MLSS concentrations that are substantially different from the previous day should be recollected and reanalyzed.

In summary, the *RAS* flow

- Returns settled solids from the clarifier to the activated sludge basin;
- In conjunction with WAS, makes possible independent control of the HRT and SRT;
- Influences the clarifier SLR in combination with influent flow, MLSS concentration, and clarifier surface area;
- Controls the clarifier solids detention time and blanket depth;

The *RAS* influences the SLR to the clarifier and controls blanket depth.

- Controls the concentration of solids in the RAS within limits set by how well the MLSS settles and compacts;
- Can only affect the MLSS concentration in the activated sludge basins if MLSS is being stored in the clarifier blankets;
- Does not affect overall treatment time, but can affect time per pass; and
- Does not appreciably raise or lower the water level in the activated sludge basins.

RETURN ACTIVATED SLUDGE/WASTE ACTIVATED SLUDGE CONCENTRATIONS

Obtaining accurate estimates of the RAS and WAS concentrations can be difficult because sampling locations are not representative, pumps may operate intermittently, grab samples don't necessarily reflect changing conditions: composite sampling may change the solids characteristics, and the RAS flow to influent flow ratio may be changing. Fortunately, if the influent and RAS flows are both known, it is possible to calculate the RAS concentration from the MLSS concentration. Compaction will be limited by the MLSS characteristics. Always verify calculation results by comparing them to the settling and compaction behavior of the MLSS in the *settleability test*.

The *settleability test* measures the settling rate and compaction for a sample of MLSS. The test is conducted by placing a sample of MLSS in a transparent container. Over a 30-minute period, the MLSS is observed as it settles and forms a blanket. The volume of the settled sludge is recorded and used in subsequent calculations.

CALCULATION METHOD

The maximum RAS concentration achievable by the clarifier can be estimated from the settleometer test results. A detailed procedure for conducting a settleometer test is presented later in this chapter. For the test, a sample of MLSS is collected from the end of the activated sludge basin and is allowed to settle for 30 minutes. The settleometer mimics how the MLSS settles in the clarifier. A blanket should form in the bottom of the settleometer, as shown in Figure 8.30. After 30 minutes, the volume of the settled sludge is recorded. If the MLSS is allowed to remain in the clarifier for 30 minutes, then it should compact to the same degree as the MLSS in the settleometer. If the MLSS concentration is known, then the settled sludge concentration (SSC) is easily calculated using the following formula:

$$SSC_{30} = \left(\frac{\text{Initial Settleometer Volume, mL}}{SSV_{30}}\right) \times \text{MLSS mg/L} \tag{8.6}$$

Where:

SSC_{30} = Settled sludge concentration after 30 minutes and
SSV_{30} = Settled sludge volume after 30 minutes.

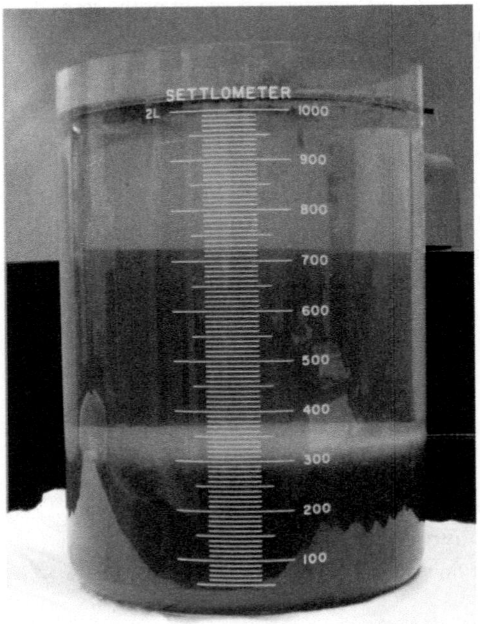

Figure 8.30 Settleometer Test (Reprinted with permission by Indigo Water Group)

Often, the MLSS will be in the clarifier for less than 30 minutes and will not achieve maximum compaction. This is especially true for WRRFs that must remove ammonia because they tend to use higher RAS flowrates. The actual WRRF RAS concentration may be estimated from the following equation. The influent flow and RAS flow must be in the same units.

$$X_{RAS} = \left(\frac{Q_{inf}}{Q_{RAS}} + 1\right) \times X_{MLSS} \tag{8.7}$$

Where:

X_{RAS} = Return activated sludge concentration,
Q_{inf} = Influent flowrate,
Q_{RAS} = Return activated sludge flowrate, and
X_{MLSS} = Mixed liquor suspended solids concentration.

CALCULATION EXAMPLE

The influent flow to a WRRF is 600 m^3/d (0.16 mgd) at 9:00 a.m. and increases to 1500 m^3/d (0.40 mgd) at 2:00 p.m. The MLSS concentration is 2800 mg/L. Assuming that the blanket in the clarifier remains below 0.6 m (2 ft) over the course of the day, calculate the RAS concentration at both 9:00 a.m. and 2:00 p.m. Assume that the RAS pump rate is fixed at 900 m^3/d. The SSV at 30 minutes is 420 mL.

Step 1—Find the RAS concentration at 9:00 a.m.

$$X_{RAS} = \left(\frac{Q_{inf}}{Q_{RAS}} + 1\right) \times X_{MLSS}$$

$$X_{RAS} = \left(\frac{600\ m^3/d}{900\ m^3/d} + 1\right) \times 2800\ mg/L$$

$$X_{RAS} = (0.67 + 1) \times 2800\ mg/L$$

$$X_{RAS} = 4667\ mg/L$$

Step 2—Find the RAS concentration at 2:00 p.m.

$$X_{RAS} = \left(\frac{Q_{inf}}{Q_{RAS}} + 1\right) \times X_{MLSS}$$

$$X_{RAS} = \left(\frac{1500\ m^3/d}{900\ m^3/d} + 1\right) \times 2800 \frac{mg}{L}$$

$$X_{RAS} = (1.7 + 1) \times 2800\ mg/L$$

$$X_{RAS} = 7467$$

Step 3—Compare the calculated RAS concentrations to the settleometer.

$$SSC_{30} = \left(\frac{\text{Initial Settleometer Volume, mL}}{SSV_{30}}\right) \times \text{MLSS mg/L}$$

$$SSC_{30} = \left(\frac{1000,\ mL}{420\ mL}\right) \times 2800\ mg/L$$

$$SSC_{30} = 6667\ mg/L$$

In this example, the settleometer result indicates that the maximum RAS concentration achievable by the clarifier is 6667 mg/L. The RAS ratio calculation estimated a higher RAS concentration at 7467 mg/L. Because the sludge isn't settling and compacting well, the RAS concentration at 2:00 p.m. will be 6667 mg/L and not 7467 mg/L, assuming the MLSS remains in the clarifier for 30 minutes. There will be times, however, when a slow-settling sludge continues to compact in the settleometer after 30 minutes. Operators should be mindful that turning the RAS down to exceedingly low levels in an attempt to further thicken and compact the solids in the clarifier blanket is counterproductive and can cause effluent quality to deteriorate.

Many WRRFs have the ability to set the RAS flow as a fixed percentage of the influent flow. This practice is called *flow pacing*. In this situation, the RAS concentration will remain relatively constant over the course of a day so long as the percentage of solids in the clarifier remains low relative to the percentage of solids in the activated sludge basin, that is, the clarifier isn't building a blanket. Flow pacing can be difficult to maintain during certain times of the day because of pumping limitations. In some cases, the RAS pump(s) can't be turned down far enough and, as a result, the percentage of RAS to influent flow increases. This causes the RAS concentration to decrease. In other cases, the RAS pumping capacity becomes the limiting factor. For example, if the influent flow to a facility peaks at 2000 m^3/d, the RAS pump(s) have a capacity of only 1000 m^3/d. For this facility, the percentage of RAS that may be returned during peak hour influent flows will only be 50%. In this situation, the clarifier may begin to accumulate solids and build a blanket. When influent flows decrease, the RAS flow percentage will increase again and solids will be removed from the clarifier more quickly and the blanket (if any) will be drawn back down. It is common and normal for the blanket depth in a secondary clarifier to rise and fall over the course of a day. The goal is to keep the blanket below 0.6 m (2 ft) whenever possible.

Some WRRFs do not have the ability to flow pace their RAS pump(s). In this situation, the operator may decide to either 1) manually adjust the RAS flow multiple times over the course of a day or 2) decide to select a fixed flowrate for the entire day. When the RAS flow is fixed, the solids will be removed from the clarifier faster during some parts of the day and slower during others. As a result, the RAS concentration will also vary over the course of the day.

Calculation 8.7 may also be used to determine the RAS flowrate required to achieve a particular RAS concentration. Maximizing the RAS concentration in the clarifier reduces the volume of water transferred to solids handling processes. Keep in mind that although low RAS flowrates can produce higher RAS concentrations, sludge compaction is limited and leaving the sludge in the clarifier for extended periods of time can degrade final effluent quality. To find the RAS flowrate that produces a desired RAS concentration, first conduct a settleometer test and calculate the SSC according to eq 8.6. Then, use eq 8.6 to calculate the RAS flowrate.

CALCULATION EXAMPLE

The SSC after 30 minutes of settling is 11 250 mg/L. Find the RAS flow that produces this concentration when the influent flow is 5100 m^3/d (1.35 mgd) and the MLSS concentration is 2900 mg/L.

Step 1—Enter known information into the formula.

$$X_{RAS} = \left(\frac{Q_{inf}}{Q_{RAS}} + 1\right) \times X_{MLSS}$$

$$11\,250\frac{\text{mg}}{\text{L}} = \left(\frac{5100\frac{\text{m}^3}{\text{d}}}{Q_{RAS}} + 1\right) \times 2900\ \text{mg/L}$$

Step 2—Rearrange the equation to solve for Q_{RAS}.

$$\frac{11\,250\text{ mg/L}}{2900\text{ mg/L}}=\frac{\left(\frac{5100\text{ m}^3/\text{d}}{Q_{RAS}}+1\right)\times 2900\text{ mg/L}}{2900\text{ mg/L}}$$

$$3.88=\frac{5100\text{ m}^3/\text{d}}{Q_{RAS}}+1$$

$$3.88-1=\frac{5100\text{ m}^3/\text{d}}{Q_{RAS}}+1-1$$

$$2.88=\frac{5100\text{ m}^3/\text{d}}{Q_{RAS}}$$

$$(2.88)(Q_{RAS})=\frac{5100\text{ m}^3/\text{d}}{Q_{RAS}}\times Q_{RAS}$$

$$(2.88)(Q_{RAS})=5300\text{ m}^3/\text{d}$$

$$\frac{(2.88)(Q_{RAS})}{2.88}=\frac{5300\text{ m}^3/\text{d}}{2.88}$$

$$Q_{RAS}=1840\text{ m}^3/\text{d}$$

Rearranging equations takes a little patience and a few simple rules that must be followed. 1) Simplify first by following the order of operations. First, do things inside of **P**arentheses, next is **E**xponents, then **M**ultiply and **D**ivide, and **A**dd and **S**ubtract. A helpful way to remember the order of operations is "**P**lease **E**xcuse **M**y **D**ear **A**unt **S**ally". Equations should be solved from left to right while still following the order of operations. A division bar acts like a parenthesis. 2) You can do anything you want to most equations; adding a number, for instance, but 3) Whatever is done, must be done to both sides of the equation. Finally, to move something from one side of the equation to the other, do the opposite. For the simple equation $3 + X = 5$, the 3 is being added to the X. To move it to the other side, subtract 3 to get X by itself. Then, subtract 3 from the other sides too. Now, $X = 2$. These same principles apply no matter how complicated the algebra problem is you are trying to solve.

SAMPLING LOCATIONS FOR RETURN ACTIVATED SLUDGE/WASTE ACTIVATED SLUDGE

Ideally, samples will be collected from well-mixed locations that are representative of the system as a whole. Splitter boxes, joiner boxes, flumes, and pump discharges all make excellent sampling locations for determining solids concentrations because they are well mixed. Figure 8.31 shows two possible configurations for RAS and WAS withdrawal from a clarifier. In the first scenario, the RAS and WAS pumps are pulling from the same pipeline. There may be a hopper or pit at the bottom of the clarifier where the pipeline enters the tank. The RAS pump operates continuously and the RAS concentration can be predicted using Formula 8.7. The WAS pump operates intermittently, but because it pulls from the same pipeline as the RAS pump, the WAS and RAS concentrations will be the same.

In the second scenario, the RAS and WAS pumps do not share a pipeline. The RAS pipeline withdraws from the clarifier at a higher elevation than the WAS pipeline. The WAS concentration is often higher than the RAS concentration because the MLSS has had more time to settle and compact before reaching the WAS withdrawal point. It becomes more difficult to estimate the WAS concentration using Formula 8.7. This disparity can be exaggerated by intermittent operation of the WAS pump. If the WAS pump only operates for 10 or 15 minutes out of each hour, then grit and heavier particles tend to accumulate in the WAS hopper while lighter floc is removed and recycled through the RAS line. If a grab sample is collected from the WAS pump immediately after it starts up, these heavier materials may fool an operator into

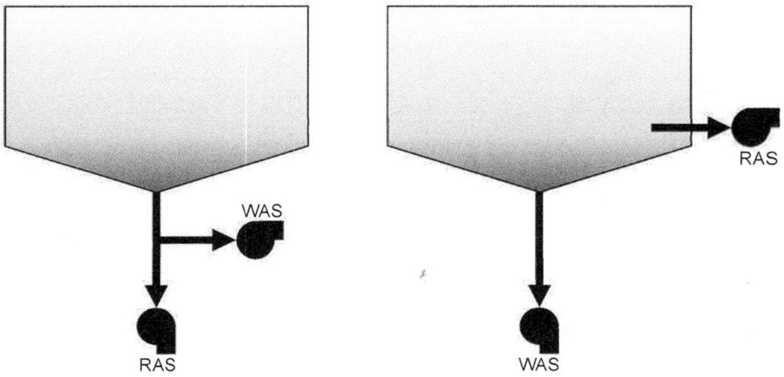

Figure 8.31 Return Activated Sludge and WAS Withdrawal Points from a Secondary Clarifier (Reprinted with permission by Indigo Water Group)

believing that the WAS concentration is much higher than it really is. As soon as the hopper has been emptied, the WAS concentration will become about the same as the RAS concentration. For this reason, a series of grab samples should be collected at the beginning, middle, and end of a pumping cycle to measure the true WAS concentration. If the WAS concentration estimate is incorrect, the MCRT/SRT calculations will also be incorrect.

Case Study—Figure 8.32 illustrates a RAS/WAS pump gallery where all three pumps are manifolded together. For this particular WRRF, any combination of clarifiers can be placed into service with one or more RAS pumps. Note that the pump in the middle can serve as either a RAS or WAS pump. An actuating valve on the discharge side of this pump allows RAS to either be returned to the activated sludge process or be diverted to the aerobic digester as WAS. Because all of the pumps are manifolded together, a sample collected from any one pump should be comparable to any other pump. The WAS and RAS concentrations will be equal.

The design engineer provided two sample taps, shown as red triangles in Figure 8.32, that were placed at opposite ends of the manifold. Operations staff diligently flushed the line prior to collecting samples for process control. Mixed liquor suspended solids concentrations measured at either of these sample taps were extremely erratic and were often lower in concentration than the MLSS in the activated sludge basins. It was clear that the sampling points were not representative. It was theorized that the MLSS was settling out at the ends of the pipe. Never knowing the true WAS concentration made it impossible for the operators to use MCRT/SRT control.

The facility operators installed three new sampling taps, shown as yellow triangles in Figure 8.32. By placing the sample taps downstream of the pumps, they ensured that the RAS would be well mixed prior to sample collection. Measured RAS concentrations have been consistent since the sampling points were moved. Return activated sludge results agree with the concentrations predicted using the RAS flow to influent flow ratio calculation discussed earlier.

WASTE ACTIVATED SLUDGE

To maintain a stable treatment process, MLSS must be removed on a regular schedule. Mixed liquor suspended solids can be removed from the bottom of the clarifier, from the RAS line, or directly from the activated sludge basin. In Figure 8.1, the WAS or WAS line is shown pulling MLSS from the bottom of the secondary clarifier, which is the most common configuration. The advantage of pulling MLSS from the bottom of the clarifier or RAS line instead of the activated sludge basin is that the RAS will have a higher concentration in milligrams per liter. This reduces the total volume that must be removed and processed. The advantage of wasting MLSS directly from the activated sludge basin is that the operator does not

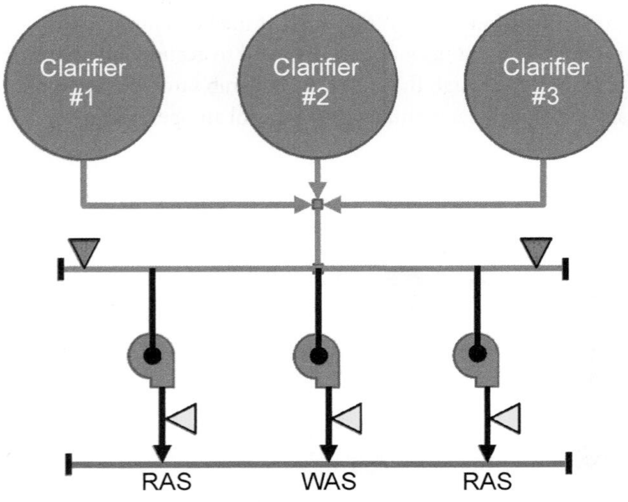

Figure 8.32 Typical RAS/WAS Pump Gallery Design (Reprinted with permission by Indigo Water Group)

need to know the concentration of MLSS to calculate the number of cubic meters or gallons to remove. A percentage of the MLSS is removed each day to achieve a target *sludge age*. If a 10-day sludge is desired, the operator simply wastes 10% of the total basin volume each day. If a 20-day sludge age is desired, the operator removes 5% of the total basin volume each day. Some early activated sludge process designs were based on hydraulic control, that is, removing solids directly from the activated sludge basin, but the desire to send thicker sludge to the solids handling side of the WRRF has limited its use.

Sludge age is the amount of time a typical microorganism cycles between the activated sludge basin and the secondary clarifier before it is removed from the system.

Waste activated sludge flow, along with environmental conditions such as water temperature and availability of BOD, nutrients, and oxygen, influences the process biology and level of treatment achieved. Slower growing microorganisms, including the nitrifying bacteria and some filaments, can only remain in the treatment process if the MLSS is held long enough for them to reproduce. Waste activated sludge determines how long the MLSS stays in the system and, therefore, helps to determine which types of microorganisms will be present. The presence or absence of filaments will influence how fast the sludge settles in the clarifier. Waste activated sludge also determines the MLSS concentration. Higher wasting rates leave fewer solids in the system. The amount of WAS removed from the system each day is determined by calculating sludge age. Sludge age calculations and the relationships between sludge age and other operational parameters are explained in detail in the next few sections.

In summary, the *WAS*

- Determines sludge age,
- Controls mass of total solids in the activated sludge basin,
- Influences sludge quality/settleability,
- Influences growth rate of the microorganisms,
- Influences the types of microorganisms that have an opportunity to proliferate, and
- Influences effluent quality.

The *WAS* determines sludge age and influences sludge quality.

TEST YOUR KNOWLEDGE

1. **Return activated sludge is higher in concentration than the MLSS in the activated sludge basins.**
   - ☐ True
   - ☐ False

2. **Slow-settling sludges may continue to settle and compact in the settleometer test after 30 minutes.**
   - ☐ True
   - ☐ False

3. **The RAS is a fixed percentage of the influent flow. The clarifier blanket depth is less than 0.6 m (2 ft). The influent flow to the WRRF is increasing. What is happening to the clarifier SLR?**
   a. Held constant by constant RAS percentage
   b. Decreasing with increasing influent flow
   c. Held constant by blanket depth
   d. Increasing with increasing influent flow

4. **Increasing the RAS flowrate will**
   a. Decrease the RAS concentration
   b. Decrease the SLR to the secondary clarifier
   c. Increase the WAS concentration
   d. Increase sludge age

5. **The RAS pumping rate was increased and the concentration of MLSS in the activated sludge basin increased. What must be true?**
   a. Before the increase, MLSS was settling in the activated sludge basins.
   b. After the increase, the SLR to the clarifier was too high.
   c. Before the increase, much of the MLSS was in the clarifier blanket.
   d. After the increase, increased exposure to BOD_5 increased the MLSS.

6. **Waste activated sludge influences all of the following EXCEPT**
   a. Sludge age
   b. Sludge settleability
   c. Sludge concentration
   d. Sludge blanket depth

7. **An activated sludge system does not currently remove ammonia. To retain nitrifying bacteria and increase their numbers, which of the following actions could be taken?**
   a. Increase return activated sludge
   b. Decrease sludge wasting
   c. Increase HDT
   d. Decrease clarifier SLR

8. The influent flow to the activated sludge basin is 8.7 ML/d (2.3 mgd). The RAS flow is 63% of the influent flow at 5.48 ML/d (1.45 mgd). If the MLSS concentration is 2600 mg/L, what is the expected RAS concentration?
   a. 4000 mg/L
   b. 4200 mg/L
   c. 6700 mg/L
   d. 7200 mg/L

9. After settling for 30 minutes in the settleometer, the SSV is 280 mL. What is the SSC if the starting MLSS concentration was 1850 mg/L?
   a. 5180 mg/L
   b. 6607 mg/L
   c. 7400 mg/L
   d. 9250 mg/L

10. The RAS concentration was calculated using flow data and the MLSS concentration. The calculation predicts a RAS concentration of 9000 mg/L. Filamentous bacteria are preventing the MLSS from settling and compacting well in the settleometer test. After 30 minutes, the SSC is only 6200 mg/L. What is the maximum RAS concentration achievable in the clarifiers if the sludge is allowed to remain in the clarifier for 30 minutes?
    a. 3100 mg/L
    b. 6200 mg/L
    c. 7600 mg/L
    d. 9000 mg/L

11. When solids are removed from the clarifier faster than they are added to the clarifier
    a. Blanket depth will increase
    b. Most of the MLSS will be in the clarifier
    c. Blanket depth will decrease
    d. Treatment efficiency will decrease

12. The RAS flow is currently set at 50% of influent flow. If the RAS flow is decreased to 30%, the operator should expect to see
    a. Reduced overall treatment time
    b. Rat holing of the clarifier sludge blanket
    c. Decreased solids loading to the clarifier
    d. Increase in RAS concentration

13. Mixed liquor suspended solids with a concentration of 1800 mg/L was placed into a settleometer and allowed to settle for 30 minutes. The initial sludge volume was 2000 mL and the 30-minute sludge volume was 280 mL. Find the SSC.
    a. 6429 mg/L
    b. 7143 mg/L
    c. 12 857 mg/L
    d. 14 286 mg/L

14. The return activated sludge flow is set at 70% of influent flow. If the MLSS concentration is 2600 mg/L, what is the anticipated RAS concentration assuming solids are not being stored in the clarifier blanket?
    a. 6300 mg/L
    b. 3700 mg/L
    c. 7500 mg/L
    d. 9000 mg/L

15. The operator knows from the settleometer test that the SSC at 30 minutes is 10 000 mg/L. The initial MLSS concentration was 2500 mg/L. If the operator wants to achieve the thickest RAS concentration for this sludge, but does not want to leave the RAS in the clarifier any longer than required, what should the RAS flowrate be when the influent flow is 4200 m^3/d (1.1 mgd)?
    a. 1400 m^3/d (0.37 mgd)
    b. 2500 m^3/d (0.66 mgd)
    c. 2800 m^3/d (0.74 mgd)
    d. 5100 m^3/d (1.35 mgd)

SLUDGE AGE

The definition of sludge age has evolved over the years as our understanding of the process has increased. Sludge age is the amount of time the solids or MLSS spend moving between the activated sludge basin(s) and secondary clarifier(s) before they are removed from the system either through the WAS line or by escaping into the final effluent. In the case of an SBR, treatment and clarification take place in the same tank. Sludge age affects every aspect of activated sludge process operation including: MLSS concentration and quality; sludge settling characteristics; types of microorganisms present in the process; F/M; treatment capacity; ability to remove ammonia, nitrite, nitrate; and phosphorus, oxygen demand, oxygen transfer efficiency, sludge production, and final effluent quality. *Sludge age should be the primary process control variable for any activated sludge system.* It should be adjusted as needed to optimize sludge settleability, minimize energy use, and meet final effluent limits. In locations that experience seasonal changes in influent water temperature, sludge age should be adjusted seasonally to compensate for increased or decreased microbial activity. The sludge age selected will dictate whether or not an activated sludge process will be able to meet treatment goals for effluent TSS, BOD, and ammonia.

Sludge age is a broad term that includes mean cell residence time (MCRT), solids retention time (SRT), and aerobic SRT.

Sludge age is a broad term that may refer to any of the following more specific terms: Gould sludge age (GSA), MCRT, SRT, or $SRT_{aerobic}$. Aerobic SRT is used by operators and design engineers to ensure the system has adequate sludge age for nitrification. Depending on the sludge age definition in use, the mass of

MLSS in the aerated zones, unaerated zones, and clarifier blankets may or may not be included in process control calculations, as shown in Table 8.4.

GOULD SLUDGE AGE

Gould sludge age is no longer in common use for reasons discussed below, although it is still referenced in many publications and operator math books.

Gould sludge age is defined as the mass of solids in the activated sludge process (MLSS) divided by the mass of TSS in the influent wastewater (Gould, 1953). In the early days of activated sludge process design, Gould and others knew that sludge younger than 1 to 2 days wouldn't flocculate and settle well and that the mass of sludge in the process needed to be 3.5 to 4.0 times the weight of the suspended solids of the incoming wastewater for this to occur (Gould, 1953). To put this in perspective, if the influent TSS concentration was 250 mg/L, then the concentration of solids in the activated sludge basin needed to be about 875 mg/L. Modern activated sludge processes operate with MLSS concentrations between 1000 and 5000 mg/L.

There are no clear advantages to using GSA and many disadvantages. Gould sludge age puts the operator in a position of trying to predict and match a constantly changing influent TSS load and does not give a clear mechanism for calculating how many solids to remove from the process each day. Sudden changes in influent load may not be noticed quickly enough for operational changes to be made. As a result, chronic over or under-removal of solids is likely.

Gould sludge age is sensitive to small changes in the wasting rate and doesn't work well for wastewaters that contain a large percentage of soluble BOD (Stall and Sherrard, 1978). For domestic wastewater, the ratio of influent *CBOD to TSS* is typically between 0.82 and 1.43 and they are often nearly equal (EnviroSim Associates, 2005; WEF et al., 2010). Gould was focused on the influent solids concentration because he thought that activated sludge worked only by flocculating and settling solids. He didn't understand that the influent BOD was food for the bacteria in the process. For domestic wastewater, this assumption worked reasonably well most of the time because TSS and CBOD were approximately equal. For wastewater from breweries and other types of food processing, much of the CBOD is soluble and was missed by Gould's equation. Gould's equation didn't work for these types of wastewater. A modification to the GSA calculation was proposed by the Water Environment Federation at one time that used influent BOD instead of influent TSS, but was never in common use. The inverse of the *modified GSA* equation is the F/M.

If the *CBOD to TSS* ratio is 0.85, this means the CBOD concentration is equal to 85% of the TSS concentration.

When you take the inverse of a fraction, you flip it upside down. The inverse of the fraction $\frac{3}{4}$ is $\frac{4}{3}$.

The *modified GSA* was calculated by taking the mass of MLSS in the activated sludge basins and dividing it by the mass of BOD in the influent. If you flip the calculation around and divide influent BOD by MLSS, you get the F/M. Neither GSA nor F/M work well for process control.

MEAN CELL RESIDENCE TIME

By 1955, many of the tools of modern-day process control were already in common use. Settleometer tests were being done in 1-L graduated cylinders and GSA calculations were performed daily at many facilities. Wasting rates were adjusted daily to achieve a target GSA (Torpey and Chasick, 1955). It was widely recognized that sludge age needed to be controlled within a narrow range to maintain sludge settleability and effluent quality. If the sludge age was too low, the sludge did not flocculate and compact well. If the sludge age was too high, the growth of filamentous bacteria caused sludge bulking and eventual process failure (Torpey and Chasick, 1955). Despite these advances, many activated sludge processes had difficulty controlling their effluent TSS and BOD. This was partly due to the reactive nature of the GSA and to its sensitivity to small changes in WAS rate.

Table 8.4 Solids Included in the Sludge Age Calculations

Sludge Age Definition	Influent TSS	Anaerobic and Anoxic Zones	Aerated Zones	Clarifier Blanket	Waste Solids
Gould sludge age (GSA)	Yes	Yes	Yes	Maybe	No
Mean cell residence rime (MCRT)	No	Yes	Yes	Yes	Yes
Solids retention time (SRT)	No	Yes	Yes	No	Yes
Aerobic solids retention time ($SRT_{aerobic}$)	No	No	Yes	No	Yes

Because the yield—*mass of MLVSS produced* per mass of BOD—is nearly constant for domestic wastewater, the concentration of MLVSS in the activated sludge basins will automatically increase or decrease to match the influent BOD load.

The first example of modern MCRT process control was published in 1958 by Garrett, who called it *hydraulic control*. Garrett's previous work showed that there is a relatively uniform relationship between the amount of BOD removed by activated sludge and the *amount of MLVSS produced* (Garrett and Sawyer, 1952). Recall the discussion on yield in Chapter 5 where about 0.42 kg of MLVSS are produced for every 1 kg of COD in the influent or roughly 0.67 kg of MLVSS for every 1 kg of BOD (Metcalf and Eddy, Inc./ AECOM, 2014). For domestic wastewater, the influent COD to BOD ratio is typically between 1.8 and 2.2 (EnviroSim Associates, 2005; WEF et al., 2010). Because the amount of MLVSS produced each day is directly related to the mass of BOD in the influent, Garrett and others proposed that the sludge age and growth rate could be controlled simply by controlling the amount of waste sludge removed from the system each day (Garrett, 1958). If the influent BOD load increases, more MLVSS will be produced and the MLVSS concentration will increase. Removing the same percentage of sludge from the system each day results in a constant sludge age. Long-term operation of a full-scale facility using this idea demonstrated that controlling the wasting rate also controlled the balance between the influent wastewater BOD load and MLVSS concentration and directly controlled effluent BOD concentrations (Garrett, 1958). In other words, controlling MCRT also controls the MLVSS concentration, F/M, and effluent quality.

Definition

The term *MCRT* was coined by Pearson (1966) and was later appropriated by Jenkins and Garrison (1968) for activated sludge process control. Mean is another word for average. Mean cell residence time is the average amount of time a bacterial cell spends in the activated sludge process. By using this specific terminology, a distinction is made between the TSS used in the GSA calculation and the biological solids in the MLVSS. Mean cell residence time was originally defined by Jenkins and Garrison as the total mass of volatile suspended solids (VSS) in the activated sludge process divided by the mass of the VSS leaving the system. Mass may be calculated in kilograms or pounds. The calculation includes the solids from all basins regardless of whether or not they are aerated as well as the solids in the clarifier blanket. Originally, the MCRT calculation included only MLVSS, but modern usage of MCRT uses MLSS rather than MLVSS. For most domestic WRRFs, the ratio of MLVSS to MLSS is fairly constant over long periods of time as long as 1) sludge age is well controlled and 2) the WRRF does not receive water treatment facility residuals or 3) add substantial quantities of inert material through chemical addition. The near constant, slowly changing ratio observed in many activated sludge processes allows for the use of MLSS, a simpler laboratory test, in place of MLVSS.

Calculation Method

Mean cell residence time is calculated using the following equation:

$$\text{MCRT, days} = \frac{\text{Activated Sludge Basin Mass} + \text{Clarifier Mass}}{\text{Waste Activated Sludge Mass} + \text{Effluent Solids Mass}} \tag{8.8}$$

Mass may be calculated as either kilograms or pounds. In a well-functioning activated sludge process, the mass of effluent solids should be small compared to the mass of solids present in the activated sludge basin, clarifier blanket, and WAS. The effluent solids can often be ignored and not be included in the calculation. When effluent TSS should be included in the calculation is discussed later in this chapter.

International Standard units for calculating mass:

$$\text{kg} = \frac{\left[\left(\text{Concentration, } \frac{\text{mg}}{\text{L}}\right)(\text{Volume, m}^3)\right]}{1000} \tag{8.9}$$

Alternatively, use dimensional analysis:

$$\left.\frac{X\ \text{mg}}{\text{L}}\right|\left.\frac{1\ \text{g}}{1000\ \text{mg}}\right|\left.\frac{1\ \text{kg}}{1000\ \text{g}}\right|\left.\frac{1000\ \text{L}}{1\ \text{m}^3}\right|\left.\frac{X\ \text{m}^3}{}\right| = \text{kg} \tag{8.10}$$

U.S. customary units for calculating mass:

To find pounds per day, use the flowrate in million gallons instead of volume in million gallons,

$$\text{Pounds} = \left(\text{Concentration, } \frac{\text{mg}}{\text{L}}\right)(\text{Volume, mil. gal})\left(8.34 \frac{\text{lb}}{\text{mil. gal}}\right) \quad (8.11)$$

Alternatively, use dimensional analysis:

$$\frac{X\text{ mg}}{\text{L}}\left|\frac{1\text{ g}}{1000\text{ mg}}\right|\left|\frac{1\text{ kg}}{1000\text{ g}}\right|\left|\frac{2.204\text{ lb}}{1\text{ kg}}\right|\left|\frac{3.785\text{ L}}{1\text{ gal}}\right|\left|\frac{1\,000\,000\text{ gal}}{1\text{ mil. gal}}\right|\left|X\text{ mil. gal}\right| = \frac{\text{lb}}{\text{mil. gal}} \quad (8.12)$$

To maintain an MCRT of 10 days, for example, the operator must remove one-tenth (10%) of the total mass of solids from the process each day. To maintain an MCRT of 20 days, the operator would remove one-twentieth (5%) of the total mass each day. Note that the influent flow to the WRRF does not affect the percentage of WAS removed each day. Mass or weight of solids removed is not the same as the volume or gallons removed. Because the WAS concentration can change frequently, it is critically important that the mass calculations be performed regularly.

Calculation Example

Find the MCRT for a treatment facility with the following characteristics:

Activated sludge basin volume = 7 570 m³ (2 mil. gal)

MLSS concentration = 2800 mg/L

Clarifier volume = 1514 m³ (0.4 mil. gal)

Clarifier solids concentration = 600 mg/L

WAS flowrate = 303 m³ (0.08 mgd)

RAS and WAS concentration = 7400 mg/L

Effluent TSS concentration = negligible

It can be helpful to draw a picture of the problem you are trying to solve. Then, label the picture with the information available. This can help define the problem better and allow you to see which pieces of information go together.

Solution with International Standard units:

Step 1—Find the mass of solids in the activated sludge basin.

$$\text{Activated Sludge Basin kg} = \frac{\left[\left(\text{Concentration, } \frac{\text{mg}}{\text{L}}\right)(\text{Volume, m}^3)\right]}{1000}$$

$$\text{Activated Sludge Basin kg} = \frac{\left[\left(2800 \frac{\text{mg}}{\text{L}}\right)(7570\text{ m}^3)\right]}{1000}$$

$$\text{Activated Sludge Basin kg} = 21\,196$$

Step 2—Find the mass of solids in the clarifier blanket.

$$\text{Clarifier Blanket kg} = \frac{\left[\left(\text{Concentration, } \frac{\text{mg}}{\text{L}}\right)(\text{Volume, m}^3)\right]}{1000}$$

$$\text{Clarifier Blanket kg} = \frac{\left[\left(600\frac{\text{mg}}{\text{L}}\right)(1514 \text{ m}^3)\right]}{1000}$$

$$\text{Clarifier Blanket kg} = 908.4$$

Step 3—Find the mass of solids wasted.

$$\text{WAS}\frac{\text{kg}}{\text{d}} = \frac{\left[\left(\text{Concentration, } \frac{\text{mg}}{\text{L}}\right)\left(\text{WAS Flow, } \frac{\text{m}^3}{\text{d}}\right)\right]}{1000}$$

$$\text{WAS}\frac{\text{kg}}{\text{d}} = \frac{\left[\left(7400\frac{\text{mg}}{\text{L}}\right)\left(303\frac{\text{m}^3}{\text{d}}\right)\right]}{1000}$$

$$\text{WAS}\frac{\text{kg}}{\text{d}} = 2242$$

Step 4—Solve for MCRT.

$$\text{MCRT, days} = \frac{\text{Activated Sludge Basin Mass} + \text{Clarifier Mass}}{\text{Waste Activated Sludge Mass} + \text{Effluent Solids Mass}}$$

$$\text{MCRT, days} = \frac{21\ 196 \text{ kg} + 908.4 \text{ kg}}{2242\frac{\text{kg}}{\text{d}} + 0}$$

$$\text{MCRT, days} = \frac{22\ 104 \text{ kg}}{2242\frac{\text{kg}}{\text{d}}}$$

$$\text{Mean Cell Residence Time, days} = 9.86$$

Solution with U.S. customary units:

Step 1—Find the mass of solids in the activated sludge basin.

$$\text{Activated Sludge Basin Pounds} = \left(\text{Concentration, } \frac{\text{mg}}{\text{L}}\right)(\text{Volume, mil. gal})\left(8.34\frac{\text{lb}}{\text{mil. gal}}\right)$$

$$\text{Activated Sludge Basin Pounds} = \left(2800\frac{\text{mg}}{\text{L}}\right)(2 \text{ mil. gal})\left(8.34\frac{\text{lb}}{\text{mil. gal}}\right)$$

$$\text{Activated Sludge Basin Pounds} = 46\ 704$$

Step 2—Find the mass of solids in the clarifier blanket.

$$\text{Clarifier Pounds} = \left(\text{Concentration, } \frac{\text{mg}}{\text{L}}\right)(\text{Volume, mil. gal})\left(8.34\frac{\text{lb}}{\text{mil. gal}}\right)$$

$$\text{Clarifier Pounds} = \left(600\frac{\text{mg}}{\text{L}}\right)(0.4\text{ mil. gal})\left(8.34\frac{\text{lb}}{\text{mil. gal}}\right)$$

$$\text{Clarifier Pounds} = 2002$$

Step 3—Find the mass of solids in the WAS.

$$\text{WAS}\frac{\text{lb}}{\text{d}} = \left(\text{Concentration, } \frac{\text{mg}}{\text{L}}\right)(\text{WAS Flow, mgd})\left(8.34\frac{\text{lb}}{\text{mil. gal}}\right)$$

$$\text{WAS}\frac{\text{lb}}{\text{d}} = \left(7400\frac{\text{mg}}{\text{L}}\right)(0.08\text{ mil. gal})\left(8.34\frac{\text{lb}}{\text{mil. gal}}\right)$$

$$\text{WAS}\frac{\text{lb}}{\text{d}} = 4937$$

Step 4—Solve for MCRT.

$$\text{MCRT, days} = \frac{\text{Activated Sludge Basin Mass} + \text{Clarifier Mass}}{\text{Waste Activated Sludge Mass} + \text{Effluent Solids Mass}}$$

$$\text{MCRT, days} = \frac{46\,704\text{ lb} + 2002\text{ lb}}{4937\frac{\text{lb}}{\text{d}} + 0}$$

$$\text{MCRT, days} = \frac{48\,706\text{ lb}}{4937\frac{\text{lb}}{\text{d}}}$$

$$\text{MCRT, days} = 9.86$$

When the target MCRT is known, the calculation is done in reverse to find the total mass and gallons that should be wasted each day. The operator determines that an MCRT of 9.9 days is too low and wishes to increase the MCRT to 12 days to maintain stable nitrification. Calculate the new wasting rate in gallons per day using the same information given in the previous example.

Solution with International Standard units:

Step 1—Find the total mass to be wasted.

$$\text{MCRT, days} = \frac{\text{Activated Sludge Basin Mass} + \text{Clarifier Mass}}{\text{WAS Mass} + \text{Effluent Solids Mass}}$$

$$12\text{ days} = \frac{21\,196\text{ kg} + 908.4\text{ kg}}{\text{WAS}\frac{\text{kg}}{\text{d}} + 0}$$

$$12\text{ days} = \frac{22\,104.4\text{ kg}}{\text{WAS}\frac{\text{kg}}{\text{d}}}$$

$$\text{WAS}\frac{\text{kg}}{\text{d}} = \frac{22\,104.4\text{ kg}}{12\text{ d}}$$

$$\text{WAS}\frac{\text{kg}}{\text{d}} = 1842$$

Step 2—Find the cubic meters of WAS to remove from the system each day.

$$\text{WAS}\frac{\text{kg}}{\text{d}} = \frac{\left[\left(\text{Concentration, }\frac{\text{mg}}{\text{L}}\right)\left(\text{WAS Flow, }\frac{\text{m}^3}{\text{d}}\right)\right]}{1000}$$

$$1842\frac{\text{kg}}{\text{d}} = \frac{\left[\left(7400\frac{\text{mg}}{\text{L}}\right)\left(\text{WAS Flow, }\frac{\text{m}^3}{\text{d}}\right)\right]}{1000}$$

$$\text{WAS Flow, }\frac{\text{m}^3}{\text{d}} = 248.9$$

Solution with U.S. customary units:

Step 1—Find the total mass to be wasted.

$$\text{MCRT, days} = \frac{\text{Activated Sludge Basin Mass} + \text{Clarifier Mass}}{\text{WAS Mass} + \text{Effluent Solids Mass}}$$

$$12 \text{ days} = \frac{46\,704 \text{ lb} + 2002 \text{ lb}}{\text{WAS}\frac{\text{lb}}{\text{d}}}$$

$$12 \text{ days} = \frac{48\,706 \text{ lb}}{\text{WAS}\frac{\text{lb}}{\text{d}}}$$

$$\text{WAS}\frac{\text{lb}}{\text{d}} = 4059$$

Step 2—Find the million gallons of WAS to remove from the system each day.

$$\text{WAS}\frac{\text{lb}}{\text{d}} = \left(\text{Concentration, }\frac{\text{mg}}{\text{L}}\right)(\text{WAS Flow, mgd})\left(8.34\frac{\text{lb}}{\text{mil. gal}}\right)$$

$$4059\frac{\text{lb}}{\text{d}} = \left(7400\frac{\text{mg}}{\text{L}}\right)(\text{WAS Flow, mgd})\left(8.34\frac{\text{lb}}{\text{mil. gal}}\right)$$

$$\text{WAS Flow, mgd} = 0.066$$

The U.S. Environmental Agency (U.S. EPA) created a graphical tool in 1977 to assist operators with daily sludge age calculations. Figure 8.33 and Figure 8.34 may be used in lieu of the calculations on the previous page(s) to estimate the wasting rate required to maintain a target sludge age. Keep in mind that calculations will be more accurate. This tool may be used with either International Standard units or U.S. customary units of measurement. The graphs are based on the SRT calculation rather than the MCRT calculation. Solids retention time does not include the solids in the secondary clarifier. Directions for using the tool are included on the figures. For daily use, consider copying Figure 8.33 and laminating it. Then, use a ruler and dry erase marker to find the wasting rate.

Advantages and Disadvantages

The MCRT, SRT, and $\text{SRT}_{\text{aerobic}}$ calculations are all based on the growth rates of the microorganisms within the activated sludge process. Unlike the GSA, each of these equations accounts for the transformation of influent BOD into MLVSS during treatment and they each provide a simple method for calculating the mass in kilograms or pounds of waste solids to remove from the process each day. Instead of trying to predict the incoming load, the operator needs only to maintain a constant percentage of solids removal from the process.

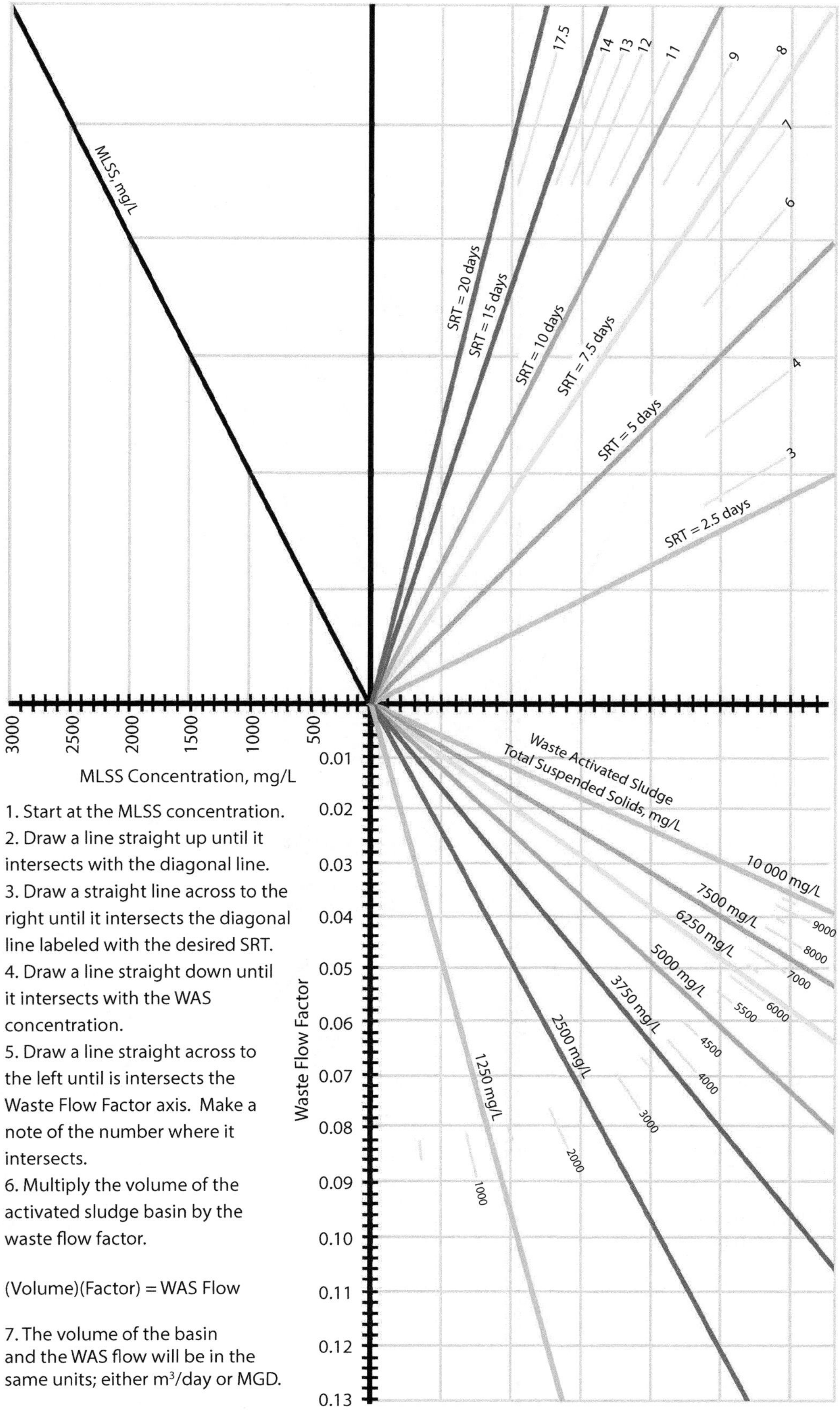
MLSS, mg/L
SRT = 20 days
SRT = 15 days
SRT = 10 days
SRT = 7.5 days
SRT = 5 days
SRT = 2.5 days
17.5
14
13
12
11
9
8
7
6
4
3
3000
2500
2000
1500
1000
500
MLSS Concentration, mg/L
Waste Flow Factor
0.01
0.02
0.03
0.04
0.05
0.06
0.07
0.08
0.09
0.10
0.11
0.12
0.13
Waste Activated Sludge
Total Suspended Solids, mg/L
10 000 mg/L
9000
7500 mg/L
8000
6250 mg/L
7000
5000 mg/L
6000
5500
3750 mg/L
4500
4000
2500 mg/L
3000
2000
1250 mg/L
1000
1. Start at the MLSS concentration.
2. Draw a line straight up until it intersects with the diagonal line.
3. Draw a straight line across to the right until it intersects the diagonal line labeled with the desired SRT.
4. Draw a line straight down until it intersects with the WAS concentration.
5. Draw a line straight across to the left until is intersects the Waste Flow Factor axis. Make a note of the number where it intersects.
6. Multiply the volume of the activated sludge basin by the waste flow factor.
(Volume)(Factor) = WAS Flow
7. The volume of the basin and the WAS flow will be in the same units; either m³/day or MGD.

Figure 8.33 Graphical Tool for Estimating Wasting Rate for a Given Target SRT (U.S. EPA, 1977)

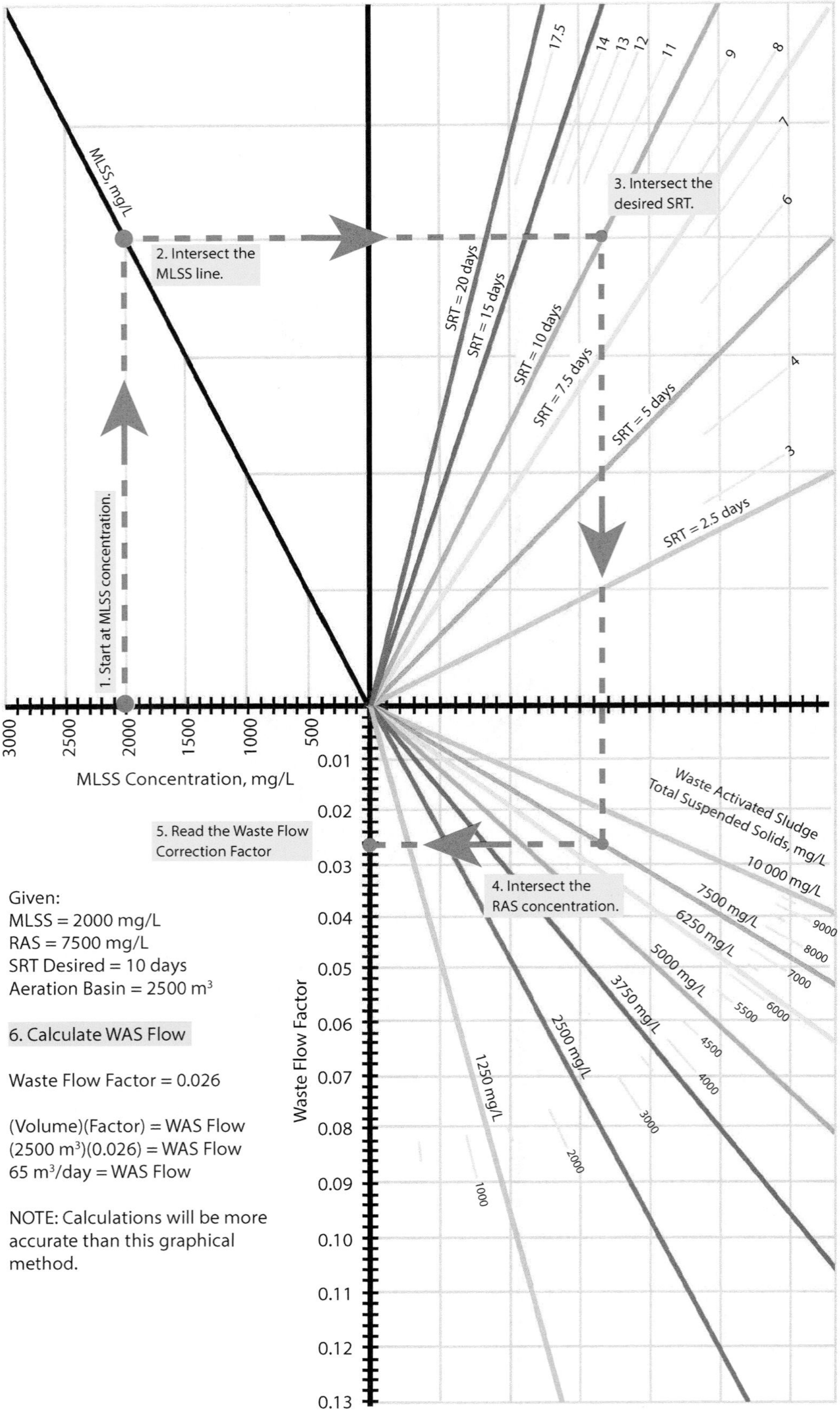

Figure 8.34 Example Calculation Using U.S. EPA Graphical Tool (U.S. EPA, 1977)

When BOD loading to the process increases or decreases, the amount of new microorganisms produced will automatically increase or decrease according to the yield. The operator does not need to monitor or predict the influent load. If the same percentage of total mass is removed from the system each day, then the mass of solids remaining in the system will adjust to the influent load automatically and the ratio of influent BOD to MLVSS (F/M) will remain relatively constant (Ekama, 2010). If the organic load to the process gradually increases, then the concentration of MLVSS will gradually increase to match the load. The converse is also true. Maintaining a constant sludge age can also reduce process upsets from sudden increases or spikes in organic loading rate.

Another advantage of using MCRT or SRT rather than the GSA or $SRT_{aerobic}$ is that much of the early research conducted on activated sludge processes were conducted using MCRT or SRT. For example, studies to determine which sludge age ranges are likely to support the growth of different types of filamentous bacteria reference MCRT or SRT. Early researchers were not very consistent about including or not including the solids in the clarifier blanket in their calculations. Continuing to use either MCRT or SRT allows for a comparison of current operating conditions with past research on filaments and helps operations staff know when a filament bloom might be likely to occur.

Most domestic WRRFs experience *diurnal* variations in their influent flow with the peak hour flow and load arriving between late morning and early afternoon. Exceptions include WRRFs with *flow equalization basins* or those with large service areas that dampen out the peak hour flow. When the peak flow enters the WRRF, it, along with the RAS flow, pushes MLSS from the activated sludge basins into the secondary clarifiers. This can cause the clarifier blankets to increase during portions of the day at some facilities. Because MCRT includes the solids in all basins, whether they are aerated or not, and in the clarifier blanket, grab samples can be collected from each location without regard to the time of day taken. Because all locations are monitored, the total sludge inventory will be accounted for.

Diurnal is a repeating pattern that occurs over a 24-hour period.

Flow equalization basins help to even out the flow and load to a WRRF. During periods of high flow or load, the basins are allowed to fill. Water is pumped out at a constant rate. The water in the tank rises and falls over a 24-hour period.

A potential disadvantage of using MCRT, SRT, or $SRT_{aerobic}$ for process control is that each of these calculations requires accurate and precise sampling and analysis of the MLSS in the activated sludge basins and WAS. The RAS/WAS concentrations often change during the day and it can be difficult to obtain representative samples from an intermittently operating WAS pump. Accurate measurement of the MLSS and WAS concentrations requires training and specialized laboratory equipment such as an analytical balance, drying oven, vacuum pump, dessicator, and/or solids meter, which may be beyond the economic reach of some very small systems. Centrifuge spin analysis may also be used to estimate MLSS and WAS concentrations; however, this method is less accurate. In the case of MCRT, it may not be possible to obtain a representative sample of the clarifier blanket if the clarifier is covered by panels or a dome. For systems that are unable to sample their clarifiers, SRT or $SRT_{aerobic}$ may be more appropriate.

Control of MCRT/SRT may be difficult to fully implement for systems that aren't fully staffed. Ideally, operators will set a target MCRT/SRT and then collect and analyze samples each day for the MCRT/SRT calculation. Wasting rates should then be adjusted accordingly to maintain constant MCRT/SRT. Systems that are not fully staffed may not be able to collect the necessary data or perform the calculation daily. There are ways to compensate for less frequent MCRT/SRT calculations, including using a 3-, 5-, or 7-day running average MCRT/SRT. Using a running average helps to minimize the effects of a single bad sampling event or a missing data point. A running average also provides a more realistic value for changes to wasting rates.

SOLIDS RETENTION TIME

Definition

The terms *SRT, MCRT*, and *sludge age* were used interchangeably throughout the early days of activated sludge and, in many cases, clarifier solids were included or not in both calculations in different publications (Stall and Sherrard, 1978; U.S. EPA, 1977). Today, MCRT is understood to include the solids in all activated sludge basins and the clarifier blanket, whereas the SRT includes only the solids in the activated sludge basins. It is nearly impossible to pinpoint exactly when the terms MCRT and SRT took on their currently accepted definitions, but it appears that the first official definition of SRT was in the 1974 U.S. EPA manual, *Process Design Manual for Upgrading Existing Wastewater Treatment Plants*. This manual defined SRT as the mass of MLSS under aeration divided by the mass removed in the WAS and effluent each day (eq 8.13). Here, solids are used in place of "mean cell" because all of the solids in the activated sludge basin are being included in the calculation, not just the volatile solids. This may seem like a fine point to make; however, as will be discussed later in this chapter, inert solids (sand, grit, eggshells, fibers, etc.) can be a substantial percentage of the MLSS and they do not contribute to treatment. As late as 1977, there was considerable debate among researchers, engineers, and operators regarding whether or not the solids in the clarifier blanket should be included in the sludge age calculation (U.S. EPA, 1977). This debate continues today. There are multiple arguments for and against including the clarifier solids in the calculations, as shown in Table 8.5. There is a move in the wastewater community to standardize on SRT rather than MCRT, particularly among design engineers. Many operator certification exams and state programs continue to use MCRT.

Calculation Method

The calculation method for SRT is almost identical to the calculation for MCRT. The only difference is that the solids in the clarifier blanket are no longer included.

$$\text{SRT, days} = \frac{\text{Aeration Basin Mass}}{\text{WAS Mass} + \text{Effluent Solids Mass}} \tag{8.13}$$

Mass may be calculated as either kilograms or pounds. In a well-functioning activated sludge process, the mass of effluent solids should be small compared to the mass of solids present in the activated sludge basin, clarifier blanket, and WAS. The effluent solids can often be ignored and not be included in the SRT calculation. Refer to the section on MCRT for example calculations.

Table 8.5 Inclusion of Clarifier Blanket Solids in the Sludge Age Calculation

Arguments Against Inclusion	Arguments for Inclusion
Collecting a representative sample from a clarifier blanket is not always possible. Clarifiers with covers or those in buildings with low ceilings can be difficult to access with a long core sampler.	Clarifier blanket depth can be estimated with a hand-held TSS meter or by lowering a modified core sampler into the clarifier on a string. Once the blanket depth is known, the mass of solids in the clarifier can be calculated if the blanket concentration is assumed to be equal to the RAS concentration.
Microorganisms in the clarifier blanket do not have access to food and oxygen. They do not contribute to treatment while in the clarifier.	Microorganisms in the clarifier blanket are only in the blanket for a short period of time each cycle. They are exposed to both food and oxygen most of the time and do contribute to overall treatment capacity.
Assuming the clarifier blankets are maintained below 2 ft, the percentage of solids in the clarifier is insignificant compared to the total inventory. Ignoring these solids won't significantly change the results of the sludge age calculation. When blanket levels are fairly constant, the error in the sludge age calculation will also be constant.	For facilities that carry deeper sludge blankets in the clarifier, experience poor sludge settleability, or that have large clarifiers relative to their activated sludge basins, the portion of solids in the clarifier blanket can be significant. Ignoring the solids in these situations can have a dramatic effect on the results of the sludge age calculation for some facilities. To be clear, deep sludge blankets in clarifiers are not recommended because this practice can degrade effluent quality and affect MLSS behavior.
Design engineers use process kinetics to design and size the activated sludge basins. These calculations only include the solids in the basins and not in the clarifier.	The kinetics equations used to size the activated sludge basins predict the total mass of solids produced. This is the yield discussed in Chapter 2. Although some of those solids may be in the clarifier temporarily, they are still included in the design equations even though engineers often assume 100% of the sludge inventory remains in the activated sludge basins.
Loading parameters (kg BOD /m^3 or lb BOD/ 1000 cu ft) for the activated sludge process were originally developed using only the MLSS in the activated sludge basin and not the clarifiers.	Again, kinetics dictate the mass of MLSS produced per mass of BOD or COD received. Some of the MLSS may temporarily be in the clarifier, but it was produced in response to the influent load.

Advantages and Disadvantages

The advantages of using SRT for process control are the same as those listed for MCRT control in the previous section of this chapter and in Table 8.5. One additional advantage of using SRT over MCRT is that there is no need to collect and analyze samples from the clarifier blankets. This reduces the total amount of time spent on sampling and analysis and can be helpful for smaller facilities with limited staff.

A disadvantage of SRT control over MCRT control is that sampling may not always account for a majority of the *solids inventory*. If an operator goes out to take a grab sample at 7:00 a.m. every morning and uses that data to calculate total mass of MLSS in the activated sludge basins, they can be confident that a majority of the MLSS has been accounted for by that sample. There should be a very small amount of MLSS in the clarifier blankets because peak flow, for most WRRFs, will not arrive until later in the day. Recall that the influent flow and RAS flow push MLSS from the activated sludge basins into the clarifiers. On a day when multiple interruptions occur—early morning safety meeting, Alex calls in sick, and a water main break requires an emergency response—the 7:00 a.m. grab sample may not be collected until after peak hour flow starts entering the WRRF. Now, the concentration of MLSS in the activated sludge basin may be slightly lower because there are more solids (potentially) sitting in the clarifier blanket. When the operator does the SRT calculation, the mass to waste will naturally be lower for that day and sludge age will increase for that day. Although the differences in the MLSS concentration and SRT calculation result shouldn't be large, they can create more variation in process control charts and may cause the operator to make unnecessary process changes. If, on the other hand, the operator collects grab samples from the activated sludge basins and clarifiers at about the same time (a few minutes apart), then 100% of the *sludge inventory* has been accounted for regardless of the time of day the sample is collected or the influent flow regimen. When clarifiers are operated to keep the sludge blanket below 2 ft at all times, the differences between the MCRT and SRT calculation results should be small regardless of the time of day samples are collected for most facilities. An exception to this rule is facilities that have large clarifiers relative to their activated sludge basins. Composite samplers may be used to collect samples of MLSS, RAS, and WAS, which can help to eliminate inaccuracies associated with grab samples.

Sludge inventory refers to the mass of MLSS in the activated sludge process.

AEROBIC SOLIDS RETENTION TIME

Design engineers use $SRT_{aerobic}$ to determine the size of the aerated basins for facilities that must remove ammonia. The bacteria responsible for converting ammonia to nitrite and nitrate, the nitrifying bacteria, grow much slower than the bacteria responsible for removing BOD, especially at colder water temperatures. The nitrifiers are strict aerobes and cannot function without an adequate supply of DO. Biochemical oxygen demand removal, on the other hand, can take place under anaerobic, anoxic, and aerated conditions. The characteristics of the nitrifying bacteria focus the design and operation of nitrifying activated sludge processes on the amount of aerobic activated sludge basin volume or time. Some WRRFs use $SRT_{aerobic}$ for process control, whereas others calculate both the total SRT and the $SRT_{aerobic}$.

Definition

The aerobic SRT is defined as the total SRT multiplied by the ratio of the total activated sludge basin volume to the aerated volume (Argaman and Brenner, 1986). The aerated volume is further defined as any basin or zone containing DO at concentrations greater than 0.2 mg/L. The $SRT_{aerobic}$ calculation does not include any MLSS that may be in anaerobic zones or basins, anoxic zones or basins, or in the clarifier blanket. For an SBR-type activated sludge process, the $SRT_{aerobic}$ includes only the amount of time when the DO concentration is above 0.2 mg/L. The concentration of 0.2 mg/L was selected as the cutoff point between aerobic and anoxic conditions because this is the concentration of DO found to inhibit denitrification (Knowles 1982). Chapter 5 gives more in-depth definitions of anaerobic, anoxic, and aerated conditions.

Calculation Method

For an activated sludge process with multiple basins or zones:

$$SRT_{aerobic}\text{, days} = \text{Solids Retention Time}_{Total}\left(\frac{\text{Volume}_{aerobic}}{\text{Volume}_{total}}\right) \quad (8.14)$$

For an SBR:

$$\text{SRT}_{\text{aerobic}},\ \text{days} = \text{Solids Retention Time}_{\text{Total}}\left(\frac{\text{Time}_{\text{aerobic}}}{\text{Time}_{\text{total}}}\right) \quad (8.15)$$

Calculation Example

Find the $\text{SRT}_{\text{aerobic}}$ for a treatment facility with the following characteristics:

Total activated sludge basin volume = 7570 m^3 (2 mil. gal)

Aerated basin volume = 4921 m^3 (1.3 mil. gal)

Anoxic basin volume = 2650 m^3 (0.7 mil. gal)

MLSS concentration = 2800 mg/L

WAS flowrate = 303 m^3/d (0.08 mgd)

WAS concentration = 7400 mg/L

Effluent TSS concentration = negligible

Solution with International Standard units:

Step 1—Find the mass of solids in the activated sludge basin.

$$\text{Activated Sludge Basin kg} = \frac{\left[\left(\text{Concentration, }\frac{\text{mg}}{\text{L}}\right)(\text{Volume, m}^3)\right]}{1000}$$

$$\text{Activated Sludge Basin kg} = \frac{\left[\left(2800\frac{\text{mg}}{\text{L}}\right)(7570\text{ m}^3)\right]}{1000}$$

$$\text{Activated Sludge Basin kg} = 21\,196$$

Step 2—Find the mass of solids wasted.

$$\text{WAS}\frac{\text{kg}}{\text{d}} = \frac{\left[\left(\text{Concentration, }\frac{\text{mg}}{\text{L}}\right)\left(\text{WAS Flow, }\frac{\text{m}^3}{\text{d}}\right)\right]}{1000}$$

$$\text{WAS}\frac{\text{kg}}{\text{d}} = \frac{\left[\left(7400\frac{\text{mg}}{\text{L}}\right)\left(303\frac{\text{m}^3}{\text{d}}\right)\right]}{1000}$$

$$\text{WAS}\frac{\text{kg}}{\text{d}} = 2242$$

Step 3—Solve for SRT.

$$\text{SRT, days} = \frac{\text{Activated Sludge Basin Mass}}{\text{WAS Mass} + \text{Effluent Solids Mass}}$$

$$\text{SRT, days} = \frac{21\ 196\ \text{kg}}{2242\dfrac{\text{kg}}{\text{d}} + 0}$$

$$\text{SRT, days} = \frac{21\ 196\ \text{kg}}{2242\dfrac{\text{kg}}{\text{d}}}$$

$$\text{SRT, days} = 9.45$$

Step 4—Find the aerobic SRT.

$$\text{SRT}_{\text{aerobic}}\text{, days} = \text{Solids Retention Time}_{\text{Total}}\left(\frac{\text{Volume}_{\text{aerobic}}}{\text{Volume}_{\text{total}}}\right)$$

$$\text{SRT}_{\text{aerobic}}\text{, days} = 9.45\ \text{days}\left(\frac{4921\ \text{m}^3}{7570\ \text{m}^3}\right)$$

$$\text{SRT}_{\text{aerobic}}\text{, days} = 6.14\ \text{days}$$

Solution with U.S. customary units:

Step 1—Find the mass of solids in the activated sludge basin.

$$\text{Activated Sludge Basin Pounds} = \left(\text{Concentration, } \frac{\text{mg}}{\text{L}}\right)(\text{Volume, mil. gal})\left(8.34\frac{\text{lb}}{\text{mil. gal}}\right)$$

$$\text{Activated Sludge Basin Pounds} = \left(2800\frac{\text{mg}}{\text{L}}\right)(2\ \text{mil. gal})\left(8.34\frac{\text{lb}}{\text{mil. gal}}\right)$$

$$\text{Activated Sludge Basin Pounds} = 46\ 704$$

Step 2—Find the mass of solids in the WAS.

$$\text{WAS}\frac{\text{lb}}{\text{d}} = \left(\text{Concentration, } \frac{\text{mg}}{\text{L}}\right)(\text{WAS flow, mgd})\left(8.34\frac{\text{lb}}{\text{mil. gal}}\right)$$

$$\text{WAS}\frac{\text{lb}}{\text{d}} = \left(7400\frac{\text{mg}}{\text{L}}\right)(0.08\ \text{mil. gal})\left(8.34\frac{\text{lb}}{\text{mil. gal}}\right)$$

$$\text{WAS}\frac{\text{lb}}{\text{d}} = 4937$$

Step 3—Solve for SRT.

$$\text{SRT, days} = \frac{\text{Activated Sludge Basin Mass}}{\text{WAS Mass} + \text{Effluent Solids Mass}}$$

$$\text{SRT, days} = \frac{46\ 704\ \text{lb}}{4937\dfrac{\text{lb}}{\text{d}} + 0}$$

$$\text{SRT, days} = \frac{46\ 704\ \text{lb}}{4937\dfrac{\text{lb}}{\text{d}}}$$

$$\text{SRT, days} = 9.46$$

Step 4—Find the $SRT_{aerobic}$.

$$SRT_{aerobic} = \text{Solids Retention Time}_{Total}\left(\frac{\text{Volume}_{aerobic}}{\text{Volume}_{total}}\right)$$

$$SRT_{aerobic} = 9.46 \text{ days}\left(\frac{1.3 \text{ mil. gal}}{2 \text{ mil. gal}}\right)$$

$$SRT_{aerobic} = 6.15 \text{ days}$$

The SRT calculation done here uses the same WRRF data as the MCRT calculation in the previous section. In this example, the SRT was 9.45 days while the MCRT was 9.86 days. Excluding the solids in the clarifier blanket made very little difference in the sludge age calculation because the percentage of sludge in the clarifier blanket was small.

Advantages and Disadvantages

There are no real disadvantages of $SRT_{aerobic}$ other than those already given in the section on SRT. The advantages of using $SRT_{aerobic}$ for process control are the same as those listed for MCRT and SRT control earlier in this chapter and in Table 8.5. Water resource recovery facilities that are required to remove ammonia may calculate $SRT_{aerobic}$ to ensure that there is enough aerated volume and/or time to support full and stable nitrification. For activated sludge facilities that were designed with separate basins or zones for anaerobic, anoxic, and aerated conditions, the $SRT_{aerobic}$ calculation was done by the design engineer and the ratio of $SRT_{aerobic}$ to SRT_{total} is fixed. Although the operators are unable to change the ratio in this case, they are able to indirectly control the $SRT_{aerobic}$ by controlling SRT_{total}. For activated sludge facilities that practice on/off aeration and for SBR, the ratio of $SRT_{aerobic}$ to SRT_{total} is controlled in time rather than in space. In these situations, the operator has more direct control over the $SRT_{aerobic}$ and can adjust it to almost any percentage of the total. For additional discussion on minimum $SRT_{aerobic}$ required for nitrification, refer to the section of this chapter on selecting a target sludge age.

TEST YOUR KNOWLEDGE

1. **In simplest terms, sludge age is simply the length of time it takes to drain an activated sludge basin.**
   - ☐ True
   - ☐ False

2. **Because yield, the mass of MLVSS produced per mass of BOD, is nearly constant for domestic wastewater, the MLSS concentration will increase and decrease with the influent BOD load even when sludge age is held constant.**
   - ☐ True
   - ☐ False

3. **This method is recommended for activated sludge process control.**
   a. Sludge age
   b. Constant MLSS mass
   c. Constant F/M
   d. RAS control

4. **Gould sludge age was developed in the 1920s and was based on the idea that**
   a. Activated sludge is primarily a physical–chemical treatment process.
   b. Activated sludge is primarily a biological treatment process.
   c. Influent soluble BOD is critical to activated sludge performance.
   d. Influent TSS is transformed in the activated sludge basin to MLSS.

5. **Gould sludge age is calculated by dividing the mass of MLSS in the activated sludge basins by**
   a. Mass of WAS leaving the system
   b. Mass of influent TSS
   c. Mass of influent BOD
   d. Mass of RAS in the return

6. **In the term *MCRT*, mean is another word for**
   a. Aggressive
   b. Average
   c. Acclimated
   d. Aerodynamic

7. **Controlling the amount of WAS leaving the system each day controls the amount of MLVSS available to treat the incoming organic load. Therefore, controlling sludge age also controls**
   a. Growth rate
   b. Yield
   c. Influent TSS
   d. Return rates

8. **Sludge age is held constant. If the influent BOD concentration increases, the MLSS concentration will**
   a. Remain about the same
   b. Increase proportionally to the load
   c. Decrease in percent volatile solids
   d. Equal the influent BOD time sludge age

9. **The desired MCRT is 10 days. If MLSS is removed directly from the activated sludge basin, what percentage of the MLSS must be removed each day?**
   a. 5%
   b. 10%
   c. 20%
   d. 25%

10. **Under what condition will maintaining a constant MLSS concentration approximate constant sludge age?**
    a. Volume of WAS removed each day is constant.
    b. RAS is a fixed percentage of influent flow
    c. Effluent solids remain below 30 mg/L.
    d. Influent loads are nearly constant.

11. **One advantage of using MCRT over SRT is**
    a. Additional samples are required
    b. 100% of the sludge inventory is accounted for
    c. Difficulty in sampling clarifier sludge blankets
    d. Enables sludge storage in the clarifiers

12. **Estimate the desired wasting rate from Figure 8.33 given the following information: MLSS concentration is 2000 mg/L, target sludge age is 7.5 days, WAS concentration is 6250 mg/L, and the activated sludge basin holds 2000 m^3 (0.53 mgd).**
    a. 50 m^3 (0.013 mgd)
    b. 160 m^3/d (0.044 mgd)
    c. 320 m^3/d (0.088 mgd)
    d. 88 m^3/d (0.023 mgd)

13. **The SRT is currently at 6 days. The target SRT is 8 days. The operator should**
    a. Increase the WAS
    b. Decrease the WAS
    c. Increase the RAS
    d. Decrease the RAS

14. **Mean cell residence time and SRT are both calculated by dividing the mass of sludge in the system by the mass of sludge __________.**
    a. In the influent
    b. In the effluent
    c. Leaving the system
    d. In the digester

15. **An activated sludge process contains 8167 kg (18 000 lb) of MLSS in the activated sludge basins and another 2269 kg (5000 lb) in the secondary clarifiers. If the operator removes 907 kg (2000 lb) of MLSS in the WAS each day, what is the MCRT?**
    a. 3.6 days
    b. 4 days
    c. 9 days
    d. 11.5 days

16. **Given the following information, find the SRT.**

    Activated sludge basin volume: 11 355 m^3 (3 mil. gal)
    MLSS concentration: 2500 mg/L
    Wasting rate: 436 m^3/d (80 gpm)
    WAS/RAS concentration: 8000 mg/L
    a. 5 days
    b. 8 days
    c. 11 days
    d. 14 days

WHEN TO INCLUDE EFFLUENT TOTAL SUSPENDED SOLIDS

The Clean Water Act set *secondary treatment standards* for BOD_5 and TSS removal. For mechanical facilities like activated sludge processes, the final effluent concentrations for BOD_5 and TSS can't exceed 30 mg/L for the 30-day average concentration or 45 mg/L for the 7-day average concentration (40 CFR § 133.102). These limits represent the expected level of effluent quality attainable. Many, if not most, activated sludge processes perform much better than this and routinely discharge BOD and TSS concentrations below 10 mg/L. For a WRRF that is meeting the secondary treatment standards, the mass of TSS leaving in the final effluent should be very small compared to the mass in the activated sludge basin(s) and clarifier(s). Table 8.6 compares the effects of including, or not including, the effluent TSS in both the SRT and MCRT calculations for a conventional activated sludge system. Facility data are given in Figure 8.35 of this chapter. For this example, an influent flow of 30 280 m^3/d (8 mgd) was used for an HRT in the activated sludge basin of 6 hours at average daily flow.

The *secondary treatment standards* for mechanical treatment facilities, including activated sludge processes, limit effluent TSS and BOD_5 concentrations to less than 30 mg/L for the monthly average and 45 mg/L for the weekly average.

Table 8.6 Effect of Effluent TSS on MCRT and SRT Calculations

Effluent TSS Concentration and Mass	MCRT, Days		SRT, Days	
	Effluent TSS Included	Effluent TSS Not Included	Effluent TSS Included	Effluent TSS Not Included
15 mg/L 454.2 kg (1001 lb)	8.2	9.9	7.9	9.5
30 mg/L 908.4 kg (2001.6 lb)	7.0	9.9	6.7	9.5
45 mg/L 1362.6 kg (3002.4 lb)	6.1	9.9	5.9	9.5
60 mg/L 1816.8 kg (4003.2 lb)	5.5	9.9	5.2	9.5
100 mg/L 3028.0 kg (6672.0 lb)	4.2	9.9	4.2	9.9

Effluent TSS will have a bigger effect on the SRT and MCRT calculations with increasing sludge age and as effluent TSS concentrations increase.

When the effluent TSS is below 15 mg/L, the effect of *effluent TSS* on the sludge age calculations is 1.7 days. In this case, effluent TSS could be safely left out of the SRT or MCRT calculation for most facilities. The section on how to select a target SRT/MCRT of this chapter discusses minimum sludge ages required to maintain stable nitrification at different water temperatures. It is possible for a facility to nitrify well at an SRT of 9.5 days, but be unable to nitrify at an SRT of 7.9 days. In this situation, the effluent TSS concentration could have a huge effect on whether or not the WRRF is able to continue meeting its discharge permit limits. When determining whether or not to include effluent TSS in the MCRT or SRT calculation, operators must take permit limits into consideration as well as the need to select a sludge age that provides an adequate buffer for errors in sampling, analysis, and calculations. Operators may also choose to leave effluent TSS out of their calculations and consistently operate with a slightly higher sludge age to compensate. With this strategy, facilities with effluent ammonia limits should keep careful watch on effluent ammonia concentrations. If they increase even slightly, the sludge age may need to be increased. As effluent TSS increases and the facility is no longer in compliance, the effluent TSS has a larger and larger effect on the SRT and MCRT calculations. As sludge age increases beyond 10 days, the effect of effluent TSS also increases because less sludge is being removed from the system overall and the mass of solids leaving in the final effluent may be nearly the same as the mass of solids leaving in the WAS.

Larger facilities often have daily sampling requirements and will collect a sample for effluent TSS every day. If the data are available, it should be included in the MCRT or SRT calculation for greater accuracy. Including the effluent TSS routinely will provide a more accurate record over time, especially during

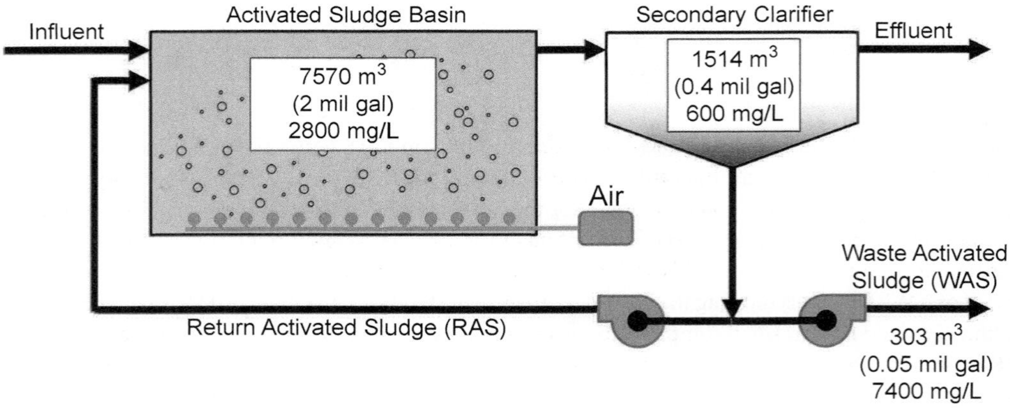

Figure 8.35 Process Data for MCRT and SRT Calculations (Reprinted with permission by Indigo Water Group)

periods of process upset. Some facilities may wish to exclude effluent TSS from their daily process calculations for regulatory reasons. Generally, if samples of the final effluent are collected and analyzed more frequently than is required by the discharge permit, the sample results must be included in the monthly monitoring report to the state or U.S. EPA. Many states exempt samples collected solely for process control purposes if they are taken in a different manner or location than the permit requires and/or if they are analyzed by a non-U.S. EPA approved method. A sample collected from the secondary clarifier effluent may be interpreted by a regulatory agency as representative of the final effluent even though it is collected upstream of disinfection because secondary treatment is already complete. Collecting grab samples for process control can complicate permit reporting because they may not be representative of average day operation, but a regulatory agency may wish to have the results included in the monthly monitoring report. *Be sure to check with your regulatory agency and report all sample results as required.*

Some activated sludge facilities have settling ponds or tertiary filters that remove solids from the process without recycling them back to the activated sludge basins. It may be necessary to include these solids in the SRT and MCRT calculations as well.

POTENTIAL EFFECTS OF FOAMING EVENTS ON SLUDGE AGE

During a foaming event, it is possible for a significant portion of the sludge inventory to be entrained in the foam. Foam solids are not easily quantified and aren't typically included as part of the total sludge inventory. As a result, during a foaming event, the sludge age may be higher than the sludge age calculations indicate. In extreme foaming events, the foam may overflow the activated sludge basins and end up outside of the basins on the ground and walkways. The loss of solids decreases sludge age; however, because this lost foam is almost impossible to measure, it may not be accounted for in sludge age calculations. In this case, the sludge age may be much lower than the sludge age calculations indicate.

WASTING FREQUENCY

The SRT or MCRT should be calculated at least once per day and wasting rates adjusted accordingly to maintain a constant or near constant SRT/MCRT. Wasting over longer periods of time, 10 minutes out of every hour for example, is preferable because this practice maintains a more constant SRT/MCRT, which, in turn, maintains sludge settleability. Mass wasting of solids should be avoided because this does not control SRT/MCRT and can cause sudden changes in sludge settleability. As a general rule, the mass of solids wasted from the system each day should not change by more than about 15% over the previous day. This is kilograms or pounds removed, not gallons.

Wasting frequency and duration at some facilities may be limited by staffing or by pumping capacity. For example, if the maximum WAS pumping rate is 5 m^3/min and the operator needs to waste 4000 m^3/d to maintain a desired SRT, then wasting will have to be spread out over 800 minutes (13.3 hours). Package facilities are small, typically skid-mounted or trailer type, activated sludge facilities that are often used to treat wastewater from small areas like an individual business, trailer park, camp, or subdivision. Package facilities often lack automatic controls and may only be staffed a few days a week. Fortunately, many package facilities do not have ammonia limits in their discharge permits, so tight control over SRT is less important than it is for larger facilities. Still, the SRT must be controlled well enough for the MLSS to flocculate, settle, and compact well. Wasting should be done at as high a frequency as possible—daily being ideal and several days each week being less ideal.

In the early days of activated sludge, extended aeration *activated sludge processes* were allowed to "self-waste" into the final effluent. This operational "technique" is documented in several U.S. EPA manuals and operator training guides. Self-wasting occurs when the MLSS concentration increases to the point where the clarifier is no longer capable of settling those solids. The blanket gradually builds in the clarifier, and eventually MLSS flows out over the weirs and into the final effluent. This is referred to as "blowing the blanket". Self-wasting will continue until the MLSS concentration decreases to a point where the clarifier becomes once again capable of settling. Repeat the cycle. A self-wasting activated sludge process would never be allowed today because operating in this way does not meet the secondary treatment standards for BOD_5 and TSS. Facilities that are staffed only one or two days a week may be self-wasting when the operator is not there. Owners should be encouraged to staff package facilities and other small systems adequately so that intentional wasting can be done, at a minimum, several days each week.

An *activated sludge process* should never be allowed to self-waste by losing solids over the clarifier weir to the final effluent in excess of the discharge permit limits.

Example #1—The influent flow and load to a WRRF is relatively constant with few changes from day to day. The operations staff is targeting an SRT of 5 days because this SRT has historically produced the best-settling MLSS and the lowest turbidity effluent. The SRT is calculated every day, Monday through Friday. The operator then sets the wasting pump to run for a certain number of minutes out of each hour over a 24-hour period. The WAS pump is controlled by a computer. The number of minutes is adjusted each day to match the WAS volume from the SRT calculation. The WRRF is not staffed on the weekends. On Saturday and Sunday, the operator uses the average wasting rate for the week because they won't be there to adjust the cycles.

Example #2—A smaller WRRF does not have the ability to program their wasting pump and is only staffed during the day. The influent flow and load are fairly constant from day to day. The operator knows from past experience that they will need to waste between 300 and 800 m^3 each day. Wasting for a few minutes of every hour would require the facility operator to manually turn the pump on and off each time. Instead, the operator elects to waste from the system twice each day. In the morning, they collect samples so they can perform the SRT calculation. At the same time, they start the WAS pump and waste 300 m^3 while they analyze the process control samples and perform the SRT calculation. The SRT calculation indicates that an additional 375 m^3 of WAS should be removed to maintain the target SRT. Just before going home toward the end of the shift, the operator starts the WAS pump and removes the additional WAS.

WHEN TO USE MEAN CELL RESIDENCE TIME, SOLIDS RETENTION TIME, OR AEROBIC SOLIDS RETENTION TIME

Whether a facility chooses to use $SRT_{aerobic}$, SRT, or MCRT for process control depends, in part, on how their process is configured and whether ammonia removal is required by the discharge permit (Figure 8.36). An SBR with an effluent ammonia limit will find $SRT_{aerobic}$ useful for controlling the amount of air-on time in a treatment cycle as will facilities with the ability to independently control the airflow to different portions of their activated sludge basins or to operate some basins as either anoxic or aerobic. For a facility with small activated sludge basins and large clarifiers, MCRT may be more appropriate because even when the clarifier sludge blanket is below 0.6 m (2 ft) of depth, a significant portion of the MLSS may be in the clarifier. Facilities with large activated sludge basins, like oxidation ditches, may prefer to use SRT because it requires less sampling and analysis and, even when the clarifiers contain several feet of blanket, the majority of the MLSS will be in the activated sludge basin.

Example #1—An extended aeration oxidation ditch activated sludge facility was designed for an HRT of 24 hours. The influent flow is 9462.5 m^3 (2.5 mgd) and the oxidation ditch volume is 9462.5 m^3 (2.5 mil. gal). There are two secondary clarifiers that are 18.3 m (60 ft) in diameter and 3.65-m (12-ft) deep. The MLSS concentration is 3000 mg/L and the RAS concentration is 7800 mg/L. The clarifier blanket depth is 1.5 m (5 ft). Calculate the percentage of MLSS in the oxidation ditch and clarifier. Assume that the RAS concentration and the sludge blanket concentration are approximately equal.

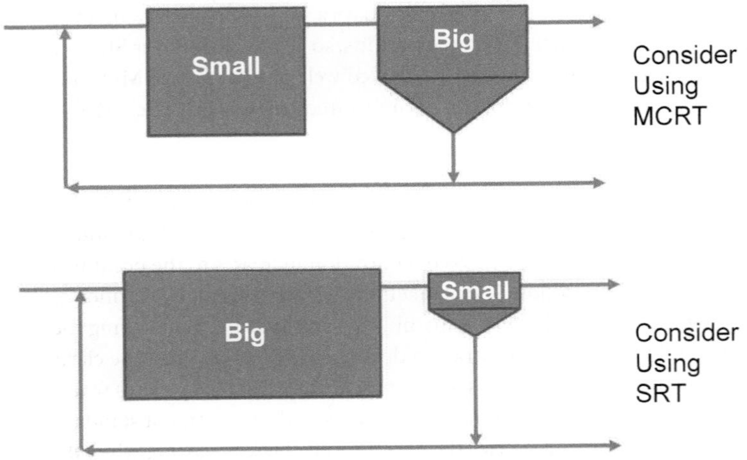

Figure 8.36 Activated Sludge Basin Size Versus Clarifier Size (**Reprinted with permission by Indigo Water Group**)

Step 1—Find the mass of solids in the activated sludge basin.

$$\text{Activated Sludge Basin kg} = \frac{\left[\left(\text{Concentration, } \frac{\text{mg}}{\text{L}}\right)(\text{Volume, m}^3)\right]}{1000}$$

$$\text{Activated Sludge Basin kg} = \frac{\left[\left(3000\frac{\text{mg}}{\text{L}}\right)(9462.5\text{ m}^3)\right]}{1000}$$

$$\text{Activated Sludge Basin kg} = 28\,388$$

Step 2—Find the volume of the clarifier.

$$\text{Volume} = 0.785(\text{Diameter}^2)(\text{Depth})(2\text{ Clarifiers})$$

$$\text{Volume} = 0.785(18.3\text{ m})^2(3.65\text{ m})(2\text{ Clarifiers})$$

$$\text{Volume} = 1919\text{ m}^3$$

Step 3—Find the mass of solids in the clarifier blanket.

$$\text{Clarifier kg} = \frac{\left[\left(\text{Concentration, } \frac{\text{mg}}{\text{L}}\right)(\text{Volume, m}^3)\right]}{1000}\left(\frac{\text{Blanket Depth}}{\text{Clarifier Depth}}\right)$$

$$\text{Clarifier kg} = \frac{\left[\left(7800, \frac{\text{mg}}{\text{L}}\right)(1919\text{ m}^3)\right]}{1000}\left(\frac{1.5\text{ m}}{3.65\text{ m}}\right)$$

$$\text{Clarifier kg} = 6152$$

Step 4—Find the percentage of solids in the clarifier blanket.

$$\text{Percent} = \frac{\text{Part}}{\text{Whole}} \times 100$$

$$\text{Percent} = \frac{6\,152\text{ kg}}{28\,387.5\text{ kg} + 6152\text{ kg}} \times 100$$

$$\text{Percent} = 17.8\%$$

Because the activated sludge basin is large compared to the clarifiers, most of the MLSS will be in the basin even when the clarifier blanket is deep. These solids will not have a big effect on the SRT or MCRT calculation. Keep in mind that the blanket depth should be maintained below 2 ft at all times. For this situation, a facility may choose to use SRT for process control.

Example #2—A conventional activated sludge facility was designed for an HRT of 6 hours. Influent flow is 9462.5 m³ (2.5 mil. gal). The activated sludge basin holds 2366 m³ (0.625 mil. gal). There are two secondary clarifiers that are 24.4 m (80 ft) in diameter and 3.65-m (12-ft) deep. The MLSS concentration is 3000 mg/L and the RAS concentration is 7800 mg/L. The clarifier blanket depth is 0.6 m (2 ft). In this situation, the total mass of MLSS in the activated sludge basin is 7097 kg and the mass of MLSS in the clarifier blankets is 4374 kg. Even though the blanket depth is only 0.6 m (2 ft) in each of the clarifiers, 38% of the total sludge inventory is in the clarifier. For this situation, a facility should use MCRT for process control.

Example #3—An activated sludge treatment facility receives alum sludge from its water treatment facility. The water facility discharge is somewhat irregular, with no discharge at all on some days and large volumes discharged on other days. The MLSS concentration in the activated sludge basins can suddenly increase when alum sludge is entering the facility. As a result, the operations staff has a difficult time calculating sludge age. In this case, the operators should select either SRT or MCRT for process control, but use MLVSS rather than MLSS in the calculation. The alum sludge does not contribute to the growth of the microorganisms and passes through the process unchanged. The total MLSS measurement includes this inert material in the sludge age calculation, which could result in accidental overwasting and loss of nitrification. A WRRF receiving water treatment facility residuals may also want to base adjustments to the sludge wasting rate on the 5- or 7-day running average SRT or MCRT.

Example #4—An activated sludge treatment facility has multiple basins in each treatment train. The first basin is anaerobic, the second is anoxic, the third is a swing basin and can be either anoxic or aerated, and the last is fully aerated. The operator must decide whether to operate the third basin as an additional anoxic basin or if it should be fully aerated. In this scenario, the operator should calculate the $SRT_{aerobic}$ to determine if there is sufficient aerobic sludge age to support full nitrification when the swing basin is operated in anoxic mode. The $SRT_{aerobic}$ should also be calculated for all SBRs and for facilities practicing on/off aeration.

SECTION EXERCISE

1. An activated sludge process has three activated sludge basins that hold 11.4 ML (3 mil. gal) each. There are also three secondary clarifiers that hold 3.8 ML (1 mil. gal). This facility is not required to remove ammonia. If one activated sludge basin and two secondary clarifiers are in service, which sludge age calculation is the most appropriate?
2. A small activated sludge facility can only control their wasting pump by manually turning it on and off. This means the operator must walk across the site each time they want to waste. The same operator also operates the water treatment facility and must frequently be away from the WRRF in the middle of the day. If the operator collects samples of the MLSS and RAS early in the morning, the laboratory can have results back by 2 p.m. each day. What can this operator do to make sure the pump wastes enough each day without overwasting?
3. An operator of an extended aeration activated sludge process typically operates with a 25-day SRT. Should the operator include the effluent TSS in their calculations?

SECTION EXERCISE SOLUTIONS

1. MCRT. The volume of the clarifiers in service is comparable in size to the volume of the activated sludge basin.
2. Collect samples first thing in the morning and deliver them to the lab. At the same time, waste one-third to one-half of the amount of WAS that was wasted from the previous day. When the lab results come back, calculate the SRT or MCRT and waste the remaining pounds.
3. Yes. As sludge ages increase, wasting goes down and effluent TSS starts to have a bigger impact on the calculation result.

TEST YOUR KNOWLEDGE

1. **Wasting should be done all at once at the end of each week rather than daily to minimize the number of visits to the treatment facility.**
   - ☐ True
   - ☐ False
2. **Solid retention time calculations and wasting rate adjustments (if needed) should be done**
   a. Daily
   b. Weekly
   c. Every 2 weeks
   d. Monthly
3. **On Monday, Tuesday, and Wednesday, the operator wasted 1000 kg (2205 lb) from the system each day. The current SRT is 12 days, but the target sludge age is 8 days. What is the maximum amount of sludge the operator can waste on Thursday while staying within the 15% rule?**
   a. 880 kg (1874 lb)
   b. 1150 kg (2536 lb)
   c. 5000 kg (11 025 lb)
   d. 15 000 kg (33 069 lb)

4. An activated sludge facility uses large quantities of ferric chloride in a downstream process to precipitate phosphorus. The precipitated phosphorus and iron, now iron sludge, is removed with tertiary filters. When the filters are cleaned, iron sludge is recycled back to the activated sludge process. How might the operator compensate for the changes in total MLSS concentration caused by these nonbiological solids?

a. Increase blower output to increase available oxygen
b. Decrease sludge age to remove solids faster
c. Switch to constant mass for process control
d. Substitute MLVSS for MLSS in the SRT calculation

5. All of the following may affect the accuracy of an SRT calculation EXCEPT

a. Accuracy of WAS flow meter
b. Lack of a blanket in the clarifier
c. Foaming in the activated sludge basin
d. High-effluent suspended solids

6. Solids in the clarifier blanket should be included in the sludge age calculation when

a. The MLSS is settling and compacting well.
b. During peak wet weather events.
c. Clarifier sludge blankets are deep.
d. Clarifiers are small compared to the basins.

7. Before collecting grab samples from the final effluent for process control calculations, operators should

a. Check with their regulatory agency
b. Verify that the effluent is clean
c. Reduce RAS and WAS flows
d. Remove algae from clarifier weirs

8. An activated sludge process has extremely large basins with a combined HDT of 28 hours. The clarifiers are relatively small and have an HDT of only 2 hours. The operator maintains the blankets below 3 ft of depth. Which of the following calculations for sludge age is most appropriate for this facility?

a. GSA
b. MCRT
c. SRT
d. $SRT_{aerobic}$

9. A conventional activated sludge process has an HRT of 6 hours. There are three clarifiers online with a combined detention time of 4 hours. Which of the following sludge age calculations is most appropriate for this system?

a. GSA
b. MCRT
c. SRT
d. $SRT_{aerobic}$

FOOD-TO-MICROORGANISM RATIO

DEFINITION

The F/M is defined as the mass of BOD or COD added to the activated sludge process each day divided by the mass of MLVSS in the process. *It is important to note that the F/M equation always uses MLVSS rather than MLSS.* The volatile portion of the MLSS is a good approximation of the mass of live microorganisms in the process, but also includes dead microorganisms and organic debris. Food-to-microorganism is literally the amount of "food" available per microorganism in the process. The F/M is also the inverse of the modified GSA equation. With the modified GSA equation, sludge age is calculated by taking the mass of solids in the activated sludge basin and dividing by the mass of influent BOD. The F/M calculation simply flips the modified GSA on its head. The F/M is directly related to the growth rate of the microorganisms in the process. Recall from Chapter 5 that the microorganisms in secondary treatment processes are almost never growing at their maximum rate. That means every time available resources are increased, the growth rate should also increase.

Activated sludge operational modes are defined, in part, by the range of the F/M, as shown in Table 8.2. The F/M for conventional activated sludge processes is accepted to be between 0.2 and 0.4 kg of BOD/kg of MLVSS·d (0.2 and 0.4 lb of BOD/d/lb of MLVSS) and between 0.05 and 0.15 kg of BOD/kg of MLVSS·d (0.05 and 0.15 lb of BOD/d/lb of MLVSS) for extended aeration systems (U.S. EPA, 1974, 1978b).

CALCULATION METHOD

The F/M is calculated according the following formula:

$$F/M = \frac{F}{M} = \frac{\text{kg or lb of BOD applied per day}}{\text{kg or lb MLVSS inventory}} \qquad (8.16)$$

CALCULATION EXAMPLE

A WRRF consists of screening, grit removal, primary clarification, conventional activated sludge, and disinfection. The influent flow is 4000 m^3/d (1.06 mgd) and has a BOD_5 concentration of 300 mg/L. The primary clarifier removes 30% of the influent BOD5 and 40% of the influent TSS. The activated sludge basins have a combined volume of 1600 m^3 (0.42 mgd) and an MLSS concentration of 3200 mg/L. The MLSS is 80% volatile. Calculate the F/M.

Solution with International Standard units:

Step 1—Find the mass of BOD_5 in the influent.

$$\frac{300\ \text{mg}}{\text{L}}\left|\frac{1\ \text{g}}{1000\ \text{mg}}\right|\left|\frac{1\ \text{kg}}{1000\ \text{g}}\right|\left|\frac{1000\ \text{L}}{1\ \text{m}^3}\right|\left|\frac{4000\ \text{m}^3}{\text{d}}\right| = 1200\ \text{kg/d}$$

Step 2—Find the mass of BOD_5 going to the activated sludge process. If the primary clarifier is removing 30% of the influent BOD, then 70% is going to the activated sludge process.

$$(\text{Total Mass})(\text{Percentage}) = \text{Mass to Activated Sludge Process}$$
$$(1200\ \text{kg})(100\% - 30\%) = \text{Mass to Activated Sludge Process}$$
$$840\ \text{kg} = \text{Mass to Activated Sludge Process}$$

Step 3—Find the mass of MLSS in the activated sludge process.

$$\frac{3200\ \text{mg}}{\text{L}}\left|\frac{1\ \text{g}}{1000\ \text{mg}}\right|\left|\frac{1\ \text{kg}}{1000\ \text{g}}\right|\left|\frac{1000\ \text{L}}{1\ \text{m}^3}\right|\left|\frac{1600\ \text{m}^3}{}\right| = 5120\ \text{kg}$$

Step 4—Find the mass of MLVSS in the activated sludge process.

$$(\text{Total Mass})(\text{Percentage}) = \text{Volatile Mass}$$
$$(5120\ \text{kg})(80\%) = 4096\ \text{kg}$$

Step 5—Calculate the F/M.

$$F/M = \frac{F}{M} = \frac{\text{kg or lb of BOD applied per day}}{\text{kg or lb MLVSS inventory}}$$
$$F/M = \frac{F}{M} = \frac{840\ \text{kg BOD applied per day}}{4096\ \text{kg MLVSS inventory}}$$
$$F/M = \frac{F}{M} = 0.21\ \text{per day}$$

Based on the F/M alone, this is likely a conventional activated sludge process.

Advantages and Disadvantages

Using the F/M for process control suffers from many of the same disadvantages of the GSA calculation and are discussed earlier in this chapter. The most obvious drawbacks to the F/M are that it puts the operator in a position of both monitoring and predicting the influent load and that it does not provide a clear mechanism for determining how much MLSS should be removed from the process each day by wasting. As will be discussed in the process control section of this chapter, the F/M is dependent on the MCRT/SRT. A WRRF may select a target MCRT/SRT designed to bring the F/M within a desired range. If the F/M is too high or too low, the MLSS may not settle well (U.S. EPA, 1977). Although many WRRFs calculate the F/M regularly, it should not be used as the sole or primary method of process control (Ekama, 2010).

Early activated sludge process control focused on trying to maintain enough sludge mass or inventory to effectively treat the incoming organic load. An earlier version of Manual of Practice No. 11 (WEF, 1996) explained it this way: "*A good way to understand the importance of [maintaining the optimal amount of activated sludge solids in the system to treat the BOD in the influent wastewater] is to consider the amount of BOD in the untreated wastewater as "food", and the activated sludge solids as "microorganisms" that will eat the food in one sitting. If there are not enough microorganisms available to eat all the food in one sitting, there will be leftovers. Leftovers means that effluent BOD is higher than desired".* We now know that the number of microorganisms available is determined by the influent organic load and the yield (Chapter 5). We also know that the mass of MLSS in the system is controlled by the wasting rate, which is determined by the desired sludge age. Setting the sludge age automatically sets the F/M. Finally, if the sludge age is long enough to support flocculation and compaction so that solids separation can occur in the secondary clarifier, there should be a large enough population of microorganisms in the process to completely remove the available BOD.

The F/M can be useful for predicting *denitrification rates* as described in the chapter on nutrient removal. The rate, or speed, of denitrification is directly proportional to the amount of available BOD and the F/M. In an activated sludge process with both pre-anoxic and post-anoxic zones like the one shown in Figure 8.37, denitrification rates may be as much as 10 times faster in the pre-anoxic zone than the post-anoxic zone because of the availability of excess BOD.

The *denitrification rate* is how fast the heterotrophic bacteria will convert nitrate to nitrogen gas. The conversion requires about 4 mg/L of BOD_5 for every 1 mg/L of NO_3-N converted to nitrogen gas.

VOLUMETRIC LOADING RATE

The volumetric loading rate is the mass of BOD_5 per volume of activated sludge basin. It does not include the volume in the clarifier. Volumetric loading rate is expressed as kilograms of BOD_5 per cubic meter of activated sludge basin per day (kg/m^3·d) or pounds of BOD_5 per day per 1000 cubic feet (lb BOD_5/d/1000 cu ft). It is primarily used by design engineers, but can be helpful for determining when to remove an activated sludge basin from service or to place another into service. The lower the volumetric loading rate, the more likely it is that the activated sludge process will remove ammonia in addition to BOD_5.

PERCENT REMOVAL

Percent removal is useful for assessing activated sludge performance under different operating conditions. Composite samples of the influent and effluent will give information about average performance. Grab samples pulled from the same locations may be used to determine performance under peak load or other conditions. Percent removal may be calculated for BOD_5, $CBOD_5$, ammonia, nitrate, phosphorus, or other parameters. Removal percentage is calculated according to the following formula:

$$\text{Removal, \%} = \left(\frac{\text{In} - \text{Out}}{\text{In}}\right) \times 100 \quad (8.17)$$

Example Calculations for Percent Removal

An activated sludge process has an influent BOD_5 concentration of 240 mg/L. A sample taken from the secondary clarifier effluent contains 12 mg/L of BOD_5. Find the percent removal.

$$\text{Removal, \%} = \left(\frac{\text{In} - \text{Out}}{\text{In}}\right) \times 100$$

$$\text{Removal, \%} = \left(\frac{240 - 12}{240}\right) \times 100$$

$$\text{Removal, \%} = 95\%$$

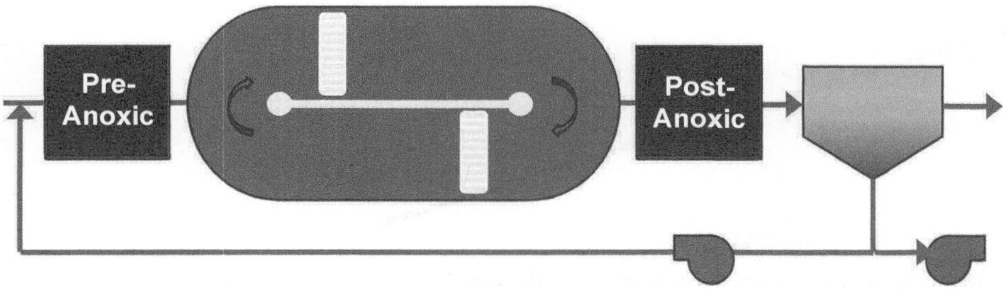

Figure 8.37 Pre-Anoxic and Post-Anoxic Zones in an Oxidation Ditch-Type Activated Sludge Process (Reprinted with permission by Indigo Water Group)

TEST YOUR KNOWLEDGE

1. Setting a target SRT or MCRT also determines the F/M.
   - ☐ True
   - ☐ False
2. The activated sludge basin volume is 30 000 m³ (7.9 mil. gal). If the MLSS concentration is 2200 mg/L and is 80% volatile, how many kilograms (pounds) of MLVSS is in the basin?
   a. 52 800 kg (115 959 lb)
   b. 66 000 kg (144 949 lb)
   c. 82 500 kg (181 867 lb)
   d. 97 000 kg (231 788 lb)
3. A WRRF receives 817 kg (1800 lb) of BOD_5. If the primary clarifier removes 35% of the incoming BOD_5, what is the load to the activated sludge process?
   a. 286 kg (630 lb)
   b. 531 kg (1170 lb)
   c. 1256 kg (2769 lb)
   d. 2234 kg (5142 lb)
4. A conventional activated sludge system that performs only BOD_5 removal typically has an F/M in the range of
   a. 0.05 to 0.15
   b. 0.2 to 0.4
   c. 0.4 to 0.6
   d. 0.8 to 1.2
5. Disadvantages of using F/M for process control include
   a. Helps maintain constant microorganism growth rate
   b. Requires influent ammonia monitoring for nitrfying facilities
   c. Provides a clear mechanism for determining wasting
   d. Forces operator to monitor and predict the influent load

RELATIONSHIPS BETWEEN PROCESS VARIABLES

Sludge age should be the primary control variable for all activated sludge systems (Ekama, 2010; Jenkins and Garrison, 1968; Pretorius et al., 2016; U.S. EPA, 1977). Once a target sludge age has been selected, it will dictate the wasting rate; MLSS concentration; MLVSS concentration; F/M; and MLSS growth rate, microbiology, and sludge settleability. It is not possible to control more than one of these variables at a time. Burchett and Tchobanoglous (1974) noted that "*It is evident that the new growth of microorganisms per day in the activated sludge system is related to the F/M and the MCRT. Second, and more important, it may be concluded that if one of the two variables is controlled at a given level, the other will seek and find its associated level.*" Figure 8.38 is a simplistic representation of the relationships between these important process variables. In reality, the relationships between the variables are more nuanced, but the basic relationships depicted here remain true. In Figure 8.38, the sludge age is increasing along with the MLSS concentration. The percent of MLVSS, F/M, and wasting rate are all decreasing. If the operator decided to lower the sludge age, then the MLSS would go down with it and all of the variables on the right side (percent MLVSS, F/M, and WAS rate) would increase.

WHY DOES MLSS INCREASE WITH INCREASING SRT?

As sludge age is increased, fewer kilograms (pounds) of MLSS are removed from the system in the WAS each day. This is the definition of sludge age. For a 10-day sludge age, one-tenth (10%) of the solids are removed each day. For a 5-day sludge age, one-fifth (20%) of the solids are removed each day. Removing fewer kilograms (pounds) of MLSS causes the MLSS concentration (mg/L) in the activated sludge basins to increase.

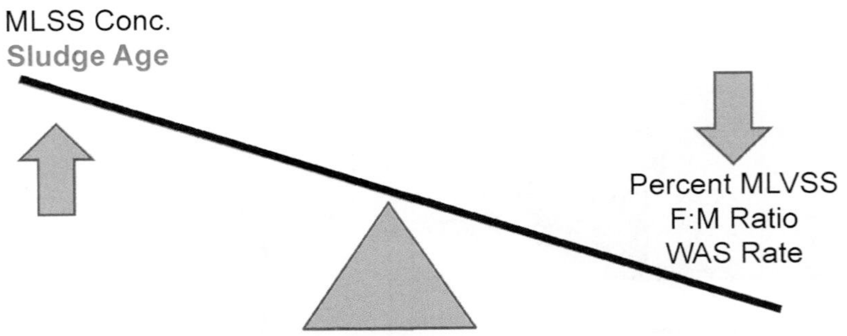

Figure 8.38 Relationships Between Activated Sludge Process Control Variables (Reprinted with permission by Indigo Water Group)

WHY DOES THE RATIO OF MLVSS TO MLSS DECREASE WITH INCREASING SRT?

The MLSS consist of both organic and inorganic (inert) material. The live microorganisms constitute the "active" portion of the MLSS and are the fraction of MLSS that provides treatment. The MLVSS represent the active microorganisms. We assume that all organic solids in the MLSS are live, active microorganisms even though dead microorganisms and organic debris are also measured in the MLVSS test. The MLSS also contain inert material that came in with the influent wastewater. Sand, grit, and eggshells can't be broken down by the microorganisms. This material passes through the process unchanged. The total MLSS added each day will be the sum of the new bacteria produced (yield) plus non-biodegradable organics (leftovers from endogenous respiration) plus inert material.

$$\text{Total MLSS Added} = \text{New Cells Grown} + \text{Existing Cells} + \text{Inert Material} \quad (8.18)$$

Figure 8.39 illustrates how biodegradable materials are used and changed through biological treatment. Biochemical oxygen demand entering the system is used by the bacteria for both cellular maintenance and to build more bacteria. The mass of new bacteria produced is fixed by the yield (0.4 kg MLVSS/kg COD or 0.67 kg MLVSS/kg BOD), as discussed in Chapter 5. At the same time, the amount of BOD required to maintain all of the existing bacteria increases as the number of active bacteria increases. In other words, a larger fraction of the influent BOD is going toward maintenance of existing cells and a smaller percentage of the total cells in the system will have enough energy to reproduce. The growth rate of the MLVSS as a whole decreases.

As the sludge ages and there is less food available per bacteria, some bacteria will be lost through endogenous respiration as they consume their own cell mass and exopolymer in an attempt to survive. Some bacteria will lyse or break open and become food for others. Still others will be lost through predation as higher life forms such as stalked ciliates, which become more prevalent at older sludge ages. The MLVSS concentration and the number of viable bacteria will begin to plateau as shown in Figure 8.40. The MLVSS concentration doesn't tell the whole story. Both dead bacteria and non-biodegradable organic material are also volatile. By one estimate, about 23% of the biological solids (MLVSS) produced are relatively inert and, therefore, accumulate in the system (McCarty and Brodersen, 1962). It may appear as though the number of live bacteria is increasing with longer and longer sludge ages, but because the yield is fixed, what is really happening is that inert material is building up in the activated sludge basins. It isn't any more possible to build new bacteria without influent *BOD* than it is for people to gain weight without eating.

It isn't possible to grow more bacteria without BOD_5. The MLSS concentration will continue to increase with older and older sludge ages, but the increase is an accumulation of dead microorganisms, organic debris, and inert material.

Sand, grit, and eggshells and some organic compounds can't be broken down by the bacteria. This inert fraction of the influent wastewater builds up in the activated sludge basins at the same rate each day. Think about adding 100 kg (or pounds) of sand to the activated sludge process on day one. On day two, 100 more kilograms are added. On day three, another 100 kg are added. If the sludge age is 20 days, this means that each day 100 additional kilograms of inert material will be added and only 5% (5 kg) will be removed. After 20 days, a whopping 1900 kg of inert material will have accumulated in the process. The amount of inert material goes up in a straight line as shown in Figure 8.40. At the same time, the MLVSS concentration levels off because the microorganisms grow slower when less food is available (lower F/M). The combination of effects causes the total solids concentration (MLSS) to increase while the percentage of volatile solids (MLVSS) decreases. For wastewaters that contain larger amounts of inert material, the effect will be more dramatic. A healthy activated sludge system that is not adding chemicals will typically have an MLVSS to MLSS ratio in the 0.75 to 0.85 range (U.S. EPA, 1977). The MLSS in an aerobic or anaerobic digester, on the other hand, may only be 60% volatile. The percent volatile solids in the *digester* is even

Chemicals such as ferric chloride may be added for odor control or to precipitate phosphorus. Alum is also used for phosphorus precipitation. These inorganic chemicals build up in the process and contribute to the non-volatile portion of the MLSS.

A *digester* is a solids-handling process used to break down organic material and reduce the total mass of sludge. A brief overview of digesters is included in Chapter 1.

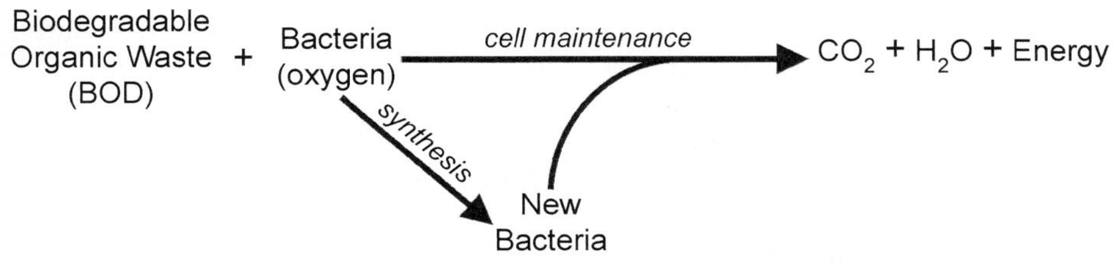

Figure 8.39 Conversion of BOD to MLVSS (Reprinted with permission by Indigo Water Group)

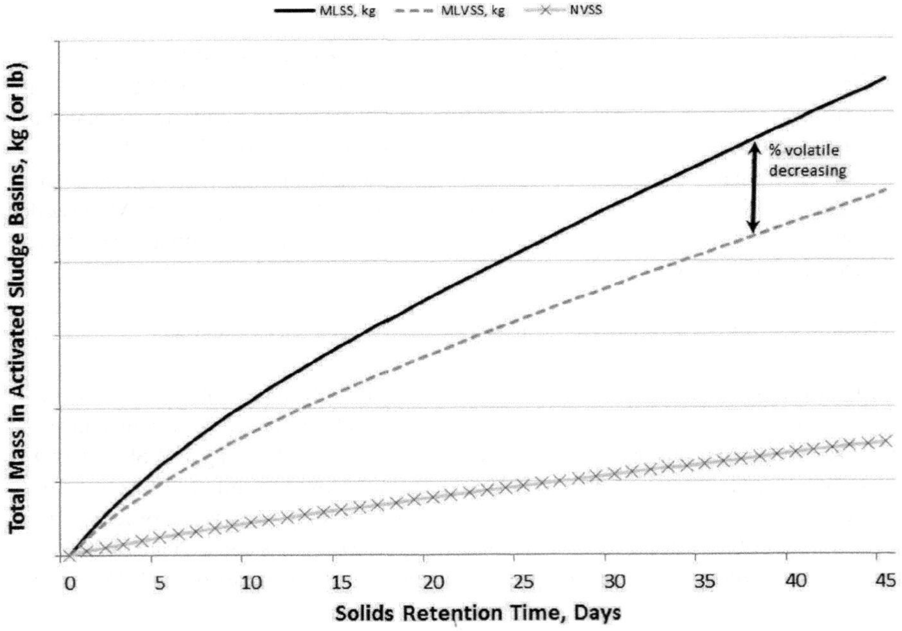

Figure 8.40 Changes to MLSS, MLVSS, and Non-Volatile Solids with Increasing SRT (Reprinted with permission by Indigo Water Group)

lower because the bacteria are breaking down by endogenous respiration, but the inert solids pass through the digester unchanged.

IS MAINTAINING A CONSTANT MLSS CONCENTRATION THE SAME AS CONSTANT SLUDGE AGE?

Not necessarily. Mixed liquor suspended solids concentration increases with increasing sludge age, but without more information, it is impossible to predict the MLSS concentration just from sludge age. For many wastewater facilities, the influent load will be different on the weekends than on weekdays and may vary seasonally as well. The organic load can go up and down as can the amount of inert material. For operators to keep a constant MLSS concentration under changing influent conditions, they must increase and decrease the amount of solids they are removing from the system each day. If the percentage of mass removed each day changes, then sludge age is also changing. For some systems with very consistent influent flows and loads, maintaining constant MLSS can approximate constant sludge age.

Example #1—The City of Bellingham's Post Point WRRF (Washington) uses in-basin TSS probes to monitor the MLSS and WAS concentrations. The WAS flow is automatically adjusted to maintain an operator-selected SRT. At one point, the operators noticed that the MLSS concentration had dropped below its typical range. The operators assumed the in-basin MLSS probe had malfunctioned or was out of calibration. After reviewing their influent data, they discovered that the influent BOD load had decreased. As a result, the MLSS concentration had automatically decreased to maintain a constant SRT (Pretorius et al., 2016).

Example #2—A metropolitan city WRRF receives alum residuals from their water treatment facility. The mass of alum residuals varies considerably from day to day and can cause the MLSS concentration to change by as much as 500 mg/L in a single day. The WRRF uses constant MLSS for process control rather than constant SRT. On Monday, the MLSS concentration had increased to 3200 mg/L. The operator increased the wasting rate to drop the MLSS concentration back to their target of 3000 mg/L. On Tuesday, the water treatment facility discharged an unknown quantity of alum residuals that caused the MLSS concentration to increase to 3700 mg/L. The operator increased the wasting rate again to bring the MLSS concentration back to 3000 mg/L. On Monday, the percent MLVSS was 80%, but, on Tuesday, had fallen to 76% volatile because of the alum addition. On Wednesday, the operator collected samples for effluent ammonia and was surprised to discover that the effluent ammonia concentration had increased from less than 1 mg/L to 22 mg/L. Why did this happen? A minimum SRT is required to maintain stable nitrification.

When the operator increased wasting on Tuesday, they removed too many of the biological solids and dropped the SRT below the required minimum. The MLSS concentration stayed constant, but the SRT dropped dramatically.

Example #3—A resort community has more visitors in the winter than in the summer. Each visitor contributes to the amount of load and flow received by the WRRF. The WRRF uses the constant MLSS mass method for process control and has a target MLSS concentration of 2500 mg/L. Past experience has shown that this concentration of MLSS produces a well-settling sludge and good effluent quality. Because the organic load is higher in the winter, more MLSS is produced and the operator must waste more to keep the MLSS concentration at the target of 2500 mg/L. In the summer when the organic load decreases, the operator wastes less as less sludge is produced. Although the MLSS concentration was kept constant year-round, the SRT is not constant. It is higher in the summer and lower in the winter. This WRRF often loses the ability to remove ammonia for short periods during the winter when the SRT drops too low.

CAN I OPERATE WITH CONSTANT MLSS INSTEAD OF CONSTANT SLUDGE AGE?

A constant MLSS concentration or constant MLSS mass approach is appealing because it is simple to understand, reduces the amount of sampling and process control required, and is less dependent on accurate sampling of the WAS concentration. Many activated sludge systems operate effectively using constant MLSS. When influent loading and water temperature are consistent, maintaining a constant MLSS can approximate maintaining a constant SRT. For this to be true, both of the following conditions must be true:

- Influent BOD (kilograms or pounds) is constant and
- Influent TSS (kilograms or pounds) is constant.

A disadvantage of this control method is its secondary emphasis on SRT. Caution must be used with this control technique where widely varying BOD concentrations are experienced. For example, a situation may exist in which the secondary influent BOD to a particular activated sludge process suddenly increases by 50% and stabilizes at this level. Because of the increased food supply, the amount of activated sludge produced would also increase, and the MLSS levels in the activated sludge basin would climb. Observing this, the operator who is holding a fixed MLSS level naturally would increase the wasting rate, trying to cut down the solids level to its typical range. By reducing the solids level to a normal value, the operator is decreasing the SRT. This could cause serious problems. Overwasting can cause a loss of nitrification, change sludge settling characteristics, and affect overall effluent quality. The constant MLSS approach should not be used by WRRFs that are required to remove ammonia.

Some facilities successfully use the constant mass approach for process control. It can work well for facilities that 1) are not required to remove ammonia, 2) have consistent influent flows and loads, and 3) are operating well below their design capacity.

WHY DOES F/M DECREASE WITH INCREASING SRT?

If the MLSS concentration is increasing and the amount of BOD entering the facility remains about the same, then the amount of food per microorganism will decrease. Yield and decay are discussed in Chapter 5.

Solids retention time is related to the F/M by the following equation (U.S. EPA, 1974):

$$\text{SRT, days} = \frac{1}{(\text{Y})\left(\dfrac{\text{F}}{\text{M}}\right)(\text{BOD Removal Efficiency}) - k_\text{d}} \tag{8.19}$$

Where:

Y = Yield,
F/M = Food-to-microorganism ratio, and
k_d = Decay coefficient.

As SRT increases, the F/M will decrease proportionally. The converse is also true. Referring to Table 8.7, for an SRT in the range of 20 to 30 days, the F/M will be considerably lower (0.05 to 0.15 kg BOD/kg MLVSS·d) than that for a system with an SRT in the 5- to 15-day range (0.2 to 0.4 kg BOD/kg MLVSS·d). The activated sludge process should be controlled by setting the target SRT or MCRT and allowing the F/M

Table 8.7 Design and Operating Parameters for the Activated Sludge Process (U.S. EPA, 1974; 1978b)

Process Modification	Solids Retention Time, days	F/M, kg BOD Applied/kg MLSS·d (lb/d/lb)	Volumetric Loading, kg $BOD_5/m^3 \cdot d$ (lb BOD_5/d/1000 cu ft)	MLSS, mg/L	Hydraulic Detention Time, hr
Conventional (plug flow)	5 to 15	0.2 to 0.4	20 to 40	1 500 to 3000	4 to 8
Extended aeration	20 to 30*	0.05 to 0.15	10 to 25	3000 to 6000	18 to 36
Step-feed	3 to 7	0.2 to 0.4	40 to 60	2000 to 3500	3 to 5
Pure oxygen	3 to 20	0.25 to 1.0	150 to 200	4000 to 7000	2 to 4

*Excessively long SRTs are no longer recommended because of the growth of filamentous bacteria and deterioration of final effluent quality that can occur at long SRTs.

to self-adjust. It is not possible to control both of these variables at the same time. Selection of an MCRT or SRT is discussed later in this chapter.

THE DESIGN CONDITION

Almost every WRRF operations and maintenance manual or drawing set contains a chart listing the design parameters for MCRT or SRT, MLSS concentration, and F/M. This is the design point. Many operators frustrate themselves in an attempt to meet all three conditions simultaneously or they may attempt to maintain a fixed mass (kilograms or pounds) of MLSS in the activated sludge basins each day regardless of changing influent wastewater characteristics. The in-facility operating conditions will never exactly match the design point for several reasons. First, the design point can only be reached when the WRRF is operating at its design flow and load. Most discharge permits require that a facility be in the planning stages for the next expansion when either the maximum month flow or load exceeds 80% of the design capacity and to be under construction when they reach 95%. Because WRRFs are designed for peak loads and flows, they will have excess capacity under average operating conditions.

Second, few WRRFs monitor their influent for volatile and non-volatile suspended solids. Without this information, it is impossible for the design engineer to predict the total MLSS concentration. Imagine two different activated sludge processes. The first receives 200 kg a day of TSS that is 85% volatile. This means there are 30 kg of non-volatile (inert) solids building up in the basins each day. The other process receives 200 kg of TSS that is 50% volatile. This means there are 100 kg of non-volatile (inert) solids building up in the basins each day. Even if both of these WWRFs maintain the same sludge age, their MLSS concentrations will be quite different. The design engineer made a reasonable estimate of how much inert material was in the influent wastewater. If the estimate doesn't match the actual concentration in the influent exactly, then the MLSS concentration won't match the design point exactly either.

The MLSS concentration can also be influenced by recycle streams from solids handling processes, chemical addition, and receipt of water treatment facility residuals. Facilities that practice chemical phosphorus removal generally use alum or ferric in large doses downstream of the activated sludge basins. These chemicals are often recycled as part of filter backwash or other processes and end up in the activated sludge basins.

SUMMARY OF PROCESS CONTROL STRATEGIES

Table 8.8 summarizes the various process control strategies currently in use along with the advantages and disadvantages of each. As noted by Pretorius et al. (2016), facilities sometimes abandon MCRT/SRT control of their activated sludge processes in favor of less-reliable control parameters such as attempting to maintain a constant MLSS concentration or F/M. Reasons for using constant MLSS or a constant F/M in place of MCRT/SRT control include adherence to older process control strategies, staffing limitations, difficulty in obtaining representative samples, lack of on-site laboratory facilities, a lack of understanding with respect to the advantages of MCRT/SRT control, and successful operation using either constant MLSS or a constant F/M. However, MCRT/SRT control offers the best control of the activated sludge process and should be used whenever practicable (Ekama, 2010; Jenkins and Garrison, 1968; Pretorius, et al., 2016; U.S. EPA, 1977). Constant MLSS and constant F/M control strategies should not be used at WRRFs that are required to remove ammonia.

Table 8.8 Advantages and Disadvantages of Common Process Control Strategies

Process Control Method	Advantages	Disadvantages
Sludge age—including MCRT, SRT, and $SRT_{aerobic}$	■ Widely used with reliable results. ■ Based on growth kinetics. ■ Accounts for transformation of influent BOD and TSS into biological solids (MLSS). ■ Automatically adjusts MLSS concentration to match incoming BOD load, thereby simultaneously controlling MLSS concentration and F/M. ■ Operator does not need to predict or match influent conditions. ■ Clear mechanism for calculating wasting rates. ■ Target sludge age controls presence or absence of different types of microorganisms in the process. ■ Self-stabilizing. ■ Produces excellent sludge quality when the target MCRT/SRT is within a reasonable range. ■ When kept in the correct range for a given water temperature, maintains stable nitrification and minimizes filament growth.	■ Requires accurate measurement of MLSS and WAS solids. ■ May be difficult to obtain representative samples needed for calculations. ■ Staffing limitations in small systems may make sludge age control difficult to fully implement. ■ Control charts are recommended along with calculation of running averages to minimize the influence of bad data points or days when wasting is not performed.
MCRT	■ Accounts for the total sludge inventory by including both the activated sludge basins and clarifier solids. ■ Preferred for systems with large clarifiers and smaller activated sludge basins. ■ Historic data for sludge ages that promote filament growth are based on MCRT and SRT.	Requires more data collection than either SRT or $SRT_{aerobic}$.
SRT_{total}	■ Less data collection required than for MCRT. ■ Preferred for systems with small clarifiers and large activated sludge basins. ■ Historic data for sludge ages that promote filament growth are based on MCRT and SRT.	■ Assumes solids are not stored in the secondary clarifier blanket. For systems with large clarifiers relative to the activated sludge basins, this may not be true even when blankets are below 0.6 m (2 ft). ■ Historic data for predicting the growth of filamentous bacteria are based on MCRT or SRT_{total} and not $SRT_{aerobic}$. Calculating $SRT_{aerobic}$ only may result in excessively long sludge ages, growth of filaments, and subsequent deterioration of sludge settleability.
$SRT_{aerobic}$	■ Used by design engineers for basin sizing. ■ Ensures there is adequate aerobic basin time or space to support stable nitrification. ■ May be used to adjust anoxic versus aerated zone size within basins. ■ May be used to adjust on/off cycle times for SBRs.	
Constant MLSS (mass or concentration)	■ Simple to understand. ■ Minimum amount of laboratory data required. ■ Capable of producing good settling sludge and effluent quality when influent loads are consistent. ■ Can approximate SRT control for some systems. ■ Best suited for WRRFs that are not required to nitrify.	■ Constant MLSS does not hold sludge age constant, which may result in variable sludge quality (poor settleability) and uncontrolled loss of nitrification. ■ No clear mechanism for calculating wasting rates. ■ If the influent load is not constant, the SRT and F/M can fluctuate. ■ Filamentous bacteria may be easily retained when the unmonitored SRT exceeds the SRT required for filament growth. ■ More prone to foaming and bulking.

(continues)

Table 8.8 Advantages and Disadvantages of Common Process Control Strategies (*Continued*)

Process Control Method	Advantages	Disadvantages
Constant F/M	■ Some argue that constant F/M has an advantage in WRRFs, such as those that receive large quantities of industrial wastewater, where influent characteristics can change rapidly. ■ Should produce a consistently settling sludge. ■ Early work in activated sludge microbiology is based on F/M.	■ Requires accurate measurement of MLSS, MLVSS, WAS, and influent BOD. ■ Operator must predict and match changing influent conditions. ■ Necessarily uses 5-day-old data in the case of BOD or more than 3-hour-old data in the case of COD. ■ Reactive—responds to past conditions ■ Requires a significant amount of laboratory work for influent and process monitoring. ■ No clear mechanism for calculating wasting rates. ■ Chasing F/M may drastically alter sludge age. ■ May not maintain stable nitrification.

TEST YOUR KNOWLEDGE

1. **Sludge age and F/M can be controlled independently.**
   - ☐ True
   - ☐ False

2. **Operators may increase the total mass of live microorganisms in their process by increasing sludge age.**
   - ☐ True
   - ☐ False

3. **The SRT is currently 14 days. The operator increases the wasting rate to lower the SRT to 10 days. Which of the following variables will decrease?**
   a. F/M
   b. Percentage of MLVSS
   c. MLSS concentration
   d. BOD load

4. **Mixed liquor suspended solids concentrations increase with increasing SRT, but the percentage of MLVSS decreases because**
   a. Live microorganisms are wasted faster than inert material.
   b. Inert material accumulates in the process unchanged.
   c. Floc sizes increase at older SRTs and keep grit suspended.
   d. More BOD_5 per microorganism.

5. **The influent BOD load is unchanged. Solids retention time is increasing. What must be true?**
   a. Growth rate is decreasing.
   b. Yield is increasing.
   c. MLSS concentration is decreasing.
   d. Wasting rate is increasing.

6. **The SRT is currently 12 days. If the SRT is decreased to 8 days, what will happen to the F/M?**
   a. It will increase by 50%.
   b. It will remain the same.
   c. It will increase.
   d. It will decrease.

7. **Maintaining a constant MLSS concentration as a process control strategy approximates constant SRT when**
   a. F/M is between 0.2 and 0.4 kg BOD_5/kg MLVSS (lb/lb).
   b. The WRRF is below 50% of its rated capacity.
   c. MLSS does not change more than 200 mg/L per day.
   d. Influent BOD_5 and TSS loads are nearly constant.

8. **This is almost never measured for WRRF influent, but it can have a big effect on the MLSS concentration.**
   a. BOD_5
   b. TSS
   c. Ammonia
   d. VSS

9. **Match the process control strategy to its primary disadvantage over other methods.**

a. MCRT	1. Operator must predict influent loads
b. SRT	2. Requires more data collection
c. $SRT_{aerobic}$	3. Assumes no solids in the clarifier blanket
d. F/M	4. If influent loads vary, SRT and F/M will also vary
e. Constant MLSS	5. Growth of filamentous bacteria tied to total SRT

Process Control for Activated Sludge Basins

As discussed earlier, the purpose of biological treatment is to convert biodegradable organic material in the influent wastewater into MLVSS and to flocculate non-biodegradable particulate material. The primary goal of activated sludge processes is to grow larger, heavier colonies of bacteria that can then be efficiently separated from the treated water by gravity. The flocculated mass of bacteria and non-biodegradable solids is collectively called the *MLSS*. Uptake and conversion of BOD is relatively rapid and is often complete within 90 minutes of the influent wastewater entering the activated sludge basin. Bioflocculation takes much longer and requires that the solids (MLSS) remain in the process for many days. Secondary goals for activated sludge may include the removal of ammonia, nitrite, nitrate, and/or phosphorus.

Refer back to Table 8.7 for the operating ranges for various process control variables for a few of the most common operating modes in domestic WRRFs. These are not the only operating modes available. Keep in mind that the numerical ranges given in Table 8.7 are functional definitions and that it is entirely possible for a single, well-functioning activated sludge process to have some variables fall within more than one definition. These functional definitions are from the 1960s and 1970s and reflect operating practices and research conducted during that time. The SRT range for extended aeration activated sludge is quite long at 20 to 30 days. Today, we know that sludge ages as long as 20 days can encourage the growth of filamentous bacteria, which can affect sludge settleability, and we know that much younger sludge ages are capable of producing an effluent low in ammonia, nitrite, and nitrate. Don't worry if your activated sludge process doesn't fit neatly into one of these definitions. Guidance on how to select the right sludge age for your facility is given later in this chapter.

SELECTING A TARGET MEAN CELL RESIDENCE TIME, SOLIDS RETENTION TIME, OR AEROBIC SOLIDS RETENTION TIME

Conventional activated sludge processes generally have MCRTs or SRTs between 5 and 15 days, whereas extended aeration activated sludge processes have MCRTs or SRTs as long as 30 days. Selecting the right MCRT, SRT, or $SRT_{aerobic}$ for a facility should be based on achieving the best possible sludge settling characteristics while meeting other process goals, such as low effluent ammonia, total nitrogen, and phosphorus limits. If the sludge is too young or too old, process goals will not be met.

FLOCCULATE, SETTLE, AND COMPACT

The optimum MCRT, SRT, or $SRT_{aerobic}$ for a given treatment facility will be the one that produces an MLSS that flocculates, settles, compacts, and produces a final effluent that meets the limitations of the discharge permit (Wahlberg, 2016). The ability of the MLSS to flocculate, settle, and compact is measured by conducting daily settleometer tests and calculating the *sludge volume index (SVI)*. Both procedures are discussed in detail later in this chapter. If the sludge flocculates well, then it will be capable of agglomerating loose bacteria and particulate matter to leave a crystal-clear *supernatant*, as shown in Figure 8.41 (a). If a sludge does not flocculate well, the supernatant may be turbid or cloudy, as shown in Figure 8.41 (b). A turbid supernatant may not meet the 30 mg/L TSS limit in most discharge permits and will be more difficult to disinfect.

The *SVI* calculation gives the volume that 1 g of settled sludge occupies in the settleometer test after 30 minutes of settling. It is expressed in milliliters per gram. The SVI allows operators to compare settleometer test results from one day to the next. Without this calculation, operators would not know if the reason the SSV was higher today than last week is because the MLSS concentration increased (more sludge) or because the MLSS was settling and compacting differently.

Research has shown that for bioflocculation to occur, a minimum SRT of 2 days is needed and that at SRTs between 0.25 and 2 days, free bacteria dominate rather than flocs (Maharajh, 2010). Free bacteria and smaller floc particles do not settle. Many activated sludge design manuals set the minimum SRT for various system types at 3 days (Metcalf and Eddy, Inc./AECOM, 2014; WEF et al., 2010). Figure 8.42 illustrates some early work by Sayigh and Malina, Jr. (1978), correlating BOD removal to SRT. Better than 85% BOD removal was achieved for SRTs greater than 2 days when the water temperature was at least 10 °C. For colder water, the minimum SRT required for good flocculation was longer than 3 days. Good BOD removal can occur even when the sludge does not flocculate well.

The *supernatant* is the upper liquid layer in a settleometer test or treatment process.

Young Sludge Bulking

When a new activated sludge process is started up, it is common to see white foam with very poor settling. As the sludge gets a bit older, the MLSS flocs begin to take shape and become heavy and dense enough to begin to settle. However, young sludge floc is often open in structure because of the very fast growth rate of the floc formers. As a result, young sludge floc can be very light and fluffy, which causes it to settle slowly. In older manuals, this condition is sometimes referred to as young sludge "bulking", with the floc described

Figure 8.41 Evidence of MLSS Flocculating Properties (Reprinted with permission by Indigo Water Group)

as "straggler floc". Bulking occurs in any MLSS with an SVI greater than 200 mL/g. West (1974) described straggler floc as "small, almost transparent, very light, fluffy, buoyant sludge particles 3.2 to 6.4 mm (one-eighth to one-quarter inch) in diameter". Once the SRT reaches 2 to 3 days, the floc structure becomes denser and settleability improves. In summary, for an activated sludge process to flocculate, settle, compact, and produce an effluent that meets permit limits for BOD_5 and TSS, the SRT must be at least 2 days.

Old Sludge Bulking

Excessively long SRTs can also result in bulking sludge results from either the formation of "pin floc" or an abundance of filamentous bacteria. West (1973) defined pin floc as very small, compact floc that was typically less than 0.8 mm (1/32 in.) in diameter. Pin floc may be observed hanging suspended without obvious movement upward or downward throughout a moderately turbid secondary clarifier. In the

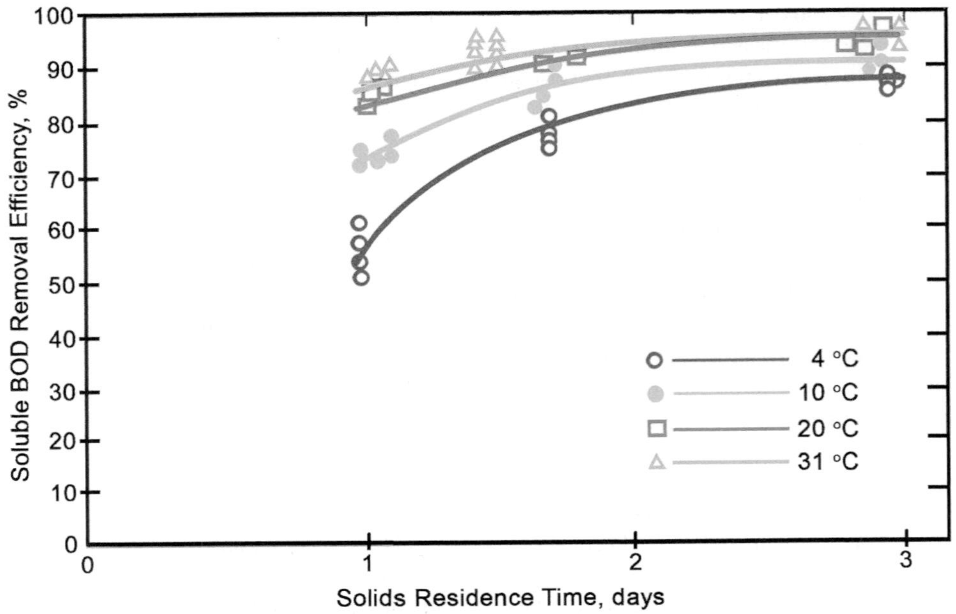

Figure 8.42 Soluble BOD Removal Efficiency Versus SRT (Sayigh and Malina, Jr., 1978)

settleometer test, pin floc is characterized by rapidly settling, discrete sludge particles instead of a well flocculated blanket. Pin floc is associated with high SRT, low F/M operation. High SRT, low F/M operation can cause the floc formers to go into endogenous respiration where they consume their external polymers and cell mass in order to survive. The loss of exopolymer can break up the floc particles.

High SRTs encourage the growth of certain types of filamentous bacteria (Jenkins et al., 2004). Old sludge filaments include *M. parvicella*, Type 0041, Type 0675, Type 1851, and Type 0803 (Jenkins et al., 2004). Note that filaments can proliferate under a number of different environmental conditions and that many are able to proliferate at relatively short SRTs (Jenkins et al., 2004). Filamentous bacteria can cause the floc structure to be more open and lacy and less dense. In extreme circumstances, the filaments can extend so far beyond the floc particle that they connect floc particles together. This is called *bridging*. An example of bridging is shown in Figure 8.43. When filaments are present in abundance, the MLSS particles will settle more slowly and won't compact as well at the bottom of the clarifier. The goal becomes to keep the MCRT or SRT long enough to flocculate, but short enough to prevent the growth of old sludge filaments and the formation of pin floc.

Populations of higher life forms, including rotifers and stalked ciliates, often decrease at longer SRTs. As the bacterial growth rate slows down, there are fewer and fewer free bacteria to support the growth of the other microorganisms that feed on them. Like other predator–prey relationships, when the number of available prey decreases, then the number of predators quickly comes down as well.

Sludge Settleability

The SSV after 5 minutes of settling in the settleometer test indicates how fast the floc is settling. For the first 5 minutes, the floc is essentially in free fall with few collisions between floc particles to slow the flocs down. This is the free settling zone (Eckenfelder and Melbinger, 1957). The 5-minute settled sludge volume (SSV_5) is affected by water temperature, floc particle size and structure, and the presence or absence of filamentous bacteria. Many facilities see a strong seasonal pattern in which the SSV_5 is higher in the winter and lower in the summer simply because of changes in water temperature. Floc particles are close to the same density as water. When the water gets colder, it becomes more dense and the difference in density between the floc and the surrounding water is less pronounced. As a result, the settling velocity decreases. The settling velocity can also be affected by the solids concentration. As the MLSS concentration increases, the settling velocity decreases because the density of the fluid increases and there are more collisions between particles (Eckenfelder and Melbinger, 1957). This can be observed by comparing how samples

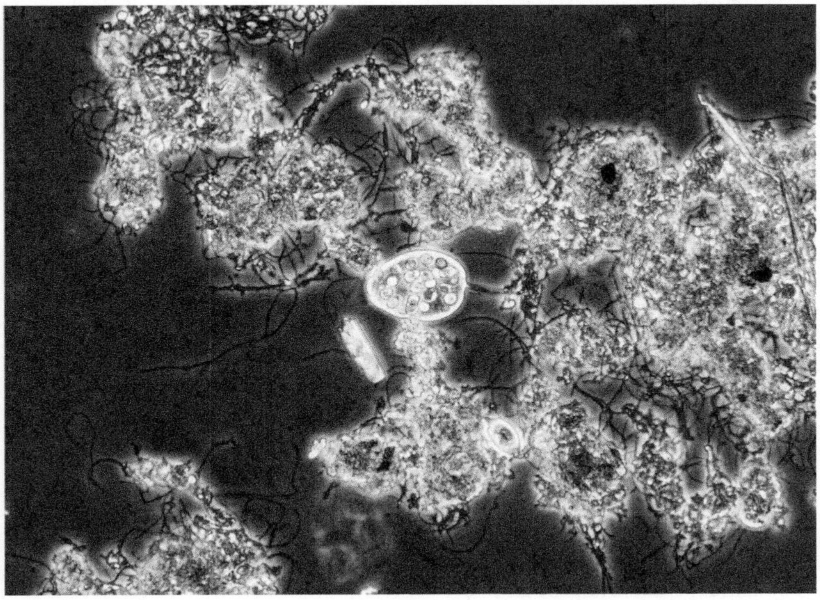

Figure 8.43 Filamentous Bridging Between Floc Particles (Reprinted with permission by Indigo Water Group)

of MLSS and RAS perform in the settleometer. The RAS will always settle at a slower rate—higher SSV_5—than the MLSS. If filaments are present and extend beyond the edges of the floc particles or, worse, create bridges between floc particles, this can create drag and slow down the settling velocity. If stalked ciliates and other predator organisms are present, they will scavenge free bacteria and help to produce a clear supernatant. Developing a sludge that flocculates and settles to leave behind a clear supernatant requires that the operator control the types and numbers of microorganisms present by adjusting sludge age.

Sludge Compaction

The 30-minute SSV in the settleometer indicates how well the floc is compacting. For slow-settling sludges, it may take longer than 30 minutes for compaction to begin. A series of U.S. EPA pamphlets produced in the 1970s recommended leaving the MLSS in the clarifier for 60 minutes or longer to achieve maximum sludge compaction (West, 1973a, 1973b, 1978). Leaving the MLSS in the clarifier for extended periods of time is no longer recommended because it can contribute to high effluent suspended solids from *deflocculation* and denitrification in the clarifier blanket (Ekama, 2010).

Deflocculation is when floc particles break apart into smaller floc particles.

The SVI calculation takes the SSV_{30} and divides it by the MLSS concentration in milligrams per liter (mg/L). The calculation result is in units of milliliters per gram (mL/g). In other words, how much space does one gram of solids occupy after being allowed to settle and compact for 30 minutes. Generally, the lower the SVI, the better the sludge quality. However, for the SVI calculation to be meaningful, the MLSS concentration must be within a reasonable range. Otherwise, the results cannot be compared in a meaningful way to literature values. A detailed discussion of the SVI calculation can be found in the section on Data Collection, Sampling, and Analysis in this chapter. It has long been accepted that an SVI below 70 mL/g can leave behind a turbid supernatant and that SVIs greater than 200 mL/g represent bulking sludge that isn't compacting well (Palm et al., 1980; Trygar, 2010). Water resource recovery facilities that practice enhanced biological phosphorus removal (EBPR) use chemically enhanced primary treatment, receive water facility residuals, and/or add other flocculants to their treatment process often report SVIs lower than 70 mL/g. How well a sludge compacts is generally dictated by the presence or absence of filamentous bacteria. Higher concentrations of filaments will prevent the MLSS particles from compressing together much in the same way that it is difficult to stuff unwound Christmas tree lights back into a box for storage. For the sludge to compact well, the operator must control the number of filaments, many of which can be controlled by maintaining an appropriate sludge age.

MAINTAIN MIXED LIQUOR SUSPENDED SOLIDS CONCENTRATIONS WITHIN A REASONABLE RANGE

Activated sludge MLSS concentrations are limited on the low end by bioflocculation and on the high end by both 1) the ability to transfer enough oxygen into the activated sludge basins to support the microorganisms and 2) the ability of the secondary clarifiers to flocculate, settle, and compact those solids. For most systems, this effectively limits the MLSS concentration to a range of 1000 to 4000 mg/L, although some types of activated sludge systems can operate successfully outside of this range. For the MLSS to flocculate, there needs to be a minimum number of collisions between floc particles. Collisions help the floc particles combine into larger particles and sweep up particulates from the influent wastewater. Generally, an MLSS concentration of at least 1000 mg/L is needed for enough collisions to occur. If the MLSS concentration falls too low, an operator can reduce the number of activated sludge basins in service.

As the MCRT/SRT increases, so does the MLSS concentration. As the MLSS concentrations increase, it becomes more and more difficult to dissolve oxygen into the wastewater, which increases energy usage. Eventually, a point is reached where so much air is being blown into the activated sludge basin that the resulting turbulence breaks the flocs apart, the MLSS settleability deteriorates, and effluent turbidity increases. High MLSS concentrations increase the SLR to the secondary clarifier. If the SLR gets too high, the clarifier won't be able to settle the solids.

WASHOUT MEAN CELL RESIDENCE TIME/SOLIDS RETENTION TIME—INFLUENCING BIOLOGY AND SETTLEABILITY

Most of the bacteria in the system reproduce rapidly and have generation times of 1 to 3 hours, with a few types reproducing as rapidly as once every 10 minutes (Brock and Madigan, 1991). For these bacteria, the MCRT/SRT must simply be long enough for the MLSS to flocculate to keep them in the system. Some bacteria, like the nitrifying bacteria and some filamentous bacteria, take much longer to reproduce. Some ciliates reproduce once every 4 to 5 hours (WEF, 2017). Like all ecological systems, higher predators and more complex organisms take longer to reproduce and can only develop after their prey are well established.

The washout-MCRT or washout-SRT is the minimum MCRT/SRT required to keep a particular organism in the system. Microorganisms are constantly entering the facility in the influent. In fact, many of the microorganisms in the activated sludge process are soil organisms (WEF, 2017). If the MCRT/SRT is too low, these organisms will be removed from the system as fast as they enter and a stable population won't develop. Although the term *washout* is used, the microorganisms are removed from the system through wasting. An operator performing a microscopic examination of their activated sludge may see the occasional nematode, for example, but unless the MCRT/SRT is old enough to keep that nematode in the system long enough for it to reproduce, the numbers of nematodes will not increase. This is true for every organism that can live, thrive, and survive in the activated sludge process, including the nitrifying bacteria, filamentous bacteria, and the macrofauna discussed earlier. The target MCRT/SRT selected by the operator will determine which microorganisms have an opportunity to develop stable populations. Other environmental conditions will also affect the biology of the activated sludge and the organisms present, including organic and nitrogen loading rates, DO concentrations, pH, alkalinity, and temperature. If other required conditions aren't met—food supply, oxygen, and so forth,—then even having a long enough sludge age may not produce a stable population of the desired organism. In practice, select a sludge age that is two to three times higher than the washout sludge age to keep a desired organism in the MLSS.

The following are some example estimated washout SRTs for different microorganisms at 20 °C:

- *Nocardioforms* = 1.5 days (Jenkins et al., 2004)
- Ammonia oxidizing bacteria (AOB) = 1.6 days (U.S. EPA, 2010)
- Phosphate accumulating organisms (PAOs) = 3 to 4 days (U.S. EPA, 2010)
- Nitrite oxidizing bacteria (NOB) = 2 days (U.S. EPA, 2010)
- *Sphaerotilus natans* = 3 to 7 days (Jenkins et al., 2004)
- *Microthrix parvicella* = Greater than 10 days (Jenkins et al., 2004)

Nocardioforms are a group of bacteria that includes *Nocardia*, *Gordona*, *Mycobacterium*, and other bacteria in the actinomycetes family. Earlier textbooks and manuals refer only to *Nocardia*.

What these washout SRTs tell us is that if an activated sludge process has a total SRT of 6 days and the in-basin water temperature is 20 °C, we would expect to find stable populations of AOB and NOB, but not *S. natans* or *M. parvicella*. If the process had an anaerobic zone and the correct ratio of influent BOD to phosphorus, we would also expect to find PAOs. The SRT by itself does not guarantee a stable population if other necessary conditions are not also met. Similarly, nocardioforms may or may not be present. Some filaments can be controlled simply by maintaining an SRT below the washout SRT for that filament. Old sludge filaments include *M. parvicella*, Type 0041, Type 0675, Type 1851, and Type 0803 (Jenkins et al., 2003). *M parvicella* is known for causing foaming and bulking episodes, especially during winter operating conditions, in WRRFs that must remove ammonia year-round. Figure 8.44 shows ranges of MCRTs for different filaments. Note that for many filaments like nocardioforms and *S. natans*, the washout SRT is so low that it isn't possible to waste them out of the process.

Washout SRT is affected by temperature. For every 10 °C drop in water temperature, the growth rate of the bacteria decreases by 50% and the washout SRT doubles. Growth rates for floc-forming and filament-forming bacteria are similarly affected. U.S. EPA (2010) created Figure 8.45 using a combination of calculated theoretical washout SRTs and experimental data by Hellinga et al. (1998). Data labeled as "H" are experimental and data labeled as "M" are calculated from AOB and NOB growth rates. At a water temperature of 20 °C, the washout SRT for the AOBs is approximately 1.6 days and the washout SRT for the NOBs is approximately 2.0 days. To maintain a stable population and to avoid accidental loss of these bacteria resulting from accidental overwasting, the target SRT would need to be two to three times as long or between 4 and 6 days. If the water temperature drops to 10 °C, the washout SRT for the AOBs increases to 3 to 5 days. This means a target SRT of between 10 and 15 days is needed to maintain stable nitrification. For a WRRF with winter water temperatures of 10 °C and summer water temperatures of 20 °C, the target SRT will need to move between 6 and 15 days to maintain nitrification. If the AOB and NOB are limited in

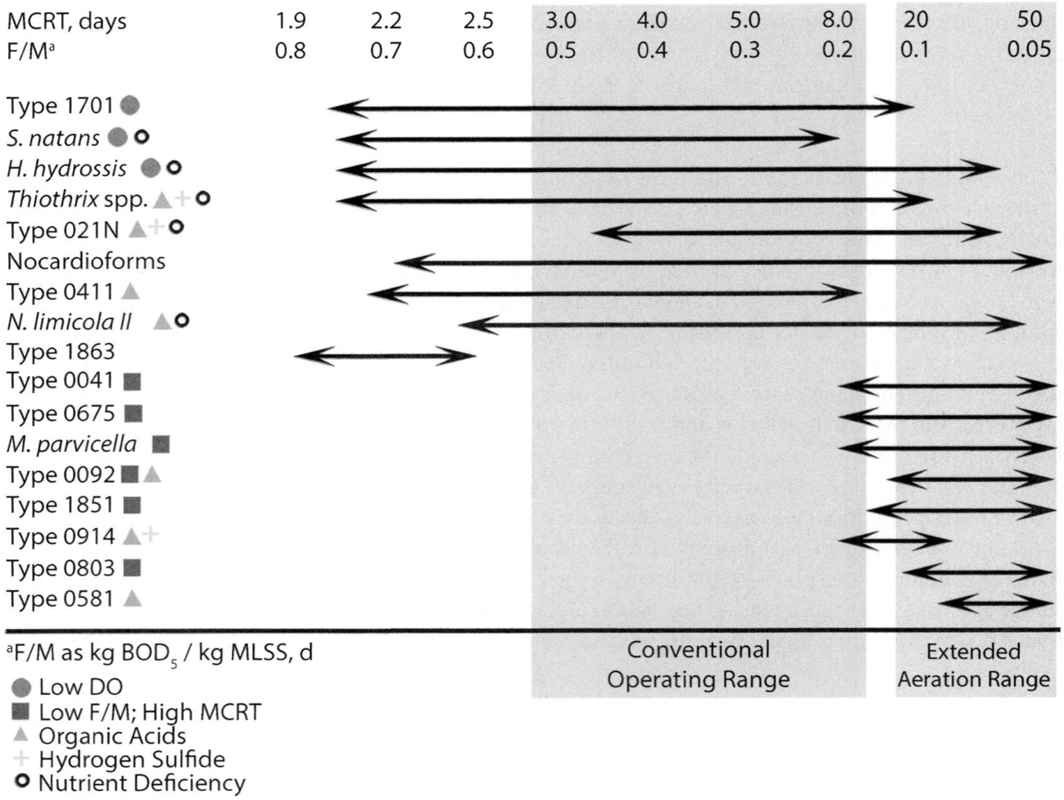

Figure 8.44 Washout SRTs for Various Filamentous Bacteria (modified from *Manual of the Causes and Control of Activated Sludge Bulking, Foaming, and Other Solids Separation Problems* by Jenkins, et al., Copyright © 2003. Reproduced by permission of Taylor and Francis Group, LLC, a division of Informa plc.)

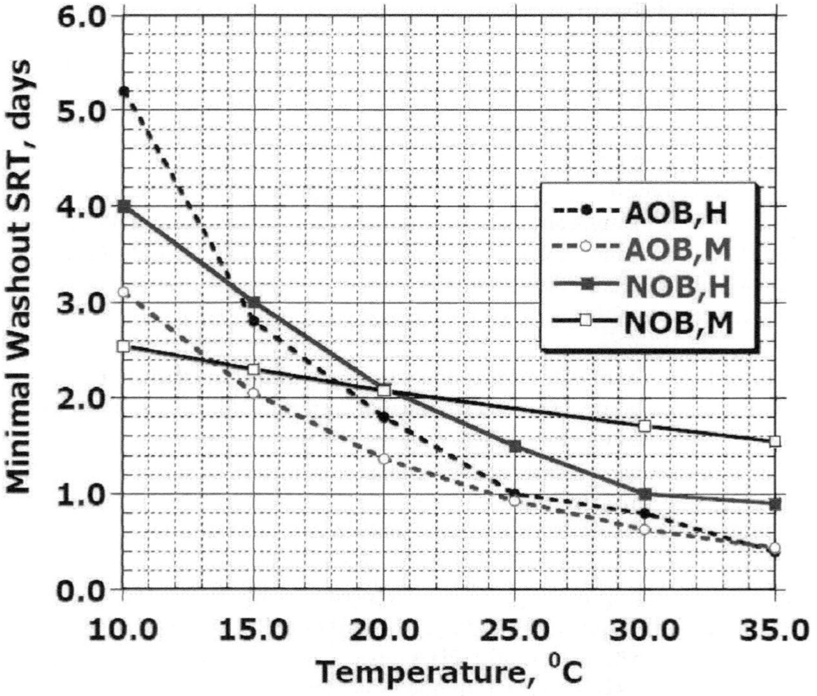

Figure 8.45 Washout SRTs for AOB and NOB at Different Water Temperatures (U.S. EPA, 2010)

some other way as well, for example low DO concentrations, then longer SRTs may be needed than those predicted by Figure 8.45.

MEET EFFLUENT AMMONIA LIMITS

The "right" sludge age for a facility depends on whether or not ammonia removal is required to meet permit limits. For WRRFs that are not required to nitrify, the MCRT/SRT only needs to be long enough to support bioflocculation and produce a low turbidity effluent. Generally, the lower the MCRT/SRT, the less expensive it will be to operate the facility as fewer kilograms (pounds) of MLSS will be under aeration. The longer the MLSS is aerated, the more energy is required to operate the blowers. On the other hand, total sludge production decreases slightly with increasing sludge age as a greater portion of the incoming BOD is used for cell maintenance instead of growth. For WRRFs that are required to nitrify, the $SRT_{aerobic}$ must be long enough to support stable nitrification, but short enough to keep filaments under control and maintain good sludge settleability. Because the washout $SRT_{aerobic}$ for the nitrifying bacteria depends on water temperature, many WRRFs that remove ammonia will need to adjust their target $SRT_{aerobic}$ seasonally.

Example #1—An oxidation ditch activated sludge process experiences changes in water temperature from 14 °C in January to 23 °C in July. The WRRF has effluent ammonia limits of 1.2 mg/L most months. The operator of this facility records the water temperature in the ditch every day and then selects a target SRT based on water temperature. As a result, the SRT is very gradually increased and decreased over the year to compensate for seasonal changes. Prior to implementation of seasonal SRT control, this facility experienced severe bulking and foaming episodes during the winter with SVIs in excess of 600 mL/g. After implementation of seasonal SRT control, they have operated for three successive winters without bulking or foaming.

Example #2—A three-stage, plug flow activated sludge process discharges all of its treated effluent to a storage reservoir. Discharges from the reservoir must meet stringent ammonia limits, but the reservoir can store more than 180 days of effluent. This gives the operators at this facility considerable flexibility to experiment with SRT. If the SRT is dropped too low for a short period of time and nitrification is lost, any ammonia discharged will be diluted below the permit limits by the reservoir. The activated sludge process has baffles separating the anaerobic, anoxic, and oxic zones that trap foam in the basins. As a result, the WRRF can have outbreaks of severe *M. parvicella* foaming at relatively low SRTs. The chief facility operator has been pushing the SRT progressively lower to determine the lower boundary of safe operation. In the summer, when water temperatures were 22 °C, they were able to maintain nitrification with an SRT of only 5 days. In the winter, when water temperatures dropped to 12 °C, they were able to maintain nitrification with an SRT of 11.5 days. If the SRT fell to 11 days, effluent ammonia concentrations would suddenly increase. This facility monitors effluent ammonia daily and increases its SRT if effluent ammonia increases.

SUMMARY ON SELECTING A TARGET MEAN CELL RESIDENCE TIME, SOLIDS RETENTION TIME, OR AEROBIC SOLIDS RETENTION TIME

The following are things to keep in mind when selecting a target sludge age:

- A minimum sludge age is needed to support bioflocculation even when nitrogen removal is not a requirement. If the sludge age is too low, the MLSS flocs that form won't be dense enough to settle in the clarifier. Use settleometer results to determine the best achievable settling characteristics at the lowest sludge age;
- Longer sludge ages are necessary when ammonia removal is required by the discharge permit. Target sludge ages should be based on a combination of water temperature, sludge settleability, and effluent ammonia concentrations;
- Suboptimal conditions—low pH, low DO, and insufficient alkalinity—affect the growth rate of the nitrifying bacteria and may require longer SRTs;
- As sludge age and MLSS concentrations increase, aeration costs increase;
- Longer sludge ages will encourage the growth of undesirable organisms like *Microthrix parvicella,* which may result in foaming and settling issues. Stay below the washout sludge age for these organisms whenever possible. Note that not all filaments are associated with longer SRTs;
- Excessively long sludge ages will cause the MLSS to go into endogenous respiration. Floc size will decrease as exopolymer is consumed as food. Settling may be very fast with turbid effluent; and
- Check your operating data. Is there a past sludge age that produces the best SVI?

TEST YOUR KNOWLEDGE

1. An effluent soluble BOD_5 concentration below 10 mg/L is an indication of well-flocculated MLSS.
   - ☐ True
   - ☐ False

2. Sludge bulking can result when the sludge age is either too high or too low.
   - ☐ True
   - ☐ False

3. Conventional and extended aeration are the two most common operating modes for domestic facilities. Check the box to indicate which of these operating modes has a higher HDT, SRT, F/M, or volumetric loading rate.

	Conventional	Extended Air
Detention time, hours		
SRT, days		
F/M kg BOD_5/kg MLVSS		
Volumetric loading rate		

4. The minimum sludge age needed to produce well-flocculated MLSS is around
   a. 2 to 3 days
   b. 5 to 15 days
   c. 15 to 30 days
   d. Greater than 30 days

5. Large, dense, well-flocculated MLSS should
   a. Produce low effluent ammonia concentrations
   b. Contain an abundance of filaments
   c. Remove particulates from the water as they settle
   d. Leave a cloudy supernatant behind

6. White foam is associated with
   a. Poorly flocculating activated sludge
   b. *Beggiatoa* filaments
   c. SRTs lower than 2 days
   d. F/M greater than 0.4

7. For filamentous bacteria to proliferate in activated sludge, environmental conditions must support their growth and
   a. Sludge age must be long enough
   b. DO must be low
   c. Volatile fatty acids are present
   d. Grease concentrations are high

8. How fast a floc particle settles in the secondary clarifier is influenced by all of the following EXCEPT
   a. Water temperature
   b. Presence of filaments
   c. Solids concentration
   d. DO concentration

9. Some facilities track their SSVs after 5 minutes because this is a good measure of sludge settling velocity. Why do many facilities see the SSV_5 decrease in summer and increase in winter?
   a. Loading changes with the seasons
   b. Water temperature affects settling velocity
   c. Typically more filaments in the summer
   d. Higher MLSS concentrations in summer

10. Which MLSS will tend to settle the slowest?
    a. MLSS of 3000 mg/L, few filaments present
    b. MLSS of 5000 mg/L, many filaments present
    c. MLSS of 3000 mg/L, many filaments present
    d. MLSS of 5000 mg/L, few filaments present

11. If MLSS has an SVI greater than 200 mL/g, it is defined as
    a. Well flocculated
    b. Filament free
    c. Well compacted
    d. Bulking

12. The maximum MLSS concentration for most activated sludge systems is limited by
    a. Bioflocculation
    b. Basin size
    c. Clarifier capacity
    d. Predation

13. The washout SRT for a newly discovered microorganism is 8 days at 20 °C. To keep this organism in the system, the target SRT should be at least
    a. 8 days
    b. 12 days
    c. 16 days
    d. 28 days

14. To prevent the growth of the filament *Sphaerotilus natans* in activated sludge, low DO conditions must be avoided and the SRT should be kept below about (assume a water temperature of 20 °C [68 °F])
    a. 3 days
    b. 9 days
    c. 15 days
    d. 21 days

SETTING A TARGET DISSOLVED OXYGEN CONCENTRATION

Selection of a target DO concentration in the aerated zones or basins should be based on 1) maintaining enough mixing energy to keep the MLSS in suspension, 2) providing enough oxygen to remove BOD_5, 3) preventing the growth of low DO filaments, 4) supporting nitrification if required by the discharge permit, and 5) minimizing operational costs by saving energy. The following sections discuss recommended DO concentrations for achieving these goals.

PREVENTING THE GROWTH OF LOW DISSOLVED OXYGEN FILAMENTS

For decades, operators have been taught to target a minimum DO concentration of 2 mg/L in the activated sludge basins. U.S. EPA recommended keeping DO concentrations between 2 and 4 mg/L in their 1973 operations pamphlets (West, 1973a, 1973b). This concentration range was thought to be the safe operating range that would prevent the growth of low DO filaments and minimize sludge bulking. Today, many systems operate successfully at much lower DO concentrations without experiencing bulking sludge due to the growth of low DO filaments because of the way they are operated. This section describes the conditions necessary for low DO filaments to proliferate.

For the microorganisms in the activated sludge process to grow at their maximum rate, they need access to both oxygen and food. The structure of floc particles creates a situation in which the oxygen concentration inside the floc will always be lower than the oxygen concentration outside of the floc. Figure 8.46 is a simple schematic of a floc particle. The microorganisms on the outside of the floc can easily access oxygen and food. As long as the concentrations of both are available in excess—the microorganisms have access to more than they can immediately use—they will grow at their maximum rate. Growth is not limited. Excess oxygen and fuel will pass further into the floc. The microorganisms on the inside of the floc don't have free access. They can only access the oxygen and fuel that penetrates into the floc. That means it has to get past the bacteria on the outside of the floc. A gradient is established where the oxygen and fuel concentrations are highest on the outside of the floc and are gradually decreasing through to the center of the floc. For larger flocs, the interior of the floc will be anoxic rather than aerobic, as shown in Figure 8.46. The lack of resources inside the floc causes the microorganisms there to grow slower. They can't achieve their maximum growth rates and growth is limited. Lau et al. (1984) concluded that the rapid growth of the floc formers under high DO conditions prevented both food and DO from reaching the interior of the floc particles. In other words, the apparent decrease in the growth rate of *S. natans* occurs because competition for food between the floc formers and *S. natans*—not because DO is inhibiting growth.

At the beginning of this chapter, the competition theory between filament formers and floc formers was discussed. Essentially, some filament formers can achieve their maximum growth rates at much lower oxygen and fuel concentrations than the floc formers. When the DO concentrations are low, low DO

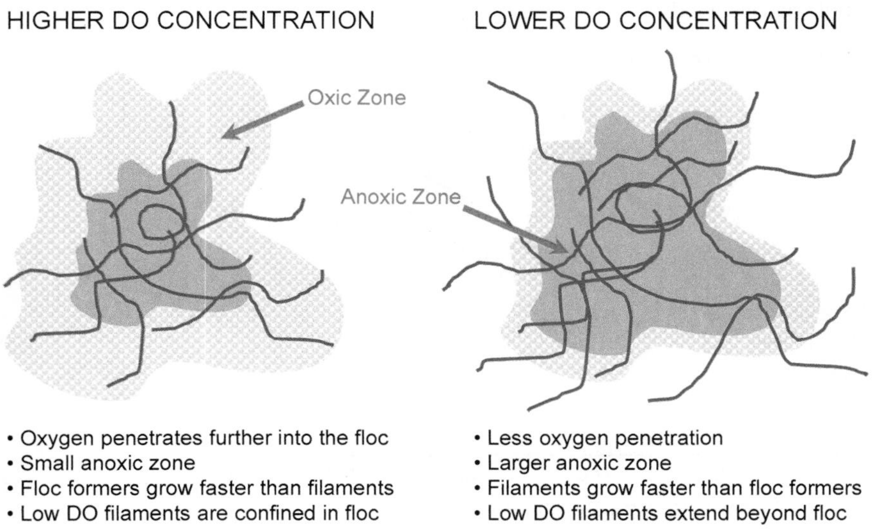
HIGHER DO CONCENTRATION
LOWER DO CONCENTRATION
Oxic Zone
Anoxic Zone
• Oxygen penetrates further into the floc
• Small anoxic zone
• Floc formers grow faster than filaments
• Low DO filaments are confined in floc
• Less oxygen penetration
• Larger anoxic zone
• Filaments grow faster than floc formers
• Low DO filaments extend beyond floc

Figure 8.46 Growth of Low DO Filaments (Reprinted with permission by Indigo Water Group)

filaments like *Sphaeotilus natans* and *H. hydrossis* can keep growing at their maximum rate, but the floc formers slow down. If the oxygen concentration is low enough, the low DO filaments will grow faster than the floc formers. The result is filaments that extend beyond the edges of the floc. The million-dollar question for operators is how low can the DO concentration go before low DO filamentous bulking occurs? The answer to this question has huge implications for overall operating costs.

The 2- to 4-mg/L DO recommendation from U.S. EPA was based on data collected from multiple full-scale activated sludge facilities, most of which were operated in conventional mode for F/M. It wasn't until 1980 that the first controlled experiments were conducted to determine the effect of DO concentration on sludge settleability. Palm et al. (1980) conducted a series of experiments in which they operated laboratory-scale, continuous-flow activated sludge systems. Two side-by-side systems were operated at the same F/M, but different DO concentrations. The DO concentrations were gradually lowered in one of the systems until low DO sludge bulking occurred. *Sphaerotilus natans* was identified as the filament responsible for bulking. The results of these experiments are shown in Figures 8.47 and 8.48. Figure 8.47 is based on COD data. Figure 8.48 is based on BOD data. Chemical oxygen demand was converted to BOD using a conversion factor established for the wastewater used in these experiments (Palm et al., 1980). Blocks in red show when low DO bulking occurred and *Sphaerotilus natans* was present. Blocks in green show when low DO bulking did not occur at the same F/M. The areas highlighted in blue in Figure 8.47 and gray in Figure 8.48 are the safe operating zones where low DO bulking does not occur. Palm et al. (1980) were able to demonstrate the following:

- The amount of DO uptake is directly related to the amount of COD or BOD_5 removed. High F/M and high DO on the outer portion of the floc encourage rapid growth. Higher numbers of rapidly growing bacteria results in a lot of competition for the available oxygen. As a result, the DO at the floc surface is used up quickly and less DO penetrates into the floc. A low DO or anoxic condition occurs in the center of the floc particle. The larger the low DO zone, the more likely it becomes for low DO filaments like *H. hydrossis* to proliferate. The low DO filaments also have access to increased amounts of BOD and can grow at their maximum rate;
- Some filaments can grow faster than floc formers under low DO conditions. *Sphaerotilus natans* was the dominant low DO filament in these experiments;
- Higher DO concentrations are needed at higher F/M to prevent bulking;
- Dissolved oxygen of 1.5 to 2 mg/L is sufficient to prevent low DO bulking when the F/M is less than 0.35 kg BOD_5/kg MLVSS (0.5 to 0.6 kg COD/kg MLVSS). This agrees with U.S. EPA's earlier recommendation and with research by Wilen and Balmer (1999);

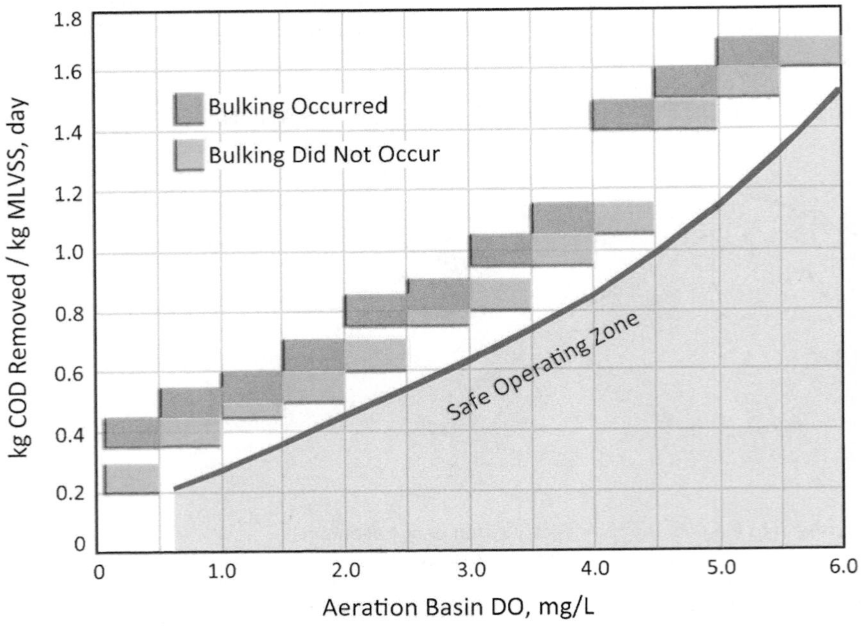

Figure 8.47 Incidence of Low DO Bulking Versus F/M Based on COD (modified from Palm et al., 1980)

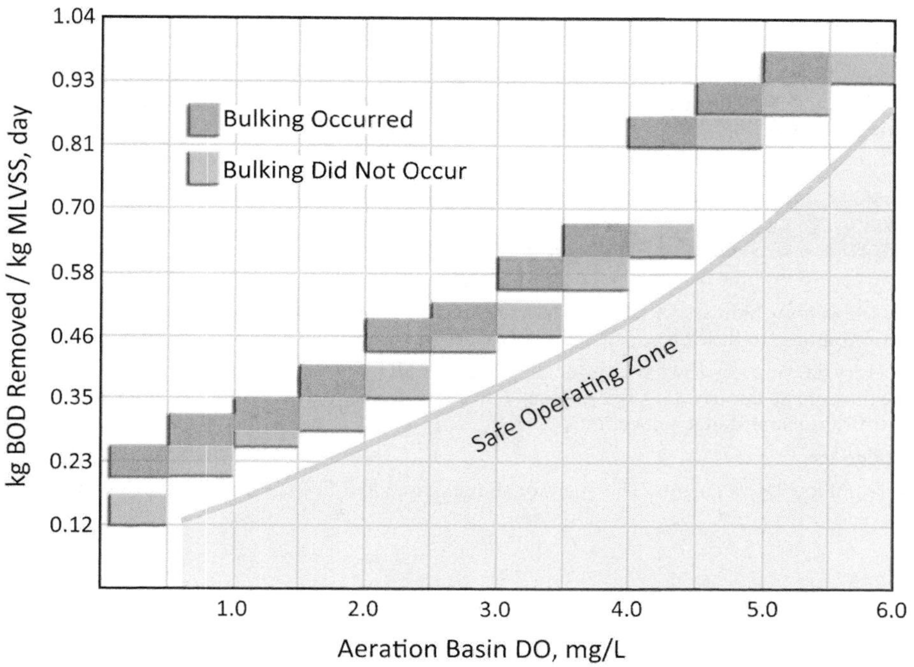

Figure 8.48 Incidence of Low DO Bulking Versus F/M Based on BOD_5 (assumes a COD to BOD ratio of 1.6 for domestic wastewater. The COD to BOD ratio can vary from one WRRF to another, especially if there are large industrial dischargers in the service area. Experiments used primary effluent [settled domestic wastewater] rather than raw wastewater). (Modified from Palm et al., 1980).

- Lower DO concentrations can prevent low DO bulking at lower F/M. The graphs shown in Figures 8.47 and 8.49 imply that DO could be taken all the way to zero, but the lowest concentration tested varied between 0.1 and 0.5 mg/L during the test. Dissolved oxygen concentrations lower than 0.5 mg/L are not recommended; and
- Low DO bulking can be cured simply by increasing the DO concentration; however, the time required to produce a stable, non-bulking sludge following a bulking episode is 2 to 3 sludge ages. If the SRT was 10 days, then the system will require 20 to 30 days of operation at a higher DO concentration for bulking to subside. It takes time to flush the filaments from the system.

MAXIMIZING NITRIFICATION RATES

The nitrifying bacteria need higher DO concentrations than the other floc-forming bacteria. Generally, a DO concentration of 2 mg/L is recommended to support maximum nitrification rates (Carberry and Englande [Eds.], 1983). There is a critical DO concentration, above which there isn't a significant increase in how fast the nitrifying bacteria convert ammonia to nitrite and nitrate. Below the critical DO concentration, the nitrifying bacteria can be inhibited. The conversion of ammonia to nitrite and nitrate slows down and becomes less efficient. If the DO concentration gets too low, nitrification will stop. For additional information on the DO requirements for the nitrifying bacteria, refer to the chapter on Nutrient Removal.

SECTION EXERCISE

1. A ski resort has an activated sludge process. The SRT is currently at 12 days and the MLSS concentration is 1800 mg/L. The resort has been open since September, but Christmas break is coming up in a few weeks, which is when they have the most visitors. The operator is concerned because the influent BOD_5 load will increase when all of the visitors show up. The operators don't think they have enough MLSS to maintain an F/M of 0.2 kg BOD_5/kg MLVSS·d (0.2 lb BOD_5/lb MLVSS·d). The operator decides to decrease their wasting rate so they can build up their inventory before the new BOD_5 arrives. Was this the right decision? Why or why not?
2. An activated sludge facility must remove ammonia year-round. Winter water temperatures are expected to drop to about 10 °C. Currently, water temperatures are near 20 °C. How should the operator make adjustments to the process over the next several months?

3. An operator decides to gradually lower the sludge age to flush filamentous bacteria out of the treatment process. When the sludge age reached 8.5 days, the effluent ammonia started to increase. Ammonia removal is not required. What should the operator do?
4. The DO in the activated sludge basins is 2.5 mg/L and the F/M is approximately 0.3 kg BOD_5/kg MLVSS·d. Looking at a sample of MLSS under the microscope, the operator sees filaments. The operator decides to turn up the DO, but nothing changes. How come nothing changed? What should the operator have done?

SECTION EXERCISE SOLUTIONS

1. This was not the right decision. It isn't possible to grow biomass without food for the microorganisms. If constant SRT or MCRT is maintained, the microorganisms will grow in response to the BOD_5 load and the F/M will automatically be in the right range.
2. Record the water temperature in the basin every day or every few days. Gradually adjust the SRT or MCRT by increasing or decreasing wasting, but never more than 10% increase or decrease per day. This will gradually move the facility between summer and winter operating conditions while maintaining nitrification and sludge settleability.
3. The operator may continue to lower the sludge age.
4. Nothing changed because the filaments were not low DO filaments. The operator should have identified the filament type before taking action.

TEST YOUR KNOWLEDGE

1. **When DO concentrations are low, certain types of filaments grow faster than the floc-forming bacteria because**
   a. DO is toxic to filament formers
   b. Floc-former growth slows down
   c. Filaments are protected by floc formers
   d. Sludge age increases
2. **If the F/M is 0.4 kg BOD_5/kg MLVSS, what DO concentration will prevent the growth of low DO filaments?**
   a. 0.5 mg/L
   b. 1 mg/L
   c. 2 mg/L
   d. 3 mg/L
3. **An activated sludge facility is not required to remove ammonia. The facility's F/M is only 0.2 kg BOD_5/kg MLVSS. What is the most likely outcome of maintaining a DO concentration of 1.0 mg/L in the activated sludge basins?**
   a. Growth of low DO filaments
   b. Floc-forming bacteria dominate
   c. Sludge bulking
   d. Low effluent ammonia
4. **An activated sludge process is experiencing extreme foaming and settleability is poor. Reviewing records, the chief facility operator discovers that the sludge age has increased over the last week to 16 days. Operators know from experience that the foaming subsides as long as the SRT is well controlled below 12 days. They make the necessary adjustments to bring the SRT below 12 days. How long will it take for the process to completely recover and stabilize at the new sludge age?**
   a. Less than 12 days
   b. Between 24 and 36 days
   c. Between 36 and 50 days
   d. Greater than 75 days
5. **Which of the following filaments grows in response to low DO conditions?**
   a. *S. natans*
   b. Type 021N
   c. *M. parvicella*
   d. Nocardioforms
6. **The SVI has increased over the last week from 120 to 180 mL/g. A microscopic evaluation reveals many *H. hydrossis* filaments. The operator should**
   a. Decrease sludge age
   b. Decrease DO concentrations
   c. Increase sludge age
   d. Increase DO concentrations

Process Variables for Secondary Clarifiers

Secondary clarifiers have both hydraulic capacity and solids handling capacity. In most cases, except in the case of low MLSS concentrations, the SLR will determine how many clarifiers need to be in service and where the RAS flowrate should be set.

INFLUENT FLOW

Influent flow can affect clarifier operation and performance because of its inherent variability. The secondary clarifiers are connected hydraulically to the influent flow. Even though variability in influent flow is *attenuated* by the processes between the headworks and secondary clarifiers, clarifier performance is still influenced by influent flow. The typical response to increased flow to a secondary clarifier is to increase the sludge blanket depth, which, in turn, can cause the effluent TSS concentrations to increase. Extreme changes in flowrate can increase the blanket depth until it crests the clarifier weir and causes major solids loss.

Attenuated means the effect is reduced. The variability in influent flow decreases as the water flows through the treatment process.

Flow variability has two components: diurnal and meteorological. Diurnal changes are the natural variations in flow that occur over a 24-hour period (Chapter 2). Diurnal flow changes contribute a flow peaking factor of approximately 1.5 and so typically do not impose any significant hydraulic disturbance on the clarifiers. Meteorological or seasonal changes can cause significant changes in influent flow according to rain and snow frequency and intensity, which can be elevated in the summer or the winter depending upon geography and climate. Snowmelt can also contribute to high influent flows. The level of *infiltration and inflow* (I/I) can elevate the flow peaking factor significantly, particularly in small communities. The net effect is that seasonal changes in flowrate can cause significant increases in effluent TSS concentrations.

Inflow and infiltration is groundwater and stormwater that enters the collection system through cracks, submerged manhole covers, missing clean out covers, and other openings. Inflow and infiltration is water that should not be entering the collection system.

The hydraulic effect of influent flow changes induced by the above range of operating conditions can be estimated by calculating the SOR, as discussed in the next section, and observing clarifier responses to changes in SOR. Flow changes can also affect clarifier SLR.

SURFACE OVERFLOW RATE

The SOR is a measure of the hydraulic loading to the secondary clarifier. It is calculated by taking the influent flow and dividing it by the clarifier surface area, as show in eq 8.20. Unlike the SLR calculation, SOR does not include RAS flow. It is expressed as cubic meters of flow per square meter per day ($m^3/m^2{\cdot}d$) or gallons per day per square foot (gpd/sq ft). Metric units may also be liters per day per square meter ($L/m^2{\cdot}d$), which are the units used in the formula sheet at the front of this book and on the ABC operator certification exams. Calculate SOR with one of the following formulas.

In International Standard units:

$$\text{SOR}, \frac{\text{m}^3}{\text{m}^2 \cdot \text{d}} = \frac{\text{Influent Flow, m}^3/\text{d}}{\text{Area, m}^2} \quad (8.20)$$

In U.S. customary units:

$$\text{SOR}, \frac{\text{gpd}}{\text{sq ft}} = \frac{\text{Influent Flow, gpd}}{\text{Area, sq ft}} \quad (8.21)$$

CALCULATION EXAMPLE

Find the SOR for a secondary clarifier given the following information:

Influent flow = 28.4 ML/d (7.5 mgd)

RAS flow = 14.2 ML/d (3.75 mgd)

MLSS concentration = 3500 mg/L

Clarifier surface area = 729 m^2 (7850 sq ft)

Number of clarifiers = 2

Solution with International Standard units:

Step 1—Convert the influent flow from megaliters per day to cubic meters per day. Note that the RAS flow is not included in the SOR calculation.

$$\frac{28.4\ \text{ML}}{\text{d}}\left|\frac{1\,000\,000\ \text{L}}{1\ \text{ML}}\right|\left|\frac{1\ \text{m}^3}{1000\ \text{L}}\right| = 28\,400\ \text{m}^3/\text{d}$$

Step 2—Find the total surface area of the clarifiers.

$$2\ \text{clarifiers}\left|\frac{729\ \text{m}^2}{1\ \text{clarifier}}\right| = 1458\ \text{m}^2$$

Step 3—Calculate the SOR.

$$\text{SOR}, \frac{\text{m}^3}{\text{m}^2\cdot\text{d}} = \frac{\text{Influent Flow, m}^3/\text{d}}{\text{Area, m}^2}$$

$$\text{SOR}, \frac{\text{m}^3}{\text{m}^2\cdot\text{d}} = \frac{28400\ \text{m}^3/\text{d}}{1458\ \text{m}^2}$$

$$\text{SOR}, \frac{\text{m}^3}{\text{m}^2\cdot\text{d}} = 19.5$$

Compare the result to the design criteria in Table 8.36 of 16.3 to 28.6 $\text{m}^3/\text{m}^2\cdot\text{d}$. The clarifiers in this example are operating within the desired range.

Solution with U.S. customary units:

Step 1—Convert the influent flow from million gallons per day to gallons per day.

$$\frac{7.5\ \text{mil. gal}}{\text{d}}\left|\frac{1\,000\,000\ \text{gal}}{1\ \text{mil. gal}}\right| = 7\,500\,000\ \text{gpd}$$

Step 2—Find the total surface area of the clarifiers.

$$2\ \text{clarifiers}\left|\frac{7850\ \text{sq ft}}{1\ \text{clarifier}}\right| = 15\,700\ \text{sq ft}$$

Gravity and RAS help pull particles toward the bottom of the clarifier while *SOR* pushes up toward the surface of the clarifier. The velocity of SOR is almost always smaller than the downward pull of gravity and RAS.

Step 3—Calculate the *SOR*.

$$\text{SOR}, \frac{\text{gpd}}{\text{sq ft}} = \frac{\text{Influent Flow, gpd}}{\text{Area, sq ft}}$$

$$\text{SOR}, \frac{\text{gpd}}{\text{sq ft}} = \frac{7\,500\,000\ \text{gpd}}{15\,700\ \text{sq ft}}$$

$$\text{SOR}, \frac{\text{gpd}}{\text{sq ft}} = 478$$

Compare the result to the design criteria in Table 8.3 of 400 to 700 gpd/sq ft. The clarifiers in this example are operating within the desired range.

Surface overflow rate is important because the sum of the particle downward velocities resulting from solids settling and removal of RAS from the bottom of the clarifier must be greater than SOR for successful liquid–solids separation. It was not until much later that, after much field testing and observations, it became clear that this downward velocity is almost always greater than the SOR.

Clarifiers rarely fail because the *SOR* is pushing solids out (clarification failure) over the clarifier weirs. Clarifiers almost always fail because more solids are being put into them than can be settled to the bottom and removed in the RAS (thickening failure) (Wahlberg, 1996). Modeling of clarifiers and the documentation of full-scale clarifier experience have suggested that SOR alone is not a major contributor to poor clarifier performance; rather, SLR has been identified as a more important criterion that controls secondary clarifier performance (Parker et al., 2001). In many cases, the observation that high effluent TSS concentrations have been induced by high SOR values has been tempered by the discovery that other design and operational constraints have contributed to poor clarifier performance, including hydraulic factors such as poor inlet/outlet design, inadequate RAS return capacity at peak flow conditions, inadequate detention time for flocculation in the center well, and denitrification caused by excess detention time of sludge in the blanket.

WEIR OVERFLOW RATE

The weir overflow rate is the amount of flow per linear meter (foot) of weir. Like SOR, the weir overflow rate is related to the influent flow and does not include RAS flow. Although rarely calculated for process control purposes, some states and other regulatory agencies limit the weir overflow rates to clarifiers during design. The weir overflow rate measures the velocity of the water as it passes over the weirs. If the velocity is too high, particles may be pulled up and out of the clarifier. Operators may wish to calculate the weir overflow rate in conjunction with SOR and SLR when determining how many clarifiers should be in service. The weir overflow rate may be calculated with the following equations:

In International Standard units:

$$\text{Weir Overflow Rate, } \frac{\text{m}^3}{\text{m}\cdot\text{d}} = \frac{\text{Influent Flow, m}^3/\text{d}}{\text{Weir Length, m}} \tag{8.22}$$

In U.S. customary units:

$$\text{Weir Overflow Rate, } \frac{\text{gpd}}{\text{ft}} = \frac{\text{Influent Flow, gpd}}{\text{Weir Length, ft}} \tag{8.23}$$

The weir length for a rectangular clarifier is easily found by measuring the length of the weir, or in the case of multiple weirs, adding together the lengths of all the weirs. The weir length for a circular clarifier is typically equal to the circumference of the clarifier. Inboard and inset launders, however, will have a smaller diameter than the clarifier. When finding the weir length, the difference in diameter must be taken into account.

The formula for the circumference of a circle is

$$\begin{gathered}\text{Circumference} = (\pi)(\text{diameter})\\ \text{or}\\ \text{Circumference} = 2(\pi)(\text{radius})\end{gathered} \tag{8.24}$$

Use a value of 3.14159 for π (pi).

Consider the clarifier with an inset launder shown in Figure 8.49. The clarifier diameter is 30.5 m (100 ft). If the launder is inset 2 m (6.6 ft) from the outside edge of the clarifier, then the diameter of the outer weir will be 26.5 m (86.8 ft). The inset distance must be subtracted from both sides of the clarifier to find the new diameter. If the launder is 0.6-m (2-ft) wide, then the diameter of the inner launder will be 25.3 m (82.8 ft). To find the total weir length, use formula 8.24 to find the length of each weir. Then, add them together.

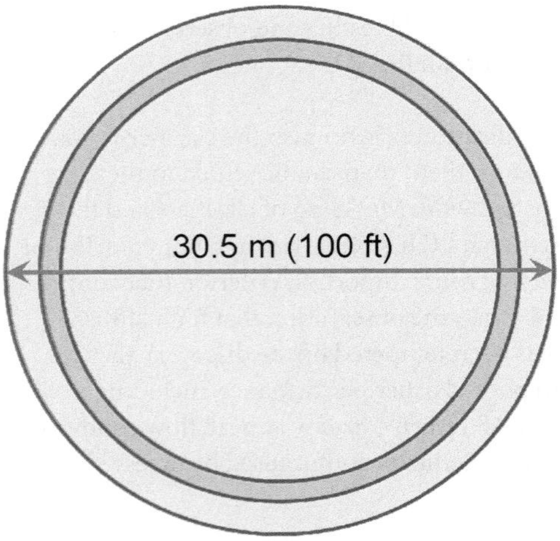

Figure 8.49 Overhead View of Clarifier with Inset Launder (Reprinted with permission by Indigo Water Group)

SOLIDS LOADING RATE

The SLR is a measure of the total solids loading to the secondary clarifier. It is calculated from the mass of solids flowing to the clarifier divided by the total surface area. Both the influent flow and the RAS flow push MLSS from the activated sludge basin into the clarifier. The SLR is expressed as kilograms per square meter per day (kg/m²·d) or pounds per square foot per day (lb/d/sq ft). The SLR may be calculated using the following formulas.

In International Standard units:

$$\text{SLR}, \frac{\text{kg}}{\text{m}^2 \cdot \text{d}} = \frac{\text{Solids Applied, kg/d}}{\text{Surface Area, m}^2} \tag{8.25}$$

In International Standard units for calculating mass:

$$\text{kg/d} = \frac{\left[\left(\text{Influent Flow}, \frac{\text{m}^3}{\text{d}} + \text{RAS Flow}, \frac{\text{m}^3}{\text{d}}\right)\left(\text{Concentration}, \frac{\text{mg}}{\text{L}}\right)\right]}{1000} \tag{8.26}$$

Note that the RAS flow must be added to the influent flow.

In U.S. customary units:

$$\text{SLR}, \frac{\text{lb}}{\text{sq ft} \cdot \text{d}} = \frac{\text{Solids Applied, lb/d}}{\text{Surface Area, sq ft}} \tag{8.27}$$

Note that the RAS flow must be added to the influent flow.

In U.S. customary units for calculating mass:

$$\text{lb/d} = (\text{Influent flow, mgd} + \text{RAS flow, mgd})\left(\text{Concentration}, \frac{\text{mg}}{\text{L}}\right)\left(8.34 \frac{\text{lb}}{\text{mil. gal}}\right) \tag{8.28}$$

CALCULATION EXAMPLE

Find the surface overflow rate for a secondary clarifier given the following information:

Influent flow = 28.4 ML/d (7.5 mgd)

RAS flow = 14.2 ML/d (3.75 mgd)

MLSS concentration = 3500 mg/L

Clarifier surface area = 729 m^2 (7850 sq ft)

Number of Clarifiers = 2

Solution with International Standard units:

Step 1—Convert the influent flow and RAS flow from megaliters per day to cubic meters per day.

$$\left.\frac{28.4\text{ ML}}{\text{d}}\right|\frac{1\,000\,000\text{ L}}{1\text{ ML}}\left|\right|\frac{1\text{ m}^3}{1000\text{ L}}\Bigg| = 28\,400\text{ m}^3/\text{d}$$

$$\left.\frac{14.2\text{ ML}}{\text{d}}\right|\frac{1\,000\,000\text{ L}}{1\text{ ML}}\left|\right|\frac{1\text{ m}^3}{1000\text{ L}}\Bigg| = 14\,200\text{ m}^3/\text{d}$$

Step 2—Find the mass of solids entering the clarifiers.

$$\text{kg/d} = \frac{\left[\left(\text{Influent Flow, }\frac{\text{m}^3}{\text{d}} + \text{RAS Flow, }\frac{\text{m}^3}{\text{d}}\right)\left(\text{Concentration, }\frac{\text{mg}}{\text{L}}\right)\right]}{1000}$$

$$\text{kg/d} = \frac{\left[\left(28\,400\frac{\text{m}^3}{\text{d}} + 14\,200\frac{\text{m}^3}{\text{d}}\right)\left(3500\frac{\text{mg}}{\text{L}}\right)\right]}{1000}$$

$$\frac{\text{kg}}{\text{d}} = 149\,100$$

Step 3—Find the total surface area of the clarifiers.

$$\left.\frac{2\text{ clarifiers}}{}\right|\frac{729\text{ m}^2}{1\text{ clarifier}}\Bigg| = 1458\text{ m}^2$$

Step 4—Find the SLR.

$$\text{SLR, }\frac{\text{kg}}{\text{m}^2\cdot\text{d}} = \frac{\text{Solids Applied, kg/d}}{\text{Surface Area, m}^2}$$

$$\text{SLR, }\frac{\text{kg}}{\text{m}^2\cdot\text{d}} = \frac{149\,100\text{ kg/d}}{1458\text{ m}^2}$$

$$\text{SLR, }\frac{\text{kg}}{\text{m}^2\cdot\text{d}} = 102$$

Compare the result to the design criteria in Table 8.3 of 100 to 150 kg/m^2·d. The clarifiers in this example are operating at the low end of range.

Solution with U.S. customary units:

Step 1—Find the mass of solids entering the clarifiers.

$$\text{lb/d} = (\text{Influent flow, mgd} + \text{RAS flow, mgd})\left(\text{Concentration}, \frac{\text{mg}}{\text{L}}\right)\left(8.34\frac{\text{lb}}{\text{mil. gal}}\right)$$

$$\text{lb/d} = (7.5\text{ mgd} + 3.75\text{ mgd})\left(3500\frac{\text{mg}}{\text{L}}\right)\left(8.34\frac{\text{lb}}{\text{mil. gal}}\right)$$

$$\text{lb/d} = 328\ 388$$

Step 2—Find the total surface area of the clarifiers.

$$\frac{2\text{ clarifiers}}{} \left| \frac{7850\text{ sq ft}}{1\text{ clarifier}} \right| = 15\ 700\text{ sq ft}$$

Step 3—Find the SLR.

$$\text{SLR}, \frac{\text{lb}}{\text{sq ft}\cdot\text{d}} = \frac{\text{Solids Applied, lb/d}}{\text{Surface Area, sq ft}}$$

$$\text{SLR}, \frac{\text{lb}}{\text{sq ft}\cdot\text{d}} = \frac{328\ 388\text{ lb/d}}{15\ 700\text{ sq ft}}$$

$$\text{SLR}, \frac{\text{lb}}{\text{sq ft}\cdot\text{d}} = 20.9$$

Compare the result to the design criteria in Table 8.3 of 20 to 30 lb/d/sq ft. The clarifiers in this example are operating at the low end of the range.

IMPACT OF SLUDGE SETTLEABILITY ON SOLIDS LOADING RATE

Table 8.3 gives ranges of SLR for secondary clarifiers. The ranges can be a little misleading though because how many solids the clarifier can effectively settle depends on both the solids concentration AND how well/fast those solids settle. When the sludge settles poorly, it takes longer for it to reach the bottom of the clarifier. It can't be removed until it reaches the bottom. This limits the capacity of the clarifier. If solids are added to the clarifier faster than they are removed, the clarifier will, eventually, fill with sludge and the blanket will go over the weir. Conversely, when the sludge is settling really well, it becomes possible to increase the SLR.

There are a few different tools available to help predict the solids handling capacity of secondary clarifiers. They are helpful for running "what if" scenarios before making a process change. Most are mathematical models built with Excel spreadsheets. These tools contain simple graphics that help the user understand if their clarifier is underloaded or overloaded and where the RAS flowrate should be set for optimum performance. The Water Environment Research Foundation/Clarifier Research Technical Committee (CRTC) Protocol (Wahlberg, 2001) provides guidance on how to develop one of these models: State Point Analysis. Building the spreadsheet and collecting data is time consuming, but is more accurate than other approaches.

Daigger (1995) and Daigger and Roper (1985) developed a convenient clarifier operating diagram (Figure 8.50). The diagram should be used with caution as models don't perfectly reflect real world situations.

How it works. The clarifier operating point can be located on the diagram by using two of the following operating parameters: actual SLR, underflow rate, or RAS solids concentration. If all three are known,

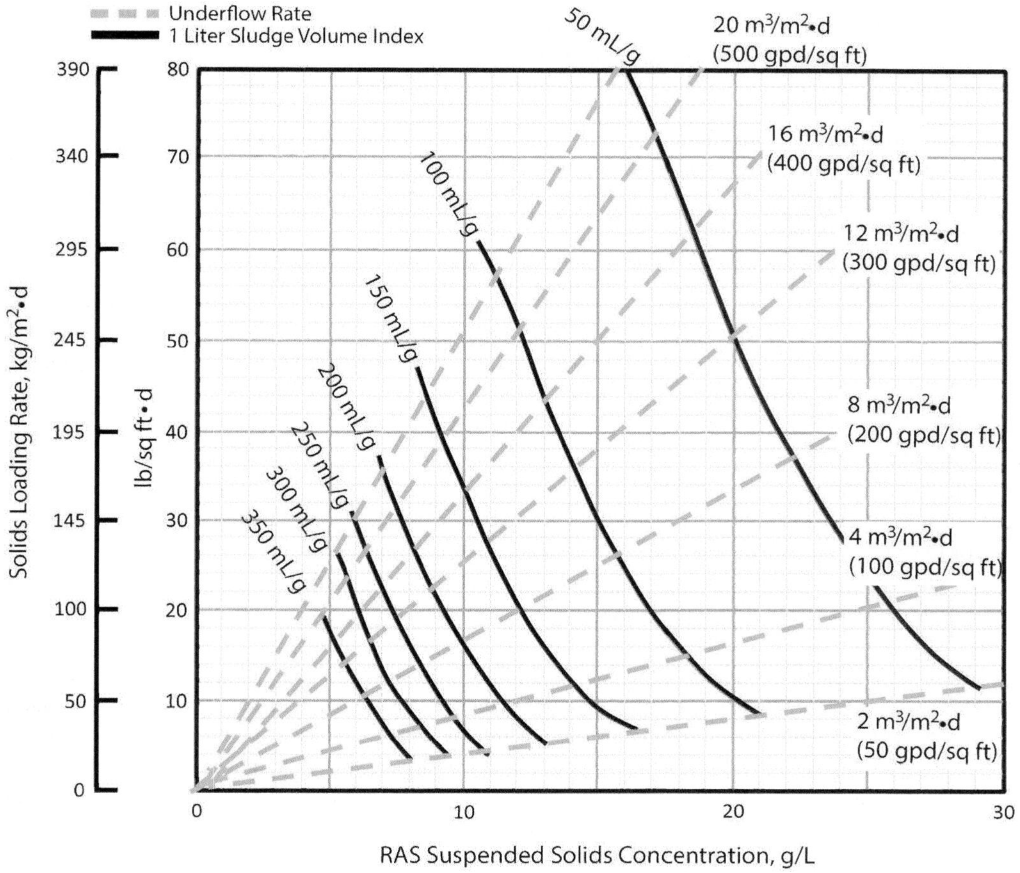

Figure 8.50 Daigger Clarifier Operating Chart (Modified from Daigger, 1995)

then the third piece of data can be used as a check on the other two. The underflow rate is the RAS flowrate divided by the surface area of the clarifier. It is calculated in the same way as SOR, but instead of using influent flow, RAS flow is used. In Figure 8.50, the underflow rate lines are shown as dashed, green lines. One additional piece of data will be needed: the current SVI. Most WRRFs use a 2-L Mallory-type settleometer to determine SVI. The data used to create the operating chart were collected with a 1-L graduated cylinder. For the greatest accuracy when using the chart, use a 1-L graduated cylinder for the settleometer test.

Example:

A secondary clarifier has the following characteristics:

- SLR = 107.4 kg/m²·d (22 lb/d/sq ft)
- RAS concentration = 12 000 mg/L
- Underflow rate = 9 m³/m²·d (220 gpd/sq ft)
- Settleometer test conducted in a 1-L graduated cylinder has an SVI of 150 mL/g

Step 1—Find the RAS concentration on the *x*-axis and draw a line straight up across the graph.

Step 2—Find the SLR on the *y*-axis and draw a line straight across the graph.

Step 3—Mark the location where the two lines cross.

If either the SLR or RAS concentration is not available, find the dashed green line that most closely matches the current underflow rate. Mark the location where it crosses the line drawn in step 1 or step 2.

The place where the two lines intersect is the current operating point for the clarifier. Lastly, find the heavy black line that most closely matches the calculated SVI for the 1-L settleometer test. In this example, the 150-mL/g line.

- If the operating point is below and to the left of the current SVI line, the clarifier is operating below the maximum SLR for that SVI. The clarifier should be able to settle the solids.
- If the operating point falls on top of the current SVI line, the SLR is operating at its limit. This condition is referred to as "critically loaded".
- If the operating point is above and to the right of the current SVI line, the clarifier is overloaded and failure is likely. If the clarifier remains overloaded, the blanket will increase until it flows over the weirs.

The operating point in Figure 8.51 is between the SVI lines for 100 mL/g and 150 mL/g. If sludge settleability deteriorates for some reason and SVI increases over 150 mL/g, then operational changes will have to be made to decrease the SLR.

RETURN ACTIVATED SLUDGE

Return activated sludge return is needed to transfer sludge from the secondary clarifiers to the activated sludge basins. Typically, RAS return rates are expressed as a percentage of flow to the activated sludge basins. For low SRT systems with low MLSS concentrations, RAS flow is typically 25 to 50% of influent flow. For higher SRT systems with high MLSS concentrations and systems that remove ammonia, RAS flow is on the order of 75 to 120% of influent flow.

As seen above, RAS is an integral part of the definition of SLR. Changing the RAS flowrate will influence the clarifier SLR. Increasing the RAS flowrate will draw down the sludge blanket and transfer sludge back to the activated sludge basin. If a substantial portion of the MLSS was in the clarifier blanket, this action will increase the MLSS concentration in the activated sludge basin and, therefore, increase the clarifier SLR. Lowering the RAS flowrate will have the opposite effect; sludge will progressively accumulate in the bottom of the clarifier, allowing more time for thickening and increasing the sludge blanket solids concentration.

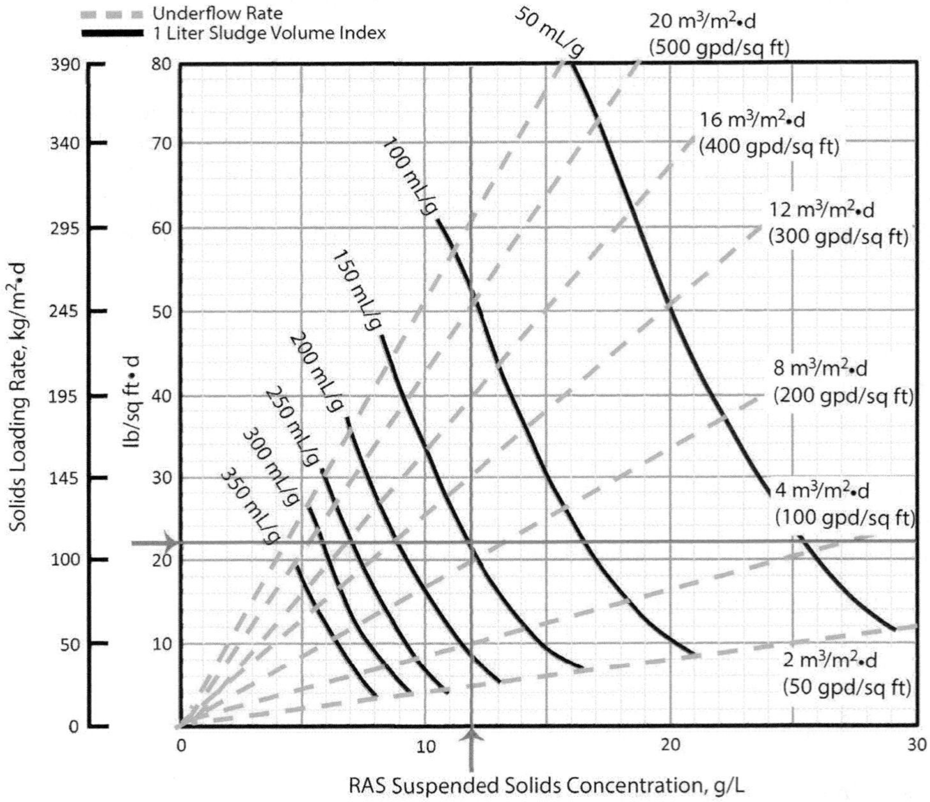

Figure 8.51 Example Using Daigger Clarifier Operating Chart (Daigger, 1995)

The net effect is to transfer more solids to the clarifier and the MLSS concentration will slowly decline, reducing the clarifier SLR. Changes to the MLSS concentration in the activated sludge basin will be small unless 1) solids are being stored in the clarifier blanket or 2) the clarifiers are large compared to the activated sludge basins. Typically, sludge blanket depths are maintained between 0.16 and 0.7 m (0.5 and 2 ft) from the bottom of the clarifier.

There is no optimum recommended RAS concentration or flowrate because RAS flowrate is correlated with so many other parameters that are varied in the operation of an activated sludge system.

TEST YOUR KNOWLEDGE

1. **This process variable is typically what limits clarifier capacity.**
   a. SOR
   b. Weir loading rate
   c. RAS flowrate
   d. SLR

2. **Find the SOR for a clarifier that is 19.8 m (65 ft) in diameter when the influent flow is 7.57 ML/d (2 mgd) and the RAS is set at 60% of influent flow.**
   a. 9.8 $m^3/m^2 \cdot d$ (241 gpd/sq ft)
   b. 24.6 $m^3/m^2 \cdot d$ (603 gpd/sq ft)
   c. 14.7 $m^3/m^2 \cdot d$ (362 gpd/sq ft)
   d. 39.3 $m^3/m^2 \cdot d$ (965 gpd/sq ft)

3. **The RAS flow is increased from 50% of influent flow to 65% of influent flow. Treatment time in the activated sludge basins**
   a. Decreases with increasing RAS flow
   b. Should only be adjusted in 10% increments
   c. Remains the same regardless of RAS flow
   d. Increases with increasing RAS flow

4. **Find the SOR for a clarifier that is 36.6 m (120 ft) in diameter when the influent flow is 26.5 ML/d (7 mgd).**
   a. 14.7 $m^3/m^2 \cdot d$ (348 gpd/sq ft)
   b. 25.2 $m^3/m^2 \cdot d$ (619 gpd/sq ft)
   c. 36.3 $m^3/m^2 \cdot d$ (892 gpd/sq ft)
   d. 56.7 $m^3/m^2 \cdot d$ (1396 gpd/sq ft)

5. **A rectangular secondary clarifier has a total weir length of 18.3 m (60 ft). The influent flow is 5 ML/d (1.32 mgd). Find the weir overflow rate.**
   a. 205 $m^3/m \cdot d$ (16 500 gpd/ft)
   b. 234 $m^3/m \cdot d$ (18 857 gpd/ft)
   c. 273 $m^3/m \cdot d$ (22 000 gpd/ft)
   d. 328 $m^3/m \cdot d$ (26 400 gpd/ft)

6. **A secondary clarifier that is 16.8 m (55 ft) in diameter is equipped with an inset launder. The outermost weir is 0.75 m (2.5 ft) from the outer wall of the clarifier. What is the diameter of the outer weir?**
   a. 15.3 m (50 ft)
   b. 16.1 m (52.5 ft)
   c. 16.8 m (55 ft)
   d. 18.3 m (60 ft)

7. **A WRRF receives 11.4 ML/d (3 mgd) of influent flow. The MLSS concentration is 3000 mg/L and the RAS concentration is 7600 mg/L. Find the total mass going to the clarifier if the RAS is set at 65% of influent flow.**
   a. 46 170 kg (101 331 lb)
   b. 56 430 kg (123 849 lb)
   c. 90 249 kg (198 909 lb)
   d. 143 245 kg (314 386 lb)

8. **Find the SLR to a clarifier if the mass of MLSS going to the clarifier is 20 000 kg (44 080 lb) and the surface area is 117 m^2 (1260 sq ft).**
   a. 53 $kg/m^2 \cdot d$ (10.8 lb/hr/sq ft)
   b. 92 $kg/m^2 \cdot d$ (18.8 lb/hr/sq ft)
   c. 117 $kg/m^2 \cdot d$ (23.9 lb/hr/sq ft)
   d. 171 $kg/m^2 \cdot d$ (35.0 lb/hr/sq ft)

9. **If the SLR is too high**
   a. Blanket depth will increase
   b. RAS thickness will increase
   c. Blanket depth will decrease
   d. RAS thickness will decrease

10. **Last week, the SVI was 120 mL/g. This week, the SVI increased to 210 mL/g. The operator may need to**
    a. Increase the RAS flowrate
    b. Stop adding polymer to the clarifier
    c. Remove a clarifier from service
    d. Reduce the SLR

11. **A clarifier has a SLR of 171 $kg/m^2 \cdot d$ (35 lb/d/sq ft). The RAS concentration is 13 000 mg/L. Use the diagram in Figure 8.50 to determine which of the following statements is true.**
    a. The clarifier can settle sludge with an SVI up to 150 mL/g.
    b. The underflow rate is between 12 and 16 $m^3/m^2 \cdot d$ (300 and 400 gpd/sq ft)
    c. If the current, operating SVI is 85 mL/g, the clarifier will be overloaded.
    d. Because the operating point does not fall directly on a line, the chart cannot be used.

Process Control for Secondary Clarifiers

Process control goals for secondary clarifiers include

- Maintaining the blanket below 0.7 m (2 ft),
- Preventing denitrification in the clarifier blanket (floating sludge),
- Preventing settled sludge from becoming septic, and
- Keeping RAS flowrates as low as possible while still meeting the other goals.

MAINTAINING MINIMAL BLANKETS

Maintaining the blanket depth below 0.7 m (2 ft) begins in the activated sludge basins by providing an environment that produces dense, compact sludge with few filaments. As discussed earlier, adjusting the RAS flowrate will increase or decrease blanket depth and also change the RAS concentration. Many facilities flow pace the RAS flow to the influent flow. The objective here is to keep the ratio constant over a 24-hour period so that the RAS concentration and blanket depths are also fairly constant. Return activated sludge concentrations will fluctuate to some extent as the MLSS concentration in the activated sludge basin goes up and down in response to the influent BOD_5 load. Facilities may be limited by how far down they can turn their RAS pumps at low flow and/or by their maximum pumping capacities at peak flow. Other facilities, especially small systems, use a "set-it-and-forget-it" approach by selecting a RAS flowrate that allows the blanket to build during peak flow and decrease after peak flow has passed. The disadvantage to a set-it-and-forget-it approach is that it could allow blankets to rise too much in some circumstances like storm events. Storm events can increase I/I in systems with combined sewers or collection systems that are not in great condition. When blankets are deep, there is a risk of losing solids over the weirs and into the final effluent.

There will come a time in nearly every operator's career when the MLSS is settling poorly and a blanket begins to build in the secondary clarifier. In response, the operator will increase the RAS flowrate to pull solids out of the clarifier faster. Two different things could happen. Either the blanket will decrease or, paradoxically, the blanket could go up even faster. Understanding why this happens is critical to secondary clarifier process control. Look back at eqs 8.25 through 8.28. These four equations have been combined into two larger equations below, one for International Standard units and one for U.S. customary units.

In International Standard units for calculating SLR:

$$\text{SLR}, \frac{\text{kg}}{\text{m}^2\cdot\text{d}} = \frac{\left[\left(\text{Influent Flow}, \frac{\text{m}^3}{\text{d}} + \text{RAS Flow}, \frac{\text{m}^3}{\text{d}}\right)\left(\text{Concentration}, \frac{\text{mg}}{\text{L}}\right)\right]}{(1000)(\text{Surface Area of Clarifier, m}^2)} \tag{8.29}$$

In U.S. customary units for calculating SLR:

$$\text{SLR}, \frac{\text{lb}}{\text{d}\cdot\text{sq ft}} = \frac{\left[(\text{Influent Flow, mgd} + \text{RAS Flow, mgd})\left(\text{Conc.}, \frac{\text{mg}}{\text{L}}\right)\left(8.34\frac{\text{lb}}{\text{mil. gal}}\right)\right]}{(\text{Surface Area of Clarifier, sq ft})} \tag{8.30}$$

Combining the equations makes it easier to see the relationships between the different process variables. If either the influent flow or the RAS flow is increased, then the SLR will also increase. If the MLSS concentration increases, then the SLR will also increase. If the clarifier surface area is increased, the SLR decreases. We know from our previous discussion that the mass of solids the clarifier can settle is related to how well (fast) the sludge is settling and compacting. When the sludge is settling and compacting poorly, the allowable SLR goes down. If the SLR has already been exceeded, that is more solids are coming into the clarifier than can be settled and withdrawn, then increasing the RAS flowrate will make the problem worse. Why? Because increasing the RAS flowrate increases the SLR on a clarifier that is already overloaded. Sometimes, when the sludge is settling poorly, the correct response is to turn the RAS down rather than up.

There are four different things an operator can change to adjust the SLR and improve clarifier performance. Looking at eqs 8.29 and 8.30, it is clear that while the operators have no control over the influent flow to the WRRF, they can manipulate 1) RAS flowrate, 2) MLSS concentration, and 3) clarifier surface area (number of clarifiers in service). Manipulating the RAS flowrate and increasing the number of clarifiers are easily accomplished and will have immediate effects.

Reducing the MLSS concentration in the short term can be done by placing another activated sludge basin (or multiple basins) into service. If the same mass (kilograms or pounds) of MLSS is spread over a greater volume, then the concentration will decrease and so will the SLR. Extra basins may not be available. Reducing MLSS concentrations in the long term requires adjusting the SRT. As the SRT is decreased, the MLSS concentration will also decrease; however, this long-term solution may take weeks to reach an MLSS concentration that the clarifiers can settle and compact. Facilities that nitrify may not be able to reduce their SRT without losing nitrification.

The fourth variable is less obvious. It is the settleability of the MLSS as measured by the SVI. If the SVI can be improved, then the allowable SLR goes up. In other words, the clarifier will be able to settle more solids. The SVI can be drastically improved in the short term by adding a chemical to the MLSS to improve flocculation. Polymer, ferric chloride, or alum may be added upstream of the secondary clarifier. If poor settling occurs because of filamentous bacteria, chlorine or polyaluminum chloride (PACl) may be added to kill the filaments and reduce their numbers. Recommended dosages are included in the troubleshooting section of this chapter. Chemical usage can be expensive and it increases sludge production. It should be used only as a last resort. Many facilities choose to add chemicals for a few weeks out of the year when their flows are highest or when their sludge settleability is poorest. This is a fine strategy that can be more cost-effective than building more clarifiers. When chemical addition becomes chronic, it is time to evaluate the process control strategy that is producing such poorly settling sludge and/or to think about building more clarifiers.

PREVENTING DENITRIFICATION AND SEPTICITY

If the sludge is allowed to sit in the clarifier too long, denitrification can occur and/or the sludge may become septic. Denitrification is the conversion of nitrate to nitrogen gas. Many of the bacteria in MLSS will automatically perform this conversion if 1) DO concentrations fall below approximately 0.5 mg/L, 2) BOD_5 is either available in the surrounding water or the bacteria contain absorbed BOD_5 that they have not finished digesting, and 3) excess nitrite and nitrate is available (typically greater than 5 mg/L as N combined). Denitrification in the clarifier can be a huge problem because the nitrogen gas bubbles produced can attach to the floc and carry it back to the surface.

A small amount of denitrification will make the surface of the clarifier appear as if someone had sprinkled fireplace ashes across the surface. In fact, it is called *ashing* when the clarifier looks this way. As more nitrogen gas is produced, more settled sludge will return to the surface and can appear as an almost uniform scum layer (Figure 8.52) or as large clumps rising in an otherwise clear clarifier (Figure 8.53). Sometimes, it can be difficult to determine if scum and floating floc are from denitrification or if they are simply straggler flocs that are too light and fluffy to settle. Key indicators that it is denitrification are the appearance of gas bubbles on the surface, especially behind the rake arm as it moves through the blanket, and sludge that appears on the effluent side of the scum baffle. Both conditions are shown in Figure 8.52. There is only one way for sludge to end up on the effluent side of the scum baffle; it went underneath the baffle and came back up on the other side.

It isn't possible, most of the time, to put enough air into the MLSS to maintain aerobic conditions in the sludge blanket. If you have access to fresh activated sludge, collect a sample and take it back to the laboratory. Shake it for several minutes to get the oxygen concentration as high as possible. Then, use a DO meter to measure the DO concentration continuously over several minutes. Most of the time, the DO concentration will drop rapidly and be completely gone before 10 minutes have passed. Because the MLSS will remain in the clarifier long enough to settle and compact, it is unlikely there will be much residual DO in the blanket. Recall from Chapter 5 that as soon as the facultative heterotrophic bacteria run out of oxygen, they will begin to use nitrite and nitrate if they are available.

Denitrification requires 1) excess BOD_5 stored in the microorganisms, 2) excess nitrite and/or nitrate, and 3) time for the denitrification reaction to take place. Denitrification, if it is taking place, will be worst

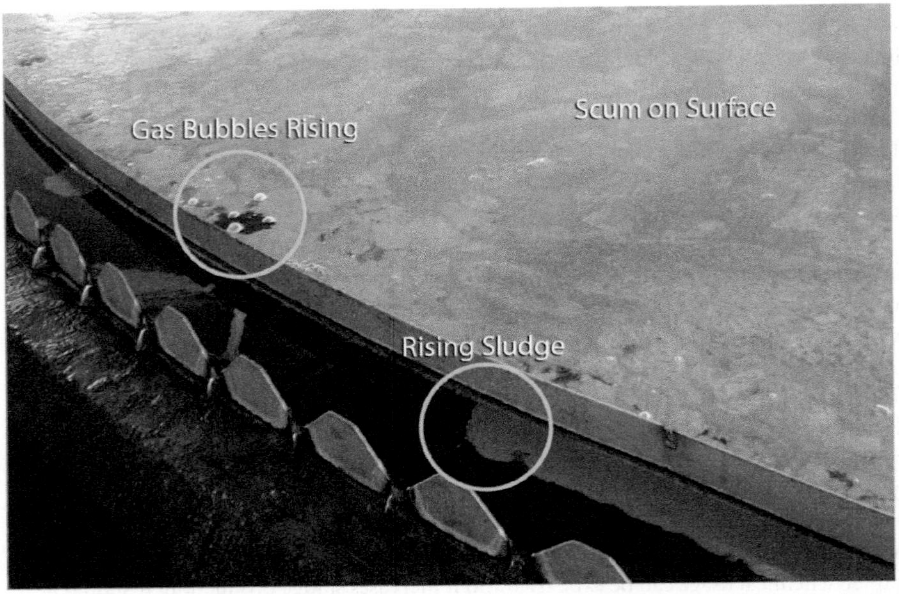

Figure 8.52 Denitrification in a Secondary Clarifier (Reprinted with permission by Indigo Water Group)

during peak loading to the facility when the F/M is highest. If denitrification is suspected, look for bubbles behind the rake arm during peak flow and load. If one of these three variables can be eliminated, then denitrification can be prevented. Excess BOD_5 can be reduced by increasing the SRT. Nitrite and nitrate may be eliminated either by reducing the SRT to prevent nitrification from happening in the first place or by forcing denitrification to take place in the activated sludge basin rather than the clarifier (Chapter 9). Reducing one or more of these variables may not prevent denitrification completely, but it can slow it down enough that it no longer poses a problem for clarifier operation. If possible, denitrify in the activated sludge basins before the MLSS is transferred to the secondary clarifiers. If denitrification continues to cause problems in the clarifier or if upstream denitrification is not an option, increase the RAS flowrate to remove the settled sludge before it has enough time to denitrify.

Figure 8.53 Clumps of Sludge Rising in a Secondary Clarifier (Reprinted with permission by Indigo Water Group)

MINIMIZING RETURN ACTIVATED SLUDGE FLOWRATES

Return activated sludge flowrates should be kept as low as possible while meeting all other operational goals. Excessive pumping reduces the RAS concentration and can affect solids handling processes. Pumping costs money in terms of both electricity and in pump wear and tear. If the RAS can be set as low as 40% of influent flow while keeping the sludge blanket below 0.7 m (2 ft) and preventing denitrification, then it should be set at that point. If it can be set even lower while meeting all process control and effluent quality goals, then consider setting it at that point. Facilities that remove ammonia often find that RAS rates as high as 125% of influent flow are necessary to prevent denitrification in the clarifier blanket. When selecting a RAS rate, be sure it will produce the desired results at low, average, and high influent flowrates.

SECTION EXERCISE

1. The MLSS concentration is 2800 mg/L and the SRT is 12 days. The in-basin water temperature is 17 °C. After 30 minutes in the settleometer, the MLSS volume decreases from 1000 mL down to 225 mL. Find the SSV in the settleometer test.
2. The RAS rate is flow paced and is set at 85% of the influent flow. What is the predicted RAS concentration?
3. The operator hates decanting the digester and wants to send as little water as possible to the digester. How low can they turn down the RAS based on how well the sludge is settling and compacting in the clarifier?
4. After a week of operating with very low RAS rates, sludge settleability begins to deteriorate. Blankets are building in the secondary clarifiers. At first, when the operators turned up the RAS flowrate, the blankets went back down. Now, it doesn't seem to matter how high or low they turn the RAS, the blankets just keep going up. If they don't get this fixed in a few hours, the sludge blanket will go over the weirs. What should they do right away? What should they do in the long run?
5. Things are back to normal thanks to the operator's quick thinking. Water temperatures have increased and the operator notices clumps of floating sludge in the secondary clarifier. What is happening and how can the operator stop it?

SECTION EXERCISE SOLUTIONS

1.

$$SSC_{30} = \left(\frac{\text{Initial Settleometer Volume, mL}}{SSV_{30}}\right) \times \text{MLSS mg/L}$$

$$SSC_{30} = \left(\frac{1000\text{ mL}}{225\text{ mL}}\right) \times 2800\text{ mg/L}$$

$$SSC_{30} = 12\,444\text{ mg/L}$$

2.

$$X_{RAS} = \left(\frac{Q_{inf}}{Q_{RAS}} + 1\right) \times X_{MLSS}$$

$$X_{RAS} = \left(\frac{1}{0.85} + 1\right) \times 2800\text{ mg/L}$$

$$X_{RAS} = 6094\text{ mg/L}$$

3. They can turn the RAS down as low as 29% of influent flow.

$$X_{RAS} = \left(\frac{Q_{inf}}{Q_{RAS}} + 1\right) \times X_{MLSS}$$

$$12\,444\frac{\text{mg}}{\text{L}} = \left(\frac{1}{Q_{RAS}} + 1\right) \times 2800\frac{\text{mg}}{\text{L}}$$

$$4.44 = \frac{1}{Q_{RAS}} + 1$$

$$0.29 = Q_{RAS}$$

4. Short-term solutions include 1) adding polymer or some other coagulant, 2) placing another activated sludge basin(s) in service, 3) placing another clarifier(s) in service. Long-term solutions include lowering the SRT by increasing the wasting rate.
5. Denitrification is taking place in the clarifier blanket. If the facility does not have to remove ammonia, the SRT can be lowered to stop nitrification. If the facility does have to remove ammonia, the RAS flowrate can be increased to remove the sludge from the clarifier before denitrification can occur.

TEST YOUR KNOWLEDGE

1. **Return activated sludge flowrates should be kept as low as possible while meeting treatment goals to**
   a. Maximize treatment time in the activated sludge basin
   b. Build a deep sludge blanket to squeeze the water from the RAS
   c. Minimize potential for denitrification to take place
   d. Avoid sending excess water to solids handling processes
2. **One advantage of flow pacing the RAS pumps to the WRRF influent flow is**
   a. Variable diurnal blanket depth
   b. Nearly constant RAS concentration
   c. Maximizes RAS concentration at low flow
   d. Solids may be stored in the clarifier
3. **Sludge settleability has deteriorated over the last 2 weeks and the clarifiers are building deep blankets. The operator turns up the RAS flowrate to remove the sludge from the clarifier faster, but instead of going down, the blankets get even deeper. This is because**
   a. RAS pumps were not turned up enough
   b. RAS pumps need time to catch up, then the blanket will drop
   c. Increasing RAS made the overloading problem worse
   d. SOR is too low
4. **The secondary clarifier at a small facility is overloaded. There are no other clarifiers available. A short-term solution may be to**
   a. Adjust SVI with chemicals
   b. Increase wasting to reduce SRT
   c. Build additional clarifiers
   d. Increase RAS flow
5. **Gas bubbles generated in the sludge blanket may cause**
   a. Short-circuiting
   b. Ashing or clumping
   c. Air binding
   d. Ratholing
6. **A key indicator that denitrification is taking place in a circular secondary clarifier blanket is**
   a. RAS concentration decreases with increasing pump speed
   b. Flocculation center well is full of heavy, thick foam
   c. Sludge collecting between scum baffle and weir
   d. Scum on clarifier surface turns from brown to green

Operation

DAILY OPERATION

After normal operation has been established, perform daily observation of the following items:

- Visually check equipment at least once per shift to ensure that it is functioning properly. Listen for any unusual noises because they may signal abnormal conditions;
- If multiple treatment trains are in service, verify flow is evenly distributed between basins and clarifiers;
- Visually inspect the surface of the activated sludge basins. If abnormal conditions such as foaming, unmixed areas, or turbulent areas are observed, refer to the troubleshooting section of this chapter;
- Monitor the DO concentrations at several points along the length of aerated basins;
- Record the activated sludge basin water temperature. For facilities that are required to remove ammonia, in-basin water temperature will influence selection of the target sludge age;
- Collect a sample from the end of the activated sludge basin and conduct a settleometer test. Calculate SVI. Record the SSV at 5 minutes and 30 minutes. Graph SVI, SSV_5, and SSV_{30} with control charts. Calculate the 30-minute SSC. Compare result to actual RAS/WAS concentration. If settleometer is compacting much better than the clarifier, consider adjusting the RAS flowrate;
- Use a composite sampler (sludge judge) to monitor blanket depths in the secondary clarifiers at least once per shift and preferably several times per shift. Be sure to measure during peak flow. If blanket depths are greater than 0.7 m (2 ft), consider adjusting the RAS flowrate;
- Collect and analyze all samples needed to calculate sludge age. This may include grab or composite samples of MLSS, WAS, and the clarifier blanket as well as the average daily flowrate for the facility. Calculate sludge age and record result;

- Compare current sludge age to target sludge age. Adjust wasting rates as needed to maintain target sludge age; and
- Operators may also wish to conduct a microbiological examination and F/M calculations weekly.

STARTUP OF AN ACTIVATED SLUDGE PROCESS

The primary objective of startup is to develop a well-flocculated mass of MLSS with good settling properties as quickly as possible. If possible, plan to bring activated sludge basins online during warmer weather. Mixed liquor suspended solids grows faster during warmer weather and startup time will be reduced. The following procedures for initial startup of an activated sludge process were created by combining and modifying startup procedures from *Start-Up of Municipal Wastewater Treatment Facilities* (U.S. EPA, 1973) and *Design Manual: Fine Pore Aeration Systems* (U.S. EPA, 1989).

There are two approaches to starting up an activated sludge process: growing all of the MLSS on-site using influent wastewater or importing some MLSS from another facility to seed the process. When importing MLSS, be sure seed sludge is good quality with few filaments and a low SVI. To determine the minimum MLSS concentration needed for startup of a single activated sludge basin, the design MLSS concentration is multiplied by two factors (eq 8.31). Using COD instead of BOD allows for rapid testing throughout startup. For facilities receiving domestic wastewater, convert COD to BOD by dividing by 2. The actual conversion factor will vary from one facility to another, but using a factor of 2 is close enough for assessing startup conditions.

$$\text{Design MLSS}, \frac{\text{mg}}{\text{L}} \times \left(\frac{\text{Startup Flow to Basin}}{\text{Design Flow of Basin}}\right) \times \left(\frac{\text{Current BOD}_5, \frac{\text{mg}}{\text{L}}}{\text{Design BOD}_5, \frac{\text{mg}}{\text{L}}}\right) = \text{Minimum Startup MLSS, mg/L} \quad (8.31)$$

Equation 8.31 is for bringing a single activated sludge basin; online, however, the number of basins started will depend on the influent flow. For example, if the new facility has a capacity of 19 ML/d (5 mgd) and four activated sludge basins, then the design capacity per basin will be the design flow divided by the number of basins. In this case, each basin is designed to receive 4.75 ML/d (1.25 mgd). If the influent flow is currently 7.6 ML/d (2 mgd), then two basins will be needed at startup. Always use the next higher number of basins if the flow doesn't divide evenly. If multiple basins will be brought online at the same time, multiply the result from calculation 8.31 by the following factor:

$$\frac{\text{Flow to basin(s) to be started / number or volume of basins}}{\text{Design flow to basin(s) / total number or volume of basins}} \quad (8.32)$$

The minimum startup MLSS concentration is the concentration that should be achieved before any sludge is wasted from the system. It does not need to be adjusted for temperature. The minimum startup MLSS is proportioned to the current flow and BOD concentration to help ensure the starting SRT and F/M are within the correct range for the design. After the first activated sludge basin is online and operating normally, other basins can be brought online in the same way by using seed sludge wasted from the first basin. It is important to cycle the MLSS through the secondary clarifier during startup and return it through the RAS line. Running the process normally influences the floc structure and helps to ensure the formation of round, dense floc. During startup and normal operation, microorganisms that don't flocculate and settle to the bottom of the clarifier won't be returned to the activated sludge basin. Instead, they leave through the clarifier overflow.

INSPECTION AND PRETESTING

Before starting up the system, all equipment and basins should be inspected and tested to ensure that

- All debris is removed from the basins and piping systems;
- All gates and valves have been checked for smoothness of operation and are in the closed position;
- Basin and clarifier weirs are level;

- Froth control system (if available) is in good condition. Sprayers should be open without blockages. Test each sprayer nozzle to see that it is securely fastened;
- If mechanical aerators are used, they should be rotated first by hand to ensure proper alignment and smoothness of operation. The mounting should be carefully inspected to ensure it is fastened securely. The motor should be lubricated properly according to the manufacturer's recommendations. All electric motors should be jogged to verify that the wiring is connected properly and that the motors are turning in the right direction;
- If diffused air is used, check to make sure diffusers are installed in accordance with the manufacturer's specifications. Tube (sock) diffusers should be tightened and oriented properly, gaskets and o-ring seals should be elastic and properly seated, the piping system should be level, and bolts and other hardware should be properly adjusted;
- The inspection of the diffused air delivery system includes
    - Checking the air filter and condensation trap,
    - Checking the air lines for leaks,
    - Checking valves for proper and smooth operation,
    - Inspecting the blower for proper lubrication, clearances, and safety guards,
    - Inspecting the coupling from the motor to the blower for proper alignment, and
    - Inspecting air and pressure gauges for proper operation and calibration;
- Follow the manufacturer's specifications in feeding air to the diffuser system before the diffusers become submerged. Always feed at least the manufacturer's minimum recommended airflow rate per diffuser to prevent backflow of wastewater through the diffusers and into the air distribution piping;
- Fill the basin to a level of about 30 cm (1 ft) above the diffusers. Service water, if available, is preferable to wastewater or MLSS for the initial filling. Use caution during the early stages of filling to prevent the force of incoming wastewater from damaging the air diffusion system or its supports;
- Operate the aeration system for a few hours. During this time, inspect all air piping and diffusers for leaks. Make repairs as needed. Recheck gates and valves for proper seating. Motors should be inspected for vibration, noise, and overheating;
- Piping systems should be filled with water and inspected for leaks; and
- Startup, inspection, and testing of clarifiers are discussed in Chapter 4—Primary Treatment.

STARTING WITH IMPORTED MIXED LIQUOR SUSPENDED SOLIDS

The use of seed sludge from another facility is the most reliable and fastest means of startup. When available, enough seed sludge should be placed into the activated sludge basin to result in a minimum MLSS concentration of 500 mg/L when the basin is filled with wastewater.

Step 1—At the end of inspection and testing, there should be enough water in the activated sludge basins to cover the diffusers and the aeration system should be operating. Verify that the DO concentration in the basin is at least 2 mg/L. If the aeration system is not still operating, restart it before proceeding with step 2. The aeration system will keep the MLSS in suspension and prevent fouling of the diffusers in fine-bubble systems.

Step 2—Fill the activated sludge basin with influent wastewater.

Step 3—Add seed sludge to the basin and continue aerating. The MLSS concentration in the basin should be at least 500 mg/L.

Step 4—Begin adding influent or primary effluent to the basin at approximately 10% of the design flow for the basin.

Step 5—When the basin is full, open the gates to allow flow into the secondary clarifiers.

Step 6—Increase influent or primary effluent flow by adding another 10% of the design flow each day.

Step 7—Start the clarifier mechanism (Chapter 4) and open the gates or valves downstream of the clarifier.

Step 8—Start the RAS pumps when the clarifiers are at least 50% full.

Step 9—When the minimum MLSS startup concentration has been reached, begin wasting from the basin each day. Determine the wasting volume by selecting a target SRT and calculating the mass (kilograms or pounds) of WAS to remove each day.

STARTING WITH INFLUENT WASTEWATER

If seed sludge is not available, influent may be used.

Step 1—At the end of inspection and testing, there should be enough water in the activated sludge basins to cover the diffusers and the aeration system should be operating. Verify that the DO concentration in the basin is at least 2 mg/L. If the aeration system is not still operating, start it before proceeding with step 2.

Step 2—Begin startup by filling the activated sludge basin with influent, bypassing the primary clarifiers (if present). Raw influent will contain more microorganisms than primary effluent.

Step 3—After the basins are filled, close the influent gate or valve. Aerate the wastewater for approximately 8 hours.

Step 4—Turn off the aerators and allow the mixture to settle for 30 to 60 minutes.

Step 5—Open the influent and effluent gates or valves to allow fresh wastewater to enter the basin. Add influent or primary effluent to the basin at approximately 10% of the design flow for the basin. This will displace some of the existing volume to be pushed out and into the secondary clarifier.

Step 6—Use the RAS pumps to return settled solids to the activated sludge basin.

Step 7—Repeat steps 3 through 6 until the MLSS concentration reaches at least 500 mg/L. Increase influent or primary effluent flow by adding another 10% of the design flow each day. The basin should be receiving 100% of its design flow after 10 days.

Step 8—Place the activated sludge process into continuous flow operation. This means water should enter and leave the process continuously. Return activated sludge pumps should be operating to return settled sludge to the activated sludge basins. Measure the DO concentration often to verify that it remains above 2 mg/L.

Step 9—When the minimum MLSS startup concentration has been reached, begin wasting from the basin each day. Determine the wasting volume by selecting a target SRT and calculating the mass (kilograms or pounds) of WAS to remove each day.

FOAMING AND TIME TO REACH STABLE OPERATION

During startup when the MLSS concentration is low, the activated sludge basins may experience severe foaming. Startup foam is white to gray and collapses easily (Figure 8.54). This type of foam typically goes away by the time the SRT reaches 2 to 3 days. The MLSS should stabilize within 7 to 10 days, producing a well-flocculated sludge that settles rapidly and produces a clear supernatant. Return activated sludge has very little odor, but what odor is there is best described as slightly musty and similar to soil or mulch. During startup, effluent ammonia concentrations will be nearly equal to the influent ammonia concentration. Nitrification may take another week to become fully established. The final effluent may contain high concentrations of BOD_5 and TSS for the first several days to a week. Chemical coagulants, such as alum or ferric chloride, may be added during startup to aid with settling. Be sure to comply with the conditions of the discharge permit at all times, including startup, and to maintain communication with your state regulatory agency or U.S. EPA.

REMOVING AN ACTIVATED SLUDGE BASIN FROM SERVICE

If an activated sludge basin must stand idle for more than 2 weeks, it should be drained and thoroughly cleaned. The following procedure for removing an activated sludge basin from service is from the *Design Manual: Fine Pore Aeration Systems* (U.S. EPA, 1989).

Figure 8.54 White Foam During Startup (Reprinted with permission by Indigo Water Group)

Groundwater pressure relief valves should open automatically to allow groundwater into the tank when groundwater levels are higher than the bottom of the tank. When the tank is full, the pressure from the water in the tank holds these valves closed. It is possible for groundwater to exert enough upward pressure on a tank to push it out of the ground or at least shift it out of place. Small amounts of pressure spread over a large area, like the bottom of a tank, can exert tremendous amounts of force. Force = Pressure × Area

Step 1—Stop the wastewater and return sludge flows to the basin. If the basin is equipped with diffused air, continue to feed air to the diffusers at or above the manufacturer's recommended minimum rate.

Step 2—Open drain lines and start drain pumps if necessary. Continue to feed air to the system until the water level is below the diffusers and the diffusers have been washed off. Monitor and adjust the airflow rate as the water level falls.

Step 3—Wash down the basin walls while the basin is draining. Material that dries onto the walls will be difficult to remove later. Wash down equipment as it is exposed.

Step 4—Once the basin is drained, verify that groundwater pressure relief valves are operational.

Step 5—Wash down basin walls and floor, air piping, diffusers, and mechanical aerators to avoid odor problems. Material that has accumulated on or under the diffuser should be removed before it can dry.

Step 6—For above-freezing conditions, tanks equipped with diffusers should be refilled with enough clean water to cover the diffusers with 1 m (3 ft) of water (U.S. EPA, 1989). If plastic piping is present, add clean water to cover the exposed piping. This will provide additional protection against UV light exposure and excessive temperature changes. Continue to feed air at a flowrate equal to or greater than the manufacturer's recommended minimum. An algaecide may be added to prevent algae growth.

Step 7—For freezing conditions, more water may be needed to protect other normally submerged piping. Although the air being fed will typically prevent serious ice damage, if an ice layer does form, do not drain the water from the basin. Falling ice can cause serious damage.

Feeding air to an empty basin can be costly in the long term. As long as the basin is filled with relatively clean water, a less costly alternative is to shut off airflow completely and allow the air distribution system to fill with water. If the basin will be out of service for an extended period of time, the diffusers should be removed and stored according to the manufacturer's recommendations.

TEST YOUR KNOWLEDGE

1. **An activated sludge process may be started up using seed sludge from a nearby anaerobic digester.**
   - ☐ True
   - ☐ False

2. **Water temperature in the activated sludge basins should be measured daily in facilities that are required to remove ammonia.**
   - ☐ True
   - ☐ False

3. **An operator has just finished the daily walkthrough of the activated sludge process. He or she decides to increase the RAS flow slightly. What did the operator observe that resulted in this decision?**
   a. MLSS not compacting well in the settleometer test
   b. Clarifier blankets deeper than 0.7 m (2 ft)
   c. Sludge age is much higher than the target
   d. RAS concentration is too low

4. **During startup, when seed sludge is not available, influent wastewater is preferred over primary effluent because**
   a. Higher BOD concentrations in influent
   b. Particulate matter is needed for floc formation
   c. Primary clarifier not yet placed into service
   d. Greater populations of microorganisms

5. **Mixed liquor suspended solids should not be wasted from the process for the first time until**
   a. Minimum startup concentration has been reached
   b. MLSS concentration begins to overload the clarifier
   c. Supernatant BOD and TSS are below 10 mg/L
   d. Rotifers and stalked ciliates are observed

6. **A primary clarifier with four groundwater pressure relief valves is taken out of service. If two of the valves fail to open with high groundwater, what is the likely outcome?**
   a. Sump pump may be needed to remove groundwater
   b. Clarifier may move or be pushed upward
   c. Clarifier may overflow as water backs up into it
   d. Operator will not be able to reach backflush

Data Collection, Sampling, and Analysis

This section summarizes some of the many process control tests that are critical for operating a successful, well controlled activated sludge process including visual inspection, clarifier blanket depth, settleability, SVI, and oxygen uptake rate (OUR). For complete, step-by-step laboratory procedures, refer to *Basic Laboratory Procedures for the Operator-Analyst*, 5th edition, that was published by WEF in 2012. Other tests needed for process control include TSS, total volatile suspended solids, pH, and DO. Facilities that are required to remove ammonia, nitrite, nitrate, and/or phosphorus will also need to analyze for alkalinity, ammonia, nitrite, nitrate, and phosphorus. Recommended minimum sampling frequencies are shown in Table 8.9.

VISUAL INSPECTION OF THE ACTIVATED SLUDGE BASIN

The surface of the activated sludge basin should be free of debris, floatable material, and foam. For aerated zones or basins, the surface of the activated sludge basin should demonstrate even mixing throughout the basin with a minimal amount of light, tan to brown foam that shifts easily making a honeycomb-type pattern. Foam coverage should be less than 5% of the total surface area. An example activated sludge basin is shown in Figure 8.55. A small amount of foam will always be present because of surfactant production by the bacteria in the process; however, it should dissipate quickly. Larger amounts of foam or foam that is difficult to move or break apart is indicative of an abnormal condition.

MIXING PATTERNS

Daily walkthroughs should be conducted once per shift when possible. Activated sludge basins should be inspected for evidence of even, complete mixing. A well-mixed, aerated basin will have a similar appearance to the basin in Figure 8.55, with an ever-changing pattern of bubbles and light foam across the surface. Anoxic and anaerobic zones and basins will be mixed with subsurface mixers. These basins will not exhibit the bubble pattern shown in Figure 8.55, but should be smooth and clear throughout the basin without clear differences in color or solids density from one area of the basin to another. Areas with foam buildup or clear water evident at the surface suggest a lack of mixing that should be investigated. Figure 8.56 shows

Table 8.9 Recommended Minimum Sampling and Analysis Frequencies for Activated Sludge

Sampling Location	Analysis[a]	Frequency[b]	Sample Type
Activated sludge basin influent	pH	Daily	Grab
	BOD	Weekly	Composite
	TSS	Weekly	Composite
	Total Kjelhdahl nitrogen[c]	Monthly	Grab
	Ammonia	Monthly	Grab
	Alkalinity	Monthly	Grab
Activated sludge basin	DO	Daily (continuous)	In situ
	Temperature	Daily (continuous)	In situ
Activated sludge basin effluent	TSS (mixed liquor suspended solids)	Daily	Grab
	Settleability	Daily	Grab
	pH	Weekly	Grab
	Microscopic	Weekly	Grab
RAS	TSS	Daily	Grab
	Flow	Daily	Totalizer
WAS	TSS	Daily	Grab
	Flow	Daily	Totalizer
Secondary clarifier	Sludge blanket	Daily	In situ
Secondary clarifier effluent	BOD	Weekly	Composite
	TSS	Weekly	Composite
	Ammonia[c]	Monthly	Composite
	Nitrate[c]	Monthly	Grab
	Nitrite[c]	Monthly	Grab
	Total phosphorus[c]	Monthly	Composite
	pH	Daily	Grab
Facility effluent	Turbidity	Daily	Grab
	Fecal coliform[d]	Daily	Grab
	Chlorine residual[d]	Daily	Grab

[a]May not show all analysis required; check with permit requirements.
[b]Sampling frequencies may need to be increased for process control or permit requirements.
[c]May not be required if nutrients (ammonia, phosphorus, etc.) are not a concern.
[d]May not be required if disinfection is not required; check with discharge permit.

areas of clear water within the activated sludge basin caused by turning off the diffused air to this portion of the basin. Without mixing, the MLSS rapidly begins to settle.

EVIDENCE OF BROKEN HEADERS AND DIFFUSERS

Dead spots and non-uniform mixing patterns typically indicate a clogged diffuser or that the diffuser header valves need adjustment to balance air distribution in the basin. Broken headers or missing, torn, and damaged diffusers can often be identified by a boiling effect that is localized to one area of the aerated zone or basin. This may be accompanied by an inability to maintain DO levels in the activated sludge basin. In the case of multiple basins being fed from a single blower, the basins without broken headers and/or diffusers may have lower DO concentrations because the air from the blower will flow predominantly to the broken header as this is the path of least resistance.

For activated sludge basins with air diffusers, a DO profile should be made every 1 to 6 months or whenever the flow pattern is changed. The air distribution should be adjusted to maintain a DO concentration of no less than 0.5 mg/L throughout the aerated portion of the activated sludge basin.

Figure 8.55 Surface of an Aerated Activated Sludge Basin in Clifton, Colorado (Reprinted with permission by Indigo Water Group)

FOAM COLOR INDICATORS

Activated sludge basins should be free of foam. The absence or presence of foam should be noted on daily walkthroughs along with a written record of foam color, thickness, and an estimation of the percentage of basin coverage. More detailed information on foaming and sludge settleability problems can be found in the troubleshooting section. In all cases, the presence of foam on an activated sludge basin should be investigated and the root cause determined, ideally before 30% of the basin is covered with foam.

Figure 8.56 Clear Water Demonstrates Solids Separation and a Lack of Mixing (Reprinted with permission by Indigo Water Group)

Three filamentous organisms can cause activated sludge foaming: nocardioforms, *Microthrix parvicella*, and Type 1863 (Jenkins et al., 2004). Nocardioforms and *M. parvicella* can produce a stable, viscous, brown foam on the surface of the activated sludge basins that can carry over to the secondary clarifiers and the final effluent. Type 1863 can produce a white-gray foam that collapses easily when the F/M is high and the SRT is below 2 days (Environmental Leverage, 2003). Foaming can range from a nuisance to a serious problem. In cold weather, this foam may freeze solid and have to be manually removed. In warm weather, it can generate odors. Unchecked, it can overflow tanks and create a slippery mess on walkways, as shown in Figure 8.57.

DEPTH OF BLANKET

As the MLSS settles in the secondary clarifier, a layer of settled sludge called the "blanket" accumulates on the floor. Ideally, the blanket will be less than 1-m (2.57-ft) deep from the floor of the clarifier to the top of the blanket; however, the depth of blanket can vary depending on the time of day and amount of flow coming into the WRRF. This is especially true in WRRFs that do not flow pace the RAS flow to influent flow. Blanket depth in clarifiers without flat floors will also vary depending on where it is measured.

APPLICATION

The depth of blanket test measures the depth of the blanket in the secondary clarifier and provides the following information:

- Enables the operator to determine the total mass of MLSS in the clarifier blanket. This information may be included in the MCRT calculation for process control;
- Alerts the operator to an increase in the sludge blanket in the final clarifier, allowing the operator to take corrective action to prevent solids from washing out over the weirs; and
- Enables the operator to determine the clarifier sludge detention time.

The depth of blanket test is also referred to as the *clarifier core sample* in some texts.

APPARATUS AND MATERIALS

Sludge blankets are typically measured with a core sampler, which is also known as a *sludge judge*. Blanket depth may also be measured using a hand-held TSS meter or an ultrasonic blanket detector; however, although these methods do provide information on blanket depth, they do not allow for collection of a

Figure 8.57 Activated Sludge Foaming (Reprinted with permission by Indigo Water Group)

core sample, which is needed to determine the total mass of MLSS in the clarifier. Core samplers are made from clear, 19-mm (0.75-in.) plastic pipe and typically consist of three sections that are 1.5-m (5-ft) long each. The sections screw together to form a single, 4.5-m (15-ft) long sampler as shown in Figure 8.58. Samplers are marked at 0.3-m (1-ft) intervals for easy determination of blanket depth and have a check valve at the bottom of the 4.5-m (15-ft) assembled sampler. The check valve allows the operator to lower the entire assembly into the clarifier, close the valve, and then withdraw the sampler with a core sample of the clarifier contents inside.

SAMPLE COLLECTION

Core samples should be taken at a point that is representative of the clarifier contents. For rectangular clarifiers, the core sample should be collected at a point halfway down the length of the clarifier. The sample must be taken at least 0.3-m (1-ft) before or after the cross collector. For circular clarifiers with floors that slope downward toward the center, as shown in Figure 8.59, the core sample should be collected one-third of the distance from the outside wall of the clarifier to the center, as shown in the figure below. Collecting the sample at this location compensates for the conical bottom of the clarifier. Recall that the formula for finding the volume of a cone uses the volume of a cylinder divided by 3. If the sample is collected too close to the outer edge, the mass of MLSS in the clarifier will be underestimated. If the sample is collected too close to the center of the clarifier, the total mass will be overestimated. For flat-bottom clarifiers, the positioning of the core sampler is not as important because the blanket depth should be the same across the clarifier. For clarifiers that slope from a high point in the center down toward a collection ring on the outside of the clarifier, the core sampler should be positioned one-third of the way from the center of the clarifier to compensate for the cone created by the sloped floor.

For circular clarifiers, the core sample should be collected when the rake arm is at a 90-deg angle relative to the sample location, as shown in Figure 8.59. The motion of the rake arm disturbs the sludge blanket. Taking the sample when the rake arm is at a 90-deg angle ensures that the blanket has had an opportunity to resettle and is representative of the general conditions at the bottom of the clarifier.

Figure 8.58 Sludge Judge Demonstration (Reprinted with permission by Indigo Water Group)

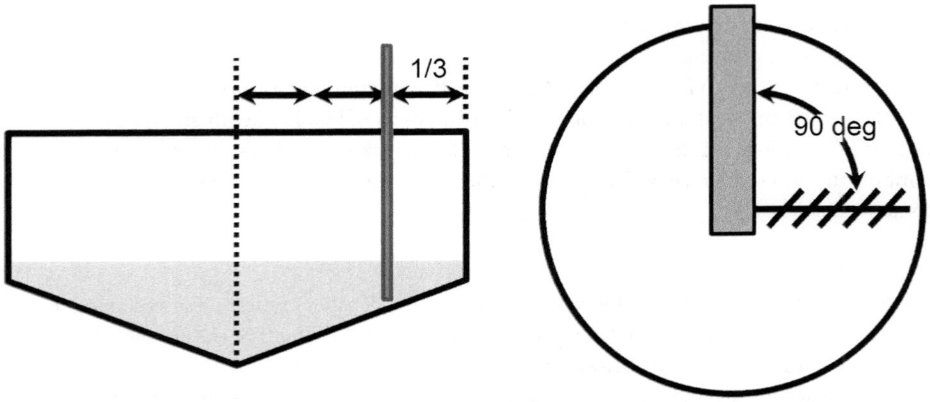

Figure 8.59 Ideal Sampling Location for Clarifier Core Sample (Reprinted with permission by Indigo Water Group)

PROCEDURE

Step 1—Position the core sampler at the clarifier railing approximately one-third of the way in from the outer edge of the clarifier for center slope circular clarifiers or one-half of the way across the length for rectangular clarifiers. Take care to keep the core sampler at least 0.3 m (1 ft) away from the cross collector in rectangular clarifiers.

Step 2—Slowly lower the core sampler to the bottom of the clarifier, taking note of the position of the rake arm in circular clarifiers and the flights in rectangular clarifiers.

Step 3—Once on the bottom, give a quick jerk to close the check valve on the bottom of the sampler.

Step 4—Slowly remove the sampler from the clarifier. Note the depth of the blanket and record it on the bench sheet. The core sampler will sometimes have a fuzzy layer where the interface between the settled sludge and clear supernatant is not well defined. The operator will need to make their best judgment on blanket depth in this situation.

Step 5—Place the bottom end into a bucket or other sampling container that can hold the entire contents of the core sampler.

Step 6—Release the entire sampler contents into the bucket.

Step 7—Mix the contents of the sample container and collect a subsample for TSS analysis.

Step 8—Repeat for each clarifier.

SPECIAL VARIATIONS OR APPLICATIONS

The depth-of-blanket test may be used as a troubleshooting tool as well as for routine process control.

- Blanket profile—This is used to assess the best location for routine measurements of average depth of blanket.
- Balancing flow to the final clarifiers—Some facilities have poor flow splits to multiple final clarifiers and it becomes necessary to adjust flows based on blanket measurements because of the lack of flow-measurement devices. With multiple units and "equal" flow splits, blankets should be within approximately 0.3 m (1 ft) of each other. For a balanced system, any variation in depth of blanket between clarifiers over 0.6 m (2 ft) is attributable to some type of hydraulic imbalance. The imbalance may be attributable to uneven loading to the clarifiers or uneven sludge withdrawal from the clarifiers. Note that for facilities with multiple clarifiers of different sizes, flow from the activated sludge basins should be proportioned according to the sizes of the clarifiers so that each clarifier has the same SLR (kilogram per square meter per day or pounds per square foot per day) and hydraulic loading rate (cubic meters per square meters per day or gallons per day per square foot).

SETTLEABILITY

GENERAL DESCRIPTION

The main purpose of the settleometer test is to give the operator an idea of how the sludge is settling in the secondary clarifier. The settleometer test simulates how activated sludge settles in the secondary clarifier. The settleometer test should be performed at least daily at activated sludge facilities. For larger facilities, the settleometer test should be performed once per shift. Information that should be recorded for long-term trend tracking includes the SSV at 5 minutes, SSV at 30 minutes, and the time if and when the blanket "popped". The SSV_5 indicates how well the MLSS is settling whereas the SSV_{30} measures how well the MLSS is compacting. The analyst may wish to refer to Method 2710 C of *Standard Methods* (APHA et al., 2017).

APPARATUS AND MATERIALS

Three pieces of equipment are necessary to run settleability tests: a settleometer, a paddle, and a timer. The equipment are described as follows:

- Mallory settleometers—three 2-L Mallory settleometers are suggested, one Mallory settleometer for the regular test, the second for dilution, and the third for any special test that may need to be run such as an extended-rise test or multiple dilutions. If only one Mallory settleometer is available, tests may be run in series. If Mallory settleometers are not available, 2-L beakers or similar devices may be substituted. A standard graduated cylinder should not be used because the narrow cylinder reduces settling velocity caused by excess friction with the walls. The base of the settleometer, where the wall meets the floor, should be square and not rounded;
- Paddle—a wide paddle is needed to gently stir, then quiet, the mixed liquor in the settleometer at the beginning of the test. The paddle should be made of smooth acrylic or a similar material, with a width slightly less than the diameter of the settleometer; and
- Timer—the timer should be capable of sounding an alarm at the end of 5-minute intervals.

SAMPLE COLLECTION, PRESERVATION, AND HOLDING TIMES

It is important that mixed liquor samples collected for the settleability test are representative of mixed liquor flowing out of the activated sludge basin and into the secondary clarifier. Typically, this sample is collected either in a weir overflow box or the secondary clarifier influent splitter structure or at a similar point prior to its entry into secondary clarifiers. If the mixed liquor comes from two or more activated sludge basins, the analyst should be sure to sample the mixed liquor after the sources have combined. However, it is also a good idea to periodically measure settleability in each individual activated sludge basin to check for differences in sludge condition between the basins. Finally, it is important to remember to collect the mixed liquor samples and not the scum and foam that may be present on the mixed liquor.

The necessity for collecting mixed liquor samples in wide-mouthed containers and starting the settleometer testing as soon as possible cannot be overemphasized. Indeed, testing should begin within 10 minutes of collection. In addition, shaking and agitation of the samples should be minimized, especially during transportation to the testing site.

Procedure

Test conditions and sampling techniques can strongly influence test results. As such, the test should be set up in a location that is free from vibration and direct sunlight. If possible, the temperature should not vary a great deal from one test to another. The following steps should be followed for the settleability procedure:

Step 1—Mix the mixed liquor sample carefully, then pour it into the settleometer carefully and rapidly with the least possible amount of additional aeration or turbulence. Stir the settleometer contents gently with a wide paddle to ensure a thorough mixing. Then, stop all movement of the mixed liquor with the paddle;

Step 2—As the paddle is smoothly and carefully removed from the settleometer, start the timer that was previously set for 5-minute intervals. Record the settled sludge volumes on the bench sheet (Figure 8.60). It is important to not leave the settleometer or the analyst will miss out on some of the most important information to be gained from the settleometer test;

Settleometer Data Sheet

Location: _______________

Method: _______________

Date: _____________________________________
Sample ID: _______________________________
Analyst: _____________________________________
Time of Test: _________________ ATC (%) _______

$$SSC = \frac{(ATC) \times (1000)}{SSV}$$

Time	SSV, mL/L	SSC, %
0	1000	
5		
10		
15		
20		
25		
30		
35		
40		
45		
50		
55		
60		

Observations in the first 5 minutes:

Floc
- ☐ Granular
- ☐ Fluffy
- ☐ Compact
- ☐ Feathery

Particle Size
- ☐ Large
- ☐ Medium
- ☐ Small

Bio-solids/ Supernate Interface
- ☐ Well Defined
- ☐ Ragged

Supernatant
- ☐ Clear
- ☐ Cloudy

Straggler Floc
- ☐ Yes
- ☐ No

Observation after 30 minutes:
- ☐ Crisp/Sharp Edges
- ☐ Sponge-like
- ☐ Fluffy/Feathery
- ☐ Homogeneous

Date: _____________________________________
Sample ID: _______________________________
Analyst: _____________________________________
Time of Test: _________________ ATC (%) _______

$$SSC = \frac{(ATC) \times (1000)}{SSV}$$

Time	SSV, mL/L	SSC, %
0	1000	
5		
10		
15		
20		
25		
30		
35		
40		
45		
50		
55		
60		

Observations in the first 5 minutes:

Floc
- ☐ Granular
- ☐ Fluffy
- ☐ Compact
- ☐ Feathery

Particle Size
- ☐ Large
- ☐ Medium
- ☐ Small

Bio-solids/ Supernate Interface
- ☐ Well Defined
- ☐ Ragged

Supernatant
- ☐ Clear
- ☐ Cloudy

Straggler Floc
- ☐ Yes
- ☐ No

Observation after 30 minutes:
- ☐ Crisp/Sharp Edges
- ☐ Sponge-like
- ☐ Fluffy/Feathery
- ☐ Homogeneous

Figure 8.60 Settleometer Bench Sheet (WEF, 1994)

Step 3—During the initial 5 minutes, the analyst should observe the visual characteristics of the settling sludge. A conscientious operator will critically observe how the sludge particles come together while forming the blanket. He or she will also observe whether the sludge compacts slowly and uniformly while squeezing clear liquid from the sludge mass or whether tightly knotted sludge particles are simply falling down through the turbid effluent. The operator will observe how much and what type of straggler floc, if any, remains in the supernatant liquor above the main sludge mass;

Step 4—The importance of conscientious, perceptive observation during the first 5 minutes cannot be overemphasized. During these important 5 minutes, the operator will acquire additional insight to sludge character and quality and will be in a much better position to evaluate what the settleometer test reveals;

Step 5—At the end of the first 5 minutes, the analyst should read and record the volume of the settleometer occupied by settled sludge. This is the most important reading taken during the test;

Step 6—The analyst should continue to read and record settled SSV at 5-minute intervals for the first 30 minutes;

Step 7—The analyst should continue taking SSV readings until SSV changes become very small from reading to reading. If a very slow settling sludge (i.e., SSV greater than 900) is being tested, readings at 90, 120, 150, 180, and 240 minutes may be required; and

Step 8—After the last reading has been taken, the sample should be allowed to stand for several hours more. The analyst should record the time when previously settled sludge starts to swell and/or rise to the surface of the settleometer. It is important to wash the settleometer and paddle with mild soap and to thoroughly rinse and dry it after the test.

SPECIAL VARIATIONS OR APPLICATIONS

There are essentially three variations of the test, all of which have been alluded to previously. The first is the rise test, which indicates that denitrification might be occurring in the secondary clarifier. The second is the dilution test for determining sludge age. The third test is to compare the settleometer supernatant to the secondary clarifier effluent, which is useful for troubleshooting clarifier performance.

Rise Test

After conducting the standard 30-minute settleometer test, allow the beaker with the sample to set for several hours and observe the time that the sludge rises. This is an indication of denitrification. If the sludge rises too quickly (e.g., less than 90 minutes), observe the depth of sludge in the secondary clarifier and the sludge detention time. The lower the RAS (underflow) flowrate, the higher the detention time. If this condition is occurring, pay close attention to the sludge blanket level in the core sampler. See if it is rising or expanding.

Dilution Test

The dilution test is conducted to determine whether there are too many solids (hindered settling) or too few solids in the aeration tank. It is important to rely on other tests, including visual observations and microbial tests, to make a judgment on necessary process control adjustments. The analyst should run three settleometer tests at the same time. Undiluted MLSS should be placed in the first beaker, 50% MLSS and 50% secondary effluent in the second, and 25% MLSS and 75% secondary effluent in the third. The analyst should conduct 60-minute settleabilities, prepare settling curves, and compare slopes. If the slopes are the same or parallel, this probably indicates settling problems caused by filamentous sludge. If an increase in settling rate occurs with the diluted samples (greater slope or rise over run), the MLSS is probably too high and wasting should be increased to lower the SRT. The increase should not be drastic, but incremental each day. The analyst should try gradually increasing the wasting over a 3-day period until the desired SRT and settling rate is achieved.

Comparison Test

The settleometer is a perfect clarifier without currents, wind, or temperature differences. The MLSS in the settleometer settles without interference. The performance of the settleometer can be compared to the performance of the clarifier. This is done by collecting two samples: a grab sample at the end of the activated sludge basin for the settleometer test and a grab sample of secondary clarifier effluent. The settleometer test is run after flocculating the MLSS sample. At the end of the test, the supernatant is collected from a sample port on the side of the settleometer. The supernatant sample and the secondary clarifier sample are both analyzed for TSS. If the clarifier is performing well, then the TSS numbers from both samples should be about the same. If the clarifier is not performing well, then the TSS in the clarifier sample will be higher than the TSS in the supernatant.

This testing protocol was developed by Water Environment Research Federation (Wahlberg, 2001). The complete protocol can be found in WERF Project 00-CTS-1: WERF/CRTC Protocols for Evaluating Secondary Clarifier Performance.

CALCULATIONS

- Observation of settling for the first 5 minutes,
- Settling rate (volume of settled sludge versus time), and
- Calculation of the 30-minute SSC.

$$SSC_{30} = \left(\frac{\text{Initial Settleometer Volume, mL}}{SSV_{30}}\right) \times \text{MLSS mg/L} \qquad (8.33)$$

Where:

SSC_{30} = Settled sludge concentration after 30 minutes and
SSV_{30} = Settled sludge volume after 30 minutes.

The SSC_{30} is the maximum achievable sludge concentration in the secondary clarifier assuming the solids remain in the clarifier blanket for 30 minutes or less. In practice, the RAS concentration will often be significantly lower than the SSC_{30} concentration because the solids are removed from the clarifier faster than 30 minutes. Occasionally, the measured RAS concentration may be higher than the calculation SSC_{30} concentration. This may be caused by denitrification in the settleometer "puffing up" the blanket or by leaving the solids in the clarifier blanket for longer than 30 minutes. The RAS concentration observed in the field is dependent on the RAS flow to influent flow ratio and can be calculated from the following equation.

$$X_{RAS} = \left(\frac{Q_{influent}}{Q_{RAS}} + 1\right) \times X_{MLSS} \tag{8.34}$$

Where:

X_{RAS} = Return activated sludge flow,
$Q_{influent}$ = Influent flowrate,
Q_{RAS} = Return activated sludge flowrate, and
X_{MLSS} = Mixed liquor suspended solids concentration.

The operator–analyst should be mindful that turning the RAS down to exceedingly low levels in an attempt to further thicken and compact the solids in the clarifier blanket is counterproductive and can cause effluent quality to deteriorate. The clarifier cannot compact the solids further than the concentration predicted by the settleometer.

- Time to rise; and
- If the dilution test is being conducted, comparison of settling rates, which provides information to evaluate the age of the sludge.
- If the flocculated/dispersed suspended solids test is being conducted, comparison of supernatant TSS or turbidity to secondary clarifier effluent TSS or turbidity, which provides information to evaluate the cause of elevated effluent TSS or turbidity.

INTERPRETATION OF RESULTS

The settleometer test is arguably one of the most useful process control tests for activated sludge. Much information can be gained from a properly conducted and interpreted test. Three data points should be tracked and plotted on control charts: the SSV_5, SSV_{30}, and the time, if any, when denitrification causes the settled sludge to rise to the top of the settleometer. Figure 8.61 shows settleometer data from a typical activated sludge process with the SSV_5, SSV_{30}, and SSV_{60} plotted over three and a half years.

The SSV_5 is an indicator of sludge settling characteristics (e.g., how fast is the particle moving through the water column). For the first 5 minutes of the test, the solids are essentially in free-fall with few collisions occurring between particles. The primary variables controlling how fast the particles settle in the first 5 minutes are the density of the particles relative to the density of the water and the presence or absence of filamentous bacteria. For activated sludge facilities that do not practice EBPR, a typical floc particle is only slightly more dense than the surrounding water. Enhanced biological phosphorus removal floc is generally denser than non-EBPR floc. As water cools, its density increases. In Figure 8.61, the SSV_5 is shown in blue. For this facility, the summer month in-basin water temperature may be as high as 20 °C (68 °F); however, in the winter months, the in-basin water temperature is typically 10 °C (50 °F). As a result, there is a strong seasonal pattern to the SSV_5, with settling rates decreasing in the winter months (higher SSV_5) and increasing in the summer (lower SSV_5) even when excess filaments are not present. Note that despite the large seasonal fluctuations in the SSV_5, the SSV_{30} remained fairly constant. For this facility, settling characteristics changed seasonally, but compaction characteristics did not. The SSV_{30} is increasing slightly in the winter months because the SRT has been increased to maintain stable nitrification. The higher SRT increases the total amount of MLSS in the system and the MLSS concentration. As a result, the MLSS is taking up more space in the settleometer blanket.

The presence of filamentous bacteria can also cause the SSV_5 increase, especially when bridging between floc particles occurs. Filaments create drag, which causes the sludge particles to settle more slowly. In Figure 8.61, there is a sudden increase in both the SSV_5 and SSV_{30} in December 2007. This sudden increase is characteristic of an increase in the numbers of filamentous bacteria. A microscopic examination should be done to confirm the presence or absence of filaments followed by staining to identify which filament is dominant.

The SSV_{30} is an indicator of sludge compaction. As the sludge particles settle in the settleometer and in the clarifier, they pile up on one another and the sludge concentration in the blanket increases. There is a limit to the amount of compaction achievable in both the settleometer and the clarifier. Compaction is limited, in part, because floc particles are negatively charged. Floc particles repel one another in much the same way as two ends of magnets can. In Figure 8.61, there is some additional compaction after 30 minutes, as evidenced by the slight differences between the SSV_{30} and SSV_{60}; however, compaction slows as the SSC increases. For very slow-settling sludges, compaction may continue for longer than 60 minutes. Practically speaking, the SSV_{30} represents the maximum achievable compaction in the secondary clarifier.

Example: The MLSS concentration is 2800 mg/L and the SSV after 30 minutes is 240 mL. The RAS is flow paced to the influent flow and is currently set at 50% of influent flow. Calculate the maximum achievable RAS concentration as predicted by the settleometer test. Then, calculate the actual RAS concentration from facility operating data.

$$SSC_{30} = \left(\frac{\text{Initial Settleometer Volume, mL}}{SSV_{30}}\right) \times \text{MLSS mg/L}$$

$$SSC_{30} = \left(\frac{1000\text{ mL}}{240\text{ mL}}\right) \times 2800\text{ mg/L}$$

$$SSC_{30} = 11\,667\text{ mg/L}$$

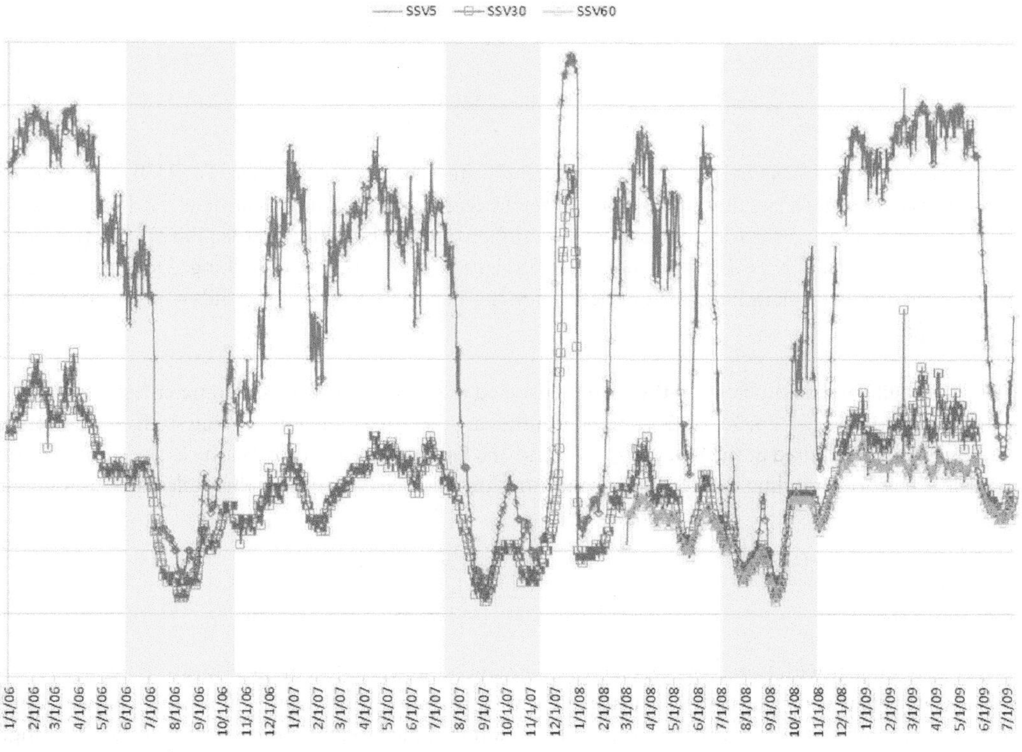

Figure 8.61 Typical SSV_5, SSV_{30}, and SSV_{60} Results (Reprinted with permission by Indigo Water Group)

The concentration of the sludge in the settleometer blanket after 30 minutes of settling is 11 667 mg/L. Sludge quality indicates that if the solids are left in the clarifier for 30 minutes, then the RAS concentration will be 11 667 mg/L. Slight differences may occur in field measurements because of inaccuracies in flow measurement. What is actually being achieved in the field?

$$X_{RAS} = \left(\frac{Q_{influent}}{Q_{RAS}} + 1\right) \times X_{MLSS}$$

$$X_{RAS} = \left(\frac{1\ m^3}{0.5\ m^3} + 1\right) \times 2800\ mg/L$$

$$X_{RAS} = 8400\ mg/L$$

SLUDGE VOLUME INDEX

GENERAL DESCRIPTION

Sludge volume index is defined as the volume of sludge in milliliters (cubic inches) occupied by 1 g (0.04 oz) of activated sludge after settling for 30 minutes. This index relates the 30-minute SSV from the settleometer process control test to the concentration of solids in the sample that the settleometer test was performed upon. The SVI will help the operator evaluate the settling characteristics of the activated sludge as the concentration of the solids in the system change. All WRRFs have unique sludge characteristics based on their design, specific influent characteristics, and operational parameters. Therefore, an SVI that represents a good settling sludge at one facility is generally not comparable to SVIs at other facilities. This is especially true if the facility is treating industrial or other unusual wastes. The reader may want to refer to Method 2710 D of *Standard Methods* (APHA et al., 2017).

APPLICATION

Operators often develop an SVI number from the settleometer test. The SVI is calculated using the SSV_{30} from a 2000-mL Mallory settleometer. The units for SVI are milliliters per gram (mL/g). Sludge volume index is determined as follows:

$$SVI, \frac{mL}{g} = \frac{SSV_{30} \times 1000\frac{mg}{g}}{MLSS\frac{mg}{L}} \tag{8.35}$$

Sludge volume indexes between 75 and 150 mL/g indicate a good settling sludge. The SVIs may be lower than this range for facilities that utilize alum or ferric, receive water facility treatment residuals, or perform biological phosphorus removal. Increasing SVI numbers means that sludge settleability is deteriorating. When the SVI reaches 200 mL/g, the sludge (MLSS) is technically "bulking". Bulking sludge may be caused by a variety of things, including nutrient deficiency, filamentous bacteria, and solids concentrations that are too high.

Using the Mallory settleometer and the aforementioned test procedures, the settleometer test is run and the SSV for 5, 10, 15, 20, 25, 30, 40, 50, and 60 minutes are recorded. For slow-settling sludge, additional readings may be required at the 90-, 120-, 150-, and 180-minute points or for as long as the sludge continues to settle. The corresponding settled sludge concentration (SSC) at each time is then calculated as follows:

$$SSC, mg/L = \frac{MLSS, \frac{mg}{L} \times 1000\ mL/L}{SSV, mL} \tag{8.36}$$

As SSV gets smaller with time, SSC gets larger as the sludge concentrates in the bottom of the settleometer. In an old, overoxidized sludge that settles fast, maximum compaction occurs within about 30 minutes. A medium-settleability sludge will typically compact to maximum within 1 hour. A young, underoxidized or filamentous sludge may not compact maximally until 2 to 3 hours of settling time. As noted previously, sludge is rarely left in the secondary clarifier blanket for longer than 30 minutes.

The settleometer test can be used to develop additional information about the system. If the settleometer is allowed to sit idly after the settling test has been completed, it can be determined whether or not significant nitrification, and the potential for denitrification, is occurring. As the settled sludge sits in the settleometer, the microorganisms run out of oxygen quickly; they begin to denitrify, converting nitrite and nitrate to nitrogen gas. If the combined concentration of nitrite and nitrate is high enough, sufficient nitrogen gas will be released to float the settled blanket or at least split the blanket and send some portion of it back to the top of the settleometer, as shown in Figure 8.62.

INTERFERENCES

It is important for the analyst to note that the SVI calculation is only valid for MLSS concentrations with a normal range of 1000 to 4000 mg/L for activated sludge. Outside of this range, the SVI calculation can give meaningless results. For example, a MLSS with a concentration of 8000 mg/L and a 30-minute SSV of 950 mL will have an SVI of 119 mL/g. If one only considered the SVI number, one would think that the sludge was settling well, which is not the case. Conversely, this sludge is hardly settling at all. When creating daily log sheets for the settleometer and SVI tests, operators are encouraged to record the MLSS concentration, SSV, and SVI. These numbers are all meaningless separate from one another.

REQUIRED DATA

- Thirty-minute settling volume in milliliters per liter from the settleometer test and
- Mixed liquor suspended solids concentrations in milligrams per liter for the same sample used in the settleometer test.

CALCULATIONS

The following equation should be used to calculate SVI:

$$\text{SVI}, \frac{\text{mL}}{\text{g}} = \frac{\text{SSV}_{30} \times 1000 \dfrac{\text{mg}}{\text{g}}}{\text{MLSS} \dfrac{\text{mg}}{\text{L}}} \tag{8.37}$$

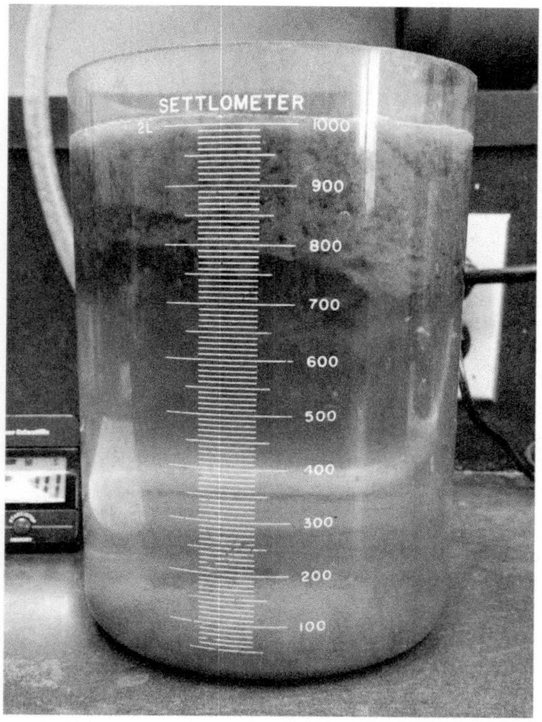

Figure 8.62 Denitrification in the Settleometer Test (Reprinted with permission by Indigo Water Group)

ACTIVATED SLUDGE MICROSCOPIC EXAMINATION

GENERAL DESCRIPTION

Changes in the MLSS, particularly the bacterial population, can be influenced quickly by pH, temperature, availability of food, and other variables. A basic microscopic observation may be performed to observe changes in floc formation, the presence of filamentous bacteria, or the sudden presence of dispersed or free bacteria. It's easy for someone with basic microscopic observation understanding to recognize the difference between a dense or weak floc. Bacteria can be found as individual cells in the bulk water (free bacteria), as filaments (growing in long chains), or, ideally, as flocculated particles (floc). The dominant form often has a direct correlation to how well the wastewater treatment system is performing.

Because identification of most bacteria requires more advanced techniques, using higher life forms commonly found in wastewater biomass such as protozoa and metazoa makes a simple microscopic observation much easier. Both these groups of organisms may be used as indicators of stable or changing operation. An operator who can determine what the organism present means to the overall system performance will have a better chance of understanding or preventing significant upsets. The difficulty becomes the understanding and transfer of this knowledge from one facility to another. Not all facilities will have a wide variety of higher life forms, but may still perform well. Operators should perform microbiological exams when the facility is operating well and become familiar with what is "normal" for their process. Weekly microscopic exams are sufficient when the process is performing well.

APPLICATION

In the past, many manuals and books would state that the presence of crawler ciliates combined with a few other types indicated a healthy biomass. Over the past 15 years of microscopic observations, we have begun to see more variations in these indicators. The best organisms are often the organisms you see when your facility is running at its best. Therefore, observing your biomass when the system is running well will help provide a good baseline of what is common.

Another key in using higher life forms as indicators is diversity. The more diverse your population, the healthier and more stable your MLSS will be. Protozoa and metazoa are orders of magnitude larger than bacteria and can be easily observed microscopically. The majority of these organisms are impacted by BOD_5 and ammonia concentrations, sludge age, and toxic shocks. Many higher life forms flourish and decline in certain environmental conditions. They also can be found in different forms depending on their surrounding conditions. As such, the appearance or lack of these organisms provides a basis for understanding how well the microlife is performing. For example, the majority of protozoa reproduce through binary fission, resulting in two daughter cells. Therefore, the presence and abundance of daughter cells may be a function of ideal environmental conditions.

One example is the presence of detached stalked ciliates. Chances are the environmental conditions have changed, and the stalked ciliate is detaching to find a more favorable environment. If an operator has taken a fresh sample of the MLSS and placed it under the microscope with no delay, it is a real-time example of what is happening in the basin at that time. If you observe an entire slide showing detached stalked ciliates (teletrochs) or empty stalks, it very likely the biomass is experiencing a shock to the system, and further testing or analysis is needed.

An important point to recognize is that some facilities may never have a large diversity or, in some instances, no protozoans or metazoans, yet their treatment system runs very well. Every facility is its own environmental system with various environmental and operating conditions. The presence or absence of higher life forms will not tell you exactly what is wrong with the system, but does suggest something is happening and a further look at the operation may be needed.

INTERFERENCES

- Dirty or scratched slide or cover slip and
- Air bubbles trapped underneath the cover slip.

APPARATUS AND MATERIALS

- Microscope, preferably phase contrast;
- MLSS sample;
- Slide;
- Cover slip; and
- Dropper.

SAMPLE COLLECTION, PRESERVATION, AND HOLDING TIMES

Collect a fresh, well-mixed, representative sample of mixed liquor from the discharge end of the activated sludge basin. Samples of foam and scum may also be collected. If the concentration of the mixed liquor is quite high, the sample should be diluted. If you choose to dilute the sample, always use the same dilution if you plan to do routine counts. Samples should be analyzed as soon as practicable after collection.

PROCEDURE

Step 1—Obtain a clean cover slip and slide.

Step 2—Use a wide-mouth pipet to pick up the sample. A long-tipped eye dropper should not be used because the small opening can break up the floc. Allow one drop of sludge from the pipet to drop in the middle of the clear area of the glass slide.

Step 3—Pick up the cover slip by two corners. Do not touch the cleaned area. Place one edge of the cover slip onto the microscope slide and drag it along the slide toward the drop of MLSS.

Step 4—As soon as the cover slip touches the drop of MLSS, allow the cover slip to fall onto the glass slide.

Step 5—Place the glass slide on the microscope stage.

Step 6—Use the 10× objective to focus the slide. Then, turn to the 20× objective. Using a lower magnification for the examination will make it difficult to see the smaller flagellates and using a larger magnification (40× or more) will narrow your field of view.

Step 7—Scan the slide using 5 passes. Start at the top left corner of the cover slip. Moving down the cover slip, count and record the number of each type of protozoa. When the edge of the cover slip is reached, the first pass is complete. Additional information that may be recorded includes

- Abundance of filaments;
- Floc structure (no filaments, open floc structure, bridging);
- Notes on floc shape (firm or weak, round and compact, or irregular and diffuse); and
- Average floc diameter.

Step 8—Move over to the right slightly and, this time, move up the cover slip keeping a running total of the number of each type of protozoa. This will complete the second pass.

Step 9—Move up and down the slide until 5 passes have been completed. Record the running total number of each protozoa type. If you encounter colonies of stalked ciliates, count or estimate the total number of heads. For the most representative results, scan three or more slides and average the number of each protozoan type.

A sample worksheet for use in microscopic examination is presented in Figure 8.63. The microscopic exam form has had many variations. The key point in any exam is understanding and qualifying the four most important points in any wastewater biomass analysis: floc, filaments, higher life forms, and bulk water. As long as the form has these four areas covered, the operator will be able to provide a detailed exam of the biomass quality. Modifications can be made to focus on particular areas, but this form is a good description of an overall analysis.

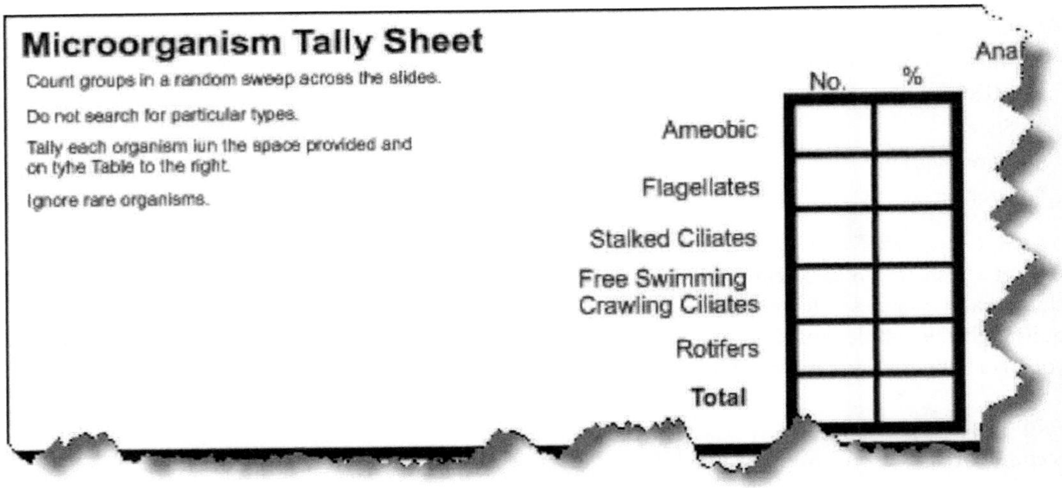

Microorganism Tally Sheet

Count groups in a random sweep across the slides.

Do not search for particular types.

Tally each organism iun the space provided and on tyhe Table to the right.

Ignore rare organisms.

Ana

	No.	%
Ameobic		
Flagellates		
Stalked Ciliates		
Free Swimming Crawling Ciliates		
Rotifers		
Total		

Figure 8.63 Microorganism Tally Sheet (WEF, 2012)

A sample form for microscopic analysis can be downloaded from the U.S. EPA NEPIS website. Search for U.S. EPA document number EPA/625/8-87/012, *Summary Report: The Causes and Control of Activated Sludge Bulking and Foaming*. The form is on page 46 and 47.

CALCULATIONS

Calculate relative dominance of each organism and groups of organisms.

- Count and record the total number of protozoa in each category.
- Next, calculate the average number of each organism from all three slides.
- Calculate the percentage of total population by dividing the average number of each organism by the total number of organisms counted and multiply by 100.
- Group all the free-swimming ciliates together and combine the percentage.
- Calculate the percentage of metazoa.
- Amoeba and flagellates can be grouped together in some cases because they both may be associated with young sludge and incomplete treatment. The presence of both of these organisms has also been found in instantaneous loading increases not related to sludge age.

OXYGEN UPTAKE RATE AND SPECIFIC OXYGEN UPTAKE RATE

GENERAL DESCRIPTION

The OUR test and resulting respiration rate estimate how active the microorganisms are in the activated sludge process. The activity of the microorganisms is related to the amount of oxygen the organisms consume. The OUR test measures how much oxygen a sample of activated sludge consumes over a specific time period. The rate of oxygen uptake can vary because of either microorganism activity or the number of microorganisms. A large number of microorganisms will use more oxygen than a smaller number of microorganisms even when they are respiring (*breathing*) at the same rate. The specific oxygen uptake rate (SOUR) test takes the OUR result and divides it by the concentration of MLVSS in the activated sludge process. This normalizes the result to the amount of oxygen consumed by 1 g of MLVSS per hour and allows operators to compare results from one day to the next even though the MLVSS concentration may be changing. Using MLVSS in the calculation relates oxygen consumption to the mass of active microorganisms in the basin instead of the total MLSS, which includes inert, non-respiring material. The SOUR test is also referred to as the *respiration rate*.

APPLICATION

The OUR test can be used to determine the aerobic biological activity wherever the sample is taken, for example, the tail-end of aeration tank (most common) or along the aeration tank to track treatment progression, to test for toxic effects of influent wastewater, and to demonstrate that digested biosolids are stabilized. It can also be used to determine the required time through the aeration tank for the microorganisms to stabilize the influent BOD and ammonia loads and become "activated", that is, reach endogenous respiration rate. Endogenous respiration rate is the rate of bioactivity when the microorganism has stabilized all stored food and the rate of growth is less than the rate of cellular degradation, that is, the cells are starving.

APPARATUS AND MATERIALS

- Calibrated DO meter;
- Two BOD bottles or two 1-L plastic bottles;
- 7.6-cm (3-in.) section of 22-mm (7/8-in.) outer-diameter plastic pipe that has been shaped on both ends to fit into the mouth of a BOD bottle. The pipe will be used to transfer MLSS between BOD bottles and should fit snuggly. Length is approximate;
- A self-stirring BOD bottle probe and/or separate magnetic stirring device (magnetic device recommended in addition to the self-stirring BOD bottle probe because fast-settling sludge may still settle with only the bottle probe mixer); and
- Stop watch or other timing device.

Other data that will be required for the SOUR calculation include the following:

Volatile suspended solids in grams per liter (ounces per gallon) of the activated sludge sample the OUR test was performed upon. For operational purposes, a TSS meter can be used for the MLSS determination and last week's volatility percentage can be used for estimating the actual MLVSS value. Note that the SOUR calculation for aerobic digesters for the purposes of meeting the vector attraction reduction requirement given in the 503 Regulations of 1.5 mg/L of DO per gram of biosolids per hour uses total solids and not VSS (U.S. EPA, 2001) and varies from the following procedure.

PROCEDURE

Step 1—Collect a fresh sample of activated sludge either from the tail end of the aeration tank or from other locations within the process to measure changes through the process.

NOTE: Perform the test within 15 minutes of collecting the sample for accurate results. Discard the sample and obtain a fresh sample if holding more than 15 minutes.

Step 2—Fill a 300-mL BOD bottle with the activated sludge sample and insert a 7.6-cm (3-in.) section of shaped plastic pipe.

Step 3—Insert the other end of the pipe into an empty BOD bottle. Pour the sample back and forth between the bottles until the DO is approximately 5 to 6 mg/L. Altitude may affect the maximum DO achievable. Do not shake the sample to aerate it because that can cause floc shear, which may expose new floc surfaces making them available for oxygen uptake. Increasing floc surface area can cause SOUR results to differ from actual conditions in the activated sludge basin.

NOTE: If BOD bottles and/or the fitted plastic pipe are not available, two 1-L bottles may be used instead.

Step 4—Remove the transfer pipe and fill one of the BOD bottles to overflowing. Some bubbles will gather at the top. Tilt the bottle and/or tap the sides of the bottle to dislodge air bubbles, then place the bottle into a Koozie or other material to help maintain constant temperature as shown in Figure 8.64.;

Step 5—Use a spatula to work the bubbles out of the sample. It is important to note that the activated sludge sample that remains after filling the BOD bottle can be used to perform the VSS test in step 12 if a meter is not being used. Alternatively, excess MLSS not used in the bottle may be analyzed for TSS;

Step 6—Insert the DO meter probe in the BOD bottle and begin stirring. Turn the DO meter to a 0 to 10 scale;

Step 7—Wait approximately 30 to 60 seconds for the DO meter reading to stabilize. It is important to note that the indicator needle or readout should be constantly dropping during this procedure;

Step 8—Record the DO concentration and temperature of the sample in 30- or 60-second intervals. Stop the test before the DO concentration reaches 1.0 mg/L. If the temperature varies by more than 0.1 °C from the start to the end of the test, the test should be done again with a fresh sample. Results are not valid when

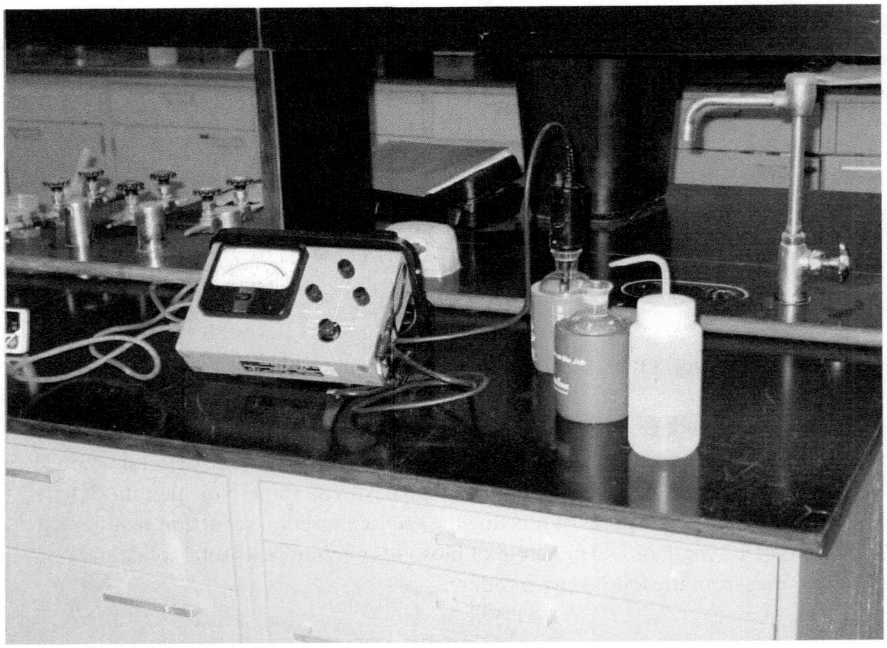

Figure 8.64 Insulated BOD Bottles for the OUR/SOUR Test (Reprinted with permission by Dave Flowers)

temperature is changing because OUR is affected greatly by temperature and changes in OUR may be primarily attributable to temperature changes and not bioactivity;

Step 9—Graph the results by plotting DO in milligrams per liter on the vertical axis and time in minutes on the horizontal axis (Figure 8.65);

Step 10—Draw a straight line connecting the majority of the points. Extend the line so that it crosses the horizontal and vertical axes. It is important to note that the operator may discover that the line is not straight at the beginning and end of the test. The straight line is drawn to negate the effects of these curves; just the straight-line section is used. At the beginning, false values caused by undissolved bubbles can cause interference. The end of the curve flattens because of the limitation of the DO meter;

Step 11—Determine the slope of the line in milligrams per liter of oxygen per minute. The easiest way to do this is to use the points where the line crosses the axes. Divide the milligrams per liter of oxygen by the time in minutes;

Step 12—If not using a TSS meter and the latest volatility percentage, determine the VSS content of the activated sludge sample. Express VSS concentration in grams per liter (g/L). This is done by dividing milligrams per liter of VSS by 1000; and

Step 13—Enter values in the equation and calculate the respiration rate. The OUR is expressed as milligrams of oxygen per hour per gram of VSS. Highly organically loaded industrial treatment systems may have such high OURs that the test sample has to be diluted with secondary clarifier effluent to slow the OUR enough to get readable values that can be graphed. At least three points in a straight line are required to establish the line.

CALCULATIONS

There are two methods commonly used for calculating OUR. Both methods should arrive at the same numerical results. In both instances, the respiration rate worksheet and graph will be used.

The first method requires drawing a straight line through the greatest number of points graphed. The line will cross both the horizontal (time) and vertical (milligrams per liter) axes of the graph. The values at

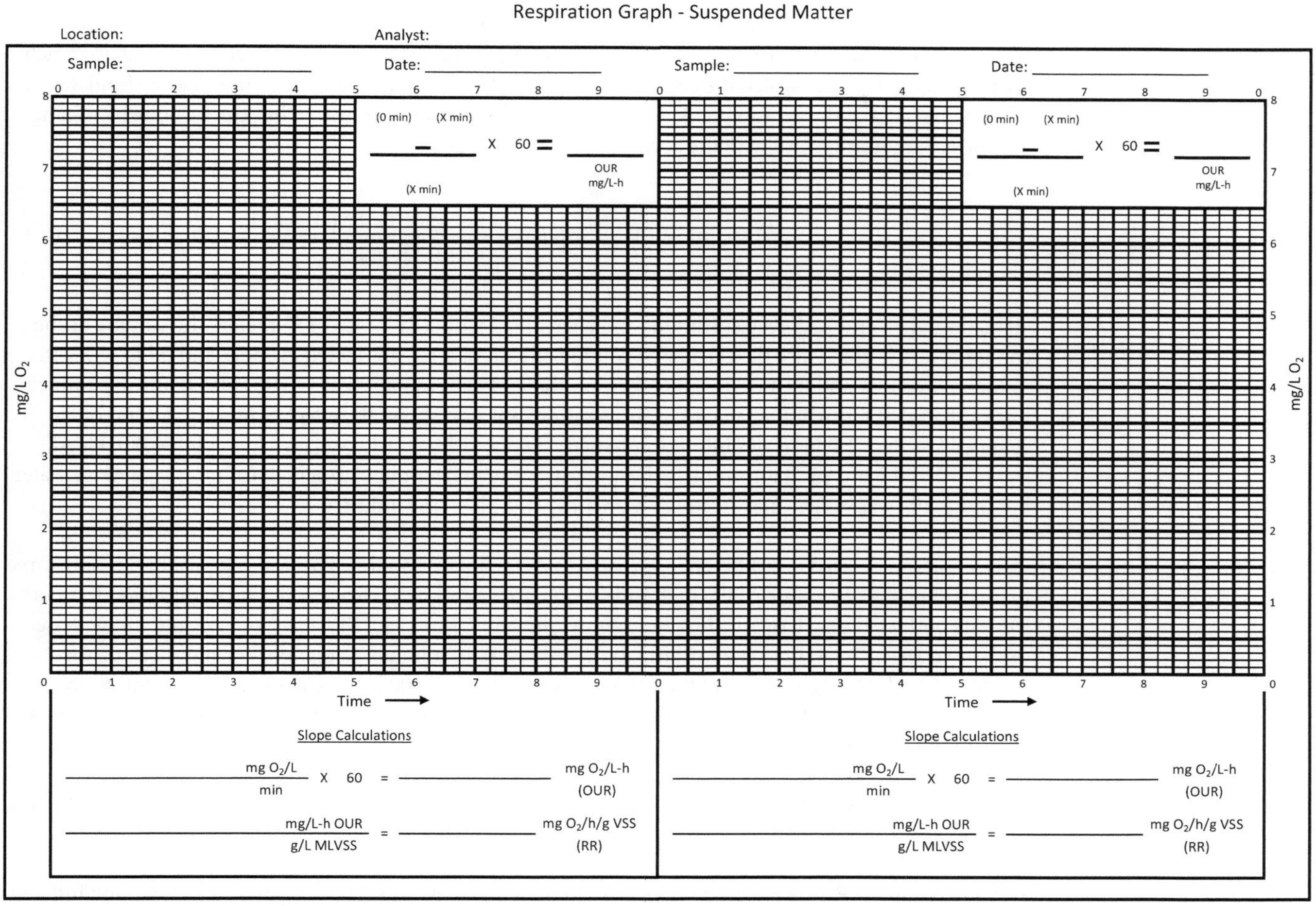

Figure 8.65 Respirometry Worksheet (Reprinted with permission by Keith Radick)

these two points are used in the OUR calculation. The milligrams per liter of oxygen value is divided by the minutes value (mg/L/min) and multiplied by 60 minutes per hour to derive milligrams per liter per hour (mg/L/h).

The second method is applicable when using the second or right-hand side of the graph to do an OUR on an aerobic digester or extended aeration system. These typically flat lines may not conveniently cross the horizontal axis. The straight line through the majority of the points should still be drawn. The analyst should pick a point on the line as time 0 and enter the milligrams per liter of oxygen value in the calculation box in the upper part of the graph. The analyst should also pick a second point on the line as the end time and enter it. The points should be at least 5 minutes apart. Then, divide by the time span between the two points and multiply by 60 minutes per hour to derive the milligrams per liter per hour OUR, as follows:

$$\text{OUR (mg/L}\cdot\text{h)} = \frac{\left(\text{DO}\ \frac{\text{mg}}{\text{L}}\ \text{at time 2} - \text{DO}\ \frac{\text{mg}}{\text{L}}\ \text{at time 1}\right)}{\text{Time at point 0} - \text{time at point 2}} \times 60\ \text{min/h} \tag{8.38}$$

$$\text{SOUR (mg/g}\cdot\text{h)} = \frac{\text{OUR(mg/L}\cdot\text{h)}}{\text{MLVSS(g/L)}} \tag{8.39}$$

The respiration rate worksheet example can be used to perform respiration rate and OUR calculations as shown in Figure 8.65.

OXYGEN UPTAKE RATE TEST VARIATION—STABILIZATION TEST

To this point, OUR testing has been limited to the end of the aeration tank; however, a different approach is called the "fed" test in which the correct amount of influent is mixed with return sludge to simulate the OUR at the head of the aeration tank. This test is best done in a respirometer, but can be done by mixing the correct amount of RAS with the correct amount of influent:

$$\text{RAS, mL} = \text{MLSS, mg/L} \times \frac{\text{Volume of Test Material}}{\text{RAS, mg/L}} \tag{8.40}$$

Here, the MLSS and RAS concentrations are determined with a calibrated TSS meter. The volume of the container is one liter. Place the calculated volume of RAS in a 1-L graduated cylinder and fill to 1 L with influent, then mix and fill a BOD bottle for aeration and determination of OUR. This OUR will be much higher than that taken at the tail end of the aeration tank and shows the effect of feeding. A ratio of fed to unfed less than 1.0 indicates inhibition/toxicity; between 1.0 and 2.0 suggests low soluble BOD load; between 2 and 5 is typical for municipal wastewater; and greater than 5.0 is high for municipal wastewater, but industrial wastewater may be much higher depending on the industry.

Fed testing can be used in a different approach, that is, stabilization test. This test is typically run only three to four times a year because of the high time requirement, but its value is great. The stabilization test is a fed test conducted on a larger scale, where approximately 7.5 to 15 L (2 to 4 gal) of RAS and influent are mixed in a 19-L (5-gal) bucket according to the fed test RAS volume calculation. For instance, if a total volume of 7.5 L (2 gal) is used and the MLSS concentration is 2000 mg/L and the RAS concentration is 8000 mg/L, then, according to formula 8.40, the volume of RAS would be 1.8 L (0.5 gal) mixed with 5.7 L (1.5 gal) of influent.

Step 1—Pour the RAS into the bucket and start aerating with two to four aquarium aerators with diffusers or compressed air line with diffusers.

Step 2—While the RAS is aerating, collect a grab sample of 3.8 to 15 L (1 to 4 gal) of aeration tank influent and mix well with the aerated RAS.

Step 3—Immediately fill a 300-mL BOD bottle with the mixture from the bucket and aerate until the DO concentration is at least 5 mg/L. Facilities located at high elevations may have difficulty getting to 5 mg/L of DO.

Step 4—Run the OUR test.

Step 5—When the test is complete, empty the 300 mL of RAS/influent mixture from the BOD bottle back into the bucket and continue aerating for an hour, trying to keep a DO approximately the same as that in the aeration tank.

Step 6—After an hour, sample again and run another OUR. Do this every hour until the OUR no longer changes. In a high SRT system, after the first hour, testing may be done every 2 hours for 4 hours, then every 4 hours until endogenous respiration is reached. The time it takes to reach endogenous respiration is the stabilization time. Figure 8.66 shows the OUR curve for a typical stabilization test.

OXYGEN UPTAKE RATE TEST VARIATION—ACTIVATED SLUDGE BASIN PROFILE

In a plug flow-type activated sludge basin, a series of OUR tests may be conducted along the length of the basin. Typical sampling locations may be at the inlet, one-quarter down the length of the basin, in the middle of the basin, three-quarters down the length of the basin, and at the outlet. There are five sampling locations total in this example, however; an operator may add additional sampling points as needed to fine-tune the activated sludge basin. Results are then graphed to produce a diagram similar to the one shown in Figure 8.67 below. In this example, endogenous respiration is reached a point midway down the activated sludge basin. The operator has an opportunity to reduce aeration in the second half of the basin to reduce energy costs. Take care to keep DO concentrations high enough to prevent the growth of some types of filamentous bacteria.

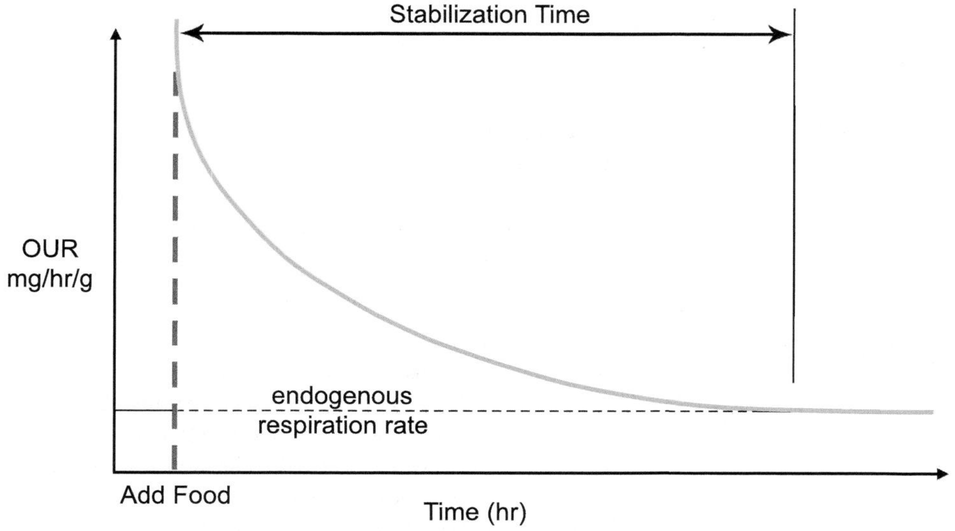

Figure 8.66 Typical Stabilization Time Plot (Reprinted with permission by Ronald Schuyler)

SIGNIFICANCE OF RESPIRATION RATE OR SPECIFIC OXYGEN UPTAKE RATE AND STABILIZATION TIME

The SOUR can be related to typical values for different activated sludge processes. For instance, a conventional activated sludge process or biological phosphorus removal process using an SRT ranging from about 3 to 6 days might expect an SOUR between 12 and 20 mg/L/h/g, whereas a process loaded more in the range of extended aeration or nitrifying activated sludge process using an SRT from about 8 to 15 days might expect an SOUR between 3 and 12 mg/L/h/g. An aerobic digester will typically see SOUR values below 2 mg/L/h/g. Highly organic-loaded industrial processes may run well above 20 mg/L/hr/g. These are typical ranges, but you may find that your process runs well at values outside these ranges. Higher SOUR values indicate a less-stabilized wastewater with a higher organic content.

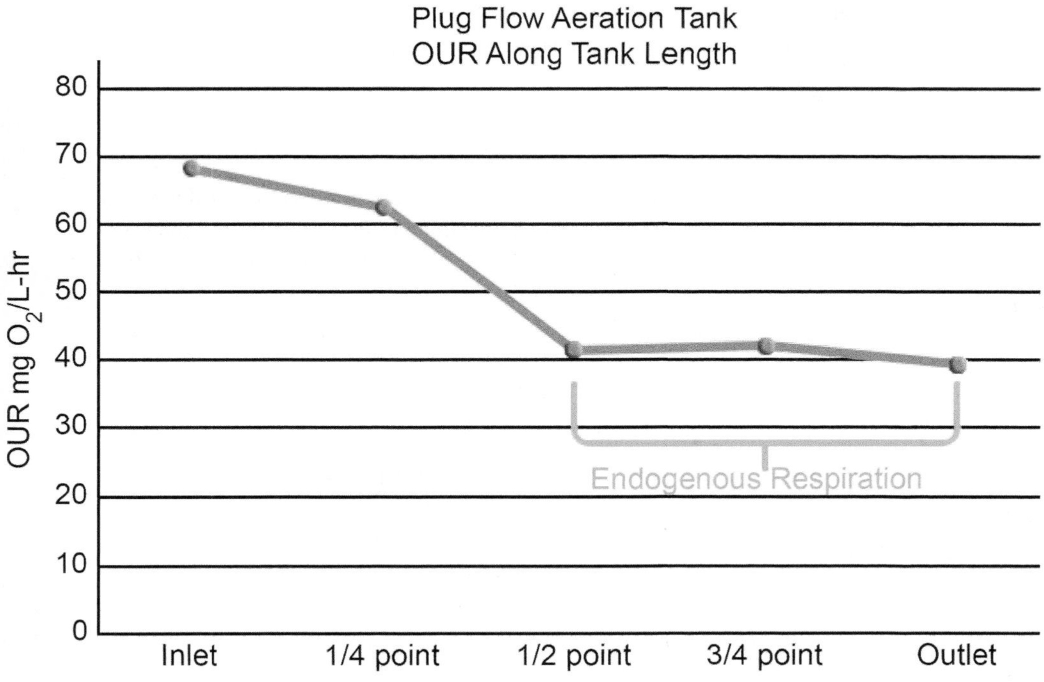

Figure 8.67 Oxygen Uptake Rate Profile of an Activated Sludge Basin (Reprinted with permission by Dave Flowers)

TEST YOUR KNOWLEDGE

1. **An operator notices that one area of the activated sludge basin is particularly turbulent. The most likely cause is**
   a. Damaged or missing diffuser
   b. Clogged air header
   c. Blower output is too high
   d. Water level is too low

2. **An activated sludge basin is covered with large amounts of crisp, white foam. This indicates**
   a. Presence of *Nocardia*
   b. Startup conditions
   c. Low organic loading rate
   d. Nutrient deficiency

3. **When collecting a clarifier core sample, the sample is collected when the rake arm is at a 90-deg angle to the sampling location for this reason.**
   a. Avoids tangling the sludge judge and clarifier mechanism
   b. Allows the sludge blanket to resettle after being disturbed by the rake arm
   c. Ensures that the lowest possible blanket depth is sampled
   d. Prevents the sludge blanket from getting too thick to sample

4. **An activated sludge process has three identical secondary clarifiers. Sludge blanket depths in each are 1 m, 1.2 m, and 3 m. What is the most likely cause for the differences?**
   a. Sludge is being removed faster from the third clarifier.
   b. The RAS pump on the first clarifier has failed.
   c. Uneven flow distribution feeding the clarifiers.
   d. WAS is only being removed from the second clarifier.

5. **In the settleometer test, the SSV at 30 minutes is an indicator of**
   a. Sludge settling velocity
   b. Sludge compaction
   c. Sludge concentration
   d. Sludge toxicity

6. **This type of container should be used for the settleometer test.**
   a. Graduated cylinder
   b. 1-L beaker
   c. 5-gal bucket
   d. Mallory settleometer

7. **When collecting samples for the settleometer test**
   a. Shake vigorously to ensure they are well mixed
   b. Minimize shaking and agitation
   c. Place samples on ice and cool to 4 °C
   d. Add 2 mL of nitric acid as a preservative

8. **Settled sludge may rise back to the top of the settleometer test for this reason.**
   a. Breakdown of solids and production of methane gas
   b. Warming during the test releases dissolved gases
   c. Denitrification and production of nitrogen gas
   d. Hydraulic currents in the settleometer

9. **Two settleometer tests are run side by side. The first settleometer contains undiluted MLSS at 4000 mg/L. The second settleometer contains 50% MLSS and 50% secondary clarifier effluent for a concentration of 2000 mg/L. Both settleometer tests are started at the same time. After 5 minutes, the SSV in the first settleometer is 900 mL but 750 mL in the second. What must be true?**
   a. Filaments are causing poor settleability in the first test.
   b. Hindered settling is taking place in the first settleometer.
   c. Wasting should be decreased.
   d. The RAS should be increased.

10. **The 30-minute settled volume is 320 mL and the MLSS concentration is 2950 mg/L. Find the SVI in milliliters per gram.**
    a. 108 mL/g
    b. 222 mL/g
    c. 320 mL/g
    d. 512 mL/g

11. **The MLSS concentration at the start of the settleometer test was 2500 mg/L. After settling for 30 minutes, the SSV is 230 mL. What is the settled sludge concentration?**
    a. 2500 mg/L
    b. 7813 mg/L
    c. 10 870 mg/L
    d. 22 609 mL/g

12. **The MLSS concentration in the activated sludge basin is 3200 mg/L. The influent flow is 2 mgd and the RAS is flow paced at 60% of the influent flow. What is the RAS concentration?**
    a. 4160 mg/L
    b. 5120 mg/L
    c. 3870 mg/L
    d. 8530 mg/L

13. **The 5-minute SSV is dependent on**
    a. Water temperature
    b. Presence of filaments
    c. MLSS concentration
    d. All of the above

14. The primary difference between an OUR and a SOUR test is

a. OUR is conducted at 25 °C and SOUR at 22 °C.

b. SOUR corrects for solids concentration.

c. OUR testing is used with anaerobic digesters.

d. SOUR is a shorter test.

15. The SOUR test measures

a. Blower and diffuser efficiency

b. DO probe fouling

c. How fast microorganisms are consuming oxygen

d. Effluent soluble BOD_5

16. Mixed liquor volatile suspended solids are used for OUR and SOUR testing rather than MLSS because

a. The MLSS analysis takes longer.

b. The MLVSS test is more accurate.

c. The MLVSS test measures live microorganisms.

d. The MLSS test does not include inert solids.

17. The SOUR for a particular WRRF is typically between 5 and 8 mg O_2/L·h. When the operator tested it at 3:00 p.m. today, the SOUR had increased to 12 mg O_2/L·h. What is the most likely cause for the increase?

a. Toxic or inhibitory substance entered the activated sludge basin

b. SRT dropped below the minimum required to support nitrification

c. Increased BOD_5 loading to the activated sludge basin

d. Number of filamentous bacteria doubled

18. A clarifier is operating with a SOR of 8 $m^3/m^2 \cdot d$ (200 gpd/sq ft) and the RAS concentration is 9000 mg/L. Using the chart in Figure 8.50, find the maximum theoretical SVI the MLSS can have without overloading the clarifier.

a. 150 mL/g

b. 200 mL/g

c. 250 mL/g

d. 300 mL/g

Maintenance

Most of the maintenance for activated sludge processes is associated with the pumps, aeration devices, and clarifier mechanism. Chapters in the forthcoming *Wastewater Treatment Fundamentals II — Solids Handling and Support Systems* are devoted to pumps and aeration systems. Clarifier maintenance was discussed in Chapter 4. A typical preventive maintenance schedule is included in Table 8.10. A site-specific schedule should be developed for each WRRF by consulting the service manuals that were provided with each piece of equipment.

Troubleshooting

Operational problems commonly experienced in the activated sludge process are discussed here. Because it is impossible to include all of the problems that may occur in the operation of the process, only the most common problems that are encountered and that can be corrected with operational changes are included.

Operators are encouraged to exercise consistent process control using SRT as the primary control mechanism as described in previous sections. Process control can be differentiated from troubleshooting by where the attention is focused. Process control sets future goals and defines allowable tolerances and deviations from those goals. Corrective actions are taken when goals are not being met. For example, a facility may set a process control goal of a 10-day SRT with an allowable deviation of plus or minus 1.5 days. When the SRT drops to 6 days, the WAS rate is reduced to increase SRT. Process control is proactive. Troubleshooting identifies a current, undesirable condition and then looks backward to find the cause. It is reactive. Having a good, consistent process control program in place should minimize situations from developing where troubleshooting becomes necessary. Sudden changes in influent characteristics such as loading, flow, temperature, or the presence of inhibitory compounds can still cause process upsets even when a facility has an established process control program.

To effectively troubleshoot a problem, the operators must first break the problem into three major categories: hydraulic issue, mechanical problem, or process problem. Once the correct category has been determined, it becomes possible to determine the probable cause and select one or more corrective measures. Corrective measures should restore the process to full efficiency with the least adverse effect on the final effluent quality and at the lowest cost. To do this, a thorough knowledge of the facility's activated sludge process and the way it fits into the overall WRRF operation is needed. When troubleshooting an activated sludge process, it is always recommended that one corrective action be implemented at a time and that the process be allowed to adapt to the change before making another. It typically requires two to three SRTs for the full effects of a corrective action to be seen when correcting a problem with sludge quality.

Table 8.10 Preventive Maintenance Schedule (Typical)

Activated Sludge System Preventive Maintenance (Typical)	Daily	Weekly	Monthly	Quarterly	Biannual	Annual	Comments
Aeration blower							
Maintain proper lubricant level		W					
Lubricate motor roller bearings							1.5 months
Check for abnormal noises and vibration				Q			
Check that air filters are in place and not clogged				Q			
Check motor bearing rise temperature				Q			
Check motor for voltage and frequency variations				Q			
Check that all covers are in place and secure				Q			
Rotate blower operation				Q			
Lubricate motor ball bearings				Q			
Check that electrical connections are tight				Q			
Check wiring integrity				Q			
Lubricate motor sleeve bearing					B		Or 2000 hours
Inspect and clean rotor ends, windings, and blades					B		
Check that electrical connections are tight and corrosion is absent					B		
Change blower bearing oil						A	
Relubricate after checking flexible couplings							As needed
Oxygen dissolution system							
Lubricate motor					B		
Change gear drive oil					B		
Lubricate flex coupling						A	
Inspect gear tooth pattern wear, shaft and bearing end play alignment, bolting, and seal condition						A	Or 2500 hours
Fine-bubble diffusers							
Check biological reactor surface pattern	D						
Check air mains for leaks	D						
Check and record operating pressure and airflow	D						
Purge water moisture from distribution piping			W				
Bump diffuser system			W				
Drain biological reactor						A	
Remove excess solids that may accumulate						A	
Clean diffusers						A	

(continues)

Table 8.10 **Preventive Maintenance Schedule (Typical) (*Continued*)**

Activated Sludge System Preventive Maintenance (Typical)	Daily	Weekly	Monthly	Quarterly	Biannual	Annual	Comments
Check that retaining rings are in place and tight						A	
Check that fixed and expansion joint retaining rings are tight						A	
Secondary Clarifier							
Remove trash and debris	D						
Test torque-control line switches		W					
Test torque overload alarm		W					
Verify torque scale pointer moves		W					
Check drive unit for accumulated condensation		W					
Check drive oil level and quality		W					
Check drive overload response controls						A	
Inspect entire mechanism above and below water line						A	
Inspect and tighten all nuts and bolts							Regular interval
Inspect lubrication for torque-overload protection device, per manufacturer's instructions							For example, every 18 months or 500 cycles
Return activated sludge pumps (centrifugal)							
Lubricate pump bearings				Q			Or 2000 hours
Lubricate motor bearings						A	
Waste activated sludge pumps (centrifugal)							
Lubricate pump bearings				Q			Or 2000 hours
Lubricate motor bearings						A	

Troubleshooting requires several skills, including investigative thinking and logical use of all tools available to help solve a problem. When troubleshooting, it is important to not perform numerous changes at once because that may result in further problems. If multiple corrective changes are made at the same time and the process does improve, it will be difficult, if not impossible, to identify which process change was the most effective. It is also important to realize that the activated sludge system is a biological process and, as such, most responses will not be noticed immediately. It may take days or weeks before a process improvement can be seen.

When a problem or situation arises, it is important to first evaluate the problem and determine if it fits into one or a combination of the following areas:

- Hydraulic,
- Mechanical, or
- Process.

Hydraulic problems may be caused by several conditions. Hydraulic overload, or surge, may be caused by rainstorms or snowmelt. This is especially true if the treatment facility is served by *combined sewers*. A sudden increase in the influent flow will decrease the HDT in the activated sludge basins. It will also push more solids from the activated sludge basins into the clarifiers, which can cause the blanket depth to increase. Decreased treatment time can affect the ability of the process to remove ammonia. Hydraulic

Combined sewers collect both stormwater and domestic wastewater. They are common in older cities, but are no longer constructed. Instead, separate piping systems are used for stormwater and wastewater.

overload may occur if there are not enough tanks in service. Smaller facilities are more likely to experience operational difficulties because of hydraulic overloading because they typically have fewer basins and excess capacity than larger facilities. A hydraulic surge through the clarifiers can stir up the sludge blanket and carry solids up and over the weirs. This may be detected by observing the area around the clarifier weirs for billowing clouds of solids or signs that solids are being washed out of the clarifier.

Some WRRFs have flow equalization basins that allow them to control how fast influent flows into the WRRF and/or to store influent flow temporarily.

Some facilities are able to manage high peak-hour flows by diverting excess influent to flow equalization tanks. The water is returned to the treatment process during periods of low influent flow. Other facilities have the ability to switch between plug flow and step-feed operational modes in their activated sludge process. Step-feed operation uses multiple influent addition points along the length of the activated sludge basin. This dilutes the MLSS, which reduces the SLR to the secondary clarifiers and reduces the likelihood that MLSS will be washed out of the clarifier and over the weirs. At the same time, step-feed operation concentrates the MLSS near the influent end of the activated sludge basins.

Another hydraulic problem that can occur in the activated sludge process is an imbalance of flow to either the activated sludge basins or secondary clarifiers. It is common engineering practice to design redundant process units to be *hydraulically equivalent*, although exceptions do occur, particularly at retrofitted facilities with limited space available for expansion. Operators should strive to maintain equal flow distribution to redundant, equally sized process units whenever possible. If basins or clarifiers are different sizes, flow should be split proportionally so each basin or clarifier is loaded at the same rate. If flow is distributed to the activated sludge basins unevenly, then the influent organic load will also be distributed unevenly. If flow is distributed unevenly to the secondary clarifiers, one or more clarifiers will build a blanket while the remaining clarifiers do not. Maintaining equitable flow distribution typically involves calibrating and adjusting hydraulic control gates and weirs. Hydraulic imbalance issues can also occur if internal flows such as RAS are not evenly dispersed.

Hydraulically equivalent means the processes are designed to process the same amount of flow. If a facility has three secondary clarifiers, they are typically all the same depth and diameter and can process the same amount of flow.

The second category is mechanical problems. A mechanical problem with a clarifier collector mechanism, RAS pump, low-pressure air blower, or mechanical aerator could result in an effluent violation if it is not noticed and corrected immediately. For example, if a return sludge pump fails and this goes unnoticed, solids will not be removed from a secondary clarifier. This, in turn, could result in the clarifier filling up with solids until the solids eventually flow over the weirs. Troubleshooting of aeration systems and pumps are included in their respective chapters. That information is not duplicated here.

The third category the operator must be aware of is process problems. Process problems are typically the most difficult to identify and correct. Facility walkthroughs should be done several times each shift to check for anything unusual. A facility that maintains good process control records and *control charts* can often recognize warning signs that will alert the operator of an impending problem. Process parameters will fluctuate up and down around an average value. The fluctuations reflect inaccuracies in sampling and measurements as well as the natural variability in the process. They should be random, that is, some days higher and some days lower than the average value. In general, three or more data points moving in the same direction should alert an operator to a potential change in the process that may require a corrective action. For the activated sludge process, operators should track in-basin DO concentrations, influent flows and loads, SRT, settleometer results, and SVI. Equipment records should include blower discharge pressure, energy usage, maintenance records, and notes about unusual noises or smells. Again, good facility records will help enable an operator to identify and correct problems. Operators should use their *supervisory control and data acquisition (SCADA)* system, if available, to produce trend charts so they can visually observe trends in process conditions to determine if the process is functioning properly. Another excellent process troubleshooting mechanism is microscopic examination of the activated sludge. A weekly microscopic examination of a properly prepared and stained sample of the MLSS can help operators determine if filamentous organisms are present in excess. Staining can help identify the type of filament, which, in turn, will help determine the best course of action to correct the problem.

Control charts are graphs of process control data. A control chart should include both upper and lower control limits to help operators identify when corrective actions are needed.

SCADA stands for *supervisory control and data acquisition*. It is a computer-based system that is used to control equipment in the facility and to collect and archive data from monitoring devices. SCADA may track equipment run time, blower discharge pressure, activated sludge basin DO concentrations, influent flow, and a host of other parameters.

Within this section and the troubleshooting guides that follow are lists of solutions that include

- Increasing or decreasing aeration,
- Balancing influent and return sludge flows,

- Adjusting return rates,
- Adjusting wasting rates,
- Adding chemical settling aids,
- Adding supplemental nutrients,
- Chlorinating return sludge or mixed liquor,
- Reducing or controlling recycle flows, and
- Rerouting recycled flows.

Some solutions produce quick results, some solutions produce changes only after many days, and some can be drastic with potentially adverse side effects. The troubleshooting information presented herein is typically oriented toward solutions that minimize the potential for overcorrection and adverse side effects. The stronger the recommended solution, the more judgment and experience that will be required in applying it.

In most cases, aeration adjustments (increasing or decreasing DO), flow balancing, and RAS flowrate adjustments will produce quick, mild results. Typically, the effect of altering SRT takes longer and is less reversible. Applying polymers or metal salts (such as alum or ferric chloride) is typically effective quickly, but can be expensive and may increase sludge wasting requirements. Adding chemicals may also affect other processes such as solids handling facilities. Correcting nutrient deficiencies can produce fast, positive results, but requires purchasing chemicals and chemical feed equipment. Determining what is causing the nutrient deficiencies is also an important form of troubleshooting. For example, is it an industrial or commercial user discharging a waste that is causing the nutrient deficiency issue? If it is, how can pretreatment measures help the situation? Reducing or controlling recycle flows produces quick, positive effects, but may require changes in the operation of solids treatment processes or the addition of offline storage or treatment facilities.

If operating problems cannot be solved in-house, help should be obtained from others such as operators of other facilities, state operators' associations, state and federal regulatory agencies, operations consultants, equipment manufacturers, and local colleges. Some problems are complex and difficult to solve because there may be several contributing factors. In these cases, only qualified individuals with firsthand knowledge and experience should be considered. Second guessing by well-intentioned people can waste time and money, divert attention from the correct solution, and possibly make matters worse.

Before making a change, collect as much data as possible and experiment with possible solutions in the laboratory. If a mistake is made, the results can be easily discarded and another solution tried. If a significant mistake is made within the facility, it is not as easily corrected.

When attempting to solve a problem, avoid treating only the symptoms. For example, chlorinating return sludge for a filamentous bulking problem will temporarily solve the problem, but the underlying cause (such as a nutrient deficiency or too low DO) must be corrected or the problem will return. In some cases, the underlying cause cannot be readily corrected because of design deficiencies or budgetary restraints and it may be necessary to settle for treating the symptom (e.g., chlorinating return sludge) until the underlying problem can be permanently solved.

TEST YOUR KNOWLEDGE

1. **Process control consists of setting goals and defining allowable tolerances and deviations from those goals.**
   - ☐ True
   - ☐ False

2. **For best results when troubleshooting, operators should take multiple corrective actions at the same time.**
   - ☐ True
   - ☐ False

3. **After lunch, an operator is conducting a walkthrough of their facility and they notice billowing clouds of sludge at the edges of the secondary clarifiers. What is the most likely cause?**
   a. Solids overloading
   b. Hydraulic surge
   c. Denitrification
   d. RAS rate too low

4. A facility has one clarifier that is 16.8 m (55 ft) in diameter and another that is 22.9 (75 ft) in diameter. How should the flow be split so that each clarifier has the same hydaulic and SLR?

a. 35% to the smaller clarifier
b. 54% to the larger clarifier
c. 73% to the smaller clarifier
d. 73% to the larger clarifier

5. Classify the action taken by whether it is process control or troubleshooting.

a. SRT target set at 10 days
b. Microscopic examination done to determine cause of foaming
c. RAS flowrate adjusted to keep clarifier blanket below 3 ft
d. Blower output increased to maintain DO of 2 mg/L
e. Second clarifier placed into service to reduce SOR
f. Adding nutrients to correct sludge bulking

AERATION SYSTEM TROUBLESHOOTING

Mixed liquor must be aerated so that aerobic microorganisms will get enough oxygen to remain active and healthy. In addition, activated sludge basin contents must be mixed to bring microorganisms into contact with all of the organic matter in the wastewater being treated. The goal is to provide enough aeration to keep the floc in suspension, but not so much that the resulting turbulence breaks up the floc particles. Mixing in the activated sludge basin can typically be checked by observing the turbulence on the basin's surface. Surface turbulence should be reasonably uniform throughout the basin; violent turbulence should not be present.

A sudden increase in DO concentration when operation of the aeration system has not changed may be a sign that something toxic has entered the aeration basin. Essentially what is happening is that the microorganisms have stopped using as much DO as they were using previously. Heavy metals, bacteriocides, and some organic compounds can be toxic to the microorganisms. They may be inhibited (temporarily slowed down) or killed. A sudden decrease in DO concentration indicates the opposite. The microorganisms are using more oxygen than they were before. This may be because the BOD loading to the WRRF has increased or because more BOD is being returned in sidestreams from other treatment processes. When ammonia is converted to nitrite and nitrate by the nitrifying bacteria, it consumes 4.6 kg O_2/kg NH_3-N and can account for 30% or more of total oxygen use. Check nitrite and nitrate concentrations leaving aerated zones and basins. If nitrification is occurring and is not desired, reduce SRT (if possible) to stop nitrification.

Airflow rates are expressed in cubic meters per hour (m^3/h) or normal cubic meters per hour (Nm^3/h). U.S. customary units are cubic feet per minute (cfm) or standard cubic feet per minute (scfm). A cubic meter or foot of air at sea level contains a certain number of molecules of nitrogen, oxygen, and other gasses. If that cubic meter or foot of air is placed into a flexible bag and brought up to Denver, Colorado, the mile-high city, it will be under less atmospheric pressure. As a result, it will expand and take up more space. Heating the air will also cause it to expand. The number of gas molecules has not changed, but the volume has changed. A normal cubic meter (standard cubic foot) of air is the number of gas molecules that take up one cubic meter (foot) of space at sea level (1 atmosphere of pressure) and at 20 °C (68 °F). The term *Nm^3/h* (scfm) is used to standardize the amount of gas available even when pressure and temperature change.

DIFFUSED AIR SYSTEMS

Dead spots and non-uniform mixing patterns will typically indicate a clogged diffuser or that the diffuser header valves need adjustment to balance air distribution with the basin. Air will follow the path of least resistance. If a diffuser is broken or missing, there will be an open path for the air to move through. This can cause large bubbles to come up in one area of the tank while little mixing is seen in the rest of the tank. Violent turbulence typically indicates a broken pipe or diffuser. Photographs of both violent and uniform turbulence in an activated sludge basin are shown in Figure 8.68.

Some probable causes of non-uniform aeration include the following:

- Air rates too high or low for proper operation of the diffuser;
- Valves needing adjustment or balancing of the air distribution;
- Diffusers needing cleaning or repair (masses of air rising over the location of the diffuser blowoff legs, if so equipped, typically indicate that the diffusers need cleaning or replacing or that there could be a line break); and
- Diffusers not at the same elevation. Less water pressure on some diffusers increases the airflow through those diffusers.

The following measures can be tried, when applicable, to correct aeration problems:

- Adjust diffuser header valves to balance air distribution and eliminate dead spots;
- Adjust airflow rate to maintain the DO concentration in the desired range. Mixing requirements of an activated sludge basin are typically 1.2 to 1.8 $Nm^3/m^2{\cdot}h$ (30 to 40 scfm/1000 cu ft) (U.S. EPA, 1989). This is the amount of air needed to keep the floc particles in suspension. Fine-bubble diffusers

(a)

(b)

Figure 8.68 Activated Sludge Basin Surface Turbulence: (a) Violent Turbulence and (b) Uniform Turbulence

typically require 1.3 to 2.2 $Nm^3/m^2 \cdot h$ (0.07 to 0.12 scfm/sq ft) of basin surface. However, *oxygen-transfer* requirements must also be met in the activated sludge basins for either coarse- or fine-bubble diffuser systems; and

- Regularly check, clean, and level the diffusers. Diffusers should be cleaned typically every 6 months to 1 year to maintain good aeration performance.

Oxygen transfer is the movement of oxygen from the air bubbles into the water. The oxygen becomes DO after transfer.

Some probable causes of inadequate aeration and/or low DO concentrations include the following:

- Increased BOD loading,
- Nitrification occurring in the activated sludge basins,
- Mechanical equipment limitations,
- Plugged diffusers,

- Blower not delivering design airflow,
- Leaks in aeration piping preventing air from reaching basins, and
- Poor oxygen transfer.

The following measures can be tried, when applicable, to correct aeration problems:

- Check the blower intake filters. Replace dirty or clogged filters;
- Check the blower discharge pressure. Increased discharge pressure without an increase in volume of air can be caused by
  - Fouled diffusers,
  - Partially closed air control valve, and
  - Clogged air filter downstream of blower.
- Check the blower preventive maintenance requirements, service record, design capacity, and present output, if possible. The blower may need repair.

JET AERATORS AND ASPIRATOR AERATORS

Occasionally, debris may become lodged inside one or more of the jets. Backflushing is designed to temporarily reverse the flow of MLSS through the jets to dislodge any trapped materials. In a typical backflush sequence, the MLSS pumps are turned off. The blower continues to run, which eventually causes both the air and liquid headers to fill with air. After approximately 1 minute, the backflush valve is opened, releasing a rush of air. The backflush valve is located at the far end of the header opposite from the MLSS pumps. Water pressure from the basin simultaneously pushes MLSS rapidly through the jets and into the liquid header. The sudden reversal of flow pushes debris out of the nozzles and into the liquid header. Jet aeration manufacturers recommend regular backwashing as part of daily operations to prevent blockages from forming.

MECHANICAL AERATION SYSTEMS

As with diffused air systems, inadequate mixing or aeration problems can occur with mechanical aeration systems. Mechanical aeration systems can either be surface splash-type aerators (Figure 8.69) or submerged mixers with or without draft tubes (Figure 8.70). Additionally, mechanical aerators can be subjected to surging and impeller fouling problems.

Inadequate Mixing and Aeration

A DO profile of the activated sludge basin can be made by taking multiple measurements throughout the basin with a hand-held meter to check the adequacy of mixing and aeration. Dissolved oxygen should be greater than 0.5 mg/L throughout the basin. Completely mixed activated sludge basins should have

Figure 8.69 Surface Aerator (Reprinted with permission by Kenneth Schnaars)

Figure 8.70 Mechanical Aerator with Draft Tube (Reprinted with permission by Kenneth Schnaars)

approximately the same DO concentration throughout. Long, narrow basins often have lower DO concentrations at the inlet end. The following are some probable causes of inadequate mixing and aeration:

- Aerator speed is too slow,
- Surface aerator impeller submergence is inadequate,
- Impeller is fouled with rags or ice,
- Aerator is undersized, and
- Mechanical aerator flooding on submerged mechanical aerator with draft tubes and air sparges.

The following measures can be tried, when applicable, to correct for inadequate mixing or aeration:

- Increase aerator speed if the aerator has a two-speed motor that has been operating on slow speed.
- If the aerator is connected to a variable frequency drive (VFD) unit, increase the VFD output to increase the speed of the aerator motor.
- Increase aerator submergence. In most activated sludge basins, the water level is set by the downstream secondary clarifier weir elevation and cannot be changed. In this case, the only way to increase submergence is to physically move the aerator by remounting it to the platform. Rotors in oxidation ditches can often be raised or lowered by adjusting their position. If possible, raise the basin effluent weir setting, which will raise the liquid level and submerge the surface aerator blade. Increasing the blade depth in the water will result in reaching the maximum allowable impeller submergence and increase the oxygen transfer into the water. Do not exceed the maximum allowable submergence given in the manufacturer's operation and maintenance manual because it could overload the motor;
- Remove rags that have caught on the blades. If rags are a problem, check the headworks for screening efficiency;
- Remove ice. This typically a problem with surface aerators; and
- For a submerged mechanical mixer with air sparger, reduce the airflow rate or turn the aerator on high speed.

Hydraulic Surging and Flooding

Mechanical aerators can be subjected to hydraulic surging. Surging occurs when the impeller submergence is below the manufacturer's recommendation and a wave pattern is established in the activated sludge basin that causes the impeller to be alternately exposed to critical over- and under-submergence.

The aerator motor will be intermittently overloaded and may shut down. Surging is more likely to occur during low-flow conditions such as when an entirely new facility or additional activated sludge basin is put online.

The following measures can be tried for surface aerators, when applicable, to correct a surging problem:

- If the surging occurs during no-flow conditions, check the activated sludge basin effluent weir for leaks, especially if it is adjustable;
- Raise or lower the activated sludge basin effluent weir and/or lower the impeller to reach proper submergence;
- If the water level cannot be changed, physically relocate the surface aerator by remounting it to the platform to achieve the desired level of submergence;
- Reduce the number of activated sludge basins in service, if possible, to increase flow and, therefore, submergence in the remaining basins; and
- Investigate the use of baffles, inlet draft tubes, and flow-straightening vanes to eliminate wave action.

Mechanical mixers with draft tubes use a downward flow pattern. Flooding occurs when the flow of the column of water that typically travels downward in the aerator draft tube reverses and starts flowing upward. This condition typically occurs if the buoyancy of the entrained air is greater than the aerator can force downward. This can also occur if the mechanical aerator is on slow speed and the airflow through the sparger is at a high rate. When an aerator is operating in a flooded condition, it reduces the oxygen-transfer efficiency and could damage the aerator. A flooded condition should be corrected as soon as possible. Care must be taken when making the correction however, because equipment damage could occur if the column of water is forced to reverse direction too quickly. Typically, the equipment manufacturer recommends that, during a severe case of flooding, the aerator be stopped and the air turned off. After several minutes, when the flow of the water through the draft tube stops, the aerator can be restarted and a correct flow pattern established. The air can then be restarted and regulated to the correct flowrate. Sometimes, if the flooding condition is not severe, operators may either reduce the airflow rate or increase the mixer speed without stopping the air or mixer. Operator experience will determine if adjustments can be made without a complete shutdown of the air and/or mixer.

Impeller Fouling

Mechanical surface aerators can be subjected to impeller-fouling problems caused by rags and ice. Rags and ice interfere with aeration and mixing, may damage the gear box, and may cause a motor overload. Rags are most likely to accumulate in facilities without primary clarifiers. Icing on surface aerators may result from heat losses in upstream tanks such as aerated grit chambers and primary clarifiers, quantity of water dispersed into the atmosphere (spray), impeller and shroud design, and, of course, climate.

Rags and ice must be removed manually. Ice must be completely removed or it may cause a significant imbalance. Ice formation may be reduced by installing metal shrouds above the aerators to keep the spray in the basin.

TEST YOUR KNOWLEDGE

1. **The surface of an aerated activated sludge basin is smooth at one end, while the other end has large bubbles coming up in one corner. What is the most likely cause?**
   a. Air rates too high at one end of the basin
   b. Broken or missing diffuser in corner of basin
   c. Valves need adjusting to balance air distribution
   d. Diffusers are not at the same elevation

2. **To remove a blockage from a jet aeration system nozzle**
   a. Backflush by filling the header with air
   b. Increase pump discharge pressure
   c. Recirculate MLSS at increased rate
   d. Insert a wire coat hanger and pull

3. **An activated sludge basin equipped with mechanical aerators has DO concentrations of 3 mg/L near the aerators, but only 0.25 mg/L at the edges of the basin. The operator should**
   a. Decrease aerator speed
   b. Lower the tank weir setting
   c. Decrease aerator submergence
   d. Increase aerator speed

4. **Flooding occurs in a mechanical aerator equipped with a draft tube when**
   a. Tank weir is too low
   b. Draft tube flow direction is reversed
   c. Aerator is operating at high speed
   d. Draft tube detached from aerator

5. **The DO concentration in the activated sludge basin increases from 2 to 6 mg/L over a 2-hour period. What is the most likely cause?**
   a. Increased BOD loading
   b. Toxic material in influent
   c. Nitrification occuring
   d. Aerator flooding

SLUDGE QUALITY PROBLEMS

Sludge quality problems fall into two major categories: foaming and bulking. The different types of foam observed in activated sludge processes are summarized in Table 8.11. Three filamentous organisms can cause activated sludge foaming: *Nocardia* spp., *Microthrix parvicella*, and Type 1863 (rare) (Jenkins et al., 2003; Richard, 1989). *Bulking* is defined as any sludge having an SVI greater than 200 mL/g. Most filaments will cause bulking if they are present in large enough numbers. Exceptions include *Nocardia spp.* and Type 1863, which cause foaming, but not bulking (Richard, 2003). Non-filamentous bulking includes the young sludge bulking (straggler floc) and old sludge bulking (pin floc) that were discussed in the process control section. Non-filamentous bulking also includes slime bulking because of nutrient deficiency. Slime bulking may or may not generate foam. Figure 8.71, a troubleshooting decision tree, is included in the next two pages of this chapter.

ACTIVATED SLUDGE BASIN FOAMING

The presence of some foam (or froth) on the surface of the activated sludge basin is normal. Typically, in a well-operated process, a small percentage of the reactor will be covered with a layer of light-tan foam in a loose pattern that moves and changes with aeration. A typical biological reactor is shown in Figure 8.72. Under certain operating conditions, foam can become excessive and can degrade operations. Foaming can

Table 8.11 Foam Types Commonly Observed in Activated Sludge Systems

Foam Type	Indicates
Thin, light tan foam that disperses easily. Ten to 25% of basin surface.	Normal operating condition
Thin, white to gray foam. Easily dispersed.	Type 1863 filament. SRT < 4 days
Stiff, white and/or billowing foam.	Startup conditions High F/M and/or SRT < 4 days Detergents and surfactants Industrial spill
Thick, scummy, and dark brown foam. Few to no filaments present in the foam.	Long SRT (old sludge) Low F/M (underloaded for MLVSS mass)
Thick, greasy, and chocolate brown. May form a crust on the surface.	*M. parvicella*
Thick, greasy, dark tan.	Nocardioforms
Very dark brown to black.	Septicity Industrial discharge
Gray, volcanic or pumice-like.	Recycling of fines from solids handling or septage receiving
Sticky, viscous, and gray.	Nutrient deficiency

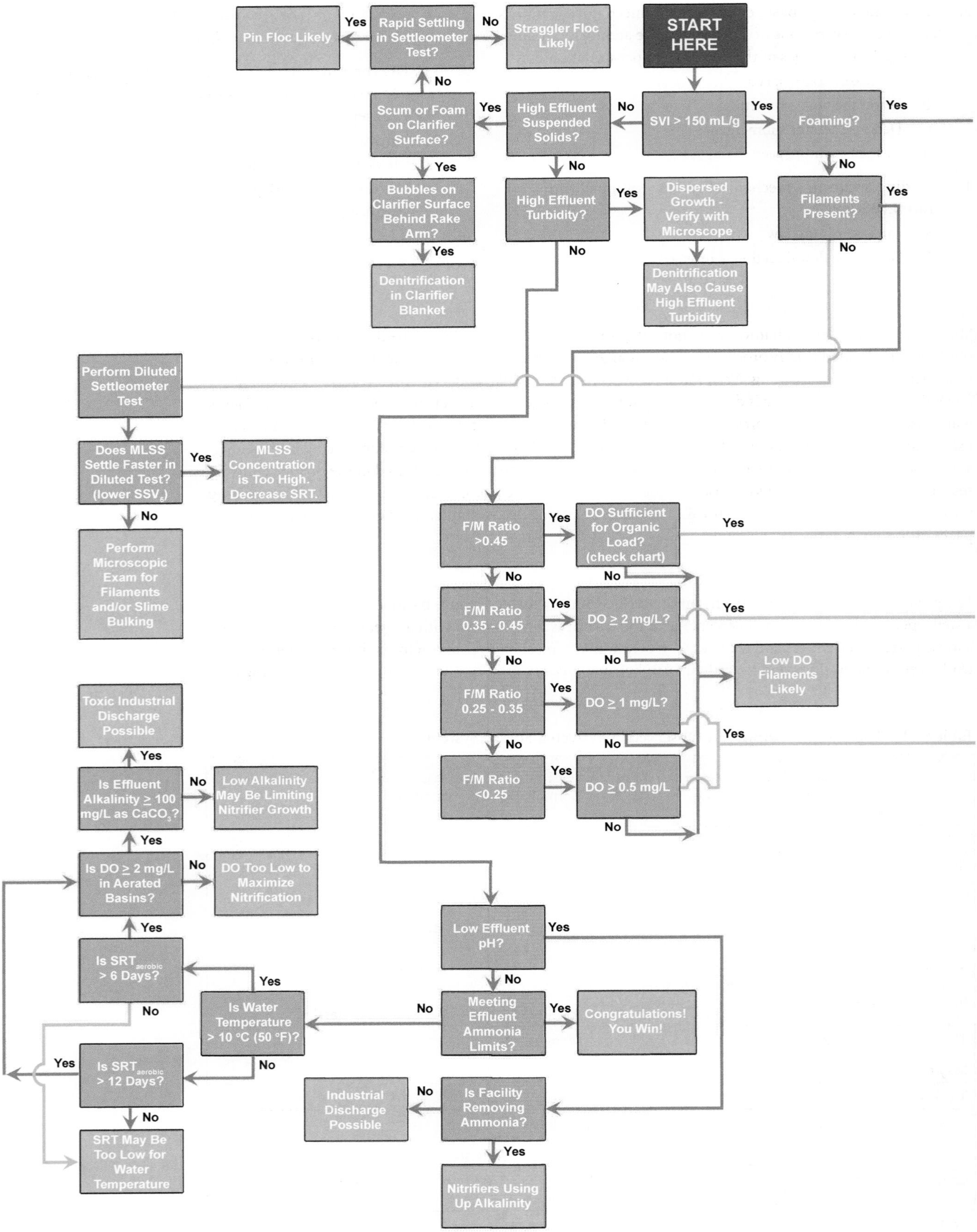

Figure 8.71 Troubleshooting Decision Tree for Activated Sludge (Reprinted with permission by Indigo Water Group)

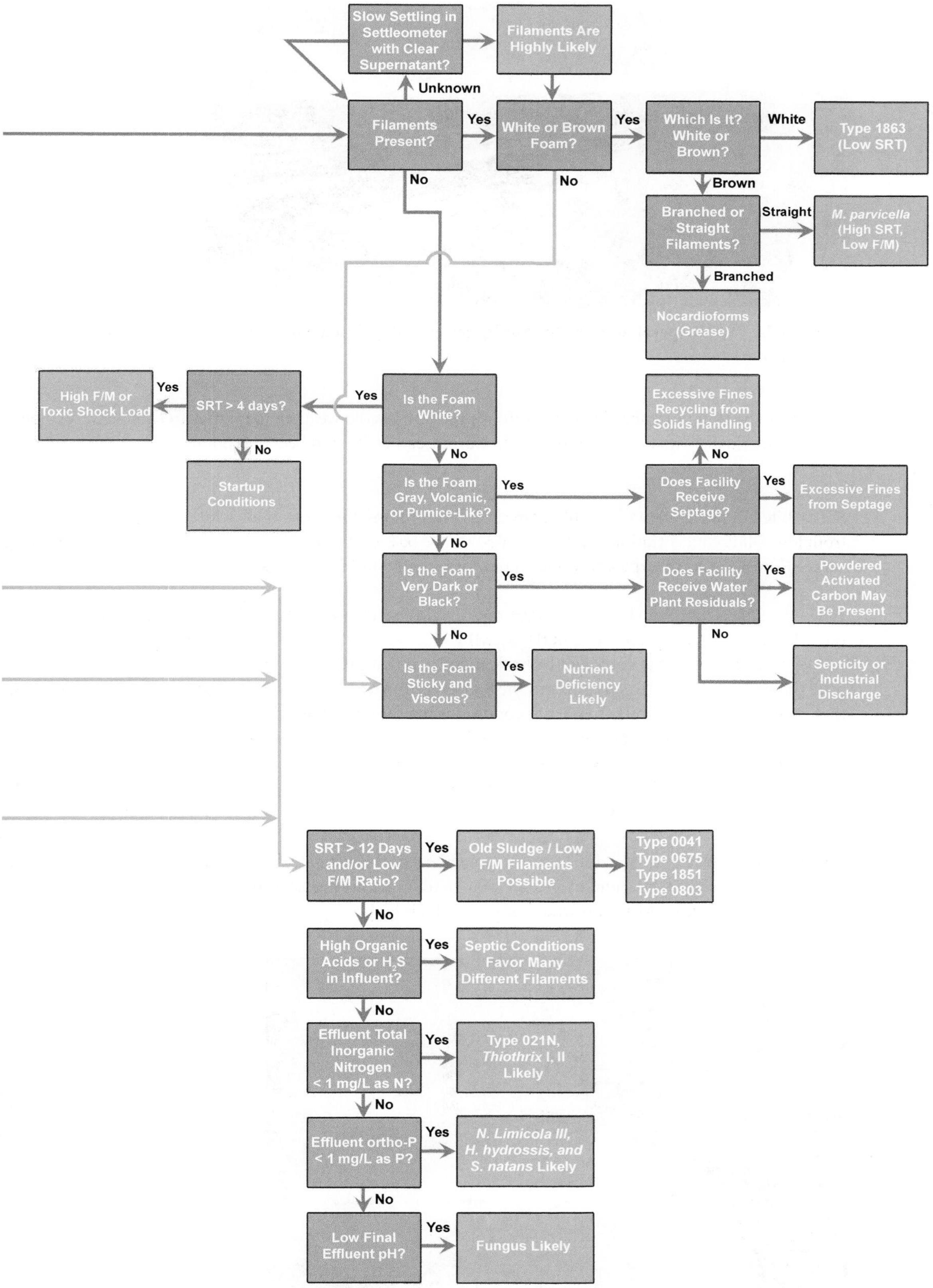

Figure 8.71 (*Continued*)

Figure 8.72 Well-Operated Activated Sludge System (Reprinted with permission by Indigo Water Group)

range from being a nuisance to being a serious problem. In cold weather, foam may freeze and have to be removed manually with a pick and shovel. In warm weather, it often becomes odorous.

White to Gray Foam

Stiff, white billowing foam can result from several conditions, but most of all, white foam problems result from low solids concentrations in the activated sludge basins. White billowing foam could indicate a young sludge (low SRT) and is typically found in new or recently started facilities (Figure 8.54 and Figure 8.73).

When the SRT is low, the MLSS concentration may also be low (low activated sludge solids inventory) and the F/M may be higher than 0.4 kg BOD_5/kg MLVSS (0.4 lb BOD_5/lb MLVSS). The foam may consist of un-degraded proteins that could not be broken down at that SRT. This type of foam typically goes away once the startup window has passed and the SRT exceeds 3 to 4 days. A sudden increase in organic loading that creates conditions similar to startup conditions may also result in white foam. White foam may also be caused by some types of poorly biodegradable household or industrial detergents; however, modern detergents have largely eliminated this issue (Jenkins et al., 2003). White billowing foam may also appear if there was an industrial spill that was toxic to microorganisms. This is accompanied by poor settling and solids washing out from the system.

The filamentous bacteria Type 1863 produces a white-gray foam that collapses easily when the F/M is high and the SRT is below 2 days, and may contribute to foaming during startup (Environmental Leverage, 2003). Type 1863 may also appear when influent FOG is high.

Figure 8.73 Billowing, White Foam on an Activated Sludge Basin (Reprinted with permission by Kenneth Schnaars)

If foam, especially billowing, white foam, is allowed to build up excessively, it can be blown by the wind onto walkways and other structures and create hazardous working conditions. It can also create an unsightly appearance, cause odors, and carry pathogenic microorganisms.

Some probable causes of white foam include the following:

- Low SRT and MLSS concentration resulting from process startup;
- Low MLSS for current organic loading, as caused by excessive sludge wasting or high organic load from an industry (often occurring after low loading periods such as weekends and early mornings);
- Low SRT resulting from unintentional loss of activated sludge biomass caused by
  - Excessive or shock hydraulic loads,
  - High sludge blanket in secondary clarifier resulting in solids washout,
  - Activated sludge not being returned to the activated sludge basin,
  - Poor sludge settleability, and
  - Mechanical deficiencies in the secondary clarifier, such as leaking seals or open drain valves or collector mechanism failure. Check to see if the torque alarm switches have tripped and/or if the shear pin is broken.
- Improper distribution of the wastewater and/or RAS flows to multiple activated sludge basins causing excessive organic loading to one or more basins. Sequencing batch reactor processes can easily receive the same amount of flow while receiving drastically different BOD_5 loads;
- Industrial loading that was toxic to the biomass; and
- Presence of filamentous bacteria Type 1863.

The following measures can be tried, when applicable, to correct this type of foaming problem:

- Verify that return sludge is flowing to the activated sludge basin. Maintain sufficient return rates to keep the sludge blanket in the lower quarter of the clarifier, preferably between 0.3 and 0.9 m (1 and 3 ft) from the bottom;
- Increase the SRT by decreasing or stopping the sludge wasting for a few days (MLSS will increase and the F/M will decrease concurrently);
- Control the airflow rates or mechanical aerator submergence to maintain DO concentrations needed to support the current F/M and/or nitrification in the activated sludge basin. During startup, DO concentrations higher than 3 mg/L may be needed for the first several days;
- Consider seeding the process with activated sludge from a well-operating facility;
- Actively enforce sewer use ordinances to avoid process upsets and deterioration of the facility effluent by controlling the type of industrial wastewater entering the treatment facility;
- Adjust weirs or valves in splitter boxes to maintain the proper distribution (proportional to basin volumes) of flows to multiple activated sludge basins and secondary clarifiers. The construction of some type of flow-distribution structure may be required;
- If flow meters are not provided, visually check the return sludge flow and compare the sludge blanket levels in each clarifier, the return sludge suspended solids concentration from each clarifier, and the MLSS concentration in each activated sludge basin. The corresponding measurements should be nearly equal if the wastewater, activated sludge basin effluent, and return flows are being properly distributed and the MLSS and RAS from different treatment trains are mixed;
- Inspect the secondary clarifier mechanism to see if it has failed or if the gasket around the center column has failed. Repair as needed;
- If the activated sludge basins have water sprays for froth control, operate the water sprays when there is danger of the foam being blown onto walkways or other structures. White, billowing foam works better with a fine spray; and
- If there are no water sprays, consider using defoamants or improvising a froth-control system using effluent water.

Excessive Brown Foams

These foams are associated with facilities operating at longer SRTs and/or low organic loading ranges. Thick, scummy, dark-brown foam with few or no filaments indicates an old sludge (long SRT). This type of foam can result in additional problems in the clarifier by building up behind influent baffles and in the flocculating center well as well as on the surface of the clarifier. This creates a scum disposal problem, as shown in Figure 8.74.

Figure 8.74 Secondary Clarifier Scum Issues

Facilities designed to remove ammonia may have low-to-moderate amounts of rich chocolate-brown foam caused by the filamentous bacteria *M. parvicella*, especially during the winter months where longer sludge ages are necessary to maintain nitrification. *M. parvicella* foam can become so thick that it develops a dry crust on top (Figure 8.75). Facilities with nocardioforms will have strong, greasy, dark-tan foam (Figure 8.76). With both nocardioforms and *M. parvicella*, there will be more filaments in the foam than in the MLSS—10 to 100 times as much (Richard, 2003).

Opportunistic pathogens and organisms that can cause disease in individuals that are already sick, have a weak immune system, or are undergoing chemotherapy. Healthy individuals are not typically infected by opportunistic pathogens.

Several species of nocardioforms and *Mycobacterium* have been isolated from activated sludge foams that can be *opportunistic human pathogens*. Activated sludge foaming should not be taken lightly because of possible harmful health effects. Samples of the foam and mixed liquor should be microscopically examined to determine if the foaming is, in fact, attributable to filamentous growths (Jenkins et al., 2003; Richard, 1989).

Nocardioform growth is typically associated with warmer water temperatures; FOG; and longer SRTs (typically more than 9 days), although it has been frequently encountered at SRTs of 2 days. Foam, however,

Figure 8.75 *Microthrix parvicella* Foam (Reprinted with permission by Indigo Water Group)

Figure 8.76 Nocardioforms Foam (Reprinted with permission by Indigo Water Group)

typically has a longer SRT than the underlying MLSS, and calculations of retention time may be incorrect (Richard, 1989). Washout SRT experiments at the Sacramento Regional Wastewater Treatment Plant saw nocardioform populations decrease to non-detectable levels when the SRT was 2.2 days at 16 °C (60.8 °F) (Cha et al., 1992).

Nocardioform foaming seems to involve the hydrophobic (water-repellant), waxy nature of the nocardioform cell wall, which tends to cause flotation under aeration (Richard, 1989). In other words, nocardioforms repel water to some extent. Nocardioform cells in the mixed liquor concentrate in the foam and continue to float even after they die. The bubbles produced by aeration carry the filaments to the surface where they collect in the foam. Facilities prone to nocardioform foaming often receive oil and grease wastes (e.g., from restaurants without grease traps), have poor or no primary scum removal, recycle scum rather than remove it from the facility, and have activated sludge basin configurations such as submerged wall cutouts or effluent gates that trap foam.

The best way to deal with nocardioform foaming is to prevent the conditions that encourage nocardioform growth. Once established, nocardioform foaming can be extremely difficult to eliminate because

- The foam is difficult to break up with water sprays and foam supression systems,
- The foam typically does not respond to chemical antifoamants,
- Some species of nocardioforms have low washout SRTs that prevent systems that are required to remove ammonia from wasting it from the system,
- The foam has a longer SRT than the MLSS because it is not wasted in the WAS, and
- Solids retention times may be limited by nitrification requirements.

M. parvicella growth is associated with colder water temperatures (12 to 15 °C), low DO, and high SRTs (low F/M). Its growth is encouraged when operating with anoxic, anaerobic, and intermittently aerated zones or basins. These are used in activated sludge processes that remove ammonia, nitrite, nitrate, and phosphorus. Foaming is especially problematic during the winter months when longer sludge ages are necessary to maintain nitrification. It is possible for *M. parvicella* to cause sludge bulking without causing foaming.

The following are some probable causes of this type of foaming problem:

- Biological reactor being operated at excessively long SRT;
- Buildup of a high MLSS concentration as a result of insufficient sludge wasting. This condition could unintentionally occur when seasonal (winter to summer) wastewater temperature change results in greater microorganism activity and, consequently, greater sludge production;

- Operating in the sludge reaeration mode. A heavy, dark-brown, greasy foam is normal in the sludge reaeration tank, particularly if the process is being operated at a low F/M; and
- Improper wasting control program.

The following measures can be tried, when applicable, to correct foaming problems caused by filamentous bacteria:

- Gradually decrease SRT by increasing the wasting rate by 10% more mass each day;
  - Facilities with effluent ammonia concentration limits will be limited in how far the SRT may be safely reduced. Generally, an $SRT_{aerobic}$ of 4 to 6 days is needed to maintain full nitrification when the water temperature in the basin is at 20 °C (68 °F) and 10 to 15 days when the water temperature is at 10 °C (50 °F).
  - Facilities with effluent ammonia concentration limits should monitor effluent ammonia concentrations while decreasing the SRT. If effluent ammonia begins to increase, the SRT must be increased. If the DO concentration is below 2 mg/L, increasing the DO concentration may increase nitrification efficiency and may allow for a lower SRT.
- Implement a better program for controlling the wasting of activated sludge; and
- If the foam contains filaments, remove it from the surface of the water and send it to the solids-disposal facilities.
  - If possible, waste foam directly from the surface of the activated sludge basins. Foam may be pushed toward one end of a basin or channel with a spray of water. Some facilities are equipped with surface wasting facilities that will allow foam to be diverted into an empty basin or channel where it can be removed.
  - Some activated sludge basins have internal walls and baffles that can trap foam. In some activated sludge systems, the MLSS flows to the secondary clarifiers through submerged pipes—known as *subsurface withdrawal*—that can also trap foam in the basins. Once foam begins to accumulate in a basin, it acts as an incubator for more filament growth. Foam can quickly go from minimal basin coverage to overflowing the walls if left unchecked. Do not wait until the basin is completely covered with foam before beginning to address the problem. It is preferable to define an operating condition, 30% basin coverage for example, when corrective action will begin.
  - A vac-truck may be used to remove foam from the surface of the activated sludge basin and deliver it to a landfill for disposal. This can be very helpful for basin designs that trap foam. This is a classic case of treating the symptom rather than the disease. It would be better to find out what is causing the problem and correct it rather than resorting to the vac-truck.
  - Ensure that the foam is not recycled back to the WRRF. If the nocardioforms are pumped to aerobic or anaerobic digesters, it could cause foaming in the digesters. Recycle streams from digesters and dewatering equipment can re-seed the activated sludge basin with filaments, sometimes setting up a vicious cycle in which foam cannot be eliminated. Ideally, send foam to a landfill. Do not haul foam to another WRRF for disposal.
  - When foam is removed from activated sludge basins, include the solids removed in the foam in the waste sludge calculations. During normal operation, the amount of solids removed with the foam is too small to matter. However, during heavy foaming, as much as 10% of WAS solids may be removed with the foam.
- If filaments appear, try to identify the type and cause of the filament (refer to the section titled, "Bulking Sludge—Filamentous Microorganisms Present", for further information);
- Implement a better program for controlling the wasting of activated sludge; and
- If the foam contains filaments, remove it from the surface of the water and send it to the solids-disposal facilities. Ensure that the foam is not recycled back to the WRRF.

Very Dark or Black Foam

The presence of a very dark or black foam indicates either insufficient aeration, which results in anaerobic/septic conditions, or that industrial wastes such as dyes and inks are present. This type of foam is often accompanied by a sour or portable toilet odor. Very dark or black MLSS without foaming or odor may be caused by the discharge of powdered activated carbon by an upstream drinking water treatment facility. The following measures can be tried, when applicable, to correct this type of foaming problem:

- Increase aeration and
- Investigate the source of industrial wastes to determine if appearance is resulting from dyes or inks.

Viscous Gray Foam

Viscous bulking and foaming is associated with nutrient deficiency. This is also called *slime bulking*. When microorganisms don't have enough nitrogen, phosphorus, or some other essential nutrient to continue growing and reproducing, they produce excess exopolymer instead. Recall from the discussion in Chapter 5 that heterotrophic bacteria gain energy from organic compounds. When the bacteria have everything they need to grow and reproduce, about 0.6 kg (6 lb) of new bacteria are produced for every kilogram (pound) of BOD_5 entering the activated sludge process. Some of the organic matter leaves the bacteria as carbon dioxide and water. When nutrients are not available, the bacteria will continue to take in organic material. They must have energy to continue living. However, without essential nutrients, they are unable to make the proteins they need to build more of themselves. As a result, the bacteria end up with a lot of excess carbon that has to go somewhere. Exopolymer is produced instead of more bacteria. The exopolymer has about the same density as the surrounding water and causes the floc particles to settle more slowly. In extreme cases, the sludge may not settle at all. Nutrient-deficient activated sludge may also produce a sticky, viscous foam.

If a sample of nutrient-deficient MLSS is treated with India ink stain and examined under the microscope, excess exopolymer will appear as white blobs. Healthy floc will appear solid black. India ink can be purchased in any art store and in many grocery and convenience stores. It is made from fine carbon black particles suspended in water. The carbon black particles can penetrate into healthy floc, but are unable to penetrate through accumulations of exopolymer. In extreme cases, the sample may look like a bunch of fish or frog eggs with a single bacteria or clump of bacteria surrounded by white exopolymer.

To correct nutrient deficiency, the operator must determine which nutrient is limiting the growth of the microorganisms and then add it continuously to the process. It is generally accepted that a ratio of 100 to 5 to 1 of BOD_5 to N to P is needed to support good floc formation. This means that if the influent BOD_5 concentration is 170 mg/L, there must be at least 8.5 mg/L of nitrogen and 1.7 mg/L of phosphorus. Other nutrients, like iron, may also be growth limiting.

Volcanic or Pumice-Like Foam

This type of foam is a result of recycling excessive amounts of fine solids from solids handling processes, especially anaerobic digesters. Fines may be recycled from digesters and from thickening and dewatering operations. Fines may also be introduced with hauled septic system waste. The fines don't flocculate well and may create both scum and foaming problems as well as increasing effluent turbidity. To correct the problem, operators must focus their attention on improving digester supernatant quality and improving solids capture in dewatering operations.

TEST YOUR KNOWLEDGE

1. **Match the foam color to its most likely cause.**

1. Thin, white-to-gray foam	a. Old sludge
2. Billowing, white foam	b. Low F/M
3. Greasy, dark tan foam	c. Nutrient deficiency
4. Thick, greasy, chocolate-brown foam with crust	d. Septic conditions
5. Dark or black foam	e. Type 1863
6. Sticky, viscous foam	f. Recycle of fines
7. Volcanic or pumice-like foam	g. Nocardioforms
8. Thick, scummy dark-brown foam	h. *M. parvicella*

2. **When starting up a new facility, how long does it typically take for startup foam to go away?**
   a. About 24 hours
   b. 2 to 4 days
   c. 5 to 7 days
   d. 2 weeks

3. **A storm event caused influent flow to the WRRF to triple. After the event, white foam is observed on the surface of the activated sludge basins and the RAS concentration has decreased. Why?**
   a. Solids loss from clarifier lowered SRT
   b. Toxic compound in stormwater flow
   c. RAS return flowrate too high
   d. MLSS pH dropped below 6.5

4. **Brown foam should be removed from activated sludge basin and clarifier surfaces to**
   a. Improve overall facility appearance
   b. Prevent flooding of mechanical aerators
   c. Keep the wind from blowing it onto walkways
   d. Reduce numbers of filamentous bacteria

5. **An activated sludge system is required to remove ammonia. The water temperature is currently 15 °C and the $SRT_{aerobic}$ is 24 days. Thick, greasy, chocolate brown foam covers the surface of the activated sludge basins and has completely filled the clarifier floc wells. The operator should**
   a. Chlorinate the RAS
   b. Manually remove foam and reduce SRT
   c. Decrease DO concentrations in the basins
   d. Increase clarifier rake arm speed

6. **The MLSS gradually turned from light tan to almost black over several hours. Dissolved oxygen concentrations are normal and excess foaming is not present. No sidestreams were entering the basin during this time. What could have caused the color change?**
   a. Septic conditions
   b. Sludge recycle from digester
   c. Powdered activated carbon
   d. Type 021N filament present

7. **The surface of the activated sludge basin and clarifier are covered with brown foam. Microscopic examination shows there are almost no filaments in the foam. What must be true?**
   a. *M. parvicella* is causing the foaming.
   b. The SRT is long and the F/M is low.
   c. Nocardioforms are causing the foaming.
   d. Denitrification is occurring in the clarifier.

8. **This foam-forming filamentous bacteria tends to cause problems during colder weather.**
   a. Nocardioform
   b. Type 021N
   c. Type 1863
   d. *M. parvicella*

9. **A WRRF is required to remove ammonia. The activated sludge basins are covered with thick, greasy, brown foam. The water temperature in the basin is currently 10 °C. How low can the operator drop the aerobic SRT before they risk losing nitrification?**
   a. 4 days
   b. 6 days
   c. 9 days
   d. 15 days

10. **A domestic WRRF receives a majority of its wastewater from a fruit processing and canning factory. The WRRF has an ongoing issue with slimy, viscous, gray foam. To fix the problem, they need to**
    a. Remove foam with a vac-truck
    b. Increase DO concentrations in the basin
    c. Add missing nutrients
    d. Add polymer to help settle fines

11. **Excessive fines from septage receiving or solids handling processes generate a pumice-like foam because**
    a. Fines don't flocculate well
    b. Promotes the growth of Type 1947
    c. Polymer in recycle causes foaming
    d. Vital nutrients are missing

ACTIVATED SLUDGE BULKING

Bulking sludge is defined as an MLSS with an SVI greater than 200 mL/g. Sludge bulking may be caused by dispersed growth, the presence of filaments, poor floc structure, and nutrient deficiency.

Dispersed growth occurs when the MLSS fails to flocculate. The microorganisms aren't producing enough exopolymer for the flocs to form properly. The presence of a cloudy supernatant above a slowly settling sludge in the settleometer test indicates dispersed-growth bulking and either improper organic loading, overaeration, or toxics (Figure 8.77). Dispersed growth often occurs for a brief period during startup and after a toxic shock or heavy organic load (increased F/M). Dispersed growth is not well understood.

The presence of clear supernatant above slowly settling sludge that compacts poorly (high SVI) indicates that settling is being hindered by the presence of filamentous microorganisms. Filaments grow in response to specific operating conditions. Filaments can also be recycled from solids-handling processes. The presence of filamentous microorganisms is corrected by improving the treatment environment by adjusting the SRT, correcting the DO concentration in the biological reactor, correcting a pH condition, eliminating excessive organic acids, or adding nutrients such as nitrogen, phosphorus, and/or iron.

The following are some probable causes of bulking problems:

- If filamentous microorganisms are present (filamentous bulking)
  - Excessively long SRT;
  - Low-DO concentrations in biological reactors;
  - Insufficient nutrients;

Figure 8.77 Dispersed Growth in a Settleometer Test (Reprinted with permission by Indigo Water Group)

- Improper pH—either too low or widely varying;
- Warm wastewater temperature;
- Widely varying organic loading;
- Industrial wastes with high BOD and low nutrients (N and P) such as simple sugars or carbohydrates (e.g., whey waste);
- High influent sulfide concentrations that cause the filamentous microorganism, *Thiothrix*, to grow and produce filamentous bulking. (Note: If high sulfides are present in the wastewater, the nodules in the *Thiothrix* filament will glow after the sample is stained. If the nodules in the *Thiothrix* filament are not glowing after staining, then it could be an indication of nutrient deficiency);
- Large amounts of filaments present in influent wastewater or recycle streams; and
- Insufficient soluble BOD_5 gradient (as found in completely mixed biological reactors).

- If few or no filamentous microorganisms are present (dispersed-growth bulking)
  - Excessively high or low SRT,
  - Improper organic loading—either too high or too low F/M,
  - Overaeration,
  - Nutrient deficiency, and
  - Toxic material.

The first step in analyzing a bulking sludge condition is to examine the mixed liquor under a microscope.

Filamentous Microorganisms Present

The first step in reducing the number of filaments is to identify the filament through staining and microscopic examination. Once the filament has been identified, the condition that encourages its growth can be eliminated. For example, *M. parvicella* is associated with longer SRTs whereas *H. hydrossis* is associated with low DO. Identifying the filament can often identify the underlying condition. Table 8.1 at the beginning of this chapter lists the different filaments that commonly occur in activated sludge and the conditions that promote their growth. All WRRFs with activated sludge processes should have a copy of *Manual on the Causes and Control of Activated Sludge Bulking, Foaming, and Other Solids Separation Problems* (Jenkins et al., 2003). If the facility staff does not have the equipment or expertise, outside help should be used. Proper identification requires *Gram and/or Neisser staining* and a good microscope capable of 1000× magnification. Use filament identification charts to determine which filament or filaments are present and

Gram stains and Neisser stains are two types of stains that may be applied to properly prepared activated sludge samples on a microscope slide. Filamentous bacteria may be identified by whether they are Gram positive (purple), Gram negative (pink), Neisser negative (straw colored), or Neisser positive (dark blue to purple/black). It is nearly impossible to properly identify most filaments without staining.

the most likely cause for their proliferation. Many filaments look alike and cannot be positively identified without staining. Guessing is counterproductive because it will often result in the wrong corrective action being taken.

Nocardioforms and *M. parvicella* are both strongly Gram positive and will appear purple after Gram staining (Jenkins et al., 2003). *Nocardia* is a short, branched filament (Figure 8.78), whereas *M. parvicella* is a long, coiled filament without branching (Figure 8.79). Nocardioforms are typically 5- to 30-mm long (Jenkins et al., 2003). *M. parvicella* filaments are typically 50- to 200-mm long (Jenkins et al., 2003). These two filaments are easily identified, especially when combined with an abundance of brown foam on the activated sludge basins.

There is typically one type of filament that is present in large quantities and one or more other filaments present in lower quantities. For example, *Sphaerotilus natans* is often dominant with smaller amounts of *Haliscomenobacter hydrossis* in a system suffering from low DO. Because the growth of many filaments is associated with more than one adverse condition (low DO, low F/M, etc.), it can be very helpful to identify multiple filaments.

If a moderate to large number of filamentous microorganisms are present, one or more of the following corrective actions should be performed to eliminate the condition that is supporting the growth of the filament(s):

- Many filaments proliferate at higher SRTs. Many filaments grow slowly and need longer SRTs to have time to reproduce and accumulate. Verify current operating conditions and confirm that SRT is within the desired range.
- Measure the DO concentration at various locations throughout the activated sludge basin. Many WRRFs have multiple zones within their activated sludge basins or separate basins to provide anaerobic, anoxic, and aerated conditions. This section applies to the aerated zones and basins only. For SBRs, this section applies to the aerated phase of the cycle.
    - If the typical DO concentration throughout the basin is less than 0.5 mg/L, there may be insufficient DO in the activated sludge basin. Increase aeration until DO increases to 1.5 to 3 mg/L throughout the tank. Some facilities do operate at DO concentrations of 0.5 mg/L at the effluent end of the basin after satisfying their nitrification requirements.

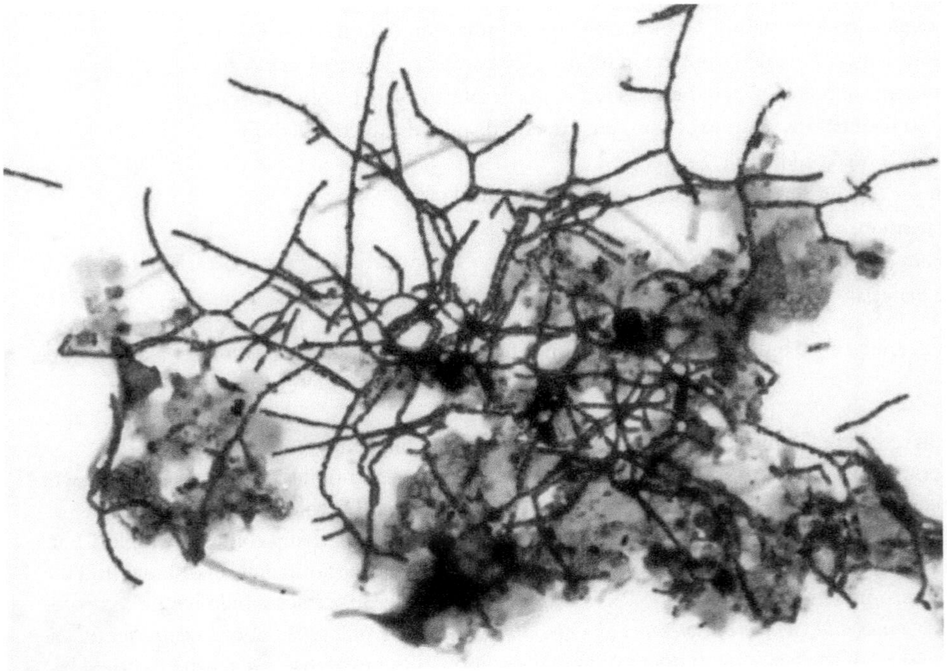

Figure 8.78 Nocardia with Floc After Gram Staining (Reprinted with permission by Toni Glymph-Martin)

Figure 8.79 ***Microthrix parvicella*** **with Floc After Gram Staining (Reprinted with permission by Toni Glymph-Martin)**

- If DO concentrations are nearly zero in some parts of the basin, but are higher in other locations, the air distribution system on a diffused air system may be off-balance or the diffusers in an area of the basin may need to be cleaned. Balance the air system and clean the diffusers. If a mechanically aerated system is used, increase aerator speed, or raise the overflow weirs, or increase the air if the mechanical aerator is also provided with an air sparger.
- If the DO concentration is low only at the head of the tank, which is being operated in the plug flow pattern, consider changing to a *step-feed* or complete-mix flow pattern or using tapered aeration, if possible. A low DO concentration only at the head of the basin may not be a problem as long as adequate reaction time is available in the rest of the basin.

Step-feed activated sludge adds influent or primary effluent at multiple locations along the length of the basin to distribute the load. Step-feed is discussed near the end of this chapter.

- Measure the pH within the activated sludge basins. If pH in the basin is less than 6.5 or is widely varying, settleability may be affected because bacteria that settle readily are inhibited. Chronic low pH operation can also encourage the growth of fungi in the MLSS and affect settleability. Basin pH may be affected by influent wastewater characteristics and by biological processes taking place within the activated sludge process.
  - If the raw wastewater pH is less than 6.5 and the water supply pH is higher, the low pH problem may be a result of industrial waste, and a survey should be conducted to identify the industry discharging low-pH water. The best procedure, if possible, is to prevent the problem by controlling it at the source. A good industrial waste monitoring and enforcement program will avoid many difficulties in this area.
  - Nitrification, the conversion of ammonia to nitrite and nitrate, destroys alkalinity and can reduce the biological reactor pH. The nitrification process can be inhibited at pH levels below 7.0 standard units (SU) and could cease if the pH goes below 6.5 SU. If this is the cause of the problem, raise the MLSS pH by adding a base such as sodium bicarbonate ($NaHCO_3$), caustic soda ($NaOH$), lime [$Ca(OH)_2$], or magnesium hydroxide [$Mg(OH)_2$] at the head of the biological reactor. Facilities with anaerobic digesters should not use chemicals containing magnesium because magnesium can cause or increase *struvite* formation in anaerobic digesters. If nitrification is required, the MLSS pH should be kept above 7.0 SU and held as constant as possible to encourage an acceptable rate of nitrification. Be sure that the treatment system is not shocked by high pH levels or overdosed with chemical to pH levels greater than 9 SU.
  - When determining the amount of $NaHCO_3$, $NaOH$, $Ca(OH)_2$, or $Mg(OH)_2$ needed for pH adjustment, obtain and weigh (or measure) a small amount (1 to 2 g) of the actual chemical to be used in the treatment process. While stirring a measured sample, add small increments of the chemical until the pH is approximately 7.2 SU. Then, weigh (or measure) the portion not used to determine the amount used in the titration.
  - Adding $NaHCO_3$, $NaOH$, $Ca(OH)_2$, or $Mg(OH)_2$ to raise pH is expensive; therefore, chemical usage must be carefully monitored so excessive chemical is not fed into the process. Mixed liquor settleability should be closely monitored to observe changes to ensure that the chemical added is effective. If no improvement occurs within 2 to 4 weeks and nothing else in the process has changed in the meantime, then chemical addition should be stopped.

Struvite, magnesium ammonium phosphate, is a nuisance crystal that can form in anaerobic digesters. The chemical formula for struvite is $MgNH_4PO_4$. Struvite can coat the insides of pipelines, pumps, digester mixers, and other equipment. It is difficult to remove, so conditions that encourage its formation should be avoided.

- A uniform low concentration of soluble BOD, as found in completely mixed basins, is considered by some to encourage the growth of certain types of filamentous microorganisms. These filaments compete with floc formers to cause low F/M bulking. One technique suggested to control low F/M filaments is the use of a non-aerated (anoxic or anaerobic selector) mixing zone for the wastewater and RAS. Anoxic selectors can help to control nocardioforms, *M. parvicella,* and other filaments (Parker et al., 2014). Anaerobic selectors for EBPR also control nocardioforms (Parker et al., 2014). In the mixing zone, a high growth rate environment is provided to promote the growth of floc-forming organisms at the expense of filamentous organisms.
- Domestic wastewater is not typically nutrient limited. However, if a facility is receiving industrial discharge, such as from a canning company, a nutrient-deficient waste could occur.
  - Measure the concentrations of N and P in the final effluent. If the sum of ammonia, nitrite, and nitrate-nitrogen in the final effluent is greater than 1 mg/L as N, then nitrogen should not be limiting. Similarly, if the soluble orthophosphate in the final effluent is greater than 0.5 mg/L as P, phosphorus should not be limiting. Samples should be collected at the end of the activated sludge basin and before any chemicals are added to precipitate phosphorus.
  - If nutrient limitations are suspected, calculate the ratios of BOD_5 to nitrogen (use total Kjeldahl nitrogen expressed as N), BOD_5 to phosphorus (expressed as P), and BOD_5 to iron (expressed as Fe). Generally, a proper nutrient balance in an activated sludge system is 100:5:1 (BOD:N:P). In general, anhydrous ammonia is used to add nitrogen, trisodium phosphate is used to add phosphorus, and ferric chloride is used to add iron. The ratio of BOD_5 to required nutrients changes with SRT. For example, high MCRTs produce less sludge and result in lower nitrogen and phosphorus requirements. Typically, 3 to 5 mg/L of nitrogen (3 mg/L at high SRTs and 5 mg/L at low SRTs) and 1 mg/L of phosphorus is needed for every 100 mg/L of BOD_5.

If there are insufficient nutrients in the wastewater, three things can happen:

1. Filamentous bacteria will predominate or take over the MLSS;
2. Organic materials will only be partially converted to end products; that is, there will be inefficient BOD_5 or COD removal; and
3. Slime bulking and foaming may occur.

If excess nutrients are added to the wastewater to overcompensate for a nutrient deficiency, a large fraction of these nutrients may not be incorporated to the MLSS and may, therefore, pass into the effluent. Nutrients, if required, should be added at the head of the activated sludge basin. Mixed liquor settleability should be carefully observed to see if it is improving. If settleability improves, the nutrient dose can be reduced by 5% per week until settleability decreases. Then, the dose should be increased by 5% and the settleability observed. Reducing the dosage gradually will help the operator determine the optimum dosage.

Nutrients are expensive and should be carefully applied. Nutrient dosage should be increased as BOD concentrations increase, which takes into account the effects of additional microorganism growth that will occur. If settleability does not readily improve, nutrient dosing should be continued until the actual problem is identified and solved because the problems that are causing poor settleability may be interrelated.

Controlling Filamentous Bulking

A practical approach to control filamentous bulking includes the following:

- Confirming that the problem is indeed caused by filaments,
- Identifying the filaments involved using Gram and Neisser stains,
- Obtaining assistance from outside experts to identify filaments if needed, and
- Determining the specific set of appropriate remedies (both short- and long-term) for the filaments involved, as follows:
  - Short-term solutions involve treating the symptoms: changing influent feed points, changing RAS flowrates, adding settling aids, adding chemicals to destroy filaments; and foam removal; and
  - Long-term solutions involve treating the cause: changing SRT, changing aeration rates, controlling mixed liquor pH, controlling influent septicity and volatile fatty acids, and adding nutrients.

Many filaments can be reduced or eliminated by adding chlorine. Chlorination exposes activated sludge to chlorine levels that damage filaments extended from the floc surface while leaving microorganisms within the floc largely untouched. Chlorine dose and the frequency at which organisms are exposed to chlorine are the two most important parameters. The chlorine dose should be based on the solids inventory in the process (biological reactors plus clarifiers). Jenkins et al. (2003) recommend dosing at least three times daily at the following concentrations:

- 2 to 3 kg Cl_2/1000 kg MLSS·d (2 to 3 lb Cl_2/d/1000 lb MLSS) to maintain an existing SVI and prevent the growth of new filaments;
- 5 to 6 kg Cl_2/1000 kg MLSS·d (5 to 6 lb Cl_2/d/1000 lb MLSS) applied for several days may be used to destroy existing filaments and reduce SVI to a manageable level; and
- 10 to 12 kg Cl_2/1000 kg MLSS·d (10 to 12 lb Cl_2/d/1000 lb MLSS) as a shock dose to bring a badly bulking system under control. This dosage will damage the floc particles and is likely to increase effluent turbidity and TSS.

Chlorine at these dosages may not work if 1) the MLSS concentration is very high or 2) the wastewater contains other compounds that use up the chlorine before it can kill/damage the filaments (Flippin and Gill, 2009). In these cases, enough chlorine should be added to keep the chlorine residual at or above 1 mg/L throughout the activated sludge basin for 30 minutes, three times a day (Flippin and Gill, 2009). *Jar testing* should be done to determine the dose needed.

When *chlorine* is added to wastewater, it combines with organic material, metals, hydrogen sulfide, and other compounds. In the process, it is converted from chlorine (Cl_2) to chloride ion (Cl^-). Chlorine is a disinfectant and can be measured as chlorine residual. Chloride is spent, can no longer oxidize, and cannot be measured as chlorine residual. The total amount of chlorine added is the dose. The chlorine consumed is the demand and the remaining chlorine is the residual. Dose = Demand + Residual

Chlorine addition is similar to chemotherapy treatments for cancer. It can kill filaments, but will also kill other microorganisms. Filamentous and floc-forming bacteria do not seem to significantly differ in their chlorine susceptibility (Richard, 1989). If the dose it too high, chlorine can cause high effluent turbidity, poor flocculation, and loss of nitrification. Chlorine addition should be used only as a last resort and should be stopped as soon as sludge settleability improves. One approach is to set a target SVI (or other measure of settleability) for satisfactory operation of secondary clarifiers and solids processing units. The target SVI can be selected based on past operating experience or by using the chart in Figure 8.50. Chlorine is then added only when the target SVI is significantly and consistently exceeded (Jenkins et al., 2003). This approach to filament control treats the symptom, not the underlying problem.

Location of the chlorine application point is critical. The point should be located where there is excellent mixing and where the sludge is concentrated (Richard, 1989). Common application points include

Jar testing uses small samples of wastewater and different doses of a chemical to figure out the best dose to achieve a desired result.

- Return activated sludge pump intake or discharge,
- Directly into the activated sludge basin at each aerator,
- Directly onto accumulated foam on the activated sludge basin or clarifier surface, and
- Internal mixed liquor recycle pump intake or discharge.

Return activated sludge chlorination is not very effective for controlling nocardioforms (Jenkins et al., 2003). Nocardioforms remain mostly within the floc and won't be exposed to the chlorine. A better approach is to spray the surface of the activated sludge basin with a fine spray of chlorine solution. Several facilities have successfully controlled nocardioforms using a chlorine dose of 0.5 to 1.0 mg/L Cl_2 directly to the foam (Jenkins et al., 2003). The dosage is based on the flowrate to the basin.

Control tests should be performed during chlorination to assess the effects of chlorine on both filamentous and floc-forming organisms. The tests should measure settleability (e.g., SVI), effluent quality (turbidity), and activated sludge quality (microscopic examination) (Jenkins et al., 2003). An adequate chlorine dose should start to improve settleability within 1 to 3 days (Richard, 1989). A turbid, milky effluent and a reduction in BOD removal are signs of overchlorination, although a small increase in effluent suspended solids and BOD concentrations are normal during chlorination for bulking control. The microscopically visible effects of chlorine on filaments include (in correct order) the following:

- Intracellular sulfur granules (if present) disappear,
- Cells deform and cytoplasm shrinks, and
- The filaments break up and dissolve.

Chlorine does not destroy the *sheathed* filaments, which will continue to cause poor sludge settling until they are wasted from the system. Chlorination should be stopped when only empty sheaths remain, and not continued until the SVI falls (Richard, 1989). Adding chlorine beyond this point may overchlorinate.

Some filamentous bacteria have a clear, protective outer covering called a *sheath*. The sheath is similar to a tube that the bacteria grow within. When sheathed filaments are chlorinated, the cells may die, but the sheath remains intact.

Because the effects of chlorine on filaments can be detected microscopically before settleability improves, microscopic examinations can provide an early indication of filament control.

Nocardioforms may also be controlled by adding a low dose (0.5 to 1.5 mg/L) of a high-weight, cationic polymer (Shao et al., 1997 and Jolis et al., 2006). It is thought that the polymer helps keep the nocardioforms in the floc and out of the foam so they can be removed by wasting. In full-scale testing at one WRRF, foaming was eliminated in fewer than 10 days (Jenkins et al., 2003). In another facility, polymer addition only worked some of the time (Jolis et al., 2006). Polymer should be added at a well-mixed location like an MLSS splitter box or the RAS pump intake.

The best growth conditions for *M. parvicella* include water temperatures below 15 °C, longer SRTs, and F/M values below 0.1 kg BOD_5/kg MLVSS·d (0.1 lb BOD_5/d/lb MLVSS) (Jenkins et al., 2004). *M. parvicella* may be controlled by reducing SRT, avoiding low DO conditions (<1 mg/L) in aerated basins, and manually removing trapped foam from basin surfaces. Chlorine is not effective for controlling *M. parvicella*. Adding PACl to the activated sludge process can be effective in controlling *M. parvicella*, but may take several weeks to achieve results. It should be noted that not all forms of PACl will work, but some facilities have had success using PAX-14. Using a chemical to solve a problem is expensive. There is the initial expense of purchasing the chemical, but also staff time for handling it, and increased sludge volume and disposal costs in the case of PACl. It is recommended that process changes, such as adjusting the SRT, be the first troubleshooting option an operator attempts.

Although chlorine and PACl are effective for controlling some types of filamentous bacteria, neither chemical will satisfactorily control poor settling caused by viscous activated sludge (nutrient deficiency) or dispersed-growth.

For both nocardioforms and *M. parvicella*, foam trapping and recycling are the primary reasons some facilities fail to eliminate these filaments (Jolis et al., 2006). Surface removal of foam can help break the cycle. Removed foam should be removed from the facility completely. Foam sent to digesters and dewatering processes reseeds the MLSS when filaments are returned in recycle streams.

In International Standard units:

$$\text{Feed Rate, }\frac{\text{kg}}{\text{d}} = \frac{\left(\text{Dosage, }\frac{\text{mg}}{\text{L}}\right)\left(\text{Flow, }\frac{\text{m}^3}{\text{d}}\right)}{(\text{Purity, Decimal Percentage})(1000)} \tag{8.41}$$

In U.S. customary units:

$$\text{Feed Rate, ppd} = \frac{\left(\text{Dosage, }\frac{\text{mg}}{\text{L}}\right)(\text{Flow, mgd})\left(8.34\frac{\text{lb}}{\text{mil. gal}}\right)}{\text{Purity, \% expressed as a decimal}} \tag{8.42}$$

Filamentous Organisms Not Present

If few or no filamentous microorganisms are present, check the SRT and/or F/M to determine if the system is operating at a higher or lower-than-normal range. Young sludge bulking or old sludge bulking can occur if the system is operated at an excessively low or excessively high SRT. Adjust the SRT as needed. The change should be reflected in the sludge settleability over a period of 2 to 3 SRTs.

The amount of turbulence and DO in the biological reactor are also important. In most facilities, DO concentrations greater than 4.0 mg/L indicate that excess air is being used, and the aeration rate should be reduced to lower the DO concentrations. Refer to the process control discussion on setting a target DO concentration. If the facility uses fine-bubble diffusers or pure oxygen, higher DO concentrations can be safely maintained without breaking apart the floc or filling it with microbubbles and causing it to float.

Excessive turbulence (i.e., overaeration) in the activated sludge basin may hinder MLSS floc formation and result in pinpoint floc being carried over with the clarifier effluent.

Toxics such as industrial wastes may also cause dispersed-growth buildup.

SECONDARY CLARIFIERS

Operational problems with secondary clarifiers can be grouped into six main categories: sludge quality, equipment malfunction, hydraulic problems, solids overload, biological activity within the clarifier, and temperature and density currents.

SLUDGE QUALITY

Clarifier performance should be compared against settleometer performance as further described in the section on Data Collection, Sampling, and Analysis. If the clarifier effluent and settleometer supernatant are of comparable quality for TSS and/or turbidity, then the clarifier is performing as well as it can with the sludge quality that is being produced in the biological reactor. Pinpoint floc, straggler floc, and filamentous bulking are fundamentally sludge quality problems that must be solved by making adjustments in the biological reactors to improve sludge settleability. The 30-minute settling test provides useful information that operators must consider when running this test. Operators should not just fill the settleometer and record the reading after 30 minutes. This test is a useful troubleshooting tool for determining how the activated process is operating because the test represents some of the conditions that are occurring in the secondary clarifier. When running this test, the operators must observe the following:

- Is the supernatant (water above the sludge) clear or cloudy?
- How well is the sludge settling?
- Are the floc particles well-formed or are they small and granular or non-existent?
- Does a large portion of the sludge rise to the surface in a short period of time?
- Is there ashing or greasy layer on the surface of the water?
- Is there a large amount of pin floc in the water?
- Are there any unusual smells such as gasoline or other substances?
- Is there any color in the water?

This list represents some observations operators must consider when running the 30-minute settling test. If an issue exists, then determine what course of action must be taken to troubleshoot and correct the problem. For example, a cloudy supernatant may indicate that the facility received a toxic material, the process is organically overloaded, or there is another issue.

Pinpoint Floc

The appearance of small, dense, pinpoint floc particles suspended in the clarifier is a common problem seen in WRRFs operating near the lower end of the loading range (between conventional and extended aeration). This problem is typically related to a sludge that settles rapidly, but lacks good flocculation characteristics.

Some probable causes of pinpoint floc include the following:

- Operating at excessively long SRTs;
- The process is being operated at an F/M near or in the extended aeration range, resulting in old sludge with poor floc formation characteristics; and
- Excessive turbulence (overaeration) in the biological reactor shears floc formations.

The following measures can be taken, when applicable, to correct the problem:

- If the sludge settling characteristics observed during the mixed liquor settleability test indicate a sludge settling too rapidly with poor floc formation, the clarifier effluent quality can be improved by gradually lowering the SRT. If nitrification is required, caution must be used to avoid decreasing nitrification by wasting too much sludge;

- If the settleability test indicates good settling with clear supernatant above the settled sludge, check for proper aeration and mixing in the biological reactor. If the average DO concentration in the reactor is more than 4 mg/L, consider reducing aeration rate; and
- If there are facilities for feeding settling aids such as alum, ferric chloride, or polymer, determine the required dose and start feeding a settling aid as a temporary measure while correcting the source of the problem.

Straggler Floc

The appearance of small, almost transparent, very light, fluffy, buoyant sludge particles rising to the clarifier surface near the effluent weirs and discharging over the weirs is a problem often seen when the MLSS concentration is too low. When this type of straggler floc occurs while the secondary effluent is otherwise exceptionally clear, and particularly if it prevails even during relatively low SORs, the problem is typically related to a low MCRT (high F/M). At some facilities, the floc is particularly noticeable during the early morning hours. In many facilities, this problem is present, but has no significant effect on effluent quality. Straggler floc is not a concern unless it causes an effluent quality problem.

The following are some probable causes of this problem:

- The activated sludge basin is being operated at an MLSS concentration that is too low. (This would typically occur during process startup until the proper MLSS concentration is reached.) A WAS rate that is too high will result in low SRT, low MLSS, and a high F/M; and
- Sludge is being wasted on a batch basis during the early morning hours, resulting in a shortage of microorganisms to handle the daytime organic loading.

The following measures can be taken, when applicable, to correct the problem:

- Decrease the WAS rate to increase the SRT and MLSS concentration.
- If wasting sludge on a batch (or intermittent) basis, avoid wasting when the BOD load is increasing. Also, increase the RAS rate as peak flows start. All of the organisms are needed at this time to handle the daily increase in organic loading. Decrease the RAS rate at night.

Hydraulic Overload

A secondary clarifier can be hydraulically overloaded because of two different situations and it is important that when operators are troubleshooting these problems they recognize each situation. Overloading can result from excessive flow or unevenly distributed flow between multiple units. The first hydraulic overload situation would be if the facility is subjected to excessive I/I or combined sewer flows. In this situation, all secondary clarifiers could be washing out, which is a facility-level problem. Washout appears as rolling clouds of MLSS at the edge of the clarifier (Figure 8.80). The operators must focus on the flows entering the facility. The second hydraulic overload situation would occur when only one secondary clarifier is washing out solids. In this situation, the operators must focus on what is occurring for that one clarifier. Either that clarifier has a higher SLR than the other clarifiers or sludge is not being removed from the bottom of it at the same rate.

Check the hydraulic loading on each clarifier by either measuring the flow to each clarifier or by estimating the flow balance between multiple units as indicated by the depth of flow over the weirs in each clarifier. Differences in depth of flow, however, will be hard to measure unless the differences in flow are excessive. If the headloss to all clarifiers is the same, the clarifier overflow weirs should all be set at the same elevation. If not, the weirs should be adjusted to equalize the headloss to them.

The problem should be approached as follows:

- Determine if flows are being distributed equally to biological reactors and secondary clarifiers. Weirs at the effluent end of the reactors must be adjusted to equally distribute the flow to each secondary clarifier if these two processs are paired together. Weirs, valves, or gates on the RAS distribution lines must be adjusted to properly distribute RAS flows. Weirs, weir gates or gates at clarifier distribution structures must be adjusted to uniformly distribute the forward flow into each secondary clarifier;
- Inspect the clarifier effluent weirs to make sure that the flow is flowing over them equally;

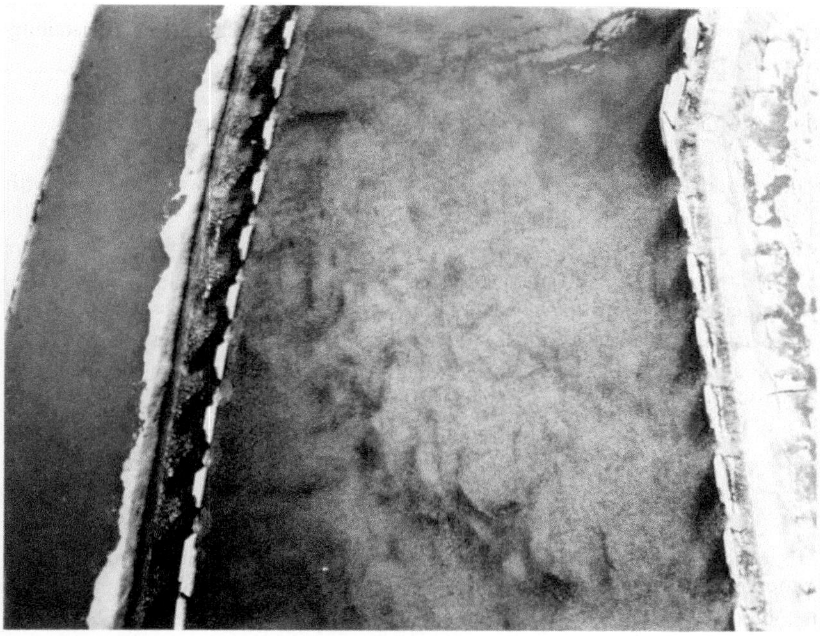

Figure 8.80 Solids Washout from High Flows

- After checking and correcting weir elevations where needed, determine the clarifier SOR at both average and peak flows. Compare the actual calculated SOR with the design value. If the actual overflow rate exceeds the design value, then the clarifiers are hydraulically overloaded and additional clarifiers may need to be placed into operation, if available. Also, perform an SVI test to check settling characteristics. If results are satisfactory and all clarifiers are operating, facility expansion may be needed.

Solids Overloading

Clarifier solids overloading occurs when the SLR is too high. The clarifier SLR is reduced when the MLSS is settling poorly. For a complete discussion on SLR and settleability, review the Process Control for Secondary Clarifiers section of this chapter. Adjusting the return rate should quickly verify whether a solids overloading problem exists. If the return rate is increased and the blanket drops, the problem is not solids overloading and increasing the return rate has helped. If an increased return rate raises the blanket and keeps it up, the clarifier is probably overloaded with solids and higher return rates will only aggravate the problem. The return rate can then be reduced by 20%, but only if the blanket level, effluent and MLSS concentrations, and mixed liquor settleability are closely watched. The sludge blanket in clarifiers should be measured after 30 minutes and again after 60 minutes to check the effect of changing the RAS flowrate. It should then be checked three times per shift to ensure that the blanket does not build up to an excessive depth. This may only work for a short time because storing solids in the clarifier is not effective with high blankets. If a reduced return rate raises the blanket, other options can be tried, as described later. In most cases, increasing the sludge return rate from a secondary clarifier is only a short-term solution. The solids will quickly re-enter the secondary clarifier and cause the sludge blanket to rise. To reduce the sludge blanket in the secondary clarifiers, the solids loading rate must be decreased.

Temperature

Temperature currents occasionally occur in large, deep clarifiers in colder climates, most often in the spring and fall. A temperature profile of the clarifier will identify the presence of temperature currents. The temperature probe on a DO meter is an excellent tool for this procedure. To make the profile, measure and record the temperature at the head, one-quarter, one-half, three-quarters, and tail end of a rectangular or square clarifier, or at quarter points across a circular clarifier. At each point, measure the temperature at the surface and quarter points down to the reactor bottom. Be careful that the temperature probe and wires do not get entangled in the sludge-collection equipment.

If deeper temperatures are consistently cooler by 1 to 2 °C (2 to 4 °F) or more, this is an indication that temperature currents are present. If temperature differences are caused by surface warming from the sun, reducing the number of clarifiers online or rigging temporary covers over the clarifiers may help. Installing baffles may improve the settling by breaking up currents and stopping turbulence.

Temperature also affects the settleability of the activated sludge. Colder temperatures are commonly associated with poorer sludge settleability and vice versa for warmer temperatures. This relationship has been attributed to temperautre-induced changes in the density of water; the density of water increases with decreasing temperature over the temperature ranges typical for WRRFs.

Clumping or Rising Sludge

Occasionally, sludge that has otherwise good settling characteristics may rise to the clarifier surface. This can occur even if settling aids such as polymer or alum are being fed as a coagulant. Floating clumps of solids and fine rising bubbles will be observed. This is caused by denitrification in the sludge blanket. Denitrification is discussed in the section on Process Control for Secondary Clarifiers. Examples of denitrification and rising sludge are shown in Figure 8.52 and Figure 8.53. A particularly severe denitrification episode is shown in Figure 8.81. The following measures can be tried, when applicable, to correct the sludge clumping problem:

- Increase the RAS rate to reduce detention time of sludge in the clarifiers. A periodic measurement of the clarifier sludge blanket depth will help determine the proper return rate;
- Where a suction-type sludge collector is used, check that all suction tubes are flowing freely with a fairly consistent suspended solids concentration. Some of the suction tubes may be improperly adjusted or plugged, resulting in coning in some areas and a sludge blanket buildup in other areas;
- Where possible, increase the sludge collector speed;
- Reduce the number of clarifiers online to shorten the detention time. Be sure not to exceed the hydraulic loading rate or SLR limitations of the clarifiers;
- If nitrification is not required, gradually decrease the SRT to stop nitrification. Initially, the solids inventory should be decreased by 20 to 30% over 1 week, and then the process operation must be observed to determine if nitrite and nitrate levels are dropping. If, after approximately 2 weeks, they have not dropped to below 5 mg/L, reduce the inventory by another 20 to 30%. As a guide, calculate the SRT and compare it to published values for nitrification. In very warm waters, it may be necessary to drop the SRT to approximately 2 days and MLVSS to less than 1000 mg/L;
- In severe cases, and if nitrification is not required, chlorinate the RAS at low rates to stop nitrification; and
- When nitrification is required, gradually decrease the WAS rate to increase the SRT and lower the F/M to be sure nitrification is complete and soluble BOD concentrations are low.

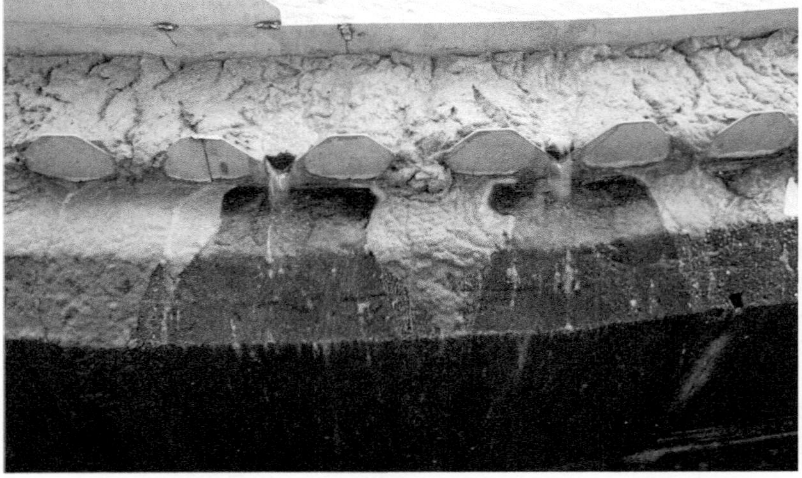

Figure 8.81 Solids Loss from Denitrification in a Secondary Clarifier (**Reprinted with permission by Indigo Water Group**)

TEST YOUR KNOWLEDGE

1. Sludge with an SVI of 225 mL/g would be considered bulking sludge.
   - ☐ True
   - ☐ False

2. This condition results when the MLSS fails to flocculate and leaves behind a turbid supernatant.
   a. Zoogleal formations
   b. Dispersed growth
   c. Pin floc
   d. Straggler floc

3. A sample is collected from the end of an activated sludge process for a settleometer test. After 30 minutes, the supernatant is crystal clear, but the sludge has only settled to a volume of 500 mL. What is the most likely cause?
   a. Pin floc
   b. High DO shear
   c. Filaments
   d. Dispersed growth

4. The first step to solving a filament problem is to
   a. Increase the wasting rate
   b. Raise DO concentrations to 3 mg/L
   c. Adjust sludge age
   d. Identify the filament by staining

5. Which two chemicals may be added to activated sludge for filament control?
   a. Ferric chloride and PACl
   b. Chlorine and PACl
   c. PACl and sodium chlorate
   d. Chlorine and ferric chloride

6. This term is used to describe the total amount of a chemical, such as chlorine, to the treatment process.
   a. Dose
   b. Demand
   c. Residual
   d. Ion

7. When should chlorine addition to an activated sludge process should be stopped?
   a. After 3 days of dosing
   b. When only empty sheaths remain
   c. SVI reaches 225 mL/g
   d. MLSS has deflocculated

8. Which of the following chemicals is most effective for controlling *M. parvicella*?
   a. Ferric chloride
   b. Chlorine
   c. Polyaluminum chloride
   d. Sodium chlorate

9. Chlorine is often applied directly to nocardioform foam rather than in the RAS line because
   a. Filaments concentrate in the foam
   b. Easier to ensure complete coverage
   c. Few facilities have ports on their RAS lines
   d. RAS exerts too much demand

10. If the MLSS concentration gets too low, this condition may result.
   a. Pinpoint floc
   b. Straggler floc
   c. Rapid settling
   d. Slow settling

11. If a secondary clarifier is already overloaded with solids, turning up the RAS flowrate will
   a. Reduce the blanket depth
   b. Decrease the SLR
   c. Increase the blanket depth
   d. Decrease overall treatment time

Safety Considerations

Activated sludge is aerated by mechanical aerators or compressed air. Aeration equipment may introduce bacteria and other microbes into the atmosphere. Practice good personal hygiene by wearing gloves and washing hands frequently. Do not eat or smoke in the WRRF except in designated areas away from the treatment process.

Falls into an aerated activated sludge basin may result in drowning because the mixed liquor is saturated with air, which reduces the buoyancy of a human body. If possible, stop aeration while working over any activated sludge basin. If it is necessary to work over a basin, wear a lifeline and life jacket. Have standby help in the immediate area with harnesses, life jackets, and lifelines. Because of reduced buoyancy, workers in aerated basins may need special life jackets.

Aeration may splash greasy material and foam on walkways and algal slime can form. Remove it frequently so walk areas are safe. Do not leave tools and other items where they may create a hazard. Walkways should have handrails and kick plates. Remove handrails only one section at a time—the minimum necessary for the immediate job—and replace them as soon as the job is complete.

Aeration equipment may produce high noise levels, but noise exposure should not exceed an 8-hour time-weighted average of 85 decibels. Wear ear protectors when working around blowers. Wear dust protection when changing or cleaning air filters.

CHAPTER SUMMARY

THEORY OF OPERATION	■ Microorganisms convert organic solids and soluble organic matter into new microorganisms and waste products. ■ Process may remove BOD_5, NH_3-N, NO_3-N, and PO_4-P. ■ Microorganisms grow suspended in the wastewater as colonies called *floc*. Floc particles are slightly more dense than water. ■ Bioflocculation combines floc particles and inert solids to form particles that settle in a secondary clarifier. ■ Flocculated mass is called the mixed liquor suspended solids (MLSS) or mixed liquor volatile suspended solids (MLVSS). ■ MLSS settles as a mass or layer with all particles eventually falling at the same speed as smaller particles catch up to larger particles. Interface is clear water above and sludge below. In primary clarifiers, particles settled individually at their own speeds with very little flocculation occurring.
OVERVIEW	■ Basic system consists of an activated sludge basin where floc is exposed to influent or primary effluent and air followed by settling in a secondary clarifier. ■ Air provides both oxygen and mixing. ■ Return activated sludge (RAS) pipe returns settled sludge from the bottom of the clarifier to the front of the process. ■ Waste activated sludge (WAS) pipe removes excess sludge and transfers it to solids handling. ■ Accelerates natural treatment process by building up a large population of microorganisms. ■ The amount of time the solids spend in the system (sludge age) is independent of the amount of time the water spends in the system. ■ A sequencing batch reactor (SBR) combines treatment and clarification in the same basin. ■ Anaerobic, anoxic, and aerated zones are used to create environments in which phosphorus, nitrite, nitrate, and ammonia may be removed. ■ Many different microorganisms are in floc. Different organisms flourish or die out because of the environmental conditions. ■ Goal is to balance floc formers and filament formers so particles will flocculate, settle, and compact. ■ Protozoa and metazoan can be useful indicators of environmental conditions in the process.
DESIGN PARAMETERS	■ Activated sludge basin design based primarily on sludge age. ■ Volumetric loading rate (mass of BOD_5 per volume of basin) and food-to-microorganism (F/M) are included in operating definitions. ■ High-rate, conventional, and low-rate designs. ■ Within these categories, there are many different process variations. ■ Clarifiers have hydraulic capacity and solids handling capacity. ■ Hydraulic capacity is the design surface overflow rate (SOR). ■ Solids loading rate (SLR) is the mass of MLSS flowing into the clarifier per area (square meters or square feet) of clarifier surface. ■ Clarifier design based on solids loading rate (SLR) and surface overflow rate (SOR). SLR is usually more important.
EXPECTED PERFORMANCE	■ Conventional and low-rate designs meet secondary treatment standards for BOD_5 and TSS. ■ High-rate systems are similar to roughing trickling filters and may remove as little as 50% of incoming BOD_5. ■ Possible to achieve <1 mg/L NH_3-N, <10 mg/L NO_3-N, and <1 mg/L PO_4-P, depending on configuration and loading rate.

EQUIPMENT

- Activated sludge basins may be concrete or steel. Minimum of two basins is preferred.
- Flow-through may be complete mix or plug-flow. Also have basins in series, batch operation, and oxidation ditch.
- Selectors increase F/M temporarily to inhibit filament growth. Small basin gets all of the RAS and influent. May be anaerobic, anoxic, or aerated. Do not work on all filament types.
- Rectangular secondary clarifiers similar to primary clarifiers.
- Circular secondary clarifiers are similar to primary clarifiers. Differences include
  - Peripheral feed, peripheral take-off flow path.
  - Typically have flocculating center well outside of smaller center well. Reflocculates MLSS in the clarifier to decrease effluent TSS and turbidity.
  - May have interior baffles: McKinney, Stamford.
  - Brushes and covers used on weirs and launders for algae control.
  - Like primary clarifiers, sludge collection mechanisms include multiple scraper blade and spiral plows. Two additional sludge removal mechanisms available: organ pipe and Tow-Bro. Both use hydraulic suction to vacuum up clarifier blanket. Hydraulic suction removes sludge faster than flights and scrapers.
- Aeration systems provide air and mixing.
  - Diffused air (fine or coarse-bubble) functions similarly to an aquarium air stone.
  - Jet aeration combines aeration and pumping. Recirculated MLSS and air forced through nozzle assembly for mixing and oxygen transfer.
  - Surface splash mechanical aerators agitate MLSS and create turbulence for oxygen transfer. May or may not have a draft tube. Horizontal rotors push flow forward, usually in an oxidation ditch, and lift drop MLSS for oxygen transfer.
  - Submerged mixers keep MLSS in suspension in anoxic and anaerobic zones.
- RAS and WAS systems typically use centrifugal pumps that may be connected to the clarifier sludge hopper directly or to a RAS/WAS wet well. Screw pumps have also been used for RAS. RAS pumps operate continuously at 25 to 125% of influent flow. May be fixed rate or flow paced to influent. Many different pump types used for WAS.
- Recirculation pumps move MLSS between zones within basins. May pump as much as 400% of influent flow.

PROCESS VARIABLES

Activated Sludge Basin

- Hydraulic detention time
- Volumetric loading rate
- WAS rate (kilograms or pounds per day)
- WAS 1) determines sludge age, 2) controls the mass of total solids in the activated sludge process, and 3) indirectly controls the growth rate of the microorganisms in the process.
- WAS, along with environmental conditions such as water temperature and availability of BOD, nutrients, and oxygen, influences the process biology, sludge settleability, and level of treatment achieved.
- The amount of WAS removed from the system each day is determined by selecting a target sludge age and calculating the mass of MLSS to remove each day.
- Sludge age is the average amount of time an average microorganism cycles between the activated sludge basin and clarifier before being removed from the system.

Secondary Clarifier

- RAS flowrate returns settled solids to the activated sludge basin. RAS and WAS together make it possible to independently control HDT and solids detention time in the activated sludge process.
- RAS controls the clarifier solids detention time and blanket depth.
- RAS concentration can be calculated if the MLSS concentration, influent flow, and RAS flow are known.
- Decreasing the RAS flowrate increases the RAS concentration until maximum compaction has been reached.
- Influent flow
- Surface overflow rate (SOR)
- Weir overflow rate (WOR)
- RAS flowrate, along with influent flow, MLSS concentration, and clarifier surface area, determine the solids loading rate (SLR) to the clarifier.
- SLR is limited by sludge settling characteristics. The better the sludge settles, the more mass (kilograms or pounds) the clarifier will be able to settle.

PROCESS VARIABLES	■ Sludge age may be defined as the MCRT, SRT, or $SRT_{aerobic}$ ■ MCRT is preferred for systems with smaller activated sludge basins. SRT is preferred for systems with larger basins. $SRT_{aerobic}$ is helpful for systems that must remove ammonia. ■ Sludge age is the primary control variable for activated sludge. Once selected, sludge age will determine the sludge wasting rate, MLSS concentration, MLSS growth rate, F/M, microbiology, and sludge settleability. It is not possible to control more than one of these variables at a time. ■ When sludge age increases, the MLSS will also increase and the percentage of MLVSS, F/M, and wasting rate all decrease. ■ Sludge age is the primary control variable, but microorganism growth rates are also affected by availability of resources (BOD_5, DO, and NH_3-N, for example) and water temperature in the basin. ■ Food-to-microorganism ratio (F/M) is the amount of BOD_5 available per equal amount of MLVSS. F/M is directly related to the growth rate of the microorganisms.	■ Several simple models available to estimate the maximum SLR for a particular SVI. The Daigger Clarifier Operating Chart uses two pieces of information to determine if a clarifier is overloaded (SLR too high) or underloaded. Locate the operating point on the graph using the actual SLR, actual RAS concentration, and underflow rate. Only two of these variables are needed to use the chart.
PROCESS CONTROL	■ Different operating modes combined ranges of SRT, F/M, loading rate, and HDT. ■ The "right" SRT is the one that produces a MLSS that will flocculate, settle, compact, and meet treatment goals. ■ Washout SRT is the minimum SRT a microorganism needs to reproduce and build a stable population in the MLSS. If the SRT is shorter than the washout SRT, that microorganism won't proliferate. ■ Washout SRT along with environmental conditions determines which microorganisms are able to live, thrive, and survive in the process. To eliminate an undesirable microorganism, one or more environmental conditions must be changed. ■ Select an SRT based on 1) water temperature, 2) treatment goals, 3) washing out undesirable microorganisms. ■ Waste small amounts often to minimize fluctuations in SRT. ■ DO concentrations should be matched to the F/M to prevent the growth of low DO filaments. Nitrification requires a DO of 2 mg/L and will typically be higher than the DO needed for filament control.	■ Maintain blankets below 0.7 m (2 ft) whenever possible. ■ Blanket depth increases when the clarifier is overloaded (SLR is too high). ■ Four variables may be adjusted to influence the clarifier SLR: • RAS flowrate • MLSS concentration • Clarifier surface area • SVI (add chemicals) ■ Pacing the RAS flow to the influent flow results in a nearly constant RAS concentration and maintains blanket depth. ■ Remove settled sludge from the clarifier before denitrification or septicity brings settled sludge back to the clarifier surface. ■ Minimize RAS flow rates to save money while meeting all other process goals.

PERSISTENT MYTHS	■ It is possible to build biomass without BOD_5 to prepare for a future load. FALSE. ■ Turning up the RAS flowrate decreases overall treatment time. FALSE. ■ The RAS can change the MLSS concentration in the activated sludge basin. MOSTLY FALSE. This is only true when a large percentage of the solids are being stored in the clarifier blanket. ■ Turning up the RAS flow changes the water level in the activated sludge basins. MOSTLY FALSE. Levels go up by fractions of a centimeter (inch).
DAILY OPERATION	■ Perform daily walkthroughs to inspect equipment and basin surfaces. ■ Monitor water temperature and DO concentrations. Facilities that nitrify should also monitor pH and alkalinity. ■ Perform settleometer tests at least daily. Record SSV_5, SSV_{30}, and calculate SSC_{30} and SVI. Compare SSC_{30} to actual RAS concentration. ■ Use a sludge judge to monitor clarifier blanket depths. ■ Adjust RAS flowrates as needed. ■ Collect and analyze samples needed to calculate sludge age. Compare current sludge age to target sludge age. Adjust wasting rate accordingly.
DATA COLLECTION, SAMPLING, AND ANALYSIS	■ Visual inspection of basin and clarifier surfaces. ■ Sludge judge sampling of the clarifier blanket. Collect the sample two-thirds of the way in from the outer edge of the clarifier if the clarifier bottom slopes toward a central hopper. ■ Conduct settleometer test in a 2-L Mallory settleometer for best results. ■ Perform a weekly microscopic examination or more often if sludge settleability changes or other unusual condition is noted. ■ OUR and SOUR tests may be used to look for evidence of toxicity and to establish a baseline of normal operating conditions.
TROUBLESHOOTING	■ Operators are encouraged to review the Troubleshooting Decision Tree Chart. Although the chart is not all-inclusive, it summarizes the majority of the troubleshooting section of this chapter. ■ Operational problems in activated sludge systems may be categorized as mechanical, process, or hydraulic. ■ Mechanical problems include various types of mechanical failures. ■ Hydraulic problems include hydraulic surges caused by a sudden increase in influent flow or uneven distribution of flow to multiple basins or clarifiers. ■ Process problems include foaming, filamentous bulking, old sludge and young sludge bulking, dispersed growth, straggler floc, pin floc, low pH, and denitrification in the clarifier blanket. Be sure to know the causes of each of these common ailments and how to interpret foam colors.
VARIATIONS ON A THEME	■ There are many different activated sludge processes in common usage. They can be grouped according to their flow pattern (how water moves through the basin) and their operational mode. ■ The different processes may have some, one, or no anaerobic and anoxic zones or internal recycle pumps. These different components can be put together in different ways. The key thing to remember is that all activated sludge processes follow the same basic principles. The reactions that take place in an anoxic zone should be the same in a step-feed type process as they are in a SBR.

CHAPTER EXERCISE

The community down the road has had a wastewater treatment pond system for the past 40 years, but their new discharge permit has a compliance schedule for meeting new, low-effluent ammonia limits. To comply with the new permit, they've built a brand new activated sludge treatment facility. You've been brought on board to help the current chief facility operator, Sarah, optimize this new activated sludge facility because of your expertise in activated sludge. Details for the new facility are given below. You'll need to help Sarah decide everything from what her sludge age should be to how many clarifiers should be in service. The answers from one question may be used in the following questions. Be sure to check your answers as you go along.

New activated sludge facility

Design flow = 22.7 ML/d (6 mgd)

Current influent flow = 14.3 ML/d (4.2 mgd)

Design BOD_5 = 250 mg/L

Design MLSS = 3000 mg/L

Activated sludge basins = 3, each basin holds 7570 m^3 (2 mil. gal). This volume includes an anoxic zone that holds 2271 m^3 (0.6 mil. gal). Two basins are in service.

Secondary Clarifiers = 4, each clarifier is 21.3 m (70 ft) in diameter.

Surface area for a single secondary clarifier is 356.1 m^2 (3846.5 sq ft).

1. The first thing you and Sarah have to do is select a target sludge age and decide how you'll calculate it. What pieces of information do you need to select a target sludge age?

    ____

2. Information in hand, use the chart in Figure 8.45 to find the washout SRT for the nitrifying bacteria and write it down below.

    ____

3. Apply an appropriate safety factor and set your $SRT_{aerobic}$. ______________________________

4. How much higher does the total SRT need to be than the $SRT_{aerobic}$? Think about the definitions for SRT_{total} and $SRT_{aerobic}$. Calculate the SRT_{total}. ______________________________

5. Now that the target sludge age is known, the wasting rate can be calculated. If the MLSS concentration is currently 2700 mg/L, how many kilograms (pounds) of MLSS should be removed today?

    ____

6. During startup, all of the clarifiers were brought online to prevent the loss of fluffy, new sludge and to make sure all of the equipment was functioning properly. If a SOR of 16.3 $m^3/m^2 \cdot d$ (400 gpd/sq ft) is used, how many clarifiers are needed now?

    ____

7. The MLSS is settling well and the clarifier blankets remain below 0.7 m (2 ft) throughout the day. You head back to your own facility. A few days later, Sarah calls in a panic because the entire activated sludge basin is covered with brown foam and it is filling up the clarifier center well. What other pieces of information do you need before you can help Sarah?
8. After talking with Sarah, you recommend two or three different corrective actions. What are they?
9. Sarah followed your first two pieces of advice, but the need to remove ammonia didn't allow her to adjust SRT. The problem is getting worse and now the clarifiers are building blankets. She is worried that solids are going to end up in the final effluent. Give Sarah at least three different things to try that might improve her situation.

CHAPTER EXERCISE SOLUTIONS

1. Treatment goals, water temperature, and the relative sizes of the activated sludge basins and clarifiers. The influent wastewater is 15 °C (59 °F). From the initial facility description, we know this facility must nitrify. The SRT will need to be long enough for nitrification. The clarifiers are small compared to the activated sludge basins, so the SRT calculation is a good choice. You'll also want to calculate the $SRT_{aerobic}$ because nitrification is required.
2. The highest washout SRT for the nitrifying bacteria at 15 °C is 3 days.
3. An $SRT_{aerobic}$ between 6 and 9 days is needed. Set the target $SRT_{aerobic}$ at 9 days.
4. If the $SRT_{aerobic}$ is 9 days, then the total SRT will need to be higher. Because the $SRT_{aerobic}$ is a fraction of the SRT_{total}, use the ratio of aerobic basin volume to total basin volume to determine the SRT_{total}.

   Step 1—Total Basin Volume − Anoxic Volume = Total Volume

   7570 m^3 (2 mil. gal) − 2271 m^3 (0.6 mil. gal) − 5299 m^3 (1.4 mil. gal)

   Step 2—Aerobic Fraction = Aerobic Volume/Total Volume

   Aerobic Fraction = 5299 m^3 (1.4 mil. gal)/7570 m^3 (2 mil. gal)

   Aerobic Fraction = 0.7

   Step 3—$SRT_{aerobic} = (SRT_{total})(\text{Aerobic Fraction})$

   $9 \text{ days} = (SRT_{total})(0.7)$

   $12.9 \text{ days} = SRT_{total}$

   Round up and set the target SRT to 13 days.

5. First, find the total mass of activated sludge inventory (pounds or kilograms). Then, use the SRT equation to find the mass that needs to be wasted.

   In International Standard units:

$$\text{kg} = \frac{\left(\text{MLSS}, \frac{\text{mg}}{\text{L}}\right)(\text{Basin volume, m}^3)}{1000}$$

$$\text{kg} = \frac{\left(2700\frac{\text{mg}}{\text{L}}\right)(2 \text{ basins})(7570 \text{ m}^3)}{1000}$$

$$\text{kg} = 40\ 878$$

$$\text{SRT} = \frac{\text{Mass in activated sludge basin}}{\text{Mass wasted}}$$

$$13 \text{ days} = \frac{40\ 878 \text{ kg}}{\text{Mass wasted}}$$

$$\text{Mass wasted} = 3145 \text{ kg}$$

In U.S. customary units:

$$\text{lb} = \left(\text{MLSS}, \frac{\text{mg}}{\text{L}}\right)(\text{Basin volume, mil. gal})(8.34)$$

$$\text{lb} = \left(2700\frac{\text{mg}}{\text{L}}\right)(2 \text{ basins})(2 \text{ mil. gal})(8.34)$$

$$\text{lbs} = 90\ 072$$

$$\text{SRT} = \frac{\text{Mass in activated sludge basin}}{\text{Mass wasted}}$$

$$13 \text{ days} = \frac{90\ 072}{\text{Mass wasted}}$$

$$\text{Mass wasted} = 6929 \text{ lb}$$

6. The SOR and influent flow are both set. Use the SOR equation to find the area needed for the current flowrate. Then, divide the area needed by the surface area of one clarifier.

$$\text{SOR} = \frac{\text{Flow}}{\text{Area}}$$

$$400\frac{\text{gpd}}{\text{sq ft}} = \frac{4\ 200\ 000 \text{ gpd}}{\text{Area}}$$

$$\text{Area} = 10\ 500 \text{ sq ft}$$

Each clarifier has 3846.5 sq ft of surface area (10 500 sq ft ÷ 3846.5 sq ft = 2.7 clarifiers needed). Round up to three clarifiers in service.

7. You need to know how well the sludge is settling and, if a microscope is available, if filaments are present and what they look like. You asked the right questions and Sarah tells you that her 5-minute SSV is more than 700 mL, but the supernatant is crystal clear. The microscope slide looks like a plate of spaghetti. You might also ask if the clarifiers are building blankets.
8. Based on Sarah's descriptions, it's certain she has an *M. parvicella* problem. You recommend getting some PACl to treat the foam, getting the city vac-truck out to remove foam from the surface, and checking to see if her sludge age is too long.
9. Sarah needs to lower the SLR to the secondary clarifier. She can reduce the RAS flowrate, put another activated sludge basin into service, put another clarifier into service, or add polymer to improve settleability.

References

ACR Publications (2011) *Pumps and Pumping.* Available online at acrp.com (accessed Nov 10, 2017).

American Public Health Association; American Water Works Association; Water Environment Federation (2017) *Standard Methods for the Examination of Water and Wastewater,* 23rd ed.; American Public Health Association: Washington, D.C.

Argaman, Y.; Brenner, A. (1986) Single-Sludge Nitrogen Removal: Modeling and Experimental Results. *J. Water Pollut. Control Fed.*, **58** (8), 853–860.

Burchett, M. E.; Tchobanoglous, G. (1974) Facilities for Controlling the Activated Sludge Process by Mean Cell Residence Time. *J. Water Pollut. Control Fed.,* **46** (5), 973–979.

Brock, T. D.; Madigan, M. T. (1991) *Biology of Microorganisms,* 6th ed.; Prentice-Hall: Englewood Cliffs, New Jersey.

Carberry, J. B.; Englande, A. J. (Eds.) (1983) *Sludge Characteristics and Behavior*, Nato Science Series E; Springer: The Netherlands.

Cha, D. K.; Jenkins, D.; Lewis, W. P.; Kido, W. H. (1992) Process Control Factors Influencing *Nocardia* Populations Activated Sludge. *Water Environ. Res.*, **64,** 37–43.

Daigger, G. T. (1995) Development of Refined Clarifier Operating Diagrams Using an Updated Settling Characteristics Database. *Water Environ. Res.*, **67,** 95.

Daigger, G. T.; Roper, R. E., Jr. (1985) The Relationship Between SVI and Activated Sludge Settling Characteristics, *J. Water Pollut. Control Fed.*, **62,** 676.

Eckenfelder, W.; Melbinger, N. (1957) Settling and Compaction Characteristics of Biological Sludges: I. General Considerations. *J. Water Pollut. Control Fed.*, **29** (10), 1114–1122.

Ekama, G. (2010) The Role and Control of Sludge Age in Biological Nutrient Removal Activated Sludge Systems. *Water Sci. Technol.*, **61** (7), 1645–1652.

Environmental Leverage (2003) *Type 1863*. http://www.environmentalleverage.com/Type%201863.htm (accessed Nov 8, 2017).

EnviroSim Associates, Ltd. (2005) *Influent Specifier (Raw) 2_2.xls*. Spreadsheet for fractionating wastewater into components required for Biowin modeling; EnviroSim Associates, Ltd.: Hamilton, Ontario, Canada.

Garrett, M. T., Jr.; Sawyer, C. N. (1952) Kinetics of Removal of Soluble BOD by Activated Sludge; *Proceedings of the 7th Industrial Waste Conference*; Purdue University, West Lafayette, Indiana; p 51.

Garrett, M. T., Jr. (1958) Hydraulic Control of Activated Sludge Growth Rate. *Sewage Ind. Wastes*, **30** (3), 253–261.

Gould, R. H. (1953) *Sewage Disposal Problems in the World's Largest City*; Paper presented at the 25th Anniversary Meeting, Federation of Sewage and Industrial Wastes Associations; New York, Oct 6–9.

Great Lakes-Upper Mississippi River Board (2014) *Recommended Standards for Wastewater Facilities: Policies for the Design, Review, and Approval of Plans and Specifications for Wastewater Collection and Treatment Facilities*, Health Research, Inc., Health Education Services Division: Albany, New York.

Hellinga, C.; Schellen, A. A. J. C.; Mulder, J. W.; Loosedrecht, M. C. M. (1988) The SHARON Process: An Innovative Method for Nitrogen Removal from Ammonium-Rich Wastewater. *Water Res.*, **37** (9), 135–142.

Jenkins, D.; Garrison, W. E. (1968) Control of Activated Sludge by Mean Cell Residence Time. *J. Water Pollut. Control Fed.*, **40** (11), 1905–1919.

Jenkins, D., Richard, M. G.; Daigger, G. T. (2003) *Manual of the Causes and Control of Activated Sludge Bulking, Foaming, and Other Solids Separation Problems*, 3rd ed.; CRC Press: Boca Raton, Florida.

Jolis, D.; Ho, C. F.; Pitt, P. A.; Jones, B. M. (2006) Mechanism of Effective Nocardioform Foam Control Measures for Non-Selector Activated Sludge Systems. *Water Environ. Res.*, **78,** 920–929.

Knowles, R. (1982) Denitrification. *Microbiol. Rev.*, **46** (1), 43–70.

Lau, A.; Strom, P.; Jenkins, D. (1984) The Competitive Growth of Floc-Forming and Filamentous Bacteria: A Model for Activated Sludge Bulking. *J. Water Pollut. Control Fed.*, **56** (1), 52–61.

Lou, I. C.; de los Reyes III, F. L. (2008) Clarifying the Roles of Kinetics and Diffusion in Activated Sludge Filamentous Bulking. *Biotechnol. Bioeng.*, **101,** 327–336.

Maharajh, N. (2010) Effect of Feed Rate and Solids Retention Time (SRT) on Effluent Quality and Sludge Characteristics in Activated Sludge Systems Using Sequencing Batch Reactors. M.S. Thesis, Environmental Science and Engineering, Virginia Polytechnic Institute and State University, Blacksburg, Virginia.

Metcalf and Eddy, Inc./AECOM (2014) *Wastewater Engineering Treatment and Resource Recovery*, 5th ed.; Tchobanoglous, G., Burton, F., Stensel, D.H., Abu-Orf, M., Bowden, G., Pfrang, W., Eds.; McGraw-Hill: New York.

McCarty, P.; Brodersen, C. (1962) Theory of Extended Aeration Activated Sludge. *J. Water Pollut. Control Fed.*, **34** (11), 1095–1103.

Palm, J. C.; Jenkins, D.; Parker, A. D. (1980) Relationship Between Organic Loading, Dissolved Oxygen Concentration, and Sludge Settleability in the Completely-Mixed Activated Sludge Process. *J. Water Pollut. Control Fed.,* **52** (10), 2484–2506.

Parker, D.; Bratby, J.; Esping, D.; Hull, T.; Kelly, R.; Melcer, H.; Merlo, R.; Pope, R.; Shafer, T.; Wahlberg, E.; Witzgall, R. (2014) A Critical Review of Nuisance Foam Formation and Biological Methods for Form Management or Elimination in Nutrient Removal Facilities. *Water Environ. Res.*; **86** (6), 483–506.

Parker, D. S.; Kaufman, W. J.; Jenkins, D. (1971) Physical Conditioning of Activated Sludge. *J. Water Pollut. Control Fed.,* **43**, 1817–1833.

Parker, D. S.; Kinnear, D. J.; Wahlberg, E. J. (2001) Review of Folklore in Design and Operation of Secondary Clarifiers. *J. Environ. Eng.*, **127** (6), 476–484.

Pearson, E. A. (1966) Kinetics of Biological Treatment. *Special Lecture Series, Advances in Water Quality Development*; University of Texas: Austin, Texas.

Pretorius, C.; Appleton, R.; Walker, S.; Jorgensen, E.; Stevensen, B.; Bateman, A. L. (2016) Why SRT Control Is Not More Widely Practiced and Three Ways to Fix It: Making SRT Control a Reliable Tool for Operators to Use. *Proceedings of the 89th Annual Water Environment Federation Technical Exhibition and Conference* [CD-ROM]; New Orleans, Louisiana, Sept 24–28; Water Environment Federation: Alexandria, Virginia.

Richard, M. (1989) *Activated Sludge Microbiology;* Water Environment Federation: Alexandria, Virginia.

Richard, M. (2003) *Microbiological and Chemical Testing for Troubleshooting Lagoons.* http://www.lagoons online.com/trouble-shooting-wastewater-lagoons.htm (retrieved Aug 18, 2017; accessed Nov 10, 2017).

Sayigh, B. A.; Malina, J. F., Jr. (1978) Temperature Effects on the Activated Sludge Process. *J. Water Pollut. Control Fed.,* **50** (4), 678–687.

Shao, Y. J.; Starr, M.; Kaporis, K.; Kim, H. S.; Jenkins, D. (1997) Polymer Addition as a Solution to *Nocardia* Foaming Problems. *Water Environ. Res.,* **69**, 25.

Stall, R. T.; Sherrard, J. H. (1978) Evaluation of Control Parameters for the Activated Sludge Process. *J. Water Pollut. Control Fed.*, **50** (3) 450–457.

Tekippe, R. (1984) Activated Sludge Circular Clarifier Design Considerations. *Proceedings of the 57th Annual Water Pollution Control Federation Technical Exposition and Conference*; New Orleans, Louisiana, Sept 30–Oct 5; Water Environment Federation: Alexandria, Virginia.

Torpey, W. N.; Chasick, A. H. (1955) Principles of Activated Sludge Operation. *Sewage Ind. Wastes,* **27** (11), 1217–1233.

Trygar, R. (2010) Sludge Volume Index is a Valuable Measure of Sludge Settleability Characteristics and Can Be Monitored to Help Prevent Process Problems. *Treat. Plant Operator.* http://www.tpomag.com/editorial/2010/03/what-the-heck-is-svi (accessed Nov 10, 2017).

U.S. Environmental Protection Agency (1973) *Start-Up of Municipal Wastewater Treatment Facilities*; EPA 430/9-74-008; U.S. Environmental Protection Agency Office of Water Program Operations: Washington, D.C.

U.S. Environmental Protection Agency (1974) *Process Design Manual for Upgrading Existing Wastewater Treatment Plants*; EPA-625/1-71-004A; U.S. Environmental Protection Agency, Technology Transfer: Washington, D.C.

U.S. Environmental Protection Agency (1977) *Process Control Manual for Aerobic Biological Wastewater Treatment Facilities*; U.S. Environmental Protection Agency: Washington, D.C.

U.S. Environmental Protection Agency (1978a) *Field Manual for Performance Evaluation and Troubleshooting at Municipal Wastewater Treatment Facilities*; U.S. Environmental Protection Agency: Washington, D.C.

U.S. Environmental Protection Agency (1978b) *Design of Wastewater Treatment Facilities Major Systems*; EPA-430/9-79-008; U.S. Environmental Protection Agency: Washington, D.C.

U.S. Environmental Protection Agency (1989) *Design Manual: Fine Pore Aeration Systems;*U.S. Environmental Protection Agency, Office of Research and Development, Center for Environmental Research Information, Risk Reduction Engineering Laboratory: Cincinnati, Ohio.

U.S. Environmental Protection Agency (2010) *Nutrient Control Design Manual*; EPA-600/R-10-100; U.S. Environmental Protection Agency: Washington, D.C.

Wahlberg, E. J. (1996) Activated Sludge Solids Inventory Control Using the State Point Concept. In *Enhancing the Design and Operation of Activated Sludge Plants*; Central States Water Environment Association: Madison, Wisconsin.

Wahlberg, E. (2001) *Project 00-CTS-1: WERF/CRTC Protocols for Evaluating Secondary Clarifier Performance, 2001 Final Report;* Water Environment Research Foundation: Alexandria, Virginia.

Wahlberg, E. J. (2016) Personal communication.

Wahlberg, E. J.; Augustus, M.; Chapman, D. T.; Chen, C.; Esler, J. K.; Keinath, T. M.; Parker, D. S.; Tekippe, R. J.; Wilson, T. E. (1994) Evaluating Activated Sludge Clarifier Performance Using the CRTC Protocol: Four Case Studies. *Proceedings of the 67th Annual Water Environment Federation Technical Exposition and Conference*, Chicago, Illinois, Oct 15–19; Water Environment Federation: Alexandria, Virginia.

Water Environment Federation (1994) *Basic Activated Sludge Process Control, Problem-Related Operations Based Education*; Water Environment Federation: Alexandria, Virginia.

Water Environment Federation (1996) *Operation of Municipal Wastewater Treatment Plants*, 5th ed.; Manual of Practice No. 11; Water Environment Federation: Alexandria, Virginia.

Water Environment Federation (2005) *Clarifier Design*, 2nd ed.; Manual of Practice No. FD-8; Water Environment Federation: Alexandria, Virginia.

Water Environment Federation (2012) *Basic Laboratory Procedures for the Operator–Analyst*, 5th ed.; Water Environment Federation: Alexandria, Virginia.

Water Environment Federation (2016) *Operation of Water Resource Recovery Facilities;* Manual of Practice No. 11; 7th ed.; Water Environment Federation: Alexandria, Virginia.

Water Environment Federation (2017) *Wastewater Biology: The Microlife*, 3rd ed.; Water Environment Federation: Alexandria, Virginia.

Water Environment Federation; American Society of Civil Engineers; Environmental & Water Resources Institute (2010) *Design of Municipal Wastewater Treatment Plants,* 5th ed.; WEF Manual of Practice No. 8/ASCE Manuals and Reports on Engineering Practice No. 76; Water Environment Federation: Alexandria, Virginia.

Water Environment Federation; American Society of Civil Engineers; Environmental and Water Resources Institute (2018) *Design of Water Resource Recovery Facilities*; 6th ed., WEF Manual of Practice No. 8/ ASCE Manuals and Reports on Engineering Practice No. 76; Water Environment Federation: Alexandria, Virginia.

West, A. W. (1973a) *Operational Control Procedures for the Activated Sludge Process—Part I Observations and Part II Control Tests;* U.S. Environmental Protection Agency, Office of Water Program Operations: Cincinnati, Ohio.

West, A. W. (1973b) *Operational Control Procedures for the Activated Sludge Process—Part IIIA Calculation Procedures;* U.S. Environmental Protection Agency, Office of Enforcement and General Counsel: Cincinnati, Ohio.

West, A.W. (1974) Operational Control Procedures for the Activated Sludge Process: Part I—Observations, Part II— Control Tests; U.S. EPA Office of Water Program Operations, National Waste Treatment Center: Cincinnati, Ohio.

West, A. W. (1978) *Updated Summary of the Operational Control Procedures for the Activated Sludge Process;* U.S. Environmental Protection Agency, Operational Technology Branch National Training and Operational Technology Center: Cincinnati, Ohio.

Wilen, B.-M.; Balmer, P. (1999) The Effect of Dissolved Oxygen Concentration on the Structure, Size, and Size Distribution of Activated Sludge Flocs. *Water Res.,* **33** (2), 391–400.

Suggested Readings

Alleman, J.; Prakasam, T. (1983) Reflections on Seven Decades of Activated Sludge History. *J. Water Pollut. Control Fed.,* **55** (5), 436–443.

Bartle, H.; Revis, C. (2014) Controlling *Microthrix parvicella* with Foam Control and Ammonia-SRT. *Proceedings of the 87th Annual Water Environment Federation Technical Exhibition and Conference* [CD-ROM]; New Orleans, Louisiana, Sept 29–Oct 1; Water Environment Federation: Alexandria, Virginia.

Facility Operations Assistance Section of the New York Department of Environmental Conservation (n.d.) *Using State Point Analysis to Maximize Secondary Clarifier Performance.* https://www1.maine.gov/dep/water/wwtreatment/state_point_article.pdf (accessed Nov 10, 2017).

Jenkins, D., Richard, M. G.; Daigger, G. T. (2003) *Manual of the Causes and Control of Activated Sludge Bulking, Foaming, and Other Solids Separation Problems,* 3rd ed.; Lewis Publishers: London.

Water Environment Federation (2012) *Basic Laboratory Procedures for the Operator–Analyst*, 5th ed.; Water Environment Federation: Alexandria, Virginia.

CHAPTER 9

Nutrient Removal

Introduction

Fundamental concepts in biological treatment were discussed in Chapter 5, including an introduction to two groups of specialized microorganisms: nitrifying bacteria and phosphate-accumulating organisms (PAOs). This chapter goes into more detail and discusses the wastewater characteristics and process variables that impact these microorganisms and that affect the ability of water resource recovery facilities (WRRFs) to remove ammonia, nitrite, nitrate, and phosphorus. Chemical phosphorus removal is also included in this chapter. Those who are unfamiliar with nutrient removal should start by reading the take-home points at the end of each section and the chapter summary. Nutrient removal might be the most complex area of treatment facility operation and can be confusing because of the amount of chemistry and math required. There is a tremendous amount to know about biological and chemical nutrient removal. This introductory chapter can't and won't cover everything. The Water Environment Federation's (WEF's) Manual of Practice No. 37, *Operation of Nutrient Removal Facilities* (2013), contains additional information and case studies.

LEARNING OBJECTIVES

Upon completing this chapter, you will be able to

- Predict the fate of different nitrogen and phosphorus species during biological and chemical treatment;
- Interpret chemical equations, predict the composition of ionic compounds, convert moles to milligrams per liter (mg/L), and calculate chemical dosages from balanced chemical equations;
- Explain the steps involved in nitrification, denitrification, and enhanced biological phosphorus removal (EBPR) and identify the microorganisms responsible for each transformation;
- List the stoichiometric requirements for and products of nitrification, denitrification, EBPR, and chemical phosphorus removal;
- Identify the most important process control parameters for nitrification, denitrification, chemical phosphorus removal, and EBPR;
- Describe the effects of environmental variables (pH, dissolved oxygen [DO], etc.) on each biological process;
- Evaluate process control data to determine whether nitrification, denitrification, or EBPR will be inhibited;
- Demonstrate how to manipulate process variables to maximize nitrification and denitrification rates;
- Determine best time of day and flowrate for adding recycle stream flows from solids-handling processes to minimize effluent nitrogen and phosphorus concentrations;
- Calculate alkalinity requirements for nitrification and chemical phosphorus removal;
- Calculate stoichiometric doses of metal salts required for phosphorus removal and estimate actual dose based on treatment objectives; and
- Select a metal salt and addition point based on treatment objectives.

Chemical Symbols and Formulas

Ions commonly found in wastewater are defined in Table 9.1. Chemical symbols and formulas are shown next to their written names throughout this chapter. Those that are used throughout the chapter are listed here.

Table 9.1 Important Ions in Wastewater Treatment

Cations	Anions
Aluminum ion (Al^{+3})	Bicarbonate ion (HCO_3^-)
Ammonium ion (NH_4^+)	Chloride ion (Cl^-)
Calcium ion (Ca^{+2})	Carbonate ion (CO_3^{-2})
Hydrogen ion (H^+)	Hydroxide ion (OH^-)
Ferric ion (Fe^{+3})	Nitrite ion (NO_2^-)
Ferrous ion (Fe^{+2})	Nitrate ion (NO_3^-)
Magnesium ion (Mg^{+2})	Phosphate ion (PO_4^{-3})
Sodium ion (Na^+)	Sulfate ion (SO_4^{-2})

(aq)—Used in a chemical formula to indicate that a chemical is dissolved in aqueous (water) solution.

(g)—Used in a chemical formula to indicate that a chemical is a gas.

(s)—Used in a chemical formula to indicate that a chemical is a solid.

Al^{+3}—Aluminum ion

$Al_{12}Cl_{12}(OH)_{24}$—Polyaluminum chloride, varies by manufacturer

$Al_2(SO_4)_3$—Aluminum sulfate (alum)

$Al_nCl_{(3n\text{-}m)}(OH)_m$—Generic formula for polyaluminum chloride

$C_5H_7O_2N$—Biomass (simple formula)

$C_{10}H_{19}O_5N$—Generic formula for wastewater BOD

$C_{60}H_{87}O_{23}N_{12}P$—Biomass (more complete formula)

$CaCl_2$—Calcium chloride

$Ca(OH)_2$—Calcium hydroxide

$CaCO_3$—Calcium carbonate

$CaMg(CO_3)_2$—Dolomitic lime

CaO—Calcium oxide (lime)

CH_3OH—Methanol

CH_3COOH—Acetate

Cl_2—Chlorine

CO_2—Carbon dioxide

CO_3^{-2}—Carbonate ion

Fe^{+2}—Ferrous iron ion

Fe^{+3}—Ferric iron ion

$Fe_2(SO_4)_3$—Ferric sulfate

Fe_2S_3—Ferric sulfide

$Fe_3(PO_4)_2$—Ferrous phosphate

$FeCl_3$—Ferric chloride

FeS—Ferrous sulfide

$FeSO_4$—Ferrous sulfate

H^+—Hydrogen ion (acid)

H_2O—Dihydrogen monoxide (water)

H_3PO_4—Phosphoric acid

HCO_3^-—Bicarbonate ion

HNO_2—Nitrous acid

KCl—Potassium chloride

$Mg(OH)_2$—Magnesium hydroxide

$MgNH_4PO_4$—Magnesium ammonium phosphate (struvite)

$MgSO_4$—Magnesium sulfate

N_2—Nitrogen gas

Na_2CO_3—Sodium carbonate (soda ash)

$NaAlO_2$—Sodium aluminate

NaCl—Sodium chloride

$NaHCO_3$—Sodium bicarbonate (baking soda)

NaHOCl—Sodium hypochlorite

NH_3—Ammonia (free ammonia)

NH_3-N—Ammonia-nitrogen (as N)

NH_4OH—Ammonium hydroxide

NH_4^+—Ammonium ion

NO_2^-—Nitrite ion

NO_2-N—Nitrite nitrogen (as N)

NO_3^-—Nitrate ion

NO_3-N—Nitrate nitrogen (as N)

O_2—Oxygen

OH^-—Hydroxide ion (base)

PO_4^{-3}—Phosphate ion

PO_4-P—Phosphate-phosphorus

Nitrogen and Phosphorus in Wastewater

The nutrients, nitrogen and phosphorus, were introduced in Chapter 1 and were discussed further in Chapter 2. Both are essential for life. In wastewater, nitrogen occurs in four basic forms: organic nitrogen, ammonia, nitrite, and nitrate (Figure 9.1). Between 60 and 70% of the total nitrogen in domestic influent is *ammonia-nitrogen* (NH_3-N) and the remaining 30 to 40% is organic nitrogen (Metcalf and Eddy, Inc./ AECOM, 2014). Organic nitrogen includes nitrogen that is part of proteins, carbohydrates, and other organics. The ammonia and organic nitrogen are measured together with the total Kjeldahl nitrogen (TKN) test. Domestic wastewater typically contains between 23 and 69 mg/L TKN, of which 14 to 41 mg/L is NH_3-N (Metcalf and Eddy, Inc./AECOM, 2014). Nitrite and nitrate are not typically present unless they are being added to the collection system for odor control or they are being discharged by an industrial user. Small amounts of nitrite and nitrate discharged into the collection system are typically used up by facultative bacteria long before the wastewater reaches the WRRF.

Nitrogen species are typically expressed as nitrogen (N). *Ammonia-nitrogen* is NH_3-N, nitrite-nitrogen is NO_2-N, and nitrate-nitrogen is NO_3-N. Phosphorus compounds are typically expressed as phosphorus (P). See Chapter 2 for an explanation on how to convert between mg/L and mg/L as N or mg/L as P.

During biological treatment, most of the influent organic nitrogen is converted to ammonia-nitrogen (NH_3-N); however, a small amount of the organic nitrogen (typically 1 to 3 mg/L as N) can't be broken down by the microorganisms during treatment (WEF, 2018). The *recalcitrant* nitrogen can still be removed

Recalcitrant means that the compound is difficult to break down. This material can often be broken down given enough time, but not within the time it spends in a typical secondary treatment process. *Inert* means the material can't be broken down at all by the microorganisms.

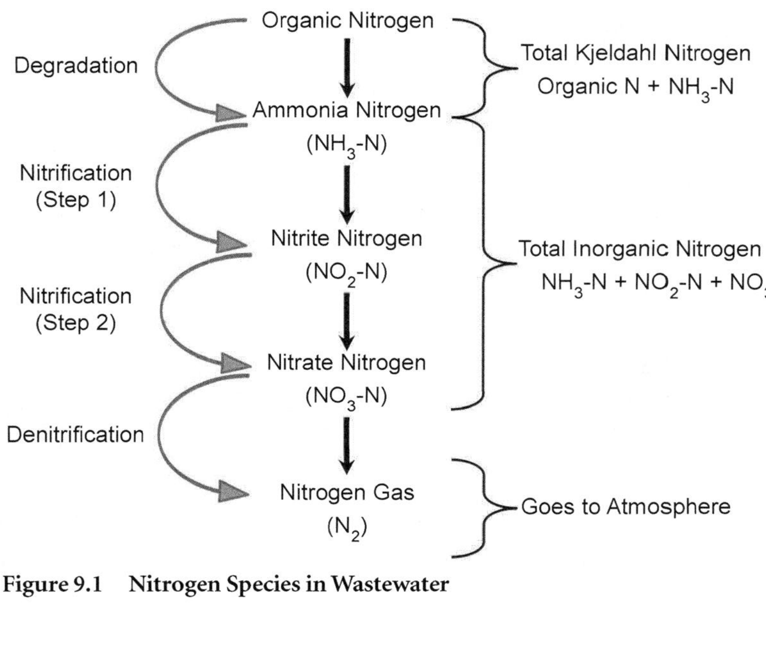

Figure 9.1 Nitrogen Species in Wastewater

by settling if it is particulate, but soluble recalcitrant organic nitrogen passes through to the final effluent. Some of the nitrogen that enters the WRRF will be incorporated to biomass or will be removed by settling, but most of it will pass into the final effluent unless the WRRF uses specialized microorganisms, the nitrifiers and denitrifiers, to remove it.

In wastewater, phosphorus occurs in three basic forms: phosphate, polyphosphate, and organic phosphorus (Table 9.2). Phosphate, also called *orthophosphate*, is typically approximately 50% of the total influent phosphorus. Orthophosphate is also called *reactive phosphate*. Orthophosphate typically exists in wastewater as either $H_2PO_4^-$ or HPO_4^{-2}, with the first form being dominant below pH 8.3 (U.S. EPA, 2010). Polyphosphates, also called *condensed phosphates*, consist of multiple phosphate molecules that are bound together chemically in long chains. Polyphosphates are used in many applications, including food additives and corrosion inhibitors. Approximately 33% of total influent phosphorus is polyphosphates. The remaining 15% of influent phosphorus is organic phosphorus. This includes phosphorus that is part of fats, proteins, and other organic molecules. Orthophosphate and polyphosphate are both soluble, but can also be *adsorbed* onto the surfaces of particles. When solids are removed by filtration, the orthophosphates and polyphosphates adsorbed to their surfaces will also be removed (Gu et al., 2011). Organic phosphorus may be soluble, colloidal, or particulate. During biological treatment, polyphosphates and most soluble organic phosphates are converted to orthophosphate (Figure 9.2).

Adsorb means to adhere to the surface. *Absorb* means to be taken up. For example, dirt adsorbs to the outside of a car, whereas water is absorbed into a sponge.

Influent total phosphorus concentrations average between 6 and 8 mg/L as P for domestic wastewater (U.S. EPA, 2010). Concentrations may be higher for municipalities with low per-person water usage and lower for municipalities with high per-person water usage. Contributions from industrial users can also increase influent phosphorus concentrations. Some of the influent phosphorus is incorporated to biomass during biological treatment; however, most of the influent phosphorus passes through the treatment process, ending up in the final effluent unless the WRRF takes additional steps to remove it through biological and/or chemical treatment (Figure 9.2).

Small amounts of soluble, organic nitrogen and soluble, organic phosphorus can't be utilized by the microorganisms or be removed by gravity in the clarifier. They pass through the treatment process unchanged and end up in the final effluent. Soluble, nonbiodegradable organic nitrogen concentrations are typically between 1 and 3 mg/L as N for domestic WRRFs, but the concentration will depend on the types of dischargers in the service area. Recalcitrant phosphorus is typically less than 0.05 mg/L as P.

Mechanisms for Nitrogen and Phosphorus Removal

Bacteria cells contain between 12 and 16% nitrogen and between 2 and 5% phosphorus by dry weight. As they grow and reproduce in the treatment process, nitrogen, phosphorus, and other trace nutrients are incorporated to the biomass. The process of incorporating nutrients to biomass is called *assimilative uptake*. During secondary treatment, 15 to 30% of the influent total nitrogen will be incorporated to biomass (U.S. EPA, 2010) and approximately 1 mg/L P will be assimilated for every 100 mg/L of 5-day biochemical oxygen demand (BOD_5) removed (WEF, 2013). Nutrients are removed from the treatment process when excess biomass is removed either by settling (ponds), sloughing (fixed film), or wasting (activated sludge). The excess biomass is typically digested and may be dewatered elsewhere in the treatment facility. Excess water returned from solids handling processes will return some ammonia and phosphorus to the liquid stream side of the treatment facility and can make it appear as though less nitrogen and phosphorus are being removed.

The connections between the liquid stream and solid stream sides of the WRRF are discussed in Chapter 1 and are shown in Figure 1.3.

Table 9.2 Forms of Phosphorus in Wastewater

	Phosphate	Polyphosphate	Organic phosphate
Chemical form	Orthophosphate	Condensed phosphates	Organically bound (part of proteins)
	Reactive phosphate		
Location	Soluble/may be adsorbed to particles	Soluble/may be adsorbed to particles	Soluble, colloidal, or particulate

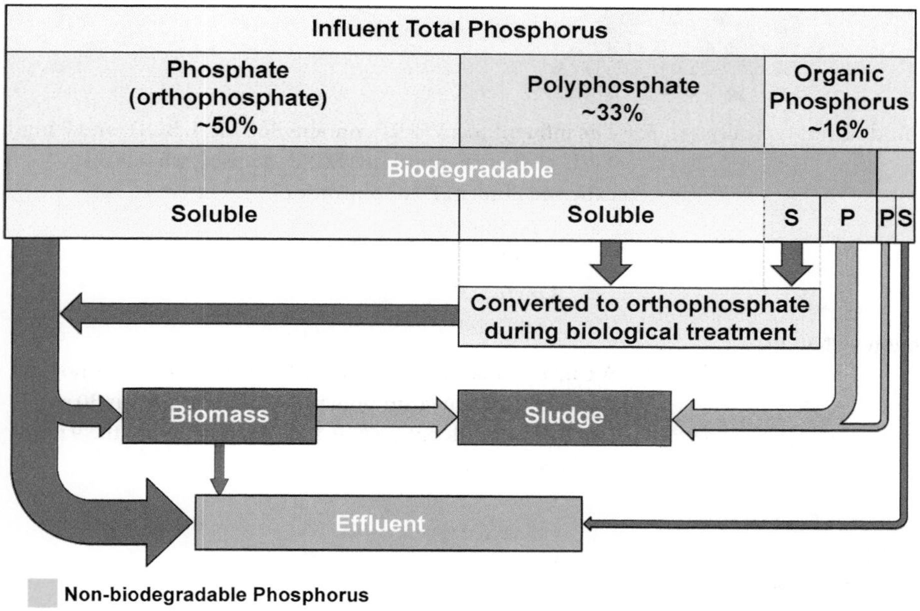

Figure 9.2 Phosphorus in Wastewater Treatment (S = soluble, P = particulate; this figure shows the fate of phosphorus in secondary treatment processes like activated sludge unless specific steps are taken to remove it) (Reprinted with permission by Indigo Water Group)

For reasons discussed in Chapter 1, many WRRFs have or will have effluent limitations for ammonia, nitrite, nitrate, and phosphorus that cannot be met through assimilative uptake alone. Other methods must be used to achieve low effluent limits. Ammonia can be converted into nitrite and nitrate by the nitrifying bacteria. Effluent ammonia-nitrogen (NH_3-N) concentrations less than 1 mg/L NH_3-N are achievable. Nitrate may be converted to nitrogen gas by a variety of facultative, heterotrophic bacteria. The nitrogen gas is returned to the atmosphere. Nitrate removal is limited by the amount of BOD_5 available. *Total inorganic nitrogen* (TIN) limits as low as 5 mg/L N can be met through biological nitrification and denitrification.

Total inorganic nitrogen is the sum of the ammonia-, nitrite-, and nitrate-nitrogen.

Phosphorus may be removed biologically by the PAOs. This process is called *EBPR* to distinguish it from assimilative uptake. Total phosphorus (TP) limits of 1 mg/L can be reliably met with EBPR, although final effluent TP concentrations down to 0.3 mg/L have been achieved at some WRRFs. Phosphorus can also be *precipitated* through chemical addition. Alum, ferric chloride, or lime is added to the wastewater where they combine with phosphorus to form a solid. The precipitate is removed by settling or filtration. Chemical phosphorus removal can meet effluent limits as low as 0.03 mg/L TP. Chemical and biological phosphorus removal methods are often used in combination. Facilities typically use biological phosphorus removal to reduce P concentrations below 1 mg/L as P followed by chemical precipitation at or after the secondary clarifiers. The pros and cons of different chemical addition points are discussed later in this chapter. Achieving effluent phosphorus concentrations below 0.5 mg/L as P depends on getting good solids removal in the secondary clarifiers. Solids that escape into the final effluent contain both nitrogen and phosphorus. Many facilities are equipped with tertiary filters to ensure solids capture.

Phosphorus precipitation is the conversion of soluble phosphorus present in the wastewater as orthophosphate ion (PO_4^{-3}) to an insoluble chemical compound.

Some ammonia and phosphorus are removed during biological treatment by being incorporated to biomass. Nitrification converts ammonia to nitrite and nitrate. Denitrification converts nitrate to nitrogen gas. Total inorganic nitrogen concentrations below 5 mg/L as N are achievable. Phosphorus may be removed biologically or chemically. Total phosphorus concentrations below 0.05 mg/L as P are achievable with chemical addition. To meet low effluent N and P limits, effluent total suspended solids (TSS) must also be low.

TEST YOUR KNOWLEDGE

1. **Which of the following nitrogen compounds is not typically found in influent wastewater?**
   a. Ammonia
   b. Nitrite
   c. Organic nitrogen
   d. TKN

2. **Nitrate is sometimes added to the collection system to**
   a. Control odors
   b. Precipitate phosphorus
   c. Remove BOD
   d. Increase biomass

3. **The wastewater from a WRRF contains 2 mg/L of organic nitrogen, 5 mg/L of nitrate-nitrogen, and 1 mg/L of ammonia-nitrogen. How much TKN is leaving the facility?**
   a. 1 mg/L as N
   b. 2 mg/L as N
   c. 3 mg/L as N
   d. 7 mg/L as N

4. **After biological treatment is complete, most of the remaining phosphorus will be present as**
   a. Orthophosphate
   b. Polyphosphate
   c. Organic phosphate
   d. Biomass

5. **The influent to a WRRF contains 300 mg/L BOD_5 and 7 mg/L TP. If neither chemical nor EBPR are used, what is the expected effluent P concentration?**
   a. 1 mg/L as P
   b. 2 mg/L as P
   c. 4 mg/L as P
   d. 7 mg/L as P

6. **A conventional activated sludge treatment process is not designed to remove ammonia. The influent TKN is 30 mg/L and the final effluent TKN is 21 mg/L. Where did the 9 mg/L of TKN end up?**
   a. Released to atmosphere
   b. Converted to ammonia
   c. Precipitated with alum
   d. Incorporated to biomass

7. **The laboratory reports the following TKN and ammonia concentrations on a filtered and unfiltered sample of influent wastewater:**

Unfiltered TKN as N	36.0 mg/L
Filtered TKN as N	26.0 mg/L
Ammonia-nitrogen (NH_3-N)	22.0 mg/L

Calculate the concentrations of particulate, soluble, and total organic nitrogen.

Chemistry Review

More detailed information and examples beyond what are presented in this section may be found by searching YouTube for Khan Academy. Khan Academy is a free education service with hundreds of short videos that teach math, chemistry, biology, and other topics.

Some knowledge of chemistry is needed to fully understand nutrient removal and to calculate chemical dosages. Ultimately, operators must be able to figure out or look up the desired chemical reaction, determine how many moles of chemical are needed, convert from moles to mg/L, and set the chemical feed rate that achieves the calculated concentration. Operators must also understand why side reactions occur and how to adjust chemical dosages accordingly. By the end of this section, you should be able to read chemical equations and convert between moles and milligrams per liter. We'll begin by defining a few terms.

Atoms—Atoms are the basic building blocks of nature. Atoms can't be broken down into smaller pieces without losing their unique properties. For example, chlorine forms a green-yellow gas that reacts with sodium to form sodium chloride, or ordinary table salt. These are two properties of chlorine. Atoms contain protons (positively charged particles) and neutrons (particles without a charge) in their center and electrons (negatively charged particles) that orbit around the center.

Element—An element is a particular type of atom. Elements are defined by the number of protons they have. For example, all carbon atoms have 6 protons, all nitrogen atoms have 7 protons, and all oxygen atoms have 8 protons. The periodic table contains over a hundred boxes. Each box contains a letter or combination of letters (Figure 9.3). Each box symbolizes a different element. The most common elements in wastewater treatment include: aluminum (Al), calcium (Ca), carbon (C), chlorine (Cl), hydrogen (H), iron (Fe), magnesium (Mg), nitrogen (N), oxygen (O), phosphorus (P), and sulfur (S).

Atomic Number—The number of protons in an atom. Atomic numbers are specific to each type of element. If an atom has 8 protons, it is oxygen. If an atom has 6 protons, it is carbon. The atomic numbers are shown in the periodic table along with the chemical symbol for the element (Figure 9.4).

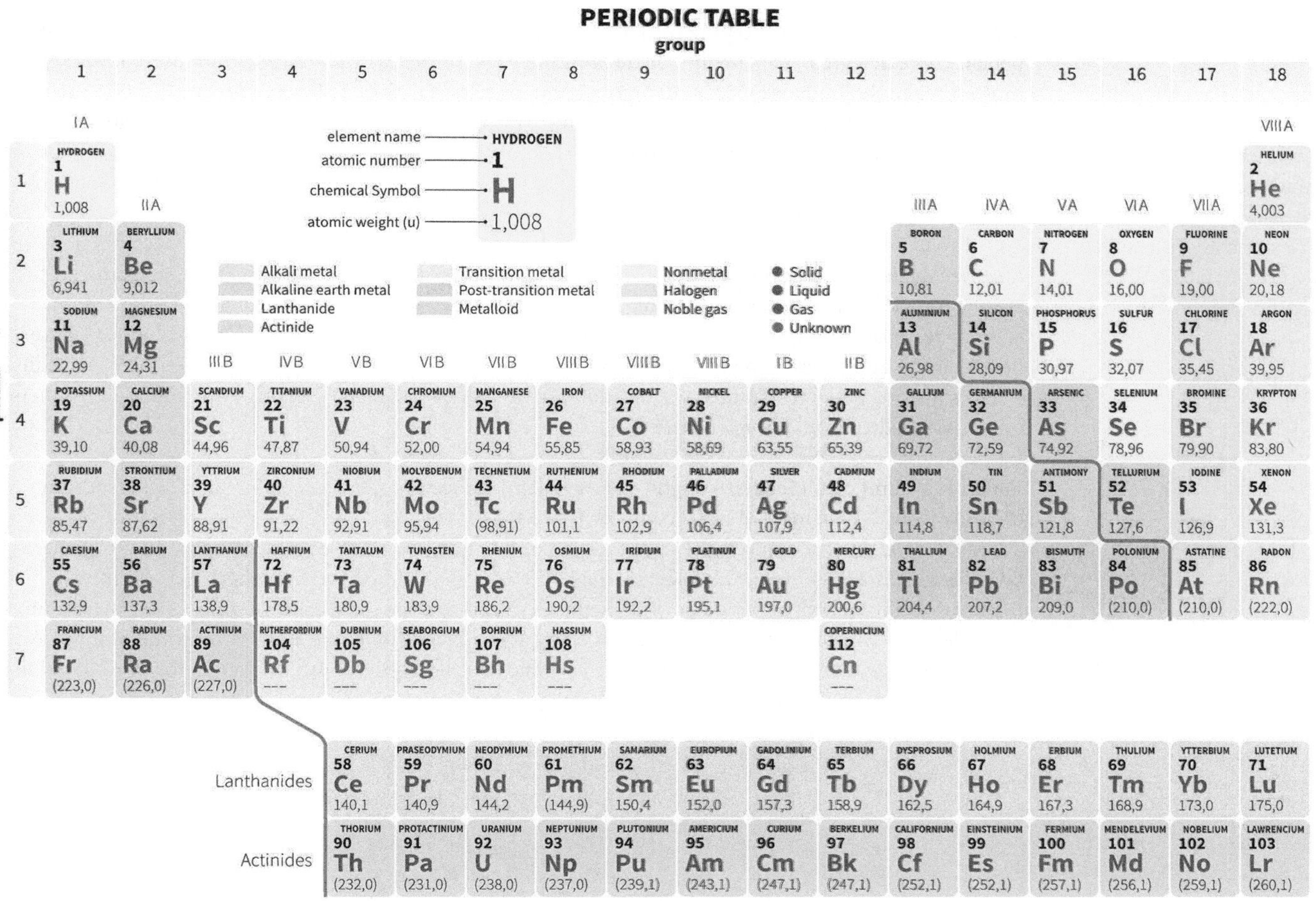

Figure 9.3 Periodic Table (Getty Images)

Molecules and Compounds—Molecules are combinations of two or more atoms that are chemically bound together. Atoms of some elements, like sulfur, are sometimes found in nature by themselves, unattached chemically to other atoms. A single sulfur atom is not a molecule. The following elements exist in nature as combinations of two of the same type of atom: bromine (Br), iodine (I), nitrogen (N), chlorine (Cl), hydrogen (H), oxygen (O), and fluorine (F). This is why DO is always shown as O_2 and nitrogen gas is shown as N_2. Putting the chemical symbols of these molecules together makes the word "BrINClHOF", which makes it easier to remember which elements form pairs with themselves. The BrINClHOF elements react with many other elements too, not just themselves. Molecules made from two or more different elements are called *compounds*. Water (H_2O) is a compound. All compounds are molecules, but not all molecules are compounds.

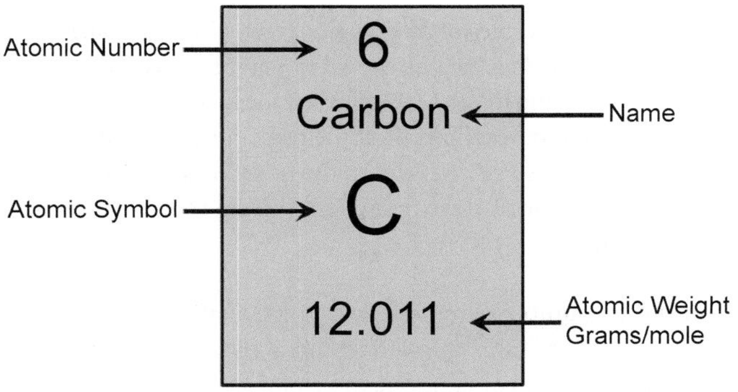

Figure 9.4 Periodic Table Information for Carbon (Reprinted with permission by Indigo Water Group)

Avogadro is a famous Italian scientist. He conducted many experiments with different elements to figure out how many atoms reacted with one another to produce new compounds. His calculations resulted in Avogadro's constant: 6.022×10^{23} atoms or molecules in a mole.

Mole—Moles are a way of keeping track of large numbers of atoms or molecules. Shoes come in a pair, pencils come in a gross, paper comes in a ream, and eggs come in a dozen. Atoms and molecules come in moles. One mole contains 6.022×10^{23} atoms or molecules. That's 602 200 000 000 000 000 000 000—a whole lot! Moles help us keep track of the number of atoms and molecules needed to react with other atoms and molecules in chemical reactions. For example, 1 mole of calcium chloride ($CaCl_2$) contains 1 mole of calcium and two moles of chloride. Chemical reactions and moles are used in this chapter to calculate chemical dosing rates.

Atomic Weight—The atomic weight is how much one mole of atoms weighs. For different elements, the atomic weight is equal to the number of protons plus the average number of neutrons. Carbon atoms contain 6 protons and typically have 6 neutrons. The atomic weight of carbon is 12.011 g/mole (Figure 9.4). Some carbon atoms have 7 neutrons, which is why the atomic weight isn't exactly 12 g/mole. This means that if someone patiently counted out 602 200 000 000 000 000 000 000 carbon atoms and weighed them on a balance, they would weigh 12.011 g.

Formula Weight or Molecular Weight—How much 1 mole of molecules weighs. Table salt is sodium chloride (NaCl). One mole of NaCl contains 1 mole of sodium (Na^+) and one mole of chloride (Cl^-). To find the weight of a mole of NaCl, the weights of the individual elements are simply added together. Sodium weighs 23.0 g/mole and chlorine weighs 35.5 g/mole. Adding their weights together gives the formula weight for sodium chloride of 58.5 g/mole. Molecules are often more complicated than sodium chloride. They can contain multiple types of elements and more than one atom of the same element. An example is ferric chloride, $FeCl_3$. Look at the periodic table and find the atomic weight for iron. To get the formula weight for $FeCl_3$, add together the weight of one iron atom and three chlorine atoms.

$$Fe + Cl + Cl + Cl = FeCl_3$$
$$56 \text{ g/mole} + 35.5 \text{ g/mole} + 35.5 \text{ g/mole} + 35.5 \text{ g/mole} = FeCl_3$$
$$162.5 \text{ g/mole} = FeCl_3$$

Ions—Ions are atoms or molecules that have a charge. Cations are positively charged. Anions are negatively charged. Ions are created when atoms and molecules gain or lose electrons. Chloride ion has one more electron than protons, which results in a −1 charge. Hydrogen ion has one proton and no electrons, which results in a +1 charge. Many of the inorganic compounds that are important in wastewater treatment will be present as ions in the water (Table 9.1).

Knowing the charges on common ions can help predict what molecules may be formed in a chemical reaction. For example, magnesium has a +2 charge whereas chloride has a –1 charge. The final compound must be neutral–zero charge. When magnesium and chloride are combined, the final compound will be $MgCl_2$ because two chlorine atoms are needed to neutralize the +2 charge on magnesium. When ferrous iron ions (+2 charge) are combined with phosphate ions (–3 charge), the quantity of each ion has to be adjusted so the final compound is neutral. In this case, three ferrous iron atoms will have a combined charge of +6 and two phosphate ions will have a combined charge of –6. The final compound is $Fe_3(PO_4)_2$. The phosphate ion (PO_4^{-3}) is shown in parenthesis because the outside subscript 2 applies to the entire ion.

Chemical Reactions—Chemical reactions occur when molecules of one type interact with molecules of another type to form something new. A chemical equation, like the one shown in Figure 9.5, is used to show how the molecules react with one another. The starting molecules are called the *reactants* and the ending molecules are called the *products*. The full-sized numbers in front of the reactants and products tell how many moles of each type of molecule are needed. The subscripts within the molecules tell how many moles of each element are needed to form the molecule. The full-sized numbers in front of the reactants and products may be changed to balance the equation.

Chemical reactions must be balanced. This means the number of moles of each type of atom must be the same on both sides of the chemical equation. The product in this case, ferric sulfide (Fe_2S_3), contains two moles of iron (Fe). The reactant, ferric chloride ($FeCl_3$), only contains one mole of iron (Fe). This means that 2 moles of $FeCl_3$ will be needed to produce 1 mole of Fe_2S_3. When balancing a chemical equation, balance the oxygen and hydrogen atoms last. Water (H_2O) can be added as a product if needed. Most of the

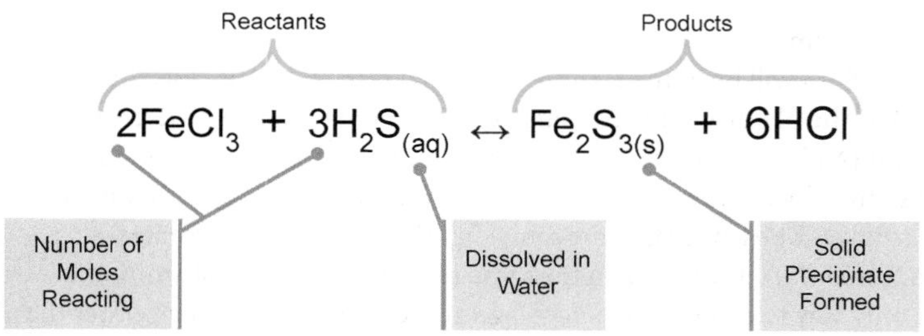

Figure 9.5 Chemical Reactions and Stoichiometry (Reprinted with permission by Indigo Water Group)

time, whole numbers are used to balance equations; however, formulas with large, complex compounds sometimes use fractions or decimals for balancing. Sometimes, the subscripts (g), (aq), or (s) are used to show if a reactant or product is a gas, dissolved in water (aqueous), or a solid.

Stoichiometry—Stoichiometry is the ratio of moles of reactants to moles of products needed to complete the reaction. In the example in Figure 9.5, two moles of ferric chloride react with three moles of hydrogen sulfide to produce one mole of ferric sulfide. The mole ratio between ferric chloride and ferric sulfide is 2 to 1.

Equilibrium—*Equilibrium* means balance. See the arrow in the center between the reactants and products in Figure 9.5? This indicates that the reaction can go either way depending on environmental conditions like pH, temperature, and the presence of other compounds. Many of the chemical reactions that are important in wastewater treatment are in equilibrium. If the arrow in the center of this equation only pointed to the right, it would mean that the chemical reaction is not in equilibrium and that when the reaction is complete, very few reactants, if any, will remain.

When a chemical reaction is in equilibrium, the percentages of reactants and products will be the same for a particular set of environmental conditions. For example, ammonia exists in wastewater as both ammonium ion (NH_4^+) and ammonia (NH_3), as shown in Figure 9.6. Ammonia (NH_3) is also called *free ammonia*. The percentages of ammonium ion (NH_4^+) and ammonia (NH_3) depend on pH and

Nitrification is required by many discharge permits because the un-ionized form of ammonia (NH_3) is toxic at the very low concentration of 0.02 mg/L. The un-ionized form is also called the *free ammonia.*

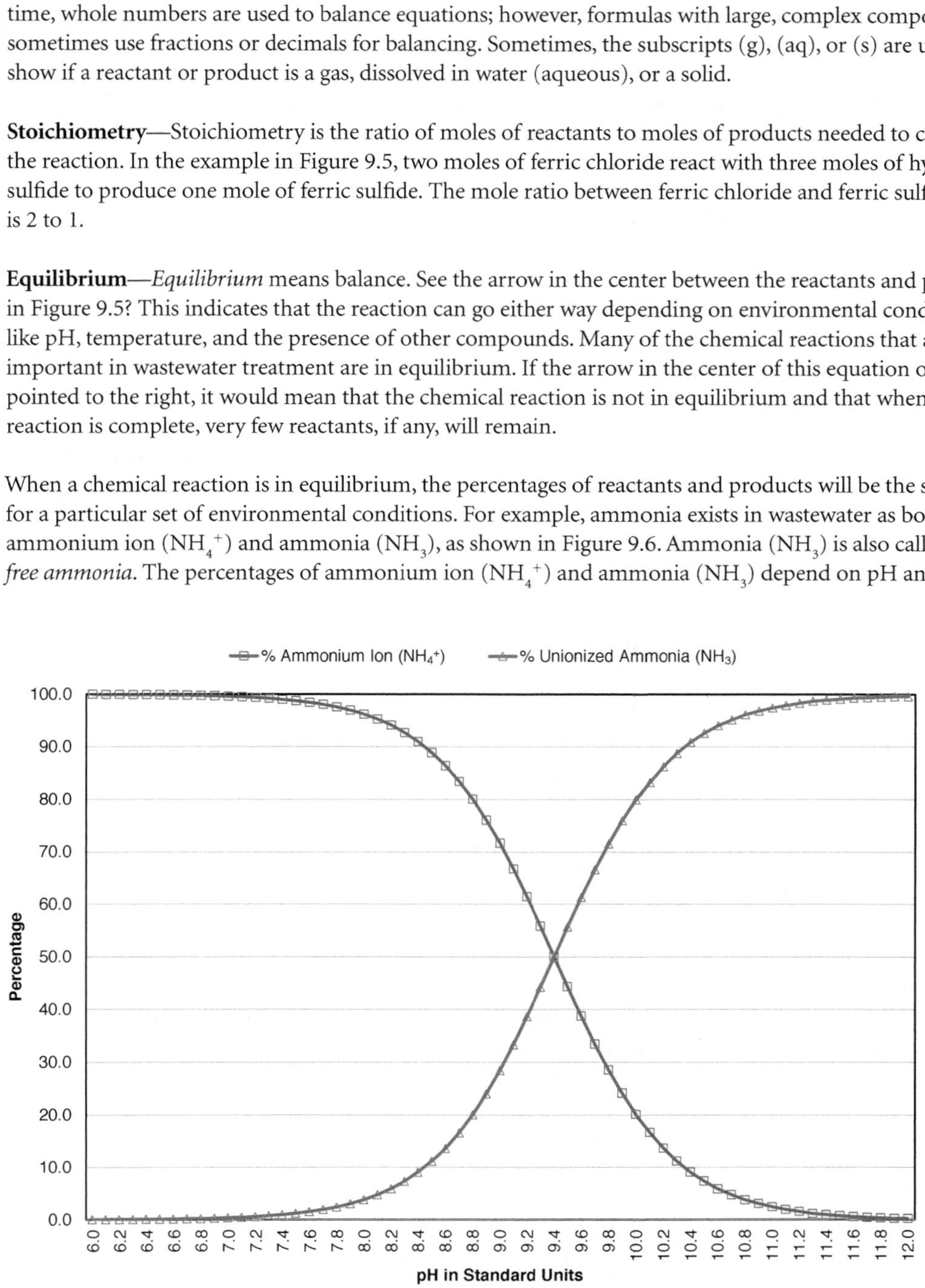

Figure 9.6 Equilibrium between Ammonium Ion and Ammonia (Reprinted with permission by Indigo Water Group)

temperature (Figure 9.6). At the equilibrium point, 50% of the total ammonia will be in the form of ammonium ion (NH_4^+) and the other 50% will be in the form of ammonia (NH_3). In this case, the equilibrium point is at about pH 9.4. If the pH drops, the percentage of ammonia (NH_3) will decrease and the percentage of ammonium ion (NH_4^+) will increase. If either ammonia (NH_3) or ammonium ion (NH_4^+) is removed from the system, whichever form is present in excess will automatically convert into the other form so that the percentages of each remain the same. In other words, the percentages of ammonium ion (NH_4^+) and ammonia (NH_3) will remain the same regardless of the total ammonia concentration ($NH_3 + NH_4^+$). The only way to change the percentage of each form is to change either the pH or temperature. Similar equilibrium graphs exist for chlorine, alkalinity (carbonate and bicarbonate), and hydrogen sulfide, although the equilibrium points are different for each.

Side reactions are undesired chemical reactions that take place in addition to the desired reaction. For example, when alum is added to precipitate phosphorus, some of the alum will react with water and other ions instead of phosphorus.

Converting from Moles to Milligrams per Liter—Chemical reactions use moles, but almost everything else in wastewater treatment uses concentrations like milligrams per liter or percent. Operators should be able to convert between moles and milligrams per liter to calculate chemical dosing rates. The actual dosage needed is typically higher than the calculated dosage because of *side reactions.* It is possible to use a trial-and-error approach to find the correct chemical dosage, but being able to calculate a starting point saves time and energy and helps to optimize chemical use over the long term.

The following steps should be followed to convert between moles and milligrams per liter:

- Look up or create the desired chemical equation;
- Ensure the chemical equation is balanced. If not, balance the equation;
- Calculate the formula weight for the chemical that will be added;
- Calculate the formula weight for the target compound; and
- Convert moles to using dimensional analysis.

Example: Determine the dose of ferric chloride needed to react with 5 mg/L of hydrogen sulfide in the WRRF influent. Use the stoichiometry from Figure 9.5.

Step 1—Calculate the formula weights for ferric chloride and hydrogen sulfide. Use the periodic table (Figure 9.3) to find the atomic weights for each element.

$$FeCl_3 = (1)(56\text{ g Fe/mole}) + (3)(35.5\text{ g Cl/mole}) = 162.5\text{ g/mole}$$

$$H_2S = (2)(1\text{ g H/mole}) + (1)(32\text{ g S/mole}) = 34\text{ g/mole}$$

Dimensional analysis is a method of converting one set of units into another set of units. For example, converting from cubic meters per day to gallons per day or from moles to grams.

Step 2—Convert moles to with *dimensional analysis.*

$$\frac{2\text{ mole } FeCl_3}{1\text{ mole } H_2S}\left|\frac{162.5\text{ g } FeCl_3}{1\text{ mole } FeCl_3}\right|\left|\frac{1\text{ mole } H_2S}{34\text{ g } H_2S}\right| = 9.56\text{ g } FeCl_3 \text{ per g } H_2S$$

Because this is a weight ratio, the units can be almost anything as long as they are still a weight or a concentration. For example, 9.56 g $FeCl_3$ per of H_2S could also be milligram per milligram, pound per pound, or milligram per liter per milligram per liter. The weight ratio remains the same. If the units needed to be liters (gallons), we would need to know how much volume 9.56 g of $FeCl_3$ and 1 g of H_2S occupy and create a new ratio.

Step 3—Multiply by the influent concentration to determine total dose needed. Remember that the actual dose may need to be slightly higher because of side reactions.

$$\frac{5\frac{mg}{L}H_2S}{}\left|\frac{9.56\frac{mg}{L}FeCl_3}{1\frac{mg}{L}H_2S}\right| = 47.8\frac{mg}{L}FeCl_3$$

TEST YOUR KNOWLEDGE

1. **Which of the following terms refers to a basic building block of nature that can't be broken down further without losing its essential properties?**
   a. Atom
   b. Element
   c. Compound
   d. Molecule

2. **Elements are defined by**
   a. Chemical properties
   b. Number of neutrons
   c. Chemical symbol
   d. Number of protons

3. **The atomic number is the number of protons, but the atomic weight is**
   a. Sum of protons, neutrons, and electrons
   b. Grams per molecule of diatom
   c. Sum of protons and neutrons
   d. Number of neutrons only

4. **Molecules differ from compounds in this way.**
   a. Compounds contain two or more atoms of the same element.
   b. Molecules are made from at least three different elements.
   c. Molecules must contain either oxygen or hydrogen.
   d. Compounds are made from at least two different elements.

5. **Find the formula weight for calcium carbonate ($CaCO_3$). Calcium weighs 40 g/mole, carbon weighs 12 g/mole, and oxygen weighs 16 g/mole.**
   a. 68 g/mole
   b. 84 g/mole
   c. 100 g/mole
   d. 204 g/mole

6. **Find the formula weight for calcium nitrate [$Ca(NO_3)_2$]. The atomic weights are calcium = 40 g/mole, nitrogen = 14 g/mole, and oxygen = 16 g/mole.**
   a. 102 g/mole
   b. 116 g/mole
   c. 164 g/mole
   d. 204 g/mole

7. **What is the chemical formula for ferric chloride?**
   a. $FeCl_2$
   b. $FeCl_3$
   c. FeOHCl
   d. $FeOCl_2$

8. **Sodium ion has a +1 charge. Nitrite ion has a −1 charge. What is the chemical formula for sodium nitrite?**
   a. $Na(NO_2)_2$
   b. Na_3NO_2
   c. $NaNO_2$
   d. Na_2NO_2

9. **If ferrous ions (Fe^{+2}) are combined with phosphate ions (PO_4^{-3}), the chemical formula for the resulting compound would be**
   a. Fe_2PO_4
   b. $Fe_3(PO_4)_2$
   c. $FePO_4$
   d. $Fe_6(PO_4)_2$

10. **Chemical reactions must**
    a. Be balanced
    b. Generate heat
    c. Produce acid
    d. Consume alkalinity

11. **Which pH range will have the highest percentage of ammonia (NH_3) when the total ammonia concentration is 20 mg/L as N?**
    a. <5
    b. 5 to 7
    c. 7 to 9
    d. >9

Biological Nitrification

Chapter 5 introduced the nitrifying bacteria. These bacteria use ammonia (NH_3) or nitrite (NO_2^-) as an energy source and use the inorganic carbon compounds carbonate (CO_3^{-2}) and bicarbonate (HCO_3^-) as their carbon sources instead of carbonaceous biochemical oxygen demand (CBOD). Carbonate and bicarbonate are two components of alkalinity that dominate in wastewater when the pH is greater than about 6.5 standard units (SU).

There are a variety of different bacteria that can convert ammonia (NH_3-N) to nitrite (NO_2-N) including *Nitrosomonas, Nitrosococcus, Nitrosospira*, and others. They are collectively referred to as *ammonia oxidizing bacteria* (AOB). *Nitrosomonas* is the most common AOB in domestic wastewater treatment systems (WEF, 2015). Another group of bacteria, the nitrite oxidizing bacteria (NOB), convert nitrite (NO_2^-) to nitrate

Nitrogen species (ammonia, nitrite, and nitrate) are typically shown as their ions in chemical equations and in general discussion. Laboratory results are shown "as nitrogen" and do not include the charge symbol for the ion. For example, nitrite (NO_2^-) becomes mg/L *nitrite-nitrogen* (NO_2-N). Chapter 2 includes examples showing how to convert.

1 mg/L NH_3 is equivalent to 0.824 mg/L NH_3-N. 1 mg/L NO_2^- is equivalent to 0.304 mg/L NO_2-N (30.4% nitrogen by weight).
1 mg/L NO_3^- is equivalent to 0.226 mg/L NO_3-N.

(NO_3^-). The NOB include *Nitrobacter, Nitrospira, Nitrococcus,* and *Nitrospina. Nitrobacter* and *Nitrospira* are the most common NOBs in domestic wastewater treatment systems (WEF, 2015). Figure 9.7 shows the sequential steps for converting ammonia (NH_3) to nitrate (NO_3^-).

The NOB grow faster than the AOB when water temperatures are below 25 °C (77 °F). Most WRRFs in the United States operate between 10 and 22 °C (55 and 72 °F). However, because the AOB produce the nitrite (NO_2^-) needed by the NOB, the AOB must be present before the NOB can appear. As soon as nitrite (NO_2^-) is produced by the AOB, it is used by the NOB. *Nitrite-nitrogen* concentrations (NO_2-N) rarely exceed 1 mg/L as N in a well-functioning nitrification process (Muirhead and Appleton, 2008). An abnormal increase in nitrite (NO_2^-) concentrations is commonly referred to a "nitrite lock". Nitrite lock occurs when the AOB outperform the NOB. Nitrite lock is discussed in more detail in the process control section.

NITRIFICATION STOICHIOMETRY

The proportion of reactants and products involved in this process of biochemical reactions is referred to as "stoichiometry." Understanding stoichiometry is important because it defines the basic inputs and outputs for each of the steps in the process and can determine which of these inputs will limit the reaction. Stoichiometry can be thought of as making the list of parts needed to build the end products. Equation 9.1 shows the stoichiometry for converting 1 mole of ammonia (NH_3) to nitrite (NO_2^-). The first portion of the equation shows the equilibrium reaction between ammonium ion (NH_4^+) and ammonia (NH_3). Recall from Chapter 5 that it requires more energy to move a charged molecule across a cell membrane than to move a neutral molecule across. The AOBs use NH_3 (no charge) rather than NH_4^+ (charged) for this reason. Ammonia (NH_3) is combined with oxygen (O_2). These bacteria are referred to as "obligate aerobes" because they grow only when DO is available. The reaction produces nitrite (NO_2^-), water (H_2O), and some hydrogen ions (H^+). The reaction will also produce some new AOBs, approximately 0.15 g of volatile suspended solids (VSS) for every 1 g of ammonia-nitrogen (NH_3-N) converted to nitrite-nitrogen (NO_2-N) (U.S. EPA, 2010).

$$NH_4^+ \leftrightarrow NH_3 + H^+ + 1.5O_2 \rightarrow NO_2^- + H_2O + 2H^+ \tag{9.1}$$

The stoichiometry reveals some information about the AOBs that is important for process control. Because the AOBs use NH_3 instead of NH_4^+, the chemical equilibrium between these two forms will determine how much NH_3 is available for use. As the pH decreases, the percentage of NH_3 available decreases rapidly (Figure 9.6). This means that it will be harder for the AOBs to access it. Low pH inhibits or slows down AOB growth by restricting the supply of ammonia (NH_3). The stoichiometry shows that 1.5 moles of oxygen will be needed to convert 1 mole of NH_3 to NO_2^-. A lack of oxygen will also inhibit or slow down the growth of the AOBs. Think back to the factory analogy used in Chapter 5. The speed of the AOB/NOB assembly line depends on having enough ammonia, nitrite, and DO. Without these raw materials, the assembly line can't produce new bacteria. If one or both is lacking, the assembly line will slow down. Finally, 2 moles of hydrogen ions (H^+) are generated. Hydrogen ions are acid. The AOBs don't produce nitrite ion (NO_2^-). They produce nitrous acid (HNO_2). The AOBs can produce enough acid to lower the pH in the treatment process. To prevent a pH drop, the treatment process must have enough alkalinity (buffering capacity) to *neutralize* the acid. The AOBs and NOBs also need alkalinity as a source of inorganic carbon to grow and reproduce. There must be enough alkalinity in the wastewater to support growth and neutralize the acid.

pH measures the hydrogen ion concentration. The higher the hydrogen ion concentration, the lower the pH will be. Chapter 2 defines pH and alkalinity.

Neutralize, in this case, means to make something ineffective. It can also mean a compound, molecule, or atom that does not have a charge. Ammonium ion (NH_4^+) is positively charged, but ammonia (NH_3) is neutral.

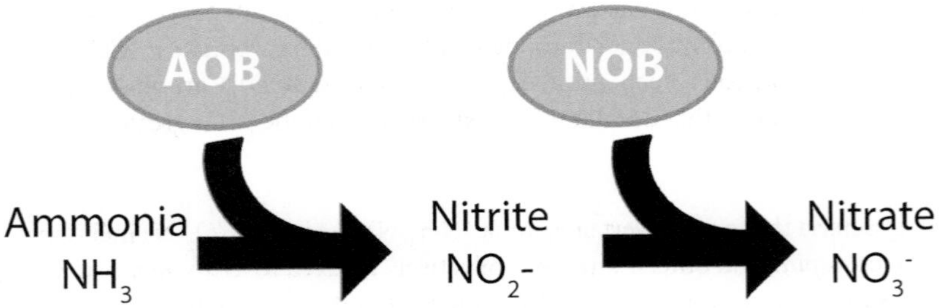

Figure 9.7 Nitrification Sequence (Reprinted with permission by Indigo Water Group)

Equation 9.2 shows the stoichiometry for converting 1 mole of nitrite (NO_2^-) to nitrate (NO_3^-). The NOBs need another 0.5 mole of oxygen (O_2). The NOBs are obligate aerobes just like the AOBs. The reaction produces some new NOBs, approximately 0.05 mg VSS/mg NO_2-N converted to NO_3-N (U.S. EPA, 2010).

$$NO_2^- + 0.5O_2 \rightarrow NO_3^- \tag{9.2}$$

Equations 9.1 and 9.2 have been simplified and don't include the new AOBs and NOBs produced by growth or the intake of inorganic carbon. The two equations are combined in eq 9.3 to show the overall stoichiometry along with carbon usage and biomass production. *Biomass* is shown in eq 9.3 as $C_5H_7O_2N$. The stoichiometry of the overall equation reveals another interesting piece of information: very few new AOBs and NOBs will be produced per mole of ammonia (NH_3) converted to nitrate (NO_3^-). The yield is low compared to the yield for bacteria that can use CBOD as their carbon and energy source. Recall from Chapter 5 that the yield for heterotrophic bacteria is 0.4 mg VSS/mg chemical oxygen demand (COD) (0.6 mg VSS/mg BOD). Why is this important? The growth rate of the nitrifying bacteria is much slower—10 to 20 times slower—than bacteria using CBOD and oxygen. For wastewater processes to remove ammonia, the biomass has to stay in the system long enough for the nitrifying bacteria to grow and reproduce.

Biomass is often shown chemically as $C_{60}H_{87}O_{23}N_{12}P$, which is slightly more accurate. Take a moment to calculate the formula weight for biomass and then find the percentage of N and P by weight. The formula weight is 1374 g/mole. It is 12.2% nitrogen and 2.3% phosphorus by weight.

$$\begin{aligned} &NH_4^+ \leftrightarrow NH_3 + H^+ + 1.863\ O_2 + 0.098\ CO_2 \rightarrow \\ &0.0196\ C_5H_7O_2N + 0.98\ NO_3^- + 1.88\ CO_2 + H_2O + 1.88\ H^+ \end{aligned} \tag{9.3}$$

Based on the stoichiometry of the overall reaction, 2 moles of oxygen are required to convert 1 mole of ammonia to nitrate, which is equivalent to 4.57 g O_2/g NH_3-N oxidized.

$$\frac{1\ \text{mole}\ NH_3}{1\ \text{mole N}}\left|\frac{1\ \text{mole N}}{14\ \text{g}}\right|\left|\frac{2\ \text{moles}\ O_2}{1\ \text{mole}\ NH_3}\right|\left|\frac{32\ \text{g}\ O_2}{1\ \text{mole}\ O_2}\right| = 4.57\ \text{g}\ O_2/\text{g}\ NH_3 - N$$

Compare this to the amount of oxygen used by heterotrophic bacteria to process CBOD. Heterotrophic bacteria only need 1 g O_2/g CBOD. A facility that removes ammonia will use 30 to 50% more oxygen overall than a facility that only removes CBOD. For activated sludge processes, nitrification substantially increases the cost of treatment resulting from the increased electrical cost of running blowers.

During nitrification, 2 moles of hydrogen ion (acid) are formed. Carbonate ions (CO_3^{-2}) in the wastewater, one component of alkalinity, react with the hydrogen ions (H^+) to prevent them from accumulating and lowering the pH. One carbonate ion (CO_3^{-2}) can neutralize two hydrogen ions (H^+). For historic reasons, laboratories report alkalinity as milligrams per liter of *calcium carbonate* ($CaCO_3$). One mole of alkalinity as $CaCO_3$ neutralizes 2 moles of hydrogen ions (H^+). Converting from moles to grams reveals that 7.14 g of alkalinity as $CaCO_3$ are needed for every 1 g of ammonia (NH_3-N) converted to nitrate (NO_3-N).

Calcium has a +2 charge and *carbonate* has a −2 charge. Because hydrogen has a +1 charge, 1 mole of carbonate ion can absorb 2 moles of hydrogen ions.

$$\frac{2\ \text{moles}\ H^+}{1\ \text{mole}\ NH_3}\left|\frac{1\ \text{mole}\ NH_3}{1\ \text{mole N}}\right|\left|\frac{1\ \text{mole N}}{14\ \text{g N}}\right|\left|\frac{1\ \text{mole}\ CaCO_3}{2\ \text{moles}\ H^+}\right|\left|\frac{100\ \text{g}\ CaCO_3}{1\ \text{mole}\ CaCO_3}\right| = \frac{7.14\ \text{g}\ CaCO_3}{\text{g}\ NH_3 - N}$$

The consumption of alkalinity during nitrification may decrease the pH of the wastewater if there isn't enough alkalinity available to neutralize all of the acid that is produced. Table 9.3 summarizes the typical oxygen and alkalinity relationships of the nitrification process.

Table 9.3 Key Nitrification Relationships

Parameter	Relationship
Oxygen use	4.57 mg O_2 per mg NH_3-N converted to NO_3-N
Alkalinity consumption	7.14 mg alkalinity as $CaCO_3$ per mg NH_3-N converted to NO_3-N
Biomass produced	0.10 mg new cells per mg NH_3-N converted to NO_3-N

Nitrification is a two-step process. The AOBs convert ammonia (NH_3) to nitrite (NO_2^-) and the NOBs convert nitrite (NO_2^-) to nitrate (NO_3^-). The process consumes 4.57 mg O_2 and 7.14 mg alkalinity for every 1 mg NH_3-N converted to NO_3-N. The AOBs and NOBs grow slower than other microorganisms.

TEST YOUR KNOWLEDGE

1. **Nitrification is a three-step process involving two groups of bacteria.**
   - ☐ True
   - ☐ False

2. **The NOB obtain their energy from nitrite and their carbon from alkalinity.**
   - ☐ True
   - ☐ False

3. **Carbonate and bicarbonate ions are two components of alkalinity in wastewater.**
   - ☐ True
   - ☐ False

4. **The AOB grow faster than the NOB when the water temperature is below 25 °C (77 °F).**
   - ☐ True
   - ☐ False

5. **In a well-functioning nitrification process, the effluent nitrite-nitrogen should be less than 1 mg/L as N.**
   - ☐ True
   - ☐ False

6. **The AOB obtain their energy from**
   a. Ammonium ion (NH_4^+)
   b. DO (O_2)
   c. Alkalinity as $CaCO_3$
   d. Ammonia (NH_3)

7. **pH affects the AOB by**
   a. Binding to the cell membrane
   b. Reducing the availability of NH_3
   c. Interfering with oxygen uptake
   d. Increasing alkalinity

8. **At pH 7 and 20 °C (68 °F), the total ammonia concentration was reported as 20 mg/L as N. What is the free or un-ionized ammonia concentration?**
   a. Less than 0.2 mg/L
   b. 10 mg/L
   c. 13 mg/L
   d. 20 mg/L

9. **Nitrifying bacteria convert ammonia to nitrate under**
   a. Anaerobic conditions
   b. Anoxic conditions
   c. Aerobic conditions
   d. Reducing conditions

PROCESS VARIABLES FOR NITRIFICATION

Process variables that are important for nitrification include temperature, aerobic solids retention time (SRT), organic loading rate (OLR), DO, pH, alkalinity, nitrogen loading patterns, and the presence of inhibitory compounds.

WASTEWATER TEMPERATURE

Ammonia oxidizing bacteria and NOB are particularly sensitive to temperature. Nitrification occurs in wastewater temperatures from 4 to 45 °C (39 to 113 °F), with an optimum rate occurring at about 30 °C (86 °F) (U.S. EPA, 1993). However, most WRRFs operate with water temperatures between 10 and 20 °C (50 and 68 °F). The nitrification rate doubles for every 8 to 10 °C increase (14.4 to 18 °F increase) in water temperature.

AEROBIC SOLIDS RETENTION TIME

For activated sludge processes, the most important process control variable for removing ammonia is the aerobic solids retention time ($SRT_{aerobic}$). The $SRT_{aerobic}$ must be long enough to allow the nitrifying bacteria to reproduce. If the $SRT_{aerobic}$ is too short, then any nitrifying bacteria entering the treatment process will be flushed out of the system as quickly as they enter. Nitrification should be thought of as an "all or nothing" operation. Either there will be enough time for the nitrifying bacteria to reproduce and build up a stable population, or there won't be. It isn't possible to grow just enough nitrifying bacteria to produce a final effluent with an ammonia concentration of 10 mg/L, for example. How long the $SRT_{aerobic}$ must be depends on water temperature. As water temperature decreases, the $SRT_{aerobic}$ must be increased to compensate for slower growth rates. Be careful not to increase it too much though as this can encourage the growth of filamentous bacteria, cause foaming in the activated sludge basins, increase the mixed liquor suspended solids (MLSS) concentration, and lead to sludge settleability problems.

For activated sludge processes, the SRT and $SRT_{aerobic}$ should be adjusted gradually as the temperature in the activated sludge basins changes seasonally. Making small changes to match water temperature each week or two will ensure a smooth transition between summer and winter operation. The theoretical minimum $SRT_{aerobic}$ is about 2 days at 20 °C (68 °F) and about 5 days at 10 °C (55 °F) (U.S. EPA, 2010). Water resource recovery facilities should not be operated at the minimum $SRT_{aerobic}$, but should use a safety factor to account for variability in several factors including activated sludge basin configuration, influent nitrogen load, temperature, wasting schedules, and toxic compounds. Safety factors between 2 and 3.5 are typically used; however, the safety factor used at a particular facility should be based on operational experience. A completely mixed activated sludge basin might require a safety factor of 2 or more, whereas a plug flow basin that is broken into multiple compartments may be able to use a smaller safety factor (WEF, 2018). Chapter 8 includes a lengthy discussion on selecting SRT and $SRT_{aerobic}$.

For fixed-film treatment processes like trickling filters and rotating biological contactors (RBCs), operators have limited control over how long the biomass remains in the treatment process. Decreasing distributor arm rotation speed in trickling filters and decreasing the rotational speed in RBCs can reduce biofilm shear, which may increase biofilm thickness and biomass retention time to some degree. Nitrification in fixed-film processes is more dependent on the OLR and nitrogen loading rate than on biomass retention time.

ORGANIC LOADING RATE

High OLRs favor the growth of heterotrophic bacteria over the nitrifying bacteria. In fixed-film systems like trickling filters and RBCs, high OLRs cause the heterotrophic bacteria to grow rapidly. These bacteria overtake and bury the slower growing nitrifying bacteria deeper in the biofilm. Once buried, the nitrifiers aren't able to get enough oxygen and ammonia and their growth rates decrease. In general, the AOB and NOB won't become a significant portion of the biofilm until the soluble BOD_5 is less than 20 mg/L or the five-day CBOD ($CBOD_5$) is less than 20 mg/L (WEF, 2013). Reducing the organic loading slows down the growth of the heterotrophs and prevents them from overtaking the nitrifiers. The same phenomenon is seen in highly loaded activated sludge processes like pure oxygen systems. The heterotrophs outcompete the nitrifying bacteria for oxygen, so little or no nitrification can take place.

For biofilm processes, $CBOD_5$ removal and nitrification are generally sequential. In other words, $CBOD_5$ removal occurs first followed by nitrification. Before nitrification can take place, competition from the heterotrophs for oxygen and space on the media surface must be reduced. This can be accomplished by removing $CBOD_5$ in another process upstream of the fixed-film process or within the fixed-film process itself. The first stage or stages of a fixed-film process can be used to remove $CBOD_5$ followed by ammonia removal in the later stages after the $CBOD_5$ concentration has been reduced. Specifically:

- An OLR of 7.3 to 9.8 g $CBOD_5/m^2 \cdot d$ (1.5 to 2.0 lb $CBOD_5$/d/1000 sq ft) (media surface) will reduce $CBOD_5$ to less than 20 mg/L so nitrification can begin, and
- An ammonia-nitrogen loading rate of 1.0 to 2.0 g TKN/$m^2 \cdot d$ (0.2 to 0.4 lb TKN/d/1000 sq ft) (media surface area) will reduce the ammonia-nitrogen rate to less than 2.0 mg/L NH_3-N (WEF, 2013).

These values are approximations for several different fixed-film processes (U.S. EPA, 1993). The actual performance of any particular system will depend on the type and quantity of media in use, wastewater characteristics, and other important factors, including temperature, DO concentration, and pH.

For activated sludge systems, the $SRT_{aerobic}$ is the most important process control variable for maintaining nitrification. The minimum $SRT_{aerobic}$ needed is strongly temperature dependent and should be adjusted seasonally based on the water temperature in the activated sludge basins. For fixed-film processes, the most important process control variable is the OLR. Ammonia oxidizing bacteria and NOB won't be major components of a biofilm until the soluble BOD_5 is less than 20 mg/L.

DISSOLVED OXYGEN CONCENTRATION

Nitrifiers are obligate aerobes, meaning they are only able to grow under aerobic conditions although they can survive in anaerobic or anoxic conditions. It is generally accepted that nitrification is not limited at DO concentrations greater than 2.0 mg/L. The end of Chapter 5 includes examples of kinetic calculations that demonstrate the effect of DO and ammonia concentrations on nitrifier growth rate.

Multiple studies have shown that the NOB are more sensitive to low DO conditions than the AOB (U.S. EPA, 2010). The heavy black line in Figure 9.8 shows the effect of DO concentration on the effluent nitrite-nitrogen (NO_2-N) concentration when the water temperature is 20 °C (68 °F). Here, a shorter SRT (~5 days) can be used to keep the effluent ammonia-nitrogen (NH_3-N) below 1 mg/L as N. As DO concentrations decrease, nitrite-nitrogen concentrations increase. This indicates that the AOB are unaffected by the lower DO concentration, but the NOB are affected. The NOB are less active and are growing slower than the AOB under these conditions. The thin, dotted line in Figure 9.8 shows the effect of DO on effluent nitrite-nitrogen (NO_2-N) concentrations when the water temperature is 10 °C (55 °F) and a longer sludge age (~12 days) is used. In this case, the NOB seem to be unaffected by the low DO concentration. What actually happens is that the longer SRT causes the AOB activity and growth rate to decrease until it matches or is lower than the NOB. When the AOB and NOB growth rates are matched, nitrite-nitrogen (NO_2-N) does not accumulate.

For activated sludge systems, the DO concentration that begins to depress nitrification rates depends, in part, on the MLSS concentration, mixing, and floc size and density (U.S. EPA, 2010). High MLSS concentrations and/or poor mixing can prevent oxygen from reaching all of the floc particles (uneven distribution). When flocs are larger and denser, oxygen isn't able to penetrate as deeply into individual flocs. It is used up by the bacteria in the outer portions of the floc. In these situations, higher DO concentrations may be needed to keep effluent ammonia-nitrogen (NH_3-N) concentrations below 1 mg/L as N. Lower DO concentrations may be acceptable when the MLSS concentration is lower, the MLSS is well-mixed, and when floc sizes are less dense or smaller. The *half-saturation coefficient* (K_s) for DO in a completely mixed activated sludge process is 0.47 mg/L (Manser et al., 2005). This implies that some activated sludge processes should be able to completely nitrify with DO concentrations as low as 1 mg/L. Operators may wish to experiment with different DO concentrations, gradually increasing and decreasing them while monitoring effluent ammonia concentrations, to determine the optimum DO concentrations for their facilities.

Recall from Chapter 5 that the K_s is the concentration that results in exactly half the maximum growth rate.

PH AND ALKALINITY

Nitrification rates decrease rapidly as the pH drops below 6.8 pH SU and are highest when the pH is between 7.5 and 8.0 SU (U.S. EPA, 2010). Influent wastewater often falls within this range; however, nitrification produces acid. For every 1 mg/L of ammonia-nitrogen (NH_3-N) converted to nitrate-nitrogen

Wastewater at pH 6 contains 10 times more hydrogen ions than wastewater at pH 7.

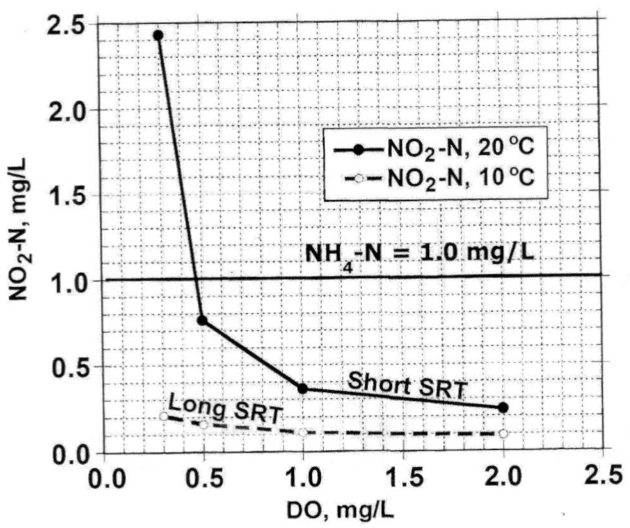

Figure 9.8 Effect of DO on Nitrite Accumulation (U.S. EPA, 2010)

(NO_3-N), 7.14 mg/L of alkalinity as $CaCO_3$ will be consumed. If enough alkalinity is consumed, the pH of the wastewater will decrease. A minimum effluent alkalinity of at least 50 mg/L and preferably 100 mg/L as $CaCO_3$ should be maintained. Below 50 mg/L as $CaCO_3$, the pH may decrease rapidly from around 7 to below 5 SU. Remember, pH is on a log-scale. Increases in hydrogen ion concentration may not change the pH very much. Alkalinity, however, can change substantially. For WRRFs that nitrify, alkalinity and pH measurements are both needed for process control.

Alkalinity primarily comes from the drinking water used in the service area, but is also added to the water as it is used. Anaerobic breakdown of BOD_5 in the collection system also generates alkalinity. Alkalinity in water supplies varies between 50 and 400 mg/L. Communities that rely on well water tend to have higher alkalinity concentrations in their influent than communities that use reservoirs, lakes, and rivers as their drinking water supply. Many WRRFs must add supplemental alkalinity to support nitrification.

Low pH is problematic because it reduces the percentage of ammonia (NH_3) available and increases the percentage of ammonium ion (NH_4^+). It becomes more difficult for the AOB to obtain their preferred energy source. Secondly, the nitrifying bacteria rely on inorganic carbon—alkalinity—as their carbon source. Systems that operate consistently at pH values between 6.5 and 7 can acclimate or get used to low pH conditions. It can sometimes be better to operate at a slightly depressed pH all the time than to add chemicals for pH adjustment that cause the pH to fluctuate.

Denitrification converts nitrate to nitrogen gas and generates alkalinity. Facilities can often reduce or eliminate chemical addition for pH control simply by denitrifying.

Example Calculation: The influent to a WRRF contains 32 mg/L of ammonia-nitrogen (NH_3-N) and 250 mg/L of total alkalinity as $CaCO_3$. The total TKN is unknown, but is assumed to be 30% of the total nitrogen in the influent. Assume the organic nitrogen will end up in the waste sludge and does not need to be converted to nitrate-nitrogen (NO_3-N). How much alkalinity is needed to completely nitrify? Is there enough alkalinity in the influent wastewater or will supplemental alkalinity be needed?

Solution: For every milligram per liter of NH_3-N converted to NO_3-N, 7.14 mg/L of alkalinity as $CaCO_3$ is needed. Additional alkalinity, at least 50 mg/L as $CaCO_3$, is needed to prevent a pH drop.

$$\left.\frac{32\frac{mg}{L}NH_3-N}{}\right|\left.\frac{7.14\frac{mg}{L}\text{alkalinity as }CaCO_3}{1\frac{mg}{L}NH_3-N}\right| = 228.48\frac{mg}{L}\text{alkalinity as }CaCO_3$$

Alkalinity needed for nitrification = 228.48 mg/L as $CaCO_3$

Alkalinity needed in final effluent = 50 mg/L as $CaCO_3$

Total alkalinity needed = 278.48 mg/L as $CaCO_3$

The operator will need to add supplemental alkalinity. Operators should refer to Appendix G in Chapter 20 of *Operation of Water Resource Recovery Facilities* (MOP 11; WEF, 2016) for an example calculation for the amount of $CaCO_3$ as solid to add to the wastewater.

NITROGEN LOADING PATTERN

Provided that environmental conditions (pH, DO, alkalinity) do not limit the growth of the nitrifying bacteria, the quantity or mass of AOB and NOB that grow in the system will be a function of the ammonia load. It is important to understand that the nitrifying bacteria cannot store ammonia or nitrite internally (WEF, 2013). Heterotrophic bacteria store carbon and energy as glycogen, poly-β-hydroxybutyrate, and other compounds. This helps them survive and continue growing when BOD_5 is not immediately available. The nitrifying bacteria can only grow and reproduce when ammonia and oxygen are both available. At the same time, how quickly they can use ammonia and oxygen is limited by their maximum growth rate. If they are already growing at their maximum rate, the extra ammonia can't be used. To take advantage of *transient* increases in ammonia loads, the nitrifiers must first increase their numbers, which they do very slowly. As a result, the nitrifying bacteria are poorly equipped to take advantage of peak-hour ammonia loading. Remember: it isn't possible to grow more bacteria without load. The numbers of nitrifying

Conditions that change rapidly are "*transient*".

bacteria in the treatment process will be representative of something closer to the average ammonia loading condition rather than the peak-loading condition. When high concentrations of ammonia pass through the process for short periods of time, the nitrifiers can't grow quickly enough to take advantage of it.

Ammonia interacts with chlorine during disinfection and can interfere with disinfection efficiency. These interactions are described in the disinfection chapter.

Recall from Chapter 2 that the influent flows and loads for domestic WRRFs follow a strong diurnal pattern (Figure 9.9). Flow and loads decrease at night when people are sleeping and increase during the day when they are at work or play. Influent ammonia concentrations follow a similar pattern. Figure 9.9 shows 3 days of influent flows and ammonia concentrations collected at a small WRRF. These data were used to calculate the hourly and average daily ammonia loads in kilograms per day (pounds per day) (Figure 9.10). The flat blue line has been arbitrarily set at 130% of the average daily ammonia load for this example. Between 9:00 a.m. and 2:00 p.m., the hourly ammonia load is greater than 130% of the average daily ammonia load. Excess ammonia may pass through to the final effluent during this time. Grab samples collected from the effluent may contain several milligrams per liter of NH_3-N, whereas samples collected at other times of the day would show less than 1 mg/L of NH_3-N.

Some WRRFs are equipped with flow equalization basins that help even out flows and loads to the treatment process. Influent flows and loads may also be equalized within the activated sludge basins. For example, an oxidation ditch with a hydraulic detention time (HDT) of 16 to 24 hours dampens the effect of peak-hour loads. This is one reason oxidation ditches and other extended aeration activated sludge processes are able to consistently produce extremely low effluent ammonia concentrations, often below 0.2 mg/L as N. Activated sludge processes with short detention times and some fixed-film processes are more likely to produce higher effluent ammonia concentrations during peak ammonia loading.

HYDRAULIC DETENTION TIME

Hydraulic detention time and *hydraulic residence time* (HRT) are two terms used to describe the amount of time it takes to fill a vessel, drain a vessel, or the average amount of time water spends within a vessel. The Association of Boards of Certification (ABC) uses HDT in their formula sheets and operator certification exams. Water Environment Federation manuals and engineering texts use HRT.

Hydraulic detention time in activated sludge processes does not have a big impact on nitrification. Although most activated sludge basins are designed for an HDT of at least 6 hours, it has been proven that it is possible to achieve effluent ammonia concentrations below 1 mg/L as N with an HDT as short as 3 hours (WEF, 2018). As long as the $SRT_{aerobic}$ is long enough and other environmental conditions (low DO, pH, alkalinity) aren't limiting growth, the HDT has little effect. Keep in mind that increasing or decreasing the return activated sludge (RAS) flowrate does not change the overall treatment time (Chapter 8). Increasing the RAS flowrate will affect the amount of time per pass, but the wastewater and the nitrifiers will spend the same amount of time overall under aerobic conditions when the RAS flowrate is high as when it is low.

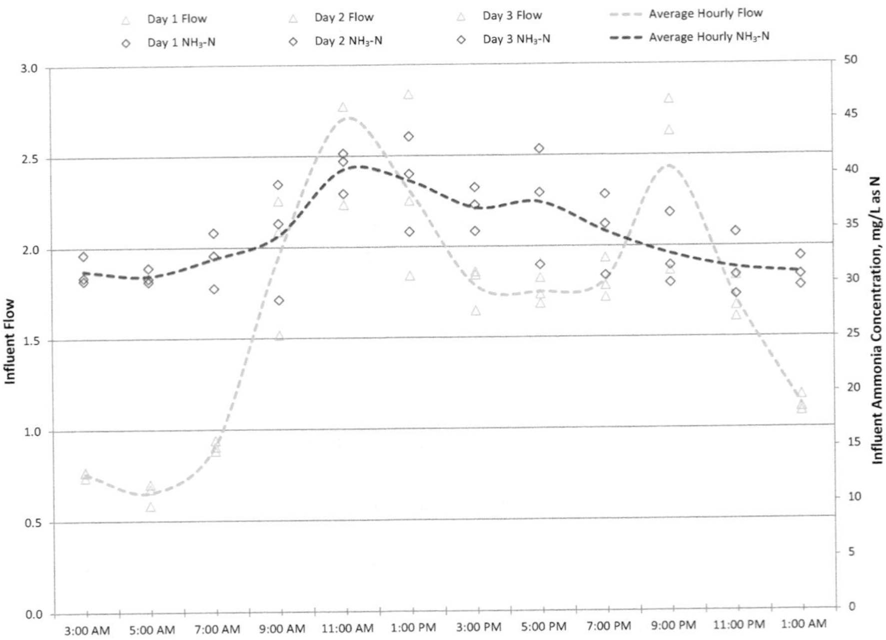

Figure 9.9 Diurnal Flow and Ammonia-Nitrogen Concentrations (Reprinted with permission by Indigo Water Group)

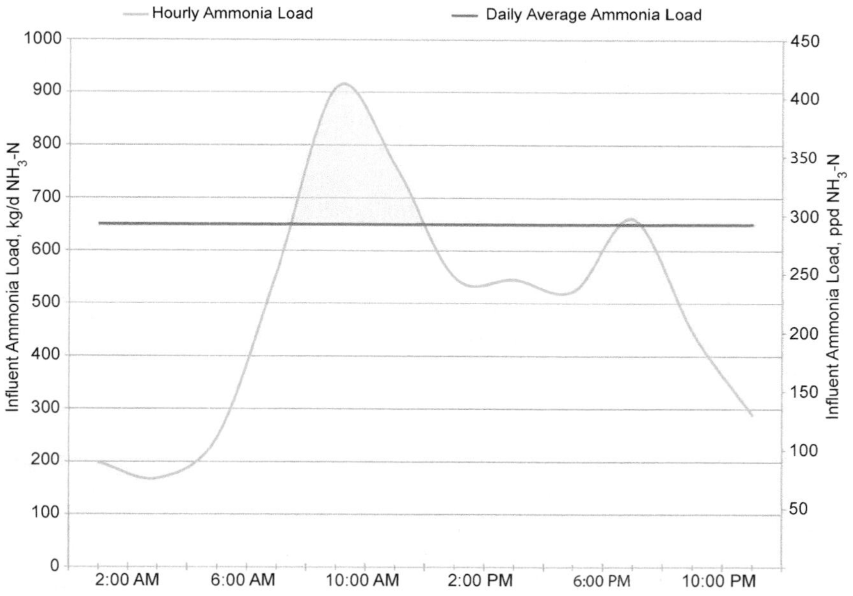

Figure 9.10 Diurnal Ammonia Load, kg/d (ppd) (Reprinted with permission by Indigo Water Group)

If deep blankets are allowed to build in the secondary clarifier, it can reduce both the aerobic HDT and the $SRT_{aerobic}$. Keep blanket depths below 0.7 m (2 ft) for best results.

Hydraulic detention time can become a factor in step-feed activated sludge processes. With step feed, the wastewater is introduced at several locations along the length of the activated sludge basin (Figure 9.11). When influent wastewater or primary effluent is introduced toward the end of the activated sludge basin, detention times may decrease to an hour or less for that portion of the flow. Effluent ammonia concentrations may increase as a result.

INHIBITORY SUBSTANCES

Nitrifiers are particularly susceptible to inhibition from a variety of organic and inorganic substances. They are particularly susceptible to wide fluctuations in the concentration of inhibitory substances, but

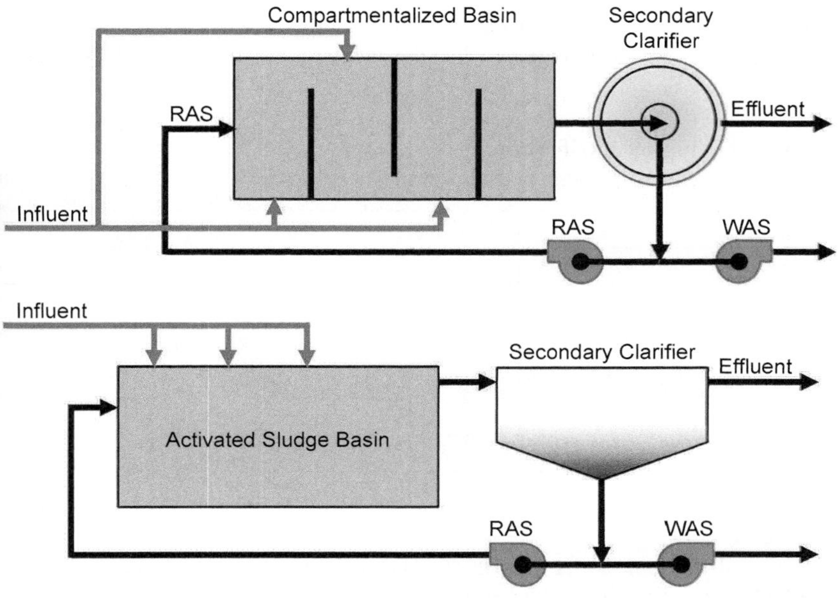

Figure 9.11 Step-Feed Activated Sludge (Reprinted with permission by Indigo Water Group)

may exhibit only minor effects if these substances are present in low concentrations that are consistently applied to the system. More than 180 organic compounds and heavy metals have been shown to inhibit nitrification, a few of which are listed in Table 9.4. A sudden loss of nitrification that occurs without a process control change could indicate the presence of a toxic or inhibitory substance.

Nitrification can be inhibited when the DO concentration falls below 2 mg/L. Nitrite oxidizing bacteria are affected more by low DO than the AOB, which can cause nitrite to accumulate in some situations. Alkalinity is needed for nitrifier growth and to neutralize acid. Excess alkalinity of 50 to 100 mg/L is needed to prevent pH drop. Peak ammonia loads cannot always be fully processed by the AOB and NOB because of their slow growth rates.

PROCESS CONTROL FOR NITRIFICATION

Process control for nitrification consists of maintaining the biomass in the treatment process long enough to build a stable population of nitrifiers, providing the resources (ammonia, oxygen, and alkalinity) the nitrifying bacteria need to live and reproduce, reducing competition from heterotrophic bacteria, and avoiding inhospitable conditions like low pH. The AOB and NOB are constantly entering the treatment process with the influent wastewater. Operators need only to provide an environment that encourages their growth.

- For activated sludge systems: selecting an $SRT_{aerobic}$ based on water temperature;
- For fixed-film processes: minimizing organic loading rates;
- Monitoring the influent and effluent for the following parameters:
  - Ammonia,
  - Nitrite,
  - Nitrate,
  - Alkalinity,
  - pH; and
- Monitoring within the process for DO and pH. Adjust as needed.

MANAGING RECYCLE STREAMS

Sludge from primary and secondary clarifiers is typically sent to the solids handling side of the WRRF for additional processing. This may include thickening, digestion, and/or dewatering. Sludge digestion breaks down the organic material and releases ammonia and phosphorus. In aerobic digestion, the nitrifying bacteria will convert the ammonia to nitrate within the digester. The nitrate can then be converted to nitrogen gas by facultative bacteria in the aerobic digester when the digester is not being aerated. Typically,

Table 9.4 Substances Reported to Inhibit Nitrification

Substance	Inhibitory Concentration	Substance	Inhibitory Concentration
Nickel[a]	0.25 mg/L	Methanol	160 mg/L
Chromium[a]	0.25 mg/L	Sulphides	5 mg/L
Copper[a]	0.10 mg/L	2-Bromophenol	0.35
Ammonia[b]	1000 mg/L as N for AOB 20 mg/L as N for NOB	3-Clorophenol	0.20
Cadmium	14.3 mg/L	Flavonoids	0.01
Chlorine	1 mg/L	Tannin	0.01
Hydrogen sulfide	50 mg/L	Thiourea	1

[a]Metcalf and Eddy, Inc./AECOM (2014)
[b]Inhibitory concentrations of total ammonia at pH 7 and 20 °C. The un-ionized form (NH_3) causes toxicity. The percentage of NH_3 and NH_4^+ is dependent on pH and temperature (U.S. EPA, 1993).

the amount of nitrogen returned from aerobic digestion and sludge dewatering is low. In anaerobic digestion, ammonia accumulates as more and more sludge is broken down. Without oxygen, ammonia can't be converted to nitrate. It is not unusual for the supernatant of an anaerobic digester to contain 800 to 2000 mg/L of NH_3-N (WEF et al., 2018). Anaerobic digester supernatant and water removed during thickening and dewatering operations that is returned to the liquid stream side of the WRRF can increase the ammonia load to the liquid side of the WRRF by up to 50% (WEF et al., 2018).

Water returned from centrifuges is called *centrate* and water returned from filter presses is called *filtrate.*

Care should be taken to avoid excessive loads from sidestreams that overwhelm the secondary treatment process. Recall that the nitrifying bacteria are limited in how much ammonia they can convert to nitrate. Excess ammonia will pass through to the final effluent, potentially causing a discharge permit violation. Aeration systems are also limited in how much oxygen they can provide. Without adequate oxygen, the nitrifiers will be inhibited and unable to convert ammonia at their maximum rate. In addition, the alkalinity in anaerobic digester supernatant is typically in the range of 2500 to 5000 mg/L as $CaCO_3$ (WEF, 2016). Recall that 1 mg/L of NH_3-N needs 7.14 mg/L of alkalinity as $CaCO_3$. High ammonia return streams often do not contain enough alkalinity to process the ammonia load. Supplemental alkalinity may be needed.

High ammonia sidestreams should be returned gradually to distribute the load and prevent spikes in the effluent ammonia concentration. Ideally, sidestreams should be managed to return ammonia when the influent ammonia loads are the lowest. Look back at Figure 9.10. Returning the sidestream ammonia load when the influent load is lowest will 1) flatten out the diurnal loading pattern, 2) raise the average, 3) increase the nitrifying bacteria population, and 4) reduce the likelihood of ammonia passing through to the final effluent. For most facilities, this means high ammonia sidestreams should be returned at night when many WRRFs are lightly staffed or unstaffed. Thickening and dewatering tend to be done from mid-morning to late afternoon when the influent ammonia load to the WRRF is typically highest. To solve the scheduling problem, many facilities are equipped with holding tanks that collect digester supernatant and other sidestreams during the day. Water is pumped from the holding tank to a point upstream of the secondary treatment process. The water level in the holding tank will increase and decrease over a 24-hour period.

NITRITE LOCK

Under certain environmental conditions, such as a poorly established or inhibited NOB population, nitrite (NO_2^-) concentrations will increase (Muirhead and Appleton, 2008). A poorly established NOB population may occur for a brief period during startup of a new facility, but can also occur in facilities that move into and out of nitrification. This is often seen in activated sludge processes where the sludge age fluctuates over and under the minimum needed to maintain nitrification. Nitrite oxidizing bacteria are more sensitive to environmental conditions than the AOB. Nitrite lock may be caused by low DO, low alkalinity or low pH, high temperature, and the presence of toxic or inhibitory substances (Muirhead and Appleton, 2008).

Nitrite lock is problematic for two reasons: nitrite is extremely toxic to fish and nitrite reacts with chlorine and hypochlorite. Chlorine gas (Cl_2) and sodium hypochlorite (NaHOCl) react with nitrite (NO_2^-), converting it to nitrate (NO_3^-). As shown below in eqs 9.4 and 9.5, 1 mg/L of nitrite-nitrogen (NO_2-N) entering a chlorine disinfection process can consume 5 mg/L of chlorine residual. If operators are not aware that nitrite lock is occurring, there may not be enough chlorine residual to adequately disinfect.

$$Cl_2 + H_2O \rightarrow HOCl + HCl \tag{9.4}$$

$$HOCl + NO_2^- \rightarrow NO_3^- + HCl \tag{9.5}$$

$$\left.\frac{1\text{ mole }Cl_2}{1\text{ mole HOCl}}\right|\left.\frac{1\text{ mole HOCl}}{1\text{ mole }NO_2^-}\right\|\left.\frac{1\text{ mole }NO_2^-}{1\text{ mole N}}\right\|\left.\frac{1\text{ mole N}}{14\text{ g N}}\right\|\left.\frac{70\text{ g }Cl_2}{1\text{ mole }Cl_2}\right| = 5.07\frac{\text{g }Cl_2}{\text{g }NO_2-\text{N}}$$

When chlorine or sodium hypochlorite are added to water, hypochlorous acid (HOCl) is produced. Hypochlorous acid breaks down further to hypochlorite ion (OCl^-). The percentage of hypochlorous acid and hypochlorite ion formed is dependent on temperature and pH.

ALKALINITY ADDITION

To avoid pH reduction, a residual alkalinity of at least 50 mg/L as $CaCO_3$ is needed. When alkalinity supplementation is required, there are several chemicals that may be used (Table 9.5). Which chemical is used at a particular WRRF depends on a number of criteria, including availability, cost, storage and feed requirements, safety considerations, and operator preference. Sodium hydroxide, calcium hydroxide, magnesium hydroxide, and lime are strong bases with pH higher than 10. They can easily cause chemical burns to eyes and skin and burn through clothing. Handling these chemicals requires personal protective equipment and training.

Sodium Carbonate

Sodium carbonate or soda ash is typically available as a dry product; therefore, it must be dissolved on-site. Dry chemical feeders may be used to add sodium carbonate wherever the wastewater is well-mixed. The maximum pH level of a sodium carbonate solution is approximately 12.

Sodium Bicarbonate

Sodium bicarbonate use is not widespread because other chemicals are easier to handle and cost less. It is, however, one of the few chemicals available that can increase alkalinity without causing large increases in pH. Sodium bicarbonate is typically available as a dry product and must be dissolved on-site. Addition is similar to sodium carbonate. The maximum pH level of a sodium bicarbonate solution is 8.3. Accidental overdosing is unlikely to cause a discharge permit violation for pH or shock the microorganisms in a secondary treatment processes.

Sodium Hydroxide

Sodium hydroxide, commonly called *caustic* or *caustic soda*, is used for alkalinity supplementation because of its ease of handling. Sodium hydroxide is delivered as a liquid and can be added with a chemical metering pump. Sodium hydroxide is often not the lowest-cost chemical for alkalinity addition; however, it is easy to use and annual maintenance costs for caustic storage and feed systems are lower than for other chemicals. Many utility operators believe that the ease of handling of sodium hydroxide far outweighs the higher cost of the chemical. Sodium hydroxide may be purchased at a 50%, by weight, solution strength or, in some locations, it is available at 20 or 25%, by weight, solution strengths. It is classified as a strong base and can raise the pH much higher than expected when overdosed. An automatic system to control chemical dose based on pH can be used to prevent overdosing and optimize chemical usage. Once the liquid temperature drops below 12.8 °C (55 °F), sodium hydroxide will begin to *crystallize* out of solution. *Heat-tracing systems* may be used on the storage tank and piping system components to prevent crystallization. It is difficult to redissolve crystallized sodium hydroxide. If sodium hydroxide is delivered to the site at 50% strength and then diluted on-site with facility water or potable water, *scaling* will occur at the point of mixing. The local pH will rise well above pH 10 and calcium carbonate will form immediately.

Most chemicals are more soluble in warm liquids than in cold. As the temperature of the liquid decreases, the amount of dissolved chemical that can remain dissolved also decreases. The excess chemical precipitates or *crystallizes* and appears as chunks of solid chemical.

Heat-tracing systems use electricity with a wiring system that is wrapped around piping and containers to help maintain a minimum desired temperature. They are similar in function to an electric blanket.

Scaling is a buildup of calcium carbonate ($CaCO_3$) on pipes and other surfaces. The pH change caused by adding sodium hydroxide (NaOH) reduces the $CaCO_3$ solubility and causes it to precipitate.

Quicklime

Before being used, calcium oxide, or quicklime, must be slaked. Slaking is a process where the calcium oxide is mixed with water and allowed to react. Slaker operating temperatures are typically between 120 and 180 °C (248 and 356 °F). The elevated temperatures increase the amount of scaling that can form. Quicklime is typically cheaper than hydrated lime. However, slaking is a labor-intensive process.

Lime *slaking* refers to mixing calcium oxide (CaO) with water to form calcium hydroxide [$Ca(OH)_2$]. Slaking generates a great quantity of heat.

Table 9.5 Chemicals Commonly Used to Increase Alkalinity

Chemical	Common Name	Chemical Formula	mg $CaCO_3$ Alkalinity Added per mg Chemical
Sodium bicarbonate	Baking soda	$NaHCO_3$	0.6
Sodium carbonate	Soda ash	Na_2CO_3	0.9
Calcium magnesium dicarbonate	Dolomite	$CaMg(CO_3)_2$	1.1
Sodium hydroxide	Caustic/caustic soda	NaOH	1.3
Calcium hydroxide	Slaked lime	$Ca(OH)_2$	1.4
Magnesium hydroxide	Milk of magnesia/brucite	$Mg(OH)_2$	1.7
Calcium oxide	Quicklime	CaO	1.8

Calcium Hydroxide

Calcium hydroxide, or hydrated lime, has been *slaked* by the manufacturer and is sold as a dry material. Hydrated lime must be slurried before use. The *slurry* tank may be susceptible to scaling depending on the hardness of the makeup water used. Either *potable* or non-potable water may be used for slaking lime. Hard water will cause more scaling and precipitation than soft water. Calcium hydroxide is typically prepared in a 3 to 5%, by weight, slurry. A small fraction of the calcium hydroxide will dissolve and raise the pH to approximately 12, which is high enough to be dangerous to operators. Calcium hydroxide should be added at a point of turbulence to ensure sufficient mixing.

A *slurry* is a mixture of fine particles in water. Lime slurry contains undissolved lime particles and has a consistency like fine mud.

Potable means drinkable. Potable water is drinking water. Non-potable water may be final effluent or drinking water that was previously used elsewhere in the facility and is no longer drinkable.

Magnesium Hydroxide

Magnesium hydroxide is increasingly being used in biological nutrient removal (BNR) processes because it will not raise the pH above approximately 10.5, which correlates to its precipitation point range of pH 10.2 to 10.5. Magnesium hydroxide also dissolves more slowly than lime and can act as time-release chemical addition. However, adding magnesium compounds to facilities with anaerobic digesters can cause *struvite* formation in the anaerobic digesters and their associated equipment and piping systems as well as in solids dewatering equipment. Magnesium hydroxide is generally sold as a slurry, although a dry product can also be purchased. Scaling will begin above pH 8.

Struvite is the common name for magnesium ammonium phosphate ($MgNH_4PO_4$). Struvite crystallizes and forms scale within anaerobic digesters, dewatering equipment, and associated equipment. Magnesium is usually the missing or limiting ingredient for struvite formation.

1) High ammonia recycle streams must be managed to prevent ammonia from passing through to the final effluent. Ideally, return high ammonia loads gradually during low flow. 2) Increasing nitrite levels (nitrite lock) occurs when there aren't enough NOB or the NOB are inhibited in some way. Nitrite lock interferes with chlorine disinfection. 3) Alkalinity addition may be needed to prevent pH drop.

TEST YOUR KNOWLEDGE

1. **The operator of an activated sludge process knows from past experience that the water temperature in the aeration basin will drop from 22 °C (72 °F) to 15 °C (59 °F) between August and December. What process change should be made?**
   a. Decrease the concentration with water temperature
   b. Increase the $SRT_{aerobic}$ gradually based on water temperature
   c. Increase the SRT and MLSS concentration to build biomass before November
   d. Wait until mid-October and then double the SRT

2. **The minimum $SRT_{aerobic}$ needed for nitrification at 10 °C (50 °F) is 5 days. If a safety factor of 2.5 is used, what is the target $SRT_{aerobic}$?**
   a. 2.5 days
   b. 5 days
   c. 12.5 days
   d. 25 days

3. **An RBC process consists of four wheels operated in series. Where in the process is nitrification most likely to take place?**
   a. First wheel
   b. Second wheel
   c. Third wheel
   d. Fourth wheel

4. **A trickling filter receives secondary effluent that normally contains less than 15 mg/L of soluble BOD_5 and TSS. The trickling filter effluent normally contains less than 1 mg/L of NH_3-N. A process upset causes settled solids to escape the clarifier and enter the trickling filter. Which of the following is most likely to occur?**
   a. Effluent ammonia concentrations will be unaffected.
   b. Effluent ammonia concentrations will increase.
   c. Effluent alkalinity concentrations will decrease.
   d. Effluent alkalinity concentrations will be unaffected.

5. **A trickling filter receives dilute primary effluent containing 75 mg/L BOD_5 and 22 mg/L of NH_3-N. The effluent from the trickling filter contains 5 mg/L BOD_5 and <1 mg/L NH_3-N. Where in the trickling filter did nitrification begin?**
   a. In the outer, aerobic layers of the biofilm
   b. Where DO concentrations reached 7 mg/L
   c. Where soluble BOD_5 decreased to <15 mg/L
   d. About two-thirds from the top of the media bed

6. **An activated sludge process with an SRT of 6 days receives an abnormally high ammonia load. The blowers are unable to deliver enough air and the DO concentration in the basins drops to 1 mg/L. Which of the following may occur next?**
   a. Nitrifying bacteria will die.
   b. Effluent ammonia will decrease.

c. Average floc size will increase.
d. Chlorine consumption will increase.

7. **How much alkalinity will be needed to convert 20 mg/L ammonia-nitrogen (NH_3-N) to nitrate-nitrogen (NO_3-N)? There must be 100 mg/L of alkalinity as $CaCO_3$ remaining in the final effluent.**
   a. 91 mg/L
   b. 143 mg/L
   c. 192 mg/L
   d. 243 mg/L

8. **When supernatant from an anaerobic digester is returned to the activated sludge process in the middle of the day**
   a. Effluent ammonia concentrations may increase.
   b. Alkalinity addition will be required to prevent pH drop.
   c. Nitrite-lock will occur and increase chlorine demand.
   d. Nitrifier growth rates will increase to meet the load.

9. **This process control parameter is the least important in sustaining nitrification in activated sludge processes.**
   a. Temperature
   b. SRT
   c. DO
   d. HDT

10. **An activated sludge facility operates with a 25-day SRT in the winter. As soon as the water temperature drops below 16 °C, the SVI increases and foam accumulates on the activated sludge basins. The final effluent contains less than 1 mg/L NH_3-N. The chief facility operator begins to chlorinate the RAS to control filaments and improve settling. After 3 days, the effluent NH_3-N concentration increases to 22 mg/L. What must be true?**
    a. Chlorine concentration in the RAS too high
    b. $SRT_{aerobic}$ is at least 50% of the total SRT
    c. SRT is not long enough to support nitrification
    d. Oxygen levels were depleted by chlorine addition

SECTION EXERCISE

This exercise includes several short troubleshooting scenarios. Using what you know about nitrification, determine the cause of the problem, and suggest a solution. Write down your answer before looking at the answer key.

Scenario #1—A small activated sludge system was left unattended over a long weekend. The process was running well and the operators thought it would be fine for a few days. The process is almost all manually controlled and requires operations staff to manually turn pumps and blowers on and off. A computer system would send out an alert if a piece of equipment stopped running. After designating an on-call person to respond to alarms, they went home for the long weekend. When the operators returned on Tuesday, all of the equipment was functioning correctly, but not everything was as it should be. The first thing they noticed is that all of their blowers were running at maximum capacity, but the DO in the aerated activated sludge basins was barely above 1 mg/L. There is floating sludge in the secondary clarifiers and a settleometer test pops after only 15 minutes. The only piece of information you have at the moment is that the WRRF is only designed to remove BOD_5 and TSS, not ammonia. What happened while the operators were away? What is causing the low DO in the aerated basins and what should the operators do to fix the problem?

Scenario #2—A trickling filter removes BOD_5 in the upper half and converts ammonia to nitrate in the lower half. For the past 2 days, the influent flow has been higher than normal and the ammonia concentration in the trickling filter effluent has increased. The operator checks the pH, alkalinity, and DO. The pH is 7.4 and alkalinity is 120 mg/L as $CaCO_3$ in the filter effluent. Dissolved oxygen is 7.8 mg/L. What should the operator check next and why?

Scenario #3—A ski resort has a small activated sludge treatment process that is designed to nitrify, but not denitrify. The facility is outdoors and temperatures often drop to subzero temperatures. To keep the autosampler and autosampler intake line from freezing, the operators wrapped the intake line and the sample carboy with an electric blanket. This makes it difficult to pull the sample carboy in and out of the autosampler, so the autosampler is rarely cleaned. The operators learned in a recent training class that lower sludge ages result in lower operational costs. Under pressure from management to reduce costs, they begin reducing their sludge age. Eventually, the sludge age reaches 3 days. Even though the water is only 12 °C (53.6 °F), the composite samples from the autosampler are still showing $<$1 mg/L ammonia-nitrogen (NH_3-N) in the final effluent. How can this be? What do you know must be true? Can a facility with cold water nitrify completely with a sludge age of only 3 days? Why are the samples from the autosampler showing $<$1 mg/L NH_3-N? Perhaps it is the laboratory? How could the operator check the laboratory results?

Scenario #4—A WRRF sends the waste sludge from their activated sludge process to an aerobic digester. Aerobic digesters are very similar to extended aeration activated sludge processes except that they don't receive influent or primary effluent. To survive, the heterotrophic bacteria consume their own cell mass in a process called *endogenous respiration*. Heterotrophic bacteria that run out of stored food eventually break open and become food for the surviving bacteria in the process. As organic matter in the digester breaks down, it releases ammonia and phosphorus. Historically, the DO in the aerobic digester was kept near 0.5 mg/L. In an attempt to get better sludge destruction, the operators increased the DO concentration to 2 mg/L. After about a week, the pH in the digester fell to 5.2. Why did this happen?

Scenario #5—A trickling filter facility has watched anxiously as their effluent ammonia concentration has gradually crept up over the last 18 months from <1 mg/L NH_3-N to more than 6 mg/L NH_3-N. The service area consists of homes and small businesses, but does not have any manufacturing or other industrial discharges. The population is stable. In fact, the BOD_5 load to the facility has not changed in the last 3 years and influent flow has actually decreased by almost 30%. The effluent from the trickling filter has a pH of 6.7 and a DO concentration of 8.2 mg/L. What should the operator check next? If your guess is correct, what corrective action should the facility take?

SECTION EXERCISE SOLUTIONS

Scenario #1—With no one there to waste, the sludge age increased over the 3-day weekend. The activated sludge process began to nitrify, which increased the oxygen demand. Because the facility was not designed for nitrification, the blowers were not large enough to supply as much oxygen needed. This caused the DO to decrease. To fix the problem, the operators should lower the sludge age by increasing the wasting rate.

Scenario #2—The operator should check the influent wastewater for two things: BOD_5 load and inhibitory compounds. If the BOD_5 load is increasing, then the soluble BOD_5 won't be less than 25 mg/L until further down in the trickling filter. This leaves less space on the media for the nitrifying bacteria. Inhibitory compounds like heavy metals and organics could also be inhibiting nitrification. The operator should collect a sample and send it for analysis.

Scenario #3—What is true: the nitrifying bacteria grow slowly and need longer ages when the water is cold. Are there any places where the wastewater is warm? Yes, in the autosampler carboy and tubing. This facility can't be nitrifying, but the autosampler could be. To check the laboratory, collect grab samples from the influent and effluent of the WRRF and send them for analysis. In this true story, the autosampler had developed a nitrifying biofilm that was converting ammonia to nitrate inside the sample carboy and tubing.

Scenario #4—The nitrifying bacteria don't need BOD_5 to convert ammonia to nitrate. As the solids in the digester break down, more and more ammonia is released. Previously, the low DO concentration prevented the nitrifying bacteria from converting very much of the ammonia to nitrate. When the DO was increased to 2 mg/L, the nitrifying bacteria were able to grow at a faster rate. The acid produced from nitrification lowered the pH.

Scenario #5—The operator should check the effluent alkalinity and check the influent for inhibitory compounds. In this case, the effluent alkalinity had dropped to only 35 mg/L as $CaCO_3$. The influent BOD_5 and ammonia load remained the same, but decreased water usage in the service area means less alkalinity is available. To correct the problem, the WRRF should add alkalinity to the trickling filter influent.

Biological Denitrification

Denitrification is the conversion of nitrate to nitrogen gas. Recall from Chapter 5 that facultative heterotrophic bacteria are capable of using nitrate as a substitute for oxygen. When BOD_5 is processed using oxygen, the waste products are carbon dioxide (CO_2) and water (H_2O). When BOD_5 is processed using nitrate, the waste products are nitrogen gas (N_2), carbon dioxide (CO_2), and water (H_2O). Because nitrogen gas is not very soluble in water, bubbles of the gas form in the wastewater. They travel to the water surface and are released to the atmosphere. The atmosphere is naturally 78% nitrogen, so the nitrogen gas (N_2) released from the treatment process does not harm the environment. Denitrification in secondary clarifiers can carry settled sludge in the clarifier blanket back to the surface of the clarifiers.

Many different species of bacteria are capable of denitrification. For denitrification to take place, three conditions must be met. 1) Organic carbon must be available. This is typically measured as BOD_5 or COD. Generally, the higher the BOD_5 concentration, the faster denitrification will occur. 2) Dissolved oxygen concentrations must be low. Oxygen is more efficient for the bacteria than nitrate. For this reason, if DO is available, denitrification won't occur. The bacteria will continue to use oxygen until it runs out. Dissolved oxygen concentrations as low as 0.3 mg/L can prevent denitrification (U.S. EPA, 1993). 3) There must be sufficient HDT for denitrification to take place.

DENITRIFICATION STOICHIOMETRY

The overall reaction depends on the carbon source used. The stoichiometry for three different carbon sources is shown in eqs 9.6 through 9.8 (Metcalf and Eddy, Inc./AECOM, 2014). Note that cell growth is not included in these simplified reactions.

Wastewater ($C_{10}H_{19}O_3N$)

$$C_{10}H_{19}O_3N + 10NO_3^- \rightarrow 5N_2 + 10CO_2 + 3H_2O + NH_3 + 10OH^- \quad (9.6)$$

Methanol (CH_3OH)

$$5CH_3OH + 6NO_3^- \rightarrow 3N_2 + 5CO_2 + 7H_2O + 6OH^- \quad (9.7)$$

Acetate (CH_3COOH)

$$5CH_3COOH + 8NO_3^- \rightarrow 4N_2 + 10CO_2 + 6H_2O + 8OH^- \quad (9.8)$$

Even though all three of the carbon sources are different chemically, 1 mole of hydroxide ion (OH^-) is produced for every 1 mole of nitrate ion (NO_3^-) converted to nitrogen gas (N_2). Hydroxide is a base and another component of alkalinity. One mole of hydroxide ions (OH^-) neutralizes 1 mole of hydrogen ions (H^+). Nitrification produced 2 moles of H^+, which consumed 7.14 mg/L of alkalinity as $CaCO_3$ for every 1 mg/L of ammonia-nitrogen (NH_3-N). Denitrification can neutralize half of the H^+ produced and recover half of the alkalinity or 3.57 mg/L of alkalinity for every 1 mg/L of nitrate-nitrogen (NO_3-N) converted to nitrogen gas (N_2). The stoichiometry can be verified by converting moles of hydroxide ion into moles of alkalinity as $CaCO_3$ as shown below. Recovering lost alkalinity through denitrification can often reduce or eliminate the need for alkalinity addition. Not all of the alkalinity consumed during nitrification can be regained because denitrification only produces half of the alkalinity consumed by nitrification.

The +2 charge on the calcium ion reacts with the −1 charge on the hydroxide ion. One mole of calcium reacts with two moles of hydroxide ion to neutralize the charge.

$$\frac{1 \text{ mole } OH^-}{1 \text{ mole } NO_3^-} \left| \frac{1 \text{ mole } CaCO_3}{2 \text{ moles } OH^-} \right| \left| \frac{100 \text{ g } CaCO_3}{1 \text{ mole } CaCO_3} \right| \left| \frac{1 \text{ mole } NO_3^-}{1 \text{ mole N}} \right| \left| \frac{1 \text{ mole N}}{14 \text{ g N}} \right| = \frac{3.57 \text{ g } CaCO_3}{\text{g } NO_3 - N}$$

Nitrate is taking the place of oxygen in each of the chemical reactions (eqs 9.6, 9.7, and 9.8). Facilities that denitrify recover some of the oxygen used for nitrification. Every 1 mg/L nitrate-nitrogen (NO_3-N) denitrified is equivalent to 2.86 mg/L of DO. This is more than 60% of the oxygen needed for nitrification. Similarly, every 1 mg/L nitrite-nitrogen (NO_2-N) is equivalent to 1.73 mg/L of DO. Denitrifying, even when not required by the discharge permit, can reduce overall energy costs.

Denitrification also results in new bacterial cells. The cell yield depends on the carbon source. Some carbon sources contain more energy than others. For example, if methanol is the carbon source, then the cell yield is about 0.5 mg VSS per 1.0 mg of nitrate-nitrogen (NO_3-N) removed. Put another way, the yield is about 0.2 to 0.3 mg VSS per mg COD used (Metcalf and Eddy, Inc./AECOM, 2014). If domestic wastewater is the source, then the cell yield is approximately 1.5 mg VSS per 1.0 mg of nitrate-nitrogen (NO_3-N) removed. Put another way, the yield is 0.30 mg VSS/mg COD used (Metcalf and Eddy, Inc./AECOM, 2014). Compare this to the yield when the same microorganisms use DO: 0.4 mg VSS/mg COD used. Another advantage of denitrification is that the cell yield per kilogram (pound) of BOD_5 removed is lower than when oxygen is used. Fewer waste biosolids are produced in denitrifying facilities. Table 9.6 summarizes the key denitrification relationships.

Table 9.6 Key Denitrification Relationships

Parameter	Relationship
Oxygen	Below 0.3 mg/L to prevent inhibition
Alkalinity produced	3.57 mg alkalinity as $CaCO_3$ per mg NO_3-N
Biomass produced	0.30 mg new cells per mg COD used or 1.5 mg new cells per mg NO_3-N

Denitrification converts nitrate (NO_3-N) to nitrogen gas (N_2). Many different types of bacteria can denitrify. The process requires organic carbon and produces 3.57 mg of alkalinity as $CaCO_3$ for every 1 mg/L nitrate (NO_3-N) denitrified. Sludge yield is slightly lower for facilities that denitrify.

TEST YOUR KNOWLEDGE

1. **Nitrogen gas produced during denitrification**
   a. Is discharged with the final effluent.
   b. Ends up in the atmosphere.
   c. Combines with organic material.
   d. Improves sludge settling in clarifiers.

2. **For denitrification to occur, which of the following conditions must exist?**
   a. Excess alkalinity must be present.
   b. pH must be lower than 6.5
   c. DO concentration is high.
   d. Organic carbon is available.

3. **Dissolved oxygen can inhibit denitrification at concentrations as low as**
   a. 0.3 mg/L
   b. 0.8 mg/L
   c. 1.3 mg/L
   d. 1.8 mg/L

4. **Activated sludge processes that denitrify**
   a. Consume more alkalinity than those that don't.
   b. Generate less sludge than those that don't.
   c. Often have difficulty maintaining neutral pH.
   d. Increase nitrogen loading to solids handling processes.

5. **For every 1 mg/L NO_3-N converted to nitrogen gas, this much alkalinity is recovered.**
   a. 3.57 mg/L as $CaCO_3$
   b. 7.14 mg/L as $CaCO_3$
   c. 7.48 mg/L as $CaCO_3$
   d. 8.34 mg/L as $CaCO_3$

6. **Denitrification of 1 mg/L of NO_3-N recovers the equivalent of _________ mg/L of DO in the treatment process.**
   a. 1.73 mg/L
   b. 2.86 mg/L
   c. 3.57 mg/L
   d. 7.14 mg/L

PROCESS VARIABLES FOR DENITRIFICATION

The most important process variables for denitrification are the availability of organic carbon, scarcity of DO, and process configuration. Denitrification is much less sensitive to pH than nitrification; however, denitrification rates are lower below pH 6.0 and above pH 8.0 (U.S. EPA, 1993).

BIOCHEMICAL OXYGEN DEMAND OR CHEMICAL OXYGEN DEMAND

The amount of nitrate that can be removed depends on how much organic carbon is available. A general rule is that 4 mg/L of influent or primary effluent BOD_5 (approximately 8 mg/L COD) is needed for every 1 mg/L of nitrate-nitrogen (NO_3-N) (Barth et al., 1968). Facilities with primary clarifiers may not have enough BOD_5 remaining in the primary clarifier effluent to achieve low effluent nitrate-nitrogen (NO_3-N) limits. Ammonia, being soluble, passes through the primary clarifier to the secondary treatment process. The end result is a primary clarifier effluent that often does not contain 4 mg/L of BOD_5 for every 1 mg/L of nitrate-nitrogen (NO_3-N). The actual amount of BOD_5 needed will depend on the carbon source used and other operating conditions, including maintaining anoxic conditions.

A variety of supplemental carbon sources have been used for denitrification, including methanol, molasses, waste beer, cheese whey, and other food wastes. When a supplemental carbon source is required, local industries should be surveyed to determine if a high COD waste product is available. This not only provides a low-cost supplement to the WRRF, it can reduce operating costs for the industry. Many facilities add methanol (CH_3OH) to increase the amount of BOD_5 available for denitrification. Unlike industrial wastes that must be analyzed regularly to determine the BOD_5 or COD content before using them, methanol is a consistent product that does not change from one chemical delivery to the next. Between 2.7 and 3.3 mg of methanol (CH_3OH) are needed for every 1 mg/L of nitrate-nitrogen (NO_3-N) (U.S. EPA, 2010). Methanol has a COD of 1.5 mg COD/mg CH_3OH.

Methanol is flammable and can be dangerous to ship, store, and handle. Many WRRFs are switching to nonflammable, sometimes proprietary, chemicals as carbon sources to provide a better work environment for their operators. MicroC is one propriety chemical that blends methanol with glycerol and other ingredients to reduce flammability. Acetic acid (vinegar) and sodium acetate are also used as supplemental carbon sources.

DISSOLVED OXYGEN AND ANOXIC CONDITIONS

Anoxic conditions are defined as containing nitrate, but no oxygen. In biofilm systems, anoxic conditions are created when DO does not penetrate through the entire biofilm. Activated sludge systems create anoxic zones by providing separate, unaerated tanks with mixers or by isolating space within a tank with walls or curtains (Figure 9.12). In oxidation ditches and long, narrow activated sludge basins, anoxic zones can be created simply by aerating only some areas. As the wastewater moves away from the aerated areas, the DO is rapidly used up by microorganisms and the wastewater becomes anoxic. Sequencing batch reactors create anoxic conditions by cycling the air on and off while continuing to mix the wastewater. Facilities that were not originally designed to denitrify also use this method. In both biofilm systems and activated sludge processes, nitrate must first be generated by nitrification in aerobic areas and then be transferred to the anoxic areas. In fixed-film processes, this is accomplished by recycling flow to the front of the treatment process. There are many different variations of activated sludge processes that arrange anoxic and aerated zones in different ways to achieve total nitrogen removal. For the activated sludge system shown in Figure 9.12, an internal mixed liquor recycle (IMLR) pump is pulling nitrified wastewater that is full of nitrate from the end of the aerated zone back to the beginning of the anoxic zone.

SPECIFIC DENITRIFICATION RATE

The specific denitrification rate is how fast the nitrate (NO_3^-) is converted into nitrogen gas (N_2) by the bacteria in the treatment process. The speed of the reaction is limited by the amount of organic carbon (BOD_5 or COD) and nitrate that are available. Recall from Chapter 5 that the bacteria in secondary

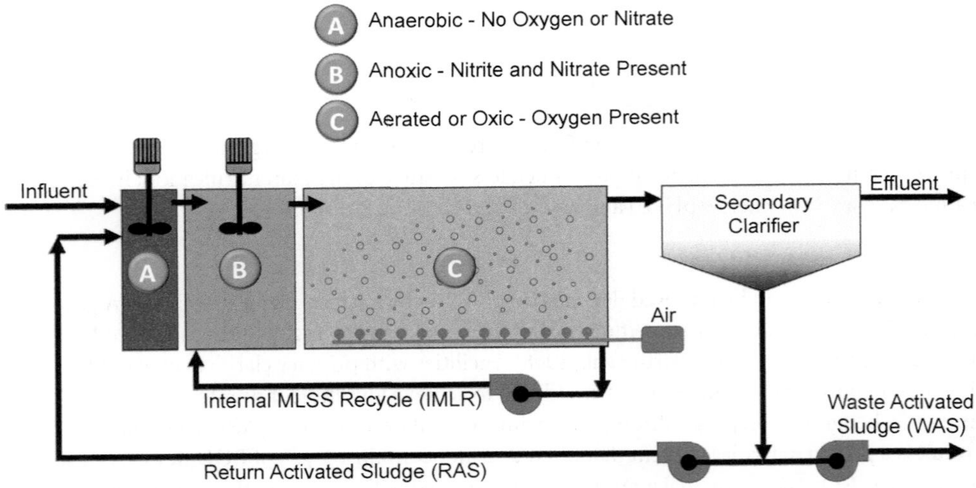

Figure 9.12 Anaerobic, Anoxic, and Oxic Zones in an Activated Sludge Process (A^2/O Process) (Reprinted with permission by Indigo Water Group [modified from U.S. EPA, 1993])

treatment processes rarely have enough resources to grow at their maximum rate. When additional resources are added, the bacteria grow faster and use more resources. Most of the time, the bacteria run out of organic carbon (BOD_5 or COD) long before they run out of nitrate (NO_3^-). This is because the influent wastewater has a limited supply of BOD_5 and because nitrification has to take place before denitrification. Much of the influent BOD_5 will be used by the bacteria in the aerobic portions of activated sludge processes and biofilm systems.

The *food-to-microorganism ratio* (F/M) is equal to the mass of organic load (pound or kilograms) entering the activated sludge basin divided by the mass of mixed liquor volatile suspended solids (MLVSS) (pounds or kilograms).

Figure 9.13 plots the specific denitrification rate versus the *food-to-microorganism ratio* (F/M). Notice that the denitrification rate increases with increasing F/M. Also note that when the influent contains a greater percentage of readily biodegradable (soluble) BOD_5, the denitrification rate keeps going up in an almost straight line. When less of the BOD_5 is readily biodegradable, the denitrification rate starts to slow down as the F/M gets higher and higher. The graph tells us that the availability of BOD_5 is controlling the denitrification rate, not the availability of nitrate (NO_3^-).

Most activated sludge systems are operated as either conventional, with an F/M range of 0.2 to 0.4 g BOD_5/g MLVSS or as extended aeration, with an F/M range of 0.05 to 0.15 g BOD_5/g MLVSS. These typical operating ranges for F/M are based on average operating conditions. Keep in mind that the influent BOD_5 changes during the day, so the F/M also changes and is sometimes higher and sometimes lower. Why is this important? Denitrification is more efficient during peak loading. This is why denitrification is more likely to occur in secondary clarifier blankets during the middle of the day at most WRRFs than any other time.

Chapter 8 includes a discussion and photographs of denitrification in secondary clarifiers.

As the wastewater moves through an activated sludge or fixed-film process, the BOD_5 is used up by the microorganisms. The F/M decreases from beginning to end. Why is this important? Denitrification will be most efficient where the F/M is highest. Many activated sludge processes have a separate anoxic zone at the beginning of the process like the one shown in Figure 9.12. Influent BOD_5 or primary effluent and RAS are combined there. By keeping the anoxic zone small and separated, dilution of the influent wastewater (or primary effluent) is minimized and the F/M is kept as high as possible. Anoxic zones with high F/M have another benefit. They inhibit the growth of some types of filamentous bacteria.

An activated sludge process may include pre-anoxic and post-anoxic zones (Figure 9.14). Denitrification rates for pre-anoxic zones in full-scale systems have been reported from 0.04 to 0.42 g NO_3-N/g MLVSS·d (Metcalf and Eddy, Inc., 2003). Denitrification rates for post-anoxic zones are much lower at 0.01 to

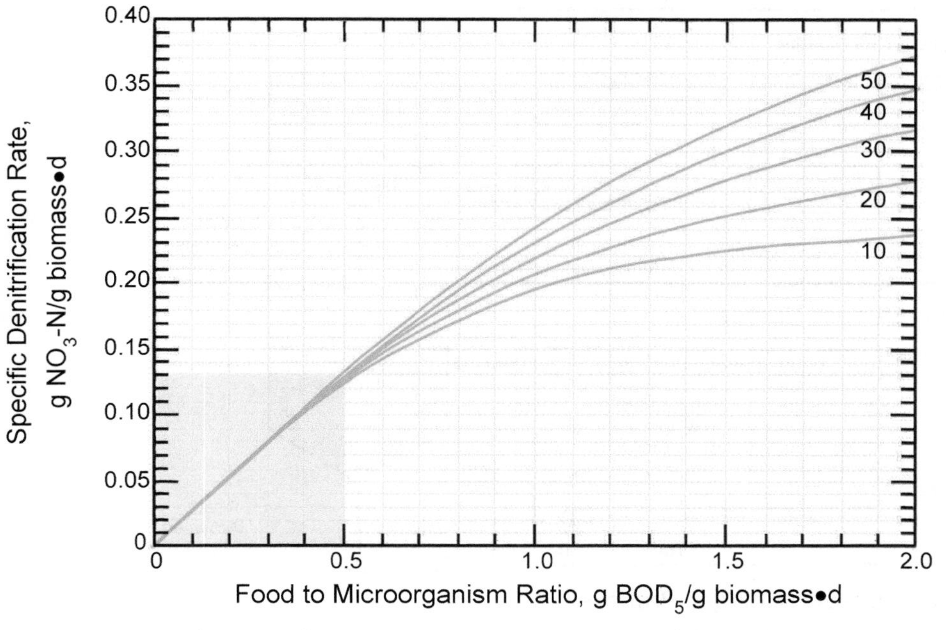

Figure 9.13 Specific Denitrification Rate at 20 °C as a Function of the Food-to-Microorganism Ratio and the Percentage of Readily Biodegradable BOD to Total BOD (Adapted from Metcalf and Eddy, 2003. Copyright © 2003 by McGraw-Hill Education, reprinted with permission.)

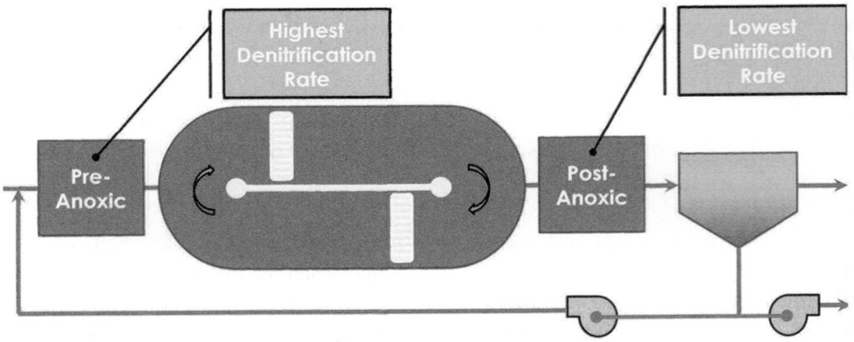

Figure 9.14 Pre-Anoxic and Post-Anoxic Zones (Reprinted with permission by Indigo Water Group)

0.04 g NO_3-N/g MLVSS·d (Metcalf and Eddy, Inc., 2003). The high F/M in the pre-anoxic zone boosts denitrification rates up to 10 times the post-anoxic zone rate. Supplemental carbon is often added to the post-anoxic zone to increase denitrification rates, especially when facing extremely low discharge limits. Operators must take care not to overdose at this point in the treatment process because it could cause a discharge permit violation for BOD_5.

INTERNAL MIXED LIQUOR RECYCLE RATIO

One of the earliest proposed activated sludge processes combining nitrification and denitrification was the Wuhrmann process in the early 1960s (Figure 9.15a). The front of the process was aerobic and the back end of the process was anoxic (U.S. EPA, 1993). Wuhrmann reasoned that because nitrification had to happen before denitrification could take place, putting the aerobic zone at the front of the process

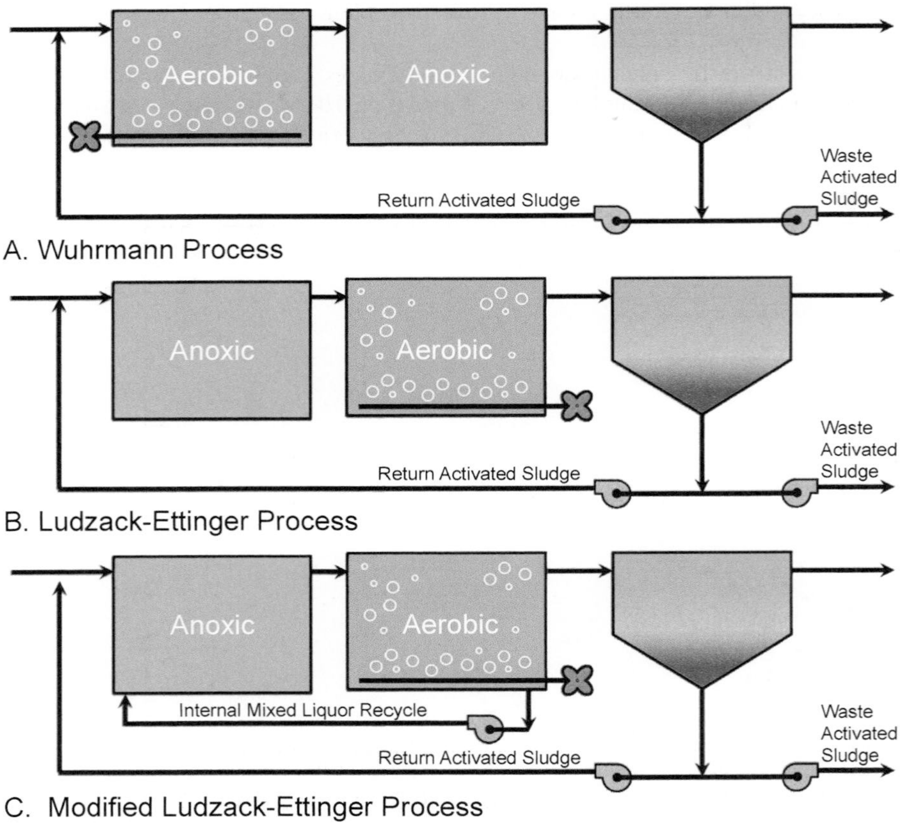

Figure 9.15 Variations on the Activated Sludge Process: Wuhrmann and Ludzack-Ettinger (Reprinted with permission by Indigo Water Group [modified from U.S. EPA, 1993])

made the most sense. Wuhrmann's design didn't denitrify as well as later designs because by the time the wastewater reached the anoxic zone, very little BOD_5 remained. Some denitrification occurred, but slowly. Around the same time, Ludzack and Ettinger proposed a similar process, but switched the order of the aerobic and anoxic zones (Figure 9.15b). The Ludzack-Ettinger process places influent (or primary effluent) BOD_5 where it is needed most—in the anoxic zone. Look carefully at the diagram in Figure 9.15b. The only way for nitrate to get from the aerobic zone back to the anoxic zone is through the RAS line. The potential for high denitrification rates exists, but not enough nitrate is recycled even when pumping the RAS at 125% of influent flow. Ten years later, the first modified Ludzack-Ettinger (MLE) process was proposed. An internal pumping system was added to recycle flows from the end of the aerobic zone back to the beginning of the anoxic zone (Barnard, 1973). The internal recycle doesn't move MLSS through the secondary clarifier, so it can be adjusted from very low percentages to very high percentages without affecting secondary clarifier performance. This modification allows the MLE process to reliably achieve very low concentrations of effluent nitrate-nitrogen (NO_3-N) provided there is enough BOD_5 available in the influent (or primary effluent). The MLE process is one of the most common activated sludge configurations for BNR. Activated sludge systems with a single anoxic zone will typically achieve effluent total nitrogen concentrations of <10 mg/L as N (U.S. EPA, 1993). To reach even lower concentrations, a second anoxic zone or tertiary denitrification process is needed.

The percentage of nitrogen removed will depend on 1) the IMLR pumping rate, 2) the RAS pumping rate, and 3) the availability of organic carbon. Assuming there is plenty of organic carbon available, it is possible to remove up to 85% of the *oxidizable total nitrogen* in the influent. If the IMLR is set equal to the influent flow, 50% of the influent total oxidizable nitrogen can be removed. If the IMLR is set equal to 200% of the influent flow, 67% of the total influent nitrogen can be removed. At 300% IMLR, 75% removal is achieved and at 400% IMLR, 80% removal is possible. The RAS also returns nitrate to the anoxic zone and can increase overall removal (Table 9.7). If the RAS is equal to 50% of the influent flow and the IMLR is equal to 200% of the influent flow, the total nitrogen removal increases slightly from 67 to 70%. Increasing the RAS to 100% of influent flow can push removal rates even higher to 75%.

Oxidizable nitrogen is organic nitrogen and ammonia that can be broken down and converted into nitrite and nitrate. Practically speaking, the oxidizable nitrogen is the same as total Kjeldahl nitrogen (TKN). Domestic wastewater contains a small amount of organic nitrogen that can't be broken down in the treatment process.

It's tempting to turn the IMLR and RAS pumps as high as they will go in the hopes of better performance, but the percentage of nitrogen removed won't continue to increase. Higher pumping rates increase wear and tear on the pumps as well as energy usage and operating costs. Wastewater recycled from the end of the aerated zone contains oxygen. The higher the IMLR pumping rate, the more oxygen is returned to the anoxic zone. This is known as "oxygen poisoning". Remember, the bacteria will always use DO first. Returning oxygen in the RAS uses up BOD_5 in the anoxic zone and inhibits denitrification. Most WRRFs use an IMLR that is 200 to 400% of the influent flow. Some specialized activated sludge processes like oxidation ditches and membrane treatment systems use IMLR pumping rates as high as 6 or 7 times the influent flow. Figure 9.16 provides a typical example of expected effluent nitrate concentration for three different starting TKN concentrations.

TEMPERATURE

Denitrification performance, as with other biological processes, is affected by changes in temperature, that is, denitrification rates will increase or decrease as temperatures rise and fall. However, denitrification performance is not as limited to temperature as nitrification. Therefore, with temperature as the variable, denitrification performance will continue as long as the nitrification process produces nitrate.

Table 9.7 Theoretical Oxidized Nitrogen Removal Performance (U.S. EPA, 1993)

IMLR, % of Influent Flow	RAS at 50% of Influent Flow	RAS at 100% of Influent Flow
0	34	50
50	50	60
100	60	67
200	70	74
300	78	80
400	82	85

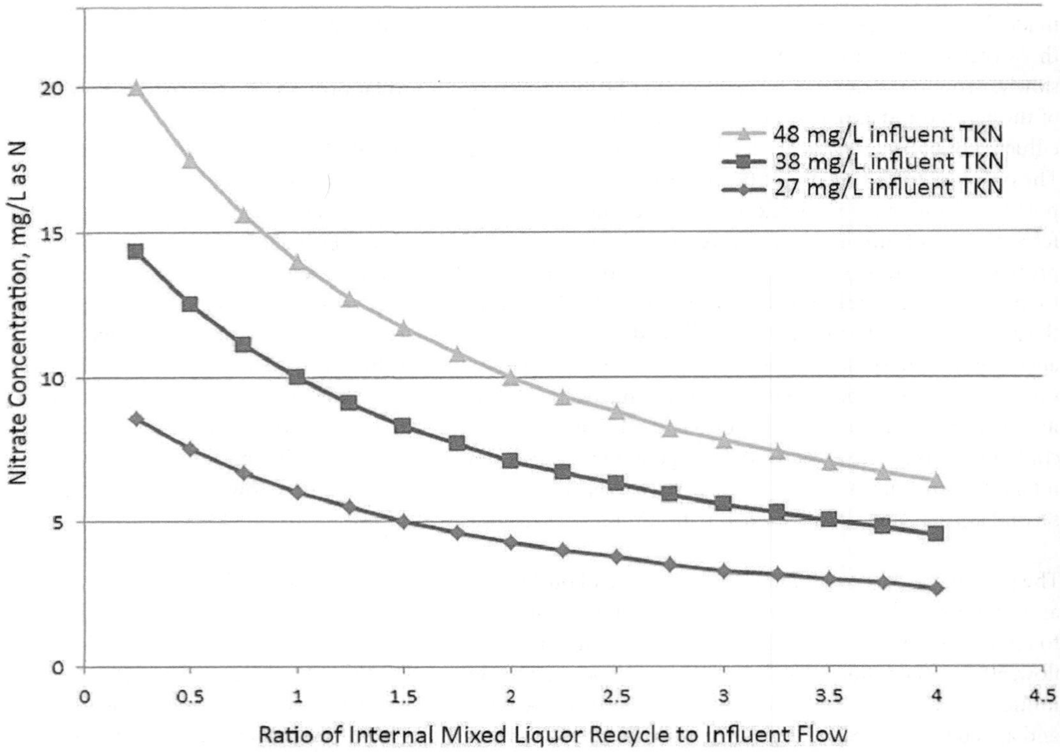

Figure 9.16 Effluent Nitrate Concentrations Versus IMLR Ratio (RAS at 50% of influent flow, influent BOD_5 = 250 mg/L, 2 mg/L non-oxidizable TKN) (Reprinted with permission by Indigo Water Group and Joel Rife)

Approximately 4 mg/L of BOD_5 is needed to convert 1 mg/L NO_3-N to nitrogen gas (N_2). During denitrification, the bacteria almost always run out of BOD_5 before running out of nitrate. As BOD_5 concentrations increase, the denitrification rate increases proportionally until the bacteria are growing at their maximum rate. Operators can add BOD_5 to the process to increase the amount of nitrate removed.

TEST YOUR KNOWLEDGE

1. About this much BOD_5 is needed to convert 1 mg/L of nitrate-nitrogen (NO_3-N) to nitrogen gas (N_2).
   a. 1 mg/L
   b. 4 mg/L
   c. 7 mg/L
   d. 10 mg/L

2. Primary clarifiers can affect denitrification by
   a. Converting influent total nitrogen to nitrate-nitrogen
   b. Poisoning the anoxic zone with oxygen
   c. Changing the BOD_5 to nitrate-nitrogen ratio
   d. Increasing particulate BOD_5 to the secondary process

3. A WRRF needs to add organic carbon for denitrification. The operators are concerned about flammability and want something they don't have to test before each use. Which of the following carbon sources is best suited to their requirements?
   a. Methanol
   b. Waste beer
   c. Cheese whey
   d. Acetic acid

4. Which of the following must occur first?
   a. Convert ammonia to nitrite
   b. Convert nitrite to nitrate
   c. Convert nitrate to nitrogen gas
   d. Convert nitrogen gas to nitrite

5. **Denitrification rates will be highest at this location.**
   a. Bottom of a trickling filter
   b. Pre-anoxic zone
   c. Secondary clarifier blanket
   d. Post-anoxic zone

6. **The F/M is 0.25 g BOD_5/g MLVSS at 10:00 a.m. If the F/M increases to 0.50 g BOD_5/g MLVSS at 2:00 p.m., the denitrification rate will**
   a. Decrease by 50%
   b. Remain the same
   c. Approximately double
   d. Be limited by nitrate

7. **Denitrification is most likely to occur in the secondary clarifier blanket when**
   a. Influent flows and loads are low.
   b. Primary clarifier performs better than average.
   c. Methanol is added to tertiary filters.
   d. It is the middle of the day.

8. **The IMLR**
   a. Removes sludge from the clarifier
   b. Returns nitrate to the anoxic zone
   c. Recycles nitrified effluent in an RBC
   d. Wastes sludge to the digester

9. **The influent to an MLE activated sludge process contains 20 mg/L nitrate-nitrogen (NO_3-N) and the effluent contains 12 mg/L NO_3-N. If the internal recycle is increased from 200 to 400% of influent flow, effluent nitrate-nitrogen remains the same. What must be true?**
   a. Too much oxygen recycled to the anoxic zone
   b. Alkalinity has dropped below 50 mg/L as $CaCO_3$
   c. Inadequate supply of BOD_5 for more denitrification
   d. Upstream nitrification process has failed

PROCESS CONTROL FOR DENITRIFICATION

Process control for denitrification in fixed-film processes is limited to adjusting the amount of nitrified effluent that is recycled back to the front of the process. Process control for denitrification in activated sludge systems consists of

- Ensuring the anoxic zone(s) remain anoxic;
- Adjusting the IMLR flow;
- Adjusting the RAS flow;
- Adjusting the recycle flow in a fixed-film process;
- Adding supplemental BOD_5 (if needed); and
- Monitoring the influent and effluent for the following parameters:
  - Oxidation–reduction potential (ORP),
  - Nitrite, and
  - Nitrate.

MAINTAINING ANOXIC CONDITIONS

Minimizing the amount of DO transferred to the anoxic zone ensures that the bacteria will use nitrate to process BOD_5. Dissolved oxygen must be used up before denitrification can begin, which means less time is spent using nitrate. Dissolved oxygen use in the anoxic zone consumes valuable BOD_5 and reduces the amount left for denitrification. If supplemental carbon is being added to support denitrification, it will be consumed using DO first, resulting in increased chemical costs. Potential sources of DO in the anoxic zone include the influent wastewater (or primary effluent), transfer from the air, entrainment during mixing, RAS, and IMLR. Influent DO should be minimal as it is consumed in the collection system. However, pumping stations, especially those equipped with screw pumps, can increase DO concentrations. This effect is more pronounced when the water is colder and can hold more dissolved gases. Many WRRFs include pumping stations immediately before or after the headworks. Raising the water up at the beginning of the treatment process often allows it to flow by gravity through to the final effluent. Preliminary and primary treatment can also add DO as the wastewater cascades over weirs and moves through open channels.

Cold water can contain more dissolved gas than warm water. Think about opening a can of soda that has been sitting in a hot car all day. What happens? The soda shoots out of the can under pressure as most of the carbon dioxide gas is released all at once. The same soda pulled from a cold refrigerator contains the same amount of carbon dioxide, but it remains in the soda.

The most significant transfer of oxygen is from the IMLR. If the DO concentrations are high at the end of the aerated zone, more DO will be pumped back to the anoxic zone through the IMLR. Minimizing the DO by managing the aeration conditions at the IMLR pump intake is important. Ideally, the aerated zone will be long and narrow with multiple aeration devices along the length that are independently controlled. The goal is to maximize DO concentration at the influent end of the basin where demand is highest, but decrease the concentration as the wastewater flows toward the outlet. The DO at the influent end may be 2 to 3 mg/L, but the DO at the effluent end may be 0.5 to 1 mg/L (Figure 9.17). Some WRRFs use tapered aeration: more diffusers at the influent end and fewer at the effluent end. As a WRRF nears its design capacity, preserving effective anoxic zone capacity becomes increasingly important to enable optimum denitrification performance. Take care to keep the DO concentration high enough to prevent the growth of low DO filaments.

Review Chapter 8 for guidance on setting target DO concentrations in activated sludge processes.

OXIDATION–REDUCTION POTENTIAL

Anoxic zones in activated sludge processes are typically monitored using ORP. *Oxidation* means to add oxygen, lose hydrogen, or lose electrons. *Reduction* means to remove oxygen, add hydrogen, or gain electrons. Oxidation and reduction always occur together with one chemical substance being oxidized, while another is reduced. If the two chemical reactions are physically separated, but joined by a wire, the electrons lost from one chemical substance will travel down the wire to the other chemical substance. The difference in electrical charge between the two points is the electrical potential, also known as *voltage*. Hence the term, *ORP*. Oxidation–reduction potential is measured in millivolts (mV). There are 1000 mV in one volt (V). Essentially, ORP measures the difference in potential (charge) between all the different chemical substances in wastewater. Wastewaters with higher concentrations of oxidized substances like nitrate, nitrite, and DO tend to have positive ORPs. Wastewaters with higher concentrations of reduced substances like hydrogen sulfide (or a lack of oxidized substances) tend to have negative ORPs.

Think of the IMLR pump as recycling MLSS, nitrate, DO, and ORP. As the IMLR pump rate increases, DO and ORP will also increase in the anoxic zone.

Oxidation–reduction potential changes when DO is present and when it isn't. It is useful for monitoring conditions in anaerobic, anoxic, and aerobic processes. In general, ORP values lower than −200 mV indicate anaerobic conditions, values between −50 and +50 mV indicate anoxic conditions, and values higher than +50 mV indicate aerobic conditions. Typically, the ORP in an anoxic reactor must be below +50 mV to trigger denitrification. These ranges are based on an ORP probe with a silver/silver chloride reference electrode and a saturated potassium chloride (KCl) fill solution. The actual ORP reading depends on the specific reference electrode being used and can be affected by water quality. Facilities that use ORP for process control should determine which ORP ranges indicate good operating conditions for their processes. Oxidation–reduction potential measurements are typically taken *in situ* by ORP probes mounted directly in the activated sludge basins.

When measurements are taken directly from the treatment process instead of transporting them to a laboratory, they are "in place" or *in situ* measurements.

Oxidation–reduction potential probes look very similar to pH probes. Combined ORP probes contain two measurement devices: a primary electrode and a reference electrode. The primary electrode is submerged in the sample—in this case, wastewater. The reference electrode is submerged inside the probe in a standard solution (fill solution). The measurement from the primary electrode is compared to the reference electrode by the electronics in the meter to produce an ORP reading in millivolts. Different ORP probes produce different ORP readings for the same solution, based on the specific reference electrode and fill solution used. Typical reference electrodes include standard hydrogen, saturated calomel electrode, or silver/silver chloride electrode. A solution showing 0 mV with a hydrogen electrode would read the approximate values shown in Table 9.8 for other probe types. Most ORP probes use a silver wire in *saturated* KCl solution as the

Saturated means the solution is the highest possible concentration for a particular temperature and pressure. Hot tea can dissolve more sugar than cold tea. The point where no more sugar can be dissolved into the tea is the sugar saturation point.

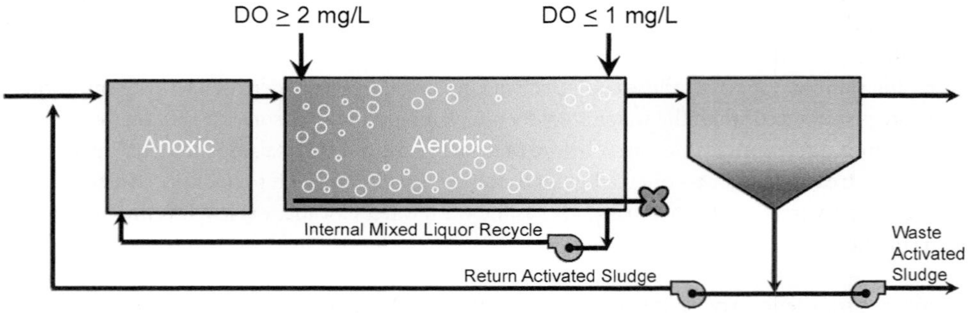

Figure 9.17 Tapered Aeration to Minimize Recycle of DO (Reprinted with permission by Indigo Water Group)

Table 9.8 Oxidation–Reduction Potential Reference Electrodes

Reference Electrode	ORP (mV)
Standard hydrogen electrode	0
Saturated calomel electrode	−241
Silver/silver chloride, 1 molar KCl*	−236
Silver/silver chloride, 4 molar KCl*	−200
Silver/silver chloride, saturated KCl*	−197

*A 1-molar solution contains 1 mole/L.

reference electrode and a platinum wire as the primary electrode (Spencer, 2012). The key point to take away from this discussion is that ORP readings can be quite different from one probe to the next. Don't worry if the result from the ORP probe in the laboratory doesn't exactly match the result from the ORP probe in the treatment process. If the in-process probe is a silver/silver chloride electrode with 4 molar KCl fill solution and the laboratory probe is a saturated calomel electrode, the readings produced by placing each probe in the exact same water will differ (in theory) by −41 mV. In practice, the difference can be much larger.

Different types of microbial activity occur within their own ranges of ORP. Table 9.9 lists typical ORP ranges using a silver/saturated potassium chloride reference electrode, which is the most commonly used ORP probe. An ORP probe may be useful in determining the type of biological activity likely to be occurring at any point in the treatment process.

Wastewater flows continuously through most activated sludge processes. Oxidation–reduction potential probes are typically mounted in both the anaerobic and anoxic zones. Many WRRFs also have ORP probes at the beginning and end of the aerated zone. These probes should reflect the ORP values shown in Table 9.9, although the readings will increase and decrease during the day. Sequencing batch reactors change operating conditions over time instead of space. A single ORP probe mounted in an SBR may be exposed to anaerobic, anoxic, and aerated conditions as the SBR works through a complete treatment cycle. Figure 9.18 is a representation of how ORP, pH, and DO concentrations are expected to change during a typical SBR treatment cycle (Basu et al., 2006). When the SBR is aerobic and air is being supplied to the MLSS, the ORP reading is increasing. The increase indicates that oxygen is being added to the process. The BOD_5 and ammonia are becoming oxidized into biomass, carbon dioxide (CO_2), water, nitrite (NO_2^-), and nitrate (NO_3^-). Wastewater pH is decreasing slightly resulting from nitrification. Next, the air is turned off and the DO concentration decreases rapidly. The ORP reading decreases with the DO concentration. Less oxygen in the water means there is less oxidation potential. Once the DO concentration is below approximately 0.3 mg/L, the ORP readout decreases rapidly. The decrease indicates that nitrate is being reduced to nitrogen gas. Figure 9.19 shows in-basin monitoring data collected from an SBR operating with two different F/M (Han et al., 2011). Notice how the ORP decreases sharply halfway through the anoxic cycle in the lower graph (B). This is the "nitrate knee". The knee indicates the end of denitrification either

Table 9.9 Oxidation–Reduction Potential Ranges for Various Biochemical Reactions*

Microbial Activity	Approximate ORP (mV)
Carbon oxidation (conversion of BOD_5 to biomass, carbon dioxide, and water with oxygen)	+50 to +200
Polyphosphate accumulation	+50 to +250
Nitrification	+150 to +350
Denitrification	−50 to +50
Polyphosphate release	−40 to −175
Volatile fatty acid (VFA) formation	−40 to −200
Sulfide formation	−50 to −250
Methane formation	−200 to −400

*Values are for a silver/silver chloride electrode with saturated potassium chloride fill solution.

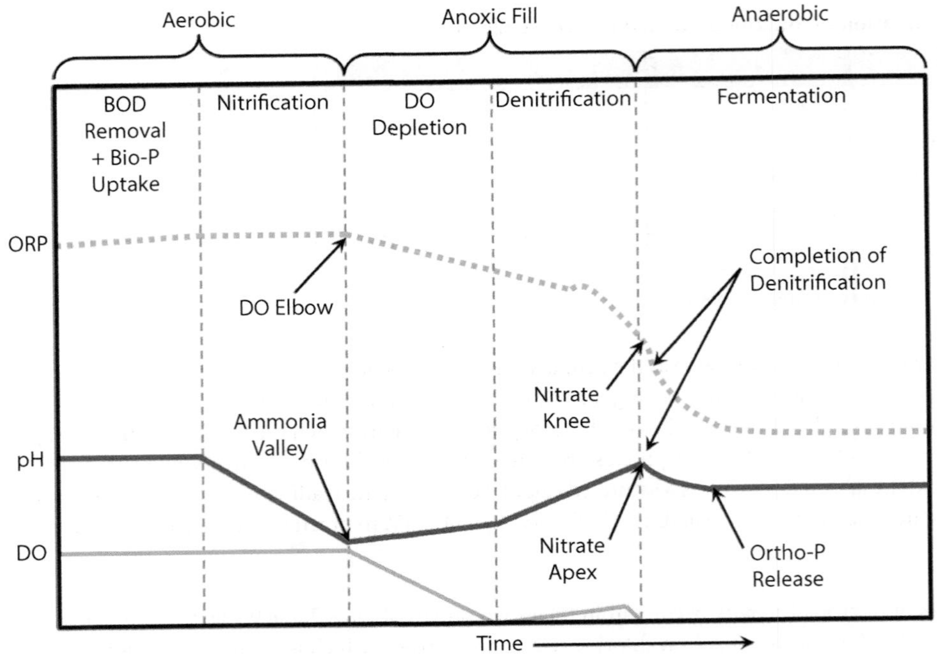

Figure 9.18 Theoretical ORP, pH, and DO Profiles in an SBR (adapted from Basu et al., 2006)

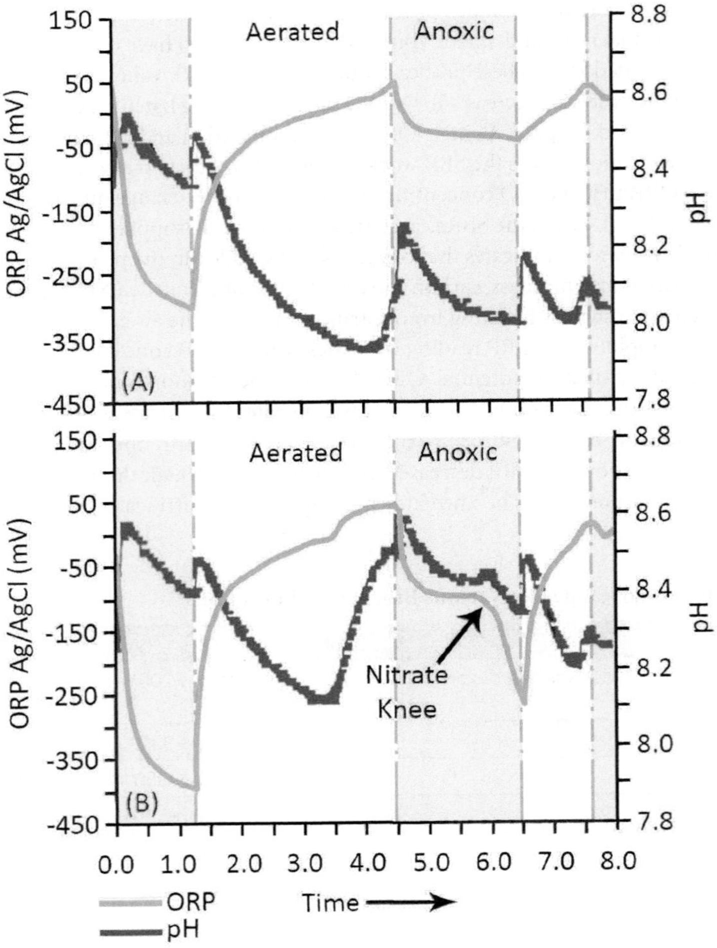

Figure 9.19 Operating Data from a Sequencing Batch Reactor (adapted from Han et al., 2011)

because the bacteria ran out of BOD_5 or nitrite and nitrate. The upper graph (A) does not have a strong nitrate knee. Because of the higher F/M, the bacteria did not run out of BOD_5.

Oxidation–reduction potential measurements are affected by temperature. Because the effect isn't easy to correct for, ORP probes typically do not incorporate temperature sensors and do not adjust the ORP result. Rather, the operator must make adjustments to how the ORP reading is interpreted. Changes in the hydrogen ion (H^+) concentration will also change ORP readings. Fortunately, in most well-operating treatment processes, the pH does not change very much from day to day.

Oxidation–reduction potential is used extensively in chlorine disinfection to monitor the chlorine concentration in the water.

Process control for denitrification includes keeping the anoxic zone anoxic and ensuring there is enough BOD_5 available to support the reaction.

TEST YOUR KNOWLEDGE

1. **An activated sludge process has an anaerobic zone, anoxic zone, and aerated zone. Mixed liquor suspended solids are pumped from the end of the aerated zone to the beginning of the anoxic zone. The recycle flow is equal to 200% of the influent flow. The RAS is set to 75% of the influent flow. The biggest source of DO in the anoxic zone is**
   a. Influent
   b. Primary effluent
   c. IMLR
   d. RAS
2. **An activated sludge basin has a DO concentration of 2 mg/L. The ORP is**
   a. Between −250 mV and −400 mV
   b. Between −250 mV and −175 mV
   c. Between −50 mV and +50 mV
   d. Between +50 mV and +200 mV
3. **Which of the following statements about ORP is true?**
   a. ORP increases at the end of denitrification.
   b. ORP is affected by water quality and temperature.
   c. The fill solution for an ORP probe is unimportant.
   d. In-basin ORP probes should match laboratory ORP results.
4. **The primary effluent going to an activated sludge process has a BOD_5 concentration of 140 mg/L and an ammonia-nitrogen concentration of 55 mg/L. If the discharge permit limit for nitrate-nitrogen is 5 mg/L, what must be true?**
   a. Supplemental carbon source may be needed.
   b. ORP will decrease through the aerated zone.
   c. IMLR should be set to 400% of influent flow.
   d. Sludge age should be less than 5 days.
5. **This marks the end of denitrification when ORP is used to monitor an SBR.**
   a. DO elbow
   b. Ammonia valley
   c. Nitrate knee
   d. Fermentation

SECTION EXERCISE

This exercise includes several short troubleshooting scenarios. Using what you know about denitrification, determine the cause of the problem, and suggest a solution. Write down your answer before looking at the answer key.

Scenario #1—An activated sludge process consists of an anoxic zone followed by an aerobic zone. An IMLR line connects the end of the aerobic zone to the beginning of the anoxic zone. The influent TKN is 50 mg/L as N. When the internal recycle is set to 200% of the influent flow, the facility can reduce nitrate-nitrogen to 14 mg/L NO_3-N. Increasing the recycle ratio to 300 or 400% does not decrease the effluent nitrate concentration. What is the most likely cause? How can your suspicion be confirmed? What can the operator of this facility do to meet an effluent nitrate-nitrogen limit of 8 mg/L NO_3-N?

Scenario #2—An SBR activated sludge facility uses ORP for process control. The aerated time is about two-thirds of the total cycle time and the anoxic time is about one-third. The SBR effluent nitrate-nitrogen (NO_3-N) is currently 13 mg/L as N. To meet the requirements of the discharge permit, it must be below 8 mg/L as N. The influent BOD_5 to NO_3-N ratio is 5 to 1. Oxidation–reduction potential readings during the aerated phase range from +100 to +250 mV and drop to +40 mV during the anoxic phase. What can the operator do to improve nitrate-nitrogen removal?

SECTION EXERCISE SOLUTIONS

Scenario #1—It appears that the denitrifying bacteria have run out of BOD_5 before running out of nitrate-nitrogen (NO_3-N). This could be confirmed by sampling the WRRF influent for BOD_5 and TKN. If the ratio of BOD_5 to TKN is less than 3, the denitrifying bacteria are probably running out of BOD_5. Monitoring could also be done at the end of the anoxic zone. If the BOD_5 to NO_3-N ratio is less than 4 to 1, the process is probably BOD_5 limited. To meet a nitrate-nitrogen limit of 8 mg/L NO_3-N, the operator will need to add a supplemental carbon source like methanol to the anoxic zone.

Scenario #2—The operator should increase the amount of anoxic time. The SBR is not achieving anoxic conditions as evidenced by the ORP readings. Denitrification occurs between −50 and +50 mV. Increasing the duration of the anoxic cycle will give the bacteria an opportunity to use of residual oxygen and force the switch to nitrate.

Chemical Phosphorus Removal

Removal of phosphorus via chemical precipitation has been a common practice at utilities for decades. Chemical phosphorus removal is conventionally done by adding common *metal salts*, primarily aluminum sulfate [$Al_2(SO_4)_3$] and ferric chloride ($FeCl_3$). The aluminum ion (Al^{+3}) or iron (Fe^{+3}) metal ions combine with orthophosphate (PO_4^{-3}). The new compounds formed are not very soluble in water and instead "come out of solution" to form solid particles that can be removed by gravity settling or filtration. Only about 50% of the phosphorus entering the WRRF is orthophosphate (PO_4^{-3}). Fortunately, most of the condensed phosphate and organic phosphate will be converted into orthophosphate during biological treatment. Depending on where chemical addition is done—headworks, primary clarifier, or secondary clarifier—the percentage of total phosphorus that can be removed changes.

Chemically speaking, a *salt* is a chemical compound formed when an ion with a positive charge combines with an ion that has a negative charge. Sodium chloride, table salt, is formed from sodium ion (Na^+) and chloride ion (Cl^-). Ferric chloride and aluminum sulfate are also salts.

Chemical precipitation of phosphorus is often the phosphorus removal method of choice for small- and medium-size facilities. Much of the reason for its popularity is its simplicity from both a process and an operational perspective, resulting in reliable and predictable performance. This is particularly the case when an existing WRRF has to meet new phosphorus limits and the existing WRRF isn't easily modified for enhanced biological phosphorus removal. Because EPBR can sometimes have periods of inferior performance or upsets, chemical phosphorus removal is often provided as a backup method or as a polishing step to meet extremely low effluent limits.

Ever wonder why ferric chloride isn't called iron chloride? The Latin word for iron is *ferrum*. The chemical symbol for iron is Fe. The chemical symbol for lead, Pb, also comes from Latin. The Latin word for lead is *plumbum*.

Observe a precipitation reaction. Household ammonia, a cleaning solution, contains ammonium hydroxide (NH_3OH). Epsom salt, used to treat muscle aches and sunburns, is magnesium sulfate ($MgSO_4$). Both chemicals are soluble in water. When they are mixed together, they combine according to the following chemical reaction to produce magnesium hydroxide [$Mg(OH)_2$]:

$$2NH_3OH + MgSO_4 \rightarrow Mg(OH)_{2(s)} + 2NH_4^+ + SO_4^{-2} \qquad (9.9)$$

Magnesium hydroxide is not very soluble in water, so it comes out of solution as a white precipitate. To try this experiment at home, first dissolve some Epsom salt in about 23 mL (1.5 tablespoons) of water. For best results, dissolve as much Epsom salt as possible. If solid Epsom salt remains in the bottom of the container, pour the supernatant only into a new container. Use a clear container for this experiment to make it easier to see the precipitate as it forms. Then, add about 46 mL of ammonia (3 tablespoons) to the Epsom salt solution. Be careful not to breathe in the ammonia fumes and be sure to wear gloves and wash your hands after use. Concentrated ammonia can cause chemical burns to eyes and skin. Watch carefully. The solution should become slightly cloudy. The precipitate will eventually settle to the bottom of the jar.

CHEMICAL PHOSPHORUS REMOVAL STOICHIOMETRY

It is important to understand the basic chemical reactions that drive chemical phosphorus removal. In general, cations of the following metals can be used for the precipitation of phosphorus (orthophosphates) from wastewater:

- Aluminum,
- Iron,
- Calcium, and
- Magnesium.

Chemicals containing calcium and magnesium are much less commonly used than aluminum and iron compounds. Calcium oxide (CaO), also known as lime, is used by some facilities to precipitate phosphorus. However, lime precipitation generates more sludge for disposal than either iron or aluminum compounds. Lime is difficult to handle operationally and can be dangerous to operators. Lime precipitation works best when the pH is greater than 9, so the pH of the treated wastewater must be adjusted with acid addition prior to discharge. For all of these reasons, the use of lime for phosphorus removal is in decline. Therefore, there will be no further discussion of lime precipitation. The following sections will discuss the different chemicals available, the precipitation process, and sample calculations.

ALUMINUM SALTS

There are several aluminum compounds that are used in the wastewater industry for phosphorus removal, including aluminum sulfate, polyaluminum chloride (PACl), and sodium aluminate. The most widely used chemical is aluminum sulfate, commonly known as *alum*, although sodium aluminate is gaining popularity because of its ease of handling and because it does not have as large an effect on pH as alum or ferric. The strength and composition of sodium aluminate and PACl products vary considerably from one supplier to another. Before deciding which chemical and which supplier to use, operators should perform jar tests and compare costs. Jar testing consists of several side-by-side tests, typically conducted in a laboratory (Figure 9.20). Several large beakers are filled with wastewater from the treatment process. The beakers are dosed with different concentrations of a chemical and mixed to determine which concentration works best. The solids produced in a jar test can be measured to estimate sludge production. Jar testing won't match field conditions exactly, but it does provide a reasonably good estimate.

Regardless of the chemical used, the active component is aluminum ion (Al^{+3}), which combines with orthophosphate ions (PO_4^{-3}) to form aluminum phosphate ($AlPO_4$). The two ions have equal, but opposite charges. In eq 9.10, one mole of Al^{+3} combines with one mole of PO_4^{-3} to form one mole of $AlPO_4$.

$$Al^{+3} + PO_4^{-3} \rightarrow AlPO_{4(s)} \qquad (9.10)$$

Figure 9.20 Jar Testing Device (Reprinted with permission by Indigo Water Group)

The mole ratios from eq 9.10 may be converted into chemical dosing concentrations in milligrams per liter. Because the atomic weight of aluminum is 27 g/mole and the atomic weight of phosphorus is 31 g/mole, the weight ratio of aluminum to phosphorus in eq 9.10 is 0.87 to 1, as shown in the following calculation. This is the stoichiometric dose.

$$\frac{1 \text{ mole } Al^{+3}}{1 \text{ mole P}} \left| \frac{27 \text{ g } Al^{+3}}{1 \text{ mole } Al^{+3}} \right\| \frac{1 \text{ mole P}}{31 \text{ g P}} = 0.87 \text{ g } Al^{+3}/\text{g P}$$

In reality, precipitation of phosphorus with aluminum and iron salts is a more complex process. When these chemicals are added to wastewater, they combine with water molecules to form other precipitates like aluminum and iron hydroxides. A few examples of these types of chemical reactions with either aluminum (Al^{+3}) or iron ions (Fe^{+2} or Fe^{+3}) reacting with hydroxide ions (OH^-) from water are shown in eqs 9.11 through 9.13. There are many other more complex reactions with water molecules and other ions that take place as well. Notice that all three end products in eqs 9.11 through 9.13 are solids. They precipitate along with the phosphorus to form a chemical floc. The floc helps capture smaller particles as it settles and removes phosphorus in the process.

$$Al^{+3} + 3OH^- \rightarrow Al(OH)_{3(s)} + 3H^+ \quad (9.11)$$

$$Fe^{+2} + 2OH^- \rightarrow Fe(OH)_{2(s)} + 2H^+ \quad (9.12)$$

$$Fe^{+3} + 3OH^- \rightarrow Fe(OH)_{3(s)} + 3H^+ \quad (9.13)$$

Iron atoms can have either a +2 or a +3 charge: divalent or trivalent. Iron with a +2 charge is ferrous iron. Iron with a +3 charge is ferric iron.

Something else to notice about eqs 9.11 through 9.13 is that the aluminum and iron salts are consuming hydroxide ions (OH^-) and producing hydrogen ions (H^+). Hydroxide is one component of *alkalinity*. Hydrogen ions are acid. This acidity consumes valuable alkalinity, reducing the solution's buffering capacity. Where aluminum is utilized, the alkalinity consumed is 5.8 mg as $CaCO_3$ per mg of Al^{+3} applied, and where iron is applied, it is 2.7 mg as $CaCO_3$ per mg of Fe^{+3} applied.

Alkalinity is buffering capacity or the ability to neutralize acid without a big change in pH. A decrease in alkalinity doesn't always correspond to a decrease in pH.

Domestic wastewater typically has an influent phosphorus concentration between 6 and 8 mg/L as P. The metal salt dose is expressed as the moles of metal added (Me_{dose}) per mole of soluble phosphorus (P_{ini}) and is written as Me_{dose}/P_{ini}, where "ini" in the subscript refers to the initial phosphorus concentration. Aluminum or iron doses of 1.5 to 2.0 Me_{dose}/P_{ini} are typically needed to reduce the soluble phosphorus concentration to 1 mg/L as P. Doses in this range should remove 80 to 98% of soluble phosphorus. Reaching very low effluent total phosphorus concentrations requires much higher doses. To meet an effluent discharge limit of 0.10 mg/L as P, the ratio needed might be as high as 6 or 7 Me_{dose}/P_{ini}. Polyaluminum chloride and sodium aluminate often require even higher chemical doses.

Why does the dose needed increase when trying to reach lower and lower phosphorus concentrations? Imagine a bag of marbles. Most of the marbles in the bag are clear to symbolize water molecules. A small percentage of the marbles are green to symbolize phosphate. The marbles are mixed together inside the bag. If a single marble is randomly drawn from the bag, it is more likely to be a clear marble (water) than a green marble (phosphate). As green marbles are removed from the bag, it becomes less and less likely that the next randomly drawn marble will be green. In the beginning, every third marble might be green. Once most of the green marbles have been removed, it might require 10 or 15 or 20 draws from the bag to produce a single green marble. The same type of thing is happening when aluminum or iron is added to the wastewater. Aluminum and iron prefer to react with phosphate, which gradually reduces the phosphate concentration. Aluminum and iron will also react with water. As the phosphate concentration decreases, more and more of the metal salt ends up reacting with water molecules instead. The metal dose (Me_{dose}/P_{ini}) increases exponentially for lower and lower soluble phosphorus concentrations.

Aluminum Sulfate (Alum)

Alum is available as a dry, powdery chemical commonly known as *filter alum*. Dry alum has an average chemical formula of $Al_2(SO_4)_3 \cdot 14H_2O$. The water molecules attached to the chemical formula for aluminum sulfate indicate that there is a certain amount of water in the structure of the dry alum crystal under normal conditions. Chemical crystals that contain water in this way are said to be hydrated. However, alum for wastewater treatment is commonly delivered as a 49% solution. This indicates that 49 parts of dry alum are combined with 51 parts of water by weight. By convention, the alum dose applied in the treatment process is expressed in equivalent units of dry alum (including water of crystallization) per volume of water (wastewater). Properties of alum and other aluminum-based chemicals are summarized in Table 9.10.

The reaction of alum with phosphate can be described by the following equation (U.S. EPA, 2010):

$$Al_2(SO_4)_3 \cdot 14H_2O + 2H_3PO_4 \rightarrow 2AlPO_4 + 3H_2SO_4 + 18H_2O \qquad (9.14)$$

One mole of aluminum sulfate contains 2 moles of aluminum. On the other side of the equation, 2 moles of aluminum phosphate ($AlPO_4$) are created for every 1 mol of aluminum sulfate added. If the concentration of phosphate is known, the alum dose needed to precipitate it can be calculated.

Example: Determine the stoichiometric dose of aluminum sulfate (alum) needed to react with 7 mg/L of phosphorus. Use the stoichiometry from eq 9.14.

Step 1—Calculate the formula weight for aluminum sulfate. Use the periodic table (Figure 9.3) to find the atomic weights for each element.

Element	Atomic Weight g/mole		Number of Atoms		Total Weight
Aluminum	27	×	2	=	54
Sulfur	32	×	3	=	96
Oxygen (Alum)	16	×	12	=	192
Hydrogen (Water)	1	×	28	=	28
Oxygen (Water)	16	×	14	=	224
					594 g/mole

Step 2—Convert moles to milligrams per liter with dimensional analysis.

$$\frac{1 \text{ mole } Al_2SO_4 \cdot H_2O}{2 \text{ moles P}} \left| \frac{594 \text{ g } Al_2SO_4 \cdot H_2O}{1 \text{ mole } Al_2SO_4 \cdot H_2O} \right\| \frac{1 \text{ mole P}}{31 \text{ g P}} = 9.6 \text{ g } Al_2SO_4 \cdot H_2O \text{ per g P}$$

Table 9.10 Properties of Aluminum-Based Chemicals Used for Phosphorus Precipitation (WEF, 2013)

Name	Aluminum Sulfate, Dry	Aluminum Sulfate, 49% Solution*	Sodium Aluminate, Anhydrous	Polyaluminum Chloride
Form	Dry	Liquid	Powder	Liquid
Common name	Filter alum	Alum	None	PACl
Chemical formula	$Al_2(SO_4)_3 \cdot 14H_2O$	$Al_2(SO_4)_3 \cdot 14H_2O$	$NaAlO_2$	Varies – $Al_{12}Cl_{12}(OH)_{24}$
Formula weight, g/mole	594	594	82	Varies by manufacturer
Commercial strength, by weight	17% Al_2O_3	8.3% Al_2O_3	41 – 46% Al_2O_3	4.9 – 26.7% Al_2O_3
Aluminum metal contents, % by weight	9.1	4.4	33	Varies by manufacturer
Alkalinity	Consumes	Consumes	Generates	Depends on formulation.
pH change	Potential decrease	Potential decrease	Potential increase	Typically no change. Depends on formulation.
Weight ratio of dry chemical for stoichiometric precipitation of P	9.6	9.6	2.64	Varies

*49% alum solution has a dry alum [$Al_2(SO_4)_3 \cdot 14H_2O$] content of 0.647 kg/L (5.4 lb/gal) and an aluminum metal content of 0.059 kg/L (0.492 lb/gal).

Because this is a weight ratio, the units can be almost anything as long as they are still a weight or a concentration. For example, 9.6 g $Al_2(SO_4)_3 \cdot H_2O$ per g of P could also be mg/mg, lb/lb, or mg/L per mg/L. The weight ratio remains the same.

Step 3—Multiply by the influent concentration to determine total dose needed. Remember that the actual dose may need to be slightly higher because of side reactions.

Chapter 10 includes sample calculations for turning a desired dose in milligrams per liter into a chemical feed rate in kilograms per day (pounds per day) or milliliters per minute.

$$\frac{7\ \frac{mg}{L}\ P}{} \left| \frac{9.6\ \frac{mg}{L}\ Al_2SO_4 \cdot H_2O}{1\ \frac{mg}{L}\ P} \right| = 67.2\frac{mg}{L}\ Al_2SO_4 \cdot H_2O$$

This is the minimum concentration of alum that will be needed in the wastewater. In practice, the actual dose will need to be 1.5 to 2 times higher to account for the alum used up in side reactions when there is at least 1 mg/L P remaining in the wastewater (eq 9.11). Remember that the dose needed increases dramatically when trying to achieve lower and lower phosphorus concentrations. The steps outlined above may be used with any aluminum or iron containing compound as long as the stoichiometry (balanced chemical equation) is known. Weight ratios for aluminum and iron compounds are listed in Tables 9.10 and 9.11.

A *polymer* is a larger chemical compound made by linking together several smaller chemical compounds. Polymers are like beads in a necklace. Each individual bead is a separate chemical compound. The beads may all be the same type or there may be several different types of beads linked together. In polymers, the individual beads are called *monomers*.

Polyaluminum Chloride

An alternative aluminum-based chemical available for phosphorus removal is PACl. It is a *polymer* with a general composition of $Al_nCl_{(3n-m)}(OH)_m$. Instead of numbers, notice that the letters *n* and *m* are used to show that the polymers vary in length, but that the ratios between aluminum (Al^{+3}), chloride (Cl^-), and hydroxide (OH^-) are fixed. Polyaluminum chloride was originally used in drinking water treatment to improve solids removal. In activated sludge processes, PACl has been used to limit the growth of the filamentous bacteria *Microthrix parvicella*. Polyaluminum chloride products vary from one manufacturer to the next and may include additives that change the way the polymer reacts chemically. Polyaluminum chloride use generally does not depress wastewater pH, although the effect on pH will depend on the exact composition.

Trihydrate translates to three waters. It refers to the three water molecules in the chemical formula.

Sodium Aluminate

Sodium aluminate can also serve as a source of aluminum. The chemical formula for sodium aluminate is $Na_2Al_2O_4$ or $NaAlO_2$. One commercial form of sodium aluminate is granular *trihydrate*, which may be written as $Na_2O \cdot Al_2O_3 \cdot 3H_2O$. The formula weight of the granular trihydrate is 215 g/mole. The trihydrate form contains approximately 46% Al_2O_3 and is 25% aluminum by weight. It is also sold as a solution in various strengths. In contrast to alum, which reduces wastewater pH, a rise in pH may occur when sodium aluminate is added to wastewater. The reaction between sodium aluminate and phosphorus is shown in eq 9.15 below. The hydroxide ions (OH^-) produced increase alkalinity and may also increase pH. Excess aluminum ions (Al^{+3}) will still react with hydroxide ions (OH^-) in the water and consume alkalinity, but,

Table 9.11 Properties of Iron-Based Chemicals Used for Phosphorus Precipitation (WEF, 2013)

	Ferric Chloride	Ferric Chloride*	Ferric Sulfate	Ferrous Chloride	Ferrous Sulfate
Form	Dry	37% solution	Dry	Dry	Dry
Formal name	Iron (III) chloride	Iron (III) chloride	Iron (III) sulfate	Iron (II) chloride	Iron (II) sulfate
Chemical formula	$FeCl_3$	$FeCl_3$	$Fe_2(SO_4)_3$	$FeCl_2$	$FeSO_4 \cdot 7H_2O$
Formula weight, g/mole	162.5	162.5	400	126.8	278
Commercial strength, by weight	100%	35 to 45%	100%	20 to 25%	55 to 58%
Iron metal contents, % by weight	34.5	12.8	28	79	20
Weight ratio of dry chemical for stoichiometric precipitation of P	5.24	5.24	6.45	6.14	7.35 (13.5 with attached water)

*37% ferric chloride solution has a specific density of 1.36 kg/L (11.4 lb/gal), a dry $FeCl_3$ content of 0.504 kg/L (4.2 lb/gal), and an iron metal content of 0.173 kg/L (1.44 lb/gal).

in this case, more hydroxide will be generated than consumed. Sodium aluminate in a dry form or in solutions of various strengths was also originally developed for drinking water treatment applications and is typically used when additional alkalinity is required.

$$Na_2O \cdot Al_2O_3 + 2PO_4^{-3} \rightarrow 2AlPO_{4(s)} + 2NaOH + 6OH^- \tag{9.15}$$

IRON SALTS

The most common iron compounds used for phosphorus precipitation from wastewater are ferric chloride ($FeCl_3$) and, sometimes, ferric sulfate [$Fe_2(SO_4)_3$]. Ferric compounds are sometimes added in the collection system, during preliminary treatment (headworks), or at the primary clarifier for odor control. Iron combines with sulfide ion (S^{-2}) to form an insoluble precipitate just as it combines with phosphate ion (PO_4^{-3}) according to the following reaction:

$$2FeCl_3 + 3H_2S \rightarrow Fe_2S_{3(s)} + 6HCl \tag{9.16}$$

Hydrogen sulfide has a strong rotten egg smell. Other odor-causing sulfide compounds include dimethyl disulfide and methyl mercaptan. Both smell like rotten cabbage. Iron compounds are also added to anaerobic digesters to reduce concentrations of hydrogen sulfide in *biogas*. Ferric chloride ($FeCl_3$) is often the chemical of choice for WRRFs because it can be used for multiple purposes.

Biogas is a mixture of methane and carbon dioxide. The methane in biogas is identical to natural gas. The name *biogas* indicates that it is produced through microbial activity.

The chemistry of phosphorus precipitation with iron compounds is similar to that of aluminum compounds. Table 9.11 includes general information for some of the different iron-based chemicals that are used to precipitate phosphorus. *Divalent* ferrous (Fe^{+2}) compounds, primarily in the form of spent pickle liquor, are sometimes used as an inexpensive alternative when locally available. Pickling is a cleaning method that uses strong acids to remove rust, scale, and impurities from iron and steel. The used acid, called *pickle liquor*, contains high concentrations of iron as well as other metals. If pickle liquor is used to precipitate phosphorus, it is important to make sure that the concentrations of other metals, if present, won't cause problems with biosolids disposal or discharge permit compliance.

Divalent refers to an atom or ion with either a −2 or a +2 charge. Trivalent atoms or ions have a −3 or +3 charge and monovalent atoms and ions have a −1 or +1 charge.

Ferrous compounds react with orthophosphate, but don't precipitate as well as ferric compounds. When used, they should be added directly to an aerated activated sludge basin. The oxygen in the basin will convert the ferrous ion (Fe^{+2}) to ferric ion (Fe^{+3}). The conversion from Fe^{+2} to Fe^{+3} will increase the amount of oxygen needed in the activated sludge basin. Ferrous ions should never be added to the final clarifier for phosphorus removal because excess or unreacted ferrous ions will carry over into the disinfection process. For facilities that disinfect with chlorine, ferrous ions will combine with chlorine and convert it to chloride ion. Chlorine is a disinfectant, but chloride is not. For facilities that disinfect with UV light, excess iron (ferric or ferrous) will absorb UV light, which reduces the amount of light available for disinfection. Adding ferric or ferrous iron at the final clarifier can also increase effluent TSS as a result of the metal hydroxide floc that is formed.

Ferric ion (+3 charge) combines with phosphate ion to form ferric phosphate, also called iron III phosphate, as follows:

$$Fe^{+3} + PO_4^{-3} \rightarrow FePO_{4(s)} \tag{9.17}$$

Ferrous ion (+2 charge) combines with phosphate ion to form ferrous phosphate, also called iron II phosphate, as follows:

$$3Fe^{+2} + 2PO_4^{-3} \rightarrow Fe_3(PO_4)_{2(s)} \tag{9.18}$$

Iron ions react with water molecules and other ions in the wastewater in the same way that the aluminum ions reacted to form a chemical floc (eqs 9.12 and 9.13). Iron also reacts with sulfides to form a precipitate (eq 9.16). Again, the amount of iron needed to precipitate phosphorus will be higher than what is predicted by eqs 9.17 and 9.18 because some of the iron will be used up in these side reactions. As with adding aluminum compounds, the required dose increases exponentially when trying to achieve effluent P concentrations below 1 mg/L as shown in Figure 9.21 below.

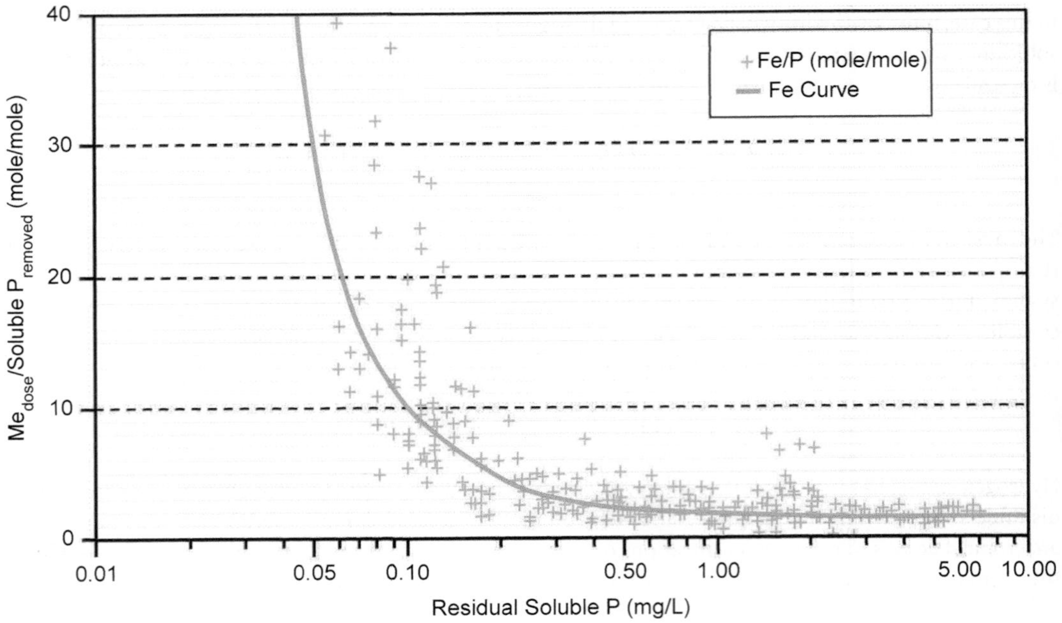

Figure 9.21 Chemical P Removal Ferric Iron Dose Curve (ratio of iron [Fe^{+3}] dose to phosphorus removed as a function of residual soluble orthophosphate concentration) (Luedecke et al., 1987, and data from the Blue Plains Advanced Wastewater Treatment Plant, Washington, D.C.)

The mole ratios from eqs 9.17 and 9.18 can be converted into chemical dosing concentrations in milligrams per liter. Iron has an atomic weight equal to 56 g/mole and phosphorus has an atomic weight of 31 g/mole. For trivalent iron (Fe^{+3}, ferric), this works out to 1.8 g of Fe^{+3} needed to precipitate 1 g of P.

$$\frac{1 \text{ mole } Fe^{+3}}{1 \text{ mole P}} \left| \frac{56 \text{ g } Fe^{+3}}{1 \text{ mole } Fe^{+3}} \right\| \frac{1 \text{ mole P}}{31 \text{ g P}} = 1.8 \text{ g } Fe^{+3}/\text{g P}$$

Ferrous iron (Fe^{+2}) reacts with phosphate ion a bit differently because the charge on the iron ions no longer matches the charge on the phosphate ions (eq 9.18). Ferrous iron has +2 charge and phosphate has a −3 charge. For the final compound to be balanced and neutral (no charge), the reaction requires 3 moles of ferrous iron and 2 moles of phosphate. For divalent iron (Fe^{+2}, ferrous), this works out to 2.71 g of Fe^{+2} needed to precipitate 1 g of P.

$$\frac{3 \text{ moles } Fe^{+2}}{2 \text{ moles P}} \left| \frac{56 \text{ g } Fe^{+2}}{1 \text{ mole } Fe^{+2}} \right\| \frac{1 \text{ mole P}}{31 \text{ g P}} = 2.71 \text{ g } Fe^{+2}/\text{g P}$$

Ferric Chloride

Ferric chloride is a dark brown, oily liquid that contains between 35 and 45% ferric chloride ($FeCl_3$), by weight. It is extremely acidic with a pH of 2 when diluted to 1% strength (U.S. EPA, 1987). Ferric chloride ($FeCl_3$) stains concrete and other materials a characteristic orange. Ferric chloride ($FeCl_3$) reacts with phosphate according to the following reaction (U.S. EPA, 2010).

$$FeCl_3 \cdot 6H_2O + H_2PO_4^- + 2HCO_3^- \rightarrow FePO_4 + 3Cl^- + 2CO_2 + 8H_2O \quad (9.19)$$

The stoichiometric weight ratio of ferric ion (Fe^{+3}) to phosphate ion (PO_4^{-3}) was calculated in the previous section at 1.8 g Fe^{+3}/P. The amount of ferric chloride ($FeCl_3$) that must be added is higher than the weight ratio of Fe^{+3}/P because of the added weight from the chlorine atoms. To find the weight ratio of ferric chloride ($FeCl_3$) to phosphorus, first calculate the formula weight for ferric chloride ($FeCl_3$). Then, use dimensional analysis to find the weight ratio of ferric chloride to phosphorus, as shown below. This is the stoichiometric dose.

$$\frac{1 \text{ mole FeCl}_3}{1 \text{ mole P}}\left|\frac{162.5 \text{ g FeCl}_3}{1 \text{ mole FeCl}_3}\right|\left|\frac{1 \text{ mole P}}{31 \text{ g P}}\right| = 5.24 \text{ g FeCl}_3/\text{g P}$$

Ferrous Sulfate

Ferrous sulfate, the primary ingredient in pickle liquor, is commercially available as a dry powder with the chemical formula $FeSO_4 \cdot 7H_2O$. The water molecules attached to the chemical formula for ferrous sulfate indicate that there is a certain amount of water in the structure dry crystal under normal conditions. This industrial waste product can be used for phosphorus precipitation, but isn't as efficient as using ferric chloride or alum. The primary component of pickle liquor is ferrous sulfate, $FeSO_4$. Ferrous sulfate reacts with phosphate according to the following reaction (Ghassemi and Recht, 1971):

$$3FeSO_4 + 2H_2PO_4 + HCO_3^- \rightarrow Fe_3(PO_4)_{2(s)} + 3SO_4^{-2} + CO_2 + H_2O \quad (9.20)$$

The stoichiometric dose may be calculated from eq 9.20 and the formula weight for ferrous sulfate ($FeSO_4$) as shown below:

$$\frac{3 \text{ moles FeSO}_4}{2 \text{ moles P}}\left|\frac{152 \text{ g FeSO}_4}{1 \text{ mole FeSO}_4}\right|\left|\frac{1 \text{ mole P}}{31 \text{ g P}}\right| = 7.35 \text{ g FeSO}_4/\text{g P}$$

Ferrous iron has a +2 charge, whereas ferric iron has a +3 charge. When ferrous sulfate is added to an aerobic process, most of the ferrous iron (Fe^{+2}) is converted to ferric iron (Fe^{+3}) by DO. It then reacts as described in the previous section on ferric chloride ($FeCl_3$). Some of it remains as ferrous iron. The precipitation products end up being a mix of ferric phosphate ($FePO_4$), ferrous phosphate [$Fe_3(PO_4)_2$], ferric hydroxide [$Fe(OH)_3$], ferrous hydroxide [$Fe(OH)_2$], and other iron hydroxides. The precipitate is made up of small, fine particles that don't settle very well compared to the precipitate produced by adding ferric chloride or alum. The reaction isn't as efficient, which means higher chemical doses are often needed when using ferrous sulfate in place of ferric chloride. Ferrous sulfate, being an industrial waste product, can contain other heavy metals and contaminants, which have the potential to impact both the final effluent and biosolids quality. Still, if pickle liquor is readily available from a nearby industry, it can be more cost-effective than purchasing other chemicals.

Ferrous iron behaves differently when added to an anaerobic process like an anaerobic digester. With no oxygen present, the ferrous iron (Fe^{+2}) remains in the ferrous form. Ferric iron (Fe^{+3}) that is added to an anaerobic process will be reduced to ferrous iron (Fe^{+2}). Ferrous iron (Fe^{+2}) combines preferentially with sulfide (S^{-2}) to form ferrous sulfide (FeS) and then combines with phosphate (PO_4^{-3}) to form ferrous phosphate [$Fe_3(PO_4)_2$]. Ferrous phosphate is commonly known as *vivianite* and precipitates as a dark green to blue crystal. Many WRRFs actively add iron compounds to anaerobic digesters to bind up sulfide and phosphate. Sulfide removal improves the quality of the digester gas and reduces corrosion problems from hydrogen sulfide (H_2S) within the digester and its associated equipment. Phosphate is removed to help prevent the formation of another crystal, magnesium-ammonium-phosphate ($MgNH_4PO_4$). The common name for this compound is *struvite*. Struvite precipitation is problematic for operations when it occurs in piping and solids handling equipment and can be very difficult to remove. Purposely creating vivianite reduces the soluble phosphate concentration in the digester and reduces struvite formation. Ferrous or ferric chloride is recommended for treating headworks, primary clarifiers, anaerobic digesters, and solids-handling equipment because adding ferrous sulfate ($FeSO_4$) can cause more hydrogen sulfide (H_2S) to form, not less. If ferrous chloride or ferric chloride is added to the WRRF headworks or primary clarifier, it will react with any sulfides that may be present before reacting with phosphate. The presence of sulfides will increase chemical consumption and sludge generation.

DRINKING WATER TREATMENT RESIDUALS

Alum and ferric chloride are widely used in drinking water treatment to remove impurities. Alum or ferric is dosed in excess to create chemical floc, which agglomerates colloids, bacteria, and other contaminants. Chemical sludge produced during drinking water treatment is referred to as *residuals*. Many municipalities discharge their drinking water residuals to the collection system where it combines with and removes phosphorus. Ferric chloride sludge also removes hydrogen sulfide. Discharging water treatment residuals to the collection system can reduce or eliminate the need for chemical addition at the WRRF. However, the residuals must be discharged frequently to maintain a steady supply in the WRRF influent. Often, drinking water facilities discharge residuals intermittently and/or discharge widely varying amounts of residuals from one day to the next. Water resource recovery facilities should coordinate with upstream drinking water facilities to make use of this resource when possible.

ROLE OF POLYMERS

Organic polymers do not remove or precipitate soluble phosphorus on their own to any significant extent. However, their use in addition to an aluminum- or iron-based compound can help precipitation products flocculate. In other words, polymer aids in combining smaller particles to form bigger, denser particles. The use of polymer can be particularly effective when low-level phosphorus levels must be met. A typical polymer dose when used as a coagulant aid is 0.1 to 0.25 mg/L (U.S. EPA, 1987). A minimum of 10 seconds lag time (HDT) is recommended between the point of metal addition (aluminum or iron) and the polymer injection point.

Aluminum and iron salts precipitate phosphorus. Aluminum sulfate (alum) and ferric chloride are the most commonly used. The stoichiometric doses are based on the balanced chemical reactions. Actual dosages needed are 1.5 to 2 times higher to reduce soluble phosphorus to 1 mg/L as P. Chemical doses increase exponentially when achieving phosphorus concentrations below 1 mg/L as P. Aluminum and iron addition consumes alkalinity and may cause pH to decrease. Polymer use can increase solids capture.

TEST YOUR KNOWLEDGE

1. **Calcium oxide (lime) is the most commonly used chemical for phosphorus precipitation.**
   - ☐ True
   - ☐ False

2. **Orthophosphate and combined phosphates may be removed through chemical precipitation with ferric chloride or aluminum sulfate.**
   - ☐ True
   - ☐ False

3. **The chemical formula for alum, $Al_2(SO_4)_3 \cdot 14H_2O$, indicates that water molecules make up part of its crystal structure.**
   - ☐ True
   - ☐ False

4. **The chemical symbol for iron is**
   a. In
   b. Ir
   c. Fe
   d. Pb

5. **Which of the following chemicals is expected to produce the greatest quantity of sludge when used to precipitate phosphorus?**
   a. Ferric chloride
   b. Calcium oxide
   c. Aluminum sulfate
   d. Ferrous sulfate

6. **A WRRF has low influent alkalinity and must remove ammonia and phosphorus. Which of the following chemicals would be the best choice for precipitating phosphorus?**
   a. Ferric chloride
   b. Ferrous sulfate
   c. Sodium aluminate
   d. Aluminum sulfate

7. **The stoichiometric dose of aluminum ion to phosphorus is 0.87 g Al^{+3}/g P. If the influent total phosphate concentration is 10 mg/L as P and contains 50% orthophosphate, calculate the stoichiometric dose for removable P.**
   a. 4.35 mg/L
   b. 8.7 mg/L
   c. 12.6 mg/L
   d. 17.4 mg/L

8. **Currently, the WRRF is required to meet an effluent phosphorus limit of 1.5 mg/L as P. The new discharge permit lowers the limit to 0.05 mg/L. The amount of chemical needed to meet the new limit will**
   a. Increase by 33%
   b. Approximately double
   c. Increase by 75%
   d. Increase exponentially

9. **When aluminum sulfate reacts with water, these are formed.**
   a. Insoluble phosphates
   b. Hydroxide compounds
   c. Divalent ions
   d. Insoluble sulfides

10. **How many kilograms (pounds) of ferric chloride does 100 kg (pounds) of a 35% ferric chloride solution contain?**
   a. 17.5
   b. 35
   c. 65
   d. 70

11. **This is one reason a WRRF may decide to use ferric chloride instead of alum for phosphorus precipitation.**
   a. More expensive
   b. Control filaments
   c. Steel mill nearby
   d. Odor control

12. **Compared to ferric chloride, ferrous sulfate is**
   a. Less efficient at precipitating phosphorus
   b. Better at reducing hydrogen sulfide
   c. More flexible with respect to addition points
   d. Available at higher concentrations

13. **A new and improved iron-based chemical, Iron Blast, is being touted by the local sales representative. Given the following information, find the stoichiometric dose for this new chemical as milligram per liter Iron Blast per milligram per liter P.**

   Chemical formula – $Fe(NO_3)_3$
   1.8 g Fe^{+3}/g P, 2.71 g Fe^{+2}/g P
   Fe = 56 g/mole, N = 14 g/mole, O = 16 g/mole, and P = 31 g/mole

   a. 3.8 mg/L
   b. 7.8 mg/L
   c. 6.8 mg/L
   d. 14 mg/L

14. **Vivianite is formed in anaerobic digesters when ____________ is added to the digester.**
   a. Ferric chloride
   b. Ferrous chloride
   c. Either ferric or ferrous chloride
   d. Neither ferric nor ferrous chloride

15. **When either alum or ferric chloride is added to wastewater**
   a. Hydrogen sulfide is precipitated.
   b. pH levels increase.
   c. Sludge production decreases.
   d. Alkalinity is consumed.

SLUDGE PRODUCTION

Aluminum and iron phosphates and hydroxides add to the total mass of sludge produced and will increase solids loading to digesters and thickening and dewatering equipment. Precipitation chemistry is complicated and is heavily influenced by water chemistry. The best way to estimate how much additional sludge will be produced is historic operating data or jar testing if facility data are not available. A good general rule estimate is that about 10 g of chemical sludge will be produced per gram of P removed by chemical addition (U.S. EPA, 2010).

ALKALINITY

Aluminum and iron addition consumes alkalinity (eqs 9.11, 9.12, and 9.13). One mole of aluminum ion (Al^{+3}) reacts with 3 moles of water (H_2O) and produces 2 moles of hydrogen ions (H^+). On a weight ratio basis, every 1 mg/L of alum [$Al_2(SO_4)_3 \cdot 14H_2O$] added will react to produce 0.26 mg/L of insoluble aluminum hydroxide [$Al(OH)_3$] while consuming 0.5 mg/L of alkalinity as $CaCO_3$. Every 1 mg/L of ferric chloride ($FeCl_3 \cdot 6H_2O$) produces approximately 0.4 mg/L of ferric hydroxide [$Fe(OH)_3$] and consumes 0.56 mg/L of alkalinity as $CaCO_3$. While this may not sound like a big reduction in alkalinity, high doses of alum and ferric can have a big impact.

Consider a utility with the following characteristics:

Influent BOD_5 = 300 mg/L
Influent Ammonia (NH_3-N) = 25 mg/L as N
Influent Phosphorus = 8 mg/L as P
Influent Alkalinity = 280 mg/L as $CaCO_3$.

The WRRF must meet an effluent ammonia-nitrogen limit of 2 mg/L as N, but typically removes ammonia to nondetectable levels. Ferric chloride ($FeCl_3$) will be used to precipitate phosphorus. The WRRF must also reduce phosphorus below 1 mg/L as P. Will there be enough alkalinity in its influent to support both objectives?

Step 1—Determine how much alkalinity is needed to support nitrification.

$$\frac{25\frac{mg}{L}NH_3-N}{} \left| \frac{7.14\frac{mg}{L}CaCO_3}{1\frac{mg}{L}NH_3-N} \right| = 178.5\frac{mg}{L}\text{alkalinity consumed}$$

Step 2—Estimate the dose of ferric chloride needed. The weight ratios for alum and ferric chloride were determined previously and can be found in Table 9.10 and 9.11. The system needs to get P below 1 mg/L, so a concentration of 7.5 mg/L will be used to estimate the stoichiometric dose.

$$\frac{7.5\frac{mg}{L}P}{} \left| \frac{5.24\frac{mg}{L}FeCl_3}{1\frac{mg}{L}P} \right| = 39.3\frac{mg}{L}FeCl_3\text{ stoichiometric dose}$$

Step 3—The stoichiometric dose must be multiplied by a factor to account for iron used up by side reactions. To reduce phosphorus down to 1 mg/L as P, a factor of 1.5 to 2 is reasonable. In practice, a larger factor may be needed depending on water chemistry. Only jar testing or full-scale testing can determine the actual dose required.

$$\left(39.3\frac{mg}{L}FeCl_3\right)(1.5) = 58.6\frac{mg}{L}FeCl_3\text{ estimated field dose}$$

Step 4—Estimate the alkalinity consumption from ferric chloride addition.

$$\frac{58.6\frac{mg}{L}FeCl_3}{} \left| \frac{0.56\frac{mg}{L}\text{alkalinity as }CaCO_3}{1\frac{mg}{L}FeCl_3} \right| = 33\frac{mg}{L}\text{alkalinity consumed}$$

Step 5—Estimate the amount of alkalinity remaining in the final effluent.

$$\begin{array}{l} 280\text{ mg/L influent alkalinity as }CaCO_3 \\ -178.5\text{ mg/L alkalinity as }CaCO_3\text{ consumed by nitrification} \\ -33\text{ mg/L alkalinity as }CaCO_3\text{ consumed by ferric chloride addition} \\ \hline 68.5\text{ mg/L alkalinity as }CaCO_3\text{ remaining} \end{array}$$

The alkalinity remaining in the final effluent is below the minimum recommended to sustain nitrification. The operator at this facility may need to add supplemental alkalinity or consider denitrifying to recover some of the alkalinity that was lost. If chemical addition is done after the secondary clarifier, the alkalinity concentration is acceptable as long as the pH is still within the range allowed by the WRRF's discharge permit.

pH RANGE

Phosphorus removal efficiency is highly dependent on pH. If the pH is too low, the aluminum and iron phosphates become more soluble and the precipitate dissolves back into solution. The same phenomenon can be observed by placing a shower head or coffee pot encrusted with calcium carbonate scale in vinegar. The low pH of the vinegar dissolves the calcium carbonate and removes the scale. If the pH is too high, more aluminum and iron is shifted into forming hydroxides, which reduces phosphate precipitation. The goal from an operations standpoint is to maintain the pH within an optimal range that maximizes phosphorus precipitation.

The optimum pH range depends on the chemical used. In Figure 9.22, the shaded gray area highlights the optimum pH range for aluminum sulfate [$Al_2(SO_4)_3$]. Aluminum sulfate (alum) works best when the pH is kept between 5.5 and 6.5, but performs reasonably well up to about pH 7. As the pH increases beyond 7, the percentage of phosphorus removed decreases quickly. Because alum use results in a small pH depression and most existing treatment systems operate at a near neutral pH, adding alum almost automatically results in a pH within the optimum range. Sometimes, it is necessary to add excess alum to bring the pH within the desired range. Sodium aluminate ($Na_2Al_2O_4$) has the same preferred pH range for phosphorus precipitation, but increases alkalinity and possibly pH when added. It is used primarily for systems that already don't have enough alkalinity.

Ferric chloride ($FeCl_3$) works best when the pH is kept between 3.5 and 5, but performs reasonably well up to about pH 6.5. In Figure 9.22, the shaded orange area highlights the pH range where ferric phosphate ($FePO_4$) solubility is the lowest. Outside of this range, the soluble phosphorus concentrations increase. Ferrous phosphate [$Fe_3(PO_4)_2$] is least soluble at pH 8; however, this precipitation reaction primarily takes place in anaerobic environments.

MIXING AND CONTACT TIME

Orthophosphate removal significantly improves if vigorous mixing is provided at the point of chemical addition. The benefits of mixing seem to be greater at a higher pH and are more noticeable with alum than ferric (Sagberg et al., 2006; Szabo et al., 2008). Extending the contact time between the precipitated chemical floc and wastewater allows additional soluble phosphates to adsorb onto the outside of the floc particles and increases removal efficiency. Consequently, chemical addition points should be at a turbulent (well-mixed) location as far upstream from where the solids are removed (i.e., clarifier or filter) as practical. Mixing can be increased by adding a mechanical mixer or air *sparger*.

A *sparger* blows air into the water to mix and aerate. Spargers are typically located on or above the water surface.

CHEMICAL ADDITION POINTS

Aluminum and iron salts may be added at many different locations throughout the treatment process, as shown in Figure 9.23. Precipitation that takes place at any location upstream of the primary clarifiers is called *pre-precipitation*. Co-precipitation describes chemical addition upstream of biological treatment so

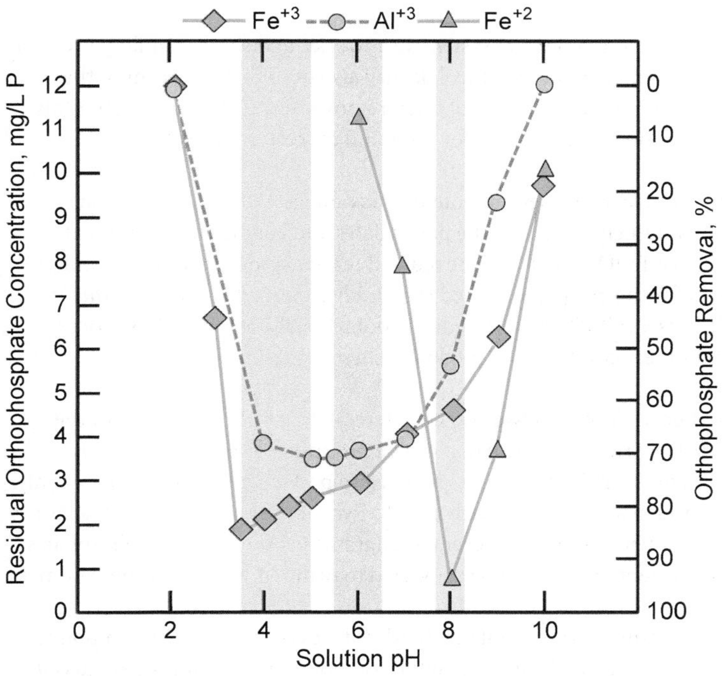

Figure 9.22 Soluble Phosphate Removal Versus Wastewater pH (the optimum pH for ferrous [Fe^{+2}] precipitation is 8, but this primarily occurs in the absence of oxygen) (adapted from Ghassemi and Recht, 1971)

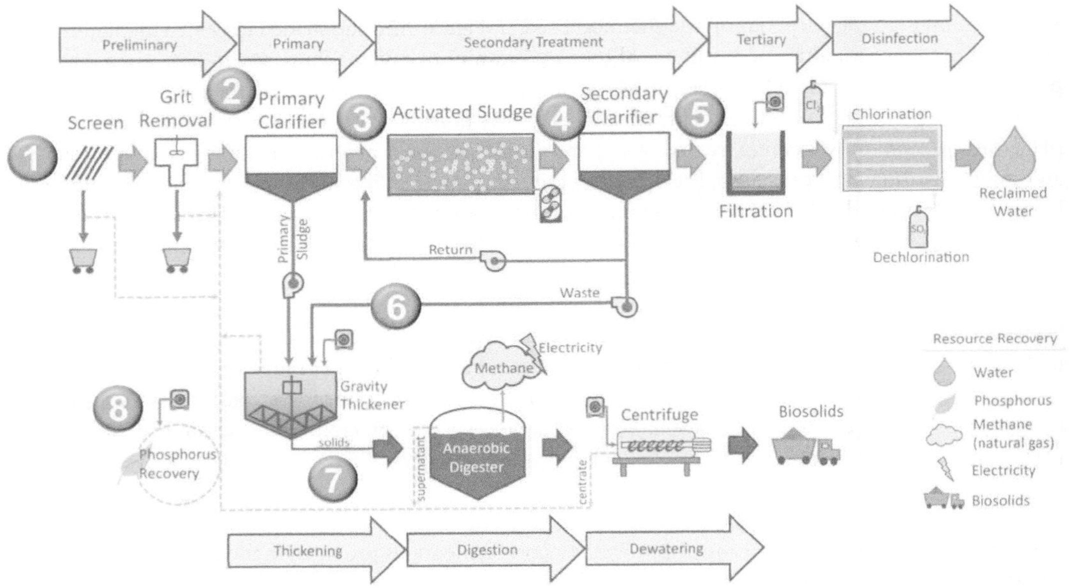

Figure 9.23 Chemical Addition Locations for Phosphorus Removal: ■ 1. In the collection system or facility headworks; ▲■ 2. Upstream of the primary clarifier; ▲ 3. Upstream of the secondary treatment process; ▲ 4. Upstream of the secondary clarifier; ▲ 5. Between the secondary clarifier and a tertiary filter; ■ 6. Directly to the feed for solids thickening equipment; ▲■ 7. Directly to anaerobic digesters; or ▲ 8. Directly to supernatant, centrate, and filtrate recycle streams. (■ = Odor and corrosion control, ▲ = Phosphorus precipitation) (Reprinted with permission by Indigo Water Group)

the precipitate is removed with the biological sludge. Chemical addition done downstream of the secondary clarifier is "post-precipitation". Although chemical addition points are often determined during design and construction of the facility, operators are often able to choose between multiple possible addition points and are sometimes able to modify or move addition points.

Ferric chloride ($FeCl_3$) is typically added to the collection system, headworks, and solids handling processes primarily to precipitate sulfides for odor and corrosion control. It may also be added upstream of the primary clarifier for the same reason. Dosing for odor control often removes some phosphorus, but this isn't the primary goal. Aluminum compounds are not used for odor and corrosion control.

Removing phosphorus in the primary clarifiers has several potential advantages. Aluminum or iron compounds added to the primary clarifier help flocculate the particulates and colloids in the influent wastewater. This increases the amount of BOD_5 and TSS removed and reduces the load to the secondary treatment process. Chemical addition for this purpose is called *chemically enhanced primary treatment (CEPT)*. If ferric or ferrous chloride is used, CEPT can also reduce odors. Finally, because phosphorus concentrations are higher here than in the secondary effluent, lower chemical doses can be used.

Phosphorus removal in primary clarifiers is limited in two ways. First, recall that only about 50% of the phosphorus entering a domestic WRRF is orthophosphate that can be precipitated. Second, primary clarifiers can only remove settleable material. This means precipitated and particulate phosphorus can only be removed if the particles are large enough and dense enough to sink. Even if the best possible phosphorus removal is achieved, the primary effluent will still contain several milligrams per liter of P. Polyphosphates that pass through to the secondary treatment process will be converted to orthophosphate during treatment.

Chemical precipitation in the primary clarifiers has several big disadvantages over other addition points. If too much BOD_5 is removed, denitrification may be limited in the secondary treatment process. Recall that ammonia, which makes up 60 to 70% of influent nitrogen, is soluble and can't be removed by the primary clarifier. If not enough phosphorus remains in the primary effluent, downstream biological treatment processes could be nutrient limited. The third potential problem is that it removes alkalinity before nitrification. If too much alkalinity is removed, the pH can drop and nitrification in the secondary treatment process may be inhibited. Finally, chemical precipitation in the primary clarifiers increases both primary sludge production and overall facility sludge production more than any other addition point.

For activated sludge processes, aluminum or ferric compounds may be added upstream of the activated sludge basins, between the activated sludge basin and secondary clarifier, or after secondary clarification. The first two addition points (3 and 4 in Figure 9.23) are attractive because (1) total phosphate concentrations in the final effluent may be reduced below 1 mg/L P, (2) sludge settling characteristics are often improved, (3) effluent turbidity may be reduced, (4) tertiary filters are typically not required, and (5) simple operation. Many facilities add alum or ferric upstream of activated sludge processes for these reasons. On the other hand, co-precipitation consumes alkalinity that could be used to support nitrification. It increases the mass of MLSS in the process and decreases the percentage of MLVSS. This is because only a portion of the chemical sludge that is generated each day is removed in the waste activated sludge (WAS). The rest is returned to the activated sludge basins where it builds up in proportion to sludge age. This increases the solids loading rate to the secondary clarifiers. Adding chemical phosphorus removal to an existing WRRF may require additional activated sludge basins to keep the MLSS concentration down or additional clarifiers. The recycled chemical sludge contains phosphorus. Normally, MLVSS is between 1.5 and 2.5% phosphorus by weight (0.02 g P/g MLVSS). For facilities that remove phosphorus chemically, the MLSS typically contain 4 or 5% phosphorus by weight. Meeting low effluent phosphorus limits becomes more difficult as a result. Consider a facility with an effluent total phosphorus limit of 0.5 mg/L as P and a final effluent TSS concentration of 12 mg/L. Effluent TSS is escaped MLSS. Without chemical addition, the TSS likely contains about 1.5% phosphorus and the amount of phosphorus leaving with the TSS is only 0.24 mg/L as P. With chemical addition, the TSS may contain 5% phosphorus, which means the particulate P in the final effluent is 0.6 mg/L as P—higher than the discharge permit limit. There will also be some soluble phosphorus in the final effluent. Reducing effluent TSS becomes critically important when trying to meet very low limits. Chemical addition after the secondary clarifier eliminates all of these issues.

Chemical addition for fixed-film processes is typically done upstream of the primary clarifier or upstream of the secondary clarifier (U.S. EPA, 1976). Often, a combination of primary and secondary clarifier dosing is used. Metal salt addition directly to a fixed-film process does not seem to cause any operational problems, although the biofilm may slough more than usual. However, because of the short HDT and lack of mixing, high degrees of phosphorus removal cannot be achieved (U.S. EPA, 1976).

Facilities that must meet discharge permit limits of 0.5 mg/L as P or less typically add chemicals after the secondary clarifier (location 5 in Figure 9.23). Because biological treatment converts organic and condensed phosphates to orthophosphate, nearly all of the phosphorus remaining can be precipitated. Settling out the majority of the solids prior to chemical addition helps reduce chemical usage. Clarification is followed by some type of tertiary filtration. There are many different filter types, but they all perform the same essential task: removing TSS from the final effluent. As a filter operates, it collects solids. In this case, the solids consist of biological solids (MLSS or biofilm) and chemical floc. Eventually, the filter reaches capacity and must be cleaned through backwashing. Backwashing essentially operates the filter in reverse, forcing clean effluent through the filter to remove accumulated solids. The backwashed solids may be sent to the solids handling side of the WRRF. The water used to clean the filter is typically returned to the front of the treatment process. Tertiary filtration increases operational complexity and operating costs.

When used, the polymer addition point should be as far downstream as practicable from the point where the metal salts are added. This will allow the floc to continue adsorbing phosphorus for as long as possible. Polymer and other chemicals should only be added where the wastewater is well mixed or where mixing can be added.

Chemical precipitation depends on the chemical dose, pH, mixing, and contact time. Each metal salt has an optimum pH range where the metal phosphate precipitate produced is least soluble. Stay below pH 6.5 for the most efficient chemical usage. Iron salts also remove sulfide. The addition point selected will affect phosphorus removal, sludge generation, and downstream processes.

TEST YOUR KNOWLEDGE

1. **Jar testing indicates that a dose of 76 mg/L of alum is needed to precipitate phosphorus. How much alkalinity will be consumed?**
   a. 38 mg/L as $CaCO_3$
   b. 43 mg/L as $CaCO_3$
   c. 76 mg/L as $CaCO_3$
   d. 152 mg/L as $CaCO_3$

2. **Chemical precipitation of phosphorus in aerobic processes works best when the pH is**
   a. Acidic
   b. Neutral
   c. Basic
   d. Equilibrated

3. **Ferric phosphate ($FePO_4$) is least soluble at pH**
   a. 2
   b. 5
   c. 6
   d. 9

4. **Chemical addition should only be done in locations where the wastewater is**
   a. Septic
   b. Aerated
   c. Well mixed
   d. Quiescent

5. **A WRRF is able to add alum in three different locations: primary clarifier outlet, activated sludge inlet channel, and the activated sludge outlet box. To maximize contact time for phosphorus precipitation, where should ferric chloride be added?**
   a. Directly to activated sludge basin
   b. Primary clarifier outlet
   c. Activated sludge inlet channel
   d. Activated sludge outlet box

6. **One reason alum may be added upstream of the primary clarifier is to**
   a. Precipitate hydrogen sulfide
   b. Convert polyphosphates
   c. Remove ammonia
   d. Improve clarifier performance

7. **Adding chemicals for phosphorus precipitation after the secondary clarifier is advantageous for this reason.**
   a. Separate sludge handling equipment needed
   b. Increases fine particulates in the final effluent
   c. Prevents alkalinity loss from affecting nitrification
   d. Prevents conversion of ferrous iron to ferric iron

8. **A WRRF must meet an effluent phosphorus limit of 0.10 mg/L total phosphorus. What must be true?**
   a. Chemicals are added only upstream of the primary clarifier.
   b. Effluent suspended solids are removed by filtration.
   c. A combination of biological and chemical removal is required.
   d. MLSS total phosphorus concentration is less than 0.5%

9. **The effluent TSS concentration is 10 mg/L. The MLSS is 4% phosphorus by weight. The influent contains 0.5 mg/L as P of soluble, nonbiodegradable condensed phosphates. Assuming all other phosphorus has been removed, what is the concentration of total phosphorus in the final effluent?**
   a. 0.4 mg/L as P
   b. 0.9 mg/L as P
   c. 4.0 mg/L as P
   d. 4.5 mg/L as P

Biological Phosphorus Removal

The PAOs are a specialized subgroup of facultative heterotrophic bacteria. When these bacteria are cycled between anaerobic and either anoxic or aerated conditions, they will pick up and store excess phosphorus in an anoxic or aerated environment. This process is called "luxury uptake" because the bacteria are incorporating more phosphorus than they need for growth. The PAOs are able to store up to 40% of their dry weight as phosphorus. The overall phosphorus content of the MLVSS, considering other microorganisms, is in the range of 6 to 15%. Phosphorus is removed from the wastewater when excess biomass is removed from the system.

Understanding why luxury uptake occurs begins with understanding what happens to CBOD under anaerobic conditions. In an anaerobic secondary treatment process, some of the CBOD is broken down through fermentation by anaerobic bacteria into soluble CBOD and simpler organic molecules called volatile fatty acids (VFAs). Volatile fatty acids are chemical compounds that contain five carbons or less such as acetic acid (2 carbons), propionic acid (3 carbons), and butyric acid (4 carbons). Acetic acid is more commonly known as vinegar. Volatile fatty acids are a preferred source of carbon and energy by heterotrophic bacteria, including the PAOs, because these compounds are easily absorbed into the bacteria.

The PAOs have a logistical problem—when they are in anaerobic conditions, they are exposed to VFAs, but without oxygen or nitrite or nitrate present, they can't access them. Picking up "fuel" takes energy, which is unavailable to most facultative heterotrophs in the absence of oxygen, nitrite, and nitrate.

The PAOs have a creative solution to this problem. They build a chemical battery from phosphorus. Most cells, including human cells, use a compound called *adenosine triphosphate* (ATP) to store energy. Adenosine triphosphate consists of three phosphate molecules chemically bound together. The chemical bond between phosphorus atoms stores energy until the cell needs it. Adenosine triphosphate is continuously formed and consumed within cells as energy is stored and recovered. The PAOs take ATP to the next level and form an energy-rich compound called *polyphosphate*, which strings together large numbers of phosphate molecules. Polyphosphate (poly-P) is a chemical battery. When the PAO needs energy, it breaks the chemical bonds between phosphorus molecules to recover the stored energy and temporarily releases phosphorus. Glycogen, a form of stored sugar, is consumed in the process. This strategy allows the PAOs to pick up and store VFAs when they are in anaerobic conditions. Because they have a battery pack, they are able to out-compete other types of bacteria for the available VFAs.

Cycling the PAOs between anaerobic and anoxic or aerated conditions is required for luxury uptake to occur. Without exposure to an anaerobic zone, there is no incentive for the PAOs to uptake phosphorus. Figure 9.24 presents the process schematic and concentration profiles for orthophosphate and soluble $CBOD_5$. Figure 9.25 shows a schematic representation of the biochemical processes involved.

When the PAOs are under anaerobic conditions, they use energy stored in polyphosphate to pick up VFAs while simultaneously consuming glycogen. The VFAs are then combined to create an internal storage product called *poly-β-hydroxybutyrate* (PHB). The lack of oxygen or nitrite or nitrate prevents the PAOs from consuming the VFAs immediately. One way to think about the anaerobic zone is like a buffet line. The PAOs heap their plates with tasty VFAs, but they cannot begin eating until they get back to their seats in the anoxic and aerated zones. Stored PHB increases while the amounts of stored polyphosphate and glycogen decrease. As the phosphorus battery is broken down, orthophosphate is released into the surrounding wastewater. This causes the orthophosphate concentration in the wastewater to increase. Concentrations may reach 70 to 90 mg/L as P in the anaerobic zone for typical domestic wastewater. Notice that Figure 9.24 shows increasing orthophosphate concentrations and decreasing soluble $CBOD_5$ as the wastewater moves through the anaerobic zone.

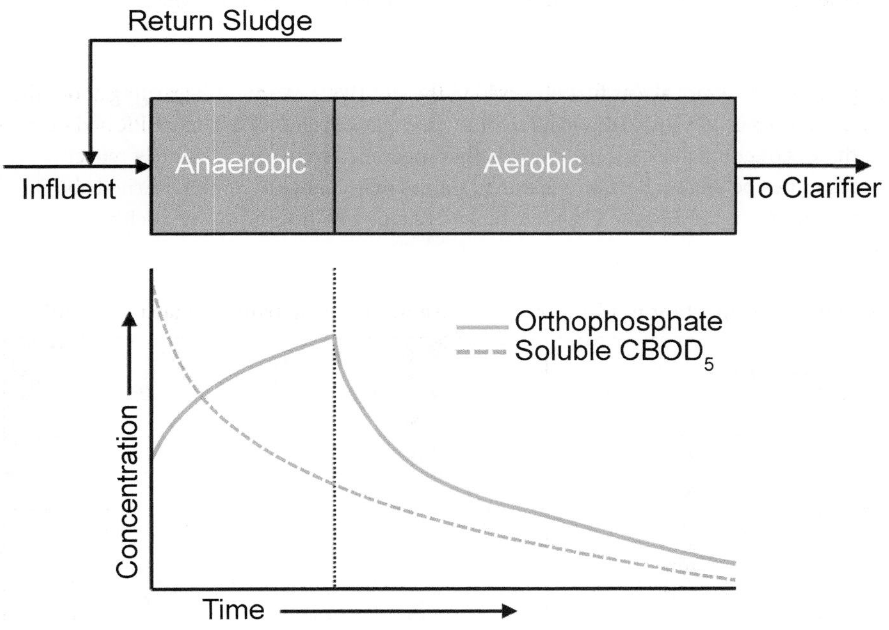

Figure 9.24 Typical Concentration Profiles for Soluble $CBOD_5$ and Orthophosphate in an EBPR Process (U.S. EPA, 1987)

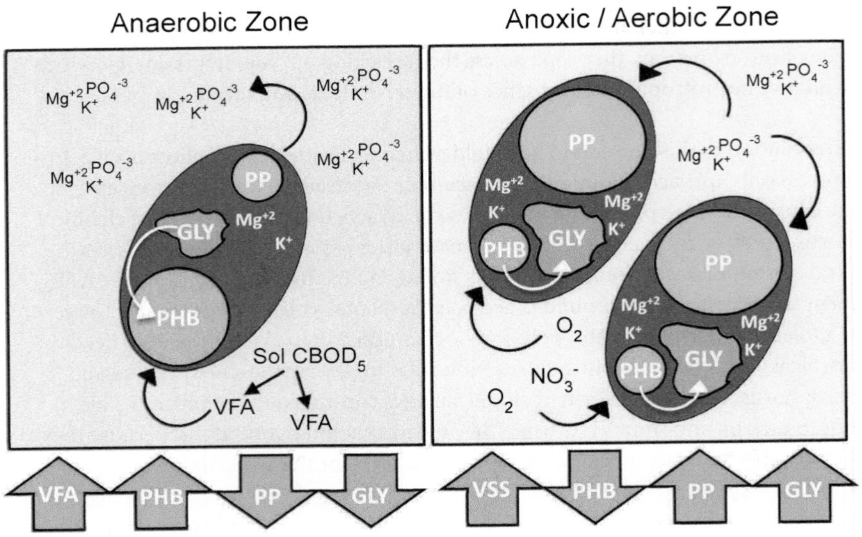

Figure 9.25 Schematic Representation of EBPR (Reprinted with permission by Indigo Water Group)

Recall that it is difficult for bacteria to move charged ions across their cell membranes. Phosphate is large and has a −3 charge. The PAOs overcome this problem by moving a magnesium ion (Mg^{+2}) and a potassium ion (K^{+}) across the cell membrane with the phosphate ion (PO_4^{-3}). The result is a complex or group of ions that is neutral (zero charge). Domestic wastewater typically contains an abundance of both magnesium and potassium; however, EBPR has been limited by a lack of these ions in some industrial wastewaters. In the anaerobic zone, magnesium and potassium are released with phosphorus. In the anoxic and aerobic zones, all three ions are taken up by the PAOs.

When the PAOs move into either anoxic or aerobic conditions, they immediately begin to consume PHB and rebuild their polyphosphate battery. Glycogen is also replenished. The PAOs use the carbon and energy from PHB to grow and reproduce, thereby increasing their numbers. Each new daughter cell takes up orthophosphate to build its own chemical battery. Even though orthophosphate is being continuously released and reabsorbed, the growth of new PAOs means that, overall, phosphate is removed from the passing wastewater. When MLSS is wasted out of the process, phosphorus is removed as well.

STOICHIOMETRY

Enhanced biological phosphorus removal was first observed in the late 1960s. We are still learning about the microorganisms responsible. As of July 2017, three different microorganisms important to EBPR had been identified. However, there is ongoing discussion about whether their current names accurately reflect the genus and species that they belong to. Be aware that their names may change as more is learned about them. Microorganisms important in EBPR are listed in Table 9.12 along with some of their defining

Table 9.12 Microorganisms Important in EBPR (created with information from Barnard et al., 2017a)

	Tetrasphaera	*Accumulibacter*	*Competibacter + Defluviicoccus*
Produces VFAs	Yes	No	No
Consumes VFAs	Possibly	Yes	Yes
Stores PHB	No	Yes—anaerobic	No
Stores glycogen	Yes—anaerobic	Yes—aerobic	Yes—anaerobic
Cycles poly-P	Yes	Yes	No
Denitrifies	Yes	No	Yes
Anoxic uptake of P	Possibly	No	No
Anaerobic zone, mV	Below −300 mV	−50 to −200 mV	−50 to −200 mV

characteristics. *Candidatus Accumulibacter* is the classic PAO organism whose biology is described in the previous section. *Tetrasphaera* also accumulates orthophosphate and stores it as poly-P in the aerobic zone and may also be able to accumulate P under anoxic conditions. Like *Accumulibacter*, *Tetrasphaera* uses the energy from stored phosphorus to pick up organic compounds and create internal storage products. Unlike *Accumulibacter*, *Tetrasphaera* prefers complex organic compounds over VFAs and will produce VFAs when the ORP in the anaerobic zone is less than −300 mV. Both organisms are important for removing phosphorus. Operating anaerobic zones to reach and maintain very low ORP readings (below −300 mV) may encourage the growth of both organisms (Barnard et al., 2017). The third group of organisms is the GAOs, or glycogen-accumulating organisms. The GAOs include *Competibacter* and *Defluviicoccus*. The GAOs compete with the POAs for VFAs in the anaerobic zone, but they don't accumulate and remove phosphorus. When the GAOs become the dominant group in a process, EBPR can fail.

In the anaerobic zone, approximately 0.5 mg P is released for every mg VFA (as COD) that is sequestered and turned into PHB.

PROCESS VARIABLES FOR ENHANCED BIOLOGICAL PHOSPHORUS REMOVAL

INFLUENT CHARACTERISTICS

The most important process variable for EBPR is the ratio of organic carbon to phosphorus. The amount of VFAs and organic material that can be fermented into more VFAs that are available will determine whether or not EBPR can take place. Volatile fatty acids naturally form in the collection system and in the anaerobic zone as a result of fermentation. The important thing to know from an operational perspective isn't necessarily how many VFAs are in the influent, but how many can be generated in the anaerobic zone. Measuring the readily biodegradable $CBOD_5$ is a better predictor of EBPR performance. This measurement includes both VFAs and organic compounds that can become VFAs under the right conditions.

The minimum ratios typically used to determine whether or not a facility will be able to remove phosphorus by EBPR are listed in Table 9.13. These ratios refer to the activated sludge basin influent and should account for recycle loads and removals in primary clarifiers. Typically, recycle flows represent additional phosphorus load, whereas primary clarifiers removed $CBOD_5$. As a result, primary effluent will contain lower organic carbon-to-phosphorus ratios compared to influent.

If adequate VFAs cannot be produced through fermentation in the collection system and/or the anaerobic zone, they may be generated in a couple of different ways within the treatment process. The first option is to operate the primary clarifier with a deep, anaerobic sludge blanket. Primary clarifiers operated in this way are called *active primaries*. Primary sludge will break down anaerobically in the sludge blanket, releasing VFAs that will be carried up and out of the clarifier and into the anaerobic zone. Care should be taken with an active primary not to get the primary sludge so thick that it prevents easy movement of the sludge collection mechanism or that it can't be pumped out of the clarifier. Some facilities have offline fermenters where primary sludge is held in a separate tank to ferment. The supernatant is then returned to the anaerobic zone. Return activated sludge may also be fermented in a similar way. If none of these solutions can be implemented, acetic or propionic acid may be purchased and added to the anaerobic zone as a VFA supplement or chemical precipitation may be used to remove residual phosphorus.

Table 9.13 Minimum Substrate to Phosphorus Requirements for EBPR (WEF, 2013)

Measurement	Organic Carbon-to-Phosphorus Ratio	Remarks
$CBOD_5$	25:1	Provides a rough and initial estimate. Uses data that are typically required by discharge permits.
Soluble $CBOD_5$	15:1	Better indicator than $CBOD_5$.
COD	45:1	More accurate than CBOD. Not measured by all facilities.
VFA	7:1 to 10:1	More accurate than COD. Involves specialized laboratory analysis.
Readily biodegradable COD	15:1	Most accurate. Measures VFA formation potential. Accounts for VFA formation in the anaerobic zone. Requires specialized laboratory tests.

ANAEROBIC ZONE OXIDATION–REDUCTION POTENTIAL

The anaerobic zone performs two critical functions.

1. Volatile fatty acid uptake: Allows PAOs to take up and store VFAs as PHB. This is a rapid reaction requiring approximately 15 to 45 minutes to complete when anaerobic conditions are established at an ORP of around −200 mV (Jeyanayagam and Downing, 2017).
2. Volatile fatty acid production: Under "deep" anaerobic conditions (ORP of −300 mV or less), the fermentable readily biodegradable $CBOD_5$ is converted to VFAs and is available to the PAOs. The reaction kinetics are slower and can take as long as 90 to 120 minutes (Jeyanayagam and Downing, 2017).

To achieve consistently low ORP readings, the anaerobic zone must be protected from both DO and nitrite/nitrate. Sources of DO include turbulence, screw pumps or airlift pumps, free fall over weirs, RAS, and backflow from the anoxic or aerobic zone. Turbulence may be minimized by decreasing mixer speeds in the anaerobic zone. With slow speed top entry mixers, the mixing energy can be reduced to 2 W/m^3 (0.08 hp/1000 cu ft) (Barnard et al., 2017). The goal is to minimize turbulence at the surface that could pull more oxygen into the wastewater. Minimizing DO and nitrate return in the RAS requires optimization of nitrification and denitrification as well as tapered aeration, which was discussed earlier in this chapter. Nitrate and DO recycle can be reduced further by only passing a fraction of the RAS through the anaerobic zone and diverting the rest directly to the anoxic zone (Barnard et al., 2017).

pH

Low pH in either the anaerobic zone or anoxic/aerobic zones can inhibit EBPR. The pH should be kept between 6.5 and 8 in the anaerobic zone and above pH 6.9 in the anoxic/aerobic zones. Outside of these ranges, EPBR efficiency declines rapidly.

SOLIDS RETENTION TIME

The washout SRT for the PAO is lower than it is for the AOB and NOB. A facility that has a well-flocculated MLSS should have an SRT long enough to maintain a stable population of PAOs provided there is an adequate supply of VFAs. Enhanced biological phosphorus removal systems have been shown to operate well at SRTs greater than 3 days.

PROCESS CONTROL FOR ENHANCED BIOLOGICAL PHOSPHORUS REMOVAL

Process control for EBPR consists of

- Ensuring an adequate supply of VFAs,
- Protecting the anaerobic zone,
- Maintaining a strongly negative ORP in the anaerobic zone,
- Maximizing solids capture,
- Minimizing recycle loads,
- Avoiding secondary release of phosphorus, and
- Minimizing competition from GAOs.

SECONDARY RELEASE

Secondary release is the term for release of poly-P without the uptake of VFAs. The phosphorus released will not be taken up again in the aerobic zone and will result in elevated effluent phosphorus concentrations. Table 9.14 shows some of the potential causes of secondary phosphorus release. All are caused by one of two fundamental problems: the PAOs are exposed to anaerobic conditions in a place where VFAs are not available or they run out of their PHB storage product before returning to the anaerobic zone. Secondary release may be caused by overly long aerobic, anoxic, or anaerobic HDTs. Long SRTs can also result in secondary release.

MINIMIZING COMPETITION

Phosphate-accumulating organisms are not the only organisms that can access VFAs in the anaerobic zone. Glycogen-accumulating organisms, which have a similar metabolism to PAOs, can also uptake and store VFAs. They use glycogen as their energy source in the anaerobic zone and don't uptake phosphorus. Because GAOs compete with PAOs for the VFAs without contributing to P removal, their dominance can cause poor EBPR performance.

Table 9.14 Location and Cause of Secondary Phosphorus Release (WEF, 2013)

Environment/Operation	Reason for Secondary Phosphorus Release
Anaerobic zone	VFA depletion because of oversized anaerobic zone
Anoxic zone	Nitrate depletion because of oversized anoxic zone (becomes anaerobic)
Second anoxic zone	Lack of nitrates resulting in anaerobic conditions
Aerobic zone	Long SRT leading to cell death
Final clarifier	Septic conditions caused by deep sludge blanket and low RAS flowrate
RAS piping	Septic conditions from long detention time in pipe
Sludge storage	Septic conditions from poorly or unaerated sludge storage
Aerobic digestion	Cell death and phosphorus release
Anaerobic digestion	Cell death and phosphorus release

Factors responsible for PAO-GAO competition include carbon source composition, temperature, and pH. Conditions that promote GAO proliferation generally include wastewater warmer than 20 °C (68 °F) and low pH. Both of these conditions are more likely to occur with industrial rather than municipal wastewater (Jeyanayagam and Downing, 2017). The GAOs appear to compete better when the VFAs in the anaerobic zone are 100% acetic acid or 100% propionic acid. When the anaerobic zone contains a mixture of different VFAs, the PAOs tend to dominate. For facilities that must add supplemental carbon, a blend of VFAs is preferred (Jeyanayagam and Downing, 2017).

For EBPR to take place, the PAOs must be cycled between anaerobic and anoxic/aerobic conditions. A mixture of different types of VFAs must be available to prevent competition from GAOs. For best performance, keep the ORP in the anaerobic zone below −300 mV.

Process Configurations

There are many variations on the activated sludge process that contain more or fewer anaerobic, anoxic, and aerated basins/zones switched around in different ways. Some have a single internal recycle line or none at all. Others have multiple recycle lines connecting different types of basins/zones. Several examples of commonly used configurations are shown in Figures 9.26 through 9.29 below. Engineers and operators have tweaked the designs in an attempt to achieve the lowest effluent nitrogen and phosphorus concentrations possible. The important thing to remember is that even though the processes look different, the same fundamental principles apply to all of them. Each process tries to minimize the amount of oxygen recycled to the anoxic zone; keep oxygen, nitrite, and nitrate out of the anaerobic zone; and make the most efficient use of influent (or primary effluent) BOD_5.

The *A²/O* process adds an anaerobic zone to the MLE process (Figure 9.26). The A²/O can attain total phosphorus concentrations below 1 mg/L as P and total nitrogen concentrations as low as 8 mg/L as N (WEF et al., 2018). Notice that IMLR is returned to the anoxic zone and not the anaerobic zone. For the anaerobic zone to remain anaerobic, it must be protected from recycle of nitrite and nitrate as well as DO. The internal recycle rate typically ranges from 100 to 400% of the influent flow.

The *A²/O* name comes from the three zones: **A**naerobic, **A**noxic, **O**xic. *Oxic* means to contain oxygen. *Oxic* and *aerobic* are used interchangeably.

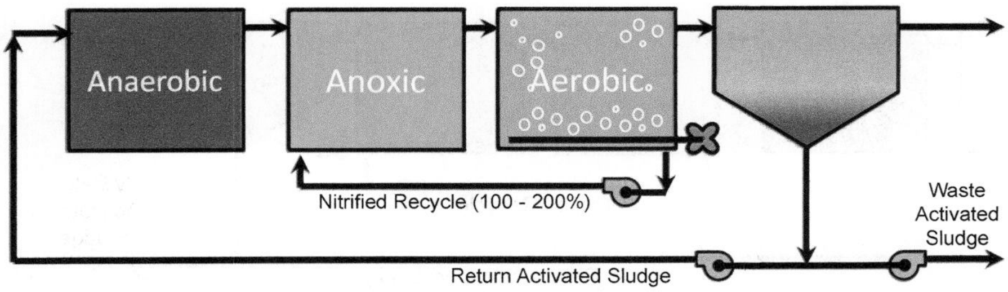

Figure 9.26 A²O Process (Three Stage BNR) (Reprinted with permission by Indigo Water Group [modified from U.S. EPA, 1993])

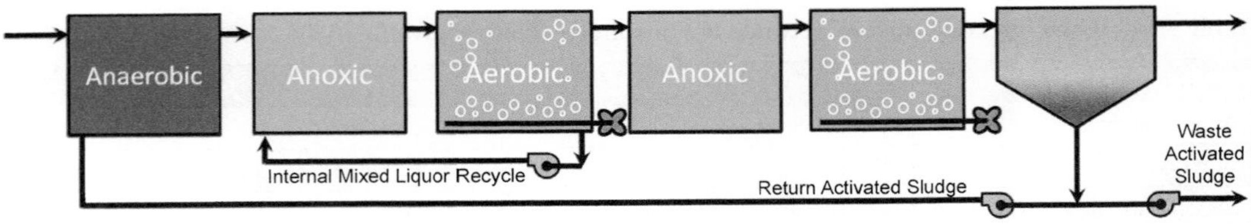

Figure 9.27 Modified Five-Stage Bardenpho Process (**Reprinted with permission by Indigo Water Group** [**modified from U.S. EPA, 1993**])

The Bardenpho process, named for Dr. James Barnard, may be configured with four stages or five. The only difference between the two configurations is that the five-stage Bardenpho includes an anaerobic zone, as shown in Figure 9.27. Most of these facilities have required supplemental chemical addition (metal salts and/or carbon) to meet effluent phosphorus limits of less than 1.0 mg/L as P (WEF et al., 2018).

Researchers at the University of Cape Town (UCT) developed the process shown in Figure 9.28. The original UCT process contained a single anoxic zone. The modified UCT process contains two separate anoxic zones. Both the RAS and aerated tank contents are recycled to the second anoxic zone, and the contents of the anoxic zone are then recycled to the anaerobic zone. This recycle sequence decreases the chance of introducing residual nitrate to the anaerobic zone. It may seem odd, but the RAS will have a higher nitrate concentration than the anoxic zone. Nitrate returned to the anoxic zone is diluted by the influent wastewater and is rapidly consumed by microorganisms. The internal recycle can be controlled to maintain near zero nitrates in the anoxic zone. Although the process effectively eliminates nitrate recycle to the anaerobic zone, the MLSS concentration is lower in the anaerobic zone than in the rest of the process.

The Johannesburg process, shown in Figure 9.29, was also developed in South Africa. The distinguishing feature of this process is the addition of an anoxic zone ahead of the anaerobic zone. This pre-anoxic zone does not receive any influent flow. Its purpose is to hold the RAS long enough for any remaining nitrates to be converted to nitrogen gas. The microorganisms use stored carbon compounds like glycogen to denitrify. Influent is fed directly to the anaerobic zone to take advantage of the VFAs generated in the collection system to fuel EBPR.

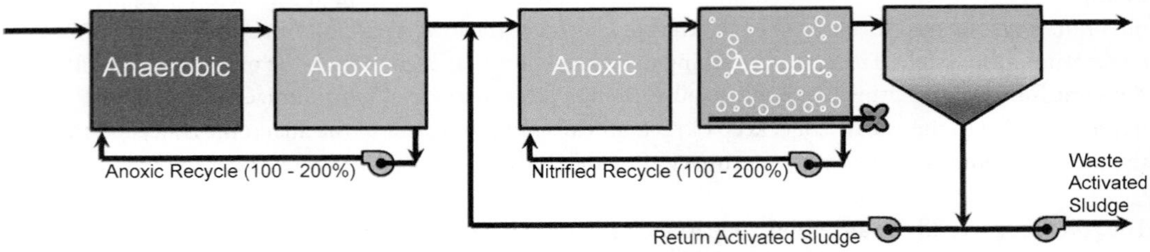

Figure 9.28 Modified UCT Process (**Reprinted with permission by Indigo Water Group** [**modified from U.S. EPA, 1993**])

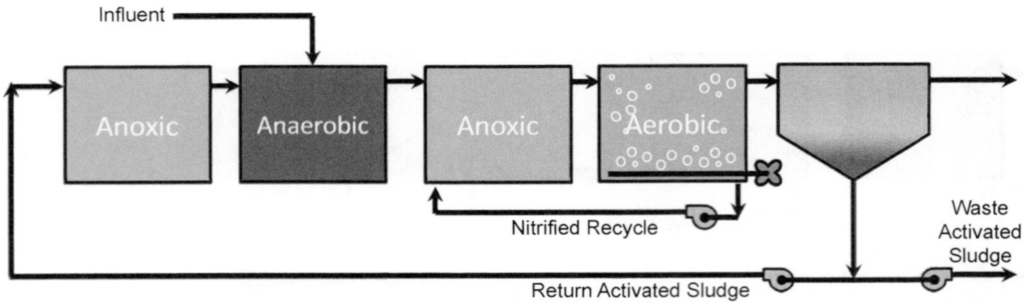

Figure 9.29 Johannesburg Process (**Reprinted with permission by Indigo Water Group** [**modified from U.S. EPA, 2010**])

TEST YOUR KNOWLEDGE

1. Luxury uptake of phosphorus occurs
   a. In the anaerobic zone
   b. When the ORP is −250 mV
   c. In the presence of nitrate or oxygen
   d. When the SRT is less than 3 days

2. Volatile fatty acids include
   a. Glucose
   b. Acetic acid
   c. Sulfuric acid
   d. Carbohydrates

3. Which of the following biochemical reactions take place in the anaerobic zone?
   a. Consumption of PHB
   b. Storage of glycogen
   c. Release of phosphorus
   d. Secondary release

4. In an EBPR process, where will the liquid phase phosphorus concentration be the highest?
   a. Anaerobic zone
   b. Anoxic zone
   c. Aerated zone
   d. Secondary clarifier

5. Growth and reproduction of PAOs primarily takes place
   a. During VFA uptake
   b. When PHB is consumed
   c. During phosphorus release
   d. When glycogen is consumed

6. Phosphorus that is removed from an EBPR process
   a. Volatilizes to the atmosphere
   b. Combines with aluminum or iron
   c. Is sequestered in the WAS
   d. Occurs only under anoxic conditions

7. This organism is capable of both fermentation and luxury uptake of phosphorus.
   a. *Accumulibacter*
   b. *Tetrasphaera*
   c. *Competibacter*
   d. *Defluviicoccus*

8. About how long does it take for the PAOs to complete uptake of VFAs when they are in the anaerobic zone?
   a. Less than 10 minutes
   b. 15 to 45 minutes
   c. 90 to 120 minutes
   d. Longer than 3 hours

9. An extended aeration activated sludge process is operated with an SRT of 15 days in the winter. The new supervisor decides to increase the SRT to 35 days. How will this decision affect final effluent quality?
   a. Phosphorus concentrations may increase.
   b. Ammonia concentrations will decrease.
   c. Nitrite concentrations will increase.
   d. Orthophosphorus concentrations may decrease.

10. Phosphorus-accumulating organisms use VFAs in the anaerobic zone to do which of the following?
    a. Grow
    b. Gain energy
    c. Convert to internal storage products
    d. Reduce phosphorus

11. Which one of these is an indicator of good EBPR performance?
    a. High phosphorus content of MLSS
    b. Low BOD_5 concentration in the effluent
    c. Increased oxygen consumption in aerobic zone
    d. Negative ORP readings in the anoxic zone

12. Which ions are released in the anaerobic zone and taken up with phosphorus in the anoxic/aerated zones of an EBPR system?
    a. Magnesium and calcium
    b. Sodium and chloride
    c. Magnesium and potassium
    d. Bicarbonate and hydrogen

CHAPTER SUMMARY

	Nitrogen	Phosphorus
INFLUENT PARAMETERS	■ Expressed as N ■ TKN = NH_3-N + Organic Nitrogen ■ TIN = NH_3-N + NO_2-N + NO_3-N ■ TN = TKN + NO_2-N + NO_3-N ■ 23 to 69 mg/L as N in influent	■ Expressed as P ■ Orthophosphate and condensed phosphates are soluble. Organic phosphate may be soluble, colloidal, or particulate. ■ 6 to 8 mg/L as P in influent typical ■ Polyphosphate and organic phosphates converted to orthophosphate during biological treatment
REMOVAL METHODS	■ Assimilative uptake for biomass growth ■ 15 to 30% N removed by assimilative uptake ■ Biological nitrification, NH_3-N < 1 mg/L ■ Biological denitrification, NO_3-N < 5 mg/L	■ Assimilative uptake for biomass growth ■ Approximately 1 mg/L P removed by assimilative uptake for every 100 mg/L BOD_5 ■ Chemical phosphorus removal, P < 0.03 mg/L ■ EBPR < 1.0 mg/L P ■ Methods often used in combination

	Nitrification	Denitrification
STOICHIOMETRY	■ Nitrification is a two-step process. The AOB convert ammonia to nitrite and the NOB convert nitrite to nitrate. ■ Converting 1 mg/L of NH_3-N to NO_3-N consumes 4.57 mg/L of DO and 7.14 mg/L of alkalinity as $CaCO_3$. ■ The AOB and NOB grow 10 to 20 times slower than heterotrophic bacteria. ■ Only 0.10 g of new bacteria (volatile suspended solids) are produced per g of NH_3-N converted to NO_3-N.	■ Denitrification converts nitrate to nitrogen gas. ■ Many different facultative heterotrophic bacteria can denitrify. ■ DO concentrations as low as 0.3 mg/L can inhibit denitrification. ■ Converting 1 mg/L of NO_3-N to N_2 requires 4 mg/L of influent BOD_5 and generates 3.57 mg/L of alkalinity as $CaCO_3$. ■ Nitrate produced is the equivalent of 2.86 mg/L of DO when recycled to an anoxic zone. ■ Methanol and other organic compounds increase the amount of nitrate removed. Between 2.7 and 3.3 mg/L of methanol is needed for every 1 mg/L NO_3-N. ■ Denitrification reduces sludge yield. ■ Yield depends on carbon source.

PROCESS VARIABLES	■ The $SRT_{aerobic}$ is the most important process control variable for activated sludge processes. ■ SRT is highly temperature dependent. ■ OLR is the most important variable for fixed-film processes. ■ Soluble BOD_5 must be less than 15 mg/L for nitrification in biofilms. ■ DO concentrations of 2 mg/L help ensure maximum nitrification rates in activated sludge processes. ■ Low DO may cause nitrite lock. ■ Nitrification rates are highest between pH 7.5 and 8.0, but nitrifiers can acclimate to lower pH. ■ pH inhibits nitrification by limiting the availability of un-ionized ammonia (NH_3). ■ A minimum alkalinity concentration of 50 mg/L as $CaCO_3$, and preferably 100 mg/L, is needed to prevent pH drop. ■ AOB can't store ammonia and don't grow quickly enough to take advantage of transient ammonia loads.	■ Denitrification is limited by the availability and type of organic carbon. ■ Anoxic tanks or zones contain nitrate, but no oxygen. ■ Nitrate is produced in areas with aerobic conditions and must be returned to areas with anoxic conditions. ■ The specific denitrification rate (SDNR) is how fast denitrification occurs. ■ SDNR depends on the organic carbon to nitrate ratio. ■ The higher the F/M, the faster denitrification will occur. ■ Denitrification is faster with soluble BOD than particulate. ■ Pre-anoxic zones denitrify faster than post-anoxic zones because more organic carbon is available. ■ Activated sludge processes with a single anoxic zone or basin can achieve total effluent nitrogen of <10 mg/L as N or up to 85% removal of the total oxidizable nitrogen in the influent. ■ Increasing the IMLR will increase nitrate removal when organic carbon is available.
PROCESS CONTROL	■ Retain biomass in the process long enough to allow the nitrifying bacteria to reproduce and build a stable population. ■ Use safety factor to prevent moving above and below the minimum $SRT_{aerobic}$ required to prevent nitrite accumulation. ■ Reduce competition from heterotrophic bacteria by minimizing organic loading rates. ■ Ensure adequate supplies of ammonia, DO, and alkalinity. ■ Manage recycle streams to keep ammonia loading as consistent as possible.	■ Add supplemental BOD to maintain a BOD to NO_3-N ratio of at least 4 : 1. ■ For activated sludge processes, increase the IMLR to decrease effluent NO_3-N. ■ Recycle ratios typically less than 400% of influent flow. ■ Ensure the anoxic zone remains anoxic by minimizing DO recycled in the IMLR and RAS. ■ Keep anoxic zone ORP between −50 mV and +50 mV. ■ ORP readings are dependent on electrode type and fill solution used. ■ For fixed-film processes, increasing recycle flow increases denitrification.

	Chemical Precipitation	Enhanced Biological Phosphorus Removal
STOICHIOMETRY	■ Metal salts combine with orthophosphate ion to form insoluble compounds that come out of solution (precipitate). ■ Aluminum sulfate (alum), ferric chloride, and sodium aluminate are the most commonly used chemicals for phosphorus precipitation. ■ Weight ratios are calculated from balanced chemical equations. The stoichiometric dose to precipitate of 1 mg/L of phosphate as P requires • 0.87 mg/L aluminum ions, or • 1.8 mg/L ferric iron ions, or • 9.6 mg/L alum, or • 2.64 mg/L sodium aluminate, or • 5.24 mg/L ferric chloride. ■ Side reactions consume aluminum and iron ions by forming hydroxides. ■ To reduce effluent P to 1 mg/L, 1.5 to 2 times the stoichiometric dose is typically needed. ■ To meet ultra-low effluent P limits, 6 to 7 times the stoichiometric dose may be needed. ■ Chemical precipitation increases sludge production.	■ MLSS is cycled between anaerobic and anoxic/aerobic conditions to promote the growth of PAOs. ■ *Accumulibacter* and *Tetrasphaera* perform luxury uptake of phosphorus. ■ Phosphate stores energy as poly-P. ■ Under anaerobic conditions, poly-P is used to take up VFAs. ■ VFAs are stored as poly-β-hydroxybutyrate (PHB). ■ Magnesium and potassium ions neutralize phosphate ions and make it easier to pass them through the cell membrane. ■ Fermentation occurs in the anaerobic zone. ■ *Tetrasphaera* performs fermentation when ORP is below −300 mV. ■ Luxury uptake occurs in the anoxic and aerobic zones.
PROCESS VARIABLES	■ Aluminum and iron addition consumes alkalinity (mg/L as $CaCO_3$). • 1 mg/L alum consumes 0.5 mg/L. • 1 mg/L ferric consumes 0.56 mg/L. ■ Supplemental alkalinity may be needed. ■ Sodium aluminate increases alkalinity. ■ Wastewater pH affects the amount of precipitate formed. Optimum pH range depends on the chemical used. ■ Alum = 5.5 to 6.5 SU ■ Ferric chloride = 3.5 to 5.0 ■ Both work well up to pH 6.5. ■ Ferrous iron converts to ferric iron in aerobic processes. Precipitation in anaerobic processes is best at pH 8. ■ Improve P removal with rapid chemical mixing and longer contact times. ■ Iron compounds are added at different locations to achieve different process objectives including odor and corrosion control, improvement of clarifier performance, reducing struvite formation, and precipitating phosphorus. ■ Aluminum compounds are used to remove phosphorus and improve clarifier performance.	■ The organic carbon to P ratio is the most important process control variable for EBPR. ■ Organic carbon must be available as VFAs or readily biodegradable $CBOD_5$. ■ VFAs are produced in the collection system and anaerobic zone. ■ VFAs may also be produced by fermenting sludge in an active primary or other location. ■ pH below 6.5 in the anaerobic zone inhibits EBPR. ■ pH below 6.9 in the anoxic/aerobic zones inhibits EBPR. ■ A minimum SRT of 3 days is needed to maintain PAOs in an activated sludge process.

PROCESS CONTROL		
PROCESS CONTROL	■ Perform jar tests followed by full-scale testing to confirm dosing calculations. ■ Add chemicals as far upstream as possible from the precipitate removal to maximize contact time. ■ Add polymer as far downstream of metal salt addition as possible. ■ Control dosages upstream of the primary clarifier to prevent • excessive BOD_5 removal, • excessive alkalinity consumption, and • inhibition of nitrification or denitrification in the secondary treatment process. ■ Minimize effluent TSS.	■ Process control for EBPR includes • Ensuring an adequate supply and variety of VFAs, • Preventing DO and nitrate from reaching the anaerobic zone, • Keeping the ORP below −300 mV in the anaerobic zone, • Minimizing TSS in the final effluent, • Managing recycle loads, • Avoiding secondary release of P, and • Minimizing competition from GAOs.

References

Barnard, J. (1973) Biological Denitrification. *J. Water Pollut. Control Fed.*, **72**, 705–720.

Barnard, J. L.; Dunlap, P.; Steichen, M. (2017a) A Comprehensive Theory to Understand All Biological Phosphorus Removal Observations. *Proceedings of the Water Environment Federation Nutrient Symposium*; Fort Lauderdale, Florida, June 12–14; Water Environment Federation: Alexandria, Virginia.

Barnard, J. L.; Dunlap, P.; Steichen, M. (2017b) Rethinking the Mechanisms of Biological Phosphorus Removal. *Water Environ. Res.*, **89**, 2043–2054.

Barth, E.; Brenner, R.; Lewis, A. R. (1968) Chemical Biological Control of Nitrogen and Phosphorus in Wastewater Effluent. *J. Water Pollut. Control Fed.*, **40**, 2054.

Basu, S.; Pilgram, S.; Keck, D.; Painter, C. (2006) ORP and pH Based Control of SBR Cycles for Nutrient Removal from Wastewater. *Proceedings of the 79th Annual Water Environment Federation Technical Exhibition and Conference* [CD-ROM]; Dallas, Texas, Oct 21–25; Water Environment Federation: Alexandria, Virginia.

Ghassemi, M.; Recht, H. (1971) *Phosphate Precipitation with Ferrous Iron;* U.S. Environmental Protection Agency, Office of Research and Monitoring: Canoga Park, California.

Gu, A.; Liu, L.; Neethling, J.; Stensel, H.; Murthy, A. S. (2011) Treatability and Fate of Various Phosphorus Fractions in Different Wastewater Treatment Processes. *Water Sci. Technol.*, **63** (4), 804–810.

Han, Z.; Zhu, J.; Ding, Y.; Wu, W.; Chen, Y.; Zhang, R.; Wang, L. (2011) Effect of Feeding Strategy on the Performance of Sequencing Batch Reactor with Dual Anoxic Feedings for Swine Wastewater Treatment. *Water Environ. Res.*, **83**, 643–649.

Jeyanayagam, S.; Downing, L. (2017) Converting to Biological Phosphorus Removal Can Be Cost-Effective for Ohio Plants. *Buckeye Bull.*, **90** (2), 64–68.

Luedecke, C.; Hermanowicz, S.; Jenkins, D. (1987). Precipitation of Ferric Phosphate in Activated Sludge: A Chemical Model and Its Verification. *Water Sci. Technol.*, **21**, 325–338.

Manser, R.; Gujur, W.; Siegrist, A. H. (2005) Consequences of Mass Transfer Effects on the Kinetics of Nitrifiers. *Water Res.*, **39** (19), 4633–4642.

Metcalf and Eddy, Inc./AECOM (2014) *Wastewater Engineering Treatment and Resource Recovery,* 5th ed; McGraw-Hill Education: New York.

Metcalf and Eddy, Inc. (2003) *Wastewater Engineering Treatment and Reuse,* 4th ed.; McGraw-Hill: New York.

Muirhead, W.; Appleton, R. (2008) Operational Keys to Nitrite Lock. *Water Pract.,* **2** (4).

Sagberg, P.; Ryfors, P.; Berg, K. (2006) 10 Years of Operation of an Integrated Nutrient Removal Treatment Plant: Ups and Downs; Background and Water Treatment. *Water Sci. Technol.,* **83**.

Spencer, M. (2012) *The Everyman's Guide to the Miraculous but Misunderstood ORP Sensor.* April 18, 2012. Retrieved June 25, 2017, from Water Analytics, Inc.: http://www.wateranalytics.net/sites/default/files/Aquametrix%20ORP%20White%20Paper.pdf (accessed Nov 1, 2017).

Szabo, A.; Takacs, I.; Murthy, S.; Daigger, G.; Licsko, I.; Smith, D. (2008) The Significance of Design and Operational Variables in Chemical Phosphorus Removal. *Water Environ. Res.,* **80**, 407.

U.S. Environmental Protection Agency (1976) *Process Design Manual for Phosphorus Removal;* EPA 625/1-76-001a; U.S. Environmental Protection Agency, Technology Transfer, Great Lakes National Program Office: Chicago, Illinois.

U.S. Environmental Protection Agency (1987) *Design Manual: Phosphorus Removal;* EPA 625/1-87/001; U.S. Environmental Protection Agency Center for Environmental Research Information, Water Engineering Research Laboratory: Cincinnati, Ohio.

U.S. Environmental Protection Agency (1993) *Manual: Nitrogen Control;* EPA 625/R-93/010; U.S. Environmental Protection Agency, Office of Research and Development, Office of Water: Washington, D.C.

U.S. Environmental Protection Agency (2010) *Nutrient Control Design Manual;* EPA 600/R-10/100; U.S. Environmental Protection Agency, Office of Research and Development, National Risk Management Research Laboratory—Water Supply and Resources Division: Cincinnati, Ohio.

Water Environment Federation (2013) *Operation of Nutrient Removal Facilities;* Manual of Practice No. 37; Water Environment Federation: Alexandria, Virginia.

Water Environment Federation (2015) *Shortcut Nitrogen Removal—Nitrite Shunt and Deammonification;* Water Environment Federation: Alexandria, Virginia.

Water Environment Federation (2016) *Operation of Water Resource Recovery Facilities,* 7th ed.; Manual of Practice No. 11; Water Environment Federation: Alexandria, Virginia.

Water Environment Federation (2018) *Activated Sludge and Nutrient Removal,* 3rd ed.; Water Environment Federation: Alexandria, Virginia.

Water Environment Federation; American Society of Civil Engineers; Environmental and Water Resources Institute (2018) *Design of Water Resource Recovery Facilities,* 6th ed.; WEF Manual of Practice No. 8/ASCE Manuals and Reports on Engineering Practice No. 76; Water Environment Federation: Alexandria, Virginia.

Suggested Readings

Spencer, M. (2012) *The Everyman's Guide to the Miraculous but Misunderstood ORP Sensor.* April 18, 2012. Retrieved June 25, 2017, from Water Analytics, Inc.: http://www.wateranalytics.net/sites/default/files/Aquametrix%20ORP%20White%20Paper.pdf (accessed Nov 1, 2017).

U.S. Environmental Protection Agency (1987) *Design Manual: Phosphorus Removal;* EPA 625/1-87/001; U.S. Environmental Protection Agency Center for Environmental Research Information, Water Engineering Research Laboratory: Cincinnati, Ohio.

U.S. Environmental Protection Agency (1993) *Manual: Nitrogen Control;* EPA 625/R-93/010; U.S. Environmental Protection Agency, Office of Research and Development, Office of Water: Washington, D.C.

U.S. Environmental Protection Agency (2010) *Nutrient Control Design Manual;* EPA 600/R-10/100; U.S. Environmental Protection Agency, Office of Research and Development, National Risk Management Research Laboratory—Water Supply and Resources Division: Cincinnati, Ohio.

Water Environment Federation (2013) *Operation of Nutrient Removal Facilities;* Manual of Practice No. 37; Water Environment Federation: Alexandria, Virginia.

CHAPTER 10

Disinfection

Introduction

Waterborne diseases arise from the contamination of water by any dozens of potential *pathogens* including viruses, bacteria, and protozoa. The pathogens of greatest concern are *enteric* bacteria, viruses, and parasites. Diseases that can be spread via bacterial contamination include salmonellosis (e.g., typhoid and paratyphoid fevers), cholera, gastroenteritis from enteropathogenic *Escherichia coli (E. coli)*, and shigellosis (bacillary dysentery). Infectious viruses include the hepatitis virus, poliovirus, Coxsackie viruses A and B, eschoviruses, reoviruses, and adenoviruses (U.S. EPA, 1986). Parasitic diseases include giardiasis and amoebic dysentery.

A *pathogen* is a virus, bacteria, or protozoan capable of causing illness in humans.

Enteric refers to microorganisms that live in the intestines of humans and animals.

Disinfection kills or damages potentially infectious microorganisms. Damaged microorganisms are incapable of reproducing and can't cause disease. Disinfection is not the same as sterilization. Sterilization kills or inactivates all living things. Disinfection inactivates microorganisms or reduces their numbers to safe levels where infection becomes unlikely. This chapter will describe the most commonly used disinfection practices including gaseous chlorine (Cl_2), sodium hypochlorite (NaOCl), calcium hypochlorite [$Ca(ClO)_2$], and UV irradiation.

LEARNING OBJECTIVES

Upon completing this chapter, you will be able to

- Compare and contrast the goal of disinfection versus sterilization;
- Explain the concept of indicator organisms and list six characteristics that make an ideal indicator organism;
- Compare and contrast testing methodologies for indicator organisms;
- Discuss the effect of temperature and pH on chlorine chemistry;
- Predict the effect of various chemical components, for example, nitrite and ammonia, on chlorine residual;
- Understand the difference between combined, free, and total chlorine residual (TCR);
- Explain why understanding the breakpoint curve is important for facilities that nitrify;
- Manipulate contact time (CT) or residual to achieve a desired level of inactivation of indicator organisms;
- Adjust chlorine dose to meet discharge permit limits for indicator organisms;
- Label all components of different types of disinfection equipment and describe the functions of each;
- Inspect, operate, and maintain chlorination, dechlorination, and UV disinfection equipment;
- Place disinfection equipment into service and remove it from service safely;
- Calculate chemical feed rate, dose, or flow given two of the three variables;
- Determine required chemical feed pump settings in milliliters per minute (mL/min);
- Discuss the mechanism behind UV inactivation of genetic material;
- Explain how effluent quality can affect UV disinfection efficiency; and
- Troubleshoot common mechanical and process control problems.

Chemical Symbols and Formulas

$Ca(ClO)_2$—Calcium hypochlorite

$Ca(OH)_2$—Calcium hydroxide

$CaCO_3$—Calcium carbonate

Cl^-—Chloride ion

Cl_2—Chlorine gas

Cl_2O—Chlorine monoxide

Fe^{+0}—Solid iron metal

Fe^{+2}—Ferrous iron ion

Fe^{+3}—Ferric iron ion

H^+—Hydrogen ion

H_2O—Water

H_2S—Hydrogen sulfide

H_2SO_3—Sulfurous acid

H_2SO_4—Sulfuric acid

HOCl—Hypochlorous acid

I^-—Iodide ion

I_2—Iodine

KCl—Potassium chloride

KI—Potassium iodide

Mn^{+2}—Manganous ion

Mn^{+4}—Manganic ion

N_2—Nitrogen gas

Na^+—Sodium ion

$Na_2S_2O_3$—Sodium thiosulfite

$Na_2S_2O_5$—Sodium metabisulfite

Na_2SO_3—Sodium sulfite

Na_2SO_4—Sodium sulfate

NaCl—Sodium chloride

$NaHSO_3$—Sodium bisulfite

$NaHSO_4$—Sodium bisulfate

NaOCl—Sodium hypochlorite

NaOH—Sodium hydroxide

NCl_3—Trichloramine

NH_2Cl—Dichloramine

NH_3—Ammonia

$NHCl_2$—Monochloramine

NO_2^-—Nitrite ion

NO_2-N—Nitrite-Nitrogen (Nitrite as N)

NO_3^-—Nitrate ion

OCl^-—Hypochlorite ion

OH^-—Hydroxide ion

SO_2—Sulfur dioxide

SO_4^{-2}—Sulfate ion

Purpose and Function

Secondary treatment processes like activated sludge can remove up to 95% of waterborne pathogens from raw wastewater. Tertiary treatment can remove even more. Nonetheless, more pathogen inactivation or removal is typically needed to protect public health and the environment.

Chlorine is the most common method of disinfecting wastewater in the United States. A survey in 2002 suggested that although chlorine gas was the primary disinfectant used by water resource recovery facilities (WRRFs) in 1979 (94.9%), chlorine gas usage had dropped to only 76.5% of all WRRFs by 1996 (Connell, 2002). Some facilities switched from gaseous chlorine to liquid sodium hypochlorite or UV radiation. Chlorine works by destroying proteins and inactivating critical *enzymes* needed for microbial life.

Enzymes are organic molecules that help different chemical reactions take place. They are important for making new proteins, breaking down compounds to obtain energy, and producing new daughter cells.

Sufficiently damaged microorganisms are unable to reproduce and can't self-repair. If they can't reproduce, they cannot cause illness.

Chlorine is toxic to aquatic life and cannot be discharged into the environment in high concentrations. Water resource recovery facilities that use chlorine or chlorine compounds for disinfection will have concentration limits for chlorine in their discharge permits. Depending on the amount of dilution available at the discharge point, limits may be as low as 0.05 mg/L. Once disinfection is complete, excess chlorine is typically neutralized by adding sulfur dioxide (SO_2) gas or one of several different sulfite salts. In the process of chlorination, chlorine compounds act as both a disinfectant and oxidant. An oxidant is a chemical that adds oxygen to another substance, loses hydrogen, or loses electrons in a chemical reaction. In the process of dechlorination, sulfur dioxide and sulfite salts act as reducing and dechlorinating agents. Reducing agents remove oxygen from another substance, gain hydrogen, or gain electrons in a chemical reaction.

Alternative disinfectants, such as UV light, ozone, and chloramines, are also used in WRRFs. Ultraviolet disinfection has been gaining in popularity with approximately 21% of all WRRFs using it in 2007 (Washington State University, 2017). Often, these alternative methods are selected because the WRRF has extremely low effluent chlorine limits, there are concerns over chlorine safety, or there is a need to avoid forming disinfection byproducts (DBPs). When chlorine is added to water, some of it combines with dissolved organic compounds to form DBPs. Disinfection byproducts include the trihalomethanes, haloacetic acids, and nitrosodimethylamine. Some of these compounds are known, or are suspected, to elevate the risk of cancer in humans. Because of this, DBPs have been regulated in drinking water since 1998. Some states, like California, are beginning to regulate DBPs in treated effluent. Many drinking water treatment facilities pull water from streams, rivers, and lakes that receive WRRF effluent. Longer rivers, like the Mississippi, have many WRRFs and drinking water facilities along their length. In the arid western United States, indirect and direct potable reuse projects provide safe, clean drinking water from recycled wastewater. Managing disinfection practices to minimize DBP generation helps to protect the water supply.

Ultraviolet disinfection uses UV light to damage the *deoxyribonucleic acid (DNA)* of microorganisms in the final effluent. Microorganisms that are sufficiently damaged can no longer make some enzymes and proteins. They are unable to reproduce and will eventually die. Ultraviolet light is generated by specialized light bulbs that are similar in appearance to the fluorescent light bulbs used in most offices and stores. Ultraviolet bulbs are commonly called *lamps*. The lamps contain a small amount of mercury. When electricity is applied to these lamps, the mercury inside absorbs some of the energy and then re-releases it as UV light. For the light to be effective, it must reach the DNA within the microorganisms. The water must be relatively free of substances that absorb or scatter UV light including turbidity, total suspended solids (TSS), and iron. DNA that receives more UV light will be damaged more than DNA that receives less.

Deoxyribonucleic acid (DNA) is one type of genetic molecule used by viruses and cells. These molecules contain instructions to build every protein a virus or cell might need to live and reproduce. If the instructions are sufficiently damaged, the viruses and cells are not able to reproduce and will eventually die.

TEST YOUR KNOWLEDGE

1. **Disinfection kills or inactivates all of the microorganisms in wastewater.**
   - ☐ True
   - ☐ False
2. **Chlorine works by destroying proteins and inactivating critical enzymes so microorganisms can't reproduce.**
   - ☐ True
   - ☐ False
3. **Why is disinfection of WRRF effluent required?**
   a. To protect public health and the environment
   b. To decrease the potential for biological growth
   c. To reduce nutrient loading
   d. To completely eliminate all microorganisms
4. **Facilities that use chlorine for disinfection**
   a. Add sodium to neutralize chlorine and make it less toxic prior to discharge
   b. Have concentration limits for chlorine in their discharge permits
   c. Are required to maintain an effluent residual of at least 0.05 mg/L
   d. Discharge greater numbers of microorganisms than facilities using UV disinfection

5. **One reason a WRRF may decide to use UV irradiation instead of chlorine for disinfection might be**
   a. Chlorine is extremely expensive.
   b. Concerns over chlorine safety.
   c. Frequent power outages.
   d. Safety concerns with UV irradiation.

6. **These compounds can form when chlorine reacts with dissolved organic compounds:**
   a. Chloramines
   b. Organoozones
   c. Disinfection byproducts
   d. Sulfite salts

Indicator Organisms

Many different pathogens are potentially present in wastewater including viruses, bacteria, and protozoa. Testing for each type of pathogen requires large sample volumes and can be expensive because multiple test methods are needed. Testing for individual pathogens can be difficult because pathogens, when present, are typically a tiny percentage of the total number of microorganisms. In addition, reliable testing methods exist for some waterborne pathogens, but not for others (WHO, 2003). Small numbers of pathogens can be easily missed. Think about the service area and all of the persons served by the WRRF. Each one of them contributes millions of bacteria to the sewer each day, but only a small number of persons in the service area are likely to be sick with a *waterborne illness* at any given time. Doesn't it make more sense to look for the types of bacteria that are plentiful instead of the relatively rare pathogens? Running multiple tests for different types of pathogens is like looking for needles in a haystack. Instead of looking for needles buried in haystacks, look for haystacks. They are easier to find. This is the *indicator organism* concept. The haystacks, in this case, are the indicator organisms and the needles are the pathogens. Indicator organisms indicate that *fecal contamination* ***has*** occurred and that pathogens ***may*** be present. That doesn't mean pathogens will be present every time an indicator organism is present. Pathogens may not be present at all. It simply means there is a higher likelihood that pathogens are there along with the indicator organism.

Waterborne illnesses may be spread by contact with contaminated water. Cold viruses are spread through the air and don't survive long in wastewater. Cholera, on the other hand, spreads rapidly via contaminated water.

Indicator organisms have several key properties that make them useful for predicting whether or not pathogens associated with waterborne illness are likely to be present. Pathogens are also referred to as the *target organisms*. Indicator organisms are

Fecal contamination means that fecal material or wastewater containing fecal material is present. *E. coli* is a bacterium that lives in the intestines of warm blooded animals. If *E. coli* is present, it likely came from feces or wastewater contaminated with feces.

- Universally present in the feces of humans and warm-blooded animals;
- Present in larger numbers than the pathogen(s);
- Easier to test for than the pathogen(s);
- Closely associated with the pathogen(s);
- Not *ubiquitous*; and
- Removed by wastewater treatment similarly to waterborne pathogens. If the indicator organism is removed, killed, or inactivated during treatment, it is assumed the target organisms (pathogens) have also been removed, killed, or inactivated.

Ubiquitous means something is common or found nearly everywhere.

The most commonly used indicator organisms for wastewater treatment are total coliforms, fecal coliforms, *E. coli*, and enterococci. These microorganisms are not, with *few exceptions*, harmful to humans. They live in *symbiosis* with us in our intestines. In exchange for a warm, food-rich environment, they break down substances that humans can't break down on their own, produce vitamin K, and crowd out other microorganisms that could make us sick. Total coliforms include all of the fecal coliforms, *Klebsiella*, and a few other types of bacteria. It is the broadest, most inclusive, category. Because it includes some types of bacteria that are associated with the environment and not with fecal contamination, testing for total coliforms can sometimes indicate that fecal contamination is present when it is not. Fecal coliforms are a subset of the total coliforms. They are more tightly associated with fecal contamination because these bacteria live in the guts of warm-blooded animals. *E. coli* and enterococci are specific species of bacteria that are also fecal coliforms. One way to compare these three groups of microorganisms is with football. The draft pool includes all of the players that might be chosen to be part of a preseason team and players that will not become part of a team. After preseason is over, several players will be removed to reduce the size of the team. Total coliforms are the draft pool. Fecal coliforms are the preseason team and both *E. coli* and enterococci will become part of the final team. Each grouping is more specific than the last and is more tightly associated with the spread of disease. This is why the U.S. Environmental Protection Agency (U.S. EPA) has recommended that *E. coli* and enterococci be used to assess how well disinfection is working. Discharge permits may contain limits for one or more indicator organism.

Symbiosis is a mutually beneficial relationship.

E. coli are not normally pathogenic, but one strain of *E. coli*—O157 : H7—can produce toxins that can cause severe illness in humans. *E. coli* O157 : H57 sometimes enters the food supply through contaminated cheese and meat, particularly hamburger, resulting in food poisoning.

Laboratory results for indicator organisms are expressed as either the number of colony forming units (#CFUs) per 100 milliliters or as the most probable number (MPN) per 100 milliliters. The measurement units reflect differences in the two most commonly used testing methods. Individual bacteria are too small to be seen with the naked eye. An analyst looking through a microscope can't easily tell the difference between live and dead bacteria or between fecal coliforms, *E. coli*, and other types of bacteria. Most bacteria look the same. To get around these problems, testing methods increase the number of indicator organisms to make them easier to count and/or measure their activity in some way.

Some testing methods filter wastewater samples to capture individual bacteria on the filter paper. The filter paper and bacteria are then transferred to a petri plate containing a layer of *agar* in the bottom. The agar contains ingredients that allow the indicator organisms to grow, while suppressing the growth of other microorganisms. Different testing methods use different ingredients. The petri plate is placed upside down (agar side up) in an incubator for 24 hours. This helps keep individual bacteria isolated and prevents water from accumulating on the agar. During this time, the indicator organisms multiply rapidly, eventually forming piles that are large enough to be seen without a microscope (Figure 10.1). These piles are called *colonies*. Colonies form wherever indicator organisms land on the filter paper. The plate in Figure 10.1 shows blue fecal coliform colonies. The agar contains a blue dye that is taken up by the bacteria. By counting the number of colonies, it is possible to estimate the number of indicator organisms in the original sample. The method is not exact. Sometimes, two or more indicator organisms will end up on the filter paper so close together that they grow to form a single colony. Some indicator organisms won't grow fast enough to produce colonies. This is why the term *colony forming unit* (CFU) is used when reporting results.

Agar is a gelatin that is extracted from seaweed. It provides a surface for the bacteria to grow on as well as organic matter and nutrients.

The number of indicator organisms may also be estimated using the MPN method. Instead of filtering the sample, the bacteria are allowed to grow suspended in test tubes containing broth media like the one shown in Figure 10.2. The broth contains organic matter and nutrients. It also contains substances that only the indicator organisms can consume, such as *lactose* and 4-methylumbelliferyl-beta-D-glucuronide (MUG). It may also contain *pH indicator*. When total coliforms consume lactose, they produce acid and gas. A small, upside down glass test tube submerged in the test vial captures the gas. In Figure 10.2, the gas bubble seen in the tube indicates a positive result for fecal coliforms. *E. coli* is able to break down MUG into byproducts that glow blue when exposed to a black light. Testing methods rely on these properties to determine if a broth-filled test tube contains indicator organisms. For example, a tube that produced both gas and acid, but does not glow under a black light would be considered positive for total coliforms, but negative for *E. coli*.

Lactose is a sugar found in milk. Media that allow the growth of some organisms, but not others are called *selective media*.

pH indicator is a compound that changes color depending on the pH. In the MPN test, a color change tells the analyst that acid was produced by the microorganisms.

In one version of the MPN test, a large number of broth tubes (15 to 20) are prepared containing different amounts of wastewater. They are then placed into a warm incubator for approximately 24 hours. During this time, the bacteria reproduce rapidly. The analyst then records the number of tubes at each dilution

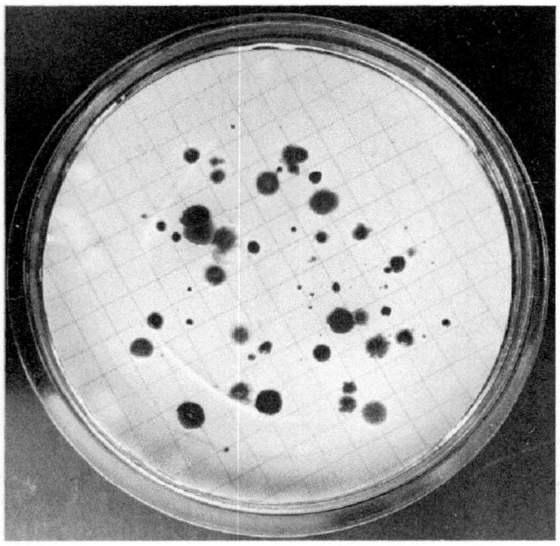

Figure 10.1 Fecal Coliform Colonies

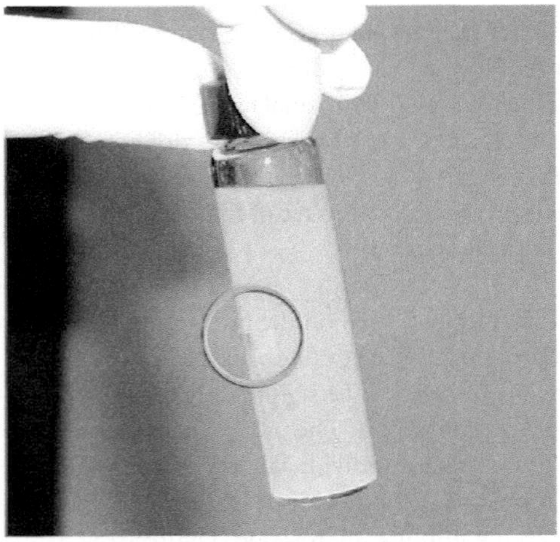

Figure 10.2 Gas Bubble Indicating a Positive Test Result in the MPN Test (Reprinted with permission by Indigo Water Group)

that tested positive. The Idexx Quanti-Tray™, shown in Figure 10.3, is another version of the MPN test. Each well in the tray will be clear (negative result) or yellow (positive for total coliforms) at the end of the 24-hour incubation period. Wells that are positive for *E. coli* glow blue when exposed to a black light. The analyst records the number of large and small wells that tested positive. For both versions of the MPN test, statistical tables are then used to determine the MPN of indicator organisms in the original sample.

Imagine a bag containing 100 marbles. Ten of the marbles are red, but the rest are white. If the contents of the bag were evenly divided among 100 containers, it would be very easy to count the number of containers with red marbles and get an exact count of red marbles. Now imagine if the marbles were divided among only 25 containers. Some containers will have a single red marble, some will have more than one red marble, and most won't have any. The MPN test methods can only tell us how many containers have at least one red marble in them, but can't tell us how many red marbles each container holds. Statistics are used to estimate the most *likely* number of red marbles in total based on the number of containers with at least one red marble. Most probable is another way of saying most likely. The math required for these types of statistical calculations is complicated. Instead of calculating the MPN each time, tables of MPN values are used instead. Once an analyst has counted the number of positive tubes or wells, the result is looked up in a statistical table to find the corresponding MPN. Laboratory results for MPN-based methods are reported as MPN/100 mL.

There are several different test methods, summarized in Table 10.1, for both membrane filtration plate counts and for MPN. All of these methods are based on the concepts presented in the previous

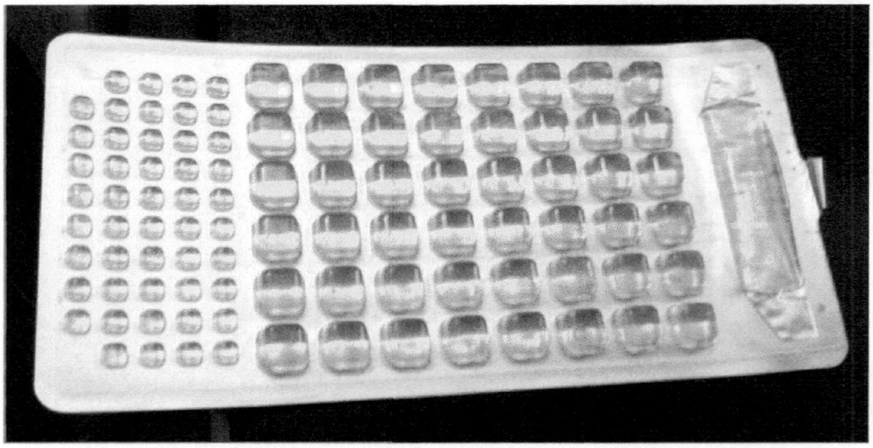

Figure 10.3 Idexx Quanti-Tray System (Reprinted with permission by Indigo Water Group)

Table 10.1 Testing Methods for Different Indicator Organisms (40 CFR, Part 136, §136.3)

Indicator Organism	Method Type	Method Sources
Total coliforms	MPN	U.S. EPA (1978c) page 114 Standard Methods 9221 B-2006
	Membrane Filter	U.S. EPA (1978c) page 108 Standard Methods 9222 D-2006
Fecal coliforms	MPN	U.S. EPA (1978c) page 132 U.S. EPA Method 1680 and 1681 Standard Methods 9221 C E-2006 Colilert-18 (Idexx Laboratories, Inc.)
	Membrane Filter	U.S. EPA (1978c) page 124 Standard Methods 9222 D-2006
E. coli	MPN	Standard Methods 9221 B.2-2006 / 9221 F-2006 Standard Methods 9223 B-2004 Colilert, Colilert-18 (Idexx Laboratories, Inc.)
	Membrane Filter	U.S. EPA Method 1603 mColiBlue-24 (Hach Company)

paragraphs. Several excellent resources exist that contain detailed instructions on how to conduct each test, including U.S. EPA Manual 600/8-78/017, *Microbiological Methods for Monitoring the Environment, Water, and Wastes*, which is available as a free download from the National Service Center for Environmental Publications website at https://nepis.epa.gov.

TEST YOUR KNOWLEDGE

1. **If the final effluent from a WRRF contains 120 CFU/100 mL of *E. coli*, pathogens are present.**
   - ☐ True
   - ☐ False

2. **Reliable test methods exist for all of the pathogens potentially present in wastewater.**
   - ☐ True
   - ☐ False

3. **Indicator organisms**
   a. Are ubiquitous (found everywhere)
   b. Require multiple testing methods
   c. Are associated with fecal contamination
   d. Indicate pathogens are present

4. **Limits on which of the following microorganisms are typically included in discharge permits as an indicator of adequate disinfection?**
   a. *Klebsiella*
   b. *E. coli*
   c. Cholera
   d. *Cryptosporidium*

5. **Put these microorganisms in order from least specific (most inclusive group) to most specific.**
   a. Fecal coliforms
   b. *E. coli*
   c. Total coliforms

6. **The final effluent result for *E. coli* was reported as 350 MPN/100 mL. What must be true?**
   a. The sample was filtered during analysis.
   b. The fecal coliform result must be less than 350 MPN/100 mL.
   c. Results include *Klebsiella* and other indicator organisms.
   d. A statistical table was used to estimate the number.

7. **A vial from the MPN test is cloudy from bacterial growth, shows gas production, and glows blue under a black light. Which of the following statements is true?**
   a. The test is positive for fecal coliforms.
   b. The test is positive for total coliforms.
   c. The test is positive for total coliforms and *E. coli*.
   d. The test is positive for total coliforms and fecal coliforms.

Chlorine Disinfection

Trade percent (%) = (g/L available chlorine) / 10

Weight percent is the weight of chemical divided by the total weight of solution. A container containing 2 kg (lb) of lime and 98 kg (lb) of water is 2% lime by weight.

Sometimes, brand names become nearly equivalent to the type of product. For example, it is rare for someone to ask for a facial tissue. Instead, they ask for a Kleenex®. A cotton swab is almost universally called a Q-tip® and a coke can mean anything from an actual Coca-Cola® to nearly any type of soft drink.

This section on chlorine theory contains many chemical equations. Don't panic! These equations should not be memorized. Focus on definitions, how chlorine is consumed or transformed in the reactions, and how environmental conditions affect which chemical species are present.

Chlorine is available as pure gaseous chlorine (Cl_2), liquid sodium hypochlorite (NaOCl), and solid calcium hypochlorite [$Ca(ClO)_2$]. Gaseous chlorine is 100% pure chlorine. Sodium hypochlorite is more commonly referred to as *bleach*. The bleach used in a WRRF typically contains 10 to 15% (*trade percent*) hypochlorite, whereas household bleach contains between 5.25 and 8% (trade percent) hypochlorite. Trade percent is not the *weight percent* of the solution, but is the industry's terminology for the strength of a hypochlorite solution compared to a similar quantity of pure chlorine. One percent trade strength is equivalent to 1 g of chlorine as chlorine gas (Connell, 2002). Trade percent is also referred to as the *percentage of available chlorine*. Calcium hypochlorite may be purchased as a solid and is available in powder, granular, disk, briquette, or tablet form, depending on the supplier. Calcium hypochlorite is often referred to as *powder bleach* or *High Test Hypochlorite (HTH®)*. HTH® is only one of several brand names for calcium hypochlorite. Calcium hypochlorite contains 65 to 70% calcium hypochlorite, by weight. In other words, one kilogram (pound) of HTH contains 0.65 to 0.70 kg (lb) of chlorine.

Chlorine may also be generated on-site from raw materials. On-site sodium hypochlorite generation has been done since the 1930s. This process uses a salt or brine solution and electric power to generate chlorine. Sodium chloride (NaCl) is the most commonly used salt. The salt is dissolved to make a brine solution, which is then diluted and passed across electrodes powered by a low-voltage current. The process produces hypochlorite ranging from 0.8 to 12.5% (trade percent) in solution. On-site generation is appealing because the only chemical delivered is salt, which can be stored indefinitely. On-site generation will not be discussed further in this chapter.

THEORY OF OPERATION

The forms of chlorine primarily used for wastewater disinfection are chlorine gas (Cl_2), sodium hypochlorite (NaOCl), and calcium hypochlorite [$Ca(OCl)_2$]. Each form of chlorine reacts similarly with water to form hypochlorous acid (HOCl) according to the reactions shown in eqs 10.1 through 10.3. Hypochlorous acid (HOCl) breaks down further to form hydrogen ions (H^+) and hypochlorite ions (OCl^-), as shown in eq 10.4. Recall from the chemistry section in Chapter 9 (Nutrient Removal) that the double-sided arrows in each of these chemical reactions means that they are in equilibrium. The percentages of each chemical form present depend on the pH and temperature of the water. If pH or temperature change, the equilibrium will shift from one side of the equation to the other and the amount of each chemical species will change.

$$\begin{gathered} Cl_{2(g)} + H_2O \leftrightarrow HOCl + HCl \\ \text{Chlorine gas} + \text{Water} \leftrightarrow \text{Hypochlorous acid} + \text{Hydrochloric acid} \end{gathered} \tag{10.1}$$

$$\begin{gathered} NaOCl + H_2O \leftrightarrow HOCl + NaOH \\ \text{Sodium hypochlorite} + \text{Water} \leftrightarrow \text{Hypochlorous acid} + \text{Sodium hydroxide} \end{gathered} \tag{10.2}$$

$$\begin{gathered} Ca(OCl)_2 + H_2O \leftrightarrow HOCl + Ca(OH)_2 \\ \text{Calcium hypochlorite} + \text{Water} \leftrightarrow \text{Hypochlorous acid} + \text{Calcium hydroxide} \end{gathered} \tag{10.3}$$

$$\begin{gathered} HOCl \leftrightarrow H^+ + OCl^- \\ \text{Hypochlorous acid} \leftrightarrow \text{Hydrogen ion} + \text{Hypochlorite ion} \end{gathered} \tag{10.4}$$

Notice that when chlorine gas (Cl_2) is added to water, the end products in eq 10.1 are hypochlorous acid (HOCl) and hydrochloric acid (HCl). The acid produced consumes 1.4 mg/L of alkalinity as calcium carbonate ($CaCO_3$) for every 1 mg/L of chlorine (Cl_2) added. If either the amount of starting alkalinity is low or a large amount of chlorine is added, chlorine gas addition can cause pH to decrease. For most WRRFs, the chlorine dose added will be less than 20 mg/L and the effect on pH will be negligible. Some WRRFs may see a noticeable pH drop when effluent alkalinity is less than about 80 mg/L as $CaCO_3$ prior to chlorine gas addition. When sodium hypochlorite (NaOCl) or calcium hypochlorite [$Ca(OCl)_2$] are added to water, hydroxide compounds are produced (eqs 10.2 and 10.3). Hydroxide (OH^-) is a base. This adds alkalinity and tends to increase pH.

The equilibrium between hypochlorous acid (HOCl) and hypochlorite ion (OCl^-) is strongly pH dependent, as shown in Figure 10.4. When the pH is near 7.6, the percentages of hypochlorous acid (HOCl) and hypochlorite ion (OCl^-) are equal. The equilibrium point is where the green and blue lines cross in Figure 10.4. Notice that it is located at pH 7.6. The equilibrium point moves slightly to the right as the water gets colder (higher pH) and moves slightly to the left (lower pH) as the water gets warmer. This effect is barely noticeable in domestic wastewater because water temperatures tend to remain with a narrow range between about 10 and 20 °C (50 and 68 °F). When the pH is greater than 8.5, hypochlorous acid (HOCl) is less than 10% of the total whereas hypochlorite ion (OCl^-) is more than 90% of the total. When the pH is less than 6.5, the equilibrium shifts to more than 90% hypochlorous acid (HOCl) and less than 10% hypochlorite ion (OCl^-). Below pH 2, hypochlorous acid (HOCl) reverts back to chlorine gas (Cl_2). Hypochlorous acid (HOCl) is a more effective disinfectant than hypochlorite ion (OCl^-). It is thought that hypochlorite ion (OCl^-) is a poor disinfectant because its negative charge prevents it from easily crossing the cell membrane. Because of this, it takes less time to achieve the same amount of disinfection at low pH than at high pH.

DOSE, DEMAND, AND RESIDUAL

In addition to forming hypochlorous acid (HOCl) and hypochlorite ion (OCl^-), chlorine reacts with other compounds present in the wastewater such as hydrogen sulfide (H_2S), ferrous iron (Fe^{+2}), manganese (Mn^{+2}), nitrite (NO_2^-), and organic compounds. Equations 10.5 through 10.8 illustrate the reactions between hypochlorous acid (HOCl), nitrite (NO_2^-), ferrous iron (Fe^{+2}), and manganous ion (Mn^{+2}). These chemical reactions use up the oxidizing power of chlorine, converting it from hypochlorous acid (HOCl) into chloride ion (Cl^-). Hypochlorous acid (HOCl) and hypochlorite ion (OCl^-) are strong oxidizers and good disinfectants. The chloride ion (Cl^-) that remains after chemical reactions are complete is spent. It has given up its oxidizing ability and is no longer available as a disinfectant.

$$\begin{gathered} H_2S + 4HOCl \rightarrow H_2SO_4 + 4H^+ + 4Cl^- \\ \text{Hydrogen sulfide} + \text{Hypochlorous Acid} \rightarrow \text{Sulfuric acid} + \text{Hydrogen ions} + \text{Chloride ions} \end{gathered} \tag{10.5}$$

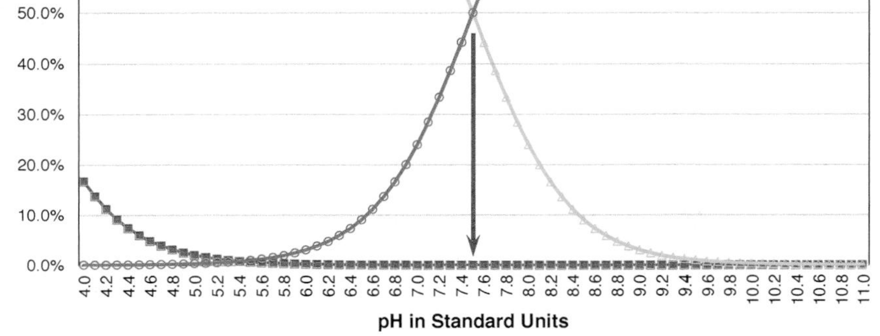

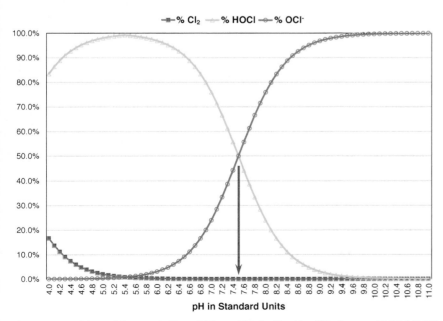

Figure 10.4 Hypochlorous Acid and Hypochlorite Ion Equilibrium at 20 °C (68 °F) (Equilibrium point at 20 °C is pH 7.58. Equilibrium point at 0 °C is pH 7.82 and the equilibrium point at 30 °C is pH 7.50. Calculated from equations and constants in Morris, 1966). (Reprinted with permission by Indigo Water Group)

$$\begin{array}{c} HOCl + 2Fe^{+2} + H^{+} \rightarrow 2Fe^{+3} + Cl^{-} + H_2O \\ \text{Hypochlorous acid} + \text{Ferrous iron} + \text{Hydrogen Ion} \rightarrow \text{Ferric iron} + \text{Chloride} + \text{Water} \end{array} \quad (10.6)$$

$$\begin{array}{c} HOCl + Mn^{+2} + H^{+} \rightarrow Mn^{+4} + Cl^{-} + H_2O \\ \text{Hypochlorous acid} + \text{Manganous ion} + \text{Hydrogen ion} \rightarrow \text{Manganic ion} + \text{Chloride} + \text{Water} \end{array} \quad (10.7)$$

$$\begin{array}{c} HOCl + NO_2^{-} \rightarrow NO_3^{-} + Cl^{-} + 2H^{+} \\ \text{Hypochlorous acid} + \text{Nitrite} \rightarrow \text{Nitrate} + \text{Chloride} + \text{Hydrogen ion} \end{array} \quad (10.8)$$

The amount of chlorine required, calculated from eqs 10.5 through 10.8, to react with each of these inorganic compounds is given in Table 10.2. These are concentration or weight ratios. For example, if the final effluent contains 1 mg/L of ferrous iron (Fe^{+2}), it will consume 0.6 mg/L of chlorine (Cl_2). Equation 10.9 gives an example of how to convert the *mole ratios* from eqs 10.5 through 10.8 into concentrations that can be measured in the field. In eq 10.8, 1 mole of hypochlorous acid (HOCl) reacts with 1 mole of nitrite (NO_2^-). In practice, operators will measure the concentration of hypochlorous acid (HOCl) present as chlorine (Cl_2). In other words, the testing method will give a result as milligrams per liter Cl_2 even when sodium or calcium hypochlorite is being added. Similarly, laboratory results for nitrite (NO_2^-) are typically reported as nitrite-nitrogen (NO_2-N) or "as N". To convert from moles to grams, the formula weights for chlorine (Cl_2) and nitrite as N (NO_2-N) are used so the resulting ratio is in the same units as the laboratory results. For every 1 mg/L of nitrite-nitrogen (NO_2-N), 5.1 mg/L of chlorine will be consumed.

A chemistry review is included at the beginning of Chapter 9—Nutrient Removal.

In a well-functioning nitrification process, nitrite-nitrogen concentrations should be very low, less than 0.25 mg/L as N. However, a process upset or onset of nitrification can cause nitrite-nitrogen concentrations to increase.

$$\frac{1 \text{ mole } Cl_2}{1 \text{ mole } NO_2^-} \cdot \frac{71 \text{ g}}{1 \text{ mole } Cl_2} \cdot \frac{1 \text{ mole } NO_2^-}{1 \text{ mole N}} \cdot \frac{1 \text{ mole N}}{14 \text{ g}} = \frac{5.07 \text{ g } Cl_2}{\text{g } NO_2\text{-N}} \quad (10.9)$$

The amount of chlorine added to the wastewater is the *dose*. The chlorine that is used up (consumed) in chemical reactions is the chlorine *demand*. The *residual* is the amount of chlorine remaining after the demand has been satisfied. Dose, demand, and residual are related according to eq 10.10. The residual chlorine concentration can be measured using a variety of test methods. The dose can be estimated if the concentration and volume of chemical being added are known. The demand can only be calculated if both the dose and residual are known. Demand cannot be measured.

$$\begin{array}{c} \text{Dose} = \text{Demand} + \text{Residual} \\ \text{or} \\ \text{Chlorine Added} = \text{Chlorine Used} + \text{Chlorine Remaining} \end{array} \quad (10.10)$$

CHLORAMINE FORMATION

Ammonia (NH_3) reacts with chlorine in water to form chloramines. In this process, chlorine can be converted to form any one of the three chlorine and ammonia compounds—monochloramine (NH_2Cl), dichloramine ($NHCl_2$), and trichloramine (NCl_3). Chlorine atoms take the place of one or more hydrogen atoms in the ammonia molecule. There are several routes chloramine formation may take depending on pH and other factors. For purposes of simplification, eqs 10.11 through 10.13 may be used to describe the

Table 10.2 Chlorine Required for Inorganic Reactions (AWWA, 2006)

Compound	Weight Ratio of Chlorine to Compound
Ferrous Iron (Fe^{+2})	0.6
Manganese (Mn^{+2})	1.3
Nitrite (NO_2^-)	1.5
Nitrite as N (NO_2-N)	5.1
Hydrogen sulfide (H_2S) to Sulfur (S)	2.1
Hydrogen sulfide (H_2S) to Sulfate (SO_4^{-2})	6.2

formation of mono-, di-, and trichloramine. All three steps may take place simultaneously and compete with one another. Organic chloramines and disinfection byproducts may also be formed when chlorine reacts with proteins and other organic nitrogen compounds.

$$\begin{gathered} NH_3 + HOCl \rightarrow NH_2Cl + H_2O + H^+ + Cl^- \\ \text{Ammonia} + \text{Hypochlorous acid} \rightarrow \text{Monochloramine} + \text{Water} + \text{Hydrogen ions} + \text{Chloride} \end{gathered} \tag{10.11}$$

$$\begin{gathered} NH_2Cl + HOCl \rightarrow NHCl_2 + H_2O + H^+ + Cl^- \\ \text{Monochloramine} + \text{Hypochlorous acid} \rightarrow \text{Dichloramine} + \text{Water} + \text{Hydrogen ions} + \text{Chloride} \end{gathered} \tag{10.12}$$

$$\begin{gathered} NHCl_2 + HOCl \rightarrow NCl_3 + H_2O + H^+ + Cl^- \\ \text{Dichloramine} + \text{Hypochlorous acid} \rightarrow \text{Trichloramine} + \text{Water} + \text{Hydrogen ions} + \text{Chloride} \end{gathered} \tag{10.13}$$

In each of these equations, some of the chlorine (Cl) from hypochlorous acid (HOCl) is combined with ammonia (NH_3) to form chloramines and some is converted to chloride (Cl^-). The chloride (Cl^-) is spent and is no longer available as a disinfectant. The chlorine attached to chloramines, however, is still able to disinfect. Chloramines are referred to as the *combined available residual chlorine.* Chloramines take longer to react with proteins and enzymes than hypochlorous acid (HOCl) or hypochlorite ion (OCl^-) and take longer to disinfect.

FREE CHLORINE RESIDUAL VERSUS TOTAL CHLORINE RESIDUAL

Chlorine residual may be measured using a variety of different test methods. Each method measures either the free available chlorine (FAC) residual or the TCR. The FAC residual consists of both hypochlorous acid (HOCl) and hypochlorite ion (OCl^-). It does not include chloramines. The TCR includes chloramines, HOCl, and OCl^-. Neither the free nor the TCR includes chloride ion (Cl^-). The free residual concentration may be the same as the TCR (equal to), but it can never be higher than the TCR. Water resource recovery facilities that use chlorine for disinfection are typically required to monitor for TCR.

BREAKPOINT CHLORINATION

Figure 10.5 shows two dose and chlorine residual plots, one with no chlorine demand and the other with chlorine demand. The water with chlorine demand contains some ammonia and other compounds that react with chlorine. The x-axis (horizontal) shows the chlorine-to-ammonia ratio. The chlorine dose increases from left to right across the graph. The y-axis (vertical) shows the TCR concentration. The TCR concentration increases from the bottom of the graph to the top. When there is no chlorine demand, the TCR measurement increases in exact proportion to the amount of chlorine added. For example, if 1 mg/L of chlorine is added, 1 mg/L of TCR will be measured. This one-to-one ratio is shown as the dashed green line in Figure 10.5.

When ammonia and other compounds are present, the graph becomes more complicated. The breakpoint curve, shown as a solid blue line in Figure 10.5, shows how the TCR concentration changes in response to the chlorine dose added. The breakpoint curve contains three distinct parts: 1) initial demand and creation of combined residual, 2) destruction of combined residual, and 3) post break point. In the first zone on the far left side of the graph, chlorine combines with ammonia to form mono, di, and tri-chloramine. Chlorine is also reacting with other compounds in the water like iron and manganese. In this first zone, all of the TCR residual is combined residual. Free residual chlorine can't exist because it immediately reacts with ammonia and other substances. Notice that the TCR concentration is not increasing in a one-to-one ratio with dose. This is because some of the chlorine applied is being combined with ammonia, but some is reduced to chloride ion (eqs 10.11 through 10.13). Chlorine that is reacting with other compounds (iron, manganese, nitrite, and other compounds) is also being reduced to chloride ion. Because of demand, the residual is less than the dose.

As the chlorine dose continues to increase, the amount of excess chlorine eventually overwhelms the amount of ammonia. Examine Figure 10.5 and find the hump that marks the end of zone one and the

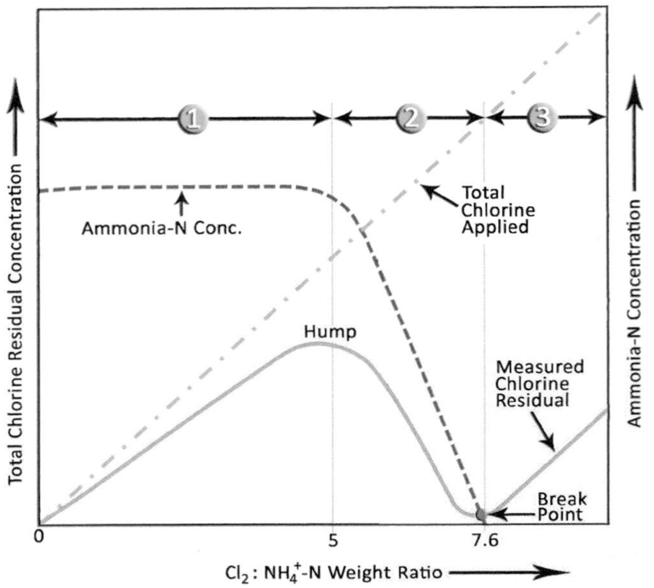

Figure 10.5 Breakpoint Chlorination (adapted from U.S. EPA [1978b])

Stoichiometry is a written chemical equation that shows how many moles of one substance are required to react with another substance in a chemical reaction.

beginning of zone two. What is happening to the TCR concentration before and after the hump? What is happening to the chlorine dose? The chloramines (combined residual) formed in the first zone are being chemically burned away in the second zone. The *stoichiometry* of this reaction is shown in eq 10.14. While this reaction can take place to a limited extent in zone one, the frequency increases greatly when the chlorine to ammonia-nitrogen ratio reaches approximately 5 to 1. In other words, approximately 5 mg/L of chlorine (Cl_2) is needed for every 1 mg/L of ammonia-nitrogen (NH_3-N) before the reaction shown in eq 10.14 is the dominant reaction. The end products are chloride ion (Cl^-) and nitrogen gas (N_2). The ammonia concentration decreases throughout zone two (Figure 10.5). Neither end product is a disinfectant. This is why the TCR concentration is decreasing in zone two even though the chlorine dose is increasing.

$$\begin{gathered} 2NH_3 + 3HOCl \rightarrow N_{2(g)} \uparrow + 3H_2O + 3H^+ + 3Cl^- \\ \text{Ammonia} + \text{Hypochlorous acid} \rightarrow \text{Nitrogen gas} + \text{Water} + \text{Hydrogen ions} + \text{Chloride} \end{gathered} \tag{10.14}$$

See the red dot on the graph? That is the break point. The break point marks the point where all of the combined residual that ***can*** be chemically burned away ***has*** been burned away. Notice that the break point curve (blue line) doesn't return all the way back to zero residual. Some combined residual often remains. This is the irreducible combined residual, which consists primarily of organic chloramines that don't provide much disinfection. Equation 10.14 predicts that a chlorine to ammonia-nitrogen ratio of 7.6 to 1 is needed to reach the break point. In practice, ratios of 10 : 1 or more may be needed because chlorine reacts with other compounds in addition to ammonia. The break point is also the point with the lowest residual chlorine concentration. Past the break point, in zone three, the chlorine residual measurement increases in direct proportion to the amount of dose added—just like it did in the zero-demand residual plot. If 2 mg/L of chlorine is added past the break point, the measured residual will go up by 2 mg/L. The demand has been satisfied. Chlorine added after the break point is free chlorine residual.

Here's what's interesting about the break point graph: it is possible to get the exact same result for the TRC analysis at three different points. Examine the example break point graph in Figure 10.6. See the three points labeled with yellow circles? For this example wastewater, all three points have a TRC concentration of 2 mg/L even though the chlorine dose is different at all three points. The free residual chlorine concentrations also differ. There can't be any free chlorine to the left of the break point. At locations 1 and 2, the TRC is 2 mg/L and the free chlorine residual is 0 mg/L. At location 3, the TRC is still 2 mg/L, but the free chlorine residual is 1.5 mg/L. Why is that? Remember the combined residual that couldn't be chemically

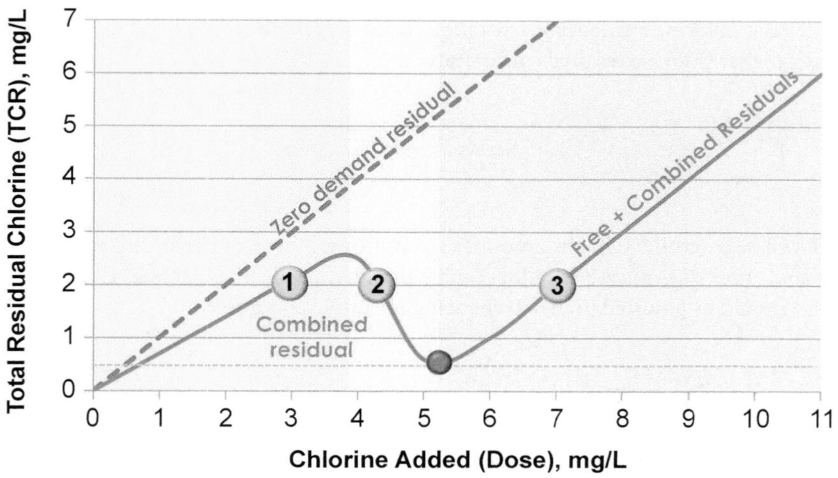

Figure 10.6 Total and Free Chlorine Residual and the Break Point Curve (adapted from U.S. EPA [1978b])

burned away? See it below the break point? The TRC is equal to the combined residual plus the free chlorine residual. In this example, the TRC at point 3 is equal to 2 mg/L and the combined residual is 0.5 mg/L. The free chlorine residual is the difference between the total and combined residual or 1.5 mg/L.

Wastewater systems typically operate to the left of the hump. Here, the chlorine to ammonia ratio is less than 5 to 1 and chloramines are formed, but not destroyed. As influent flows and loads to the WRRF change throughout the day, the ammonia concentration in the final effluent may also change. This can cause the effluent to swing back and forth across the break point. Imagine a WRRF that uses flow pacing to match the volume of sodium hypochlorite added to the amount of flow received. Early in the morning, the chlorine dose added is 5 mg/L and the effluent ammonia concentration is 0.2 mg/L. The TCR concentration is 4.6 mg/L and the effluent is well disinfected. The ratio of chlorine dose to ammonia-nitrogen is 10 to 1 and the facility is operating to the right of the break point. Peak flow begins to enter the WRRF around 10:00 a.m. and by 1:00 p.m. the effluent ammonia concentration has increased to 1.5 mg/L as N. Now, the chlorine to ammonia-nitrogen ratio is 3.3 to 1. The facility is operating to the left of the break point. This means that for some portion of the day, the facility was operating at or very near the break point. Remember, the break point is the lowest TCR concentration. If the TCR gets too low, the effluent may not be disinfected. Operators should adjust the chlorine dose to remain to the far left of the break point in zone one (Figure 10.5) and avoid break point chlorination. Operating within zone one can also reduce chemical usage and operating costs.

Drinking water facilities operate to the right of the break point. This is why wastewater operators measure TRC and drinking water operators (mostly) measure free residual chlorine.

Break point chlorination is not used in conventional wastewater treatment to remove ammonia because of the large quantities of chlorine required and the potential for harming the environment by accidentally discharging high concentrations of chlorine.

CONTACT TIME

Chlorine doesn't inactivate microorganisms instantaneously. It takes time for chlorine to react with enzymes, proteins, and other critical components and damage them beyond repair. How much time is needed for disinfection depends on several factors, including

- Chlorine concentration,
- Contact time,
- Water temperature,
- Water pH,
- Mixing, and
- Microorganism type.

The first two factors, chlorine concentration and CT, are related to one another according to eq 10.15. The amount of CT required is set according to the level of disinfection required. Then, either the chlorine residual or the CT can be manipulated to achieve the desired CT. For example, if the CT required to achieve disinfection is 10 mg/L·min and the chlorine residual concentration is 2 mg/L, then 5 minutes will be needed to disinfect. Alternatively, a chlorine residual of 0.5 mg/L could be used for 20 minutes to achieve the same CT. A very low chlorine residual concentration requires more time to achieve the same amount of disinfection as a higher chlorine residual concentration.

$$CT, \frac{mg}{L} \cdot min = \left(\text{Chlorine Residual}, \frac{mg}{L} \right) (CT, min) \quad (10.15)$$

For most WRRFs, the CT will be determined by the amount of flow moving through the facility and the size of the basin where disinfection takes place. Operators can't typically control either of these. Chlorine dose can be controlled and should be adjusted to match changing operating conditions.

DECHLORINATION

Dechlorination of wastewater is typically accomplished by adding reducing agents that react with the excess chlorine to either produce a chlorine-free wastewater effluent or one below 0.05 mg/L chlorine. The most common reducing agents are those associated with sulfur dioxide (SO_2) and sulfite salts. The reactions of these compounds with chlorine are illustrated in eqs 10.16 through 10.20. Although chlorine (Cl_2) is shown in each equation, similar reactions take place with hypochlorous acid (HOCl). Notice that one of the end products in every reaction is chloride ion (Cl^-). Facilities may choose to use one sulfur compound over another based on cost and ease of handling.

$$\begin{gathered} SO_2 + Cl_2 + 2H_2O \rightarrow H_2SO_4 + 2H^+ + 2Cl^- \\ \text{Sulfur dioxide} + \text{Chlorine} + \text{Water} \rightarrow \text{Sulfuric Acid} + \text{Hydrogen ions} + \text{Chloride} \end{gathered} \quad (10.16)$$

$$\begin{gathered} Na_2SO_3 + Cl_2 + H_2O \rightarrow Na_2SO_4 + 2H^+ + 2Cl^- \\ \text{Sodium sulfite} + \text{Chlorine} + \text{Water} \rightarrow \text{Sodium sulfate} + \text{Hydrogen ions} + \text{Chloride} \end{gathered} \quad (10.17)$$

$$\begin{gathered} NaHSO_3 + Cl_2 + H_2O \rightarrow NaHSO_4 + 2H^+ + 2Cl^- \\ \text{Sodium bisulfite} + \text{Chlorine} + \text{Water} \rightarrow \text{Sodium bisulfate} + \text{Hydrogen ions} + \text{Chloride} \end{gathered} \quad (10.18)$$

$$\begin{gathered} Na_2S_2O_5 + 2Cl_2 + 3H_2O \rightarrow 2NaHSO_4 + 4H^+ + 4Cl^- \\ \text{Sodium metabisulfite} + \text{Chlorine} + \text{Water} \rightarrow \text{Sodium bisulfate} + \text{Hydrogen ions} + \text{Chloride} \end{gathered} \quad (10.19)$$

$$\begin{gathered} Na_2S_2O_3 + Cl_2 + H_2O \rightarrow Na_2SO_4 + S + 2H^+ + 2Cl^- \\ \text{Sodium thiosulfate} + \text{Chlorine} + \text{Water} \rightarrow \text{Sodium sulfate} + \text{Sulfur} + \text{Hydrogen ions} + \text{Chloride} \end{gathered} \quad (10.20)$$

A small amount of sodium thiosulfate is added to sample bottles when collecting samples for microbiological analysis. It is important to dechlorinate the sample when it is collected so disinfection does not continue after collection.

Overdosing of sulfite compounds can consume most or all of the oxygen in the final effluent and the receiving stream. Operators must be careful to add just enough sulfite during dechlorination to eliminate the chlorine residual. Stoichiometric weight ratios for different sulfite compounds are presented in Table 10.3. In practice, slightly higher doses are used to ensure all of the chlorine residual is neutralized.

Table 10.3 Sulfite Compound to Chlorine Residual Ratios Required for Neutralization

Sulfite Compound	Amount Needed to Neutralize 1 mg/L of Chlorine Residual (Cl_2)
Sulfur dioxide	0.9
Sodium bisulfite	1.47
Sodium sulfite	1.78
Sodium metabisulfite	1.34

TEST YOUR KNOWLEDGE

1. Trade percent is the same thing as the percentage of available chlorine.
   - ☐ True
   - ☐ False

2. The chlorine concentration of all three forms of chlorine (gas, sodium hypochlorite solution, and calcium hypochlorite) is expressed as trade percent.
   - ☐ True
   - ☐ False

3. Dose is equal to the demand minus the residual.
   - ☐ True
   - ☐ False

4. Chlorine that has reacted with ammonia to form chloramines is still able to disinfect.
   - ☐ True
   - ☐ False

5. Chlorine and hypochlorites break down in water to form
   a. Hypochlorous acid and hypochlorite ion
   b. Hypochlorous acid and hydrochloric acid
   c. Hypochlorite ion and hydrogen ion
   d. Hydrochloric acid and chloride ion

6. The final effluent from a WRRF has extremely low alkalinity. Which of the following chemicals would NOT be a good choice for disinfection because it may cause the effluent pH to decrease?
   a. Chlorine gas
   b. Calcium hypochlorite
   c. Sodium hypochlorite
   d. Hypochlorous acid

7. Calcium chloride is used for disinfection. Which of the following statements is true?
   a. Alkalinity will be consumed and pH may decrease.
   b. Hypochlorous acid and hydrochloric acid will be produced.
   c. Alkalinity will be produced and pH may increase.
   d. Sodium hydroxide and hypochlorous acid will be produced.

8. At what pH will the percentage of hypochlorous acid be highest during chlorine disinfection?
   a. 6.5
   b. 7.0
   c. 7.5
   d. 8.0

9. The pH in the final effluent from a lagoon treatment facility rises during the day and falls at night. The chlorine dose is held constant throughout the day. The operator may need to
   a. Add alkalinity to minimize pH changes in the middle of the day
   b. Heat the final effluent to shift the chemistry toward hypochlorous acid
   c. Increase the dose to compensate for the shift toward hypochlorite ion
   d. Decrease the dose because of increased microorganism kill at high pH

10. Calculate the chlorine demand using the following data.

    —Raw water flow is 1.9 ML/d (0.50 mgd)
    —Chlorinator dose is 3.0 mg/L
    —Chlorine residual is 1.8 mg/L

    a. 0.8 mg/L
    b. 1.2 mg/L
    c. 3.0 mg/L
    d. 4.8 mg/L

11. Which of the following chemical species will NOT be measured in the TCR test?
    a. Hypochlorous acid
    b. Hypochlorite ion
    c. Chloride ion
    d. Monochloramine

12. The final effluent from a WRRF typically contains less than 0.25 mg/L of nitrite-nitrogen (NO_2-N). An upset in the activated sludge process has caused the NO_2-N concentration to increase to 3 mg/L. How will this affect a chlorine disinfection process?
    a. The nitrite-nitrogen will consume an additional 15 mg/L of chlorine.
    b. The nitrite-nitrogen will shift the equilibrium toward hypochlorite ion.
    c. The nitrite-nitrogen will not affect disinfection efficiency in any way.
    d. The nitrite-nitrogen will decrease chlorine demand by 2.75 mg/L.

13. When the chlorine to ammonia-nitrogen (NH_3-N) ratio is greater than 8 to 1, what must be true?
    a. Only monochloramine will form.
    b. Free chlorine residual will be present.
    c. TCR is maximized.
    d. Disinfection cannot occur.

14. Wastewater operators must be familiar with the break point curve because
    a. The ratio of chlorine to ammonia in the effluent can change during the day.
    b. Accidental operation at or near the break point can result in disinfection failure.
    c. Operating to the right of the break point consumes excess chemical and increases operating costs.
    d. All of the above.

15. The chlorine residual concentration, and ability to disinfect, are lowest

a. To the left of the break point
b. At the hump, before combined residual destruction
c. At the break point
d. To the right of the break point

16. Which of the following statements about break point chlorination is false?

a. It is possible to read the same TCR concentration at more than one point on the break point curve.
b. The combined chlorine residual is completely eliminated at the break point when enough chlorine is added.
c. Wastewater operators sometimes operate to the right of the break point where free chlorine residual is measured.
d. The TRC concentration includes the combined residual and the free residual.

17. An operator is able to adjust both the chlorine dose and the CT at their facility. If the CT is doubled and all other parameters remain the same, the operator could

a. Decrease the dose by half to achieve the same CT
b. Increase the dose by half to achieve the same CT
c. Decrease the dose by 30% to achieve the same CT
d. Increase the dose by 30% to achieve the same CT

18. When sulfur dioxide or sulfite salts are added to water containing chlorine

a. Chlorine is converted to hypochlorous acid.
b. Chlorine is oxidized to chlorohydrogen ions.
c. Chlorine is reduced to chloride.
d. Chlorine is converted to hypochlorite ion.

DESIGN PARAMETERS

Design parameters include chlorine dose and CT. The chlorine dose required depends on effluent quality. Total suspended solids, ammonia, hydrogen sulfide, iron, manganese, and other substances can increase chlorine demand. Table 10.4 includes chlorine doses used by engineers when sizing chemical addition and storage facilities. These doses are for domestic wastewater that meets secondary treatment standards for 5-day biochemical oxygen demand and TSS. In practice, the chlorine dose should be adjusted to meet discharge permit limits for total coliforms, fecal coliforms, or *E. coli*.

Contact time requirements vary from state to state. In general, contact chambers are designed to provide at least 15 minutes of CT at the design peak hour flowrate (GLUMRB, 2014). Some states may also require a minimum of 30 minutes of CT at the design average flowrate. Because most facilities operate at less than 80% of their design flowrates, the actual contact time will be higher.

EXPECTED PERFORMANCE

Ultimately, performance depends on the CT and effluent quality. Chlorine disinfection systems are capable of achieving *E. coli* concentrations below 2 CFU/100 mL ($<$2 MPN/100 mL). Discharge permit limits for *E. coli* (or other indicator organisms) are facility specific. U.S. EPA recommends that *E. coli* concentrations in recreational waters remain below 126 CFU/100 mL (geometric mean) and 410 CFU/100 mL (statistical threshold value) (U.S. EPA, 2012). The statistical threshold value for *E. coli* states that 90% of all sample results must be below 410 CFU/100 mL.

The geometric mean is a special method of averaging numbers together. It is used in statistics to reduce the influence of very large and very small numbers in a set of data. The geometric mean is used when averaging results for *E. coli* and other bacteria because they reproduce exponentially. One bacterium becomes two, two become four, and so on. An effluent sample that is analyzed immediately after being taken will have fewer bacteria in it than one that is allowed to sit for several hours before analysis. The geometric mean may be calculated using eq 10.21, where each X represents a single result and n is the total number of results being averaged.

Table 10.4 Chlorine Dose Guidance (GLUMRB, 2014)

Type of Treatment	Chlorine Dose*
Trickling filter	10 mg/L
Activated sludge	8 mg/L
Tertiary filtration	6 mg/L
Nitrified effluent	6 mg/L

*Dosage varies with effluent quality. Guideline only.

$$\text{Geometric Mean} = [(X_1)(X_2)(X_3)(X_4)(X_5)(X_n)]^{1/n} \quad (10.21)$$

Example Calculation

Find the geometric mean for a set of *E. coli* results: 2300, 70, 55, 6, 300, and 32 CFU/100 mL.

$$\text{Geometric Mean} = [(X_1)(X_2)(X_3)(X_4)(X_5)(X_n)]^{1/n}$$
$$\text{Geometric Mean} = [(2300)(70)(55)(6)(300)(32)]^{1/6}$$
$$\text{Geometric Mean} = [510\,048\,000\,000]^{1/6}$$
$$\text{Geometric Mean} = 89 \text{ CFU/100 mL}$$

An alternative method for finding the geometric mean is to 1) take the log of each number, 2) average the log results together, and 3) take the inverse log of the average (10^x button on the calculator). Both calculation methods will give the same result.

In general, the more efficiently a WRRF is operated, the easier it will be to disinfect the effluent. Any failure to provide adequate treatment upstream will increase the level of pathogens and the chlorine requirements. High solids content increases the chlorine requirement, as does the soluble organic load.

The geometric mean is used when reporting *E. coli*, fecal coliform, total coliform, and other indicator organism results on discharge monitoring reports.

TEST YOUR KNOWLEDGE

1. **Contact time will be higher at peak hour flow than at average daily flow.**
   - ☐ True
   - ☐ False
2. **The chlorine dose should be adjusted**
   a. To reduce *E. coli* below 2 CFU/100 mL
   b. To meet discharge permit limits
   c. To match the design dose in the facility design manual
   d. To comply with Ten States Standards
3. **The chlorine disinfection system is designed to provide 15 minutes of CT at peak hour flow. Which of the following statements must be true?**
   a. Only one of the chlorine contact chambers needs to be in service.
   b. Chlorine doses must be increased to meet the minimum CT during average flow.
   c. There is enough CT to satisfy the design criteria in all states.
   d. Most of the time, the CT will be greater than 15 minutes.
4. **U.S. EPA recommends that the *E. coli* concentration in recreational waters be kept, on average, below**
   a. 2 CFU/100 mL
   b. 53 CFU/100 mL
   c. 126 CFU/100 mL
   d. 410 CFU/100 mL
5. **Calculate the geometric mean for the following *E. coli* results: 2, 35, 1800, 45, 160.**
   a. 62
   b. 76
   c. 98
   d. 408

EQUIPMENT

Gas chlorination and dechlorination facilities generally consist of chlorination and dechlorination chemicals, chemical storage containers, chlorinators, sulfonators, *injectors*, and a gas room equipped with storage area, scales, gas detectors, piping, valves, and safety equipment. Facilities using 68-kg (150-lb) cylinders will also have cylinder restraints and either tank- or wall-mounted chlorinators with vacuum regulators. Facilities using 907-kg (1-ton) containers are equipped with cradles to hold the containers, trunnions for rotating containers into position, and an overhead crane. Chlorine and sulfur dioxide gas are stored in containers. Most of the chemical is in liquid form, while some is in the gas form. Either liquid or gas may be used, depending on the type of system. Liquid chlorine systems have evaporators to convert liquid chlorine to gaseous chlorine upstream of the chlorinator. Gas rooms constructed after 1988 have scrubbers to capture and neutralize chlorine gas leaks. Gas rooms constructed prior to 1988 may or may not have scrubbers. The equipment is fairly reliable, easy to operate, and familiar to many WRRF staff. Hypochlorination facilities typically consist of either sodium or calcium hypochlorite, liquid sulfite compounds, chemical storage containers, and metering pumps.

Injectors are also called *ejectors* or *eductors*, depending on the manufacturer.

CHLORINATION AND DECHLORINATION CHEMICALS

Elemental Chlorine

The chemical symbol for chlorine is Cl. Chlorine almost always exists as molecules with two chlorine atoms each (Cl_2). At room temperature, chlorine (Cl_2) is a greenish-yellow gas. When cooled to −33.97 °C (−29.15 °F) at 101.325 kilopascals (kPa) (1 atmosphere, 14.7 psi) of pressure, it condenses to form a clear, amber-colored liquid (The Chlorine Insititute, 2014). Additional physical properties of chlorine are given in Table 10.5.

Boiling is the conversion of a liquid to a gas.

Chlorine gas is 2.5 times heavier than air. In the event of a chlorine leak, the gas will collect near the floor and in the lowest areas. In liquid form, chlorine is approximately 1.5 times heavier than water. Liquid chlorine *boils* at −33.97 °C (−29.15 °F) and vaporizes rapidly. This means that as soon as the pressure over the liquid is released, the chlorine turns to gas almost instantaneously. One volume of liquid chlorine yields approximately 460 volumes of gas. Thus, 1 kg (2.204 lb) of liquid chlorine produces enough gas to fill 0.31 m^3 (5.4 cu ft). The 460-times expanded volume contains the highest concentration of chlorine gas possible after all the liquid has been vaporized. In practice, chlorine gas will continue expanding until it fills the room; however, the concentration will be diluted as the volume occupied increases. Chlorine has a strong, characteristic odor: the smell of bleach. The *odor threshold* for chlorine gas is between 0.08 and 0.4 ppmv.

The *odor threshold* is the lowest concentration of something that can be detected by smell. Odor thresholds vary considerably from one person to the next for the same chemical.

ppmv stands for parts per million per volume of air.

Chlorine is classified as either dry or wet. It is important to know the difference because wet chlorine is extremely corrosive to most metals, whereas dry chlorine is not. Dry chlorine does not contain *free water*. Small amounts of water may be present in dry chlorine; however, the water is dissolved within the liquid chlorine and cannot be seen. The solubility of water in chlorine is temperature dependent (Table 10.6). For example, chlorine that contains 130 mg/L of water would be considered wet at 10 °C (50 °F), but dry at 20 °C (68 °F). Most metals are not affected by dry chlorine in the gas or liquid state when the temperature is below 232 °C (450 °F). However, dry chlorine reacts with titanium metal and ignites carbon steel when temperatures exceed 232 °C (450 °F). Wet chlorine contains more water than can be dissolved in the liquid chlorine. Free water floats on top of liquid chlorine because it is less dense than water. Recall that when chlorine mixes with water, hypochlorous and hydrochloric acids are formed. Because of this, wet chlorine is corrosive to most metals.

Free water is water that can be seen separated from the liquid chlorine. Because liquid chlorine is heavier than water, free water floats on the surface of liquid chlorine.

Precautions should be taken to keep chlorine and chlorine equipment dry. Containers, valves, and piping should be closed and capped when not in use to keep water out (The Chlorine Insititute, 2014). Water from the atmosphere can *condense* on any surface, including inside valves and exposed piping.

Condense means to turn from a vapor (gas) into a liquid. A glass filled with an ice-cold drink can pull water from the atmosphere and condense. Water droplets form on the outside of the glass.

Each phase of chlorine service requires a different type of piping. Chlorine liquid or gas pressure piping is typically 18-mm (0.75-in.) or 25-mm (1-in.) carbon steel, seamless schedule 80 material. Chlorine

Table 10.5 Physical Properties of Chlorine (Connell [2002] based on data from The Chlorine Institute [1997])

Property	Value
Boiling point	−33.97 °C (−29.15 °F) at 101.325 kPa (1 atm)
Freezing point (melting point)	−100.98 °C (−149.76 °F) at 101.325 kPa (1 atm)
Gas density	3.213 kg/m^3 (0.2006 lb/cu ft) at standard conditions[a]
Gas color	Greenish-yellow
Odor	Bleach
Specific gravity (gas)	2.485[b] at standard conditions[a]
Liquid density at 0.0 °C (32 °F)	1467 kg/m^3 (91.56 lb/cu ft)
Liquid color	Amber
Specific gravity (liquid)	1.467[c]

[a]Standard conditions are defined as 0 °C (32 °F) and 101.325 kPa (1 atm) (14.696 psi).
[b]Density of air is 1.2929 kg/m^3 (0.0807 lb/cu ft).
[c]Ratio of the density of chlorine at 0 °C (32 °F) to water at 4 °C (39 °F).

Table 10.6 Solubility of Water in Liquid Chlorine (data points taken from The Chlorine Institute [2014])

Temperature	Solubility of Water in Chlorine, mg/L*
0 °C (32 °F)	80
5 °C (41 °F)	100
10 °C (50 °F)	120
15 °C (59 °F)	150
20 °C (68 °F)	190
25 °C (77 °F)	240

*Above this concentration, free water will exist and the chlorine will be wet.

gas under vacuum or up to 41.3 kPa (6 psig) may use plastic pipe or tubing constructed from polyvinyl chloride (PVC), chlorinated polyvinyl chloride (CPVC), acrylonitrile-butadiene-styrene (ABS), polyethylene (PE), polypropylene (PP), or fiberglass-reinforced polyester (FRP) materials. These are all different types of plastics. Chlorine solution may use any of the materials used for chlorine gas plus rubber hose and steel-lined pipe. Because of the corrosive nature of chlorine solutions, steel and iron pipe must not be used unless they have a corrosion-resistant lining.

The primary difference between *PVC* and *CPVC* is that PVC can be used at temperatures up to 60 °C (140 °F), whereas CPVC can be used at temperatures up to 93 °C (200 °F). Above these temperatures, the material begins to soften (PVC Pipe Supplies, 2017).

Commercial chlorine is classified as a nonflammable, toxic, compressed gas. Although classified as nonflammable, chlorine will support combustion under certain conditions. Many materials that burn in air (oxygen) will also burn in a chlorine atmosphere (The Chlorine Insititute, 2014). For this reason, piping and other equipment should not be welded or heated without properly evacuating and purging the chlorine first (The Chlorine Insititute, 2014). Chlorine is highly reactive and, under specific conditions, can rapidly react with and oxidize many compounds and elements, including grease. Equipment and piping must be cleaned prior to use to remove any oils (The Chlorine Insititute, 2014). Chemicals and equipment should never be stored in gas rooms.

Although engineers design piping systems, operators and maintenance staff often make repairs and sometimes replace existing lines. Knowledge of acceptable materials and safe working practices are essential for maintaining chemical systems.

Chlorine is only slightly soluble in water, with a maximum solubility at 100 kilopascals (kPa) (1 atm) of approximately 10 000 mg/L at 9.6 °C (49.3 °F). The solubility of chlorine, like all gases, decreases with increasing temperature.

Hypochlorites

Hypochlorites are *salts* of hypochlorous acid. Sodium hypochlorite (NaOCl) is the only liquid hypochlorite form in current use. Calcium hypochlorite [$Ca(OCl)_2$] is the predominant solid form. As shown in eqs 10.1 through 10.4, hypochlorite compounds react with water similarly to elemental chorine to produce hypochlorous acid and hypochlorite ion.

Chemically speaking, *salts* are compounds made from combining positively charged ions (cations) and negatively charged ions (anions). Sodium hypochlorite is made from sodium (Na^+) and hypochlorite ($OHCl^-$) ions.

Typically, commercial or industrial-grade hypochlorite solutions have strengths of 12% (trade percent) and can be purchased in the 10 to 15% (trade percent) range. Other strengths are available, such as 5.25% (trade percent), commonly referred to as *household bleach*. The physical properties of different strength hypochlorite solutions are given in Tables 10.7 and 10.8.

Table 10.7 Physical Properties of Sodium Hypochlorite Solutions (The Chlorine Institute, 2000)

Property	Weight (%)				
	5	7.5	10	12.5	15
Specific gravity at 20 °C (68 °F)	1.076	1.109	1.142	1.175	1.206
Density at 20 °C (68 °F), kg/L	1.076	1.109	1.142	1.175	1.206
Density at 20 °C (68 °F), lb/cu ft	67.14	69.20	71.26	73.32	75.25
Density at 20 °C (68 °F), lb/gal	9.01	9.25	9.53	9.80	10.06

Table 10.8 Hypochlorite Solution Strengths (The Chlorine Institute, 2000)

	Density		Availability chlorine (weight percent)	
Trade Percent	kg/L	lb/gal	g/L	lb/gal
1	1.1014	8.45	10	0.083
5	1.070	8.92	50	0.416
10	1.139	9.50	100	0.833
12	1.168	9.74	120	0.999
13	1.183	9.86	130	1.08
15	1.211	10.1	150	1.25

Hypochlorite solutions are defined by many different terms as identified in the following equations:

$$\text{Trade Percent (\%)} = \frac{\left(\text{Available Chlorine, } \frac{g}{L}\right)}{10} \tag{10.22}$$

$$\text{Weight Percent (\%)} = \frac{\text{Trade percent}}{\text{Specific Gravity}} \tag{10.23}$$

$$\text{Pounds per gallon}\left(\frac{lb}{gal}\right) = \frac{g}{L} \times 0.0083 \tag{10.24}$$

All sodium hypochlorite solutions are unstable. Heat, light, storage time, and impurities, such as iron, accelerate product degradation. Higher strength hypochlorite solutions degrade faster than lower strength solutions. Sodium hypochlorite solution should not be stored for more than 60 days. Because strength decreases over time, the volume of hypochlorite solution required to achieve the same dose increases with storage time. Hypochlorites are destructive to wood and corrosive to many metals. Hypochlorite solutions give off chlorine monoxide gas (Cl_2O) (The Chlorine Institute, 2000).In the presence of moisture, these vapors will corrode metals such as iron and copper.

The solid form (calcium hypochlorite) is available in powder, granular, disk, briquette, or tablet form. These forms come in drums of 22.7 kg (50 lb) or 45.5 kg (100 lb). The strength of calcium hypochlorite is nominally 65% minimum calcium hypochlorite by weight. The remainder (35%) is water and inert materials. Typical strengths range to 70%, depending on the supplier. Calcium hypochlorite is classified as an oxidizer by the Department of Transportation (DOT).

The solid form (calcium hypochlorite) is unstable under normal atmospheric conditions. Reactions may occur spontaneously with numerous chemicals, including turpentine, oils, water, and paper. Therefore, calcium hypochlorite should be stored in a dry location and only be used with equipment that is free of organics. Serious fire and explosion hazards exist when using this material.

Sulfur Dioxide

In the gaseous state, sulfur dioxide is colorless, with a suffocating, pungent odor similar to a freshly struck match. It is approximately 2.3 times heavier than air. Like chlorine gas, sulfur dioxide gas will collect near the floor when released. Liquid sulfur dioxide is supplied as a pressurized, colorless, liquefied gas. The solubility of sulfur dioxide gas in water is approximately 17 times greater than the solubility of chlorine gas. In solution, sulfur dioxide reacts with water to form sulfurous acid (H_2SO_3), which breaks down further to sulfite ion (SO_3^{-2}). Additional characteristics of sulfur dioxide are listed in Table 10.9.

Metal is *galvanized* by coating it with zinc. The zinc protects the metal against corrosion.

Sulfur dioxide is a reactive chemical and, in the presence of moisture, will react with common metals such as iron and copper. For wet sulfur dioxide gas, stainless steel (316 or Alloy 20) can be used. Some plastics may also be used with gaseous sulfur dioxide including Teflon™, PVC, and CPVC. Dry sulfur dioxide (liquid or gas) is not corrosive to steel and most other common metals. However, *galvanized* metals

Table 10.9 Physical Properties of Sulfur Dioxide (Compressed Gas Association, 2017)

Property	Value
Boiling point	−10.0 °C (−14.0 °F) at 101.325 kPa (1 atm)
Freezing point (melting point)	−75.5 °C (−104.6 °F) at 101.325 kPa (1 atm)
Gas density	2.927 kg/m^3 (0.1810 lb/cu ft) at standard conditions[a]
Gas color	Colorless
Specific gravity (gas)	2.2638[b] at standard conditions[a]
Liquid density at 21.1 °C (70 °F)	1379 kg/m^3 (86.06 lb/cu ft)
Liquid color	Colorless
Specific gravity (liquid)	1.436[c]

[a]Standard conditions are defined as 0 ° (32 °F) and 101.325 kPa (1 atm) (14.696 psi).
[b]Density of air is 1.2929 kg/m^3 (0.0807 lb/cu ft).
[c]Ratio of the density of sulfur dioxide at 0 °C (32 °F) to water at 4 °C (39 °F).

should not be used. Because sulfur dioxide does not burn or support combustion, there is no danger of fire or explosion.

Sulfite Salts and Solutions

There are many forms of sulfite salts and solutions available for use as dechlorination agents. The choice of using a salt or solution is typically made based on availability and cost. Of all the sulfite salts and solutions available, sodium bisulfite ($NaHSO_3$) is used most frequently in wastewater dechlorination. Sodium sulfite (Na_2SO_3) is available as a crystalline solid in several grades. The purity of the industrial grade is satisfactory for most wastewater treatment applications. Use of sodium sulfite requires mixing with water to form a solution of approximately 20%. Sodium bisulfite ($NaHSO_3$) is supplied as a 39% solution in two grades—pure and technical. The technical grade is typically the choice for WRRFs. Sodium bisulfite solutions are yellow in color with a pungent sulfur dioxide gas odor. These solutions are highly corrosive. Sodium metabisulfite is a solid, white, crystalline product that emits the odor of sulfur dioxide gas. All three compounds produce the same active ion, sulfite (SO_3^-), when dissolved in water. Sulfite salts are typically more expensive than sulfur dioxide. Sodium metabisulfite is more stable than sodium sulfite and sodium bisulfite.

TEST YOUR KNOWLEDGE

1. **Chlorine equipment and piping must be free of oil and grease.**
   - ☐ True
   - ☐ False

2. **As water temperatures increase, the amount of chlorine gas that can be dissolved in the water also increases.**
   - ☐ True
   - ☐ False

3. **Wet chlorine contains more than 150 mg/L of water.**
   - ☐ True
   - ☐ False

4. **Chlorine gas ______________ than air.**
   a. Contains more water
   b. Is heavier than
   c. Contains less water
   d. Is lighter than

5. **This is an effective disinfectant that is less hazardous than chlorine or calcium hypochlorite.**
   a. Sodium hypochlorite
   b. Potassium hydroxide
   c. Hydrochloric acid
   d. Sodium hydroxide

6. **Sodium hypochlorite may ________ with time unlike gaseous chlorine.**
   a. Degrade
   b. Crystalize
   c. Volatilize
   d. Strengthen

7. **Sodium hypochlorite can have a short shelf life. Concentration will be affected by all but**
   a. Temperature
   b. Time
   c. Concentration
   d. Pressure

8. **The term *wet chlorine* refers to**
   a. Sodium hypochlorite solutions
   b. Damp calcium hypochlorite powder
   c. Liquid chlorine containing free water
   d. Blends of calcium hypochlorite and water

9. **This type of piping material should never be used with chlorine solutions (sodium hypochlorite)**
   a. PVC
   b. Steel or iron
   c. PE
   d. FRP

10. **A hypochlorite solution that is 15% (trade percent) strength contains this much chlorine:**
   a. 1.5 g/L
   b. 15 g/L
   c. 150 g/L
   d. 1500 g/L

11. **While chlorine smells like bleach, sulfur dioxide smells like**
   a. Rotten eggs
   b. A freshly struck match
   c. Cabbage
   d. Decaying grass

12. **Match the chemical to its color.**

Chlorine gas	yellow
Chlorine liquid	colorless
Sulfur dioxide gas	white
Sodium bisulfite solution	greenish-yellow
Sodium metabisulfite	amber

CHLORINE CONTAINERS

Most facilities use chlorine and sulfur dioxide in 68-kg (150-lb) cylinders or 908-kg (1-ton) containers, although 45-kg (100-lb) cylinders are also available. Some larger facilities may use railroad tank cars. Railroad tank cars are available in capacities ranging from 14 500 to 81 000 kg (16 to 90 tons). The weights given for cylinders, containers, and rail cars refer to the weights of chlorine and do not include the weights of the containers. The shipping and handling of chlorine gas is regulated by DOT for land movement and the U.S. Coast Guard for movement on rivers, lakes, and oceans.

Cylinders

Chlorine and sulfur dioxide cylinders are seamless steel containers designed according to DOT specifications. Each cylinder is equipped with a Chlorine Institute valve located at the top of the cylinder (Figure 10.7). By law, only 80% of the cylinder volume is used when filled with liquid chlorine. The unused 20% is to allow for expansion of the liquid chlorine resulting from any temperature increases experienced after filling. Cylinders are moved, stored, and used in the upright position with the cylinder valve at the top. The bonnet covers the valve and protects it from damage when the cylinder is not in use. Cylinders should never be moved, even for short distances, without the bonnet securely in place.

Under normal operating conditions, most of the chlorine or sulfur dioxide within the cylinder will be in liquid form. Some of the liquid will be converted to gas and fill the space above the liquid. More gas is produced at higher temperatures. The gas exerts pressure on the inside of the cylinder. Greater amounts of gas exert greater amounts of pressure. Figure 10.8 shows the relationship between chlorine gas pressure and temperature. At 16 °C (60 °F), the gas pressure in the cylinder will be 495.66 kPa (71.89 *psig*). Notice that the pressure increases rapidly with increasing temperature. The pressure can become great enough to rupture the cylinder.

The acronym *psig* stands for pounds per square inch gauge. This is the amount of pressure above atmospheric pressure. A gauge that measures in psig will give the same result whether it is at sea level or on top of a mountain.

Each cylinder valve contains a fusible plug located below the valve seat (Figure 10.9). The fusible plug is a safety device. It is designed to melt when the chlorine temperature reaches between 70 and 74 °C (158 and 165 °F). At this temperature, chlorine expands to fill the vessel to 100% of its volume, generating a pressure of 2075 kPa (301 psig). When the fusible plug melts, the chlorine is able to escape and the pressure is relieved. While this causes a chlorine leak, it prevents the cylinder from exploding. Look carefully at Figure 10.9. Notice that the fusible plug is below the metal to metal seal (the metal to metal contact may also be referred to as a needle valve). The fusible plug can melt to relieve pressure regardless of whether the cylinder is in use (valve open) or not (valve closed). The threads at the bottom of the cylinder valve screw directly into the top of the gas cylinder, as shown in Figure 10.7. The fusible plugs used with sulfur dioxide cylinders are the same as the ones used with chlorine cylinders.

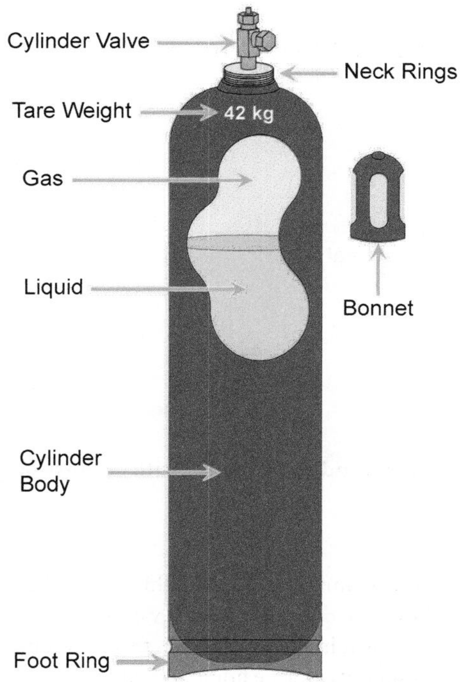

Figure 10.7 Chlorine Cylinder, 68 kg (150 lb) (Reprinted with permission by Indigo Water Group)

Ton Containers

Ton containers are large, horizontal vessels that can store 907 kg (1 ton) of chlorine. Ton containers are moved, stored, and used in a horizontal position (Figure 10.10). Two valves are located on one end of the container. The valves used on ton containers are similar to those used on cylinders, but have a larger opening and do not contain fusible plugs. The valves have a domed, protective cover that prevents damage during shipment and storage. Covers can be seen resting on the floor in front of the containers in Figure 10.10. The cover is removed only when the containers are placed into service.

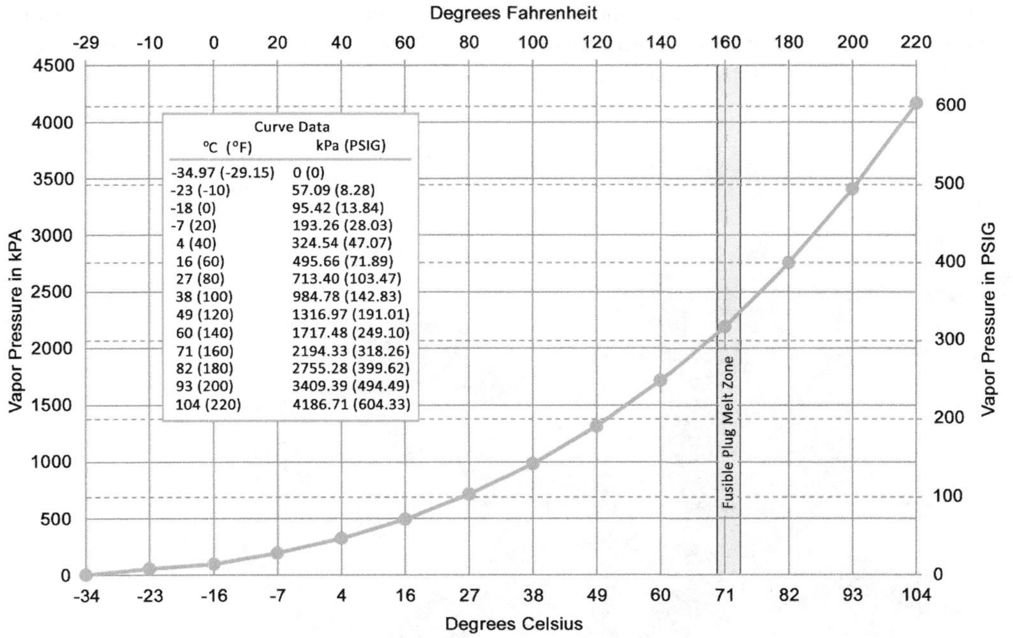

Figure 10.8 Chlorine Vapor Pressure Versus Temperature (adapted from The Chlorine Institute, 1997)

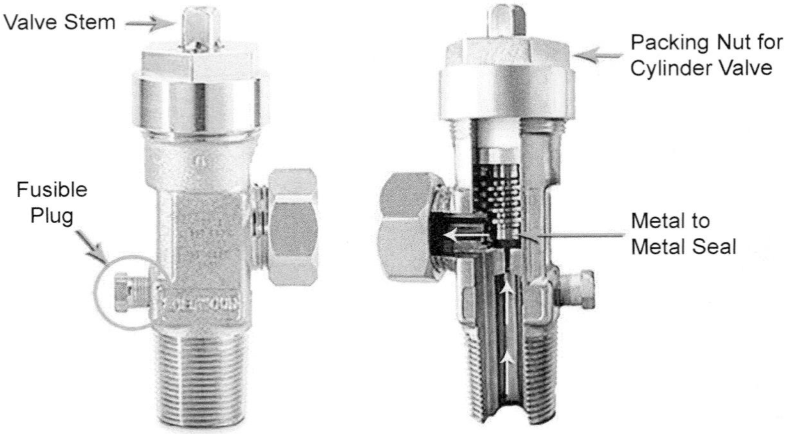

Figure 10.9 Cylinder Valve and Fusible Plug for a 68-kg (150-lb) Cylinder (Reprinted with permission by Sherwood Valve)

Containers rest inside a cradle that prevents them from rolling away. Trunnions, wheels within the cradle, allow the container to be rotated within the cradle. Containers should be positioned so the valves are located one above the other in a straight line up from the floor, as shown in Figure 10.11. In this position, the upper valve will supply chlorine gas and the bottom valve will supply chlorine liquid. An eduction tube is connected to each valve as shown in Figure 10.11. The tube ensures that gas (top valve) or liquid (bottom valve) is always available to the valve even as the container empties. If the container is rotated so the two valves change positions, the top valve will still draw gas and the bottom gas will still draw liquid. Sometimes, an eductor tube breaks off inside a container. If this occurs, gas or liquid may come out of both valves, depending on how full the container is at the time. This can also happen if the valves are not aligned vertically.

Each container has six fusible plugs with three located on each end of the container (Figure 10.12). Because ton containers are so large, it is possible for a portion of the container to reach 70 to 74 °C (158 to 165 °F) while the rest of the container remains cool. Placing fusible plugs in multiple locations helps ensure that the entire container is monitored. The plugs are positioned at 120 deg to one another on the face of each end near the outer edge of the container. The fusible plugs on a ton container look different than the smaller fusible plugs on a 68-kg (150-lb) cylinder, but they function the same way (Figure 10.13). Instead of being threaded into the valve, they are threaded directly into the ton container. The center of the plug melts to relieve pressure within the container.

Figure 10.10 Ton Containers (Reprinted with permission by Armando Mendoza)

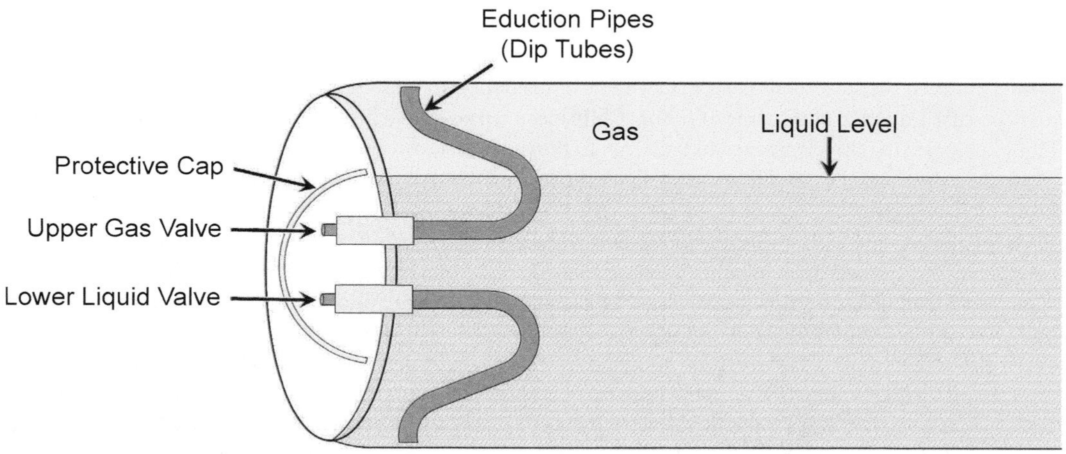

Figure 10.11 Ton Container with Eduction Pipes (Reprinted with permission by Indigo Water Group)

Figure 10.12 Fusible Plugs and Protective Cap on a Ton Container (Reprinted with permission by Indigo Water Group)

Figure 10.13 Fusible Plugs for Ton Containers and Cylinders (Reprinted with permission by Sherwood Valve)

Tank Cars and Trucks

Tank cars and tank trucks have an external steel shell and an inner steel shell, like a very large thermos bottle. Between the steel shells is 100 mm (4 in.) of insulation to maintain the liquid chlorine temperature and pressure provided at the time of filling. Chlorine is removed as a liquid from tank trucks and rail cars. Liquid chlorine fed directly from rail tank cars to the process is common practice at larger facilities. Tank trucks have been used to feed liquid directly to the process or to on-site storage tanks.

Each car is equipped with two liquid valves and two gas valves. A safety relief valve provides overpressure protection and is set to open at either 1550 or 2584 kPa (225 or 375 psig), depending on the tank car. The liquid lines contain excess flow valves ahead of each liquid-discharge valve to stop the flow of chlorine and prevent any further discharge in the event of a downstream line break. Three different capacities of excess flow valves are available and are designed to seat (close) at flows of 3178, 4540, and 6810 kg/h (7000, 10 000, and 15 000 lb/hr). Although the tank cars have gas valves, they are not used for removing chlorine gas. The gas valves are used when increased chlorine pressure is needed to aid in removal of liquid chlorine. Pressurized, dry air is added to the tank car through the gas valve to increase the car pressure.

According to DOT regulations, tank cars must always be attended during unloading and no more than one tank may be connected to an unloading facility at the same time. Any deviations must receive an exemption from the DOT. Because tank cars are supported on the rail-car wheels by heavy-duty springs, connections to tank cars must be flexible enough to accommodate a rise in the tank car elevation as chlorine is unloaded.

TEST YOUR KNOWLEDGE

1. **Fusible plugs are designed to melt at 70 to 74 °C (158 to 165 °F) to relieve gas pressure.**
   - ☐ True
   - ☐ False
2. **Fusible plugs only work when the cylinder valve is fully closed.**
   - ☐ True
   - ☐ False
3. **Rail cars may be used to supply either chlorine gas or chlorine liquid directly to the process.**
   - ☐ True
   - ☐ False
4. **A 68-kg (150-lb) chlorine cylinder typically contains this much chlorine:**
   a. 45 kg (100 lb)
   b. 54 kg (120 lb)
   c. 68 kg (150 lb)
   d. 85 kg (188 lb)
5. **Chlorine cylinders and containers are filled to a maximum of ________ of their volume.**
   a. 70%
   b. 80%
   c. 90%
   d. 100%
6. **Which of the following chlorine cylinders is likely to have the highest gas pressure?**
   a. Cylinder stored at 10 °C (50 °F)
   b. Cylinder containing 75% liquid chlorine
   c. Cylinder located outdoors on a hot day
   d. Cylinder with a closed metal to metal seal
7. **A 907-kg (1-ton) chlorine container has this many fusible plugs:**
   a. 1
   b. 3
   c. 6
   d. 9
8. **An operator connects a chlorine line to the bottom valve of a 907-kg (1-ton) container expecting to remove liquid chlorine. Instead, chlorine gas comes out through the lower valve. They switch the line to the upper valve, but still pull chlorine gas instead of liquid. What is the most likely reason for both valves to produce gas?**
   a. Eductor tube has broken off within the container.
   b. Container is positioned so both valves are below the liquid level.
   c. Container is less than 50% full of liquid chlorine.
   d. Excessive heat has converted all of the liquid chlorine to gas.

GAS CHLORINE SYSTEMS

All gaseous chlorine systems take advantage of the fact that chlorine boils and is converted from a liquid to a gas at −34 °C (−29 °F). Within the chlorine cylinder or container, the amount of liquid present versus the amount of gas present is dependent on temperature. Remember: boiling requires heat. More liquid is converted to gas as temperature increases, which increases gas pressure (Figure 10.8). The withdrawal of chlorine gas from cylinders and containers must have sufficient pressure to move the gas at a required rate. This means that there must be sufficient heat available internally and externally to vaporize the liquid chlorine and maintain gas pressure. As gas is removed, more of the liquid will vaporize to satisfy the liquid/gas equilibrium. Vaporization draws heat from the remaining liquid and from the cylinder or container, which cools the remaining liquid and cylinder or container, reduces the vaporization rate, and decreases gas pressure. This drop in temperature and pressure can slow, or even stop, the gas withdrawal rate.

When water droplets (condensation) begin to form on the outside surface of the cylinder or container, it indicates that the liquid temperature in the cylinder or container has reached the *dew point* of the surrounding air. Typically, condensation will only appear at or below the internal liquid level. When the liquid temperature reaches 0 °C (32 °F), the condensed moisture on the cylinder or container walls will turn to ice. Because ice acts as an insulator, ice formation must be avoided. Ice reduces the transfer of heat from the air through the cylinder or container wall and into the liquid. If the ice is not removed or prevented from forming in the first place, there will not be enough heat to continue to boil the liquid chlorine. Chlorine gas pressure will drop and the feed rate will slow, eventually stopping (Figure 10.8). This phenomenon is referred to as "freezing" the cylinder or container, but the chlorine and equipment don't actually freeze. This is impossible because chlorine freezes at −101 °C (−150 °F). The cylinder or container simply won't get that cold. The chlorine remains liquid; there just isn't enough gas pressure for gas flow to continue.

The *dew point* is the temperature at which the air begins to release water. Warm air holds more water than cool air. As the air cools, the water is released. This is what happens when a glass filled with ice "sweats" and forms water droplets on the outer surface.

A persistent myth is that a maximum of 18 kg/d (40 lb/d) may be withdrawn from a 68-kg (150-lb) chlorine cylinder and a maximum of 180 kg/d (400 lb/d) may be withdrawn from a 907-kg (1-ton) container. Although the Chlorine Institute advises that the dependable, continuous discharge rate from a cylinder is approximately 0.8 kg/h (1.75 lb/hr) or up to 19.2 kg/d (42.0 lb/d) and 6.8 kg/h (15.0 lb/hr) or 163.2 kg/d (360.0 lb/d) for 907-kg (1-ton) containers, experience shows that higher feed rates are possible. For the recommended, reliable feed rates, the Chlorine Institute model assumes an ambient temperature of 21.1 °C (70 °F) and a *backpressure* of 241 kPa (35 psig). If either heat transfer can be increased or the backpressure reduced, the amount of gas that can be removed from a cylinder or container without icing will increase. Cylinder feed rates can be doubled if air circulation is provided. Circulation minimizes condensation, prevents icing, and helps with heat transfer. With directly mounted vacuum regulators, gas feed rates of 45.6 kg/d (100 lb/d) are attainable from cylinders. It is possible to achieve a 227 kg/d (500 lb/d) feed rate from 907-kg (1-ton) containers using modern equipment designs and by providing air circulation around the container.

Backpressure is the pressure that the chlorine gas must discharge against.

The continuous withdrawal rate of sulfur dioxide gas is slower than chlorine. The rate is approximately 0.91 kg/h (2.0 lb/hr) from 68-kg (150-lb) cylinders whereas the rate from 907-kg (1-ton) containers is 11.3 kg/h (25.0 lb/hr) at 21.1 °C (70 °F). The withdrawal rate is slower for sulfur dioxide because it doesn't generate as much gas pressure within the cylinder or container as chlorine at the same temperature. Contrary to the procedures for handling chlorine, direct heat in the form of a heating blanket can be applied to sulfur dioxide cylinders and containers. However, the blankets must be controlled so that the maximum temperature of the blanket does not exceed 37.8 °C (100.0 °F). Care must also be taken to ensure that the heat from the blanket does not melt the fusible plug(s).

Pressure-Based Gas Feed Systems

Pressure-based chlorine gas systems rely on the gas pressure produced within the container to push the gas down the line and into the water being treated. These systems were in common use prior to 1960. A typical pressure system consists of the chlorine cylinder or container, pressure control regulators, gas control valve, flow indicator, and check valve (Figure 10.14). The check valve at the injection point prevents water from backing into the gas line. For a pressure system to work, the pressure in the gas line must be greater than the backpressure of the water being treated. Injecting chlorine gas into a pressurized water line or at some depth in a basin or channel requires that the gas pressure be higher than the backpressure created by the water. Operating under pressure made these systems dangerous. A small pinhole leak anywhere in the system could fill a room with chlorine gas. Pressure systems still exist, but have largely been abandoned in favor of safer vacuum-based systems.

Most people think of boiling as adding heat. Although this is one way to boil a liquid, it isn't the only way. Think of the liquid molecules as having a certain amount of energy. When heat is added, each molecule has more energy. Eventually, the molecules gain enough energy that they can push against the pressure from the atmosphere and escape, becoming a gas. The amount of energy needed at atmospheric pressure is called the *boiling point*. Another way to boil a liquid is to remove the pressure that is holding the molecules together as a liquid. If the pressure gets low enough, the existing energy of the liquid molecules will be enough energy to escape and form a gas, even though additional heat wasn't added. The drop in pressure required to boil a liquid at room temperature is called the *vapor pressure*.

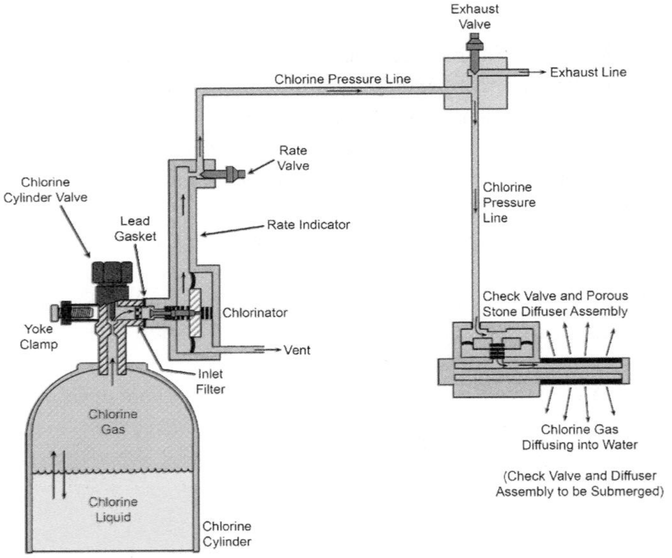

Figure 10.14 Pressure-Operated Gas Feed System (Courtesy of De Nora Water Technologies, Inc.)

A *check valve* only allows flow in one direction. If flow stops or begins to go backward through the valve, it closes to stop the flow.

The water pressure at the bottom of a column of water 0.3048-m (1-ft) high is 2.985 kPa (0.433 psi). The deeper the water, the greater the pressure.

Lead washers have traditionally been used for dry chlorine gas; however, various types of PTFE and asbestos fiber washers are also available (The Chlorine Institute, 2017). Washers may only be used once, regardless of material.

Systems with wall-mounted chlorinators that do not use cylinder- or container-mounted vacuum regulators maintain chlorine gas pressure between the cylinder or container and the chlorinator. A leak in this region will allow chlorine gas to escape.

Monel is a metal alloy made from 60 to 70% nickel, 25 to 35% copper, iron, manganese, carbon, and silicon. It is highly acid resistant and is often used in chemical handling.

Vacuum-Based Chlorine Gas Systems

Vacuum based chlorination systems consist of a chlorinator and an injector/ejector/eductor (name varies with manufacturer) (Figure 10.15). Chlorinators may be mounted directly on the chlorine cylinder or container, as shown in Figure 10.16, or on the wall. The first direct tank-mounted chlorinator was developed and patented in 1960. Cylinder-mounted chlorinators connect directly to the cylinder valve. An adjustable yoke holds the chlorinator in place (Figure 10.16). A *lead or polytetrafluoroethylene (PTFE) washer* placed between the chlorinator and the cylinder valve creates an inert, gas-tight seal when properly compressed. When the chlorinator is mounted to the wall, an ancillary valve and vacuum regulator are typically installed on the cylinder or container, as shown in Figure 10.17. The vacuum regulator converts gas pressure to vacuum by limiting the amount of gas allowed to leave the cylinder or container. This practice keeps all gas piping in the equipment room under vacuum, which improves safety and permits the use of plastic piping, such as schedule 80 PVC, from the regulator to the feeding equipment and ejector. Pressure systems require steel piping. With cylinder- or container-mounted chlorinators, the vacuum regulator is an integral part of the chlorinator. A flexible piece of tubing called a "pigtail" connects the vacuum regulator to the chlorinator. If a vacuum regulator is not present, the pigtail connects the container or cylinder to the chlorinator via the ancillary valve. Pigtails may be made of *Monel*-lined copper or clear plastic. In general, pressure systems use Monel-lined copper and vacuum systems use clear plastic pigtails.

The injector/ejector/eductor is located at the point where chlorine gas is added to the water being treated. The injector creates a vacuum that pulls chlorine gas from the chlorinator and into the water being treated. If a leak develops anywhere between the chlorinator and the injector, air will be pulled into the line. Loss of vacuum causes the chlorinator to close, stopping the flow of gas from the cylinder or container.

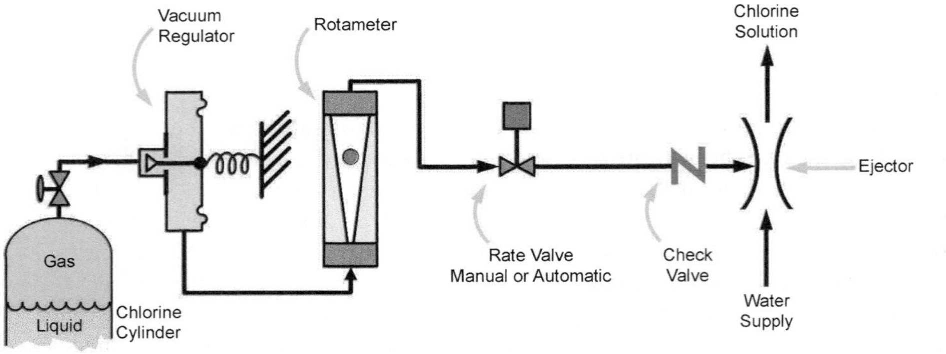

Figure 10.15 Vacuum-Operated Gas Feed System (Courtesy of De Nora Water Technologies, Inc.)

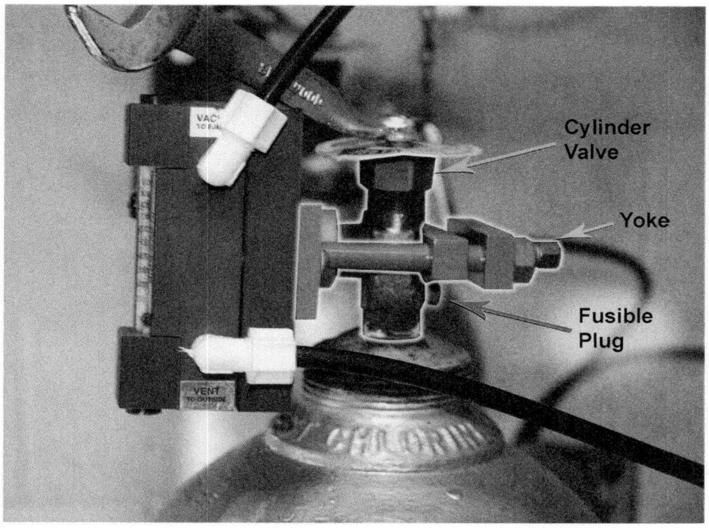

Figure 10.16 Chlorinator Mounted Directly on Cylinder (Reprinted with permission by Larry Thomas)

The injector continues to operate, evacuating the remaining chlorine gas and pulling room air into the gas line. Chlorine gas does not escape into the room, making vacuum-based systems much safer than pressure-based systems. A vacuum regulator combined with a wall-mounted chlorinator provides the same level of protection.

Tank-mounted chlorinators consist of three main components: a diaphragm valve, rotameter, and needle valve. The diaphragm valve is a large, flexible disk with a spring device mounted at the center to hold it closed (Figures 10.18 through 10.20). Diaphragm valves use force to open and close. Force is equal to pressure multiplied by area. A large area with a small amount of pressure can exert the same amount of force as a small area with a large amount of pressure. The pressure within the chlorine container is relatively high at around 586 kPa (85 psi) at 21.1 °C (70 °F). This pushes against the diaphragm and holds it closed (Figure 10.21). Downstream of the chlorinator, a slight vacuum is created by the *venturi*, which is negative pressure. In order for the diaphragm to flex open and allow gas to enter the chlorinator, the force on the downstream side (vacuum line) must be greater than the force on the upstream side (chlorine container).

Try making a *venturi* with a piece of rubber tubing and your kitchen sink. Stretch the tubing over the end of the faucet, allowing the tubing to dangle into the sink. Then, cut a small hole in the tubing just below the end of the faucet. Turn the water on as high as it will go. Place your finger over the hole. Can you feel the vacuum? Turn the water up and down and observe what happens to the strength of the vacuum.

Figure 10.17 Wall-Mounted Chlorinator with Vacuum Regulator on Cylinder: (A) Vacuum regulator, (B) Wall-mounted chlorinator, and (C) Scale (Reprinted with permission by Indigo Water Group)

Figure 10.18 Chlorinator/Sulfonator Diaphragm (Reprinted with permission by Lyle Leubner of North Little Rock Wastewater Utility)

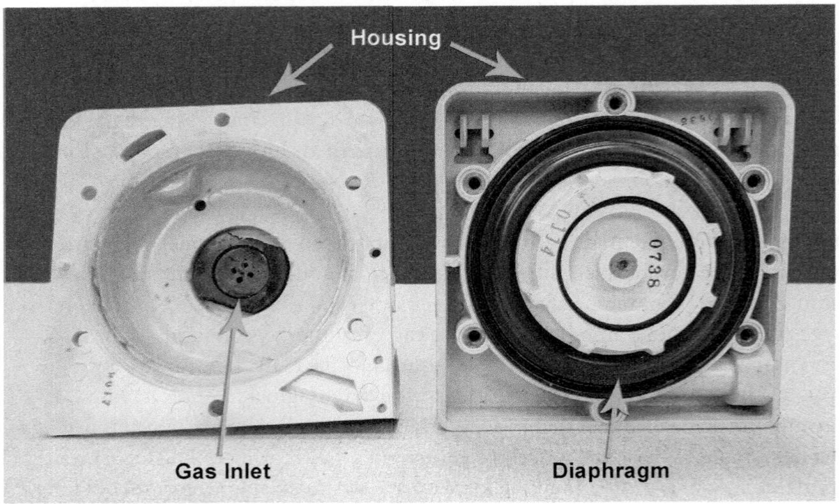

Figure 10.19 Chlorinator Components Showing Diaphragm (Reprinted with permission by Lyle Leubner of North Little Rock Wastewater Utility)

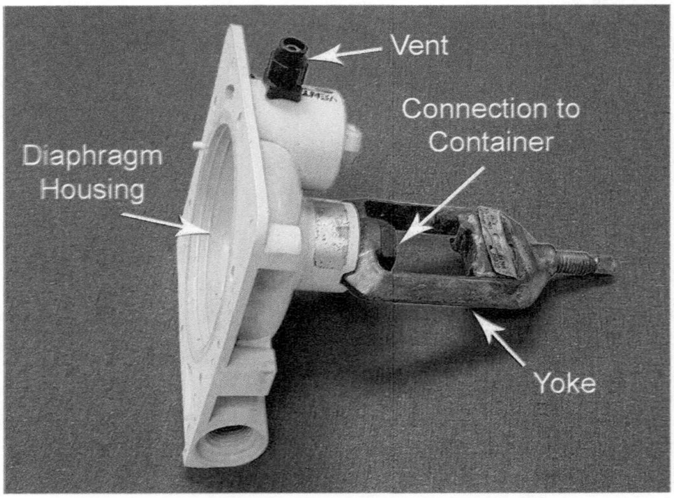

Figure 10.20 Chlorinator Components with Yoke for Direct Cylinder Mounting (Reprinted with permission by Lyle Leubner of North Little Rock Wastewater Utility)

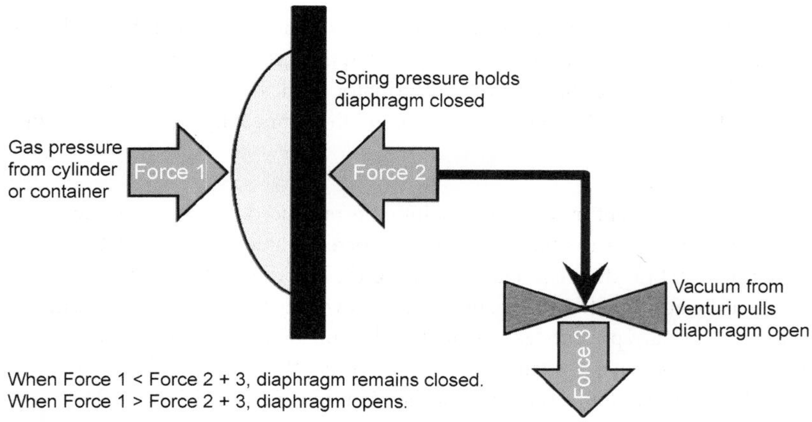

Figure 10.21 Opposing Forces (Reprinted with permission by Indigo Water Group)

This is accomplished by exposing a large portion of the diaphragm on the downstream side and a small portion on the upstream side. That way, a small amount of force downstream can pull the diaphragm open. The spring mechanism provides some additional force on the diaphragm. If a leak develops and vacuum is lost, the diaphragm slams shut, preventing chlorine gas from leaving the container.

The rotameter is a device that measures gas flowrate. The rotameter looks like a clear rectangle in Figure 10.22, but actually has a slight V shape—narrower at the bottom and wider on top. Rotameters may also be wall mounted separately from the chlorinator. As gas flows through the rotameter, a ball is pushed upward (Figure 10.22). The height of the ball in the rotameter corresponds to the gas flowrate or feed rate. Most rotameters indicate the gas flow in kilograms per day (pounds per day) of chlorine fed. A needle valve is used to control the amount of gas passing through the rotameter. The valve is adjusted by turning the knob on the chlorinator that is located just above the rotameter (Figure 10.22). Gas may leave the rotameter either through the upper gas line, which leads to the injector, or through the lower gas line. The lower gas line is a vent that ends at a gas scrubber or outside the gas room.

The injector has two components: a check valve to prevent water from backing into the gas line and a venturi, which creates the vacuum needed to pull open the diaphragm in the chlorinator and allow chlorine gas to enter the gas line. The venturi is essentially a section of pipe that quickly tapers from a larger diameter to a much smaller diameter and then reopens to the original diameter. Water is pumped through the venturi (Figure 10.15). When it passes through the narrowest point of the venturi, called the "throat", the velocity of the water increases. Velocity is equal to flow per area. If the diameter of the pipe is decreased

Figure 10.22 Rotameter (Feed Rate Indicator) (Reprinted with permission by Indigo Water Group)

from 2.54 cm (1 in.) to 1.25 cm (0.5 in.), the cross-sectional area decreases by a factor of 4 and the velocity increases by a factor of 4. By making the throat very narrow compared to the rest of the pipe, very high velocities can be achieved. The chlorine gas line runs from the chlorinator to the throat of the venturi. The high velocity of the water moving through the throat creates suction on the connecting line—the vacuum needed to open the diaphragm.

After the chlorine gas and water mix at the venturi, a chlorine solution is produced that is sent to the chlorine application point. Operators must ensure that they have the correct water pressure and flowrate to the venturi. Otherwise, the venturi won't produce enough vacuum for the chlorinator to operate properly. The throat diameter may be adjusted to increase or decrease flow velocity, which will influence the amount of vacuum produced. Typically, pressure gauges are installed upstream and downstream of the venturi to allow operators to monitor water pressure. It is important that the facility staff (operation and maintenance [O&M]) check with the manufacturer and/or the manufacturer's O&M manual to understand the correct requirements for the water pressure and flowrate.

Gas feeders for either chlorine or sulfur dioxide are similar in design and construction, but use different materials for the gas being fed. The vacuum regulator component of the chlorinator is made of ABS as its main component. Sulfonators use PVC. Metallic components for chlorinators are silver or a tantalum alloy. Sulfonators use stainless steel or tantalum alloy components. They are not interchangeable.

Gas Manifolds

When the vacuum regulator in a chlorination system is not mounted directly on the cylinder or container, a manifold is required. Gas manifolds consist of flexible copper connectors lined with Monel, typically with isolation valves, and a rigid pipe section made of carbon steel with one or more drip legs (Figures 10.23 and 10.24). Drip legs may also be located at the chlorinator. Drip legs collect liquid chlorine that may enter the manifold or be created in the manifold by reliquefaction. A heater at the bottom of the drip leg converts the liquid chlorine to gas. Pressure manifolds typically include heat tracing of the pressure lines, the use of pressure-reducing valves (PRVs), and insulation of the pressure line. These measures help ensure that the pressure piping is always as warm as or warmer than the cylinder or container to help prevent reliquefaction. Pressure lines typically slope from the chlorinator downward toward the cylinder or container. Any chlorine that condenses back to liquid will run toward the cylinder or container instead of toward the chlorinator. This is very important because the chlorinator cannot tolerate exposure to liquid chlorine. Even a small amount of liquid chlorine can damage the chlorinator.

Automatic Switchover Devices

Automatic switchover devices allow the system to automatically switch from an empty cylinder or container to a full one. The technology differs somewhat from one manufacturer to the next. Two cylinders or containers, each equipped with vacuum regulators, are connected to the switchover device. After

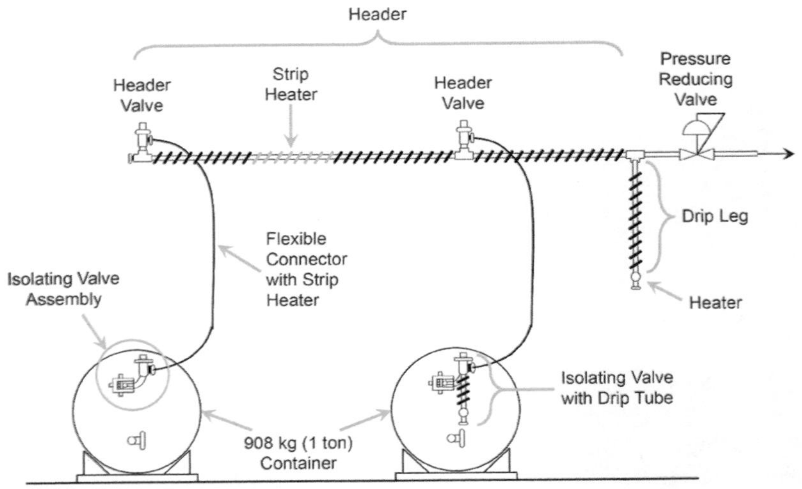

Figure 10.23 Ton Containers Manifolded Together to Increase Gas Withdrawal

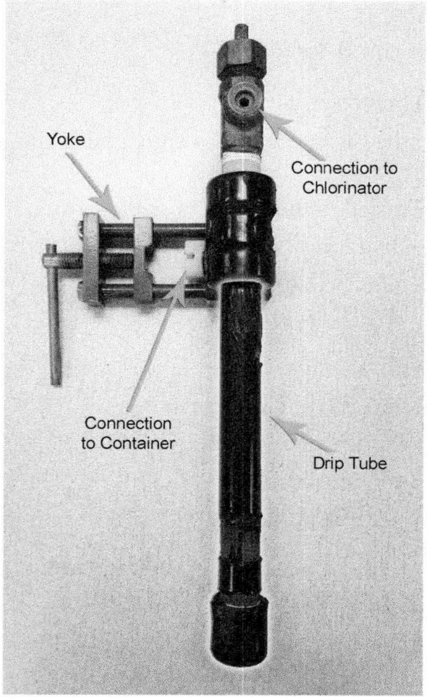

Figure 10.24 Drip Leg (Reprinted with permission by Lyle Leubner of North Little Rock Wastewater Utility)

connection, both cylinder or container valves are opened. Chlorine or sulfur dioxide gas is flowing from both cylinders or containers to the switchover device. The switchover device contains a detent-type lockout and diaphragm valve similar to the one inside a chlorinator or sulfonator. The lockout device keeps the valve on the second cylinder or container closed until it is needed. Gas is allowed to flow from the first cylinder or container (Wallace & Tiernan, 2017). When the first cylinder or container is empty, the system vacuum increases to a higher-than-normal level. The increased vacuum pulls against the detent and pulls it open (Wallace & Tiernan, 2017). This allows gas from the second cylinder or container to flow through the switchover device while continuing to pull the remaining gas from the first cylinder or container.

Automatic Shutoff Devices

Automatic valve operators mount directly on the cylinder or ton-container valve. The valve operator only closes and cannot open any cylinder or container valve. Activation of the valve can be from contact closures such as a leak, fire, or earthquake, or a manual action taken by an operator at a remote location. Valve operators are pneumatically (gas pressure) or electrically controlled. Six connections can be operated from one console.

TEST YOUR KNOWLEDGE

1. **The amount of chlorine in the liquid versus gas state in a cylinder, container, or rail car is dependent on temperature.**
   - ☐ True
   - ☐ False

2. **The maximum amount of chlorine that may be withdrawn from a 68-kg (150-lb) cylinder is 18 kg/d (40 lb/d).**
   - ☐ True
   - ☐ False

3. **Heat blankets may be applied to sulfur dioxide cylinders and containers, but not to chlorine vessels.**
   - ☐ True
   - ☐ False

4. **Liquid chlorine is used instead of chlorine gas when slower withdrawal rates are required.**
   - ☐ True
   - ☐ False

5. **Chlorine gas is withdrawn from a 68-kg (150-lb) cylinder. Condensation is observed on the bottom half of the cylinder. What must be true?**
   a. The chlorine within the cylinder is beginning to freeze.
   b. The cylinder is approximately half full of liquid chlorine.
   c. If the withdrawal rate is decreased, ice may form.
   d. There is too much air movement around the cylinder.

6. **Ice formation on cylinders and containers should be avoided because**
   a. It can damage the cylinder or container wall.
   b. It may interfere with the function of the fusible plug.
   c. It prevents heat from moving into the cylinder or container.
   d. It increases the gas withdrawal rate and generates pressure.

7. **Which of the following types of chlorine systems is the safest?**
   a. Gas pressure system
   b. Cylinder-mounted chlorinator
   c. Wall-mounted chlorinator without vacuum regulator
   d. Liquid chlorine system

8. **This piece of equipment is used in combination with a wall-mounted chlorinator to keep the entire system under vacuum.**
   a. Vacuum regulator
   b. Venturi
   c. Eductor tube
   d. PTFE washer

9. **Chlorine injectors function based on**
   a. Pressure
   b. Coriolis effect
   c. Vacuum
   d. Plug flow

10. **A chlorine system consists of a wall-mounted chlorinator and cylinder-mounted vacuum regulators. A leak develops between the chlorinator and the venturi. What is the most likely outcome?**
    a. Gas room will fill with chlorine gas.
    b. Chlorinator diaphragm will close.
    c. Venturi operation will stop.
    d. Rotameter ball will remain elevated.

11. **For a venturi to function correctly:**
    a. The opening between the gas line and the venturi must be closed.
    b. The throat must be adjusted to the widest possible setting.
    c. Chlorine dose and residual must be greater than 10 mg/L.
    d. Pressure and flowrate must be high enough to produce a vacuum.

12. **Manifolds are equipped with drip legs to**
    a. Protect the chlorinator from liquid chlorine
    b. Increase heat transfer and withdrawal rates
    c. Reduce the need for sulfonator blankets
    d. Replace the pigtail-type connectors

13. **The water flowrate through the venturi in a gas eductor has decreased below the manufacturer's recommended minimum flowrate. What is the most likely outcome?**
    a. Gas flowrate will increase.
    b. Vacuum will be lost.
    c. Reliquificiation of chlorine.
    d. Outgassing to atmosphere.

14. **Chlorine gas systems that operate under pressure must use this type of connector:**
    a. Clear plastic pigtail
    b. Monel-lined copper pigtail
    c. Schedule 80 PVC
    d. Aluminum spiral wound PVC

Liquid Chlorine Systems

For each 907-kg (1-ton) container, gas withdrawal rates are limited to 8 to 10 kg/hr (400 to 500 lb/d) for chlorine and 4.5 to 5 kg/hr (225 to 250 lb/d) for sulfur dioxide. Higher feed rates of either chemical require gas manifolding. When there isn't enough floor space to allow for manifolding, liquid chlorine may be withdrawn instead. Liquid chlorine may be withdrawn from either 907-kg (1-ton) containers or rail cars. In general, liquid withdrawal is unaffected by temperature as long as there is sufficient pressure within the container or rail car. In other words, there is no real upper limit on the liquid withdrawal rate. Because chlorine is fed to the chlorine feed equipment (chlorinators) as a gas, liquid chlorine must be converted to the gas form after withdrawal. This is accomplished with chlorine vaporizers. A vaporizer is also referred to as an *evaporator*.

A chlorine or sulfur dioxide vaporizer consists of an inner chamber where the liquid chlorine or sulfur dioxide is introduced and an outer chamber that contains heated water (Figure 10.25). The water bath is maintained at a high temperature by immersion heaters. The vaporizer has controls for water level and temperature and is equipped with cathodic protection. Chlorine or sulfur dioxide gas is released from the liquid in the inner chamber and exits the vaporizer through a superheated baffle and into the system piping. An automatically controlled PRV regulates downstream pressure. A rupture-disk-protected relief valve is designed to burst and release gas when the gas pressure gets too high. It prevents an explosion from overpressure. Gas flows through a gas strainer to the gas feeder (chlorinator or sulfonator).

The operation of chlorine or sulfur dioxide vaporizers does not increase the pressure of the chemical in either the liquid or gas portion of the system. However, because the vaporizer itself operates under pressure, all vaporizers are designed to meet the American Society of Mechanical Engineers (ASME) for pressure vessels.

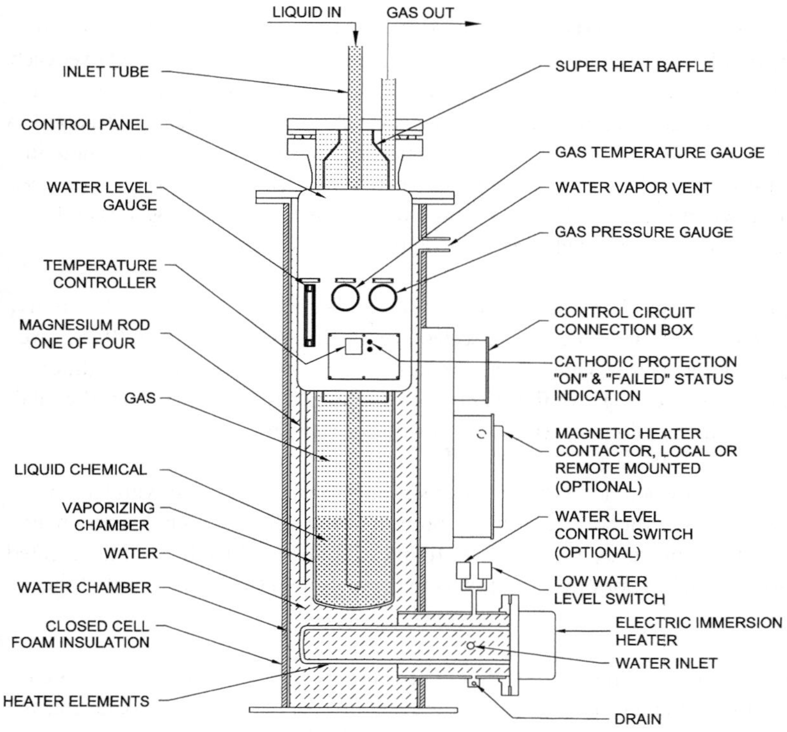

Figure 10.25 Vaporizer (Courtesy of De Nora Water Technologies, Inc.)

There is no chlorine or sulfur dioxide liquid level control in the inner chamber. The vaporizer level is maintained by the feed rate, which is set by the downstream gas feeder (chlorinator or sulfonator). Liquid carryover can occur whenever the feed rate exceeds the ability of the vaporizer to provide sufficient heat. It will also occur if there is a buildup of debris in the chamber that affects heat transfer. The appearance of moisture or frost on the external walls of downstream piping is indicative of liquid carryover and must be corrected quickly. If it is not corrected, the chlorinator or sulfonator may be damaged.

All vaporizers are equipped with several devices for safe operation, including the following:

- Pressure-reducing and shutoff valve—the PRV closes on low water temperature, low water level, or loss of power. It also helps prevent liquid carryover because the pressure drop that occurs as gas moves through the valve generates heat. The extra heat helps to vaporize any liquid chlorine or sulfur dioxide that may be present.
- Pressure-relief valve—the ASME code requires a pressure-relief device to protect against overpressurization of the system.
- Temperature control—the water bath temperature is controlled in a narrow range sufficient to vaporize the chlorine or sulfur dioxide. Typically, the water temperature is kept below 82.2 °C (180.0 °F). A high water temperature will shut off the power to the water heaters and shut the system down.
- High-pressure switch—the high-pressure switch disconnects the vaporizer heater power supply and provides an alarm contact in the event of high system pressure.
- Water-level switch—a low water level in the water bath will shut the system down. The switch also controls a valve in the makeup water line to maintain the water at a desired level.
- *Cathodic protection*—protects against vaporizing chamber and water bath tank corrosion. Donor electrodes are submerged in the water bath and an ammeter monitors the cathodic protection's performance.

Chlorine and Sulfur Dioxide Gas Rooms

Chlorine and sulfur dioxide gas rooms are always at ground level and have the same basic layout. Having the gas rooms at ground level prevents storage of cylinders in low-lying areas where leaks could cause gases to accumulate and eliminates the need to maneuver cylinders and containers up and down stairs. Gas storage and use areas should be dedicated rooms. In these facilities, nothing should be stored, and no work should be performed that is not related directly to handling chlorine or sulfur dioxide.

For corrosion to occur, all four parts of the galvanic cell must be present. They are 1) anode, 2) cathode, 3) metal, and 4) a conduction solution, typically water. Electrons move from the anode to the cathode. This causes the metal at the anode to become ionized. For example, the iron in an iron pipe does not typically have a charge. This is shown chemically as Fe^{+0} or Fe. When part of the iron pipe becomes an anode, the iron atoms give up electrons. This transforms the iron metal at the anode from Fe^{+0} to Fe^{+2} or Fe^{+3}. Once the iron has a charge, it becomes soluble in the passing water and can be dissolved. Charged iron atoms at the anode will also react with oxygen and carbon dioxide in the water to form rust (iron oxide) and scale. The electrons move from the anode to the cathode, where they participate in other chemical reactions. Scale builds up at the cathode. The result is that some areas of the pipe form pits while other areas form bumps and chimneys. To stop corrosion, at least one of the four components of the galvanic cell must be eliminated. This is nearly impossible to do in many wastewater environments. Instead, *cathodic protection* is used. Cathodic protection provides a piece of sacrificial metal, typically zinc or magnesium, to create a new anode for the galvanic cell. Zinc and magnesium give up electrons more readily than iron, steel, and the other metals that cathodic protection is used to protect. Corrosion still occurs, but is moved to a different location—the zinc or magnesium rods. In the case of the evaporator, the anode corrodes so the rest of the evaporator does not.

Plan and profile views of a typical gas room are shown in Figures 10.26 and 10.27, respectively. Notice in Figure 10.26 that full and empty chlorine containers are stored separately. Signs should be placed over each storage area to designate which containers are full and empty. Cylinders and containers may also be physically tagged to further identify which ones contain chlorine or sulfur dioxide. Labeling ensures that first responders are able to quickly identify full containers. Additionally, a wall separates the cylinders in use from those in storage. Chlorine gas rooms must have leak detectors that alarm whenever the chlorine concentration goes above 1 ppmv. The gas room shown in Figure 10.26 has two gas sensors, one in the operating area and one in the storage area. Multiple sensors may be needed depending on the size of the room.

Cylinders and containers are typically placed on a scale when in use. Smaller scales, like the one shown in Figure 10.17, accommodate two cylinders at a time. Larger scales, like the one shown in Figure 10.28, have a balance-beam-type arrangement and a large plate on one end of the beam that supports the weight of multiple cylinders or containers. The tare weight of the cylinders or containers is subtracted from the total weight so the operator knows how many kilograms (pounds) of chlorine remain.

Gas cylinders must be restrained, even when not in use. A chain strong enough to hold the cylinder when full should be placed about two-thirds of the way up from the bottom of the cylinder. A single chain may be used to restrain multiple cylinders; however, it is safer to restrain each one individually. Facilities located in earthquake zones must also restrain the bottoms of cylinders to prevent them from sliding out of the upper restraint.

One-ton containers should be located in cradles with trunnions. Trunnions are necessary for both stored and in-use containers, to allow the container to be rotated in the event of a leak, so that the leaking area is at the top of the container, resulting in the release of gas rather than liquid. Trunnions are also necessary to permit rotating the container, so that the valves on the 900-kg container are vertical. Hold-down chains for each container, both stored and in use (including those on scales), are recommended to prevent container movement, especially in earthquake-prone areas. A properly designed 900-kg-container-handling facility will also have an overhead monorail hoist and motorized trolley of at least 1800-kg (2-ton) capacity. Slow-speed cranes typically are used to prevent jerky movements of containers.

Gas rooms should be kept at 20 °C (68 °F) at all times. A room that is too cold may reduce the gas feed rate. A room that is too warm can cause pressure to build up in gas cylinders and containers. Gas cylinders and containers should always be stored at temperatures equal to or greater than the chlorinator to prevent reliquefaction of chlorine gas in gas lines and manifolds. Never use a space heater in a gas room or apply heat directly to a cylinder or container.

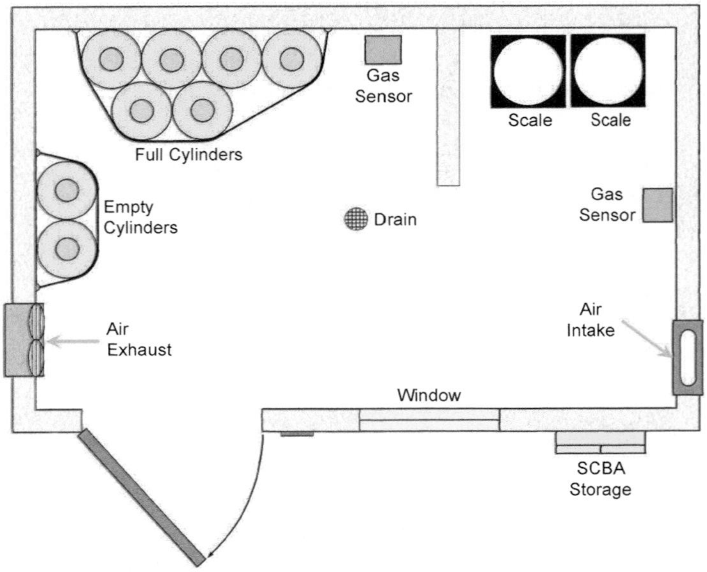

Figure 10.26 Plan View of a Typical Chlorine Gas Room

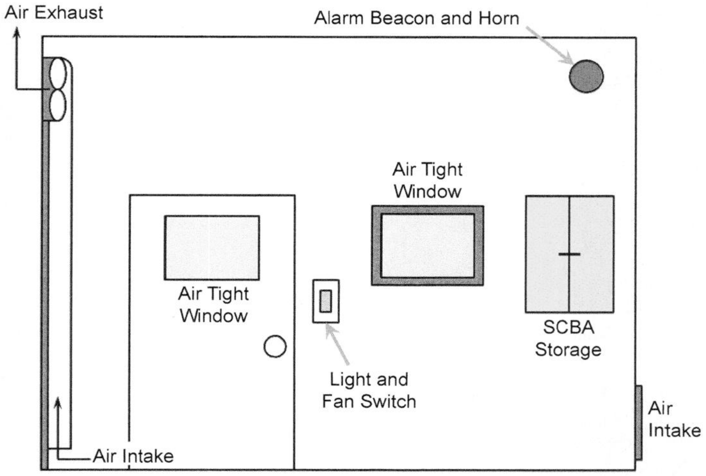

Figure 10.27 Profile View of a Typical Chlorine Gas Room

The room or building must be gas tight to contain leaks. Notice in Figures 10.27 and 10.29 that the doors open out and each has a clear, glass window at the top. This reduces the need for entry and allows the operator to monitor the feed rate at a glance as well as check for obvious leaks without entering the room. Gas room entrances must have Occupational Safety and Health Administration (OSHA)-approved warning signs showing that the room contains chlorine or sulfur dioxide. Signs should also be posted on any other exposed sides of the gas room. The light switch, fan controls, and safety equipment are located outside of the gas room. This allows operators to access them in the event of a leak without entering the gas room. Safety equipment includes *self-contained breathing apparatus (SCBA)* and the appropriate chlorine container repair kit. Supplied air is also acceptable. Repair kits are specific to the container type: 68-kg (150-lb) cylinders use an A kit, 907-kg (1-ton) containers use a B kit, and rail cars use a C kit. The kits contain tools and patches for repairing chlorine and sulfur dioxide leaks.

An *SCBA* consists of tanks of air, air hoses, and a face mask. A self-contained underwater breathing apparatus or SCUBA gear for diving may be a more familiar form of this type of equipment. Though not identical, they are similar. Supplied air systems provide air via a fan or compressor and hose connected to a face mask.

Chlorine gas and sulfur dioxide gas are both heavier than air. Air exhaust vents are located near the floor where these gases will collect in the event of a leak. The exhaust system should exchange all of the air in the room once every 3 to 4 minutes. In other words, 15 to 20 air exchanges per hour. Air exchanges remove gas that may be present resulting from leaks. They also reduce corrosion of equipment, piping, and valves.

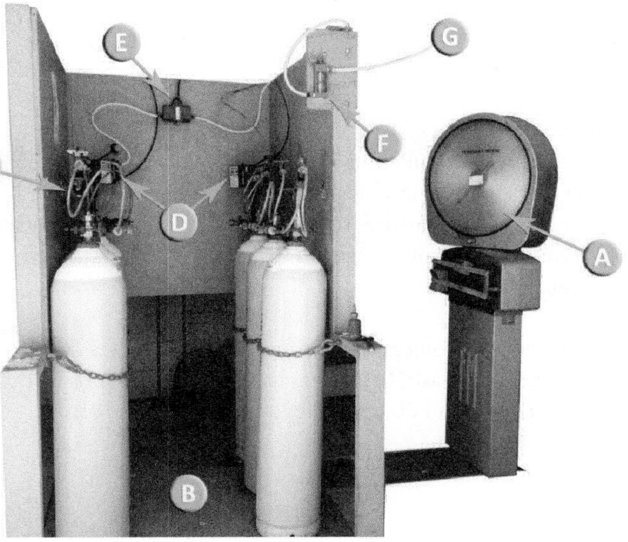

Figure 10.28 Gas Room with Manifolded Cylinders on Floor Scale: (A) Balance beam scale readout, (B) Balance floor, (C) Pigtail connector, (D) Chlorinators, (E) Automatic changeover device, (F) Rotometer, and (G) Gas line to injector/ejector/eductor (Reprinted with permission by Indigo Water Group)

Figure 10.29 Exterior of a Chlorine Gas Room at a Small WRRF (Reprinted with permission by Indigo Water Group)

In the United States, there are three major organizations that produce model fire codes: The International Fire Code Institute (California), The Building Officials and Code Administrators (Illinois), and The Standard Fire Prevention Code (Alabama) (De Nora Water Technologies, 2015). In addition, there is the National Fire Protection Association (Massachusetts). States and municipalities decide which set of standards to follow.

Scrubbers

Gas rooms may be equipped with scrubbers to remove chlorine from the air before it is discharged to the environment. Building and fire codes have had a strong effect on the use of scrubbers in chlorine gas facilities. These codes, particularly noticeable in California where the Uniform Fire Code (UFC) was developed, required that a scrubber be provided to contain all of the gas potentially produced from the largest cylinder or container in use. Prior to 1988 when Article 80 of the UFC was enacted, whether or not a scrubber was required depended largely on state and local requirements. Under Article 80, scrubbers are required for all chlorine gas rooms unless they meet at least one of the following criteria:

- A single 68-kg (150-lb) cylinder,
- Up to two 68-kg (150-lb) cylinders that are stored in either a gas cabinet OR a room with a 1-hour fire rating OR a room with fire sprinklers, or
- Up to four 68-kg (150-lb) cylinders may be stored if the room has both a 1-hour fire rating and fire sprinklers (De Nora Water Technologies, 2015).

The major types of scrubbers depend on the activation of a system that creates a vacuum capable of passing all chlorine liquid or gas leaks through a scrubber containing sodium hydroxide solution. Because chlorine is more soluble in sodium hydroxide (caustic) than in water, all of the released gas could be easily absorbed and the leak of gas or liquid would be contained. Scrubbing systems are typically installed outside of the gas room because of their size. Automatic scrubber activation is typically by a signal from a gas detector, although it can be activated manually.

When there are no leaks, the systems will sit dormant for long periods of time. It is often necessary to operate the scrubber under test conditions on a periodic basis. Each manufacturer's instructions will provide details on the recommended testing, frequency, and duration. The scrubbing system caustic solution may require periodic changing.

Chlorine Gas System Equipment Summary

Figures 10.30 through 10.34 summarize the different chlorine and sulfur dioxide gas system configurations. Automatic changeover systems, injectors/ejectors/eductors, and other ancillary equipment are not shown for simplicity. Table 10.10 summarizes the functions of major components of gas and liquid chlorine systems.

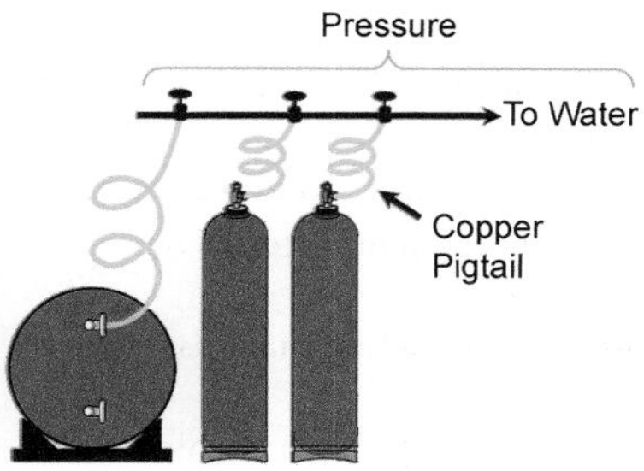

Figure 10.30 **Pressure Gas System (Reprinted with permission by Indigo Water Group)**

Pressure System

- Gas only
- Containers or cylinders connected directly to manifold
- Drip legs
- Gas pressure provides force to push chlorine into the water
- Most dangerous
- Leaks go to atmosphere
- No chlorinator, sulfonators, vacuum regulator, or ejector

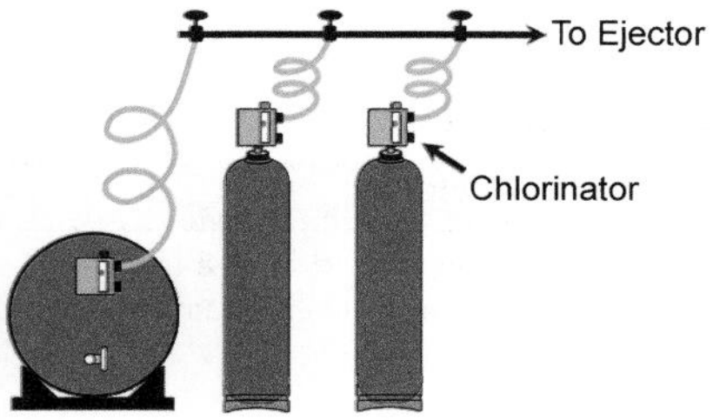

Figure 10.31 **Cylinder- or Container-Mounted Chlorinator (Reprinted with permission by Indigo Water Group)**

Cylinder- or Container-Mounted Chlorinator/Sulfonator

- Gas only
- Chlorinator or sulfonator mounted directly to cylinder or container
- 100% vacuum system
- Safest configuration
- Loss of vacuum (leak) shuts system down
- Uses chlorinator/sulfonator and ejector

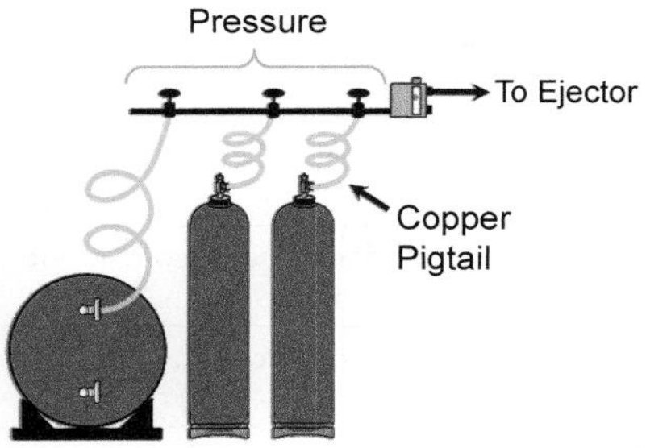

Figure 10.32 **Wall-Mounted Chlorinator with Gas Pressure (Reprinted with permission by Indigo Water Group)**

Wall-Mounted Chlorinator/ Sulfonator with Pressurized Pigtails and Manifold

- Gas only
- Partially pressurized
- Leaks upstream of chlorinator/sulfonators go to atmosphere
- Loss of vacuum (leak) downstream of chlorinator/sulfonator shuts system down
- Uses chlorinator/sulfonator and ejector

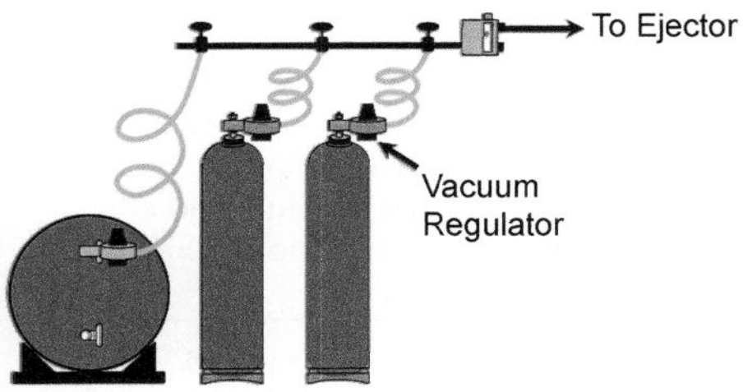

Figure 10.33 **Wall-Mounted Chlorinator with Vacuum Regulator (Reprinted with permission by Indigo Water Group)**

Wall-Mounted Chlorinator/Sulfonator with Vacuum Regulator on vessel

- Gas only
- 100% vacuum system
- Safest configuration
- Loss of vacuum (leak) shuts system down
- Uses chlorinator/sulfonator and ejector

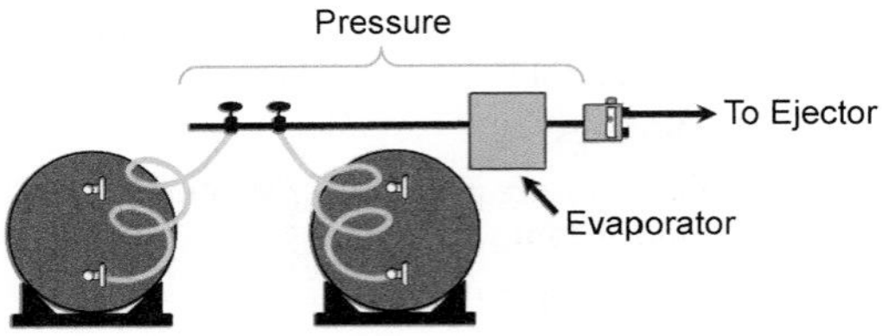

Figure 10.34 Liquid Chlorine System (Reprinted with permission by Indigo Water Group)

Liquid Chlorine System

- Liquid chlorine/sulfur dioxide drawn from containers
- Evaporator converts liquid to gas
- Liquid withdrawal not affected by temperature
- Greater gas production than gas only systems
- Pressure system upstream of chlorinator/sulfonators
- Vacuum system downstream of chlorinator/sulfonators
- Uses ejector

Table 10.10 Gas Chlorination System Components and their Uses (WEF, 2016)

Components	Features	Uses
Container hoist	Up-down and forward-reverse push buttons; at least a 18 015-kg (2-ton) capacity.	Move 907-kg (1-ton) containers to and from a truck loading dock and to place and remove containers from a weight scale.
Container or cylinder scale	Weight tare bar	To set the tare weight of a specific container or cylinder. Each 68-kg (150-lb) cylinder or 907-kg (1-ton) container has a tare weight stamped on it.
	Readout faceplate or digital dial indicating pounds of chlorine remaining	To illustrate how much chlorine has been used by subtracting past and present weights.
	Trunnions	To allow all 907-kg (1-ton) containers to be rotated while on the scale to properly position valves. Also, if a leak occurs, trunnions allow a container to be rotated so the leak is gas side up.
	Low-weight alarm	To alert operators that little chlorine remains so chlorine cylinders or containers can be replaced.
Automatic changeover system and pipeline device	Pigtail and yoke	To connect 68-kg (150-lb) or 907-kg (1-ton) container to manifold.
	Pressure switch	To monitor chlorine pressure in-line from scale to evaporator or chlorinator. When pressure is low, automatically switches from empty cylinder or container to full cylinder or container.
	Motor-operated valves	To isolate empty cylinder or container and introduce chlorine from full cylinder or container to facilitate automatic switchover without interrupting chlorination.
	Rupture disk and expansion chamber	To protect the pipeline by allowing rupture disk to break and fill expansion chamber when chlorine (especially liquid) is trapped between two closed valves. Pressure switches typically activate an alarm before and after the rupture disk is broken, based on the calibrated pressure to break the rupture disk.
Evaporator	Electric heater, steam, or hot water types	To convert liquid chlorine into gaseous chlorine for metering in a chlorinator and dissolution by an injector. Evaporators are necessary in facilities using more than 680 kg (1500 lb/d) of chlorine.
	Water bath and heat exchanger	To provide the heat needed to convert liquid chlorine into gas. Typically, immersion heaters in an external heat exchanger are thermostatically controlled to automatically maintain a specified water-bath temperature.

(continues)

Table 10.10 Gas Chlorination System Components and their Uses (WEF, 2016) (*Continued*)

Components	Features	Uses
Evaporator (*continued*)	Magnesium rods	To protect the water bath by allowing sacrificial rods to take up the corrosion.
	Discharge PRV	To reduce the pressure of the outlet gas conveyed to the chlorinator. Also, to shut off chlorine when at least one of the following evaporator conditions occurs: 1) low gas temperature, 2) low water bath level, 3) an evaporator power failure.
Chlorinator		To accurately meter chlorine gas for dissolution in an injector and inject solution to the treated effluent.
	V-notch vacuum plug	To limit the maximum capacity of chlorine gas that can be drawn through a chlorinator.
	Rotameter tube	To provide a metered amount of chlorine gas. Rotameters can be adjusted manually or automatically.
	Electric positioner	To automatically adjust the rotameter based on 4- to 20-mA signals from an effluent flow meter or chlorine residual analyzer.
	Vacuum regulator	To reduce the pressure of chlorine gas to 101.325 kPa (1 atm, 14.7 psi). Chlorine gas can then be conveyed through PVC pipes.
	Pressure and vacuum gauges	To monitor the pressure of chlorine gas before regulation, and the vacuum created by the injector design and setting.
Chlorine residual analyzer	Loss of vacuum alarm	To activate an alarm when low vacuum conditions occur.
	Sample pump, detector cell, and electronics	To continuously monitor the effluent chlorine residual so it can be charted for trending purposes. Various analyzers are available to measure free or total chlorine residual.
Redox (ORP)	Submersible sensor, probe cleaning system electronics	To continuously control the chlorine dose based on demand. Dose rates are set to achieve a predetermined level of disinfection. Redox will work in either free or total chlorine environments.
Gas injector	Also called an *ejector* or *eductor*, depending on the manufacturer. Features a water inlet, gas inlet, solution outlet, and venturi with variable throat.	To dissolve chlorine gas into solution by creating a vacuum with a variable throat orifice. This vacuum draws the chlorine gas through the chlorinator, to the injector, and then discharges the solution to the application point.
Chlorine gas leak detector	Electromechanical cell, variable ppm detection range, and test push button	To monitor chlorine gas leakage in the scale, evaporator, chlorinator, and injector, and activate an alarm when a leak occurs.
Safety equipment	SCBA	To enter an area where chlorine is leaking. Wearer breathes from the air tank via the face mask and regulator.
	Emergency repair kits	To provide essential components for temporary repair of chlorine leaks.
	Room ventilation system	To purge escaped chlorine gas from lower elevations of the chlorination area. Ventilation systems are activated from an outside switch prior to anyone entering the room.

SODIUM AND CALCIUM HYPOCHLORITE

Typically, the application of sodium or calcium hypochlorite solution to the treated effluent is done by a set of chemical-feed pumps that are referred to as *hypochlorinators*. The basic components are a storage reservoir or mixing tank for the hypochlorite solution; a metering pump, which consists of a positive-displacement pumping mechanism, motor or solenoid, and feed-rate-adjustment device; and an injection device. Depending on the size of the system, a plastic or fiberglass vessel is used to hold a low-strength hypochlorite solution. Hypochlorite solutions are corrosive to metals commonly used in the construction of storage tanks. Feeding of calcium hypochlorite requires a mixing device, typically a motorized propeller or agitator located in the tank. Also in the tank is a foot-valve and suction strainer connected to the suction inlet of the hypochlorinator. The strainer prevents undissolved chunks of calcium hypochlorite from entering the chemical feed pump. Mixing takes place in a separate tank. For both sodium and calcium hypochlorite systems, the pump suction intake should be kept above the storage tank bottom to avoid feeding sediment into the pump.

Figure 10.35 shows a sodium hypochlorite addition system at a small WRRF. The hypochlorite tank, valves, and metering pumps are located above spill pallets to catch drips and spills. For very small systems like this one, operators often dilute the more concentrated sodium hypochlorite delivered to the facility in a smaller tank. This is sometimes referred to as a "day tank" because it typically only holds enough diluted hypochlorite solution for a few days of operation.

Most installations of calcium hypochlorite use the compressed (tablet) forms available. A tablet feeder, shown in Figure 10.36, holds a stack of tablets. Tablet feeders are also called "erosion feeders". Water flows through the bottom of the tablet feeder, gradually dissolving the bottommost tablets. As these tablets dissolve, new ones drop in to take their place. Warmer water will dissolve the tablets faster than cold water. Water with lower total dissolved solids will dissolve the tablets faster than water with high total dissolved solids.

The water that flows through the tablet feeder may be a small portion of the total flow being treated or all of the flow. If only a portion of the flow goes through the tablet feeder (sidestream treatment), the concentrated hypochlorite solution produced is either gravity fed or pumped to the water being treated. With calcium hypochlorite powder, there can be sediment in the tanks after mixing, that is, inert material that does not dissolve. Because of this, mixing tanks will need to be cleaned periodically to remove it.

Decomposition of hypochlorite can cause pressure buildup in piping systems and has been known to rupture plastic pipe. Because contact with metals will accelerate the decomposition of hypochlorite and emit

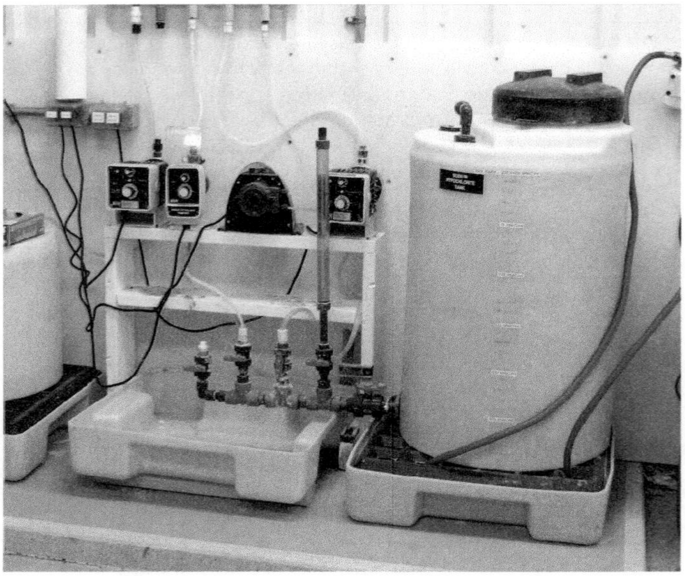

Figure 10.35 Sodium Hypochlorite Storage Tank and Metering Pumps (**Reprinted with permission by Indigo Water Group**)

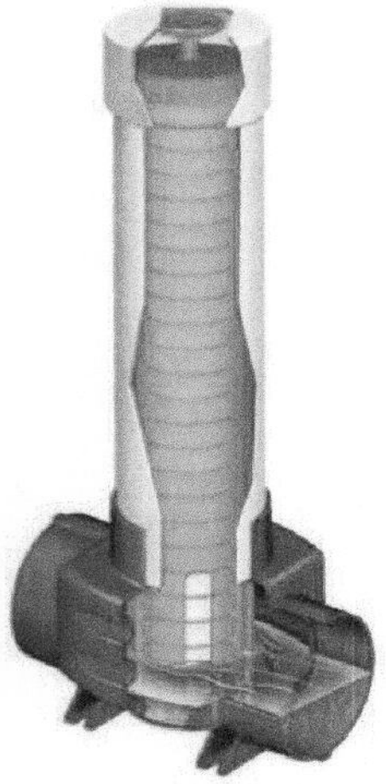

Figure 10.36 Tablet Feeder (Reprinted with permission by Norweco)

oxygen gas, components in the piping system, such as valves, gauges, and pressure switches, must be made of or be protected by compatible materials such as PVC, CPVC, and PP. Air-release valves are typically installed at the highest points in hypochlorite piping systems. Gas production within the storage tank can *air-bind* some types of metering pumps. Peristaltic pumps do not suffer from air-binding.

For pumps to function, they must be able to move liquid through the pump. When air accumulates inside centrifugal pumps, the pump rotates freely in the air instead of pushing liquid. The pump may also overheat.

Sodium hypochlorite solutions can leach through piping seams, connections, fittings, and valves. These leaks are unsightly. Their appearance can be exacerbated by the presence of sodium hydroxide (caustic) in sodium hypochlorite solutions. The caustic reacts with carbon dioxide in the air and leaves large, white deposits at the leak site. Scaling can occur when high alkalinity water is used as dilution water for hypochlorite solutions. The calcium in the water used for dilution and the high pH create the scaling problem. Scaling is problematic because it can interfere with operation of valves, it narrows the diameter of pipes, and pieces can break away to clog pumps and fittings.

MIXING EQUIPMENT

Mixing ensures that all of the water being treated is exposed to the same concentration of chlorine. Without proper mixing, some water might not have high enough chlorine residual concentrations to adequately disinfect. In the case of dechlorination, mixing brings hypochlorous acid and hypochlorite ion into direct contact with the sulfite ions. A wide variety of equipment may be used to mix chemicals with the final effluent prior to entering the chlorine contact chamber. A mixing device in a small-diameter pipeline may be as simple as having the chemical feed pipeline tap into the effluent pipeline at a 90-deg angle. The chemical feed must be at a higher pressure than the effluent pipeline to force the chemical solution into the flow. Another option is to inject chemical solution through a diffuser located inside a pipe. A diffuser is a smaller diameter piece of pipe with multiple holes. Because chemical is being added across the pipeline diameter, this provides better mixing than the single tap injection point. A third option is to use an in-line static mixer. Static mixers are designed to create turbulence inside the pipe. They typically have one or more obstacles for the water to flow around and through. Chemicals may also be added into small tanks equipped with some type of mechanical mixer.

CHORINE CONTACT CHAMBER

Chlorine disinfection takes place in either a chlorine contact chamber or pipeline. Contact chambers are long and narrow. Length-to-width ratios of 40 to 1 or greater are preferred (Metcalf & Eddy, Inc./AECOM, 2014). The goal is to maximize the chlorine contact time and reduce the potential for short-circuiting. By making the contact chamber long and narrow, water moves through it similarly to the way water moves through a pipe—from beginning to end with little or no mixing from one end to the other. This type of flow pattern is commonly referred to as plug flow. Contact chambers are typically built folded back on themselves in a serpentine pattern, as shown in Figure 10.37. This folded footprint reduces construction costs by taking advantage of common walls. Baffles reduce the potential for short-circuiting and prevent solids from settling out and collecting in the corners. A photograph of a contact chamber at a 3.9-ML/d (1-mgd) facility is shown in Figure 10.38. Two or more contact chambers are typically present so one can be taken offline for cleaning and maintenance while maintaining disinfection through the remaining contact chambers.

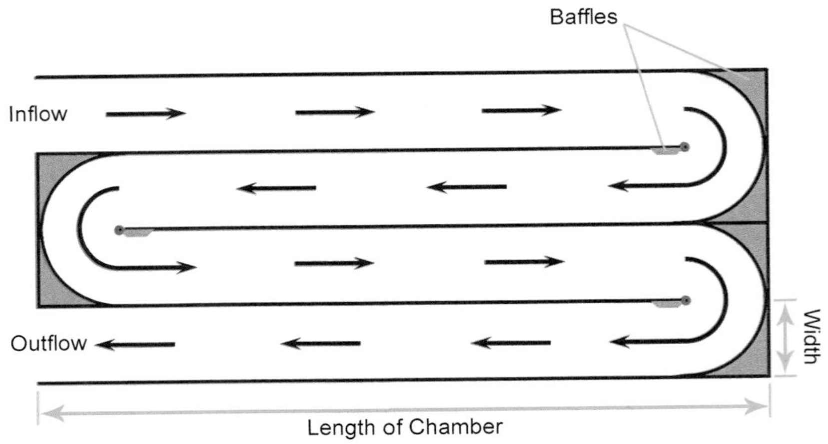

Figure 10.37 **Aerial View of Serpentine Chlorine Contact Chamber** (Marske and Boyle, 1973)

Figure 10.38 **Chlorine Contact Chamber with Slide Gates** (Reprinted with permission by Indigo Water Group)

TEST YOUR KNOWLEDGE

1. When withdrawing chlorine or sulfur dioxide liquid from a cylinder or container, this piece of equipment is needed to convert the liquid to a gas:
   a. Vacuum regulator
   b. Vaporizer
   c. Eductor tube
   d. Rupture disk

2. The tubing downstream of a chlorine evaporator is covered with a light frost. This indicates that
   a. The temperature within the evaporator is too high.
   b. Air circulation around the chlorine container is too low.
   c. Moisture is escaping from around the burst disk.
   d. The liquid withdrawal rate exceeds the evaporation rate.

3. To prevent metal corrosion from occurring within evaporators
   a. Zinc or magnesium rods create a sacrificial anode.
   b. Anti-corrosion chemicals are added to the interior chamber.
   c. Electrical current is applied across the two chambers.
   d. Very pure water is used within the outer chamber.

4. Gas storage rooms may be used to house tools and spare parts
   a. When tools must be kept nearby and other storage is unavailable.
   b. Only during the summer when the doors may be propped open.
   c. Never. Gas storage and use areas should be dedicated rooms.
   d. At any time. There are no restrictions on gas room storage.

5. **When placing a new sulfur dioxide cylinder on a scale, the scale starting weight should be set to**
   a. The tare weight of the cylinder
   b. The tare weight of the cylinder plus 68 kg (150 lb)
   c. The manufacturer's recommended default
   d. The weight of the cylinder minus the tare weight

6. **Gas cylinders should be restrained**
   a. 1/4 of the way up from the bottom
   b. 1/3 of the way up from the bottom
   c. 1/4 of the way down from the top
   d. 1/3 of the way down from the top

7. **In the event of a liquid chlorine leak from a 907-kg (1-ton) container**
   a. Safely rotate the container so gas escapes instead.
   b. Use the monorail hoist to remove the container from the room.
   c. Obtain a patch from a class B kit to repair the liquid leak.
   d. The liquid will freeze rapidly, stopping the leak.

8. **Safety equipment should be located**
   a. Next to the exhaust vent
   b. Outside the gas room
   c. In the administration building
   d. Between the empty and full containers

9. **Match the equipment to its function.**

1. Rupture disk	a. Connect cylinders and containers to manifolds
2. Magnesium rods	b. Reduces gas pressure
3. Pigtail and yoke	c. Relieves pressure when chlorine is trapped between closed valves
4. Trunnions	d. Uses water velocity to create a vacuum to pull chlorine
5. Vacuum regulator	e. Allow 907-kg (1-ton) containers to be easily rotated.
6. Venturi	f. Sacrificial anodes in evaporator water bath. Prevent corrosion.

10. **The pump suction intake on a hypochlorite system should be placed**
    a. As close to the tank bottom as possible
    b. Near the top of the tank
    c. Above the bottom of the tank
    d. About mid-depth in the tank

11. **A centrifugal pump is used to transfer sodium hypochlorite solution to a diffuser in a chlorine contact chamber. Over time, the flow output of the pump has gradually decreased even though the pump speed has remained the same. What could be the problem?**
    a. pH is too low
    b. Air binding
    c. Corrosion of pump housing
    d. Blockage

12. **Sodium hypochlorite leaks through piping seams**
    a. May be detected with carbonate solution
    b. Are the result of high alkalinity
    c. Are indicative of internal scaling
    d. Appear as large, white deposits

13. **Water moves through the chlorine contact chamber as plug flow to**
    a. Reduce the overall size of the chamber
    b. Minimize chlorine contact time
    c. Reduce construction costs
    d. Reduce opportunities for short-circuiting

14. **A WRRF has two chlorine contact chambers. Each is designed to provide 15 minutes of CT at the peak hour design flow. The facility currently receives about 35% of its design flow. What must be true?**
    a. One contact chamber does not provide enough CT at the current flow.
    b. Both contact chambers should be placed into service.
    c. Only one contact chamber is needed at the current influent flow.
    d. A higher chlorine residual concentration is needed.

PROCESS VARIABLES

Chlorine disinfection efficacy is determined primarily by *CT* and chlorine residual concentration. These two variables are multiplied together to calculate the CT (eq 10.15). Contact time has units of milligrams per liter per minute (mg/L·min). As discussed earlier in this chapter, it is possible achieve the same level of disinfection with a low concentration of chlorine over a long period of time or vice versa as long as the CT remains the same. Wastewater operators have almost no control over the DT in the disinfection pipeline or chlorine contact chamber other than being able to place additional contact chambers into service. They can control the chlorine dose and residual.

Chlorine *CT* is the same as detention time (DT) and hydraulic retention time. Detention time calculation examples may be found in earlier chapters.

How much CT is needed depends on the effluent characteristics. Cold water is harder to disinfect than warm water because the chemical reaction is slower. For every 10 °C decrease (18 °F decrease) in water temperature, the amount of time needed to get the same amount of disinfection doubles. Colder water may require an increase in the chlorine dose and residual compared to warm water with the same contact time. Chlorine gas is less soluble in warm water than in cold water. Warm water may make it difficult to produce highly concentrated sodium hypochlorite solution. Operators can compensate for lower concentrations by increasing the volume of hypochlorite solution added. Industrial facilities are affected by chlorine gas solubility limits more than domestic facilities. High pH water is more difficult to disinfect than low pH water because the chemical equilibrium shifts. At high pH, there is more hypochlorite ion (OCl^-) and less hypochlorous acid (HOCl). Hypochlorous acid is the better, more efficient disinfectant. As a result, WRRFs with *high pH* effluent may need a higher chlorine dose to achieve the same level of disinfection as a facility with lower pH effluent.

How high is high pH? Take a moment and look back at Figure 10.4. When the pH is 8 or higher, more than 80% of the chlorine residual will be in the hypochlorite ion (OCl^-) form.

The goal is to keep the chlorine residual at a high enough level to achieve the necessary CT to disinfect while minimizing overdosing, which drives up operational costs. The target CT can be estimated by multiplying the suggested chlorine feed rates given in Table 10.4 by 15 minutes. However, this is a *very* rough estimate because pH, temperature, and demand will affect how much CT is actually needed. Chlorine demand is affected by substances in the effluent that consume chlorine, including TSS, organic matter, ammonia, nitrite, iron, and manganese. Poor process performance upstream of disinfection will increase demand. The only way for operators to know if the chlorine residual is in the right range is to collect samples of the final effluent and analyze them for the indicator organism(s) in their discharge permits. This should be done during different times of day, under multiple flow conditions, and on different days of the week so the operator can be confident that disinfection goals are being met at all times. If indicator organism levels remain low, it may be possible to reduce the chlorine dose. If indicator organism levels increase, the dose may also need to be increased.

Case Study—A small WRRF located in a rural community has only one operator. The operator is required to collect samples for their discharge permit compliance once a week. Because the laboratory is far away and the operator is pressed for time on permit sampling days, he or she always grabs the effluent sample at 7:00 a.m. The sample is taken by courier to the state-certified laboratory. The *E. coli* result was always below 10 CFU/100 mL and the permit limit was 2000 CFU/100 mL. The operator began to gradually reduce the chlorine feed rate. Eventually, he or she didn't even need to add sulfur dioxide at the outfall because the chlorine residual was gone by the time the effluent was halfway through the contact chamber. One day, the state inspector arrived and took a grab sample of the effluent at 2:00 p.m. The *E. coli* result came back at more than 12 000 CFU/100 mL. Both the inspector and the operator were surprised because the WRRF had a long history of extremely low *E. coli* results. Why did this happen? First, the chlorine feed rate was adequate at 7:00 a.m., but didn't provide enough disinfecting power when peak flow began entering the WRRF at 10:00 a.m. Second, the increased influent flow and load increased the amount of TSS and ammonia in the final effluent, which increased demand and lowered residual.

FEED RATE CALCULATION

The feed rate calculation is used to determine the mass of chlorine or hypochlorite in kilograms per day or pounds per day. It is useful for determining how much chlorine or hypochlorite will be used over a period of time, which helps operators determine when to place chemical orders and to make long-term budgets. The feed rate calculation can also be used for troubleshooting. Notice that the feed rate calculations are based on chlorine dose and not residual (eqs 10.25 and 10.26).

In International Standard units:

$$\text{Feed Rate}, \frac{\text{kg}}{\text{d}} = \frac{\left(\text{Dosage}, \frac{\text{mg}}{\text{L}}\right)\left(\text{Flowrate}, \frac{\text{m}^3}{\text{d}}\right)}{(\text{Purity, \% expressed as a decimal})(1000)} \tag{10.25}$$

In U.S. customary units:

$$\text{Feed Rate}, \frac{\text{lb}}{\text{d}} = \frac{\left(\text{Dosage}, \frac{\text{mg}}{\text{L}}\right)(\text{Flowrate, mgd})\left(8.34\frac{\text{lb}}{\text{mil. gal}}\right)}{(\text{Purity, \% expressed as a decimal})} \tag{10.26}$$

EXAMPLE CALCULATIONS FOR FEED RATE

Example #1—A small WRRF disinfects with gaseous chlorine. The lead operator needs to know how much chlorine is used, on average, each day. The average daily flow for the facility is 9.46 ML/d (2.5 mgd). Historically, the chlorine dose has averaged 12 mg/L. How many 68-kg (150-lb) cylinders will be needed each month? Assume a typical month has 30 days.

In International Standard units:

Step 1—Convert the effluent flowrate from ML/d to m³/d.

$$\frac{9.46\ \text{ML}}{\text{d}}\left|\frac{1000\ \text{m}^3}{1\ \text{ML}}\right| = 9460\frac{\text{m}^3}{\text{d}}$$

Step 2—Determine the feed rate in kg/d using eq 10.25.

$$\text{Feed Rate}, \frac{\text{kg}}{\text{d}} = \frac{\left(\text{Dosage}, \frac{\text{mg}}{\text{L}}\right)\left(\text{Flowrate}, \frac{\text{m}^3}{\text{d}}\right)}{(\text{Purity, \% expressed as a decimal})(1000)}$$

$$\text{Feed Rate}, \frac{\text{kg}}{\text{d}} = \frac{\left(12\frac{\text{mg}}{\text{L}}\right)\left(9460\frac{\text{m}^3}{\text{d}}\right)}{(1)(1000)}$$

$$\text{Feed Rate}, \frac{\text{kg}}{\text{d}} = 114$$

Step 3—Determine how many cylinders are needed for 30 days.

$$\frac{114\ \text{kg}}{\text{d}}\left|\frac{30\ \text{days}}{1\ \text{month}}\right|\left|\frac{1\ \text{cylinder}}{68\ \text{kg}}\right| = 50\frac{\text{cylinders}}{\text{month}}$$

In U.S. customary units:

Step 1—Determine the feed rate in lb/d using eq 10.26.

$$\text{Feed Rate}, \frac{\text{lb}}{\text{d}} = \frac{\left(\text{Dosage}, \frac{\text{mg}}{\text{L}}\right)(\text{Flowrate, mgd})\left(8.34\frac{\text{lb}}{\text{mil. gal}}\right)}{(\text{Purity, \% expressed as a decimal})}$$

$$\text{Feed Rate}, \frac{\text{lb}}{\text{d}} = \frac{\left(12\frac{\text{mg}}{\text{L}}\right)(2.5\ \text{mgd})\left(8.34\frac{\text{lb}}{\text{mil. gal}}\right)}{(1)}$$

$$\text{Feed Rate}, \frac{\text{lb}}{\text{d}} = 250.2$$

Chlorine gas is 100% pure. In decimal form, 100% is simply 1.

Step 2—Determine how many cylinders are needed for 30 days.

$$\frac{250.2\ \text{lb}}{\text{d}}\left|\frac{30\ \text{days}}{1\ \text{month}}\right|\left|\frac{1\ \text{cylinder}}{150\ \text{lb}}\right| = 50\frac{\text{cylinders}}{\text{month}}$$

Example #2—A medium-sized WRRF uses 15% sodium hypochlorite solution to disinfect. The influent flow averages 83.27 ML/d (22 mgd) and the facility uses an average chlorine dose of 8 mg/L. How many liters (gallons) of hypochlorite solution will be needed each day? A 15% hypochlorite solution has a density of 1.20593 g/cm^3 (10.06 lb/gal).

In International Standard units:

Step 1—Convert the effluent flowrate from ML/d to m^3/d.

$$\frac{83.27\text{ ML}}{\text{d}}\left|\frac{1000\text{ m}^3}{1\text{ ML}}\right| = 83\,270\frac{\text{m}^3}{\text{d}}$$

Step 2—Determine the feed rate in kg/d using eq 10.25.

$$\text{Feed Rate, }\frac{\text{kg}}{\text{d}} = \frac{\left(\text{Dosage, }\frac{\text{mg}}{\text{L}}\right)\left(\text{Flowrate, }\frac{\text{m}^3}{\text{d}}\right)}{(\text{Purity, \% expressed as a decimal})(1000)}$$

$$\text{Feed Rate, }\frac{\text{kg}}{\text{d}} = \frac{\left(8\frac{\text{mg}}{\text{L}}\right)\left(83\,270\frac{\text{m}^3}{\text{d}}\right)}{(0.15)(1000)}$$

$$\text{Feed Rate, }\frac{\text{kg}}{\text{d}} = 4441$$

Step 3—Convert from kilograms to liters.

$$\frac{4441\text{ kg}}{\text{d}}\left|\frac{1000\text{ g}}{1\text{ kg}}\right|\left|\frac{\text{cm}^3}{1.20593\text{ g}}\right|\left|\frac{1\text{ mL}}{1\text{ cm}^3}\right|\left|\frac{1\text{ L}}{1000\text{ mL}}\right| = 3683\frac{\text{L}}{\text{d}}$$

In U.S. customary units:

Step 1—Determine the feed rate in pounds per day using eq 10.26.

$$\text{Feed Rate, }\frac{\text{lb}}{\text{d}} = \frac{\left(\text{Dosage, }\frac{\text{mg}}{\text{L}}\right)(\text{Flowrate, mgd})\left(8.34\frac{\text{lb}}{\text{mil. gal}}\right)}{(\text{Purity, \% expressed as a decimal})}$$

$$\text{Feed Rate, }\frac{\text{lb}}{\text{d}} = \frac{\left(8\frac{\text{mg}}{\text{L}}\right)(22\text{ mgd})\left(8.34\frac{\text{lb}}{\text{mil. gal}}\right)}{(0.15)}$$

$$\text{Feed Rate, }\frac{\text{lb}}{\text{d}} = 9786$$

Step 2—Convert from pounds to gallons.

$$\frac{9786\text{ lb}}{\text{d}}\left|\frac{1\text{ gal}}{10.06\text{ lb}}\right| = 973\frac{\text{gal}}{\text{d}}$$

Example #3—The operator of a 28.39-ML/d (7.5-mgd) WRRF dutifully calculates the amount of sodium hypochlorite solution their facility should be consuming each day. They also measure the amount of hypochlorite solution remaining in their storage tanks each day. At the end of a month, the operator notices that the facility is consuming more hypochlorite than predicted by their calculations—3541 kg/d (7806 lb/d). The operator suspects that the effluent flow meter is not accurate. An off-site laboratory analysis confirms that the sodium hypochlorite solution strength is 15% available chlorine. It has not degraded or lost strength since delivery. The current dose is 18 mg/L. Use the available information and the feed rate calculations to work backwards and find flowrate.

In International Standard units:

Step 1—Determine the flowrate in m³/d using eq 10.25.

$$\text{Feed Rate}, \frac{\text{kg}}{\text{d}} = \frac{\left(\text{Dosage}, \frac{\text{mg}}{\text{L}}\right)\left(\text{Flowrate}, \frac{\text{m}^3}{\text{d}}\right)}{(\text{Purity, \% expressed as a decimal})(1000)}$$

$$3541\frac{\text{kg}}{\text{d}} = \frac{\left(18\frac{\text{mg}}{\text{L}}\right)\left(\text{Flowrate}, \frac{\text{m}^3}{\text{d}}\right)}{(0.15)(1000)}$$

$$3541\frac{\text{kg}}{\text{d}} = (0.12)\left(\text{Flowrate}, \frac{\text{m}^3}{\text{d}}\right)$$

$$29\,520 = \text{Flowrate}, \frac{\text{m}^3}{\text{d}}$$

The calculated flowrate is higher than the measured flowrate. The operator decides to calibrate the flow meter.

In U.S. customary units:

Step 1—Determine the flowrate in million gallons per day using eq 10.26.

$$\text{Feed Rate}, \frac{\text{lb}}{\text{d}} = \frac{\left(\text{Dosage}, \frac{\text{mg}}{\text{L}}\right)(\text{Flowrate, mgd})\left(8.34\frac{\text{lb}}{\text{mil. gal}}\right)}{(\text{Purity, \% expressed as a decimal})}$$

$$7806\frac{\text{lb}}{\text{d}} = \frac{\left(18\frac{\text{mg}}{\text{L}}\right)(\text{Flowrate, mgd})\left(8.34\frac{\text{lb}}{\text{mil. gal}}\right)}{(0.15)}$$

$$7806\frac{\text{lb}}{\text{d}} = (1000.8)(\text{Flowrate, mgd})$$

$$7.8 = \text{Flowrate, mgd}$$

The calculated flowrate is higher than the measured flowrate. The operator decides to calibrate the flow meter.

Example #4—The chlorine residual must be reduced below 0.05 mg/L before the final effluent may be discharged to the receiving stream. The chlorine dose is 15 mg/L and a chlorine residual of 1.2 mg/L remains at the end of the chlorine contact chamber. Dechlorination will be done using sulfur dioxide gas. If the effluent flow averages 11.73 ML/d (3.1 mgd), how much sulfur dioxide will be needed each day?

In International Standard units:

Step 1—Convert the effluent flowrate from ML/d to m³/d.

$$\frac{11.73\text{ ML}}{\text{d}}\left|\frac{1000\text{ m}^3}{1\text{ ML}}\right| = 11\,730\frac{\text{m}^3}{\text{d}}$$

Step 2—Determine the feed rate in kg/d using eq 10.25.

$$\text{Feed Rate, }\frac{\text{kg}}{\text{d}} = \frac{\left(\text{Dosage, }\frac{\text{mg}}{\text{L}}\right)\left(\text{Ratio of }\frac{SO_2}{Cl_2}\right)\left(\text{Flowrate, }\frac{\text{m}^3}{\text{d}}\right)}{(\text{Purity, \% expressed as a decimal})(1000)}$$

$$\text{Feed Rate, }\frac{\text{kg}}{\text{d}} = \frac{\left(1.2\frac{\text{mg}}{\text{L}}\right)\left(0.9\frac{SO_2}{Cl_2}\right)\left(11\,730\frac{\text{m}^3}{\text{d}}\right)}{(1)(1000)}$$

$$\text{Feed Rate, }\frac{\text{kg}}{\text{d}} = 12.7$$

The ratios needed for different sulfite compounds are given in Table 10.3.

Sulfur dioxide is 100% pure chemical. This is shown as a 1 in decimal form.

In U.S. customary units:

Step 1—Determine the flowrate in mgd using eq 10.26.

$$\text{Feed Rate, }\frac{\text{lb}}{\text{d}} = \frac{\left(\text{Dosage, }\frac{\text{mg}}{\text{L}}\right)\left(\text{Ratio of }\frac{SO_2}{Cl_2}\right)(\text{Flowrate, mgd})\left(8.34\frac{\text{lb}}{\text{mil. gal}}\right)}{(\text{Purity, \% expressed as a decimal})}$$

$$\text{Feed Rate, }\frac{\text{lb}}{\text{d}} = \frac{\left(1.2\frac{\text{mg}}{\text{L}}\right)\left(0.9\frac{SO_2}{Cl_2}\right)(3.1\text{ mgd})\left(8.34\frac{\text{lb}}{\text{mil. gal}}\right)}{(1)}$$

$$\text{Feed Rate, }\frac{\text{lb}}{\text{d}} = 27.9$$

CHEMICAL FEED PUMP SETTING CALCULATION

The rotometer used with chlorine and sulfur dioxide gas feeders displays the chemical feed rate in kilograms per day or pounds per day directly. The chemical feed pumps used with hypochlorite solutions do not. Instead, operators must calculate the feed pump setting in milliliters per minute. If the desired feed rate in kilograms per day or pounds per day is already known, a unit conversion may be used to convert the feed rate into milliliters per minute. To perform the conversion, operators must know the feed chemical density. The chemical density may be found on the safety data sheet (SDS) available from the chemical supplier and in Tables 10.7 and 10.8.

In International Standard units:

$$\frac{\text{Dose, kg}}{\text{d}}\left|\frac{1\text{ d}}{1440\text{ min}}\right|\left|\frac{1\text{ L}}{\text{Density, kg}}\right|\left|\frac{1000\text{ mL}}{1\text{ L}}\right| = \text{Feed Pump Setting, }\frac{\text{mL}}{\text{min}} \quad (10.27)$$

In U.S. customary units:

$$\frac{\text{Dose, lb}}{\text{d}}\left|\frac{1\text{ d}}{1440\text{ min}}\right|\left|\frac{\text{gal}}{\text{Density, lb}}\right|\left|\frac{3.785\text{ L}}{1\text{ gal}}\right|\left|\frac{1000\text{ mL}}{1\text{ L}}\right| = \text{Feed Pump Setting, }\frac{\text{mL}}{\text{min}} \quad (10.28)$$

The chemical feed pump setting may also be calculated using dose in milligrams per liter as shown in eqs 10.29 and 10.30. For these equations, the chemical density must be in units of g/cm³ or mg/mL. The example calculations include examples of how to convert density into the necessary units.

In International Standard units:

$$\text{Feed Pump Setting}, \frac{\text{mL}}{\text{min}} = \frac{\left(\text{Flow}, \frac{\text{m}^3}{\text{d}}\right)\left(\text{Dose}, \frac{\text{mg}}{\text{L}}\right)}{\left(\text{Density}, \frac{\text{g}}{\text{cm}^3}\right)(\text{Active Chemical}, \%)\left(1440\frac{\text{min}}{\text{d}}\right)} \quad (10.29)$$

In U.S. customary units:

$$\text{Feed Pump Setting}, \frac{\text{mL}}{\text{min}} = \frac{(\text{Flow, mgd})\left(\text{Dose}, \frac{\text{mg}}{\text{L}}\right)\left(3.785\frac{\text{L}}{\text{gal}}\right)\left(1\,000\,000\frac{\text{gal}}{\text{mil. gal}}\right)}{\left(\text{Density}, \frac{\text{mg}}{\text{mL}}\right)(\text{Active Chemical}, \%)\left(1440\frac{\text{min}}{\text{d}}\right)} \quad (10.30)$$

EXAMPLE CALCULATIONS FOR CHEMICAL FEED PUMP SETTING

An 18.93-ML/d (5-mgd) WRRF uses a 12% available chlorine sodium hypochlorite solution to disinfect. The desired dose at the head of the chlorine contact chamber is 9 mg/L. The manufacturer's data sheets show the density of the hypochlorite solution as 1.168 kg/L (9.74 lb/gal). Find the chemical feed pump setting in milliliters per minute.

In International Standard units:

Step 1—Convert density from kg/L to g/cm³.

$$\frac{1.168\text{ kg}}{\text{L}}\left|\frac{1000\text{ g}}{1\text{ kg}}\right|\left|\frac{1\text{ L}}{1000\text{ mL}}\right|\left|\frac{1\text{ mL}}{1\text{ cm}^3}\right| = 1.168\frac{\text{g}}{\text{cm}^3}$$

Step 2—Convert flow from ML/d to m³/d.

$$\frac{18.93\text{ ML}}{\text{d}}\left|\frac{1000\text{ m}^3}{1\text{ ML}}\right| = 18\,930\frac{\text{m}^3}{\text{d}}$$

Step 3—Find the chemical feed pump setting.

$$\text{Feed Pump Setting}, \frac{\text{mL}}{\text{min}} = \frac{\left(\text{Flow}, \frac{\text{m}^3}{\text{d}}\right)\left(\text{Dose}, \frac{\text{mg}}{\text{L}}\right)}{\left(\text{Density}, \frac{\text{g}}{\text{cm}^3}\right)(\text{Active Chemical}, \%)\left(1440\frac{\text{min}}{\text{d}}\right)}$$

$$\text{Feed Pump Setting}, \frac{\text{mL}}{\text{min}} = \frac{\left(18\,930\frac{\text{m}^3}{\text{d}}\right)\left(9\frac{\text{mg}}{\text{L}}\right)}{\left(1.168\frac{\text{g}}{\text{cm}^3}\right)(0.12)\left(1440\frac{\text{min}}{\text{d}}\right)}$$

$$\text{Feed Pump Setting}, \frac{\text{mL}}{\text{min}} = 844$$

In U.S. customary units:

Step 1—Convert density from lb/d to mg/mL.

$$\frac{9.74\ \text{lb}}{\text{gal}}\left|\frac{1\ \text{gal}}{3.785\ \text{L}}\right|\left|\frac{1\ \text{L}}{1000\ \text{mL}}\right|\left|\frac{1\ \text{kg}}{2.204\ \text{lb}}\right|\left|\frac{1000\ \text{g}}{1\ \text{kg}}\right|\left|\frac{1000\ \text{mg}}{1\ \text{g}}\right| = 1167.57\frac{\text{mg}}{\text{mL}}$$

Step 2—Find the chemical feed pump setting.

$$\text{Feed Pump Setting}, \frac{\text{mL}}{\text{min}} = \frac{(\text{Flow, mgd})\left(\text{Dose}, \frac{\text{mg}}{\text{L}}\right)\left(3.785\frac{\text{L}}{\text{gal}}\right)\left(1\ 000\ 000\frac{\text{gal}}{\text{mil. gal}}\right)}{\left(\text{Density}, \frac{\text{mg}}{\text{mL}}\right)(\text{Active Chemical, \%})\left(1440\frac{\text{min}}{\text{d}}\right)}$$

$$\text{Feed Pump Setting}, \frac{\text{mL}}{\text{min}} = \frac{(5\ \text{mgd})\left(9\frac{\text{mg}}{\text{L}}\right)\left(3.785\frac{\text{L}}{\text{gal}}\right)\left(1\ 000\ 000\frac{\text{gal}}{\text{mil. gal}}\right)}{\left(1167.57\frac{\text{mg}}{\text{mL}}\right)(0.12)\left(1440\frac{\text{min}}{\text{d}}\right)}$$

$$\text{Feed Pump Setting}, \frac{\text{mL}}{\text{min}} = 844$$

TEST YOUR KNOWLEDGE

1. The water temperature has decreased from 20 °C (68 °F) to 10 °C (50 °F). If 20 minutes of CT were needed before, how many minutes of CT are needed now?
   a. 10 minutes
   b. 20 minutes
   c. 30 minutes
   d. 40 minutes

2. This type of water is the most difficult to disinfect:
   a. Cold, low pH
   b. Warm, low pH
   c. Cold, high pH
   d. Warm, high pH

3. Experience has shown that a chlorine residual of 2 mg/L provides adequate disinfection with 30 minutes of CT. If the CT is reduced to 10 minutes because of a wet weather event, how much chlorine residual will be needed to achieve the same level of disinfection?
   a. 2 mg/L
   b. 4 mg/L
   c. 6 mg/L
   d. 8 mg/L

4. Chlorine gas is used to disinfect the final effluent. If the dose is 13.5 mg/L and the effluent flow is 9.46 ML/d (2.5 mgd), how much chlorine will be used each day?
   a. 1.28 kg/d (2.81 lb/d)
   b. 1.83 kg/d (4.05 lb/d)
   c. 115 kg/d (252 lb/d)
   d. 128 kg/d (281 lb/d)

5. A WRRF disinfects with 12% sodium hypochlorite solution. The effluent flow averages 37.85 ML/d (10 mgd). If the dose is set at 6 mg/L, how much sodium hypochlorite solution will be needed each day?
   a. 18.9 kg/d (41.7 lb/d)
   b. 227 kg/d (500 lb/d)
   c. 1670 kg/d (3740 lb/d)
   d. 1892 kg/d (4170 lb/d)

6. Given the following information, find the chemical feed pump setting in mL/min. Effluent flowrate = 87.06 ML/d (23 mgd), chlorine dose = 7.5 mg/L, hypochlorite strength = 12% available chlorine, hypochlorite density = 1.168 g/cm^3 (1167.57 mg/mL).
   a. 1617 mL/min
   b. 3235 mL/min
   c. 6470 mL/min
   d. 7213 mL/min

7. The chlorine dose at the head of the contact chamber is 5.7 mg/L. At the end of the contact chamber, the chlorine residual is 2.2 mg/L. If the flowrate is 4.54 ML/d (1.2 mgd), how many kilograms (pounds) per day of sulfur dioxide will be needed?
   a. 9 kg/d (22 lb/d)
   b. 26 kg/d (57 lb/d)
   c. 36 kg/d (79 lb/d)
   d. 72 kg/d (180 lb/d)

PROCESS CONTROL

There are several ways to control the feed rate of chlorine gas or hypochlorite solutions: manual control, automatic-flow proportioning or open-loop control, automatic residual or closed-loop control, or automatic compound-loop control, which combines flow and residual signals to vary the feed rate. Flow proportioning is sometimes referred to as *feed-forward control*, and residual control is sometimes referred to as *feedback control*.

Flow pacing is based on the concept that varying the chlorine feed rate in proportion to flowrate will provide adequate quantities of chlorine at any flow. A typical arrangement found in some WRRFs involves the use of a flow control device to measure the secondary clarifier effluent flow and sending the flow signalto an automatic controller, which effectively controls the pacing of the sodium hypochlorite or chlorine-solution dosing pump. However, in wastewater, this is not always correct. The chlorine demand will vary with flow and can vary independently of flow, depending on the constituents in it. For example, changes in the effluent ammonia and nitrite concentrations will increase or decrease chlorine demand.

Residual control involves varying the chlorine feed rate based on the deviation of concentration from a setpoint on a controller. In other words, how different is the measured concentration from the target concentration. For systems in which the flowrate is nearly constant on a daily basis or is strictly seasonal, this type of control system works well. For systems in which the flowrate varies often and demand is variable also, residual control may not be as effective as flow pacing, because residual control systems do not react well to large variations in flowrate over short periods of time. These methods can be applied to control based on free chlorine residual, total chlorine residual, or oxidation–reduction potential (ORP).

Compound-loop control provides the ability to use both flow and residual input to control chlorine feed. Flow is the primary driver, wheras residual is used to trim the chemical feed. When the setpoint in a compound-loop control system is further controlled automatically, the configuration is referred to as *cascade control*. Cascade control requires the use of another analyzer downstream of the compound-loop control analyzer. Output from the cascade-control analyzer is used to regulate the compound-loop controller.

The choice of control strategy for a particular installation is based on regulatory requirements, existing facilities, wastewater treatment system design, economics, cost-effectiveness, and required system maintenance. The more complicated the selected control system, the more likely it is that service requirements will be more exacting and, therefore, will require more training and an increase in the skill level of operating personnel.

OXIDATION–REDUCTION POTENTIAL

Also referred to as *high-resolution redox* (HRR), ORP is based on millivolt readings (potential) created across an electrode. These millivolt readings are generated by adding chlorine to water and by other ions in the water. Higher voltage readings correspond to higher concentrations of chlorine in solution. The system responds directly to the balance between the oxidizers (chlorine) and the reducers (contaminants). The ORP hardware includes a sensing electrode and a reference electrode, which are immersed in the water to measure the electrical potential of the oxidizers present. This potential is then displayed in millivolts. After chlorination, ORP levels may vary from 380 to 750 mV. The reduction potential increases with increases in sulfite concentrations. After sulfonation (addition of sulfite ions), ORP levels can vary from 190 to 250 mV. Although ORP does not measure residuals of chlorine or sulfite directly, it does provide an indication of the level. The ORP system monitors and controls the chlorine feed based on immediate changes in chemical demand rather than on the residual.

Every wastewater effluent will have a different ORP baseline. The ORP baseline of untreated water changes as the concentration of various ions changes, ultimately reaching equilibrium. If this reading happens to be 200 mV, then the ORP contributed by the chlorine will be added to this number. This ORP baseline is typically determined after the system has been in operation for at least 24 hours and operating conditions are normal. When equilibrium has occurred, the system is standardized via a titrimetric chlorine residual procedure. Care must be taken when standardizing the available chlorine residual to the ORP measurements to establish a consistent relationship necessary to meet the current National Pollutant Discharge Elimination System discharge requirements.

To allow some conversion of free chlorine to chloramine, most manufacturers recommend placing the ORP probe approximately 5 to 10 minutes downstream of the chlorine application point. Oxidation–reduction potential measurements are much simpler to obtain and the ORP cell is easier to maintain than either the membrane or amperometric chlorine analyzer cells. It has been observed that there is a marked difference in the ORP millivolt curves between free chlorine residuals and chloramine residuals. Therefore, HRR can be calibrated to control in either a free or a total chlorine environment.

MANUAL CONTROL

Manual control of chlorine feed is the simplest strategy and, because of its simplicity, may often be the most effective method. A manual-control system requires less maintenance and operator expertise than any form of automatic control. The basic chlorinator feeds chlorine at a predetermined constant feed rate, which is changed by the operator as required. In the case of hypochlorination, the chemical-dosing rate is controlled at the desired level by manually adjusting the pump stroke length and/or the motor speed. Manual control has a low capital cost, but it is prone to either overfeeding or underfeeding of chlorine and, therefore, excessive dechlorination, insufficient disinfection, or overchlorination. Manual-control systems are used where flowrate and demand are fairly constant. An example of where this may be appropriate is the discharge from a pond treatment system.

SEMIAUTOMATIC CONTROL

The same equipment used in manual-control systems can often be used to partially automate the operation of a system.

One such system is on–off control. The chlorinator can be turned on and off automatically in response to a signal, such as a wet-well level or pump activation. The feeder can be turned on and off by controlling any of the following: booster pump operation, a *solenoid* valve in the water supply line to the injector or ejector, or a solenoid valve located in the gas vacuum line between the rate-control valve and the injector.

A *solenoid* is a coil of wire that acts like a magnet when electrical current is applied. Solenoids are used to open and close valves and switches.

Another option for semiautomatic control is a method known as *band control*. In this technique, two chlorine gas-flow metering tubes are used in the gas vacuum line in conjunction with two vacuum line solenoid valves and a chlorine residual analyzer. The analyzer, or a recorder receiving a signal from the analyzer, is equipped with two setpoint alarm contacts. The contacts are preset at points of maximum and minimum levels of residual. These contacts activate the vacuum line solenoid valves as follows:

- When the residual is lower than the low setpoint, both valves are open;
- When the residual is higher than the high setpoint, both valves are closed; and
- When the residual is between the two setpoints, one valve is closed.

The above control actions can also be achieved in the case of hypochlorinators, by turning the dosing pumps on or off in response to the signal from the controller.

FLOW-PROPORTIONAL CONTROL

In most WRRFs, flow is variable, and it is not possible or practical to construct equalization basins. Therefore, control of chlorination feed is often set in proportion to flow, which enables the ratio to be varied by adjusting the dosage. In this strategy, a flow signal is transmitted from a flow meter to the chlorinator, where an automatic valve opens or closes, depending on the signal level (typically *4- to 20-mA direct current*) from the flow meter. If flow increases by 20%, the chlorine feed rate also increases by 20%. The goal is to maintain a constant chlorine residual in the contact chamber.

A *4- to 20-mA* signal is an electronic signal that increases or decreases in proportion to something else. For example, if a valve opens fully—100%—the signal will be 20 mA. If the valve is fully closed, the signal will be 4 mA.

Flow-proportional chlorinators are sized by establishing a design dosage for the chlorine feed rate in grams per hour (pounds per day). This design sets the maximum wastewater flow and maximum chlorine flow dosage at a 1-to-1 ratio. This means that, at a 10% signal from the wastewater flow meter, the chlorinator flow meter will read 10% of full scale; at 90% of the wastewater flowrate, the chlorine flow meter will read 90%; and so on. Flow-proportional controllers have a dosage-control adjustment that allows the operator to vary the design dosage ratio from 10 to 1 turndown to 1 to 4 turn up.

Flow measurement for hypochlorinator control commonly is performed with a flow meter, and the signal is recorded continuously and sent by a transmitter (flow indicator transmitter) to a *programmable logic controller* (*PLC*). The controller sends a signal to the speed controller of the dosing pump, to deliver the desired amount of hypochlorite solution in proportion to the effluent flow. The mathematical relationship between the output (pumping rate of hypochlorite solution) in response to an input (effluent flow) signal is programmed into the PLC, such that the controller maintains the dosing rate in proportion to the flow. The setpoint for the controller is the proportion of flow between the input and the output, which it tries to maintain.

A *PLC* is essentially a simple computer that takes a signal (reading) from a piece of equipment like a chlorine analyzer or flow meter and translates it into a signal for another piece of equipment like a solenoid on a valve. That signal may cause a valve to open or close to some degree. Programmable logic controllers are used throughout WRRFs for many different purposes.

RESIDUAL CONTROL

One of the criteria for successful operation of residual control systems is to minimize lag time. Lag time consists of the following four primary components:

- Time required for the flow to pass from the injection point to the sampling point,
- Time required for the flow to pass from the sample point to the analyzer (including the speed of response of the analyzer),
- Time required for the chlorine gas or chlorine solution to reach the diffuser, and
- Response speed of the control valve in the chlorinator.

The first component of lag time depends on the flowrate and distance between the injection point and sample point. To limit lag time, this distance should be minimized. For optimal control, the sample point should be located at a distance that corresponds to an approximately 90-second travel time at maximum flowrate from the injection point.

The distance from the sample point to the analyzer and speed of response of the analyzer can be minimized also, by locating the analyzer as close as possible to the contact chamber and sample point. The use of a dedicated analyzer produces a consistent, rapid reading, because recalibration is unnecessary, and sample-line cleaning is not required. The sample line should be cleaned periodically with high concentrations of chlorine to remove any buildup of algae and slime. Also, the sample line should maintain a sample velocity of approximately 3 m/s (10 ft/sec) to minimize lag time and provide for plug flow.

When the injector and control valve are located at the diffuser site, any change in chlorine feed called for by the control signal is sensed rapidly because the distance to the point of addition is minimized. The speed of response of the control valve in the chlorinator is relatively insignificant compared with the other three factors. This discussion does not suggest that the chlorination reaction takes 90 seconds or that contact for a minimum of 30 minutes is unnecessary. Rather, the implication is that, after 90 seconds, chlorination reactions are sufficiently complete to be measured for control purposes, and, if the control system is functioning properly, the chlorine residual after 30 minutes of contact may be obtained by manual sampling and analysis.

If manual sampling is insufficient to provide an adequate safeguard against improper disinfection, an additional dedicated residual analyzer can be installed at a sample point at least 30 minutes' detention downstream of the injection point. The use of dedicated analyzers allows continuous measurement and control, and the additional cost is negligible compared with the benefit derived. With a typical feedback-control loop when using a hypochlorinator for the residual chlorine concentration, the chlorine analyzer probe transmits a signal to a PLC. The controller then sends a signal to pace the dosing pump. This adjusts the dosing rate so that the desired residual chlorine concentration, which is the setpoint of the controller, is maintained.

COMPOUND-LOOP CONTROL

A compound-loop system consists of two interlocking control loops—the flow loop and the residual loop. These loops can be interlocked in one of several ways. One way is to have the signal from a residual controller vary the dosage adjustment of the chlorinator. Alternatively, signals from the analyzer and flow meter can be sent to a multiplier, which then sends a composite signal corresponding to combined measured feedback to the chlorinator. The chlorinator then injects the appropriate amount of chlorine based on this mass signal.

An advantage of the compound-loop controller is that it can introduce damping to the system, enabling the system to avoid uncontrollable swings in residual concentration that result from excess lag time. The multiplier type of compound-loop controller has the advantage of being able to react quickly to rapid

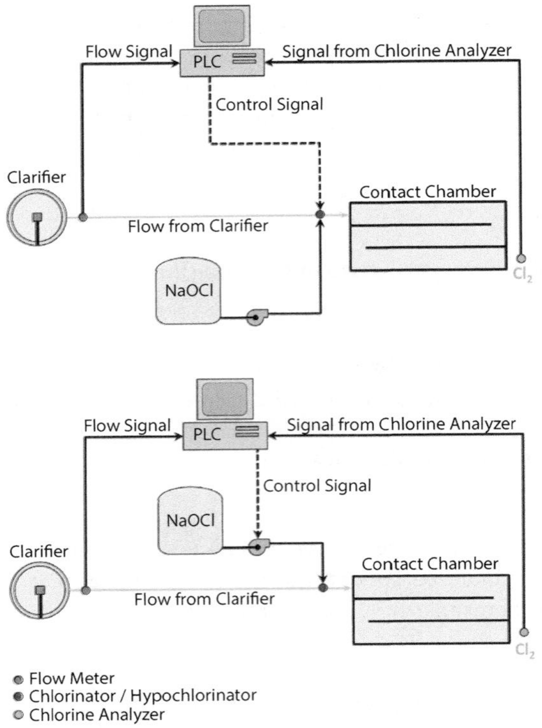

Figure 10.39 Compound-Loop Control (Reprinted with permission by Indigo Water Group)

changes in residual (and, therefore, demand). In WRRFs, a compound-loop system not only provides more accurate control, but, equally important, it can save costs by minimizing the overfeeding of chlorine by operators and reducing the amount of chlorine and chlorine byproducts in the effluent. Figure 10.39 represents the compound-control loop for hypochlorite-solution dosing based on the effluent flow and residual chlorine concentrations.

CASCADE CONTROL

In cascade-control systems, additional instrumentation, measurement, and input to the control scheme are used. In this system, an additional chlorine-residual analyzer is provided to sample the contact chamber. Located at or near the contact-chamber discharge, this second analyzer provides input to the control scheme by varying the setpoint of the compound-loop analyzer, thereby providing an additional damping effect on the variation of chlorine residual. The cascade-control loop for dosing hypochlorite solution is presented in Figure 10.40.

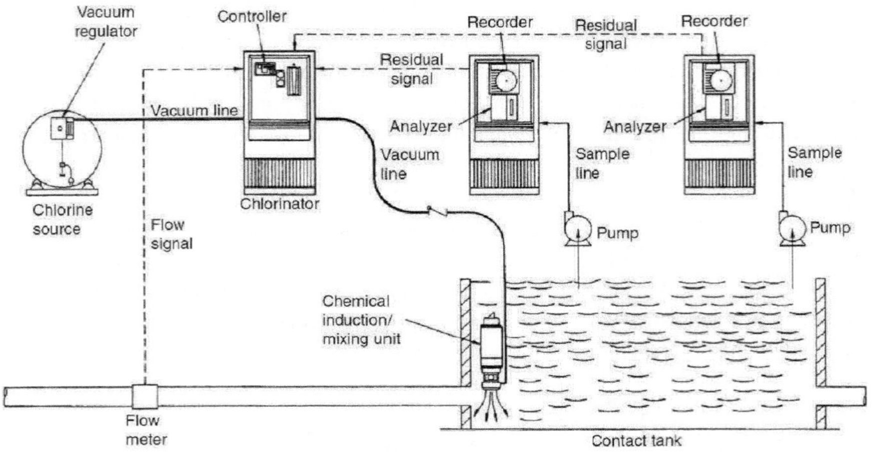

Figure 10.40 Cascade Control (WEF et al., 2018)

TEST YOUR KNOWLEDGE

1. **Flow pacing the chlorine dose to the effluent flow will always maintain a constant chlorine residual.**
   - ☐ True
   - ☐ False
2. **The ORP is higher after chlorination and lower after dechlorination.**
   - ☐ True
   - ☐ False
3. **Residual control adjusts the chlorine dose by measuring residual either directly or with ORP.**
   - ☐ True
   - ☐ False
4. **The simplest control strategy for chlorine disinfection is**
   a. Manual
   b. Semiautomatic
   c. Flow-proportional
   d. Cascading loop
5. **The electronic signal for an electronically controlled valve currently reads 4 mA. The valve is**
   a. Fully closed
   b. 4% open
   c. 40% open
   d. Fully open
6. **One advantage of compound-loop control is**
   a. Slow reaction time
   b. Ability to manually adjust
   c. Steady chlorine residual
   d. Simplicity of operation

OPERATION

DAILY OPERATION

Operators should follow these recommendations when receiving new chlorine cylinders, containers, and tank cars from the supplier:

- Record the number stamped on each cylinder, container, or tank car when it is received. Label them as full and store them separately from empties.
- Check each cylinder or container as follows:
  - Place a hand on the outside of the cylinder or container. If it is excessively warm, it could indicate that the cylinder or container has been contaminated with water. Hot cylinders and containers should be returned to the supplier and not be used.
  - Remove the protective bonnet or cover and make sure the main valves are tightly closed.
  - Make sure the threaded caps and valve seats do not have deep scratches. These are potential sources of leaks when the cylinder or container is put into service. Cylinders and containers with damaged threads or valves should be returned to the supplier.
  - Slowly and slightly tighten the valve packing nut(s).
  - Check each valve for leaks using the vapor from a 17.7% (20 Baume) ammonium hydroxide solution. A white precipitate will form in the air where the ammonium vapor reacts with chlorine gas. Never apply liquid ammonia directly to a chlorine leak or fitting. Any cylinders or containers with leaks should be repaired (if possible) and returned to the supplier.
  - Replace the threaded caps and protective bonnets or covers.
- Keep the bonnet or cover in place until the cylinder or container is positioned on the scale. Secure cylinders by chaining two-thirds of the way up from the bottom. Strap 907-kg (1-ton) containers into cradles.
- Allow new cylinders and containers to rest for 24 hours before placing them into service. This ensures that all of the liquid chlorine is consolidated at the bottom of the vessel and prevents small droplets of liquid from entering the gas lines.

Startup and shutdown procedures for chlorination systems should follow a specific sequence (especially larger systems that include evaporators).

STARTUP

The following directions detail one possible sequence of steps for placing a chlorine or sulfur dioxide system into service. Procedures will vary depending on type of equipment, vacuum or pressure, and whether gas or liquid chlorine is being used. When approaching a gas cylinder or container, operators should assume that the valve is open and gas or liquid is flowing. While there is no requirement for operators to wear an SCBA when changing chlorine cylinders and containers, it is highly recommended. Safety equipment should be easily accessible.

1. If an empty container or cylinder is being replaced, first verify that there is no gas passing through the chlorinator. The rotameter ball should be at the bottom of the rotameter and the gas flow indicator on the chlorinator should indicate no gas flow.
2. Turn the cylinder or container valve clockwise to close it. There should already be a wrench in place on the cylinder or container valve (Figure 10.41). Wrenches are left in place so the cylinder or container can be closed quickly in an emergency. Verify again that there is no gas passing through the chlorinator. For systems with manifolds, verify there is no gas pressure in the manifold before proceeding.

Specialized wrenches are used with cylinders and containers. The twisted shaft provides a flat surface to push against when changing out equipment. Wrenches are left in place when cylinders and containers are in use.

3. To remove either a vessel-mounted chlorinator, sulfonator, or ancillary valve, loosen the tightening nut on the back of the chlorinator yoke. Be careful not to stress or kink the gas and vent lines. Once the chlorinator, sulfonator, or ancillary valve has been removed, the lead washer will be visible at the connection point.
4. Replace the protective valve cover and mark the cylinder or container as empty.
5. Use a hand truck (dolly) or hoist to position a new, full cylinder or container onto the scale. Never roll a cylinder on its edge. Keep protective covers on cylinders and containers until they are in place and are ready to be connected. Rotate 907-kg (1-ton) containers so the top and bottom valves are perpendicular to the floor. Gently tighten the valve to ensure it is closed before removing the threaded valve cap.
6. Use a pocket knife or utility knife to remove the lead washer from the chlorinator, sulfonator, or vacuum regulator. The washer should have deep grooves like the ones shown in Figure 10.42 because the lead is compressed during installation. These grooves make it impossible to reuse washers to form a new, gas-tight seal. Always use new, approved washers to prevent leaks.
7. Before replacing the lead washer, make sure the surface of the valve and the surface of the chlorinator, sulfonator, or vacuum regulator are free of debris and clean. Place the lead washer into position.
8. Place the chlorinator, sulfonator, or ancillary valve onto the cylinder or container. Hand-tighten the yoke to hold it in place. Continue tightening with a wrench until the lead washer can be seen squeezing out from between the cylinder or container valve and the chlorinator, sulfonator, or ancillary valve.
9. Place the wrench on the valve and open the cylinder or container by one-quarter turn and quickly reclose it.
10. Use a 20 Baume (17.7%) ammonia solution to check for leaks around all threads and connections. Do not apply the ammonia solution directly to the equipment. The ammonia vapor is all that is needed. If the ammonia vapor comes into contact with any chlorine gas, a white cloud of smoke will appear.

Place a piece of sponge in the bottom of a squeeze bottle filled one-quarter to one-third full of ammonia solution. The sponge will absorb the ammonia and help prevent accidental release of ammonia liquid while allowing ammonia vapor to escape.

11. If there is a leak, remove the chlorinator and install a new washer. If no leaks are found, open the cylinder or container one-quarter turn. Never open the cylinder or container by more than one-quarter turn. Always leave the wrench on the valve stem. It is important to be able to close the cylinder or container quickly in an emergency.
12. With the valve open, reset the gas indicator and adjust the feed knob to bring the rotometer ball up to the desired dose (kilograms per day or pounds per day).
13. Open the appropriate valves at the point of solution application.

Figure 10.41 Angled Wrench on Cylinder Valve (Reprinted with permission by Indigo Water Group)

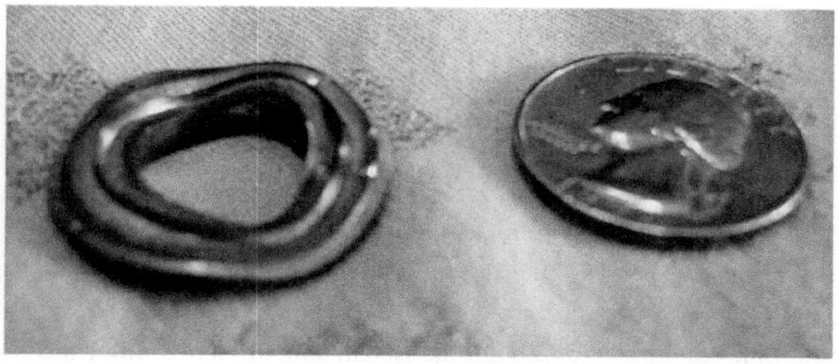

Figure 10.42 Used Lead Washer (Reprinted with permission by Indigo Water Group)

14. Activate water flow at the gas injector to create the required vacuum. Typically, a vacuum of at least 17 to 24 kPa (5 to 7 in. of Hg) is required at the chlorinator for proper chlorinator operation.
15. Adjust the net weight scales to the correct weight.
    Note: Steps 16 and 17 apply only to systems with evaporators.
16. Activate the evaporator heat-exchanger system to attain the required operating temperature, typically 71 to 82 °C (160 to 180 °F).
17. Open the evaporator inlet and outlet valves.
18. Activate the chlorinator, alarm systems, and (if applicable) electric positioner.
19. Activate the chlorine gas-leak detectors.
20. Open the chlorinator inlet and outlet valves.
21. Activate the automatic changeover system (if applicable).
22. While checking for leaks with fumes from an aqueous ammonia squeeze bottle, slowly open the main chlorine manifold and tank valves no more than one turn (some WRRFs operate successfully with tank valves open only one-quarter turn). Repair any leak immediately.
23. Activate the chlorine-residual analyzer.
24. Monitor all chlorination-system devices for proper operation (as described in the following sections on "Troubleshooting Checklist and Process Control" and "Routine Operations").

SHUTDOWN

Some WRRFs do not require year-round chlorination, and so seasonal shutdown of the chlorination process is necessary. The following steps should be taken to shut down the entire chlorination system:

1. Close the container or cylinder main valves.
   Note: Allow all chlorination devices to operate until chlorinator gas pressure is zero and the rotameter has stopped bouncing. Then, proceed with the following steps:
2. Close the main manifold valves at the scale.
3. After a few minutes, close the water-supply valve at the injector.
4. Close the valve at the point of application for the solution (if this is a check valve, then it may remain open).
5. Deactivate the evaporator heat-exchange system.
6. Close the evaporator inlet and outlet valves.
7. Deactivate the chlorinator controls.
8. Close the chlorinator inlet and outlet valves.
9. Deactivate the automatic changeover system, chlorine-residual analyzer, and chlorine-gas leak detector.
10. Flush all systems that handle liquid or gaseous chlorine with dry nitrogen and maintain a nitrogen pad on equipment. It is important not to trap chlorine liquid between two closed valves. Be sure the entire system has been evacuated.
    Note: If chlorine containers or cylinders remain on-site, then the leak detector should remain operational.
11. If necessary, remove chlorine containers or cylinders from the scale and store them properly (consult the chlorine supplier about long-term storage).

MAINTENANCE

A comprehensive maintenance program is critical to WRRF safety and process efficiency. It should include equipment inspections at a frequency greater than or equal to the manufacturer's recommendations. There should be enough equipment and process redundancy to facilitate cleaning and repairs without loss of disinfection efficiency. Frequently inventory accessories for chlorine gas systems and keep at least the following items on hand:

- New, approved gaskets;
- A squeeze bottle one-half full of 17.7% (20 Baume) liquid ammonium hydroxide solution;
- Appropriate wrenches for 68-kg (150-lb) cylinders and 907-kg (1-ton) containers;
- At least two SCBAs;
- An emergency repair kit for the appropriate container type, and, if the system contains a manifold; and
- One new pigtail connector tube long enough to connect the cylinder or container to the manifold. (Note: Must be clean and dry before using.)

Complete documentation of tasks and analyses plus a good recordkeeping system are also essential to ensure that data are recorded for historical trends and comparison purposes. Some key elements of a chlorination system preventive maintenance program are listed below.

GAS-RELATED EQUIPMENT

- Chlorinators and Sulfonators
  - Weigh chlorine and sulfur dioxide containers and cylinders on a scheduled basis. Consult chlorine supplier for details.
  - Inspect and clean the stem, spring, passageway, trap, and filter on vacuum chlorine regulators.
  - Check the operation of automatic changeover systems by closing the cylinder or container main valves on in-service tanks.
  - Inspect and replace pigtail connectors on a scheduled basis. A crinkling sound coming from a Monel-lined copper pigtail indicates that the Monel is wearing out and the pigtail should be replaced. Clear plastic pigtails should be replaced when the plastic is discolored. Replace pigtail connectors immediately if kinking or other damage is visible.
  - Clean out the chlorinator rotameter when deposits are noticed inside the glass rotameter tube or if the float sticks.
  - Clean the chlorinator v-notch while cleaning the rotameter.
  - Lubricate the chlorinator's electric-positioner motor shaft with a few drops of non-detergent oil and a scheduled basis.
- Liquid Chlorine Equipment
  - Lubricate evaporator heat-exchanger recirculation pump motor with non-detergent oil according to the manufacturer's recommendations.
  - Inspect the cathodic-protection anode magnesium rods inside evaporators on a scheduled basis. Replace rods as recommended by the manufacturer.
  - Inspect and clean the strainer, stem, spring, passageway, and (if applicable) trap and filter on PRVs.
- Safety Equipment
  - Test the alarm circuitry of the chlorine leak detector on a scheduled basis.
  - Certify and hydrotest SCBA units.
  - Check and adjust the leak detector electrolyte level and sensor drip rate.
  - Flush the leak detector with distilled water.
  - Check the operability of the leak detector air fan.
  - Check for leakage at all o-rings and leak detector's site glass.
  - If applicable, replace the cartridge filter in the cap of the leak detector electrolyte reservoir on a scheduled basis.

SODIUM HYPOCHLORITE

- Clean the pump and effluent water strainers regularly. Wear rubber gloves and goggles.
- Regularly inspect the spring-loaded backpressure valve and the balls and seats of pump valves. Operate the piping valves regularly.
- Rotate the use of hypochlorite feed pumps and operate them briefly at all speeds.
- Sodium hypochlorite can release gas bubbles that accumulate in the suction piping to the feed pumps, causing air binding. Maintain a flooded suction or air-release mechanism to ensure proper operation.
- Maintain the proper oil level in the pump reservoirs.

- Occasionally, pump a 5% solution of hydrochloric acid to prevent carbonate scale from accumulating in the pump. Flush the pump with water before and after using the acid.
- Inspect and, as necessary, repair the storage-tank linings at least once per year. Be sure to adhere to confined space entry protocols when entering tanks.

CHLORINE CONTACT CHAMBER

A major cause of effluent disinfection permit violations is reintroduction of microorganisms to the effluent during the effluent-disinfection process. Sources of microorganisms include

- Microbial films/slimes that will grow on the walls and floor of the contact chamber,
- Settled particles on the floor of the contact chamber,
- Animals in the contact chamber,
- Bird defecation in the contact chamber,
- Windborne debris falling into the contact chamber, and
- Regrowth between the end of the contact chamber and the sampling location.

The contact chamber should be dewatered and cleaned at least once each year.

Discourage birds from frequenting the contact chamber area by placing wires over the tops of the handrails to stop perching. Wires may also be stretched across the contact chambers to limit access to the water surface.

TEST YOUR KNOWLEDGE

1. **A chlorine cylinder or container that is hot to the touch may be contaminated with internal moisture.**
   - ☐ True
   - ☐ False
2. **Cylinders may be rolled on their edges for transport so long as the distance traveled is less than 10 ft.**
   - ☐ True
   - ☐ False
3. **Cylinders and containers should be allowed to rest for 24 hours after delivery before being placed into service.**
   - ☐ True
   - ☐ False
4. **Lead washers may be reused if they are compressed sufficiently to form new grooves.**
   - ☐ True
   - ☐ False
5. **There is a federal law requiring SCBAs to be worn when changing chlorine cylinders and containers.**
   - ☐ True
   - ☐ False
6. **A gas cylinder valve should only be opened ____________ turn, so it can be closed quickly in an emergency.**
   a. One-quarter
   b. One-half
   c. One full turn
   d. Completely open
7. **Chlorine leaks can be located quickly by**
   a. Applying a dilute ammonia solution directly to a chlorine valve or fitting
   b. Using the vapor only from a squirt bottle filled with ammonia solution
   c. Spreading a thin layer of soapy water over the suspected leak and watching for bubbles
   d. Using a hand-held sulfur dioxide monitor calibrated for chlorine release
8. **One method of confirming that there is no gas flowing through a chlorinator or sulfonator is to**
   a. Check the downstream system pressure
   b. Confirm the scale weight equals tare weight
   c. Apply ammonia vapor at the valve stem
   d. Verify the ball is at the bottom of the rotameter
9. **Clear plastic pigtails should be replaced when**
   a. A crinkling sound is heard.
   b. They become discolored.
   c. A new cylinder or container is used.
   d. Manifolds are backflushed.
10. **Deposits that form within the glass rotameter may**
    a. Cause the float to become stuck
    b. Be removed with sodium hydroxide
    c. Interfere with pigtail operation
    d. Obscure the feed rate markings

11. This chemical may be used to remove calcium carbonate scale from a chemical-metering pump pumping sodium or calcium hypochlorite solution.

a. Sodium hydroxide
b. Sulfuric acid
c. Hydrochloric acid
d. Magnesium hydroxide

12. Chlorine contact chambers should be drained and cleaned at least

a. Weekly
b. Monthly
c. Annually
d. Every 2 years

TROUBLESHOOTING

Table 10.11 presents a troubleshooting guide for common problems associated with chlorination systems. Refer to manufacturers' manuals for troubleshooting specific pieces of equipment.

SAFETY CONSIDERATIONS

CHLORINE GAS

Perhaps the most substantial drawback associated with the use of chlorine gas is the safety risk. Safety issues related to chlorine are documented in *Standard for Fire Protection in Wastewater Treatment and Collection Facilities* (NFPA, 2016) and by OSHA.

Chlorine is a toxic gas and is considered an acutely hazardous substance under the *Superfund Amendment and Reauthorization Act (SARA)*. A major chlorine leak at a WRRF could injure or even kill facility personnel and others. Emergency planning must consider the evacuation of personnel and people living near the facility. Because chlorine is regulated under the Emergency Planning and Community Right to Know Act (EPCRA), WRRFs using chlorine must comply with the reporting requirements and other provisions of that law. The EPCRA applies whenever 45 kg (100 lbs) or more of chlorine is on-site. In addition, releases of 4.5 kg (10 lb) of chlorine or 227 kg (500 lb) of sulfur dioxide to the environment must be reported under EPCRA. Facility management is responsible for training employees in the proper handling and safety aspects of chlorine and informing neighbors and local government agencies (e.g., fire and police departments) about the physical system. One goal of EPCRA is to ensure that first responders know how much chlorine may be on-site and where it is located. Facilities that use chlorine should consider obtaining Chlorine Institute pamphlets and safety videos (http://www.chlorineinstitute.org). Their brochures may be downloaded free of charge. The OSHA enacted the Process Safety Management Rule in 1992, which has additional emergency planning requirements for chlorine gas systems that have 681 kg (1500 lbs) of chlorine or 453 kg (1000 lbs) of sulfur dioxide on-site at any time.

The *superfund* was set up by Congress to provide funds for cleaning up environmental contamination when the responsible person or company couldn't be quickly identified or held responsible. It has been used to clean up abandoned factory sites, landfills, and mines. When the superfund was reauthorized by Congress in 1986, they added provisions in Title III (the third part of the act) to help keep track of dangerous chemicals at the point of manufacture and use. The SARA, also known as Title III, requires users of dangerous chemicals to meet requirements for community notification and emergency planning. Tracking chemicals from cradle to grave has reduced the number of new superfund sites.

Only trained and certified operators should work with chlorine. Operators should always use the "buddy system" when working with chlorine and use a SCBA or supplied air when opening valves or changing cylinders and containers. Because of the hazardous nature of chlorine gas, security measures must be in place to ensure that only authorized personnel have access to chlorine storage and equipment areas.

Extremely low levels of chlorine are undetectable by most humans; therefore, instruments are useful in detecting chlorine below 0.1 ppmv in air. The odor threshold varies with individuals. However, most people are able to detect concentrations higher than 0.2 ppmv in air.

The threshold limit value (TLV) established for chlorine gas by the American Conference of Governmental Industrial Hygienists (ACGIH) is 0.5 ppm as a time-weighted average (TWA). The TLV is accepted as the safe limit of exposure for a worker day after day without any adverse effects. In 1989, the OSHA permissible exposure level (PEL) for chlorine was set at 0.5 ppmv (8-hour average exposure) and the short-term exposure limit (STEL) was set at 1.0 ppmv (15-minute exposure). In other words, a worker can have a short-term exposure to chlorine concentrations up to 1.0 ppmv, but the TWA exposure concentration over an 8-hour workday still can't exceed 0.5 ppmv. The National Institute of Occupational Health and Safety (NIOSH) issued the immediately dangerous to life or health (IDLH) value at 10.0 ppmv.

Municipalities and utilities are not legally required to follow OSHA recommendations. Most still comply with OSHA to protect workers, limit liability, and reduce insurance costs. Some follow state guidelines, recommendations from other organizations like ACGIH, or develop their own internal standards. Internal standards are often more protective of workers.

Below 5.0 ppmv, there are no known acute or chronic health effects from chlorine exposure. Between 5.0 and 10.0 ppmv, effects of chlorine exposure (choking, coughing, watery eyes, mild skin irritation, and nose and lung irritation) are temporary. An exposure to 30 ppmv causes immediate chest pain, shortness of

Table 10.11 Troubleshooting Guide (U.S. EPA, 1978a)

Indicators/Observations	Probable Cause	Check or Monitor	Solutions
1. Low chlorine gas pressure at chlorinator.	a. Insufficient number of cylinders or containers connected to system.	a. Reduce feed rate and note if pressure rises appreciably after short period of time. If pressure increases, 1.a. is the cause.	a. Connect enough cylinders or containers to the system so that chlorine feed rate does not exceed the withdrawal rate from the cylinders or containers.
	b. Stoppage or flow restriction between cylinders or containers and chlorinators.	b. Reduce feed rate and note if icing and cooling effect on supply lines continues.	b. Disassemble chlorine header system at point where cooling begins, locate stoppage, and clean.
2. No chlorine gas pressure at chlorinator.	a. Chorine cylinders or containers empty or not connected to system.	a. Visual inspection.	a. Connect cylinders or containers or replace empties.
	b. Plugged or damaged PRV.	b. Inspect valve.	b. Repair the reducing valve after shutting off cylinder or container valves.
3. Chlorinator will not feed any chlorine.	a. Pressure-reducing valve in chlorinator is dirty.	a. Visual inspection.	a. 1. Disassemble chlorinator and clean valve stem and seat. a. 2. Precede valve with a filter-sediment trap.
	b. Chlorine cylinder hotter than chlorine control apparatus.	b. Cylinder area temperature.	b. 1. Reduce temperature in cylinder area. b. 2. Do not connect a new cylinder that has been sitting in the sun.
4. Chlorine gas escaping from chlorine PRV.	a. Main diaphragm of PRV ruptured because of either improper assembly, fatigue, or corrosion.	a. Place ammonia bottle near PRV to confirm leak.	a. 1. Disassemble valve and diaphragm. a. 2. Inspect chlorine supply system for moisture intrusion.
5. Inability to maintain chlorine feed rate without icing of chlorine system.	a. Insufficient evaporator capacity.	a. Reduce feed rate to about 75% of evaporator capacity. If this eliminates the problem, 5.a. is the cause.	
	b. External PRV cartridge is clogged.	b. Inspect cartridge.	b. Flush and clean cartridge.
6. Chlorination system unable to maintain water bath temperature sufficient to keep external PRV open.	a. Heating element malfunction.	a. Evaporator water bath temperature.	a. Remove and replace heating element.
7. Inability to obtain maximum feed rate from chlorinator.	a. Inadequate chlorine gas pressure.	a. Gas pressure.	a. Increase pressure—replace empty or low cylinders/containers.
	b. Water pump injector clogged with deposits.	b. Inspect injector.	b. Clean injector parts using hydrochloric acid. Rinse with fresh water and replace in service.
	c. Lead in vacuum relief valve.	c. Disconnect vent line at chlorinator; place hand over vent connection to vacuum relief valve, observe if this results in more vacuum and higher chlorine feed rate.	c. Disassemble vacuum relief valve and replace all springs.
	d. Vacuum leak in joints, gaskets, tubing, etc. in chlorinator system.	d. Moisten joints with ammonia solution, or put paper containing orthotolidine at each joint in order to detect leak.	d. Repair all vacuum leaks by tightening joints, replacing gaskets, replacing tubing, and/or compression nuts.

(continues)

Table 10.11 Troubleshooting Guide (U.S. EPA, 1978a) (*Continued*)

Indicators/Observations	Probable Cause	Check or Monitor	Solutions
8. Inability to maintain adequate chlorine feed rate.	a. Malfunction or deterioration of water supply pump.	a. Inspect pump.	a. Overhaul pump. If turbine pump is used, try closing down needle valve to maintain proper discharge pressure.
9. Wide variation in chlorine residual produced in effluent.	a. Chlorine flow proportion capacity inadequate to meet facility flow.	a. Check chlorine meter capacity against facility flow meter capacity.	a. Replace with higher capacity chlorinator meter.
	b. Malfunctioning automatic controls.	b. Go directly to Solutions.	b. Call manufacturer's field service personnel.
	c. Solids settled in chlorine contact chamber	c. Solids in contact chamber.	c. Clean contact chamber.
	d. Fluctuating ammonia or nitrite concentrations in final effluent.	d. Ammonia and nitrite concentrations.	d. Increase chlorine dose to maintain residual.
	e. Flow proportioning control device not calibrated correctly.	e. Check calibration.	e. Recalibrate in accordance with the manufacturer's instructions.
10. Chlorine residual analyzer recorder controller does not control chlorine residual properly.	a. Electrodes fouled.	a. Visual inspection.	a. Clean electrodes.
	b. Loop-time too long.	b. Check loop time.	b. Reduce loop time by doing the following: 1. Move injector closer to point of application. 2. Increase velocity in sample line to analyzer cell. 3. Move cell closer to sample point. 4. Move sample point closer to point of application.
	c. Poor mixing of chlorine at point of application.	c. Set chlorine feed rate at constant dosage and analyze a series of grab samples for consistency.	c. Install mixing device to cause turbulence at point of application.
	d. Rotameter tube range is improperly set	d. Check tube range to see if it gives too small or too large an incremental change in feed rate.	d. Replace with a proper range of feed rate.
11. Indicator organism count fails to meet required standards for disinfection.	a. Breakpoint chlorination occurring.	a. Check chlorine residual at different times of day.	a. Increase chlorine dose to maintain adequate residual during breakpoint chlorination.
	b. Inadequate chlorination equipment capacity.	b. Check capacity of equipment.	b. Replace equipment as necessary to provide treatment based on peak hour flow.
	c. Inadequate chlorine residual control	c. Continuously record residual in effluent.	c. Use chlorine residual analyzer to monitor and control the chlorine dosage automatically.
	d. Short-circuiting in the contact chamber.	d. Contact time.	d. Install baffling and/or mixing device in contact chamber.
	e. Solids build up in contact chamber.	e. Visual inspection.	e. Clean contact chamber to reduce solids build-up.
	f. Chlorine residual.	f. Chlorine residual, effluent ammonia and nitrite concentrations.	f. Increase contact time and/or increase chlorine feed rate.

breath, and coughing. At higher dosages, 40 to 60 ppmv, permanent lung damage can occur. Concentrations over 1000 ppmv are fatal after only a few minutes of exposure.

Sulfur dioxide is an irritant. The odor of sulfur dioxide is detectable at 0.5 ppmv. This is 2.5 times greater than chlorine. Limitations on sulfur dioxide exposure are not as restrictive as those on chlorine. Safe exposure concentration limits differ between agencies. ACGIH recommends a TLV-TWA of 2 ppmv. Both OSHA and NIOSH recommend a PEL of 5 ppmv. The IDLH concentration is 100 ppmv.

SODIUM HYPOCHLORITE

Sodium hypochlorite solutions emit fumes similar to chlorine gas. Although no STEL, PEL, or TWA have been issued by OSHA, good safety practice calls for the use of goggles, gloves, and other protective equipment when using or transferring sodium hypochlorite solutions. Access to an emergency eyewash and shower are recommended for operators handling sodium hypochlorite. As with any form of hypochlorite, the undiluted chemical can cause severe burns on the skin and clothing. It is recommended that operators working with any of the hypochlorites wear protective clothing.

CALCIUM HYPOCHLORITE

Operators who use calcium hypochlorite should wear eye protectors and dust masks when transporting the powder or mixing it with water. All areas exposed to hypochlorite should be washed thoroughly. Rubber gloves are recommended to provide hand protection. One note of caution is strongly stressed in the handling and use of calcium hypochlorite: operators must only mix or add calcium hypochlorite to water. Do not add water to calcium hypochlorite. Adding water can cause a violent chemical reaction.

Strong acids and bases react with water in rapid, exothermic reactions. Exothermic reactions generate heat. Always add acids and bases to water, never the other way around. The excess water absorbs and dissipates the heat. The rhyme "*Do like you oughta, add the acid into water*" can help you remember.

Operators and facilities should include representatives from the local fire department in chlorine awareness and safety training whenever possible. Coordination between facilities and fire departments is essential for a well-defined health and safety site plan. Including fire departments in training events is an opportunity to ensure response plans for chlorine incidents are current and fire department staff are familiar with the facility.

TEST YOUR KNOWLEDGE

1. **Most humans are able to detect extremely low levels of chlorine gas by smell.**
   - ☐ True
   - ☐ False
2. **Sulfur dioxide is more toxic to humans than chlorine gas.**
   - ☐ True
   - ☐ False
3. **The OSHA has not set an exposure limit for sodium hypochlorite solutions.**
   - ☐ True
   - ☐ False
4. **The EPCRA requires facilities with at least ___________ of chlorine gas on site to comply with reporting requirements and other provisions.**
   a. 45 kg (100 lb)
   b. 68 kg (150 lb)
   c. 272 kg (600 lb)
   d. 907 kg (1 ton)
5. **A chlorine release to the environment as small as __________ must be reported under EPCRA.**
   a. 1.0 kg (2.2 lb)
   b. 4.5 kg (10 lb)
   c. 45 kg (100 lb)
   d. 68 kg (150 lb)
6. **Under the Process Safety Management Rule, facilities with at least ___________ lb of chlorine on-site have emergency planning requirements.**
   a. 453 kg (1000 lb)
   b. 681 kg (1500 lb)
   c. 1000 kg (2204 lb)
   d. 1500 kg (3306 lb)
7. **One goal of EPCRA is to**
   a. Reduce the amount of chlorine kept on-site.
   b. Increase paperwork for utilities.
   c. Keep first responders safe.
   d. Ensure compliance with secondary containment rules.
8. **Both the ACGIH and the OSHA set the safe level of exposure to chlorine gas over an 8-hour work day at**
   a. 0.1 ppmv
   b. 0.5 ppmv
   c. 1.0 ppmv
   d. 10 ppmv

UV Disinfection

Since the late 1970s, concerns about chlorine-related storage, handling, and water quality effects have prompted some WRRFs to consider alternatives to chlorine disinfection. Ultraviolet disinfection has become the second most popular method of disinfection in the United States. A useful and important resource for operators is *UV Disinfection Guidelines* (NWRI and WRF, 2012).

THEORY OF OPERATION

In Chapter 6—Wastewater Treatment Ponds, the visible light spectrum was discussed as it relates to photosynthesis. Visible light includes only light that humans are able to see without specialized equipment, from violet (V) to red (R). It is a small part of the much larger electromagnetic spectrum, which includes everything from gamma waves to radio waves (Figure 10.43). Ultraviolet light is a band of light below violet on the electromagnetic spectrum between 10 and 400 *nanometers* (nm). Ultraviolet light is invisible to humans, although some animals and insects are able to see it. Recall from Chapter 6—Wastewater Treatment Ponds that wavelength is one way of describing different types of light. Light with a wavelength of 620 nm appears orange to us, whereas light with a wavelength of 495 nm appears blue. The wavelength of light is related to the amount of energy it contains. As wavelength decreases, the amount of energy increases proportionally. In more technical terms, wavelength is the distance from one peak to the next in a wave of light (Figure 10.44).

One *nanometer* is equal to one billionth of a meter. A sheet of paper is about 100 000-nm thick.

We don't normally think of light as energy, but that is exactly what it is. This energy can be beneficial to life, like when plants harvest light energy to live and grow. It can also be harmful if it is too intense or the exposure time is too long. Think about what happens when you spend too much time in the sun. In the short term, a sunburn may be the result. In the long term, *DNA* damage from sun exposure may eventually result in skin cancer. This is because UV light is high energy and is just the right size to interact with large molecules like DNA. Ultraviolet light is especially damaging to DNA at wavelengths between 250 and 270 nm (U.S. EPA, 1986).

DNA stands for deoxyribonucleic acid. DNA is one type of genetic molecule used by viruses and cells. These molecules contain instructions to build every protein a virus or cell might need to live and reproduce. If the instructions are sufficiently damaged, the viruses and cells are not able to reproduce and will eventually die.

Molecules of DNA are made from four different building block molecules called *base pairs*. They are: adenine (A), thymine (T), guanine (G), and cytosine (C). The four base pairs are connected together in long chains containing hundreds of thousands of base pairs each. By arranging the four base pairs in different ways, all of the instructions needed to produce the proteins, enzymes, hormones, and other molecules that cells need to live are written down. Think of the base pairs as a simple, four-letter alphabet arranged to make DNA a biochemical instruction manual for creating life. Four letters may not seem like enough to form an entire alphabet. The base pairs are read by the cell in groups of three, with each group of three forming a short code. Using groups of three instead of individual letters increases the number of possible "letters" in the alphabet to sixty-four.

Every DNA molecule contains two chains of base pairs, placed one opposite the other, to form a ladder. The ladder is referred to as the *double helix* because of the way the two strands coil and wind around one

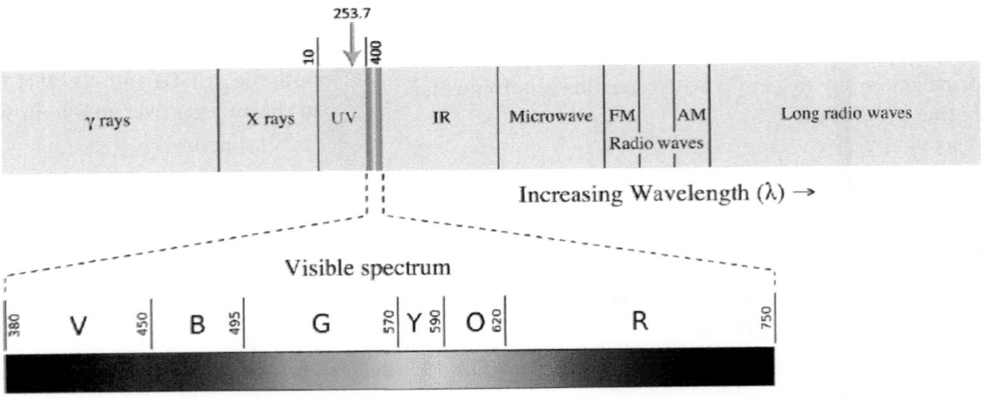

Figure 10.43 Electromagenetic Spectrum (R = red, O = orange, Y = yellow, G = green, B = blue, V = violet, UV = ultraviolet, IR = infrared) (adapted from image by Philip Ronan, Gringer [CC BY-SA 3.0 (https://creativecommons.org/licenses/by-sa/3.0)], via Wikimedia Commons)

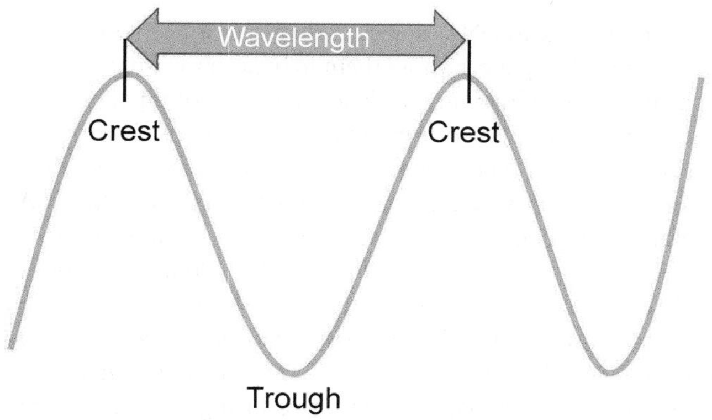

Figure 10.44 Wavelength (Reprinted with permission by Indigo Water Group)

another. An example of a DNA ladder is shown at the top of Figure 10.45. Notice that guanine (G) is always opposite cytosine (C) and adenine (A) is always opposite thymine (T). This unique matching system allows DNA to self-replicate and helps prevent information loss if a DNA molecule is damaged. When cells divide, they first have to duplicate all of their life instructions (DNA) so a copy can be passed on to each daughter cell. A special *enzyme* attaches to the DNA molecule and moves down its length, disconnecting the two halves of the ladder. Different enzymes in the cell then rebuild the opposing sides of each ladder as shown in the middle of Figure 10.45. Because there is only one perfect match for each type of base pair, the two new DNA molecules can each be reconstructed to match the original. The DNA that is damaged can often be repaired using the same method. Information in the undamaged chain is used to repair the damaged chain.

Enzymes are molecules produced by cells that help them carry out specific biological functions. Amylase is an enzyme found in saliva that converts starch into sugar.

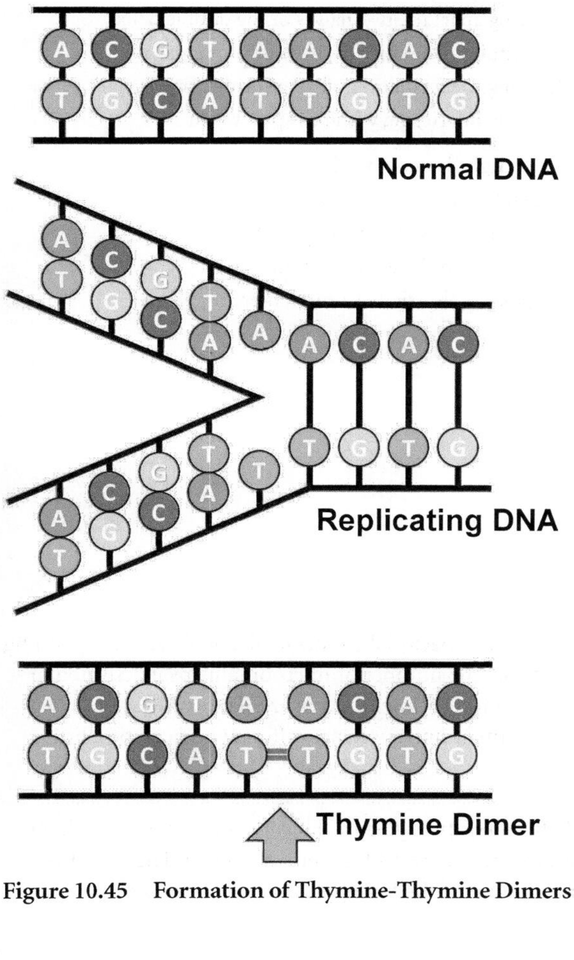

Figure 10.45 Formation of Thymine-Thymine Dimers in DNA (Bernstein, 1981)

The word *dimer* literally means two parts. It is a molecule made by joining two monomers together. A monomer is a molecule that can be chained together to form larger molecules. In this case, the monomer is a single DNA base pair. An oligomer contains a few monomers strung together and a polymer contains many monomers strung together.

The *slider* in a zipper is the piece that you pull on to zip or unzip the zipper.

Ultraviolet disinfection causes a particular type of damage to DNA molecules. In places where two thymine (T) base pairs are side-by-side in one chain, UV light can cause them to break away from the adenine (A) base pairs on the opposite chain. Instead of rejoining with adenine, the two thymine molecules fuse together to form a thymine-thymine *dimer*. Dimer formation prevents the cell from replicating. Think of a DNA molecule as a very large zipper. The teeth from one side of the zipper mesh together with the teeth from the other side. Think of the enzyme that separates the two halves of the DNA ladder before replication as the *slider* in the zipper. When all of the teeth (base pairs) in the zipper mesh together properly, the slider (enzyme) moves smoothly up and down the zipper (DNA) doing its job of splitting and rejoining the teeth (base pairs). When some of the teeth (base pairs) are damaged or aren't fitting together properly, the slider (enzyme) becomes stuck. It can't move past the damaged teeth to finish unzipping. Ultraviolet disinfection doesn't necessarily kill the microorganism, but it can damage it and create enough thymine-thymine dimers to prevent the microorganism from reproducing. Microorganisms that can't reproduce can't cause illness. These microorganisms are inactive.

Ultraviolet light is generated for disinfection using specialized light bulbs called *lamps*. They look similar to the fluorescent light bulbs found in most offices and stores, but are made from more robust materials. Instead of thin glass, they are made from clear quartz. The lamps contain mercury and mercury vapor. Electricity is passed through the UV lamp. The energy is temporarily absorbed by mercury atoms and then released. Different types of atoms absorb and release energy at different wavelengths. Mercury releases absorbed energy primarily at 254 nm, which creates the UV light needed for disinfection.

TEST YOUR KNOWLEDGE

1. **Visible light is the portion of the electromagnetic spectrum that can be seen by humans.**
   - ☐ True
   - ☐ False
2. **Some wavelengths of light can damage genetic material.**
   - ☐ True
   - ☐ False
3. **Damaged DNA can sometimes be repaired by a special enzyme within the cell.**
   - ☐ True
   - ☐ False
4. **Which of the following wavelengths of light would be most damaging to a bacterium?**
   a. 190 nm
   b. 260 nm
   c. 480 nm
   d. 700 nm
5. **Ultraviolet light disrupts DNA by**
   a. Blasting apart the two halves of the double helix
   b. Substituting adenine for thymine in bonds
   c. Causing the formation of thymine dimers
   d. Directly interfering with base pair repair
6. **Ultraviolet light is generated**
   a. When liquid mercury fluoresces
   b. As a byproduct of static electricity
   c. Only at a wavelength of 254 nm
   d. By passing electricity through mercury vapor

DESIGN PARAMETERS

Ultraviolet disinfection systems are designed to reduce active bacterial counts to allowable levels depending on the discharge permit limits. With chlorine disinfection, the number of microorganisms inactivated or killed depends on the dose (amount of chlorine added) and the time of exposure. The same is true for UV disinfection; however, the terminology and measurement units used are different and can be a little confusing.

Instead of concentration, UV systems depend on light intensity or *fluence*. Fluence is the amount of light energy delivered per area and is expressed as milliwatts per square centimeter (mW/cm^2). The term fluence is used instead of concentration because only a tiny amount of the UV applied to the water is actually absorbed by microorganisms, typically less than 1%. Fluence is also referred to as the *irradiance*. Instead of chlorine CT, the amount of UV light available for disinfection per unit of time is called the *fluence rate*. The basic equation for fluence rate or dose (eq 10.31) is similar to the CT equation used for chlorine disinfection (eq 10.32). Fluence rate (dose) is measured in *millijoules* per square centimeter (mJ/cm^2) or milliwatt·seconds per square centimeter (mW·s/cm^2). One W·s is equal to one J.

In the metric system, the basic unit of energy is the Joule, named after James Prescott Joule, a famous scientist. One Joule is equal to 0.000000278 kW·h or 1 W·s.

$$\text{Dose} = \text{Light Intensity} \times \text{time}$$

or

$$\text{Fluence Rate}, \frac{\text{mW} \cdot \text{s}}{\text{cm}^2} = \text{Fluence}, \frac{\text{mW}}{\text{cm}^2} \times \text{time} \quad (10.31)$$

$$\text{CT} = \text{Chlorine Concentration} \times \text{time} \quad (10.32)$$

There isn't a one-size-fits-all design for UV systems because the UV dose delivered depends on a variety of factors, including

Individual UV bulbs are also called *lamps*.

- Lamp type—Different types of *lamps* emit different wavelengths of UV light and vary in their effectiveness.
- Lamp UV power output—Various types of UV lamps provide different UV output per unit of lamp surface area. The higher the output, the larger the intensity at the lamp surface. In addition, the UV output of the lamp will increase with lamp length.
- Power setting of the lamp—The UV intensity is lower if the lamp is operated at less than maximum power.
- Lamp age—After the initial burn-in period, the lamp output and resultant UV intensity decrease gradually as the lamp reaches the end of its useful life.
- Quality and age of the quartz sleeve—The quartz sleeve absorbs some of the UV light, which reduces the delivered dose. Quartz UV absorption may increase as the sleeve ages.
- Fouling on the quartz sleeve—Organic or inorganic fouling can reduce the UV transmission into the wastewater.
- Transmittance of the wastewater—Transmittance is the amount of light that passes through the wastewater. Transmittance is the opposite of absorbance. As UV light passes through the wastewater, its intensity is reduced exponentially with distance from the source, because water and the substances in wastewater absorb some of the light.
- Type and size of the suspended solids—The concentration, type, and size of suspended solids will affect the UV light intensity because suspended solids can absorb and scatter light.
- Particles can shield microorganisms from UV radiation—The UV intensity received by a microorganism floating freely in the wastewater will be greater than the UV intensity received by a microorganism embedded in a particle at the same location.
- Distance from the center of the lamp—The UV intensity decreases with increasing distance from the lamp.
- Hydraulics—Dead space reduces the hydraulic detention time through a UV system. Also, without good mixing through the UV system, a microorganism may pass through the UV reactor between lamps and be exposed to a smaller UV dose.

The UV dose (fluence rate) that must be used to disinfect wastewater is either specified by a regulatory agency or is determined by the design engineer. For example, the State of Colorado's design regulation calls for a minimum fluence of 30 mJ/cm^2 for effluent from activated sludge processes and requires that design engineers assume a UV transmittance of only 65% (CDPHE WQCD, 2012). Ultraviolet system manufacturers and engineers use a combination of historic effluent data for a facility and future flow projections to determine the fluence needed.

EXPECTED PERFORMANCE

Ultraviolet disinfection can treat secondary WRRF effluents, reliably achieving bacterial levels of fewer than 200 colonies of fecal coliforms/100 mL. Many facilities achieve much lower levels and often report <2 CFU/100 mL for *E. coli* and fecal coliforms. Performance is dependent on UV light penetration through the wastewater.

TEST YOUR KNOWLEDGE

1. **Ultraviolet disinfection may use either a high dose and short CT or a low dose and longer CT to achieve the same level of disinfection.**
   - ☐ True
   - ☐ False

2. **Quartz sleeves are used with UV lamps because they do not absorb UV light.**
   - ☐ True
   - ☐ False

3. **For chlorine disinfection, CT is the product of the chlorine residual concentration and DT. Which of the following terms for UV disinfection is most like CT?**
   a. Fluence
   b. DT
   c. Fluence rate
   d. Millijoules

4. **Ultraviolet disinfection efficiency may be reduced by all of the following EXCEPT**
   a. Fouling of the quartz sleeve
   b. Decrease in effluent turbidity
   c. Aging of the UV lamp
   d. Increase in effluent TSS

5. **Which of the following statements is true?**
   a. UV light intensity decreases with distance from the lamp.
   b. UV light easily penetrates particles to inactivate microorganisms.
   c. Dead spaces within a UV reactor increase the DT.
   d. Fluence rates requirements are determined by the manufacturer.

6. **As effluent turbidity increases**
   a. Fecal coliform concentrations remain unchanged.
   b. Less UV light will be needed to disinfect.
   c. UV light penetration will also increase.
   d. *E. coli* concentrations are likely to increase.

EQUIPMENT

The UV light process has several components, each with distinct functions (Table 10.12). Ultraviolet systems may be installed in either a closed conduit similar to a large-diameter pipe or in an open channel. Nearly all UV systems in WRRFs use UV lamps inside quartz sleeves that are submerged directly into the passing wastewater. Ultraviolet systems may consist of a single UV lamp or contain multiple lamps arranged in groups called *modules*. When multiple modules are placed together to function as a single unit, the group is called a *UV bank*. Figure 10.46 shows one method of arranging UV lamps, horizontally and parallel to flow. This example includes one bank with multiple modules. Lamps may also be arranged vertically (Figure 10.47) or at an angle relative to flow (Figure 10.48). The system shown in Figure 10.47 includes two treatment trains, side-by-side, with four banks of lamps each.

Figure 10.49 shows individual lamps (A) arranged horizontally within a module. An electronic ballast (B) with a control panel (C) at the top provides power to the individual lamps. The power cord (D) connects the module to a larger control panel. Four modules can be seen in Figure 10.50, connected to a control panel that spans the UV channel. Most of the time, a UV channel will contain multiple banks in series. This allows operators to take one or more banks out of service while maintaining disinfection with the remaining banks. It also increases the total CT and fluence rate (dose). Because UV rays do not travel very far in water, the lamps are placed close together in modules, typically only a few centimeters (inches) apart.

Most UV systems used in WRRFs are equipped with some type of mechanical wiper. The wipers periodically travel the length of the lamps to help prevent biofilms from forming, algae from attaching and growing, chemical scale from depositing, and solids and other debris from accumulating. An example of a mechanical wiper system can be seen in Figure 10.49. The wipers (E) may be pneumatically or hydraulically driven. In this example, the wiper system is hydraulically driven and the hydraulic fluid lines may be seen (F). The entire wiper module moves along the module of lamps, cleaning all of the lamps at the same time. Each wiper sleeve contains a gasket, typically made of Teflon™, that acts as a squeegee (Figure 10.51). The gaskets wear out and must be periodically replaced. Depending on the size and complexity of the UV system, wipes may be manually or automatically operated. The wiping frequency for an automatic system

Table 10.12 Ultraviolet Light Disinfection System Components and their Uses (WEF et al., 2018)

Component	Features	Function
UV disinfection chamber	Long and narrow. Designed to minimize short-circuiting.	Provides sufficient DT for UV dose to take effect.
Control box	Remote signals, UV lamp ballasts	Control UV system's electrical operations and alarms.
UV lamp	Three-pin plug	Generates UV light.
	Elapsed time meter, cannot be reset	Records elapsed run time (in hours) and keeps a permanent record.
Flow control	Manually operated gates	Diverts flow to standby reactor for channel servicing.
	Counterweighted flow control (underflow) gate or downward opening gat weir	Maintains constant water level in the channel at varying flows to ensure the water level over the UV lamps is constant.
Alarm	Remote warning buzzer and visible light	Indicates overflow or low UV output.
Intensity monitor	Photocell and electronic circuit	Measures and monitors the UV energy through the water or at the lamp surface.
Indicator lights	Manufacturer specific	Indicate safe operating conditions and adverse conditions depending on color.

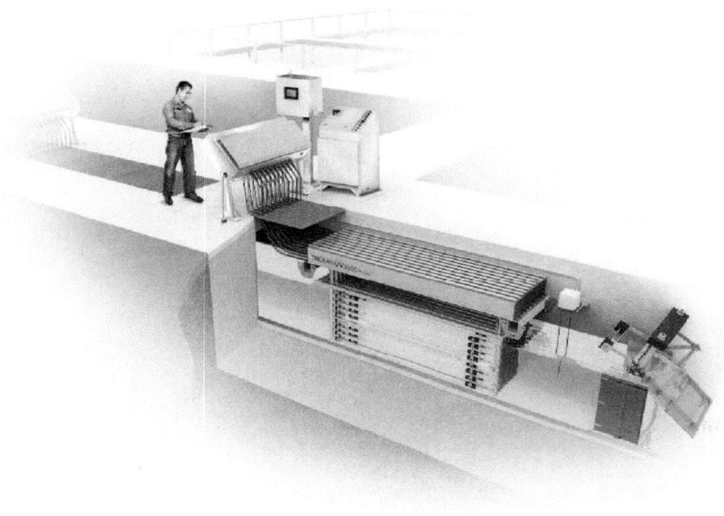

Figure 10.46 Ultraviolet Lamps Arranged Parallel to Flow (Courtesy of TrojanUV)

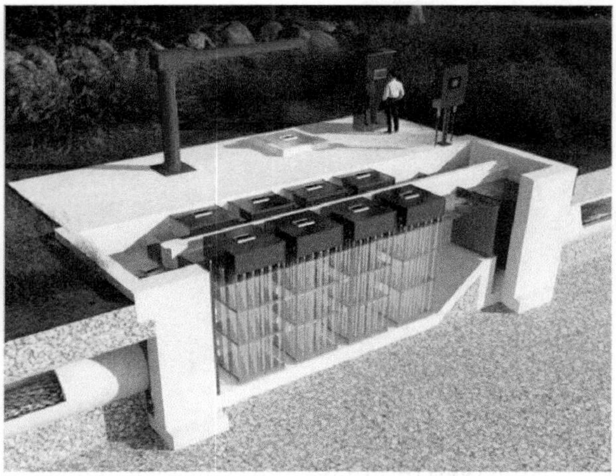

Figure 10.47 Ultraviolet Lamps Arranged Vertical to Flow (Courtesy of Glasco UV)

Figure 10.48 Ultraviolet Lamps Arranged at an Angle Relative to Flow (Courtesy of TrojanUV)

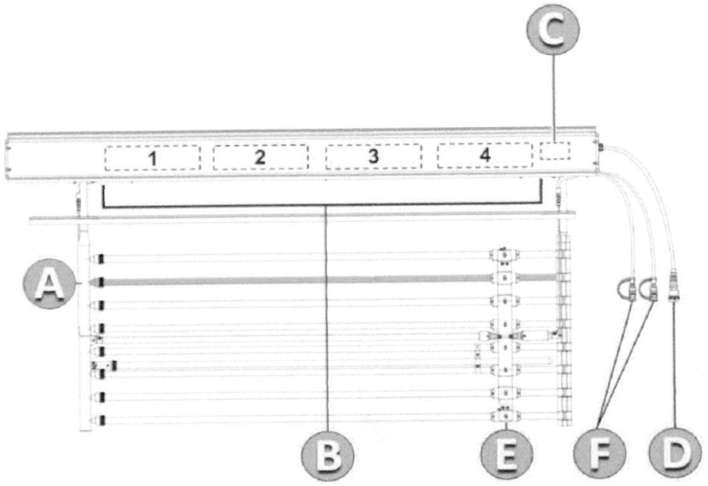

Figure 10.49 Module of UV Lamps (Courtesy of TrojanUV)

Figure 10.50 Ultraviolet Modules Connected to Control Panel (Reprinted with permission by Indigo Water Group)

Figure 10.51 Teflon™ Gasket for Cleaning UV Lamps (Reprinted courtesy of Indigo Water Group)

may be set at a fixed time interval, such as every 2 minutes, or be proportional to flow, such as one wipe for every so many liters (gallons) that pass through the system.

Even when wiper systems are in use, chemical and/or biological fouling will gradually occur on the quartz sleeves. When this happens, the lamps must be removed for a more in-depth cleaning. For small, open-channel UV systems, manual chemical cleaning and wiping by an operator may be sufficient. Cleaning racks for individual lamps or modules are typically provided (Figure 10.52). In larger open-channel systems, lamp modules or entire banks may be removed at a time and cleaned in a dip tank. In this case, a hoist or crane is needed for removing and handling the banks. Dip tanks may be filled with citric acid, phosphoric acid, commercial bathroom cleaner, or other dilute acid solution. Different manufacturers recommend different cleaning protocols. Some open-channel UV systems are equipped with air-scouring systems that release air bubbles underneath the lamps. As the air bubbles travel up and around the lamps, they scrub the quart sleeves. Air scour systems can increase the amount of time between chemical cleanings, but don't replace chemical cleanings.

Ultraviolet systems with pneumatically operated wiper systems and/or air scour cleaning require a pressurized air supply. Air may be supplied from a dedicated compressor or blower. Alternatively, air for the UV system may be provided from the facility air system provided the air quality and quantity meet the requirements of the UV manufacturer.

Figure 10.52 Cleaning Rack with UV Module (Reprinted with permission by Manuel Freyre, City of Northglenn, Colorado)

LAMP TYPES

There are three main types of mercury lamps used in disinfection applications: low-pressure, amalgam, and medium-pressure lamps. The lamps are enclosed within UV-transparent quartz sleeves to protect the lamps from exposure to water and to control lamp temperature (Figure 10.53). Gaskets at both ends of the quartz tube prevent water from reaching the lamp inside.

Low-Pressure Lamps

Germicidal means it is effective for damaging or killing microorganisms.

Monochromatic means single color. Monochromatic light is emitted at a single wavelength. *Polychromatic* means many colors. Polychromatic light is emitted at multiple wavelengths.

Low-pressure lamps are the oldest, most commonly used type of lamp. The low-pressure lamps generate mostly *germicidal, monochromatic* UV light at a wavelength of 253.7 nm. In a low-pressure mercury lamp, the mercury vapor is at equilibrium with a small liquid mercury reservoir inside the lamp. A typical low-pressure lamp contains approximately 60 mg of mercury. Lamp efficiency is a function of the vapor pressure of the mercury, which is, in turn, a function of the "cold-spot" temperature. For optimum lamp efficiency, the cold spot should be kept near 40 °C (104 °F). Lamp efficiency decreases at warmer and colder temperatures. The lamp output can, therefore, be affected by changes in water temperature.

Two standard lamp lengths are typically used in conventional disinfection systems: 0.9 m (36 in.) and 1.6 m (64 in.). Low-pressure lamps are between 25 and 30% efficient at producing UV radiation at the ideal wavelength, but their intensity is relatively low. A low-output lamp typically lasts for 8000 hours of operation; however, turning the lamps on and off frequently makes the filaments brittle and shortens lamp life.

Amalgam Lamps

Amalgam lamps decrease and control the vapor pressure of the mercury by combining it with other metals to create a solid amalgam (Figure 10.54). As a result, the mercury vapor can be maintained near its optimum value over a broader range of temperatures than a low-pressure lamp. This allows these lamps to be operated at a higher power, resulting in increased UV output for a given lamp size, while maintaining the nearly monochromatic (single light wavelength) output of the low-pressure lamp. Accordingly, amalgam lamps, also known as *low-pressure, high-output lamps*, have become increasingly popular. The amalgam lamps generally operate at higher electrical current, higher temperature, and higher total internal pressure than low-pressure lamps. A typical amalgam lamp contains approximately 120 mg of mercury, which is twice as much as a low-pressure lamp. Amalgam lamps are not affected by water temperature fluctuations, but can take up to 800 seconds to reach full power.

Medium-Pressure Lamps

The wavelengths of light emitted by a particular UV system are not operator adjustable.

Medium-pressure lamps do not use amalgams, but operate at higher power density and temperature so the mercury within them is completely vaporized. The lamps themselves may reach 800 °C (1472 °F), but the outer quartz sleeve typically remains below 60 °C (140 °F). Medium-pressure lamps contain the most mercury of all three lamp types, with a typical lamp containing 300 mg of mercury. Medium-pressure lamps are *polychromatic* and emit a broader range of UV light wavelengths, including the germicidal wavelengths between 200 and 300 nm (Figure 10.55). These lamps are less efficient than low-pressure lamps. A medium-pressure lamp typically lasts for 12 000 hours of operation.

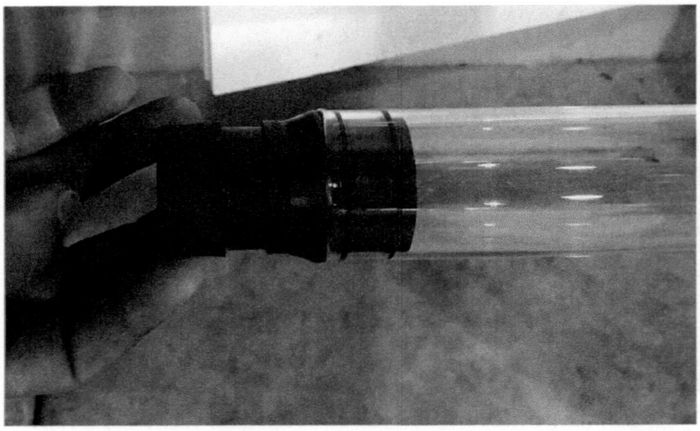

Figure 10.53 Quartz Sleeve (Reprinted with permission by Indigo Water Group)

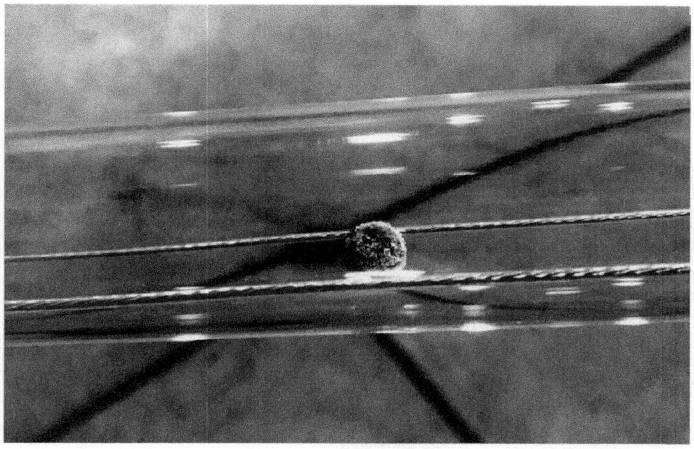

Figure 10.54 Mercury Amalgam within a Low-Pressure, High-Output Lamp (Reprinted with permission by Indigo Water Group)

ULTRAVIOLET CHAMBER

Because disinfection occurs rapidly, the UV chamber or channel is relatively small, even at large treatment facilities. Contact time may be as short as a few seconds. Turbulent flow past the lamps is necessary to ensure mixing and UV light exposure so all of the material in the effluent receives a lethal dose of UV light. A perforated plate like the one shown in Figure 10.56 is typically placed upstream of the UV system to create turbulence.

BANK/MODULE LIFTING

In larger open-channel systems, a hoist or crane may be required for removing and handling the banks. Entire banks may be removed at a time and be cleaned in a dip tank. Bank removal, or at least module removal, may also be required for lamp inspection and replacement.

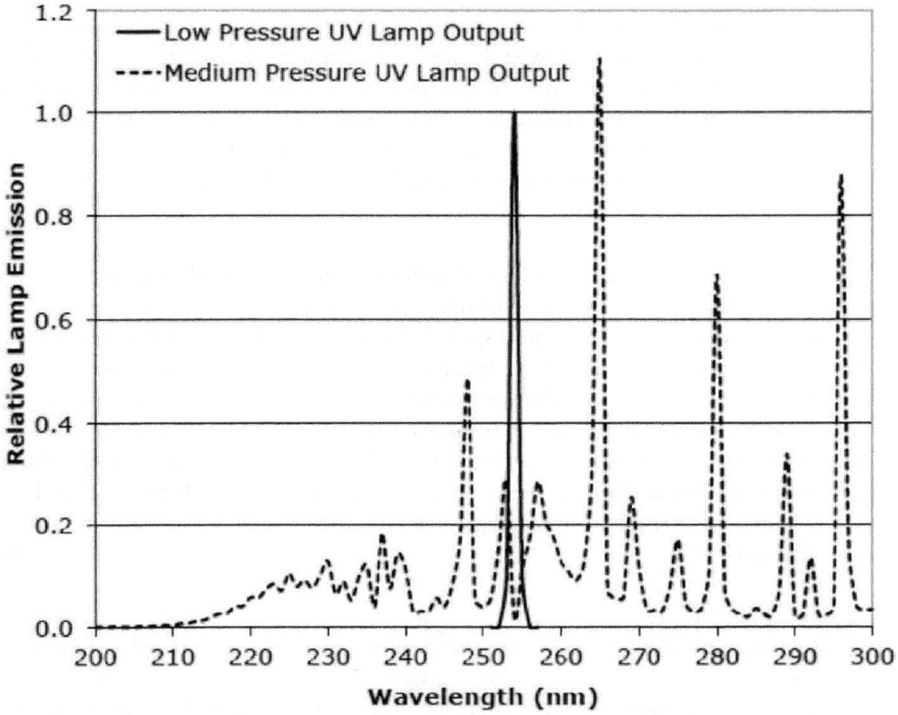

Figure 10.55 Ultraviolet Light Output from Low-Pressure and Medium-Pressure Lamps (WEF and IUA, 2015)

Figure 10.56 Ultraviolet Channel Baffle (Reprinted with permission by Indigo Water Group)

Jib cranes or davit hoists are often mounted adjacent to the UV channels. To transfer a bank to a cleaning tank, a monorail hoist system or a bridge crane is typical.

ULTRAVIOLET-INTENSITY SENSORS

The UV-intensity measurement reflects changes in lamp output related to the lamp power setting, lamp aging, quartz sleeve fouling and aging, and changes in degree Celcius (WERF, 2007). Ultraviolet manufacturers typically provide one UV-intensity sensor per system or per bank of UV lamps. The intensity sensor measures the amount of light (fluence) delivered at that location only. The sensor cannot report an average light intensity within the entire reactor or bank. Intensity sensors may be used to control the amount of energy delivered to the bulbs, which, in turn, controls the amount of UV light produced. Placement of the sensor is important. An intensity sensor located close to the lamp will respond more to changes in lamp output and sleeve fouling, but less to changes in wastewater quality. Intensity sensors located further away from the lamps will respond more to changes in water quality than lamp aging and fouling. Intensity sensors must be precisely placed if the intent is to incorporate changes to wastewater ultraviolet transmittance (UVT) in the intensity measurement. If UVT is monitored separately with a transmittance monitor, the intensity sensor may be placed more closely to the UV lamps and sensor placement is not as critical.

An *algorithm* is a mathematical model that predicts the value of one number from one or more measurements or pieces of data. A UV algorithm might predict system performance from intensity or transmittance or both. It could also predict percentage of fouling from the same information.

Some UV system manufacturers have developed dose *algorithms* for system-display, reporting, and dose-pacing purposes. The algorithms are calibrated to intensity sensor readings, with or without the inclusion of transmittance meter data. Other UV control systems use intensity sensors to report relative differences in performance or only for triggering alarms. Intensity sensors can be used to estimate lamp aging or sleeve fouling effects over time. Sensor readings are recorded for a new lamp before fouling has occurred and again over time. Intensity decline can then be used to predict when cleaning is needed or when effluent quality goals may not be met. Sensor readings can also be compared pre- and post-cleaning to estimate cleaning effectiveness. They can be compared over time on clean systems to discern the need to replace lamps and/or sleeves. Low-intensity warning/alarm setpoints can be used to warn against potentially inadequate disinfection.

A *wet weather event* is caused by excessive rain or snow entering the collection system. This causes the influent flow to the WRRF to increase substantially over a short period of time.

ULTRAVIOLET TRANSMITTANCE METERS

Ultraviolet transmittance is one of the most critical water quality parameters in determining disinfection effectiveness. Industrial discharges and *wet weather events* often significantly affect UVT. It is important to note that changes to UVT cannot be discerned by the human eye. Just because water exhibits color (e.g., a greenish tinge resulting from algae) does not indicate that the water also will exhibit a low UVT. Conversely, certain chemicals are colorless, yet have high UV-absorbing characteristics, which result in significant effects on UVT. Simply because an effluent is crystal clear does not indicate high UVT.

Ultraviolet disinfection systems may be equipped with online UVT sensors. During operation, the disinfection control system may regulate dose delivery in response to UVT in an effort to minimize energy requirements. More energy is applied to increase light intensity (dose) when needed and less is applied when it is not. Another option is to maintain the system dose required for the minimum design UVT and use online transmittance monitoring to trigger alarms.

FLOW MONITORING

Ultraviolet disinfection control systems typically use flowrate information to match the quantity of UV equipment in service, or lamp power, with the current flowrate. However, flow instrumentation is not typically provided by UV disinfection system manufacturers. Data from a facility effluent flow meter are used as an input to the disinfection control system. A 4- to 20-mA signal is typically sent from the flow meter. The data may be relayed through the facility supervisory control and data acquisition (SCADA) system or, particularly at smaller facilities, there may be a direct data connection between the effluent flow meter and the disinfection control system.

Facility operators should consider how the recorded effluent flowrate corresponds to the flow through the UV disinfection system. These flowrates may differ because of the addition or diversion of various recycle or service water streams. If the difference in flowrate is significant or highly variable, a more accurate UV disinfection flowrate could be calculated in a PLC by adding or subtracting other measured or estimated flowrates from the recorded effluent flow, with the calculation result sent by the facility SCADA system to the disinfection control system. Operators should refer to the flow meter manufacturer's manual for maintenance and calibration requirements.

LEVEL MONITORING

In open-channel UV disinfection systems, the wastewater within the channel must be maintained at a nearly constant level with little allowable fluctuation. The wastewater level is typically controlled by a mechanical counterbalanced gate or by fixed or adjustable weirs downstream of the UV reactors. Level control in open-channel UV systems is required to prevent

- Too high a water level resulting in an undisinfected layer of water passing above the lamps or flooding of the channels or nonwetted disinfection system components, such as ballasts; and
- Too low a water level resulting in UV lamps operating in the air instead of submerged, which may damage the lamps through overheating or expose staff to UV radiation.

Ultraviolet disinfection systems are typically designed to ensure that the water level will remain within the required range. However, open channels may be equipped with level monitoring instrumentation to produce alarms for protection against unexpected conditions. Discrete (e.g., float switches) or analog (ultrasonic level sensors, pressure transmitters, radar level sensors, etc.) level instrumentation may be used to monitor level in open UV channels. Low- and high-level setpoints in the disinfection control system would trigger alarms and/or equipment shutdown. Closed-vessel UV disinfection systems must remain full of water at all times and, therefore, the connecting piping systems are typically designed to ensure full flow. Operators should refer to the level instrumentation manufacturer's manual for maintenance and calibration requirements.

Flow measurement devices are discussed in Chapter 6 of Book II. Additional information may be found in *Teledyne Isco Open Channel Flow Measurement Handbook* (Teledyne Isco, 2014).

TEMPERATURE MONITORING

Most UV disinfection lamps rely on flowing wastewater for lamp cooling because of the high operating temperatures of the lamps. Medium-pressure and low-pressure high-output (amalgam) UV lamps operate at the highest temperatures. Ultraviolet disinfection systems are typically designed to ensure that the expected range of design flowrates will be sufficient for lamp cooling. However, temperature sensors may be provided by UV system manufacturers to prevent overheating. Flowrates in pumped systems could drop to zero if the pumps are not operating.

Discrete (e.g., temperature switches) or analog (temperature sensors) instrumentation may be used to monitor temperature in UV reactors. High-temperature setpoints in the disinfection control system trigger alarms and/or equipment shutdown. Operators should refer to the temperature instrumentation manufacturer's manual for maintenance and calibration requirements.

Table 10.13 Typical Alarm Conditions for UV Disinfection Systems (adapted from U.S. EPA [2006])

Alarm Name	Alarm Category	Description
Lamp age	Minor alarm	Lamp end-of-life per defined operational lamp life
UV sensor calibration	Minor alarm	UV sensor calibration interval reached
Lamp/ballast failure	Minor alarm	Failure of a single lamp or ballast
Multiple lamp/ballast failures	Significant alarm	Failure of adjacent lamps/ballast or failure of 5% of lamps/ballast in a single reactor
Low UV intensity	Significant alarm	Measured UV intensity is below design/validated value
Low UV transmittance	Significant alarm	Measured UVT is below the minimum design value
Low UV dose	Significant alarm	The calculated UV dose is below the design/validated dose
High flowrate	Significant alarm	Measured effluent flowrate exceeds the UV design capacity
High liquid level	Significant alarm	There may be an undisinfected layer of water above the lamps or potential flooding
Mechanical cleaning system failure	Significant alarm	Failure of the lamp cleaning system
Low liquid level	Critical alarm	Potential for exposed lamps or overheating
High temperature	Critical alarm	Overheating occurring

CONTROL SYSTEMS

Control systems provided by UV disinfection system manufacturers typically provide a variety of warnings, minor alarms, and significant and critical alarms. The terminology and categorization will vary between manufacturers. The warnings and alarms may be audible locally at the UV system control panel(s) or they may be transmitted to the facility SCADA system. For facilities that are not staffed around the clock, it is recommended that, at a minimum, significant/critical alarms are connected to a remote alarm system (such as an auto-dialer) to notify off-site staff of the alarm. Some systems may be configured so that UV control system will automatically shut down UV equipment upon certain critical alarms. Table 10.13 provides a list of typical UV system warnings and alarms.

TEST YOUR KNOWLEDGE

1. **Banks contain more UV bulbs than modules.**
   - ☐ True
   - ☐ False
2. **Mechanical wipers may be manually, hydraulically, or pneumatically operated.**
   - ☐ True
   - ☐ False
3. **The intensity sensor in a UV system reports the average light intensity over the entire system.**
   - ☐ True
   - ☐ False
4. **Effluents that are clear and colorless have low UV absorbance and high UV transmittance.**
   - ☐ True
   - ☐ False
5. **The ballast in a UV module**
   a. Provides power to each lamp
   b. Is intended to hold the lamps in place
   c. Adds weight to reduce vibration
   d. Always contains six lamps
6. **Mechanical wipers in UV systems**
   a. Prevent water droplets from accumulating
   b. Reduce biofilm and algae growth
   c. Automatically soak lamps in citric acid
   d. Typically contain glass fiber squeegees
7. **Match the UV system component to its function.**

a. Lamp	1. Indirect measurement of UV dose
b. Ballast	2. Cleans lamps within UV channel
c. Dip tank	3. Generates UV light
d. Quartz tube	4. Removes scale from quartz tubes
e. Intensity monitor	5. Controls lamp temperature
f. Transmittance monitor	6. Measures lamp light output
g. Air scour	7. Provides power to lamps

8. **This type of UV lamp emits light over many different wavelengths.**
   a. Low-pressure
   b. Amalgam
   c. Low-pressure, high-intensity
   d. Medium-pressure

9. **Ultraviolet lamps that emit light at only 254 nm are**
   a. Monochromatic
   b. Medium-pressure
   c. Polychromatic
   d. High-pressure

10. **Low-pressure UV bulbs typically last for 8000 hours of operation, but doing this can shorten lamp life.**
   a. Operating lamps continuously
   b. Adjusting the wavelength above 300 nm
   c. Turning lamps on and off frequently
   d. Increasing the wiper cycle frequency

11. **A mercury amalgam lamp**
   a. Generates light at multiple wavelengths.
   b. Contains about twice as much mercury as a low-pressure lamp.
   c. Requires less time to warm up than a low-pressure lamp.
   d. Is more sensitive to temperature changes than other lamps.

12. **A UV channel may contain baffles or a perforated plate**
   a. To create turbulence
   b. To keep solids in suspension
   c. To hold up the ends of UV lamps
   d. To increase the DT

13. **Manufacturers typically provide _____ UV sensors.**
   a. One per bank
   b. One per module
   c. Two per bank
   d. Two per module

14. **A WWRF receives industrial discharges that cause fluctuations in their effluent turbidity. Which of the following locations is preferred for a UV sensor that will be used to adjust lamp output?**
   a. As close to the lamp as possible
   b. Halfway between two lamps
   c. Some distance away from the lamps
   d. At the entrance to the UV channel

15. **The UV intensity reading has been gradually decreasing over the past month. What is the most likely cause?**
   a. Rags or debris caught on sensor
   b. Inorganic or biological fouling of sleeves
   c. Increased flowrates and power usage
   d. New quartz sleeves over lamps

16. **This type of meter would be most effective for determining the effectiveness of UV lamp cleaning.**
   a. Transmittance meter
   b. Thermographic sensor
   c. Intensity meter
   d. Turbidity meter

17. **The water level in a UV channel**
   a. Should be kept as high as possible to keep lamps submerged.
   b. Is typically controlled with a fixed or adjustable weir.
   c. Increases and decreases as the facility influent flow changes.
   d. May be varied from day to day to reduce algae growth.

18. **One consequence of UV lamps operating in air instead of remaining submerged might be**
   a. Operator exposure to UV light and possible skin burns
   b. Excessive cooling of the lamps and loss of radiation
   c. Undisinfected layer of water passing over the lamps
   d. Flooding of ballasts and electrical short-circuiting

19. **Effluent pumps send secondary effluent to the UV disinfection system. The effluent pump fails. The weir on the UV channel keeps the lamps submerged. If the pump remains off, what is likely to happen?**
   a. Lamps will be exposed to air.
   b. UV channel may overflow.
   c. Ballasts may short-circuit.
   d. Lamps may overheat.

20. **Of the following possible alarm conditions for a UV disinfection system, which is severe enough that it might require automatic shutdown of the UV system?**
   a. Low UV transmittance
   b. Lamp/ballast failure
   c. Low liquid level
   d. Lamp nearing end of life

PROCESS VARIABLES

Ultraviolet disinfection efficiency depends on the following:

- Lamp intensity,
- Ultraviolet light transmittance or absorbance, and
- Exposure time.

LAMP INTENSITY

The lamp's age strongly influences intensity. As lamps age, the lamp output and light intensity decrease. The UV transmittance of the quartz sleeve of the lamp decreases with time. This is called *solarizing*. Lower lamp output and solarization together cause a slow decline in lamp performance over the lamp's lifetime. Lamps should be replaced when light intensity falls below a manufacturer recommended level and/or at manufacturer recommended time intervals. For example, a manufacturer may recommend lamps be replaced when light intensity falls below 20 mW/cm^2 or after 6000 hours of run time. Lamps should never be run to failure.

Lamp output should be monitored and recorded as part of the lamp history record. Also, an adequate inventory of replacement lamps should be maintained.

ULTRAVIOLET TRANSMITTANCE OR ABSORBANCE

The effectiveness of UV disinfection depends on the light intensity and the amount of time the microorganisms are exposed. Any condition that reduces either light intensity or CT will decrease the UV disinfection system's performance. The two parameters that most affect performance are UV *transmittance* and suspended solids concentration. The UV transmittance through water, defined as the percentage of UV light intensity not *absorbed* after passing through 1 cm of water, depends on dissolved and suspended matter. Reduced transmittance lowers the intensity of the light reaching the bacteria, resulting in less inactivation. Suspended solids can lower the UV transmittance by scattering and absorbing the light; suspended solids can also reduce the inactivation level by *encapsulating* the bacteria and protecting them from exposure to UV light. Facilities using ferrous (as ferric salts) for phosphorus removal or enhanced sedimentation–coagulation also may reduce the effective performance of UV disinfection. Water clarity is not always a good indicator of UV transmittance because water that appears clear in visible light may actually absorb invisible UV wavelengths.

FLOWRATE AND CONTACT TIME

Flowrate affects the CT. Increasing the flowrate across the UV lamps decreases the CT and lowers the inactivation rate. Many WRRFs have a wide variation in flow, especially comparing average daily design and peak flow values. Most UV systems are designed to disinfect at the peak flowrate. If short-circuiting occurs in the contact channel, the contact duration will be reduced.

PROCESS CONTROL

Depending on the manufacturer, operators may be able to set minimum transmissivity or intensity settings. Setpoints should be selected to meet effluent quality goals under minimum, average, and maximum flow conditions. Determining the ideal setpoints may require some trial and error.

Ultraviolet disinfection is a significant energy consumer at WRRFs, so strategies to conserve energy are often central to operation. A key element of energy conservation is through the use of flow and dose pacing. Flow and dose pacing typically are incorporated to a UV system by the manufacturer. Flow pacing is modulating (changing) the power or number of lamps in response to wastewater flow. Dose pacing, typically considered the most energy-efficient means of operational control, is modulating the power to the UV lamps to achieve the desired dose based on three to four factors: flow, lamp power, UVT, and, in some cases, UV intensity. However, excessive lamp cycling may reduce lamp life; thus, many operators will leave lamps running for a minimum duration each time a lamp is turned on.

The controller often will be designed to cycle operational banks and channels in an attempt to age all of the lamps at a similar rate. Operators should check the programmed algorithm and verify that the system is operating within the validated range. If the disinfection permit is seasonal, then significant energy can be saved by shutting down the UV system during the non-disinfection season.

OPERATION

DAILY OPERATION

Generally, a UV disinfection system should be operated in accordance with the O&M manual provided by the manufacturer and/or design engineer. Depending on the facility, the system may be operated in flow-paced, dose-paced, or manual mode. To ensure that the UV system functions properly, a number of routine maintenance activities, as described in this chapter, should be implemented.

Typical operations tasks are summarized in Table 10.14. During normal operation, secondary or tertiary effluent is passed through the UV reactor(s) at flowrates and transmittances within the design range of the system. The system will typically be operated with monitoring systems so alarms can indicate problems with the system.

STARTUP

Before water is introduced to the UV system, a number of startup checks need to be completed. The equipment/channels should be inspected to ensure that no foreign matter is in the channel. In addition, comprehensive input/output checks need to occur during the startup activities. These checks are necessary to verify the integrity and accuracy of signals between UV reactors, that is, ancillary equipment such as gates, valves, and analyzers; local control panels; the master control panel; and any associated SCADA platform. Additionally, a check will need to be made for the status of each lamp. Lamps should be repaired or replaced as needed.

Before starting the UV disinfection system, operators should check the following items:

- The power supply to the following units:
  - Motor control center,
  - Local disconnect, and
  - Local control panel;
- The standby power system. To ensure readiness, test the UV system under standby power and periodically exercise the standby power supply;
- The normal operating positions of the valves upstream and downstream of the appropriate chambers;
- The readiness of the UV unit, including indicator lights, monitors, and gauges;
- The effluent outfall (must be open); and
- The effluent flow control gate or adjustable weir (should operate freely and automatically).

Once the prestart checks have been completed and the lamps are completely submerged, operators may move the local on–off switch to the on position.

SHUTDOWN

When taking the UV unit out of service, operators should

- Move the on–off selector switch to the off position;
- Perform lockout/tagout safety procedures to isolate individual lamp racks without affecting the rest of the operating system;
- Close the appropriate valves upstream and downstream of the chambers;
- If shutdown is a result of mechanical failure (long-term) or is for routine preventative maintenance, then open, and lockout/tagout all electrical breakers;

Table 10.14 Typical Operational Tasks for UV Disinfection Systems

Frequency	Recommended Tasks
Daily	■ Perform overall visual inspection of UV reactors. ■ Ensure system control is in automatic mode (if applicable). ■ Check control panel display for status of system components and alarm status and history. ■ Ensure that all online analyzers, flow meters, and data-recording equipment are operating within normal parameters. ■ Review effluent data.
Weekly	■ Initiate manual operation of wipers (if provided).
Monthly	■ Check lamp run time values. Consider changing lamps if operating hours exceed guaranteed life or UV intensity is low.
Semiannually	■ Check ballast cooling system for leaks and unusual noises from cooling fans. ■ Check operation of automatic and manual valves.

- Properly store UV equipment and controls; and
- Clean the channel to remove algae and other accumulated materials.

MAINTENANCE

Proper maintenance of any equipment system is required to provide efficient operations, including the following (U.S. EPA, 2006; WEF, 2006):

- Weekly—check and calibrate the online UVT analyzers;
- Monthly—check the quartz sleeves and wipers for cracks and/or leaks;
- Weekly to monthly—check the cleaning efficacy. If the UV system is not equipped with an automatic wiping system, the quartz sleeves must be cleaned manually, which can range from simple wiping by hand to in-channel cleaning or a separate tank dedicated to cleaning;
- Quarterly—test the ground fault interrupt;
- Semiannually—check the cleaning fluid reservoir (semiannually);
- When required
  - Check the accuracy of the flow meters (as recommended by manufacturer);
  - Replace lamps
    - When a lamp fails,
    - Reaches the predetermined service hours,
    - The system does not achieve the required performance,
    - Online monitoring alarm signals lamp replacement, and
    - The UV monitor indicates that the UV intensity is low;
  - Replace quartz sleeves
    - At prespecified schedule,
    - When damage or cracks are observed, and
    - When excessive fouling occurs that cannot be removed by cleaning; and
  - Replace ballasts when they fail.

Table 10.15 summarizes significant maintenance considerations for UV systems.

Drift is when the reading gradually increases or decreases. Drift occurs when the sensor can't get an accurate reading.

Intensity sensors require frequent cleaning and calibration verification to provide accurate results. Periodic cleaning prevents scaling and fouling on the UV-intensity receiving surface, and the cleaning frequency will depend on the site-specific water quality. Calibration verification may consist of maintaining a single unused sensor as a reference standard and inserting that sensor in place of operational sensors to measure *drift*. In other words, a clean sensor is kept on hand so clean sensor and dirty sensor readings can be compared. When drift of operational sensors becomes excessive, a calibration factor can be applied to the control system or those sensors can be recalibrated or replaced. Some manufacturers provide intensity sensors that may be field calibrated (typically when intensity readings are not used for dose control or verification), whereas many sensors may only be recalibrated at the factory. For reclaimed water systems, the National Water Resources Institute (NWRI) requires that intensity sensors must be calibrated at least monthly (NWRI and WRF, 2012). *The Ultraviolet Disinfection Guidance Manual for the Final Long Term 2 Enhanced Surface Water Treatment Rule* (U.S. EPA, 2006) provides detailed calibration procedures for intensity monitors. For other UV control systems that do not consider intensity readings, the sensor calibration and maintenance schedule may be determined by manufacturer's recommendations or selected by facility staff.

Ultraviolet transmittance sensors must be maintained and calibrated according to manufacturer recommendations. The NWRI and WRF guidelines (NWRI and WRF, 2012) require that water reclamation UV disinfection systems verify the accuracy of online transmittance sensors weekly by comparison with a calibrated laboratory UVT analyzer. The frequency of transmittance sensor calibration and maintenance in low-dose wastewater treatment applications will depend on whether the control system provided by the UV system manufacturer relies on UVT sensors for dose pacing. At a minimum, the cleaning, maintenance, and calibration recommendations of the sensor manufacturer should be followed.

Table 10.15 Typical Maintenance Tasks for UV Disinfection Systems

Frequency	Task or General Guideline	Action
Weekly	Check online UV transmittance analyzer calibration (if applicable).	Calibrate UV transmittance analyzer when manufacturer's guaranteed measurement uncertainty is exceeded.
Monthly	Check sleeves and wipers for leaks.	Replace sleeves, seats, O-rings, or wiper seals if damaged or leaking.
Monthly	UV intensity calibration check.	Check sensor calibration at the lamp power output used during routine operating conditions.
Semiannually	Check cleaning fluid reservoir (if applicable).	Replenish solution if the reservoir is low. Drain and replace solution if it is discolored.
Annually	Test trip ground fault interrupt (GFI).	Maintain GFI breakers in accordance with manufacturer's recommendations.
As needed	Replace lamps.	Replace lamps when any of the following conditions occur: ■ Initiation of low UV intensity after verifying that condition is caused by low lamp output. ■ Initiation of lamp failure alarm, or ■ System not achieving discharge permit requirements after exceeding guaranteed lamp run time.
Manufacturer's recommended frequency	Check flow meter calibration.	If effluent weir or flume is available, manually check depth and compare with primary device measurement. Primary device should be calibrated at recommended frequency or when measurement uncertainty is exceeded.
	Clean and calibrate transmittance monitor (if installed).	Clean and calibrate according to manufacturer's recommended frequency.
When lamps are replaced	Properly dispose of lamps.	Send spent lamps to a mercury recycling facility or return to manufacturer.
When quartz sleeves are replaced	Properly dispose of quartz sleeves.	Replace sleeves as recommended by the manufacturer when damage, cracks, or excessive fouling occurs that would impede UV intensity.

LAMP REPLACEMENT

Lamps are replaced at a frequency that typically ranges from 6 months to 3 years. Some facilities replace lamps strictly in accordance with manufacturer's recommendations, whereas other facilities only replace them when intensity decreases or indicator counts increase.

SLEEVE REPLACEMENT

Similar to lamps, sleeves are replaced at a frequency that typically ranges from 1 to 5 years. Similar to lamps, some facilities only replace sleeves when intensity decreases or indicator counts increase and, for low-pressure systems, they may retain sleeves in use for the life of the system.

BALLAST REPLACEMENT

Ballasts are replaced at a frequency that typically ranges from 3 to 8 years.

SLEEVE CLEANING

Control of lamp sleeve fouling is achieved by a variety of techniques. The following are sleeve-cleaning options that are used currently:

- Manual cleaning strategies, which require periodic removal of the sleeves for soaking in a chemical bath or manual wiping with a chemical cleaner;

- Automated online strategies, which use mechanical cleaning devices that wipe frequently and require periodic manual chemical cleaning;
- Automated chemical/mechanical cleaning systems (that may require periodic manual chemical cleaning [e.g., for areas outside the lamp's arc length and/or depending on the nature of scaling constituents]);
- Chemical removal of scale is achieved by applying a dilute acid (pH of approximately 1 to 3) to the fouled surface. Acid can be applied by either wiping individual lamps or immersing entire lamp modules. Immersion techniques are more efficient for scale removal. For large systems, module-immersion hardware is a necessity;
- Several different acid solutions have been used for chemical cleaning, including citric acid, phosphoric acid, and commercially available bathroom cleaners. Selection of an appropriate acid will depend on site-specific requirements, but disposal of spent acids should be incorporated to the decision. For large systems, the use of food-grade citric acid or phosphoric acid should be considered so that the neutralized liquid containing the spent acid can be diverted to the headworks of the WRRF (it is important to note that operators should first consider the effect of recycle streams containing spent cleaning solutions on facility performance);
- A number of physical processes can be incorporated to mitigate scaling. Introducing air bubbles at the base of a channel for short periods of time, but frequently (e.g., 10 min/d), has been performed successfully in some UV systems. This procedure will not eliminate the need for cleaning at facilities where fouling occurs, but will be effective in increasing the interval between cleanings;
- Citric acid, phosphoric acid, proprietary mixtures, and commercially available bathroom cleaners were used most commonly to clean sleeves. Typically, WRRFs should use a commercially available, inexpensive cleaning agent that is handled and disposed of easily. Materials issues (e.g., corrosion) should be considered if cleaning is to be performed in situ. A small, bench-scale, flow-through unit can be used to evaluate a number of agents by trial and error and the optimal cleaning frequency. Historical reported manual cleaning frequencies are highly site-specific and range from weekly to yearly, with a median frequency of approximately once per month (U.S. EPA, 1992). However, with the proliferation of automated cleaning systems, cleaning cycle times have increased from these levels;
- Low-pressure, low-intensity lamp cleaning in horizontal systems is typically accomplished by either bank or module removal to a mobile or dedicated cleaning station. The level of cleaning complexity can range from a drained area equipped with a holding rack, hose, and cleaning solution to automatic air sparging or an ultrasonic dip tank for large banks accessed with overhead hoists; and
- Low-pressure, low-intensity vertical lamp cleaning typically is accomplished in a similar manner to that of horizontal systems. Current options include dip tanks and an air-scouring system, which is engaged in place and under process conditions. It is used to increase the interval between chemical lamp-cleaning cycles, which can either be done in situ (isolating the channel) or by transferring the module to a dip tank.

CHANNEL CLEANING

The deposition of solids and growth of algae are known to result in permit exceedances because both solids and algae can harbor bacteria. Cleaning these nuisance items is standard practice for successful operation of UV systems. Solids accumulation and algae growth will occur in most UV reactors, with tertiary filtration reducing the amount of solids deposition. Solids accumulation and growth will also occur on the UV modules, outside of the arc length of the lamps; growth has also been observed downstream of the UV systems on the effluent weirs. Thus, to minimize the effect of solids and algae, the UV channels (and pressurized UV reactors) should be cleaned on a schedule.

TEST YOUR KNOWLEDGE

1. **This commonly used chemical in WRRFs is known to reduce UV transmittance.**
   a. Sodium hypochlorite
   b. Ferric chloride
   c. Sodium hydroxide
   d. Citric acid
2. **Two common methods for conserving energy in UV disinfection are**
   a. Flow pacing and dose pacing
   b. Supplemental chlorine addition and flow pacing
   c. Bypassing wet weather flows and adding chlorine
   d. Reducing cleaning cycles and watching television
3. **How often should UV lamp run time values be monitored?**
   a. Daily
   b. Weekly
   c. Monthly
   d. Semiannualy

4. **Ultraviolet systems should only be started after**
   a. Feed pumps have been started.
   b. Ballasts are fully de-energized.
   c. Channel covers are in place.
   d. All lamps are completely submerged.

5. **Ultraviolet lamps should be replaced**
   a. When they reach a predetermined number of service hours.
   b. When excessive fouling occurs that cannot be removed.
   c. When algae growth penetrates the quartz sleeves.
   d. When the lamp intensity falls below 90%.

6. **For reclaimed water systems, NWRI requires intensity sensors be calibrated**
   a. Daily
   b. Weekly
   c. Monthly
   d. Annually

7. **Intensity sensors must be cleaned frequently**
   a. To ensure accuracy
   b. To remove scale
   c. To prevent drift
   d. All of the above

8. **Which of the following chemicals is often used for large-scale cleaning of UV lamps?**
   a. Reagent-grade sulfuric acid
   b. Food-grade citric acid
   c. Sodium hypochlorite
   d. Ferric or ferrous chloride

TROUBLESHOOTING

Table 10.16 summarizes typical troubleshooting considerations for UV systems. Additional discussion on common areas that require troubleshooting is provided in the following paragraphs.

Table 10.16 Typical Troubleshooting Checklist for UV Disinfection Systems

Equipment	What to Check	Potential Problems	Corrective Actions
Ballasts	Surface temperature on ballasts during operation on normal utility power	Overheating resulting from poor panel ventilation	Add panel ventilation or cooling system.
	Surface temperature on ballasts during operation on standby power	Overheating resulting from power-distorting harmonics from electronic ballasts	Check power quality under various UV loads. May require addition of system or equipment to filter out harmonics.
UV system intensity meter	Lamp indicators	Burned-out lamp	Replace as needed.
		Wrong sequence	Respective lamps indicate sequence of individual components.
	Indicates chamber UV intensity	Buildup on quartz jacket	Clean routinely as necessary.
Intensity monitor	Photocell and electronic circuit, indicating meter alarm condition and pilot lights	Nonfunctional	Repair or replace.
UV lamps	Lamps	Burned out	Replace as necessary.
	Heat buildup or low pressure lamp exposure to air	Little or no flow	Increase water supply.
		Malfunction of level control system	Clean level sensors; repair or replace as necessary.
	Ground fault interrupter indicator	Broken quartz sleeve or seal system failure	Check sleeve(s) for breaks and leaks; dry contacts; replace lamp components as required.
Control box indicator lights	Amber	Low UV output	Clean chamber or replace lamp.
	Red	Poor water quality	Clean chamber or replace lamp. Check upstream processes.
Glad seal assembly	O-ring	Water leaks	Tighten the gland nut to compress O-ring, or replace.
Electrical service	AC volts, DC volts, ohms AC	Over range	Use multimeter and set ranges according to manufacturer's recommendation.
Lamp out warning system	Circuit board	Defective	Replace.
	Indicator bulb	Burned out	Replace.
	Pilot light	Burned out	Replace.

ELECTRICAL ISSUES

Issues with power and control circuits, including for the ballasts, lamps, and control systems, are common causes of UV problems. It is typically recommended that manufacturers' manuals be consulted for troubleshooting electrical issues. Proper precautions are essential; often, electrical troubleshooting requires a licensed electrician.

LOW ULTRAVIOLET TRANSMITTANCE

Low transmittance is almost always a consequence of poor performance of upstream treatment processes. Excessive solids loss, increased turbidity, and overdosing of ferric or ferrous chloride will all reduce UV transmittance. Poor UV system maintenance and inadequate cleaning will reduce UV intensity. In some instances, industrial effluents may contribute components with high UV absorbance, which can strongly affect the UVT of the combined domestic and industrial wastewater. When collecting UVT data, especially when troubleshooting low UVT, it is important to measure the parameter on both filtered and unfiltered samples to determine if dissolved constituents may be absorbing UV light (Swift et al., 2007). A number of industries have been implicated as discharging wastewater with high dissolved UV absorbance (low filtered UVT) as a result of the presence of organic compounds not readily degraded (Swift et al., 2001), including sunblock, coffee, pharmaceutical, and chemical manufacturers; centralized waste treatment facilities; and printed-circuit-board manufacturers. Several WRRFs have been documented as having potentially violated effluent disinfection standards as a result of the presence of certain organic compounds passing through the WRRF and lowering effluent UVT.

HYDRAULICS

Hydraulic issues are fairly common causes of problems with UV systems, particularly when upstream or downstream equipment or processes have changed. A common problem with open-channel systems is when the water level is too high or too low, as might occur if a level control gate is malfunctioning. Checking water level is often recommended when disinfection performance has decreased and problems with lamps, ballasts, or UVT have been ruled out.

TEST YOUR KNOWLEDGE

Examine the photographs from a poorly performing UV system. In your own words, describe two problems that can be seen and the effect each will have on effluent quality.

____

____

____

____

(Reprinted with permission by Indigo Water Group)

SAFETY CONSIDERATIONS

Safety is one of the most important items to consider in O&M of UV disinfection systems. Safety equipment provided by the manufacturer will need to be used (as described by the manufacturer) to ensure that injury does not occur.

ELECTRICAL HAZARDS

To prevent electrical hazards, all safety and operational precautions required by the current National Fire Protection Agency National Electric Code, OSHA, local electric codes, and the UV manufacturer should be followed and include the following precautions (U.S. EPA, 2006):

- Proper grounding,
- Lockout/tagout procedures,
- Use of proper electrical insulators, and
- Installation of safety cutoff switches.

ULTRAVIOLET RADIATION HAZARDS

Ultraviolet radiation can cause several safety issues like burns and damage to eyes and skin. To protect staff from injuries associated with exposure to UV light, the following precautious should be implemented:

- Although channel covers/grating are typically provided to prevent or minimize potential UV light exposure during operation of the system, warning signs regarding UV radiation should be posted;
- Eyes should never be exposed to UV radiation. It is important to not look directly into a UV lamp that is on. (It should be noted that, in closed-vessel UV systems, there is minimal exposure to UV light compared to open-channel UV systems, although caution must still be exercised if sensors are removed for inspection from sensor ports.);
- Personal protective equipment should include wraparound goggles or face shields that will absorb UV light. Protect hands, arms, and face from UV radiation by covering exposed skin;
- Potentially dangerous UV radiation exposure may result if lamps are operated in dry conditions without being submerged or properly covered. Ensure lamps are submerged to the manufacturer's recommended depth and covers are in place prior to startup and throughout system operation;
- Routine maintenance should be performed with UV lamps off, if possible; and
- Some test and maintenance procedures may require operating the system without proper shielding. If this is necessary, all persons in the area should wear protective gear and an alarm should warn persons who might be entering the area to stay out. Skin and eyes easily absorb UV radiation and are vulnerable to injury. Conjunctivitis, or welder flash, may occur within several hours. Although painful, conjunctivitis is usually temporary.

LIFTING HAZARDS

Ultraviolet modules and banks can be heavy and may require cranes or davits for lifting when performing routine maintenance. Although the design of lifting systems that are required for some UV systems should follow all local, state, and federal (e.g., National Electrical Code) guidelines, operators should also be trained on using manufacturer recommendations on removal of UV equipment. In all instances, caution should be exercised to avoid risk of injury when lifting and moving modules and banks. Cranes should only be operated by experienced staff who have received all necessary training for safely operating the cranes.

CHEMICAL HAZARDS

Cleaning solutions typically contain acids and should be handled with appropriate precautions. Cleaning solutions may contain phosphoric or other moderate (citric) to strong (hydrochloric) acids. Gloves and eye protection are necessary. Consult the relevant *SDSs* for specific recommendations on chemical handling and disposal.

Ultraviolet lamps contain mercury that can be released if the lamps are accidentally broken. Contact with mercury should be avoided. Released mercury must be contained with a mercury containment kit. Mercury disposal may fall under state and federal waste disposal rules; therefore, it is important to consult both manufacturer guidelines and local, state, and federal requirements regarding proper handling procedures and disposal for used, defective, and broken lamps. Lamps may not be sent to a municipal landfill. In some cases, disposal of released mercury can be done at the same time as UV lamp disposal.

Until 2012, *SDSs* were called material safety data sheets. These multipage documents summarize the properties of chemical substances, warn users of hazards, and provide guidance on safe handling, storage, and usage. Safety data sheets are part of the OSHA hazard communication standard.

OTHER HAZARDS

Other hazards include the following:

- Ultraviolet lamps remain hot for a considerable time after they have been turned off. Lamps should be allowed to cool off after use for at least 5 minutes for low-pressure lamps and 15 minutes for medium-pressure lamps;
- There is a risk of flooding during peak flow events in open-channel systems. Protection with interlocks to shut off electricity based on water level may be needed to protect equipment from water damage (short-circuiting, damaging equipment, and creating an electrocution hazard) if equipment is energized and is flooded;
- Tripping and slipping hazards may exist around UV channels. This is a risk especially when grates are removed; and
- Staff should take care in handing broken lamps because broken glass can pose a risk of staff being cut. Store lamps, quartz sleeves, and ballasts in separate dry areas on shelves to avoid breakage and to provide for safe, easy access. Repackage used lamps, sleeves, and ballasts for safe disposal.

Data Collection, Sampling, and Analysis

This section summarizes the different methods available for chlorine residual analysis. Analysis methods for indicator organisms are summarized in Table 10.1 at the beginning of this chapter. For complete, step-by-step laboratory procedures, refer to *Basic Laboratory Procedures for the Operator-Analyst*, 5th edition, that was published by the Water Environment Federation in 2012.

The current method of chlorine residual analysis most frequently used is the amperometric method. Some laboratories use the colorimetric technique. Colorimetric methods are frequently used when rapid field tests are required. The currently approved laboratory methods for measuring chlorine residual in wastewater are found in Part 4500-Cl of *Standard Methods*. Table 10.17 compares the various test methods.

Table 10.17 Chlorine Residual Testing Methods (Connell, 2002)

Method	Standard Methods Procedure Number	Residual Type	Interferences	Range	Comments
Iodometric I & II	4500-Cl B&C	Total	Manganese oxide, oxidizing, reducing agents, ferric, and nitrite ions	1 mg/L with start iodide or 40 mg/L with thiosulfate	Simple to perform
Amperometric	4500-Cl D	Free, combined or total, monochloramine, and dichloramine	Other halogens, chlorine dioxide, and nitrogen trichloride (trichloramine), high speed mixing	0.1 to 2 mg/L	Higher degree of training than test kits
Low-level Amperometric	4500-Cl E	Total	Same as above	10 to 100 mg/L	
N, N-diethyl-pphenylenediamine (DPD) ferrous titrimetric	4500-Cl F	Free, combined, or total	Strong oxidants, manganous oxide, copper, chromate	<0.1 mg/L	More difficult to perform, requires math calculations for different components
DPD colorimetric	4500-Cl G	Free, combined or total	Strong oxidants, color, turbidity, chromate	<0.1 mg/L	Subjective and lacks precision and accuracy as titration levels. May turn pink in the absence of chlorine.
FACTS	4500-Cl H	Free	Color, turbidity, and other oxidants	0.1 to 10 mg/L	Same as above
Iodometric electrode	4500-Cl I	Total	Oxidizing agents, manganese oxide, iodate, bromine, and cupric ions	<1 mg/L	No direct reading

SAMPLE COLLECTION

Effluent samples for indicator organisms are always grab samples and must be collected in sterile plastic or glass containers. A small amount of sodium thiosulfate powder or solution is added to each sampling container prior to sterilization. The sodium thiosulfate eliminates any chlorine residual present at the time of sampling. That way, the sample will not continue to disinfect while awaiting analysis. Sample analysis should begin as soon as possible after collection. When not analyzed immediately, samples should be cooled to less than 10 °C (50 °F). Samples for indicator organisms may be stored for a maximum of 8 hours from the time the sample is collected to the time that incubation begins.

Samples for chlorine residual are always grab samples. Acceptable sample containers are made from polyethylene or glass. Containers should be clean, but do not need to be sterilized. Chlorine residual samples must be analyzed within 15 minutes of collection.

TEST YOUR KNOWLEDGE

1. **Operators may safely look directly at UV lamps while they are operating.**
   - ☐ True
   - ☐ False
2. **Keeping UV channels covered reduces algae growth and operator exposure to UV light.**
   - ☐ True
   - ☐ False
3. **Welder flash or conjunctivitis occurs immediately after prolonged UV exposure.**
   - ☐ True
   - ☐ False
4. **Mercury containment kits should be used to clean up any mercury released from a broken UV lamp.**
   - ☐ True
   - ☐ False
5. **Samples collected for either indicator organisms or chlorine residual must be**
   a. Grab samples
   b. Composite samples
   c. Collected in sterile containers
   d. Contain sodium thiosulfate
6. **Chlorine residual samples must be analyzed**
   a. As soon as reasonably practicable.
   b. By a state certified laboratory.
   c. Within 15 minutes of collection.
   d. After cooling to less than 10 °C (50 °F).
7. **A small amount of sodium thiosulfate is added to sample bottles intended for *E. coli* analysis**
   a. As a preservative to increase hold time.
   b. To neutralize residual chlorine.
   c. To sterilize the sampling container.
   d. As a way to prevent *E. coli* from replicating.

CHAPTER SUMMARY

<table>
<tr><th></th><th>Chlorine Disinfection</th><th>UV Disinfection</th><th></th></tr>
<tr><td>PURPOSE</td><td colspan="2">■ Protect public health and the environment.
■ Inactivate pathogens.
■ Damaged pathogens can't reproduce and cause illness.</td><td>PURPOSE</td></tr>
<tr><td>INDICATOR ORGANISMS</td><td colspan="2">■ Testing for all pathogens is costly and time-consuming.
■ Indicator organisms live in the guts of warm-blooded animals. If they are present, then pathogens may also be present.
■ Look for haystacks instead of needles.
■ Indicator organisms are not pathogens.
■ Rapid detection of fecal contamination.
■ Laboratory results expressed as #CFUs/100 mL or MPN/100 mL.</td><td>INDICATOR ORGANISMS</td></tr>
<tr><td>THEORY OF OPERATION</td><td>■ Chlorine destroys proteins and critical enzymes required for pathogens to function.
■ Dechlorination neutralizes chlorine and protects the environment.
■ Chlorine is available as gaseous chlorine, liquid sodium hypochlorite, or solid calcium hypochlorite (also known as HTH).
■ All forms of chlorine produce hypochlorous acid and hypochlorite ion when added to water.
■ The percentage of each ion varies with pH and temperature.
■ Dose = Demand + Residual
■ Total Residual = Combined Residual + Free Residual
■ Ammonia combines with chlorine to form combined residual, but can also destroy chlorine residual.
■ The amount of pathogen inactivation achieved is a product of chlorine residual concentration and time (CT).
■ With dechlorination, sulfite ions reduce chlorine compounds to chloride. Sulfite becomes sulfate in the process.</td><td>■ UV light damages DNA.
■ Thymine dimers form.
■ Teeth in the DNA "zipper" are stuck together, so DNA can't be used or replicated.
■ UV radiation is generated by passing energy through mercury vapor.
■ Low-pressure and amalgam lamps produce light at 254 nm.
■ Medium pressure lamps produce light at multiple wavelengths.</td><td>THEORY OF OPERATION</td></tr>
<tr><td>DESIGN PARAMETERS</td><td>■ Dose depends on water quality.
■ Required CT varies from state to state.
■ Generally, 15 minutes of CT at peak hour design flow.
■ States may also require 30 minutes of CT at average daily design flow.</td><td>■ UV light intensity is the fluence (mW/cm^2).
■ UV light provided per unit of time is the fluence rate (mJ/cm^2 or mW·sec/cm^2).
■ Fluence rate is equal to UV dose.
■ Fluence rate depends on water quality.
■ Fluence rates required vary from one regulatory agency to the next.</td><td>DESIGN PARAMETERS</td></tr>
</table>

EXPECTED PERFORMANCE	■ Dependent on CT and effluent quality. ■ Capable of meeting <2 CFU/100 mL for *E. coli* and fecal coliforms. ■ Reported as the geometric mean.	■ Dependent on fluence rate and effluent quality. ■ Capable of meeting <2 CFU/100 mL for *E. coli* and fecal coliforms. ■ Reported as the geometric mean.
EQUIPMENT	■ Different types of systems include • Gas pressure • Gas vacuum • Combination gas pressure and vacuum • Liquid chlorine (100%) • Liquid hypochlorite solution ■ Cylinders and containers can be manifolded together for higher feed rates. ■ Pigtails connect cylinders and containers to manifolds or wall-mounted chlorinators. ■ Ancillary valve and vacuum regulator used with wall mounted chlorinators and sulfonators. ■ Evaporator used with liquid chlorine system converts liquid to gas. ■ Vacuum systems are safer than pressure and liquid chlorine and sulfur dioxide systems. ■ Hypochlorite systems consist of chemical storage tank, metering pump, and a mixing device or diffuser. ■ Chlorine contact chamber or pipeline with at least a 40 to 1 length-to-width ratio provides CT.	■ Mercury vapor lamps are grouped into modules. ■ Modules consist of one or more lamps and a ballast to provide power. ■ Modules grouped to form banks. ■ Typically two or more banks in series. ■ Lamps may be oriented vertically, horizontally, or at an angle relative to flow. ■ Lamps may be placed parallel to flow or across the flow path. ■ Wiper system prevents buildup on quartz sleeves. ■ Multiple types of sensors used to monitor and control UV light output: • UV intensity sensors • UV transmittance meters • Level and flow monitoring • Temperature monitoring ■ UV channel just wide enough to hold UV modules/banks.
CHEMICALS	■ Gaseous chlorine, sodium hypochlorite, or calcium hypochlorite for chlorination. ■ Sulfur dioxide or sulfite salts for dechlorination. ■ Chlorine and sodium hypochlorite are sold by trade percent. ■ Trade percent and percent available chlorine are equivalent. ■ Calcium hypochlorite is sold by weight percent of available chlorine. ■ Operators should be familiar with the chemical properties of each of these chemicals. ■ Dry and wet chlorine corrode different materials. ■ Whether chlorine is defined as wet depends on both water concentration and temperature. ■ Hypochlorite solutions should not be stored for >30 days.	■ Wiper systems cannot prevent all buildup on quartz sleeves. ■ Lamps may be cleaned manually or by soaking in a dip tank. ■ Dip tanks are filled with dilute phosphoric acid, citric acid, bathroom cleaner, or other dilute acid. ■ Cleaning protocols are manufacturer specific. ■ Some systems use air scour to clean lamps without removing them from the channel. ■ Air scour extends time between manual or dip tank cleanings.

CONTAINERS	■ Three size containers for chlorine gas and sulfur dioxide gas: 68 kg (150 lb), 907 kg (1 ton), and rail cars (up to 81 000 kg [90 tons]). ■ Filled to 80% capacity to allow for liquid expansion. ■ Equipped with fusible plugs to relief pressure. ■ 68-kg (150-lb) cylinders provide only gas. ■ 907-kg (1-ton) containers provide either liquid or gaseous chlorine through eduction pipes. ■ Rail cars provide only liquid.	■ Three types of UV lamps—low-pressure, amalgam, and medium-pressure. ■ Low pressure and amalgam lamps generate UV light at 254 nm. ■ Medium pressure lamps generate UV light at multiple wavelengths. ■ Lamps are encased in quartz sleeves. ■ Lamps must be completely submerged when operating.	LAMPS
CHLORINATORS/SULFONATORS	■ In combination with an injector, converts gas pressure to vacuum. ■ Rotometer indicates gas feed rate. ■ Diaphragm closes and shuts off gas supply in the event of a leak. ■ Different materials of construction for chlorine versus sulfur dioxide. ■ Direct mount or wall mounted units. ■ Vacuum regulator typically used with wall mounted chlorinators and sulfonators. ■ Lead washer forms gas tight seal between cylinder or container and chlorinator/sulfonator. ■ Venturi at the point of injection uses velocity of water through a narrow opening to create vacuum. ■ Vacuum pulls on the chlorinator diaphragm to open it, drawing gas into the water through the injector. ■ Automatic switchover devices help maintain continuous chemical addition.		
VAPORIZERS	■ Gas withdrawal limited by available heat. ■ Liquid withdrawal from cylinders and containers is not limited by ice formation. ■ Convert liquid chlorine and sulfur dioxide to gas. ■ Heated water bath boils chlorine or sulfur dioxide at a preset temperature. ■ Vaporizers do not increase system pressure.		
HYPOCHLORINATORS	■ Concentrated hypochlorite solution is stored in a plastic or fiberglass tank. ■ Smaller systems may also have a day tank of lower strength solution. ■ Chemical metering pumps transfer the hypochlorite solution to the water being treated. ■ Mixing is required before or at the beginning of the chlorine contact chamber.		
TABLET FEEDERS	■ Calcium hypochlorite tablets are stacked in a tube. ■ Water flowing through the bottom dissolves the tablets. ■ Concentrated hypochlorite solution generated.		

CONTACT TIME	■ 15 minutes of CT at design peak hour flow ■ 30 minutes of CT at design average daily flow	■ Contact times for UV systems may only be a few seconds. ■ Turbulent flow past the UV lamps ensures all of the water is exposed to UV.	CONTACT TIME
PROCESS VARIABLES	■ Chlorine efficacy is determined by CT and chlorine residual concentration. ■ Cold water is harder to disinfect than warm. ■ High pH water is harder to disinfect than low. ■ Poor performance of upstream processes increases chlorine demand. ■ Adjust CT to meet effluent goals.	■ UV efficacy is determined by UV light transmittance (dose received by the microorganisms) and CT. ■ Dose in UV systems is the fluence rate. ■ Anything that absorbs or scatters UV light will decrease disinfection efficiency. ■ Poor performance of upstream processes reduces transmittance. ■ Adjust UV fluence to meet effluent goals.	PROCESS VARIABLES
PROCESS CONTROL	■ Multiple methods to control chlorine feed rate. ■ Manual control is simplest. ■ Flow pacing adjusts chlorine dose to match flowrate. Changes in demand not taken into account. ■ Residual control may use ORP to adjust flowrate. ■ Compound loop control uses both flow and chlorine residual signals to adjust dose.	■ Flow and dose pacing used by most UV systems. ■ Power is increased or decreased to increase or decrease UV light production. ■ Transmittance and intensity measurements may be used to adjust the dose.	PROCESS CONTROL
OPERATION	■ Monitor effluent residual concentration. ■ Adjust chlorine dose to meet effluent goals. ■ Monitor cylinder and container weights daily. ■ Never reuse washers. ■ Watch for condensation or icing. Correct as needed. ■ Monitor equipment for proper operation.	■ Operate UV systems in accordance with the manufacturer's recommendations. ■ Replace lamps when transmittance limit or run time hours are reached. ■ Ensure routine maintenance is completed.	OPERATION
MAINTENANCE	■ Replace Monel-lined pigtails when a crinkling sound is heard. ■ Replace clear plastic pigtails when discolored. ■ Replace gas lines if kinking or other damage is observed. ■ Test safety equipment, alarm circuitry, and leak monitoring equipment regularly. ■ Remove hypochlorite deposits with dilute hydrochloric acid. ■ Keep debris and settled solids from accumulating in the contact chamber.	■ Check and calibrate sensors weekly. ■ Clean quartz sleeves as needed to maintain UV transmittance. ■ Monitor lamp life and replace as needed. ■ Replace ballasts when they fail. ■ Clean UV channel regularly to remove deposits and algae.	MAINTENANCE
SAFETY	■ Reporting requirements under EPCRA, also known as SARA, Title III. ■ OSHA's Process Safety Management Rule has emergency planning requirements for facilities with 681 kg (1500 lbs) of chlorine or 453 kg (1000 lbs) of sulfur dioxide on site at any time. ■ Chlorine exposure (PEL) should not exceed 0.5 ppmv on average over an 8-hour work day. ■ Short-term exposure limit (15 min) of 1 ppmv is acceptable. ■ IDLH concentration for chlorine is 10 ppmv. ■ Exposure to 30 ppmv of chlorine causes chest pain and coughing. ■ Concentration of 1000 ppmv fatal after a few minutes. ■ Sodium and calcium hypochlorite solutions can cause chemical burns. ■ Wear gloves, goggles, and other personal protective equipment to prevent exposure.	■ UV radiation can cause skin burns and eye damage. ■ Keep UV channels covered to reduce UV exposure. ■ Never look directly into an operating UV lamp. ■ Keep lamps submerged at all times when in operation. ■ Cleaning solutions can cause chemical burns. ■ Wear gloves, goggles, and other personal protective equipment to prevent exposure. ■ Mercury can be released when UV lamps are accidentally broken. ■ Clean up released mercury with a mercury containment kit.	SAFETY

References

American Water Works Association (2006) *Water Chlorination and Chloramination Practices and Principles,* 2nd ed.; AWWA Manual M20; American Water Works Association: Denver, Colorado.

Bernstein, L. A. (1981) *Biological Influences on Environmental Toxicity*. Deeds and Data, Water Pollution Control Federation, **18** (5).

Black & Veatch Corporation (2010) *White's Handbook of Chlorination and Alternative Disinfectants,* 5th ed.; Wiley & Sons: Hoboken, New Jersey.

Colorado Department of Public Health and Environment, Water Quality Control Division (2012) *State of Colorado Design Criteria for Domestic Wastewater Treatment Works*; WPC-DR-1; Colorado Department of Public Health and Environment, Water Quality Control Division: Denver, Colorado.

Compressed Gas Association (2017) *Sulfur Dioxide,* 6th ed.; CGA G-3; Compressed Gas Association: Arlington, Virginia.

Connell, G. (2002) *The Chlorination/Dechlorination Handbook*; Water Environment Federation: Alexandria, Virginia.

De Nora Water Technologies (2015, Oct) The Uniform Fire Code: Realities and Options: 005.9101.3; De Nora Water Technologies, Capital Controls: Colmar, Pennsylvania.

Great Lakes Upper Mississippi River Board of State and Provincial Public Health and Environmental Managers (2014) *Recommended Standards for Wastewater Facilities: Policies for the Design, Review, and Approval of Plans and Specifications for Wastewater Collection and Treatment Facilities*; Minnesota Department of Administration: St. Paul, Minnesota.

Marske, D. M.; Boyle, V. D. (1973) Chlorine Contact Chamber Design—A Field Evaluation. *Water and Sewage Works*, 120, 70.

Metcalf & Eddy, Inc./AECOM (2014) *Wastewater Engineering: Treatment and Resource Recovery*; McGraw Hill Education: New York.

Morris, J. (1966) The Acid Ionization Constant of HOCl from 5 °C to 35 °C. *J. Phys. Chem.*, **70** (12), 3798–3806.

National Fire Protection Association (2016) *Standard for Fire Protection in Wastewater Treatment and Collection Facilities*; NFPA 820; National Fire Protection Association: Quincy, Massachusetts.

National Water Research Institute and Water Research Foundation (2012) *Ultraviolet Disinfection Guidelines for Drinking Water and Water Reuse,* 3rd ed.; National Water Research Institute and Water Research Foundation: Fountain Valley, California.

PVC Pipe Supplies (2017) *What Are the Differences Between CPVC and PVC?* PVC Pipe Supplies. http://pvcpipesupplies.com/cpvc-vs-pvc (accessed Oct 15, 2017).

Swift, J.; Wilson, J.; Hunter, G. (2007) Implementing Local Limits for the Control of WWTP Effluent Ultraviolet Transmittance. *Proceedings of the 80th Annual Water Environment Federation Technical Exposition and Conference* [CD-ROM]; San Diego, California, Oct 13–17; Water Environment Federation: Alexandria, Virginia.

Swift, J.; Wilson, J.; Johnson, M.; Jacobsen, B. (2001) The Impact of UV-Absorbing Wastewater from Printed Circuit Board Manufacturing Facility on the Performance of a Municipal UV Disinfection System. *Proceedings of the 74th Annual Water Environment Federation Technical Exposition and Conference* [CD-ROM]; Atlanta, Georgia, Oct 13–17: Water Environment Federation: Alexandria, Virginia.

Teledyne Isco (2014) *Teledyne Isco Open Channel Flow Measurement Handbook*, 7th ed.; Teledyne Isco, Inc.: Lincoln, Nebraska.

The Chlorine Institute (1997) *The Chlorine Manual; Pamphlet I,* 6th ed.; The Chlorine Institute: Washington, D.C.

The Chlorine Institute (2000) *Sodium Hypochlorite, Safety and Handling (Pamphlet 96),* 2nd ed.; The Chlorine Institute: Washington, D.C.

The Chlorine Institute (2014) *Chlorine Basics (Pamphlet 1),* 8th ed.; The Chlorine Institute: Arlington, Virginia.

The Chlorine Institute (2017) *Gaskets for Chlorine Service (Pamphlet 95),* 5th ed., Revision 2; The Chlorine Institute: Arlington, Virginia.

U.S. Environmental Protection Agency (1978a) *Field Manual for Performance Evaluation and Troubleshooting at Municipal Wastewater Treatment Facilities*; EPA-0430/9-78-001; U.S. Environmental Protection Agency, Office of Water Program Operations: Washington, D.C.

U.S. Environmental Protection Agency (1978b) *Full-scale Demonstration of Nitrogen Removal by Breakpoint Chlorination: Environmental Protection Technology Series*; EPA-600/2-78-029; U.S. Environmental Protection Agency, Municipal Environmental Research Laboratory: Cincinnati, Ohio.

U.S. Environmental Protection Agency (1978c) *Microbiological Methods for Monitoring the Environment, Water, and Wastes*; EPA-600/8-78-017; U.S. Environmental Protection Agency: Washington, D.C.

U.S. Environmental Protection Agency (1986) *Design Manual: Muncipal Wastewater Disinfection*; EPA-625/1-86-021; U.S. Environmental Protection Agency, Office of Research and Development, Water Engineering Research Laboratory, Center for Environmental Research Information: Cincinnati, Ohio.

U.S. Environmental Protection Agency (1992) *Ultraviolet Disinfection Technology Assessment*; EPA-832/R-92-004; U.S. Environmental Protection Agency: Washington, D.C.

U.S. Environmental Protection Agency (2006) *Ultraviolet Disinfection Guidance Manual for the Final Long Term 2 Enhanced Surface Water Treatment Rule*; EPA-815/R-06-007; U.S. Environmental Protection Agency: Washington, D.C.

U.S. Environmental Protection Agency (2012) *2012 Recreational Water Quality Criteria*; EPA-820/F-12-061; U.S. Environmental Protection Agency, Office of Water: Washington, D.C.

Wallace & Tiernan (2017) *Series V-2000 Wall-Mounted Chlorinator*; Technical Data Sheet 25.056; Wallace & Tiernan: Belleville, New Jersey.

Washington State University (2017) *Emerging Technologies*. http://e3tnw.org/ItemDetail.aspx?id=13 (accessed Oct 12, 2017).

Water Environment Federation (2006) *Wastewater Disinfection Training Manual*; Water Environment Federation: Alexandria, Virginia.

Water Environment Federation (2016) *Operation of Water Resource Recovery Facilities,* 7th ed.; WEF Manual of Practice No. 11; Water Environment Federation: Alexandria, Virginia.

Water Environment Federation; American Society of Civil Engineers; Environmental and Water Resources Institute (2018) *Design of Water Resource Recovery Facilities,* 6th ed.; WEF Manual of Practice No. 8/ASCE Manuals and Reports on Engineering Practice No. 76; Water Environment Federation: Alexandria, Virginia.

Water Environment Federation; International Ultraviolet Association (2015) *Ultraviolet Disinfection for Wastewater—Low-Dose Application Guidance for Secondary and Tertiary Discharges*; Water Environment Federation: Alexandria, Virginia

Water Environment Research Foundation (2007) *Disinfection of Wastewater Effluent—Comparison of Alternative Technologies*; Report No. 04-HHE-4; Water Environment Research Foundation: Alexandria, Virginia.

World Health Organization (2003) *Guidelines for Safe Recreational Water Environments*; *Volume 1: Coastal and Fresh Waters.* http://www.who.int/water_sanitation_health/bathing/srwe1/en (accessed Jan 2015).

Suggested Readings

Arasmith, S. (1997) *O&M of Chlorine Systems*; ACR Publications, Inc.: Albany, Oregon. http://www.acrp.com.

Loge, F.; Emerick, R.; Thompson, D.; Nelson, D.; Darby, J. (1999) Factors Influencing Ultraviolet Disinfection Performance Part I: Light Penetration to Wastewater Particles. *Water Environ. Res.*, **71**, 377–381.

The Chlorine Institute (1999) *Water and Wastewater Operators Chlorine Handbook*, 1st ed.; The Chlorine Institute: Washington D.C.

The Chlorine Institute (2000) *Emergency Response Plans for Chlorine Facilities*, 5th ed.; The Chlorine Institute: Washington, D.C.

List of Acronyms

CFU	= colony forming unit
°C	= degrees Celsius
°F	= degrees Fahrenheit
μ	= specific growth rate of microorganisms, represented by Greek letter μ (pronounced mew)
μm	= micrometer = 0.001 mm
μ_{max}	= maximum growth rate of microorganisms when saturated (not limited) by substrate
μS	= microsiemens
A^2/O	= anaerobic, anoxic, oxic activated sludge process
ABC	= Association of Boards of Certification
ABS	= acrylonitrile-butadiene-styrene
ac	= acre
ACGIH	= American Conference of Governmental Industrial Hygienists
ADF	= average daily flow
aerobic	= in the presence of oxygen
AIT	= analysis indicator transmitter
anaerobic	= without oxygen, nitrite, or nitrate
anoxic	= without oxygen, but in the presence of nitrite or nitrate
AOB	= ammonia-oxidizing bacteria
ASME	= American Society of Mechanical Engineers
atm	= atmosphere
ATP	= adenosine triphosphate
AWT	= advanced wastewater treatment
BFP	= belt filter press
BNR	= biological nutrient removal
BOD	= biochemical oxygen demand
BOD_5	= biochemical oxygen demand (5-day)
BrINClHOF	= mnemonic device for remembering the elements that form two atom molecules with themselves in nature
$C_6H_{12}O_6$	= sugar
CAC	= combined available chlorine
CBOD	= carbonaceous biochemical oxygen demand
$CBOD_5$	= carbonaceous biochemical oxygen demand (5-day)
CCTV	= closed-circuit television
CFR	= Code of Federal Regulations
cfs	= cubic feet per second
CFU/100 mL	= colony forming units per 100 milliliters
CI	= Chlorine Institute
CIU	= categorical industrial user
cm	= centimeter
CO_2	= carbon dioxide
COC	= chain-of-custody
COD	= chemical oxygen demand
CPVC	= chlorinated polyvinyl chloride
CT	= contact time
cu ft	= cubic feet
CWA	= Clean Water Act
DAFT	= dissolved air flotation thickener
DBP	= disinfection byproduct
DMR	= discharge monitoring report
DNA	= deoxyribonucleic acid
DO	= dissolved oxygen
DOT	= Department of Transportation (U.S.)
DT	= detention time
EBPR	= enhanced biological phosphorus removal
effluent	= treated wastewater
EPCRA	= Emergency Planning and Community Right-to-Know Act
EPS	= extracellular polymeric substances
F/M	= food-to-microorganism ratio
FAC	= free available chlorine
FIT	= flow indicator transmitter
FOG	= fats, oils, and grease
FRP	= fiber-glass-reinforced plastic (may also mean fiber-glass-reinforced polyester)
ft	= feet
ft/sec	= feet per second
g	= gram
gal	= gallon
gpd	= gallons per day
gpd/cap	= gallons per capita per day (per person)
gpd/sq ft	= gallons per day per square foot
gpm	= gallons per minute
gpm/sq ft	= gallons per minute per square foot
GSA	= Gould sludge age
H_2O	= water
H_2S	= hydrogen sulfide
ha	= hectare
HAA	= haloacetic acid
HCO_3^{-2}	= bicarbonate ion
HDPE	= high-density polyethylene
HDT	= hydraulic detention time
HHR	= high-resolution redox
HLR	= hydraulic loading rate
HMA	= hot mix asphalt
HRT	= hydraulic retention time
HTH	= high-test hypochlorite
IDLH	= immediately dangerous to life of health
IMLR	= internal mixed liquor recycle
in.	= inch
influent	= raw wastewater entering the WRRF
JHB	= Johannesburg activated sludge process
kg	= kilogram
kg/ha·d	= kilograms per hectare per day
kg/m^3	= kilograms per cubic meter

kPa = kilopascal, a unit of pressure
K_s = half-saturation coefficient or half-saturation constant
L = liter
L/cap·d = liters per capita (per person) per day
L/d = liters per day
lb = pound
lb/ac/d = pounds per acre per day
lb/cu ft = pounds per cubic foot
lb/1000 cu ft/d = pounds per thousand cubic feet per day
lb/1000 sq ft/d = pounds per thousand square feet per day
lb/gal = pounds per gallon
LOTO = lockout/tagout
m = meter
m/s = meters per second
m^3 = cubic meter
m^3/d = cubic meters per day
$m^3/(m^2{\cdot}d)$ = cubic meters per square meters per day
MCRT = mean cell residence time
Me_{dose} = metal salt dose
mg = milligram
mg/L = milligrams per liter
mgd = million gallons per day
mil. gal = million gallons
mJ/cm^2 = millijoules per square centimeters
ML/d = megaliters per day
mL/min = milliliters per minute
MLE = modified Ludzack-Ettinger activated sludge process
MLSS = mixed liquor suspended solids
MLVSS = mixed liquor volatile suspended solids
mm = millimeter = 0.001 m
MPN/100 mL = most probable number per 100 milliliters
MSDS = material safety data sheet
MUG = 4-methylumbelliferyl-beta-D-glucuronide
mV = millivolts
mW/cm^2 = milliwatts per square centimeters
$mW{\cdot}sec/cm^2$ = milliwatt·seconds per square centimeters
N_2 = nitrogen gas
NDMA = nitrosodimethylamine
NH_3 = ammonia or free ammonia
NH_3-N = ammonia-nitrogen (as N)
NH_4^+ = ammonium ion
NH_4^+-N = ammonium-nitrogen ion (as N)
NH_4Cl = ammonium chloride
NIOSH = National Institute of Occupational Health and Safety
nm = nanometer = 0.001 μm
NO_2^- = nitrite ion
NO_2 = nitrite
NO_2-N = nitrite-nitrogen (as N)
NO_3^- = nitrate ion
NO_3 = nitrate
NO_3-N = nitrate-nitrogen (as N)
NOB = nitrite-oxidizing bacteria
NPDES = National Pollutant Discharge Elimination System
NTF = nitrifying trickling filter
NTU = nephelometric turbidity unit
NWRI = National Water Resources Institute
O&M = operations and maintenance
O_2 = oxygen
OLR = organic loading rate
ORP = oxidation–reduction potential
OSHA = Occupational Safety and Health Administration
OTR = oxygen-transfer rate
OUR = oxygen uptake rate
P = phosphorus
PACl = polyaluminum chloride, also shown as PAX
PAO = phosphate-accumulating organism
PE = polyethylene
PEL = permissible exposure level
PFRP = process to further reduce pathogens
PHB = poly-β-hydroxybutyrate
P_{ini} = initial soluble phosphorus concentration
PLC = programmable logic controller
PMDAS = mnemonic device for remembering the order of mathematical operations when solving equations—“Poly Makes Donuts After School” stands for parenthesis and powers, multiply, divide, add, and subtract
PO_4 = phosphate
PO_4^{-3} = phosphate ion
PO_4-P = phosphate as phosphorus (as P)
PP = polypropylene
ppd = pounds per day
ppm = parts per million
ppmv = parts per million per volume of air
PRV = pressure-reducing valve
psi = pounds per square inch
psig = pounds per square inch gauge
PSRP = process to significantly reduce pathogens
PTFE = polytetrafluorethylene
PVC = polyvinyl chloride
PVdC = polyvinyldiene chloride
RAS = return activated sludge
RBC = rotating biological contactor
RR = respiration rate
RW = Redwood media
S = concentration of limiting nutrient or substrate
S = elemental sulfur
SARA = Superfund Amendment and Reauthorization Act
sBOD = soluble BOD
SBR = sequencing batch reactor
SCADA = supervisory control and data acquisition
SCBA = self-contained breathing apparatus
SDS = safety data sheet (formerly material safety data sheet or MSDS)

SIU = significant industrial user
SK = flushing intensity (Spülkraft)
SLR = solids loading rate
SO_4^{-2} = sulfate ion
SOR = surface overflow rate
SOUR = specific oxygen uptake rate
SPDES = State Pollutant Discharge Elimination System
SRB = sulfate-reducing bacteria
SRT = solids retention time
$SRT_{aerobic}$ = aerobic solids retention time
SSA = specific surface area
SSC = settled sludge concentration
SSV = settled sludge volume
SSV_{30} = settled sludge volume *after 30 minutes*
SSV_5 = settled sludge volume *after 5 minutes*
STEL = short-term exposure limit
STV = statistical threshold value
SU = standard units, used for pH
SVI = sludge volume index
TDS = total dissolved solids
THM = trihalomethane
TIN = total inorganic nitrogen
TKN = total Kjeldahl nitrogen (pronounced kell-doll)
TLV = threshold limit value
TOC = total organic carbon
TP = total phosphorus
TRC = total residual chlorine
TS = total solids
TSS = total suspended solids
TVSS = total volatile suspended solids
TWA = time-weighted average
U.S. EPA = U.S. Environmental Protection Agency
UFC = Uniform Fire Code
UV = ultraviolet (radiation)
UVT = ultraviolet transmittance
vector = a transmitter of disease
VF = vertical flow media
VFA = volatile fatty acid
VFD = variable-frequency drive
VOC = volatile organic compound
VSS = volatile suspended solids
WAS = waste activated sludge
WET testing = whole effluent toxicity testing
WRF = Water Research Foundation
WRRF = water resource recovery facility (formerly wastewater treatment plant, or WWTP)
XF = crossflow media
Y = yield, g VSS/g substrate

Answer Keys

CHAPTER 1

TEST YOUR KNOWLEDGE, PAGE 4

1. False. Natural treatment systems have a limited capacity to assimilate pollutants. Urban areas concentrate a large number of people and businesses into a small area and discharge them into one or a few locations. The result is reduced dissolved oxygen, fish kills, and other adverse effects.
2. True
3. True
4. False, most come from animal sources like bacon.
5. B
6. C
7. D
8. A
9. D
10. A
11. A
12. C

TEST YOUR KNOWLEDGE, PAGE 8

1. True
2. True
3. True
4. False. The Federal penalties shown in Table 1.2, including jail time, only apply when a person knowingly and willfully falsifies, conceals, or covers up a permit violation or misrepresents a permit violation with fraudulent statements.
5. False. Permit applications and required reports must be signed by an authority state-licensed individual of responsible charge, a principal executive officer, or ranking elected official.
6. True
7. B
8. A
9. D

TEST YOUR KNOWLEDGE, PAGE 14

1. True
2. False. Trash racks have wide openings to remove trash and large debris.
3. A
4. D
5. B
6. A
7. B
8. C
9. A

TEST YOUR KNOWLEDGE, PAGE 21

1. True
2. True
3. True
4. False. The purpose of secondary treatment is to increase the size of particles remaining in the influent after screening, degritting, and primary sedimentation. The biological solids grown during secondary treatment must be separated from the treated wastewater before discharge using a secondary clarifier or other solids separation process.
5. False. Lagoons are lined to prevent contamination of underlying groundwater. They may be lined with natural clay (bentonite) or with a synthetic liner.
6. True
7. False. Trickling filters are so named because the wastewater is trickled down over the media surface. Free space between pieces of media is required for air to penetrate and provide oxygen to the microorganisms that make up the biofilm.
8. True
9. False. Disinfection reduces the numbers of bacteria and pathogens in the final effluent, but it does not sterilize the wastewater. Sterilization is the complete destruction or inactivation of all living things.
10. B
11. B
12. B
13. C
14. B
15. A
16. C
17. C
18. B
19. 1 = A, 2 = D, 3 = B, 4 = E, 5 = C
20. 1 = C, 2 = E, 3 = B, 4 = D, 5 = A

TEST YOUR KNOWLEDGE, PAGE 28

1. True
2. True
3. False. Biosolids are regulated under Section 503 of Title 40 of the Code of Federal Regulations.
4. True
5. Secondary clarifier, thickening, digestion, dewatering
6. A
7. C
8. A
9. A
10. C
11. A
12. D
13. B
14. 1 = C, 2 = D, 3 = B, 4 = E, 5 = A
15. C

CHAPTER 2

TEST YOUR KNOWLEDGE, PAGE 40

1. False. The Pretreatment Act does not distinguish between commercial and industrial users. It only defines domestic and industrial wastewater.
2. True
3. True
4. D
5. A
6. D

TEST YOUR KNOWLEDGE, PAGE 50

1. True
2. False. An influx of stormwater would likely decrease the influent temperature.
3. True
4. False. While some conductivity meters give a TDS result, it is not a true measurement. TDS is calculated by applying a factor to the conductivity measurement. Although it is possible to build a correlation between conductivity and TDS for a particular wastewater, the only way to accurately measure TDS is by drying down a sample of filtered wastewater and weighing the residue left behind.
5. C
6. B
7. D
8. A
9. B
10. A
11. B
12. C
13. B

CHAPTER 3

TEST YOUR KNOWLEDGE, PAGE 57

1. False. Headworks may contain all or none of these depending on the size and complexity of the WRRF.
2. False. Flow measurement can come before either screening or grit removal.
3. B
4. C
5. D

TEST YOUR KNOWLEDGE, PAGE 60

1. True
2. True
3. True
4. C
5. A
6. C. The amount of screened material doubles for every 13-mm (0.5-in.) reduction of clear opening size. Here, the size was reduced twice so the amount of screenings will go up by a factor of 4.
7. A
8. D = micro screen, A = fine screen, C = bar screen, B = trash rack

TEST YOUR KNOWLEDGE, PAGE 72

1. True
2. False. Mechanical cleaning reduces labor costs.
3. False. Fine screens capture more of everything than coarse screens, including organic material.
4. True
5. A
6. 1 = E, 2 = G, 3 = A, 4 = B, 5 = F, 6 = C, 7 = H, 8 = D
7. D
8. C

TEST YOUR KNOWLEDGE, PAGE 78

1. False. Combined sewers bring both sewage and stormwater to the WRRF. After a storm, expect accumulated debris to be flushed from the collection system and into the WRRF. Cleaning frequency should increase.
2. True
3. B
4. B
5. A = 3, B = 1, C = 2, D = 6, E = 5, F = 4, G = 7 (B, C, A, F, E, D, G)
6. A

TEST YOUR KNOWLEDGE, PAGE 80

1. False. The volume of grit removed is highly dependent on the activities in the service area, condition of the collection system, type of collection system, and other factors.
2. True
3. C
4. A
5. D
6. B

TEST YOUR KNOWLEDGE, PAGE 89

1. False. Velocity is calculated by dividing the flowrate by the cross-sectional area. Alternatively, velocity can also be calculated by dividing the distance traveled by time.
2. True
3. C. To solve this problem, divide the current velocity by the desired velocity. The current velocity is 0.8 m/s (2.6 ft/sec). The maximum desired velocity is 0.3 m/s (1 ft/sec). 0.8 ÷ 0.3 = 2.7 basins in service. The minimum desired velocity is 0.24 m/s (0.8 ft/sec). 0.8 ÷ 0.24 = 3.3 basins in service.
4. D
5. A
6. C
7. B
8. D

TEST YOUR KNOWLEDGE, PAGE 93

1. True
2. False. The pump run time should be coordinated with both the time needed to fill the hopper and the pumping time needed to empty the hopper.
3. False. The mechanism should be started as soon as the flights are covered with water. Waiting until the basin is full will result in excessive amounts of grit accumulating around the flights. When finally started, the mechanism may not be able to move at all.
4. B
5. A
6. C
7. B

TEST YOUR KNOWLEDGE, PAGE 98

1. True
2. False. Septage tends to be higher strength than domestic wastewater.
3. True
4. False. Grease trap waste is considered industrial and commercial waste, not domestic. It is regulated under 40 CFR Part 257.
5. C
6. B
7. D

CHAPTER 4

TEST YOUR KNOWLEDGE, PAGE 107

1. False. The clarifier can only remove the settleable portion of TSS and BOD_5.
2. D
3. B

TEST YOUR KNOWLEDGE, PAGE 109

1. True
2. False. The particle with more surface area will settle slower, not faster.
3. False. Compaction is complete after about 2 hours.
4. C
5. D
6. B
7. C
8. A

TEST YOUR KNOWLEDGE, PAGE 111

1. False. The percent removal depends on the amount of particulate and soluble BOD_5 entering the clarifier.
2. True
3. True
4. B
5. D
6. B

TEST YOUR KNOWLEDGE, PAGE 113

1. True
2. True
3. False. Most clarifiers have 0.5 to 0.7 m (1.5 to 2.0 ft) of freeboard.
4. C
5. B
6. A

TEST YOUR KNOWLEDGE, PAGE 115

1. True
2. False. The feedwell reduces inlet velocity.
3. B
4. A

TEST YOUR KNOWLEDGE, PAGE 121

1. False. Weirs for circular clarifiers are typically v-notch weirs.
2. True
3. False. The drive is typically mounted on the center platform.
4. D
5. C
6. A
7. B

TEST YOUR KNOWLEDGE, PAGE 126

1. True
2. True
3. True
4. B
5. D
6. C
7. B
8. C
9. A

TEST YOUR KNOWLEDGE, PAGE 132

1. False. Flow is longitudinal from one end to the other.
2. True
3. True
4. B
5. D
6. D
7. A
8. B
9. C
10. A
11. D
12. C
13. C

TEST YOUR KNOWLEDGE, PAGE 140

1. True
2. False. *Surface overflow rate* is another name for velocity specific to clarifiers.

3. 4805.6 m^3 (169 560 cu ft)
4. 1.8 hours
5. 28.8 $m^3/m^2 \cdot d$ (707.7 gpd/sq ft)
6. D
7. A
8. A
9. C
10. D
11. B

TEST YOUR KNOWLEDGE, PAGE 148

1. True
2. True
3. 1400 kg/d (3083 lb/d)
4. 63.7 m^3/d (16 828 gpd)
5. 46.4 minutes
6. B
7. C
8. A
9. D
10. B

TEST YOUR KNOWLEDGE, PAGE 151

1. True
2. False. Recycle flows should be returned when they have the least effect. Returning during peak hour flow will increase the SOR, possibly to unacceptably high levels.
3. C
4. D
5. B

TEST YOUR KNOWLEDGE, PAGE 155

1. False. Some sludge collection mechanisms have water-lubricated bearings. These mechanisms should only be started when they are submerged.
2. True
3. False. Sludge left in lines will continue to break down. Gasses produced may build pressure and can damage equipment and lines.
4. C
5. A
6. D
7. B

TEST YOUR KNOWLEDGE, PAGE 158

1. True
2. B
3. A
4. C

TEST YOUR KNOWLEDGE, PAGE 161

1. A
2. C
3. B
4. D

CHAPTER 5

TEST YOUR KNOWLEDGE, PAGE 180

1. True
2. False. A microscope is needed to see individual bacteria.
3. False. Some material is biodegradable, but takes longer than 5 days to break down and will not be measured by the 5-day test.
4. False. Liquid is absorbed, not adsorbed, into a sponge.
5. C
6. B
7. C
8. D
9. B
10. A

TEST YOUR KNOWLEDGE, PAGE 186

1. True
2. True
3. True
4. False. Metazoa are larger than bacteria and protozoa.
5. B
6. D
7. B
8. C
9. A

TEST YOUR KNOWLEDGE, PAGE 194

1. True
2. False. Autotrophic bacteria cannot use CBOD. They obtain carbon from dissolved carbonate.
3. False. Autotrophic bacteria use energy-poor fuel sources like ammonia.
4. True
5. C
6. B
7. 1 = C, 2 = B, 3 = A
8. C
9. B
10. C
11. A
12. B
13. C
14. C
15. B

TEST YOUR KNOWLEDGE, PAGE 202

1. True
2. True
3. False. The growth rate will be limited by the number of bacteria in the process.
4. False. Each substrate has a minimum concentration needed to support the fastest possible growth of the bacteria. Multiple substrates may be limiting. For example, nitrifying heterotrophic bacteria could lack both oxygen and ammonia.
5. True

6. C
7. B
8. D
9. B
10. D
11. A

CHAPTER 6

TEST YOUR KNOWLEDGE, PAGE 211

1. False. Not all pond systems have screening and grit removal.
2. True
3. False. Ponds do not remove nutrients consistently.
4. C
5. D
6. A
7. D

TEST YOUR KNOWLEDGE, PAGE 214

1. C
2. B
3. B
4. C
5. A
6. B
7. 1 = C, 2 = F, 3 = A, 4 = E, 5 = B, 6 = D
8. B

TEST YOUR KNOWLEDGE, PAGE 221

1. A
2. C
3. B
4. A
5. D
6. D
7. B
8. C
9. A
10. C
11. B
12. A
13. B
14. C
15. C

TEST YOUR KNOWLEDGE, PAGE 225

1. B
2. C
3. C

TEST YOUR KNOWLEDGE, PAGE 227

1. D
2. B
3. C
4. A

TEST YOUR KNOWLEDGE, PAGE 238

1. False. Actual detention times are shorter because of incomplete mixing.
2. False. Dike slopes are not as steep on the outside of the pond. Steep slopes are hard to mow.
3. True
4. B
5. C
6. 1 = C, 2 = F, 3 = E, 4 = A, 5 = B, 6 = G, 7 = D
7. A
8. D
9. A
10. C
11. B
12. D
13. D
14. B
15. C
16. A
17. D

TEST YOUR KNOWLEDGE, PAGE 246

1. True
2. True
3. False. Oxygen saturation concentrations decrease with decreasing barometric pressure (elevation).
4. C
5. C
6. B
7. A
8. B
9. D
10. A
11. C

TEST YOUR KNOWLEDGE, PAGE 252

1. False. Aquashade and other dyes that block sunlight are not approved by U.S. EPA for use in wastewater treatment ponds.
2. True
3. B
4. B
5. D
6. A

TEST YOUR KNOWLEDGE, PAGE 258

1. C
2. A
3. A
4. D
5. False. Regulations change over time. A chemical applied last season or last week may no longer be on the approved list.
6. True

CHAPTER 7

TEST YOUR KNOWLEDGE, PAGE 270

1. True
2. True
3. False. Trickling filters are not submerged. The wastewater trickles over the media.
4. D
5. C
6. A

TEST YOUR KNOWLEDGE, PAGE 279

1. True
2. False. Substances move from areas of high concentration to areas of low concentration.
3. False. EPS increases with biofilm age.
4. A
5. B
6. C
7. A
8. B
9. D
10. B
11. D
12. C
13. C

TEST YOUR KNOWLEDGE, PAGE 284

1. True
2. True
3. False. As loading increases, as with a roughing filter, the percent removal decreases and the BOD concentration in the effluent increases.
4. C
5. C
6. B
7. A
8. B
9. C

TEST YOUR KNOWLEDGE, PAGE 287

1. False. Many trickling filters are hydraulically driven and do not have motors.
2. True
3. A
4. C
5. D
6. B

TEST YOUR KNOWLEDGE, PAGE 294

1. A = 3, B = 5, C = 1, D = 2, E = 4
2. C
3. D
4. B
5. A

TEST YOUR KNOWLEDGE, PAGE 302

1. True
2. False. The bearings at the top of the mast are better protected from corrosion.
3. B
4. D
5. A
6. A
7. D
8. C
9. A

TEST YOUR KNOWLEDGE, PAGE 312

1. False. It is the mass of BOD_5 per VOLUME of trickling filter media.
2. True
3. D
4. A
5. A
6. D
7. C
8. A
9. D
10. B. The volume of the trickling filter is 9679.8 m^3 (343 359 cu ft) and the total load of BOD going to the trickling filter is 4086 kg (9007.2 lb). The recycle flow should not be included in this calculation.
11. C
12. B

TEST YOUR KNOWLEDGE, PAGE 317

1. False. Water will fall through the tower. Heat will be transferred from the hot air to the cooler water. This makes the air inside the trickling filter cooler than the outside air.
2. True
3. True
4. B
5. B
6. D
7. B

TEST YOUR KNOWLEDGE, PAGE 321

1. True
2. False. Taking grit basins and primary clarifiers out of service can increase the amount of grit, grease, scum, and organic load going to the trickling filter. Excess grit can cause media plugging and grease and scum accumulation, especially in rock media trickling filters. Increasing the organic loading rate will increase the amount of biofilm produced and may require an adjustment of both the recycle ratio and distributor speed.
3. True
4. B
5. A
6. C
7. C
8. A

TEST YOUR KNOWLEDGE, PAGE 324

1. B
2. C. Should also check to see if distributor orifices are blocked.
3. B. A hooked wire may also be used.
4. A
5. D
6. C
7. A

TEST YOUR KNOWLEDGE, PAGE 333

1. B
2. C
3. C
4. A
5. D
6. B
7. A

TEST YOUR KNOWLEDGE, PAGE 341

1. True
2. False. Metals expand when heated and contract when cooled.
3. True
4. B
5. C
6. A
7. B
8. C
9. B
10. C
11. A

TEST YOUR KNOWLEDGE, PAGE 348

1. False. Loading is calculated as mass applied per surface area.
2. False. Ammonia removal requires more time.
3. B
4. C
5. B
6. C
7. B

TEST YOUR KNOWLEDGE, PAGE 350

1. B
2. C
3. B
4. D
5. A
6. C

TEST YOUR KNOWLEDGE, PAGE 354

1. B
2. C
3. A
4. D
5. A
6. C
7. B

CHAPTER 8

TEST YOUR KNOWLEDGE, PAGE 368

1. True
2. True
3. B
4. D
5. C
6. A

TEST YOUR KNOWLEDGE, PAGE 375

1. True
2. False. MLSS particles settle at the same speed as a blanket.
3. C
4. D
5. A
6. B
7. B
8. B
9. A
10. C
11. B

TEST YOUR KNOWLEDGE, PAGE 379

1. False. Too many filaments will cause settling problems in the clarifier.
2. A = 4, B = 1, C = 7, D = 2, E = 6, F = 3, G = 5
3. C
4. D
5. A = 3, B = 5, C = 1, D = 2, E = 4
6. A

TEST YOUR KNOWLEDGE, PAGE 381

1. B
2. A
3. B
4. C
5. A

TEST YOUR KNOWLEDGE, PAGE 398

1. True
2. True
3. False. The oxygen must be transferred from the air to the water to be dissolved.
4. D
5. B
6. C
7. D
8. B
9. C
10. A
11. B
12. D
13. D
14. A
15. B
16. A

17. B
18. D
19. B
20. C

TEST YOUR KNOWLEDGE, PAGE 407

1. True
2. True
3. D
4. A
5. C
6. D
7. B
8. C. [(2.3 mgd ÷ 1.45 mgd) + 1] × 2600 mg/L = 6724 mg/L, round to 6700 mg/L
9. B. (1000 mL ÷ 280 mL) × 1850 mg/L = 6607 mg/L
10. B. The clarifier can't perform better than the settleometer.
11. C
12. D
13. C. (2000 mL ÷ 280 mL) × 1800 mg/L = 12 857 mg/L
14. A. To solve, assume the influent flow is 1 m^3/d (or 1 mgd) and set the RAS flow to 0.7 m^3/d (or 0.70 mgd). Then, RAS concentration = [(1.0 ÷ 0.7) + 1] × 2600 mg/L = 6314 mg/L
15. A. 10 000 mg/L = [(1.0 ÷ RAS flow) + 1] × 2500 mg/L. Rearrange the equation to solve for RAS flow.

TEST YOUR KNOWLEDGE, PAGE 422

1. False. Sludge age is how long the average sludge particle remains in the system before being wasted.
2. True
3. A
4. A
5. B
6. B
7. A
8. B
9. B
10. D
11. B
12. D
13. B
14. C
15. D
16. B. First, find the mass of MLSS in the activated sludge basin. This is 28 387.5 kg (62 550 lb). Then, find the mass of WAS. This is 3488.3 kg (7686 lb). For U.S. customary units, it is necessary to convert the wasting rate in gpm to mgd first. 80 gpm = 0.1152 mgd. Then, divide the mass of MLSS by the mass of WAS to find SRT.

TEST YOUR KNOWLEDGE, PAGE 428

1. False. Wasting should be done at least daily.
2. A
3. B
4. D
5. B
6. C
7. A
8. C
9. B

TEST YOUR KNOWLEDGE, PAGE 432

1. True
2. A. First, find the total mass of MLSS. Then, multiply by 0.8 (80%) to find the mass of MLVSS.
3. B. If the clarifier removes 35%, then 65% of the load goes through to the activated sludge process.
4. B
5. D

TEST YOUR KNOWLEDGE, PAGE 438

1. False. Setting the sludge age also sets the F/M.
2. False. Yield is fixed. It isn't possible to grow more microorganisms without more load.
3. C
4. B
5. A
6. C
7. D
8. D
9. A = 2, B = 3, C = 5, D = 1, E = 4

TEST YOUR KNOWLEDGE, PAGE 446

1. False. BOD uptake can be complete even if the MLSS is not flocculating.
2. True
3. Detention time = extended air is longer, SRT = extended air is longer, F/M = conventional is higher, volumetric loading rate = conventional is higher
4. A
5. C
6. C
7. A
8. D
9. B
10. B
11. D
12. C
13. C
14. B

TEST YOUR KNOWLEDGE, PAGE 450

1. B
2. C
3. B
4. B
5. A
6. D

TEST YOUR KNOWLEDGE, PAGE 459

1. D
2. B. The surface area of the clarifier is 308 m^2 (3317 sq ft)
3. C

4. B. The surface area of the clarifier is 1052 m^2 (11 304 sq ft)
5. C
6. A. The distance from the outer wall to the weir must be subtracted from both sides.
7. B. The RAS concentration is extra information not needed to solve the problem.
8. D
9. A
10. D
11. B

TEST YOUR KNOWLEDGE, PAGE 464

1. D
2. B
3. C
4. A
5. B
6. C

TEST YOUR KNOWLEDGE, PAGE 469

1. False. Seed sludge should come from a well-functioning activated sludge basin.
2. True
3. B
4. D
5. A
6. B

TEST YOUR KNOWLEDGE, PAGE 490

1. A
2. B
3. B
4. C
5. B
6. D
7. B
8. C
9. B
10. A
11. C
12. D
13. D
14. B
15. C
16. C
17. C
18. B

TEST YOUR KNOWLEDGE, PAGE 495

1. True
2. False. Take one corrective action at a time.
3. B
4. A
5. a, c, d, and e are all process control. b and f are troubleshooting.

TEST YOUR KNOWLEDGE, PAGE 500

1. B
2. A
3. D
4. B
5. B

TEST YOUR KNOWLEDGE, PAGE 509

1. 1 = E, 2 = B, 3 = G, 4 = H, 5 = D, 6 = C, 7 = F, 8 = A
2. B
3. A
4. D
5. B
6. C
7. B
8. D
9. D
10. C
11. A

TEST YOUR KNOWLEDGE, PAGE 521

1. True
2. B
3. C
4. D
5. B
6. A
7. B
8. C
9. A
10. B
11. C

CHAPTER 9

TEST YOUR KNOWLEDGE, PAGE 540

1. B
2. A
3. C
4. A
5. C. 1 mg/L of P is removed for every 100 mg/L BOD_5.
   300 mg/L BOD_5 = 3 mg/L P
6. D
7. The total organic nitrogen is the difference between the TKN and the ammonia-nitrogen:

   $$\text{Organic nitrogen} = 36.0 - 22.0 = 14.0 \text{ mg/L N}$$

   All of the ammonia is soluble. Therefore, the soluble organic nitrogen is the difference between filtered TKN and the ammonia concentration.

   $$\text{Soluble organic nitrogen} = 26.0 - 22.0 = 4.0 \text{ mg/L N}$$

   The particulate organic nitrogen is the difference between the total organic nitrogen and the soluble organic nitrogen.

   $$\text{Particulate organic nitrogen} = 14.0 - 4.0 = 10.0 \text{ mg/L N}$$

TEST YOUR KNOWLEDGE, PAGE 545

1. A
2. D
3. C
4. D
5. C
6. C
7. B
8. C
9. B
10. A
11. D

TEST YOUR KNOWLEDGE, PAGE 548

1. False. Nitrification is a two-step process.
2. True
3. True
4. False. The NOB grow faster than the AOB when water temperatures are below 25 °C.
5. True
6. D
7. B
8. A
9. C

TEST YOUR KNOWLEDGE, PAGE 557

1. B
2. C
3. D
4. B
5. C
6. D
7. D
8. A
9. D
10. A

TEST YOUR KNOWLEDGE, PAGE 561

1. B
2. D
3. A
4. B
5. A
6. B

TEST YOUR KNOWLEDGE, PAGE 566

1. B
2. C
3. D
4. A
5. B
6. C
7. D
8. B
9. C

TEST YOUR KNOWLEDGE, PAGE 571

1. C
2. D
3. B
4. A
5. C

TEST YOUR KNOWLEDGE, PAGE 580

1. False. Calcium oxide (lime) usage is less common than both aluminum and iron salts.
2. False. Only orthophosphate can be chemically precipitated.
3. True
4. C
5. B
6. C
7. A
8. D
9. B
10. B
11. D
12. A
13. B
14. C
15. D

TEST YOUR KNOWLEDGE, PAGE 586

1. A
2. A
3. B
4. C
5. B
6. D
7. C
8. B
9. B

TEST YOUR KNOWLEDGE, PAGE 593

1. C
2. B
3. C
4. A
5. B
6. C
7. B
8. B
9. A
10. C
11. A
12. C

CHAPTER 10

TEST YOUR KNOWLEDGE, PAGE 603

1. False. Sterilization kills or inactivates all microorganisms. Disinfection reduces their numbers to safe levels.
2. True

3. A
4. B
5. B
6. C

TEST YOUR KNOWLEDGE, PAGE 607

1. False. Pathogens may or may not be present.
2. False. Many pathogens do not have reliable test methods to determine their presence.
3. C
4. B
5. C. Total coliforms → A. fecal coliforms → B. *E. coli*
6. D
7. C

TEST YOUR KNOWLEDGE, PAGE 615

1. True
2. False. Calcium hypochlorite is expressed as a weight ratio of chlorine.
3. False. Dose is equal to demand PLUS residual.
4. True
5. A
6. A
7. C
8. A
9. C
10. B
11. C
12. A
13. B
14. D
15. C
16. B
17. A
18. C

TEST YOUR KNOWLEDGE, PAGE 617

1. False. More flow is moving through the facility at peak hour flow, which reduces the amount of time it spends in the process.
2. B
3. D
4. C
5. A

TEST YOUR KNOWLEDGE, PAGE 621

1. True
2. False. Solubility decreases with increasing temperature.
3. False. How much water chlorine can contain before it is considered "wet" depends on temperature.
4. B
5. A
6. A
7. D
8. C
9. B
10. C
11. B
12. Chlorine gas = greenish-yellow, Chlorine liquid = amber, Sulfur dioxide gas = colorless, Sodium bisulfite solution = yellow, Sodium metabisulfite = white

TEST YOUR KNOWLEDGE, PAGE 626

1. True
2. False. The fusible plug is located below the metal to metal seal and can open regardless of the valve position.
3. False. Although rail cars are equipped with gas lines, they are used for pressurizing the car rather than removing chlorine gas.
4. C
5. B
6. C
7. C
8. A

TEST YOUR KNOWLEDGE, PAGE 633

1. True
2. False. Although this is the recommended reliable withdrawal rate, higher rates may be obtained by minimizing backpressure and increasing air circulation.
3. True
4. False. Liquid chlorine is used when gas cannot be withdrawn from the container fast enough to meet demand.
5. B
6. C
7. B
8. A
9. C
10. B
11. D
12. A
13. B
14. B

TEST YOUR KNOWLEDGE, PAGE 644

1. B
2. D
3. A
4. C
5. D
6. D
7. A
8. B
9. 1 = C, 2 = F, 3 = A, 4 = E, 5 = B, 6 = D
10. C
11. B
12. D
13. D
14. C

TEST YOUR KNOWLEDGE, PAGE 652

1. D

2. C
3. C
4. D
5. D
6. B
7. A

TEST YOUR KNOWLEDGE, PAGE 657

1. False. Changes in effluent quality can cause changes in the chlorine residual concentration even when the chlorine feed is flow paced.
2. True
3. True
4. A
5. A
6. C

TEST YOUR KNOWLEDGE, PAGE 661

1. True
2. False. A handtruck (dolly) should be used no matter how short the distance.
3. True
4. False. Lead washers may never be reused.
5. False. There is no requirement to wear an SCBA during changeout.
6. A
7. B
8. D
9. B
10. A
11. C
12. C

TEST YOUR KNOWLEDGE, PAGE 665

1. False. Most humans can't detect low levels of chlorine, which is why gas detectors are needed in gas rooms.
2. False. Sulfur dioxide is less toxic than chlorine.
3. True
4. A
5. B
6. C
7. C
8. B

TEST YOUR KNOWLEDGE, PAGE 668

1. True
2. True
3. True
4. B
5. C
6. D

TEST YOUR KNOWLEDGE, PAGE 670

1. True
2. False. Quartz sleeves absorb some UV light.
3. C
4. B
5. A
6. D

TEST YOUR KNOWLEDGE, PAGE 678

1. True. Banks are groups of modules.
2. True
3. False. The intensity sensor can only report the light intensity at the sensor location.
4. False. Clear, colorless effluents may contain UV absorbing substances.
5. A
6. B
7. A = 3, B = 7, C = 4, D = 5, E = 6, F = 1, G = 2
8. D
9. A
10. C
11. B
12. A
13. A
14. C
15. B
16. C
17. B
18. A
19. D
20. C

TEST YOUR KNOWLEDGE, PAGE 684

1. B
2. A
3. C
4. D
5. A
6. C
7. D
8. B

TEST YOUR KNOWLEDGE, PAGE 686

The control panel shows multiple lamps are burned out. This creates a hydraulic hole. Water passing through these areas is not being disinfected. The image of the lamps shows an excessive accumulation of solids and large solids may be seen between the lamps. This reduces transmissivity and reduces disinfection.

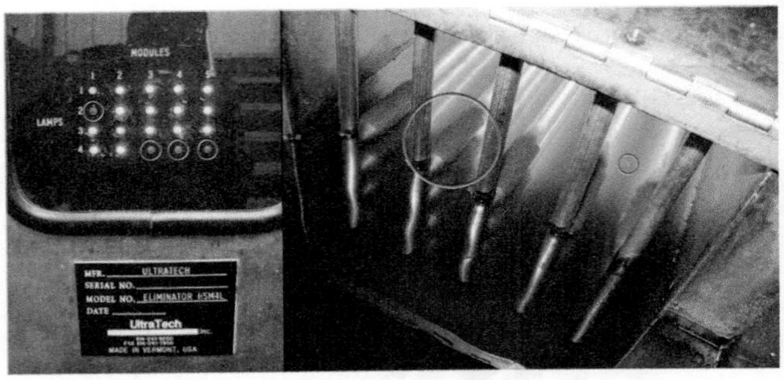

(Reprinted with permission by Indigo Water Group)

TEST YOUR KNOWLEDGE, PAGE 689

1. False. Looking directly at a UV lamp may cause conjunctivitis or welder flash.
2. True
3. False. Like a bad sunburn, the effects of welder flash may take several hours to appear.
4. True
5. A
6. C
7. B

Periodic Table

PERIODIC TABLE

group

element name → HYDROGEN
atomic number → 1
chemical Symbol → H
atomic weight (u) → 1,008

- Alkali metal
- Alkaline earth metal
- Lanthanide
- Actinide
- Transition metal
- Post-transition metal
- Metalloid
- Nonmetal
- Halogen
- Noble gas
- ● Solid
- ● Liquid
- ● Gas
- ● Unknown

period	1	2	3	4	5	6	7	8	9	10	11	12	13	14	15	16	17	18
	IA	IIA	IIIB	IVB	VB	VIB	VIIB	VIIIB	VIIIB	VIIIB	IB	IIB	IIIA	IVA	VA	VIA	VIIA	VIIIA
1	HYDROGEN 1 H 1,008																	HELIUM 2 He 4,003
2	LITHIUM 3 Li 6,941	BERYLLIUM 4 Be 9,012											BORON 5 B 10,81	CARBON 6 C 12,01	NITROGEN 7 N 14,01	OXYGEN 8 O 16,00	FLUORINE 9 F 19,00	NEON 10 Ne 20,18
3	SODIUM 11 Na 22,99	MAGNESIUM 12 Mg 24,31											ALUMINIUM 13 Al 26,98	SILICON 14 Si 28,09	PHOSPHORUS 15 P 30,97	SULFUR 16 S 32,07	CHLORINE 17 Cl 35,45	ARGON 18 Ar 39,95
4	POTASSIUM 19 K 39,10	CALCIUM 20 Ca 40,08	SCANDIUM 21 Sc 44,96	TITANIUM 22 Ti 47,87	VANADIUM 23 V 50,94	CHROMIUM 24 Cr 52,00	MANGANESE 25 Mn 54,94	IRON 26 Fe 55,85	COBALT 27 Co 58,93	NICKEL 28 Ni 58,69	COPPER 29 Cu 63,55	ZINC 30 Zn 65,39	GALLIUM 31 Ga 69,72	GERMANIUM 32 Ge 72,59	ARSENIC 33 As 74,92	SELENIUM 34 Se 78,96	BROMINE 35 Br 79,90	KRYPTON 36 Kr 83,80
5	RUBIDIUM 37 Rb 85,47	STRONTIUM 38 Sr 87,62	YTTRIUM 39 Y 88,91	ZIRCONIUM 40 Zr 91,22	NIOBIUM 41 Nb 92,91	MOLYBDENUM 42 Mo 95,94	TECHNETIUM 43 Tc (98,91)	RUTHENIUM 44 Ru 101,1	RHODIUM 45 Rh 102,9	PALLADIUM 46 Pd 106,4	SILVER 47 Ag 107,9	CADMIUM 48 Cd 112,4	INDIUM 49 In 114,8	TIN 50 Sn 118,7	ANTIMONY 51 Sb 121,8	TELLURIUM 52 Te 127,6	IODINE 53 I 126,9	XENON 54 Xe 131,3
6	CAESIUM 55 Cs 132,9	BARIUM 56 Ba 137,3	LANTHANUM 57 La 138,9	HAFNIUM 72 Hf 178,5	TANTALUM 73 Ta 180,9	TUNGSTEN 74 W 183,9	RHENIUM 75 Re 186,2	OSMIUM 76 Os 190,2	IRIDIUM 77 Ir 192,2	PLATINUM 78 Pt 195,1	GOLD 79 Au 197,0	MERCURY 80 Hg 200,6	THALLIUM 81 Tl 204,4	LEAD 82 Pb 207,2	BISMUTH 83 Bi 209,0	POLONIUM 84 Po (210,0)	ASTATINE 85 At (210,0)	RADON 86 Rn (222,0)
7	FRANCIUM 87 Fr (223,0)	RADIUM 88 Ra (226,0)	ACTINIUM 89 Ac (227,0)	RUTHERFORDIUM 104 Rf ---	DUBNIUM 105 Db ---	SEABORGIUM 106 Sg ---	BOHRIUM 107 Bh ---	HASSIUM 108 Hs ---				COPERNICIUM 112 Cn ---						

Lanthanides	CERIUM 58 Ce 140,1	PRASEODYMIUM 59 Pr 140,9	NEODYMIUM 60 Nd 144,2	PROMETHIUM 61 Pm (144,9)	SAMARIUM 62 Sm 150,4	EUROPIUM 63 Eu 152,0	GADOLINIUM 64 Gd 157,3	TERBIUM 65 Tb 158,9	DYSPROSIUM 66 Dy 162,5	HOLMIUM 67 Ho 164,9	ERBIUM 68 Er 167,3	THULIUM 69 Tm 168,9	YTTERBIUM 70 Yb 173,0	LUTETIUM 71 Lu 175,0
Actinides	THORIUM 90 Th (232,0)	PROTACTINIUM 91 Pa (231,0)	URANIUM 92 U (238,0)	NEPTUNIUM 93 Np (237,0)	PLUTONIUM 94 Pu (239,1)	AMERICIUM 95 Am (243,1)	CURIUM 96 Cm (247,1)	BERKELIUM 97 Bk (247,1)	CALIFORNIUM 98 Cf (252,1)	EINSTEINIUM 99 Es (252,1)	FERMIUM 100 Fm (257,1)	MENDELEVIUM 101 Md (256,1)	NOBELIUM 102 No (259,1)	LAWRENCIUM 103 Lr (260,1)

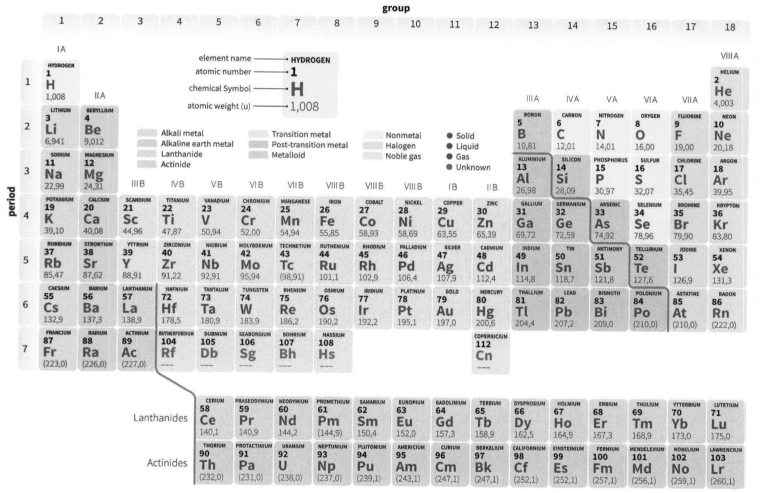

Index

D

R

Ready to Become a PO?

1 **Visit**
Go to www.ProfessionalOperator.org to find information about eligibility, complete your profile, and apply to become a Professional Operator.

2 **Test**
Once approved, you can choose from one of hundreds of testing centers to schedule your certification exam.

3 **Certify**
Upon successfully passing your exam, you'll be certified as a premier water professional, having earned your PO.

Benefits of Earning Your Professional Operator Certification

- PO Designation – Privilege to include PO after your name in all communications
- Certificate Package – You'll receive a PO certificate, suitable for framing
- PO Pin – Wear it with pride (included with your PO Certificate Package)
- POWER Event – Be recognized on an industry stage
- Digital Badge – Share your credentials in real time online
- Business Cards & Merchandise – Access to special PO-branded items
- Connect with other POs – Stay in touch with fellow POs via social media

www.ProfessionalOperator.org